Lecture Notes in Computer Science 10827

Commenced Publication in 1973
Founding and Former Series Editors:
Gerhard Goos, Juris Hartmanis, and Jan van Leeuwen

More information about this series at http://www.springer.com/series/7409

Jian Pei · Yannis Manolopoulos
Shazia Sadiq · Jianxin Li (Eds.)

Database Systems for Advanced Applications

23rd International Conference, DASFAA 2018
Gold Coast, QLD, Australia, May 21–24, 2018
Proceedings, Part I

Springer

Editors
Jian Pei
Simon Fraser University
Burnaby, BC
Canada

Yannis Manolopoulos
Aristotle University of Thessaloniki
Thessaloniki
Greece

Shazia Sadiq
University of Queensland
Brisbane, QLD
Australia

Jianxin Li
University of Western Australia
Crawley, WA
Australia

ISSN 0302-9743   ISSN 1611-3349 (electronic)
Lecture Notes in Computer Science
ISBN 978-3-319-91451-0   ISBN 978-3-319-91452-7 (eBook)
https://doi.org/10.1007/978-3-319-91452-7

Library of Congress Control Number: 2018942340

LNCS Sublibrary: SL3 – Information Systems and Applications, incl. Internet/Web, and HCI

This Springer imprint is published by the registered company Springer Nature Switzerland AG
The registered company address is: Gewerbestrasse 11, 6330 Cham, Switzerland

Preface

It is our great pleasure to present the proceedings of the 23rd International Conference on Database Systems for Advanced Applications (DASFAA). DASFAA 2018 is an annual international database conference, which showcases state-of-the-art R&D activities in database systems and their applications. It provides a forum for technical presentations and discussions among database researchers, developers, and users from academia, business, and industry.

DASFAA 2018 was held on the Gold Coast, Australia, during May 21–24, 2018. The Gold Coast is a coastal city in the state of Queensland, 66 km (41 mi) from the state capital Brisbane. With a population of 638,090 (2016), the Gold Coast is the sixth largest city in Australia. It is a major tourist destination with its sunny subtropical climate and has become widely known for its surfing beaches, high-rise-dominated skyline, theme parks, nightlife, and rainforest hinterland. It is also the major film production hub for Queensland. The Gold Coast will host the 2018 Commonwealth Games.

This year we introduced a Senior Program Committee (SPC) at DASFAA. The SPC comprised 12 distinguished leaders in the area of database systems and advanced applications: Amr El Abbadi, UC Santa Barbara, USA; K. Selcuk Candan, Arizona State University, USA; Lei Chen, Hong Kong University of Science and Technology, Hong Kong; Chengfei Liu, Swinburne University of Technology, Australia; Nikos Mamoulis, University of Ioannina/University of Hong Kong, Hong Kong; Kyuseok Shim, Seoul National University, Korea; Michalis Vazirgiannis, Ecole Polytechnique Paris, France; Xiaokui Xiao, Nanyang Technological University, Singapore; Xiaochun Yang, Northeastern University, China; Jeffrey Xu Yu, Chinese University of Hong Kong, Hong Kong; Xiaofang Zhou, University of Queensland, Australia; and Aoying Zhou, East China Normal University, China. We are grateful for the role played by the SPC and acknowledge that the SPC provided a significant level of support and expert advice in the efficient paper-reviewing process that resulted in an excellent selection of papers.

We received 360 submissions, each of which was assigned to at least three Program Committee (PC) members and one SPC member. The thoughtful discussion on each paper by the PC with facilitation and meta-review provided by the SPC resulted in the selection of 83 full research papers (acceptance ration of 23%). In addition, we included 21 short papers, six industry papers, and eight demo papers in the program. This year the dominant topics for the selected papers included learning models, graph and network data processing, and social network analysis, followed by text and data mining, recommendation, data quality and crowd sourcing, and trajectory and stream data. Selected papers also included topics relating to network embedding, sequence and temporal data processing, RDF and knowledge graphs, security and privacy, medical data mining, query processing and optimization, search and information retrieval, multimedia data processing, and distributed computing. Last but not least, the conference program included keynote presentations by Dr. C. Mohan (IBM Almaden Research Center, San Jose, USA), Prof. Xuemin Lin (UNSW, Sydney, Australia), and Prof. Yongsheng Gao (Griffith University, Brisbane, Australia).

Four workshops were selected by the workshop co-chairs to be held in conjunction with DASFAA 2018: the 5th International Workshop on Big Data Management and Service (BDMS 2018); the 5th International Symposium on Semantic Computing and Personalization (SeCoP 2018); the Second International Workshop on Graph Data Management and Analysis (GDMA 2018); and the Third Workshop on Big Data Quality Management (BDQM 2018). The workshop papers are included in a separate volume of the proceedings also published by Springer in its *Lecture Notes in Computer Science* series.

We are grateful to the general chairs, Yanchun Zhang, Victoria University, and Rao Kotagiri, University of Melbourne, all SPC members, PC members and external reviewers who contributed their time and expertise to the DASFAA 2018 paper reviewing process. We would like to thank all the members of the Organizing Committee, and many volunteers, for their great support in the conference organization. Special thanks go to the DASFAA 2018 local Organizing Committee chair, Junhu Wang (Griffith University), for his tireless work before and during the conference. Many thanks to the authors who submitted their papers to the conference. Lastly we acknowledge the generous financial support from Griffith University, Destination Gold Coast, and Springer.

March 2018

Shazia Sadiq
Jian Pei
Yannis Manolopoulos

Organization

General Co-chairs

Yanchun Zhang   Victoria University, Australia
Rao Kotagiri   University of Melbourne, Australia

Program Committee Co-chairs

Jian Pei   Simon Fraser University, Canada
Yannis Manolopoulos   Aristotle University of Thessaloniki, Greece
Shazia Sadiq   The University of Queensland, Australia

Industrial/Practitioners Track Co-chairs

Yu Zheng   Urban Computing Group, Microsoft Research, China
Qing Liu   Data61, CSIRO, Australia

Demo Track Co-chairs

Sebasitian Link   University of Auckland, New Zealand
Chaoyi Pang   NIT, Zhejiang University, China

Workshop Co-chairs

Chengfei Liu   Swinburne University of Technology, Australia
Lei Zou   Peking University, China

Tutorial Chair

Yoshiharu Ishikawa   Nagoya University, Japan

PhD Consortium Chair

Zhiguo Gong   University of Macau, China

Panel Co-chairs

Sean Wang   Fudan University, China
Sven Hartman   Clausthal University of Technology, Germany

Proceedings Chair

Jianxin Li   The University of Western Australia, Australia

Publicity Co-chairs

Ji Zhang   University of Southern Queensland, Australia
Xin Wang   Tianjin University, China
Shuo Shang   KAUST, Saudi Arabia

Local Organization Co-chairs

Junhu Wang   Griffith University, Australia
Bela Stantic   Griffith University, Australia
Alan Liew   Griffith University, Australia

DASFAA Liaison officer

Kyuseok Shim   Seoul National University, South Korea

Web Master

Xuguang Ren   Griffith University, Australia

Senior Program Committee Members

Amr El Abbadi   UC Santa Barbara, USA
K. Selcuk Candan   Arizona State University, USA
Lei Chen   Hong Kong University of Science and Technology,
                                  SAR China
Chengfei Liu   Swinburne University of Technology, Australia
Nikos Mamoulis   University of Ioannina/University of Hong Kong,
                                  SAR China
Kyuseok Shim   Seoul National University, Korea
Michalis Vazirgiannis   Ecole Polytechnique Paris, France
Xiaokui Xiao   Nanyang Technological University, Singapore
Xiaochun Yang   Northeastern University, China
Jeffrey Xu Yu   Chinese University of Hong Kong, SAR China
Xiaofang Zhou   University of Queensland, Australia
Aoying Zhou   East China Normal University, China

Program Committee

Alberto Abello   Universitat Politècnica de Catalunya, Spain
Akhil Arora   Indian Institute of Technology, India
Jie Bao   Independent

Sabrina De Capitani Vimercati	University of Milan, Italy
Bin Wang	NEU, China
Jianmin Wang	Tsinghua University, China
Wei Wang	National University of Singapore
Xin Wang	Tianjin University, China
John Wu	Berkeley Lab, USA
Xiaokui Xiao	Nanyang Technological University, Singapore
Xike Xie	University of Science and Technology of China, China
Jianlian Xu	Hong Kong Baptist University, SAR China
Xiaochun Yang	Northeastern University, China
Yu Yang	Simon Fraser University, Canada
Hongzhi Yin	The University of Queensland, Australia
Man Lung Yiu	The Hong Kong Polytechnic University, SAR China
Ge Yu	Northeastern University, China
Jeffrey Xu Yu	The Chinese University of Hong Kong, SAR China
Yi Yu	National Institute of Informatics, Japan
Ye Yuan	NEU, China
Fuzheng Zhang	Microsoft
Wenjie Zhang	The University of New South Wales, Australia
Ying Zhang	University of Technology, Australia
Zhengjie Zhang	Yitu Technology
Bolong Zheng	Aalborg University, Denmark
Kai Zheng	University of Science and Technology of China, China
Yu Zheng	JD Finance
Aoying Zhou	East China Normal University, China
Xiangmin Zhou	RMIT University, Australia
Xiaofang Zhou	The University of Queensland, Australia
Yongluan Zhou	University of Copenhagen, Denmark
Hengshu Zhu	Baidu Inc.
Yuanyuan Zhu	Wuhan University, China
Andreas Zuefle	George Mason University, USA

Additional Reviewers

Al-Baghdadi, Ahmed	Bellandi, Valerio	Ceh Varela, Edgar
Alserafi, Ayman	Benkrid, Soumia	Ceravolo, Paolo
Alves Peixoto, Douglas	Berkani, Nabila	Chen, Chen
Anisetti, Marco	Bilalli, Besim	Chen, Jinpeng
Ardagna, Claudio	Bioglio, Livio	Chen, Lu
Askar, Ahmed	Cao, Xin	Chen, Xiaoshuang
Banerjee, Prithu	Casagranda, Paolo	Chen, Xilun
Behrens, Hans	Castelltort, Arnaud	Cheng, Yu

Chondrogiannis,
 Theodoros
Cong, Zicun
Cui, Yufei
Dellal, Ibrahim
Dian, Ouyang
Du, Boxin
Du, Dawei
Du, Xingzhong
Feng, Kaiyu
Feng, Shi
Feng, Xing
Frey, Christian
Fu, Xiaoyi
Galhotra, Sainyam
Galicia Auyon, Jorge
Gan, Junhao
Garg, Yash
Gianini, Gabriele
Gkountouna, Olga
Gong, Qixu
Gu, Yu
Guo, Long
Guo, Shangwei
Guo, Tao
Gurukar, Saket
Hao, Yifan
Hewasinghage, Moditha
Hu, Jiafeng
Hu, Xia
Huang, Jun
Huang, Shengyu
Huang, Xiangdong
Huang, Zhipeng
Imani, Maryam
Jovanovic, Petar
Kang, Jian
Kang, Rong
Kefalas, Pavlos
Khan, Hina
Khouri, Selma
Lai, Longbin
Lei, Mingtao
Li, Guorong
Li, Hangyu
Li, Huan

Li, Huayu
Li, Jingjing
Li, Liangyue
Li, Lin
Li, Mao-Lin
Li, Pengfei
Li, Xiaodong
Li, Xinsheng
Li, Xiucheng
Liang, Yuan
Liu, Qing
Liu, Sicong
Liu, Weiwei
Liu, Wu
Liu, Yiding
Luo, Siqiang
Ma, Chenhao
Ma, Yujing
Mao, Jiali
Mattheis, Sebastian
Mesmoudi, Amin
Moscato, Vincenzo
Munir, Rana Faisal
Mustafa, Ahmad
Nadal, Sergi
Nei, Wendy
Nelakurthi, Arun
Nelakurthi, Arun Reddy
Nie, Tiezheng
Pande, Shiladitya
Pang, Junbiao
Paraskevopoulos, Pavlos
Peng, Hao
Peng, Jinglin
Pham, Nguyen Tuan Anh
Pham, Tuan Anh
Piantadosi, Gabriele
Qin, Chengjie
Qin, Dong
Rai, Niranjan
Rakthanmanon, Thanawin
Ren, Weilong
Roukh, Amine
Ruan, Sijie
Sarwar, Raheem
Shan, Caihua

Shao, Yingxia
Sharma, Vishal
Song, Shaoxu
Sperlì, Giancarlo
Su, Li
Sun, Haiqi
Tang, Bo
Tao, Hemeng
Tiakas, Eleftherios
Tzouramanis, Theodoros
Vachery, Jithin
Varga, Jovan
Vassilakopoulos, Michael
Wang, Hanchen
Wang, Hongwei
Wang, Kai
Wang, Li
Wang, Qinyong
Wang, Shuhui
Wang, Sibo
Wang, Wei
Wang, Weiqing
Wang, Zhefeng
Wen, Lijie
Xiao, Chuan
Xu, Cheng
Xu, Jianqiu
Xu, Tong
Xu, Wenjian
Xu, Xing
Xue, Zhe
Yan, Jing
Yavanoğlu, Uraz
Zhang, Ce
Zhang, Fan
Zhang, Jilian
Zhang, Liming
Zhang, Pengfei
Zhang, Si
Zhao, Kaiqi
Zhao, Weijie
Zhou, Qinghai
Zhou, Yao
Zhu, Lei
Zhu, Zichen

Contents – Part I

Social Network Analytics

Sequence and Temporal Data Processing

Trajectory and Streaming Data

Contents – Part II

Learning Models

Multimedia Data Processing

Distributed Computing

Network Embedding

Enhancing Network Embedding with Auxiliary Information: An Explicit Matrix Factorization Perspective

Junliang Guo, Linli Xu$^{(\boxtimes)}$, Xunpeng Huang, and Enhong Chen

Anhui Province Key Laboratory of Big Data Analysis and Application,
School of Computer Science and Technology,
University of Science and Technology of China, Hefei, China
{guojunll,hxpsola}@mail.ustc.edu.cn, {linlixu,cheneh}@ustc.edu.cn

Abstract. Recent advances in the field of network embedding have shown the low-dimensional network representation is playing a critical role in network analysis. However, most of the existing principles of network embedding do not incorporate auxiliary information such as content and labels of nodes flexibly. In this paper, we take a matrix factorization perspective of network embedding, and incorporate structure, content and label information of the network simultaneously. For structure, we validate that the matrix we construct preserves high-order proximities of the network. Label information can be further integrated into the matrix via the process of random walk sampling to enhance the quality of embedding in an unsupervised manner, i.e., without leveraging downstream classifiers. In addition, we generalize the Skip-Gram Negative Sampling model to integrate the content of the network in a matrix factorization framework. As a consequence, network embedding can be learned in a unified framework integrating network structure and node content as well as label information simultaneously. We demonstrate the efficacy of the proposed model with the tasks of semi-supervised node classification and link prediction on a variety of real-world benchmark network datasets.

1 Introduction

The rapid growth of applications based on networks has posed major challenges of effective processing of network data, among which a critical task is network data representation. The primitive representation of a network is usually very sparse and suffers from overwhelming high dimensionality, which limits its generalization in statistical learning. To deal with this issue, network embedding aims to learn latent representations of nodes on a network while preserving the structure and the inherent properties of the network, which can be effectively exploited by classical vector-based machine learning models for tasks including node classification, link prediction, and community detection, etc. [1,3,6,10].

Recently, inspired by the advances of neural representation learning in language modeling, which is based on the principle of learning the embedding

© Springer International Publishing AG, part of Springer Nature 2018
J. Pei et al. (Eds.): DASFAA 2018, LNCS 10827, pp. 3–19, 2018.
https://doi.org/10.1007/978-3-319-91452-7_1

vector of a word by predicting its context [12,13], a number of network embedding approaches have been proposed with the paradigm of learning the embedding vector of a node by predicting its neighborhood [3,17,20]. Specifically, latent representations of network nodes are learned by treating short random walk sequences as sentences to encode structural proximity in a network. Existing results demonstrate the effectiveness of the neural network embedding approaches in the tasks of node classification, behavior prediction, etc.

However, existing network embedding methods, including DeepWalk [17], LINE [20] and node2vec [3], are typically based on structural proximities only and do not incorporate other information such as node content flexibly. In this paper, we explore the question whether network structure and auxiliary properties of the network such as node content and label information can be integrated in a unified framework of network embedding. To achieve that, we take a matrix factorization perspective of network embedding with the benefits of natural integration of structural embedding and content embedding simultaneously, where label information can be incorporated flexibly.

Specifically, motivated by the recent work [9] that explains the word embedding model of Skip-Gram Negative Sampling (SGNS) as a matrix factorization of the words' co-occurrence matrix, we build a co-occurrence matrix of structural proximities for a network based on a random walk sampling procedure. The process of SGNS can then be formulated as minimizing a matrix factorization loss, which can be naturally integrated with representation learning of node content. In addition, label information can be exploited in the process of building the co-occurrence matrix to enhance the quality of network embedding, which is achieved by decomposing the context of a node into the structure context generated with random walks, as well as the label context based on the given label information.

Our main contributions can be summarized as follows:

- We propose a unified framework of Auxiliary information Preserved Network Embedding with matrix factorization, abbreviated as *APNE*, which can effectively learn the latent representations of nodes, and provide a flexible integration of network structure, node content, as well as label information without leveraging downstream classifiers.
- We verify that the structure matrix we generate is an approximation of the high-order proximity of the network known as rooted PageRank.
- We extensively evaluate our framework on four benchmark datasets and two tasks including semi-supervised classification and link prediction. Results show that the representations learned by our proposed method are general and powerful, producing significantly increased performance over the state of the art on both tasks.

2 Related Work

2.1 Network Embedding

Network embedding has been extensively studied in the literature [1]. Recently, motivated by the advances of neural representation learning in language modeling, a number of embedding learning methods have been proposed based on the Skip-Gram model. A representative model is DeepWalk [17], which exploits random walk to generate sequences of instances as the training corpus, followed by utilizing the Skip-Gram model to obtain the embedding vectors of nodes. Node2vec [3] extend DeepWalk with sophisticated random walk schemes. Similarly in LINE [20] and GraRep [2], network embedding is learned by directly optimizing the objective function inspired from the Skip-Gram model. To further incorporate auxiliary information into network embedding, many efforts have been made. Among them, TADW [22] formulates DeepWalk in a matrix factorization framework, and jointly learns embeddings with the structure information and preprocessed features of text information. This work is further extended by HSCA [25], DMF [24] and MMDW [21] with various additional information. In SPINE [4], structural identities are incorporated to jointly preserve local proximity and global proximity of the network simultaneously in the learned embeddings. However, none of the above models jointly consider structure, content and label information in a unified model, which are the fundamental elements of a network [1]. Recently, TriDNR [16] and LANE [5] tackle this problem both through implicit interactions between the three elements. TriDNR leverages multiple skip-gram algorithms between node-word and word-label pairs, while LANE first constructs three network affinity matrices from the three elements respectively, followed by executing SVD on affinity matrices with additional pairwise interactions. Although empirically effective, these methods do not provide a clear objective articulating how the three aspects are integrated in the embeddings learned, and is relatively inflexible to generalize to other scenarios.

In contrast to the above models, we propose a unified framework to learn network embeddings from structure, content and label simultaneously. The superiority of our framework is threefold: (a) the structure matrix we generate contains high-order proximities of the network, and the label information is incorporated by explicitly manipulating the constructed matrix rather than through implicit multi-hop interactions [5,16]; (b) instead of leveraging label information through an explicitly learned classifier (e.g., SVM [21], linear classifier [24] and neural networks [6,23]) whose performance is not only related to the quality of embeddings but also the specific classifiers being used, we exploit the label information without leveraging any downstream classifiers, which enables the flexibility of our model to generalize to different tasks; (c) while most of the above models only consider text descriptions of nodes, we use raw features contained in datasets as the content information, which is more generalized to various types of networks in real world such as social networks.

2.2 Matrix Factorization and Word Embedding

Matrix Factorization (MF) has been proven effective in various machine learning tasks, such as dimensionality reduction, representation learning, recommendation systems, etc. Recently, connections have been built between MF and word embedding models. It is shown in [8] that the Skip-Gram with Negative Sampling (SGNS) model is an Implicit Matrix Factorization (IMF) that factorizes a word-context matrix, where the value of each entry is the pointwise mutual information (PMI) between a word and context pair, indicating the strength of association. It is further pointed out in [9] that the SGNS objective can be reformulated in a representation learning view with an Explicit Matrix Factorization (EMF) objective, where the matrix being factorized here is the co-occurrence matrix among words and contexts.

In this paper, we extend the matrix factorization perspective of word embedding into the task of network embedding. More importantly, we learn the network embedding by jointly factorizing the structure matrix and the content matrix of the network, which can be further improved by leveraging auxiliary label information. Different from most existing network embedding methods based on matrix factorization, which employ either trivial objective functions (F-norm used in TADW) or traditional factorization algorithms (SVD used in GraRep) for optimization, we design a novel objective function based on SGNS in our framework. Furthermore, the proposed method is general and not confined to specific downstream tasks, such as link prediction [3] and node classification [17], and we do not leverage any classifiers either.

3 Network Embedding with Matrix Factorization

In this section, we propose a novel approach for network embedding based on a unified matrix factorization framework, which consists of three procedures as illustrated in Fig. 1. We follow the paradigm of treating random walk sequences as sentences to encode structural proximities in a network. However, unlike the EMF objective for word embedding where the matrix to factorize is clearly defined as the word-context co-occurrence matrix, for network embedding, there is a gap between the random walk procedure and the co-occurrence matrix. Therefore, we start with proposing a random walk sampling process to build a co-occurrence matrix, followed by theoretical justification of its property of preserving the high-order structural proximity in the network, based on which we present the framework of network embedding with matrix factorization.

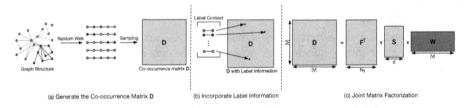

(a) Generate the Co-occurrence Matrix D   (b) Incorporate Label Information   (c) Joint Matrix Factorization

Fig. 1. The overall procedure of our framework (*APNE*). Different colors indicate different labels of nodes. (Color figure online)

Algorithm 1. Sampling the general co-occurrence matrix

Input: The transition matrix $\boldsymbol{P}$, window size l
Output: Co-occurrence matrix $\boldsymbol{D}$
1: Sample random walks C based on $\boldsymbol{P}$
2: **for** every node sequence in C **do**
3:    Uniformly sample (i, j) with $|i - j| < l$
4:    $D_{v_i, v_j} = D_{v_i, v_j} + 1$
5: **end for**

3.1 High-Order Proximity Preserving Matrix

Given an undirected network $G = \{V, E\}$ which includes a set of nodes V connected by a set of edges E, the corresponding adjacency matrix is $\boldsymbol{A}$, where $A_{i,j} = w_{i,j}$ indicates an edge with weight $w_{i,j}$ between the i-th node v_i and the j-th node v_j. And we denote the transition matrix of G as $\boldsymbol{P}$, where $P_{i,j} = \frac{w_{i,j}}{\sum_{k=1}^{|V|} w_{i,k}}$. Next, a list of node sequences C can be generated with random walks on the network.

Given C, we can generate the co-occurrence matrix $\boldsymbol{D}$ of G with the n-gram algorithm. The procedure is summarized in Algorithm 1. In short, for a given node in a node sequence, we increase the co-occurrence count of two nodes if and only if they are in a window of size l.

Next we show that the co-occurrence matrix generated by Algorithm 1 preserves the high-order structural proximity in the network with the following theorem:

Theorem 1. *Define the high-order proximity $\boldsymbol{S}$ of the network G as*

$$S^l = \sum_{k=1}^{l} \boldsymbol{P}^k$$

where l denotes the order of the proximity as well as the window size in Algorithm 1. Then, under the condition that the random walk procedure is repeated enough times and the generated list of node sequences C covers all paths in the network G, we can derive that according to [22]:

$$l \cdot \boldsymbol{D}^{nor} = \boldsymbol{S}^l \tag{1}$$

where l is the window size in Algorithm 1, and the matrix $\boldsymbol{D}^{nor}$ denotes the expectation of row normalized co-occurrence matrix $\boldsymbol{D}$, i.e., $D_{i,j}^{nor} = \mathbb{E}[\frac{D_{i,j}}{\sum_{k=1}^{|V|} D_{i,k}}]$.

Note that the (i, j)-th entry of the left side of Eq. (1) can be written as $\mathbb{E}[\frac{D_{i,j}}{\sum_{k=1}^{|V|} D_{i,k}/l}]$, which is the expected number of times that v_j appears in the left or right l-neighborhood of v_i.

To investigate into the structural information of the network encoded in the co-occurrence matrix $\boldsymbol{D}$, we first consider a well-known high-order proximity of a network named rooted PageRank (RPR) [19], defined as $\boldsymbol{S}^{\text{RPR}} = (1 - \beta_{\text{RPR}})(\boldsymbol{I} -$

$\beta_{\mathrm{RPR}}\boldsymbol{P})^{-1}$, where $\beta_{\mathrm{RPR}} \in (0,1)$ is the probability of randomly walking to a neighbor rather than jumping back. The (i,j)-th entry of $\boldsymbol{S}^{\mathrm{RPR}}$ is the probability that a random walk from node v_i will stop at v_j in the steady state, which can be used as an indicator of the node-to-node proximity. $\boldsymbol{S}^{\mathrm{RPR}}$ can be further rewritten as:

$$\boldsymbol{S}^{\mathrm{RPR}} = (1 - \beta_{\mathrm{RPR}})(\boldsymbol{I} - \beta_{\mathrm{RPR}}\boldsymbol{P})^{-1} \tag{2}$$

$$= (1 - \beta_{\mathrm{RPR}}) \sum_{k=0}^{\infty} \beta_{\mathrm{RPR}}^k \boldsymbol{P}^k \tag{3}$$

We next show that for an undirected network, where $\boldsymbol{P}$ is symmetric, the row normalized co-occurrence matrix $\boldsymbol{D}^{nor}$ is an approximation of the rooted PageRank matrix $\boldsymbol{S}^{\mathrm{RPR}}$.

Theorem 2. *When l is sufficiently large, for $\boldsymbol{D}^{nor}$ defined as $\boldsymbol{D}^{nor} = \frac{1}{l}\sum_{k=1}^{l} \boldsymbol{P}^k$, and $K = \lfloor -\frac{\log l(1-\beta_{RPR})}{\log \beta_{RPR}} \rfloor$, the ℓ-2 norm of the difference between $\boldsymbol{D}^{nor}$ and $\boldsymbol{S}^{RPR}$ can be bounded by K:*

$$\|\boldsymbol{S}^{RPR} - \boldsymbol{D}^{nor}\|_2 \leq 2 - 2\beta_{RPR}^{K+1} \tag{4}$$

Proof of Theorem 2. Here we omit the superscript of $\boldsymbol{S}^{\mathrm{RPR}}$ and the subscript of β_{RPR} in the proof for simplicity. Substituting (2) and reformulating the left side of (4) we have:

$$\|\boldsymbol{S} - \boldsymbol{D}^{nor}\|_2 = \left\| (1-\beta) \sum_{k=0}^{\infty} \beta^k \boldsymbol{P}^k - \frac{1}{l} \sum_{k=1}^{l} \boldsymbol{P}^k \right\|_2$$

$$= \left\| (1-\beta) \sum_{k=0}^{l} \beta^k \boldsymbol{P}^k - \frac{1}{l} \sum_{k=0}^{l} \boldsymbol{P}^k + \frac{1}{l} + (1-\beta) \sum_{k=l+1}^{\infty} \beta^k \boldsymbol{P}^k \right\|_2$$

$$\leq \left\| \sum_{k=0}^{l} \boldsymbol{P}^k \left[(1-\beta)\beta^k - \frac{1}{l} \right] \right\|_2 + (1-\beta) \left\| \sum_{k=l+1}^{\infty} \beta^k \boldsymbol{P}^k \right\|_2 + \frac{1}{l}$$

$$\leq \sum_{k=0}^{l} \lambda_{\max}^k \left| (1-\beta)\beta^k - \frac{1}{l} \right| + (1-\beta) \sum_{k=l+1}^{\infty} \beta^k \lambda_{\max}^k + \frac{1}{l}$$

where $\lambda_{\max}$ is the largest singular value of matrix $\boldsymbol{P}$, which is also the eigenvalue of $\boldsymbol{P}$ for the reason that $\boldsymbol{P}$ is symmetric and non-negative. Note that $\boldsymbol{P}$ is the transition matrix, which is also known as the Markov matrix. And it can be easily proven that the largest eigenvalue of a Markov matrix is always 1, i.e., $\lambda_{\max} = 1$. We eliminate the absolute value sign by splitting the summation at $K = \lfloor -\frac{\log l(1-\beta)}{\log \beta} \rfloor$, then we have:

Algorithm 2. Sampling general co-occurrence matrix with structure and label context

---

Input: The transition matrix P, labeled nodes L, parameters m, l, d
Output: Co-occurrence matrix D

1: Sample random walks C of length d based on P
2: **for** every node sequence in C **do**
3:     Uniformly sample (i, j) with $|i - j| < l$
4:     $D_{v_i,v_j} = D_{v_i,v_j} + 1$
5: **end for**
6: **for** $k = 1$ to m **do**
7:     Uniformly sample a node v_i in L
8:     Uniformly sample a node v_j with the same label as node v_i
9:     $D_{v_i,v_j} = D_{v_i,v_j} + 1$
10: **end for**

$$\|\boldsymbol{S} - \boldsymbol{D}^{nor}\|_2 \le \sum_{k=0}^{K} \left[(1-\beta)\beta^k - \frac{1}{l} \right] + \sum_{k=K+1}^{l} \left[\frac{1}{l} - (1-\beta)\beta^k \right]$$

$$+ (1-\beta) \sum_{k=l+1}^{\infty} \beta^k + \frac{1}{l}$$

$$= 1 - \beta^{K+1} - \beta^{K+1}(1 - \beta^{l-K}) + \frac{l - 2K}{l} + \beta^{l+1}$$

$$= 1 - 2\beta^{K+1} + 2\beta^{l+1} + \frac{l - 2K}{l}$$

Note that when l is sufficiently large, according to the definition of K, we have $K \ll l$. Given $\beta \in (0, 1)$, we can derive:

$$\|\boldsymbol{S} - \boldsymbol{D}^{nor}\|_2 \le 2 - 2\beta^{K+1}.$$

□

With Theorem 2, we can conclude that the normalized co-occurrence matrix $\boldsymbol{D}^{nor}$ we construct is an approximation of the rooted PageRank matrix S^{RPR} with a bounded ℓ-2 norm.

Note that in TADW [22] and its follow-up works [24,25] which also apply matrix factorization to learn network embeddings, the matrix constructed to represent the structure of a network is $\frac{P+P^2}{2}$, which is a special case of $\boldsymbol{D}^{nor}$ when $l = 2$. As comparison, we construct a general matrix while preserving high-order proximities of the network with theoretical justification.

3.2   Incorporating Label Context

Apparently, the co-occurrence value between node v_i and context v_c indicates the similarity between them. A larger value of co-occurrence indicates closer proximity in the network, hence higher probability of belonging to the same class. This intuition coincides with the label information of nodes. Therefore, with the benefit of integer values in $\boldsymbol{D}$, label information can be explicitly incorporated in the procedure of sampling $\boldsymbol{D}$ to enhance the proximity between nodes, which can additionally alleviate the problem of isolated nodes without co-occurrence in structure, i.e., we consider isolated nodes through label context instead of structure context.

Specifically, we randomly sample one node among labeled instances, followed by uniformly choosing another node with the same label and update the corresponding co-occurrence count in $\boldsymbol{D}$. As a consequence, the co-occurrence matrix $\boldsymbol{D}$ captures both structure co-occurrence and label co-occurrence of instances. The complete procedure is summarized in Algorithm 2, where m is a parameter controlling the ratio between the structure and label context.

In this way, while preserving high-order proximities of the network, we can incorporate supervision into the model flexibly without leveraging any downstream classifiers, which is another important advantage of our method. By contrast, most existing methods are either purely unsupervised [22] or leveraging label information through downstream classifiers [21,24].

3.3 Joint Matrix Factorization

The method proposed above generates the co-occurrence matrix from a network and bridges the gap between word embedding and network embedding, allowing us to apply the matrix factorization paradigm to network embedding. With the flexibility of the matrix factorization principle, we propose a joint matrix factorization model that can learn network embeddings exploiting not only the topological structure and label information but also the content information of the network simultaneously.

Given the co-occurrence matrix $\boldsymbol{D} \in \mathbb{R}^{|V| \times |V|}$ and the content matrix $\boldsymbol{F} \in \mathbb{R}^{N_f \times |V|}$, where $|V|$ and N_f represent the number of nodes in the network and the dimensionality of node features respectively. Let d be the dimensionality of embedding. The objective here is to learn the embedding of a network G, denoted as the matrix $\boldsymbol{W} \in \mathbb{R}^{d \times |V|}$, by minimizing the loss of factorizing the matrices $\boldsymbol{D}$ and $\boldsymbol{F}$ jointly as:

$$\min_{\boldsymbol{W},\boldsymbol{S}} MF(\boldsymbol{D}, \boldsymbol{F}^T \boldsymbol{S} \boldsymbol{W}) \qquad (5)$$

where $MF(\cdot, \cdot)$ is the reconstruction loss of matrix factorization which will be introduced later, and $\boldsymbol{S} \in \mathbb{R}^{N_f \times d}$ can be regarded as the *feature embedding* matrix, thus $\boldsymbol{F}^T \boldsymbol{S}$ is the *feature embedding* dictionary of nodes.

By solving the joint matrix factorization problem in (5), the structure information in $\boldsymbol{D}$ and the content information in $\boldsymbol{F}$ are integrated to learn the network embeddings $\boldsymbol{W}$. This is inspired by Inductive Matrix Completion [14], a method originally proposed to complete a gene-disease matrix with gene and disease features. However, we take a completely different loss function here in light of the word embedding model of SGNS with a matrix factorization perspective [9].

We first rewrite (5) in a representation learning view as:

$$\min_{\boldsymbol{W},\boldsymbol{S}} \sum_i MF(\boldsymbol{d}_i, \boldsymbol{F}^T \boldsymbol{S} \boldsymbol{w}_i) \tag{6}$$

where $MF(\cdot,\cdot)$ is the representation loss functions evaluating the discrepancy between the ith column of $\boldsymbol{D}$ and $\boldsymbol{F}^T \boldsymbol{SW}$. $\boldsymbol{F}^T \boldsymbol{S}$ is the feature embedding dictionary, and the embedding vector of the ith node, $\boldsymbol{w}_i \in \mathbb{R}^d$, can be learned by minimizing the loss of representing its structure context vector $\boldsymbol{d}_i$ via the feature embedding $\boldsymbol{F}^T \boldsymbol{S}$.

We then proceed to the objective of factorizing the co-occurrence matrix $\boldsymbol{D}$ and the content matrix $\boldsymbol{F}$ jointly, denoted as $MF(\boldsymbol{d}_i, \boldsymbol{F}^T \boldsymbol{S} \boldsymbol{w}_i)$. We follow the paradigm of explicit matrix factorization of the SGNS model and derive the following theorem according to [9]:

Theorem 3. *For a node i in the network, we denote $Q_{i,c}$ as a pre-defined upper bound for the possible co-occurrence count between node i and context c. With the equivalence of Skip-Gram Negative Sampling (SGNS) and Explicit Matrix Factorization (EMF) [9], the representation loss $MF(\cdot,\cdot)$ can be defined as the negative log probability of observing the structure vector $\boldsymbol{d}_i$ given i and $\boldsymbol{F}^T \boldsymbol{S}$ when $Q_{i,c}$ is set to $k\frac{\#(i)\#(c)}{|D|} + \#(i,c)$. To be more concrete,*

$$MF(\boldsymbol{d}_i, \boldsymbol{F}^T \boldsymbol{S} \boldsymbol{w}_i) = -\sum_{c\in|V|} \log P(d_{i,c}|\boldsymbol{f}_c^T \boldsymbol{S} \boldsymbol{w}_i)$$

where $\boldsymbol{f}_c \in \mathbb{R}^{N_f}$ is the c-th column of the content matrix $\boldsymbol{F}$, i.e., the feature vector of node c, $\#(i,c)$ is the co-occurrence count between node i and c, $\#(i) = \sum_{c\in|V|}\#(i,c)$, $\#(c) = \sum_{i\in|V|}\#(i,c)$, $|D| = \sum_{i,c\in|V|}\#(i,c)$ and k is the negative sampling ratio.

Based on Theorem 3, we can derive:

$$MF(\boldsymbol{D}, \boldsymbol{F}^T \boldsymbol{SW}) \triangleq \sum_{i=1}^{|V|} MF(\boldsymbol{d}_i, \boldsymbol{F}^T \boldsymbol{S} \boldsymbol{w}_i)$$

$$= -\sum_{i=1}^{|V|}\sum_{c=1}^{|V|} \log P(d_{i,c}|\boldsymbol{f}_c^T \boldsymbol{S} \boldsymbol{w}_i)$$

Finally, we can formulate the objective of the joint matrix factorization framework with parameters $\boldsymbol{W}$ and $\boldsymbol{S}$ as:

$$L(\boldsymbol{W},\boldsymbol{S}) = MF(\boldsymbol{D}, \boldsymbol{F}^T \boldsymbol{SW})$$

$$= -\sum_{i=1}^{|V|}\sum_{c=1}^{|V|} \log P(d_{i,c}|\boldsymbol{f}_c^T \boldsymbol{S} \boldsymbol{w}_i) \tag{7}$$

3.4 Optimization

To minimize the loss function in (7) which integrates structure, label and content simultaneously, we utilize a novel optimization algorithm leveraging the alternating minimization scheme (ALM), which is a widely adopted method in the matrix factorization literature.

First we derive the gradients of (7) as:

$$\frac{\partial L(\boldsymbol{W},\boldsymbol{S})}{\partial \boldsymbol{S}} = \frac{\partial MF(\boldsymbol{D},\boldsymbol{F}^T\boldsymbol{S}\boldsymbol{W})}{\partial \boldsymbol{S}}$$

$$= \sum_{i\in|V|} -\boldsymbol{d}_i\boldsymbol{w}_i^T + \mathbb{E}_{\boldsymbol{d}_i'|\boldsymbol{F}^T\boldsymbol{S}\boldsymbol{w}_i}[\boldsymbol{d}_i']\boldsymbol{w}_i^T$$

$$= \boldsymbol{F}(\mathbb{E}_{\boldsymbol{D}'|\boldsymbol{F}^T\boldsymbol{S}\boldsymbol{W}}\boldsymbol{D}' - \boldsymbol{D})\boldsymbol{W}^T$$

$$\triangleq \mathrm{grad}_{\boldsymbol{S}}$$

$$\frac{\partial L(\boldsymbol{W},\boldsymbol{S})}{\partial \boldsymbol{W}} = \boldsymbol{S}^T\boldsymbol{F}(\mathbb{E}_{\boldsymbol{D}'|\boldsymbol{F}^T\boldsymbol{S}\boldsymbol{W}}\boldsymbol{D}' - \boldsymbol{D})$$

$$\triangleq \mathrm{grad}_{\boldsymbol{W}}$$

We denote $\mathrm{grad}_{\boldsymbol{W}}$ and $\mathrm{grad}_{\boldsymbol{S}}$ as the gradients of $\boldsymbol{W}$ and $\boldsymbol{S}$ in the loss function (7) respectively. Note that the expectation $\mathbb{E}_{\boldsymbol{d}_i'|\boldsymbol{F}^T\boldsymbol{S}\boldsymbol{w}_i}$ can be computed in a closed form [9] as:

$$\mathbb{E}_{d_{i,c}'|\boldsymbol{f}_c^T\boldsymbol{S}\boldsymbol{w}_i}[d_{i,c}'] = Q_{i,c}\sigma(\boldsymbol{f}_c^T\boldsymbol{S}\boldsymbol{w}_i) \tag{8}$$

where $\sigma(x) = \frac{1}{1+e^{-x}}$ is the sigmoid function.

The algorithm of Alternating Minimization (ALM) is summarized in Algorithm 3. The algorithm can be divided into solving two convex subproblems (starting from line 3 and line 6 respectively), which guarantees that the optimal solution of each subproblem can be reached with sublinear convergence rate with a properly chosen step-size [15]. One can easily show that the objective (7)

Algorithm 3. ALM algorithm for generalized explicit matrix factorization

Input: Co-occurrence matrix $\boldsymbol{D}$, content matrix $\boldsymbol{F}$, ALM step-size μ and maximum number of outer iterations I
Output: Node embedding matrix $\boldsymbol{W}$, feature embedding dictionary $\boldsymbol{S}$
1: Initialize $\boldsymbol{W}$ and $\boldsymbol{S}$ randomly
2: **for** $i = 1$ to I **do**
3:    **repeat**
4:      $\boldsymbol{W} = \boldsymbol{W} - \mu \cdot \mathrm{grad}_{\boldsymbol{W}}$
5:    **until** Convergence
6:    **repeat**
7:      $\boldsymbol{S} = \boldsymbol{S} - \mu \cdot \mathrm{grad}_{\boldsymbol{S}}$
8:    **until** Convergence
9: **end for**

descents monotonically. As a consequence, Algorithm 3 will converge due to the lower bounded objective function (7).

The time complexity of one iteration in Algorithm 3 is $O((nnz(\boldsymbol{F})d|V|)^2)$, where $nnz(\boldsymbol{F})$ is the number of non-zero elements in $\boldsymbol{F}$. For datasets with sparse node content, e.g., Cora, Citeseer, Facebook, etc., we implement $\boldsymbol{f}_c^T \boldsymbol{S}$ in Eq. (8) efficiently as a product of a sparse matrix with a dense matrix, which reduces the complexity from $O(|V|N_f d)$ to $O(nnz(\boldsymbol{F})d)$.

4   Experiments

The proposed framework is independent of specific downstream tasks, therefore in experiments, we test the model with different tasks including link prediction and node classification. Below we first introduce the datasets we use and the baseline methods that we compare to.

Datasets. We test our models on four benchmark datasets. The statistics of datasets are summarized in Table 1. For the node classification task, we employ datasets of Citation Networks [18], where nodes represent papers while edges represent citations. And each paper is described by a one-hot vector or a TFIDF word vector. For the link prediction task, we additionally include a social network dataset Facebook [7]. This dataset consists of 10 ego-networks from the online social network Facebook, where nodes and edges represent users and their relations respectively. Each user is described by users' properties, which is represented by a one-hot vector.

Baselines. For both tasks, we compare our method with network embedding algorithms including DeepWalk [17], node2vec [3], TADW [22] and HSCA [25]. For the node classification task, we further include DMF [24], LANE [5] and two neural network based methods, Planetoid [23] and GCN [6]. To measure the performance of link prediction, we also evaluate our method against some popular heuristic scores defined in node2vec [3]. Note that we do not consider TriDNR [16] as a baseline for the reason that they use text description as node content in citation networks, while in social networks such as Facebook, there is no natural text description for each user, which prevents TriDNR from generalizing to various types of networks. In addition, as MMDW [21] and DMF [24] are both semi-supervised variants of TADW with similar performance in our setting, we only compare our model with DMF for brevity.

Table 1. Dataset statistics

Dataset	# Classes	# Nodes	# Edges	# Feature
Citeseer	6	3327	4732	3703
Cora	7	2708	5429	1433
Pubmed	3	19717	44338	500
Facebook	-	4309	88234	1283

Experimental Setup. For our model, the hyper-parameters are tuned on the Citeseer dataset and kept on the others. The dimensionality of embedding is set to 200 for the proposed methods. In terms of the optimization parameters, the number of iterations is set to 200, the step-size in Algorithm 3 is set to $1e - 7$. The parameters in Algorithm 2 are set in consistency with DeepWalk, i.e., walk length $d = 40$ with window size $l = 5$. We use *APNE* to denote our unsupervised model of network embedding where the co-occurrence matrix is generated by Algorithm 1, and *APNE+label* denotes the semi-supervised model which uses Algorithm 2 to incorporate label context into the co-occurrence matrix. Unless otherwise specified, in all the experiments, we use one-vs-rest logistic regression as the classifier for the embedding based methods[1].

4.1 Semi-supervised Node Classification

We first consider the semi-supervised node classification task on three citation network datasets. To facilitate the comparison between our model and the baselines, we use the same partition scheme of training set and test set as in [23]. To be concrete, we randomly sample 20 instances from each class as training data, and 1000 instances from all samples in the rest of the dataset as test data.

The experimental results are reported in Table 2. In the comparison of unsupervised models, the proposed *APNE* method learns embeddings from the network structure and node content jointly in a unified matrix factorization framework. As a consequence, *APNE* outperforms notably on all datasets with improvement from 10.1% to 34.4%. Compared with TADW and HSCA, which both incorporate network topology and text features of nodes in a matrix

Table 2. Accuracy of semi-supervised node classification (in percentage). Upper and lower rows correspond to unsupervised and semi-supervised embedding methods respectively.

Method	Citeseer	Cora	Pubmed
DeepWalk	41.5	67.3	66.4
node2vec	47.2	69.8	70.3
TADW	54.0	72.0	41.7
HSCA	47.7	65.4	41.7
APNE	**72.6**	**79.3**	**81.5**
DMF	65.5	58.5	59.3
LANE	60.3	65.2	-
Planetoid	67.3	73.4	76.7
GCN	70.3	**81.5**	79.0
APNE+label	**72.8**	79.6	**82.1**

---

[1] Code available at https://github.com/lemmonation/APNE.

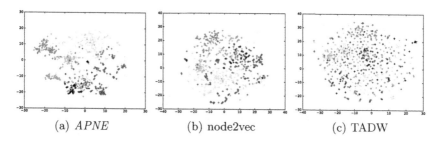

(a) *APNE*   (b) node2vec   (c) TADW

Fig. 2. t-SNE visualization of embeddings on Cora

factorization model simultaneously, our method is superior in the following: (a) the matrix we construct and factorize represents the network topology better as proven in Sect. 3.1; (b) the loss function we derive from SGNS is tailored for representation learning.

Meanwhile, in the comparison of semi-supervised methods, the proposed *APNE* model outperforms embedding based baselines significantly, illustrating the promotion brought by explicitly manipulating the constructed matrix rather than implicitly executing multi-hop interactions. In addition, LANE suffers from extensive complexity both in time and space, which prevents it from being generalized to larger networks such as Pubmed. Although being slightly inferior to GCN on the Cora dataset, considering that *APNE* is a feature learning method independent of downstream tasks and classifiers, the competitive results against the state-of-the-art CNN based method GCN justify that the node representations learned by *APNE* preserve the network information well.

In general, the proposed matrix factorization framework outperforms embedding based baselines and performs competitive with the state-of-the-art CNN based model, demonstrating the quality of embeddings learned by our methods to represent the network from the aspects of content and structure. Between the two variants of our proposed framework, *APNE* and *APNE+label*, the latter performs consistently better on all datasets, indicating the benefits of incorporating label context.

We further visualize the embeddings learned by our unsupervised model *APNE* and two unsupervised embedding-based baselines on the Cora dataset with a widely-used dimension reduction method t-SNE [11], and results are shown in Fig. 2. One can observe that different classes are better separated by our model, and nodes in the same class are clustered more tightly.

In order to test the sensitivity of our framework to hyper-parameters, we choose different values of the negative sampling parameter k in Theorem 3 and the number of iterations of label context sampling m in Algorithm 2 and evaluate the model on Citeseer on the node classification task.

For each pair of parameters, we repeat the experiments 10 times and compute the mean accuracy. Results are shown in Fig. 3. The horizontal axis represents different values of m. And $m = 0$ represents results when the model is purely unsupervised, otherwise results are from semi-supervised models. The vertical

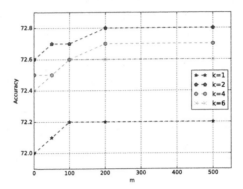

Fig. 3. Parameter effect of *APNE* on the node classification task (in percentage)

axis is the classification accuracy on Citeseer. Clearly, increasing m brings a boost of the performance of the model, as we infer in Sect. 3.2. This justifies the effectiveness of the approach we propose to incorporate the label context. In addition, the performance of the proposed models with different values of k is relatively stable.

4.2   Link Prediction

We further test our unsupervised model on the link prediction task. In link prediction, a snap-shot of the current network is given, and we are going to predict edges that will be added in the future. The experiment is set up as follows: we first remove 50% of existing edges from the network randomly as positive node pairs, while ensuring the residual network connected. To generate negative examples, we randomly sample an equal number of node pairs that are not connected. Node representations are then learned based on the residual network. While testing, given a node pair in the samples, we compute the cosine similarity between their representation vectors as the edge's score. Finally, Area Under Curve (AUC) score and Mean Average Precision (MAP) are used to evaluate the consistency between the labels and the similarity scores of the samples.

Results are summarized in Table 3. As shown in the table, our method *APNE* outperforms all the baselines consistently with different evaluation metrics. We take a lead of topology-only methods by a large margin, especially on sparser networks such as Citeseer, which indicates the importance of leveraging node features on networks with high sparsity. Again, we consistently outperform TADW and HSCA which also consider text features of nodes.

The stable performance of our proposed *APNE* model on different datasets justify that embeddings learned by jointly factorizing the co-occurrence matrix D and node features F can effectively represent the network. More importantly, the problem of sparsity can be alleviated by incorporating node features in a unified framework.

Table 3. Results of link prediction

Method	Citeseer		Cora		Pubmed		Facebook	
	AUC	MAP	AUC	MAP	AUC	MAP	AUC	MAP
Common neighbor	0.567	0.781	0.616	0.797	0.561	0.778	0.797	0.882
Jaccard's coefficient	0.567	0.782	0.616	0.795	0.561	0.776	0.797	0.877
Adamic adar	0.560	0.780	0.617	0.801	0.561	0.778	0.798	0.885
Preferential attachment	0.675	0.721	0.679	0.705	0.863	0.852	0.675	0.675
DeepWalk	0.656	0.725	0.734	0.793	0.721	0.781	0.891	0.914
node2vec	0.502	0.731	0.723	0.790	0.728	0.785	0.888	0.911
TADW	0.914	0.936	0.854	0.878	0.592	0.620	0.909	0.921
HSCA	0.905	0.928	0.861	0.885	0.632	0.660	0.926	0.917
APNE	**0.938**	**0.940**	**0.909**	**0.910**	**0.925**	**0.916**	**0.956**	**0.949**

4.3 Case Study

To further illustrate the effectiveness of *APNE*, we present some instances of link prediction on the Cora dataset. We randomly choose 2 node pairs from all node samples and compute the cosine similarity for each pair. Results are summarized in Table 4. The superiority of *APNE* is obvious in the first instance, where TADW gives a negative correlation to a positive pair. For this pair, although the first paper is cited by the second one, their neighbors do not coincide. As a consequence it is easy to wrongly separate these two nodes into different categories if the structure information is not sufficiently exploited.

As for the second instance, both papers belong to the Neural Networks class but not connected in the network. Specifically, the first paper focuses on H-Infinity methods in control theory while the second paper is about recurrent neural networks, and there exist papers linking these two domains together in the dataset. As a consequence, although these two nodes can hardly co-occur in random walk sequences on the network, their features may overlap in the dataset. Therefore, the pair of nodes will have a higher feature similarity than the topology similarity. Thus by jointly considering the network topology and the node features, our method gives a higher correlation score to the two nodes that are disconnected but belong to the same category.

Table 4. Two randomly chosen node pairs from Cora dataset

Title	Same class	Connected	Cosine similarity	
			APNE	TADW
A cooperative coevolutionary approach to function	$\checkmark$	$\checkmark$	**0.471**	−0.002
Multi-parent reproduction in genetic algorithms				
A class of algorithms for identification in H_∞	$\checkmark$	×	**0.158**	−0.129
On the computational power of neural nets				

5   Conclusion

In this paper, we aim to learn a generalized network embedding preserving structure, content and label information simultaneously. We propose a unified matrix factorization based framework which provides a flexible integration of network structure, node content, as well as label information. We bridge the gap between word embedding and network embedding by designing a method to generate the co-occurrence matrix from the network, which is actually an approximation of high-order proximities of nodes in the network. The experimental results on four benchmark datasets show that the joint matrix factorization method we propose brings substantial improvement over existing methods. One of our future directions would be to apply our framework to social recommendations to combine the relationship between users with the corresponding feature representations.

Acknowledgements. This research was supported by the National Natural Science Foundation of China (No. 61673364, No. U1605251 and No. 61727809), and the Fundamental Research Funds for the Central Universities (WK2150110008).

References

1. Cai, H., Zheng, V.W., Chang, K.C.C.: A comprehensive survey of graph embedding: problems, techniques and applications. preprint arXiv:1709.07604 (2017)
2. Cao, S., Lu, W., Xu, Q.: GraRep: Learning graph representations with global structural information. In: Proceedings of the 24th ACM International on Conference on Information and Knowledge Management, pp. 891–900. ACM (2015)
3. Grover, A., Leskovec, J.: node2vec: scalable feature learning for networks. In: Proceedings of the 22nd ACM SIGKDD International Conference on Knowledge Discovery and Data Mining, pp. 855–864. ACM (2016)
4. Guo, J., Xu, L., Chen, E.: SPINE: structural identity preserved inductive network embedding. arXiv preprint arXiv:1802.03984 (2018)
5. Huang, X., Li, J., Hu, X.: Label informed attributed network embedding. In: Proceedings of the Tenth ACM International Conference on Web Search and Data Mining, pp. 731–739. ACM (2017)
6. Kipf, T.N., Welling, M.: Semi-supervised classification with graph convolutional networks. arXiv preprint arXiv:1609.02907 (2016)
7. Leskovec, J., Krevl, A.: SNAP datasets: stanford large network dataset collection, June 2014. http://snap.stanford.edu/data
8. Levy, O., Goldberg, Y.: Neural word embedding as implicit matrix factorization. In: Advances in Neural Information Processing Systems, pp. 2177–2185 (2014)
9. Li, Y., Xu, L., Tian, F., Jiang, L., Zhong, X., Chen, E.: Word embedding revisited: a new representation learning and explicit matrix factorization perspective. In: IJCAI 2015, pp. 3650–3656 (2015)
10. Liu, L., Xu, L., Wangy, Z., Chen, E.: Community detection based on structure and content: a content propagation perspective. In: IEEE International Conference on Data Mining (ICDM), pp. 271–280. IEEE (2015)
11. Maaten, L.V.D., Hinton, G.: Visualizing data using t-SNE. J. Mach. Learn. Res. 9(Nov), 2579–2605 (2008)

12. Mikolov, T., Chen, K., Corrado, G., Dean, J.: Efficient estimation of word representations in vector space. arXiv preprint arXiv:1301.3781 (2013)
13. Mikolov, T., Sutskever, I., Chen, K., Corrado, G.S., Dean, J.: Distributed representations of words and phrases and their compositionality. In: Advances in Neural Information Processing Systems, pp. 3111–3119 (2013)
14. Natarajan, N., Dhillon, I.S.: Inductive matrix completion for predicting gene-disease associations. Bioinformatics **30**(12), i60–i68 (2014)
15. Nesterov, Y.: Introductory Lectures on Convex Optimization: A Basic Course, vol. 87. Springer, Heidelberg (2013). https://doi.org/10.1007/978-1-4419-8853-9
16. Pan, S., Wu, J., Zhu, X., Zhang, C., Wang, Y.: Tri-party deep network representation. Network **11**(9), 12 (2016)
17. Perozzi, B., Al-Rfou, R., Skiena, S.: DeepWalk: online learning of social representations. In: Proceedings of the 20th ACM SIGKDD International Conference on Knowledge Discovery and Data Mining, pp. 701–710. ACM (2014)
18. Sen, P., Namata, G., Bilgic, M., Getoor, L., Galligher, B., Eliassi-Rad, T.: Collective classification in network data. AI Mag. **29**(3), 93 (2008)
19. Song, H.H., Cho, T.W., Dave, V., Zhang, Y., Qiu, L.: Scalable proximity estimation and link prediction in online social networks. In: Proceedings of the 9th ACM SIGCOMM Conference on Internet Measurement Conference. ACM (2009)
20. Tang, J., Qu, M., Wang, M., Zhang, M., Yan, J., Mei, Q.: LINE: large-scale information network embedding. In: Proceedings of the 24th International Conference on World Wide Web, pp. 1067–1077. ACM (2015)
21. Tu, C., Zhang, W., Liu, Z., Sun, M.: Max-margin DeepWalk: discriminative learning of network representation. In: IJCAI 2016, pp. 3889–3895 (2016)
22. Yang, C., Liu, Z., Zhao, D., Sun, M., Chang, E.: Network representation learning with rich text information. In: IJCAI 2015, pp. 2111–2117 (2015)
23. Yang, Z., Cohen, W., Salakhutdinov, R.: Revisiting semi-supervised learning with graph embeddings. arXiv preprint arXiv:1603.08861 (2016)
24. Zhang, D., Yin, J., Zhu, X., Zhang, C.: Collective classification via discriminative matrix factorization on sparsely labeled networks. In: Proceedings of the 25th ACM International on Conference on Information and Knowledge Management, pp. 1563–1572. ACM (2016)
25. Zhang, D., Yin, J., Zhu, X., Zhang, C.: Homophily, structure, and content augmented network representation learning. In: IEEE 16th International Conference on Data Mining (ICDM), pp. 609–618. IEEE (2016)

Attributed Network Embedding with Micro-meso Structure

Juan-Hui Li[1], Chang-Dong Wang[1(✉)], Ling Huang[1], Dong Huang[2],
Jian-Huang Lai[1,3], and Pei Chen[1]

[1] School of Data and Computer Science, Sun Yat-sen University, Guangzhou, China
sysuLiJuanHui@163.com, changdongwang@hotmail.com, huanglingh1@hotmail.com,
stsljh@mail.sysu.edu.cn, chenpei@mail.sysu.edu.cn
[2] College of Mathematics and Informatics, South China Agricultural University,
Guangzhou, China
huangdonghere@gmail.com
[3] XinHua College, Sun Yat-sen University, Guangzhou, China

Abstract. Recently, network embedding has received a large amount of attention in network analysis. Although some network embedding methods have been developed from different perspectives, on one hand, most of the existing methods only focus on leveraging the plain network structure, ignoring the abundant attribute information of nodes. On the other hand, for some methods integrating the attribute information, only the lower-order proximities (e.g. microscopic proximity structure) are taken into account, which may suffer if there exists the sparsity issue and the attribute information is noisy. To overcome this problem, the attribute information and mesoscopic community structure are utilized. In this paper, we propose a novel network embedding method termed Attributed Network Embedding with Micro-Meso structure (ANEM), which is capable of preserving both the attribute information and the structural information including the microscopic proximity structure and mesoscopic community structure. In particular, both the microscopic proximity structure and node attributes are factorized by Nonnegative Matrix Factorization (NMF), from which the low-dimensional node representations can be obtained. For the mesoscopic community structure, a community membership strength matrix is inferred by a generative model from the linkage structure, which is then factorized by NMF to obtain the low-dimensional node representations. The three components are jointly correlated by the low-dimensional node representations, from which an objective function can be defined. An efficient alternating optimization scheme is proposed to solve the optimization problem. Extensive experiments have been conducted to confirm the superior performance of the proposed model over the state-of-the-art network embedding methods.

Keywords: Network embedding · Node attribute
Microscopic proximity structure · Mesoscopic community structure

© Springer International Publishing AG, part of Springer Nature 2018
J. Pei et al. (Eds.): DASFAA 2018, LNCS 10827, pp. 20–36, 2018.
https://doi.org/10.1007/978-3-319-91452-7_2

1 Introduction

Network embedding aims to learn a low-dimensional node representation that reflects the inherent properties of a network, which plays a key role in many network analysis tasks such as visualization, node classification, link prediction, and entity retrieval [1–7]. In particular, it can well address the sparsity issue associated with network structure. Another benefit is that by transforming the topological linkage structure of network into the low-dimensional node representations, the node-interdependence is implicitly encoded into the node representations, from which both the large-scale distributed computing models (e.g., MapReduce) and off-the-shelf machine learning methods (e.g., node classification) can be directly applied.

Some recent methods propose to exploit different network structural properties to enhance network embedding, yet they mostly, if not all, ignore an important and inherent property of the network, i.e., the node attributes, which generally contains rich semantically meaningful information, such as user attributes in social network and paper titles in citation network [8,9]. Although some attempts have been made to preserve node attributes in network embedding [10,11], yet they attempt to infer low-dimensional node representations from the lower-order proximities, e.g., first- and second-order proximities [12–14], or the higher-order proximities [15,16]. With only the the microscopic structure preserved, they may still suffer when the attribute is noisy and the network has the sparsity issue. To cope with this problem, Wang et al. [17] further proposed the modularized non-negative matrix factorization (M-NMF) method to incorporate the mesoscopic community structure into network embedding, which encodes versatile organizational and functional properties of the network. However, the abundant attribute information is lost. It remains an open problem how to simultaneously incorporate the microscopic proximity structure, the mesoscopic community structure, and the node attributes into a unified and unsupervised network embedding framework.

In this paper, we propose a novel network embedding method termed Attributed Network Embedding with Micro-meso structure (ANEM) that triply preserves the microscopic proximity structure, the mesoscopic community structure and the node attributes. For the microscopic proximity structure, both the first- and second-order proximities of nodes are considered, which are summed together to form proximity matrix. To preserve the mesoscopic community structure, the recently proposed generative model termed BigCLAM [18] is used to infer the community membership strength matrix from the linkage structure. While the node attributes are characterized by a matrix with rows indicating the node ids and columns indicating the attribute dimensionality. By introducing other three matrices, these three components are factorized into the low-dimensional node representations under the framework of Nonnegative Matrix Factorization (NMF) [19], so that the learned results aggregate the information of the three components in a seamless way. NMF is used here since it generally models the generation of directly observable variables from the hidden variables [20], which coincides with our goal to learn the node representations. Under such a scheme, the three components are jointly correlated by the low-dimensional node representations, from which an objective function can be defined. An efficient alternating optimization scheme is proposed to solve the optimization problem. Extensive experiments conducted on six real-world datasets show that the proposed ANEM method outperforms most of the state-of-the-art network embedding methods in the tasks of node classification and clustering.

2   Related Work

Recently, some network embedding methods have been developed from different perspectives. Perozzi et al. [1] proposed a DeepWalk algorithm, where the truncated random walks are deployed to generate the node sequences, which are further fed into a neural language model (Skip-gram) [21] to produce the latent node embeddings. Thereafter, Tang et al. [13] proposed a large-scale information network embedding method called LINE that preserves both the first-order and second-order proximity. In [15], Cao et al. developed a GraRep model, which defines different loss functions to capture different k-step local relational information for different k, and then obtains the global representation by integrating the learned representations from each loss function. In [22], the proposed Node2Vec model generates the node sequence by balancing the breadth-first sampling and the depth-first sampling, and then learns the node representations through maximizing the likelihood of preserving network neighborhoods of nodes. To capture the highly non-linear network structure and preserve the global and local structures, Wang et al. [14] designed a Structural Deep Network Embedding (SDNE) model, where the multiple layers of non-linear functions are utilized. By preserving the asymmetric transitivity through approximating the higher-order proximity, Ou et al. [23] proposed a High-Order Proximity preserved Embedding (HOPE) model. Recently, Wang et al. [17] developed a M-NMF model which combines the mesoscopic community structure and the microscopic proximity structure simultaneously for learning low-dimensional node representations. However, most of the aforementioned methods only utilize the microscopic structure, with less methods combing both the microscopic structure and the mesoscopic community structure (e.g., [17]). In addition, rich information is lost by ignoring the valuable node attributes.

Some recent efforts have been made in integrating the node attributes information to learn the low-dimensional representations. In [24], the proposed TADW (Text-Associated DeepWalk) model is extended from DeepWalk [1] by means of combing the text features of each node via the matrix factorization. Thereafter, in [10], the proposed TriDNR model aggregates the inter-node relationships, node-content correlation, and label-content correspondence to learn the optimal node representation. In [25], the network embedding is aggregated from the ID embedding and the attribute embedding, both of which are learned through the multi-layer neural network. In [26], Li et al. proposed the property preserved algorithm through jointly optimizing the topology-derived objective function and the property-derived function. Unlike most of the existing unsupervised algorithms, Huang et al. [11] proposed the LANE framework by incorporating the label information into the attributed network embedding while preserving their correlations. However, most of these methods fail to capture the information of mesoscopic community structure.

Although the above methods work well in incorporating one or two of the three components (i.e., the microscopic proximity structure, mesoscopic community structure and node attributes), they fail to integrate all the information from these three components, resulting in less discriminative embeddings. To the best of our knowledge, there is still a lack of network embedding methods that can triply preserve microscopic proximity structure, the mesoscopic community structure and the node attributes in an unsupervised manner.

3   The Proposed Model

Let $\mathcal{G} = (\mathcal{V}, \mathcal{E}, \mathbf{D})$ denote an attributed network consisting of n nodes, where $\mathcal{V}$ denotes the set of nodes, $\mathcal{E}$ denotes the edge set and $\mathbf{D} \in \mathbb{R}^{n \times m}$ denotes the node attribute matrix with m being the dimensionality of the node attributes. Let $\mathbf{A} \in \mathbb{R}^{n \times n}$ denote the adjacency matrix. The goal is to learn the low-dimensional node representation denoted as $\mathbf{U} \in \mathbb{R}^{n \times d}$ that can comprehensively reflect the inherent properties of the attributed network, where d is the dimensionality of the representation vectors.

The main idea of ANEM is presented in Fig. 1. As shown in this figure, there is an attributed network consisting of 9 nodes associated with the node attributes. Similar nodes in this network belong to the same community. In the attributed network embedding, three inherent components, i.e., the microscopic proximity structure, the mesoscopic community structure and the node attributes, are preserved by NMF in a unified model, which is solved by the alternating optimization scheme to obtain the embedding representations $\mathbf{U}$. In Fig. 1, nodes 1 and 2 have similar properties in the original attributed network which results in similar node representation vectors [0.53 0.27 0.64 0.5] and [0.56 0.30 0.4 0.55]. In what follows, the proposed Attributed Network Embedding with Micro-meso structure (ANEM) model will be introduced in detail.

3.1   Modeling the Microscopic Proximity Structure

For the microscopic proximity structure, both the first-order and second-order proximities [13] are considered simultaneously. Specifically, the first-order proximity is the local pairwise proximity between two nodes. It describes the proximity of two linked nodes, i.e., $A_{ij} > 0$ indicates the first-order proximity between node i and j, otherwise, their first-order proximity is 0. However, the first-order proximity information observed in real world is only a small proportion, which leads to the sparse linkage structure. For the nodes without edges, the proximity information is lost even though they are intrinsically very similar to each other.

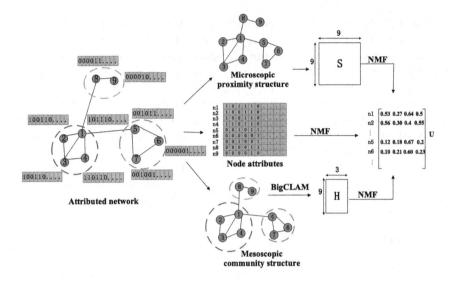

Fig. 1. Illustration of ANEM.

A remedy is to considering their common neighbors, i.e., the nodes sharing similar neighbors tend to be similar to each other.

Formally, the first-order proximity is denoted by $\mathbf{S}^{(1)} \in \mathbb{R}^{n \times n}$, and as discussed above, we have $\mathbf{S}^{(1)} = \mathbf{A}$. Let $\mathcal{N}_i = [\mathbf{S}_{i1}^{(1)}, \ldots, \mathbf{S}_{in}^{(1)}]$ denote the first-order proximity of node i with other nodes. Then the second-order proximity is defined as [13]

$$\mathbf{S}_{ij}^{(2)} = \frac{\mathcal{N}_i \mathcal{N}_j}{||\mathcal{N}_i||\,||\mathcal{N}_j||}. \tag{1}$$

To preserve both the first-order proximity and the second-order proximity, the proximity matrix $\mathbf{S} \in \mathbb{R}^{n \times n}$ is defined as follows,

$$\mathbf{S} = \mathbf{S}^{(1)} + \eta \mathbf{S}^{(2)} \tag{2}$$

where η is a balancing parameter controlling the importance of the second-order proximity, and following [17], we set $\eta = 5$. Under the framework of Nonnegative Matrix Factorization (NMF) [19], the proximity matrix $\mathbf{S}$ can be decomposed into a nonnegative basis matrix $\mathbf{M} \in \mathbb{R}^{n \times d}$ and the nonnegative node representation matrix $\mathbf{U} \in \mathbb{R}^{n \times d}$ [19]:

$$\min_{\mathbf{M}, \mathbf{U}} ||\mathbf{S} - \mathbf{M}\mathbf{U}^T||_F^2 \quad \text{s.t.} \quad \mathbf{M} \geq 0, \mathbf{U} \geq 0. \tag{3}$$

3.2   Modeling the Mesoscopic Community Structure

For preserving the mesoscopic community structure, the BigCLAM [18] generative model is used to infer a community membership strength matrix, which is a recently proposed but popular method for community detection in networks.

In the previous work [17], the classical modularity is used to model the community structure, which however leads to a very sparse binary membership indicator matrix. On the contrary, the BigCLAM model more generally allows generating the memberships with strengths, i.e., each community attracts its member nodes depending on the value of $\mathbf{H} \in \mathbb{R}^{n \times c}$, whose element $\mathbf{H}_{ur}$ in row $\mathbf{H}_u$ indicates the probability of node u belonging to community r. Parameter c is set according to the community number we have already known in the dataset. Specifically, to find the most likely membership matrix, the maximization of the likelihood function is designed as follows:

$$\max_{\mathbf{H}} \sum_{(u,v) \in \mathcal{E}} \log(1 - \exp(-\mathbf{H}_u \mathbf{H}_v^T)) - \sum_{(u,v) \notin \mathcal{E}} \mathbf{H}_u \mathbf{H}_v^T \quad \text{s.t.} \quad \mathbf{H} \geq 0. \quad (4)$$

By introducing an auxiliary nonnegative matrix $\mathbf{C} \in \mathbb{R}^{c \times d}$, which is a community representation matrix, $\mathbf{U}_u \mathbf{C}_r^T$ can be considered as a description of the propensity that node u belongs to community r. As the membership matrix $\mathbf{H}$ encodes the probabilities of nodes belonging to the communities, $\mathbf{UC}^T$ should be as closely consistent as possible with $\mathbf{H}$. To this end, the following minimization is designed:

$$\min_{\mathbf{U},\mathbf{C}} ||\mathbf{H} - \mathbf{UC}^T||_F^2 \quad \text{s.t.} \quad \mathbf{U} \geq 0, \mathbf{C} \geq 0. \quad (5)$$

3.3 Modeling the Node Attributes

The nodes in the network contain rich semantically meaningful information, which plays a key role in network embedding [10,11,25,27–29]. Let $\mathbf{D} \in \mathbb{R}^{n \times m}$ denote the node attributes, where the i-th row $\mathbf{D}_i$ is the attribute vector of node i. By introducing a nonnegative basis matrix $\mathbf{N} \in \mathbb{R}^{m \times d}$, we use the NMF framework to approximate the node attribute matrix $\mathbf{D}$, which gives rise to the following objective function:

$$\min_{\mathbf{N},\mathbf{U}} ||\mathbf{D}^T - \mathbf{NU}^T||_F^2 \quad \text{s.t.} \quad \mathbf{N} \geq 0, \mathbf{U} \geq 0. \quad (6)$$

3.4 The Overall Objective Function

As can be seen, each of the three components defined above involves the low-dimensional node representation $\mathbf{U}$. To make $\mathbf{U}$ contain the information from the microscopic proximity structure, the mesoscopic community structure and the node attributes, the consensus relationship among these three components should be established. To this end, the overall objective function is designed as follows,

$$\min_{\mathbf{M},\mathbf{U},\mathbf{H},\mathbf{C},\mathbf{N}} L = ||\mathbf{S} - \mathbf{MU}^T||_F^2 + \alpha ||\mathbf{H} - \mathbf{UC}^T||_F^2$$

$$- \beta \left(\sum_{(u,v) \in \mathcal{E}} \log(1 - \exp(-\mathbf{H}_u \mathbf{H}_v^T)) - \sum_{(u,v) \notin \mathcal{E}} \mathbf{H}_u \mathbf{H}_v^T \right) + \gamma ||\mathbf{D}^T - \mathbf{NU}^T||_F^2$$

$$\text{s.t.} \quad \mathbf{M} \geq 0, \mathbf{U} \geq 0, \mathbf{H} \geq 0, \mathbf{C} \geq 0, \mathbf{N} \geq 0 \quad (7)$$

where α, β, γ are the parameters that adjust the contributions of each component, the effect of which will be analyzed in our experiments. Notice that NMF is used for the three components since it generally models the generation of directly observable variables from the hidden variables [20], which coincides with our goal to learn the node representations.

4   Optimization

By using the alternating optimization scheme, the objective function Eq. (7) can be decomposed into five subproblems, i.e. updating one variable when fixing the remaining four variables.

Updating M: Updating $\mathbf{M}$ when fixing the other variables leads to the standard NMF optimization [19], which is implemented as follows,

$$\mathbf{M} \leftarrow \mathbf{M} \odot \frac{\mathbf{SU}}{\mathbf{MU}^T\mathbf{U}} \tag{8}$$

Updating U: Updating $\mathbf{U}$ when fixing the other variables leads to the joint NMF optimization [30], which is implemented as follows,

$$\mathbf{U} \leftarrow \mathbf{U} \odot \frac{\mathbf{S}^T\mathbf{M} + \alpha\mathbf{HC} + \gamma\mathbf{DN}}{\mathbf{U}(\mathbf{M}^T\mathbf{M} + \alpha\mathbf{C}^T\mathbf{C} + \gamma\mathbf{N}^T\mathbf{N})} \tag{9}$$

Updating C: Updating $\mathbf{C}$ when fixing the other variables leads to the standard NMF optimization, which is implemented as follows,

$$\mathbf{C} \leftarrow \mathbf{C} \odot \frac{\mathbf{H}^T\mathbf{U}}{\mathbf{CU}^T\mathbf{U}} \tag{10}$$

Updating N: Updating $\mathbf{N}$ when fixing the other variables leads to the standard NMF optimization, which is implemented as follows,

$$\mathbf{N} \leftarrow \mathbf{N} \odot \frac{\mathbf{D}^T\mathbf{U}}{\mathbf{NU}^T\mathbf{U}} \tag{11}$$

Updating H: Updating $\mathbf{H}$ when fixing the other variables leads to the following optimization problem:

$$\min_{\mathbf{H}} \ \alpha||\mathbf{H} - \mathbf{UC}^T||_F^2 - \beta\left(\sum_{(u,v)\in\mathcal{E}} \log(1 - \exp(-\mathbf{H}_u\mathbf{H}_v^T)) - \sum_{(u,v)\notin\mathcal{E}} \mathbf{H}_u\mathbf{H}_v^T\right)$$
$$\text{s.t.} \quad \mathbf{H} \geq 0. \tag{12}$$

By rewriting the above objective function into the following objective function w.r.t. each row of $\mathbf{H}$

$$L(\mathbf{H}_u) = \alpha||\mathbf{H}_u - (\mathbf{UC}^T)_u||^2 - \beta\left(\sum_{(u,v)\in\mathcal{E}} \log(1 - \exp(-\mathbf{H}_u\mathbf{H}_v^T)) - \sum_{(u,v)\notin\mathcal{E}} \mathbf{H}_u\mathbf{H}_v^T\right)$$
$$\text{s.t.} \quad \mathbf{H}_u \geq 0 \tag{13}$$

we can obtain the gradient of the variable H_u as follows,

$$\nabla L(\mathbf{H}_u) = 2\alpha(\mathbf{H}_u - (\mathbf{U}\mathbf{C}^T)_u) - \beta\left(\sum_{v\in\mathcal{N}(u)} \mathbf{H}_v \frac{\exp(-\mathbf{H}_u\mathbf{H}_v^T)}{1 - \exp(-\mathbf{H}_u\mathbf{H}_v^T)} - \sum_{v\notin\mathcal{N}(u)} \mathbf{H}_v\right)$$

$$= 2\alpha(\mathbf{H}_u - (\mathbf{U}\mathbf{C}^T)_u)$$
$$- \beta\left(\sum_{v\in\mathcal{N}(u)} \mathbf{H}_v \frac{\exp(-\mathbf{H}_u\mathbf{H}_v^T)}{1 - \exp(-\mathbf{H}_u\mathbf{H}_v^T)} - \left(\sum_v \mathbf{H}_v - \mathbf{H}_u - \sum_{v\in\mathcal{N}(u)}\mathbf{H}_v\right)\right)$$

$$= 2\alpha(\mathbf{H}_u - (\mathbf{U}\mathbf{C}^T)_u)$$
$$- \beta\left(\sum_{v\in\mathcal{N}(u)} \mathbf{H}_v \frac{1}{1 - \exp(-\mathbf{H}_u\mathbf{H}_v^T)} - \left(\sum_v \mathbf{H}_v - \mathbf{H}_u\right)\right) \qquad (14)$$

By applying the gradient descent algorithm, $\mathbf{H}_u$ can be updated as follows,

$$\mathbf{H}_u \leftarrow \mathbf{H}_u - \lambda\nabla L(\mathbf{H}_u) \qquad (15)$$

where λ is the learning rate, which is set $\lambda = 40$ uniformly in this paper.

The above five variables are randomly initialized and then iteratively updated until the number of iterations reaches the predefined maximum number of iterations (i.e., 100 here) or the relative difference of the objective function in two adjacent iteration steps, i.e.

$$\left|\frac{L \text{ in the current iteration} - L \text{ in the last iteration}}{L \text{ in the last iteration}}\right| \qquad (16)$$

is smaller than 10^{-3}.

4.1 Complexity Analysis

The time complexity of ANEM mainly depends on the matrix computation in the update procedure, namely the updating rules in Eqs. (8), (9), (10), (11) and (15). For these five equations, by introducing the average number of the neighbors of each node denoted by p, the computational time complexities are $O(n^2d+d^2n)$, $O(n^2d+ncd+nmd+d^2n+d^2c+d^2m)$, $O(cnd+d^2n)$, $O(nmd+d^2n)$ and $O(n^2dc + npc + n^2c)$, respectively. Since in most cases, $d, c < n$, the major computation of ANEM is in Eq. (15). Therefore, the overall computational time complexity of ANEM is $O(n^2dc)$.

5   Experiments

5.1   Experimental Settings

Six publicly available networks with node attributes are used in our experiments, which are four subnetworks from the WebKB network[1], the Terrorist Attack network [31], and the Citeseer network [32].

---

[1] http://www.cs.cmu.edu/~webkb/.

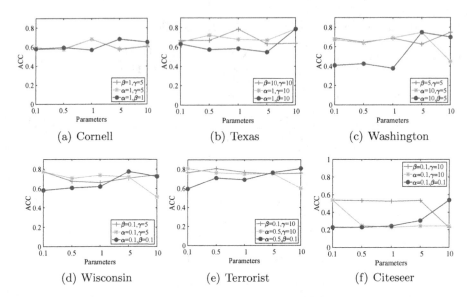

Fig. 2. Parameter analysis of α, β, γ: the classification task in terms of ACC.

1. **WebKB:** This dataset contains the hypertexts collected from various universities, i.e. Cornell, Texas, Washington and Wisconsin, each taken as a subnetwork. It consists of 877 webpages and 1608 edges, where Cornell has 195 webpages and 304 edges, Texas has 187 webpages and 328 edges, Washington has 230 webpages and 446 edges, and Wisconsin has 265 webpages and 530 edges. Each webpage is associated with a 1703-dimensional attribute vector. Moreover, each subnetwork is divided into 5 communities according to the following labels: Course, Student, Faculty, Project and Staff.
2. **Terrorist Attack:** This dataset is the affiliation network classified into 6 communities from the Profile in Terror project. It consists of 1293 terror attacks and 3172 links. Each attack is associated with 106-dimensional 0–1 vector indicating the attributes that are present and the attributes that are absent. We briefly use Terrorist to represent this dataset in the following paper.
3. **Citeseer:** This dataset is a citation network consisting of 3312 scientific publications and 4732 links, where each publication is associated with a 3703-dimensional 0–1 word feature vector. The publications are from distinct research areas which are divided into 6 classes.

The performance of network embedding is evaluated on the tasks of both node classification and clustering where the k-nearest neighbor (KNN) classifier and the spectral clustering based on the normalized cut are used respectively. In the network embedding, the default value of the dimensionality d is set to be 100. In the KNN classification task, k is set to be 3, 4, 5, 6, 7, and 80% of the learned node representations are randomly selected as the training data with class labels, with the remaining nodes as the testing data without class labels.

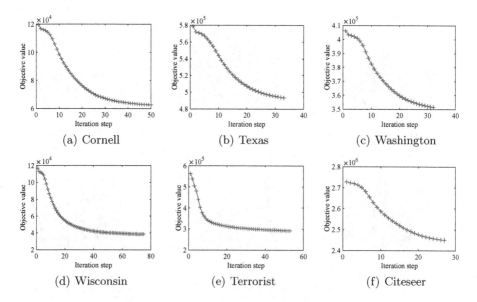

Fig. 3. Convergence analysis on the six datasets.

In the clustering task, the similarity matrix is built from the k-nearest neighbor graph [33] with k set to be 3, 4, 5, 6, 7. The classification results and the clustering results are evaluated by comparing the obtained class labels with the ground-truth class labels, in terms of both accuracy (ACC) [34] and Purity. Higher ACC and Purity values indicate better classification and clustering performance.

5.2   Parameter Analysis

We first analyze the impact of the parameters α, β, γ on the performance of our ANEM model in terms of ACC with each parameter varying from {0.1, 0.5, 1, 5, 10}. Figure 2 shows the ACC results by varying one parameter when fixing the other two parameters. Taking the results shown in Fig. 2(b) as an example, when fixing the values of parameters $\beta = 10$ and $\gamma = 10$, ANEM has achieved the best result when $\alpha = 1$ on the Texas dataset. Similarly, when $\beta = 10$ and $\gamma = 10$, the corresponding best results are achieved. Therefore, the best values of the three parameters are {$\alpha = 1, \beta = 10, \gamma = 10$} for the Texas dataset. For the remaining datasets, the values of α, β, γ are determined using the same evaluation strategy as Texas. The reason for using the different values of α, β, γ for different datasets is due to the diverse impact of the microscopic proximity structure, the mesoscopic community structure, and the node attributes in different networks. Hence, in the following experiments, the values of α, β, γ are set when the best result is obtained on each dataset separately.

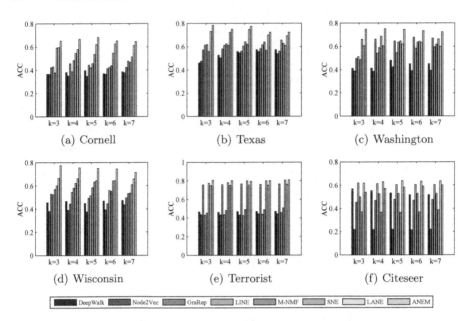

Fig. 4. Comparison results: the classification task in terms of ACC.

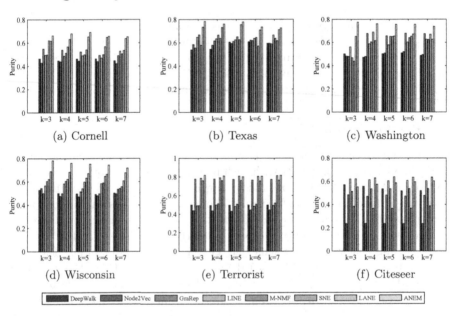

Fig. 5. Comparison results: the classification task in terms of Purity.

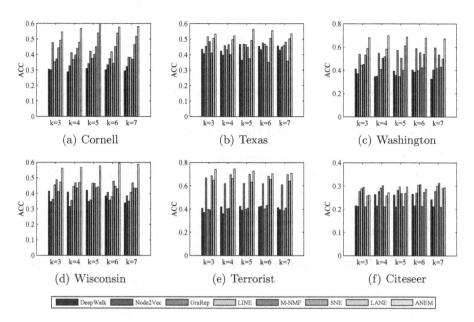

Fig. 6. Comparison results: the clustering task in terms of ACC.

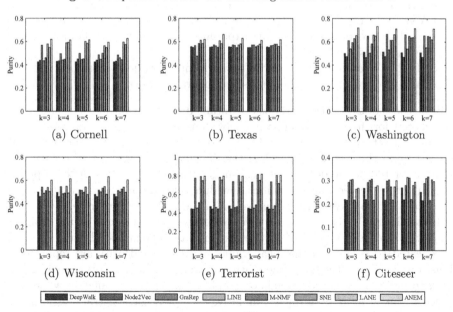

Fig. 7. Comparison results: the clustering task in terms of Purity.

5.3 Convergence Analysis

To evaluate the convergence property of our model, the objective value as a function of the iteration step on each of the six datasets is reported in Fig. 3.

In this figure, we can see that on most of the datasets, the algorithm tends to converge when the iteration number is larger than 20, and converges to a stable value when the iteration number reaches 30. In particular, on the Citeseer dataset as shown in Fig. 3(f), the algorithm converges relatively faster and reaches the stable state when the iteration number is larger than 20.

5.4 Comparison Results with the Existing Algorithms

In this section, comparison experiments are conducted to compare the performance of our model with seven state-of-the-art network embedding algorithms.

1. DeepWalk [1]: It combines the truncated random walks based on the topological structure and the Skip-gram to produce the node embeddings.
2. Node2Vec [22]: It preserves the neighborhoods of nodes based on the topological structure to generate node embeddings.
3. GraRep [15]: It utilizes the topological structure to define different loss functions and learns the node embeddings from the loss functions.
4. LINE [13]: The first-order and second-order proximities of the topological structure are both preserved to learn the node embeddings.
5. M-NMF [17]: Both the microscopic topological structure and the mesoscopic community structure are preserved to learn the node embeddings.
6. LANE [11]: The information of the network topology, node attributes and the labels are preserved. In our experiment, the version without the label information is used.
7. SNE [25]: The information of the network topology, node attributes are preserved through utilizing the multi-layer neural network.

All the codes of the above methods are obtained from the authors' websites. The parameters for these seven compared algorithms are set in such a way that either the default settings suggested by the authors are utilized or they are tuned by trials to find the best settings. And the dimensionality d is set in such a way that either 100 or the default settings suggested by the authors. After applying these network embedding algorithms, low-dimensional node representations can be obtained respectively. In the classification task, the node representations are fed into the KNN classifier with 80% randomly selected nodes as the training data while the remaining 20% as the testing data. The cross-validation process is repeated 10 times and the mean values of ACC and Purity are reported in Figs. 4 and 5 respectively. In the clustering task, the node representations are used to build the k-nearest neighbor graph as the similarity matrix where the spectral clustering is performed. Similarly, we repeat the spectral clustering 10 times, and the average values of ACC and Purity are reported in Figs. 6 and 7 respectively.

Overall, compared with the existing methods, the proposed ANEM method exhibits the best performance on most of the datasets in terms of both ACC and Purity. In particular, on the Washington and Wisconsin datasets, ANEM has obtained significantly higher ACC values than the second best algorithm in

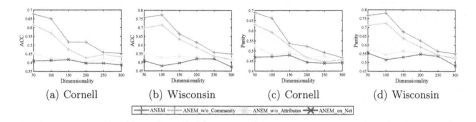

(a) Cornell   (b) Wisconsin   (c) Cornell   (d) Wisconsin

Fig. 8. Comparison results of ANEM and its variations: the classification task in terms of ACC and Purity on the Cornell and Wisconsin datasets.

both the classification and clustering tasks. Among the seven existing algorithms, the M-NMF method incorporates the community structure, while LANE and SNE both account for the node attributes. Compared with the M-NMF method, ANEM achieves 53.6% and 64.8% improvement in terms of ACC and Purity respectively on the Washington dataset in the 3nn classification task, and 38.7% and 25.8% in the spectral clustering task when $k = 4$. For the two algorithms which consider the node attributes, i.e., the LANE and SNE methods, LANE shows relatively better performance on most of the datasets. Compared with the LANE method, ANEM achieves 16.5% and 13.4% improvement in terms of ACC and Purity respectively on the Wisconsin dataset in the 3nn classification task, and 21.1% and 25.0% in the spectral clustering task when $k = 4$. Overall, the comparison results have demonstrated the superior performance of our model, i.e., confirming the effectiveness of incorporating all the three components.

5.5   Comparison Results with Its Variations

To investigate the effectiveness of incorporating the community structure and node attributes, we compare network embedding performance of ANEM and its variations, i.e., ANEM_w/o_Community, ANEM_w/o_Attributes and ANEM_on_Net on the Cornell and the Wisconsin dataset, by means of the KNN classification. The first two variations omit the information of the community structure and node attributes respectively, while the third one only leverages the network structure. The k in the KNN classification is fixed to 3, and the dimensionality d of the node representations varies from {50, 100, 150, 200, 250, 300}. The average ACC and Purity values of 10 times cross-validation process are presented in Fig. 8.

Experimental results in Fig. 8 show that without the community structure and node attributes, ANEM_on_Net achieves the worst result, which confirms the effectiveness of preserving the community structure and node attributes information in network embedding. As d becomes larger, the performance of ANEM and its variations degenerates. When $d > 150$, ANEM_w/o_Attributes has similar performance as ANEM_on_Net. On both of the Cornell and Wisconsin datasets, ANEM_w/o_Community achieves higher ACC and Purity values than the ANEM_w/o_Attributes, which demonstrates the more significant impact of preserving the node attributes. Overall, the proposed ANEM method generates the best results compared with all its variations, further confirming the necessity of incorporating all the three components.

6 Conclusions

Network embedding has attracted an increasing amount of attention in recent years. However, it remains an open challenge to incorporate the microscopic proximity structure, the mesoscopic community structure and the node attributes for the network embedding. To this end, we developed a novel Attributed Network Embedding with Micro-meso structure (ANEM) method, triply preserving the first- and second-order proximities, community membership strength matrix generated by BigCLAM from linkage structure and the information about node attributes. By jointly correlating these three components, an overall objective function is designed, leading to an alternating optimization under the NMF framework. Extensive experiments conducted on the six publicly available attributed networks show that ANEM achieves superior performance on both of the node classification and clustering tasks over the state-of-the-art network embedding methods.

Acknowledgments. This work was supported by NSFC (61502543 & 61602189), Guangdong Natural Science Funds for Distinguished Young Scholar (2016A030306014), the PhD Start-up Fund of Natural Science Foundation of Guangdong Province, China (2016A030310457), and Tip-top Scientific and Technical Innovative Youth Talents of Guangdong special support program (2016TQ03X542).

References

1. Perozzi, B., Al-Rfou, R., Skiena, S.: Deepwalk: online learning of social representations. In: KDD, pp. 701–710 (2014)
2. Yang, D., Wang, S., Li, C., Zhang, X., Li, Z.: From properties to links: deep network embedding on incomplete graphs. In: CIKM, pp. 367–376 (2017)
3. Li, C., Li, Z., Wang, S., Yang, Y., Zhang, X., Zhou, J.: Semi-supervised network embedding. In: Candan, S., Chen, L., Pedersen, T.B., Chang, L., Hua, W. (eds.) DASFAA 2017. LNCS, vol. 10177, pp. 131–147. Springer, Cham (2017). https://doi.org/10.1007/978-3-319-55753-3_9
4. Li, H., Wang, H., Yang, Z., Odagaki, M.: Variation autoencoder based network representation learning for classification. In: Proceedings of ACL 2017, Student Research Workshop, pp. 56–61 (2017)
5. Cavallari, S., Zheng, V.W., Cai, H., Chang, K.C.C., Cambria, E.: Learning community embedding with community detection and node embedding on graphs. In: CIKM (2017)
6. Lai, Y.A., Hsu, C.C., Chen, W.H., Yeh, M.Y., Lin, S.D.: Preserving proximity and global ranking for node embedding. In: NIPS, pp. 5261–5270 (2017)
7. Wang, H., Zhang, F., Hou, M., Xie, X., Guo, M., Liu, Q.: SHINE: signed heterogeneous information network embedding for sentiment link prediction. In: WSDM (2018)

8. Burger, J.D., Henderson, J., Kim, G., Zarrella, G.: Discriminating gender on Twitter. In: Proceedings of the Conference on Empirical Methods in Natural Language Processing, pp. 1301–1309 (2011)

9. Pennacchiotti, M., Popescu, A.M.: Democrats, republicans and starbucks afficionados: user classification in Twitter. In: KDD, pp. 430–438 (2011)

10. Pan, S., Wu, J., Zhu, X., Zhang, C., Wang, Y.: Tri-party deep network representation. In: IJCAI, pp. 1895–1901 (2016)

11. Huang, X., Li, J., Hu, X.: Label informed attributed network embedding. In: WSDM, pp. 731–739 (2017)

12. Tenenbaum, J.B., De Silva, V., Langford, J.C.: A global geometric framework for nonlinear dimensionality reduction. Science 290(5500), 2319–2323 (2000)

13. Tang, J., Qu, M., Wang, M., Zhang, M., Yan, J., Mei, Q.: Line: large-scale information network embedding. In: WWW, pp. 1067–1077 (2015)

14. Wang, D., Cui, P., Zhu, W.: Structural deep network embedding. In: KDD, pp. 1225–1234 (2016)

15. Cao, S., Lu, W., Xu, Q.: GraRep: learning graph representations with global structural information. In: CIKM, pp. 891–900 (2015)

16. Ribeiro, L.F., Saverese, P.H., Figueiredo, D.R.: Struc2vec: learning node representations from structural identity. In: KDD, pp. 385–394 (2017)

17. Wang, X., Cui, P., Wang, J., Pei, J., Zhu, W., Yang, S.: Community preserving network embedding. In: AAAI, pp. 203–209 (2017)

18. Yang, J., Leskovec, J.: Overlapping community detection at scale: a nonnegative matrix factorization approach. In: WSDM, pp. 587–596 (2013)

19. Lee, D.D., Seung, H.S.: Algorithms for non-negative matrix factorization. In: NIPS, pp. 556–562 (2001)

20. Lee, D.D., Seung, H.S.: Learning the parts of objects by non-negative matrix factorization. Nature 401(6755), 788 (1999)

21. Mikolov, T., Chen, K., Corrado, G., Dean, J.: Efficient estimation of word representations in vector space. arXiv preprint arXiv:1301.3781 (2013)

22. Grover, A., Leskovec, J.: Node2vec: scalable feature learning for networks. In: KDD, pp. 855–864 (2016)

23. Ou, M., Cui, P., Pei, J., Zhang, Z., Zhu, W.: Asymmetric transitivity preserving graph embedding. In: KDD, pp. 1105–1114 (2016)

24. Yang, C., Liu, Z., Zhao, D., Sun, M., Chang, E.Y.: Network representation learning with rich text information. In: IJCAI, pp. 2111–2117 (2015)

25. Liao, L., He, X., Zhang, H., Chua, T.S.: Attributed social network embedding. arXiv preprint arXiv:1705.04969 (2017)

26. Li, C., Wang, S., Yang, D., Li, Z., Yang, Y., Zhang, X., Zhou, J.: PPNE: property preserving network embedding. In: Candan, S., Chen, L., Pedersen, T.B., Chang, L., Hua, W. (eds.) DASFAA 2017. LNCS, vol. 10177, pp. 163–179. Springer, Cham (2017). https://doi.org/10.1007/978-3-319-55753-3_11

27. Huang, X., Li, J., Hu, X.: Accelerated attributed network embedding. In: SDM, pp. 633–641 (2017)

28. Li, J., Dani, H., Hu, X., Tang, J., Chang, Y., Liu, H.: Attributed network embedding for learning in a dynamic environment. arXiv preprint arXiv:1706.01860 (2017)

29. Huang, X., Song, Q., Li, J., Hu, X.B.: Exploring expert cognition for attributed network embedding. In: WSDM (2018)

30. Akata, Z., Thurau, C., Bauckhage, C.: Non-negative matrix factorization in multimodality data for segmentation and label prediction. In: 16th Computer Vision Winter Workshop (2011)

31. Zhao, B., Sen, P., Getoor, L.: Event classification and relationship labeling in affiliation networks. In: Proceedings of the Workshop on Statistical Network Analysis (SNA) at the 23rd International Conference on Machine Learning (ICML) (2006)
32. Liu, L., Xu, L., Wangy, Z., Chen, E.: Community detection based on structure and content: a content propagation perspective. In: ICDM, pp. 271–280 (2015)
33. Von Luxburg, U.: A tutorial on spectral clustering. Stat. Comput. **17**(4), 395–416 (2007)
34. Cai, D., He, X., Han, J., Huang, T.S.: Graph regularized nonnegative matrix factorization for data representation. IEEE Trans. Pattern Anal. Mach. Intell. **33**(8), 1548–1560 (2011)

An Efficient Exact Nearest Neighbor Search by Compounded Embedding

Mingjie Li[1(✉)], Ying Zhang[1], Yifang Sun[2], Wei Wang[2], Ivor W. Tsang[1], and Xuemin Lin[2]

[1] Centre for Artificial Intelligence, University of Technology Sydney, Sydney, Australia
Mingjie.Li@student.uts.edu.au, {ying.zhang,ivor.tsang}@uts.edu.au
[2] The University of New South Wales, Sydney, Australia
{yifangs,weiw,lxue}@cse.unsw.edu.au

Abstract. Nearest neighbor search (NNS) in high dimensional space is a fundamental and essential operation in applications from many domains, such as machine learning, databases, multimedia and computer vision. In this paper, we first propose a novel and effective distance lower bound computation technique for Euclidean distance by using the combination of linear and non-linear embedding methods. As such, each point in a high dimensional space can be embedded into a low dimensional space such that the distance between two embedded points lower bounds their distance in the original space. Following the *filter-and-verify* paradigm, we develop an efficient exact NNS algorithm by pruning candidates using the new lower bounding technique and hence reducing the cost of expensive distance computation in high dimensional space. Our comprehensive experiments on 10 real-life and *diverse* datasets, including image, video, audio and text data, demonstrate that our new algorithm can significantly outperform the state-of-the-art exact NNS techniques.

1 Introduction

Nearest neighbor search (NNS) aims to find a point in a reference database which has the smallest distance to a given query point. It is a fundamental and significant operation in many domains, such as machine learning, multimedia databases and computer vision. In these applications, each object is usually represented by a point in a high dimensional space. For instance, by utilizing the deep learning technique [14], an image can be embedded into a point in a 4096-dimensional space. Then, for a given image (i.e., a high dimensional point), NNS can be used to identify the most similar one within an image database. In this paper, we focus on the Euclidean distance which is one of the most popular distance metrics widely used in NNS applications.

It is commonly believed that the computation of the exact NNS in high dimensional space *in the worst case* is very expensive due to the *curse of dimensionality* [11]. In recent years, instead of finding the exact nearest neighbors, an

© Springer International Publishing AG, part of Springer Nature 2018
J. Pei et al. (Eds.): DASFAA 2018, LNCS 10827, pp. 37–54, 2018.
https://doi.org/10.1007/978-3-319-91452-7_3

enormous amount of research effort has been attracted to the problem of approximate nearest neighbor search (ANNS), which circumvents the curse of dimensionality by the trade-off between the search time and search accuracy. Representative algorithms include data-independent hashing methods (e.g., LSH [11] and SRS [23]), learning to hash methods (e.g., AGH [17] and OPQ [6]), tree-based methods (e.g., Optimized KD-tree [22] and FLANN [20]) and neighborhood graph-based methods (e.g., KGraph [4] and SW [18]). Readers may refer to [15] for a comprehensive performance evaluation on the ANNS algorithms.

Despite of the hardness of exact NNS, thanks to the fact that the *intrinsic dimensionality* of the real-life high dimensional data is usually much lower [1,15], it is still feasible to develop efficient and practical *exact* NNS algorithms for high dimensional real-life data. A variety of exact NNS algorithms have been proposed in the literature. Some of them are based on tree structures such as KD-tree [2], iDistance [12], and cover-tree [3]. As reported in [10], they cannot scale to high dimensional space due to the poor performance of tree structure in high dimensional space.

OST [16], FNN [10] and HB+ [5] are three most recent exact NNS algorithms in high dimensional Euclidean space. OST proposes an orthogonal search tree using the PCA basis obtained from the data set, where the tree depth corresponds to the total number of PCA dimensions used for projection. Particularly, the search tree is constructed in a similar way to KD-tree [2], where the data points are recursively partitioned based on their sorted projection values in each individual PCA dimension. Exact NN search can be conducted on this search tree, where the difference between the query projection and each node is used as distance lower bound for NN candidate pruning, and then the distance verification is executed if the pruning fails. FNN obtains the distance lower bounds between query and data points by nonlinearly embedding the dataset into a 2-dimensional space according to two important statistics (mean and variance) of the coordinate values of all dimensions. The performance can be enhanced by partitioning the dimensions into t disjoint groups and compute the statistics for each group, where each point is embedded to a $2t$-dimensional space. Then NNS can be conducted following the filter-and-verify paradigm by using the distance lower bounds between query and data points in the embedded space. HB+ is a newly proposed method which is an extension of HB [21]. The key idea of HB is to devise distance lower bound between a query and a data point by exploiting separating hyperplanes between the query point and the corresponding cluster of the data point. The complexity of the lower bound computation is $O(k^2 d)$ where k and d denote the number of clusters and the dimensionality, respectively. This seriously limits the search performance of HB in high dimensional space. HB+ alleviates this issue by accelerating the lower bound computation. Nevertheless, as shown in our empirical study, the search performance of HB+ is not competitive to FNN and OST.

In this paper, we propose a new embedding technique for Euclidean space to devise distance lower bound, which is essential for exact NNS. In a nutshell, we embed data points in the d dimensional space $\mathcal{R}^d$ into a $m_1 + m_2$ dimensional space, denoted by $\mathcal{E}^{m_1+m_2}$ $(m_1 + m_2 \ll d)$. The first m_1 dimensions of $\mathcal{E}^{m_1+m_2}$ is a *linear* embedding of data points in $\mathcal{R}^d$, and the last m_2 dimensions are obtained by the *non-linear* embedding method. Specifically, we choose a subspace $\mathcal{S}^{m_1}$ of the d-dimensional Euclidean space $\mathcal{R}^d$ with dimensionality m_1, and the remaining dimensions form the other subspace $\mathcal{S}^{d-m_1}$. Then each point p in $\mathcal{R}^d$ can be embedded into a $m_1 + m_2$ dimensional Euclidean space where the coordinate values of the first m_1 dimensions come from $\mathcal{S}^{m_1}$ and the coordinate values of the last m_2 dimensions are from m_2 times space partitioning on $\mathcal{S}^{d-m_1}$. We show that the distance between two points in the embedded space is a lower bound of their distance in the original space.

We also theoretically show that the optimal subspace $\mathcal{S}^{m_1}$ which maximizes the linear embedding part of distance lower bound can be achieved by PCA. It is well-known that the Euclidean distance is preserved under any orthogonal transformation [7]. Thus we first apply PCA on the data points in the original space $\mathcal{R}^d$ to obtain a t-dimensional truncated PCA $(t < d)$, and choose the first m_1 $(m_1 \ll t)$ dimensions as the optimal subspace $\mathcal{S}^{m_1}$, and then we apply m_2 $(m_2 \ll t)$ times space partitioning on subspace $\mathcal{S}^{t-m_1}$ to achieve the last m_2 non-linear embedded dimensions. Following the *filter-and-verify* paradigm, we develop an efficient exact NNS algorithm for high dimensional data by pruning candidates using the distance lower bound obtained by our embedding technique and hence reducing the expensive cost of distance verification in high dimensional space. We show in the empirical study that the proposed method can provide a much better search performance compared with OST, FNN and HB+.

Contributions. Our contributions are summarized as follows.

- We propose a new embedding technique, combining the linear and non-linear methods, for devising Euclidean distance lower bound. We also show that the linear embedding part of the distance lower bound can be optimized by the PCA technique.
- We develop an efficient *exact* NN search algorithm following the *filter-and-verify* paradigm, by leveraging a distance lower-bounding in the embedded low-dimensional space.
- Our comprehensive experiments on 10 real-life high dimensional data demonstrate the effectiveness and efficiency of our proposed techniques. Our algorithm can significantly outperform the state-of-the-art exact NN search techniques.

Road Map. The rest of this paper is organized as follows. Section 2 presents some preliminaries for the exact NNS problem. Section 3 describes our embedding method and the distance lower bound. Our exact NNS algorithm is presented in Sect. 4. Section 5 shows the experimental results and Sect. 6 concludes the paper.

2   Preliminaries

Before delving into the details of our techniques, we briefly review the knowledge used in our algorithm and then introduce the problem definition. Notations frequently used in this paper are summarized in Table 1.

Table 1. Summary of notations

Notation	Definition
$\mathcal{R}^d$	An Euclidean space (d dimensions)
$\mathcal{S}^{m_1}$	A subspace of $\mathcal{R}^d$ (m_1 dimensions)
$\mathcal{P}^t$	A t dimensional truncated PCA ($t < d$)
$\mathcal{E}^{m_1+m_2}$	An embedded space ($m_1 + m_2$ dimensions)
p	A data point in original space $\mathcal{R}^d$
p_i	The i-th coordinate value of p
p'	The projection point of p in subspace $\mathcal{S}^{m_1}$
p^*	The embedded point of p in space $\mathcal{E}^{m_1+m_2}$
$\|qp\|$	the Euclidean distance between point q and p
$\overrightarrow{pq}$	The vector from point p to point q

In this paper, we consider the Euclidean distance as our distance metric. Throughout the paper, we use "space" to denote the "Euclidean space". Let p denote a point in the space $\mathcal{R}^d$ with d dimensions, and we use p_i to represent its coordinate value in the i-th dimension.

Definition 1. *The Euclidean distance between two d-dimensional points q and p in the space $\mathcal{R}^d$, denoted as $\|pq\|$, is computed as $\|pq\| = \sqrt{\sum_{i=1}^{d}(q_i - p_i)^2}$.*

Given any two points p and q in $\mathcal{R}^d$, we use pq to represent the line segment between p and q. By $\overrightarrow{pq}$, we denote the vector from p to q. We use $\|pq\|$ and $\|\overrightarrow{pq}\|$ to denote the length of pq and $\overrightarrow{pq}$, i.e., the Euclidean distance between p and q. By $\angle abc$, we denote the angle between ab and bc.

Problem Definition. Let D denote a dataset with n data points in a d dimensional Euclidean space $\mathcal{R}^d$. For *exact* **NNS**, given a query point q, the exact nearest neighbor of q is the point in D that has the smallest distance to q.

3   Embedding and Distance Lower Bound

In this section, we first introduce the motivation of our distance lower bound technique, followed by our embedding method, linear and non-linear embedding. Then we formally show the correctness of our distance lower bound within the embedded space, followed by its optimization. Finally, we remark that the above techniques can be applied to the PCA space of $\mathcal{R}^d$ to achieve better performance.

3.1 Motivation

Suppose the original space $\mathcal{R}^3 = \{w_1, w_2, w_3\}$ is a 3-dimensional space as shown in Fig. 1. Given a query point q and a data point p, we use q' and p' to denote their orthogonal projections on a one-dimensional subspace $\mathcal{S}^1 = \{w_1\}$. Now we will show how to devise a distance lower bound for $\|pq\|$ based on $\|p'q'\|$, $\|qq'\|$, and $\|pp'\|$.

Let o be the point in $\mathcal{R}^d$ generated by moving q along the direction of $\overrightarrow{q'p'}$ with distance $\|q'p'\|$. It is trivial to see that qo is parallel to $q'p'$, and $\|qo\| = \|q'p'\|$, thus $\|op'\| = \|qq'\|$. It is easy to know that qo is perpendicular to op, i.e., $\angle qop$ is a right angle. According to the Pythagorean Theorem, we have $\|pq\|^2 = \|qo\|^2 + \|op\|^2$. By Triangle Inequality, we have $\|op\| \geq |\,\|op'\| - \|pp'\|\,|$. Then we have the following distance lower bound for $\|pq\|$:

$$\|pq\|^2 \geq \|p'q'\|^2 + (\|pp'\| - \|qq'\|)^2. \tag{1}$$

Let q^* (resp. p^*) be a 2-dimension point with coordinate values $q_1^* = q_1$ (resp. $p_1^* = p_1$) and $q_2^* = \|qq'\|$ (resp. $p_2^* = \|pp'\|$). Recall that q_1 denotes the coordinate value of point q on the 1st dimension. By doing this, we introduce another dimension to the subspace $\mathcal{S}^1$, resulting in a 2-dimensional embedded space, denoted by $\mathcal{E}^2$. The 1st dimension of $\mathcal{E}^2$ directly comes from $\mathcal{S}^1$, while the 2nd dimension is the distance between the point and its projection on the subspace $\mathcal{S}^1$. The points p and q from $\mathcal{R}^3$ are mapped to p^* and q^* in the embedded space $\mathcal{E}^2$. Then inequality 1 can be re-written as follows:

$$\|pq\|^2 \geq (p_1 - q_1)^2 + (\|pp'\| - \|qq'\|)^2 = \|p^*q^*\|^2. \tag{2}$$

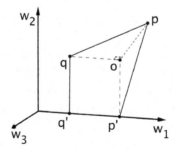

Fig. 1. Motivation of distance lower bound

Inequality 2 implies that the distance between p and q in the embedded space (i.e., $\|p^*q^*\|$) is always no larger than their distance in the original space (i.e., $\|pq\|$).

In the following three subsections, we formally define our embedding method and then prove the correctness of the distance lower bound in the embedded space, followed by the optimization of the distance lower bound.

3.2 Embedding Method

Given the Euclidean space $\mathcal{R}^d = \{w_1, w_2, \ldots, w_d\}$, where $w_1, w_2, \ldots, w_d$ are the orthonormal basis of $\mathcal{R}^d$. We use $\mathcal{S}^{m_1}$ to denote a subspace of $\mathcal{R}^d$ with m_1 dimensions ($m_1 \ll d$), and the remaining dimensions form another subspace $\mathcal{S}^{d-m_1}$. Then, we partition the $\mathcal{S}^{d-m_1}$ into m_2 ($m_2 \ll d$) disjoint subspaces, denoted by $\mathcal{S}^{n_1}, \mathcal{S}^{n_2}, \ldots$, and $\mathcal{S}^{n_{m_2}}$, respectively, where $n_1 + n_2 + \ldots + n_{m_2} = d - m_1$. Then the subspace $\mathcal{S}^{m_1}$ and $\mathcal{S}^{n_1}$ form a new subspace, denoted by $\mathcal{S}^{m_1+n_1}$. Accordingly, $\mathcal{S}^{m_1}$ and $\mathcal{S}^{n_2}$ form the second subspace $\mathcal{S}^{m_1+n_2}$. Consequently, we generate m_2 new subspaces, denoted by $\mathcal{S}^{m_1+n_1}, \mathcal{S}^{m_1+n_2}, \ldots$, and $\mathcal{S}^{m_1+n_{m_2}}$, respectively. They share a common subspace $\mathcal{S}^{m_1}$.

The embedded space regarding these m_2 subspaces is a $(m_1 + m_2)$-dimensional space, denoted by $\mathcal{E}^{m_1+m_2}$. Then for each point $p \in \mathcal{R}^d$, we use $\hat{p}^1$, $\hat{p}^2, \ldots$, and $\hat{p}^{m_2}$ to denote its corresponding projections on the subspaces $\mathcal{S}^{m_1+n_1}$, $\mathcal{S}^{m_1+n_2}, \ldots$, and $\mathcal{S}^{m_1+n_{m_2}}$, respectively. Then for each $\hat{p}^i$, where $i = 1, 2, \ldots, m_2$, we use p'^i to denote its projection on subspace $\mathcal{S}^{m_1}$. We directly denote p' as the projection of p on $\mathcal{S}^{m_1}$, it is trivial that $p' = p'^1 = p'^2 = \ldots = p'^{m_2}$. By p^*, we denote the corresponding point of p in the embedded space $\mathcal{E}^{m_1+m_2}$, where $p_j^* = p_j'$ for $1 \le j \le m_1$ and $p_{m_1+k}^* = \|\hat{p}^k p'^k\|$ for $1 \le k \le m_2$. Note that $\|\hat{p}^k p'^k\|$ is the distance between $\hat{p}^k$ and its projection on the subspace $\mathcal{S}^{m_1}$. In other words, the embedded space $\mathcal{E}^{m_1+m_2}$ contains two parts: linear embedding (the first m_1 dimensions) and non-linear embedding (the last m_2 dimensions).

3.3 Correctness of Our Distance Lower Bound

In this subsection, we formally show that the distance between two points within the embedded space $\mathcal{E}^{m_1+m_2}$ can serve as the lower bound of their distance in original $\mathcal{R}^d$.

Theorem 1. *Given an Euclidean space $\mathcal{R}^d$ with an orthonormal basis $\{w_1, w_2, \ldots, w_d\}$, the subspace $\mathcal{S}^{m_1} = \{w_1, w_2, \ldots, w_{m_1}\}$ and the corresponding embedded space $\mathcal{E}^{m_1+m_2}$, we have $\|p^* q^*\| \le \|pq\|$ where p and q are two points in $\mathcal{R}^d$ while p^* and q^* are their embedded points in $\mathcal{E}^{m_1+m_2}$.*

Proof. (1) When $m_2 = 1$, there is no space partitioning on $\mathcal{S}^{d-m_1}$. Taking Fig. 1 for example, we let q' and p' be the projections of q and p on $\mathcal{S}^{m_1}$, respectively. We define o as the point in $\mathcal{R}^d$ such that $\overrightarrow{q'q} = \overrightarrow{p'o}$. Then we can have $\overrightarrow{qo} = \overrightarrow{q'p'}$ and $\overrightarrow{op} = \overrightarrow{p'p} - \overrightarrow{p'o}$. It is trivial that $\overrightarrow{q'p'}$ is within subspace $\mathcal{S}^{m_1}$, and $\overrightarrow{p'o}$ and $\overrightarrow{p'p}$ are both within the subspace $\mathcal{S}^{d-m_1}$. Since $\mathcal{S}^{m_1}$ is orthogonal to $\mathcal{S}^{d-m_1}$, $\overrightarrow{qo}$ and $\overrightarrow{op}$ are perpendicular to each other. According to the Pythagorean Theorem and Triangle Inequality, we can have $\|pq\|^2 \ge \|q'p'\|^2 + (\|q'q\| - \|p'p\|)^2$.

Now consider p^* and q^*, which are the embedded points of p and q in the embedded space $\mathcal{E}^{m_1+1}$. By definition, we have $p^* = (p', \|pp'\|)$ and $q^* = (q', \|qq'\|)$, then we have:

$$\|p^* q^*\|^2 = \sum_{i=1}^{m_1} (p_i' - q_i')^2 + (\|pp'\| - \|qq'\|)^2 = \|q'p'\|^2 + (\|q'q\| - \|p'p\|)^2 \le \|pq\|^2.$$

Hence, when $m_2 = 1$, the Theorem 1 holds. This case is equivalent to the inequality lemma in [16].

(2) When $m_2 = 2$. For $m_2 = 1$, we have $\|q'p'\|^2 + (\|q'q\| - \|p'p\|)^2 \leq \|pq\|^2$. Since $\|q'p'\|^2$ is directly from the subspace $\mathcal{S}^{m_1}$ of $\mathcal{R}^d$, we can have:

$$\sum_{i=m_1+1}^{d} (p_i - q_i)^2 \geq (\|q'q\| - \|p'p\|)^2. \tag{3}$$

As $m_2 = 2$, we generate 2 new subspaces $\mathcal{S}^{m_1+n_1}$ and $\mathcal{S}^{m_1+n_2}$, respectively, where $n_1 + n_2 = d - m_1$. For each new subspace, we can have that:

$$\|\hat{q}^1\hat{p}^1\|^2 \geq \|q'p'\|^2 + (\|\hat{q}^1q'\| - \|\hat{p}^1p'\|)^2, \tag{4}$$

$$\|\hat{q}^2\hat{p}^2\|^2 \geq \|q'p'\|^2 + (\|\hat{q}^2q'\| - \|\hat{p}^2p'\|)^2. \tag{5}$$

The Eqs. 4 and 5 can be further re-written as:

$$\sum_{i=m_1+1}^{m_1+n_1} (p_i - q_i)^2 \geq (\|\hat{q}^1q'\| - \|\hat{p}^1p'\|)^2, \tag{6}$$

$$\sum_{i=m_1+n_1+1}^{d} (p_i - q_i)^2 \geq (\|\hat{q}^2q'\| - \|\hat{p}^2p'\|)^2. \tag{7}$$

After combining the Eqs. 6 and 7, we can have:

$$\sum_{i=m_1+1}^{d} (p_i - q_i)^2 \geq (\|\hat{q}^1q'\| - \|\hat{p}^1p'\|)^2 + (\|\hat{q}^2q'\| - \|\hat{p}^2p'\|)^2. \tag{8}$$

Finally, we can have:

$$\|pq\|^2 \geq \|q'p'\|^2 + (\|\hat{q}^1q'\| - \|\hat{p}^1p'\|)^2 + (\|\hat{q}^2q'\| - \|\hat{p}^2p'\|)^2 = \|p^*q^*\|^2. \tag{9}$$

Therefore, when $m_2 = 2$, the Theorem 1 holds.

(3) When $m_2 \geq 3$. Based on the cases of $m_2 = 1$ and $m_2 = 2$, it is easy to prove that the Theorem 1 holds for $m_2 \geq 3$, so we omit it here.

3.4   Optimization of Our Distance Lower Bound

In this subsection, we discuss the optimization of our distance lower bound within the embedded space $\mathcal{E}^{m_1+m_2}$. Our distance lower bound is from the combination of linear embedding and non-linear embedding, so the optimization consists of two parts.

For the linear case, given a dataset with n points in space R^d, we aim to find a subspace $\mathcal{S}^{m_1}$ of R^d that is able to maximize the average square distance of all pairwise projected data points on $\mathcal{S}^{m_1}$. This optimization objective is given as follows:

$$Maximize \quad \frac{1}{n^2} \sum_{i=1}^{n} \sum_{j=1}^{n} d_{ij}^2. \tag{10}$$

where the d_{ij} denotes the distance between projected point i and projected point j on $\mathcal{S}^{m_1}$. The optimization of the above objective leads to our another important theorem, which is given as follows. The proof is similar to the Multidimensional scaling [24].

Theorem 2. *Given a dataset X with n points in space R^d, the subspace $\mathcal{S}^{m_1} = \{w_1, ..., w_{m_1}\}$ ($m_1 \ll d$) which can maximize the average square distance of all pairwise projected data points on $\mathcal{S}^{m_1}$, is the m_1-dimensional PCA of the dataset X.*

Proof. We assume that the dataset X is already centered, i.e., $\sum_{i=1}^{n} x_l^i = 0$, *for all* $l = 1, ..., d$, where the x_l^i denotes the l-th coordinate value of the i-th point. Then we can have that the dataset $\tilde{X}$, which is the orthogonal projection of dataset X on $\mathcal{S}^{m_1}$, is also centered, as $\sum_{i=1}^{n} w_l^T x^i = w_l^T \sum_{i=1}^{n} x^i = 0$, *for all* $l = 1, ..., m_1$. Let $\tilde{x}^i$ denote the projection of data point i on $\mathcal{S}^{m_1}$, then

$$\sum_{i=1}^{n} \tilde{x}_l^i = 0, for\ all\ l = 1, ..., m_1. \tag{11}$$

Let b_{ij} denote the inner product between $\tilde{x}^i$ and $\tilde{x}^j$. Since $||\tilde{x}^i - \tilde{x}^j||^2 = (\tilde{x}^i)^T \tilde{x}^i + (\tilde{x}^j)^T \tilde{x}^j - 2(\tilde{x}^i)^T \tilde{x}^j$, we can have:

$$d_{ij}^2 = b_{ii} + b_{jj} - 2b_{ij}. \tag{12}$$

The Eq. 11 leads to

$$\sum_{i=1}^{n} b_{ij} = \sum_{i=1}^{n} \sum_{l=1}^{m_1} \tilde{x}_l^i \tilde{x}_l^j = \sum_{l=1}^{m_1} \tilde{x}_l^j \sum_{i=1}^{n} \tilde{x}_l^i = 0, for\ all\ j = 1, ..., n. \tag{13}$$

With a notation $T = \sum_{i=1}^{n} b_{ii}$, by Eqs. 12 and 13, we have:

$$\sum_{i=1}^{n} d_{ij}^2 = T + nb_{jj}, \quad \sum_{j=1}^{n} d_{ij}^2 = T + nb_{ii}. \tag{14}$$

Hence, our optimization objective can be rewritten as:

$$\frac{1}{n^2} \sum_{i=1}^{n} \sum_{j=1}^{n} d_{ij}^2 = \frac{1}{n^2} \times 2nT = 2 \times \frac{1}{n} \sum_{i=1}^{n} b_{ii} = 2 \times \frac{1}{n} \sum_{i=1}^{n} (\tilde{x}^i)^T \tilde{x}^i. \tag{15}$$

PCA can be defined as the orthogonal projection of the data onto a lower dimensional linear space, such that the variance of the projected data is maximized [9].

In other words, a m_1-dimensional PCA of dataset X is a subspace that can maximize the average square distance of the projected data points to their central point. As the projected data points are centered, our optimization objective is equivalent to that of PCA, Theorem 2 holds.

In terms of the non-linear case, after the $\mathcal{S}^{m_1}$ is decided, we aim to choose a space partitioning way for $\mathcal{S}^{d-m_1}$, which can tighten the partial distance lower bound contributed by the non-linear embedding as much as possible. However, as the subspace partitioning is quite computationally expensive, in this paper, we set $m_2 = 2$ and choose a simple partitioning way for $\mathcal{S}^{d-m_1}$. Specifically, we partition the $\mathcal{S}^{d-m_1}$ into 2 disjoint but consecutive subspaces with bases $\{w_{m_1+1}, ..., w_{m_1+\lfloor (d-m_1)/2 \rfloor}\}$ and $\{w_{m_1+\lfloor (d-m_1)/2 \rfloor +1}, ..., w_d\}$, respectively. Our empirical study shows that this partitioning way already gives us a good performance for all datasets.

3.5   Using PCA Technique

Theorem 1 holds for any Euclidean linear space with an orthonormal basis. According to Theorem 2, we choose the t-dimensional truncated PCA of dataset in $\mathcal{R}^d$ for embedding ($t < d$). In contrast to full PCA (i.e., the d-dimensional PCA), choosing a truncated PCA with smaller dimensions can save more preprocessing time, and the Theorem 1 still holds in a t-dimensional PCA space. Since the Euclidean distance is preserved under any orthogonal transformation [7], we first conduct PCA on the original dataset to achieve the t-dimensional PCA space, denoted by $\mathcal{P}^t$, and then first m_1 ($m_1 \ll t$) dimensions (components) are used as the subspace $\mathcal{S}^{m_1}$ for the linear embedding while the remaining $t - m_1$ dimensions are partitioned into m_2 ($m_2 \ll t$) subspaces for non-linear embedding. Such that our embedded space $\mathcal{E}^{m_1+m_2}$ is generated by the compound of the linear and non-linear embedding, on which the NN search will be conducted.

4   Our Exact NNS Algorithm

This section presents our exact NNS algorithm. We first show the key idea of the algorithm, followed by the detailed implementation of the algorithm and finally the performance analysis.

4.1   Motivation

Given the embedded space $\mathcal{E}^{m_1+m_2}$, a straightforward implementation is to randomly choose a point and compute its distance to the query q as the distance threshold λ. Then for each remaining point, we compute its distance lower bound w.r.t q (i.e., the distance under $\mathcal{E}^{m_1+m_2}$), and the point will be safely excluded if its lower bound is larger than λ. Otherwise, we can compute its true distance to q and update λ. Finally, the point contributes to λ is the exact NN of q. There are two shortcomings of this implementation: (i) it may take many steps to find a good distance threshold λ, and (ii) we need to compute the distance lower

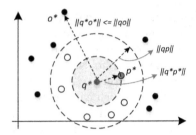

Fig. 2. Motivation of exact NNS using embedded space

bounds for all data points, which is still expensive even the dimensionality of the embedded space is low.

To address the above issues, we seek help from the multi-dimensional index techniques which can efficiently support exact NNS and range search on relatively low dimensional space (e.g., 8–20 dimensions). Regarding the example shown in Fig. 2, we suppose the data points in the embedded space $\mathcal{E}^{m_1+m_2}$ are organized by a multi-dimensional index. Then, we can identify the true NN of the query q by issuing one NN search and one range search in the embedded space, together with the verification of the survived candidates. Specifically, we first find the nearest neighbor of q^* (i.e., embedding of q in $\mathcal{E}^{m_1+m_2}$) in the embedded space, which is p^* in our example. Then we take $\|pq\|$ as the initial distance threshold λ, and issue a range query from q^* with radius $\|pq\|$. Clearly, we only need to access points within the range because the distance lower bounds of other points are already larger than the current distance threshold. For instance, we do not need to access the point o^* in the example of Fig. 2. By doing this, we can find a good distance threshold for better pruning performance and avoid computing distance lower bounds for many points outside of the range search. Detailed implementations will be introduced in the following subsection.

Remark 1. In this paper, we choose cover tree [3] as the multi-dimensional index structure for the embedded data points due to its good performance under our settings. Note that we use the cover tree to organize the low dimensional embedded points in $\mathcal{E}^{m_1+m_2}$, not the high dimensional data points in $\mathcal{R}^d$. It is already reported in [10] that the performance of cover tree in high dimensional space is not competitive compared with FNN.

4.2   Exact NNS Algorithm

In Algorithm 1, we illustrate the details of the exact nearest neighbor search (NNS) algorithm by using our embedding technique. In our implementation, the dataset in original space $\mathcal{R}^d$ is represented by D, and the data points in PCA space $\mathcal{P}^t$, denoted by $\tilde{D}$, are computed off-line on the D. We set the first m_1 dimensions of the embedded space $\mathcal{E}^{m_1+m_2}$ as the first m_1 dimensions of the PCA space $\mathcal{P}^t$, and the last m_2 dimensions of $\mathcal{E}^{m_1+m_2}$ are from m_2 times space partitioning on $\mathcal{P}^{t-m_1}$. The corresponding points in the embedded space can also be pre-computed, denoted by D^*. We use a cover tree [3] to organize the embedded points, which can efficiently support NN search and range search when $m_1 + m_2$ is small.

Algorithm 1. Exact NN Search $(D, \tilde{D}, D^*, q)$

Input   : D: data points in original space $\mathcal{R}^d$,
           $\tilde{D}$: data points in PCA space $\mathcal{P}^t$,
           D^*: data points in embedded space $\mathcal{E}^{m_1+m_2}$,
           q: query point in space $\mathcal{R}^d$
Output: r : the nearest neighbor of q in D

1 $\tilde{q} \leftarrow$ transfer q from space $\mathcal{R}^d$ to PCA space $\mathcal{P}^t$;
2 $q^* \leftarrow$ the embedded point of q in $\mathcal{E}^{m_1+m_2}$;
3 $p^* \leftarrow$ the nearest neighbor of q^* in D^*;
4 $p \leftarrow$ the corresponding point of p^* in D;
5 $min\_dist := \|pq\|^2$; $r \leftarrow p$;
6 $C \leftarrow$ data points $\{o^*\}$ in D^* with $\|q^*o^*\| \leq \|pq\|$;
7 **for** *each data point* $o^* \in C$ **do**
8   **if** $\|q^*o^*\|^2 \geq min\_dist$ **then**
9         $continue$;
10     $dist := \|q^*o^*\|^2 - \Delta$;
11     $\tilde{o} \leftarrow$ the corresponding points of o^* in $\tilde{D}$;
12     $is\_rejected \leftarrow false$;
13   **for** $j := (m_1 + 1) \rightarrow t$ **do**
14        $dist := dist + (\tilde{q}_j - \tilde{o}_j)^2$;
15   **if** $dist \geq min\_dist$ **then**
16           $is\_rejected \leftarrow true$;
17           $break$;
18   **if** $is\_rejected = false$ **then**
19   distance verification in D;
20   **if** $dist < min\_dist$ **then**
21        $min\_dist := dist$; $r \leftarrow o$;

22 **return** r

At Lines 1–2, the query point q is mapped to the PCA space and q^* is its corresponding point in the embedded space. Line 3 retrieves the nearest neighbor of q^*, denoted by p^*, by issuing NN search on the cover tree. Note that p^* is the embedded point of p from D. In the algorithm, we use r and min_dist to denote the current closest point of q and its threshold (i.e., squared Euclidean distance), which are initialized at Line 5. Then we issue the range search on the cover tree with centre at q^* and radius $\|pq\|$. The data points within the search range are kept in the set C for further processing (Line 6).

Lines 7–21 incrementally verify the candidate data points in C. If a data point o cannot be pruned based on its distance lower bound (i.e., $\|q^*o^*\|$) and the current distance threshold min_dist, we need to compute its distance to q. To reduce the verification cost, we do not directly compute the distance between o and q in original space $\mathcal{R}^d$. We have $\tilde{p}_i = p_i^*$ for $1 \le i \le m_1$ because the first m_1-dimensions of embedded space are from the first m_1-dimensions of PCA space $\mathcal{P}^t$. Thus, we can reuse the computation of the distance lower bound $\|q^*o^*\|$ by ignoring the contribution of the last m_2 dimensions, denoted as Δ, at Line 10. As such, we can accumulatively compute the distance in PCA space $\mathcal{P}^t$ and immediately terminate the distance computation when the distance computed so far already exceeds the distance threshold (Lines 13–17). If the verification in PCA space $\mathcal{P}^t$ fails, we conduct the distance verification in original space $\mathcal{R}^d$ (Lines 18–19). Meanwhile, we update the distance threshold as well as the corresponding data point (Lines 20–21). The nearest neighbor will be returned after all candidate points are explored. Our empirical study shows that at most cases, the verification in $\mathcal{R}^d$ does not happen.

Remark 2. Algorithm 1 can be easily extended to support k nearest neighbor search (kNNS). Instead of NN search, we issue a kNN search at Line 3 and p^* is the k-th nearest neighbor of q^* in the embedded space. Then the distance threshold is the distance of the k-th closest data point to q seen so far.

4.3   Performance Analysis

The dominant cost of the pre-processing phase is the PCA computation. In the literature, many research efforts have been devoted to develop efficient PCA computation, and various exact and approximate algorithms have been proposed [8,13,19]. In our implementation, we use the popular randomized PCA computing algorithm [8], which takes $O(nd \log t)$ time to compute the PCA space $\mathcal{P}^t$ ($t < d$). In this paper, the t is set to be 60 for all datasets. It takes $O(ndt)$ time to transform data points from space $\mathcal{R}^d$ to the PCA space $\mathcal{P}^t$. Regarding the embedded data points in $\mathcal{E}^{m_1+m_2}$ ($m_1 + m_2 \ll t$), we simply take the first m_1 dimensions of the data points in $\mathcal{P}^t$ (i.e., their projections in the subspace $\mathcal{P}^{m_1}$). The computation of the last m_2 dimensions (i.e., the distances from partitioned subspace to the subspace $\mathcal{P}^{m_1}$) takes time $O(ndm_2)$. The embedded data points are organized by cover tree with construction time $O(c^6 n \log n)$, where c is related to the intrinsic dimensionality of the $m_1 + m_2$ dimensional embedded data [3].

The dominant cost of NN search comes from the pruning (Line 3–6) and verification (Line 7–21). The pruning phase consists of NN search and range search on the embedded data. The NN search and range search costs on the cover tree are bounded by $O(c^{12} \log n)$ and $O(c^{12} l \log n)$, respectively, where l is the number of points within the range search [3]. Note that, in the worst case, it takes $O(n(m_1+m_2))$ time to compute distance lower bounds for all data points. Regarding the verification phase, it contains the verifications in PCA space $\mathcal{P}^t$ and original space $\mathcal{R}^d$, respectively, and it takes $O(t + d)$ time in the worse case to compute the distance for each survived candidate point.

5   Experiment

In this section, we report and analyze the experimental results.

5.1   Experimental Setting

Algorithms. We choose the following exact algorithms for conducting L_2-distance based exact NN search on high dimensional data for comparison.

- LNL is the proposed algorithm. The abbreviation stands for **Linear** and **Non-Linear** embedding. The cover tree used in our algorithm is from the cover tree source code[1].
- OST is a PCA-based exact NNS method proposed in [16]. We implement OST and make its performance as good as possible.
- FNN [10] is the state-of-the-art method for exact NN search in high dimensional data. The source code of FNN is public available[2].
- HB+ is a cluster-based exact NNS algorithm proposed in [5]. The source code is obtained from authors.

Datasets. We use 10 real-life datasets with different types, including image data (**Cifar**[3], **Deep**[4], **Sun**[5], **Gist**[6], **MNIST**[7], **Nusw**[8], and **Trevi**[9]), audio data (**Audio**[10]), text data (**GoogleNews**[11]), and video data (**Youtube**[12]); they are also widely used in prior research literature to evaluate nearest neighbor query performance. Table 2 summarizes these datasets. For each dataset, after the deduplication, we randomly select 200 data points and reserve them as the query points.

Implementation Details. All algorithms are implemented in standard C++ and compiled with G++ with -O3 in Linux. All experiments are performed on a machine with Intel Xeon 3.33GHz CPU and Redhat Linux System, with 32G main memory. We report the average NN search time, average kNN search time and preprocessing time.

---

[1] http://hunch.net/~jl/projects/cover_tree/cover_tree.html.
[2] http://research.yoonho.info/fnnne.
[3] http://www.cs.toronto.edu/~kriz/cifar.html.
[4] https://yadi.sk/d/I_yaFVqchJmoc.
[5] http://groups.csail.mit.edu/vision/SUN/.
[6] http://corpus-texmex.irisa.fr.
[7] http://yann.lecun.com/exdb/mnist/.
[8] http://lms.comp.nus.edu.sg/research/NUS-WIDE.htm.
[9] http://phototour.cs.washington.edu/patches/default.htm.
[10] http://www.cs.princeton.edu/cass/audio.tar.gz.
[11] https://code.google.com/archive/p/word2vec/.
[12] http://www.cs.tau.ac.il/~wolf/ytfaces/index.html.

Table 2. Dataset summary

Name	n	d	Type
Audio	53,387	192	Audio
Cifar	50,000	512	Image
Deep	1,000,000	256	Image
GoogleNews	2,999,800	300	Text
Sun	79,106	512	Image
Gist	982,677	960	Image
MNIST	69,000	784	Image
Trevi	99,900	4096	Image
Nusw	268,643	500	Image
Youtube	346,194	1,770	Video

5.2   Exact Nearest Neighbor Search

In this subsection, we compare the performance of OST [16], FNN [10], HB+ [5] and the proposed LNL on the exact NNS task.

Parameter Settings. For OST, FNN and HB+ algorithms, we set their experimental parameters according to the suggestion by authors.

Next, we determine the value of m_1 and m_2 in LNL by conducting experiments on **Audio** dataset. The other datasets follow the similar trends. Note that the parameter t of the truncated PCA space $\mathcal{P}^t$ is set to be 60 for all datasets. We firstly investigate the impact of m_1 on NN search time with m_2 set to be 2. Figure 3(a) shows the trade-off between m_1 and the average search time of LNL. Then, we study the influence of m_2 on NN search time with $m_1 = 8$, which can be seen in Fig. 3(b). In our experiments, we choose $m_1 = 8$ and $m_2 = 2$ for all the datasets to avoid manually tuning, as this setting does achieve a relatively good performance over all the datasets.

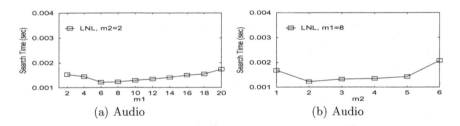

(a) Audio   (b) Audio

Fig. 3. Search time with respect to m_1 and m_2

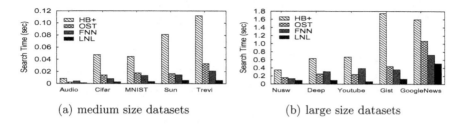

(a) medium size datasets   (b) large size datasets

Fig. 4. Comparison of search time on all datasets

Efficiency of NN Search. Figure 4 reports the searching performance of the four algorithms on all datasets. It is clear that LNL outperforms the other three methods on all the 10 datasets, especially on the high dimensional datasets, such as **Gist**, **Trevi** and **Youtube**. On most datasets, LNL can achieve 2 to 6 times faster than OST and FNN, and 7 to 20 times faster than HB+. For example, LNL needs on average 0.0032 s to process a query on **MNIST** dataset while OST, FNN and HB+ need 0.0178 s, 0.0135 s and 0.04497 s, respectively. The high efficiency of NN search achieved by LNL is because of two reasons. On one hand, LNL tends to select a point that is closer to the query as the start point in NN search. On the other hand, for the same points, LNL usually has a much tighter distance lower bound than the other three algorithms. Further more, the range query allows LNL to avoid computing distance lower bound for all data points, which can reduce a certain amount of searching cost. Note that both of LNL and OST use PCA technique, however, OST individually use the PCA on a search tree without the consideration of the coherence between PCA components. Hence, OST is not competitive to the proposed LNL.

Comparison with respect to k. Figure 5 shows the trends of the performance of four algorithms for kNN search with increasing k on six datasets. Although they are all not sensitive to k, especially when k is large (e.g., ≥ 20), LNL has the most stable performance when k changes. This is because the pruning power of LNL decreases slower than that of other three methods when k increases. We can observe that the proposed LNL still performs the best among the other tree algorithms on the kNN search task.

Pre-processing Time. We show the pre-processing time for all four algorithms in Fig. 6. FNN has the smallest preprocessing time, as it does not need to construct any index but only need to compute the mean and variance of the elements for each partition in the vector space. HB+ takes the highest pre-processing time because it needs to construct the clusters which is quite time consuming. The preprocessing time of LNL and OST are similar, since both of them need the PCA computation and index structure. Note that although LNL requires more pre-processing time than FNN, it is still faster than HB+ and is practical for this *off-line* procedure. For example, LNL can process one million 960-dimensional data points (e.g., **Gist**) in 3 minutes. Considering of the excellent search performance of LNL, it offers an attractive tradeoff between *off-line* and *online* processing times.

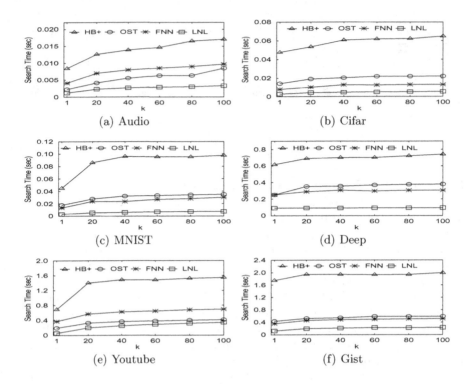

Fig. 5. Comparison with respect to k

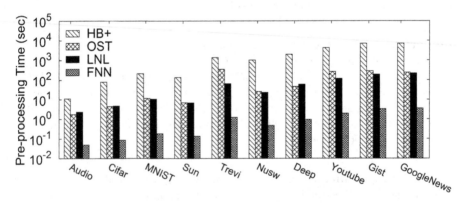

Fig. 6. Pre-processing time

6   Conclusion

In this paper, we investigate the problem of exact nearest neighbour search (NNS) in high dimensional space. We design a new embedding method which can embed the high dimensional points into a low dimensional space. Then an efficient exact NNS algorithm is developed following the *filter-and-verify* paradigm based on the embedded data. Extensive experiments on 10 real-life datasets with various types demonstrate that our method significantly outperforms the state-of-the-art exact NNS techniques in high dimensional space.

Acknowledgement. Ying Zhang is supported by ARC DE140100679 and DP170103710. Wei Wang is supported by ARC DP170103710, and D2DCRC DC25002 and DC25003. Ivor W. Tsang is supported by ARC grant FT130100746, DP180100106, and LP150100671. Xuemin Lin is supported by NSFC 61672235, DP170101628 and DP180103096.

References

1. Amsaleg, L., Chelly, O., Furon, T., Girard, S., Houle, M.E., Kawarabayashi, K., Nett, M.: Estimating local intrinsic dimensionality. In: KDD, pp. 29–38 (2015)
2. Bentley, J.L.: Multidimensional binary search trees used for associative searching. Commun. ACM **18**(9), 509–517 (1975)
3. Beygelzimer, A., Kakade, S., Langford, J.: Cover trees for nearest neighbor. In: Proceedings of the 23rd International Conference on Machine Learning, pp. 97–104. ACM (2006)
4. Dong, W., Charikar, M., Li, K.: Efficient k-nearest neighbor graph construction for generic similarity measures. In: WWW, pp. 577–586 (2011)
5. Feng, X., Cui, J., Liu, Y., Li, H.: Effective optimizations of cluster-based nearest neighbor search in high-dimensional space. Multimedia Syst. **23**(1), 139–153 (2017)
6. Ge, T., He, K., Ke, Q., Sun, J.: Optimized product quantization for approximate nearest neighbor search. In: Proceedings of the IEEE Conference on Computer Vision and Pattern Recognition, pp. 2946–2953 (2013)
7. Golub, G.H., Van Loan, C.F.: Matrix Computations, vol. 3. JHU Press, Baltimore (2012)
8. Halko, N., Martinsson, P.-G., Tropp, J.A.: Finding structure with randomness: probabilistic algorithms for constructing approximate matrix decompositions. SIAM Rev. **53**(2), 217–288 (2011)
9. Hotelling, H.: Analysis of a complex of statistical variables into principal components. J. Educ. Psychol. **24**(6), 417 (1933)
10. Hwang, Y., Han, B., Ahn, H.-K.: A fast nearest neighbor search algorithm by nonlinear embedding. In: 2012 IEEE Conference on Computer Vision and Pattern Recognition (CVPR), pp. 3053–3060. IEEE (2012)
11. Indyk, P., Motwani, R.: Approximate nearest neighbors: towards removing the curse of dimensionality. In: Proceedings of the Thirtieth Annual ACM Symposium on Theory of Computing, pp. 604–613. ACM (1998)
12. Jagadish, H.V., Ooi, B.C., Tan, K.-L., Yu, C., Zhang, R.: iDistance: an adaptive B+-tree based indexing method for nearest neighbor search. ACM Trans. Database Syst. (TODS) **30**(2), 364–397 (2005)
13. Jolliffe, I.T.: Principal component analysis and factor analysis. In: Jolliffe, I.T. (ed.) Principal Component Analysis, pp. 150–166. Springer, New York (2002). https://doi.org/10.1007/0-387-22440-8_7
14. Krizhevsky, A., Sutskever, I., Hinton, G.E.: Imagenet classification with deep convolutional neural networks. In: Advances in Neural Information Processing Systems, pp. 1097–1105 (2012)

15. Li, W., Zhang, Y., Sun, Y., Wang, W., Zhang, W., Lin, X.: Approximate nearest neighbor search on high dimensional data - experiments, analyses, and improvement (v1.0). CoRR, abs/1610.02455 (2016)
16. Liaw, Y.-C., Leou, M.-L., Wu, C.-M.: Fast exact k nearest neighbors search using an orthogonal search tree. Pattern Recogn. **43**(6), 2351–2358 (2010)
17. Liu, W., Wang, J., Kumar, S., Chang, S.: Hashing with graphs. In: ICML, pp. 1–8 (2011)
18. Malkov, Y., Ponomarenko, A., Logvinov, A., Krylov, V.: Approximate nearest neighbor algorithm based on navigable small world graphs. Inf. Syst. **45**, 61–68 (2014)
19. Martinsson, P.-G., Rokhlin, V., Tygert, M.: A randomized algorithm for the decomposition of matrices. Appl. Comput. Harmonic Anal. **30**(1), 47–68 (2011)
20. Muja, M., Lowe, D.G.: Scalable nearest neighbor algorithms for high dimensional data. IEEE Trans. Pattern Anal. Mach. Intell. **36**(11), 2227–2240 (2014)
21. Ramaswamy, S., Rose, K.: Adaptive cluster distance bounding for high-dimensional indexing. IEEE Trans. Knowl. Data Eng. **23**(6), 815–830 (2011)
22. Silpa-Anan, C., Hartley, R.: Optimised KD-trees for fast image descriptor matching. In: IEEE Conference on Computer Vision and Pattern Recognition, CVPR 2008, pp. 1–8. IEEE (2008)
23. Sun, Y., Wang, W., Qin, J., Zhang, Y., Lin, X.: SRS: solving c-approximate nearest neighbor queries in high dimensional euclidean space with a tiny index. Proc. VLDB Endow. **8**(1), 1–12 (2014)
24. Torgerson, W.S.: Multidimensional scaling: I. Theory and method. Psychometrika **17**(4), 401–419 (1952)

BASSI: Balance and Status Combined Signed Network Embedding

Yiqi Chen[1], Tieyun Qian[1(✉)], Ming Zhong[1], and Xuhui Li[2]

[1] School of Computer Science, Wuhan University, Wuhan, China
{yiqic16,qty,clock}@whu.edu.cn
[2] School of Information Management, Wuhan University, Wuhan, China
lixuhui@whu.edu.cn

Abstract. Signed social networks have both positive and negative links which convey rich information such as trust or distrust, like or dislike. However, existing network embedding methods mostly focus on unsigned networks and ignore the negative interactions between users. In this paper, we investigate the problem of learning representations for signed networks and present a novel deep network structure to incorporate both the balance and status theory in signed networks. With the proposed framework, we can simultaneously learn the node embedding encoding the status of a node and the edge embedding denoting the sign of an edge. Furthermore, the learnt node and edge embeddings can be directly applied to the sign prediction and node ranking tasks. Experiments on real-world social networks demonstrate that our model significantly outperforms the state-of-the-art baselines.

Keywords: Signed network embedding · Balance theory
Status theory

1 Introduction

In recent years, online signed networks are proliferating rapidly. For example, the consumer review sites like Epinions allow members decide whether to trust each other; news websites such as Slashdot allow users to tag each other as friends or foes. In the signed networks, the relations between entities convey rich information and are signed positively or negatively. Signed network are useful in many applications like recommendation, advertisement, and community detection. Among which, sign prediction and ranking nodes constitute the foundations for sign network analysis. The task of sign prediction [9,18] is to infer signs of existing links. Links with positive signs may show trust and agreement, while negative links may represent the distrust and disagreement. It is essential to identify the positive or negative relation between two users. For example, when recommending friends for a user u in a social network, it is required not to list u's foes in the candidates. Sign prediction can be viewed as a sub-task of link prediction [15] but has received very limited attentions [3,9,17,18] by now.

The original version of this chapter was revised: the acknowledgement section was updated. The correction to this chapter is available at https://doi.org/10.1007/978-3-319-91452-7_61

J. Pei et al. (Eds.): DASFAA 2018, LNCS 10827, pp. 55–63, 2018.
https://doi.org/10.1007/978-3-319-91452-7_4

Ranking nodes aims to find out how important a node is based on its reputation [2,8]. The solution to this problem is traditionally developed in the area of webpage ranking, and the PageRank [2] and HITS [8] are two classic algorithms. More recently, ranking in signed network has aroused great research interests due to its wide applications like targeted advertisement or trust network construction. However, most of existing approaches [7,12,16] are simple modifications of PageRank and HITS and cannot deal with signed links directly. Hence new approaches are desired to include the impacts from both positive and negative links.

We follow this line of work and consider the problem of sign prediction and node ranking in signed networks. Our approach is motivated by balance [6] and status [5] theory. We build on the work of SiNE [18], which considers balance theory as "users should sit closer to their friends than their foes" and adopts a similarity based function to measure such a relation. We go beyond a single usage of balance theory and propose a novel framework which incorporates the balance and status theory into a unified deep learning framework.

2 The Proposed BASSI Model

2.1 Balance Theory and Status Theory

Balance and status are two fundamental theories in social science. **Balance theory** is originally defined for undirected networks to model the relations of likes and dislikes. It implies that "the friend of my friend is my friend" and "the enemy of my enemy is my friend" [6]. **Status theory** [5,10] is proposed to represent the social status of the people in directed networks, where the status may denote the relative ranking, prestige, or skill level. For example, a positive/negative link from a to b denotes "b has higher/lower status than a". The status relation should be transitive, which means that "a person respected me should be respected by my subordinate". In order to illustrate the balance and status theory in the directed sign network, we adopt the triangle representations [10,15] and list the possibilities of 12 triangles in Fig. 1.

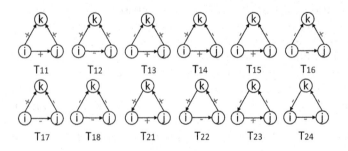

Fig. 1. Twelve types of signed triangles.

For example, for T_{15} in Fig. 1, balance theory indicates that the sign of edge e_{ij} should be a "+" (j is i's friend) given that k is an enemy of i and k is an enemy of j. Similarly, for T_{22} in Fig. 1, status theory suggests that the sign of edge e_{ij} should be a "−" (the status of i is higher than that of j) given that i has higher status than k and k has higher status than j.

In real-world signed networks, Leskovec et al. found that more than 0.9 of triangles satisfy balance [9] and status theory [10], respectively. Hence in the following, we investigate how to incorporate the balance and status theory in a unified framework.

2.2 Modeling Balance Theory

We now take the triangle T_{15} in Fig. 1 as an example to show the detail of modeling balance theory. As we illustrated in the previous section, given that $e_{jk} = -$ and $e_{ik} = -$, we have $e_{ij} = +$ if we follow the rule of "the enemy of my enemy is my friend". Mathematically, taking the sign of three edges together, our goal is to maximize the following objective function.

$$J^b_{T_{15}} = P(+|e_{ij}) + P(-|e_{jk}) + P(-|e_{ik}), \tag{1}$$

where $J^b_{T_{15}}$ is the overall probability for predicting the sign of three edges in T_{15} guided by balance theory. Taking all triangles in the network into consideration, we can define the objective function J_{bal} for triangles satisfying balance and use a loss function L_{bal} to measure the difference between the observation and the prediction as:

$$L_{bal} = L(J_{bal}) = \sum_{t \in T_{sam}} L(J^b(t)) = \sum_{t \in T_{sam}} L^b_\triangle(t) = \sum_{t \in T_{sam}} \lambda^b_{ij} L^b_{ij} + \lambda^b_{ik} L^b_{ik} + \lambda^b_{jk} L^b_{jk} \tag{2}$$

where $L^b_\triangle$ denotes the loss for sampled triangles, T_{sam} is the set of triangles satisfying balance theory and $\lambda^b_{ij}, \lambda^b_{ik}, \lambda^b_{jk}$ are the indicator function denoting whether an edge is missing from the sampled triangle.

L^b_{ij} in Eq. 2 is used to measure the difference between the predicted value $P(+|e_{ij})$ for the sign of the edge e_{ij} and the ground truth value y_{ij}, and it can be defined using a cross-entropy loss function. Hence we have:

$$L^b_{ij} = -y_{ij} \log P(+|e_{ij}) - (1 - y_{ij}) \log(1 - P(+|e_{ij})) \tag{3}$$

Similarly we can define the loss for other edges and get L^b_{ik}, L^b_{jk} in Eq. 2.

2.3 Modeling Status Theory

We continue to use the triangle T_{15} in Fig. 1 to show the detail. We denote the status ranking score of node v_i, v_j, v_k as R_i, R_j, R_k respectively. Given that $e_{ij} = +$ and $e_{ik} = -$ in T_{15}, we have the status relationships as $R_i < R_j$ and $R_i > R_k$ if we follow the rule of "the person respected by me should have higher status than me". Mathematically, taking the status relationships of two edges together, our goal is to minimize the following objective function.

$$J_{T_{15}}^s = Q(R_i < R_j, R_i - R_j) + Q(R_i > R_k, R_i - R_k), \tag{4}$$

where $J_{T_{15}}^s$ is the overall distances from status relationships $R_i < R_j$ and $R_i > R_k$, and Q is the function to measure the distances between true status relationship $(R_i < R_j)$ and predicted status relationship value $(R_i - R_j)$. Taking all triangles in the network into consideration, we can define the objective function J_{sta} for the triangles satisfying status theory and use a loss function L_{sta} to measure the difference between the observation and the prediction as:

$$L_{sta} = L(J_{sta}) = \sum_{t \in T_{sam}} L(J^s(t)) = \sum_{t \in T_{sam}} L_\triangle^s(t) = \sum_{t \in T_{sam}} \lambda_{ij}^s L_{ij}^s + \lambda_{ik}^s L_{ik}^s \tag{5}$$

L_{ij}^s in Eq. 5 is used to measure the difference between the predicted status relationship value $R_i - R_j$ from the edge e_{ij} and the "ground truth" value q_{ij}, and it can be defined using a square loss function. Hence we have:

$$L_{ij}^s = Q(R_i < R_j, R_i - R_j) = (q_{ij} - (R_i - R_j))^2 \tag{6}$$

Similarly we can define the loss for other edges and get L_{ik}^s, L_{ki}^s in Eq. 5. Since we do not have the ground truth value q_{ij} measuring status relationship between v_i and v_j, we define the following boundary function to calculate the value of q_{ij}.

$$q_{ij} = \begin{cases} max(R_i - R_j, \gamma), & i \to j : - \\ min(R_i - R_j, -\gamma), & i \to j : + \end{cases} \tag{7}$$

where γ is a threshold of status relationship value.

2.4   BASSI Model

Based on the mathematically modeled balance and status theory, we now propose a novel BASSI model to combine two theories into a unified deep network. The whole objective function of BASSI can be written as:

$$L_{all} = L_{bal} + L_{sta} + \lambda_{reg} L_{reg} \tag{8}$$

where $L_{reg} = ||\Theta||_2^2$ is the L_2 regularizer for all parameters Θ in the neural network and λ_{reg} is the corresponding weighing factor ($\lambda_{reg} = 0.0001$ in our experiments). With the objective function L_{bal} and L_{sta} given above, our task now is to find a function f to measure the probability $P(+|e_{ij})$ or $P(-|e_{ij})$ of the sign of an edge e_{ij} and a function g to get the ranking score R_i of a node v_i. Motivated by recent advances in deep learning which has been proven to be powerful in learning nonlinear representations, we propose a deep neural network to learn the embedding of nodes and edges as well as the function f and g. Figure 2 shows the architecture of BASSI.

The input to the framework is the set of triplets $(\mathbf{x_i}, \mathbf{x_j}, \mathbf{x_k})$ denoting the embedding of the nodes in triangles extracted from the signed network.

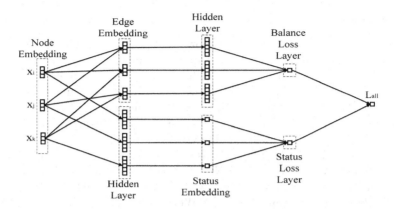

Fig. 2. Architecture of BASSI model

The upper part is the embedding process based on balance theory, where we start from the node embedding layer to optimize the relationships in a triangle in next layers. More formally, we define the probability $P(+|e_{ij})$ of the sign of an edge e_{ij} as:

$$P(+|e_{ij}) = f(\mathbf{x_i}, \mathbf{x_j}) = f(\mathbf{e_{ij}}) = \sigma(\mathbf{e_{ij}W_1} + \mathbf{b_1}), \tag{9}$$

where σ is a sigmoid function, $\mathbf{W_1}$ and $\mathbf{b_1}$ is the weight and bias in the second layer of the upper part, respectively, and $\mathbf{e_{ij}}$ is the embedding of an edge e_{ij} and defined as:

$$\mathbf{e_{ij}} = \mathbf{x_i W_{eh}} + \mathbf{x_j W_{et}} + \mathbf{b_e}, \tag{10}$$

where $\mathbf{W_{eh}}$ and $\mathbf{W_{et}}$ are the weights and $\mathbf{b_e}$ the bias in the first layer of the upper part.

The lower part is the embedding process based on status theory, where we begin with the node embedding layer to model the relationships between nodes in the point of view of ranking scores. Formally, we define the following ranking score function g for a node v_i as:

$$R_i = g(v_i) = \sigma((\mathbf{x_i W_2} + \mathbf{b_2})\mathbf{W_3} + \mathbf{b_3}), R_i \in (0,1), \tag{11}$$

where $\mathbf{W_2}$, $\mathbf{b_2}$, and $\mathbf{W_3}$, $\mathbf{b_3}$ are the weights and biases in the first and second layer of the lower part.

To train our BASSI model, we simply adopt the stochastic gradient descent approach [1] to update the parameters in the neural network and get the representations for nodes $\mathbf{X}$ in the signed network.

3   Experimental Evaluation on Sign Prediction

We first conduct the sign prediction experiment to check whether our embeddings improve the performance of signed network analysis [18].

3.1 Settings

We conduct experiments on two well known and publicly available signed social network datasets. Slashdot [10] is a technology-related news website known for its specific user community. Users are allowed to tag each other as friends or foes. Epinions [10] is an online social network of a general consumer review site. Members of the site can decide whether to "trust" each other.

For this prediction experiment, we use four state-of-the-art baselines: Deep-Walk [11], LINE [13], FExtra [9], and SiNE [18]. We randomly select 80% edges as training set and the remaining 20% as the test set. After we get node embeddings from DeepWalk, LINE or SiNE, we follow the way in [4] to get edge features through four operators (average, Hadamard, weighted-L1, weighted-L2) and then choose the best one to display. With the learnt edge features, we train a logistic regression classifier on training set and use it to predict the edge sign in test set. We run with different random seed for 5 times to get the average scores. We set all the embedding dimension 20 as SiNE [18] does. For the main hyper-parameters in BASSI, we set $epoch = 10$, status difference gap $\gamma = 0.5$, missing third edge learning rate $\eta = 0.01$. For baselines, we follow the default settings in their paper.

3.2 Results for Sign Prediction

We report the average auc, macro-F1 and micro-F1 as evaluation metrics as those in [9,18]. Table 1 shows the results. Scores in bold denote the highest performance among all methods.

Table 1. Results for sign prediction on Slashdot and Epinions

Dataset	Metric	DW	LINE	SiNE	FExtra	BASSI
Slashdot	auc	0.7703	0.5470	0.8357	0.8866	**0.9041**
	macro-F1	0.6005	0.4364	0.7174	0.7391	**0.7906**
	micro-F1	0.7775	0.7742	0.8260	0.8465	**0.8601**
Epinions	auc	0.8170	0.5704	0.7822	0.9446	**0.9503**
	macro-F1	0.6141	0.4605	0.6154	0.8070	**0.8714**
	micro-F1	0.8693	0.8533	0.8627	0.9213	**0.9387**

We can observe that on both datasets our BASSI model significantly outperforms all baselines in terms of three metrics. (1) DeepWalk and LINE are the worst, showing that it is not suitable to directly apply unsigned network embedding methods to this problem. (2) SiNE performs better than LINE and DeepWalk in most of cases while it is worse than FExtra and BASSI. This can be due to that SiNE ignores sign's direction information as it is designed for undirected signed network. (3) FExtra is the second best, but it cannot compete with BASSI. For example, BASSI has the macro-F1 score of 0.8714, while that for FExtra is 0.8070, showing a 7.98% increase. The reason can be that the embedding process in BASSI can capture more complex latent relations than the feature-engineering approach FExtra.

4   Experimental Evaluation on Node Ranking

Based on the status theory in signed network, the ranking score of each node can be seen as its status. To demonstrate the property of such ranking scores in real-world signed social network, we design a status comparison experiment. Specifically, we simplify the comparison procedure used in the [12,14] as follows: we take the test edges as the ground truth, and compare the status between two adjacent nodes by their ranking scores. For example, $v_i \xrightarrow{+(-)} v_j$ can be transformed into $R_i < (>)R_j$ based on status theory. In this way, we can measure how the ranking scores generated by different methods are consistent with the ground truth.

We conduct the status comparison experiments on the Epinions and Slashdot datasets. We use the following six baselines: Prestige [20], PageRank [2], Exp [16], PageRank [2], MPR [12], MHITS [19] and Troll-Trust [19]. We use 80% edges as training set and 20% edges as test set and use accuracy an the evaluation metric as that in [12,19]. For Troll-Trust, we test different combinations of $\beta = [0.01,$ 0.1, 0.2, 0.5, 0.9] and $\lambda_1 = [0.1, 0.5, 1.0, 5.0, 10.0, 100.0]$ and choose the best results on Slashdot and Epinions. For other baselines, we follow the setting of Troll-Trust [19]. For our BASSI model, we simply get the ranking scores from the model trained in the sign prediction task. The results are shown in Table 2.

Table 2. Accuracy for status comparison on Slashdot and Epinions

Dataset	Method						
	Prestige	PageRank	Exp	MPR	MHITS	Troll-Trust	BASSI
Slashdot	0.4619	0.6273	0.5920	0.5815	0.5518	0.5915	**0.7345**
Epinions	0.5134	0.6515	0.6457	0.6503	0.5883	0.6424	**0.7270**

It is clear that our BASSI model preserves the status property significantly better than any other ranking methods in both networks. This could be attributed to that BASSI unifies two theories in a framework, and it benefits from the balance theory to learn better status representations for the nodes.

5   Conclusion

In this paper, we propose a novel BASSI (BAlance and Status combined SIgned network embedding) model to learn the node and edge representations in social networks. In particular, we first define two new objective functions to mathematically modeling the balance theory and status theory. We then design a deep neural structure to combine two theories in a unified framework. Based on the deep network, we learn the node embedding and edge embedding denoting the status of a node and the sign of an edge, which can be directly used in node ranking and sign prediction tasks. We conduct extensive experiments on real world networks. The results demonstrate that our model significantly outperforms the state-of-the-art baselines.

In the future, we plan to investigate how the learnt representations can be used in other applications like community detection or recommendation. We are also interested in making connections between our model and the social properties of real world networks.

Acknowledgement. The work described in this paper has been supported in part by the NSFC projects (61572376, 91646206), the 111 project (B07037), and Natural Science Foundation of Hubei Province under Grant No. 2018CFB616.

References

1. Bottou, L.: Stochastic gradient learning in neural networks. Proc. Neuro-Nımes **91**(8) (1991)
2. Brin, S., Page, L.: Reprint of: the anatomy of a large-scale hypertextual web search engine. Comput. Netw. **56**(18), 3825–3833 (2012)
3. Chiang, K.Y., Natarajan, N., Tewari, A.: Exploiting longer cycles for link prediction in signed networks. In: Proceedings of CIKM, pp. 1157–1162 (2011)
4. Grover, A., Leskovec, J.: node2vec: Scalable feature learning for networks. In: Proceedings of SIGKDD, pp. 855–864 (2016)
5. Guha, R., Kumar, R., Raghavan, P., Tomkins, A.: Propagation of trust and distrust. In: Proceedings of WWW, pp. 403–412. ACM (2004)
6. Heider, F.: Attitudes and cognitive organization. J. Psychol. **21**(1), 107–112 (1946)
7. de Kerchove, C., Van Dooren, P.: The PageTrust algorithm: how to rank web pages when negative links are allowed? In: Proceedings of SDM, pp. 346–352. SIAM (2008)
8. Kleinberg, J.M.: Authoritative sources in a hyperlinked environment. JACM **46**(5), 604–632 (1999)
9. Leskovec, J., Huttenlocher, D., Kleinberg, J.: Predicting positive and negative links in online social networks. In: Proceedings of WWW, pp. 641–650. ACM (2010)
10. Leskovec, J., Huttenlocher, D., Kleinberg, J.: Signed networks in social media. In: Proceedings of SIGCHI, pp. 1361–1370. ACM (2010)
11. Perozzi, B., Al-Rfou, R., Skiena, S.: Deepwalk: online learning of social representations. In: Proceedings of SIGKDD, pp. 701–710 (2014)
12. Shahriari, M., Jalili, M.: Ranking nodes in signed social networks. SNAM **4**(1), 172 (2014)
13. Tang, J., Qu, M., Wang, M., Zhang, M., Yan, J., Mei, Q.: Line: large-scale information network embedding. In: Proceedings of WWW, pp. 1067–1077 (2015)
14. Tang, J., Chang, S., Aggarwal, C., Liu, H.: Negative link prediction in social media. In: Proceedings of ICDM, pp. 87–96. ACM (2015)

15. Tang, J., Chang, Y., Aggarwal, C., Liu, H.: A survey of signed network mining in social media. ACM Comput. Surv. (CSUR) **49**(3), 42 (2016)
16. Traag, V.A., Nesterov, Y.E., Van Dooren, P.: Exponential ranking: taking into account negative links. In: Bolc, L., Makowski, M., Wierzbicki, A. (eds.) SocInfo 2010. LNCS, vol. 6430, pp. 192–202. Springer, Heidelberg (2010). https://doi.org/10.1007/978-3-642-16567-2_14
17. Wang, S., Aggarwal, C., Tang, J., Liu, H.: Attributed signed network embedding. In: Proceedings of CIKM, pp. 137–146. ACM (2017)
18. Wang, S., Tang, J., Aggarwal, C., Chang, Y., Liu, H.: Signed network embedding in social media. In: Proceedings of SDM, pp. 327–335. SIAM (2017)
19. Wu, Z., Aggarwal, C.C., Sun, J.: The troll-trust model for ranking in signed networks. In: Proceedings of ICDM, pp. 447–456. ACM (2016)
20. Zolfaghar, K., Aghaie, A.: Mining trust and distrust relationships in social web applications. In: Proceedings of ICCP, pp. 73–80. IEEE (2010)

Recommendation

Geographical Relevance Model for Long Tail Point-of-Interest Recommendation

Wei Liu[1,5], Zhi-Jie Wang[1,5(✉)], Bin Yao[2,5,6], Mengdie Nie[1,5], Jing Wang[3],
Rui Mao[4,6], and Jian Yin[1,5(✉)]

[1] School of Data and Computer Science, Sun Yat-Sen University, Guangzhou, China
hugh.wei@foxmail.com, {wangzhij5,issjyin}@mail.sysu.edu.cn,
niemengdie@icloud.com
[2] Shanghai Jiao Tong University, Shanghai, China
yaobin@cs.sjtu.edu.cn
[3] Neusoft Institute Guangdong, Foshan, China
jingyun_wj@163.com
[4] Shenzhen University, Shenzhen, China
mao@szu.edu.cn
[5] Guangdong Key Laboratory of Big Data Analysis and Processing,
Guangzhou, China
[6] Guangdong Province Key Laboratory of Popular High Performance Computers,
Shenzhen, China

Abstract. Point-of-Interest (POI) recommendation plays a key role in people's daily life, and has been widely studied in recent years, due to its increasingly applications (e.g., recommending new restaurants for users). One of important phenomena in the POI recommendation community is the *data sparsity*, which makes deep impact on the quality of recommendation. Existing works have proposed various models to alleviate the bottleneck of the data sparsity, and most of these works addressed this issue *from the user perspective*. To the best of our knowledge, few attention has been made to address this issue *from the POI perspective*. In this paper, we observe that the "long tail" POIs, which have few checkins and have less opportunity to be exposed, take up a great proportion among all the POIs. It is interesting and meaningful to investigate the long tail POI recommendation *from the POI perspective*. To this end, this paper proposes a new model, named GRM (geographical relevance model), that expands POI profiles via relevant POIs and employs the geographical information, addressing the limitations of existing models. Experimental results based on two public datasets demonstrate that our model is effective and competitive. It outperforms state-of-the-art models for the long tail POI recommendation problem.

Keywords: Long tail · Relevance model · Geographical information
Point-of-interest recommendation

© Springer International Publishing AG, part of Springer Nature 2018
J. Pei et al. (Eds.): DASFAA 2018, LNCS 10827, pp. 67–82, 2018.
https://doi.org/10.1007/978-3-319-91452-7_5

1   Introduction

Nowadays, location-based services are widely used in our daily life [16, 17, 30, 45, 46]. For instance, Yelp and Meituan can help individuals discover favourite foods, shopping malls, hotels, etc. Foursquare and Webchat can assist people in discovering fun places, friends' footprint, etc. Almost all these applications incorporate the function of Point-of-Interest (POI) recommendation (which is similar to *trip recommendation* [36]). Particularly, POI recommendation can help these applications understand individuals' favourites more deeply, and so it has the potential to provide personalized services for individuals. Besides, it can also help merchants to solicit more potential customers.

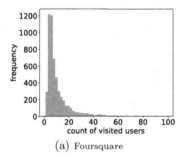

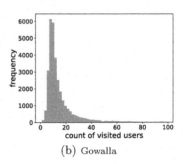

(a) Foursquare   (b) Gowalla

Fig. 1. Histogram of the visiting frequency of all POIs (a) distribution on Foursquare. (b) distribution on Gowalla.

In past decades, there are numerous papers studying the problem of POI recommendation [3, 11]. For example, some works used the rich context information (e.g., geographical information [7, 8, 10, 18, 20, 47] and content information [2, 5, 11]) to address data sparsity in POI recommendation. Some works (e.g., [12]) discussed the diversity of POI recommendation. In the existing literature, most of works focused on POI recommendation *from the user perspective* (e.g., [1–3, 5]). That is, prior works mainly concentrated on developing effective solutions to recommend POIs for individuals (i.e., users). In addition, most of models/methods (e.g., [1, 19]) developed in prior works are inclined to recommend "popular" POIs for users. As such, the *POIs with less check-ins* have less chance to be exposed, incurring a lot of "long tail" POIs. For example, Fig. 1 shows the distribution of check-ins on public datasets "Foursquare and Gowalla" (other details are listed in Table 1). From this figure, we can see that a few POIs' records are more than 40, yet a majority of POIs have less than 10 records. Clearly, the distribution of check-ins on POIs is **extremely skewed**. This observation is consistent with the findings in recent works [14, 15, 17]. Note that, the frequently visited POIs mostly belong to popular locations, and existing models/systems usually tend to recommend these "popular" POIs. Naturally, this leads most of unpopular POIs left out (i.e., the long tail POIs appear). The above phenomenon is usually called the *long tail phenomenon*.

Table 1. Details of two datasets

Dataset	#User	POI	#Check-in	#Avg.check-in	#Avg.visited	Density
Foursquare	2,321	5,596	194,108	83.6	34.64	1.49%
Gowalla	10,162	24,250	456,988	44.97	18.8	0.18%

Essentially, existing works (e.g., [35,37]) have already addressed the long tail phenomenon by recommending long tail POIs to users. By doing so, it is benefit for users (e.g., exploring the fresh environment) and also for merchants (e.g., the promotion of niche POIs contributes to the increase of revenue). Furthermore, it is worth noting that the good quality of POI recommendation includes not only the accuracy but also other ingredients such as serendipity and novelty. In this regard, recommending long tail POIs to users are also helpful for improving the recommendation quality. It can be seen that, the essence of their works is also from the user perspective, although they recommended "long tail" POIs instead of "popular" POIs. Specifically, in this paper we attempt to improve the recommendation quality by finding/recommending the best users for each long tail POI. For clarity, we call it **long tail POI recommendation**. To the best of our knowledge, few attention has been made to investigate the long tail POI recommendation. It is not hard to understand that, this idea should be especially effective for addressing the long tail phenomenon, since it can allow each (current) long tail POIs to match n best users, significantly alleviating the long tail POIs.

Although the idea above seems to be powerful, there are still many challenges needing to be addressed. (i) For long tail POI recommendation, the recommendation accuracy is also deeply troubled by *data sparsity*, it could incur the learning insufficient, when the recommender learns user's preference features and POIs' attributive features. (ii) The data structure of user-POI and check-in records is usually extremely skewed. For example, as shown in Table 1, the quantity of POIs is much more than the quantity of users, and the number of average check-ins of POIs is much less. It could be hard to characterize the features of POIs.

To alleviate various challenges and address the long tail recommendation problem, we propose a geographical relevance model (GRM) that employs two key ideas: (i) expanding POI profiles via relevant POIs to obtain candidate users, and (ii) developing the "mixed" similarity that employs the geographical information to further remedy the data sparsity. Our model builds a statistical relevance model for each long tail POI, which enables us to recommend target users for long tail POIs conveniently and effectively.

To summarize, our main contributions are as follows:

- We formally define the long tail recommendation problem from the POI perspective.
- We present geographical relevance model to address the long tail POI recommendation problem.

- We conduct extensive experiments to verify the effectiveness of our proposed model. Experimental results consistently demonstrate that our proposed model is effective and competitive.

The rest of this paper is organized as follows. In Sect. 2, we review the related work and define the long tail POI recommendation problem formally. In Sect. 3, the geographical relevance model is proposed, and the experimental results are discussed in Sect. 4. Finally, we conclude the paper in Sect. 5.

2   Preliminaries

In this section, we first review previous works most related to ours (Sect. 2.1), and then define our problem formally (Sect. 2.2).

2.1   Related Work

In recent years, *POI recommendation* has attracted much attention due to various applications [9,24,25,27,28,33,39–41,44]. In existing works, many researches focused on improving the recommendation *accuracy*. There are many representative methods such as the memory-based *collaborative filtering* (CF) method [8,20,21], the model-based CF methods [7,23], the weighted matrix factorization based methods [19], etc. Particularly, recent studies have also attempted to adapt implicit feedback data to improve the recommendation *quality* [1,11,24,25].

It is well known that *data sparsity* makes deep impact on the recommendation quality [7,25]. To cope with this issue, prior works have proposed many effective techniques by utilizing various context information. The main context information used in the POI recommendation community includes: (i) Geographical information [1,7,8,19,20,27,28]. For example, Zhang and Chow [8] suggested to depict individual's personalized geographical distribution, instead of using a universal distribution for all individuals; the personalized *geographical distribution information* is utilized to improve the recommendation accuracy. Ye et al. [20] used the power law distribution to characterize geographical *clustering* [13] phenomenon, and they utilized the characterized *geographical clustering information* to improve the recommendation accuracy. (ii) Content information [5,10,11,29,30]. For instance, Wang et al. [10] exploited a general model to characterize individual's interest and the crowd preference to help user explore unacquainted environments. He et al. [11] proposed an efficient method for POI recommendation, which learns individual's category preference first (by a listwise ranking approach), then selects POI candidates in the recommended category (by the spatial influence and the category ranking influence). (iii) Temporal information [24–26,31,32]. For example, Yuan et al. [26] and Gao et al. [31] split check-ins into different time bins, and then learnt the temporal pattern of individual's preference. Feng et al. [24] and Liu et al. [25] utilized the sequential relationship between two check-ins to recommend POIs for users. (iv) Social relationship [2,6,14,20,35]. For example, Zhang and Chow [2] aggregated the check-in frequency of a user's friends on the POI, and modelled the social check-in frequency as a power-law distribution. And (**v**) trajectory information [4,22,34]. For example, Wang *et al.* [34] utilized a gravity model to estimate spatial influence using trajectory information.

Among the works mentioned above, the ones highly related to ours could be [1,7,8,19,20,27,28], since both these works and ours utilize the geographical information. Particularly, the work closest to our could be [8], since both their model and ours (i) belong to the collaborative filtering based methods; (ii) utilize the geographical information, e.g., some similar probability density functions are employed to estimate the spatial relationship. Nevertheless, our work is different from their work in several aspects at least: (i) their work directly utilized similar neighbourhoods as the candidates, but the long tail POI has few "check-ins" users, it is inaccurate to directly employ similar users. Therefore, to remedy data sparsity, our work utilize a relevance model to expand POI's profile; (ii) their work utilized the probability density estimation for user-POI spatial relationship, whereas we utilize the kernel density estimation to calculate the spatial relationship between POIs; and (iii) their work employed the spatial relationship to represent part of user's rating on POI, yet we utilize the spatial relationship to expand target POI's neighbours and to remedy the data sparsity.

Besides the works mentioned above, another type of works [35,37] are also highly related to ours, since both their papers and ours discuss the long tail issue in POI recommendation. For example, Yin *et al.* [35] represented the user-item information with undirected edge-weighted graph, and extended Hitting Time algorithm to help users find their favourite long tail items. Valcarce *et al.* [37] proposed an item relevance model to help vendors get rid of long tail products. Nevertheless, these models/methods cannot efficiently work for our problem, since their works addressed the long tail issue (in POI recommendation) from the user perspective. Instead, in this paper we focus on the long tail issue (in POI recommendation) from the POI perspective. That is, unlike the traditional POI recommendation, here we are interested in recommending users for long tail POIs, instead of recommending long tail POIs for users. We would like to point that, the work in [48] mentioned the inverted version of the classic POI recommendation problem. Nevertheless, the focus of their paper is still to address how to suggest POIs to users with a better quality. Hence, our work is essentially different from theirs.

2.2 Problem Definition

As discussed before, most of existing works for POI recommendation from the user perspective. That is, they focused on the recommendation task—finding POIs for users. Instead, in this paper we are interested in the inverted version of the classic recommendation task. That is, we want to find best users for each long tail POI. Addressing this problem is helpful for alleviating the long tail phenomenon, as mentioned earlier. In what follows, we formally formulate our problem.

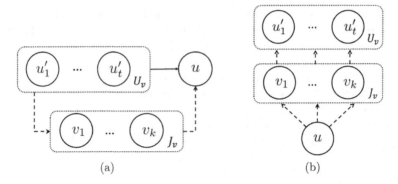

Fig. 2. Illustration of GRM. (a) Relevance model R_v for POI v. (b) Derivation of the estimating process.

Denote by U the set of (all) users, V the set of (all) POIs, and V' the set of long tail POIs, where $V' \subset V$. Similar to [20], let C be the user-POI matrix that represents the relationship between each user and POI. The check-in activity of a user u at a POI v is denoted as $c_{u,v}$. Note that, $c_{u,v}$ could only be 1 or 0. When $c_{u,v} = 1$, it means u has a check-in at v in the past; otherwise, u has no check-in at v. It is easy to understand that each POI contains some "history" visitors. We use $U_v = \{u'_1, ..., u'_t\}$ to denote the set of visitors, which can be viewed as the *POI profile*. Based on U_v, our task is to compute the probability $p(u|U_v)$ for each user u, and then recommend top-n users for the corresponding long tail POI v. For ease of discussion, we use a list, L_v^n, to represent/store the ranked top-n users for a POI v. Note that, the key challenge is to compute $p(u|U_v)$, since the long tail POI has few visitors. In the next section, we show how to address this problem in detail.

3   Geographical Relevance Model

We first describe the basic idea of our model, and then show how to obtain the important elements used in our model.

▶ **Basic idea.** The recommendation can be considered as a process of expanding POI profiles with candidate users, as shown in Fig. 2(a) (see the upper part). Here $U_v = \{u'_1, ..., u'_t\}$ is the set of users who visited the POI v; meanwhile, it also represents the profile of POI v, which shall be used to estimate the probability of a user u, $p(u|U_v)$.

To expand POI profiles, typically, one can directly employ similar users with visitors in POI profiles. Yet, for the long tail POI, it has few check-in records. Thus, directly obtaining similar users on the corresponding long tail POI is difficult for traditional collaborative filtering models. However, via relevance model R_v, the computation of the user probability on the profile of POI v can be estimated by the set of relevant POIs J_v. To understand, Fig. 2(a) (see the lower part) shows the set of relevant POIs J_v corresponding to the relevance model R_v, for a long tail POI v. The computation problem of $p(u|U_v)$ is transformed into $p(u|R_v)$. Note that, throughout this paper we assume users recorded in POI's profile are independent from each other, but dependent on the users recorded in the profiles of relevant POIs.

▶ **Computing the probability** $p(u|R_v)$. To this step, one can easily verify that the long tail POI recommendation can be roughly considered as the problem of computing the probability of a user u under the relevance model of long tail POI v. In what follows, we show how to obtain $p(u|R_v)$ for each user $u \in U$. This step is especially important, since it serves as a vital ingredient of our model. At a high level, our technique for estimating the probability is inspired by [38], in which a relevance-based Language Models is developed for pseudo-relevance feedback. Specifically, the derivation process is as follows:

- Since the probability of next user u (sampled from the relevance model) relies on the set of users $U_v = \{u'_1, ..., u'_t\}$ who visited the POI v, one can have $p(u|R_v) \approx p(u|u'_1, ...u'_t)$. By applying the definition of conditional probability, one can have $p(u|u'_1, ...u'_t) = \frac{p(u, u'_1, ...u'_t)}{p(u'_1, ...u'_t)}$. Note that, the denominator in the fraction can be ignored, since it remains constant for the same POI v. So, we have

$$p(u|R_v) \approx p(u|u'_1, ...u'_t) = \frac{p(u, u'_1, ...u'_t)}{p(u'_1, ...u'_t)} \propto p(u, u'_1, ...u'_t) \quad (1)$$

- To estimate the relevance model of POI v, R_v, we pick a user u given the prior probability $p(u)$. The sampling probability of users $u'_1, ...u'_t$ depends on user u, as shown in the lower part of Fig. 2(b). To estimate the conditional probability $p(u'|u)$, we sample a user $u' \in U_v$ from the relevant POI's v_j distribution with probability $p(u'|v_j)$, as depicted in the upper part of Fig. 2(b). The formula is shown as follows.

$$p(u, u'_1, ...u'_t) = p(u) \prod_{u' \in U_v} p(u'|u) \approx p(u) \prod_{u' \in U_v} \sum_{v_j \in J_v} p(u'|v_j) p(v_j|u) \quad (2)$$

- By Bayes' Theorem, we have $p(v_j|u) = \frac{p(u|v_j)p(v_j)}{p(u)}$. By combing Formula 2, we have $p(u, u'_1, ...u'_t) \approx p(u) \prod_{u' \in U_v} \sum_{v_j \in J_v} p(u'|v_j) \frac{p(u|v_j)p(v_j)}{p(u)}$. Further, by combing Formula 1, we have

$$p(u|R_v) \propto p(u) \prod_{u' \in U_v} \sum_{v_j \in J_v} p(u'|v_j) \frac{p(u|v_j)p(v_j)}{p(u)} \quad (3)$$

- We next show how to obtain each element in Formula 3. We can see that there are four main elements: $p(u|v_j)$, $p(u'|v_j)$, $p(u)$, and $p(v_j)$. We consider the uniform distribution for $p(u)$ and $p(v_j)$ in this paper (i.e., $p(u) = \frac{1}{|U|}, p(v_j) = \frac{1}{|V|}$). To compute $p(u|v_j)$, the maximum likelihood estimation of a multinomial distribution over the check-ins can be used for this task. However, this method may suffer from the data sparsity. In our paper, we use Absolute Discounting (AD) [42] to smooth the maximum likelihood estimation

$p(u|v_j) = \frac{c_{u,v_j}}{\sum_{u' \in U_{v_j}} c_{u',v_j}}$ with the user probability in the collection. In brief, we use AD to subtract the same constant, δ, from the count of all the seen check-ins. The selection of an appropriate δ will be detailed in experiment. Then, a count (proportional to the probability in the collection) is added to each user. Then, we have

$$p(u|v_j) = \frac{max(c_{u,v_j} - \delta, 0) + \delta|U_{v_j}|p(u|C)}{\sum_{u' \in U_{v_j}} c_{u',v_j}} \tag{4}$$

where $p(u|C)$ is computed as follows:

$$p(u|C) = \frac{\sum_{j \in V} c_{u,v_j}}{\sum_{u' \in U, v_j \in V} v_{u',v_j}} \tag{5}$$

Note that, $p(u'|v_j)$ can be obtained using the same method above.

To this step, it remains to explain how to obtain J_v (i.e., relevant POIs), although we briefly describe it in Fig. 2(a). Note that, J_v is a critical component in our model. We next address how to obtain it in detail.

▶ **Computing relevant POIs J_v.** The key of point for obtaining J_v is the computation of the similarity of v_i and v_j. To achieve this, our method consists of several steps: (i) computing the general similarity of v_i and v_j (in our paper, we use the cosine similarity); (ii) computing the spatial similarity; (iii) combing the "mixed" similarity by combing the above two types of similarities. We next discuss each step.

- Since the similar POIs are the approximation of the real relevant POIs, we call the similar POIs with respect to POI v are the pseudo-relevant POIs in the relevance model R_v. In our paper, we employ kNN algorithm to compute the set of pseudo-relevant POIs based on the check-ins, according to the *pairwise similarity* [43]. Here, we utilize cosine similarity, which yields better results in our experiments. The cosine similarity s between two POIs v_i and v_j is as follows:

$$s(v_i, v_j) = \frac{\sum_{u \in U_{v_i} \cap U_{v_j}} c_{u,v_i} c_{u,v_j}}{\sqrt{\sum_{u \in U_{v_i}} c_{u,v_i}^2 \sum_{u' \in U_{v_j}} c_{u',v_j}^2}} \tag{6}$$

- It is easy to understand that, the normalized distance between two POIs can be used as the spatial similarity. However, the spatial similarity is not linear relationship with distance. To obtain POIs' spatial similarity from the distance information and reflect their non-linear relationship, we use the kernel estimation method, which is a non-parametric way to estimate the probability density function of a random variable. The spatial similarity $sp(v_i, v_j)$ is computed as

$$sp(v_i, v_j) = \frac{1}{\sqrt{2\pi}h} e^{-\frac{(d_{v_i v_j})^2}{2h^2}}, \tag{7}$$

where $d_{v_i v_j}$ is the spatial distance between v_i and v_j. The method to select the bandwidth h will be introduced in Sect. 4.2.

- To make our model comprehensive and robust, following prior works [7,8,20] we integrates multiple similarity functions into one. Specifically, we here integrate the general similarity and the spatial similarity. For clarity, we denote by $s_m(v_i, v_j)$ the "mixed" similarity, which is computed as follows:

$$s_m(v_i, v_j) = (1 - \alpha)s(v_i, v_j) + \alpha \cdot sp(v_i, v_j)$$
$$0 \le \alpha \le 1, \tag{8}$$

where α is a parameter used to balance the weight. Since the above formula is a linear combination of different factors, there is an extra variable α to be inferred. One can easily understand that, when facing new datasets, the variable may need to re-adjust. To address this trouble, our proposed method utilizes the proportion of two factors in exponential space to replace the extra variable. This way, it can avoid this extra variable, as shown below.

$$s_m(v_i, v_j) = \frac{\exp(s(v_i, v_j))}{Z} s(v_i, v_j) + \frac{\exp(sp(v_i, v_j))}{Z} sp(v_i, v_j)$$
$$Z = \exp(s(v_i, v_j)) + \exp(sp(v_i, v_j)). \tag{9}$$

4   Performance Evaluation

In this section, we first cover the experimental settings including datasets, evaluation metrics and benchmark models, and then discuss the experimental results.

4.1   Experimental Settings

In the experiments, we use two real datasets to evaluate our proposed method. One is **Foursquare**[1], and another is **Gowalla**[2]. The Foursquare dataset is made of 2,321 users on 5,596 POIs from August 2010 to July 2011. In contrast, the Gowalla dataset is made of 10,162 users on 24,250 POIs from February 2009 to October 2010. Both datasets are very sparse. More details about them can be found in Table 1 (recall Sect. 1).

Since this paper deals with a novel recommendation task, i.e., an inverted version of the classic POI recommendation problem, in our experiments we consider the POIs with less than 10 visitors as the long tail POIs. We randomly select 50% of the visited users as the training set, and the rest of users as a test set. We learn user's preference to POIs from the training set, and then we recommend the best candidate users for each long tail POI. The recommendation model is to rate each users unvisited and rank them by the ratings. Then it returns the top-n users as the recommendation list (to the POI). By comparing the recommendation list and test set, we evaluate the model's accuracy. Evaluation metrics include Pre@n and Rec@n, which are computed as

---

[1] Foursquare is available at https://pan.baidu.com/s/1hrYNwJM.
[2] Gowalla is available at https://pan.baidu.com/s/1i4DgFmX.

$$\text{Pre@}n = \frac{1}{|V'|} \sum_{v=1}^{|V'|} \frac{|L_v^n \cap T_v|}{n}$$

$$\text{Rec@}n = \frac{1}{|V'|} \sum_{v=1}^{|V'|} \frac{|L_v^n \cap T_v|}{|T_v|}. \tag{10}$$

where L_v^n represents the top-n users recommended by the model for POI v, T_v represents the user set in which the users really visited POI v, and $|V'|$ is the number of long tail POIs. In our experiments, we run 5 times for each test and report the average value.

To evaluate the performance of our proposed model, we compare it against five baselines. Since no targeted model exists for the task discussed in this paper, we adapt state-of-the-art models to achieve the task. The details are as follows.

- *Popularity.* This model is a classic and naive recommender model, but is usually used to compare sophisticated models. This model chooses the most popular users for all the POIs in traditional recommendation task. In our task, it implies that, for each POI, it recommends the same set of users. For concise, we call this baseline **Pop**.
- *User-based neighborhood recommender.* They employ the classic collaborative filtering technique, which is used to compute a set of k nearest neighbors (kNN) for each user or POI [43]. The neighbourhood relationships are computed using pairwise similarities (e.g. Pearson's correlation coefficient). The recommender aims to predict the probability of the target user, based on the check-ins of the neighbourhoods. For short, we name the user-based neighborhood recommender as k**NN-UB**. For recommending users to long tail POIs, we generate a recommendation list L_v for each $v \in V'$. This list contains those users $u \in U$ with the largest prediction ratings.
- *Item-based neighborhood recommender.* It is almost the same with the above recommend, except the it is based on the item, while kNN-UB is based on the user. For short, we name it as k**NN-IB**.
- *Hitting time.* This recommender is designed for recommending long tails to users [35], which is contrary to our task. The method overcomes the data sparsity by considering the recommendation task as a random walk in a graph. We utilize this model to build an edge-weighted undirected graph, in which the nodes are POIs and users. Each rating is the weight of an edge that connects two nodes (i.e., user and POI). Given such a graph, their method computes the hitting time from POI v with respect to the target user u, which is the average number of steps that a random walker needs to take (from node v to node u). For short, we call it **HT**.
- *Rank-GeoFM.* It is a ranking-based model that learns users' preference rankings, and includes the geographical influence of neighbouring POIs [1]. This technique is one of the strongest state-of-the-art top-N recommendation model. We first use this model to estimate each user's rating on the target POI, and then get the recommendation list for each long tail POI.

4.2   Experimental Results

Figure 3(a) shows the results by varying k (the number of nearest neighbourhoods). In this set of experiments, we can see that, when k is small, the model has few relevant POIs to expand POI profiles. However, when k exceeds a threshold ($k = 40$ in Foursquare, $k = 30$ in Gowalla), the relevant POIs set introduce more noise data. At this time, it is not helpful to improve model performance by increasing k. Here the number of nearest neighborhoods k used in the rest of experiments is set to 40 for Foursquare and 30 for Gowalla, unless stated otherwise. To study the impact of bandwidth h in the spatial similarity (recall Sect. 3), we test different settings for this parameter. Figure 3(b) shows the results of Pre@5 on both datasets. We can see from the figure that GRM first increases and then decreases (for Foursquare), when h increases. Particularly, when $h = 0.2$, its performance reaches the best. One possible reason is that when the distance of two POIs is less than 0.2 km, POI's property is more likely to be similar. Yet, when the distance is more than 0.2 km, the similarity between POI's property is fluctuated drastically. The tendency on Gowalla is similar to that on Foursquare. In the rest of experiments, the parameter h is set to 0.2, unless stated otherwise. Besides, we set δ in Foursquare as 0.6, δ in Gowalla as 0.4.

As we expected, the popularity method is poor in our task (cf., Fig. 4). Interestingly, the classic neighbourhood methods (kNN-UB and kNN-IB) perform even worse than this naive strategy in most of experiments (cf., Fig. 4). It indicates that traditional neighbourhood algorithms could be unsuitable for this task. This is mainly because computing neighbourhoods for long tail POIs is difficult, due to the fact that the long tail POI has few check-ins. Essentially, this indirectly shows that the pairwise similarities, such as Pearson's correlation coefficient used in neighbourhood methods, has little help to improve the performance, when only few co-occurrences between vectors are available. On the other hand, it also implies that finding user neighbourhoods who have information about long tail POIs is even more challenging, although finding neighbourhoods for long tail POIs is also challenging.

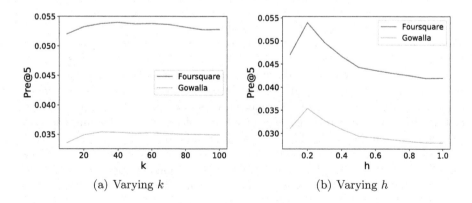

(a) Varying k (b) Varying h

Fig. 3. Results by varying k and h on two public datasets: Foursquare and Gowalla

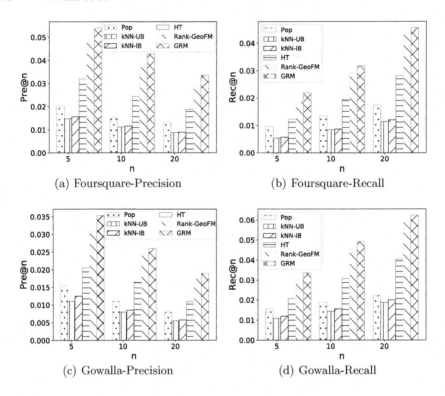

Fig. 4. Performance results by comparing GRM against others competitors.

Furthermore, we observe that the item-based approach (*k*NN-IB) performs similarly, or better than the user-based counterpart (*k*NN-UB) in all the tested scenarios (cf., Figs. 4 and 5). Note that, it has been verified [43] that, for the traditional recommendation task, the item-based approaches tend to achieve better accuracy, compared to most of methods. Fortunately, for the novel recommendation task, we observe from Fig. 4 that the item-based approaches are also not bad.

From Fig. 4, we can see that the performance of the HT method is good. This could be due to the fact that HT computes the average number of steps, for which a random walker needs to go from one node to another. This way, the model can generate recommendations for both users and POIs, because all of them are nodes in the same graph (connected by check-ins). Thus, this model does not establish any type of difference between users and POIs. The symmetry of this model between both entities is the key of point for achieving the good recommendation quality in this novel recommendation task.

Even so, the strongest baseline among these five baselines could be Rank-GeoFM, as shown in Fig. 4. We observe that, Rank-GeoFM produces scores for user and POIs. Thus, the creation of a recommendation list in this task is done by sorting users in descending order, with respect to the corresponding POI. The reason why Rank-GeoFM produces relatively good recommendation could

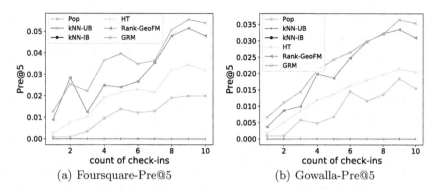

(a) Foursquare-Pre@5   (b) Gowalla-Pre@5

Fig. 5. Performance results on POIs with different sparsity.

be that, this method needs no adaptation to our task. Additionally, it utilizes a pairwise ranking method to learn user's and POI's feature vectors. Furthermore, it incorporates geographical information with the influence of nearest geographical neighbourhoods. Last but not least, this method is also symmetric with respect to users and POIs. All these properties could be responsible for the good results of this baseline method. Yet, the symmetry property could be still one of the major reasons.

At the same time, we also test each models' performance for different sparsity degrees. Based on the number of check-in records, we divide POIs into ten parts. On Foursquare, we can observe that GRM outperforms the rest of baselines consistently, except when count of check-ins equals 2, as shown in Fig. 5. The performance of some baselines varies when we change the number of visitors. Also, the strongest baseline, Rank-GeoFM, outperforms other baselines except when we deal with POIs with very few visitors. Even so, GRM could be still more favourable, since it is able to effectively deal with both the scenarios with high sparsity degree, and the uncomplicated ones.

In summary, from the above analysis, one can see that, although all the baselines are designed for dealing with the traditional recommendation task—suggesting POIs to users—the results show that some models have good performance for the novel task—recommending users to the long tail POIs. Even so, our proposed model, GRM, is still the best for this task, demonstrating its competitiveness.

5   Conclusions

In this paper, we investigated POI recommendation from the POI perspective. That is, how to recommend potential users for the long tail POIs. We formulated this task formally and designed a novel model to address this problem. Our model is based on the probabilistic collaborative filtering model. We conducted experiments to verify the performance of our proposed model. Experimental results demonstrated that it outperformed a set of representative state-of-the-art recommendation models for our proposed problem. In the future, we would like to incorporate various context information into our model to study the effect of various context information.

Acknowledgment. This work was supported by the NSFC (61472453,61672351, 61729202, 91438121, U1401256, U1501252, U1611264, U1711261, U1636210 and U1711262), the Opening Projects of Guangdong Key Laboratory of Big Data Analysis and Processing (201808), Guangdong Province Key Laboratory of Popular High Performance Computers of Shenzhen University (SZU-GDPHPCL2017), State Key Laboratory of Mathematical Engineering and Advanced Computing (2017A01), the 2016 Characteristic Innovation Project (Natural Science) of Education Department of Guangdong Province of China (2016KTSCX162), Foshan Science and Technology Bureau Project (2016AG100382), and Guangdong Pre-national project (2014GKXM054).

References

1. Li, X., Cong, G., Li, X., Pham, T., Krishnaswamy, S.: Rank-GeoFM: a ranking based geographical factorization method for point of interest recommendation. In: SIGIR, pp. 433–442 (2015)
2. Zhang, J., Chow, C.: GeoSoCa: exploiting geographical, social and categorical correlations for point-of-interest recommendations. In: SIGIR, pp. 443–452 (2015)
3. Hosseini, S., Yin, H., Zhang, M., Zhou, X., Sadiq, S.: Jointly modeling heterogeneous temporal properties in location recommendation. In: Candan, S., Chen, L., Pedersen, T.B., Chang, L., Hua, W. (eds.) DASFAA 2017. LNCS, vol. 10177, pp. 490–506. Springer, Cham (2017). https://doi.org/10.1007/978-3-319-55753-3_31
4. Shang, S., Chen, L., Jensen, C., Wen, J., Kalnis, P.: Searching trajectories by regions of interest. TKDE **29**(7), 1549–1562 (2017)
5. Yin, H., Zhou, X., Cui, B., Wang, H., Zheng, K., Nguyen, Q.: Adapting to user interest drift for POI recommendation. TKDE **28**(10), 2566–2581 (2016)
6. Shang, S., Chen, L., Wei, Z., Jensen, C., Wen, J., Kalnis, P.: Collective travel planning in spatial networks. TKDE **28**(5), 1132–1146 (2016)
7. Cheng, C., Yang, H., King, I., Lyu, M.: Fused matrix factorization with geographical and social influence in location-based social networks. AAAI **12**, 17–23 (2012)
8. Zhang, J., Chow, C.: iGSLR: personalized geo-social location recommendation: a kernel density estimation approach. In: SIGSPATIAL, pp. 334–343 (2013)
9. Shang, S., Lu, H., Pedersen, T., Xie, X.: Modeling of traffic-aware travel time in spatial networks. In: MDM, pp. 247–250 (2013)
10. Wang, W., Yin, H., Chen, L., Sun, Y., Sadiq, S., Zhou, X.: Geo-SAGE: a geographical sparse additive generative model for spatial item recommendation. In: SIGKDD, pp. 1255–1264 (2015)
11. He, J., Li, X., Liao, L.: Category-aware next point-of-interest recommendation via listwise Bayesian personalized ranking. In: IJCAI, pp. 1837–1843 (2017)
12. Chen, X., Zeng, Y., Cong, G., Qin, S., Xiang, Y., Dai, Y.: On information coverage for location category based point-of-interest recommendation. In: AAAI, pp. 37–43 (2015)
13. Shang, S., Zheng, K., Jensen, C., Yang, B., Kalnis, B., Li, G., Wen, J.: Discovery of path nearby clusters in spatial networks. TKDE **27**(6), 1505–1518 (2015)

14. Noulas, A., Scellato, S., Lathia, N., Mascolo, C.: A random walk around the city: new venue recommendation in location-based social networks. In: Social-Com/PASSAT, pp. 144–153 (2012)
15. Lichman, M., Smyth, P.: Modeling human location data with mixtures of kernel densities. In: SIGKDD, pp. 35–44 (2014)
16. Shang, S., Ding, R., Zheng, K., Jensen, C., Kalnis, P., Zhou, X.: Personalized trajectory matching in spatial networks. VLDB J. **23**(3), 449–468 (2014)
17. Bao, J., Zheng, Y., Wilkie, D., Mokbel, M.: Recommendations in location-based social networks: a survey. Geoinformatica **19**(3), 525–565 (2015)
18. Shang, S., Liu, J., Zheng, K., Lu, H., Pedersen, T., Wen, J.: Planning unobstructed paths in traffic-aware spatial networks. GeoInformatica **19**(4), 723–746 (2015)
19. Lian, D., Zhao, C., Xie, X., Sun, G., Chen, E., Rui, Y.: GeoMF: joint geographical modeling and matrix factorization for point-of-interest recommendation. In: SIGKDD, pp. 831–840 (2014)
20. Ye, M., Yin, P., Lee, W., Lee, D.: Exploiting geographical influence for collaborative point-of-interest recommendation. In: SIGIR, pp. 325–334 (2011)
21. Levandoski, J., Sarwat, M., Eldawy, A., Mokbel, M.: Lars: a location-aware recommender system. In: ICDE, pp. 450–461 (2012)
22. Shang, S., Yuan, B., Deng, K., Xie, K., Zheng, K., Zhou, X.: PNN query processing on compressed trajectories. GeoInformatica **16**(3), 467–496 (2012)
23. Zheng, V., Zheng, Y., Xie, X., Yang, Q.: Collaborative location and activity recommendations with GPS history data. In: WWW, pp. 1029–1038 (2010)
24. Feng, S., Li, X., Zeng, Y., Cong, G., Chee, Y., Yuan, Q.: Personalized ranking metric embedding for next new POI recommendation. In: IJCAI, pp. 2069–2075 (2015)
25. Liu, Q., Wu, S., Wang, L., Tan, T.: Predicting the next location: a recurrent model with spatial and temporal contexts. In: AAAI, pp. 194–200 (2016)
26. Yuan, Q., Cong, G., Sun, A.: Graph-based point-of-interest recommendation with geographical and temporal influences. In: CIKM, pp. 659–668 (2014)
27. Liu, B., Fu, Y., Yao, Z., Xiong, H.: Learning geographical preferences for point-of-interest recommendation. In: SIGKDD, pp. 1043–1051 (2013)
28. Liu, B., Xiong, H., Papadimitriou, S., Fu, Y., Yao, Z.: A general geographical probabilistic factor model for point of interest recommendation. TKDE **27**(5), 1167–1179 (2015)
29. Zhang, J., Chow, C., Zheng, Y.: ORec: an opinion-based point-of-interest recommendation framework. In: CIKM, pp. 1641–1650 (2015)
30. Gao, H., Tang, T., Hu, X., Liu, H.: Content-aware point of interest recommendation on location-based social networks. In: AAAI, pp. 1721–1727 (2015)
31. Gao, H., Tang, J., Hu, X., Liu, H.: Exploring temporal effects for location recommendation on location-based social networks. In: RecSys, pp. 93–100 (2013)
32. Liu, Y., Liu, C., Liu, B., Qu, M., Xiong, H.: Unified point-of-interest recommendation with temporal interval assessment. In: SIGKDD, pp. 1015–1024 (2016)
33. Shang, S., Chen, L., Wei, Z., Jensen, C., Zheng, K., Kalnis, P.: Trajectory similarity join in spatial networks. PVLDB **10**(11), 1178–1189 (2017)
34. Wang, Y., Yuan, N., Lian, D., Lin, L., Xie, X., Chen, E., Rui, Y.: Regularity and conformity: location prediction using heterogeneous mobility data. In: SIGKDD, pp. 1275–1284 (2015)
35. Yin, H., Cui, B., Li, J., Yao, J., Chen, C.: Challenging the long tail recommendation. PVLDB **5**(9), 896–907 (2012)
36. Shang, S., Ding, R., Yuan, B., Xie, K., Zheng, K., Kalnis, P.: User oriented trajectory search for trip recommendation. In: EDBT, pp. 156–167 (2012)

37. Valcarce, D., Parapar, J., Barreiro, Á.: Item-based relevance modelling of recommendations for getting rid of long tail products. KBS **103**(C), 41–51 (2016)
38. Lavrenko, V., Croft, W.B.: Relevance based language models. In: SIGIR, pp. 120–127 (2001)
39. Parapar, J., Bellogín, A., Castells, P., Barreiro, Á.: Relevance-based language modelling for recommender systems. IPM **49**(4), 966–980 (2013)
40. Shang, S., Lu, H., Pedersen, T.B., Xie, X.: Finding traffic-aware fastest paths in spatial networks. In: Nascimento, M.A., Sellis, T., Cheng, R., Sander, J., Zheng, Y., Kriegel, H.-P., Renz, M., Sengstock, C. (eds.) SSTD 2013. LNCS, vol. 8098, pp. 128–145. Springer, Heidelberg (2013). https://doi.org/10.1007/978-3-642-40235-7_8
41. Valcarce, D., Parapar, J., Barreiro, Á.: A study of priors for relevance-based language modelling of recommender systems. In: RecSys, pp. 237–240 (2015)
42. Zhai, C., Lafferty, J.: A study of smoothing methods for language models applied to information retrieval. TOIS **22**(2), 179–214 (2004)
43. Desrosiers, C., Karypis, G.: A comprehensive survey of neighborhood-based recommendation methods. In: Ricci, F., Rokach, L., Shapira, B., Kantor, P. (eds.) Recommender Systems Handbook, pp. 107–144. Springer, Boston (2011). https://doi.org/10.1007/978-0-387-85820-3_4
44. Cremonesi, P., Koren, Y., Turrin, R.: Performance of recommender algorithms on top-n recommendation tasks. In: RecSys, pp. 39–46 (2010)
45. Wang, Z., Wang, D., Yao, B., Guo, M.: Probabilistic range query over uncertain moving objects in constrained two-dimensional space. TKDE **27**(3), 866–879 (2015)
46. Xie, K., Deng, K., Shang, S., Zhou, X., Zheng, K.: Finding alternative shortest paths in spatial networks. TODS **37**(4), 29:1–29:31 (2012)
47. Wang, Z., Yao, B., Cheng, R., Gao, X., Zou, L., Guan, H., Guo, M.: SMe: explicit & implicit constrained-space probabilistic threshold range queries for moving objects. GeoInformatica **20**(1), 19–58 (2016)
48. Feng, S., Cong, G., An, B., Chee, Y.: POI2Vec: geographical latent representation for predicting future visitors. In: AAAI, pp. 102–108 (2017)

Exploiting Context Graph Attention for POI Recommendation in Location-Based Social Networks

Siyuan Zhang$^{(\boxtimes)}$ and Hong Cheng

The Chinese University of Hong Kong, Hong Kong, China
{syzhang,hcheng}@se.cuhk.edu.hk

Abstract. The prevalence of mobile devices and the increasing popularity of location-based social networks (LBSNs) generate a large volume of user mobility data. As a result, POI recommendation systems, which play a vital role in connecting users and POIs, have received extensive attention from both research and industry communities in the past few years. The challenges of POI recommendation come from the very sparse user check-in records with only positive feedback and how to integrate heterogeneous information of users and POIs. The state-of-the-art methods usually exploit the social influence from friends and geographical influence from neighboring POIs for recommendation. However, there are two drawbacks that hinder their performance. First, they cannot model the different degree of influence from different friends to a user. Second, they ignore the user check-ins as context information for preference modeling in the collaborative filtering framework.

To address the limitations of existing methods, we propose a *Context Graph Attention* (CGA) model, which can integrate context information encoded in different context graphs with the attention mechanism for POI recommendation. CGA first uses two context-aware attention networks to learn the influence weights of different friends and neighboring POIs respectively. At the same time, it applies a dual attention network, which considers the mutual influence of context POIs for a user and the context users for a POI, to learn the influence weights of different context vertices in the user-POI context graph. A multi-layer perceptron integrates the context vectors of users and POIs for estimating the visiting probability of a user to a POI. To the best of our knowledge, this is the first work that applies the attention mechanism for POI recommendation. Extensive experiments on two public check-in data sets show that CGA can outperform the state-of-the-art methods as well as other attentive collaborative filtering methods substantially.

Keywords: POI recommendation · Context graph attention
Neural network

1 Introduction

The prevalence of mobile devices stimulates user activities in location-based social networks (LBSNs), which are now becoming part of our daily life. People can share

© Springer International Publishing AG, part of Springer Nature 2018
J. Pei et al. (Eds.): DASFAA 2018, LNCS 10827, pp. 83–99, 2018.
https://doi.org/10.1007/978-3-319-91452-7_6

their visited Points of Interest (POIs) with friends, or search for interesting locations when they travel to a new city on LBSNs. As a result, a large volume of check-in data have been generated in LBSNs, which give rise to the POI recommendation systems that build connections between users and POIs. Different from traditional recommendation systems designed for movies and E-commerce websites, where there are explicit ratings and rich content information for each item, the challenges for POI recommendation come from the very sparse user check-in records with only positive feedback and how to integrate heterogeneous information of users and POIs.

Existing methods for POI recommendation [11,12,15,18,20] exploit the check-ins by friends of a user and visiting patterns of neighboring POIs of a POI to alleviate the data sparsity problem. The state-of-the-art methods [15,20] are based on implicit feedback models such as weighted matrix factorization and ranking-based factorization method. However, the performance of existing methods is not satisfactory for two reasons. First, they cannot model the different degree of influence from different friends to a user. Existing methods either assign equal weights to all friends [20] or determine the influence of friends with fixed heuristic values [11,21]. They ignore the fact that friends have different influence to a user when he/she checks in at different POIs. Second, few models consider the interactions between users and POIs as context information to alleviate the data sparsity problem. If we represent the user check-ins in a context graph (according to the definition in [20], a *context graph* is a graph that describes the affinity relationship between instances), the preference of a user can be modeled by his/her check-in POIs, and the visiting pattern of a POI can be modeled by its visiting users. Incorporating such a context graph between users and items into the collaborative filtering (CF) framework has shown promising improvement on movie recommendation and multimedia recommendation [3,8,10]. The challenge in exploiting the user-POI context graph is how to filter irrelevant context vertices for both users and POIs when estimating the visiting probability of a user at a POI.

To address the limitations of existing methods, we propose a **Context Graph Attention (CGA)** model to fuse different context information in LBSNs for POI recommendation. CGA is inspired by the attention mechanism adopted by many neural network based learning tasks [1,3,7,17,19]. The basic assumption of the attention mechanism is that only selective parts of input features are informative for the end machine learning tasks, which is reasonable in modeling context information for POI recommendation. CGA first builds three context graphs: a user friend context graph, a POI neighborhood context graph and a user-POI context graph to encode the heterogeneous information in an LBSN. Then it represents each user and each POI in the three context graphs with attentive latent vectors, which are the weighted sum of the embeddings of relevant context vertices in different context graphs. Two types of attention networks: context-aware attention network and dual attention network are proposed to compute the weights of different context vertices in different context graphs. The final representation for users and POIs consists of their original context-free latent vectors and the attentive latent vectors from context graphs. A multi-layer perceptron is applied on the final representations of users and POIs to estimate the visiting probabilities.

For a user, to model the influence from his/her friends, CGA computes the attentive weight from each friend using a context-aware attention network, which is a multi-layer neural network that takes characteristic of his/her friends, the target user and a candidate POI as inputs. The learned attentive weights are context aware, in the sense that they depend on the candidate POI. Similarly, for a POI, to model the influence from its neighboring POIs, CGA computes the attentive weights using another context-aware attention network, which further exploits the geographical influence measured by a distance decay function in the weight computation process. From the check-in records, CGA uses a dual attention network, which consists of two context-aware attention networks and some connections between them, to sequentially output the attentive context vectors for users and POIs. The dual attention network enables the mutual influence of two types of context vertices in the weight computation process. Thus it can further filter out irrelevant context vertices in the final attentive vectors.

In summary, we make the following contributions in this paper:

- We propose an integrated framework named Context Graph Attention (CGA) that can model heterogeneous context information with the attention mechanism in LBSNs for POI recommendation. To the best of our knowledge, this is the first attention based POI recommendation model that integrates the heterogeneous user-POI context graph for preference modeling.
- We design a context-aware attention network for the homogeneous context graph (between users and between POIs) and a dual attention network for heterogeneous context graph (between users and POIs), which can model the different and mutual influence of context vertices respectively.
- Through extensive experiments on two public check-in data sets, we show that CGA outperforms the state-of-the-art POI recommendation method as well as other attentive CF based methods for general item recommendations.

The remainder of this paper is organized as follows. We give a formal definition of our problem in Sect. 2. Section 3 introduces the proposed Context Graph Attention model. Section 4 describes details on the attentive weight computation and model learning. Section 5 reports the experiment results. We discuss related works in Sect. 6 and conclude this paper in Sect. 7.

2   Problem Description

Assume we have a user set $U = \{u_1, \ldots, u_N\}$ and a POI set $V = \{v_1, \ldots, v_M\}$. We use a user-POI matrix $Y \in \mathbb{R}^{N \times M}$ to denote the check-in records of all users in U. An entry $y_{ij} = 1$ if user u_i performed check-in at POI v_j. Otherwise, $y_{ij} = 0$. We denote the POIs visited by user u_i as V_i, and the set of users that performed check-in at v_j as U_j. The POI recommendation problem in an LBSN is defined as:

Definition 1 *(POI recommendation). Given a user $u_i \in U$, a POI set V, and the user-POI matrix Y, return a ranked list of POIs $\{v_j \in V \setminus V_i\}$, in descending order of the estimated check-in probability $\widehat{y}_{ij}$ by user u_i.*

3 The Context Graph Attention Model

3.1 Context Graphs in LBSN

We define three context graphs to encode the heterogeneous information in an LBSN. The *user friend context graph* captures the social relationships between users.

Definition 2 *(User friend context graph). The user friend context graph is denoted as $G_{UU} = (U, E_{UU})$, where a vertex $u_i \in U$ represents a user, and an edge $e = (u_i, u_k)$ indicates that u_i and u_k are friends. The context vertices of u_i in G_{UU} are denoted as $F_i = \{u_k | (u_i, u_k) \in E_{UU}\}$.*

The *POI neighborhood context graph* captures the geographical proximity between POIs.

Definition 3 *(POI neighborhood context graph). The POI neighborhood context graph is denoted as $G_{VV} = (V, E_{VV})$, where a vertex $v_i \in V$ represents a POI, and an edge $e = (v_i, v_j)$ indicates that v_j is one of the k-nearest neighbors of v_i. The context vertices of v_i in G_{VV} are denoted as $L_i = \{v_j | (v_i, v_j) \in E_{VV}\}$.*

In addition, we construct a *user-POI context graph* from the user-POI matrix Y.

Definition 4 *(User-POI context graph). The user-POI context graph is a bipartite graph denoted as $G_{UV} = (U \cup V, E_{UV})$, where an edge $e = (u_i, v_j) \in E_{UV}$ if $y_{ij} = 1$ in the user-POI matrix Y. The context vertices of u_i in G_{UV} are denoted as $P_i = \{v_k | (u_i, v_k) \in E_{UV}\}$; and the context vertices of v_j in G_{UV} are denoted as $Q_j = \{u_k | (u_k, v_j) \in E_{UV}\}$.*

The first two context graphs are homogeneous, while the user-POI context graph is heterogeneous as the vertices are of different types. To the best of our knowledge, no previous work has considered the user-POI interaction as a context graph for POI recommendation. Examples of the three context graphs are shown in Fig. 1.

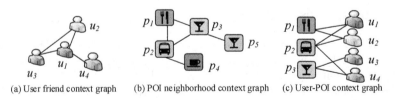

(a) User friend context graph   (b) POI neighborhood context graph   (c) User-POI context graph

Fig. 1. The three context graphs used in our model

3.2   Attentive Representations for Users and POIs in Context Graphs

Existing methods [18,20] use graph embedding based methods to model the information in context graphs. However, they assume that vertices with similar context neighbors have similar latent representations which, however, may not hold in the real scenario. To prove this, we perform data analysis on a Gowalla data set, which contains 736,148 check-ins made in California and Nevada [4,5].

In the user friend context graph G_{UU}, for each pair of friends, we calculate the cosine similarity of their check-in POI lists; in the POI neighborhood context graph G_{VV}, for each pair of neighbor POIs, we calculate the cosine similarity of the visiting user lists. We plot the cosine similarity histogram in Fig. 2. We observe that more than 20% of the friend pairs in G_{UU} have 0 cosine similarity, more than 50% have cosine similarity in $(0, 0.1]$, and only less than 10% have cosine similarity larger than 0.2. In Fig. 2(b), more than 25% of neighbor POIs have 0 cosine similarity, more than 30% have cosine similarity in $(0, 0.1]$, and only less than 10% have cosine similarity larger than 0.3. This analytical results show that the assumption in existing methods [18,20] may not hold.

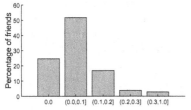 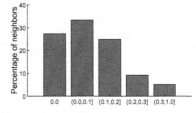

(a) Cosine similarity between friends   (b) Cosine similarity between neighbor POIs

Fig. 2. Cosine similarity between vertices in user-friend context graph and POI-neighborhood context graph

The above observation motivates us to re-consider the importance of different context vertices in the representation learning for users and POIs. We propose to learn the context representations for users and POIs in different context graphs using the attention mechanism. The *attention mechanism* was first applied in image/video recommendation [3] and neural machine translation [1], which assumes that only selective parts of the input features are informative for the end machine learning tasks. In the task of POI recommendation, applying the attention mechanism means selecting informative context vertices in the context graphs for learning an attentive representation.

Specifically, to represent u_i in G_{UU} for modeling the influence from his/her friends F_i, we first use friend embeddings $\{\boldsymbol{f}_{i,1}, \boldsymbol{f}_{i,2}, \ldots, \boldsymbol{f}_{i,|F_i|}\}$ to encode u_i's friend context vertices in G_{UU}. Then the representation of u_i in G_{UU} is defined as an attentive vector:

$$\widetilde{\boldsymbol{f}}_i = \sum_{u_{i,k} \in F_i} \alpha(i,k) \boldsymbol{f}_{i,k}, \tag{1}$$

where $\alpha(i, k)$ is the attentive weight of $\boldsymbol{f}_{i,k}$. We can model the influence of different friends by assigning suitable attentive weights, which are computed by a context-aware attention network. Similarly, after using neighborhood embeddings $\{\boldsymbol{l}_{j,1}, \boldsymbol{l}_{j,2}, \ldots, \boldsymbol{l}_{j,|L_j|}\}$ to encode context vertices in G_{VV} for v_j, we can compute an attentive vector $\widetilde{\boldsymbol{l}}_j$ to represent v_j in G_{VV} as:

$$\widetilde{\boldsymbol{l}}_j = \sum_{v_{j,k} \in L_j} \beta(j, k) \boldsymbol{l}_{j,k}, \tag{2}$$

where the attentive weight $\beta(j, k)$ is computed by another context-aware attention network.

Different from the neighborhood model in general item recommendation systems [8,10], where the preference of a user is represented by the set of rated items, we not only consider the context vertices (i.e., POIs) of a user in G_{UV}, but also consider the context vertices (i.e., users) of a POI in G_{UV} for POI recommendation. We believe that the dual context setting can further alleviate the data sparsity problem in POI recommendation. To represent u_i and v_j in G_{UV}, we first introduce two sets of context vertex embeddings: user-POI embeddings $\{\boldsymbol{p}_{i,1}, \boldsymbol{p}_{i,2}, \ldots, \boldsymbol{p}_{i,|P_i|}\}$ which are the embeddings of context vertices for u_i, and POI-user embeddings $\{\boldsymbol{q}_{j,1}, \boldsymbol{q}_{j,2}, \ldots, \boldsymbol{q}_{j,|Q_j|}\}$ which are the embeddings of context vertices for v_j in G_{UV}. Then we define two attentive vectors $\widetilde{\boldsymbol{p}}_i$ for u_i and $\widetilde{\boldsymbol{q}}_j$ for v_j to represent their context information in G_{UV} as:

$$\widetilde{\boldsymbol{p}}_i = \sum_{v_{i,k} \in P_i} \gamma(i, k) \boldsymbol{p}_{i,k}, \tag{3}$$

$$\widetilde{\boldsymbol{q}}_j = \sum_{u_{j,k} \in Q_j} \gamma(j, k) \boldsymbol{q}_{j,k}, \tag{4}$$

where $\gamma(i, k)$ and $\gamma(j, k)$ are the attentive weights for different context vertices in G_{UV}. As the context vertices of u_i and v_j can have mutual influence on each other through the edges in the context graph G_{UV}, we use a new dual attention network, which builds connections between the context-aware attention networks for two sets of context vertices P_i and Q_j, to compute the attentive weights.

Compared with the existing graph embedding based methods, which give equal weights to all context vertices [20] or using predefined, fixed edge weights [18] for different context vertices, our attentive context representation scheme enables learning the weights of different context vertices automatically to optimize the POI recommendation performance. We will describe how to compute the attentive weights from the attention networks and how to use the attentive vectors for POI recommendation in the following sections.

3.3   Solution Framework

The solution framework of our context graph attention model is shown in Fig. 3. We use the three context graphs G_{UU}, G_{VV}, and G_{UV} as inputs to our model.

To compute the visiting probability of user u_i at POI v_j, we first use the context-free user embedding $\boldsymbol{u}_i$ and POI embedding $\boldsymbol{v}_j$ to represent u_i and v_j in a latent space. Then two context-aware attention networks are applied on G_{UU} and G_{VV} to output an attentive friend vector $\widetilde{\boldsymbol{f}}_i$ for u_i and an attentive neighbor vector $\widetilde{\boldsymbol{l}}_j$ for v_j. A dual attention network is applied on G_{UV} to output an attentive user-POI vector $\widetilde{\boldsymbol{p}}_i$ for u_i and an attentive POI-user vector $\widetilde{\boldsymbol{q}}_j$ for v_j.

The final user vector of u_i is a summation on user embedding and its two attentive context vectors in G_{UU} and G_{UV}. The final POI vector of v_j is a summation on POI embedding and its two attentive context vectors in G_{VV} and G_{UV}. We then merge the user vector and the POI vector through concatenation and feed it into a multi-layer perceptron (MLP). A binary softmax layer is applied on top of the MLP layers to output the estimated visiting probability $\widehat{y}_{ij}$ of u_i at v_j. We use the cross entropy between the estimated $\widehat{y}_{ij}$ and the true entry label y_{ij} as the loss function for training parameters.

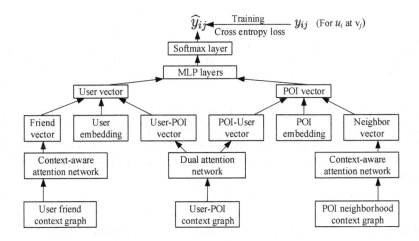

Fig. 3. The context graph attention framework

4 Attention Computation and Model Learning

In this section, we describe how to compute the attentive weights using attention networks, how to train the CGA model, and perform prediction using the model.

4.1 Context-Aware Attention for Friends and Neighbors

Compute Attentive Friend Vector. As we have discussed, different friends of a user can have different influence on him/her. Furthermore, even the same friend can have different influence to the user at different locations. For example, the check-in behavior of a user around a work place may be very similar to that of his colleagues, but the check-in behavior around his/her home location may be quite different from that of his colleagues. To model such different influences, we propose to use a context-aware attention network to output the friend vector in the user context graph. Given the user embedding $\boldsymbol{u}_i$, the POI embedding $\boldsymbol{v}_j$ and the friend embeddings $\{\boldsymbol{f}_{i,1}, \boldsymbol{f}_{i,2}, \dots, \boldsymbol{f}_{i,|F_i|}\}$, we compute the attentive score $a(i,k)$ for each friend embedding with a two-layer network as:

$$a(i,k) = \boldsymbol{w}_f^T \phi(\boldsymbol{W}_f[\boldsymbol{f}_{i,k}; \boldsymbol{u}_i; \boldsymbol{v}_j] + \boldsymbol{b}_f), \qquad (5)$$

where we first concatenate the three embeddings $\boldsymbol{f}_{i,k}, \boldsymbol{u}_i$ and $\boldsymbol{v}_j$ to form a new vector. Then we transform it to another vector by multiplying kernel $\boldsymbol{W}_f$ and adding bias $\boldsymbol{b}_f$. $\phi(x) = \max(0, x)$ is the ReLU function for activation of the first layer, $\boldsymbol{w}_f$ is the parameter for the second layer. The final weights of friends are obtained by normalizing the above attentive scores using Softmax as:

$$\alpha(i,k) = \frac{\exp(a(i,k))}{\sum_{u_{i,k'} \in F_i} \exp(a(i,k'))}. \qquad (6)$$

Then we compute the final context-aware attentive friend vector $\widetilde{\boldsymbol{f}}_i$ by Eq. (1). Compared with the self attention proposed in [3], where $a(i,k)$ is not related to $\boldsymbol{v}_j$ and $\widetilde{\boldsymbol{f}}_i$ is fixed for different $\boldsymbol{v}_j$, the context-aware attention network introduces more flexibility in modeling influences from friends and candidate POI v_j.

Compute Attentive Neighbor Vector. Existing studies on POI recommendation [12,15,16] suggested that incorporating neighborhood information of POIs can improve the recommendation performance. In order to utilize the geographical influence weighted by distance, we compute the attentive score for each neighbor embedding $b(j,k)$ as:

$$b(j,k) = \boldsymbol{w}_l^T \phi(\boldsymbol{W}_l[\boldsymbol{l}_{j,k}; \boldsymbol{v}_j; \boldsymbol{u}_i; g(j,k)] + \boldsymbol{b}_l), \qquad (7)$$

where $g(j,k) = 1/(0.5 + d(v_j, v_k))$ is a distance decay function defined in [12], $d(v_j, v_k)$ is the distance between POI v_j and its neighbor v_k. $\boldsymbol{W}_l, \boldsymbol{b}_l$ and $\boldsymbol{w}_l$ are the network parameters. After normalization, the context-aware neighbor attention is:

$$\beta(j,k) = \frac{\exp(b(j,k))}{\sum_{v_{j,k'} \in L_j} \exp(b(j,k'))}. \qquad (8)$$

Then we compute the context-aware attentive neighbor vector $\widetilde{\boldsymbol{l}}_j$ by Eq. (2).

4.2   Dual Attention for User-POI Context Graph

We show the structure of the proposed dual attention network in Fig. 4. Given the user-POI embeddings $\{\boldsymbol{p}_{i,1}, \boldsymbol{p}_{i,2}, \dots, \boldsymbol{p}_{i,|P_i|}\}$ and the POI-user embeddings $\{\boldsymbol{q}_{j,1}, \boldsymbol{q}_{j,2}, \dots, \boldsymbol{q}_{j,|Q_j|}\}$ as inputs, the dual attention network first summarizes the user-POI embeddings with a user-POI summary vector through mean pooling. Then it sequentially outputs an attentive POI-user vector and an attentive user-POI vector through two context-aware attention networks. The structure of the

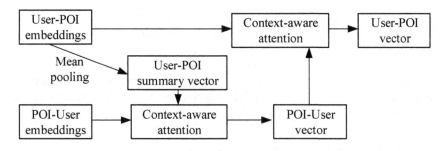

Fig. 4. The dual attention network

dual attention network is consistent with our two intuitions: (1) a user would like to check in at a POI either because the POI is similar to some of his/her visited POIs or the POI has been visited by some users with similar preference; (2) the attentive weights of context users and context POIs have mutual influence.

Context POIs Guided Context Users Attention. As we focus on recommending POIs to users, we first perform mean pooling on the user-POI embeddings and obtain the user-POI summary vector $\overline{\boldsymbol{p}}_i$ as:

$$\overline{\boldsymbol{p}}_i = \frac{1}{|P_i|} \sum_{v_{i,k} \in P_i} \boldsymbol{p}_{i,k}. \tag{9}$$

Then we compute the dual attentive scores for $\boldsymbol{p}_{i,k}$ as:

$$c(j,k) = \boldsymbol{w}_q^T \phi(\boldsymbol{W}_q[\boldsymbol{q}_{j,k}; \boldsymbol{v}_j; \boldsymbol{u}_i; \overline{\boldsymbol{p}}_i] + \boldsymbol{b}_q), \tag{10}$$

where $\boldsymbol{W}_q$ and $\boldsymbol{b}_q$ are the parameters for the first layer neural network and $\boldsymbol{w}_q$ is the parameter for the second layer. Compared with the context-aware attention for friends and neighbors in Sect. 4.1, adding user-POI summary vector $\overline{\boldsymbol{p}}_i$ in Eq. (10) helps to filter out context users of v_j that have no overlapping with context POIs of u_i in G_{UV}. After normalization, the attention for context users of v_j is:

$$\gamma(j,k) = \frac{\exp(c(j,k))}{\sum_{u_{j,k'} \in Q_j} \exp(c(j,k'))}. \tag{11}$$

The dual attention POI-user context vector $\widetilde{\boldsymbol{q}}_j$ is computed by Eq. (4).

Context Users Guided Context POIs Attention. After generating the dual-attention POI-user context vector $\widetilde{\boldsymbol{q}}_j$, we use it to guide the attention of context POIs for u_i as follows:

$$c(i,k) = \boldsymbol{w}_p^T \phi(\boldsymbol{W}_p[\boldsymbol{p}_{i,k}; \boldsymbol{u}_i; \boldsymbol{v}_j; \widetilde{\boldsymbol{q}}_j] + \boldsymbol{b}_p), \tag{12}$$

$$\gamma(i,k) = \frac{\exp(c(i,k))}{\sum_{v_{i,k'} \in P_i} \exp(c(i,k'))}, \tag{13}$$

where $\boldsymbol{W}_p$ and $\boldsymbol{b}_p$ are the first layer parameters, $\boldsymbol{w}_p$ is the second layer parameter. Incorporating attentive POI-user vector $\tilde{\boldsymbol{q}}_j$ can help to select more informative visited POIs for characterizing u_i's intention when performing check-in at v_j. Since the attentive POI-user vector $\tilde{\boldsymbol{q}}_j$ is influenced by the user-POI embedding summary vector $\bar{\boldsymbol{p}}_i$, the dual attention layer enables the mutual influence between $\{\boldsymbol{p}_{i,*}\}$ and $\{\boldsymbol{q}_{j,*}\}$. The final dual attention user-POI context vector $\tilde{\boldsymbol{p}}_i$ is computed by Eq. (3).

4.3   POI Visiting Probability Prediction

Given the user embedding $\boldsymbol{u}_i$, the attentive context vectors $\tilde{\boldsymbol{f}}_i$ and $\tilde{\boldsymbol{p}}_i$, the final user vector can be represented through summation: $\boldsymbol{uvec}_i = \boldsymbol{u}_i + \tilde{\boldsymbol{f}}_i + \tilde{\boldsymbol{p}}_i$. Similarly, the final POI vector is: $\boldsymbol{vvec}_j = \boldsymbol{v}_j + \tilde{\boldsymbol{l}}_j + \tilde{\boldsymbol{q}}_j$. To model the non-linear interactions between $\boldsymbol{uvec}_i$ and $\boldsymbol{vvec}_j$, we first merge $\boldsymbol{uvec}_i$ and $\boldsymbol{vvec}_j$ through concatenation. Then we feed the new vector $[\boldsymbol{uvec}_i; \boldsymbol{vvec}_j]$ into a multi-layer perceptron (MLP) as:

$$
\begin{aligned}
\boldsymbol{z}_1 &= \phi(\boldsymbol{W}_1[\boldsymbol{uvec}_i; \boldsymbol{vvec}_j] + \boldsymbol{b}_1), \\
\boldsymbol{z}_2 &= \phi(\boldsymbol{W}_2\boldsymbol{z}_1 + \boldsymbol{b}_2), \\
&\quad\cdots\cdots \\
\boldsymbol{z}_H &= \phi(\boldsymbol{W}_H\boldsymbol{z}_{H-1} + \boldsymbol{b}_H), \\
\hat{y}_{ij} &= \sigma(\boldsymbol{w}_y^T\boldsymbol{z}_H),
\end{aligned}
\tag{14}
$$

where $\boldsymbol{W}_h$ and $\boldsymbol{b}_h$ are the kernel and bias for the h-th layer, $\phi(x) = \max(0, x)$ is the ReLU function for activation. After H layers' non-linear transformation, we use the sigmoid function $\sigma(x) = 1/(1 + e^{-x})$ to output the final predicted visiting probability $\hat{y}_{ij}$, where $\boldsymbol{w}_y$ is the parameter in $\sigma(x)$.

4.4   Model Training

Since the user-POI matrix Y is sparse with only positive feedback $Y^+ = \{y_{ij}|y_{ij} = 1\}$, we randomly sample some unobserved entries from $Y^n = \{y_{ij}|y_{ij} = 0\}$ to form the negative instance set Y^- for training a binary classifier. The number of negative instances $|Y^-|$ is proportional to the positive instance number $|Y^+|$. We use the *binary cross-entropy loss* as the loss function for training,

$$
\begin{aligned}
\mathcal{J} &= -\log p(Y^+ \cup Y^- | \Theta) \\
&= -\sum_{y_{ij} \in Y^+} \log \hat{y}_{ij} - \sum_{y_{ij} \in Y^-} \log(1 - \hat{y}_{ij}) \\
&= -\sum_{y_{ij} \in Y^+ \cup Y^-} y_{ij} \log \hat{y}_{ij} + (1 - y_{ij}) \log(1 - \hat{y}_{ij})
\end{aligned}
\tag{15}
$$

The parameters of our model are $\Theta = \{\Theta_e, \Theta_a, \Theta_p\}$, where $\Theta_e = \{\boldsymbol{u}, \boldsymbol{v}, \boldsymbol{f}, \boldsymbol{l}, \boldsymbol{p}, \boldsymbol{q}\}$ are the parameters for embeddings, $\Theta_a = \{\boldsymbol{W}_*, \boldsymbol{w}_*, \boldsymbol{b}_*|* = f, l, p, q\}$ are the parameters for attentions, $\Theta_p = \{ \boldsymbol{w}_y, \boldsymbol{W}_h, \boldsymbol{b}_h | h = 1, 2, \ldots, H \}$ are the parameters for MLP. The CGA model can be trained efficiently by stochastic gradient descent (SGD) with minibatch Adam [9].

5 Experiments

In this section, we conduct experiments on two public check-in data sets to evaluate the performance of our proposed POI recommendation method.

5.1 Settings

Data Sets. The first is a Gowalla data set that contains 736,148 check-ins made in California and Nevada between Feb. 2009 and Oct. 2010 [4,5]. The second is a Foursquare data set made by 4,163 users living in California [18]. Both data sets contain friendship information between users. After filtering users with less than 15 visited POIs and POIs with less than 10 visited users, we get 3,172 users, 5,665 POIs and 132,634 check-ins in Gowalla. The Foursquare data set contains 1,513 users, 2,686 POIs and 55,554 check-ins after applying the same filtering condition. For each data set, we use the earliest 80% visited POIs of each user for training and the remaining 20% for testing.

Evaluation Metrics. Three metrics are used to evaluate the performance of different POI recommendation algorithms: Pre@K (precision), Rec@K (recall) and hit rate. Given the top-K returned POIs for a user, we compute the precision and recall. We do the POI recommendation for a set of test users and report the average precision and recall for all test users. To compute the hit rate, we follow the strategy in [6,20] of mixing each ground truth POI in the test data with 100 random POIs that are not visited by the user and performing ranking with different algorithms. Then we check whether the ground truth POI is in the top-K list and report the average HR@K (hit rate) values for all test POIs.

Baseline Methods. We denote our method as Context Graph Attention (**CGA**) and compare CGA with the following baselines.

- **PACE** [20]: the state-of-the-art method for POI recommendation. PACE combines MLP and context graph embedding for jointly training user embeddings and POI embeddings. The context graph embedding component is considered as the regularizer for alleviating the data sparsity problem. Since it outperforms many matrix factorization based methods such as **IRenMF** [16], **ASMF** and **ARMF** [11], we do not compare with those methods in experiments.
- **NCF** [6]: it combines MLP and general matrix factorization to model the interaction between users and items in two different latent spaces. It achieves the state-of-the-art performance on movie recommendation and picture recommendation. But it has not been tested on POI recommendation in previous works.

- **ACF** [3]: the first work that applies the attention mechanism on multimedia recommendation for pictures and short videos. We only use the item level attention component in ACF as POIs have no rich component features as pictures and short videos. Since the original ACF uses dot product to compute the ranking score between users and items, we further feed the representations of users and items produced by ACF into an MLP to output the ranking score. We denote the new variant as **ACF+MLP**.

Parameter Settings. For all the compared methods, we set the embedding size $D = 16$. The number of layers H in MLP is 3 and the size of MLP layer is $32 \rightarrow 16 \rightarrow 8$. For the attention network in ACF (ACF+MLP) and CGA, the size of attention unit d is the same as the embedding size. We choose $k = 30$ (k-nearest neighbors) in the construction of POI context graph. To train the neural network, we set the batch size to 512 and learning rate to 0.0001. For each positive entry in Y^+, we sample 5 negative instances to form Y^-.

5.2   Overall Performance

We report the Pre@K, Rec@K and HR@K of different methods on the two data sets in Figs. 5, 6 and 7, respectively. We observe that: (1) Our method CGA outperforms all the other methods on the three metrics. Compared with the second best method PACE, CGA increases the precision by 28.5%, recall by 28.9% and hit rate by 35.9% on Gowalla data set. The improvements on the Foursquare data set are 21.9% on precision, 26.3% on recall and 28.1% on hit rate. The results demonstrate the benefits of exploiting context graph attention for POI recommendation. (2) PACE has better performance than NCF, which shows the effectiveness of exploiting user and POI context graphs. (3) The performance of ACF+MLP is close to that of NCF and is worse than that of PACE, which shows that only applying the item level attention for POI recommendation is not enough due to data sparsity in check-in records. (4) ACF+MLP outperforms the original ACF, which shows that the MLP can improve the performance of the matrix factorization method with the attention mechanism.

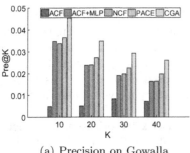
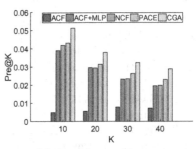

(a) Precision on Gowalla   (b) Precision on Foursquare

Fig. 5. Precision on two data sets

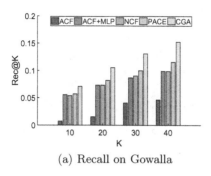

(a) Recall on Gowalla

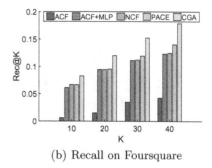

(b) Recall on Foursquare

Fig. 6. Recall on two data sets

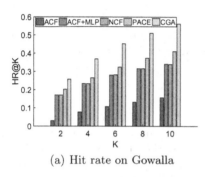

(a) Hit rate on Gowalla

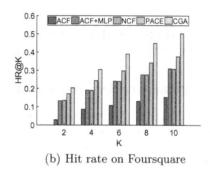

(b) Hit rate on Foursquare

Fig. 7. Hit rate on two data sets

5.3 Effectiveness of Different Attention Networks

To show the effectiveness of different attention networks, we create two variants of CGA as follows:

– **CGA_NoAtt:** it performs mean-pooling on all the context vertex embeddings in the three context graphs for both users and POIs. All context vertices in context graphs have equal weights in the final context vectors for both users and POIs, i.e., without learning the attention weights.
– **CGA_ContextAtt:** it performs context-aware attention on all the context vertex embeddings in the three context graphs. Specifically, it ignores $\overline{p}_i$ in (10) and $\widetilde{q}_j$ in (12). Thus there is no mutual influence on computing the attentive user-POI vector and POI-user vector.

The results on the two data sets are shown in Fig. 8. Three observations are made: (1) CGA_NoAtt has a higher hit rate than PACE, which shows the benefits of adding the user-POI context graph for POI recommendation; (2) CGA_ContextAtt outperforms CGA_NoAtt, which shows the effectiveness of applying the attention mechanism on modeling context graph information for users and POIs; (3) CGA has the best performance among all variants. This is

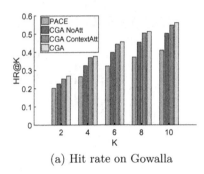
(a) Hit rate on Gowalla

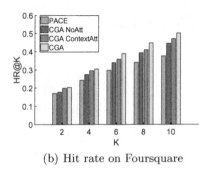
(b) Hit rate on Foursquare

Fig. 8. Hit rate on two data sets for different attention networks

because the proposed dual attention network can exploit the mutual influence of context vertices for users and POIs in the user-POI context graph.

5.4   Parameter Sensitivity Test

We evaluate the performance of CGA w.r.t. four parameters: the size of the embedding D, the size of the attention units d, the number of layers H in MLP and the number of neighbors k in POI neighborhood graph. The results on HR@10 are reported for both data sets. We make the following observations: (1) Small value of D (e.g., $D = 8$) decreases the performance of CGA, while the HR@10 becomes stable when D is larger than 16 as shown in Table 1(a). (2) The HR@10 increases when d varies from 8 to 64, but the increase is marginal as shown in Table 1(b); (3) Table 1(c) shows that MLP with $H = 3$ has significant performance improvement compared with $H = 0$, which verifies the effectiveness of using MLP to model the interaction between users and POIs; (4) The number of neighbors k has marginal influence on the performance of CGA as shown in Table 1(d). In summary, the performance of CGA depends on the size of embeddings D and the number of layers H, but is not sensitive to the size of attention units d and the number of neighbors k.

Table 1. HR@10 w.r.t. different parameters in CGA

(a) HR@10 w.r.t. D

Data set	$D=8$	$D=16$	$D=32$	$D=64$
Gowalla	0.557	0.565	**0.591**	0.582
Foursquare	0.465	**0.504**	0.502	0.503

(b) HR@10 w.r.t. d

Data set	$d=8$	$d=16$	$d=32$	$d=64$
Gowalla	0.562	0.566	**0.574**	**0.574**
Foursquare	0.475	0.482	0.488	**0.506**

(c) HR@10 w.r.t. H

Data set	$H=0$	$H=1$	$H=2$	$H=3$
Gowalla	0.504	0.530	0.550	**0.581**
Foursquare	0.406	0.457	0.460	**0.492**

(d) HR@10 w.r.t. k

Data set	$K=10$	$K=30$	$K=50$	$K=70$	$K=90$
Gowalla	0.567	0.566	0.567	**0.582**	0.581
Foursquare	0.487	0.484	0.479	0.473	**0.492**

6   Related Works

6.1   POI Recommendation

In order to alleviate the data sparsity problem in check-in records and exploit rich context in LBSN, many methods for POI recommendation have been proposed [2,15], most of which are based on a fused model that extends the matrix factorization methods with context embeddings [11–14,16,18]. In [15], four types of POI recommendation models, including matrix factorization models, poisson factor models, link-based models and hybrid models, are compared by extensive experiments on three public data sets. The experiment results show that models [12,13,16] based on implicit feedback and incorporating geographical influence have better performance than other types of models. Recently, a neural embedding method [20] that bridges collaborative filtering and semi-supervised learning with context graphs has been proposed, which outperforms the state-of-the-art matrix factorization based methods such as [11,16]. However, these context embedding methods either ignore the influence of different context vertices or set the weights of different context vertices with fixed and heuristic values. Our proposed method exploits the attention mechanism to overcome these limitations and further incorporates the user-POI context graph to improve the POI recommendation performance.

6.2   Attention Mechanism

The attention mechanism [1,3,7,17,19] has been successfully applied in many machine learning tasks such as neural machine translation [1], image/video recommendation [3] and hashtag recommendation [17]. [3] proposes a two-level self attention network to model the item-level and component-level attention for multimedia recommendation. [17] uses a co-attention network to jointly model the mutual influence of image and text information in microblog. Our proposed context graph attention model first extends the self attention network in [3] to context-aware attention network for modeling the different influence from different context vertices. Then we further extend the context-aware attention network to dual attention network inspired by the idea of co-attention network [17]. We also show that incorporating deep neural network structure like MLP can further improve the performance of attention-based collaborative filtering model for POI recommendation.

7   Conclusions

In this paper, we proposed a context graph attention (CGA) model for POI recommendation in LBSNs. CGA first models the different influence of friends and POI neighbors with a context-aware attention network. Then it uses a dual attention network to model the context information in the user-POI context graph. The learned attentive representation for different context graphs can be fed into a multi-layer perceptron for modeling non-linear interaction between users and POIs. Extensive experimental results on two public check-in data sets show that the context attention model can outperform the state-of-the-art method for POI recommendation and other attentive collaborative filtering methods by a large margin.

References

1. Bahdanau, D., Cho, K., Bengio, Y.: Neural machine translation by jointly learning to align and translate. In: ICLR (2015)
2. Bao, J., Zheng, Y., Wilkie, D., Mokbel, M.: Recommendations in location-based social networks: a survey. GeoInformatica **19**(3), 525–565 (2015)
3. Chen, J., Zhang, H., He, X., Nie, L., Liu, W., Chua, T.S.: Attentive collaborative filtering: multimedia recommendation with item-and component-level attention. In: SIGIR, pp. 335–344 (2017)
4. Cho, E., Myers, S.A., Leskovec, J.: Friendship and mobility: user movement in location-based social networks. In: SIGKDD, pp. 1082–1090 (2011)
5. Feng, S., Li, X., Zeng, Y., Cong, G., Chee, Y.M., Yuan, Q.: Personalized ranking metric embedding for next new POI recommendation. In: IJCAI. pp. 2069–2075 (2015)
6. He, X., Liao, L., Zhang, H., Nie, L., Hu, X., Chua, T.S.: Neural collaborative filtering. In: WWW, pp. 173–182 (2017)
7. Xiao, J., Ye, H., He, X., Zhang, H., Wu, F., Chua, T.-S.: Attentional factorization machines: learning the weight of feature interactions via attention networks. In: IJCAI, pp. 3119–3125 (2017)
8. Kabbur, S., Ning, X., Karypis, G.: FISM: factored item similarity models for top-N recommender systems. In: SIGKDD, pp. 659–667 (2013)
9. Kingma, D., Ba, J.: Adam: a method for stochastic optimization. In: ICLR (2015)
10. Koren, Y.: Factorization meets the neighborhood: a multifaceted collaborative filtering model. In: SIGKDD, pp. 426–434 (2008)
11. Li, H., Ge, Y., Hong, R., Zhu, H.: Point-of-interest recommendations: learning potential check-ins from friends. In: SIGKDD, pp. 975–984 (2016)
12. Li, X., Cong, G., Li, X.L., Pham, T.A.N., Krishnaswamy, S.: Rank-geoFM: a ranking based geographical factorization method for point of interest recommendation. In: SIGIR, pp. 433–442 (2015)
13. Lian, D., Zhao, C., Xie, X., Sun, G., Chen, E., Rui, Y.: GeoMF: joint geographical modeling and matrix factorization for point-of-interest recommendation. In: SIGKDD, pp. 831–840 (2014)
14. Liu, X., Liu, Y., Li, X.: Exploring the context of locations for personalized location recommendations. In: IJCAI, pp. 1188–1194 (2016)
15. Liu, Y., Pham, T.A.N., Cong, G., Yuan, Q.: An experimental evaluation of point-of-interest recommendation in location-based social networks. VLDB **10**(10), 1010–1021 (2017)
16. Liu, Y., Wei, W., Sun, A., Miao, C.: Exploiting geographical neighborhood characteristics for location recommendation. In: CIKM, pp. 739–748 (2014)
17. Zhang, Q., Wang, J., Huang, H., Huang, X., Gong, Y.: Hashtag recommendation for multimodal microblog using co-attention network. In: IJCAI, pp. 3420–3426 (2017)

18. Xie, M., Yin, H., Wang, H., Xu, F., Chen, W., Wang, S.: Learning graph-based POI embedding for location-based recommendation. In: CIKM, pp. 15–24 (2016)
19. Xiong, C., Callan, J., Liu, T.Y.: Word-entity duet representations for document ranking. In: SIGIR, pp. 763–772 (2017)
20. Yang, C., Bai, L., Zhang, C., Yuan, Q., Han, J.: Bridging collaborative filtering and semi-supervised learning: a neural approach for poi recommendation. In: SIGKDD, pp. 1245–1254 (2017)
21. Ye, M., Yin, P., Lee, W.C., Lee, D.L.: Exploiting geographical influence for collaborative point-of-interest recommendation. In: SIGIR, pp. 325–334 (2011)

Restricted Boltzmann Machine Based Active Learning for Sparse Recommendation

Weiqing Wang[1], Hongzhi Yin[1(✉)], Zi Huang[1], Xiaoshuai Sun[2], and Nguyen Quoc Viet Hung[3]

[1] University of Queensland, Brisbane, Australia
{weiqingwang,h.yin1,uqzhuang}@uq.edu.au
[2] Harbin Institute of Technology, Harbin, China
xiaoshuaisun.hit@gmail.com
[3] Griffith University, Brisbane, Australia
quocviethung1@gmail.com

Abstract. In recommender systems, users' preferences are expressed as ratings (either explicit or implicit) for items. In general, more ratings associated with users or items are elicited, more effective the recommendations are. However, almost all user rating datasets are sparse in the real-world applications. To acquire more ratings, the active learning based methods have been used to selectively choose the items (called interview items) to ask users for rating, inspired by that the usefulness of each rating may vary significantly, i.e., different ratings may bring a different amount of information about the user's tastes. Nevertheless, existing active learning based methods, including both static methods and decision-tree based methods, encounter the following limitations. First, the interview item set is predefined in the static methods, and they do not consider the user's responses when asking the next question in the interview process. Second, the interview item set in the decision tree based methods is very small (i.e., usually less than 50 items), which leads to that the interview items cannot fully reflect or capture the diverse user interests, and most items do not have the opportunity to obtain additional ratings. Moreover, these decision tree based methods tend to choose popular items as the interview items instead of items with sparse ratings (i.e., sparse items), resulting in "Harry Potter Effect" (http://bickson.blogspot.com.au/2012/09/harry-potter-effect-on-recommendations.html). To address these limitations, we propose a new active learning framework based on RBM (Restricted Boltzmann Machines) to add ratings for sparse recommendation in this paper. The superiority of this method is demonstrated on two publicly available real-life datasets.

1 Introduction

The success of retailers such as Amazon.com and Netflix has largely been attributed to the recommender system, which is the key to attracting users and

© Springer International Publishing AG, part of Springer Nature 2018
J. Pei et al. (Eds.): DASFAA 2018, LNCS 10827, pp. 100–115, 2018.
https://doi.org/10.1007/978-3-319-91452-7_7

promoting products. In most recommender systems, users' tastes are expressed as ratings for items, and the number of available ratings determines the performance of these recommender systems. However, in the real-world rating datasets, most users and items have sparse ratings and they generally follow power-law distributions [1,20,23,31,37]. To improve the recommendation accuracy for both sparse users and sparse items (i.e., sparse recommendation), more ratings are expected for them. Thus, how to obtain more ratings for sparse users and items has become a challenging and practical problem in the domain of recommender systems [3,35].

Conventional active learning techniques have been focused on smartly selecting unlabeled instances to be labeled, and they have been widely recognized as an efficient way to reduce the annotation cost [30]. Recently, active learning is used to obtain additional ratings to deal with the sparse recommendation problem [3,10,36], inspired by the observation that the usefulness of each rating varies, i.e., different ratings may bring a different amount of knowledge about the users' preferences.

The traditional active learning models in recommender systems construct the interview process with a static seed set of items selected based on measurements such as informativeness, popularity and coverage [12]. Such static methods are not personalized or adaptive because the items are selected statically in batch, and they do not consider the user responses during the interview process. It has been convincingly argued that an effective method for designing the interview process should ask the interview questions adaptively and take into account the user's responses when asking the next question in the interview process [2].

A balanced decision tree has been widely used to progressively select interview items to query the user's response in previous work [10,17,37]. They first construct a decision tree with each non-leaf node being an interview question (i.e., an item) and then query a user's preferences progressively along a path from the root node to a leaf node based on the user's response. A non-leaf node divides the users into *Lovers* and *Haters*. Although these decision tree based methods can adaptively choose the interview items, they encounter the following limitations. First, the height of the decision tree is equal to the number of interview questions for a user, and the number of interview questions k is usually limited to be very small (i.e., $k \leq 5$), therefore the whole set of interview items for all users (i.e., the non-leaf nodes in the decision tree) are rather limited. For example, even when $k = 5$, the total number of interview items in the decision tree is only 31, which means that only 31 items have the opportunity to obtain the additional ratings. Besides, decision tree based methods tend to pick up the popular items as interview items [15], resulting in "Harry Potter Effect"[1]. "Harry Potter Effect" refers to the phenomenon that recommender systems tend to be biased towards generally popular items. The popular items have already accumulated sufficient ratings, and it is unnecessary to obtain additional ratings

---

[1] http://bickson.blogspot.com.au/2012/09/harry-potter-effect-on-recommendations.html.

for them. What is urgently demanded is to obtain more ratings for items with sparse ratings.

In this paper, we focus on addressing the limitations of existing active learning-based methods and improving recommendation for both sparse users and items by adaptively constructing k, a limited number, interview questions for each user. During each interview question, given an item m, a user u is asked whether she likes m or not instead of rating m because choosing like or dislike is much easier than giving a numeric rating to an item. The recommender system refines its impression about u and then directs u to the next interview question according to her response. Based on her answers to the k interview questions, the recommender is able to more accurately predict u's preferences on the other items. However, how to get most of the user's preferences by asking k questions is a NP-hard problem due to the combinational explosion [9]. Hence, we propose a greedy strategy to iteratively choose the next interview item that can maximize the reduction of uncertainty about the user's preferences. Specifically, we adopt the entropy to measure the uncertainty of a user's preferences on all items, and greedily select the item with the biggest information gain as the interview item at each iteration. Popular items are liked by most users and have significant correlation with most of other items. From the information theory perspective, popular items tell little about the user's preferences and choosing the popular items to query the user's response does not significantly reduce the entropy of the user's preferences. Thus, our proposed entropy-based greedy method tends to query the user's response for sparse items.

Given a user u's previous response to the interviewed items, many methods (i.e., matrix factorization models) can be applied to update her preferences in the interview process. However, these methods or models need to be retrained or recomputed every time after receiving a new response from the user. To reduce the time cost in the interview process, we introduce a class of two-layer undirected graphical models [19] that generalized from Restricted Boltzmann Machine (RBM) to update user's preferences in a real-time manner. This method is able to update user' preferences without retraining during the interview process.

2   Related Work

Collaborative filtering(CF) is one of the most successful recommendation approaches and has been proven to be effective in practice. However, CF recommendation suffers from data sparsity problem, which include sparse item and sparse user problem.

Sparse item recommendation is also known as long-tail recommendation. The existing research work can be divided into two categories: the ones trading-off between recommendation accuracy and long-tail performance [26,32] and the ones targeting only long-tail recommendation [21,22]. For example, Wang et al. [26] formulate a multi-objective framework under which two contradictory objective functions are designed to describe the recommender system to recommend accurate and unpopular items, respectively. To optimize these two objective functions, a novel multi-objective evolutionary algorithm is proposed to find a trade-off solution by optimizing two objective functions simultaneously. Shi et al. [21] propose a graph model which employs an item normalization based on the global popularity and a PCA analysis to promote less popular items. However, all these existing long-tail recommender systems concentrate on making recommendation for warm-start users.

Another challenging task for recommender systems is what to recommend for users with little or no data (i.e. cold-start users). A popular approach is to use active learning to collect the most useful initial ratings [3,36].

Traditional active learning strategy in recommender systems is to pick up a static seed set. The preferences of users are efficiently achieved by a short interview during which they are asked to rate several carefully selected items in advance. The items are chosen based on several criteria, such as Popularity [16], Contention [16,17], Coverage [4] and so on.

Instead of interviewing all users with the same set of items, recent researches focus on selecting the interviewed items dynamically according to the users' response. This is known as a really exciting application of crowd-sourcing, which has been shown to be effective in a wide range of applications, and is seeing increasing use [25]. The majority of existing bootstrapping methods with dynamic seed sets are based on decision trees [5,10,37]. The users are first clustered into groups and then a classical decision tree classifier is built which recover the user clusters based on user ratings of specific items. Specifically, each non-leaf node asks for user opinions on a specific item and the user is directed to subtrees based on her answer. Finally, the preferences of the users within the same leaf node are used to generate the recommendation by the collaborative filtering. There are different ways of generating the recommendations based on the leaf node. Some of them simply use the average rating of users in the same leaf node as the prediction [5,37] while some others adopt matrix factorization (MF) to generate the recommendations [10]. As we analyzed before, the whole set of interview items are rather limited for all the methods based on decision tree and these methods also tend to acquire ratings for popular items. This argument has also been validated experimentally on two publicly available datasets in this work.

3   Problem Formulation

In this section, we present the problem definition. Hereafter, following the common symbolic notation, we will use upper case bold letters to denote matrices, lower case bold letters to denote column vectors, calligraphic letters to represent sets (e.g., $\mathcal{M}$) and non-bold letters to represent scalars.

Assume that we have an item set $\mathcal{M}$ and a user set $\mathcal{N}$. Let $\mathbf{R}$ be a binary matrix, with $r_{ij} = 1$ indicating i^{th} user likes j^{th} item and $r_{ij} = 0$ denoting i^{th} user dislikes j^{th} item. Note that, in the rest of the paper, *ratings* mean *binary ratings* with values either 0 or 1 unless specified. For a user, interviewing k items leads to a sequence of ratings $\mathbf{d} = <d_1, d_2, \ldots, d_k>$, where d_i represents the rating obtained in the i-th interview question. Given a prediction method f (i.e., a recommender model) and an evaluation metric δ, $\delta(f(\mathbf{d}))$ is the recommendation accuracy given $\mathbf{d}$.

Problem 1 (**Optimizing Top-k Interviews**). Given an item set $\mathcal{M}$, a user set $\mathcal{N}$ and the rating matrix $\mathbf{R}$, for a sparse user u, the problem of optimizing top-k interviews is the identification of a sequence $\mathbf{d} = <d_1, d_2, \ldots, d_k>$ so that $\delta(f(\mathbf{d})) > \delta(f(\mathbf{d}'))$ for any other sequence $\mathbf{d}'$.

Solving Problem 1 is challenging due to the mutual relationship between items. Validating one item can affect the uncertainty of ratings on other items. Further, finding an optimal solution requires investigation of all permutations of all subsets with size k, which is NP-hard [9].

4   Proposed Approach

4.1   Overview

To solve the Problem 1, we propose a greedy method with the heuristic that maximizes the uncertainty reduction of the user's preferences in guiding the selection of items. Specifically, given a sparse user u, we aim at selecting an item m in each iteration so that, after obtaining u's feedback on m, the uncertainty of u's preferences over the other unrated items is minimal.

For a user u, a *probabilistic rating set* is constructed as $\mathcal{P}^u = <\mathcal{M}^{\tilde{u}}, \mathbf{P}^u>$, where $\mathcal{M}^{\tilde{u}}$ stores the items that u has not rated and $\mathbf{P}^u$ is an $|\mathcal{M}^{\tilde{u}}| \times 2$ probability matrix with $p_{i,1}^u$ representing the probability of u liking the ith item and $p_{i,0}^u$ is the probability of u disliking the ith item in $\mathcal{M}^{\tilde{u}}$.

If a sparse user u has no rating data at all, $\mathcal{M}^{\tilde{u}}$ is initialized with the whole set of items at beginning. Then, the process continuously selects an item to seek for u's response. Each iteration of the process comprises two steps: (1) select the most informative item to furthest reduce the uncertainty of u's preferences based on her response in the previous interactions; (2) given u's response to the selected item, update her preferences. Note that, if u has no history data, the first item is selected randomly as there is no interaction history based on which we can make a selection. We will introduce how to select the most informative item for u and how to infer her preferences based on her ratings on interviewed items in the following two sections. To make the description of our approach clearer and easier to follow, we present the user preference learning first.

4.2   User Preference Learning

In this section, we present how to learn a user's preferences over other unrated items given her ratings on some items. Many methods can be applied, such as CF and MF. However, these methods or models need to be recomputed or retrained every time after the user rates a new item. This is very time consuming and impractical in the on-line interview process. To reduce the time cost, we introduce a two-layer undirected graphical model generalizing RBM, which has been widely used in [7,8], inspired by [19].

We use a different RBM for each user shown in Fig. 1. The hidden features are different for different users, but they share the weights because if two users have rated the same movie, their two RBMs should use the same weights between the rating for that movie and the hidden features. Before a new user comes, the RBM based model is trained over **R** to learn the weights and biases for all the users. As the corresponding weights are shared between all the users, when a new user comes, given her feedback on some items, we can predict the probability she likes/dislikes an unrated item, following [19]. In this way, we do not need to retrain the model during the interview, which makes our method efficient and scalable.

In the following part, we will concentrate on learning the gradients for the parameters of a single user-specific RBM. The final gradients with respect to the shared weight parameters can then be obtained by averaging over all users. Assume $\mathcal{M}^u$ stores the set of items that a user u has rated and **V** be a $2 \times |\mathcal{M}^u|$ observed binary indicator matrix with $v_{1i} = 1$ if the user likes the ith movie and $v_{0i} = 1$ otherwise. $h_j, j = 1, \ldots, F$ are the binary hidden features which have different values for different users. $\mathbf{W}^0$ and $\mathbf{W}^1$ are $|\mathcal{M}| \times F$ matrices are shared by all users and are used to store symmetric interaction parameters between features and ratings of items.

RBM Model. The graphical representation of RBM model is given in Fig. 1. Given visible binary ratings, we draw the hidden features according to a conditional Bernoulli distribution as follows:

$$p(h_j = 1|\mathbf{V}) = \sigma(b_j + \sum_{i \in M^u} \sum_{r \in \{0,1\}} v_{ri} \mathbf{W}^r_{ij}) \tag{1}$$

where $\sigma(x) = \frac{1}{1+e^{-x}}$ is the logistic function and b_j is the bias of jth feature.

A conditional multinomial distribution is used to model each column of the observed rating matrix **V**:

$$p(v_{ri} = 1|\mathbf{h}) = \frac{exp(b_{ri} + \sum_{j=1}^{F} h_j \mathbf{W}^r_{ij})}{\sum_{r'=0}^{1} exp(b_{r'i} + \sum_{j=1}^{F} h_j \mathbf{W}^{r'}_{ij})} \tag{2}$$

where $r \in \{0, 1\}$ and b_{ri} is the bias of rating r for the ith movie. The marginal distribution over the visible ratings **V** is inferred with Eq. 3:

$$p(\mathbf{V}) = \prod_{\mathbf{h}} \frac{exp(-E(\mathbf{V}, \mathbf{h}))}{\sum_{\mathbf{V}', \mathbf{h}'} exp(-E(\mathbf{V}', \mathbf{h}'))} \tag{3}$$

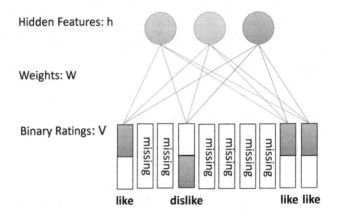

Hidden Features: h

Weights: W

Binary Ratings: V

missing missing missing missing missing missing

like   dislike   like like

Fig. 1. The graphical model of RBM

where $E(\mathbf{V}, \mathbf{h})$ is computed in the following Equation:

$$E(\mathbf{V}, \mathbf{h}) = - \sum_{i \in M^u} \sum_{j=1}^{F} \sum_{r \in \{0,1\}} \mathbf{W}_{ij}^r h_j v_{ri}$$

$$- \sum_{i \in M^u} \sum_{r \in \{0,1\}} v_{ri} b_{ri} - \sum_{j=1}^{F} h_j b_j \tag{4}$$

For the RBM, exact maximum likelihood learning is intractable. We adopt the Contrastive Divergence (CD) [6] to perform efficient model training.

Preference Learning. Given the learned parameters and the user u's existing ratings $\mathbf{V}$ on selected items, we can infer the probabilities that she likes or dislikes each unrated item. For an item i that u has not rated, the unnormalized probabilities that she likes or dislikes this item are inferred as follows:

$$p(v_{ri} = 1|\mathbf{V})$$

$$\propto \sum_{\mathbf{h}} exp(-E(v_{ri}, \mathbf{V}, \mathbf{h}))$$

$$= \Gamma_i^r \prod_{j=1}^{F} (1 + exp(\sum_{l \in M^u} \sum_{r' \in \{0,1\}} v_{r'l} W_{lj}^{r'} + v_{ri} W_{ij}^r + b_j)) \tag{5}$$

where $\Gamma_i^r = exp(v_{ri} b_{ri})$. Once we obtain unnormalized probabilities, we can perform normalization over rating $\{0, 1\}$ for each item to get probabilities $p(v_{ri} = 1|\mathbf{V})$.

The classic RBM model for collaborative filtering can only make recommendations to warm-start users. In the following section, we will present how this model is applied to make recommendations to cold-start users.

4.3 Adaptive Item Selection

To make recommendations for sparse users, we need to acquire their ratings on items in an interview way. To reduce the number of interview items, we aim to select the item that can reflect the most of the user's preferences at each interaction. The item selection method presented in this section is motivated by the observation that the validations of some items are more beneficial than others, since items are not independent, but connected by users' ratings, and a smartly chosen item will have a positive effect on the prediction of many other items. Based on this heuristic, we formally define a measure for the expected benefit of the interview of an item. This measure is the information gain from information theory. Information theory has been widely used in cold-start recommendation and crowd sourcing [9,17]. The information gain quantifies the potential benefit of knowing the true value of an unknown variable [18] (i.e., the rating of an unknown item in our case).

Since for a given user u, $p_{i,0}^u$ and $p_{i,1}^u$ for each item i that u has not rated form a distribution, i.e., $p_{i,0}^u + p_{i,1}^u = 1$, we can model each item as a random variable. Then, the overall uncertainty of u's probabilistic answer set is computed by the Shannnon entropy over a set of random variables. Specifically, the entropy of u's preferences over an item i that u has not rated is computed as follows:

$$H(u,i) = - \sum_{r \in \{0,1\}} p_{i,r}^u \times log(p_{i,r}^u) \tag{6}$$

Based on the entropy of each item in $\mathcal{M}^{\tilde{u}}$, the uncertainty of the probabilistic rating set $\mathcal{P}^u$ is computed as the sum of the entropies of all items:

$$H(\mathcal{P}^u) = \sum_{i \in \mathcal{M}^{\tilde{u}}} H(u,i) \tag{7}$$

The information gain of the selection of an item is the reduction of uncertainty of the probabilistic answer set. For each iteration of the interview process, we aim to find the item which has the largest information gain. Specifically, to decide which item to select, we assess the expected difference in the target user's preferences uncertainty before and after her response for each item, and the respective change is quantified as information gain. Following [9], we define the information gain of user u selecting an item i as follows:

$$IG(u,i) = H(\mathcal{P}^u) - H(\mathcal{P}^u|i) \tag{8}$$

where $H(\mathcal{P}^u|i)$ is a conditional variant of the entropy measure. It refers to the entropy of the probabilistic answer set $\mathcal{P}^u$, conditioned on u's input on i^{th} item. It measures the expected entropy of $\mathcal{P}^u$ given u's rating on i^{th} item [9] and it's defined as follows:

$$H(\mathcal{P}^u|i) = \sum_{r \in \{0,1\}} p_{i,r}^u \times H(\mathcal{P}^{u,i=r}) \tag{9}$$

where $\mathcal{P}^{u,i=r} = <\mathcal{M}^{\tilde{u}}\backslash i, \mathbf{P}^{u,i=r}>$ is the updated *probabilistic rating set* given u's rating r on i^{th} item. Specifically, for each item j in the set $\mathcal{M}^{\tilde{u}}\backslash i$, we infer the items $p_{j,0}^u$ and $p_{j,1}^u$ in $\mathbf{P}^{u,i=r}$ with Eq. 5.

Given a sparse user u, for each iteration of the interview process, we select the item i with maximum information gain $IG(u,i)$ as the interview item. After obtaining u's answer to the interview item, we update her probabilistic rating set $\mathcal{P}^u$, based on which the next interview item is selected. Note that, the proposed item selection method is a greedy algorithm. As Problem 1 is monotonic and submodular, this greedy algorithm yields a $(1 - 1/e) \approx 0.63$ approximation [13] to the exact algorithm.

4.4   Making Recommendation

For a user u, when the interview is done, we get u's all the ratings $\mathbf{V}$, which include both the previous existing ratings and the additional ratings obtained by the interview. For each item i that u has not rated, the unnormalized probabilities that u likes this item, which is denoted as $p(v_{1i} = 1|\mathbf{V})$, are inferred with Eq. 5. Then, the items with highest probabilities are recommended to u.

5   Experiments

In this section, we first demonstrate the experimental setting and then present the results. Note that, we focus on the sparse users and sparse items, thus, the low accuracy is common in these systems and we concentrate on the relative improvements instead of the absolute values.

5.1   Experimental Setup

Datasets. We perform experiments on two real widely used benchmark datasets: Movielens and Netflix. The statistics of these two datasets are listed in Table 1.

Following the standard setup in [5,17,37], we focus on the format of interview questions being "Do you like item m?". However, the raw rating data in both Movielens and Netflix are integers ranging from 1 to 5. Following [37], we infer a user u's binary rating r over an item m based on the raw rating a as follows:

$$r = \begin{cases} 0 & a \leq 3 \\ 1 & a > 3 \end{cases} \tag{10}$$

Data Splitting. We sample 25% users randomly as the test users to be interviewed. For each test user u, if u has more than n_{max} records, we pick up n_{max} records randomly as the training data and move the other records associated with u to the test set. If u has no more than n_{max} records, we move all her data to the test set. For the other users, all their data are stored in the training set.

Table 1. Statistics of the two datasets

	Movielens	Netflix
# of users	6,040	480,189
# of items	3,900	11,770
# of ratings	1,000,209	100,480,507
Sparsity	95.754%	98.822%

In this way, we can simulate the evaluation of recommendation for the sparse users as all the users in the test set have no more than n_{max} history records left in the training set. Additionally, to simulate the recommendation for the sparse items, the records containing $p\%$ most popular items are marked off from the test set. As a result, both the users and items in the test set are sparse. The aim of our interview process is to improve the recommendation related to these sparse users and items by adding ratings for them.

Comparison Methods. We compare with five state-of-the-art methods which can obtain additional ratings for sparse recommendations through an interview process. Our proposed method is referred as "*i*-RBM" as a short of "information theoretic RBM".

Popularity. The popularity-based method randomly chooses popular items to query the user's response.

Entropy0. This is a top performing method among the bootstrapping methods using static seed sets proposed in [17]. It picks the item to be interviewed by considering both its popularity and contention.

Note that, for both *Popularity* and *Entropy0*, CF method is applied to generate the recommendation after additional ratings are obtained through the interviewed questions.

Tree. There are many existing work exploiting decision trees to guide the interview process. We compare our method with LAL-based Decision Tree [10]. LAL is short for *Learning Active Learning*. The decision tree is constructed based on the training supervised by the existing warm-start users. For each non-leaf node, it chooses the best splitting item, namely the item that partitions the users such that the total squared prediction error is minimized. Following a root-to-leaf path, a user ultimately ends at a leaf node. She gets the recommendation based on the preferences of the other users in the same leaf node with MF method.

fMF. Different from *Tree* which separates the construction of decision tree and the prediction with MF, *fMF* [37] combines MF and tree based interview model into a single framework. It restricts the user profiles to be a function of the answers to the interview questions in the form of decision tree. An iterative optimization algorithm that alternates between tree construction and latent item profiles extraction is proposed to learn the tree and the item profiles.

i-RBM-s. To further explore the benefits brought by adaptively choosing the interview items, we also compare with a simplified version of *i*-RBM. This simplified version is referred as *i*-RBM-s, which selects k interview items at once in a batch mode based on the user's ratings in the training set. During the interview process, it acquires the user's ratings for these selected items. After the interview, it makes recommendations for this user based on both her ratings in the training dataset and the new acquired ratings in the interview with Eq. 5.

Evaluation Methods. Our proposed model can be evaluated online or offline. In the first case, the system interacts with real users through a graphical interface. This requires building a fully developed recommender system, with a large number of registered users, which is expensive and time consuming. Moreover, it is hard and impractical to compare several strategies in alternative systems [3]. Thus, the majority of the works [11,14,24] in this research area have focused on offline evaluations where pre-collected rating dataset is used to simulate users' real response during the interaction with the system.

Simulation of User Response. Many existing work [24,37] sample part of the ratings of each user in the test set. This sampled rating set is called as *answer set* and it is used to simulate the generation of responses in the interview process. Specifically, given a user u and an interviewed item m, if there is a rating associated with u and m in the answer set, this rating is used as u's response. But this method can only simulate u's response to the items in the so-called answer set, but cannot simulate her response to many items that have not been rated by her in the answer set. To deal with this limitation, we propose a more reasonable and reliable offline evaluation method to simulate the user's response during the interview process. For each test user, we construct a proxy user from the "Gods view" and this proxy user has all ratings of the test user in both training set and test set. Then, based on all these ratings, we use the trained RBM model to infer the proxy users preference distribution over all items using Eq. 5. For each interviewed item m for the test user u, her response will be assumed to be "like" if $p_{like} - p_{dis} > 0$; otherwise the answer will be assumed to be "dislike", where p_{like} and p_{dis} are the probabilities that the proxy user likes and dislikes the item m respectively. Thus, we simulate a test user's response based on her proxy user's preferences. By this way, we can simulate the test user's response to any item. Obviously, the proxy user is more close to the corresponding real user, as a real user is fully aware of her interests and preferences, which have nothing to do with ratings. Actually, ratings are just a way to reflect the real user's preferences from the recommender system's perspective. In most time, ratings can only reflect part of a real user's preferences. Note that the selection of the interview items and the final top-k recommendations are based on the test user's preferences rather than the proxy user's preferences. The test users preferences are estimated from her ratings in the training set and the newly acquired ratings during the interview process.

In this paper, one popular ranking related evaluation methodology and measurement Accuracy@n which has been widely applied in recommendation

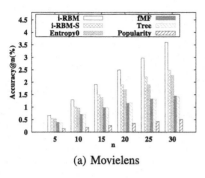
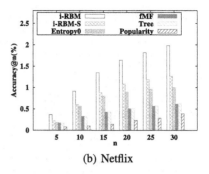

(a) Movielens   (b) Netflix

Fig. 2. Performance on both datasets

community [27–29, 33, 34] is adopted. We define the users in the test set as U_{test}. For each u in U_{test}, we get the item set S_u that u likes based on u's ratings in the test set as the ground truth. Then, we conduct the interview process for u based on the training data. For each case (u, m) where $m \in S_u$:

(1) We compute the scores for all u's unrated items in $\mathcal{M}$ with Eq. 5;
(2) We form a ranked list by ordering all items according to their scores. Let o denote the position of m within this list. The best result corresponds to the case where m precedes all the unrated items (i.e., $o = 1$);
(3) We form a top-n recommendation list by picking the n top ranked items from the list. If $o \leq n$, we have a hit (i.e., the ground truth item m is recommended to the user). The probability of a hit increases with the increasing value of n and we always have a hit when $n = |\mathcal{M}|$. The computation of Accuracy@n proceeds as Eq. 11. #hit@n is the number of hits in the test set.

$$Accuracy@n = \frac{\#hit@n}{\sum_{u \in U_{test}} |S_u|} \qquad (11)$$

5.2   Recommendation Effectiveness

Following [19], we set $F = 100$. To speed-up the training process on the Netflix dataset, we divide the data into smaller batches, each containing about 1000 users, and update the weight parameters after each batch. The model is trained for between 40 and 50 passes on the training dataset. The weights are initialized with small random values sampled from a zero-mean normal distribution with standard deviation 0.01 and updated with a learning rate of 0.01 and a weight decay of 0.001.

We evaluate the performance of i-RBM in sparse recommendation compared with the baselines. Figure 2(a) and (b) present the performance of methods on two datasets. We fix the number of interviewed questions as 5 [24, 37] for all the methods. n_{max} and p are all set to 5. We present Accuracy@n of the methods when n is set to $5, 10, 15, 20, 25, 30$ respectively. From these figures, we see that all the methods achieve higher accuracy values with the increase of n. This is

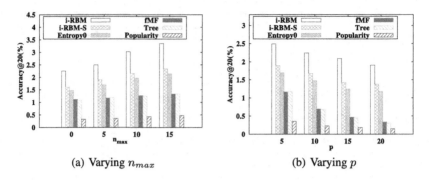

Fig. 3. The performance under varying sparse settings on movielens

because with more items recommended, it is more likely to find the items that the users like. Another observation is that i-RBM outperforms the comparison methods significantly in all settings. This shows the advantage of i-RBM. We also observe that i-RBM performs better than i-RBM-S. This demonstrates that integrating the feedback dynamically during the interview indeed improves the quality of the added ratings. Moreover, i-RBM, i-RBM-S and Entropy0 perform much better than fMF, Tree and Popularity. The reason is that i-RBM, i-RBM-S and Entropy0 select the interviewed items based on the information gain while the other three methods tend to select the popular items. What's worth noting is that Popularity performs even worse than the random recommendation ($Accuracy$@20 = 0.513%) in the sparse setting. This does make sense because Popularity tends to recommend popular items as it adds more ratings to the popular items. Thus, in recommending the long-tail items, it may perform even worse than the random recommendation which treats all the items equally.

To evaluate the performance of the methods in different sparse settings, we evaluate the performance by setting n_{max} and p to varying values on Movielens dataset. A larger n_{max} indicates a denser data set while a bigger p represents a sparser data set. Figure 3(a) and (b) report the recommendation effectiveness by varying n_{max} and p respectively. We set $p = 5$ in Fig. 3(a) and $n_{max} = 5$ in Fig. 3(b). We present Accuracy@20 in both figures. From the Fig. 3(a) and (b), we see that our proposed method i-RBM outperforms the comparison methods significantly in all the presented sparse settings. Another observation is that the accuracy of all the methods decrease with the increase of sparsity. This is consistent with our assumption that recommender systems tend to perform more poorly on sparser datasets.

5.3 Further Discussions

As we argued before, due to the limitation of the height of decision trees, the number of items which can be selected for rating elicitation are quite limited. Moreover, because of the metric used to select the splitting nodes in decision trees, the selected items for rating elicitation tend to be popular items, which

Table 2. Item diversity of i-RBM and tree

	$k = 1$	$k = 2$	$k = 3$	$k = 4$	$k = 5$
i-RBM	429	548	613	670	719
Tree	1	3	7	15	31

Table 3. Average numbers of ratings associated with the interviewed items

	$k = 1$	$k = 2$	$k = 3$	$k = 4$	$k = 5$
i-RBM	259.55	242.13	242.73	237.99	237.59
Tree	1166.00	1500.44	1144.79	980.93	717.42

Table 4. Median numbers of ratings associated with the interviewed items

	$k = 1$	$k = 2$	$k = 3$	$k = 4$	$k = 5$
i-RBM	115	125	122	123	124
Tree	1166	1166	1166	450	450

will lead to the Harry Potter Effect. To evaluate that i-RBM is able to elicit ratings for various sparse items instead of limited popular items more directly, in this section, we study both the varieties and popularities of items selected by i-RBM for ratings on Movielens. In this section, we still set $p = 5$ and $n_{max} = 5$.

Table 2 presents the numbers of different items selected for interview for i-RBM and Tree. From this table, we can see that, compared with decision tree based methods, the i-RBM is able to elicit ratings for much more diverse items.

To further validate that the items selected by i-RBM are not limited to the popular items as decision tree does, we also study the popularity of the selected items for both these two methods on Movielens in this section. We present the average numbers of history records associated with the interviewed items in Table 3 and the median numbers of history records associated with the interviewed items in Table 4. From these two tables, we can see clearly that the, compared with Tree, i-RBM tends to acquire extra ratings for the less popular items.

Tables 2, 3 and 4 jointly demonstrate that, compared with Tree, i-RBM is able to acquire additional ratings for more diverse and less popular items.

6   Conclusion

In this paper, we proposed an interactive framework to obtain additional ratings for sparse recommendation. We formally defined the problem of maximum prediction accuracy with restricted number of interview questions. To guide the selection of interview items, a greedy strategy was introduced. This strategy is motivated by the observation that the validations of some items are more beneficial than others. Based on this motivation, this greedy strategy selects the next item to be interviewed which maximizes the expected uncertainty reduction about the user's preferences. To make our framework more efficient during the interview, we introduced a class of two-layer undirected graphical models based on RBM to update user preferences. The proposed method has been proven effective on two publicly available real-life datasets.

Acknowledgment. The work described in this paper is partially supported by ARC Discovery Early Career Researcher Award (DE160100308), and ARC Discovery Project (DP170103954).

References

1. Cai, C., He, R., McAuley, J.: SPMC: socially-aware personalized Markov chains for sparse sequential recommendation. In: IJCAI, pp. 1476–1482 (2017)
2. Elahi, M., Ricci, F., Rubens, N.: Active learning strategies for rating elicitation in collaborative filtering: a system-wide perspective. ACM TIST $5(1)$, 13 (2013)
3. Elahi, M., Ricci, F., Rubens, N.: A survey of active learning in collaborative filtering recommender systems. Comput. Sci. Rev. **20**, 29–50 (2016)
4. Golbandi, N., Koren, Y., Lempel, R.: On bootstrapping recommender systems. In: CIKM, pp. 1805–1808 (2010)
5. Golbandi, N., Koren, Y., Lempel, R.: Adaptive bootstrapping of recommender systems using decision trees. In: WSDM, pp. 595–604 (2011)
6. Hinton, G.E.: Training products of experts by minimizing contrastive divergence. Neural Comput. **14**(8), 1771–1800 (2002)
7. Hinton, G.E., Osindero, S., Teh, Y.W.: A fast learning algorithm for deep belief nets. Neural Comput. **18**(7), 1527–1554 (2006)
8. Hinton, G.E., Salakhutdinov, R.: Reducing the dimensionality of data with neural networks. Science **313**(5786), 504–507 (2006)
9. Hung, N.Q.V., Thang, D.C., Weidlich, M., Aberer, K.: Minimizing efforts in validating crowd answers. In: SIGMOD, pp. 999–1014 (2015)
10. Karimi, R., Nanopoulos, A., Schmidt-Thieme, L.: A supervised active learning framework for recommender systems based on decision trees. User Model. User-Adapt. Interact. **25**(1), 39–64 (2015)
11. Kluver, D., Konstan, J.A.: Evaluating recommender behavior for new users. In: RecSys, pp. 121–128 (2014)
12. Liu, N.N., Meng, X., Liu, C., Yang, Q.: Wisdom of the better few: cold start recommendation via representative based rating elicitation. In: RecSys, pp. 37–44 (2011)
13. Nguyen, T.T., Weidlich, M., Duong, C.T., Yin, H., Nguyen, Q.V.H.: Retaining data from streams of social platforms with minimal regret. In: IJCAI (2017)
14. Park, S., Chu, W.: Pairwise preference regression for cold-start recommendation. In: RecSys, pp. 21–28 (2009)
15. Park, Y., Tuzhilin, A.: The long tail of recommender systems and how to leverage it. In: RecSys, pp. 11–18 (2008)
16. Rashid, A.M., Albert, I., Cosley, D., Lam, S.K., McNee, S.M., Konstan, J.A., Riedl, J.: Getting to know you: learning new user preferences in recommender systems. In: IUI, pp. 127–134 (2002)

17. Rashid, A.M., Karypis, G., Riedl, J.: Learning preferences of new users in recommender systems: an information theoretic approach. SIGKDD Explor. **10**(2), 90–100 (2008)
18. Russell, S.J., Norvig, P.: Artificial Intelligence - A Modern Approach. Pearson Education, London (2010). (3rd internat. edn.)
19. Salakhutdinov, R., Mnih, A., Hinton, G.E.: Restricted Boltzmann machines for collaborative filtering. In: ICML, pp. 791–798 (2007)
20. Sedhain, S., Menon, A.K., Sanner, S., Xie, L., Braziunas, D.: Low-rank linear cold-start recommendation from social data. In: AAAI, pp. 1502–1508 (2017)
21. Shi, K., Ali, K.: GetJar mobile application recommendations with very sparse datasets. In: SIGKDD, pp. 204–212 (2012)
22. Shi, L.: Trading-off among accuracy, similarity, diversity, and long-tail: a graph-based recommendation approach. In: RecSys, pp. 57–64 (2013)
23. Song, K., Gao, W., Feng, S., Wang, D., Wong, K., Zhang, C.: Recommendation vs sentiment analysis: a text-driven latent factor model for rating prediction with cold-start awareness. In: IJCAI, pp. 2744–2750 (2017)
24. Sun, M., Li, F., Lee, J., Zhou, K., Lebanon, G., Zha, H.: Learning multiple-question decision trees for cold-start recommendation. In: WSDM, pp. 445–454 (2013)
25. Tong, Y., Chen, L., Zhou, Z., Jagadish, H.V., Shou, L., Lv, W.: Slade: a smart large-scale task decomposer in crowdsourcing. TKDE (2018)
26. Wang, S., Gong, M., Li, H., Yang, J.: Multi-objective optimization for long tail recommendation. KBS **104**, 145–155 (2016)
27. Wang, W., Yin, H., Chen, L., Sun, Y., Sadiq, S.W., Zhou, X.: Geo-SAGE: a geographical sparse additive generative model for spatial item recommendation. In: SIGKDD, pp. 1255–1264 (2015)
28. Wang, W., Yin, H., Chen, L., Sun, Y., Sadiq, S.W., Zhou, X.: ST-SAGE: a spatial-temporal sparse additive generative model for spatial item recommendation. ACM TIST **8**(3), 48:1–48:25 (2017)
29. Wang, W., Yin, H., Sadiq, S.W., Chen, L., Xie, M., Zhou, X.: SPORE: a sequential personalized spatial item recommender system. In: 32nd IEEE International Conference on Data Engineering, ICDE 2016, Helsinki, Finland, 16–20 May 2016, pp. 954–965 (2016)
30. Yan, S., Chaudhuri, K., Javidi, T.: Active learning from imperfect labelers. In: NIPS, pp. 2128–2136 (2016)
31. Yin, H., Cui, B.: Spatio-Temporal Recommendation in Social Media. Springer Briefs in Computer Science. Springer, Heidelberg (2016). https://doi.org/10.1007/978-981-10-0748-4
32. Yin, H., Cui, B., Li, J., Yao, J., Chen, C.: Challenging the long tail recommendation. PVLDB **5**(9), 896–907 (2012)
33. Yin, H., Cui, B., Zhou, X., Wang, W., Huang, Z., Sadiq, S.W.: Joint modeling of user check-in behaviors for real-time point-of-interest recommendation. TIST **35**(2), 11:1–11:44 (2016)
34. Yin, H., Wang, W., Wang, H., Chen, L., Zhou, X.: Spatial-aware hierarchical collaborative deep learning for POI recommendation. TKDE **29**(11), 2537–2551 (2017)
35. Yin, H., Zhou, X., Cui, B., Wang, H., Zheng, K., Hung, N.Q.V.: Adapting to user interest drift for POI recommendation. TKDE **28**(10), 2566–2581 (2016)
36. Zhang, Z., Jin, X., Li, L., Ding, G., Yang, Q.: Multi-domain active learning for recommendation. In: AAAI, pp. 2358–2364 (2016)
37. Zhou, K., Yang, S., Zha, H.: Functional matrix factorizations for cold-start recommendation. In: SIGIR, pp. 315–324 (2011)

Discrete Binary Hashing Towards Efficient Fashion Recommendation

Luyao Liu[1]([✉]), Xingzhong Du[1], Lei Zhu[2], Fumin Shen[3], and Zi Huang[1]

[1] School of ITEE, The University of Queensland, Brisbane, Australia
{luyao.liu,x.du}@uq.edu.au, huang@itee.uq.edu.au
[2] School of Information Science and Engineering, Shandong Normal University,
Jinan, China
leizhu0608@gmail.com
[3] School of Computer Science and Engineering, UESTC, Chengdu, China
fumin.shen@gmail.com

Abstract. How to match clothing well is always a troublesome problem in our daily life, especially when we are shopping online to select a pair of matched pieces of clothing from tens of thousands available selections. To help common customers overcome selection difficulties, recent studies in the recommender system area have started to infer the fashion matching results automatically. The conventional fashion recommendation is normally achieved by considering visual similarity of clothing items or/and item co-purchase history from existing shopping transactions. Due to the high complexity of visual features and the lack of historical item purchase records, most of the existing work is unlikely to make an efficient and accurate recommendation. To address the problem, in this paper we propose a new model called Discrete Supervised Fashion Coordinates Hashing (DSFCH). Its main objective is to learn meaningful yet compact high level features of clothing items, which are represented as binary hash codes. In detail, this learning process is supervised by a clothing matching matrix, which is initially constructed based on limited known matching pairs and subsequently on the self-augmented ones. The proposed model jointly learns the intrinsic matching patterns from the matching matrix and the binary representations from the clothing items' images, where the visual feature of each clothing item is discretized into a fixed-length binary vector. The binary representation learning significantly reduces the memory cost and accelerates the recommendation speed. The experiments compared with several state-of-the-art approaches have evidenced the superior performance of the proposed approach on efficient fashion recommendation.

1 Introduction

With the rapid growth of e-commerce, traditional offline clothing sales have been moving to the online websites [33]. Facing the eyeful of clothing items available online, customers usually have limited time on fashion matching and are easy to suffering from selection difficulties. It is a very common scenario that we

© Springer International Publishing AG, part of Springer Nature 2018
J. Pei et al. (Eds.): DASFAA 2018, LNCS 10827, pp. 116–132, 2018.
https://doi.org/10.1007/978-3-319-91452-7_8

feel difficult to decide 'which trousers would fashionably match this jumper' or 'what kind of skirt would go well with this shirt'. Clothing recommendation is now a trending service provided by a number of major online shopping websites. Hand-picked fashion coordinates such as model images which are advised by the fashion insides are presented to customers to assist them choosing a better matching style. However, the hand-picked solution is usually unscalable and labor-consuming. In result, recent research efforts in the recommender system area try to infer the fashion matching results automatically for the customers [25], which has strong potential to provide considerable economic value to the existing online services.

The existing work technically provides the clothing fashion matching automatically in three steps: (1) learning representations of clothing items by high-dimensional vectors with real values based on visual features and matching tuples; (2) calculating the Euclidean distances between the matching target and complementary clothing; (3) selecting the nearest complementary clothing as the matching results [12].

Since the ability of visual aware is significantly advanced by the progress in the computer vision area, recent work [8,14,24,25,33] mainly focus on how to embed the matching relations between clothing items into the embedding vectors. Although the matching accuracy has been improved by recent studies to some extent, the fashion recommendation task still faces three challenges [1,25,33].

- *Inference efficiency.* With the sustainable growth of e-commerce, a large amount of clothing is available online at high speed nowadays. Considering that the existing work need to store a high-dimension real-value vector for each item, the persistent and temporal storage costs for inference are heavy burden due to the massive data scale. In addition, the existing work employ the Euclidean distance to calculate the nearest neighbors for each query target. Given the huge amount of clothing, the inference process would be very slow. As a result, it is necessary to develop a compact feature representation for clothing items to support high efficient and scalable fashion matching with limited storage cost.
- *Label quality.* Precise labels that represent matching relationships are important for constructing an effective learning system. In other words, a matching matrix to carry the relationships (i.e., matched, un-matched, unknown) among clothing items is the essential priori knowledge for the learning process in the recommender system. As fashion matching is subjective without a clear definition, precise matching relationships are generally achieved from fashion expertise. To the best of our knowledge, the existing datasets for fashion matching, i.e., Deep Fashion [22,23] and Amazon Product Data [26,34], construct the matching labels purely according to customers' shopping carts in single transactions. Obviously, co-purchased items cannot be guaranteed relevant or matched with each other. The matching labels generated in this way is not reliable for fashion matching supervision.

– *Fashion understanding.* Individuals may have different understanding of fashion. Fashion, from the perspective of automatic fashion matching, need to be understood by the learning over user-clothing interactions and visual features. Accordingly, how to design a better learning process to effectively capture the fashion is in high demand for personalization.

In this paper, we propose an efficient fashion recommendation method to learn meaningful yet compact representations of clothing items to capture their intrinsic visual appearances and the matching relationships. The efficiency problem in existing methods is addressed with high competitive recommendation accuracy. Specifically, we design a supervised hashing framework, called Discrete Supervised Fashion Coordinates Hashing (DSFCH), that learns discrete binary representations of clothing items from their visual content features and the matching matrix constructed based on expertise knowledge. The proposed framework guarantees that each clothing item is discretized into a fixed-length binary vector when the training stops. The discretization significantly reduces the memory cost and accelerates the inference speed. Our experiments validate that the learned binary representations effectively facilitate the fashion matching with competitive recommendation accuracy.

It is worthwhile to highlight the key contributions of our proposed method:

– we propose a supervised learning to hash framework that learns the discrete binary representations of clothing items from their visual content features and the matching matrix constructed based on expertise knowledge. An iterative optimization guaranteed with convergence is proposed to effectively solve the optimal binary representation of clothing items. The discretization can significantly reduce the memory cost and accelerate the fashion recommendation speed.
– We construct two real life fashion datasets with clothing images and professional fashion coordinates advices. These datasets are built up based on websites Netaporter[1] and Farfetch[2]. To the best of our knowledge, this is the first large-scale fashion database with professional advices for fashion recommendation.

The rest of the paper is structured as follows. Section 2 reviews the related work. Details about the proposed methodology are presented in Sect. 3. In Sects. 4 and 5, we introduce the experiments. Section 6 concludes the paper.

2   Related Work

Due to the limited space here, in this section, we only focus on the most related works on fashion recommendation and hashing techniques.

---

[1]   www.net-a-porter.com/au/.
[2]   www.farfetch.com/au/.

2.1 Fashion Recommendation

Motivated by the huge impact for e-commerce applications, fashion recommendation [10,25,33] has been receiving increasing attentions. Content-based recommender systems [15] attempt to model each user's preference toward particular types of goods. An early work [9] proposes a probabilistic topic model to learn information about coordinates from visual features by training full-body photographs from fashion magazines. The model finds reference photographs that are similar to the query image based on image content and recommends fashion items that are similar to those in the reference photograph.

Beyond exact matching between user photos and clothing images [9,10,13], recommendation systems require learn the human notions between outfit collections [30,33] and mining personal taste [4] with surrounding auxiliary information. In [25] the authors aim to model human notion of what is visually correlated by investigating a large scale dataset and affluent corresponding information. The model understands human preference more than just the visual similarity between the two. The system suggests people what not to wear and who is more fashionable.

A variety of approaches are proposed to incorporate deep learning into recommender systems [36]. A feature transformation learning [33] extends the traditional metric learning by utilizing Siamese Convolutional Neural Network (CNN) [7] architecture and projects images into a latent fashion style space to express the compatibility of outfit with the help of cross-category labels and user co-purchase data. Similarly, a recent work [12] combines fashion design and image classification by training image representations to achieve personalized fashion recommendation.

Forecasting future fashion trend is also an interesting way [1] to recommend fashion outfits before they occur. A study in [5] investigates the correlation between attributes popular in New York fashion shows versus what is seen later on the street. Another model [1] analyses fine-grained visual styles from large scale fashion data in an unsupervised manner to identify unique style signatures. The model provides a semantic description on key visual attributes to predict the future popularity of the styles.

However, existing fashion recommendation approaches still suffer from the problem of inference efficiency, label quality and fashion understanding.

2.2 Hashing

Hashing [2] is an advanced indexing technique that can achieve both high retrieval efficiency and memory saving. With binary embedding of hashing, the original time-consuming similarity computation can be substituted with efficient bit operations. Thus, the similarity search process could be greatly accelerated with constant or linear time complexity [43]. Moreover, binary representation could significantly shrink the memory cost of data samples, and thus accommodate large-scale similarity search with very limited memory. Due to these desirable advantages, hashing has been received great attention in literature [3,20,37,39,42].

According to the learning dependence on semantic labels, existing learning-based hashing methods can be categorized into two major families: unsupervised hashing [6,11,18,20,38,40] and supervised hashing [16,19,27,43]. Supervised hashing learns effective binary codes based on the supervised semantic labels. It usually achieves better performance than unsupervised hashing methods.

The inner product of binary codes play an important role on cross-modality retrieval [17] and supervised hashing [21]. As indicated by the existing studies [21, 28,29], it has been proved that code inner product can characterize the similarity of two binary hash codes in Hamming space.

3   Methodology

In this section, we will detail our proposed Discrete Supervised Fashion Coordinates Hashing (DSFCH) for efficient fashion recommendation. We develop a unified hashing learning framework. A kernelized feature embedding is employed to efficiently capture the nonlinear structure of the raw feature in original space with a single vector. An inner-product fitting model is designed to preserve the correlation between various images of clothing items into binary hash codes.

3.1   Problem Formulation

Let $X = \{x_1, x_2, \ldots, x_n\} \in \Re^{n \times d}$ represent an image representation matrix for the collection of clothing items, n is the number of data samples and d is the dimension of feature representation. As mentioned above, we aim to learn a hash function $Z(x) = sgn(F(x))$, which maps x from the original space into a Hamming space. Here $sgn(\cdot)$ is the signum function which returns 1 if $x \geq 0$, -1 if $x < 0$. We will discuss $F(x)$ in Sect. 3.2.

The projected binary codes are defined as $B = \{b_1, b_2, \ldots, b_n\} \in \{-1, 1\}^{n \times r}$, where r denotes the hash code length and $b_i^T, b_j^T \in \{-1, 1\}^{1 \times r}$ denote the i_{th}, j_{th} row of B, respectively. Formally, the hashing projection loss can be formulated as:

$$\min_{B,F} \frac{1}{2} \sum_{i=1}^{n} (b_i - F(x_i))^2 \tag{1}$$

$$s.t. \ b_i \in \{-1, 1\}^r.$$

We introduce a fashion matching matrix $S \in \{0, 1\}^{n \times n}$ to semantically guide the hash code learning process. The matrix records each pairwise similarity S_{ij} as 1 if two clothing items are correlated, and 0 if their matching relations are unknown. As mentioned above, existing studies [17,21,41] have approved that the inner product of binary codes can characterize their similarity in Hamming space. In this paper, to preserve the fashion matching relations in binary codes, we try to solve the following optimization problem:

$$\min_{B,F} \frac{1}{2} \sum_{i=1}^{n} \sum_{j=1}^{n} (S_{ij} - \frac{1}{r} b_j^T \cdot b_i)^2 + \frac{1}{2} \nu \sum_{i=1}^{n} (b_i - F(x_i))^2 \tag{2}$$

$$s.t.\ b_i, b_j \in \{-1, 1\}^r$$

where $\nu > 0$ is the parameter to balance regularization terms. Considering that the elements of S are comprised of 0 and 1, and the binary quantization loss between each b and $F(x_i)$ can be minimized by imposing the binary constraints on B. Therefore we rewrite the problem (2) as:

$$\min_{B,F} \frac{1}{2} \sum_{i=1}^{n} \sum_{j=1}^{n} C_{ij} \odot (S_{ij} - \frac{1}{r} b_j^T b_i)^2 + \frac{1}{2} \nu \sum_{i=1}^{n} (b_i - F(x_i))^2 \tag{3}$$

$$s.t.\ b_i, b_j \in \{-1, 1\}^r$$

where C_{ij} indicates the precision parameter for S_{ij}. The element-wise product "$\odot$" means we are only interested in the b_j where $S_{ij} = 1$ corresponds to each b_i, which is targeted to our application. According to [35], we set C_{ij} a higher value when $S_{ij} = 1$ than when $S_{ij} = 0$,

$$C_{ij} = \begin{cases} a, & if\ S_{ij} = 1 \\ b, & if\ S_{ij} = 0 \end{cases} \tag{4}$$

where a and b are tuning parameters satisfying $a > b > 0$. Here we follow the same settings in [35] as $a = 1, b = 0.01$.

3.2   Kernelized Feature Embedding

Large-scale real-world data contains a lot of noises which negatively affect the accuracy of the projections. Specifically, the learned hash codes will be affected unavoidably by variances, redundancies and noises [17]. It will result in in crucial representation problems of raw features. Thus we utilize RBF kernel embedding to achieve better performance [21]. The nonlinear form can be formulated as:

$$F(x) = \phi(x) \cdot H \tag{5}$$

where $\phi(x) \in \Re^{1 \times m}$ is a m-dimensional row vector obtained by the kernel mapping: $\phi(x) = [exp(\|x - a_1\|^2/\epsilon, \cdots, exp(\|x - a_m\|^2/\epsilon)]$, where $\| \cdot \|$ denotes the Frobenius norm operation, $\{a_u\}_{u=1}^{m}$ indicates the randomly selected m anchor points from the training samples and ϵ is the kernel width. The $H \in \Re^{m \times r}$ is the projection matrix which maps the original image feature into the low dimensional space. Once the kernelized feature embedding is obtained, we derive the overall objective formulation as:

$$\min_{B,F} \frac{1}{2} \sum_{i=1}^{n} \sum_{j=1}^{n} C_{ij} \odot (S_{ij} - \frac{1}{r} b_j^T b_i)^2 + \frac{1}{2} \nu \sum_{i=1}^{n} (b_i - F(x_i))^2 + \lambda \|H\|^2 \tag{6}$$

$$s.t.\ b_i, b_j \in \{-1, 1\}^r$$

where λ denotes the penalty parameter. The next step is to optimize the hash functions and find the optimal solution.

3.3   Optimization

Directly solving the minimization problem in Eq.(3) is NP-hard. Thus, we propose an iterative approach to convert this problem into a few sub-problems with each solving one variable when fixing all other variables. For each sub-problem, it is tractable and able to get the optimal solution.

Optimizing F. If B is fixed in Eq. (6), the projection matrix H is independent to other regularization terms. Therefore we can easily compute the H by solving:

$$\min_{H} \|B - \phi(X)H\|^2 + \lambda\|H\|^2 \tag{7}$$

Eq. (5) can be solved by linear regression. The optimal H can be derived as:

$$H = (\phi(X)^T\phi(X) + \lambda I)^{-1}\phi(X)^T B. \tag{8}$$

Optimizing B. It is still challenging to optimize B due to the discrete constraints in Eq. (6) which is NP-hard problem. So we try to find a closed-form solution for each single b_i by fixing all other bits $\{b_j\}_{j\neq i}^n$ during optimization. We can rewrite the each iteration step of Eq. (6) as:

$$L = \sum_{i=1}^{n}\sum_{j=1}^{n} C_{ij} \odot (S_{ij} - \frac{1}{r}b_j^T b_i)^2 + \nu \sum_{i=1}^{n}(b_i - F(x_i))^2 + \lambda\|H\|^2 \tag{9}$$

where L denotes the total loss of each loop, and it will achieve convergence after K_{th} iteration. It should be noticed that when we apply another embedded iteration to solve each single b_i of B, we relax the discrete constraint. When i is ranged from 1 to n, we calculate the partial derivatives of loss term l_i with respect to the output b_i. The partial derivation process can be written as:

$$\frac{\partial l_i}{\partial b_i} = 2\sum_{j=1}^{n} C_{ij} \odot (S_{ij} - \frac{1}{r}b_j^T b_i)\frac{\partial(-\frac{1}{r}b_j^T b_i)}{\partial b_i} + 2\nu(b_i - F(x_i))\frac{\partial b_i}{\partial b_i}$$

$$= (\sum_{j=1}^{n} C_{ij} \odot \frac{1}{r^2}S_{ij}b_j b_j^T + \nu I)b_i - (\sum_{j=1}^{n} C_{ij} \odot \frac{1}{r}S_{ij}b_j + \nu F(x_i)). \tag{10}$$

Due to Eq. (4), $C_{ij} \odot S_{ij} = S_{ij}$. Let $\nabla l_i = 0$, the optimized solution can be calculated as:

$$b_i = sgn((\sum_{j=1}^{n} \frac{1}{r^2}S_{ij}b_j b_j^T + \nu I)^{-1}(\sum_{j=1}^{n} \frac{1}{r}S_{ij}b_j + \nu F(x_i))). \tag{11}$$

We can observe that computing single bit binary codes for each data point relies on the rest of pre-learned $(n-1)$ binary codes. It is also noted that b_j^k should be selected from the previous iterative round of pre-learned B^{k-1} corresponding to each b_i. Thus, we need to learn and update b_i for n times in each iteration to obtain the final optimized B. The iteration complexity here is $O(knr + knr^3)$ where $k, r \ll n$. More importantly, we still keep the discrete constrains for B outside the embedded iteration.

Algorithm 1. *Discrete Supervised Fashion Coordinates Hashing (DSFCH)*

Input: Training data $X = \{x_1, x_2, \ldots, x_n\} \in \Re^{n \times d}$, matching matrix S, precision parameter C, code length r, number of anchor points m, maximum iteration number K, parameters λ and ν.
Output: Binary codes $B \in \{-1, 1\}^{n \times r}$, hashing projection matrix H.

Randomly select m samples $\{a_u\}_{u=1}^{m}$ from the training data and map the training data via the RBF kernel function $\phi(x)$
Initialize $B_0 \in \{-1, 1\}^{n \times r}$.
repeat
    Optimizing F:
        Calculate H using Eq.(8):

$$H = (\phi(X)^T \phi(X) + \lambda I)^{-1} \phi(X)^T B$$

    Optimizing B:
        Calculate each b_i of B using Eq.(11):

$$b_i = sgn((\sum_{j=1}^{n} \frac{1}{r^2} S_{ij} b_j b_j^T + \nu I)^{-1} (\sum_{j=1}^{n} \frac{1}{r} S_{ij} b_j + \nu F(x_i)))$$

    Calculate the loss of each iteration using Eq.(9):

$$L = \sum_{i=1}^{n} \sum_{j=1}^{n} C_{ij} \odot (S_{ij} - \frac{1}{r} b_j^T b_i)^2 + \sum_{i=1}^{n} \nu(b_i - F(x_i))^2 + \lambda \|H\|^2$$

  until Convergence

Initializing B. Obviously we should initialize B_0 to start F sub-problem before conducting the K iterations. Inspired by SH [40] and KSH [19], we tried to initialize the binary codes by thresholding spectral graph decomposition. However, the performance was unsatisfactory and considering the time consumption, we choose to use random binary codes $B_0 \in \{-1, 1\}^{n \times r}$ which is sufficient to show the effectiveness of our method.

Precision Parameter C_{ij}. In the above section, we have presented the discrete learning algorithm for each bit of hash codes. We haven't discussed the influence of the C_{ij} which is a precision parameter for rating the correlation matrix S_{ij}. Without C_{ij}, our model will compute all of the 0 labels (unknown cases) same as the ones with 1 labels, which dramatically reduces the learning effectiveness. With considering C_{ij}, we trust the labelled cases more than the unknown cases when C_{ij} is high (e.g. here we define it as $a = 1$). In addition, the parameter helps the model balance the weight of loss between matching and unknown cases (by defining $b = 0.01$ when $S_{ij} = 0$). It means the model considers the loss of 100 unknown cases as 1 trust case.

Online Recommendation. Once we get the optimized projection matrix H, for given a query q, the predicted binary code can be simply computed by a signum function on a linear embedding. The formula is $B_q = Z(x) = sgn(\phi(q) \cdot H)$. We use inner product to calculate ranking score which is formulated as:

$$score = \frac{1}{r} B \cdot B_q^T. \tag{12}$$

The ranking score will be sorted in descend order and the larger value get the better recommending priority.

4 Experimental Dataset

4.1 Fashion Dataset

As one of the key contributions of this work, two real-life fashion datasets are constructed by crawling clothing data from two well-known online shopping websites Netaporter and Farfetch. These two websites demonstrate millions of clothing images, where each item is associated with detailed descriptions such as category, brand, price, similar items, matching advice and groups of pictures taken from different views. At the current stage, more than 80,000 clothing items have been stored in our fashion database with more than 30,000 professional clothing matching suggestions, which is detailed in Table 1. In this paper, we are only focused on the clothing visual appearance and matching advice (Fig. 1).

Table 1. Statistics of the fashion dataset

# of	Netaporter	Farfetch
Items	13,190	68,563
Categories	57	174
Matching pairs	15,476	18,244

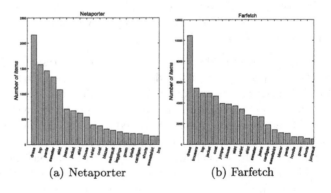

(a) Netaporter   (b) Farfetch

Fig. 1. Number of clothing in top 20 categories for both datasets.

4.2   Feature Extraction

The deep convolutional Neural Network (CNN) [31] is employed in this work to capture the visual appearance of clothing items. We extract the 4096 dimensional visual features from the second fully-connected layer (i.e. FC7). These features are used as the input of our learning model and also for the matching matrix self-augmentation (Fig. 2).

4.3   Description of Matching Matrix

The matching matrix indicates the identified matching items based on both the professional advices and the self-augmented relationships. The original clothing items are divided into different categories, such as T-shirt, pants, etc. Possible matching relationships are not limited to the items from different categories. In reality, matched pairs may from the same category, where one example is shown in Fig. 3 This fact clearly points out the difference between our work of fashion recommendation and the conventional visual similarity based clothing retrieval, where the later one is limited to finding the similar items from the same category. The initial matching matrix is constructed based on the professional advice that is provided by the websites. All these advice is hand-picked (i.e., manually generated) and obviously quite limited. Due to this, the matching matrix is very sparse (Fig. 3).

Fig. 2. Example of professional advices for matching. These three clothing match each other.

Fig. 3. Example of matched pairs from the same category "top".

4.4   Data Preprocessing

The real world data on the website contains a lot of noises such as typo, wrong labels, ambiguity of name, strange id numbers and off-line items which bring negative impact on recommendation model training. Therefore we made a lot of efforts on correcting and eliminating those noises to get pure valid pair labels. In particular, for off-line items, if the items are still stored in the image database and do have pair matchings with other items, we still save them as valid records.

In addition, we only focus on the items being labelled. Before training, we select those positive records which at least are labelled with another item. The valid dataset size of Netaporter and Farfetch are 10,793 and 18,753 respectively. For Netaporter dataset, we select 500 records as query set, the remaining 10,293 records are determined as the training and retrieval set. For Farfetch dataset, we select 1000 records as query set, the remaining 17,753 records as the training and retrieval set.

After we separate the whole data into training and testing parts, some of the records which belong to the training part will lose their pair labels due to the sparsity of the matching matrix. For example, if one record only has one matching pair which is selected into the test part by accident, this record becomes invalid. In this paper, we propose an effective self-augmentation process to alleviate the problem.

4.5 Matching Matrix Self-augmentation

Due to the sparsity of the initial matching matrix, we conduct a self-augmentation process to enrich the density of the matching relationships.

Firstly, we directly calculate the Euclidean distance of CNN features between each clothing in order to find the K-Nearest Neighbour similar items. Then we find all of the matched items for each clothing by the matching matrix. Finally we assign each matched item with the most n similar neighbours of the clothing as matching pair. In other words, if two items x_i x_j are labelled with 1 (i.e. $S_{ij} = 1$), we find K-NN samples x_{ik} and x_{jk} where $x_{ik}, x_{jk} \in X$ and $x_{ik} \neq x_i$, $x_{jk} \neq x_j$. Then assign $S_{ik,j} = S_{jk,i} = 1$. As a result, the scale of density is multiplied by n.

Intuitively, it can be understood that if a white long-sleeve shirt is labelled with a jeans and there is another white long-sleeve shirt which is super close to the previous shirt on visual content, we can infer that the second shirt is also well matching the jeans. But we do not label them crossly and it is expected that our model is able to learn those intrinsic relationships.

5 Experiment

5.1 Evaluation Metrics

We compute the AUC (Area Under the ROC curve) to evaluate the recommendation performance of each model. The AUC measures the quality of a ranking based on pairwise comparisons. Higher value of AUC means higher performance on recommendation. The AUC is formulated as:

$$AUC = \frac{1}{|\mathcal{Q}|} \sum_{(q,i,j)\in\mathcal{Q}} \delta(x_{q,i} > x_{q,j}),$$

where $\delta(\cdot)$ is an indicator function and $\mathcal{Q}$ is the fraction of the data withheld for testing. In other words, we are counting the fraction of times the model correctly ranks i higher than j.

5.2 Compared Approaches

We compare our DSFCH with four state-of-art hashing methods, including Supervised Hashing with Kernels (KSH) [21], Inter-media hashing(IMH) [32], Canonical Correlation Analysis (CCA) [6] and Supervised Discrete Hashing (SDH) [27].

5.3 Implementation Settings

In experiments, hash code length on all datasets is varied in the range of [16, 32, 64, 128] to observe the performance. (1) For Netaporter, number of anchor points $m = 500$, maximum iteration number $K = 50$, parameters $\lambda = 10^{-2}$ and $\nu = 10^{-2}$, Self-augmentation $n = 3$, query size=500 with the same initial seed. (2) For Farfetch, number of anchor points $m = 1000$, maximum iteration number $K = 50$, parameters $\lambda = 10^{-2}$ and $\nu = 10^{-2}$. Self-augmentation $n = 3$, query size=1000 with the same initial seed.

5.4 Experiment Results

We report AUC results of all compared methods in Table 2. A query example of coordinates recommendation is shown in Fig. 4. For the Netaporter dataset, we can easily find that our method DSFCH outperforms the competitors on all

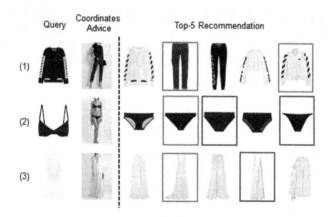

Fig. 4. Example of fashion recommendation by DSFCH with top-5 returned candidates. The returned clothing items that are the same as the recommended ones by professionals are highlighted with green frame. The items bounded in blue boxes are the same as the matched ones identified by the proposed self-augmentation process. (1) Given a query of "top", the recommended results are jeans, pants, top and jackets. The 5th is a matching advice of the query (https://www.net-a-porter.com/au/en/product/735424). The 1st and 4th are visual similar to the 5th. (2) Given a query of "bra", the recommended results are briefs. (3) Given a query of "cape", the recommended results are pants, gown and culottes.

cases. The largest improvement appears on 128 bits about 40.5% than the second best approach. For the Farfetch dataset, our method DSFCH is outperform at 64 and 128 code bits. In particular, DSFCH can achieve 0.5228 AUC on 16 bits for Farfetch if the Self-augmentation parameter setting is $n = 2$, which shows that different lengths of binary code accommodate different n. In addition, there is also an interesting finding from the experimental results that the AUC increases significantly along with the increase of the hash code length on both two datasets.

Table 2. AUC of all approaches on two datasets. The best result in each column is marked with bold.

Methods	Netaporter				Farfetch			
	16	32	64	128	16	32	64	128
KSH	0.5124	0.5016	0.4874	0.5252	**0.5495**	**0.5621**	0.5731	0.5720
IMH	0.4650	0.4650	0.4650	0.4650	0.5070	0.5070	0.5070	0.5070
CCA	0.5067	0.5070	0.4981	0.4953	0.5073	0.5046	0.5037	0.5114
SDH	0.4624	0.4796	0.5114	0.5174	0.4369	0.4931	0.4322	0.5436
DSFCH	**0.5468**	**0.5911**	**0.6622**	**0.7379**	0.4815	0.5209	**0.6262**	**0.7105**

5.5   Self-augmentation Study

We are trying to enrich the density of the matching matrix by KNN search, the testing result on 128 bits is shown below in Figure 3. We tested both Netaporter and Farfetch dataset. We found that, when the nearest neighbours number $n = 3$, the performance is peaking at highest value. Along with the number n increasing, the performance is dropping sharply (Fig. 5).

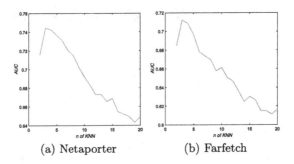

(a) Netaporter   (b) Farfetch

Fig. 5. Comparison of AUC curves for different Self-augmentation parameter n on 128 bits with two datasets.

5.6   Discrete or Not?

In practical, we compare the performance in two situations: (1) we keep the binary constraint of B in Eq. (11) during training; (2) we relax the discrete constraints to get a continuous B and threshold it at last. In most cases, we notice that discrete method is better than relaxed method. Keeping the binary constraints is getting better and better performance along with the code length increasing. Which can be understood that short code length suffers more penalty from quantization loss. All these two experiment is tested on Netaporter dataset, maximum iteration number $K = 50$, self-augmented neighbours $n = 3$, with the same query set and initial seed (Table 3).

Table 3. Comparative performance between discrete or relaxed methods on Netaporter dataset.

	Constraint	16 bits	32 bits	64 bits	128 bits
AUC	Discrete	**0.5702**	0.5898	**0.6684**	**0.7500**
	Relaxed	0.5663	**0.6004**	0.6655	0.7134

Figure 6 shows the convergence procedures in discrete and relaxed respectively. It can be seen from the figures that discrete method is much faster to get convergent than the other one. Also the convergence curve of discrete is more stable than the relaxed one.

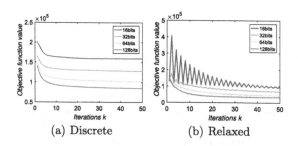

(a) Discrete   (b) Relaxed

Fig. 6. Objective function value variations with the number of iterations on discrete and relaxed methods.

5.7   Parameter Sensitivity Experiment

In practical, we notice that the parameter ν has a significant impact on learning performance, so a parameter study is applied shown below in Fig. 7. Testing range is $\{10^{-3}, 10^{-2}, 10^{-1}, 1, 10^1, 10^2, 10^3\}$ on 128 bits, argumentation $n = 3$.

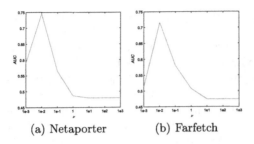

(a) Netaporter   (b) Farfetch

Fig. 7. Comparison of AUC for different ν on 128 bits with two datasets.

6 Conclusion

In this paper, we propose an effective model, dubbed as Discrete Supervised Fashion Coordinates Hashing (DSFCH), to learn meaningful yet compact visual features of clothing items, and thus support large-scale fashion recommendation. The learning process is supervised by a clothing matching matrix, which is initially constructed based on the limited pre-known matching pairs with self-augmentation. The proposed model jointly learns the intrinsic matching patterns from the matching matrix and the discrete binary representations from the images of clothing items. The binary representation significantly reduces the memory cost and accelerates the fashion recommendation. Extensive experiments have been conducted to provide comprehensive performance studies on different parameter settings. The comparisons with the-state-of-the-arts methods have evidenced the superior performance of the proposed approach for fashion recommendation.

References

1. Al-Halah, Z., Stiefelhagen, R., Grauman, K.: Fashion forward: forecasting visual style in fashion. In: ICCV, October 2017
2. Andoni, A., Indyk, P.: Near-optimal hashing algorithms for approximate nearest neighbor in high dimensions. Commun. ACM **51**(1), 117–122 (2008)
3. Andoni, A., Razenshteyn, I.: Optimal data-dependent hashing for approximate near neighbors. In: STOC, STOC 2015, pp. 793–801. ACM (2015)
4. Bracher, C., Heinz, S., Vollgraf, R.: Fashion DNA: merging content and sales data for recommendation and article mapping. CoRR abs/1609.02489 (2016)
5. Chen, K., Chen, K., Cong, P., Hsu, W.H., Luo, J.: Who are the devils wearing Prada in New York city? In: Proceedings of the 23rd ACM international conference on Multimedia, pp. 177–180. ACM (2015)
6. Gong, Y., Lazebnik, S., Gordo, A., Perronnin, F.: Iterative quantization: a procrustean approach to learning binary codes for large-scale image retrieval. TPAMI **35**(12), 2916–2929 (2013)
7. Hadsell, R., Chopra, S., LeCun, Y.: Dimensionality reduction by learning an invariant mapping. CVPR **2**, 1735–1742 (2006)

8. He, R., Packer, C., McAuley, J.: Learning compatibility across categories for heterogeneous item recommendation. In: ICDM, pp. 937–942. IEEE (2016)
9. Iwata, T., Wanatabe, S., Sawada, H.: Fashion coordinates recommender system using photographs from fashion magazines. In: IJCAI, vol. 22, p. 2262 (2011)
10. Jagadeesh, V., Piramuthu, R., Bhardwaj, A., Di, W., Sundaresan, N.: Large scale visual recommendations from street fashion images. In: SIGKDD, KDD 2014, pp. 1925–1934. ACM (2014)
11. Jiang, Q.Y., Li, W.J.: Scalable graph hashing with feature transformation. In: IJCAI, IJCAI 2015, pp. 2248–2254. AAAI Press (2015)
12. Kang, W.C., Fang, C., Wang, Z., McAuley, J.: Visually-aware fashion recommendation and design with generative image models. arXiv preprint arXiv:1711.02231 (2017)
13. Kiapour, M.H., Han, X., Lazebnik, S., Berg, A.C., Berg, T.L.: Where to buy it: matching street clothing photos in online shops. In: ICCV, pp. 3343–3351 (2015)
14. Koren, Y., Bell, R., Volinsky, C.: Matrix factorization techniques for recommender systems. Computer 42(8), 30–37 (2009)
15. Lew, M.S., Sebe, N., Djeraba, C., Jain, R.: Content-based multimedia information retrieval: state of the art and challenges. TOMM 2(1), 1–19 (2006)
16. Liong, V.E., Lu, J., Wang, G., Moulin, P., Zhou, J.: Deep hashing for compact binary codes learning. In: CVPR, pp. 2475–2483 (2015)
17. Liu, L., Lin, Z., Shao, L., Shen, F., Ding, G., Han, J.: Sequential discrete hashing for scalable cross-modality similarity retrieval. TIP 26(1), 107–118 (2017)
18. Liu, L., Zhu, L., Li, Z.: Learning robust graph hashing for efficient similarity search. In: Huang, Z., Xiao, X., Cao, X. (eds.) ADC 2017. LNCS, vol. 10538, pp. 110–122. Springer, Cham (2017). https://doi.org/10.1007/978-3-319-68155-9_9
19. Liu, W., Wang, J., Ji, R., Jiang, Y.G., Chang, S.F.: Supervised hashing with kernels. In: CVPR, pp. 2074–2081 (2012)
20. Liu, W., Mu, C., Kumar, S., Chang, S.F.: Discrete graph hashing. In: NIPS, NIPS 2014, pp. 3419–3427. MIT Press (2014)
21. Liu, W., Wang, J., Ji, R., Jiang, Y.G., Chang, S.F.: Supervised hashing with kernels. In: CVPR, pp. 2074–2081. IEEE (2012)
22. Liu, Z., Luo, P., Qiu, S., Wang, X., Tang, X.: DeepFashion: powering robust clothes recognition and retrieval with rich annotations. In: CVPR (2016)
23. Liu, Z., Yan, S., Luo, P., Wang, X., Tang, X.: Fashion landmark detection in the wild. In: ECCV (2016)
24. McAuley, J., Pandey, R., Leskovec, J.: Inferring networks of substitutable and complementary products. In: SIGKDD, pp. 785–794. ACM (2015)
25. McAuley, J., Targett, C., Shi, Q., van den Hengel, A.: Image-based recommendations on styles and substitutes. In: SIGIR, SIGIR 2015, pp. 43–52. ACM (2015)
26. McAuley, J., Yang, A.: Addressing complex and subjective product-related queries with customer reviews. In: Proceedings of the 25th International Conference on World Wide Web, pp. 625–635. International World Wide Web Conferences Steering Committee (2016)
27. Shen, F., Shen, C., Liu, W., Shen, H.T.: Supervised discrete hashing. In: CVPR, pp. 37–45 (2015)
28. Shen, F., Liu, W., Zhang, S., Yang, Y., Tao Shen, H.: Learning binary codes for maximum inner product search. In: ICCV, pp. 4148–4156 (2015)
29. Shrivastava, A., Li, P.: Asymmetric LSH (ALSH) for sublinear time maximum inner product search (mips). In: NIPS, pp. 2321–2329 (2014)
30. Simo-Serra, E., Fidler, S., Moreno-Noguer, F., Urtasun, R.: Neuroaesthetics in fashion: modeling the perception of fashionability. In: CVPR, pp. 869–877 (2015)

31. Simonyan, K., Zisserman, A.: Very deep convolutional networks for large-scale image recognition. arXiv preprint arXiv:1409.1556 (2014)
32. Song, J., Yang, Y., Yang, Y., Huang, Z., Shen, H.T.: Inter-media hashing for large-scale retrieval from heterogeneous data sources. In: SIGMOD, SIGMOD 2013, pp. 785–796. ACM (2013)
33. Veit, A., Kovacs, B., Bell, S., McAuley, J., Bala, K., Belongie, S.: Learning visual clothing style with heterogeneous dyadic co-occurrences. In: ICCV, pp. 4642–4650 (2015)
34. Wan, M., McAuley, J.: Modeling ambiguity, subjectivity, and diverging viewpoints in opinion question answering systems. In: ICDM, pp. 489–498. IEEE (2016)
35. Wang, C., Blei, D.M.: Collaborative topic modeling for recommending scientific articles. In: SIGKDD, pp. 448–456. ACM (2011)
36. Wang, H., Wang, N., Yeung, D.Y.: Collaborative deep learning for recommender systems. In: SIGKDD, pp. 1235–1244. ACM (2015)
37. Wang, J., Kumar, S., Chang, S.F.: Semi-supervised hashing for scalable image retrieval. In: CVPR, pp. 3424–3431 (2010)
38. Wang, J., Kumar, S., Chang, S.F.: Semi-supervised hashing for large-scale search. TPAMI 34(12), 2393–2406 (2012)
39. Wang, J., Xu, X.S., Guo, S., Cui, L., Wang, X.L.: Linear unsupervised hashing for ANN search in Euclidean space. Neurocomputing 171, 283–292 (2016)
40. Weiss, Y., Torralba, A., Fergus, R.: Spectral hashing. In: Koller, D., Schuurmans, D., Bengio, Y., Bottou, L. (eds.) NIPS, pp. 1753–1760. Curran Associates, Inc. (2009)
41. Xia, R., Pan, Y., Lai, H., Liu, C., Yan, S.: Supervised hashing for image retrieval via image representation learning. AAAI 1, 2156–2162 (2014)
42. Xu, H., Wang, J., Li, Z., Zeng, G., Li, S., Yu, N.: Complementary hashing for approximate nearest neighbor search. In: ICCV, pp. 1631–1638 (2011)
43. Zhang, P., Zhang, W., Li, W.J., Guo, M.: Supervised hashing with latent factor models. In: SIGIR, SIGIR 2014, pp. 173–182. ACM (2014)

Learning Dual Preferences with Non-negative Matrix Tri-Factorization for Top-N Recommender System

Xiangsheng Li[1], Yanghui Rao[1(✉)], Haoran Xie[2], Yufu Chen[1],
Raymond Y. K. Lau[3], Fu Lee Wang[4], and Jian Yin[5]

[1] School of Data and Computer Science, Sun Yat-sen University, Guangzhou, China
`raoyangh@mail.sysu.edu.cn`
[2] Department of Mathematics and Information Technology,
The Education University of Hong Kong, Tai Po, Hong Kong
[3] Department of Information Systems, City University of Hong Kong,
Kowloon Tong, Hong Kong
[4] Caritas Institute of Higher Education, Tseung Kwan O, Hong Kong
[5] Guangdong Key Laboratory of Big Data Analysis and Processing,
Guangzhou 510006, People's Republic of China

Abstract. In recommender systems, personal characteristic is possessed by not only users but also displaying products. Users have their personal rating patterns while products have different characteristics that attract users. This information can be explicitly exploited from the review text. However, most existing methods only model the review text as a topic preference of products, without considering the perspectives of users and products simultaneously. In this paper, we propose a user-product topic model to capture both user preferences and attractive characteristics of products. Different from conventional collaborative filtering in conjunction with topic models, we use non-negative matrix tri-factorization to jointly reveal the characteristic of users and products. Experiments on two real-world data sets validate the effectiveness of our method in Top-N recommendations.

Keywords: Top-N recommender system · Topic model
Matrix tri-factorization

1 Introduction

The emergence of e-commerce facilitates the development of recommender systems. In recent years, an increasing number of companies have applied recommender systems to automatically suggest products or services to their

---

The first two authors contributed equally to this work which was finished when Xiangsheng Li was an undergraduate student of his final year.

© Springer International Publishing AG, part of Springer Nature 2018
J. Pei et al. (Eds.): DASFAA 2018, LNCS 10827, pp. 133–149, 2018.
https://doi.org/10.1007/978-3-319-91452-7_9

customers [1]. Top-N recommendation is a personalized information filtering strategy which aims to identify a set of items that best fit interests and needs of users [2]. As one of the classical approaches for Top-N recommender systems, Collaborative Filtering (CF) via matrix factorization [3] assumed that users who exhibited similar preferences tend to have similar rating patterns for each product. Since the incomplete user-product rating data is leveraged only, traditional CF-based methods suffer from the issue of the sparsity of rating vectors [4]. Thus, the topic information of items have been extracted and then adopted in the collaborative topic regression (CTR) [5] for recommending scientific articles. CTR captures the semantic information from the item contents by latent Dirichlet allocation (LDA) [6], which can effectively identify the attractive characteristics of items. For example, the food quality and the environment may be a restaurant's topic preferences, but the price may be the attractive characteristics of electronic products. However, we argue that users not only are attracted by item characteristics (topic preferences), but also have their personal rating patterns. Taking ratings of a restaurant as an example, we assume that the restaurant's environment is important to user A, while user B focuses more on the food quality. Although both the environment and the food quality are critical aspects for restaurants, user A is more likely to give a lower rating than user B if the restaurant's environment is poor but has a good food quality. Thus, jointly exploiting both user preferences and product characteristics can obtain the dominant aspects in ratings to users and the dominant attributes (or aspects) of items, which is a key motivation of our research.

In this paper, we propose an approach named user-product topic model (UPTM). The key idea is to find the user preferences and attractive characteristics of products. Specifically, these preferences and characteristics are considered as a topic distribution, in which each topic value represents the level that a user prefers or a product is attracted. Then, these topic preferences are incorporated into matrix tri-factorization to model the ratings. The major process is as below: Firstly, the topic preferences of users and products are jointly extracted from the review text. Secondly, the topic information is incorporated into a non-negative matrix tri-factorization to facilitate rating prediction. Compared to the traditional bi-factorization, tri-factorization can better reveal the latent structures among products (samples) and attributes (features) [7]. Matrix tri-factorization is significant only when it cannot be transformed into bi-factorization, and this happens when certain constraints are applied to the tri-factorization [8]. Thus, we incorporate the topic preferences of users and products into user and product latent vectors and add a mapping matrix. The predicted ratings are demonstrated to improve the performance of Top-N recommendations compared to the conventional CF-based approach combined with a topic model. The main contributions of our paper are summarized as follows:

- We propose a probabilistic matrix tri-factorization model that incorporates both user and product preferences. By taking both types of preferences into account, the model extends the features of recommender systems.

- Our model is the first research to consolidate the non-negative matrix tri-factorization with topic information for recommendations. Revealing the latent aspects among users, products and reviews, our model effectively improves the quality of Top-N recommendations.

Experimental results on two real-world datasets indicate that our model can outperform baselines consistently.

2 Related Work

2.1 Matrix Factorization

Matrix factorization is a basic model for CF-based recommendations, in which products are recommended to a user based on other users with similar preferences of products. To learn joint latent space for users and products, a preliminary study developed a probabilistic matrix factorization (PMF) [9] that combined matrix factorization with probabilistic models. Recently, nonnegative matrix factorization (NMF) has been shown to be useful in CF recommendations [10,11]. NMF aims to factor a matrix X into two lower-dimension matrices and minimizes the square error between X and the approximation of X using those lower-dimension matrices. NMF is applied when certain non-negativity constraints exist, which makes the result easier to explain since it is natural to consider that users have non-negative affinities for some user communities based on their interests [12]. Guillamet et al. [13] extended the NMF to a weighted non-negative matrix factorization (WNMF) to improve the capabilities of representations. Experimental results show that WNMF achieves a great improvement in the classification accuracy compared with NMF. Ding et al. [14] provided an analysis of the relationship between bi-factorization and tri-factorization, and proposed an orthogonal non-negative matrix tri-factorization for clustering. This model is demonstrated to better capture the latent features of products and reveal hidden aspects underlying products [7]. Kang et al. [15] recently proposed a matrix completion method based on a low-rank assumption, which shows the effectiveness of matrix factorization.

2.2 Topic-Based Recommendations

To learn how users prefer products, understanding the hidden preferences for a product is quite important, such as food quality for a restaurant or price for an electronic product. Modeling these hidden factors is key to obtaining state-of-the-art performance on product recommendation system [16]. Therefore, many recommender systems rely on users feedback, which typically comes in the form of a plain-text review and a numeric score. However, in spite of the wealth of research on utilizing numeric score, the plain-text review is not well exploited.

To exploit the textual information of products, a collaborative topic regression (CTR) [5] was proposed by integrating a topic model in the matrix factorization. CTR used LDA [6] to mine the topic information from the item's text and incorporated it into PMF [9]. Compared to LDA and PMF, CTR is an appealing method in that it produces promising and interpretable results. The aforementioned method, however, did not consider the topic preference from both user and product perspectives. More specifically, CTR utilized matrix bi-factorization to capture items' attractive characteristics only. Furthermore, McAuley and Leskovec [17] proposed a model called HFT, which combines latent dimensions in rating data with topics in review text based on matrix factorization and LDA, and obtain highly interpretable textual labels for latent rating dimensions. Although HFT further mined the information under the connection between rating and review text, it also ignored the dual preference between users and items.

3   User-Product Topic Model

3.1   Problem Definition

For the reader's convenience in understanding our description of the model, we define the following terms and notations: We consider a review text as a document, which describes the evaluation of a certain product from the aspect of a certain user. Thus, an online collection consists of T documents is denoted as $\{d_1, d_2, \ldots, d_T\}$, expressed by $|U|$ users $\{u_1, u_2, \ldots, u_{|U|}\}$ over $|P|$ products $\{p_1, p_2, \ldots, p_{|P|}\}$, together with the corresponding ratings. The number of documents authored by user u is denoted as D_u while the number of documents described about product p is denoted as D_p. In particular, a document d for each user-product pair contains a sequence of N words denoted by $\{w_1, w_2, \ldots, w_N\}$ and a given rating r. A user u can make comment on several products, and a product p may be reviewed by multiple users.

The key process of our method is to find the user preferences $\theta_{u_i} \in \mathbb{R}^K$ and the attractive characteristics of products $\theta_{p_j} \in \mathbb{R}^K$, i.e., the topic preference on the aspects of users and products. Typically, user i is represented by a latent vector $u_i \in \mathbb{R}^K$ and product j by a latent vector $p_j \in \mathbb{R}^K$. The rating prediction r_{ij} that describes whether user i will like product j with the inner product between their latent representations and a mapping matrix $\boldsymbol{H}$, i.e., $u_i \boldsymbol{H} v_j^T$.

3.2   Generative Process

Our proposed approach, designated a user-product topic model (UPTM), aims to jointly learn the users' and products' latent topic vector. The motivation of designing a UPTM is to incorporate user and product topic preferences into non-negative matrix tri-factorization to factorize ratings. Typically, users will reveal their own shopping preference in the review text, and products, being reviewed by a larger number of users, can also be discovered what they attract different users in the reviews over them. Based on this phenomenon, we can assume that the topic preference of a user or a product can be exploited from the collection of reviews that this user issued or this product received. Thus, our proposed

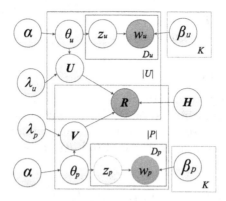

Fig. 1. Generative process of the proposed UPTM

UPTM is able to find topic preferences of users and products from review texts simultaneously. Given K topics, the generative process of UPTM, as shown in Fig. 1, is described as follows:

1. For the collection of documents expressed by user u_i:
    (a) Draw a topic distribution $\theta_{u_i} \sim$ Dirichlet (α);
    (b) For each word w_u in a document:
        i. Draw a topic assignment $z_u \sim$ Multinomial (θ_{u_i});
        ii. Draw the word $w_u \sim$ Multinomial (β_{z_u}).
2. For the collection of documents over product p_j:
    (a) Draw a topic distribution $\theta_{p_j} \sim$ Dirichlet (α);
    (b) For each word w_p in a document:
        i. Draw a topic assignment $z_p \sim$ Multinomial (θ_{p_j});
        ii. Draw the word $w_p \sim$ Multinomial (β_{p_j}).
3. For the i-th user, sample user latent vector $u_i \sim \mathcal{N}(\theta_{u_i}, \lambda_u^{-1} I_K)$.
4. For the j-th product, sample product latent vector $v_j \sim \mathcal{N}(\theta_{p_j}, \lambda_p^{-1} I_K)$.
5. For the (i, j)-th user-product pair, draw the rating

$$r_{ij} \sim \mathcal{N}(u_i \boldsymbol{H} v_j^T, c_{ij}^{-1}),$$

where $\mathcal{N}(\mu, \sigma^2)$ is a Gaussian distribution with a mean μ and a variance σ^2. Here, c_{ij} is the confidence parameter over user u_i and product p_j. If c_{ij} is larger, we trust r_{ij} more. The topic preferences on the perspective of users and products are obtained from the collection of reviews that this user issued or this product received. Typically, $\boldsymbol{H}$ is a matrix that maps dual (i.e., user and product) preferences to the rating space. The parameters λ_u and λ_p balance the proportion of users' topic preferences and ratings, and the proportion of products' topic preferences and ratings, respectively. Given that topic preferences are non-negative interests in a community [12], u_i, v_j and $\boldsymbol{H}$ are constrained to be non-negative. In this way, the objective can be considered non-negative matrix tri-factorization, which is a 3-factor decomposition of non-negative dyadic data $\boldsymbol{R} \in \mathbb{R}_+^{|U| \times |P|}$ that takes the form $R \approx UHV^T$, where $U \in \mathbb{R}_+^{|U| \times K}$, $H \in \mathbb{R}_+^{K \times K}$, and $V \in \mathbb{R}_+^{|P| \times K}$ are constrained to be non-negative matrices.

3.3   Parameter Estimation

Since it is intractable to compute the full posterior of u_i, v_j, θ_{u_i} and θ_{p_j}, we develop an expectation maximization (EM)-style algorithm for learning parameters. Maximizing the posteriors with fixed hyper-parameters is equivalent to maximizing the complete log likelihood of U, V, θ_d, H, and R given $\lambda_u, \lambda_p, \beta_u$, and β_p, as follows:

$$
\begin{aligned}
\mathcal{L} = &-\frac{\lambda_u}{2} \sum_i (u_i - \theta_{u_i})^T (u_i - \theta_{u_i}) \\
&- \frac{\lambda_p}{2} \sum_j (v_j - \theta_{p_j})^T (v_j - \theta_{p_j}) \\
&+ \sum_i \sum_{n_u} log \sum_k \theta_{ik} \beta_{k,w_{in_u}} \\
&+ \sum_j \sum_{n_p} log \sum_k \theta_{jk} \beta_{k,w_{jn_p}} \\
&- \sum_{i,j} \frac{c_{ij}}{2} (r_{ij} - u_i H v_j^T)^2,
\end{aligned}
\tag{1}
$$

where n_u and n_p denote the n-th item in the word set of the collection of documents expressed by user u, and the collection of documents over product p, respectively. A confidence parameter c_{ij} is used to determine the weight of ratings in different cases, and the Dirichlet parameter α is set to 1 by following [5]. To maximize the above likelihood function, we propose an alternating approach by coordinate ascent, i.e., by iteratively optimizing one variable while fixing the others, and repeat the procedure until convergence. The update formula of $\mathcal{L}$ with respect to U, V, and H is as follows:

$$
U_{ik} \leftarrow U_{ik} \sqrt{\frac{[C \odot RVH^T]_{ik} + \lambda_u \theta_{u_i k}}{[C \odot (UHV^T)VH^T]_{ik} + \lambda_u U_{ik}}},
\tag{2}
$$

$$
V_{jk} \leftarrow V_{jk} \sqrt{\frac{[(C \odot R)^T UH]_{jk} + \lambda_p \theta_{p_j k}}{[(C \odot (UHV^T))UH]_{jk} + \lambda_p V_{jk}}},
\tag{3}
$$

$$
H_{ij} \leftarrow H_{ij} \sqrt{\frac{[U^T (C \odot R)V]_{ij}}{[U^T (C \odot (UHV^T))V]_{ij}}},
\tag{4}
$$

where U_{ik} is the k-th item of u_i, V_{jk} is the k-th item of v_j, H_{ij} is the i-th row and the j-th column item of H, C is the confidence parameter matrix with each element c_{ij}, and $\odot$ is the element-wise product.

Equations 2 and 3 show how the parameters λ_u and λ_p affect the user and product latent factors. A larger λ_u gives rise to the influence of topic preference rather than rating information. Similarly, a larger λ_p corresponds to a larger proportion of the product topic preference compared to rating information. These update formulae are in good and consistent agreement with traditional matrix factorization methods, with two additional issues that require proofs: (1) The correctness of the converged solution, and (2) the convergence of the algorithm.

Correctness Analysis. We optimize U by fixing V and H in Eq. 1, as follows:

$$\mathcal{L}(U) = -\frac{1}{2}\|C \odot (R - (UHV^T))\|_F^2 - \frac{\lambda_u}{2}tr(G_u^T G_u)$$
$$s.t. U \geq 0,$$

where $\|\cdot\|_F$ is the Frobenius norm and G_u is the matrix $(U - \theta_u)$. The derivative of $\mathcal{L}(U)$ with respect to U is

$$\frac{\partial \mathcal{L}(U)}{\partial U} = C \odot RVH^T - C \odot (UHV^T)VH^T - \lambda_u G_u.$$

According to the Karush-Kuhn-Tucker (KKT) complementarity condition [18] of the non-negativity of U, we have

$$[C \odot RVH^T - C \odot (UHV^T)VH^T - \lambda_u G_u]_{ik} U_{ik} = 0.$$

This is the fixed-point relation that local minima must hold, and it is true that at convergence, from Eq. 2, the solution will satisfy

$$[C \odot RVH^T - C \odot (UHV^T)VH^T - \lambda_u G_u]_{ik} U_{ik}^2 = 0.$$

This is identical to the fixed-point condition because either $U_{ik} = 0$ or the left-hand term being equal to zero will make the above equation true. The correctness analysis of updating rules for V and H are similar to that of U, by separating Eq. 1 that contains V and H, respectively.

Convergence Analysis. In the following, we will demonstrate the deduction and the convergence of our updating formulas in Eqs. 2, 3, and 4. We apply the auxiliary function approach [19] and inequality Lemma [14] for the convergence analysis.

Definition 1. $Z(h, h')$ is called an auxiliary function for $F(h)$ if the conditions

$$Z(h, h') \geq F(h), Z(h, h) = F(h)$$

are satisfied [19].

Lemma 1. If Z is an auxiliary function, then F is nonincreasing [19] under the update

$$h^{(t+1)} = \arg\min_h Z(h, h^{(t)}).$$

Proof: $F(h^{(t+1)}) \leq Z(h^{(t+1)}, h^t) \leq Z(h^t, h^t) = F(h^t)$.

Note that if Z is lower bounded and we iteratively update until $F(h^{(t+1)}) = F(h^t)$, h^t becomes a local minimum of Z, which also implies the derivative $\nabla F(h^t) = 0$. The key is to find an appropriate $Z(h, h')$. □

Lemma 2. *For any matrices* $A \in \mathbb{R}_+^{n \times n}$, $B \in \mathbb{R}_+^{k \times k}$, $S \in \mathbb{R}_+^{n \times k}$, $S' \in \mathbb{R}_+^{n \times k}$, *and* A, B *being symmetric, the following inequality holds [14]:*

$$\sum_{i=1}^{n} \sum_{p=1}^{k} \frac{(AS'B)_{ip} S_{ip}^2}{S'_{ip}} \geq tr(S^T ASB).$$

Proof: It can be referred to in [14]. □

Theorem 1. *Let*

$$\mathcal{J}(U) = -\mathcal{L}(U)$$
$$= \frac{1}{2}\|C \odot (R - (UHV^T))\|_F^2 + \frac{\lambda_u}{2} tr(G_u^T G_u) \qquad (5)$$
$$\propto tr(G_u^T G_u) - tr(2C \odot RVH^T U^T)$$
$$+ tr(C \odot (UHV^T)VH^T U^T).$$

The auxiliary function of $\mathcal{J}(U)$ *is then*

$$Z(U, U')$$
$$= \lambda_u \sum_{i,k} U_{ik}^2 + \lambda_u \sum_{i,k} \theta_{u\,ik}^2$$
$$- 2 \sum_{i,k} U'_{ik} \theta_{u\,ik} (1 + \log \frac{U_{ik}}{U'_{ik}})$$
$$- 2 \sum_{i,k} (C \odot RVH^T)_{ik} U'_{ik} (1 + \log \frac{U_{ik}}{U'_{ik}}) \qquad (6)$$
$$+ \sum_{i,k} \frac{(C \odot (U'HV^T)VH^T)_{ik} U_{ik}^2}{U'_{ik}}.$$

Furthermore, this is a convex function with respect to U *and its global minimum is*

$$U_{ik} = U_{ik} \sqrt{\frac{[C \odot RVH^T]_{ik} + \lambda_u \theta_{u\,ik}}{[C \odot (UHV^T)VH^T]_{ik} + \lambda_u U_{ik}}}.$$

Proof: According to Lemma 1, it is obvious that when $U' = U$ the equality holds $Z(U, U') = \mathcal{J}(U)$. Second, the inequality $Z(U, U') \geq \mathcal{J}(U)$ also holds because the first four terms in Eq. 6 are larger than the first two terms in Eq. 5 since the inequality

$$z \geq 1 + \log(z), \forall z > 0, \qquad (7)$$

and we can set $z = U_{ik}/U'_{ik}$. Furthermore, the last term in Eq. 6 is larger than the last term in Eq. 5 in terms of Lemma 2. Therefore, according to Lemma 1, the minimum of $Z(U, U')$ fixing U' is given by

$$
\begin{aligned}
0 &= \frac{\partial Z(U, U')}{\partial U_{ik}} \\
&= 2\lambda_u U_{ik} - 2\theta_{u\,ik} \frac{U'_{ik}}{U_{ik}} \\
&\quad - 2(C \odot RVH^T)_{ik} \frac{U'_{ik}}{U_{ik}} \\
&\quad + 2\frac{(C \odot (U'HV^T)V)_{ik}U_{ik}}{U'_{ik}}.
\end{aligned}
$$

Then, to solve U_{ik}, let $U = U^{(t+1)}$ and $U' = U^{(t)}$, we obtain the updating formula of U, as shown in Eq. 2.   □

Theorem 2. *Updating U under the update formula 2 will monotonically decrease the value in Eq. 5. The updating will finally converge.*

Proof: Since $J(U)$ is lower bounded to zero, the only condition of convergence is that it is monotonically decreasing. Due to that $J(U^0) = Z(U^0, U^0) \geq Z(U^1, U^0) \geq J(U^1) \geq \cdots$, it converges.   □

Theorem 3. *Let*

$$
\begin{aligned}
J(V) &= -\mathcal{L}(V) \\
&= \frac{1}{2}\|C \odot (R - (UHV^T))\|_F^2 + \frac{\lambda_p}{2}tr(G_p^T G_p) \\
&\propto tr(G_p^T G_p - tr(2C \odot RVH^T U^T) \\
&\quad + tr(C \odot (UHV^T)VH^T U^T).
\end{aligned}
\tag{8}
$$

Updating V under the formula Eq. 3 will monotonically decrease the value $J(V)$, and finally it converges.

Proof: Since V is similar and symmetrical to U in Eq. 1, the proof of convergence is similar to that of U.   □

Theorem 4. *Let*

$$
\begin{aligned}
J(H) &= -\mathcal{L}(H) \\
&= \frac{1}{2}\|C \odot (R - (UHV^T))\|_F^2 \\
&\propto tr(-2U^T(C \odot R)VH^T) \\
&\quad + tr(U^T(C \odot (UHV^T))VH^T).
\end{aligned}
\tag{9}
$$

The auxiliary function of $\mathcal{J}(\boldsymbol{H})$ is then

$$
\begin{aligned}
&Z(\boldsymbol{H}, \boldsymbol{H}') \\
&= -2\sum_{i,j} (\boldsymbol{U}^T(\boldsymbol{C} \odot \boldsymbol{R})\boldsymbol{V})_{ij} \boldsymbol{H}'_{ij}(1 + \log \frac{\boldsymbol{H}_{ij}}{\boldsymbol{H}'_{ij}}) \\
&\quad + \sum_{i,j} \frac{(\boldsymbol{U}^T(\boldsymbol{C} \odot (\boldsymbol{U}\boldsymbol{H}'\boldsymbol{V}^T))\boldsymbol{V})_{ij} \boldsymbol{H}^2_{ij}}{\boldsymbol{H}'_{ij}}.
\end{aligned}
\tag{10}
$$

Furthermore, this is a convex function with respect to $\boldsymbol{H}$ and its global minimum is

$$
\boldsymbol{H}_{ij} = \boldsymbol{H}_{ij} \sqrt{\frac{[\boldsymbol{U}^T(\boldsymbol{C} \odot \boldsymbol{R})\boldsymbol{V}]_{ij}}{[\boldsymbol{U}^T(\boldsymbol{C} \odot (\boldsymbol{U}\boldsymbol{H}\boldsymbol{V}^T))\boldsymbol{V}]_{ij}}}.
$$

Proof: According to Lemma 1, it is obvious that when $\boldsymbol{H}' = \boldsymbol{H}$ the equality holds $Z(\boldsymbol{H}, \boldsymbol{H}') = \mathcal{J}(\boldsymbol{H})$. Second, the inequality $Z(\boldsymbol{H}, \boldsymbol{H}') \geq \mathcal{J}(\boldsymbol{H})$ also holds because the first term in Eq. 10 is larger than the first term in Eq. 9 since the inequality (Eq. 7) and we can set $z = \boldsymbol{H}_{ij}/\boldsymbol{H}'_{ij}$. Furthermore, the second term in Eq. 10 is larger than the second term in Eq. 9 in terms of Lemma 2.

Therefore, according to Lemma 1, the minimum of $Z(\boldsymbol{H}, \boldsymbol{H}')$ fixing $\boldsymbol{H}'$ is given by

$$
\begin{aligned}
0 &= \frac{\partial Z(\boldsymbol{H}, \boldsymbol{H}')}{\partial \boldsymbol{H}_{ij}} \\
&= -2(\boldsymbol{U}^T(\boldsymbol{C} \odot \boldsymbol{R})\boldsymbol{V})_{ij} \frac{\boldsymbol{H}'_{ij}}{\boldsymbol{H}_{ij}} \\
&\quad + 2\frac{(\boldsymbol{U}^T(\boldsymbol{C} \odot (\boldsymbol{U}\boldsymbol{H}'\boldsymbol{V}^T))\boldsymbol{V})_{ij} \boldsymbol{H}_{ij}}{\boldsymbol{H}'_{ij}}.
\end{aligned}
$$

Then, to solve $\boldsymbol{H}_{ij}$, let $\boldsymbol{H} = \boldsymbol{H}^{(t+1)}$ and $\boldsymbol{H}' = \boldsymbol{H}^{(t)}$, and we obtain the updating formula of $\boldsymbol{H}$, as shown in Eq. 4.   □

Theorem 5. *Updating $\boldsymbol{H}$ under the update formula 4 will monotonically decrease the value in Eq. 9. The updating will finally converge.*

Proof: Since $\mathcal{J}(\boldsymbol{H})$ is lower bounded to zero, the only condition of convergence is that it is monotonically decreasing. Due to that $\mathcal{J}(\boldsymbol{H}^0) = Z(\boldsymbol{H}^0, \boldsymbol{H}^0) \geq Z(\boldsymbol{H}^1, \boldsymbol{H}^0) \geq \mathcal{J}(\boldsymbol{H}^1) \geq \cdots$, it converges.   □

Optimization of Other Parameters. Learning θ_{u_i} and θ_{p_j} is different because they are difficult to derive. We can, however, apply Jensen's inequality to solve this problem. For θ_{u_i}, it is constrained by a low bound with respect to θ_{u_i}. We now define $q(z_{in} = k) = \phi_{ink}$ and separate the items that contain θ_{u_i}.

We then apply Jensen's inequality as follows:

$$\mathcal{L}(\theta_{u_i}) \geq -\frac{\lambda_u}{2} \sum_i (u_i - \theta_{u_i})^T (u_i - \theta_{u_i})$$
$$+ \sum_{n_u} \sum_k \phi_{ink}(log\theta_{ik}\beta_{k,w_{in_u}} - log\phi_{in_uk})$$
$$= \mathcal{L}(\theta_{u_i}, \phi_i),$$

where $\phi_i = (\phi_{ink})_{n=1,k=1}^{|D_u| \times K}$, $|D_u|$ is the number of words in the document group of this user. The optimal ϕ_{ink} satisfies $\phi_{ink} \propto \theta_{ik}\beta_{k,w_{in_u}}$.

Thus, $\mathcal{L}(\theta_{u_i}, \phi_i)$ gives the tight lower bound of $\mathcal{L}(\theta_{u_i})$. The gradient projection [20] can be applied to optimize θ_{u_i}. We can then optimize β_u as follows:

$$\beta_{kw_i} \propto \sum_d \sum_{n_u} \phi_{in_uk}1[w_{in_u} = w].$$

This is consistent with the M-step of the EM-algorithm in LDA [6]. Moreover, for θ_{p_j}, ϕ_{jn_pk}, and $\beta_k w_j$, it is similarly updated. Note, however, that in order to ensure topics θ_{u_i} and θ_{p_j} have the same semantic information, we make β_u equal to β_p. After estimating u_i, v_j, θ_{u_i}, and θ_{p_j}, the rating is predicted by $r_{ij} \approx u_i \boldsymbol{H} v_j^T$.

Complexity Analysis. Our method applies an EM-style algorithm, so the parameter estimation algorithm is implemented in an iterative manner. The efficiency is determined by the convergence and time cost per iteration. The time cost mainly comes from two parts: topic modeling and matrix tri-factorization. For topic modeling, the time complexity is $\mathcal{O}(N_{iter} \cdot (|U| \cdot |\widetilde{D_u}| + |P| \cdot |\widetilde{D_p}|) \cdot K \cdot \tilde{l})$, where N_{iter} is the number of iterations, $\widetilde{D_u}$ and $\widetilde{D_p}$ are the average number of review text of users and products, respectively. $|U|$ and $|P|$ are the number of users and products, K is the number of topic and $\tilde{l}$ is the average length of each review text. For matrix tri-factorization, the time complexity is $\mathcal{O}(N_{iter} \cdot |U|^2 \cdot |P|^2 \cdot |K|^5)$. Thus, the total time complexity is

$$\Theta = N_{iter} \cdot ((|U| \cdot |\widetilde{D_u}| + |P| \cdot |\widetilde{D_p}|) \cdot K \cdot \tilde{l} + |U|^2 \cdot |P|^2 \cdot |K|^5).$$

4 Experiments

4.1 Datasets

We employ two datasets in our experiment: IMDB [21] and Yelp2013[1]. IMDB and Yelp2013 contain users' reviews and ratings on different aspects of movies, and on different restaurants, respectively. Table 1 presents the statistical information of these datasets.

In our experiment, we split each dataset into three parts: a training set (80%), a validation set (10%), and a testing set (10%). Each model is trained on the training set and obtains its optimal parameters on the validation set. The performance is then evaluated on the testing set.

---

[1] http://www.yelp.com/dataset_challenge.

Table 1. Statistical information of datasets.

Dataset	# users	# products	# reviews	Rating scale
IMDB	1,310	1,635	84,919	1–10
Yelp2013	1,631	1,633	78,966	1–5

4.2 Comparisons and Evaluation

The baseline models are listed as follows:

- **2NMF** [12]: Non-negative matrix bi-factorization is a type of probability matrix factorization in which a rating matrix R is factorized into two matrices U and V.
- **3NMF**: Different from 2NMF, 3NMF is a non-negative matrix tri-factorization that factorizes a rating matrix into three matrices, i.e., U, V and H. Review text is not considered in this method.
- **CTR** [5]: Collaborative topic regression is a model used to perform topic modeling and collaborative filtering simultaneously.
- **RSMC** [15]: Recommender system via matrix completion is a method based on a low-rank assumption for Top-N recommendations.

We measure the performance of the proposed model and baselines by comparing *Precision* and *Recall*, as in [5]. For each user, *Precision* and *Recall* are defined as follows:

$$Precision@M = \frac{\# \ products \ the \ user \ likes \ in \ Top \ M}{M},$$

$$Recall@M = \frac{\# \ products \ the \ user \ likes \ in \ Top \ M}{Total \ number \ of \ products \ the \ user \ likes},$$

where M is the number of returned items sorted by their predicted ratings. Specifically, *Precision* evaluates the recommendation accuracy of the model while *Recall* evaluates which of the returned items were actually in each user's purchase records. The final result reported is the average precision and recall over all users.

4.3 Experimental Setting

As we mention earlier, we leveraged a validation dataset to find the optimal parameters of all models. For 2NMF and 3NMF, we employed multiplication update rules to avoid the learning rate setting, which is similar to [12]. CTR delivered good performance when $\lambda_u = 0.01$ and $\lambda_p = 0.01$, and when $a = 1$ and $b = 0.01$, where a and b are the confidence parameters c_{ij}. For RSMC, we set $\mu = 1.2 \times 10^{-3}$ and $\gamma = 1.3$ in IMDB and $\mu = 1.5 \times 10^{-3}$ and $\gamma = 1.8$ in Yelp2013. Except for RSMC, we set a common topic dimension $K = 20$, i.e., the rating matrix R is factorized into $U^{|U| \times K}$ and $V^{|P| \times K}$ in 2NMF and CTR, while

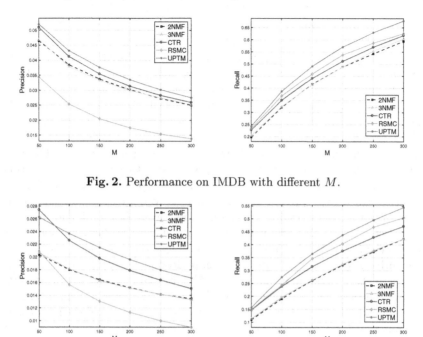

Fig. 2. Performance on IMDB with different M.

Fig. 3. Performance on Yelp2013 with different M.

it is factorized into $\boldsymbol{U}^{|U| \times K}$, $\boldsymbol{S}^{K \times K}$ and $\boldsymbol{V}^{|P| \times K}$ in 3NMF and our UPTM. The impact of topic number will be further discussed in Sect. 4.5. For our UPTM, we directly set $a = 1$, $b = 0.01$, and apply a grid search to obtain the best parameters λ_u and λ_p on the validation dataset. On IMDB, the optimal performance is achieved when $\lambda_u = 1000$ and $\lambda_p = 10$, while $\lambda_u = 100$ and $\lambda_p = 1$ obtains the best performance on Yelp2013. For evaluation, we set $M = 50, 100, 150, 200, 250$, and 300 and fix the parameters of each approach to the best results.

4.4   Performance Comparison

The overall performance of each approach on IMDB and Yelp2013 are shown in Figs. 2 and 3, respectively. For non-negative matrix factorization, 3NMF sightly outperforms 2NMF for both metrics, which shows the effectiveness of non-negative matrix tri-factorization in recommender systems. However, they both do not perform as well as other models in *Recall*. RSMC performs better than most baseline models in *Recall*, while it is the worst in *Precision* in both datasets. Our proposed model, the UPTM, outperforms 2NMF, 3NMF, RSMC and CTR on IMDB in terms of different M consistently. On Yelp2013, CTR sightly outperforms the UPTM on *Precision* when $M = 50$. However, a zero entry in the rating matrix may be due to the fact that the user is not interested in the product, which indicates that *Recall* is a more important performance measure than *Precision* on Top-N recommender systems [1,5]. This slightly worse *Precision* is unconvincing on the condition that *Recall* of the proposed UPTM improves 6.51% and 15.04% compared to CTR. On average, UPTM improves 2NMF, 3NMF, CTR and RSMC by 18.03%, 16.60%, 10.06%, and 82.06%, respectively, in terms of precision, and by 18.03%, 16.60%, 10.06%, and 6.24% in terms of *Recall*, respectively, on the dataset of IMDB. On the other dataset Yelp2013, the UPTM improves 2NMF, 3NMF, CTR, and RSMC by 28.60%, 28.58%, 6.58% and 63.41%, respectively, in terms of *Precision*, and by 37.35%, 35.27%, 14.13% and 7.51%, respectively, in terms of *Recall*.

The performance comparison shows the effectiveness of our UPTM which captures both users' and products' topic preferences. Compared to conventional non-negative matrix factorization, our model incorporates the topic information between user and products, which effectively improves the recommendation performance. Compared to CTR, our model leverages topic information on both aspect of users and products and adopts matrix tri-factorization to better reveal the latent aspects among users, products and topic features [7], which significantly improves the recommendation performance. Compared to the state-of-the-art matrix completion method, RSMC, our model also performs better, achieving the best performance in Top-N recommendations among matrix factorization methods. Note that execution times of algorithms and the performance on a small M value are also important to test the effectiveness of a recommender system, we leave these kinds of evaluations to the future work due to the limit of space.

4.5   Influence of the Number of Topics

The number of topics K is an important parameter in topic-based recommendation. We tried different values of K and the result is shown in Fig. 4. We can observe that as the cross-validation we did, $K = 20$ delivers the best performance in both data sets. Furthermore, as K gets larger, the performance gets worse. This is because too many topics over-depict the review features. Topic number below 30 is enough to depict the review features and performs well in recommendation.

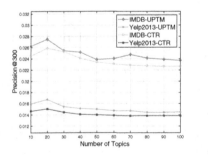 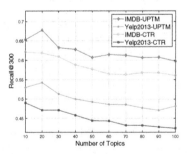

Fig. 4. *Precision* and *Recall* performance with different topic number K

We also investigate the effect of K on our model and the baseline CTR. Our model performs better over both of the evaluations. This demonstrates that considering two aspects of users and products simultaneously is superior to the one-side topic modeling. As the number of topics becomes larger, their performances get worse since the number of the documents under a specific topic come to the bottleneck. On the other hand, like LDA, if K is too small, the topics will be coarse and a lot of useful features are missing. If K is too large, some useless features may be incorporated and useful features may be confounded by those noise.

4.6   Impact of Training Data Size

A good recommender system aims to perform well even when the data is quite sparse. We examined the impact of the size of training data on each model's performance by randomly selecting $x\%$ data from the original training corpus. The values of x varied from 20 to 80, with an interval of 20. In case of coincidence, we extracted the training data 10 times and calculated the average of performance each time.

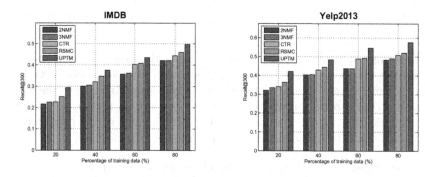

Fig. 5. The *Recall*@300 with different training data size.

To be consistent with existing evaluations [1], the *Recall*@300 performance is shown in Fig. 5. We observed that the performance of all methods increased as the size of the training data increased, and our model outperforms the baselines on both datasets.

4.7   Parameter Effect Analysis

Here, we study the effects of the parameters λ_u and λ_p on the proposed UPTM using *Recall*@300 (ref. Table 2). On IMDB, UPTM achieved the best performance when $\lambda_u = 1000$ and $\lambda_p = 10$, which indicates that product topic preference contributes more than user topic preference on this dataset. On Yelp2013, the optimal parameters for the best performance were $\lambda_u = 100$ and $\lambda_p = 1$; the contribution of product preference is consistently important in this dataset.

Table 2. The *Recall*@300 of UPTM with different λ_u and λ_p.

(a) IMDB

λ_p \ λ_u	0.01	0.1	1	10	100	1000
0.01	0.647	0.657	0.648	0.661	0.663	0.649
0.1	0.650	0.656	0.664	0.666	0.664	0.648
1	0.655	0.653	0.663	0.669	0.663	0.663
10	0.666	0.659	0.648	0.659	0.663	**0.678**
100	0.656	0.655	0.648	0.657	0.661	0.664
1000	0.657	0.656	0.659	0.666	0.647	0.668

(b) Yelp2013

λ_p \ λ_u	0.01	0.1	1	10	100	1000
0.01	0.527	0.526	0.534	0.533	0.532	0.528
0.1	0.531	0.530	0.531	0.537	0.535	0.532
1	0.527	0.538	0.532	0.537	**0.545**	0.531
10	0.531	0.529	0.532	0.529	0.541	0.531
100	0.528	0.530	0.532	0.533	0.537	0.535
1000	0.527	0.530	0.532	0.532	0.532	0.529

In addition, on both datasets, when λ_u and λ_p were both small (i.e., smaller than 1), the performance suffered, which means that both user and product topic preferences affect recommendation performance.

5 Conclusions

In this paper, we proposed a probabilistic matrix tri-factorization approach named UPTM, which applied LDA to mine the user and product topic preferences and incorporated them into the matrix factorization. We also leveraged non-negative matrix tri-factorization to factorize the rating matrix into a user latent matrix, a product latent matrix and a mapping matrix. The main conclusions of our paper are the following:

- By mining the topic preference not only from the product aspect but also from the user aspect, our UPTM was used to find a connection between a user's topic of interest and a product topic that was attractive to the user.
- The matrix factorization part of our model is based on non-negative matrix tri-factorization. By incorporating a third mapping matrix, the predicted rating was demonstrated to enhance the recommender performance.

In the future, we plan to explore the implementation of parallel calculating algorithms, which can make the proposed method scalable to large-scale datasets.

Acknowledgements. We are grateful to the anonymous reviewers for their valuable comments on this manuscript. The research has been supported by the National Natural Science Foundation of China (61502545, U1611264, U1711262), a grant from the Research Grants Council of the Hong Kong Special Administrative Region, China (UGC/FDS11/E03/16), and the Individual Research Scheme of the Dean's Research Fund 2017–2018 (FLASS/DRF/IRS-8) of The Education University of Hong Kong.

References

1. Wang, H., Wang, N., Yeung, D.Y.: Collaborative deep learning for recommender systems. In: Proceedings of the 21th ACM SIGKDD International Conference on Knowledge Discovery and Data Mining, pp. 1235–1244 (2015)
2. Linden, G., Smith, B., York, J.: Amazon.com recommendations: item-to-item collaborative filtering. IEEE Internet Comput. **7**(1), 76–80 (2003)

3. Koren, Y., Bell, R., Volinsky, C.: Matrix factorization techniques for recommender systems. Computer **42**(8), 30–37 (2009)
4. Chen, C., Zheng, X., Wang, Y., Hong, F., Lin, Z.: Context-ware collaborative topic regression with social matrix factorization for recommender systems. In: Proceedings of the 28th AAAI Conference on Artificial Intelligence, pp. 9–15 (2014)
5. Wang, C., Blei, D.M.: Collaborative topic modeling for recommending scientific articles. In: Proceedings of the 17th ACM SIGKDD International Conference on Knowledge Discovery and Data Mining, pp. 448–456 (2011)
6. Blei, D.M., Ng, A.Y., Jordan, M.I.: Latent Dirichlet allocation. J. Mach. Learn. Res. **3**, 993–1022 (2003)
7. Yang, X., Huang, K., Zhang, R., Hussain, A.: Learning latent features with infinite non-negative binary matrix tri-factorization. In: Hirose, A., Ozawa, S., Doya, K., Ikeda, K., Lee, M., Liu, D. (eds.) ICONIP 2016. LNCS, vol. 9947, pp. 587–596. Springer, Cham (2016). https://doi.org/10.1007/978-3-319-46687-3_65
8. Li, T., Ding, C.: The relationships among various nonnegative matrix factorization methods for clustering. In: Proceedings of the 6th International Conference on Data Mining, pp. 362–371 (2006)
9. Salakhutdinov, R., Mnih, A.: Probabilistic matrix factorization. In: Proceedings of Advances in Neural Information Processing Systems, pp. 1257–1264 (2007)
10. Luo, X., Zhou, M., Xia, Y., Zhu, Q.: An efficient non-negative matrix-factorization-based approach to collaborative filtering for recommender systems. IEEE Trans. Ind. Inform. **10**, 1273–1284 (2014)
11. Hernando, A., Bobadilla, J., Ortega, F.: A non negative matrix factorization for collaborative filtering recommender systems based on a Bayesian probabilistic model. Knowl.-Based Syst. **97**, 188–202 (2016)
12. Zhang, S., Wang, W., Ford, J., Makedon, F.: Learning from incomplete ratings using non-negative matrix factorization. In: Proceedings of the 2006 SIAM International Conference on Data Mining, pp. 548–552 (2006)
13. Guillamet, D., Vitrià, J., Schiele, B.: Introducing a weighted non-negative matrix factorization for image classification. Pattern Recogn. Lett. **24**, 2447–2454 (2003)
14. Ding, C., Li, T., Peng, W., Park, H.: Orthogonal nonnegative matrix t-factorizations for clustering. In: Proceedings of the 12th ACM SIGKDD International Conference on Knowledge Discovery and Data Mining, pp. 126–135 (2006)
15. Kang, Z., Peng, C., Cheng, Q.: Top-N recommender system via matrix completion. In: Proceedings of the 30th AAAI Conference on Artificial Intelligence, pp. 179–185 (2016)
16. Bennett, J., Elkan, C., Liu, B., Smyth, P., Tikk, D.: KDD Cup and workshop 2007. SIGKDD Explor. **9**, 51–52 (2007)
17. McAuley, J., Leskovec, J.: Hidden factors and hidden topics: understanding rating dimensions with review text. In: Proceedings of the 7th ACM Conference on Recommender Systems, pp. 165–172 (2013)
18. Boyd, S., Vandenberghe, L.: Convex Optimization. Cambridge University Press, Cambridge (2004)
19. Lee, D.D., Seung, H.S.: Algorithms for non-negative matrix factorization. In: Proceedings of Advances in Neural Information Processing Systems, pp. 556–562 (2001)
20. Bertsekas, D.P.: Nonlinear Programming. Athena Scientific, Belmont (1999)
21. Diao, Q., Qiu, M., Wu, C., Smola, A.J., Jiang, J., Wang, C.: Jointly modeling aspects, ratings and sentiments for movie recommendation (JMARS). In: Proceedings of the 20th ACM SIGKDD International Conference on Knowledge Discovery and Data Mining, pp. 193–202 (2014)

Low-Rank and Sparse Cross-Domain Recommendation Algorithm

Zhi-Lin Zhao[1], Ling Huang[1], Chang-Dong Wang[1(✉)], and Dong Huang[2]

[1] School of Data and Computer Science, Sun Yat-sen University, Guangzhou, China
zhaozhl7@mail2.sysu.edu.cn, huanglinghl@hotmail.com,
changdongwang@hotmail.com
[2] College of Mathematics and Informatics, South China Agricultural University,
Guangzhou, China
huangdonghere@gmail.com

Abstract. In this paper, we propose a novel Cross-Domain Collaborative Filtering (CDCF) algorithm termed Low-rank and Sparse Cross-Domain (LSCD) recommendation algorithm. Different from most of the CDCF algorithms which tri-factorize the rating matrix of each domain into three low dimensional matrices, LSCD extracts a user and an item latent feature matrix for each domain respectively. Besides, in order to improve the performance of recommendations among correlated domains by transferring knowledge and uncorrelated domains by differentiating features in different domains, the features of users are separated into shared and domain-specific parts adaptively. Specifically, a low-rank matrix is used to capture the shared feature subspace of users and a sparse matrix is used to characterize the discriminative features in each specific domain. Extensive experiments on two real-world datasets have been conducted to confirm that the proposed algorithm transfers knowledge in a better way to improve the quality of recommendation and outperforms the state-of-the-art recommendation algorithms.

Keywords: Low-rank · Sparse · Cross-domain
Recommendation algorithm

1 Introduction

Nowadays, for the sake of the benefit of businesses and the satisfaction of users, many online platforms like Amazon, Netflix and Douban use recommendation algorithms [1–3] to recommend items to users who are the most likely to be interested in them by analyzing huge amounts of data about user behaviour. More often than not, the task of recommendation algorithm is to speculate the value of missing ratings in the sparse user-item rating matrix by analyzing a few known ratings. Then some unrated items with high predicted ratings will be recommended to the target users. Collaborative Filtering (CF) [4,5] is the most widely used recommendation algorithm and one representative technology for collaborative filtering is matrix factorization.

© Springer International Publishing AG, part of Springer Nature 2018
J. Pei et al. (Eds.): DASFAA 2018, LNCS 10827, pp. 150–157, 2018.
https://doi.org/10.1007/978-3-319-91452-7_10

Over the past decade, matrix factorization has attracted an increasing amount of attention and has been applied in many areas such as machine learning [6] and data mining [7]. From the perspective of collaborative filtering [4], matrix factorization firstly extracts latent factors of users and items from the user-item rating matrix and then predicts missing ratings according to those latent factors. In practice, the rating matrix is usually quite sparse so that it is difficult to learn satisfactory latent factors of users and items which will make a great impact on the quality of rating prediction. Besides, it is a challenging task to make reliable predictions for new users due to the lack of the relevant rating data. In order to solve the above data sparsity and cold-start problems as well as improve the quality of recommendation [8], transfer learning [9] has been integrated into matrix factorization to transfer the knowledge from auxiliary data to rating data. Some typical auxiliary data are social media [10], tag information [11], user reviews [12] and product images [13].

As a special case of transfer learning, Cross-Domain Collaborative Filtering (CDCF) [14] transfers knowledge of rating data among multiple domains to make a better recommendation. It predicts ratings for all domains by learning from multiple rating matrices of different domains together. Combining all domains, we have more data to describe the features of users which can mitigate the sparsity problem. If a user does not get any ratings to items in a target domain but there are some ratings in other relative domains, we can solve the cold-start problem by learning the features of the user from other relative domains. Most existing CDCF methods [15–18] tri-factorize the rating matrix of each domain into three low dimensional matrices which represent user latent factors (or labels), codebook (or rating pattern matrix) and item latent factors (or labels) respectively. Usually, the codebook describing the relations between the clusters of users and items is shared among domains. Besides, users may appear in all domains and the user factors can also be shared, so user latent factor matrix or codebook can be viewed as a bridge to transfer knowledge. However, in the scenario where the domains are uncorrelated, the bridge can not transfer meaningful knowledge and even transfer negative knowledge which reduces the effect of recommendation. Besides, even for the same user, the performance will differ from one domain to another. So there are some domain-specific features of each user. Unfortunately, there is still a lack of methods differentiating shared features and domain-specific features adaptively.

To solve the above problems, we propose a novel cross-domain collaborative filtering algorithm termed Low-rank and Sparse Cross-Domain (LSCD) recommendation algorithm. Different from most of the existing CDCF algorithms, we decompose all the rating matrices of different domains into user and item latent feature matrices. More specifically, each user latent feature matrix consists of two parts: user-domain-shared feature matrix and user-domain-specific feature matrix. The former is used to describe the overall preferences of users among multiple domains which is modeled by a low-rank matrix. And the latter is a sparse matrix which is used to characterize the discriminative features in each domain because the expressions of some features vary from one domain to another. For example, if a user is willing to give high ratings to all items on average, we can capture this feature by analyzing the rating records of all domains. But if the user prefers movies to books, the user may give higher ratings on movie domain than book domain. So, a domain-specific feature matrix should be used to adjust shared feature matrix to fit specific domain. Therefore, if the domains are correlated, the performance can be improved because we use more rating data to learn the shared features of users. On the contrary, if the domains are uncorrelated, the performance can still be improved because we can distinguish the domain-specific features of different domains.

Extensive experiments have been conducted on two real world datasets: Amazon and MovieLens. The results show that the proposed LSCD recommendation algorithm can improve the quality of recommendation as the number of domains increases even the domains are uncorrelated and outperforms the state-of-the-art recommendation algorithms.

2   The Proposed Algorithm

We assume there are D domains in a recommendation task. The input user-item rating matrix $R_d \in \mathbb{R}^{m \times n_d}$ represents the rating relation between m users and n_d items in the dth domain. Note that the users are the same in each domain. Each entry denotes the rating of a user to an item within a certain numerical interval $[R_{min}, R_{max}]$ which will vary on different datasets. The rating will be zero if the user has not rated the item. I_d is the indicator matrix of R_d, where the value will be equal to one if the corresponding item in R_d has been rated or zero otherwise. $U \in \mathbb{R}^{m \times l}$ is the user-domain-shared feature matrix and $H_d \in \mathbb{R}^{m \times l}$ is the user-domain-specific feature matrix in the dth domain. $V \in \mathbb{R}^{n_d \times l}$ denotes the item latent feature matrix in the dth domain. Among them, $l \ll \min(m, n_d), \forall d = 1, \ldots, D$ is the number of latent features.

In the dth domain, the user latent feature matrix is the sum of the domain-shared feature matrix and domain-specific feature matrix, i.e., $U + H_d$. In the proposed algorithm, the shared features and domain-specific features can be differentiated adaptively. Based on the traditional matrix factorization algorithm, the predicted ratings of users to unrated items in the domain can be estimated by the product of users' and items' latent feature matrices,

$$P_d = (U + H_d)V_d^T. \tag{1}$$

In matrix factorization, those latent feature matrices will be learnt from rating data by minimizing the sum of squared errors between real ratings and predicted ratings. Besides, we should learn all the parameters among all domains together so that the shared features and domain specific features of users can be distinguished, and the user-domain-shared feature matrix U can be viewed as a bridge to transfer knowledge among domains. So the loss function is,

$$\frac{1}{2} \sum_{d=1}^{D} \| (R_d - (U + H_d)V_d^T) \odot I_d \|_F^2,$$

where $\| \cdot \|_F$ is Frobenius norm and $\odot$ denotes the Hadamard product.

Multi-Task Learning (MTL) [19] captures the task relationship via a shared low-rank structure, and CDCF is similar to this learning method since we want to explore the latent feature relationship among multiple domains. So the basic idea of MTL inspires us to divide the factors of users into shared part and domain-specific part. The shared part is used to model the overall preferences of users among multiple domains. Although the dimension of U is $\mathbb{R}^{m \times l}$, the number of shared features may be less than l and even equal to 0 if the domains are uncorrelated. When the domains are uncorrelated, it is not necessary to transfer knowledge because transferring negative knowledge may reduce the effect. Besides, if the domains are strongly correlated, the rank of U is at most l. So we need to find some important shared features from the l features, the number of which may be less than l. Besides, we can get a good result even the number of features is not large enough, since the most important features can be selected. Therefore, we assume U is low-rank. So users will share the same low-rank feature subspace and we should minimize rank(U) to get a low-rank structure of user-domain-shared feature matrix.

On the other hand, the preference of a user to different domains will vary slightly as discussed earlier and it can be reflected by a few features. So we use entry-wise sparse regularization term to identify those discriminative features in each domain. Because l_0-norm counts the number of nonzero elements of a solution, we should minimize $\|H_d\|_0$ to get a sparse structure of user-domain-specific feature matrix. And the matrix can be utilized to adjust domain-shared feature matrix to fit specific domain. The objective function is,

$$\mathcal{L}_O = \frac{1}{2} \sum_{d=1}^{D} \|(R_d - (U + H_d)V_d^T) \odot I_d\|_F^2 + \frac{\lambda_V}{2} \sum_{d=1}^{D} \|V_d\|_F^2$$

$$+ \lambda_U \text{rank}(U) + \lambda_H \sum_{d=1}^{D} \|H_d\|_0,$$

where λ_V, λ_U and λ_H are regularization coefficients. The ratio between λ_H and λ_U is used to control the composition's ratio of shared and domain-specific parts in user latent features. The Frobenius norm of V_d is added to prevent overfitting. However, for U and H_d, the Frobenius norms are not necessary because the low-rank and sparse regularization terms of the two variables can also do this.

Solving the above nonconvex optimization problem is NP-hard. As pointed out in [20], if the rank of U is not too large and H_d is sparse, the regularization terms $rank(U)$ and $\|H_d\|_0$ can be approximated by the tightest convex relaxation $\|U\|_*$ and $\|H_d\|_1$ respectively where $\|\cdot\|_*$ and $\|\cdot\|_1$ denote the Nuclear norm and the l_1-norm of a matrix, respectively. Therefore, the relaxed objective function is,

$$\mathcal{L} = \frac{1}{2} \sum_{d=1}^{D} \|(R_d - (U + H_d)V_d^T) \odot I_d\|_F^2 + \frac{\lambda_V}{2} \sum_{d=1}^{D} \|V_d\|_F^2 + \lambda_U \|U\|_* + \lambda_H \sum_{d=1}^{D} \|H_d\|_1.$$

The approximate projected gradient method is used to solve the above objective function. After obtaining the latent feature matrices U, H_d and V_d, the predicted rating matrix P_d of each domain can be estimated by Eq. (1).

3   Experiments

In this section, some experiments are conducted to evaluate the effectiveness of the proposed method[1].

3.1   Datasets and Evaluation Measures

In our experiments, two real-world datasets with multiple item domains are used, namely Amazon and MovieLens.

- **Amazon**[2]: This dataset is obtained from Julian McAuley [13], which contains 6,643,669 users, 2,441,053 products and 80,737,555 ratings of 24 domains from Amazon spanning Jun 1995–Mar 2013. Each record is a (user, item, rating, timestamp) tuple and the time information is not used. Four item domains are used in our experiments, i.e., Book, CD, Music and Movie.
- **MovieLens**[3]: This dataset is obtained from the Information Retrieval Group at Universidad Autónoma de Madrid, which contains 2113 users, 10,197 movies, and 855,598 ratings from MovieLens spanning 1970–2009. We use the tags of movies to classify the ratings into 18 domains and the four movie domains are used in our experiments, i.e., comedy (COM), dramatic (DRA), action (ACT), thrilling (THR) domains.

Without loss of generality, the two datasets are split randomly with 80% as the training set and 20% as the testing set. In order to evaluate the quality of the recommendation algorithms, two widely used evaluation metrics, namely Mean Absolute Error (MAE) and Root Mean Square Error (RMSE), will be used to measure the accuracy of the predicted ratings.

3.2   Comparison Experiments

We compare the results of the predicted ratings of the proposed LSCD algorithm with nine state-of-the-art recommendation algorithms, i.e., **N-CDCF-U**, **N-CDCF-I**, **MF-CDCF**, **CMF** [21], **CDTF**, **CLMF** [17], **TALMUD** [15], **CDLD** [22] and **PCLF** [23] where N-CDCF-U, N-CDCF-I, MF-CDCF and CDTF are from the same paper [16]. N-CDCF-U and N-CDCF-I are neighborhood based collaborative filtering methods computing the similarities between users and between items by cosine similarity over all items and users respectively. But the other eight algorithms (including LSCD) are matrix factorization methods. For these matrix factorization methods, we set the dimensionality of latent feature vector $l = 50$. Besides, the step size μ and the item regularization coefficient λ_V are set to 0.001 and 0.1 respectively. To be fair, all the regularization coefficients of the compared algorithms are set to 0.1. We initialize all the latent feature matrices randomly.

---

[1] Source code and datasets are available at https://github.com/sysulawliet/LSCD.

[2] http://jmcauley.ucsd.edu/data/amazon.

[3] https://grouplens.org/datasets/hetrec-2011.

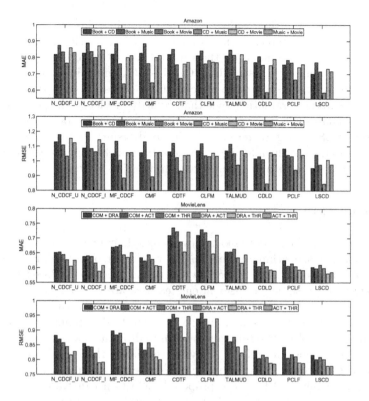

Fig. 1. Performance comparisons of different combination of two domains.

There are 11 combinations of the 4 domains. Plotting all the results in a figure will lead to chaos. Without loss of generality, we plot some typical combinations of domains. The comparison results in terms of MAE and RMSE on the two datasets over two domains are reported in Fig. 1. Six different pairs of domains are selected from each dataset respectively. The results show that the proposed LSCD algorithm outperforms the other state-of-the-art CDCF algorithms in all combinations of two domains. Generally speaking, the other five matrix factorization methods are inferior to the two neighborhood based collaborative filtering methods on the MovieLens dataset but better than the two algorithms on the Amazon dataset. The reason is that, the Amazon dataset is much sparser than the MovieLens dataset, and those matrix factorization methods can work well when data information is sparse because they predict ratings according to latent features rather than the original data. Besides, those neighborhood based collaborative filtering methods have more advantages when the rating matrix is dense since they can get a more precise user or item similarity matrix. But the proposed LSCD algorithm which is also based on matrix factorization can work well on both datasets. One reason may be that the low-rank structure of U makes it able to catch the shared feature subspace of users from few entries.

On the other hand, the results of different combinations of two domains are quite different. For example, the performance of (CD + Music) domains is better than (Book + Music) domains on the Amazon dataset. And the performance of (DRA + ACT) domains is better than (COM + ACT) domains on the MovieLens dataset. Intuitively, Music is more correlated to CD than Book and action movie is more correlated to dramatic movie than comedy movie. So the performance of correlated domains will be better for the eight recommendation algorithms. But the proposed LSCD algorithm can obtain good results even the two domains are uncorrelated since the all sparse matrices H_d can characterize those domain-specific features of users and differentiate all domains better.

4   Conclusion

In this paper, we have proposed a novel cross-domain collaborative filtering recommendation algorithm termed LSCD which can better model the latent features of users to improve the quality of rating prediction. Based on matrix factorization, we assume the user latent features are divided into shared and domain-specific parts. We use low-rank matrix to capture the shared feature subspace of users and use sparse matrix to identify discriminative features in each domain. The objective function is optimized by the approximate projected gradient method and theoretical analysis has shown the complexity and convergence of the proposed algorithm. Extensive experiments on two real-world datasets have confirmed that the proposed algorithm can transfer knowledge among domains in an even better fashion and significantly outperforms state-of-the-art recommendation algorithms.

Acknowledgments. This work was supported by NSFC (61502543 & 61602189), Guangdong Natural Science Funds for Distinguished Young Scholar (2016A030306014), the Ph.D. Start-up Fund of Natural Science Foundation of Guangdong Province, China (2016A030310457), and Tip-top Scientific and Technical Innovative Youth Talents of Guangdong special support program (2016TQ03X542).

References

1. Zheng, B., Su, H., Zheng, K., Zhou, X.: Landmark-based route recommendation with crowd intelligence. Data Sci. Eng. **1**(2), 86–100 (2016)
2. Zhao, Z.L., Wang, C.D., Wan, Y.Y., Lai, J.H.: Recommendation in feature space sphere. Electron. Commer. Res. Appl. **26**, 109–118 (2017)
3. Hu, Q.Y., Zhao, Z.L., Wang, C.D., Lai, J.H.: An item orientated recommendation algorithm from the multi-view perspective. Neurocomputing **269**, 261–272 (2017)
4. Jannach, D., Zanker, M., Felfering, A., Friedrich, G.: Recommender Systems: An Introduction. Addison Wesley Publishing, Boston (2013)
5. Zhao, Z.L., Wang, C.D., Lai, J.H.: AUI&GIV: recommendation with asymmetric user influence and global importance value. PLoS ONE **11**(2), e0147944 (2016)
6. Allab, K., Labiod, L., Nadif, M.: A Semi-NMF-PCA unified framework for data clustering. IEEE Trans. Knowl. Data Eng. **29**(1), 2–16 (2017)

7. Salakhutdinov, R., Mnih, A.: Probabilistic matrix factorization. In: NIPS, pp. 1257–1264 (2007)
8. Guo, G., Zhang, J., Thalmann, D.: Merging trust in collaborative filtering to alleviate data sparsity and cold start. Knowl.-Based Syst. **57**, 57–68 (2014)
9. Tan, B., Song, Y., Zhong, E., Yang, Q.: Transitive transfer learning. In: KDD, pp. 1155–1164 (2015)
10. Ma, H., Zhou, D., Liu, C., Lyu, M.R., King, I.: Recommender systems with social regularization. In: WSDM, pp. 287–296 (2011)
11. Wei, S., Zheng, X., Chen, D., Chen, C.: A hybrid approach for movie recommendation via tags and ratings. Electron. Commer. Res. Appl. **18**, 83–94 (2016)
12. Xin, X., Liu, Z., Lin, C.Y., Huang, H., Wei, X., Guo, P.: Cross-domain collaborative filtering with review text. In: IJCAI, pp. 1827–1834 (2015)
13. McAuley, J.J., Targett, C., Shi, Q., van den Hengel, A.: Image-based recommendations on styles and substitutes. In: SIGIR, pp. 43–52 (2016)
14. Song, T., Peng, Z., Wang, S., Fu, W., Hong, X., Yu, P.S.: Review-based cross-domain recommendation through joint tensor factorization. In: Candan, S., Chen, L., Pedersen, T.B., Chang, L., Hua, W. (eds.) DASFAA 2017. LNCS, vol. 10177, pp. 525–540. Springer, Cham (2017). https://doi.org/10.1007/978-3-319-55753-3_33
15. Moreno, O., Shapira, B., Rokach, L., Shani, G.: TALMUD: transfer learning for multiple domains. In: CIKM, pp. 425–434 (2012)
16. Hu, L., Cao, J., Xu, G., Cao, L., Gu, Z., Zhu, C.: Personalized recommendation via cross-domain triadic factorization. In: WWW, pp. 595–606 (2013)
17. Gao, S., Luo, H., Chen, D., Li, S., Gallinari, P., Guo, J.: Cross-domain recommendation via cluster-level latent factor model. In: Blockeel, H., Kersting, K., Nijssen, S., Železný, F. (eds.) ECML PKDD 2013. LNCS (LNAI), vol. 8189, pp. 161–176. Springer, Heidelberg (2013). https://doi.org/10.1007/978-3-642-40991-2_11
18. Liu, Y.F., Hsu, C.Y., Wu, S.H.: Non-linear cross-domain collaborative filtering via hyper-structure transfer. In: ICML, pp. 1190–1198 (2015)
19. Chen, J., Zhou, J., Ye, J.: Integrating low-rank and group-sparse structures for robust multi-task learning. In: KDD, pp. 42–50 (2011)
20. Candes, E.J., Li, X., Ma, Y., Wright, J.: Robust principal component analysis. J. ACM **58**(3), 1–39 (2011)
21. Singh, A.P., Gordon, G.J.: Relational learning via collective matrix factorization. In: KDD, pp. 650–658 (2008)
22. Iwata, T., Takeuchi, K.: Cross-domain recommendation without shared users or items by sharing latent vector distributions. In: AISTATS, pp. 379–387 (2015)
23. Ren, S., Gao, S., Liao, J., Guo, J.: Improving cross-domain recommendation through probabilistic cluster-level latent factor model. In: AAAI, pp. 4200–4201 (2015)

Cross-Domain Recommendation for Cold-Start Users via Neighborhood Based Feature Mapping

Xinghua Wang[1], Zhaohui Peng[1(✉)], Senzhang Wang[2], Philip S. Yu[3,4],
Wenjing Fu[1], and Xiaoguang Hong[1]

[1] School of Computer Science and Technology, Shandong University, Jinan, China
wang.xingh@foxmail.com, {pzh,hxg}@sdu.edu.cn, fuwenjing@mail.sdu.edu.cn
[2] College of Computer Science and Technology,
Nanjing University of Aeronautics and Astronautics, Nanjing, China
szwang@nuaa.edu.cn
[3] Department of Computer Science, University of Illinois at Chicago, Chicago, USA
psyu@uic.edu
[4] Institute for Data Science, Tsinghua University, Beijing, China

Abstract. Traditional Collaborative Filtering (CF) models mainly focus on predicting a user's preference to the items in a single domain such as the movie domain or the music domain. A major challenge for such models is the data sparsity problem, and especially, CF cannot make accurate predictions for the cold-start users who have no ratings at all. Although Cross-Domain Collaborative Filtering (CDCF) is proposed for effectively transferring users' rating preference across different domains, it is still difficult for existing CDCF models to tackle the cold-start users in the target domain due to the extreme data sparsity. In this paper, we propose a Cross-Domain Latent Feature Mapping (CDLFM) model for cold-start users in the target domain. Firstly, the user rating behavior is taken into consideration in the matrix factorization for alleviating the data sparsity. Secondly, neighborhood based latent feature mapping is proposed to transfer the latent features of a cold-start user from the auxiliary domain to the target domain. Extensive experiments on two real datasets extracted from Amazon transaction data demonstrate the superiority of our proposed model against other state-of-the-art methods.

Keywords: Cross-domain recommendation · Cold start
Feature mapping

1 Introduction

With the quick development of Internet and Web techniques, e-commerce has become increasingly popular. In order to help consumers find what they really desire from the massive amounts of products, recommender systems become indispensable in most e-commerce websites. Collaborative Filtering (CF) is a

J. Pei et al. (Eds.): DASFAA 2018, LNCS 10827, pp. 158–165, 2018.
https://doi.org/10.1007/978-3-319-91452-7_11

widely used technique in recommender systems due to the fact that it requires little domain-specific knowledge. Traditional CF models focus on single-domain user preference prediction and suffer from the data sparsity problem. Although Cross-Domain Collaborative Filtering (CDCF) is proposed to enrich the knowledge in the target domain by taking advantage of multi-domain ratings, and most CDCF models, e.g. TCF [1], CST [2], CMF [3] are designed to alleviate the single-domain data sparsity problem, while how to effectively recommend for the cold-start users is still not fully explored.

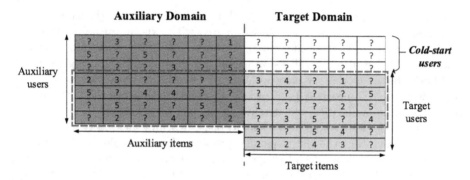

Fig. 1. Illustration of the cross-domain recommendation for cold-start users (Color figure online)

In real world, the cold-start users of an item domain may have ratings in another item domain. Thus, the problem setting about cross-domain recommendation for cold-start users studied in this paper is illustrated in Fig. 1. One can see from Fig. 1 that cold-start users only have ratings in the auxiliary domain, which is different from most previous works [1,2] that only assume the auxiliary domain data is relatively denser than the target domain data. Users who have ratings in both domains are called linked users whose rating data is marked with dashed red box in Fig. 1. It is challenging to make recommendations for the cold-start users in the target domain. First, rating matrices in different item domains are usually sparse, which makes it difficult to characterize users. Second, there is no rating data for cold-start users in the target domain, and user rating behavior and preference in different domains are varied.

To address the above challenges, we propose a Cross-Domain Latent Feature Mapping (CDLFM) model. Firstly, we handle the rating matrices in different domains separately with Matrix Factorization by incorporating User Similarities (MFUS) in order to gain domain-specific user latent features in sparse domains. Next, to transfer the knowledge of user characteristics across domains, we propose a neighborhood based gradient boosting trees method to learn the cross-domain user latent feature mapping function for each cold-start user. Due to space limit, the details of CDLFM are addressed in our technical report [4]. Our major contributions are summarized as follows:

- An improved rating matrix factorization model is proposed which is the first to consider users' similarity relationship reflected from their rating behaviors.
- A neighborhood based gradient boosting trees method is proposed for more accurately performing cross-domain latent feature mapping.
- We conduct extensive experiments on the Amazon rating data to evaluate the proposed model and make comparisons with other state-of-the-art models.

2   Related Work

The existing research related to our work mainly includes matrix factorization and cross-domain recommendation. Compared with existing rating matrix factorization models [5–7], besides taking the user rating behavior into consideration, our MFUS also capitalizes on the advantages of both the neighborhood and latent factor models by incorporating user similarities into the Matrix Factorization (MF) [5] process. TagiCoFi [7] also aims to improve the performance of MF via user similarities, but it relies on tagging information.

For cross-domain recommendation, transfer learning has been used extensively for alleviating the data sparsity problem. For the cold start problem, there have been tag-based and review-based cross-domain factorization models [8,9], and Hu [10] mentions the unacquainted world for users. EMCDR [11] and [12] try to use the Multi-Layer Perceptron (MLP) and a transformation matrix to map the user feature vector across domains, but they take all the linked users into consideration which may introduce noise. On social networks [13], Zhao [14] maps users' social networking features to another feature representation for product recommendation. In our work, no text information is available and we make cross-domain feature mapping in a more explicable way.

3   Matrix Factorization by Incorporating User Similarities

The user rating behaviors in different item domains can be quite different. Firstly, we handle the rating matrices of different domains separately. In order to better characterize users in sparse domains, we take users' rating behaviors into consideration and an improved rating matrix factorization model named MFUS (Matrix Factorization by incorporating User Similarities) is proposed.

3.1   Rating Behavior Based User Similarity Measures

Similarity Based on Common Ratings. Given two users u and v, if they have commonly rated products C_{uv}, we can compute their similarity based on their rating similarity on C_{uv}. In our experiments, we first compute the average of squared rating difference over the users' ratings on C_{uv}, and then a monotonously-decreased exponential function is used to transform the difference into a similarity value.

Similarity Based on the Estimations of Having No Interest. A user's potential preference can be also reflected by the products that he/she does not give ratings to, and we can not arbitrarily conclude that a user does not like the unrated products. We use P_{ui} to represent the probability of user u having no interest on the product i. If u dose not rate i, P_{ui} can be estimated by $P_{ui} = [1 - f_1(n_u) \times f_2(n_i)] \times [1 - f_3(n_i/n) \times f_3(n_{Hi}/n_i)]$, where $f_1(n_u) = \sqrt{1 - \frac{n_u^2}{m^2}}$, $f_2(n_i) = \sqrt{1 - \frac{n_i^2}{n^2}}$, $f_3(x) = \frac{2}{1+e^{-\sigma x}} - 1$. n and m denote the numbers of users and products in a domain. n_u and n_i represent the total rating numbers of user u and product i. n_{Hi} is the number of high ratings on product i (the high rating is 4 or 5 in our experiments). When user u rates product i, P_{ui} can be estimated from the rating score R_{ui}. For example, if $R_{ui} = 1$, $P_{ui} = 1$; if $R_{ui} = 2$, $P_{ui} = 0.8$; if $R_{ui} = 3$, $P_{ui} = 0.5$ and so on.

Given two users and the products which have not been rated by both of them, we can compute a similarity value based on the computed probability values.

Similarity Based on Rating Biases. We observe that users' rating values are usually unevenly distributed. For example, high ratings, 4 and 5, usually account for a large proportion, while low ratings hold a small proportion. We call this as the rating biases of users. Here, we adopt the idea of TF-IDF to measure the users' rating biases, and the rating bias of user u to rating score $r \in \{1, 2, 3, 4, 5\}$ can be calculated via $rf(u, r) \times log_{base}(n/uf(r))$, where $uf(r)$ represents the number of users who have given the rating score r, $rf(u, r) = n_{ur}/\left(\sum_{z=1}^{5} n_{uz}\right)$, n_{ur} represents the frequency of rating score r used in u's rating history, and $base$ is a predefined parameter ($base = 2$ in our experiments). Then we can compute the third similarity measure based on the users' rating biases.

Finally, for users u and v, the weighted average of the three similarity measures is used as the final similarity value between them.

3.2   Rating Matrix Factorization

We embed the user similarities in Sect. 3.1 into the matrix factorization model as a new regularization term and our goal is solving the following minimization problem:

$$\min_{\mathbf{U},\mathbf{V}} \frac{1}{2} \sum_{u=1}^{n} \sum_{i=1}^{m} Y_{ui} \left(R_{ui} - \mathbf{U}_{u*}\mathbf{V}_{i*}^{T}\right)^2 + \frac{\alpha}{2} tr\left(\mathbf{U}\mathbf{U}^T\right)$$
$$+ \frac{\alpha}{2} tr\left(\mathbf{V}\mathbf{V}^T\right) + \frac{\beta}{2} \sum_{u=1}^{n} \sum_{v=u+1}^{n} S_{uv} \parallel \mathbf{U}_{u*} - \mathbf{U}_{v*} \parallel^2$$

(1)

where $\mathbf{U}_{u*}$ and $\mathbf{V}_{v*}$ represent the latent features of user u and product i, $(\cdot)^T$ and $tr(\cdot)$ denote the transposition and trace of a matrix, and Y_{ui} is 1 if user u rated product i and 0 otherwise. α and β are the regularization parameter to prevent over-fitting. Equally, we transform (1) to

$$\min_{\mathbf{U},\mathbf{V}} \frac{1}{2} \sum_{u=1}^{n} \sum_{i=1}^{m} Y_{ui} \left(R_{ui} - \mathbf{U}_{u*} \mathbf{V}_{i*}^{T} \right)^{2} + \frac{\alpha}{2} tr \left(\mathbf{V} \mathbf{V}^{T} \right) + \frac{1}{2} tr \left[\mathbf{U}^{T} \left(\alpha \mathbf{I} + \beta \mathbf{L} \right) \mathbf{U} \right] \quad (2)$$

where $\mathbf{L} = \mathbf{D} - \mathbf{S}$ with $\mathbf{D}$ being a diagonal matrix whose diagonal element is $D_{uu} = \sum_{v=1}^{n} S_{uv}$, $\mathbf{S}$ is the users' similarity matrix, and $\mathbf{I}$ is an identity matrix. We apply the alternating gradient descent to optimize one column of $\mathbf{U}$ or one row of $\mathbf{V}$ at a time. If we use F to represent the objective function in (2), the gradients can be computed as follows:

$$\frac{\partial \mathrm{F}}{\partial \mathbf{U}_{*k}} = \left(\alpha \mathbf{I} + \beta \mathbf{L} \right) \mathbf{U}_{*k} - \mathbf{x}, \quad \mathbf{x} \in R^{n \times 1} \text{ with } x_u = \sum_{i=1}^{m} Y_{ui} \left(R_{ui} - \mathbf{U}_{u*} \mathbf{V}_{i*}^{T} \right) V_{ik}$$

$$\frac{\partial \mathrm{F}}{\partial \mathbf{V}_{i*}} = -\sum_{u=1}^{n} Y_{ui} \left(R_{ui} - \mathbf{U}_{u*} \mathbf{V}_{i*}^{T} \right) \mathbf{U}_{u*} + \alpha \mathbf{V}_{i*}.$$

4   Neighborhood Based Latent Feature Mapping

The proposed MFUS can learn the domain-specific latent features of users in different domains. However, for the cold-start users U_T, we can only obtain their latent features in the auxiliary domain which cannot be used directly for making recommendation in the target domain due to the different semantic meanings of latent features in different domains. However, the same user's latent features in different domains can be highly correlated. Therefore, we try to use the linked users U_L as a bridge to learn the function $\mathcal{F}$ which can map the user's latent features from the auxiliary domain to the target domain. The input of the mapping function $\mathcal{F}$ is a user's latent features in the auxiliary domain and the output is the same user's latent features in the target domain.

We adopt the Gradient Boosting Trees (GBT) method [15] to learn the mapping function $\mathcal{F}$ since it is powerful to capture higher-order transformation relationship between the input and output. Assuming the dimension of the latent features in the target domain is K_t, we can use GBT K_t times and learn the mapping function $\mathcal{F} = \left\{ f^{(k)} \left(\boldsymbol{x} \right) \right\}_{k=1}^{K_t}$, where the jth subfunction $f^{(j)} \left(\boldsymbol{x} \right)$ takes the user's latent features in the auxiliary domain as input and returns the jth mapped latent feature in the target domain. Moreover, for each cold-start user, we use the similar linked users to learn the mapping function. Thus in the last step of our model, for each user $u \in U_T$, we use $\mathbb{N}_u$ to denote the similar linked users to u with each $v \in \mathbb{N}_u$, $v \in U_L$ and $S_{uv}^a > sim$. Here, sim is a predefined similarity threshold value and S_{uv}^a is the user similarity in the auxiliary domain computed in MFUS. Latent feature pairs $\{\mathbf{U}_{v*}^a, \mathbf{U}_{v*}^t\}_{v \in \mathbb{N}_u}$, where $\mathbf{U}_{v*}^a$ and $\mathbf{U}_{v*}^t$ represent user v's latent features in the auxiliary domain and target domain, are used to learn the mapping function $\mathcal{F}_u = \left\{ f_u^{(k)} \left(\boldsymbol{x} \right) \right\}_{k=1}^{K_t}$ via GBT. According to the latent features $\mathbf{U}_{u*}^a$ and the mapping function $\mathcal{F}_u$, we can compute the user mapped latent features $\boldsymbol{u}$ in the target domain with the element $u_k = f_u^{(k)} \left(\mathbf{U}_{u*}^a \right)$. Based on the latent feature matrix of products in the target domain, the rating predictions of the cold-start user u can be computed by $\hat{\mathbf{r}} = \mathbf{V}^t \boldsymbol{u}^T$.

5   Experiments

5.1   Experiment Setup

We extract two datasets from the Amazon rating data [16]. The first extracted dataset consists of the ratings about movies and books, and the second one consists of the ratings about movies and electronic products. We first filter out the linked users and items with very small number of ratings. Besides the linked users, some active users who have given a large number of ratings in a certain domain are also included. Finally, we have 16926 linked users in the first dataset and 12004 linked users in the second one. The statistics of the two datasets are given in Table 1. RMSE and MAE are used as the evaluation metrics, and we compare our model CDLFM with the following baselines AF [1], CDCF-U [10], CDCF-I [10], CMF [3], TMatrix [12] and EMCDR [11], where EMCDR is one state-of-the-art cross-domain recommendation method for cold-start users.

Table 1. Statistics of the two datasets used for evaluation

Dataset 1	Rating value			Density
Movie	$\{1,2,3,4,5\}$	#users	17926	0.00225
		#movies	4595	
		#ratings	185421	
Book	$\{1,2,3,4,5\}$	#users	17426	0.00149
		#books	8935	
		#ratings	231564	
Dataset 2				
Movie	$\{1,2,3,4,5\}$	#users	12203	0.00307
		#movies	3625	
		#ratings	135587	
Electronics	$\{1,2,3,4,5\}$	#users	12728	0.00212
		#electronics	4302	
		#ratings	115955	

5.2   Experimental Results

Experiments with different auxiliary and target domains are denoted as MB and ME for brevity. For example, MB denotes the experiments on Dataset 1 with **M**ovies as the auxiliary domain and **B**ooks as the target domain. The dimension of latent features is set to 15 and *sim* in CDLFM is set to 0.45.

Impact of Data Density. To evaluate the impact of data density, we randomly select 50% of the total linked users as the cold-start users and construct three

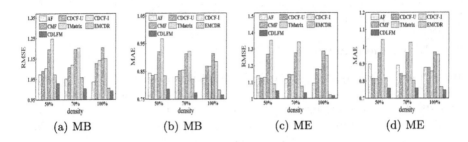

Fig. 2. Performance of methods under different data density levels

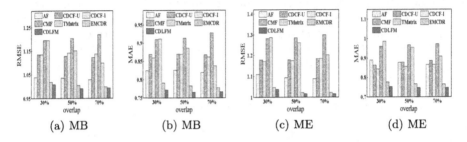

Fig. 3. Performance of methods under different overlap levels

different training sets denoted as density levels 50%, 70% and 100%. Taking the density level 70% for example, the training set consists of 70% of the total ratings in the auxiliary domain and 70% of the remaining ratings (after removing the cold-start users' ratings) in the target domain. Figure 2 report the results on different datasets and CDLFM performs best under all different data density levels. For EMCDR, MLP are learned based on all linked users which may introduce noise. Besides, from Fig. 2, one can see that the sparser the dataset is, the improvement of our model compared to EMCDR is more obvious.

In our model, MFUS can learn more accurate domain-specific latent features in sparse domains and neighborhood based GBT can learn more appropriate feature mapping function for each cold-start user. Therefore, we can predict cold-start users' latent features and preference accurately in the target domain.

Impact of the Size of Linked Users. we also experiment with three different user overlap levels, namely 30%, 50% and 70%. Taking overlap level 30% for example, we randomly select 70% of the total linked users as the cold-start users, and the remaining ratings in the dataset compose the training set. The results are reported in Fig. 3. One can see that our model achieves the best performance under all user overlap levels. Similarly, the less users overlap between two domains, the improvement of our model compared to EMCDR is more obvious.

6   Conclusions

In this paper, we present a novel model CDLFM for more effective cross-domain recommendation for cold-start users. Firstly, a new rating matrix factorization model is proposed to learn more accurate latent features of users in sparse domains. Then, a neighborhood based feature mapping method is used to learn more appropriate latent feature mapping function across domains. The experimental results demonstrate the superiority of our model.

Acknowledgements. This work is supported by NSF of China (No. 61602237, No. 61672313), 973 Program (No. 2015CB352501), NSF of Shandong, China (No. ZR2017MF065), NSF of Jiangsu, China (No. BK20171420). This work is also supported by US NSF through grants IIS-1526499, and CNS-1626432.

References

1. Pan, W., Yang, Q.: Transfer learning in heterogeneous collaborative filtering domains. Artif. Intell. **197**, 39–55 (2013)
2. Pan, W., Xiang, E.W., Liu, N.N., Yang, Q.: Transfer learning in collaborative filtering for sparsity reduction. In: AAAI, pp. 230–235 (2010)
3. Singh, A.P., Gordon, G.J.: Relational learning via collective matrix factorization. In: KDD, pp. 650–658 (2008)
4. Wang, X., Peng, Z., Wang, S., Yu, P.S., Fu, W., Hong, X.: Cross-domain recommendation for cold-start users via neighborhood based feature mapping. https://arxiv.org/
5. Salakhutdinov, R., Mnih, A.: Probabilistic matrix factorization. In: NIPS, pp. 1257–1264 (2007)
6. Koren, Y.: Factorization meets the neighborhood: a multifaceted collaborative filtering model. In: KDD, pp. 426–434 (2008)
7. Zhen, Y., Li, W.J., Yeung, D.Y.: TagiCoFi: tag informed collaborative filtering. In: RecSys, pp. 69–76 (2009)
8. Fernández-Tobías, I., Cantador, I.: Exploiting social tags in matrix factorization models for cross-domain collaborative filtering. In: CBRecSys@RecSys, pp. 34–41 (2014)
9. Song, T., Peng, Z., Wang, S., Fu, W., Hong, X., Yu, P.S.: Review-based cross-domain recommendation through joint tensor factorization. In: Candan, S., Chen, L., Pedersen, T.B., Chang, L., Hua, W. (eds.) DASFAA 2017. LNCS, vol. 10177, pp. 525–540. Springer, Cham (2017). https://doi.org/10.1007/978-3-319-55753-3_33
10. Hu, L., Cao, J., Xu, G., Cao, L., Gu, Z., Zhu, C.: Personalized recommendation via cross-domain triadic factorization. In: WWW, pp. 595–606 (2013)
11. Man, T., Shen, H., Jin, X., Cheng, X.: Cross-domain recommendation: an embedding and mapping approach. In: IJCAI, pp. 2464–2470 (2017)
12. Kazama, M., Varga, I.: Cross domain recommendation using vector space transfer learning. In: RecSys Posters (2016)
13. Wang, S., Hu, X., Yu, P.S., Li, Z.: MMRate: inferring multi-aspect diffusion networks with multi-pattern cascades. In: KDD, pp. 1246–1255 (2014)
14. Zhao, W.X., Li, S., He, Y., Chang, E.Y., Wen, J.R., Li, X.: Connecting social media to e-commerce: cold-start product recommendation using microblogging information. IEEE Trans. Knowl. Data Eng. **28**(5), 1147–1159 (2016)
15. Friedman, J.H.: Greedy function approximation: a gradient boosting machine. Ann. Statist. **29**, 1189–1232 (2000)
16. He, R., McAuley, J.: Ups and downs: modeling the visual evolution of fashion trends with one-class collaborative filtering. In: WWW, pp. 507–517 (2016)

Graph and Network Data Processing

K-Connected Cores Computation in Large Dual Networks

Lingxi Yue[1], Dong Wen[2], Lizhen Cui[1(✉)], Lu Qin[2], and Yongqing Zheng[1]

[1] School of Software Engineering, Shandong University, Jinan, China
yuelingxi@mail.sdu.edu.cn, clz@sdu.edu.cn, zhengyongqing@dareway.com.cn
[2] Centre for Artificial Intelligence, University of Technology Sydney,
Ultimo, Australia
dong.wen@student.uts.edu.au, lu.qin@uts.edu.au

Abstract. Computing k-cores is a fundamental and important graph problem, which can be applied in many areas, such as community detection, network visualization, and network topology analysis. Due to the complex relationship between different entities, dual graph widely exists in the applications. A dual graph contains a physical graph and a conceptual graph, both of which have the same vertex set. Given that there exist no previous studies on the k-core in dual graphs, we formulate a k-connected core (k-CCO) model in dual graphs. A k-CCO is a k-core in the conceptual graph, and also connected in the physical graph. Given a dual graph and an integer k, we propose a polynomial time algorithm for computing all k-CCOs. We also propose three algorithms for computing all maximum-connected cores ($MCCO$), which are the existing k-CCOs such that a $(k + 1)$-CCO does not exist. We conduct extensive experiments on six real-world datasets and several synthetic datasets. The experimental results demonstrate the effectiveness and efficiency of our proposed algorithms.

1 Introduction

Graph model has been used to represent the relationship of entities in many real-world applications, such as social networks, web graphs, collaboration networks and biological networks. Given a graph $G(V, E)$, vertices in V represent the interested entities and edges in E represent the relationship between entities. Significant research efforts have been devoted towards many fundamental problems in managing and analyzing graph data. Among them, cohesive subgraph detection has been extensively studied recently [5,9,13,17,31].

Given a graph G and an integer k, a k-core of G is a maximal connected subgraph in which each vertex has degree at least k [29]. The problem of computing k-cores draws a lot of attention [7,17,28,32] due to the elegant structural properties of k-core [29] and the linear time solution [3]. It can be applied in many areas including but not limited to community detection [11], dense subgraph discovery [2,6], graph visualization [1], and system analysis [10].

© Springer International Publishing AG, part of Springer Nature 2018
J. Pei et al. (Eds.): DASFAA 2018, LNCS 10827, pp. 169–186, 2018.
https://doi.org/10.1007/978-3-319-91452-7_12

Motivations. In many real-world applications, a single simple graph is hard to express the complex relationship between entities. [33] models a dual graph containing two complementary graphs with the same vertex set, one of which represents the physical interaction between vertices, and the other represents the conceptual interaction. They study the problem of computing the subgraph, namely DCS, which is the densest in the conceptual graph and also connected in the physical graph. However, computing the DCS in dual graphs is NP-hard. Even though an approximate solution is proposed and a relatively poorer result quality is endured in [33], the time consuming for this problem is still large and not scalable to big graphs. Additionally, they do not restrict the connectivity of DCS in the conceptual graph. The result subgraph is probably disconnected and obviously not cohesive.

Given that there exists no any research on the k-core computation in dual graphs, in this paper, we adopt the classic k-core definition to model a k-**C**onnected **CO**re (k-CCO) in dual graphs. Given a dual graph and an integer k, a k-CCO is a dual subgraph g satisfying the following three conditions: (i) the minimum degree of g is not less than k in the conceptual graph; (ii) g is connected in the conceptual graph; and (iii) g is connected in the physical graph.

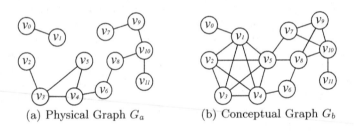

(a) Physical Graph G_a (b) Conceptual Graph G_b

Fig. 1. An example of dual graphs $G(V, E_a, E_b)$

An example of a dual graph is given in Fig. 1. Figure 1(a) is a physical graph, and Fig. 1(b) is a conceptual graph. Given an integer $k = 3$, there exists only one 3-CCO, that is the induced dual subgraph of $\{v_2, v_3, v_4, v_5\}$. The minimum degree is not less than 3, and the subgraph is connected in both two graphs.

Our k-CCO model restricts the connectivity for both two graphs and guarantees the cohesiveness of the result graph by given integer parameter k. Based on this model, we formulate two problems in this paper. Given a dual graph G and an integer k, the first problem is computing all k-CCOs in G. It offers a flexible selection for the degree constraint k and returns a subjective result for users. Similar to the DCS problem in [33], we also study a parameter-free problem, that is computing the **M**aximum-**C**onnected **CO**res ($MCCOs$) in a given dual graph G. Here, a $MCCO$ is a k-CCO in G such that there does not exist any $(k+1)$-CCO in G. The 3-CCO in the dual graph G in Fig. 1 is also a $MCCO$, since there does not exist any 4-CCO in G.

Applications. Computing k-CCOs and $MCCO$s can be applied in many areas. For example, to mine a research group, the researchers in the group should be connected in their collaboration network (physical graph), in which each edge represents two researchers have co-authored a paper. Simultaneously, each researcher should have enough neighbors in a similarity network (conceptual graph), in which each edge represents two researchers have similar research interests. In social networks, each user may have many interest labels, such as soccer, basketball, cartoon. A conceptual graph can be built by computing the interest similarity between any two users. A physical graph can be built by checking whether any two users follow each other. A social community should be connected in the physical graph, and each user in the group should have enough neighbors with similar interest.

Challenges. It is nontrivial to compute all k-CCOs. A k-$core$ in conceptual graph may be disconnected in the physical graph, and a connected component in the physical graph may conversely violate the degree constraint and connectivity constraint in the conceptual graph. For the problem of computing the $MCCO$s, let k_{max} be the maximum k in a dual graph such that a k-CCO exists. Given the solution for computing k-CCOs, the $MCCO$s can be obtained if k_{max} is known. Therefore, a main challenge in computing the $MCCO$s is computing k_{max}.

Our Approaches and Contributions. We propose a polynomial algorithm to compute all k-CCOs in dual graphs. It performs by recursively removing the vertex which violates the k-CCO definition. For the problem of computing $MCCO$s, we first follow the similar idea in computing all k-CCOs, and give a bottom-up solution. More specifically, we compute the $MCCO$s by iteratively removing all unsatisfied vertices. We also propose a top-down algorithm, which selects k_{max} following a top-down strategy and returns the k-CCOs if exist. To further improve the algorithmic efficiency, we propose a binary search algorithm for computing all $MCCO$s. The experimental results show the excellent performance of our optimized algorithm. More details can be found in Sect. 5. We summarize the main contributions in this paper as follows.

- *A k-connected core model in dual graphs.* We design a k-connected core model, which inherits the properties of classic k-$core$ model in dual graphs. To the best of our knowledge, this is the first work that studies the k-$core$ concept in dual graphs.
- *A polynomial time algorithm for computing all k-connected cores.* Given a dual graph G and an integer k, we propose a polynomial peeling-style algorithm, named KCCO, to compute all k-CCOs in G. We prove the time complexity of KCCO is $O(h \times m)$. Here, m is the number of edges in the conceptual graph, and h is a value theoretically roughly bounded by but practically much less than the number of vertices in G.

- *Three algorithms for computing the maximum-connected cores.* We give a bottom-up and a top-down algorithms for the *MCCO* computation. An optimized binary search algorithm is finally proposed to achieve significant speedup.
- *Extensive performance studies.* We conduct extensive performance studies on four synthetic graphs and six real large graphs. We also present a case study. The results demonstrate the effectiveness and efficiency of our proposed model and algorithms.

Organization. The rest of this paper is organized as follows. Section 2 introduces preliminary concept and defines the problem. Section 3 proposes an algorithm for computing all k-*CCOs*. Section 4 studies the problem of computing all *MCCOs*. Section 5 evaluates our proposed algorithms in extensive experiments. Section 6 introduces the related works, and Sect. 7 concludes the paper.

2   Preliminaries

Cores in Simple Graphs. Before studying the dual graphs, we briefly introduce several definitions and recall the problem of k-*core* computation in simple graphs. Let $G(V, E)$ be an undirected graph, where V is the set of vertices and E is the set of edges. Given a vertex u in G, we use $N_G(u)$ to denote the neighbor set of u in G, i.e., $N_G(u) = \{v \in V | (u, v) \in E\}$. The degree of a vertex u in G is denoted by $deg_G(u)$, i.e., $deg_G(u) = |N_G(u)|$. Given a vertex set S, the induced subgraph of S in G is denoted by $G[S]$, i.e., $G[s] = (S, \{(u, v) \in E | u \in S \wedge v \in S\})$. The formal definition of k-*core* is given below.

Definition 1 (K-CORE). *A k-core of graph $G(V, E)$ is a maximal connected subgraph in which each vertex has degree at least k [29].*

Definition 2 (CORE NUMBER). *The core number of a vertex u in G, denoted by $core(u)$, is the maximal number of k such that u is contained in a k-core.*

Definition 3 (DEGENERACY). *The degeneracy of a graph G, denoted by $\mathcal{D}(G)$, is the maximal number of k such that a k-core exists, i.e., $\mathcal{D}(G) = \max_{u \in V} core(u)$.*

We denote the k-*core* containing a given vertex u by $G_k(u)$, and have the following lemma.

Lemma 1. $\forall 1 \leq k < \mathcal{D}(G), G_{k+1}(u) \subseteq G_k(u)$.

Let $V_k(u)$ be the set of vertices in which each vertex v can be reached from u via a path that every vertex w in the path satisfies $core(w) \geq k$. Following lemma holds:

Lemma 2. $G_k(u) = G[V_k(u)]$.

Algorithm 1. Core-Decomposition[3]

Input: A graph $G(V, E)$
Output: The core numbers of all vertices in G

1: $G'(V', E') \leftarrow G(V, E)$;
2: **while** $V' \neq \emptyset$ **do**
3:    $k \leftarrow \min_{u \in V'} deg_{G'}(u)$;
4:    **while** $\exists u \in V', deg_{G'}(u) < k + 1$ **do**
5:       $core(u) \leftarrow k$;
6:       remove u and its incident edges from G';
7: **return** $core(u)$ for all $u \in V$;

Given the core numbers of all vertices, all k-*cores* can be easily found based on Lemma 2. The algorithm for computing all core numbers [3] is given in Algorithm 1. It performs by iteratively removing the vertex with minimum degree and its incident edges. The time complexity of Algorithm 1 is $O(m)$.

Cores in Dual Graphs. In this paper, we focus on an undirected dual graph $G(V, E_a, E_b)$, where E_a and E_b represent the edge sets in physical graph G_a and conceptual graph G_b respectively. The example of the dual graph can be found in Fig. 1. Based on the aforementioned classic k-*core* concept, we define the k-*Connected COre* (k-*CCO*) in dual graphs.

Definition 4. *Given a dual graph* $G(V, E_a, E_b)$, *a dual subgraph* $G[C]$ *is a* k-*connected core* (k-*CCO*) *if: (1)* $G_a[C]$ *is connected; (2)* $G_b[C]$ *is connected; (3)* $\forall u \in C, deg_{G_b[C]}(u) \geq k$; *and (4)* $G[C]$ *is maximal.*

Note that in the existing work [33] for computing the densest connected subgraph in dual graphs, only the connectivity in physical graph is required. This condition is insufficient to support the cohesiveness of result subgraphs, since the subgraph may be disconnected in the conceptual graph. To conquer this drawback, our k-*CCO* definition guarantees the connectivity for both physical and conceptual graphs. Based on Definition 4, we further define the *Maximum-Connected COre* (*MCCO*) below.

Definition 5. *Given a dual graph* $G(V, E_a, E_b)$, *a dual subgraph* $G[C]$ *is a maximum-connected core* (*MCCO*) *if* $G[C]$ *is a* k-*CCO, and* $(k + 1)$-*CCO does not exist.*

Definition 6 (MAXIMUM CCO NUMBER). *Given a dual graph* G, *the maximum CCO number of* G, *denoted by* $k_{max}(G)$, *is the maximum value of* k *such that a* k-*CCO exists.*

Based on Definitions 4 and 5, we formally define the two problems studied in this paper as follows.

Problem 1. Given a dual graph $G(V, E_a, E_b)$ and an integer k, find all k-*CCOs* in G.

Problem 2. Given a dual graph $G(V, E_a, E_b)$, find all *MCCOs* in G.

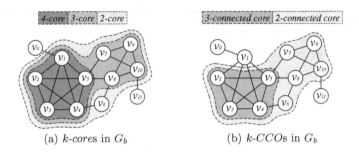

(a) k-cores in G_b (b) k-CCOs in G_b

Fig. 2. An example of k-CCOs

Example 1. We give an example of k-CCO and MCCO. The k-CCOs of the dual graph G in Fig. 1 are presented in Fig. 2(b). The k-cores of G_b are also reported as comparisons in Fig. 2(a). There does not exist a 4-CCO in G, and the MCCO of G is the 3-CCO containing v_2, v_3, v_4, and v_5. The degeneracy of G_b is 4, and the 4-core is the induced subgraph of v_1, v_2, v_3, v_4, and v_5. It is not a 4-CCO in G, since v_1 does not connect to other vertices in G_a.

3 Computing K-Connected Cores

Given an integer k, we study the problem of computing all k-CCOs in this section. We first give several lemmas about k-CCO based on Definition 4. Given a dual graph $G(V, E_a, E_b)$ and a k-CCO $G[C] \subset G$, following lemmas hold.

Lemma 3. *There exists a k-core $G_b[S]$ in G_b, such that $C \subset S$.*

Lemma 4. *There exists a connected component $G_a[H]$ in G_a, such that $C \subset H$.*

Based on Lemmas 3 and 4, we propose a peeling algorithm for computing all k-CCOs. The pseudocode is given in Algorithm 2.

Algorithm 2. Computing **K-C**onnected **CO**res (KCCO)

Input: A graph $G(V, E_a, E_b)$, and a parameter k
Output: The set $\mathbb{C}$ containing all k-CCOs in G

1: $\mathbb{C} \leftarrow \emptyset$;
2: **for each** connected component $G_a[C]$ in G_a **do**
3:    **if** $\forall u \in C, deg_{G_b[C]}(u) \geq k$ **and** $G_b[C]$ is connected **then**
4:       $\mathbb{C} \leftarrow \mathbb{C} \cup G_b[C]$;
5:    **else**
6:       **while** $\exists u \in C, deg_{G_b[C]}(u) < k$ **do**
7:          $C \leftarrow C - \{u\}$;
8:       **for each** connected component $G_b[H]$ in $G_b[C]$ **do**
9:          $\mathbb{C} \leftarrow \mathbb{C} \cup \mathsf{KCCO}(G_b[H], k)$;
10: **return** $\mathbb{C}$;

The algorithm performs by recursively removing the vertex that does not satisfy the degree constraint and the connectivity constraint in Definition 4. We compute all connected components of G_a in line 2. Lemma 4 guarantees that we will not lose any k-CCO in this step. We add $G_b[C]$ to the result set if $G_b[C]$ is connected and satisfies the degree constraint (line 3–4). Otherwise, the algorithm from line 6 to line 8 computes a k-core $G_b[H]$ of $G_b[C]$. All vertices that violate the degree constraint in $G_b[C]$ are iteratively removed from C; and for each connected component $G_b[H]$, we recursively invoke KCCO to find k-CCOs in $G_b[H]$ (line 8–9). The correctness of this step is guaranteed by Lemma 3.

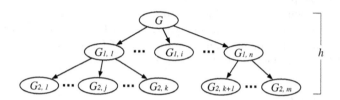

Fig. 3. DFS tree

The process of Algorithm 2 can be represented by a DFS tree as depicted in Fig. 3. Each node in the tree demonstrates an input dual graph G for the invocation of KCCO. Let h be the height of the tree. The time complexity of *Algorithm 2* is given as follows.

Theorem 1. *Given a graph $G(V, E_a, E_b)$ and an integer k, the time complexity of Algorithm 2 is $O(h|E_b|)$.*

Proof. Obtaining all connected components in line 2 of Algorithm 2 costs $O(|E_a|)$ time. Checking the degree constraint and connectivity of G_b in line 3 costs $O(|E_b|)$ time. From line 6 to line 8, Algorithm 2 also costs $O(|E_b|)$ time to remove all vertices whose degree is less than k and compute the connected components $G_b[H]$. Normally, we have $|E_a| < |E_b|$, and the time cost for each node in the DFS tree is $O(|E_b|)$, where E_b is the edge set of input conceptual graph.

Let $\mathbb{G}_l$ be the set of all input graphs on height l of DFS tree, where the height of a node is the distance from root to that node. We can find that there does not exist any vertex or edge overlap between different connected components in line 8. Given the tree height h, we have $\forall 0 \le l \le h, \sum_{G'(V', E'_a, E'_b) \in \mathbb{G}_l} |E'_b| \le |E_b|$, where $|E_b|$ is the number of edges in the original conceptual graph. Therefore, the total time complexity of Algorithm 2 is $O(h|E_b|)$.

Discussion. The time complexity of Algorithm 2 is the product of two parts:

- The first part is the tree height h. Note that in the DFS tree, the size of input graph in each node must be less than that in its parent node. Therefore, h is roughly bounded by $|V|$. However, h is much smaller than $|V|$ in practice. In our experiments, h is not larger than 5 on all datasets.

- The second part is the graph size $|E_b|$. Given that vertices violating the degree constraint are removed in line 7 of Algorithm 2, the graph size becomes small when the tree height increases. The practical performance of Algorithm 2 can be found in Sect. 5.

4   Computing Maximal-Connected Cores

We study the problem of computing all $MCCO$s in this section. A straightforward bottom-up solution BU-MCCO is first given in Subsect. 4.1. Then, we propose a top-down solution TD-MCCO in Subsect. 4.2. To further improve the algorithmic efficiency, we give a binary search algorithm, namely BIN-MCCO, in Subsect. 4.3. Our experiments demonstrate BIN-MCCO outperforms TD-MCCO and BU-MCCO. The details can be found in Sect. 5.

4.1   A Bottom-Up Approach

We give a straightforward algorithm for computing all $MCCO$s in this section. Similar to the concept of k-core, a nest property of k-CCO can be also easily obtained according to Definition 4.

Lemma 5. *Given an integer $1 < k \le k_{max}$ and a k-CCO C, there exists a $(k-1)$-CCO C' such that $C' \supseteq C$.*

Inspired by the lemma above, we propose a bottom-up algorithm, namely BU-MCCO. More specifically, we iteratively compute the k-CCOs based on computed $(k-1)$-CCOs when increasing k. The detailed pseudocode is given in Algorithm 3. We first compute all 1-CCOs in G (line 2). Then we iteratively increase k (line 6), and compute k-CCOs in $\mathbb{C}$, where $\mathbb{C}$ is the set of the previous computed $(k-1)$-CCOs (line 9). The algorithm terminates once no any k-CCO is found. The time complexity of Algorithm 3 is given as follows.

Algorithm 3. BU-MCCO

Input: A graph $G(V, E_a, E_b)$
Output: The set containing all $MCCO$s in G

1: $\mathbb{C} \leftarrow \emptyset$;
2: $\mathbb{T} \leftarrow \mathsf{KCCO}(G, 1)$;
3: $k \leftarrow 1$;
4: **while** $\mathbb{T} \ne \emptyset$ **do**
5:     $\mathbb{C} \leftarrow \mathbb{T}$;
6:     $k \leftarrow k + 1$;
7:     $\mathbb{T} \leftarrow \emptyset$;
8:     **for each** $G[C] \in \mathbb{C}$ **do**
9:         $\mathbb{T} \leftarrow \mathbb{T} \cup \mathsf{KCCO}(G[C], k)$;
10: **return** $\mathbb{C}$;

Theorem 2. *Given an input dual graph $G(V, E_a, E_b)$, the time complexity of Algorithm 3 is $O(\mathcal{D}(G_b) \times h|E_b|)$.*

Proof. KCCO costs $O(h|E_b|)$ time in line 2. The number of iterations in line 4 is at most $\mathcal{D}(G)$. Since there does not exist any overlap between any two components in $\mathbb{C}$ (line 8), the time complexity from line 4 to line 9 is $O(\mathcal{D}(G_b) \times h|E_b|)$. The total time complexity of Algorithm 4 is obtained.

4.2 A Top-Down Approach

A bottom-up solution is given in the previous section. Given that k_{max} may be very large, the time-consuming in BU-MCCO may be very large. To handle this problem, we propose a top-down algorithm, namely TD-MCCO, in this section.

Given a dual graph G, computing all $MCCO$s is equivalent to computing all $k_{max}(G)$-CCOs. We adopt a top-down strategy to select the k_{max}. An upper bound for $k_{max}(G)$ can be easily obtained according to Lemma 3:

Lemma 6. *Given a dual graph $G(V, E_a, E_b)$, $k_{max}(G) \leq \mathcal{D}(G_b)$.*

Based on Lemma 6, we propose the algorithm TD-MCCO in Algorithm 4. Core-Decomposition is invoked in line 2, and we initialize k_{max} by the graph degeneracy in line 3. For each k_{max}, the vertex set of k-*cores* of G_b is obtained in line 5 based on Lemma 3. KCCO is invoked to compute all k_{max}-CCOs in $G[C]$ (line 6). We terminate the algorithm if any k_{max}-CCO is found.

Theorem 3. *Given an input dual graph $G(V, E_a, E_b)$, the time complexity of Algorithm 4 is $O(\mathcal{D}(G_b) \times h|E_b|)$.*

Proof. The proof is similar to that for Theorem 2 and is omitted here.

Algorithm 4. TD-MCCO

Input: A graph $G(V, E_a, E_b)$
Output: The set containing all $MCCO$s in G

1: $\mathbb{C} \leftarrow \emptyset$;
2: Core-Decomposition(G_b);
3: $k_{max} \leftarrow \max_{u \in V}(core(u))$;
4: **while** $\mathbb{C} = \emptyset$ **and** $k_{max} > 0$ **do**
5:     $C \leftarrow \{u \in V | core(u) \geq k_{max}\}$
6:     $\mathbb{C} \leftarrow$ KCCO($G[C], k_{max}$);
7:     $k_{max} \leftarrow k_{max} - 1$;
8: **return** $\mathbb{C}$;

4.3 Binary Searching $MCCOs$

We propose BU-MCCO and TD-MCCO in Subsects. 4.1 and 4.2 respectively. Even though they can successfully compute all $MCCOs$ in the given dual graph G, both of them endure $\mathcal{D}(G_b)$ times of KCCO invocation in the time complexity. To conquer this drawback, we propose a binary search algorithm, namely BIN-MCCO, in this section. Similar to the conventional binary search, we maintain a lower bound $\underline{k}$ and an upper bound $\overline{k}$ of k, and attempt to find all k-CCOs in each iteration, where $k = \lfloor (\overline{k} + \underline{k})/2 \rfloor$. If no any k-CCO is found, we know there does not exist any k'-CCO for $k < k' < \overline{k}$ according to Lemma 5. In this case, we assign the upper bound by k, and continue the search. Otherwise, we assign the lower bound by k. The procedure terminates once we find all k-CCOs and $(k+1)$-CCO does not exist. The initial lower bound for k is assigned by 1, and the upper bound is assigned by $\mathcal{D}(G_b)$ based on Lemma 6. The detailed pseudocode of BIN-MCCO is given in Algorithm 5.

The core numbers for each vertex in G_b are first computed in line 1. In line 2, we set the upper bound d by $\mathcal{D}(G)+1$. This guarantees no any d-CCO exists. The subroutine BIN-Search is invoked recursively to find the $MCCOs$ (line 4). The first parameter of BIN-Search is the set of all $\underline{k}$-CCOs in G. Recall that a k-CCO must be contained in a $(k-1)$-CCO according to Lemma 5. An optimization here is that we maintain all $\underline{k}$-CCOs in BIN-Search. Instead of computing k-CCOs in the original graph, we compute k-CCOs in a smaller graph induced by $\underline{k}$-CCOs (line 9), and never lose any result. For each induced subgraph of $\underline{k}$-CCO (line 9), we prune all vertices whose core number is less than k in line 10 based on Lemma 3. Then KCCO is invoked to compute all k-CCOs. If there does not

---

Algorithm 5. BIN-MCCO

Input: A graph $G(V, E_a, E_b)$
Output: The set containing all $MCCOs$ in G

1: Core-Decomposition(G_b);
2: $d \leftarrow \max_{u \in V}(core(u)) + 1$;
3: $\mathbb{C} \leftarrow$ KCCO($G, 1$);
4: **return** BIN-Search($\mathbb{C}, 1, d$);

Procedure BIN-Search($\mathbb{C}, \underline{k}, \overline{k}$)

5: **if** $\overline{k} - \underline{k} \leq 1$ **then**
6:     **return** $\mathbb{C}$;
7: $k \leftarrow \underline{k} + (\overline{k} - \underline{k})/2$;
8: $\mathbb{T} \leftarrow \emptyset$;
9: **for each** $G[C] \in \mathbb{C}$ **do**
10:     $S \leftarrow \{u \in C | core(u) \geq k\}$;
11:     $\mathbb{T} \leftarrow \mathbb{T} \cup$ KCCO($G[S], k$);
12: **if** $\mathbb{T} = \emptyset$ **then**
13:     **return** BIN-Search($\mathbb{C}, \underline{k}, k$);
14: **else**
15:     **return** BIN-Search($\mathbb{T}, k, \overline{k}$);

exist any k-CCOs (line 12), we decrease $\overline{k}$ to k and continue searching (line 13). Otherwise, we increase $\underline{k}$ to k, and change the first parameter to the set of all k-CCOs (line 15).

Theorem 4. *Given an input dual graph $G(V, E_a, E_b)$, the time complexity of Algorithm 5 is $O(\log \mathcal{D}(G_b) \times h|E_b|)$.*

Proof. Given that there does not exist any between different $G[C]$s (line 9), the time complexity of line 9 to line 11 is $O(h|E_b|)$. Given the upper bound $\mathcal{D}(G_b)$, the total invocation of BIN-Search is bounded by $O(\log \mathcal{D}(G_b))$, and the total time complexity of Algorithm 5 is $O(\log \mathcal{D}(G_b) \times h|E_b|)$.

5   Experiments

We conduct extensive experiments to evaluate the performance of our proposed solutions. We obtain the code for DCS from the author as a comparison. All other algorithms are implemented in C++. All the experiments are conducted on a Windows Server operating system running on a machine with an Intel Xeon 2.0 GHz CPU, 32 GB 1333 MHz DDR3-RAM. The time cost for algorithms is measured as the amount of wall-clock time elapsed during the program execution.

Table 1. Statistics of real-world datasets

Datasets	$|V|$	$|E_a|$	$|E_b|$	$|E_b|/|E_a|$	$\overline{d}_b$
DBLP	40,490	203,670	400,448	1.97	9.89
Hep-TH	29,381	352,807	886,791	2.51	30.18
Epinions	49,290	487,002	729,403	1.50	14.80
CiaoDVD	14,811	40,133	124,533	3.10	8.41
Brightkite	58,228	214,078	602,836	2.82	10.35
Gowalla	196,591	950,327	1,458,456	1.53	7.42

Real-World Datasets. We evaluate the algorithms on six real graphs. The detailed statistics of these graphs are summarized in Table 1. $\overline{d}_b$ is the average degree in the conceptual graph.

We adopt a similar idea in [33] to construct the dual graphs. DBLP [30] is constructed based on the computer science bibliography *DBLP*. We select several conferences and journals in database research area. The vertices represent the authors of the published papers. An edge exists if two authors have a common paper in the physical graph, and edges in the conceptual graph are constructed by measuring the similarity between the abstracts of papers published by any two authors. Hep-TH [18] is a theory collaboration network in high energy physics area. The construction for Hep-TH is same as that for DBLP.

Epinions [23] and CiaoDVD[1] are recommendation networks. Each vertex represents a user. A physical edge exists if a user expresses a positive trust statement on the other user. To construct the conceptual graph, we calculate the correlation coefficient [22] of the common ratings between users, and connected two users by a conceptual edge if their coefficient value is larger than a threshold.

Brightkite [8] and Gowalla [8] are geosocial networks. Each vertex represents a user. The physical edges represent the friend relationship between users, and the conceptual edges are constructed based on the Euclidean distance between the locations of users.

Table 2. Statistics of synthetic datasets

	GT1	GT2	GT3	GT4
$|V|$	1×2^{20}	2×2^{20}	4×2^{20}	6×2^{20}
$|E_a|, |E_b|$	1×10^7	2×10^7	4×10^7	8×10^7

Synthetic Datasets. We adopt the same method in [33] to generate several synthetic graphs. In specific, we use the graph generator $GTgraph$[2] to construct both physical graphs and conceptual graphs. The statistics of generated graphs are summarized in Table 2.

5.1   Performance Studies on Real-World Datasets

Eval-I: Evaluating the Algorithm for Computing All k-CCOs. The time-consuming for algorithm KCCO on six real-world graphs is reported in Fig. 4. For each dataset, we select $20\% \times k_{max}, 40\% \times k_{max}, 60\% \times k_{max}, 80\% \times k_{max}$ and k_{max} as the input integer k, and present a line chart. We can find that the time cost of KCCO decreases when increasing k. This is mainly because a large number of vertices are removed when the degree constraint k is large, and the result subgraph is small.

Eval-II: Evaluating the Algorithms for Computing the $MCCO$s. The time-consuming for algorithms BU-MCCO, TD-MCCO and BIN-MCCO on six real-world graphs is reported in Fig. 5(a). Given that there exists no previous work on this problem, we give the time cost for computing DCS [33], namely DCS, as a comparison in the figure. Note that the time cost for DCS is not given in some datasets, since the procedure cannot terminate in 4 h.

As we can see from the figure, BIN-MCCO is the fastest algorithm. It costs about 13 s in Gowalla and less than 4 s in all other datasets. TD-MCCO is the second fastest algorithm in all datasets, while BU-MCCO is slightly slower than TD-MCCO. For example, in Brightkite, TD-MCCO and BU-MCCO cost about 77 s and 113 s respectively. BIN-MCCO costs about 2 s, which is almost two orders

---

[1] https://www.librec.net/datasets.html.
[2] http://www.cse.psu.edu/~kxm85/software/GTgraph/.

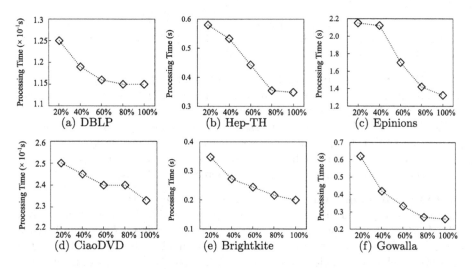

Fig. 4. Computing k-CCOs in real-world graphs

of magnitude faster than TD-MCCO and BU-MCCO. As a comparison, DCS costs over 3000 s and 750 s in DBLP and CiaoDVD respectively, while BIN-MCCO costs only about 1.3 s and 0.7 s respectively in those two datasets. The result demonstrates the high efficiency of BIN-MCCO.

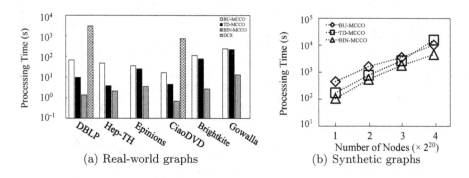

Fig. 5. Computing $MCCO$s in real-world graphs and synthetic graphs

5.2 Scalability Testing

We test the scalability of our proposed algorithms in this section. For each real-world dual graph, we randomly sample physical edges, conceptual edges and vertices respectively from 20% to 100%. When sampling physical edges, we get the incident vertices of the edges as the vertex set, and preserve the induced subgraph of this vertex set in the conceptual graph. The sampling strategy for conceptual edges is same as that for physical edges. When sampling vertices, we get the induced dual subgraph of the sampled vertices. Due to the space limitation, we only report the charts for DBLP, Epinions and Brightkite, while the results in other datasets show the similar trends.

Eval-III: Sampling Physical Edges. The running time of our proposed algorithms is reported in Fig. 6(a), (b) and(c) when sampling physical edges. We can see that BIN-MCCO is the fastest, and the time cost of all algorithms performs a slightly downward trend in all datasets. This is mainly due to the speedup of performing KCCO. In specific, when the physical edge size is large, a *k-core* in the conceptual graph is more likely to be connected in the physical graph, which means the depth of the invocation tree depicted in Fig. 3 is small.

Eval-IV: Sampling Conceptual Edges. The running time of our proposed algorithms is reported in Fig. 6(d), (e) and (f) when sampling conceptual edges. BIN-MCCO is the fastest algorithm, and the lines for BIN-MCCO in all datasets are stable. TD-MCCO is the second fastest algorithm. The time cost of TD-MCCO presents a relatively obvious increase from 20% to 100% in all datasets, and the gap between TD-MCCO and BU-MCCO decreases when edge size increases.

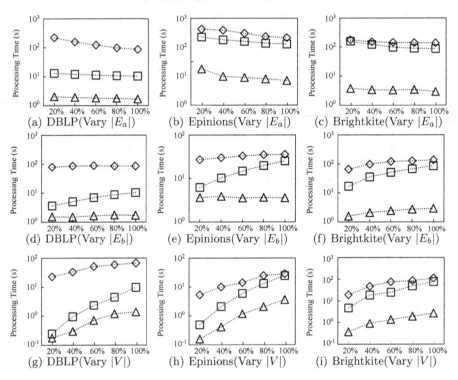

Fig. 6. Scalability testing

This is mainly because the graph degeneracy $\mathcal{D}$ of G_b increases when increasing $|E_b|$, and the gap between $\mathcal{D}$ and k_{max} increases. Therefore, more iterations in TD-MCCO are performed, and the efficiency of TD-MCCO declines.

Eval-V: Sampling Vertices. The running time of our proposed algorithms is reported in Fig. 6(g), (h) and (i) when sampling vertices. We can see that BIN-MCCO is still the fastest in all scenarios. The chart for BIN-MCCO presents a slight increase when increasing the vertex size. TD-MCCO is faster than BU-MCCO, and in some datasets, the gap between them decreases when increasing vertex size. For example, in Epinions, TD-MCCO costs about 0.5 s on 20% and reaches about 25 s on 100%. By contrast, BU-MCCO costs about 5.3 s on 20% and reaches about 29 s on 100%. The main reason is similar to that in sampling conceptual edges. From the three scalability experiments, we can see that high efficiency and stability of BIN-MCCO. The top-down solution TD-MCCO is the second fastest, while the efficiency of TD-MCCO highly depends on the graph structure, and the gap between k_{max} and $\mathcal{D}$. The bottom-up solution BU-MCCO is the slowest but performs more stable than TD-MCCO.

5.3 Performance Studies on Synthetic Datasets

Eval-VII: Evaluating the Algorithms on Synthetic Graphs. The running time for computing $MCCO$s in synthetic graphs is given in Fig. 5(b). BIN-MCCO is the fastest algorithm on all graph size. BU-MCCO has a slower increasing rate than TD-MCCO, and is even faster than TD-MCCO finally. This is mainly because the gap between k_{max} and $\mathcal{D}$ is large given a big graph size.

5.4 Effectiveness Evaluation

Eval-VII: Case Study in Gowalla. We conduct a case study to present the effectiveness of our solution. Due to the space limitation, we select a subgraph of Gowalla, and compute the $MCCO$ in this subgraph. The result is reported in Fig. 7(b). As a comparison, we also give the result of DCS in the same subgraph

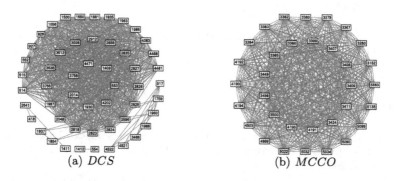

(a) DCS (b) $MCCO$

Fig. 7. The DCS and $MCCO$ in a conceptual subgraph of Gowalla

in Fig. 7(a). We can see that there exist several vertices whose degree less than three in the DCS. This demonstrates the approximate solution for DCS may generate a result with a sparse subgraph. By contrast, the degree of each vertex is not less than k_{max} in Fig. 7(b), and the result of $MCCO$ is cohesive.

6   Related Works

Computing k-core. k-core is first introduced in [29]. [3] proposes a linear time solution for core decomposition. k-core in directed graph and weighted graphs is studied in [14,15] respectively. [7] proposes a partition-based external memory algorithm for computing k-cores. [17,32] apply a semi-external model and further speed up the core decomposition algorithm for big graphs. [25] gives a distributed algorithm for core decomposition. Given that real-world graphs are highly dynamic, core number maintenance is studied in [20,28]. Locally estimating core number is studied in [26]. Several work studies k-core in different graph models, such as uncertain graphs [4], random graphs [16,21,24,27], and attribute graphs [12]. [11,19] use k-core to detect communities in the graph.

Cohesive Subgraph Detection in Dual Networks. [33] studies the cohesive subgraph problem in dual networks. An approximate algorithm is proposed for computing the densest connected subgraph in the input dual graph.

7   Conclusion

Computing k-cores is a fundamental and important graph problem. In this paper, we define the k-connected core in dual graphs. A subgraph g is a k-connected core if the minimum degree of g is at least k in the conceptual graph, and g is connected in both conceptual graph and physical graph. We propose a polynomial time algorithm for computing all k-connected cores in the dual graph. We also propose three algorithms for computing all maximum-connected cores, which are the maximum k-connected cores such that a $(k+1)$-connected core does not exist. We do extensive experiments to demonstrate the effectiveness and efficiency of our propose algorithms.

Acknowledgments. The work is supported by the National Key R&D Program (No. 2017YFB1400102, No. 2016YFB1000602), NSFC (No. 61572295), and SDNSF (No. ZR2017ZB0420).

References

1. Alvarez-Hamelin, J.I., Dall'Asta, L., Barrat, A., Vespignani, A.: Large scale networks fingerprinting and visualization using the k-core decomposition. In: NIPS, pp. 41–50 (2006)
2. Asahiro, Y., Iwama, K., Tamaki, H., Tokuyama, T.: Greedily finding a dense subgraph. J. Algorithms **34**(2), 203–221 (2000)

3. Batagelj, V., Zaversnik, M.: An o(m) algorithm for cores decomposition of networks. arXiv preprint arXiv:cs/0310049 (2003)
4. Bonchi, F., Gullo, F., Kaltenbrunner, A., Volkovich, Y.: Core decomposition of uncertain graphs. In: KDD, pp. 1316–1325 (2014)
5. Chang, L., Yu, J.X., Qin, L., Lin, X., Liu, C., Liang, W.: Efficiently computing k-edge connected components via graph decomposition. In: SIGMOD, pp. 205–216 (2013)
6. Charikar, M.: Greedy approximation algorithms for finding dense components in a graph. In: Jansen, K., Khuller, S. (eds.) APPROX 2000. LNCS, vol. 1913, pp. 84–95. Springer, Heidelberg (2000). https://doi.org/10.1007/3-540-44436-X_10
7. Cheng, J., Ke, Y., Chu, S., Özsu, M.T.: Efficient core decomposition in massive networks. In: ICDE, pp. 51–62 (2011)
8. Cho, E., Myers, S.A., Leskovec, J.: Friendship and mobility: user movement in location-based social networks. In: KDD, pp. 1082–1090 (2011)
9. Conte, A., Firmani, D., Mordente, C., Patrignani, M., Torlone, R.: Fast enumeration of large k-plexes. In: KDD, pp. 115–124 (2017)
10. da Fontoura Costa, L., Oliveira Jr., O.N., Travieso, G., Rodrigues, F.A., Boas, P.R.V., Antiqueira, L., Viana, M.P., Rocha, L.E.C.: Analyzing and modeling real-world phenomena with complex networks: a survey of applications. Adv. Phys. 60(3), 329–412 (2011)
11. Cui, W., Xiao, Y., Wang, H., Wang, W.: Local search of communities in large graphs. In: SIGMOD, pp. 991–1002 (2014)
12. Fang, Y., Cheng, R., Luo, S., Hu, J.: Effective community search for large attributed graphs. PVLDB 9(12), 1233–1244 (2016)
13. Gallo, G., Grigoriadis, M.D., Tarjan, R.E.: A fast parametric maximum flow algorithm and applications. SIAM J. Comput. 18(1), 30–55 (1989)
14. Giatsidis, C., Thilikos, D.M., Vazirgiannis, M.: D-cores: measuring collaboration of directed graphs based on degeneracy. In: ICDM, pp. 201–210 (2011)
15. Giatsidis, C., Thilikos, D.M., Vazirgiannis, M.: Evaluating cooperation in communities with the k-core structure. In: ASONAM, pp. 87–93 (2011)
16. Janson, S., Luczak, M.J.: A simple solution to the k-core problem. Random Struct. Algorithms 30(1–2), 50–62 (2007)
17. Khaouid, W., Barsky, M., Srinivasan, V., Thomo, A.: K-core decomposition of large networks on a single PC. PVLDB 9(1), 13–23 (2015)
18. Leskovec, J., Kleinberg, J., Faloutsos, C.: Graph evolution: densification and shrinking diameters. TKDD 1(1), 2 (2007)
19. Li, R.-H., Qin, L., Yu, J.X., Mao, R.: Influential community search in large networks. PVLDB 8(5), 509–520 (2015)
20. Li, R.-H., Yu, J.X., Mao, R.: Efficient core maintenance in large dynamic graphs. TKDE 26(10), 2453–2465 (2014)
21. Łuczak, T.: Size and connectivity of the k-core of a random graph. Discrete Math. 91(1), 61–68 (1991)
22. Ma, H., Zhou, D., Liu, C., Lyu, M.R., King, I.: Recommender systems with social regularization. In: Proceedings of the fourth ACM International Conference on Web Search and Data Mining, pp. 287–296 (2011)
23. Massa, P., Avesani, P.: Trust-aware recommender systems. In: RecSys, pp. 17–24 (2007)
24. Molloy, M.: Cores in random hypergraphs and Boolean formulas. Random Struct. Algorithms 27(1), 124–135 (2005)
25. Montresor, A., De Pellegrini, F., Miorandi, D.: Distributed k-core decomposition. TPDS 24(2), 288–300 (2013)

26. OBrien, M.P., Sullivan, B.D.: Locally estimating core numbers. In: ICDM, pp. 460–469 (2014)
27. Pittel, B., Spencer, J., Wormald, N.: Sudden emergence of a giantk-core in a random graph. J. Comb. Theory Ser. B **67**(1), 111–151 (1996)
28. Saríyüce, A.E., Gedik, B., Jacques-Silva, G., Wu, K.-L., Çatalyürek, Ü.V.: Streaming algorithms for k-core decomposition. PVLDB **6**(6), 433–444 (2013)
29. Seidman, S.B.: Network structure and minimum degree. Social Netw. **5**(3), 269–287 (1983)
30. Tang, J., Zhang, J., Yao, L., Li, J., Zhang, L., Su, Z.: Arnetminer: extraction and mining of academic social networks. In: KDD, pp. 990–998 (2008)
31. Wang, J., Cheng, J.: Truss decomposition in massive networks. PVLDB **5**(9), 812–823 (2012)
32. Wen, D., Qin, L., Zhang, Y., Lin, X., Yu, J.X.: I/o efficient core graph decomposition at web scale. In: ICDE, pp. 133–144. IEEE (2016)
33. Wu, Y., Jin, R., Zhu, X., Zhang,X.: Finding dense and connected subgraphs in dual networks. In: ICDE, pp. 915–926 (2015)

Graph Clustering with Local Density-Cut

Junming Shao[1]([✉]), Qinli Yang[1], Zhong Zhang[1], Jinhu Liu[1],
and Stefan Kramer[2]

[1] School of Computer Science and Engineering, Big Data Reserach Center,
University of Electronic Science and Technology of China, Chengdu 611731, China
junmshao@uestc.edu.cn
[2] Institute for Computer Science, University of Mainz, 55128 Mainz, Germany

Abstract. In this paper, we introduce a new graph clustering algorithm, called Dcut. The basic idea is to envision the graph clustering as a local density-cut problem. To identify meaningful communities in a graph, a density-connected tree is first constructed in a local fashion. Building upon the local intuitive density-connected tree, Dcut allows partitioning a graph into multiple densely tight-knit clusters effectively and efficiently. We have demonstrated that our method has several attractive benefits: (a) Dcut provides an intuitive criterion to evaluate the goodness of a graph clustering in a more precise way; (b) Building upon the density-connected tree, Dcut allows identifying high-quality clusters; (c) The density-connected tree also provides a connectivity map of vertices in a graph from a local density perspective. We systematically evaluate our new clustering approach on synthetic and real-world data sets to demonstrate its good performance.

1 Introduction

In recent years, the study of graph clustering has attracted increasing attention, and many algorithms have been developed based on different criteria, e.g., *betweenness* [12], *normalized cut (Ncut)* [23], *minimum-cut tree* [6], *modularity* [13], *synchronization* [18,21], *distance dynamics* [19], to mention a few. Although many established approaches have already achieved some success, finding the intrinsic clusters in complex networks is still a big challenge [5]. To date, most previous studies struggle to find a good graph clustering by minimizing the cuts between communities, e.g. *minimum cut* or *modularity*. However, the similarities of vertices in a community are considered only little.

In this paper, we introduce a new local density-based criterion for measuring the "goodness" of a graph clustering. The basic idea is to consider the graph clustering as a density-cut problem by removing the edges in a proposed local density-connected tree (cf. Sect. 3.2). We expect that the vertices of resulting clusters are densely connected while the vertices between clusters are sparsely linked. The "good cut" is viewed as "good", if the vertices in the same group have the same or similar topological structures, instead of only focusing on minimum cuts (e.g. *normalized cut, ratio cut*) or expected cuts (e.g. *modularity*) between

J. Pei et al. (Eds.): DASFAA 2018, LNCS 10827, pp. 187–202, 2018.
https://doi.org/10.1007/978-3-319-91452-7_13

two partitions. Our proposal is to measure similarities of vertices in and between graph clusters by constructing a local density-connected tree, where any two adjacent vertices with highest similarity (i.e. with strong edge weight and similar topological structure) are densely linked together. Based on the properties of the density-connected tree, a good graph partitioning with *density-cut* criterion is efficiently identified.

In this paper, we present a new graph clustering method *Dcut*. The major benefits of *Dcut* can be summarized as follows:

1. *Simple, yet intuitive and effective approach.* While introducing a new density-based graph clustering may appear as only a small step, it gives the overall approach a new quality. The new density-cut criterion, characterizing by the highest density of any two adjacent vertices in a local fashion, provides a more intuitive and precise way to measure the goodness of a graph clustering.
2. *High-quality clustering.* Unlike existing graph clustering approaches, *Dcut* models the local density by considering both edge weights and topological structure. As it turns out in the results section, Dcut is not only more accurate than the well-known graph clustering algorithms Ncut, Modularity, Metis, and MCL on smaller, labeled data, it also outperforms these algorithms on larger real-world data.
3. *Efficiency.* Dcut generates an intuitive and interpretable density-connected tree, which provides a connectivity map of vertices in a graph from a local density perspective. Due to the properties of the density-connected tree, *Dcut* is time efficient and easily implemented.

2 Related Work

During the past several decades, many approaches have been proposed for clustering, such as [2,8,13,17–20,22,23], etc. Here we only review the closest approaches from the literature. For detailed reviews of graph clustering, please refer to [16].

Cut-Based Graph Clustering. The minimum-cut criterion based graph clustering refers to a class of well-known techniques which seek to partition a graph into disjoint subgraphs such that the number of cuts across the subgraphs is minimized. Wu and Leahy [26] has proposed a clustering method based on such minimum cut criterion, where the cut between two subgraphs is computed as the total weights of the edges that have been removed. k–disjoint subgraphs are obtained by recursively finding the minimum cuts that bisect the existing segments. To avoid unnatural bias for partitioning out small size of subgraphs based on the minimum-cut criterion, *ratio cut* [7] has been introduced, and it uses the second smallest eigenvalue of the similarity matrix to find the suitable cut. In the same spirit, Shi and Malik [23] has proposed the *normalized cut*, to compute the cut cost as a fraction of the total edge connections to all the nodes in a graph. To optimize this criterion, a generalized eigenvalue decomposition was also used to speed up the computational time. In many cases, this class of graph clustering algorithms relying on the eigenvector decomposition of a similarity matrix (e.g. *ratio cut* and *Ncut*) is also called spectral clustering.

Modularity. Recently, modularity has been developed to measure the division of a network into communities. Unlike minimum-cut related approaches which investigate the number of edges or the total number of edge weights between two subgroups, modularity identifies a good cut by measuring the expected edges between clusters. Modularity-based graph clustering methods [13] partition a network into groups to ensure the number of edges between two groups is significantly less than the expected edges.

Multi-level Clustering. Metis is a class of multi-level partitioning techniques proposed by Karypis and Kumar [10,11]. Graph clustering starts with constructing a sequence of successively smaller (coarser) graphs, and a bisection of the coarsest graph is applied. Subsequently, a finer graph in next level is further generated based on the previous bisections. At each level, an iterative refinement algorithm such as Kernighan-Lin (KL) or Fiduccia-Mattheyses (FM) is used to further improve the bisection. A more robust overall multilevel paradigm has been introduced by Karypis and Kumar [11], which presents a powerful graph coarsening scheme. It uses the simplified variants of KL and FM to speed up the refinement without compromising the overall quality.

Markov Clustering. The Markov Cluster algorithm (MCL) [4] is a popular algorithm used in life sciences based on the simulation of (stochastic) flow in graphs. The basic idea is that dense regions in sparse graphs correspond with regions in which the number of random walks of k-length is relatively large. MCL basically identifies high-flowing regions representing the graph clusters by using the inflation parameter to separate weak and strong flow regions.

Markov Clustering. The Markov Cluster algorithm (MCL) [4] is a popular algorithm used in life sciences based on the simulation of (stochastic) flow in graphs. The basic idea is that dense regions in sparse graphs correspond to regions in which the number of random walks of length k is relatively large. MCL basically identifies high-flowing regions representing the graph clusters by using an inflation parameter to separate regions of weak and strong flow.

3   Graph Clustering Based on Density-Cut

3.1   Local Density Measure

Definition 1 (UNDIRECTED WEIGHTED GRAPH). Let $G = (V, E, W)$ be an undirected weighted graph, where V is the set of nodes, E is the set of edges and W is the corresponding set of weights. $e = \{u, v\} \in E$ indicates a connection between the nodes u and v. $w(u, v)$ represents the weight of edge e. $\forall e = \{u, v\} \in E, w(u, v) = 1$, in case of unweighted graph.

Definition 2 (NEIGHBORS OF VERTEXu). Given an undirected graph $G = (V, E, W)$, the neighborhood of a node $u \in V$ is the set $\Gamma(u)$ containing node u and its adjacent nodes.

$$\Gamma(u) = \{v \in V | \{u, v\} \in E\} \cup \{u\} \tag{1}$$

In this study, we use the well-known Jaccard coefficient [9] to quantify their local topological similarity. Generally, the more common neighbors of two adjacent nodes they have, the more similar they are. As Jaccard coefficient normalizes the number of common neighbors by the sum of the two neighborhood's size, it well captures the local connectivity density of any two adjacent nodes in a graph. Formally, the Jaccard coefficient is defined as follow.

Definition 3 (JACCARD COEFFICIENT). Given a graph $G = (V, E, W)$, the Jaccard coefficient of any two adjacent nodes u and v is defined as:

$$\rho(u, v) = \frac{|\Gamma(u) \cap \Gamma(v)|}{|\Gamma(u) \cup \Gamma(v)|} \tag{2}$$

By considering the topological structure and edge weight together, finally we define the local density of any two adjacent nodes as follows.

Definition 4 (LOCAL DENSITY). Given any two adjacent nodes u and v in the graph G, the *local density* of the two nodes u and v is defined as:

$$s(u, v) = \rho(u, v) * w(u, v) \tag{3}$$

where $w(u, v)$ is the edge weight (connection intensity) between two adjacent nodes. For unweighted graph, $w(u, v) = 1$ if u and v are connected, otherwise $w(u, v) = 0$. The local density tries to capture the structural closeness and connection intensity of two adjacent nodes simultaneously. Namely, two nodes are density connected if they share similar structure and have strong connection strength, instead of ignoring the structural difference in traditional approaches.

3.2   Density-Connected Tree

Based on the local density measure, a variant of maximum spanning tree is constructed to link all local density-connected nodes in a global way. We call the tree as density-connected tree (DCT). In this tree, all nodes of an original graph link to their most similar nodes in term of their local density, respectively. As the local density considering both edge weight/number and node structure, the density-connected tree characterizes the spatial density connectivity pattern of vertices in graphs, where the nodes with highest density are linked together (forming as a tight-knit communities), and the density of nodes linking any two different communities will be rather low. In this way, the graph clusters can be easily identified. Before we illustrate how to find a good graph clustering based on the density-connected tree, we give a brief procedure how to construct the DCT using Prim algorithm [14] (See Algorithm 1).

Algorithm 1. $T = DCT(G)$

Input: $G = (V, E, W)$
Output: T

$T = null$;
Set $\forall v \in V$ as unchecked ($v.checked = false$);
Randomly selected one node $u \in V$;
Set $u.checked = true$;
$u.connect = null$, and $u.density = null$;
$T.insert(u)$;

while $T.size < V.size$ **do**
    $maxv = -1$; $p = null$; $q = null$;
    for $i = 1$ **to** $T.size$ **do**
        $u = T.get(i)$;
        for $j = 1$ **to** $\Gamma(u).size$ **do**
            $v = \Gamma(u).get(j)$;
            if $v.checked == false$ **then**
                if $s(u,v) > maxv$ **then**
                    $maxv = s(u,v)$;
                    $p = v$;
                    $q = u$;
                end
            end
        end
    end
    $p.checked = true$;
    $p.connect = q$; $p.density = maxv$;
    $T.insert(p)$;
end

To illustrate DCT generation, Fig. 1 takes the well-known Zachary's karate club network as an example. The Karate graph data consists of 34 vertices and 78 undirected edges. Each node represents a member of the club, and each edge represents a tie between two members of the club. To construct its density-connected tree, one node is first randomly selected (e.g. node "8" in this example, see Fig. 1(b)). Next, all unchecked adjacent vertices of node "8" are viewed as the potential vertices for next insertion, namely, the nodes of "1", "2", "3" and "4". As the node "4" has the maximum similarity with node "8", it is further inserted into the tree, which directly connects it to the node "8" with the edge weight representing the similarity between the two nodes. For the next step, as there are already two vertices (node "8" and node "4") in this tree, all unchecked adjacent vertices are: $\Gamma(8) \bigcup \Gamma(4) \backslash \{4, 8\}$ (i.e. the nodes of "1", "2", "3", "13" and "14"). For the five potential vertices, the node "14" has the highest similarity with the node "4" in the constructed tree, and thereby node "14" is further inserted into the tree. Similarly, it directly connects to node "4" with the corresponding similarity. This procedure is repeated until all vertices have been inserted into the tree.

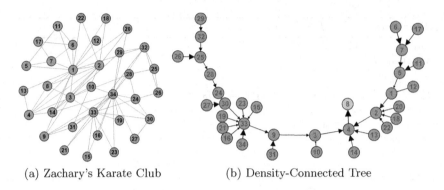

(a) Zachary's Karate Club   (b) Density-Connected Tree

Fig. 1. Illustration of construction of the density-connected tree. The thickness of the arrows in the density-connected tree indicates how densely connected two nodes are.

Theorem 1. *The density-connected tree (DCT) is unique for a given graph, if any two adjacent nodes have a distinct similarity.*

Proof. Supposing there are two separate sets R and S during the generating phase at any step, where R is the set of vertices that have been inserted into the tree, and S is the set of unchecked neighbors of vertices that are already in R. The next node (e.g. v) is selected from S, which has maximum similarity with one node (e.g. u) from R. This means for each node, it is always connected with its most similar adjacent node. Since any two adjacent nodes in the graph have a distinct similarity, the connection of nodes u and v is unique. Thus, the density-connected tree for a given graph is unique.

3.3   The Dcut Algorithm

Formally, we propose a new density-based criterion for measuring the "goodness" of a graph clustering. Instead of investigating the value of total (or normalized) edge weights connecting the two partitions, our measure computes its **local density connection** between the two partitions based on the density-connected tree. We call this measure as *density cut* (Dcut):

$$Dcut(C_1, C_2) = \frac{d(C_1, C_2)}{min(|C_1|, |C_2|)} \tag{4}$$

where C_1, C_2 are the two partitions, $d(C1, C2)$ means the corresponding density connecting the two partitions. The term of $min(|C_1|, |C_2|)$ is used to avoid the bias for partitioning out small sets of vertices.

As DCT connects all vertices without cycle, each edge connects two components of a graph. Thereby, the intuitive bipartitioning of a graph in terms of density can be easily achieved by cutting one edge in the DCT. Instead of seeking to partition an original graph such that vertices in the same partition are densely connected and the vertices across different partitions is lightly connected, Dcut allows recursively finding the optimal cut on the DCT directly.

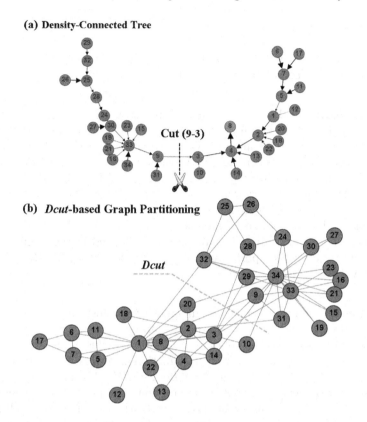

(a) Density-Connected Tree

Cut (9-3)

(b) *Dcut*-**based Graph Partitioning**

Dcut

Fig. 2. Illustration of graph clustering based on the density-connected tree.

Generally, supposing we want to partition a graph into k disjoint clusters, the *Dcut* algorithm runs in the following steps.

1. Given a graph G, compute the local density between adjacent vertices based on the node density measure (Definition 4).
2. Construct its density-connected tree (see Algorithm 1).
3. Partition the DCT by removing the edge with minimum *Dcut* value of the resulting two components.
4. Recursively repartition the segmented DCT until k components of the graph are obtained.

Figure 2 illustrates the graph clustering with density-cut criterion on the karate club network. Based on the constructed density-connected tree, the *Dcut* values for cutting all edges are computed, and the optimal cut with the minimum *Dcut* value is found between node "9" and node "3". By removing this edge from the DCT, it results in two partitions which are indicated in Fig. 2.

4   Experiments

In this section, we evaluate our proposed algorithm *Dcut* on synthetic as well as real-world data to demonstrate its benefits.

Comparison Methods. To examine the performance of *Dcut*, we compare it to two closely related graph clustering algorithms: the normalized cut criterion based graph clustering method *Ncut* [23] and the *modularity*-based graph clustering algorithm by Newman [13] (in the following named *Modularity*). In addition, we compare to two representatives of graph clustering paradigms: the well-known multi-level partitioning algorithm *Metis* by Karypis and Kumar [11] and the Markov Cluster algorithm (*MCL*) by Dongen [4]. In the experiments, *Dcut*, *Ncut* and *Metis* assume the same number of clusters K for all data sets. *MCL* takes the default inflation parameter as indicated in the original paper. All experiments have been performed on a workstation with 2.0 GHz CPU and 8.0 GB RAM.

Evaluation Measures: To compare different graph clustering algorithms with respect to effectiveness, we evaluate the clustering results in two ways. First, if class label information is available for the graph, the clustering performance is directly measured by three widely used evaluation measures: *Normalized Mutual Information (NMI)* [24], *Adjusted Rand Index (ARI)* [15] and *Cluster Purity*. All measures scale between 0 and 1 for a random or a perfect clustering result, respectively. For the graphs without ground truth, we adopt the well-known clustering coefficient proposed by Watts and Strogatz [25] as a cost function. This coefficient is a measure of the local cohesiveness that takes into account the importance of the clustered structure on the basis of the amount of triplets. Clustering results are measured by averaging the clustering coefficient of all subgraphs (clusters) obtained by different approaches.

4.1   Synthetic Data

In this section, we start with two experiments on synthetic data sets featuring various graph characteristics.

Noise Edges: First, we evaluate how well the different graph clustering algorithms can handle the additional edges in graphs, which we call noise edges. Here 20 clusters are generated, and each cluster consisting of 50 nodes are randomly interlinked with 60% intra-cluster edges. In addition to the approximately 15,000 intra-cluster edges, the number of noise edges, which are additional edges randomly added to random nodes, are present in the data varying from 2500 to 30,000. The noise is represented by inter-cluster edges being added to the data, thus, introducing inter-cluster connectivity to hamper cluster separation.

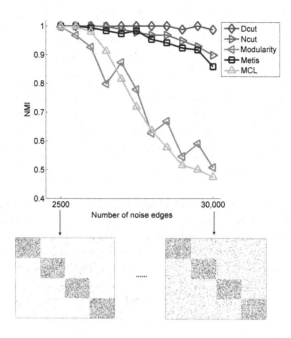

Fig. 3. Varying the number of inter-cluster edges in the data. Due to space limitations, the matrices only display 4 clusters, which is the same as in Figs. 5 and 6.

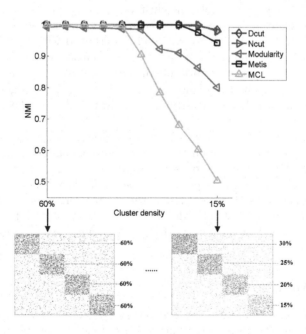

Fig. 4. Varying the densities of clusters in the data.

Table 1. Performance of different graph clustering algorithms on real-world data sets.

	College football			Politics books		
	NMI	ARI	Pur	NMI	ARI	Pur
Dcut	0.924	0.899	0.930	0.572	0.680	0.857
Ncut	0.923	0.897	0.930	0.534	0.645	0.829
Modularity	0.596	0.474	0.574	0.508	0.638	0.838
Metis	0.526	0.236	0.487	0.382	0.425	0.781
MCL	0.923	0.897	0.930	0.455	0.594	0.857

With adding more noise edges into the graph data (Fig. 3), the performance of all five approaches degrades, as measured by the normalized mutual information (NMI). *MCL* is only able to handle data with up to approximately 10,000 noise edges, and the performance starts to decrease dramatically as soon as more inter-edges are added. Like *MCL*, *Modularity* is sensitive to noise edges, which is indicated by large performance fluctuations. In contrast, the performances of *Dcut*, *Ncut* and *Metis* are more stable and robust to noise edges. *Dcut* starts to achieve relatively better results than *Ncut* and *Metis* for up to 20,000 noise edges.

Cluster Density: Next, we evaluate how the algorithms respond to a change of the intra-cluster edges of different clusters in the graph data, which we call cluster density. Here 10 clusters are first generated with 5000 inter-cluster edges, and 100 nodes in each cluster are randomly interlinked with 60% intra-cluster edges. We gradually change the number of intra-cluster edges for one cluster step by step with 5% decrease until all clusters have different densities of intra-connectivity. As a result, the highest density of intra-cluster edges in the first cluster is 60%, and the lowest density of intra-cluster edges in the last cluster is 15% (Fig. 4).

By generating the clusters with different densities in the graph data (Fig. 4), all algorithms perform well when the densities of clusters are above 50%. The performance of *MCL* begins to decrease dramatically when clusters with lower densities are included in the graph data. *Modularity* is also not able to achieve convincing results like *MCL*. As soon as the cluster density is lower than 35%, *Metis* starts to exhibit a slightly decreasing performance. Gradually changing the density of intra-cluster edges, both *Dcut* and *Ncut* achieve high clustering performance.

4.2   Real World Data

In this section, we evaluate the performances of different graph clustering algorithms on several real-world data which are all publicly available from the UCI network data repository (https://networkdata.ics.uci.edu/index.php).

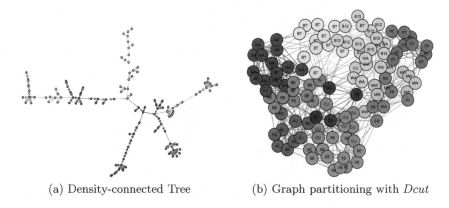

(a) Density-connected Tree   (b) Graph partitioning with *Dcut*

Fig. 5. Performance of *Dcut* on the network of American college football, where the colors of nodes indicate different graph clusters. (Color figure online)

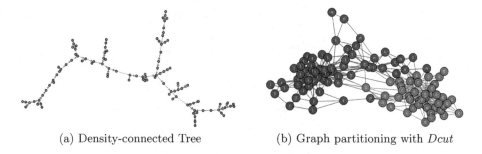

(a) Density-connected Tree   (b) Graph partitioning with *Dcut*

Fig. 6. Performance of *Dcut* on the network of books about US politics.

1. Networks with class information

To provide an objective evaluation of different graph clustering algorithms, we first investigate the networks for which the ground truth of community structure are already known. The external evaluation measures such as *NMI*, *ARI* and *purity* are applied.

American College Football: The graph data derived from the American football games of the schedule of Division I during regular season Fall 2000, where 115 vertices in the graph represent teams, and edges represent regular-season games between the two teams they connect. The teams are divided into 12 conferences containing around 8–12 teams each, and thereby the real community structure is already known.

Dcut ($K = 12$) identifies the graph clusters with a high degree of success (Fig. 5 and Table 1). Most teams are correctly grouped with the other teams in their conference with the highest cluster quality compared to the other four approaches (NMI = 0.924, ARI = 0.899, Purity = 93.0%). The good performance is due to the density-connected tree, where the most closely associated teams are densely connected together (Fig. 5(a)). *Ncut* and *MCL* also perform well,

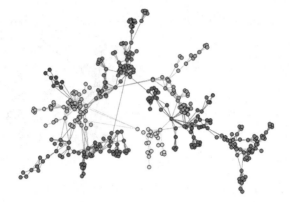

Fig. 7. Graph clustering of *Dcut* on the coauthorship network of scientists (K = 8).

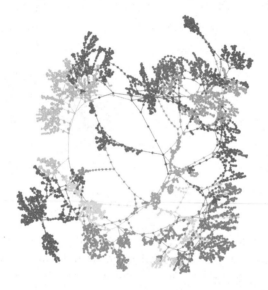

Fig. 8. Graph clustering of *Dcut* on the power grid network (K = 10).

Table 2. Evaluation of different graph clustering algorithms with clustering coefficient on real-world data sets.

Data	Dcut	Ncut	Modularity	Metis	MCL
Coauthorship network	0.1408	0.1349	0.1269	0.0984	0.1247
Power grid	0.0309	0.0255	0.0086	0.0237	0.012

and most teams are correctly grouped. For *Metis* and *Modularity*, however, it is difficult to discover the community structure. The performance of the different algorithms is summarized in Table 1.

Books About US Politics: The network consists of 105 nodes and 441 edges, which are derived from the books about US politics published around the time of the 2004 presidential election and sold by the online bookseller "Amazon.com". Edges represent frequent co purchasing of books by the same buyers. Each book is categorized as "liberal", "neutral", or "conservative" by Mark Newman based on a reading of the descriptions and reviews of the books posted on Amazon. With $K = 3$, most books can be correctly grouped by $Dcut$ with $NMI = 0.572$ (Fig. 6). Two major clusters correspond to liberal and conservative books with high cluster purity (only four books are misclustered in the two clusters), respectively. The same types of books are linked together in the density-connected tree (Fig. 6(a)). Compared with other algorithms, $Dcut$ achieves the best clustering results, as indicated in Fig. 6(b) and Table 1.

2. Networks without class information

Coauthorships in Network Science: The graph is a coauthorship network of 1589 scientists working on network theory and experiment. As the vertices of the network are not all connected, only the largest component of this network is used for graph clustering in this study. The graph clusters detected by $Dcut$ ($K = 8$) are illustrated in Fig. 7. In the plot, the obtained clusters present a high degree of scientific community structures. For comparison, the adapted clustering coefficient is applied to measure the quality of graph clusters discovered by different clustering algorithms (Table 2).

Power Grid: This network consists of 4941 vertices and 6594 edges, which represents the power grid of the Western States of the United States, compiled by Duncan Watts and Steven Strogatz. With $K = 10$, the clustering result of $Dcut$ is depicted in Fig. 8. We can observe that the power stations in each cluster show strong connections although the graph is very sparse, which results in the highest clustering coefficient of 0.031 compared to other approaches (Table 2).

Generally, $Dcut$ allows identifying a good graph clustering, and outperforms the compared algorithms on these real-world data sets, as indicated by external measures or the clustering coefficient.

4.3   Case Study

In this section, we further evaluate the performance of $Dcut$ on a case study with a protein-protein interaction (PPI) network. The objective of PPI network analysis is to investigate the structure of interaction networks of proteins, and thus provide insight into their functions that proteins in a subgraph indicating the similar molecular functions. Currently, it is an interesting and challenging problem since the functions of many proteins are unknown even for the most well-studied organisms such as yeast. Here, we use the PPI network in budding yeast, which contains 2361 proteins and 7182 interactions (http://vlado.fmf.uni-lj.si/pub/networks/data/). We analyze this interaction network with $Dcut$, and also compare its performance to the other four approaches. In the context of

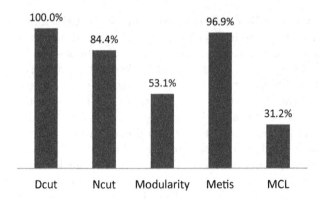

Fig. 9. Performance on protein data set. The clustering results of each algorithm are shown as a bar graph, which measures the percentage of GO enriched clusters with a significance level of <0.01.

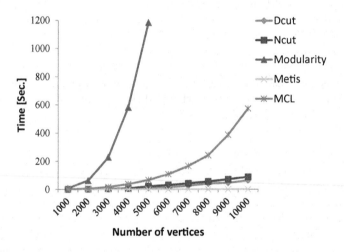

Fig. 10. The runtime of the different graph clustering algorithms.

biology, we evaluate the biological significance of obtained clusters with the help of the Gene Ontology (GO) database [1], which provides the ontology of defined terms representing functional annotations of proteins. Researchers can calculate P-value of each non-singleton cluster to demonstrate the statistical enrichment of GO molecular functions based on the hyper-geometric distribution [3].

Modularity detects 32 clusters on this network. For comparison, we set $K = 32$ for *Dcut*, *Ncut* and *Metis*. It is interesting to observe that all graph clusters detected by *Dcut* are enriched for the molecular functions (Fig. 9). For *Metis* and *Ncut*, one and five out of 32 clusters are not biologically significant for molecular functions, respectively. *Modularity* only finds 17 clusters which are enriched for the molecular functions. *MCL* generats 497 clusters, of which approximately 31% (159 out of 497) are biologically meaningful clusters.

4.4   Runtime

For runtime comparisons, we generate several synthetic data sets, where the number of clusters k varied from 10 to 100, and each cluster contains 100 nodes. Approximately 30% of the intra-cluster edges are generated, and 1% inter-cluster edges are linked. To obtain more accurate runtime results, each method processes 10 times and the times are averaged. In Fig. 10, we can observe that $Dcut$ is faster than $Modularity$. $Dcut$ is also better than MCL (approximately 8 times), and comparable with $Ncut$. However, $Dcut$ is slightly slower than $Metis$ with VLSI parallelling implementation.

5   Conclusions

In this paper, we introduce $Dcut$, a novel graph clustering algorithm. From the density point of view, the $density$-cut criterion offers a new measure to quantify the "goodness" of a graph clustering. Although $Dcut$ is a cut-based graph clustering algorithm, it largely differs from traditional cut-based algorithms such as Ncut, Min-Cut and Modularity. Except the new proposed cut criterion, one main difference appears to be that traditional cut criteria ignore the effects of topological structural similarity among nodes in the same community. Moreover, our approach allows transferring the density cut problem into a simple yet effective strategy: splitting a density-connected tree (maximum spanning tree) directly. Our extensive experiments demonstrate that $Dcut$ has many desirable properties and outperforms most state-of-art graph clustering methods.

Acknowledgments. This work is supported by the National Natural Science Foundation of China (61403062, 41601025, 61433014), Science-Technology Foundation for Young Scientist of SiChuan Province (2016JQ0007), State Key Laboratory of Hydrology-Water Resources and Hydraulic Engineering (2017490211), National key research and development program (2016YFB0502300).

References

1. Ashburner, M., Ball, C.A., Blake, J.A., Botstein, D., Butler, H., Cherry, J.M., Harris, M.A.: Gene ontology: tool for the unification of biology. Nat. Genet. **25**(1), 25 (2000)
2. Böhm, C., Plant, C., Shao, J., Yang, Q.: Clustering by synchronization. In: ACM SIGKDD International Conference on Knowledge Discovery and Data Mining, pp. 583–592 (2010)
3. Brohée, S., Faust, K., Lima-Mendez, G., Vanderstocken, G., Van Helden, J.: Network analysis tools: from biological networks to clusters and pathways. Nat. Protoc. **3**(10), 1616–1629 (2008)
4. Dongen, S.: A cluster algorithm for graphs. Technical report, Amsterdam (2000)
5. Evans, T.S.: Clique graphs and overlapping communities. J. Stat. Mech. Theory Exp. **12**, P12037 (2010)
6. Flake, G.W., Tarjan, R.E., Tsioutsiouliklis, K.: Graph clustering and minimum cut trees. Internet Math. **1**(4), 385–408 (2004)

7. Hagen, L., Kahng, A.B.: New spectral methods for ratio cut partitioning and clustering. IEEE Trans. Comput. Aided Des. Integr. Circ. Syst. **11**(9), 1074–1085 (1992)
8. Hajiabadi, M., Zare, H., Bobarshad, H.: IEDC: an integrated approach for overlapping and non-overlapping community detection. Knowl.-Based Syst. **123**, 188–199 (2017)
9. Hennig, C., Hausdorf, B.: Design of dissimilarity measures: a new dissimilarity between species distribution areas. In: Batagelj, V., Bock, H.H., Ferligoj, A., Žiberna, A. (eds.) Data Science and Classification. STUDIES CLASS, pp. 29–37. Springer, Heidelberg (2006). https://doi.org/10.1007/3-540-34416-0_4
10. Karypis, G., Kumar, V.: Multilevelk-way partitioning scheme for irregular graphs. J. Parallel Distrib. Comput. **48**(1), 96–129 (1998)
11. Karypis, G., Kumar, V.: A fast and high quality multilevel scheme for partitioning irregular graphs. SIAM J. Sci. Comput. **20**(1), 359–392 (1998)
12. Girvan, M., Newman, M.E.: Community structure in social and biological networks. Proc. Nat. Acad. Sci. **99**(12), 7821–7826 (2002)
13. Newman, M.E.: Modularity and community structure in networks. Proc. Nat. Acad. Sci. **103**(23), 8577–8582 (2006)
14. Prim, R.C.: Shortest connection networks and some generalizations. Bell Labs Tech. J. **36**(6), 1389–1401 (1957)
15. Rand, W.M.: Objective criteria for the evaluation of clustering methods. J. Am. Stat. Assoc. **66**(336), 846–850 (1971)
16. Schaeffer, S.E.: Graph clustering. Comput. Sci. Rev. **1**(1), 27–64 (2007)
17. Shao, J.: Synchronization on Data Mining: A Universal Concept for Knowledge Discovery. LAP LAMBERT Academic Publishing, Saarbrücken (2012)
18. Shao, J., He, X., Yang, Q., Plant, C., Böhm, C.: Robust synchronization-based graph clustering. In: Pei, J., Tseng, V.S., Cao, L., Motoda, H., Xu, G. (eds.) PAKDD 2013. LNCS (LNAI), vol. 7818, pp. 249–260. Springer, Heidelberg (2013). https://doi.org/10.1007/978-3-642-37453-1_21
19. Shao, J., Han, Z., Yang, Q., Zhou, T.: Community detection based on distance dynamics. In: Proceedings of the 21th ACM SIGKDD International Conference on Knowledge Discovery and Data Mining, pp. 1075–1084 (2015)
20. Shao, J., Yang, Q., Dang, H.V., Schmidt, B., Kramer, S.: Scalable clustering by iterative partitioning and point attractor representation. ACM Trans. Knowl. Discov. Data **11**(1), 5 (2016)
21. Shao, J., Wang, X., Yang, Q., Plant, C., Böhm, C.: Synchronization-based scalable subspace clustering of high-dimensional data. Knowl. Inf. Syst. **52**(1), 83–111 (2017)
22. Shao, J., Huang, F., Yang, Q., Luo, G.: Robust prototype-based learning on data streams. IEEE Trans. Knowl. Data Eng. **30**(5), 978–991 (2018)
23. Shi, J., Malik, J.: Normalized cuts and image segmentation. IEEE Trans. Pattern Anal. Mach. Intell. **22**(8), 888–905 (2000)
24. Strehl, A., Ghosh, J.: Cluster ensembles: a knowledge reuse framework for combining multiple partitions. J. Mach. Learn. Res. **3**, 583–617 (2002)
25. Watts, D.J., Strogatz, S.H.: Collective dynamics of small-worldnetworks. Nature **393**(6684), 440–442 (1998)
26. Wu, Z., Leahy, R.: An optimal graph theoretic approach to data clustering: theory and its application to image segmentation. IEEE Trans. Pattern Anal. Mach. Intell. **15**(11), 1101–1113 (1993)

External Topological Sorting in Large Graphs

Zhu Qing[1], Long Yuan[2(✉)], Fan Zhang[2], Lu Qin[3], Xuemin Lin[2], and Wenjie Zhang[2]

[1] East China Normal University, Shanghai, China
Skullpirate.qing@gmail.com
[2] The University of New South Wales, Sydney, Australia
{longyuan,lxue,zhangw}@cse.unsw.edu.au, fan.zhang3@unsw.edu.au
[3] Centre for Artificial Intelligence, University of Technology Sydney, Sydney, Australia
lu.qin@uts.edu.au

Abstract. Topological sorting is a fundamental problem in graph analysis. Given the fact that real world graphs grow rapidly so that they cannot entirely reside in main memory, in this paper, we study external memory algorithms for the topological sorting problem. We propose a contraction-expansion paradigm and devise an external memory algorithm based on the paradigm for the topological sorting problem. Our new algorithm is efficient due to the introduction of the new paradigm and can be implemented easily by using the fundamental external memory primitives. We conduct extensive experiments on real and synthesis graphs and the results demonstrate the efficiency of our proposed algorithm.

1 Introduction

Graphs have been widely used to represent the relationships of entities in a large spectrum of applications such as social networks, web search, collaboration networks, and biology. With the proliferation of graph applications, research efforts have been devoted to many problems in managing and analyzing graph data. Among them, topological sorting is a fundamental one. Given a directed graph G, topological sorting aims to compute a node numbering in which each node in G is assigned with a non-negative number such that if G contains an edge (u, v), then the assigned number of u is smaller than that of v.

Applications. Topological sorting can be used in many real-world applications:

(1) *Graph structure analysis in social network.* Topological sorting can be used to compute the node importance when analyzing the graph structure in social network [5].
(2) *Job planning and scheduling.* In job planning, the jobs are represented by nodes, and there is an edge from u to v if job u must be completed before job v can be started. Then, a topological sorting gives an order in which to perform the jobs [9].

J. Pei et al. (Eds.): DASFAA 2018, LNCS 10827, pp. 203–220, 2018.
https://doi.org/10.1007/978-3-319-91452-7_14

(3) *A key step to solve other graph problems.* Topological sorting is also an important building block for other graph algorithms, such as separator partitions of planar graphs [11], contour tree simplification [4], multi-objective shortest path computation [12].

Motivation. In the literature, there are efficient in-memory algorithm and semi-external algorithm to compute topological sorting [2,9]. An in-memory algorithm assumes that the graph is resident in main memory while a semi-external algorithm assumes that all nodes of G are kept in main memory. Nevertheless, as the sizes of many real graphs keep growing rapidly, even the nodes of a graph cannot reside entirely in main memory. For example, the Facebook social network contains 1.32 billion nodes and 140 billion edges[1]; and a sub-domain of the web graph of the EU countries contains 1.07 billion nodes and 91 billion edges[2]. In [2], the authors propose an external memory topological sorting algorithm in which the nodes of G cannot fit entirely in memory. It iteratively computes a partial topological sorting of G until the final topological sorting of G is obtained. However, the algorithm involves other complex external memory subroutines which makes it hard to implement and it cannot effectively utilize the available memory to further improve its performance, even when the memory is abundant.

Our Approach. In order to address the drawbacks of the existing solutions, we propose a new paradigm for topological sorting in this paper. Our paradigm contains two phases, namely, graph contraction phase and graph expansion phase. In graph contraction phase, we contract the nodes of the graph iteratively until all nodes of the contracted graph can fit in main memory. Then, we compute the topological sorting for the contracted graph using the efficient semi-external algorithm. In graph expansion phase, the removed nodes are added back into the graph in a reverse order of their removal and the topological sorting for the graph with the new added nodes is computed. Our new approach just uses the fundamental external memory primitives and can effectively utilize the available memory due to the contraction of the input graph and the exploiting of semi-external topological sorting algorithm.

Contributions. In this paper, we make the following contributions:

(1) *A new paradigm for topological sorting.* We investigate the drawbacks of existing semi-external and external topological sorting algorithms and propose a new contraction-expansion paradigm for the topological sorting problem, which can overcome the drawbacks of existing solutions.

(2) *A new external memory topological sorting algorithm.* Following the contraction-expansion paradigm, we devise a new external memory topological sorting algorithm. Our new algorithm just uses the fundamental external memory primitives and exploits the high efficiency of semi-external topological sorting algorithm. Besides, we also analyze the correctness and I/O complexity of our approach.

---

[1] http://newsroom.fb.com/company-info.
[2] http://law.di.unimi.it/datasets.php.

(3) *Extensive performance studies on large real and synthetic datasets.* We con-
    duct extensive performance studies using large real and synthetic graphs.
    The experimental results demonstrate the efficiency of our proposed algo-
    rithm.

2   Preliminaries

We model a directed graph as $G(V, E)$, where $V(G)$ represents the set of nodes
and $E(G)$ represents the set of directed edges in G. We denote the number of
nodes and the number of edges of G by n and m, respectively, i.e., $n = |V(G)|$ and
$m = |E(G)|$. We use $|G|$ to denote the sum of n and m, i.e., $|G| = |V(G)| + |E(G)|$.
Each node $u \in V(G)$ has a unique identity, denoted by $\text{id}(u)$. If there is a directed
edge (u, v) in G, we say u is the tail and v is head, u is an in-neighbor of v and v
is an out-neighbor of u. For each node $u \in G$, we use $\text{nbr}^-(u, G)$ and $\text{nbr}^+(u, G)$
to denote the set of u's in-neighbors and out-neighbors in G, respectively. For
a node u, the in-degree of u, denoted by $\deg^-(u, G)$, is the number of u's in-
neighbors and the out-degree of u, denoted by $\deg^+(u, G)$, is the number of u's
out-neighbors. And the degree of u, denoted by $\deg(u, G)$, is the sum of u's
in-degree and out-degree. For simplicity, we omit G from the notations if the
context is self-evident. Given a directed graph G, a path $p = (v_1, v_2, \cdots, v_k)$ is a
sequence of k nodes in $V(G)$ such that, for each $v_i (1 \le i < k)$, $(v_i, v_{i+1}) \in E(G)$.
In a directed graph, a path $(v_1, v_2, \cdots, v_k)$ forms a directed cycle if $v_1 = v_k$.

Definition 1 *(Directed Acyclic Graph).* *Given a directed graph G, G is a
directed acyclic graph (DAG) if and only if there exist no directed cycles in G.*

Definition 2 *(Topological Sorting).* *Given a DAG G, a topological sorting
of G is a node numbering in which each node is assigned with a non-negative
number such that if G contains an edge (u, v), then the assigned number of u is
smaller than that of v.*

We use Ω to denote an arbitrary node numbering and $\Omega(u)$ to denote the
number assigned to a node u by Ω. For a graph G, we use Ω_G to denote the
topological sorting of G and $\Omega_G(u)$ to denote the number assigned to a node u
by Ω_G. We also call $\Omega_G(u)$ the topological sorting number of u.

Problem Statement. In this paper, we study the problem of computing the
topological sorting for a given DAG G in an external memory model, namely,
not only G but also $V(G)$ cannot reside entirely in main memory. We use the
standard external memory model proposed in [1]. It assumes that the main
memory can only keep M elements while the remaining are kept in blocks on disk,
where one block contains B elements. Suppose one I/O access will read/write B
elements (one block) from/into disk into/from main memory. External memory
model contains two fundamental primitives: scanning N elements ($\text{scan}(N)$) and
sorting N elements ($\text{sort}(N)$). The I/O complexity of $\text{scan}(N)$ is $\Theta(\frac{N}{B})$ I/Os, and
the I/O complexity of $\text{sort}(N)$ is $O(\frac{N}{B} \cdot \log_{\frac{M}{B}} \frac{N}{B})$ I/Os.

3   Existing Solutions

3.1   Semi-external Topological Sorting Algorithm

In this section, we first introduce the semi-external topological sorting algorithm SemiTS [2], which assumes that the nodes of the graph can reside in main memory. It is based on the following property: given a graph G, a post-order traversal on a depth first search tree of G visits nodes in the reverse order of a topological sorting of G. Therefore, SemiTS computes the topological sorting in a semi-external manner by using the existing semi-external depth first search algorithm. SemiTS is shown in Algorithm 1.

Algorithm 1. SemiTS(Graph G)

1: compute a DFS tree T of G with root r using the semi-external DFS algorithm [18];
2: $\Omega_G \leftarrow \emptyset$; $i \leftarrow 1$; Stack $S \leftarrow \emptyset$;
3: PostOrder(r, T);
4: **while** $S \neq \emptyset$ **do**
5:     $u \leftarrow S.\text{pop}()$; $\Omega_G(u) \leftarrow i$; $i \leftarrow i + 1$;
6: **return** Ω_G;

7: **Procedure** PostOrder(TreeNode t, Tree T)
8: **if** $t = $ NULL **then return**;
9: **for each** child t_c of t in the corresponding DFS order **do**
10:     PostOrder (t_c, T);
11: $S.\text{push}(t)$;

SemiTS first computes a DFS tree T of G by using the state-of-the-art semi-external depth first search algorithm [18] (line 1). To facilitate the reversing of the post-order, SemiTS utilizes a stack S initializing with $\emptyset$ (line 2). Then, it conducts a post-order traversal on T (line 3). When the traversal finishes, SemiTS pops the nodes in S and assigns $\Omega_G(u)$ to u (line 4–6). Procedure PostOrder performs a post-order traversal on a DFS tree T starting from t. For a given tree node t, it first visits all its children in the corresponding DFS order (line 9–10) and then pushes t into the stack S (line 11).

3.2   External Memory Topological Sorting Algorithm

The state-of-the-art external memory algorithm, IterTS, is proposed in [2]. The main idea of IterTS is that if we have an arbitrary node numbering Ω, we can obtain the topological sorting Ω_G based on Ω by repeatedly adjusting the assigned numbers of such nodes u and v that $(u, v) \in E(G)$ but $\Omega(u) > \Omega(v)$.

For an arbitrary node numbering Ω, Algorithm 2 calls an edge $(u, v) \in E(G)$ topological sorted edge if $\Omega(u) < \Omega(v)$ and use $|\Omega|$ to denote the number of topological sorted edges for Ω. Algorithm 2 starts with an initial node numbering

Algorithm 2. IterTS(Graph G)

1: $\Omega \leftarrow$ InitOrder(G);
2: **while** $|\Omega| < m$ **do**
3:     $\Omega \leftarrow$ ImproveOrder(G, Ω);
4: **return** Ω;

5: **Procedure** InitOrder(G)
6:   compute an out-tree T of G from s;   // s is the only node in G with
      $\deg^-(s) = 0$.
7:   compute two orderings Ω_l and Ω_r by Euler Tour [8] on T;
8:   **if** $|\Omega_l| < |\Omega_r|$ **then return** Ω_r;
9: **else return** Ω_l;

10: **Procedure** ImproveOrder(G, Ω)
11:   compute an out-tree T of G from s based on Ω; // s is the only node in G with
      $\deg^-(s) = 0$.
12:   **for each** edge (u, v) following the pre-order traverse of T **do**
13:       $\Omega'(v) \to \max\{\Omega(v), \Omega'(u) + 1\}$;
14:   **for each** topological sorted edge (u, v) w.r.t Ω in increasing order according to
      $\Omega(v)$ **do**
15:       $\Omega''(v) \to \max\{\Omega'(v), \Omega''(u) + 1\}$;
16:   compute Ω_{new} by sorting nodes based on Ω'';
17: **return** Ω_{new};

---

by invoking InitOrder (line 1). After obtaining the initial Ω, it iteratively improves Ω by invoking ImproveOrder until $|\Omega|$ is m (line 2–4). For brevity, Algorithm 2 assumes G has only a single node s with $\deg^-(s) = 0$. Procedure InitOrder aims to compute a node numbering Ω with a big $|\Omega|$ heuristically. It computes an out-tree T rooted at s of G (line 6), computes two node numbering Ω_l and Ω_r through Euler Tour [8] on T (line 7) and returns the node numbering with a bigger number of topological sorted edges (line 8–9).

Procedure ImproveOrder intends to compute a new Ω_{new} based on Ω such that $|\Omega_{new}| > |\Omega|$. It first computes an out-tree T rooted at s such that for each node u, its parent in T has the maximum assigned number among all its in-neighbors in G (line 11). After that, it adjusts the assigned number for each node based on the tree edges in T and the topological sorted edges and generates two temporary node numberings Ω' and Ω'', respectively. For each tree edge (u, v) in T, it iterates them following the pre-order traverse of T and adjusts $\Omega'(v)$ by setting it as $\max\{\Omega(v), \Omega'(u) + 1\}$ (line 12–13). For each topological sorted edge (u, v) w.r.t Ω, it iterates them in the increasing order according to their $\Omega(v)$ and adjusts the assigned number of v by setting it as $\max\{\Omega'(v), \Omega''(u) + 1\}$ (line 14–15). In this way, the edges in T and the topological sorted edges w.r.t Ω are still topological sorted edge w.r.t Ω''. Then, it computes Ω_{new} by sorting the nodes based on Ω'' and returns Ω_{new} (line 16–17). Note that Algorithm 2 just shows the main idea of IterTS and omits the details about I/O issues. Given a graph G, Algorithm 2 finishes the topological sorting of G in $O(n \cdot \mathsf{sort}(|G|))$ I/Os.

3.3   Drawbacks of Existing Solutions

Regarding SemiTS, although it is significantly efficient in terms of computing topological sorting compared with IterTS [2]. It assumes that the nodes of the graph can fit in main memory, which does not satisfy our requirements. Regarding IterTS, although IterTS addresses the topological sorting problem upon the external memory model, it has the following two drawbacks: (1) Underutilization of available memory. As shown in Algorithm 2, IterTS does not take the available memory into consideration when computing Ω_G, which renders the algorithm unable to benefit from the availability of extra memory to further improve its performance, even when the memory is abundant. (2) Involvement of other complex external memory subroutines. Besides the two fundamental primitives scan and sort in external memory model, IterTS involves other complex external memory subroutines, such as, external list ranking, external Euler tour and external priority queue. All these subroutines are complex and hard to implement.

4   A New Approach

4.1   Contraction-Expansion Paradigm

As discussed in Sect. 3, SemiTS is efficient and can fully utilize the available memory, but it cannot handle the scenario that the nodes of the graph cannot be loaded in main memory. On the other hand, IterTS is an external memory solution. However, it cannot use the available memory effectively, limiting the improvement of its performance from the availability of extra memory. Motivated by this, we propose a new paradigm combining the merits of these two existing solutions. Our paradigm has two phases, namely, graph contraction phase and graph expansion phase.

In the *graph contraction phase*, we generate a list of graph $G_1, G_2, \cdots, G_k$, where $G_1 = G$, and for each $1 \leq i < k$, G_{i+1} is generated by removing a batch of nodes from G_i. The contraction phase stops when the nodes of the newly generated graph can fit in memory. In the *graph expansion phase*, after computing the topological sorting of G_k using SemiTS, the removed nodes are added back to the graph in the reverse order of their removal during the graph contraction phase, i.e., the lastly removed nodes in the graph contraction phase are firstly added back in the graph expansion phase. When a batch of nodes are added back, we compute the topological sorting for the graph with the new added nodes. More specifically, given that the topological sorting of G_k is computed using SemiTS, we compute the topological sorting of $G_{k-1}, G_{k-2}, \cdots, G_1$ in order and the topological sorting of G_i is computed based on the topological sorting of G_{i+1}. Since $G_1 = G$, the topological sorting of the original graph is obtained when the expansion phase finishes.

Advantages of Our Paradigm. Compared with the existing solutions, the advantages of our approach are twofold: (1) Regarding SemiTS, due to the introduction of contraction and expansion phases, our approach can handle scenario that the memory can not fit the nodes of the given graph. Therefore, our

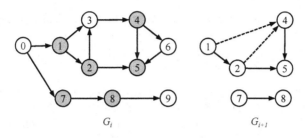

Fig. 1. Contraction phase

approach is a genuine external memory solution. (2) In contrast to IterTS, our approach can utilize the available memory effectively. In our approach, we stop the contraction phase once the available memory can fit the nodes of the contracted graph and directly use the SemiTS to compute the topological sorting of the contracted graph. In other words, the more available memory we have, the less contraction we would conduct. As a result, our approach is sensitive to the memory size and can significantly benefit from the extra memory considering the high performance of SemiTS. Besides, our contraction-expansion paradigm can be achieved by just using scan and sort, which makes our approach easy to implement.

Due to the adoption of contraction-expansion paradigm, our approach can obtain the merits of existing solutions. However, to make our paradigm practically applicable, the following issues should be addressed when designing our algorithm:

- *Contractility:* For a given G_i, the number of nodes in the contracted graph G_{i+1} should be smaller than that in G_i, i.e., $V(G_{i+1}) \subset V(G_i)$.
- *Expandability:* For a given G_{i+1} and $\Omega_{G_{i+1}}$, we should be able to compute Ω_{G_i} based on $\Omega_{G_{i+1}}$.
- *External-Memory Feasibility:* All the algorithms should be able to be implemented in an external-memory manner.

In the following, we will introduce how to address these issues one by one.

4.2   Contraction Phase

In the contraction phase, we focus on solving the contractility issue. It is trivial if we just consider the contractility alone. For example, we can arbitrarily select a batch of nodes and generate a new graph by removing those un-selected nodes. However, this approach makes satisfying the expandability requirement hard in the expansion phase. Taking the contractility and expandability into consideration simultaneously, we propose a vertex cover based contraction method, which is based on the following definition:

Definition 3 (Vertex Cover). *Given a graph G, a vertex cover of graph is a subset V' of $V(G)$ such that each edge of the graph is incident to at least one node of the set, i.e., $(u, v) \in E(G) \Rightarrow u \in V' \vee v \in V'$.*

Example 1. Consider the graph G_i in Fig. 1, the node set $\{v_1, v_2, v_4, v_5, v_7, v_8\}$ is a vertex cover of G_i, which are shown with grey shadow. And each edge of G_i is incident to at least one node in the vertex cover, for example, edge (v_0, v_1) is incident to v_1.

In the contraction phase, for an input graph G_i, we compute a vertex cover V' of G_i and consider the node in V' as $V(G_{i+1})$. The benefits of choosing a vertex cover V' of G_i as $V(G_{i+1})$ are twofold: (1) Based on Definition 3, we can easily get a vertex cover V' such that $V' \subset V(G_i)$. Therefore, the contractility is satisfied. (2) As we choose V' as $V(G_{i+1})$, for any edge $(u, v) \in E(G_i)$, either $u \in V(G_{i+1})$ or $v \in V(G_{i+1})$. Therefore, if we know $\Omega_{G_{i+1}}$, then, for a node $w \in V(G_i) \setminus V(G_{i+1})$, the relative numerical magnitudes of its in-neighbors and out-neighbors in Ω_{G_i} can be determined. As a result, we can obtain Ω_{G_i} by just considering w together with its neighbours, which satisfies the expandability requirements (the details will be discussed in Sect. 4.3).

We can obtain $V(G_{i+1})$ by computing a vertex cover V' of G_i. However, just taking the edge of G_i induced by V' as $E(G_{i+1})$ is not sufficient for the topological sorting problem. For example, in Fig. 1, if we just keep the edges induced by $\{v_1, v_2, v_4, v_5, v_7, v_8\}$ as $E(G_{i+1})$, the information that $\Omega_{G_{i+1}}(v_1)$ should be smaller than $\Omega_{G_{i+1}}(v_4)$ is lost. Therefore, we include two types of edges in $E(G_{i+1})$:

Definition 4 (Induced Edge). *Given a graph G_i and its vertex cover V', let $u, v \in V'$. If $(u, v) \in E(G_i)$, then (u, v) is an induced edge for G_{i+1}.*

Definition 5 (Contracted Edge). *Given a graph G_i and its vertex cover V', let u, v, w be three nodes in $V(G_i)$ and $u, v \in V' \wedge w \notin V'$. If $(u, w) \in E(G_i) \wedge (w, v) \in E(G_i) \wedge (u, v) \notin E(G_i)$, then (u, v) is a contracted edge for G_{i+1}.*

Example 2. Consider G_i in Fig. 1 again, the contracted graph G_{i+1} is shown on the right of Fig. 1. $V(G_{i+1})$ is the vertex cover of G_i. $E(G_{i+1})$ consists of two types of edges: the induced edges which are shown with solid line, such as (v_1, v_2), and the contracted edges which are shown with dotted line, such as (v_1, v_4).

Algorithm Design. Our contraction algorithm, Contract, is shown in Algorithm 3. It first computes $V(G_{i+1})$ by computing a vertex cover in an external-memory manner (line 1), then it computes $E(G_{i+1})$ by invoking conEdge (line 2).

Regarding computing the vertex cover, in order to reduce the number of iterations in the contraction phase, $|V_{i+1}|$ should be as small as possible. This leads to the minimum vertex cover problem which is NP-hard [9]. In the literature, [3] proposes an external memory algorithm to find a vertex cover V' with an approximation ratio $\sqrt{\Delta(G)}/2 + 3/2$, where $\Delta(G)$ is the maximum degree of G. It defines a total order $\prec$ for all nodes in the graph based on the node's degree. For each edge (u, v) in $E(G_i)$, if $u \prec v$, then u is added to V', otherwise, v is added to V'. Its I/O complexity is $O(\text{sort}(G_i))$. Since we focus on external-memory topological sorting in this paper, we just use the algorithm in [3] directly to compute the vertex cover.

Algorithm 3. Contract(Graph G_i)

1: $V(G_{i+1}) \leftarrow$ compute a vertex cover of G_i in an external-memory manner by [3];
2: $E(G_{i+1}) \leftarrow$ conEdge$(G_i, V(G_{i+1}))$;

3: **Procedure** conEdge$(G_i, V(G_{i+1}))$
4: $E^- \leftarrow$ sort $(u,v) \in E(G_i)$ by $(\text{id}(v), \text{id}(u))$;
5: $E^+ \leftarrow$ sort $(u,v) \in E(G_i)$ by $(\text{id}(u), \text{id}(v))$;
6: $E_{\text{ind}} \leftarrow (u,v) \in E(G_i)$ with $u \in V(G_{i+1})$ by sequential scan $V(G_{i+1})$ and E^+;
7: $E_{\text{ind}} \leftarrow$ sort $(u,v) \in E_{\text{ind}}$ by $(\text{id}(v), \text{id}(u))$;
8: $E_{\text{ind}} \leftarrow (u,v) \in E(G_i)$ with $u,v \in V(G_{i+1})$ by sequential scan $V(G_{i+1})$ and E_{ind};
9: $E_{\text{rem}} \leftarrow (u,v) \in E(G_i)$ with $v \notin V(G_{i+1})$ by sequential scan $V(G_{i+1})$ and E^-;
10: $E_{\text{rem}} \leftarrow (u,v, \text{nbr}^+(v, G_i))$ with $v \notin V(G_{i+1})$ by sequential scan E_{rem} and E^+;
11: **for each** edge $(u,v) \in E_{\text{rem}}$ **do**
12:    **for each** $w \in \text{nbr}^+(v, G_i)$ by sequential scan of E_{rem} **do**
13:       $E_{\text{con}} \leftarrow E_{\text{con}} \cup (u,w)$;
14: $E(G_{i+1}) \leftarrow E_{\text{ind}} \cup E_{\text{con}}$;
15: **return** $E(G_{i+1})$;

Procedure conEdge computes $E(G_{i+1})$ externally for a given G_i and $V(G_{i+1})$. As discussed above, $E(G_{i+1})$ consists of two types of edges, namely, the induced edges and the contracted edges. We denote them as E_{ind} and E_{con}, respectively. E^- and E^+ be the edges of G_i by grouping in-coming and out-going edges for each node in G_i, which can be obtained by external sorting based on $(\text{id}(v), \text{id}(u))$ and $(\text{id}(u), \text{id}(v))$, respectively (line 4–5). Here, sorting based on $(\text{id}(v), \text{id}(u))$ means when sorting edges (u,v), we sort them based on $\text{id}(v)$. If two edges have the same $\text{id}(v)$, we sort them based on $\text{id}(u)$. conEdge first constructs E_{ind} (line 6–8) and E_{con} (line 9–13), and then unions E_{con} and E_{ind} to construct $E(G_{i+1})$ (line 14–15).

To construct E_{ind}, conEdge first computes the edges (u,v) with $u \in V(G_{i+1})$ by a sequential scan of $V(G_{i+1})$ and E^+ simultaneously. When scanning $V(G_{i+1})$ and E^+, for each edge $(u,v) \in E^+$, if $u \in V(G_{i+1})$, then (u,v) is added to E_{ind} (line 6). Then, it sorts all edges $(u,v) \in E_{\text{ind}}$ based on $(\text{id}(v), \text{id}(u))$ (line 7). At last, it computes the edges (u,v) with $u,v \in V(G_{i+1})$ by a sequential scan of $V(G_{i+1})$ and E_{ind} simultaneously (line 8). conEdge constructs E_{con} based on the set of edges that will be removed from $E(G_i)$. It first computes the edges E_{rem} in which each edge (u,v) with $v \notin V(G_{i+1})$. E_{rem} can be computed by a single sequential scan of $V(G_{i+1})$ and E^- on disk (line 9). When scanning $V(G_{i+1})$ and E^-, for each edge $(u,v) \in E^-$, if $v \notin V(G_{i+1})$, then (u,v) is added to E_{rem}. After constructing E_{rem}, for each edge $(u,v) \in E_{\text{rem}}$, we compute the out-neighbors of v, which can be obtained by a single sequential scan of E_{rem} and E^+ (line 10). Then, conEdge constructs E_{con} using a single sequential scan of all edges in E_{rem} (line 11–13). In E_{rem}, for each node v removed from G_i, each of its in-neighbors u in G_i and its out-neighbors $\text{nbr}^+(v, G_i)$ are stored together in the form $(u,v, \text{nbr}^+(v, G_i))$. When scanning each removed in-coming edge (u,v) of v, the removed out-going edge (v,w) of v can be obtained in the same sequential scan of E_{rem}. Then, conEdge adds a contracted edge (u,w) into E_{ind}.

Lemma 1. *Let G_i be the input graph and G_{i+1} be the contracted graph for Algorithm 3, respectively, then, $V(G_{i+1}) \subset V(G_i)$.*

Proof. Let v be the node in G_i such that for any other node $u \in V(G_i)$, we have $u \prec v$. Based on the method in [3] to compute the vertex cover, v cannot be added into $V(G_{i+1})$, because there does not exits an edge (u, v) or an edge (v, u) with $v \prec u$. Thus, the lemma holds. □

Lemma 2. *Let G_i be the input graph and G_{i+1} be the contracted graph for Algorithm 3, respectively, assume u and v be two nodes in $V(G_{i+1})$, then $\Omega_{G_{i+1}}(u) < \Omega_{G_{i+1}}(v)$ if and only if $\Omega_{G_i}(u) < \Omega_{G_i}(v)$.*

Proof. We can prove the lemma based on the procedure of Algorithm 3 directly. □

Lemma 3. *Let G_i be the input graph and G_{i+1} be the contracted graph, the I/O complexity of Algorithm 3 is $O(\mathsf{sort}(|G_i|) + \mathsf{scan}(|G_{i+1}|))$.*

Proof. This lemma can be proved directly based on the procedure of Algorithm 3. □

4.3   Expansion Phase

In expansion phase, we aim to obtain Ω_{G_i} through the computed $\Omega_{G_{i+1}}$. Based on Definition 3, we can divided the nodes in $V(G_i)$ into three types:

- Type-I: the nodes in $V(G_i) \setminus V(G_{i+1})$ without in-neighbors in G_i, i.e., $u \in V(G_i) \setminus V(G_{i+1}) \wedge \deg^-(u, G_i) = 0$.
- Type-II: the nodes in $V(G_i) \setminus V(G_{i+1})$ with in-neighbors in G_{i+1} and the nodes in $V(G_{i+1})$ with out-neighbors in G_i, i.e., $\{u \in V(G_i) \setminus V(G_{i+1}) \wedge \deg^-(u, G_i) > 0\} \cup \{u \in V(G_{i+1}) \wedge \deg^+(u, G_i) > 0\}$.
- Type-III: the nodes in $V(G_{i+1})$ without out-neighbors in G_i, i.e., $u \in V(G_{i+1}) \wedge \deg^+(u, G_i) = 0$.

For the Type-I nodes, since they have no in-neighbors in G_i, we can obtain their topological sorting numbers in G_i by assigning their numbers first. For the Type-II nodes, we consider two types of edges: (1) $(u, v) \in E(G_i) \wedge u \in V(G_{i+1}) \wedge v \in V(G_i) \setminus V(G_{i+1})$ (2) $(u, v) \in E(G_i) \wedge u \in V(G_{i+1}) \wedge v \in V(G_{i+1})$. We use E_{exp} to denote these edges. By sorting the edges in E_{exp} based on $\Omega_{G_{i+1}}(u)$, we can observe:

Observation 1. *For the Type-II nodes of $V(G_i)$ which are also in $V(G_{i+1})$, they appear in the tail position of edges in E_{exp} and their appearing order in the tail position of edges in E_{exp} is consistent with their numerical magnitude in the topological sorting.*

Algorithm 4. Expand (Graph G_i, Graph G_{i+1}, TopologicalSorting $\Omega_{G_{i+1}}$)

1: num $\leftarrow$ 1; $v_{\text{pos}} \leftarrow$ 1; $k \leftarrow$ 1; curr $\leftarrow \emptyset$;
2: compute E^- and E^+ as line 4-5 of Algorithm 3;
3: $E_{\text{rem}} \leftarrow (u,v) \in E(G_i)$ with $v \in V(G_i) \setminus V(G_{i+1})$ by sequential scan $V(G_{i+1})$ and E^-;
4: $V_{\text{zero}}^- \leftarrow v \in V(G_i) \setminus V(G_{i+1})$ with $\deg^-(v, G_i) = 0$ by sequential scan $V(G_i) \setminus V(G_{i+1})$ and E_{rem};
5: **for each** $u \in V_{\text{zero}}^-$ **do**
6: $\quad \Omega_{G_i}(u) =$ num; num $\leftarrow$ num $+ 1$;
7: $E_{\text{ind}} \leftarrow$ compute as line 6-8 of Algorithm 3;
8: $E_{\text{rem}} \leftarrow (u,v,\text{REM})$ for each $(u,v) \in E_{\text{rem}}$; $E_{\text{ind}} \leftarrow (u,v,\text{IND})$ for each $(u,v) \in E_{\text{ind}}$;
9: $E_{\text{exp}} \leftarrow E_{\text{rem}} \cup E_{\text{ind}}$;
10: $E_{\text{exp}} \leftarrow$ sort $(u,v,\text{FLAG}) \in E_{\text{exp}}$ based on $(\Omega_{G_{i+1}}(u), \text{id}(v))$;
11: **for each** $(u,v,\text{FLAG}) \in E_{\text{exp}}$ **do**
12: $\quad$ **if** FLAG = REM **then**
13: $\quad\quad E_{v_{\text{pos}}} \leftarrow (v, v_{\text{pos}})$;
14: $\quad\quad v_{\text{pos}} \leftarrow v_{\text{pos}} + 1$;
15: $E_{v_{\text{pos}}} \leftarrow$ sort $(v, v_{\text{pos}}) \in E_{v_{\text{pos}}}$ by $(\text{id}(v), v_{\text{pos}})$;
16: $E_{v_{\text{pos}}} \leftarrow (v, v'_{\text{pos}}) \in E_{v_{\text{pos}}}$ with $v'_{\text{pos}} = \max\{v_{\text{pos}}\}$ for v by sequential scan $E_{v_{\text{pos}}}$;
17: $E_{\text{exp}} \leftarrow$ sort $(u,v,\text{FLAG}) \in E_{\text{exp}}$ based on $(\text{id}(v), \text{id}(u))$;
18: $E_{\text{exp}} \leftarrow (u,v,\text{FLAG},v'_{\text{pos}})$ by sequential scan $E_{v_{\text{pos}}}$ and E_{exp};
19: $E_{\text{exp}} \leftarrow$ sort $(u,v,\text{FLAG},v'_{\text{pos}}) \in E_{\text{exp}}$ based on $(\Omega_{G_{i+1}}(u), \text{id}(v))$;
20: **for each** $(u,v,\text{FLAG},v'_{\text{pos}}) \in E_{\text{exp}}$ **do**
21: $\quad$ **if** curr $\neq u$ **then**
22: $\quad\quad \Omega_{G_i}(u) =$ num; num $\leftarrow$ num $+ 1$; curr $\leftarrow u$;
23: $\quad$ **if** $k = v'_{\text{pos}}$ **then**
24: $\quad\quad \Omega_{G_i}(v) =$ num; num $\leftarrow$ num $+ 1$;
25: $\quad\quad k \leftarrow k + 1$;
26: **for each** u not assigned topological sorting number by sequential scan $V(G_i)$ and Ω_{G_i} **do**
27: $\quad \Omega_{G_i}(u) =$ num; num $\leftarrow$ num $+ 1$;

Observation 2. *For the Type-II nodes of $V(G_i)$ which are in $V(G_i) \setminus V(G_{i+1})$, they only appear in the head position of edges in E_{exp} and for a specific node in this case, its last appearance in E_{exp} is after the first appearances of its in-neighbors and before the first appearances of its out-neighbors in the tail position of edges in E_{exp}.*

Following these two observations, we can obtain the topological sorting of Type-II nodes in G_i based on their appearing orders in E_{exp}. For the Type-III nodes, since they have no out-neighbors in G_i, we can obtain their topological sorting numbers in G_i by handling them at last. By combining the above three cases together, we can obtain Ω_{G_i}.

Algorithm Design. Our expansion algorithm, Expand, is shown in Algorithm 4. Expand first computes the topological number for the Type-I nodes. It first computes E^- and E^+ similarly as Algorithm 3 (line 2). By a simultaneous sequential scan of $V(G_{i+1})$ and E^-, it obtains the edges $(u,v) \in E(G_i)$ with

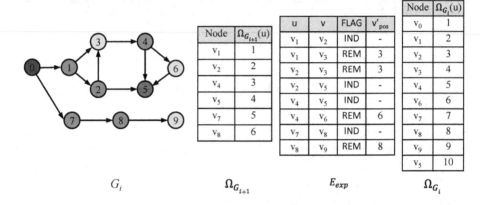

Fig. 2. Expansion phase (Color figure online)

$v \in V(G_i) \setminus V(G_{i+1})$ and stores them as E_{rem} (line 3). After that, Expand computes the nodes without in-neighbors in G_i by a simultaneous sequential scan of $V(G_i) \setminus V(G_{i+1})$ and E_{rem} and assigns topological sorting number to them (line 4–6).

Then, Expand handles the topological sorting number assignment for the Type-II nodes. To obtain their topological sorting numbers, Expand first computes the two types of edges discussed above. The edges $(u, v) \in E(G_i) \wedge u \in V(G_{i+1}) \wedge v \in V(G_i) \setminus V(G_{i+1})$ has been computed in line 3. For the edges $(u, v) \in E(G_i) \wedge u \in V(G_{i+1}) \wedge v \in V(G_{i+1})$, they can be computed similarly as Algorithm 3 (line 7). After that, Expand aims to compute the last appearance for the Type-II nodes in $V(G_i) \setminus V(G_{i+1})$ in the edge sets E_{exp} sorted by $\Omega_{G_{i+1}}(u)$. To achieve this goal, Expand first arguments the edges in E_{rem} and E_{ind} with an indicator FLAG, which contains two values: REM and IND, and combines E_{rem} and E_{ind} together in E_{exp} (line 8–9). After that, Expand sorts the edges $(u, v, \text{FLAG}) \in E_{\text{exp}}$ based on $(\Omega_{G_{i+1}}(u), \text{id}(v))$ and computes the last appearance position for the Type-II nodes in $V(G_i) \setminus V(G_{i+1})$ in E_{exp} in line 11–18. Note that for the edges with FLAG value IND, we only set their v'_{pos} as null in line 18 and sorting by $(\Omega_{G_{i+1}}(u), \text{id}(v))$ can be achieved by first augmenting the edges with $\Omega_{G_{i+1}}(u)$, we omit the details for brevity. After sorting edges $(u, v, \text{FLAG}, v'_{\text{pos}}) \in E_{\text{exp}}$ based on $(\Omega_{G_{i+1}}(u), \text{id}(v))$ (line 19), Expand sequentially scans edge $(u, v, \text{FLAG}, v'_{\text{pos}}) \in E_{\text{exp}}$ (line 20). If u is the first time to scan, it assigns the topological sorting number for u (line 21–22). If v reaches its last position, it assigns the topological sorting number for v (line 23–25).

For the Type-III nodes, Expand just conducts a simultaneous sequential scan on $V(G_i)$ and $\Omega(G_i)$ and assigns their topological sorting numbers to them (line 26–27).

Example 3. Figure 2 shows an example of expansion phase for G_i. The nodes in $V(G_i)$ are divided into three types: v_0 is the Type-I node, which is shown in blue. v_5 is the Type-III node, which is shown in aqua. The remaining nodes

---

Algorithm 5. CoExTS(Graph G)

1: $G_1 \leftarrow G$; $i \leftarrow 1$;
2: **while** $M < c \times |V(G_i)|$ **do**
3:    $G_{i+1} \leftarrow$ Contract(G_i); $i \leftarrow i + 1$;
4: $\Omega_{G_i} \leftarrow$ SemiTS (G_i);
5: **while** $i > 1$ **do**
6:    $i \leftarrow i - 1$; $\Omega_{G_i} \leftarrow$ Expand $(G_i, G_{i+1}, \Omega_{G_{i+1}})$;
7: output topological sorting Ω_G of G;

---

are Type-II nodes. The Type-II nodes in $V(G_{i+1})$, such as v_1, are shown in dark grey. The Type-II node in $V(G_i) \setminus V(G_{i+1})$, such as v_3, are shown in light grey. $\Omega_{G_{i+1}}$ shows the computed topological sorting for G_{i+1}. E_{exp} consists of two types of edges: (1) the edges (u, v) with $u \in V(G_{i+1}) \wedge v \in V(G_{i+1})$, such as (v_1, v_2), which are shown with an FLAG value IND. (2) the edges (u, v) with $u \in V(G_{i+1}) \wedge v \in V(G_i) \setminus V(G_{i+1})$, such as (v_1, v_3), which are shown with an FLAG value REM. The edges in E_{exp} are sorted by $\Omega_{G_{i+1}}(u)$. The v'_{pos} value for an IND edge is null, which is shown as $-$. The v'_{pos} value for an REM edge (u, v) is the last appearance position of v. For example, the v'_{pos} value for (v_1, v_3) is 3 as the last appearing position of v_3 in E_{exp} is 3. When computing Ω_{G_i}, we first compute the topological sorting number for Type-I nodes and v_0 is assigned with topological sorting number 1. When computing the topological number for Type-II nodes, we scan $(u, v, \mathsf{FLAG}, v'_{pos}) \in E_{\mathsf{exp}}$. If current u is different from last u, we assign the topological sorting number for u. And if current position is v'_{pos}, we assign the topological sorting number for v. At last, we assign the topological sorting number of Type-III nodes. Ω_{G_i} is shown on the right of Fig. 2.

Lemma 4. *Algorithm 4 computes Ω_{G_i} correctly.*

Proof. The correctness of Algorithm 4 can be directly derived from above analysis.   □

Lemma 5. *The I/O complexity of Algorithm 4 is $O(\mathsf{sort}(|G_i|) + \mathsf{scan}(|G_{i+1}|))$.*

Proof. This lemma can be proved directly based on the procedure of Algorithm 4.   □

4.4 Our Approach

Our complete algorithm CoExTS is shown in Algorithm 5. It follows the contraction-expansion paradigm as discussed above. For a given graph G, it iteratively contracts G until the nodes of the contracted graph G_i can be loaded in memory (line 2–3). In line 2, M represents the size of available memory size and c is a constant factor. After computing Ω_{G_i} with the semi-external algorithm SemiTS (line 4), CoExTS conducts the expansion phase in the reverse order for the graphs generated in the contraction phase (line 5–6). At last, it outputs Ω_G of the original input graph G (line 7).

Theorem 1. *Given a graph G, Algorithm 5 computes Ω_G correctly.*

Proof. The correctness of Algorithm 5 can be directly derived from above analysis.   □

Theorem 2. *Given a graph G, let $G_1, G_2, \cdots, G_k$ be the graphs generated in the contraction phase and* $\mathsf{IO_{semi}}$ *is the I/O complexity of* SemiTS*, the I/O complexity of Algorithm 5 is* $O((\sum_{i=1}^{k} \mathsf{sort}(|G_i|)) + \mathsf{IO_{semi}})$.

Proof. This theorem can be proved according to Lemmas 3 and 5 directly.   □

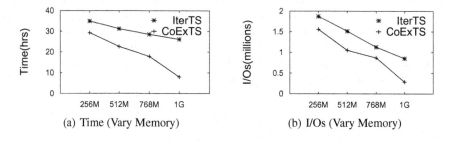

(a) Time (Vary Memory)   (b) I/Os (Vary Memory)

Fig. 3. UK-2007

5   Performance Studies

In this section, we conduct experimental studies by comparing two external memory algorithms for topological sorting, namely, IterTS (Algorithm 2) and CoExTS (Algorithm 5). All algorithms are implemented using Visual C++ and STXXL[3] and tested on a Laptop with Intel Core i7 2.8 GHz CPU and 3.5 GB memory running Windows 7.

Dataset. In our experiments, we use a real world graph and two synthetic graphs. The real world graph is UK-2007[4], which contains the webpages and their hyperlinks information in the .UK domain. The original UK-2007 consists of 105,895,908 nodes and 3,738,733,568 edges. Since the original UK-2007 contains cycles, we collapse the cycles in UK-2007 and the generated DAG contains 105,895,908 nodes and 1,238,766,251 edges. For synthetic graphs, we generate a graph by fixing the number of its nodes first and randomly add edges that cannot form cycles in the graph. We generate two synthetic graphs Random1 and Random2. Random1 contains 100M nodes with average degree 10 and Random2 contains 100M nodes with average degree 20.

Exp-1: Performance on Real Graph UK-2007. In this experiment, we compare the running time and number of I/Os of IterTS and CoExTS on UK-2007

---

[3] http://stxxl.org/.
[4] http://chato.cl/webspam/datasets/uk2007/links/.

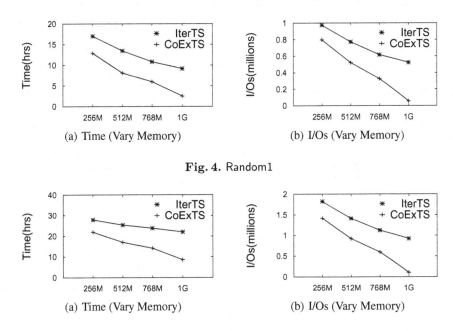

(a) Time (Vary Memory)   (b) I/Os (Vary Memory)

Fig. 4. Random1

(a) Time (Vary Memory)   (b) I/Os (Vary Memory)

Fig. 5. Random2

when we vary the size of available memory from 256M to 1G. The results are shown in Fig. 3.

Figure 3(a) shows that (1) The running time of IterTS and CoExTS decrease as the size of available memory increases and CoExTS outperforms IterTS in all cases. This is because CoExTS adopts the contraction-expansion paradigm and uses SemiTS as a subroutine and SemiTS is very efficient for topological sorting compared with CoExTS. The performance improvement by introducing SemiTS surpasses the time consumption of graph contraction and expansion. (2) the performance gap between IterTS and CoExTS increases as the size of available memory increases. This is because as the size of available memory increases, the iterations in contraction phase and expansion phase in CoExTS decrease and we can further exploit the efficiency of SemiTS. On the other hand, IterTS does not take available memory into consideration during processing. Combining these two factors together, the performance gap increases as the size of available memory increases. For the same reasons, Fig. 3(b) shows similar trends on the number of I/Os when we vary the size of available memory.

Exp-2: Performance on Synthetic Graphs. In this experiment, we compare the running time and the number of I/Os of IterTS and CoExTS on Random1 and Random2 when we vary the size of available memory from 256M to 1G. The results are shown in Figs. 4 and 5, respectively.

For the synthetic graphs, we have similar conclusions from the performance study upon UK-2007. CoExTS also outperforms IterTS in all test cases. When available memory increases, the running time and I/Os for both IterTS and

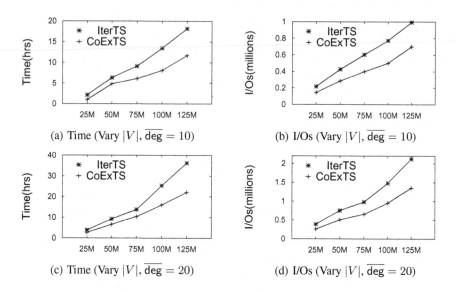

Fig. 6. Scalability

CoExTS decrease. As the size of available memory increases, it is clear that the performance gap between IterTS and CoExTS increases as well. The reason is similar as that presented Exp-1.

Exp-3: Scalability. In this experiment, we evaluate the scalability of IterTS and CoExTS. To test the scalability, we vary the number of nodes from 25M to 125M in synthetic graphs and record the running time and number of I/Os of IterTS and CoExTS, respectively. In the first test, the average degree of the synthetic graphs is 10 and the results are shown in Fig. 6(a) and (b). In the second test, the average degree of the synthetic graphs is 20 and the results are shown in Fig. 6(c) and (d). The available memory in this experiment is 512M.

As shown in Fig. 6, the running time and I/Os of both two algorithms increase as $|V|$ increases. This is because, as $|V|$ increases, IterTS needs more iterations to adjust the assigned numbers for the nodes incident to the edges which are not topological sorted edges. In the meantime, the iterations in the contraction phase and expansion phase of CoExTS also increase as $|V|$ increases. CoExTS outperforms IterTS for all cases due to the introduction of contraction-expansion paradigm and the efficiency of SemiTS. From Fig. 6, it is clear that CoExTS has a better scalability than IterTS.

6   Related Work

Topological sorting is a fundamental problem in graph analysis and has been extensively studied in the literature. For the in-memory topological sorting algorithms, [9] describes an algorithm by iteratively removing the nodes with in-degree number 0. [13] proposes a depth first search based algorithm whose idea is used in the semi-external topological algorithm. The state-of-the-art external topological sorting algorithm is proposed in [2]. We introduce it in Sect. 3 and use it as our baseline in the experiments.

Several other graph algorithms focusing on I/O efficiency are proposed in the literature. [6] describes an I/O efficient algorithm for the core decomposition problem. I/O efficient algorithm for the maximal clique enumeration problem is proposed in [7]. [15] proposes an I/O efficient algorithm for the diversified top-k clique search problem in large graphs. [3] studies three I/O efficient algorithms for vertex cover. I/O efficient algorithms for the triangle enumeration problem are presented in [10]. The I/O efficient algorithm for the k-truss problem is investigated in [14]. A semi-external algorithm for the depth first search is proposed in [18]. [16] presents three I/O efficient algorithms for edge connectivity decomposition problem. [17] studies the Steiner Maximum-Connected Components search problem in semi-external memory model.

7   Conclusion

In this paper, we study the external topological sorting algorithm in large graphs. We propose a new contraction-expansion paradigm and devise an external-memory algorithm CoExTS for the topological sorting problem. The experimental results demonstrate the efficiency of our proposed algorithm.

Acknowledgements. Long Yuan is supported by Huawei YBN2017100007. Fan Zhang is supported by Huawei YBN2017100007. Lu Qin is supported by ARC DP160101513. Xuemin Lin is supported by NSFC 61672235, ARC DP170101628, DP180103096 and Huawei YBN2017100007. Wenjie Zhang is supported by ARC DP180103096 and Huawei YBN2017100007.

References

1. Aggarwal, A., Vitter, J., et al.: The input/output complexity of sorting and related problems. Commun. ACM **31**(9), 1116–1127 (1988)
2. Ajwani, D., Cosgaya-Lozano, A., Zeh, N.: A topological sorting algorithm for large graphs. J. Exp. Algorithmics **17**, Article No. 3.2 (2012)
3. Angel, E., Campigotto, R., Laforest, C.: Analysis and comparison of three algorithms for the vertex cover problem on large graphs with low memory capacities. Algorithmic Oper. Res. **6**(1), 56–67 (2011)
4. Arge, L., Revsbæk, M.: I/O-efficient contour tree simplification. In: Dong, Y., Du, D.-Z., Ibarra, O. (eds.) ISAAC 2009. LNCS, vol. 5878, pp. 1155–1165. Springer, Heidelberg (2009). https://doi.org/10.1007/978-3-642-10631-6_116
5. Buccafurri, F., Lax, G., Nocera, A., Ursino, D.: Moving from social networks to social internetworking scenarios: the crawling perspective. Inf. Sci. **256**, 126–137 (2014)
6. Cheng, J., Ke, Y., Chu, S., Özsu, M.T.: Efficient core decomposition in massive networks. In: Proceedings of ICDE, pp. 51–62 (2011)
7. Cheng, J., Ke, Y., Fu, A.W.-C., Yu, J.X., Zhu, L.: Finding maximal cliques in massive networks. TODS **36**(4), 21 (2011)

8. Chiang, Y.-J., Goodrich, M.T., Grove, E.F., Tamassia, R., Vengroff, D.E.: External-memory graph algorithms. In: SODA, vol. 95, pp. 139–149 (1995)
9. Cormen, T.H.: Introduction to Algorithms. MIT Press, Cambridge (2009)
10. Hu, X., Tao, Y., Chung, C.-W.: Massive graph triangulation. In: Proceedings of SIGMOD, pp. 325–336 (2013)
11. Maheshwari, A., Zeh, N.: I/O-efficient planar separators. SIAM J. Comput. **38**(3), 767–801 (2008)
12. Raith, A., Ehrgott, M.: A comparison of solution strategies for biobjective shortest path problems. Comput. Oper. Res. **36**(4), 1299–1331 (2009)
13. Tarjan, R.E.: Edge-disjoint spanning trees and depth-first search. Acta Inform. **6**(2), 171–185 (1976)
14. Wang, J., Cheng, J.: Truss decomposition in massive networks. Proc. VLDB Endow. **5**(9), 812–823 (2012)
15. Yuan, L., Qin, L., Lin, X., Chang, L., Zhang, W.: Diversified top-k clique search. VLDB J. **25**(2), 171–196 (2016)
16. Yuan, L., Qin, L., Lin, X., Chang, L., Zhang, W.: I/O efficient ECC graph decomposition via graph reduction. PVLDB **9**(7), 516–527 (2016)
17. Yuan, L., Qin, L., Lin, X., Chang, L., Zhang, W.: I/O efficient ECC graph decomposition via graph reduction. VLDB J. **26**(2), 275–300 (2017)
18. Zhang, Z., Yu, J.X., Qin, L., Shang, Z.: Divide & conquer: I/O efficient depth-first search. In: Proceedings of the 2015 ACM SIGMOD International Conference on Management of Data, pp. 445–458 (2015)

Finding All Nearest Neighbors
with a Single Graph Traversal

Yixin Xu$^{(\boxtimes)}$, Jianzhong Qi, Renata Borovica-Gajic, and Lars Kulik

School of Computing and Information Systems, The University of Melbourne,
Melbourne, Australia
yixinx3@student.unimelb.edu.au,
{jianzhong.qi,renata.borovica,lkulik}@unimelb.edu.au

Abstract. Finding the nearest neighbor is a key operation in data analysis and mining. An important variant of nearest neighbor query is the all nearest neighbor (ANN) query, which reports all nearest neighbors for a given set of query objects. Existing studies on ANN queries have focused on Euclidean space. Given the widespread occurrence of spatial networks in urban environments, we study the ANN query in spatial network settings. An example of an ANN query on spatial networks is finding the nearest car parks for all cars currently on the road. We propose VIVET, an index-based algorithm to efficiently process ANN queries. VIVET performs a single traversal on a spatial network to precompute the nearest data object for every vertex in the network, which enables us to answer an ANN query through a simple lookup on the precomputed nearest neighbors. We analyze the cost of the proposed algorithm both theoretically and empirically. Our results show that the algorithm is highly efficient and scalable. It outperforms adapted state-of-the-art nearest neighbor algorithms in both precomputation and query processing costs by more than one order of magnitude.

1 Introduction

Finding the nearest neighbor is an important query in spatial databases. Its variation includes reverse nearest neighbor search [1], continuous nearest neighbor search [2], all nearest neighbor search [3] and so forth. An important variant of the nearest neighbor query, the *all nearest neighbor query*, returns the nearest neighbor of each query object over a spatial network. Despite its importance, this query has not been addressed in the research literature on spatial networks.

ANN queries have many applications. We briefly discuss two of them: (i) ridesharing and (ii) carparks. For ridesharing, the average number of daily trips using Uber reached 5.5 million in 2016 [4], which shows the importance of highly scalable and efficient ANN algorithms to match cars with riders instantly. (ii) According to a study on parking spaces of 27 districts in the United States [5], the average oversupply ratio of parking spaces to cars requiring parking is 45% among districts that have identified parking shortages. The large oversupply ratio implies that building more parking spaces is not an effective solution to

© Springer International Publishing AG, part of Springer Nature 2018
J. Pei et al. (Eds.): DASFAA 2018, LNCS 10827, pp. 221–238, 2018.
https://doi.org/10.1007/978-3-319-91452-7_15

the perceived lack of parking spaces. Instead, this study shows that there is an increasing need for real-time parking management, which is able to quickly report the locations of the nearest parking spaces for all drivers. This real-time parking management requires finding nearest neighbors (carparks) for all drivers in a road network. Both applications are examples of ANN queries in spatial networks. Figure 1 shows an example of an ANN query. Given two data objects o_1, o_2 and four query objects q_1, q_2, q_3, q_4, an ANN query is to compute the nearest data object for each query object, e.g., o_1 for q_1 and q_2, and o_2 for q_3 and q_4.

Existing studies on ANN mainly focus on the Euclidean space [3,6–11], where the distance between two points is determined by their Euclidean distance. In the real world, movements of objects are usually restricted by the underlying road network. The traveling cost between two points is not only determined by their relative positions but also affected by the route between them. Take v_7 and v_{13} in Fig. 1 as an example. The travel distance between them is much larger than their Euclidean distance because the route must make a long detour to avoid the lake. In spatial networks, the distance between two points is measured by the length of their shortest path. Data structures and heuristics used by ANN algorithms in the Euclidean space, e.g., R-tree [12] and grid-partitioning [7], are not applicable to spatial networks due to the different distance concepts. Our study fills the need for an efficient ANN algorithm in spatial networks. To the best of our knowledge, this is the first study on ANN queries in spatial networks.

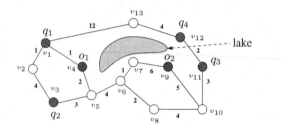

Fig. 1. An example of all nearest neighbor query.

A straightforward solution to find ANNs in spatial networks is to apply a state-of-the-art spatial network *nearest neighbor* (NN) algorithm for each query object individually. However, this solution is inefficient for large numbers of query objects. Besides, it does not scale to large networks due to high memory cost.

Applying NN algorithms straightforwardly is inefficient due to the overlap of the search regions of some query objects. For example, in Fig. 1, both the search regions of q_3 and q_4 cover the two edges between v_{10} and v_{11} and between v_9 and v_{10}, as these two edges are both on the shortest paths to their nearest neighbor the nearest neighbor of q_3 and q_4 (i.e., o_2). When the number of query objects is large, a large part of the network may be visited multiple times, thereby severely impacting query performance. Thus, an efficient ANN algorithm needs a careful design to avoid unnecessary visits.

The reason that the straightforward solution is not scalable is due to the index structures used by spatial network NN algorithms. Most of the recent spatial network NN algorithms improve their query performance by building indices during a precomputation phase. However, these indices are memory-intensive and thus do not scale to large networks. Table 1 depicts the average memory consumption of two state-of-the-art spatial network NN algorithms, G-tree [13] and IER-PHL [14], over five real-world road networks. The two algorithms consume rapidly increasing memory with the growing network size, which renders them inapplicable to large networks. To illustrate, IER-PHL requires over 64 GB memory to index networks with roughly 14 million vertices.

Table 1. Memory consumption of G-tree, IER-PHL, and VIVET over five road networks.

Road network	Number of vertices	Memory consumption		
		G-tree	IER-PHL	VIVET
Northwest US	1 million	88.4 MB	845.1 MB	9.2 MB
East US	3.6 million	339.8 MB	6.4 GB	27.5 MB
Western US	6.3 million	543.3 MB	10.4 GB	47.8 MB
Central US	14.1 million	1.5 GB	>64 GB	107.4 MB
Full US	23.9 million	2.4 GB	>64 GB	182.7 MB

We propose **VIVET** (**Vi**rtual **ve**rtex **t**raversal), a spatial network ANN algorithm that overcomes the above limitations. In the precomputation phase, VIVET runs Dijkstra's algorithm starting with a virtual vertex. The virtual vertex is created by connecting it to every data object with an edge of weight *zero* (as shown in Fig. 2). After the traversal, the shortest path from the virtual vertex to each vertex in the network is obtained. For each vertex v_i, we observe that there is always *one* data object o_j on the shortest path from the virtual vertex to v_i, and o_j is the nearest neighbor of v_i. We store the nearest neighbors of all vertices in an array N. For query processing, VIVET reports the nearest neighbor of every query object by a simple lookup to N.

VIVET significantly outperforms solutions adapted from state-of-the-art nearest neighbor algorithms described above in terms of precomputation and query cost. The precomputation of VIVET is efficient and easy to implement compared with other nearest neighbor indices because it only requires a *single traversal* over the network. Furthermore, the memory consumption of the VIVET index (the array N) is linear to the number of vertices in the network, which makes it scalable to large networks. Taking Table 1 as an example, the memory consumption of VIVET is more than an order of magnitude lower compared with the index of G-tree and two orders of magnitude lower compared with the index of IER-PHL. In query processing, VIVET refers to the array N directly to report the query results and thus outperforms the state-of-the-art NN algorithms by almost two orders of magnitude. For example, VIVET needs less than 0.02 s to answer 500,000 query objects while existing NN algorithms require more than 6 s under the same setting.

To summarize, our contributions are as follows:

- To the best of our knowledge, this is the first study on all nearest neighbor queries in spatial networks.
- We propose a simple and efficient algorithm called VIVET for ANN queries in spatial networks. VIVET is applicable to both undirected and directed networks.
- Our theoretical analysis proves the advantage of VIVET. The precomputation of VIVET requires $O((|E| + |V| + n) \log |V|)$ time and $O(|V|)$ space, where n represents the number of data objects and $|E|$, $|V|$ represent the number of edges and vertices, respectively. The overall query complexity is linear to the number of query objects m, i.e., $O(m)$.
- We conduct experiments on both real-world and synthetic data, showing that VIVET outperforms the state-of-the-art algorithms by one to two orders of magnitude in terms of query times and precomputation costs.

2 Preliminaries

We start with a few basic concepts, based on which we define the all nearest neighbor query in spatial networks.

We consider a set of n data objects $\mathcal{O} = \{o_1, o_2, \ldots, o_n\}$ and a set of m query objects $\mathcal{Q} = \{q_1, q_2, \ldots, q_m\}$. Both the data objects and the query objects are represented by points on a spatial network.

The *spatial network* is modeled by a graph $G = \langle V, E \rangle$, where V is a set of vertices and E is a set of edges. We consider both directed and undirected graphs. For ease of presentation, we assume an undirected graph by default, and will discuss how our techniques and algorithms can be adapted to directed graphs in Sect. 3.3. An edge $e_{i,j} \in E$ connects two vertices v_i and v_j in V. Such two vertices are called *adjacent vertices*. Every edge $e_{i,j}$ is associated with a *weight*, denoted by $w(e_{i,j})$, which represents the cost of traveling between v_i and v_j. A *path* between two vertices v_i and v_j is an ordered list of edges between the two vertices, denoted by $P_{i,j}$. We use $|P_{i,j}|$ to denote the number of edges in the path, and $l(P_{i,j})$ to denote the *length* of the path, which is the sum of the weights of the edges in the path. The *shortest path* between v_i and v_j is the path between them with the smallest length. This smallest length is called the *shortest path distance* between v_i and v_j, denoted by $d_n(v_i, v_j)$. We further use $d_e(v_i, v_j)$ to denote the Euclidean distance between v_i and v_j.

For simplicity, we assume that the data objects and the query objects are located at the graph vertices. This assumption can be easily met by adding vertices that represent the data objects or query objects to the graph.

Nearest Neighbor Query in a Spatial Network. Given a query object q and a set of data objects $\mathcal{O}$ in a spatial network G, a *nearest neighbor* query finds the nearest data object $o_i \in \mathcal{O}$ with the smallest shortest path distance to q, denoted by $NN(q)$:

$$NN(q) = \{o_i \in \mathcal{O} | \forall o_j \in \mathcal{O} : d_n(o_i, q) \leq d_n(o_j, q)\}$$

All Nearest Neighbor Query in a Spatial Network. Given a set of query objects $\mathcal{Q}$ and a set of data objects $\mathcal{O}$ in a spatial network G, an *all nearest neighbor* query finds the nearest data object $o_j \in \mathcal{O}$ with the smallest shortest path distance to every query object $q_i \in \mathcal{Q}$. The query answer is a set of tuples of a query object and its nearest data object, denoted by $ANN(\mathcal{Q}, \mathcal{O})$. Formally,

$$ANN(\mathcal{Q}, \mathcal{O}) = \{\langle q_i, o_j \rangle | q_i \in \mathcal{Q}, o_j \in \mathcal{O}, o_j = NN(q_i)\}$$

In Fig. 1, $ANN(Q, O) = \{\langle q_1, o_1 \rangle \langle q_2, o_1 \rangle, \langle q_3, o_2 \rangle, \langle q_4, o_2 \rangle\}$.

3   VIVET

In this section, we present our VIVET algorithm for ANN queries in spatial networks. The VIVET algorithm precomputes and stores the nearest data object of every vertex in the network. When an ANN query is issued, we simply lookup for the vertices where the query objects lie on and return the corresponding nearest data object. Next, we detail the precomputation process of VIVET, which computes the nearest neighbors for all the vertices in a spatial network with a single traversal over the network.

3.1   Precomputation

To compute the nearest neighbors for all the vertices, a straightforward method is to run a graph shortest path search algorithm such as Dijkstra's algorithm [15] starting from every vertex in the network. However, this algorithm may traverse the network too many times and access the same vertices and edges repetitively.

To avoid such repetitive computation and overlapping network traversals, we propose to traverse the network starting from a virtual vertex which connects to all data objects. The traversal will go through every vertex in the network. When the traversal reaches a vertex, the corresponding path reaching the vertex must pass a data object and this data object will be recorded as the nearest neighbor of the vertex.

We first augment the graph G with a virtual vertex v^* and connect it to every data object $o_i \in \mathcal{O}$ with a directed edge $\overrightarrow{e_{*,i}}$ of weight 0. As we assume that the data objects are all on the vertices, this process effectively connects the virtual vertex to every vertex v_i in V on which a data object lies. We denote the resulting graph as G^*, $G^* = \langle V^*, E^* \rangle$, where $V^* = V \cup \{v^*\}$ and $E^* = E \cup \{\overrightarrow{e_{*,i}} | \overrightarrow{e_{*,i}}$ connects v^* to $o_i \in \mathcal{O}\}$.

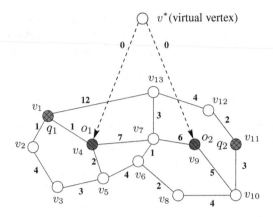

Fig. 2. An example of VIVET. (Color figure online)

We call G^* the *augmented graph*. Figure 2 illustrates such a graph. The virtual vertex v^* is connected to vertices v_4, v_9 where there are data objects o_1 and o_2. Note that even though the original graph G is undirected, the augmented graph G^* contains directed edges that only allows traveling from v^* to the vertices (data objects) in V. The directed edges here are used to ensure that the graph traversal starting from v^* will not go back to v^* (note the zero weight for the edges connecting v^* to the data objects), so as to guarantee the validity of our precomputation algorithm.

Once the augmented graph G^* is computed, we run a single-source graph shortest path algorithm (e.g., Dijkstra's algorithm) starting from the virtual vertex v^* to find the shortest path to every vertex in V. We record the data object that a path goes through. When the traversal reaches a vertex v_i, the data object on the path to v_i is recorded. We show that there is always one and only one data object on the shortest path from v^* to v_i and this data object is the nearest data object of v_i with the following two lemmas.

Lemma 1. *Given a connected graph $G = \langle V, E \rangle$ and an augmented graph $G^* = \langle V^*, E^* \rangle$ created from G, for every vertex $v_i \in V$, there must be one and only one data object on the shortest path from v^* to v_i.*

Proof. First, we prove that there must be at least one data object on the shortest path to v_i. Since G is connected, there must be a path that connects v^* to v_i by the design of the augmented graph G^*. Since v^* is only connected to the data objects, any path including the shortest path from v^* to v_i must go through at least one data object.

Next, we prove by contradiction that there is at most one data object on the shortest path to v_i. Let $\langle P = v^*, o_j, \ldots, v_i \rangle$ be the shortest path from v^* to v_i. The second vertex on by the path must be a vertex on which a data object o_j lies by design of G^*. Suppose that there is another data objects o_k in the path, i.e., $P = \langle v^*, o_j, \ldots o_k, \ldots, v_i \rangle$. Then, the distance between o_j and v_i must be larger than that between o_k and v_i, i.e., $d_n(o_j, v_i) > d_n(o_k, v_i)$. Since there is an edge that connects from v^* to every data object, there must be another path $P' = \langle v^*, o_k, \ldots, v_i \rangle$. The edge between v^* to every data object has a zero weight. Thus, the path length $l(P) = d_n(o_j, v_i) > d_n(o_k, v_i) = l(P')$, which contradicts that P is the shortest path between v^* and v_i.   □

For example, in Fig. 2, there is only one data object o_2 on the shortest path between v^* and v_{11}, which goes through v^*, o_2, v_{10}, v_{11}.

Lemma 2. *Given a connected graph $G = \langle V, E \rangle$ and an augmented graph $G^* = \langle V^*, E^* \rangle$ created from G, the data object on the shortest path from v^* to every vertex $v_i \in V$ is the nearest data object of v_i.*

Proof. The proof is similar to the second half of Lemma 1's proof and omitted.   □

Lemmas 1 and 2 guarantee the correctness of using a single-source shortest path algorithm to compute the nearest neighbor for every vertex. Any single-source graph shortest path algorithms can be used. We use Dijkstra's algorithm for its simplicity and efficiency [15].

Once the nearest neighbors of the vertices are computed, we store them as an array of vertex-NN pairs (together with the shortest path distance) for fast retrieval at query processing. We call this array the *NN array*. Table 2 illustrates the NN array built for the example shown in Fig. 2.

Table 2. VIVET index of Fig. 2.

	v_1	v_2	v_3	v_4	v_5	v_6	v_7	v_8	v_9	v_{10}	v_{11}	v_{12}	v_{13}
NN	o_1	o_1	o_1	o_1	o_1	o_1	o_2	o_1	o_2	o_2	o_2	o_2	o_2
Distance	1	2	5	0	2	6	6	8	0	5	8	10	9

Algorithm 1 summarizes the precomputation procedure of VIVET. The algorithm starts with creating the augmented graph G^* based on the spatial network G (Lines 1 to 6). Then, it initializes an array N of size $|V|$ to store the NN pairs (Line 7). The data objects located at vertices are the nearest data objects of those vertices, which yield a nearest neighbor distance of 0 (Lines 8 to 10). Next, the graph traversal starts. We use a priority queue PQ to facility the traversal (Line 11). Each element in the queue is a vertex in G^* to be visited, which is prioritized by its distance to the nearest data object computed so far in the NN array. The virtual vertex v^* is inserted into PQ to initialize the traversal (Line 12). A loop is run to keep popping out vertices from PQ (Lines 13 to 21). The vertex with the smallest distance to the nearest data object in PQ is popped out first (Line 14). When a vertex v_i is popped out and visited for the first time, vertices connected to it that have not been visited before are inserted into PQ (Lines 16 to 20). For each such vertex v_j, if the path through v_i is shorter than the existing shortest

---

Algorithm 1. Precomputation

Input : $G = \langle V, E \rangle$, object set $\mathcal{O}$
Output: NN array indexing the nearest object of every vertex $v_i \in V$.

1 create a virtual vertex v^*;
2 $E^* = E, V^* = V \cup \{v^*\}$;
3 **for** $o_i \in \mathcal{O}$ **do**
4   create a virtual edge $\overrightarrow{e_{*,o_i.vid}}$;   // $o_i.vid$ is the vertex ID of o_i
5       $w(\overrightarrow{e_{*,o_i.vid}}) = 0$;
6       $E^* = E^* \cup \{\overrightarrow{e_{*,o_i.vid}}\}$;
7 initialize an array N with size $|V|$;
8 **for** $o_i \in \mathcal{O}$ **do**
9       $N[o_i.vid].nndistance = 0$;
10      $N[o_i.vid].nnid = o_i.oid$;   // $o_i.oid$ is the object ID of o_i
11 initialize a priority queue PQ;
12 PQ.insert (v^*);
13 **while** $PQ \neq \emptyset$ **do**
14   v_i = the first element in PQ;
15   **if** v_i *has not been visited before* **then**
16   **for** *each adjacent vertex* v_j *of* v_i *that have not been visited before* **do**
17   **if** $N[j].nndistance > N[j].nndistance + w(e_{i,j})$ **then**
18                  $N[j].nndistance = N[j].nndistance + w(e_{i,j})$;
19                  $N[j].nnid = N[i].nnid$;
20   PQ.insert(v_j);
21   mark v_i as visited;
22 **return** N;

---

path to v_j, we update the distance to nearest data object of $v_j(N[j].nndistance)$ to be the nearest neighbor distance of v_i plus the weight of the edge between v_i and v_j (Line 18), and the nearest data object of $v_j(N[j].nnid)$ is updated to be that of v_i (Line 19). When PQ becomes empty, all vertices will have been visited and their nearest data objects are computed and stored in the NN array N. The array N is returned and the algorithm terminates (Line 22).

3.2 Query Processing

Once the NN array is computed, an ANN query can be processed by first locating the vertex v_j on which every query object q_i lies and then retrieving the nearest data object of v_j from the NN array, which is returned as the nearest data object of q_i. If a query object is lying on an edge, we locate both vertices of the edge and retrieve their nearest data objects. We compare the distances of the two retrieved data objects to q_i and return the closer one as the nearest data object of q_i. We omit the pseudocode of the query processing procedure for conciseness.

Continuing with the example shown in Fig. 2, where there are two query objects q_1, q_2 represented by the two red circles, q_1 is at v_1 and q_2 is at v_{11}.

The nearest neighbor of v_1 is o_1 and the nearest neighbor of v_{11} is o_2 as shown in the NN array listed in Table 2. VIVET reports o_1 and o_2 as the nearest neighbors of q_1 and q_2, respectively.

3.3   Generalizing the Algorithm

VIVET can be generalized to directed networks and to process ANN queries without precomputation with small changes.

When applied to directed networks, we need to update the traversal for single-source graph shortest path computation as follows. When a vertex is visited (Line 15 of Algorithm 1), we retrieve its inbound edges and add the vertices connected by these edges to the priority queue PQ to be visited next, i.e., we update Line 16 of Algorithm 1 to be "for every vertex v_j that has an edge pointing to v_i". We need to reverse the direction of the edges of the virtual vertex v^* such that they point from the data object vertices to v^* instead of from v^* to the data objects. By doing so, we find the shortest "reverse" paths from the data objects to the vertices in the network, which are the shortest paths from the vertices to the data objects. This approach is correct because we still use Dijkstra's algorithm graph expansion procedure, but restricting the direction of the edges to ensure that the paths found are going from the vertices to the data objects. Our experiments show similar behavior of VIVET for both undirected and directed graphs.

When processing an ANN query without the precomputed NN array, we run the single-source graph shortest path computation online. We find the shortest paths from the virtual vertex v^* to all query objects instead of all network vertices. When the shortest paths are found, the data object o_j on the shortest path to query object q_i is returned as the nearest data object of q_i. The correctness of doing so is guaranteed by Lemmas 1 and 2 above straightforwardly. Our experiments verify the efficiency of VIVET in dynamic scenarios, especially when the number of query objects is large. For example, when the network has over one million vertices and 2^{11} data objects, dynamic VIVET requires 0.5 s to answer an ANN query with 2^{16} query objects while the other state-of-the-art algorithms requires at least 0.8 s.

A *multi-source Dijkstra's algorithm* has been proposed in the literature [16] that starts by adding multiple source vertices into the priority queue PQ. The focus of [16] is to run the multi-source Dijkstra's algorithm to test the reachability of different points and find out the most time-consuming shortest path in the graph for emergency services. Another study [17] shares a similar idea and uses a multi-source shortest path approach for location privacy. To the best of our knowledge, we are the first to apply this technique for finding ANNs in spatial networks.

3.4   Algorithm Complexity

Next, we analyze the complexity of VIVET. We denote the number of data objects as n and the number of query objects as m. We also denote the numbers of vertices and edges in G^* as $|V^*|$ and $|E^*|$, which equals to $|V| + 1$ and $|E| + n$, respectively.

Precomputation. Creating the augmented graph G^* takes $O(|V| + 1 + |E| + n)$ time. The time for traversing G^* to compute the nearest data objects is determined by the time of the single-source shortest path algorithm. We use Dijkstra's algorithm, which has a time complexity of $O((|E^*| + |V^*|) \log |V^*|)$ in the worst case by using a binary heap for the priority queue PQ, which is equivalent to $O((|E| + |V| + n) \log |V|)$. Overall, the time complexity of the precomputation of VIVET is $O((|E| + |V| + n) \log |V|)$. The size of the NN array is linear to the number of vertices, i.e., $O(|V|)$.

Query Processing. The query time complexity of VIVET is linear to the number of query objects. For each query object, the nearest neighbor is computed in constant time from the NN array. Therefore, the query time complexity of VIVET is $O(m)$.

4   Experiments

We experimentally compare the performance of VIVET against the state-of-the-art NN algorithms, IER-PHL [18], G-tree [13] and INE [19]. All algorithms are implemented in C++ and run on a 64-bit virtual node with a 1.8 GHz CPU and 64 GB memory from an academic computing cloud (Nectar [20]) running on OpenStack.

4.1   Experimental Setup

We run ANN queries on real-world road network datasets as listed in Table 3, which are created for the 9th DIMACS Challenge [21]. Each undirected network has two datasets, a travel time dataset and a travel distance dataset, the edge weight of which correspond to the travel distance and the travel time between vertices, respectively. Note that the directed network *Europe* has only the travel

Table 3. Road networks.

Name	# vertices	# edges	Description
NY	264,346	733,846	New York (undirected)
COL	435,666	1,057,066	Colorado (undirected)
FLA	1,070,376	2,712,798	Florida (undirected)
NW	1,207,945	2,840,208	Northwest USA (undirected)
CAL	1,890,815	4,657,742	California & Nevada (undirected)
E	3,598,623	8,778,114	Eastern USA (undirected)
W	6,262,104	15,248,146	Western USA (undirected)
CTR	14,081,816	34,292,496	Central USA (undirected)
Europe	18,010,173	42,188,664	Europe (directed)
USA	23,947,347	58,333,344	Full USA (undirected)

time dataset. As the experimental results on the travel distance dataset are consistent with that on the travel time dataset in most cases, we focus on showing experiments on the travel time dataset due to the space limit. We use two methods to create data object sets: mapping real-world POIs and synthetically sampling. We use eight types of real-world POIs extracted from OpenStreetMap by Abeywickrama et al. [18]. We also synthetically sample vertices of the networks to be the data objects and query objects following two distributions, uniform and clustered. The uniform distribution simulates scenarios where areas with more vertices tend to have more objects, while the clustered distribution simulates scenarios where objects may be clustered in some areas. The maximum number of vertices is 50 in every cluster.

Table 4. Experiment settings.

Parameters	Values	Default
Road networks	Refer to Table 3	*NW*
Number of data objects	2^7 to 2^{16}	2^{11}
Number of query objects	2^7 to 2^{16}	2^{10}
Real-world POIs	Refer to Table 2 in [18]	*Parking*
Synthetic data objects distributions	Uniform, clustered	*Uniform*
Synthetic query objects distributions	Uniform, clustered	*Uniform*

Table 4 shows the range of variables we use in our experiments. In a default setting, we run queries on 2^{11} uniformly distributed data objects and 2^{10} uniformly distributed query objects over the network *NW*. *Park* is the default POI type in experiments using real-world POIs as data objects. We first show the algorithm performance on undirected graphs and then compare algorithms on directed graphs.

4.2 Precomputation Costs

We compare the precomputation costs of VIVET with the index-based NN algorithms IER-PHL and G-tree by measuring their time and memory consumption.

Effect of the Network Size. Figure 3a and b show the precomputation costs over different networks. All algorithms require longer time and larger memory to build indices when the network has more nodes and edges. Compared with IER-PHL and G-tree, VIVET reduces the precomputation time by two orders of magnitude and saves the memory consumption by one order of magnitude due to a single traversal over the network. Compared to the precomputation costs on the travel distance dataset as shown in Table 1, both G-tree and VIVET require consistent precomputation costs on the two datasets. IER-PHL, however, requires less memory on the travel time dataset by taking advantage of the travel

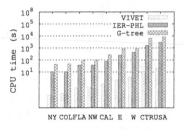

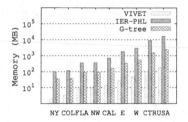

(a) Precomputation vs. networks.

(b) Memory consumption vs. networks.

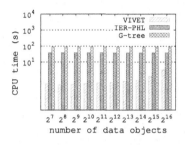

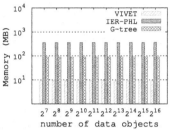

(c) Precomputation time vs. # data objects.

(d) Memory consumption vs. # data objects.

Fig. 3. Precomputation costs.

speed (geometrical length divided by travel time) to improve the effectiveness of the highway decomposition, thereby reducing the index size.

Effect of the Number of Data Objects. Figure 3c and d show the effect of the number of data objects on the precomputation costs. Varying the number of data objects has little effect on the precomputation costs of both IER-PHL and G-tree because their precomputation costs are dominated by the process of building network indices. As for VIVET, its precomputation time increases with the growing number of data objects, which is caused by the increasing number of virtual edges added in the augmented network G^*. Even though the precomputation costs of VIVET are impacted by the number of data objects, the number of data objects in real world scenarios usually lies within a reasonable range. For example, the number of parking spaces in NW is 5098 [18], which lies in the range between 2^{12} and 2^{13} as shown in Fig. 3c. The precomputation time of VIVET in this range is approximately 1.5% of that of IER-PHL and 0.5% of that of G-tree. In terms of memory consumption VIVET has a constant index size when the number of data objects changes as its index size is determined only by the number of vertices. Its index size is at least an order of magnitude smaller than those of IER-PHL and G-tree.

4.3   Query Costs

We further analyze the query performance of IER-PHL, G-tree, INE, and VIVET by comparing their query times.

Effect of the Network Size. Figure 4 shows the query times of the four algorithms on different networks. The query times of G-tree and INE increases rapidly with the growing network size due to their larger search region, while those of IER-PHL and VIVET are much less impacted by the network size. However, IER-PHL consumes large size of memory for large networks as shown in Fig. 3b. VIVET outperforms the other three algorithms by more than two orders of magnitude over all networks, which shows the efficiency and scalability of VIVET in large networks.

Effect of the Number of Data Objects. Figure 6 shows the query times of uniform and clustered data objects when varying number of data objects. VIVET again outperforms the state-of-the-art by more than two orders of magnitude. Furthermore, the query times of VIVET is unaffected by the size and distribution of data objects as it only performs a simple lookup to answer an ANN query.

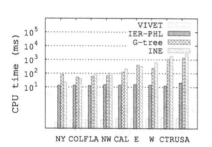

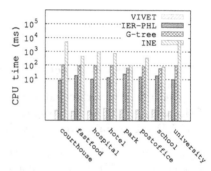

Fig. 4. Query time vs. network size.   **Fig. 5.** Query time vs. real data objects.

Effect of the Number of Query Objects. Figure 7 shows the effect of the number of query objects on the query performance. Query objects in Fig. 7a are generated following the uniform distribution while those in Fig. 7b are generated following a clustered distribution. As expected, the query times of all algorithms grow with the increasing number of query objects. VIVET is two orders of magnitude faster than the most efficient baseline IER-PHL. Furthermore, our scalability experiments show that VIVET can answer an ANN query with 10 million query objects within 0.3 s, while the other state-of-the-art algorithms require more than 2 s to answer such large number of query objects.

Real-World Object Sets. Figure 5 shows the query times of different algorithms when data objects are generated based on real-world POIs. VIVET outperforms other algorithms by more than two orders of magnitude on all types of POIs examined.

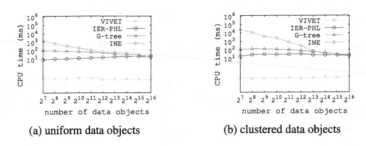

Fig. 6. Effect of the number of data objects on query time.

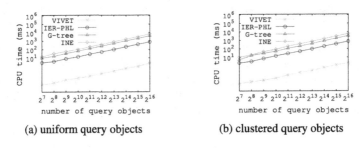

Fig. 7. Effect of the number of query objects on query time.

4.4 Experiments on Directed Graphs

Since G-tree requires undirected graphs for its graph partitioning phase while IER-PHL assumes undirected graphs for indexing, we only compare the performance of VIVET against INE on the directed network *Europe*.

Precomputation Cost. Figure 8 shows the precomputation time of VIVET when the number of data objects varies. The precomputation time required by VIVET increases slightly with increasing number of data objects, which is consistent with the experiments in undirected graphs. The memory consumption of VIVET on *Europe* is 137 MB, which remains linear to the number of vertices in the network.

Query Cost. The query times of VIVET and INE on directed network are compared in Fig. 9. VIVET outperforms INE by up to four orders of magnitude in

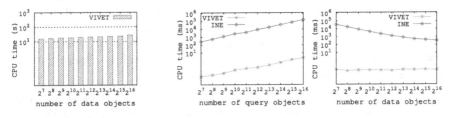

Fig. 8. Precomputation time (directed graph).

Fig. 9. Query time (directed graph).

this set of experiments. When the number of data objects increases, the query performance of INE improves due to the smaller size of the search region. However, even for dense data objects, VIVET still outperforms INE by three orders of magnitude.

5   Related Work

Nearest neighbor queries are studied extensively under various data spaces including spatial network spaces, while all nearest neighbor queries are studied mainly in the Euclidean spaces. We review studies on nearest neighbor queries in spatial networks and all nearest neighbor queries in the Euclidean space.

Nearest Neighbor Queries in Spatial Networks. A key issue in NN query processing in spatial networks lies in the high cost of computing the shortest path distance between two objects, which may require a graph traversal. Studies of NN queries in spatial networks thus explore various techniques to reduce the shortest path distance computation. Papadias et al. [19] propose two spatial network NN algorithms: *IER* and *INE*. The IER algorithm is based on the observation that the spatial network distance between two objects must be no smaller than their Euclidean distance. The INE algorithm gradually expands the search region from the query object so that the first data object reached when expanding is the query answer. Kolahdouzan and Shahabi [22] precomputes a *network Voronoi diagram* over the spatial network, which partitions the network into sections (*network Voronoi cells*). An NN query can then be answered simply by locating the network Voronoi cell containing the query object q. More recent studies use index structures to help to improve the query efficiency. The *distance browsing* algorithm [23] uses the *spatially induced linkage cognizance* (SILC) index, which stores the network shortest path distance between every pair of vertices. The *ROAD* [24] algorithm hierarchically partitions the spatial network and precomputes the spatial network distance between border vertices within every partition, where border vertices of a partition are the vertices connecting to other partitions. The *G-tree* [13] algorithm also partitions the network but differs from ROAD on the tree structures and searching paradigms.

An experimental paper [18] compares the performance of various network NN algorithms. They also proposed an algorithm named IER-PHL that combines the IER algorithm with a shortest path computation technique named *pruning highway labelling* (PHL) [14]. It is found that the query time of IER-PHL outperforms the other NN algorithms in most cases. G-tree is also competitive because it requires lower precomputation costs than IER-PHL, while ranks the second in terms of the query time. INE, on the other hand, is the most efficient algorithm when the data objects are densely distributed.

All Nearest Neighbour in Euclidean Space. ANN algorithms in the Euclidean space can be grouped into index-free and index-based algorithms. We start with the index-free algorithms. Clarkson [25] consider the case where query and data objects belong to the same set. and split the space into small cubic cells of equal size. The distance from a query object to its nearest neighbor is hence bounded by the distance to the nearest cell occupied by a data object. Vaidya [26] use a similar idea but optimize the splitting scheme. When query objects and data objects are in two different sets, the ANN query is also called the *nearest neighbor join* (NN-join) query. Xia et al. [7] propose the *Gorder* algorithm to process the ANN-join query. Gorder divides query objects and data objects into several blocks and schedules the searching order of data objects' blocks so that promising nearest neighbor candidates are visited first. Zhang et al. [8] propose a hash-based algorithm that hashes the query objects together with the data objects and divides them into buckets. For query objects in a bucket, nearest neighbors only need to be searched from data objects in the same or overlapping buckets. Chen and Chang [11] use the Hilbert curve to hash the data objects into grid cells.

Index-based ANN algorithms compute the query with a traversal over a pre-computed index structure. Böhm and Krebs [3] propose an R-tree based algorithm named *MuX*. They optimize the I/O cost by organizing input data using large pages and take advantage of a secondary search structure within pages to optimize the efficiency. Zhang et al. [8] propose two algorithms for the case where the data objects are indexed with an R-tree. Their first algorithm named *multiple nearest neighbor* (MNN) finds the nearest neighbor of every query object by computing an NN query on the R-tree of data objects. The processing order of the query objects is optimized so that close query objects can be handled consecutively. Their second algorithm named *batched nearest neighbor* (BNN) finds nearest neighbors of multiple query objects at a time. BNN first groups multiple query objects and traverses the R-tree of data objects once for finding the nearest neighbors of a group. Chen and Patel [6] use a Quad-tree variant called the *MBRQT* to index the data objects and propose a metric called *NXNDIST* (MINMAXMINDIST) to prune the search space during index traversal. Sankaranarayanan et al. [27] propose another pruning metric called *MAX-MAXDIST*. Yu et al. [9] use *iDistance* [28] as the index structure. They propose an algorithm named *iJoin* that takes advantage of the data partitioning strategy of iDistance. Emrich et al. [10] propose to index the data objects with an *SS-tree* and use trigonometric relationships to prune the search space during index traversal.

6   Conclusion

We studied all nearest neighbor queries in spatial networks and proposed a scalable and efficient algorithm named VIVET. Compared with the methods adapted from state-of-the-art nearest neighbor algorithms, VIVET reduces the precomputation and query costs by one to two orders of magnitude. The improvements are achieved via a shared computation technique that computes the nearest neighbors for all query objects at the same time with a single graph traversal. Extensive experiments using real road networks confirm the advantages of VIVET in terms of precomputation time, storage space, and query time compared to the state-of-the-art network NN algorithms.

Since existing index structures for nearest neighbor queries suffer due to large memory consumption, while VIVET effectively overcomes this limitation, it is worth to explore the applicability of the VIVET index structure on other variants of the nearest neighbor problems in spatial networks in the future. Our preliminary results already confirmed the advantage of VIVET on NN queries compared to the state-of-the-art in terms of precomputation costs and query performance.

Acknowledgment. This work is supported in part by Australian Research Council (ARC) Discovery Project DP180103332.

References

1. Safar, M., Ibrahimi, D., Taniar, D.: Voronoi-based reverse nearest neighbor query processing on spatial networks. Multimed. Syst. **15**(5), 295–308 (2009)
2. Mouratidis, K., Yiu, M.L., Papadias, D., Mamoulis, N.: Continuous nearest neighbor monitoring in road networks. In: VLDB, pp. 43–54 (2006)
3. Böhm, C., Krebs, F.: The k-nearest neighbour join: turbo charging the KDD process. Knowl. Inf. Syst. **6**(6), 728–749 (2004)
4. https://techcrunch.com/2016/07/18/uber-has-completed-2-billion-rides/
5. Weinberger, R.R., Karlin-Resnick, J.: Parking in mixed-use US districts: oversupplied no matter how you slice the pie. Transp. Res. Rec.: J. Transp. Res. Board (2537), 177–184 (2015)
6. Chen, Y., Patel, J.M.: Efficient evaluation of all-nearest-neighbor queries. In: ICDE, pp. 1056–1065 (2007)
7. Xia, C., Lu, H., Ooi, B.C., Hu, J.: GORDER: an efficient method for KNN join processing. In: VLDB, pp. 756–767 (2004)
8. Zhang, J., Mamoulis, N., Papadias, D., Tao, Y.: All-nearest-neighbors queries in spatial databases. In: SSDBM, pp. 297–306 (2004)
9. Yu, C., Cui, B., Wang, S., Su, J.: Efficient index-based KNN join processing for high-dimensional data. Inf. Softw. Technol. **49**(4), 332–344 (2007)
10. Emrich, T., Graf, F., Kriegel, H.-P., Schubert, M., Thoma, M.: Optimizing all-nearest-neighbor queries with trigonometric pruning. In: Gertz, M., Ludäscher, B. (eds.) SSDBM 2010. LNCS, vol. 6187, pp. 501–518. Springer, Heidelberg (2010). https://doi.org/10.1007/978-3-642-13818-8_35
11. Chen, H.L., Chang, Y.I.: All-nearest-neighbors finding based on the Hilbert curve. Expert Syst. Appl. **38**(6), 7462–7475 (2011)
12. Guttman, A.: R-trees: a dynamic index structure for spatial searching. In: SIGMOD, pp. 47–57 (1984)
13. Zhong, R., Li, G., Tan, K.L., Zhou, L.: G-tree: an efficient index for KNN search on road networks. In: CIKM, pp. 39–48 (2013)
14. Akiba, T., Iwata, Y., Kawarabayashi, K.I., Kawata, Y.: Fast shortest-path distance queries on road networks by pruned highway labeling. In: ALENEX, pp. 147–154 (2014)
15. Dijkstra, E.W.: A note on two problems in connexion with graphs. Numer. Math. **1**(1), 269–271 (1959)
16. Eklund, P.W., Kirkby, S., Pollitt, S.: A dynamic multi-source Dijkstra's algorithm for vehicle routing. In: ANZIIS, pp. 329–333 (1996)

17. Duckham, M., Kulik, L.: A formal model of obfuscation and negotiation for location privacy. In: Gellersen, H.-W., Want, R., Schmidt, A. (eds.) Pervasive 2005. LNCS, vol. 3468, pp. 152–170. Springer, Heidelberg (2005). https://doi.org/10.1007/11428572_10

18. Abeywickrama, T., Cheema, M.A., Taniar, D.: K-nearest neighbors on road networks: a journey in experimentation and in-memory implementation. PVLDB 9(6), 492–503 (2016)

19. Papadias, D., Zhang, J., Mamoulis, N., Tao, Y.: Query processing in spatial network databases. In: VLDB, pp. 802–813 (2003)

20. http://nectar.org.au/research-cloud/

21. http://www.dis.uniroma1.it/challenge9/

22. Kolahdouzan, M., Shahabi, C.: Voronoi-based k nearest neighbor search for spatial network databases. In: VLDB, pp. 840–851 (2004)

23. Samet, H., Sankaranarayanan, J., Alborzi, H.: Scalable network distance browsing in spatial databases. In: SIGMOD, pp. 43–54. ACM (2008)

24. Lee, K.C., Lee, W.C., Zheng, B., Tian, Y.: ROAD: a new spatial object search framework for road networks. TKDE 24(3), 547–560 (2012)

25. Clarkson, K.L.: Fast algorithms for the all nearest neighbors problem. In: FOCS, pp. 226–232 (1983)

26. Vaidya, P.M.: An O(n log n) algorithm for the all-nearest-neighbors problem. Discret. Comput. Geom. 4(1), 101–115 (1989)

27. Sankaranarayanan, J., Samet, H., Varshney, A.: A fast all nearest neighbor algorithm for applications involving large point-clouds. Comput. Graph. 31(2), 157–174 (2007)

28. Yu, C., Ooi, B.C., Tan, K.L., Jagadish, H.: Indexing the distance: an efficient method to KNN processing. In: VLDB, vol. 1, pp. 421–430 (2001)

Towards Efficient Path Skyline
Computation in Bicriteria Networks

Dian Ouyang[1], Long Yuan[2(✉)], Fan Zhang[2], Lu Qin[1], and Xuemin Lin[2]

[1] Centre for Artificial Intelligence, University of Technology Sydney,
Sydney, Australia
Dian.Ouyang@student.uts.edu.au, lu.qin@uts.edu.au
[2] The University of New South Wales, Sydney, Australia
{longyuan,lxue}@cse.unsw.edu.au, fan.zhang3@unsw.edu.au

Abstract. Path skyline query is a fundamental problem in bicriteria
network analysis and is widely applied in a variety of applications. Given
a source s and a destination t in a bicriteria network G, path skyline
query aims to identify all the skyline paths from s to t in G. In the
literature, PSQ is a fundamental algorithm for path skyline query and
is also used as a building block for the afterwards proposed algorithms.
In PSQ, a key operation is to record the skyline paths from s to v for
each node v that is possible on the skyline paths from s to t. However, to
obtain the skyline paths for v, PSQ has to maintain other paths that are
not skyline paths for v, which makes PSQ inefficient. Motivated by this,
in this paper, we propose a new algorithm PSQ$^+$ for the path skyline
query. By adopting an ordered path exploring strategy, our algorithm
can totally avoid the fruitless path maintenance problem in PSQ. We
evaluate our proposed algorithm on real networks and the experimental
results demonstrate the efficiency of our proposed algorithm. Besides,
the experimental results also demonstrate the algorithm that uses PSQ
as a building block for the path skyline query can achieve a significant
performance improvement after we substitute PSQ$^+$ for PSQ.

1 Introduction

Computing the shortest path for a given source s and a destination t in a network
is one of the most commonly used operation in online location based services,such
as Google Maps[1], Bing Maps[2]. However, using a single cost criterion is often not
sufficient in real applications. For example, in a rush hour, besides considering
the distance between s and t, people often consider the traffic congestion degree
of the path as well. In this application, people prefer to find all the paths that
could potentially be optimal under the distance or traffic congestion degree,
which leads to the path skyline query problem. Given a source s and a destination
t in a network and a set of path cost criteria, a path P_{st} is said to dominate

---

[1] https://maps.google.com/.

[2] https://www.bing.com/maps.

© Springer International Publishing AG, part of Springer Nature 2018
J. Pei et al. (Eds.): DASFAA 2018, LNCS 10827, pp. 239–254, 2018.
https://doi.org/10.1007/978-3-319-91452-7_16

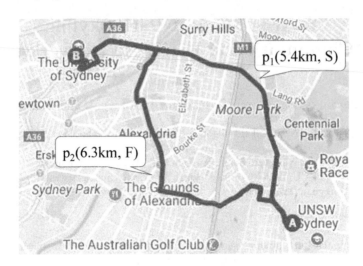

Fig. 1. Path skyline query

another path P'_{st} when P_{st} is strictly better than P'_{st} for one or more criteria while P_{st} and P'_{st} are equally good for the other criteria. A skyline path from s to t is a path that is not dominated by any other path from s to t. The path skyline query aims to find all the skyline paths from s to t.

Example 1. Figure 1 shows a path skyline query in an online location based service. Assume that a user wants find the path from the University of New South Wales (A) to The University of Sydney (B). Instead of returning the shortest path based on distance, path skyline query returns two paths p_1 and p_2 considering both the distance and the live traffic congestion. p_1 has the shortest distance between A and B but the traffic on p_1 moves slow. Contrarily, p_2 is longer than p_1 but the traffic moves faster. Path skyline query provides the flexility to the user to select the path based on his/her preference.

Application. Besides the above example, path skyline query has been adopted in a wide range of other application scenarios. For example, in the telecommunication network, the network routing problem aims to find paths that minimize the total number of links while simultaneously minimize the bandwidth [2], which can be modelled as a path skyline query problem. In bicycle trip planning, different routes are recommended based on the distance and the hardness of road conditions [21], which can also be modelled as a path skyline query problem. In earth observing satellite scheduling, path skyline query can be used to select a daily shot sequence to photograph earth landbelts [6].

Motivation. Path skyline query is a NP-hard problem [17]. In general, the number of skyline paths might increase exponentially as a function of the amount of considered cost criteria [18]. In order to keep the amount of potentially skyline paths on a moderate level and make the returned paths easier to interpret by users, in this paper, we focus on the path skyline query problem in bicriteria networks. In the literature, path skyline query in bicriteria networks has received considerable attention [8,12,15]. Among them, PSQ [8] is a fundamental algorithm for this problem and it is also a building block of the afterwards proposed algorithms for the path skyline query problem [12,15,23]. To answer the path skyline query for a source s and a destination t, PSQ traverses the nodes of G starting from s and iteratively extends the paths that are possible to lead to skyline paths from s to t. During the traversal, a key operation in PSQ is to record the skyline paths from s to v for each node v which is possible on the skyline paths from s to t. In order to achieve this, PSQ keeps all the paths from s to v which are the skyline paths hitherto as a candidate set. When a new path P_{sv} is explored, PSQ enlarges the candidate set by inserting P_{sv} into it, verifies the dominant relations among the paths in the candidate set and updates the candidate set to ensure that the paths in it are skyline paths from s to v. In this operation, although PSQ only needs to record the skyline paths from s to v, those paths that are not skyline paths are also continuously maintained. Since this operation is a basic procedure of PSQ and is performed many times in each iteration, lots of fruitless paths are maintained during the processing, which makes PSQ inefficient in terms of recording the skyline paths for each node v.

Our Approach. In order to address the drawback of PSQ, we propose a new algorithm, PSQ$^+$, for the path skyline query problem. We observe that since PSQ processes the paths from s to v disorderly, when processing a new path P_{sv}, PSQ cannot determine whether P_{sv} is a skyline path w.r.t s and v. As a result, PSQ has to keep a candidate set of skyline paths for v and verify the dominant relationship when a new path is explored. On the other hand, if we can process the paths from s to v in a certain order based on the cost criteria, we can determine whether a path is a skyline path w.r.t s and v directly. Consequently, instead of keeping the candidate set and continuously maintaining the candidate set, we are able to identify whether a path is a skyline path directly and record the exactly skyline paths for each node. Following this idea, our proposed algorithm can avoid the fruitless path maintenance problem in PSQ.

Contribution. In this paper, we make the following contributions:

(1) *Investigation of performance bottleneck of a fundamental algorithm for path skyline query.* We conduct a comprehensive literature review and investigate the performance bottleneck of a fundamental algorithm PSQ for path skyline query. Through theoretical and experimental analyses, we find that the operation of recording skyline paths for each possible nodes in PSQ restricts its efficiency.

(2) *An accurate skyline paths record method.* We propose a new method to record the skyline paths for each node which is possible on the skyline paths from s to t. Our new method can guarantee that the paths recorded for each node during the processing are exactly the skyline paths, which avoid the repetitious skyline paths maintenance in the existing solution.

(3) *Extensive performance studies on real large networks.* We conduct extensive performance studies using real large networks. The experimental results

show that PSQ$^+$ outperforms PSQ on all the networks used in our experiment. We also evaluate the afterwards proposed algorithm 2P which uses PSQ as a building block for path skyline query problem. The experimental results on the real networks demonstrate that 2P can achieve a significant performance improvement after we substitute PSQ$^+$ for PSQ.

Outline. The remainder of this paper is organized as follows. Section 2 reviews the related work. Section 3 provides the formal problem definition. Section 4 introduces a fundamental algorithm PSQ for the path skyline query problem and investigates its drawbacks. Section 5 presents our proposed algorithm. Section 6 evaluates our proposed algorithm with various real large networks in terms of efficiency and scalability. Section 7 concludes the paper.

2   Related Work

Path skyline query in bicriteria networks is a NP-hard problem [17]. [14] suggests that in practical applications we can expect to find a reasonably small number of skyline paths. In the literature, the methods to solve the path skyline query in bicriteria networks can be divided into three subclasses: labelling methods, ranking methods and two-phases methods. Labelling methods [1,3,8,11,20] maintain a set of non-dominated solutions at each node and the identified skyline paths are represented by the labels at the destination node when the labelling methods finish. In Sect. 4, we present a representative method, PSQ, in this subclass in detail. Ranking methods [19] first compute all paths having a length within a certain deviation from the length of the shortest path for one criterion. Then the skyline paths are computed based on the obtained paths by considering the other criterion. The advantage of methods in this subclass is that they do not have to compute the shortest paths from s to every other nodes in the network, but, on the other hand, it is proved to be not competitive compared with the methods in the other two subclasses, because they perform dominance test only at the destination node [15,19]. Two-phases methods [12,15,23] handle a path skyline query from s to t by adopting some pruning strategies. They contain two phases. In phase 1, they compute the so called supported skyline paths. In phase 2, they enumerate the remaining skyline paths using the labelling methods. Due to the introduction of supported skyline path and pruning strategies, two-phase methods are generally more efficient than the labelling methods and ranking methods for the path skyline query problem when the network is large and s and t are located remotely w.r.t the cost criteria in the network [15].

In the literature, there exist some works focusing on the path skyline query in multicriteria networks. [10] proposes two algorithms based on different pruning strategies to address this problem. [9] discusses continuous skyline queries in road networks. [24] studies the path skyline query on the multicriteria time-dependent uncertain networks. Surveys on existing solutions to this problem can be found in [5,7,22]. Besides, [18] study the efficient algorithms for the supported skyline path problem. Preferred path problems are studied in [4,13,16].

3 Preliminaries

A bicriteria network is a graph $G = (V, E)$ where V denote the set of nodes and E denotes the set of edges. Beside, two positive cost $\omega_1(u, v)$ and $\omega_2(u, v)$ are associated with each edge $(u, v) \in E$. We denote the number of nodes as n and the number of edges as m, i.e., $n = |V|$ and $m = |E|$. For a node $u \in V$, we use $\mathsf{nbr}(u, G)$ to denote neighbours of v. The degree of a node $u \in V$, denoted by $\deg(u, G)$, is the number of neighbors of u, i.e., $\deg(u, G) = |\mathsf{nbr}(u, G)|$. For simplicity, we omit G in the notations if the context is self-evident.

Definition 1 (Path and Cost of a Path). *Given a bicriteria network G, a path from node s to node t, denoted by P_{st}, is a sequence of nodes ($s = v_1, v_2, \cdots v_k = t$) where $(v_i, v_{i+1}) \in E$ for each $1 \leq i < k$. The cost of a path $P_{st} = (s = v_1, v_2, \cdots v_k = t)$ w.r.t the l-th cost is $\mathsf{cost}_l(P_{st}) = \sum_{i=1}^{k-1} \omega_l(v_i, v_{i+1})$, where $l \in \{1, 2\}$.*

Definition 2 (Path domination). *Given two path P_{st}^1 and P_{st}^2 in G, P_{st}^1 dominates P_{st}^2, denoted as $P_{st}^1 \prec P_{st}^2$, iff*

- $\mathsf{cost}_1(P_{st}^1) \leq \mathsf{cost}_1(P_{st}^2) \wedge \mathsf{cost}_2(P_{st}^1) < \mathsf{cost}_2(P_{st}^2)$, *or*
- $\mathsf{cost}_1(P_{st}^1) < \mathsf{cost}_1(P_{st}^2) \wedge \mathsf{cost}_2(P_{st}^1) \leq \mathsf{cost}_2(P_{st}^2)$

Definition 3 (Skyline Path). *Given a source s and a destination t in V, P_{st} is a skyline path w.r.t s and t if there exists no other path Q_{st} which dominates P_{st}.*

Definition 4 (Path Skyline Query). *Given a bicriteria network G, a source s and a destination t in V, the path skyline query w.r.t s and t, denoted by $\mathcal{PSQ}(s, t)$, identifies all the skyline paths from s to t in G.*

Problem Statement. In this paper, we study the path skyline query problem and aim to devise an efficient algorithm to find all the skyline paths for a source s and a destination t in G.

4 Existing Solution

As shown in Sect. 2, among the algorithms for path skyline query in the literature, PSQ [8] is a representative algorithm in the labelling methods and is also a building block of the two phases algorithms. Therefore, we focus on improving the efficiency of PSQ. PSQ is shown in Algorithm 1.

Given a source s and a destination t in G, PSQ enumerates all the skyline paths from s to t. To achieve this goal, for each node v which is possible on the skyline paths from s to t, PSQ uses $v.\mathsf{Skyline}$ to record the skyline paths from s to v (lines 1–2). Starting from s, PSQ traverses the nodes of G to enumerate the skyline paths in an iterative manner. Specifically, it maintains a priority queue Q (line 3) and each element (u, c_1, c_2) in Q represents an explored candidate skyline path P_{su} with the first cost as c_1 and the second cost as c_2. The priority

Algorithm 1. PSQ (Graph G, node s, node t)

1: **for each** $v \in V(G)$ **do**
2:      v.Skyline $\leftarrow \emptyset$;
3: PriorityQueue $Q \leftarrow \emptyset$;
4: Q.push$((s,0,0))$;
5: **while** $Q \neq \emptyset$ **do**
6:      $(u, c_1, c_2) \leftarrow Q$.pop();
7:      **for each** $v \in$ nbr(u) **do**
8:          cost$_1(P_{su \rightarrow v}) \leftarrow c_1 +$ cost$_1(u, v)$; cost$_2(P_{su \rightarrow v}) \leftarrow c_2 +$ cost$_2(u, v)$;
9:          **if** $P_{su \rightarrow v}$ is not dominated by any path in v.Skyline **then**
10:              insert $P_{su \rightarrow v}$ into v.Skyline; // line 9-11 maintains the skyline paths
11:              verify the dominance and eliminate all dominated paths in v.Skyline;
12:              **if** $v \neq t$ **then**
13:                  Q.push$((v, \text{cost}_1(P_{su \rightarrow v}), \text{cost}_2(P_{su \rightarrow v})))$;
14: output the skyline paths from source s to t by backtracking from t.Skyline.

function of Q is the lexicographic order of the two costs for the explored paths in Q. It first pushes $(s,0,0)$ into Q (line 4) as s is the source node. Then, it pops an element out from Q each time (line 6). For the popped element (u, c_1, c_2), it iterates each neighbor $v \in$ nbr(u) and conducts the edge relaxation for edge (u, v) (lines 7–13). It first computes the first and the second cost of path $P_{su \rightarrow v}$ (line 8). Here, $P_{su \rightarrow v}$ denotes the path from s to v through the edge (u, v). PSQ checks whether $P_{su \rightarrow v}$ is a candidate skyline path w.r.t s and v (line 9). If $P_{su \rightarrow v}$ is a candidate skyline path, PSQ inserts $P_{su \rightarrow v}$ into v.Skyline and maintains the explored candidate skyline paths from s to v by eliminating all the dominated paths in v.Skyline (lines 10–11). If v is not the destination node, element $(v, \text{cost}_1(P_{su \rightarrow v}), \text{cost}_2(P_{su \rightarrow v}))$ is pushed into Q (lines 12–13), which means $P_{su \rightarrow v}$ possibly leads to new candidate skyline paths for other nodes. The procedure terminates when Q is empty (line 5). At last, PSQ outputs the skyline paths from s to t by backtracking from t.Skyline (line 14).

Drawback of PSQ. In order to compute the skyline paths from s to t, PSQ intends to record the skyline paths w.r.t s and v for the node v which is possible on the skyline paths from s to t. To achieve this goal, PSQ computes the skyline paths w.r.t s and v gradually as follows: it keeps all the paths from s to v which are the skyline paths w.r.t s and v hitherto as a candidate set in v.Skyline. When a new path P_{sv} is explored (line 8), PSQ enlarges v.Skyline by inserting P_{sv} into it (line 10), verifies the dominant relations among the paths in v.Skyline and eliminates all the dominated paths in it (line 11). The issue of this approach is that PSQ only needs to record the skyline paths from s to v, but those paths that are not skyline paths w.r.t. s and v are also inserted and eliminated from v.Skyline repetitiously during the processing. As shown in Algorithm 1, maintaining v.Skyline is a basic procedure in PSQ and is performed for every new explored path in each iteration. As a result, lots of fruitless paths are maintained in PSQ, which makes it inefficient in terms of handling the path skyline query.

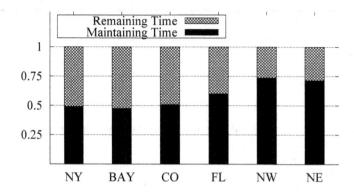

Fig. 2. Percentage of time used for the maintenance of v.Skyline in PSQ

Figure 2 shows the percentage of time used for the maintenance of v.Skyline in PSQ upon six real networks evaluated in our performance studies. We generate 1000 path skyline queries randomly for each real network and compute the average percentage of the time used for the maintenance of v.Skyline in PSQ for each query on the certain real network. As shown in Fig. 2, the time used for the maintenance of v.Skyline occupies most of the total time for the query processing in PSQ. For example, on BAY, maintaining v.Skyline occupies 47% of the total processing time and the percentage is 74% on NW. The experimental results are consistent with our aforementioned analysis.

5 Our New Approach

In order to address the drawback of PSQ, we propose a new algorithm, PSQ$^+$, for the path skyline query. Our algorithm can avoid the problem of fruitless candidate skyline paths maintenance in PSQ and guarantees that each path inserted into v.Skyline is exactly a skyline path from s to v. Before presenting our approach, we first introduce the following two lemmas:

Lemma 1. *Let P_{sv} be a skyline path from s to v in G, then $\mathsf{cost}_2(P_{sv})$ is minimum among all the second costs of the paths from s to v with the first cost as $\mathsf{cost}_1(P_{sv})$.*

Proof. We can prove this lemma by contradiction. Assume that there exists another path P'_{sv} such that $\mathsf{cost}_1(P'_{sv}) = \mathsf{cost}_1(P_{sv})$ and $\mathsf{cost}_2(P'_{sv}) < \mathsf{cost}_1(P_{sv})$. According to Definition 2, we can derive $P'_{sv} \prec P_{sv}$, which contradicts with the given condition that P_{sv} is a skyline path. Thus, the lemma holds.   □

Lemma 2. *Let P^1_{sv} and P^2_{sv} be two skyline paths from s to v in G, for a path P^3_{sv}, assume that $\mathsf{cost}_1(P^1_{sv}) < \mathsf{cost}_1(P^2_{sv}) < \mathsf{cost}_1(P^3_{sv})$ and there exists no path with the first cost in between $\mathsf{cost}_1(P^2_{sv})$ and $\mathsf{cost}_1(P^3_{sv})$, if $\mathsf{cost}_2(P^3_{sv}) < \mathsf{cost}_2(P^2_{sv})$, then $P^1_{sv} \nprec P^3_{sv}$ and $P^2_{sv} \nprec P^3_{sv}$; otherwise, $P^2_{sv} \prec P^3_{sv}$.*

Algorithm 2. PSQ$^+$ (Graph G, node s, node t, SkylinePath $\mathcal{P}$)

1: **for each** $v \in V(G)$ **do**
2:     v.Skyline $\leftarrow \emptyset$;
3: PriorityQueue $Q \leftarrow \emptyset$;
4: Q.push$((s, s, 0, 0))$;
5: **while** $Q \neq \emptyset$ **do**
6:     $(u, v, c_1, c_2) \leftarrow Q$.pop();
7:     **if** $P_{su \rightarrow v}$ is not dominated by the last path in v.Skyline **then**
8:         insert $P_{su \rightarrow v}$ into v.Skyline;
9:         **if** $v \neq t$ **then**
10:             **for each** $w \in$ nbr(v) **do**
11:                 cost$_1(P_{sv \rightarrow w}) \leftarrow c_1 +$ cost$_1(v, w)$;
12:                 cost$_2(P_{sv \rightarrow w}) \leftarrow c_2 +$ cost$_2(v, w)$;
13:                 Q.push$((v, w, \text{cost}_1(P_{sv \rightarrow w}), \text{cost}_2(P_{sv \rightarrow w})))$;
14: output the skyline paths from source s to t by backtracking from t.Skyline.

Proof. The dominance relation between P_{sv}^2 and P_{sv}^3 can be proved directly based on Definition 2. Then, we focus on the dominance relation between P_{sv}^1 and P_{sv}^3. Since P_{sv}^1 and P_{sv}^2 are two skyline paths and cost$_1(P_{sv}^1) <$ cost$_1(P_{sv}^2)$, according to Definition 2, we have cost$_2(P_{sv}^2) <$ cost$_2(P_{sv}^1)$. And if cost$_2(P_{sv}^3) <$ cost$_2(P_{sv}^2)$, we have cost$_2(P_{sv}^3) <$ cost$_2(P_{sv}^1)$. Then, we can derive that $P_{sv}^1 \not\prec P_{sv}^3$ according to Definition 2. Thus, the lemma holds.   □

According to Lemma 1, for a specific the first cost value c_1, all the paths from s to v with the first cost value as c_1 but without the minimum value of the second cost among these paths are not skyline paths. According to Lemma 2, given two skyline paths P_{sv}^1 and P_{sv}^2, for a path P_{sv}^3, we can determine the dominance relation of P_{sv}^1 and P_{sv}^3 and that of P_{sv}^2 and P_{sv}^3 by P_{sv}^2 alone if these paths are given in the increasing order based their first costs. Combining these two considerations together, we can derive that if we handle the paths from s to v in the lexicographic order of their two costs, we can determine whether a path P_{sv} is a skyline path w.r.t s and v based on the previous skyline path w.r.t s and v that has been determined. In other words, we can determine whether a path is a skyline path immediately when we handle it. As a result, we can avoid the problem of fruitless candidate skyline paths maintenance existing in Algorithm 1. The following problem is how we can handle the paths from s to v in the lexicographic order of their two costs. We can achieve this by postponing the process of v.Skyline from edge relaxation on edge (u, v) to edge (v, w). Following this idea, our improved skyline paths enumeration algorithm, PSQ$^+$, is shown in Algorithm 2.

Algorithm Design. Procedure PSQ$^+$ follows a similar framework as PSQ. v.Skyline records all the skyline paths from s to v and is initialized as $\emptyset$ (lines 1–2). PSQ$^+$ maintains a priority queue Q (line 3) and each element (u, v, c_1, c_2) in Q represents a path $P_{su \rightarrow v}$ with the first cost as c_1 and the second cost as c_2. The priority function is the lexicographic order of the two costs for the

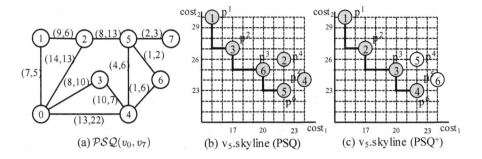

(a) $\mathcal{PSQ}(v_0, v_7)$ (b) v_5.skyline (PSQ)   (c) v_5.skyline (PSQ$^+$)

Fig. 3. PSQ and PSQ$^+$

explored paths in Q. It first pushes $(s, s, 0, 0)$ into Q (line 4) as s is the starting node. Then, it pops an element out from Q each time (line 6). For each popped element (u, v, c_1, c_2), PSQ$^+$ checks whether $P_{su \to v}$ is dominated by the last path in v.Skyline (line 7). If $P_{su \to v}$ is a skyline path, PSQ$^+$ inserts $P_{su \to v}$ into v.Skyline (line 8) and conducts the edge relaxation for each edge (v, w) where $w \in \mathsf{nbr}(v)$ (lines 10–13). For an edge (v, w), it first computes the first and the second cost of path $P_{sv \to w}$ (lines 11–12) and then pushes the element $(v, w, \mathsf{cost}_1(P_{sv \to w}), \mathsf{cost}_2(P_{sv \to w}))$ into Q (line 13). The procedure terminates when Q is empty (line 5). At last, PSQ$^+$ outputs the skyline paths from s to t by backtracking from t.Skyline (line 14). Since PSQ$^+$ can avoid the fruitless candidate skyline paths maintenance problem, it does not have explicit maintenance procedure for v.Skyline compared with PSQ.

Example 2. Consider the graph G in Fig. 3(a) and assume a query $\mathcal{PSQ}(v_0, v_7)$ on G. Figure 3(b) and (c) show the skyline paths maintained in v_5.Skyline during the query processing for $\mathcal{PSQ}(v_0, v_7)$ in PSQ and PSQ$^+$, respectively. In Fig. 3(b), x-axis represents the first cost cost_1 of a path, y-axis represents the second cost cost_2 of a path and each point represents a skyline path with its first and second costs. For example, the first and second cost of the path p^1 are 15 and 30, respectively. In Fig. 3(b), we also show the order in which PSQ processes the paths by the number shown in the point. For G, PSQ processes the paths in the order $p^1, p^4, p^2, p^5, p^6, p^3$. As shown in Fig. 3(b), when processing p^4, since only p^1 is in v_5.Skyline, PSQ considers p^4 as a skyline path and inserts it into v_5.Skyline. PSQ processes p^5 in the similar way. When processing p^6, PSQ finds that p^4 and p^5 are dominated by p^6, then, it inserts p^6 into v_5.Skyline and removes p^4 and p^5 from it. In Fig. 3(b), although p^4 and p^5 are not skyline paths, PSQ inserts them into v_5.Skyline during the processing. On the other hand, PSQ$^+$ processes the paths in the order $p^1, p^2, p^3, p^6, p^4, p^5$, which is shown in Fig. 3(c). For PSQ$^+$, when processing p^4 and p^5, p^6 has been in v_5.Skyline. Thus, p^4 and p^5 are not inserted into v_5.Skyline. As shown in Fig. 3(c), PSQ$^+$ only inserts skyline path from s to v into v.Skyline and it can avoid the fruitless candidate skyline paths maintenance problem in PSQ.

Lemma 3. *Given a path skyline query $\mathcal{PSQ}(s,t)$, for an arbitrary node v, Algorithm 2 explores the paths from s to v in line 7 in lexicographic order of their two costs.*

Proof. We can prove this by contradiction. Without loss of generality, assume that P_{sv}^1 and P_{sv}^2 are two paths from s to v with $\mathsf{cost}_1(P_{sv}^1) < \mathsf{cost}_1(P_{sv}^2)$ but Algorithm 2 explores P_{sv}^2 earlier than P_{sv}^1. Then we have two cases: (1) when Algorithm 2 explores P_{sv}^2, P_{sv}^1 is in Q. In this case, exploring P_{sv}^2 earlier than P_{sv}^1 contradicts with the property of priority queue Q. (2) when Algorithm 2 explores P_{sv}^2, P_{sv}^1 is not in Q. Let P_{su} be a subpath of P_{sv}^1 that is in Q when Algorithm 2 explores P_{sv}^2. According to $\mathsf{cost}_1(P_{sv}^1) < \mathsf{cost}_1(P_{sv}^2)$, we have $\mathsf{cost}_1(P_{su}) < \mathsf{cost}_1(P_{sv}^2)$. Based on the property of priority queue, Algorithm 2 explores P_{su} before P_{sv}^2 and extends the paths through P_{su} until P_{sv}^1 in Q. Thus, it is impossible that P_{sv}^1 is not in Q when Algorithm 2 explores P_{sv}^2. Combining these two cases together, Algorithm 2 explores P_{sv}^1 before P_{sv}^2. Similarly, we can prove the case that $\mathsf{cost}_1(P_{sv}^1) = \mathsf{cost}_1(P_{sv}^2)$ and $\mathsf{cost}_2(P_{sv}^1) < \mathsf{cost}_2(P_{sv}^2)$. Thus, Algorithm 2 explores the paths from s to v in line 7 in lexicographic order of their two costs and the lemma holds.   □

Theorem 1. *Given a path skyline query $\mathcal{PSQ}(s,t)$, for an arbitrary node v, Algorithm 2 records the skyline paths from s to v in v.Skyline.*

Proof. Without loss of generality, let P_{sv} be an arbitrary path in line 7. If P_{sv} is inserted into v.Skyline in line 8, then we have P_{sv} is not dominated by the skyline path that previously inserted into v.Skyline. According to Lemma 2, P_{sv} is not dominated by any other paths in v.Skyline. Besides, based on Lemma 3, Algorithm 2 explores the paths from s to v in line 7 in lexicographic order. Then, the paths explored after P_{sv} cannot dominate P_{sv}. Thus, Algorithm 2 stores the skyline paths from s to v in v.Skyline and the theorem holds.   □

Corollary 1. *Given a path skyline query $\mathcal{PSQ}(s,t)$, for an arbitrary node v, the paths that are not skyline paths from s to v are not maintained in Algorithm 2.*

Proof. This corollary can be proved directly based on Theorem 1 and the procedure of Algorithm 2.   □

Theorem 2. *Given a path skyline query $\mathcal{PSQ}(s,t)$, Algorithm 2 answers $\mathcal{PSQ}(s,t)$ correctly.*

Proof. Based on Theorem 1, we can derive that t.Skyline stores the skyline paths from s to t. Besides, base on Definition 3, if there is a skyline path P_{st} from s to t, then any subpath P_{sv} of P_{st} is also a skyline path w.r.t s and v, we can prove similarly as Theorem 1 that all these subpaths are inserted in v.Skyline. Since we conduct the edge relaxation for every edge (v,w) where $w \in \mathsf{nbr}(v)$ in line 10, then, all the skyline paths w.r.t s to t extending from P_{sv} will be explored in Algorithm 2, which means all the skyline paths from s to t are stored in t.Skyline when Algorithm 2 terminates. In line 14, Algorithm 2 retrieves all the skyline paths by backtracking t.Skyline, thus, Algorithm 2 computes all the skyline paths from s to t correctly and the theorem holds.   □

6   Performance Studies

In this section, we compare our proposed algorithm with other path skyline query algorithms. All experiments are conducted on a machine with an Intel Xeon 3.4 GHz CPU (8 cores) and 32 GB main memory running Linux (Red Hat Linux 4.4.7, 64bit).

Table 1. Datasets used in experiments

Name	Corresponding region	Number of nodes	Number of edges
NY	New York City	$264,346$	$733,846$
BAY	San Francisco Bay Area	$321,270$	$800,172$
COL	Colorado	$435,666$	$1,057,066$
FLA	Florida	$1,070,376$	$2,712,798$
NW	Northwest USA	$1,207,945$	$2,840,208$
NE	Northeast USA	$1,524,453$	$3,897,636$
CAL	California and Nevada	$1,890,815$	$4,657,742$
LKS	Great Lakes	$2,758,119$	$6,885,658$

Datasets. We evaluate the algorithms on eight publicly available datasets from DIMACS[3], each of which corresponds to a part of the road network in the US. In each dateset, intersections and endpoints are represented by nodes and the roads connecting these intersections or road endpoints are represented by undirected edges. For each edge (u, v) in the dateset, two positive costs are associated with (u, v), which represent the physical distance and transit time between two nodes in the network, respectively. The details of the datasets are shown in Table 1.

Algorithms. We implement and compare the following four algorithms:

- PSQ: Algorithm 1 (Sect. 4)
- PSQ$^+$: Algorithm 2 (Sect. 5)
- 2P: A representative two phase algorithm proposed in [15], which uses PSQ as a subroutine to enumerate the skyline paths.
- 2P$^+$: The two phase algorithm in which PSQ is replaced with PSQ$^+$ to enumerate the skyline paths. Except PSQ and PSQ$^+$, all the remaining parts in 2P and 2P$^+$ are the same.

All algorithms are implemented in C++ and compiled with GNU GCC 4.4.7 using optimization level 3. The time cost of the algorithm is measured as the amount of elapsed wall-clock time during the program's execution. For each test, we set the maximum running time for each test to be 200 h. If an algorithm cannot finish the query processing the in the time limit, we do not show its processing time in the figures.

---

[3] http://www.dis.uniroma1.it/challenge9/download.shtml.

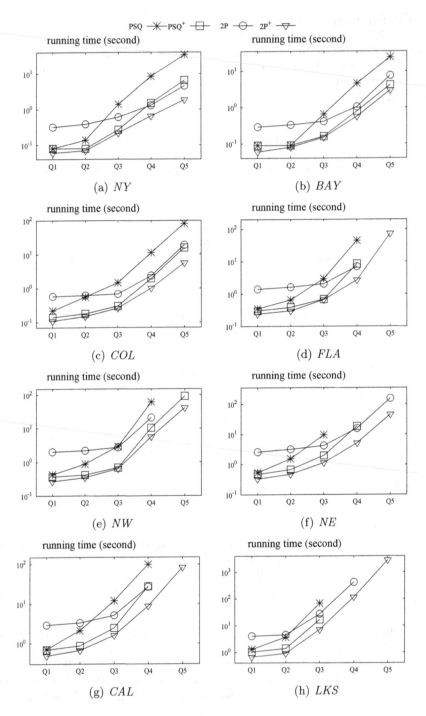

Fig. 4. Query processing time (varying query location)

Exp-1: Efficiency. In this experiment, we evaluate the efficiency of these four algorithms. Since the locations of the source node and the destination node w.r.t the cost criteria influence the efficiency of the algorithms. We generate queries for the experiments as follows: for each dataset, we generate 5 sets of queries $Q_1, Q_2, \ldots, Q_5$. To generate these query sets for a specific dataset, we first figure out the maximum distance l_{max} and the minimum distance l_{min} between any two nodes of the network w.r.t two criteria, respectively. For each criterion, let $x = (l_{max}/l_{min})^{1/10}$, and we generate 100 queries (s, t) to form Q_i for each $1 \leq i \leq 5$ such that the distance between s and t w.r.t the specific criterion is in $[x^{i-6} \cdot l_{max}, x^{i-5} \cdot l_{max}]$. In this way, we can ensure that the source node and destination node of these queries in Q_i are located more remotely than these in Q_{i-1} in the dateset w.r.t the specific criterion for all $1 < i \leq 5$. We execute all algorithms on the datasets and computes the average processing time for each query. The results are shown in Fig. 4.

Figure 4 shows that: (1) For all test cases, PSQ$^+$ outperforms PSQ and 2P$^+$ outperforms 2P. For example, on *FLA* (Fig. 4(d)), the processing time of PSQ is 44.7 s for Q_4 while that of PSQ$^+$ is 8.45 s. On the same dataset, the processing time of 2P for Q_4 is 7 s while that of 2P$^+$ is 2.67 s. On *NW* (Fig. 4(e)), both PSQ and 2P cannot finish the processing of Q_5 in the time limit while PSQ$^+$ and 2P$^+$ can finish the query processing. This is because the operation of maintaining skyline paths in PSQ and 2P is time-consuming while PSQ$^+$ and 2P$^+$ can avoid the problem of fruitless skyline maintenance completely. (2) For all evaluated four algorithms, the query processing time increases as the distance between the source node and destination node in the query increases. For example, on *COL* (Fig. 4(c)), the query processing times of Q_1 for PSQ, PSQ$^+$, 2P and 2P$^+$ are 0.22 s, 0.14 s, 0.58 s, 0.11 s, respectively. However, the query processing time of Q_5 for them are 81.8 s, 15.6 s, 18.8 s, 5.55 s. This is because as the distance between the source node and destination node increases, all the algorithms have to visit more nodes in the networks and more paths are explored during the query processing. As a result, they have to consume more time to process the query. (3) The performance gap between PSQ and PSQ$^+$ (2P and 2P$^+$) increases as the distance between the source node and destination node in the query increases. For example, on *NW* (Fig. 4(e)), regarding PSQ and PSQ$^+$, the processing time of PSQ for Q_1 is 0.43 s and that of PSQ$^+$ is 0.37 s, the gap between processing times is 0.06 s. For Q_4, the processing time of PSQ is 60.89 s and that of PSQ$^+$ is 10.5 s, the gap between processing times is 50.39 s. Regarding 2P and 2P$^+$, the processing time of 2P for Q_1 is 1.68 s and that of 2P$^+$ is 0.26 s, the gap between processing times is 1.42 s. For Q_4, the processing time of 2P is 21.1 s and that of 2P$^+$ is 5.68 s, the gap between processing times is 15.42 s. This is because with the distance between the source node and destination node in the query increases, more paths are explored, as a result, the time saved by avoiding the fruitless skyline maintenance increases as well. Therefore, the performance gap increases with the query distance increases.

Exp-2: Scalability. In this experiment, we evaluate the scalability of the four algorithms when the size of the network increases. To do this, we first divide

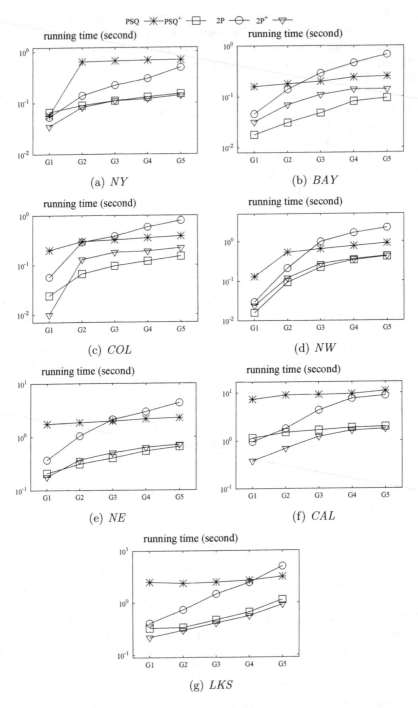

Fig. 5. Query processing time (varying dataset size)

each dataset into a 5×5 gird. Then we create five networks using the 1×1, $2 \times 2, \ldots, 5 \times 5$ grids in the middle of the dataset. These networks are donated as $G_1, G_2, \ldots, G_5$, respectively. For each $i \in [1, 4]$, it is obvious that G_i is contained in G_{i+1}. For the query set of each dataset, we generate 100 queries randomly from G_1. We execute all the algorithms on these networks and compute the average processing time for each query. The results are shown in Fig. 5. Note that G_1 of *FLA* does not contains any nodes and we cannot generate queries for it, thus, the results on *FLA* are not shown in Fig. 5.

From Fig. 5, we can observe that the average processing time for each query of all the four algorithms increases stably as the size of the network increases. This is because as the size of the network increases, more nodes and paths are explored during the query processing. As a result, the average processing time increases as the size of network increases. In the meantime, for all the test cases, PSQ$^+$ outperforms PSQ and 2P$^+$ outperforms 2P. This is also because PSQ and 2P involve the time-consuming operations for the maintenance of skyline paths while PSQ$^+$ and 2P$^+$ can completely avoid the fruitless skyline maintenance problem.

7 Conclusion

In this paper, we study the path skyline query problem in bicriteria networks. In the literature, PSQ is a fundamental algorithm for path skyline query and is also used as a building block for the afterwards proposed algorithms. We investigate the drawbacks in PSQ and propose a new algorithm, PSQ$^+$, to answer the path skyline query. By adopting an ordered path exploring strategy, our algorithm can totally avoid the fruitless path maintenance problem in PSQ. The experimental results demonstrate the efficiency of our proposed algorithms. In the further, we plan to study the path skyline query problem in multi-criteria networks.

Acknowledgments. Long Yuan is supported by Huawei YBN2017100007. Fan Zhang is supported by Huawei YBN2017100007. Lu Qin is supported by ARC DP160101513. Xuemin Lin is supported by NSFC 61672235, ARC DP170101628, DP180103096 and Huawei YBN2017100007.

References

1. Brumbaugh-Smith, J., Shier, D.: An empirical investigation of some bicriterion shortest path algorithms. Eur. J. Oper. Res. **43**(2), 216–224 (1989)
2. Clímaco, J.C., Pascoal, M.: Multicriteria path and tree problems: discussion on exact algorithms and applications. Int. Trans. Oper. Res. **19**(1–2), 63–98 (2012)
3. Climaco, J.C.N., Martins, E.Q.V.: A bicriterion shortest path algorithm. Eur. J. Oper. Res. **11**(4), 399–404 (1982)
4. Delling, D., Wagner, D.: Pareto paths with SHARC. In: Proceedings of the SEA, pp. 125–136 (2009)
5. Ehrgott, M., Gandibleux, X.: A survey and annotated bibliography of multiobjective combinatorial optimization. OR Spectr. **22**(4), 425–460 (2000)

6. Gabrel, V., Vanderpooten, D.: Enumeration and interactive selection of efficient paths in a multiple criteria graph for scheduling an earth observing satellite. Eur. J. Oper. Res. **139**(3), 533–542 (2002)
7. Garroppo, R.G., Giordano, S., Tavanti, L.: A survey on multi-constrained optimal path computation: exact and approximate algorithms. Comput. Netw. **54**(17), 3081–3107 (2010)
8. Hansen, P.: Bicriterion path problems. In: Fandel, G., Gal, T. (eds.) Multiple Criteria Decision Making Theory and Application, pp. 109–127. Springer, Berlin (1980). https://doi.org/10.1007/978-3-642-48782-8_9
9. Jang, S., Yoo, J.: Processing continuous skyline queries in road networks. In: Computer Science and Its Applications, pp. 353–356 (2008)
10. Kriegel, H.-P., Renz, M., Schubert, M.: Route skyline queries: a multi-preference path planning approach. In: Proceedings of the ICDE, pp. 261–272 (2010)
11. Martins, E.Q.V.: On a multicriteria shortest path problem. Eur. J. Oper. Res. **16**(2), 236–245 (1984)
12. Mote, J., Murthy, I., Olson, D.L.: A parametric approach to solving bicriterion shortest path problems. Eur. J. Oper. Res. **53**(1), 81–92 (1991)
13. Mouratidis, K., Lin, Y., Yiu, M.L.: Preference queries in large multi-cost transportation networks. In: Proceedings of the ICDE, pp. 533–544 (2010)
14. Müller-Hannemann, M., Weihe, K.: On the cardinality of the Pareto set in bicriteria shortest path problems. Ann. Oper. Res. **147**(1), 269–286 (2006)
15. Raith, A., Ehrgott, M.: A comparison of solution strategies for biobjective shortest path problems. Comput. Oper. Res. **36**(4), 1299–1331 (2009)
16. Sacharidis, D., Bouros, P., Chondrogiannis, T.: Finding the most preferred path. In: Proceedings of the SIGSPATIAL (2017)
17. Serafini, P.: Some considerations about computational complexity for multi objective combinatorial problems. In: Jahn, J., Krabs, W. (eds.) Recent Advances and Historical Development of Vector Optimization, pp. 222–232. Springer, Berlin (1987). https://doi.org/10.1007/978-3-642-46618-2_15
18. Shekelyan, M., Jossé, G., Schubert, M.: Linear path skylines in multicriteria networks. In: Proceedings of the ICDE, pp. 459–470 (2015)
19. Skriver, A.J.: A classification of bicriterion shortest path (BSP) algorithms. Asia-Pac. J. Oper. Res. **17**(2), 199 (2000)
20. Skriver, A.J., Andersen, K.A.: A label correcting approach for solving bicriterion shortest-path problems. Comput. Oper. Res. **27**(6), 507–524 (2000)
21. Storandt, S.: Route planning for bicycles-exact constrained shortest paths made practical via contraction hierarchy. In: ICAPS, vol. 4, p. 46 (2012)
22. Tarapata, Z.: Selected multicriteria shortest path problems: an analysis of complexity, models and adaptation of standard algorithms. Int. J. Appl. Math. Comput. Sci. **17**(2), 269–287 (2007)
23. Ulungu, E., Teghem, J., Fortemps, P., Tuyttens, D.: MOSA method: a tool for solving multiobjective combinatorial optimization problems. J. Multicriteria Decis. Anal. **8**(4), 221 (1999)
24. Yang, B., Guo, C., Jensen, C.S., Kaul, M., Shang, S.: Multi-cost optimal route planning under time-varying uncertainty. In: Proceedings of the ICDE (2014)

Answering Why-Not Questions
on Structural Graph Clustering

Chuanyu Zong[1]([✉]), Xiufeng Xia[1], Bin Wang[2], Xiaochun Yang[2], Jiajia Li[1],
Xiangyu Liu[1], and Rui Zhu[1]

[1] College of Computer Science, Shenyang Aerospace University,
Liaoning 110136, China
{zongcy,xiaxiufeng,lijiajia,liuxy}@sau.edu.cn, neuruizhu@gmail.com
[2] School of Computer Science and Engineering, Northeastern University,
Liaoning 110819, China
{binwang,yangxc}@mail.neu.edu.cn

Abstract. *Structural graph clustering* is one fundamental problem in managing and analyzing graph data. As a fast and exact density based graph clustering algorithm, pSCAN is widely used to discover meaningful clusters in many different graph applications. The problem of explaining why-not questions on pSCAN is to find why an expected vertex is not included in the specified cluster of the pSCAN results. Obviously, the pSCAN results are sensitive to two parameters: (i) the similarity threshold ϵ; and (ii) the density constraint μ, when them are not set *good* enough, some expected vertices would be missing in the specified clusters. To tackle this problem, we firstly analyze that how the parameters affect the results of pSCAN, then we propose two novel explanation algorithms to explain why-not questions on pSCAN by offering some advices on how to refine the initial pSCAN with minimum penalty from two perspectives: (i) modifying the parameter ϵ; and (ii) modifying the parameter μ. Moreover, we present some constraints to ensure the original pSCAN results are retained as much as possible in the results of refined pSCAN. Finally, we conduct comprehensive experimental studies, which show that our approaches can efficiently return high-quality explanations for why-not questions on pSCAN.

Keywords: Why-not question · pSCAN · Query refinement
Explanation

1 Introduction

In recent years, the usability study of database has attracted amount of attention in the database community. To improve the usability of database, one useful feature that needs to be supported by the database is the explanation capability for users to seek clarifications on their query results. Why-not question was proposed by Jagadish and Chapman [1], which explains why an expected tuple is missing

© Springer International Publishing AG, part of Springer Nature 2018
J. Pei et al. (Eds.): DASFAA 2018, LNCS 10827, pp. 255–271, 2018.
https://doi.org/10.1007/978-3-319-91452-7_17

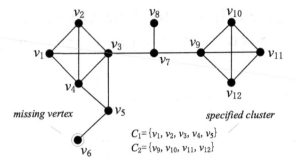

Fig. 1. An example of pSCAN, with parameters $\epsilon = 0.65$ and $\mu = 4$

in the query result. Although many efforts have been made to explain why-not questions on different queries, they are far from meeting user's requirements.

Given a graph $G = (V, E)$, structural graph clustering is to assign vertices in V to clusters and to identify the sets of hub vertices and outlier vertices as well, such that vertices in the same cluster are densely connected to each other while vertices in different clusters are loosely connected to each other. As a fast and exact structural graph clustering algorithm, pSCAN [2] is widely used in many different graph applications. pSCAN generates clusters with two parameters: (i) ϵ, a similarity threshold; (ii) and μ, a density constraint. Let $G = (V, E)$ be a social network graph, and u be a vertex in V, $N[u]$ is denoted as the *closed neighborhood* [2] of u, and $\sigma(u, v)$ is defined as the *structural similarity* between vertices u and v. Vertices u and v are *structure-similar* to each other if $\sigma(u, v) \geq \epsilon$, and u is denoted as a *core* vertex if it has at least μ neighbors that are structure-similar to it. Let $N_\epsilon[u]$ be a set of vertices whose structural similarities with u are at least ϵ; $N_\epsilon[u]$ is a cluster if it covers at least μ vertices. Then the cluster keeps growing from core vertices by including all vertices that are structure-similar to core vertices in the cluster. At last, for each vertex that is not a member of any cluster, it is called as a *hub* vertex if its neighbors belong to two or more clusters, otherwise, it is denoted as an *outlier* vertex.

As depicted in Fig. 1, we can obtain two clusters $C_1 = \{v_1, v_2, v_3, v_4, v_5\}$ and $C_2 = \{v_9, v_{10}, v_{11}, v_{12}\}$ wrt. $\epsilon = 0.65$ and $\mu = 4$. v_7 is a hub vertex since its neighbors contains v_3 and v_9, and v_8 and v_6 are outlier vertices. In many cases, however, owing to that the users lack of enough acknowledge to determine the appropriate parameters for pSCAN, the results obtained by pSCAN may not meet the requirements of users. Such as one user may feel confused when an expected vertex is missing in the specified cluster of the pSCAN result.

In Fig. 1, for example, v_5 is surprised to find that his neighbor v_6 is absent from his cluster C_1. At this point, v_5 could try to figure out an explanation for the expected vertex v_6 by relaxing at least one clustering condition (e.g., decreasing the parameter $\epsilon = 0.65$ to $\epsilon' = \sigma(v_5, v_3) = 0.61$ or decreasing the parameter $\mu = 4$ to $\mu' = 3$) to make v_6 appear in the specified cluster C_1.

Therefore, it would be very helpful to users such as v_5 if he could simply pose a single why-not question to the graph database system to seek an explanation for WHY v_6 is NOT in the same cluster with v_5. Motivated by this, we focus on the problem of effectively explaining why-not questions on pSCAN in this paper.

Through the clustering mechanism of pSCAN and the discussion about pSCAN in Fig. 1, we can easily figure out that the pSCAN results can be affected by the parameters ϵ and μ. If they are not set *good* enough, the clustering results may not satisfy the users' requirements. Hence, we can tune them to explain why-not questions on pSCAN. Moreover, one common principle for explaining why-not questions is that the original query results should be retained as much as possible in the refined query results. Hence, to ensure the quality of explanations for why-not questions on pSCAN, the refined pSCAN should not destroy the original clustering results as much as possible, i.e., those clusters in the original pSCAN results should still appear in the refined pSCAN results, and they should not be merged together with the refined ϵ' or μ'.

However, such an approach of manually seeking explanation by tuning parameters of pSCAN is rather tedious, which involves possibly many rounds of pSCAN refinement. Moreover, the users could end up over-relaxing their refined pSCAN and many more irrelevant vertices besides the expected vertex are clustered into the existed clusters.

In view of this, the main challenge of explaining why-not questions on pSCAN is how to quickly explore the effective refined parameters of pSCAN to make the missing vertex appear in the specified cluster in terms of those constraints as mentioned above. To sum up, the key contributions are as follows:

(1) We analyze the factors that affect the results of pSCAN, and propose an unified explanation framework, which consists of two modules: (i) refine the parameter ϵ; (ii) refine the parameter μ.
(2) Based on the influences of parameter ϵ on pSCAN results, we explore one novel explanation algorithm to answer why-not questions on pSCAN by modifying the parameter ϵ.
(3) Based on the influence of parameter μ on pSCAN results, we propose one effective explanation algorithm to explain why-not questions on pSCAN by modifying the parameter μ.

This paper is organized as follows. Section 2 reviews related work and presents problem statement. Section 3 presents an unified explanation framework, and analyzes the factors influencing the results of pSCAN. Section 4 proposes two novel explanation algorithms to answer why-not questions on pSCAN. Section 5 reports experimental results. We conclude the paper in Sect. 6.

2   Related Work and Problem Statement

In this section, we review related work firstly. Then we give some background definitions about pSCAN algorithm. Finally, we present the problem definition.

2.1   Related Work

Structural Graph Clustering. The SCAN approach was proposed by Xu et al. [3], it iterates through all vertices that have not been clustered to clusters. Based on the property that a vertex and its two-hop-away vertices are expected to share large parts of their neighborhoods, [4] proposed an improved approach SCAN++. Chang et al. [2] proposed a fast and exact structural graph clustering approach pSCAN in terms of three important observations.

Why-Not Questions. Huang et al. [5] explained missing tuples by presenting the users how to modify the original data. Artemis [8,9] presented users what kinds of tuples should be inserted into the original database to get the missing tuples. In addition, [6,7] proposed a minimal explaining model for SPJU queries.

   Missing tuples can be formalize as "why-not" questions, which was proposed by Chapman and Jagadish [1]. [1] answered why-not questions by identifying the "culprit" operations which can exclude expected tuples in the query statement.

   [10] proposed a new explanation strategy based on query refinement. It conquered why-not questions by automatically generating a refined query whose query results include both the original query answers and the missing answers. Another work [11] used the similar strategy to address the why-not questions on top-k queries. While Gao et al. [12] systematically explored why and why-not questions on reverse top-k queries, and Chen et al. [13] addressed missing values on spatial keyword top-k queries. Furthermore, [14] addressed the why-not questions on reverse skyline query. [15] explored the why-not questions in similar graph matching. [16] aimed to find the explanations for why and why-not questions in terms of the causality and responsibility.

2.2   Preliminary

As mentioned above, the structural neighborhood of a vertex u is denoted as $N[u] = \{v \in V | (u, v) \in E\} \cup u$. And the degree of u, denoted as $d[u]$, is the cardinality of $N[u]$ (i.e., $d[u] = |N[u]|$).

Definition 1 (Structural Similarity). *The structural similarity between vertices u and v is defined as the number of common vertices in $N[u]$ and $N[v]$ normalized by the geometric mean of their cardinalities, that is denoted as $\sigma(u,v)$:*

$$\sigma(u,v) = \frac{|N[u] \cap N[v]|}{\sqrt{d[u] * d[v]}} \tag{1}$$

   Note that, $\forall\ u, v \in V, 0 \le \sigma(u,v) \le 1$. In Fig. 1, $N[v_7] = \{v_3, v_7, v_8, v_9\}$, $N[v_3] = \{v_1, v_2, v_3, v_4, v_5, v_7\}$, thus $\sigma(v_3, v_7) = |\{v_3, v_7\}|/\sqrt{4 * 6} = 1/\sqrt{6}$.

   Let ϵ be a similarity threshold ($0 < \epsilon \le 1$), the ϵ-neighborhood of a vertex u is defined as the subset of vertices in $N[u]$ whose structural similarities with u are at least ϵ, that can be formalized as $N_\epsilon[u] = \{v \in N[u] \,|\, \sigma(u,v) \ge \epsilon\}$. Because of $\sigma(u,u) = 1$, $N_\epsilon[u]$ must include u for each $u \in V$. In Fig. 1, $N_{0.65}[v_4] = \{v_1, v_2, v_3, v_4, v_5\}$. The *similar-degree* of a vertex u with a parameter ϵ is defined as the size of the ϵ-neighborhood of vertex u. Moreover, if $|N_\epsilon[u]| \ge \mu$, u is a core vertex, otherwise, it is a *non-core* vertex. For example, $|N_{0.65}[v_3]| = |\{v_1, v_2, v_3, v_4\}| = 4$, thus, v_3 is a core vertex. While v_5 is a non-core vertex. If u belongs to a cluster and u is not a core vertex, u is denoted as a *border* vertex.

Definition 2 (Structural-Reachable). *A vertex $u \in V$ is structural-reachable from a vertex $v \in V$ wrt. the parameters ϵ and μ if there exists a path P in G (p is expressed as a sequence of vertices v_1, v_2, $v_3, \ldots, v_n$ ($n \geq 2$) in V), where $v_1 = v$, $v_n = u$ such that: (i)$|N_\epsilon[v_i]| \geq \mu$; (ii)$v_{i+1} \in N_\epsilon[v_i]$, $i \in [1, n-1]$.*

In Fig. 1, v_5 is structural-reachable from v_1 through the sequence v_1, v_4, v_5. But v_4 is not structural-reachable from v_5 since v_5 is not a core vertex. Moreover, let $u \in C$, then u is denoted as structural-reachable from the cluster C in this paper. In Fig. 1, v_5 is structural-reachable from C_1 since v_5 belongs to C_1.

Given a vertex $u \in V$, the parameters ϵ and μ, $kSN(u)$ is defined as the k-th structure-similar closed neighbor of u, then the *core-similarity* of u with μ is defined as: $csim_\mu[u] = \sigma(u, \mu'SN(u))$ ($\mu' = \mu - 1$). And the *core-density* of u with ϵ is defined as: $cden_\epsilon[u] = |N_\epsilon[u]|$. In Fig. 1, when $\mu = 4$, $csim_4[v_4] = \sigma(v_4, v_2) = 0.89$, and when $\epsilon = 0.80$, $cden_{0.80}[v_4] = |N_{0.80}[v_4]| = 5$.

The missing vertex w may be a non-core vertex or a core vertex. When w is a core vertex, if w is merged into the specified cluster C_t, all the vertices in the cluster of w can also be merged into C_t based on the clustering rational of pSCAN. To satisfy the constraints for refining pSCAN, therefore, the situation that w is a core vertex is not discussed in this paper. Based on this, let R is the original pSCAN result, and R' is the refined pSCAN result, two important constraints for refining pSCAN with refined ϵ' or μ' can be formalized as follows: (1) $\forall u \in R$, if $|N_\epsilon[u]| \geq m$, $|N_{\epsilon'}(u)| \geq \mu'$; (2) $\forall$ core vertices $u \in C_i$, $v \in C_j$ in R, $i \neq j$, $\nexists u \in C_j'$, $v \in C_j'$ in R'.

2.3 Problem Definition

To explain why-not questions on pSCAN in light of query refinement model, we formalize the pSCAN as a query $Q(\epsilon, \mu)$. Generally, there are many refined pSCAN queries for a given why-not question, which can make the missing vertex appear in the specified cluster, it is necessary to define a penalty function to compare the quality of refined pSCAN so that only the "*good*" refined pSCAN are returned as possible explanations for a missing vertex. Ideally, only the missing vertex w is assigned to the cluster C_t in the result of the refined pSCAN Q'. Any additional vertices returned in C_t and other existed clusters are considered to be irrelevant vertices that should be minimized. Therefore, the penalty function for Q' is defined to be the number of irrelevant vertices newly added to the existed clusters in R(suppose R contains k clusters), which is formalized as follows:

$$Penalty(Q') = \sum_{i=1}^{k}(|C_i'| - |C_i|), \text{ where } C_i' \text{ in } R', C_i' \supseteq C_i \qquad (2)$$

In Fig. 1, the pSCAN query is $Q(0.65, 4)$. We want to make v_6 appear in C_1. Suppose one refined pSCAN result R' is obtained by modifying the $\epsilon' = \sigma(v_5, v_3)$, which consists of two clusters $C_1' = \{v_1, v_2, v_3, v_4, v_5, v_6\}$, $C_2' = \{v_9, v_{10}, v_{11}, v_{12}\}$, then $Penalty(Q') = |C_1'| - |C_1| + |C_2'| - |C_2| = 6 - 5 + 4 - 4 = 1$.

Problem Definition. Given a pSCAN $Q(\epsilon, \mu)$ over a graph database D, the clustering result R of Q consists of k clusters $C_1, C_2, \ldots, C_k$. A why-not question on pSCAN contains an expected vertex w, and a specified cluster C_t $(1 \le t \le k)$; explaining this why-not question is to seek a refined pSCAN $Q'(\epsilon', \mu')$ by tuning the parameters ϵ or μ to obtain the new result R', such that the following three conditions should be satisfied: (1) There exists a cluster C_d' in R' which contains C_t and w; (2) The refined parameters ϵ' or μ' satisfy the constraints mentioned above; (3) The penalty $Penalty(Q')$ for Q' is minimal.

3   Influence Factors for pSCAN

In this section, firstly, we present an unified explanation framework for refining pSCAN. Then we introduce the influences of ϵ and μ on the pSCAN results.

3.1   An Explanation Framework

As mention above, the factors influencing the why-not vertex are the parameters ϵ and μ. Moreover, we can figure out that the essential reason that why a missing vertex w is absent from the specified cluster C_t is that w cannot structural-reachable from C_t with the original parameters ϵ and μ. Therefore, to make w appear in C_t, w needs to be made be structural-reachable from any core vertex in C_t by tuning the parameter ϵ or μ to the refined ϵ' or μ'. Based on this, we present an unified explanation framework as depicted in Fig. 2, which contains two modules as discussed below:

(1) **Refining** ϵ. We try to obtain a structural-reachable path from C_t to w with a maximal refined similarity threshold ϵ' by modifying the parameter ϵ.
(2) **Refining** μ. We try to explore a structural-reachable path from C_t to w with a maximal refined density constraint μ' by modifying the parameter μ.

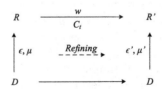

Fig. 2. Explanation framework.

Moreover, one important theorem for refining pSCAN can be obtained.

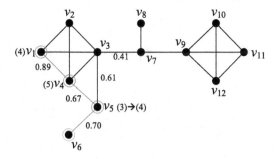

Fig. 3. Modifying the parameter ϵ.

Theorem 1. *Let one structural-reachable path* $(v_1, v_2, v_3, \ldots, v_n, v_1 \in C_t, v_n = w)$ *from* C_t *to* w *be obtained with a refined* ϵ' *or* μ', *then the path should satisfy the following conditions:* (1) $|N_{\epsilon'}[v_i]| \geq \mu'$; (2) $\sigma(v_i, v_{i+1}) \geq \epsilon'$, $i \in [1, n-1]$.

Proof. Based on Definition 2, we know that if a vertex u is structural-reachable from a vertex v, v must be a core vertex. Hence, in this path, v_i must be a core vertex and v_{i+1} must be structural-reachable from v_i with ϵ' or μ'.

3.2 The Analysis of ϵ

The pSCAN results are very sensitive to the parameter ϵ, and will make a difference when ϵ changes by just a small factor.

In Fig. 1, for example, when μ is fixed and $\epsilon = 0.65$ is modified to $\epsilon' = \sigma(v_5, v_3)$, v_5 becomes a new core vertex since its ϵ'-neighborhood covers more than $\mu = 4$ vertices as depicted in Fig. 3. And v_6 becomes a border vertex in C_1 since it is structural-reachable from v_5. In Fig. 3, the value beside a edge is the structural similarity between the vertices of this edge. The value in a parentheses beside a vertex is the similar-degree of the vertex with a similarity threshold ϵ. For example, 0.67 is the structural similarity between vertices v_4 and v_5, the similar-degree of v_4 with $\epsilon = 0.65$ is 5 as depicted in Fig. 3.

Obviously, from the above example, one important property about the parameter ϵ can be derived as follows.

Property 1. Let ϵ be decreased to ϵ', $\forall u \in V$, if $|N_\epsilon[u]| \geq \mu$, $|N_{\epsilon'}[u]| \geq \mu$.

3.3 The Analysis of μ

The pSCAN results can also be affected by the density constraint μ. That is, if the setting of μ is unreasonable, the users cannot get the expected results either.

In Fig. 1, when ϵ is fixed and μ is decreased to 2, v_5 becomes a core vertex because of $|N_{0.65}[v_5]| = 3 \geq 2$. And v_6 is structural-reachable from v_5 since $N_{0.65}[v_5]$ contains v_6. Thus, v_6 can be assigned into the designed cluster C_1 by decreasing $\mu = 4$ to $\mu' = 2$. However, the size of $N_{0.65}[v_5]$ is already 3; when $\mu' = 3$, v_6 can also be clustered into C_1 as detailed in Fig. 4.

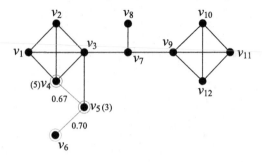

Fig. 4. Modifying the parameter μ.

Note that, after μ is decreased to μ', the original core vertices with the parameter μ are still core vertices with the refined parameter μ'.

4 Explanation Algorithms

In this section, we explain why-not questions on pSCAN by modifying the similarity threshold ϵ firstly. Then we answer why-not questions on pSCAN by modifying the density constraint μ.

4.1 Modifying the Similarity Threshold ϵ

Based on Property 1, we can make w be structural-reachable from C_t by decreasing the parameter ϵ to ϵ'. It is very obvious that the smaller the parameter ϵ' is, the more the irrelevant vertices are newly structural-reachable from existed clusters. Based on this, one property is presented as follows.

Property 2. Let $Q'_1(\epsilon'_1, \mu)$ and $Q'_2(\epsilon'_2, \mu)$ be two refined pSCAN by modifying the parameter ϵ, suppose $\epsilon'_1 < \epsilon'_2$, then $Penalty(Q'_1) > Penalty(Q'_2)$.

Hence, we need to get the maximal ϵ' which can make w be structural-reachable from C_t (denoted as ϵ'_{max}). In Fig. 3, there exist many ways to make v_6 be structural-reachable from C_1 by modifying ϵ. One way is to modify ϵ to $\epsilon' = \sigma(v_5, v_3) = 0.61$ as discussed above. The other way is to modify ϵ to $\epsilon'' = \sigma(v_3, v_7) = 0.41$, then v_6 is also structural-reachable from C_1. Based on our penalty function, we prefer to modify ϵ to equal to the bigger value $\sigma(v_5, v_3)$.

To explain missing vertex w, based on Theorem 1, we need to find a path from any core vertex u in C_t to w in G firstly, and then making the path be one structural-reachable path by modifying ϵ. Note that, the maximal similarity threshold which can make one path become one structural-reachable path is denoted as the *reachable-similairty* of this path. Let $P = v_1, v_2, \ldots, v_n$ be a structural-reachable path, then the reachable-similarity of P, denoted as $RSim(P)$, can be computed based on the following equation:

$$RSim(P) = \min_{1 \le i < n} \{csim_\mu[v_i], \sigma(v_i, v_{i+1})\} \qquad (3)$$

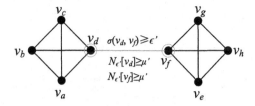

Fig. 5. An example of direct merge.

Therefore, to obtain refined ϵ'_{max} by modifying ϵ, one naive approach is to traverse all the paths from C_t to w in the graph dataset G firstly. Then making all the possible paths be structural-reachable paths based on Theorem 1 and computing the corresponding reachable-similarity for each structural-reachable path. At last, the maximal reachable-similarity among them is obtained as the refined similarity threshold ϵ'_{max} for the missing vertex w.

To speed up the process of computing ϵ'_{max}, those useless paths which cannot generate the maximal refined parameter need to be pruned as early as possible. Based on some knowledge mentioned above, we derive several important lemmas.

Lemma 1. *Given a vertex u, $u \notin C_t$, and $u \in C_o$, then the paths containing u should be pruned, i.e., when traversing the vertex u, we stop this traversal and backtrack to the previous vertex of u.*

Proof. Based on Theorem 1, except for w, those vertices in the structural-reachable paths from C_t to w must be core vertices. Hence, when the structural-reachable paths contain u, u must be a core vertex; that would lead to C_t and C_o are merged together, which violates the constraints for refining pSCAN.

Lemma 2. *Given a vertex u, if $d[u] < \mu$, the paths containing u cannot be made be structural-reachable paths by modifying ϵ and should be pruned.*

Proof. In view of Theorem 1, u must be a core vertex when using u to construct a structural-reachable path from C_t to w. However, $d[u] < \mu$ means u cannot become a core vertex, no matter how to modify the parameter ϵ. Therefore, the paths containing u should be pruned.

Lemma 3. *Let $u \in C_t$, if u also belongs to other cluster C_o as mentioned in [2], u cannot be used to construct a structural-reachable path from C_t to w.*

Proof. If $u \in C_o$, that means C_t and C_o exist overlap. When using u to construct a structural-reachable path from C_t to w, u must become a core vertex based on Theorem 1, which result in C_t and C_o are merged together, this violates the constraints for refining pSCAN. Hence, u cannot be used.

Let $w \in N[u]$, and $N[u]$ be sorted by the structural similarities between those vertices in $N[u]$ and u in ascending order, when traversing w, the remaining vertices in $N[u]$ which are not visited can be pruned since the paths containing

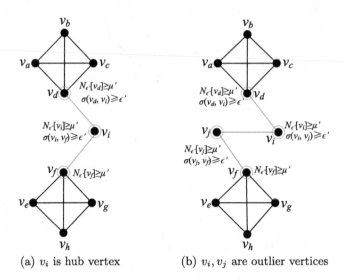

(a) v_i is hub vertex   (b) v_i, v_j are outlier vertices

Fig. 6. An example of indirect merge

those vertices cannot generate bigger refined similarity threshold. Moreover, we only need to construct structural-reachable paths from any core vertex u in C_t to w rather than constructing structural-reachable paths from all the core vertices in C_t to w. The reason is that the paths from u to w must contain the paths from all the core vertices in C_t to w since all the core vertices in C_t are connected.

As mentioned in the problem statement, after obtaining the refined parameter ϵ'_{max}, we need to check whether the refined pSCAN $Q'(\epsilon'_{max}, \mu)$ satisfies the constraints. As depicted in Property 1, the original core vertices in R are still core vertices with the decreased parameter ϵ'_{max}. From the clustering mechanism of pSCAN, we can see that two different clusters may be merged in two ways with the refined parameters ϵ' or μ'. One way is that two clusters are directly merged together as depicted in Fig. 5, vertices v_d and v_f belong to different clusters C_i and C_j, respectively. With the refined parameters ϵ' or μ', v_d and v_f are all core vertices, and v_f is structural-reachable from v_d based on $\sigma(v_d, v_f) \geq \epsilon'$. Therefore, C_i and C_j are directly merged together. The other way is that two different clusters are indirectly merged together as seen in Fig. 6. This way contains two situations as discussed below: The first situation is that two clusters are merged together based on a hub vertex as shown in Fig. 6(a), v_i is a hub vertex since its neighbors v_d and v_f belong to different clusters C_i and C_j, respectively. Vertices v_d, v_i and v_f are all core vertices wrt. the refined parameters ϵ' or μ', and $\sigma(v_d, v_i) \geq \epsilon'$ and $\sigma(v_i, v_f) \geq \epsilon'$, that means v_f is structural-reachable from v_d based on the sequence v_d, v_i, v_f. Hence, C_i and C_j are merged together. The second situation is that two clusters are indirectly merged together based on a sequence outlier vertices as depicted in Fig. 6(b), which depend on the similar principle as mentioned above.

Algorithm 1. MOEPS(ϵ, μ, w, C_t)

Input: The parameters ϵ and μ of pSCAN, a desired cluster C_t, a why-not vertex w, and a graph dataset D

Output: ϵ'_{max}

1 $c \leftarrow$ obtain one core vertex in C_t;
2 add c to S, $\epsilon'_{max} = 0$;
3 **while** S *is not empty* **do**
4     $v = S.top(), visited[v] = 1$;
5   **if** $\exists\, u \in N[v], visited[u] == 0$ and $|N[v]| \geq \mu$ and $csim_\mu[u] > \epsilon'_{max}$ and $\sigma(u,v) > \epsilon'_{max}$ **then**
6   **if** $u == w$ **then**
7           $P \leftarrow$ obtain one structural-reachable path from C_t to w;
8           $\epsilon' \leftarrow RSim(P)$;
9   **if** $\epsilon'_{max} < \epsilon'$ **then**
10             $\epsilon'_{max} = \epsilon'$;
11          $S.pop()$;
12   **else**
13   **if** $u.cID==-1$ *or* $u.cID==t$ **then**
14   add u to S;
15   **else**
16       $S.pop()$;
17 $res \leftarrow$ **CheckParaE**($\epsilon'_{max}, \mu, w, C_t$);
18 **if** $res == False$ **then**
19   ϵ'_{max}=NULL;
20 **return** ϵ'_{max};

Function **CheckParaE** is defined to check whether the refined parameter ϵ'_{max} satisfies the constraints. Firstly, we compute the ϵ'_{max}-neighborhood of core vertices in each existed clusters. If the ϵ'_{max}-neighborhood contain two core vertices belong to different existed clusters, it means the refined ϵ'_{max} does not satisfy the constraint since there exists direct merge. Otherwise, we compute the ϵ'_{max}-neighborhood of those hub vertices and outlier vertices, if two clusters are structural-reachable from each other wrt. to one or more of them, ϵ'_{max} dose not satisfy the constraints as depicted in Fig. 6. If not, ϵ'_{max} satisfies the constraints and it is an effective explanation for the missing vertex.

Based on these lemmas mentioned above, we propose one novel algorithm to explain why-not questions on pSCAN as depicted in Algorithm 1. We obtain one core vertex c in C_t firstly, and we add c to a stack S and initialize $\epsilon'_{max} = 0$ (Lines 1–2). Then we obtain the effective paths from c to w in D based on Lemmas 1 to 3, and compute the corresponding reachable-similarity of those paths based on Eq. 3 and update the value of ϵ'_{max} (Lines 3–16). Then we check whether the ϵ'_{max} satisfies those constraints for refining pSCAN by using the

Function **CheckParaE**, if not, we set ϵ'_{max} =NULL. At last, we return ϵ'_{max} (Lines 17–20).

4.2   Modifying the Density Constraint μ

As discussed in Sect. 3.3, to make w be structural-reachable from C_t, we can also decrease μ when the parameter ϵ is fixed. It is very clear that the smaller the parameter μ', the more the irrelevant vertices are assigned to the existed clusters. Based on this observation, one property about μ should be derived.

Property 3. Let $Q'_3(\epsilon, \mu'_1)$ and $Q'_4(\epsilon, \mu'_2)$ be two refined pSCAN queries by modifying the parameter μ, suppose $\mu'_1 < \mu'_2$, then $Penalty(Q'_3) > Penalty(Q'_4)$.

Clearly, many refined parameters μ' can make w be structural-reachable from C_t. We only need to obtain the maximum μ' (denoted as μ'_{max}) to minimize the irrelevant vertices newly assigned to the existed clusters based on Property 3.

Therefore, to explain missing vertex w by modifying μ based on Theorem 1, we need to find a path from any core vertex u in C_t to w firstly, and then making the path become one structural-reachable path by modifying μ. Note that, the maximal density constraint which can make one path become one structural-reachable path is defined as the *reachable-density* of this path. Let $P = v_1, v_2, \ldots, v_n$ be one structural-reachable path, then the reachable-density of P can be obtained in light of the following equation:

$$RDen(P) = \min_{1 \leq i < n} cden_\epsilon[v_i] \tag{4}$$

In view of this, to construct the structural-reachable path from C_t to w by modifying the density constraint μ, one basic approach is to traverse all the paths from C_t to w whose structural similarities between any two adjacent vertices are at least ϵ firstly. Then making all the paths be structural-reachable paths by refining μ based on the Theorem 1 and computing the corresponding reachable-density for each structural-reachable path. At last, the maximal reachable-density which can make w be structural-reachable from C_t are obtained as the refined density constraint μ'_{max} for the missing vertex w.

As mentioned above, those useless paths which cannot generate the maximal refined density constraint μ'_{max} can be pruned based on Lemmas 1 and 3. Moreover, in the process of constructing the structural-reachable paths from C_t to w, when traversing the neighbors for a current vertex, we only need to check its ϵ-neighborhood rather than its structural neighborhood. Based on this, we obtain one important lemma as follows.

Lemma 4. *Given the parameter ϵ, a vertex $u \neq w$ and $|N_\epsilon[u]| = 1$, then the paths containing u should be pruned.*

Proof. Based on Theorem 1, u should be core vertex, $|N_\epsilon[u]| = 1$ means u is a core vertex when $\mu' = 1$. But μ' should equal to or larger than 2. Therefore, u cannot be a core vertex with any μ'.

Moreover, Let $w \in N_\epsilon[u]$, and $N_\epsilon[u]$ be sorted by their similar-degree with the parameter ϵ in ascending order, when traversing w, the remaining vertices in $N_\epsilon[u]$ which are not visited can be pruned since a bigger refined density constraint cannot obtained from the paths which contain those vertices.

It's similar with refining the parameter ϵ, after obtaining the maximal refined μ'_{max}, whether μ'_{max} satisfies the constraints mentioned in problem definition is also need to be checked. Function **CheckParaM** is defined to achieve this purpose. Firstly, we compute the ϵ-neighborhood of core vertices in each existed clusters. If two core vertices with μ'_{max} belong to two different clusters are contained in one of them, μ_{max} dose not satisfy the constraints since the two clusters are merged together based on μ_{max}. Otherwise, the ϵ-neighborhood of those hub vertices and outlier vertices are obtained, if two existed clusters are structural-reachable from each other in terms of one or more of them, it means there exists indirect merge. μ'_{max} is not an effective explanation for the missing vertex.

Algorithm 2 describes how to find the optimal refined pSCAN with the minimal penalty value by modifying the parameter μ. We obtain one core vertex c in C_t firstly, and we add c to a stack S and initialize $\mu'_{max} = 0$ (Lines 1–2). Then the effective paths from c to w in D are obtained based on Lemmas 1, 3 and 4, and the corresponding reachable-density of those paths are computed based on Eq. 4 and μ'_{max} is updated (Lines 3–16). Then we check whether μ'_{max} satisfies the constraints by using the Function **CheckParaM**, if not, we set μ'_{max}=NULL. At last, we return μ'_{max} (Lines 17–20).

In my opinion, for the same why-not question, when the users trust the original parameter ϵ more than the original parameter μ, it is better to use MOMU to answer this question for them, otherwise, MOEPS is used better.

5 Experiments

In this section, we evaluate the performance of our proposed algorithms for explaining why-not questions on pSCAN using real datasets.

5.1 Experimental Setup

We conducted experiments on three real datasets[1].

- **Epinions:** Epinions is a who-trust-whom online social network of a general consumer review site Epinions.com. Members of the site can decide whether to "trust" each other. It contains 75,879 vertices and 508,837 edges.
- **Slashdot:** Slashdot is a technology-related news website know for its specific user community. The network was obtained in November 2008 and it contains 77,360 vertices and 905,468 edges.
- **Brightkite:** Brightkite was once a location-based social networking service provider where users shared their locations by checking-in. The friendship network consists of 58,228 nodes and 214,078 edges.

---

[1] http://snap.stanford.edu/data/.

Algorithm 2. MOMU(ϵ, μ, w, C_t)

Input: The parameters ϵ and μ of pSCAN, a desired cluster C_t, a why-not vertex w, and a graph dataset D

Output: μ'_{max}

1   $c \leftarrow$ obtain one core vertex in C_t;

2   add c to S, $\mu'_{max} = 0$;

3   **while** S *is not empty* **do**

4      $v = S.top(), visited[v] = 1$;

5   **if** $\exists u \in N_\epsilon[v], visited[u] == 0$ *and* $|N_\epsilon[u]| > \mu'_{max}$ *and* $|N_\epsilon[u]| > 1$ **then**

6   **if** $u == w$ **then**

7              $P \leftarrow$ obtain one structural-reachable path from C_t to w;

8              $\mu' \leftarrow RDen(P)$;

9   **if** $\mu'_{max} < \mu'$ **then**

10                 $\mu'_{max} = \mu'$;

11              $S.pop()$;

12   **else**

13   **if** $u.cID$==-1 *or* $u.cID$==t **then**

14   add u to S;

15   **else**

16          $S.pop()$;

17   $res \leftarrow$ **CheckParaM**($\epsilon, \mu'_{max}, w, C_t$);

18   **if** $res == False$ **then**

19   μ'_{max} =NULL;

20   **return** μ'_{max};

We designed different why-not questions $w_1 - w_4$ for Epinions dataset, $w_5 - w_8$ for Slashdot dataset and $w_9 - w_{12}$ for Brightkite dataset.

We studied the performance of the presented algorithms MOEPS and MOMU under various parameters, All the algorithms were implemented using GNU C++. The experiments presented below were run on a PC with 3.60 GHz CPU and 8 GB main memory with a 1 TB disk, running a Ubuntu operating system.

5.2 Performance Evaluation

Firstly, we compare the change of the execution time of the proposed algorithms under the variation of the number of vertices. In Fig. 7, as expected, the execution time of MOEPS and MOMU grow with the increase of the number of vertices. The reason is that the number of paths from the desired cluster to the missing vertex is increased. Moreover, the number of clusters, hub vertices and outlier vertices is increased, which leads to the execution time of Function **CheckParaE** and Function **CheckParaM** is increased. As depicted in Fig. 7(a), (b), and (c), the execution time of MOMU is smaller than MOEPS, which is primarily because MOEPS need to check all the paths from the desired cluster to the missing vertex,

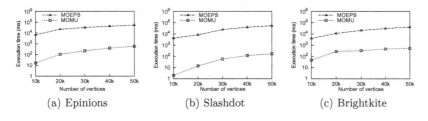

Fig. 7. Execution time of explanation algorithms under different datasizes

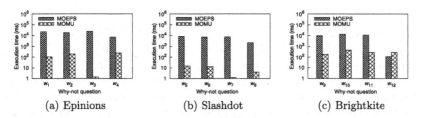

Fig. 8. Execution time of explanation algorithms under different why-not questions

while MOMU only need to check those paths whose structural similarities of all edges are at least ϵ.

Secondly, we investigate the impact of different why-not questions on the execution time of the proposed algorithms. In Fig. 8(a) and (b), for w_3 and w_7, the execution time of MOMU is much smaller than MOEPS. The principal reason is that them cannot explained by MOMU, i.e., there is no refined μ'_{max} to explain them. In Fig. 8(c), for w_{12}, the execution time of MOEPS is smaller than MOMU since it cannot be explained by MOEPS. For different why-not questions on the same dataset, the execution time of MOEPS and MOMU is different as depicted in Fig. 8, the reason is that the number of paths from the desired cluster to the missing vertex is different for the different why-not questions. For the different why-not questions on the different datasets, the execution time of MOEPS and MOMU is very different. The principal reason is the structures of datasets are different as detailed in Fig. 8.

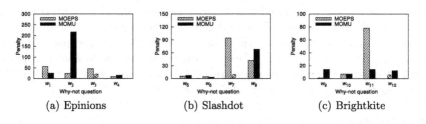

Fig. 9. Penalty value of explanation algorithms under different why-not questions

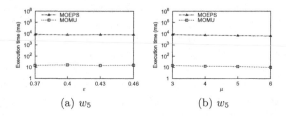

Fig. 10. Execution time of explanation algorithms with different ϵ or μ

Then we evaluate the impact of different why-not questions on the penalty of the proposed algorithms. In Fig. 9(b) and (c), for w_6, the penalty of MOEPS is larger than MOMU; for w_{12}, the penalty of MOEPS is smaller than MOMU, the main reason is that which one of the penalty of MOEPS and MOMU is the bigger is uncertain, which depend on the structures of graph datasets. Some why-not questions could not be explained since there is no structural-reachable path constructed from the desired cluster to the missing vertex by refining ϵ or μ. For example, w_3 and w_7 cannot be explained by MOMU and w_{12} cannot be explained by MOEPS, thus, the penalty of them are denoted as $\varnothing$ in Fig. 9.

Finally, we change the original parameters ϵ and μ, and verify their effects on the execution time of proposed algorithms. For w_5, when ϵ changes from 0.37 to 0.46, Fig. 10(a) shows that there is no obvious change for the execution time of MOEPS and MOMU. The main reason is that the pSCAN result do not change obviously, then the number of structural-reachable paths constructed from the desired cluster to the missing vertex for w_5 do not change obviously. The execution time of MOEPS and MOMU is decreased when μ increases from 3 to 6 as depicted in Fig. 10(b). The principal reason is that the number of effective paths from the desired cluster to the missing vertex is decreased, and the number of clusters is decreased, which result in the computation time and checking time of ϵ'_{max} and μ'_{max} are decreased.

6   Conclusion

We have developed a new explanation framework to explain why-not questions on pSCAN. And we have proposed two effective explanation algorithms, including refining the similarity threshold ϵ and refining the density constraint μ. Our experiments show that our algorithms can return high-quality explanations for why-not questions on pSCAN. In future work, we would like to analyze the join effect of the parameters ϵ and μ and to explain why-not questions on pSCAN by refining the two parameters simultaneously.

Acknowledgements. The work is supported by the National Natural Science Foundation of China (Nos. U1736104, 61572122, 61532021, 61502317, 61502316, 61702344).

References

1. Chapman, A., Jagadish, H.V.: Why not'?'. In: SIGMOD, pp. 523–534 (2009)
2. Chang, L., Li, W., Lin, X., Qin, L., Zhang, W.: pSCAN: fast and exact structural graph clustering. In: ICDE, pp. 253–264 (2016)
3. Xu, X., Yuruk, N., Feng, Z., Schweiger, T.A.J.: SCAN: a structural clustering algorithm for networks. In: KDD, pp. 824–833 (2007)
4. Shiokawa, H., Fujiwara, Y., Onizuka, M.: SCAN++: efficient algorithm for finding clusters, hubs and outliers on large-scale graphs. In: PVLDB, pp. 1178–1189 (2015)
5. Huang, J., Chen, T., Doan, A., Naughton. J.F.: On the provenance of non-answers to queries over extracted data. In: PVLDB, pp. 736–747 (2008)
6. Zong, C., Yang, X., Wang, B., Liu, C.: Minimal explanations of missing values by chasing acquisitional data. WWWJ **20**, 1333–1362 (2017). https://doi.org/10.1007/s11280-017-0438-0
7. Zong, C., Yang, X., Wang, B., Zhang, J.: Minimizing explanations for missing answers to queries on databases. In: Meng, W., Feng, L., Bressan, S., Winiwarter, W., Song, W. (eds.) DASFAA 2013. LNCS, vol. 7825, pp. 254–268. Springer, Heidelberg (2013). https://doi.org/10.1007/978-3-642-37487-6_21
8. Herschel, M., Hernández, M.A., Tan, W.C.: Artemis: a system for analyzing missing answers. In: PVLDB, pp. 1550–1553 (2009)
9. Herschel, M., Hernández, M.A.: Explaining missing answers to SPJUA queries. In: PVLDB, pp. 185–196 (2010)
10. Tran, Q.T., Chan, C.Y.: How to ConQueR why-not questions. In: SIGMOD, pp. 15–26 (2010)
11. He, Z., Lo, E.: Answering why-not questions on top-k queries. In: ICDE, pp. 750–761 (2012)
12. Liu, Q., Gao, Y., Chen, G., Zheng, B., Zhou, L.: Answering why-not and why questions on reverse top-k queries. VLDB J. **25**, 867–892 (2016)
13. Chen, L., Lin, X., Hu, H., Jensen, C.S., Xu, J.L: Answering why-not spatial keyword top-k queries via keyword adaption. In: ICDE, pp. 697–708 (2016)
14. Islam, M.S., Zhou, R., Liu, C.: On answering why-not questions in reverse skyline queries. In: ICDE, pp. 973–984 (2013)
15. Islam, M.S., Liu, C., Li, J.: Efficient answering of why-not questions in similar graph matching. TKDE **27**, 2672–2686 (2015)
16. Roy, S., Suciu, D.: A formal approach to finding explanations for database queries. In: SIGMOD, pp. 1579–1590 (2014)

SSRW: A Scalable Algorithm
for Estimating Graphlet Statistics
Based on Random Walk

Chen Yang[1], Min Lyu[1,2(✉)], Yongkun Li[1,2], Qianqian Zhao[1],
and Yinlong Xu[1,2]

[1] University of Science and Technology of China, Hefei, China
{ccyangch,qqzhao}@mail.ustc.edu.cn, {lvmin05,ykli,ylxu}@ustc.edu.cn
[2] AnHui Province Key Laboratory of High Performance Computing, Hefei, China

Abstract. Mining graphlet statistics is very meaningful due to its wide
applications in social networks, bioinformatics and information security,
etc. However, it is a big challenge to exactly count graphlet statistics
as the number of subgraphs exponentially increases with the graph size,
so sampling algorithms are widely used to estimate graphlet statistics
within reasonable time. However, existing sampling algorithms are not
scalable for large graphlets, e.g., they may get stuck when estimating
graphlets with more than five nodes. To address this issue, we propose
a highly scalable algorithm, Scalable subgraph Sampling via Random
Walk (SSRW), for graphlet counts and concentrations. SSRW samples
graphlets by generating new nodes from the neighbors of previously visited nodes instead of fixed ones. Thanks to this flexibility, we can generate
any k-graphlets in a unified way and estimate statistics of k-graphlet efficiently even for large k. Our extensive experiments on estimating counts
and concentrations of $\{4, 5, 6, 7\}$-graphlets show that SSRW algorithm is
scalable, accurate and fast.

1 Introduction

Graphlets are small connected induced subgraphs in large graphs, and they are
widely studied in many applications, such as social networks [9,16], bioinformatics [15,19], and information security [17], etc. For example, triangle counting is a
classic problem in studying social networks, which reflects the closeness between
members and can be used to make friend recommendation.

Graphlet concentration is the relative frequency of a specific graphlet in a
set of graphlets, such as the clustering coefficient. Graphlet concentration can be
used to make large graph comparison [24]. Counting graphlets is a more general
analysis task, and it also has wide applications [13], including deducing graphlet
concentrations. A naive approach to obtain graphlet statistics is exact counting,
which enumerates all connected subgraphs of a given size and then derives the
statistics by counting the number of subgraphs isomorphic to each graphlets.
However, exact counting is time-consuming, and it is inefficient even for graphs

© Springer International Publishing AG, part of Springer Nature 2018
J. Pei et al. (Eds.): DASFAA 2018, LNCS 10827, pp. 272–288, 2018.
https://doi.org/10.1007/978-3-319-91452-7_18

with moderate size. For example, it may take more than ten days to obtain the exact number of 5-graphlets of the Flickr graph with 240K nodes using ESCAPE [18], a state-of-the-art exact counting algorithm. To the best of our knowledge, there is still a lack of efficient tools to obtain the statistics of graphlets with more than five nodes in large graphs due to the extremely large number of subgraphs and the complicated structures of graphlets.

Sampling is a common approach to reduce the time complexity for estimating graphlet concentrations [8,23]. Recent works are mainly based on random walk because of its simplicity and stability, which samples k-graphlets in k consecutive steps of a random walk. However, some k-graphlets can not be sampled in such a way, and these graphlets must be sampled with other sampling methods or by utilizing the relationships between graphlets [5,7]. There are also other sampling algorithms which first sample a specific substructure in a graph, such as a path or a star, and then estimate statistics with the combinational relationship between different graphlets for unsampled graphlets [10,20,26]. Thus, existing sampling approaches do not scale well for large sized graphlets in very large graphs.

To address the scalability issue when sampling large sized graphlets, some works execute a random walk on super graphs for getting samples [4,5,23,25], where each super node in the super graph corresponds to a k-connected subgraph. Sampling on the super graph makes it easier to sample a k-subgraph than on the original graph, but the size of a k-super graph is much larger than the original graph, so populating neighbor is still inefficient when $k \geq 3$, making random walks on a super graph hard to mix quickly. Therefore, sampling with random walk on super graphs are still inefficient, especially for large graphlets.

In this paper, we propose a new algorithm, Scalable Subgraph Sampling via Random Walk (SSRW), to efficiently estimate graphlet counts and concentrations of large size. We develop a new sampling strategy for SSRW by introducing randomness to the selection of new nodes when generating a sample, say a k-subgraph sample, to ensure that the sample set contains all desired graphlets. The sampling strategy is general and thus easy to be scaled to sample subgraphs with large size. We summarize our contributions as follows.

- **A highly scalable subgraph sampling strategy.** We develop a novel strategy to quickly sample large sized subgraphs based on random walk. With our strategy, all graphlets can be sampled in a unified way. To get a subgraph sample, we generate new nodes from the neighbors of a set of previously generated nodes instead of a fixed one, which makes it possible to generate all types of subgraphs. Our strategy is simple and also easy to be scaled for graphlets with more than 5 nodes in large graphs. In addition, it is also easy to be parallelized to further improve its efficiency.

- **A framework to estimate large sized graphlet counts and concentrations.** Based on our subgraph sampling strategy, we design a new sampling algorithm, SSRW, to estimate graphlet counts and concentrations in large graphs accurately and efficiently. SSRW provides an efficient alternative for estimating large graphs, of which the exact graphlet statistics can not be derived in reasonable time.

– **Extensive experiments.** We implement our sampling algorithm for estimating the concentrations and counts of $\{4, 5, 6, 7\}$-graphlets. Extensive experiments show that our algorithm are accurate, scalable and fast. For example, for estimating 4-graphlet concentrations, the Normalized Root Mean Square Errors (NRMSEs) of 3-star and 4-clique are smaller than 0.01 and 0.02 respectively on com-Amazon with 334K nodes. Besides, our algorithm can also estimate 6 and 7-graphlet counts efficiently, which are rarely addressed before. The NRMSEs of prescribed 6-graphlets counts are smaller than 0.1 on Gnutella04 with 10.8K nodes, and are smaller than 0.15 for prescribed 7-graphlets on ca-GrQc with 4.1K nodes. We get 50× speed up when we make our algorithm running in parallel.

2   Preliminaries

2.1   Notations and Definitions

We denote a graph as $G = (V, E)$, where V, E are the sets of nodes and edges respectively. For a node $v \in V$, $N(v)$ is the set of v's neighbors, and $d(v) = |N(v)|$ is the degree of v. Given two nodes v_1 and v_2, we use (v_1, v_2) to denote the undirected edge between v_1 and v_2. A graph $G_1 = (V_1, E_1)$ is a k-subgraph of G, if $V_1 \subseteq V$, $E_1 \subseteq E$ and $|V_1| = k$. If any edge whose both ends are in V_1 is also in E_1, then we say G_1 is an induced subgraph of G.

Two graphs $G = (V, E)$ and $G' = (V', E')$ are isomorphic, denoted as $G \simeq G'$, if there is a bijection $\varphi : V \to V'$, such that for any two nodes $v_1, v_2 \in V$, there is an edge $(v_1, v_2) \in E$, if and only if there is an edge $(\varphi(v_1), \varphi(v_2)) \in E'$.

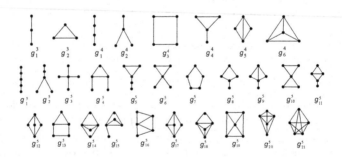

Fig. 1. All $\{3, 4, 5\}$-graphlets, where g_i^k denotes the k-graphlets of type i.

Graphlets are small, nonisomorphic and connected induced subgraphs of G. As shown in Fig. 1, there are two kinds of 3-graphlets, called as 2-path (g_1^3) and triangle (g_2^3), respectively, and 6 different 4-graphlets, as well as 21 different 5-graphlets. We note that the number of unique graphlets grows rapidly with the increase of graphlet size, and in particular, the number of 6-graphlets and 7-graphlets increase to 112 and 853, respectively. In this paper, we focus on $\{4, 5, 6, 7\}$-graphlets.

Definition of graphlet concentration: Given a graph G and a family of k-graphlets $\mathcal{G}^k = \{g_1^k, g_2^k, \ldots g_{T_k}^k\}$, let C_i^k be the number of subgraphs in G which are isomorphic to g_i^k, i.e., the count of g_i^k, and the graphlet concentration c_i^k of g_i^k is defined as $c_i^k = \frac{C_i^k}{\sum_{t=1}^{T_k} C_t^k}$ for $i = 1, 2, \ldots, T_k$.

2.2 Random Walk

A simple random walk on a graph runs as follows. It starts from an initial node in a graph, then chooses the next node from its neighbors uniformly and moves to it, and repeats this process until some stopping criteria are reached. The transition probability matrix is defined as $P = (p(i,j))_{n \times n}$,

$$p(i,j) = \begin{cases} \frac{1}{d(v_i)} & \text{if } (v_i, v_j) \in E; \\ 0 & \text{otherwise,} \end{cases}$$

where n is the number of nodes in the graph and $p(i,j)$ is the probability of transiting from v_i to v_j. Simple random walk has a unique stationary distribution π with $\pi(v) \propto d(v)$ [3], denoting the probability of node $v \in V$ being visited.

3 The Scalable Subgraph Sampling Algorithm

Our algorithm is based on a random walk running on the original graph to generate k-subgraphs as samples. Specifically, when a random walk get mixing, every node in the random walk serves as a starting node and induces a k-subgraph sample with our sampling strategy. During the sampling process, new nodes are chosen from the neighbors of multiple previously accessed nodes instead of only one fixed node. In this way, the selection of a new node is more flexible, which makes us be able to get all k-graphlets in a unified way. Finally, we correct bias according to the probability of samples and their repeat coefficients.

In the following, we first illustrate the core idea of our algorithm, i.e., the sampling strategy, then describe the method of calculating repeat coefficient α_i^k for creating an unbiased estimator, and finally summarize the framework of our algorithm called SSRW.

3.1 Sampling Strategy for k-Subgraphs

Note that all k-graphlets contain 2-path, and this implies that we could get the samples of every graphlet starting with a 2-path. Specifically, from a starting node, we first choose a 2-path by choosing the second node uniformly from the neighbors of the starting node. For $3 \leq t \leq k$, the t-th node v_t is selected from the neighbors of the previous $t-2$ nodes, i.e., all previously chosen nodes except for the first one. Basically, such a way of generating k-subgraphs by sampling a 2-path and choosing the remaining nodes uniformly at random from the neighbors of multiple nodes, helps our sampling methods expand easily and quickly in the graph G, and it is beneficial for sampling all k-graphlets in a unified way and scaling the algorithm to higher-order graphlets sampling.

Take 4-graphlets as an example to further explain the sampling process. We first select a starting node, such as a in Fig. 2(a), from a random walk of the graph at the first step, then choose the second node uniformly from a's neighbors, say b, and then continue to choose the third node uniformly from b's neighbors, say c. Now we have chosen a, b, c at the first three steps and get a 2-path $a - b - c$, as shown in Fig. 2(a). If we insist on selecting the fourth node from the neighbors of the third node (i.e., c), we can never get a 3-star, i.e., g_2^4. However, if we choose the fourth node from the neighbors of both b and c, then we can get 3-star by choosing e from b's neighbors and can also get 3-path $a - b - c - u$ by choosing u from c's neighbors. So all 4-graphlets in G can be generated.

As for 5-graphlets, the first four nodes are generated as above, the fifth node can be chosen from the neighbors of all previously visited nodes except for the starting node, and all 5-graphlets can be generated. As shown in Fig. 2(b), suppose that we have generated four nodes, a, b, c, u, we choose the fifth node from the neighbors of b, c and u. If we choose e from b's neighbors as an example, then we can get a subgraph isomorphic to g_8^5 with node set $\{a, b, c, u, e\}$.

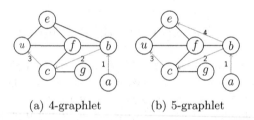

(a) 4-graphlet   (b) 5-graphlet

Fig. 2. Examples of generating 4- and 5-graphlet.

Algorithm 1. Sampling k-Subgraph

Input: starting node v_1
Output: k-tuple $(v_1, v_2, \ldots, v_k)$
 1: $v_2 \leftarrow$ Random node in $N(v_1)$;
 2: **for** i in $[3, \ldots, k]$ **do**
 3:     $v_i \leftarrow$ Random node in $N(v_2)$, $N(v_3)$, ..., or $N(v_{i-1})$;
 4: **end for**
 5: **Return** k-tuple $(v_1, v_2, v_3, \ldots, v_k)$.

Algorithm 1 shows our sampling method for k-subgraphs. To select v_t, we generate a multiset of nodes by concatenating the neighbors of $v_2, \ldots, v_{t-1}$ and uniformly select a node from the multiset. As in the example of generating 4-graphlets in Fig. 2(a), in the final step, we choose the 4-th node from the neighbors of b and c, which is a multiset $\{a, c, f, e, b, f, g, u\}$ of size $d(b) + d(c)$. Notice that node f appears twice. If we choose e from b's neighbors, we get a

3-star, and we get a 3-path (a, b, c, u) if we choose u from c's neighbors. By this way, all 4-graphlets in G are possible to be generated in a unified way. To identify from which neighbors the fourth node being selected, we use pairs to denote all possible selections. For example, in Fig. 2(a), we use $\{\langle b, a \rangle, \langle b, c \rangle, \langle b, e \rangle, \langle b, f \rangle\} \cup \{\langle c, b \rangle, \langle c, f \rangle, \langle c, g \rangle, \langle c, u \rangle\}$ to denote all choices to select the fourth node, where $\langle v_i, v_j \rangle$ means that node v_j is generated from the neighbors of v_i. We select a node as v_t from the concatenation of $N(v_2), \cdots, N(v_{t-1})$, rather than directly from $N(v_2) \cup N(v_3) \cup \ldots \cup N(v_{t-1})$, because the former helps to make unbias estimation of graphlet concentrations and counts based on the distribution of node degrees, which will be explained later in detail.

Theorem 1. *Algorithm 1 generates all k-graphlets in graph G for $k \geq 2$.*

Proof. Given a graph G and an integer $k \geq 2$, we show that every k-graphlets g_i^k is possible to be sampled by Algorithm 1 as long as it exists in G.

Given a graphlet g_i^k, since it is connected, it has a spanning tree, say T. Take a leaf node v_1 of T as the starting node of sampling. We then choose the unique neighbor of v_1 as v_2. Note that T is a connected graph with k nodes. So there are some of the remaining $k - 2$ nodes connected to $\{v_1, v_2\}$, and more specifically, they are the neighbors of v_2 since v_1 is a leaf. Then we choose one from them as v_3. Similarly, some of the remaining $k - 3$ nodes are connected to v_2 or v_3, so we can continue to choose a node from the neighbors of v_2 or v_3 as the fourth node by our sampling strategy, and so on. At each step of the sampling process, we add a new node and an edge to the sampled subgraph. So during the sampling process, we keep the sampled subgraph being connected and also make sure that the number of nodes in the sampled graph is equal to the number of edges in it plus one. So after k steps of the sampling process, the sampled subgraph is exactly T. Because T is the spanning tree of g_i^k, g_i^k is a subgraph of the induced graph of $V(T)$ (the node set of T). So g_i^k can be sampled from T, which is a possible sampling by Algorithm 1. So Algorithm 1 can generate g_i^k.

3.2   Computation of Repeat Coefficient

There may be multiple ways to sample a graphlet. The number of ways for sampling different graphlets are usually different, resulting in biases for the samples of different graphlets. To achieve unbiased estimation of graphlet concentrations and counts, we need to calculate the repeat coefficient α_i^k of each graphlet g_i^k, i.e., the number of ways being sampled.

We denote a sampling way of a k-graphlets as a tuple of $k - 1$ pairs, such as $\langle \langle v_1, v_2 \rangle, \langle v_2, v_3 \rangle, \ldots, \langle v_{k-1}, v_k \rangle \rangle$, which means v_1 is the starting node and v_i is selected from the neighbors of v_{i-1} for $2 \leq i \leq k$. For the 4-graphlets in Fig. 3, we can sample it with multiple ways, such as $\langle \langle v_1, v_2 \rangle, \langle v_2, v_3 \rangle, \langle v_3, v_4 \rangle \rangle$, or $\langle \langle v_1, v_2 \rangle, \langle v_2, v_3 \rangle, \langle v_2, v_4 \rangle \rangle$, etc.

We can compute α_i^k by firstly enumerating all sequences of k nodes, then calculating the ways of sampling g_i^k corresponding to each sequence, and at last aggregating the ways corresponding to all sequences. For example, the tailed-triangle g_4^4 (Fig. 3) has 4 nodes. There are 24 sequences of 4 nodes, and there

Fig. 3. Tailed-triangle.

are two ways to sample g_4^4 corresponding to sequence $\langle v_1, v_2, v_3, v_4 \rangle$. Since the t-th ($4 \leq t \leq k$) node v_t comes from the neighbors of any one of the second, third, ..., $(t-1)$-th node, let v_t^s be the node from whose neighbors v_t being chosen. We can denote a sampling way of a k-graphlets as a tuple of $k-1$ pairs, $\langle \langle v_1, v_2 \rangle, \langle v_2, v_3 \rangle, \langle v_4^s, v_4 \rangle, \cdots, \langle v_k^s, v_k \rangle \rangle$, which means that v_1 is the starting node, and v_i is selected from the neighbors of v_i^s for $4 \leq i \leq k$. Then the two ways corresponding to sequence $\langle v_1, v_2, v_3, v_4 \rangle$ are $\langle \langle v_1, v_2 \rangle, \langle v_2, v_3 \rangle, \langle v_3, v_4 \rangle \rangle$ and $\langle \langle v_1, v_2 \rangle, \langle v_2, v_3 \rangle, \langle v_2, v_4 \rangle \rangle$. There is no sampling ways corresponding to sequence $\langle v_3, v_1, v_2, v_4 \rangle$. We define

$$A(u,v) = \begin{cases} 1 & \text{if } (u,v) \in E; \\ 0 & \text{otherwise.} \end{cases}$$

In general, given a sequence $(w_1, w_2, \ldots, w_k)$ of k nodes for a graphlet, when a new node $w_z(z \geq 3)$ is generated, it can be derived from the neighbors of $\sum_{u=2}^{z-1} A(w_z, w_u)$ different generated nodes in our strategy. So the number of ways of generating a graphlet corresponding to sequence $(w_1, w_2, \ldots, w_k)$ is

$$f(w_1, w_2, \ldots, w_k) = A(w_1, w_2) \prod_{z=3}^{k} \left(\sum_{u=2}^{z-1} (A(w_z, w_u)) \right).$$

For example, there are $1 * 1 * (1 + 1) = 2$ ways to generate g_4^4 corresponding to sequence (v_1, v_2, v_3, v_4) as shown in Fig. 3. We enumerate all sequences of v_1, v_2, v_3, v_4, calculate corresponding ways to generate g_4^4 for each sequence, sum them up, and finally we get there are 10 different ways to generate g_4^4.

3.3   Unbiased Estimator

We use a simple random walk to generate the starting node for sampling, because it only needs a little information about the graph and usually most nodes in the graph are in the largest connected component in real-world networks. When a random walk get mixed, the probability of the appearance of node v is $P(v) = \frac{d(v)}{D}$, where D is the sum of degrees of all nodes in G.

Let $R^{(k)}$ be the set of all sequences of k nodes and $r^{(k)} \in R^{(k)}$. The sequence $r^{(k)} = (v_1, v_2, \ldots, v_k)$ is generated with probability

$$P(r^{(k)}) = \frac{1}{D} \frac{1}{d(v_2)} \frac{1}{[d(v_2) + d(v_3)]} \cdots \frac{1}{[d(v_2) + \cdots + d(v_{k-1})]}.$$

Now we define $L(r^{(k)}, g_i^k)$ as follows.

$$L(r^{(k)}, g_i^k) = \begin{cases} 1 & \text{if } s(r^{(k)}) \simeq g_i^k, \\ 0 & \text{otherwise,} \end{cases}$$

where $s(r^{(k)})$ represents the subgraph induced by $r^{(k)}$.

Theorem 2. *The unbiased estimation for C_i^k, the count of graphlet g_i^k, is $\hat{C}_i^k = \frac{1}{n\alpha_i^k} \sum_{t=1}^n \frac{L(r^{(k)}, g_i^k)}{P(r^{(k)})}$.*

Proof. Every subgraph isomorphic to graphlet g_i^k in G can be found α_i^k times in our strategy. So we have $\sum_{r^{(k)} \in R^{(k)}} L(r^{(k)}, g_i^k) = \alpha_i^k C_i^k$. By the property of expectation we have

$$E[\hat{C}_i^k] = E[\frac{1}{n\alpha_i^k} \sum_{t=1}^n \frac{L(r^{(k)}, g_i^k)}{P(r^{(k)})}] = \frac{1}{n\alpha_i^k} \sum_{t=1}^n E[\frac{L(r^{(k)}, g_i^k)}{P(r^{(k)})}]$$

$$= \frac{1}{n\alpha_i^k} \sum_{t=1}^n \left(\sum_{r^{(k)} \in R^{(k)}} \frac{L(r^{(k)}, g_i^k)}{P(r^{(k)})} P(r^{(k)}) \right) = \frac{1}{n\alpha_i^k} \sum_{t=1}^n \alpha_i^k C_i^k = C_i^k.$$

Therefore, with Law of Large Numbers, we can obtain the unbiased estimator of graphlet concentrations and we don't require any global information of the network, say D. Let $F(r^{(k)}) = d(v_2)[d(v_2) + d(v_3)] \cdots [d(v_2) + \cdots + d(v_{k-1})]$, which means $P(r^{(k)}) = \frac{1}{D} \frac{1}{F(r^{(k)})}$. By the definition of concentration, we have

$$\frac{\sum_{t=1}^n \frac{1}{\alpha_i^k} F(r^{(k)}) L(r^{(k)}, g_i^k)}{\sum_{i=1}^{T_k} \sum_{t=1}^n \frac{1}{\alpha_i^k} F(r^{(k)}) L(r^{(k)}, g_i^k)} \to c_i^k, \text{ when } n \to \infty.$$

3.4 Framework to Estimate Graphlet Counts and Bound Analysis

Based on the method and theoretical analysis in previous sections, we show our framework to estimate graphlet counts in Algorithm 2. Our algorithm starts to generate samples from starting nodes in a random walk when the random walk mixes. The k-subgraph samples for estimating k-graphlet counts are generated by Algorithm 1. Every time we get a valid k-graphlet sample g_i^k, we add corresponding reweighted coefficient $\frac{F(r^{(k)})}{\alpha_i^k} * \frac{D}{n}$ to $\hat{C}_i^k$ and finally get the estimated graphlet counts.

To speed up our algorithm, we consider to run it in parallel. Running many random walks at the same time would help us to get more samples and thus to obtain more accurate estimation. Every processing unit is chosen independently. We derive the unbiased estimation of target graphlets statistics by averaging the unbiased estimations from each processing units.

Now we discuss the relation between the number of samples and the accuracy. Define $T = \tau(\xi)$ as its ξ-mixing time for $\xi \leq \frac{1}{8}$. Let $v_1, v_2, \ldots, v_n$

Algorithm 2. SSRW for estimating graphlet counts

Input: Graph $G = (V, E)$, sample number n, graphlet size k, sum of degrees D;
Output: Estimated k-graphlet counts for of G;
1: pick s as a starting node when random walk get mixing;
2: **for** $i = 1$ to n **do**
3:     $v_1 \leftarrow s$;
4:     generate k-$\{v_1, v_2, \ldots, v_k\}$ starting from v_1 using Algorithm.1;
5:     **if** $v_1, v_2, \ldots, v_k$ are all distinct **then**
6:         $id \leftarrow$ graphlet id induced by $(v_1, v_2, ..., v_k)$;
7:         $\hat{C}_{id}^k \leftarrow \hat{C}_{id}^k + \frac{d(v_2)[d(v_2)+d(v_3)]\cdots[d(v_2)+\cdots+d(v_{k-1})]}{\alpha_{id}^k} * \frac{D}{n}$;
8:     **end if**
9:     $s \leftarrow$ random node in $N(s)$;
10: **end for**
11: **return** $(\hat{C}_1, \hat{C}_2, \ldots, \hat{C}_{T_k})$.

---

denote a n-step random walk starting from an initial distribution φ, and $\|\varphi\|_\pi = \sum_{v \in V} \varphi^2(v)/\pi(v)$, $M_i^k = \max_{r^{(k)} \in R^{(k)}} \frac{DL(r^{(k)}, g_i^k)}{P(r^{(k)})\alpha_i^k}$, we have following theorem.

Theorem 3. *There exists a bound $B = \frac{72TM_i^k \log \frac{s\|\varphi\|_\pi}{\delta}}{C_i^k \epsilon^2}$ so that when $n > B$ we have $P(|\hat{C}_i^k - C_i^k| \leq \epsilon C_i^k) \geq \delta$ for given accuracy parameter ϵ and confidence level δ.*

Where s is a const number. The detail proof is shown in [1]. So the accuracy of our algorithm is related to the properties of the graph and initial state of random walk.

4 Experiments

Our algorithm get big advance in estimating graphlet statistics with more than 5 nodes. We show our experiment results for these higher-order graphlets first. With the help of parallelization, we can get massive samples in a short time, leading to higher accuracy. We show these results in Sect. 4.3. Finally, we compare the accuracy and running time with the state-of-the-art algorithms.

4.1 Experiment Setup

The datasets used in our experiments come from Stanford Network Analysis project (SNAP) [12] and networkrepository [22], listed in Table 1. We implemented our algorithm in C++ for graphlets with less than 6 nodes. For testing 6-and 7-graphlets, we implemented our algorithm in Python with iGraph library. Our algorithm is based on simple random walk, so we choose the largest connected component (LCCs) in large graphs and remove directions and self-loop edges.

Table 1. Datasets

Graph	$|V|$	$|E|$
LiveJournal [12]	4843953	42851237
com-Amazon [12]	334862	925870
socfb-Penn94 [22]	41536	1362180
Slashdot [12]	77360	417759
Gnutella04 [12]	10876	39994
Gnutella06 [12]	8717	31525
ca-GrQc [12]	4158	13422

We use normalized root mean square error (NRMSE) to evaluate the accuracy of SSRW and other algorithms, which is defined as NRMSE $(\hat{C}_i^k)$ = $\frac{\sqrt{E(\hat{C}_i^k - C_i^k)^2}}{C_i^k}$ = $\frac{\sqrt{Var[\hat{C}_i^k] + (E[\hat{C}_i^k - C_i^k])^2}}{C_i^k}$, where $\hat{C}_i^k$ is the estimated value and C_i^k is the ground-truth.

4.2   Estimation of 6-and 7-Graphlet Statistics

To show the scalability of our algorithm, we use SSRW to estimate 6- and 7-graphlet counts. In this experiment, we choose medium sized graphs, as there is no efficient tool to calculate the exact count of 6-and 7-graphlets in large graphs.

Table 2. Performance on 6-and 7-graphlets

Graphlet id	Gnutella04 (C_i^6)	Gnutella06 (C_i^6)	ca-GrQc (C_i^7)
1	0.0511	0.0604	0.0492
2	0.0452	0.0512	0.0520
3	0.0856	0.0306	0.0893
4	0.0428	0.0429	0.1400
5	0.0410	0.0367	0.1103
6	0.0233	0.0278	0.0815
7	0.0326	0.0340	0.0701

Table 2 illustrates the accuracy of our algorithm for 6-and 7-graphlet counts estimation. We get 500K samples and repeat 1000 times to get NRMSE and choose seven of them to demonstrate. All NRMSEs are less than 0.1 for 6-graphlets and 0.15 for 7-graphlets. For getting 500K 6-graphlets samples, it takes our algorithm 44.9 s and 44.3 s on Gnutella04 and Gnutella06, respectively, while Getting 500K 7-graphlets samples needs 136 s on ca-GrQc. Parallelization will shorten running time greatly. If we sample 500K samples on Gnutella04 with 5 processing unit with every processing unit getting 100K samples, it only

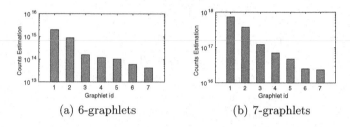

(a) 6-graphlets   (b) 7-graphlets

Fig. 4. Estimation of 6-and 7-graphlets counts on Slashdot.

needs 9.8 s. We will show the performance of parallelization in more detail in next subsection.

For showing that our algorithm also performs well in large graphs, we list the counts estimation for Slashdot with 77K nodes and 417K edges in Fig. 4. As far as we know, there is no practical tool for counting all 6-and 7-graphlets in graphs as big as Slashdot. We run 500K samples, which spends 104.8 s and 165.6 s for 6-and 7-graphlets estimation respectively. Even though we have no exact value for comparison, the unbiasedness of our algorithm and previous experiments make us believe that these values are very close to the ground truth.

4.3   Parallelization

Parallelization would help us to get massive number samples in a short time, which will improve accuracy a lot. We parallelize our algorithm with 256 CPUs, Intel Xeon system, 2.20 GHz and 512G memory.

We run parallelized SSRW to estimate the counts of 6-graphlets in Gnutella04 by varying the number of processes. Each process collects 10K samples. Figure 5(a) shows the relation of the time consuming and the number of processes we use. The average accuracy of 7 graphlets is 0.077 for 100K samples. When we perform 10 processes in parallel, which means we will have $10K \times 10 = 100K$ samples, time overhead only increases by 41.7% with the average NRMSE drops to 0.042. When we run 100 processes in parallel, the time is only doubled, but the average accuracy drops to 0.035, reduced by 50%, as shown in Fig. 5(b).

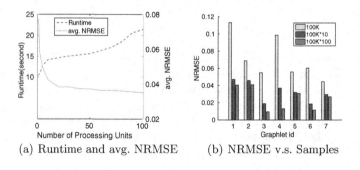

(a) Runtime and avg. NRMSE   (b) NRMSE v.s. Samples

Fig. 5. 6-graphlets counts estimation on Gnutella06.

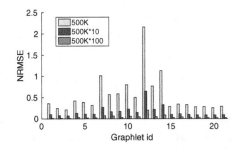

Fig. 6. Samples v.s. NRMSE on LiveJournal.

We also evaluate our parallelized algorithm on a larger graph, i.e., LiveJournal for estimating 5-graphlet counts. Figure 6 shows that the accuracy increases with the number of samples. When we run 100 processes in parallel and every thread gets 500K samples, we get 5M samples in total. Except for g_{11}^5, all NRMSEs are smaller than 0.1, and the running time is 43.9 s, while the state-of-the-art algorithm ESCAPE costs 28385 s for getting the exact value of 5-graphlet counts.

4.4 Comparison with Previous Work

To present the performance of SSRW furthermore, we compare it with the state-of-the-art algorithms Subgraph Random Walk on $G^{(2)}$ with corresponding states for a subgraph (SRW2CSS) [5] and Waddling Random Walk (WRW) [7] in the accuracy of estimating graphlet concentrations. These algorithms are mainly designed to estimate concentrations of graphlets with less than 6 nodes. The counts estimation can derive concentrations directly, so we compare the results of 4-and 5-graphlet concentrations with the above two algorithms. SRW2CSS runs on the super graph $G^{(2)}$, where each node corresponds to an edge in G. WRW runs on the original graph and designs a specific waddling protocol for those graphlets which can not be sampled in k consecutive steps of the simple random walk.

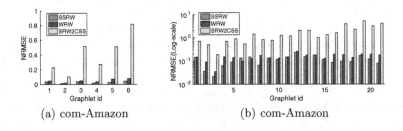

(a) com-Amazon   (b) com-Amazon

Fig. 7. NRMSE of concentration estimations for 4-and 5-graphlets.

Figure 7 shows the comparisons of the average NRMSEs of all 4-graphlets with 20K samples and all 5-graphlets with 30K samples on com-Amazon and Emai-Eron. The NRMSE is estimated over 1000 independent simulations. From

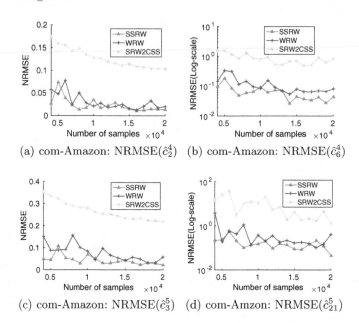

(a) com-Amazon: NRMSE($\hat{c}_2^4$)   (b) com-Amazon: NRMSE($\hat{c}_6^4$)

(c) com-Amazon: NRMSE($\hat{c}_3^5$)   (d) com-Amzon: NRMSE($\hat{c}_{21}^5$)

Fig. 8. NRMSE of concentration estimations for graphlets $(g_2^4, g_6^4, g_3^5, g_{21}^5)$.

Fig. 7, we find that SSRW and WRW are significantly more accurate than SRW2CSS for all graphlets. For 4-graphlets, SSRW outperforms SRW2CSS at least 6× and up to 18× (e.g. g_5^4 in com-Amazon), and outperforms WRW up to 1.8× (e.g. g_6^4 in com-Amazon). For 5-graphlets, SSRW outperforms SRW2CSS 10× and WRW 1.5× in general. We think that it is because SRW2CSS samples on the super graph, which induces to a much larger state space of sampling. SSRW and WRW sample on the original graph, so they can touch different parts of graph quickly and get samples more evenly from the graph.

To furthermore explore the performance of sampling algorithms on each individual graphlet, we also compare the accuracies of SSRW, WRW and SRW2CSS with different number of samples for every graphlet. Due to the limitation of pages, we only present the experimental results on dataset Com-Amazon, and the similar conclusion is derived on other datasets. In Fig. 8, we compare the three algorithms for two 4-graphlets and two 5-graphlets. It shows that SSRW outperforms the others in accuracy, converges fastest (about 5K samples) and is stable during the whole sampling process.

A recent work by Xiaowei Chen and Lui [6] performs excellently for estimating counts of 5-graphlets. We further conduct experiments to compare the accuracies of SSRW with their work. Their algorithm runs $k - 1$ consecutive steps on a random walk and Observe all Visual nodes to get k-subgraph samples under a Restricted model, so we denote their algorithm as OVR. Figure 9 shows that SSRW is significantly more accurate than OVR except for g_7^5 with the same number of samples for estimating 5-graphlet counts. The reason is that it collects too many samples in local locations.

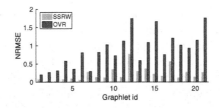

Fig. 9. NRMSE comparison between SSRW and OVR [6].

4.5 Runtime Comparison

We conduct experiments to compare the computational complexities of WRW, SRW2CSS and SSRW with C++. WRW was written in Python which limits its speed, so we rewrite it in C++. We use the algorithm provided by GUISE to identify 4- and 5- graphlets. We got SRW2CSS from its authors. All of them are running with the same configuration with 3.2 GHz Intel core i3 processor and 4 GB memory. For fair comparison, all algorithm are running in sequential. Also due to limitation of pages, we only present the results of four graphlets in Fig. 10, where we omit the loading time.

From Fig. 10, we find that although SRW2CSS obtains more samples than WRW and SSRW within the same time, its accuracy is usually lower than WRW and SSRW. SSRW performs better than WRW for estimating cliques.

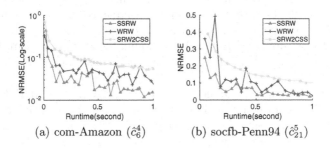

(a) com-Amazon ($\hat{c}_6^4$)   (b) socfb-Penn94 ($\hat{c}_{21}^5$)

Fig. 10. NRMSE v.s. runtime.

5 Related Work

GUISE [4] estimates graphlet concentrations based on random walk. It runs Metropolis-Hasting algorithm to perform sampling on a super graph where each super node corresponds to a connected subgraph of the original graph and two super nodes are adjacent if and only if the two corresponding subgraphs differ in only one node. When implementing Metropolis-Hasting Random Walk (MHRW) on the super graph, all subgraphs can be uniformly sampled. GUISE calculates 3, 4, 5-graphlets simultaneously.

However, GUISE is inefficient for choosing next state because of the rejections in MHRW. Some works improve GUISE, such as using a simple random walk (SRW) to avoid the rejections. It chooses next state uniformly from current state's neighbors without rejection and uses Horvitz-Thompson estimator to construct unbiased estimation. As neighbor population for k-graphlets is more difficult than $(k-1)$-graphlets, PairWise [25] and Chen et al. [5] combine consecutive states in the random walk on t-super graph to collect k-subgraphs for $t \leq k - 1$. The smaller t is, the better the algorithm performs. But they meet difficulty when $t = 1$ because consecutive k steps of random walk on G can not generate all k-graphlets when $k \geq 4$, say star g_2^4. In addition, there are much more nodes in a super graph, so the accuracy of the approaches by sampling on super graphs is worse than SSRW with the same number of samples, which is validated in our experiments.

Waddling Random Walk (WRW) [7] collects samples in the original graph with a simple random walk. When $k \geq 4$, because some k-graphlets are impossible to be generated via k consecutive steps, the authors designed a waddling protocol for such graphlets. However, WRW has not provided definitive algorithm for waddling protocol with $k > 5$. To design a wadding protocol, one should carefully study the features of all k-graphlets, which is too hard to realize when $k \geq 6$, for the number of graphlets grows rapidly. We design a flexible sampling method for SSRW, which selects the next node from the neighbors of multiple previous nodes, so SSRW is easy to be extended to sample graphlets with more nodes.

There are some works on estimating graphlet counts [10,26]. Path Sampling [10], Moss [26] estimates the counts of 4-node and 5-node graphlet counts respectively by sampling special types trees. However, these algorithms are quite complex and need pre-computation before sampling and are hard to be scaled. Chen et al. designed an algorithm based on random walk under restricted access model [6].

For exact counting of graphlets, RAGE [14], PGD [2] and ESCAPE [18] perform excellently for counting 4-graphlets and 5-graphlets. They use the relationship between the counts of different graphlets to reduce computation. But they do not perform well on dense graphs. G-tries [21] can calculate the counts of graphlets with more than 5 nodes, but it runs slowly and will return wrong answers when it runs more than 30 min.

6   Conclusion

We developed a sampling algorithm SSRW to estimate graphlet statistics in large graphs efficiently. SSRW is scalable to estimate graphlets with more than 5 nodes. Extensive experiments show that our algorithm is scalable, accurate and fast. The sampling strategy in our algorithm can also be easily applied to other scenarios, such as distribution systems for sampling subgraphs. Our sampling algorithm could serve as a basic component for solving other problems in large graphs, in which cases most of the running time is spent on identifying subgraphs.

We can further improve SSRW by developing a better subgraph identification algorithm, or using other strategy such as NBRW [11] to generate starting nodes to avoid too much repeated nodes, and we pose these as our future work.

Acknowledgements. This work is supported by NSFC (61672486, 61772484, 11671376), and Key Program of NSFC (71631006).

References

1. https://tinyurl.com/com-ssrw
2. Ahmed, N.K., Neville, J., Rossi, R.A., Duffield, N.G., Willke, T.L.: Graphlet decomposition: framework, algorithms, and applications. Knowl. Inf. Syst. **50**, 1–34 (2016)
3. Lovász, L.: Random walks on graphs: a survey. Combinatorics: Paul Erdös Is Eighty **2**(1), 1–46 (1993)
4. Bhuiyan, M.A., Rahman, M., Al Hasan, M.: Guise: uniform sampling of graphlets for large graph analysis. In: ICDM. IEEE (2012)
5. Chen, X., Li, Y., Wang, P., Lui, J.: A general framework for estimating graphlet statistics via random walk. VLDB **10**(3), 253–264 (2016)
6. Chen, X., Lui, J.C.: Mining graphlet counts in online social networks. In: ICDM. IEEE (2016)
7. Han, G., Sethu, H.: Waddling random walk: fast and accurate mining of motif statistics in large graphs. In: ICDM. IEEE (2016)
8. Hardiman, S.J., Katzir, L.: Estimating clustering coefficients and size of social networks via random walk. In: WWW. ACM (2013)
9. Holland, P.W., Leinhardt, S.: A method for detecting structure in sociometric data. Am. J. Sociol. **76**(3), 492–513 (1970)
10. Jha, M., Seshadhri, C., Pinar, A.: Path sampling: a fast and provable method for estimating 4-vertex subgraph counts. In: WWW. ACM (2015)
11. Lee, C.-H., Xu, X., Eun, D.Y.: Beyond random walk and metropolis-hastings samplers: why you should not backtrack for unbiased graph sampling. In: SIGMETRICS (2012)
12. Leskovec, J., Krevl, A.: SNAP datasets: Stanford large network dataset collection, June 2014. http://snap.stanford.edu/data
13. Lim, Y., Kang, U.: Mascot: memory-efficient and accurate sampling for counting local triangles in graph streams. In: KDD (2015)
14. Marcus, D., Shavitt, Y.: RAGE-a rapid graphlet enumerator for large networks. Comput. Netw. **56**(2), 810–819 (2012)
15. Milenkovic, T., Przulj, N.: Uncovering biological network function via graphlet degree signatures. arXiv preprint arXiv:0802.0556 (2008)
16. Mislove, A., Marcon, M., Gummadi, K.P., Druschel, P., Bhattacharjee, B.: Measurement and analysis of online social networks. In: SIGCOMM. ACM (2007)
17. Peng, W., Gao, T., Sisodia, D., Saha, T.K., Li, F., Al Hasan, M.: ACTS: extracting android app topological signature through graphlet sampling. In: 2016 IEEE Conference on Communications and Network Security (CNS), pp. 37–45. IEEE (2016)
18. Pinar, A., Seshadhri, C., Vishal, V.: ESCAPE: efficiently counting all 5-vertex subgraphs. arXiv preprint arXiv:1610.09411 (2016)

19. Pržulj, N., Corneil, D.G., Jurisica, I.: Modeling interactome: scale-free or geometric? Bioinformatics **20**(18), 3508–3515 (2004)
20. Rahman, M., Bhuiyan, M.A., Al Hasan, M.: Graft: an efficient graphlet counting method for large graph analysis. TKDE **26**(10), 2466–2478 (2014)
21. Ribeiro, P., Silva, F.: G-tries: an efficient data structure for discovering network motifs. In: Proceedings of the 2010 ACM Symposium on Applied Computing, pp. 1559–1566. ACM (2010)
22. Rossi, R.A., Ahmed, N.K.: The network data repository with interactive graph analytics and visualization. In: Proceedings of the Twenty-Ninth AAAI Conference on Artificial Intelligence (2015). http://networkrepository.com/socfb.php
23. Saha, T.K., Hasan, M.A.: Finding network motifs using MCMC sampling. In: Mangioni, G., Simini, F., Uzzo, S.M., Wang, D. (eds.) Complex Networks VI. SCI, vol. 597, pp. 13–24. Springer, Cham (2015). https://doi.org/10.1007/978-3-319-16112-9_2
24. Shervashidze, N., Vishwanathan, S., Petri, T., Mehlhorn, K., Borgwardt, K.: Efficient graphlet kernels for large graph comparison. In: Artificial Intelligence and Statistics, pp. 488–495 (2009)
25. Wang, P., Lui, J., Ribeiro, B., Towsley, D., Zhao, J., Guan, X.: Efficiently estimating motif statistics of large networks. TKDD **9**(2), 8 (2014)
26. Wang, P., Zhao, J., Zhang, X., Li, Z., Cheng, J., Lui, J.C., Towsley, D., Tao, J., Guan, X.: MOSS-5: a fast method of approximating counts of 5-node graphlets in large graphs. TKDE **30**, 73–86 (2017)

Multi-metric Graph Query Performance Prediction

Keyvan Sasani$^{(\boxtimes)}$, Mohammad Hossein Namaki, Yinghui Wu, and Assefaw H. Gebremedhin

School of EECS, Washington State University, Pullman, USA
{ksasani,mnamaki,yinghui,assefaw}@eecs.wsu.edu

Abstract. We propose a general framework for predicting graph query performance with respect to three performance metrics: execution time, query answer quality, and memory consumption. The learning framework generates and makes use of informative statistics from data and query structure and employs a multi-label regression model to predict the multi-metric query performance. We apply the framework to study two common graph query classes—reachability and graph pattern matching; the two classes differ significantly in their query complexity. For both query classes, we develop suitable performance models and learning algorithms to predict the performance. We demonstrate the efficacy of our framework via experiments on real-world information and social networks. Furthermore, by leveraging the framework, we propose a novel workload optimization algorithm and show that it improves the efficiency of workload management by 54% on average.

1 Introduction

Query performance prediction (QPP) plays an important role in database management systems. For example, it can be used to optimize workload allocation and online queries [1]. Furthermore, since QPP can be used to estimate the quality of a retrieved answer to a user's query, it can be used to prioritize search procedures, where queries with higher quality of answers are favored. Formally, given a query workload $\mathcal{W}$, a database D, and performance metrics $\mathcal{M}$ (*e.g.* response time, quality, or memory), the QPP problem is to predict $\mathcal{M}$ for each query instance in $\mathcal{W}$ over D.

This paper studies QPP for structureless graph queries that are fundamental in a wide range of applications, including knowledge and social media search. A graph query can represent a complex question that is subject to topological and semantic constraints. *Graph traversal*—e.g. regular path queries—and *pattern matching*—via subgraph isomorphism or simulation—are two commonly seen classes of graph queries. While efficient algorithms are studied to process graph queries efficiently, QPP is nontrivial for these queries. We use the following two examples to illustrate the unique challenges graph analytical workloads pose.

© Springer International Publishing AG, part of Springer Nature 2018
J. Pei et al. (Eds.): DASFAA 2018, LNCS 10827, pp. 289–306, 2018.
https://doi.org/10.1007/978-3-319-91452-7_19

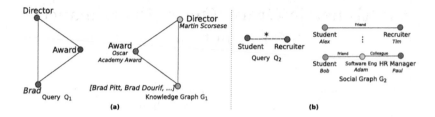

Fig. 1. Approximate graph querying, (a) knowledge search and (b) social search

Example 1. Knowledge search. Consider a query Q_1 and a portion of knowledge graph G_1 extracted from DBpedia that finds every "Brad" who worked with a "Director" and won an award [24]. This query can be represented by a graph pattern Q_1 that carries (ambiguous) keywords, with a corresponding *approximate* match as illustrated in Fig. 1(a). Each pattern node in Q_1 may have a large number of candidate matches.

As shown in this example, graph queries, unlike their relational counterparts, can be "approximate" or "structureless" [25], *i.e.* not well supported by rigid algebra and syntax. The ambiguous keyword "Brad" in this example query can lead to either "Brad Pitt", "Brad Dourif", or many other nodes in our data graph. It is often hard to exploit algebra and operator-level features (*e.g.* number of "join") [1,8] for graph matching queries—and it is exceedingly much harder for reachability queries. Furthermore, graph data is often noisy and heterogeneous. Features from data graph alone may not be reliable for QPP tasks.

Example 2. Social search. Consider a business-oriented social network in which nodes and edges represent people and their contacts, respectively. Suppose a researcher wants to know how senior students can use their connections to contact a recruiter and ask the recruiter to evaluate their resume. This question can be represented as a regular path query from a student typed person to a company recruiter. A regular path query Q_2 as shown in Fig. 1(b) asks *"which recruiters are reachable from a student with at most 2 hops utilizing only friend and colleague relations?"* on the social network G_2. Note that each person might have several positions at the same time and there might be restrictions on using connections. In this example, friend and colleague relations are allowed to be used. Imposing other constraints on the number of connections to reach the target person is also possible. The query node "Student" and "Recruiter" match {"Alex", "Bob"} and {"Tim", "Paul"}, respectively.

As illustrated in this example, while a common practice for QPP is to explore (logical and physical) query plans that are generated following a principled manner [1], this is inapplicable for approximate graph queries. A regular path may strict the edges we use to find the target which may change the query plan dynamically during its computation. In addition, deriving statistics from the graph data alone is expensive due to the sheer size of data, and the fact that the underlying graph may change over time makes the process even more complex.

In this paper, we present effective QPP methods for graph analytical workloads over multiple metrics. We develop a learning framework that solely makes use of computationally efficient query-oriented features and statistics from executed graph queries, without imposing assumptions on query syntax and algebra. Our goal is to build a general prediction framework for routinely issued, structureless graph queries. We apply the framework to design a workload optimization algorithm under bounded resources.

Contributions. Our main contributions are as follows:

- We propose a general learning framework (MGQPP) to predict *multiple* query performance metrics for various graph analytical queries. The framework employs a novel training instance generation and multi-label regression models.
- We use the framework to develop performance prediction methods for top-k queries [24], approximate matching queries [14], and general regular expression-based reachability queries [5].
- We apply MGQPP to resource intensive querying, and develop learning-based workload optimization strategies that make use of a Skyline Querying Algorithm [21] over a "query table" and extract top-k resource-bounded queries as a prioritized workload.
- We experimentally verify the efficacy of the proposed MGQPP framework over real-world graphs.

Related Work. QPP has been studied extensively, especially in the information retrieval community to either predict quality of answers [9] or resource consumption [7,19,26]. Learning techniques for QPP have been applied in relational databases for SQL workloads [22] and in semi-structured data for SPARQL queries [7,8,26]. Regression and Support Vector Machines were used to predict the performance of SPARQL queries [7], where the features are collected from SPARQL algebra and pattern [8]. A similarity metric was used to find if the incoming query is similar to one of their training data and this similarity is used as a part of their features [8]. The problem of SPARQL query execution time prediction on RDF datasets has been considered by [26]. The authors of [26] also used algebra and basic graph pattern features to train two support vector regression and k-nearest neighbor models. In contrast to the mentioned query languages, graph analytical queries are not well supported by algebra and apriori query plans. These methods are not applicable to approximate graph querying.

Efficient processing of top-k queries is an essential requirement especially when trying to manage very large data and multi access scenarios. Top-k processing techniques in relational and XML databases are surveyed in [10].

Our framework differs from the related works discussed here in several ways. (1) It considers performance as a multi-variant metric consisting of response time, answer quality, and resource consumption. Thus, it uses a multi-label regression model. (2) It does not assume any existing query plan or algebra. (3) We obtain higher prediction by introducing a diversified training instance generator using query templates of the available benchmarks [15]. (4) We study the problem of multi-performance metric graph query workload optimization using skyline algorithm which allows us to select the optimal subset of workload.

2 Problem Formulation

In this section, we formally define graph queries, performance metrics and the prediction problem which will be used later in the proposed framework.

2.1 Graph Queries

Data Graphs. We consider a labeled and directed data graph $G = (V, E, \mathcal{L})$, with node set V and edge set E. Each node $v \in V$ (edge $e \in E$) has a label $\mathcal{L}(v)$ ($\mathcal{L}(e)$) that specifies node (edge) information, and each edge represents a relationship between two nodes. In practice, $\mathcal{L}$ may specify attributes, entity types, and relation names [12].

Graph Queries. A graph analytical query Q is a graph $G_Q = (V_Q, E_Q, \mathcal{L}_Q)$. Each *query node* $u \in V_Q$ has a label $\mathcal{L}_Q(u)$ that describes the entities to be searched for (*e.g.* type, attribute values), and an edge $e \in E_Q$ between two query nodes specifies the relationship between the two entities. A match of Q in G, denoted as $\phi(Q)$, is a subgraph of G that satisfies certain matching semantics, induced by a matching relation ϕ. Specifically, each node $u \in V_Q$ has a set of matches $\phi(u)$, and each edge $e \in E_Q$ has a set of matches $\phi(e)$ [17].

Next, we define the three query classes we study under this framework.

Top-k Subgraph Queries [24]. A top-k subgraph query $Q(G, k, L)$ defines the match function ϕ as subgraph isomorphism, where the label similarity function L is derived by a set of functions drawn from a library (*e.g.* acronym, synonym, abbreviations), where each function maps nodes and edges (as ambiguous keywords) in Q to their counterparts in G. A common practice to evaluate a top-k subgraph query is to follow the Threshold Algorithm (TA) [4] that aggregates top-k tuples in relational tables.

Approximate Graph Pattern Matching [14]. A dual-simulation query $Q(G, S_V, \theta)$ relaxes the strict label equality to approximate matches of ambiguous keywords as well as the subgraph isomorphism from 1-1 bijective mapping to matching relations. The semantic has been used recently for event discovery [16,18,20]. Given a query $Q = (V_Q, E_Q, \mathcal{L}_Q)$ and a graph $G = (V, E, \mathcal{L})$, a match relation $\phi \subseteq V_Q \times V$ satisfies the following:

(1) for any node $u \in V_Q$, there is a match $v \in V$ such that $(u, v) \in \phi$ and $S_V(u, v) > \theta$, where $S_V(\cdot)$ is a similarity function over labels and θ is a threshold which assures each node has an appropriate match in an answer [24].
(2) for any $(u, v) \in \phi$ and any child (resp. parent) of u (denoted as u') in Q, there is a child (resp. parent) of v (denoted as v') in G, such that $(u', v') \in \phi$. That is, it preserves both parent and child relationships between a node u and its matches v.

Regular Path Queries [5]. Applications in traffic analysis, social analysis and Web mining often rely on queries that carry a regular expression. Similar to [5], we consider reachability queries as regular expressions.

A reachability query is defined as $Q_r = (s, t, f_e, d, \theta)$, where s and t are predicates such as node types and labels, θ is a threshold for similarity function $S_V(v, s) > \theta$ (resp. $S_V(v, t) > \theta$) for accepting each node v, and f_e is a regular expression drawn from the subclass $R ::= l \mid l^{\leq d} \mid RR$. Here, l is any potential relationship type of an edge or a wildcard _, where the wildcard _ is a variable standing for any $L(e)$; d is a user-specified positive integer that determines the maximum allowed hops from a source match s to a target match t. That is, $l^{\leq d}$ denotes the closure of l by at most d occurrences; and the operation of RR denotes the concatenation of two regular expressions. The query finds all pairs of source match v_s and target match v_t, where v_s matches s and v_t matches t via ϕ, and there exists a path ρ from v_s to v_t with a label (concatenated edge labels) that can be parsed by f_e.

2.2   Performance Metrics

We focus on multi-metric QPP for graph analytical queries. We consider the following metrics: (1) response time $t(Q, G, \mathcal{A})$, the time needed by algorithm $\mathcal{A}$ to return answers to query Q in graph G; (2) quality $q(Q, G, \mathcal{A})$, the highest quality score of the answers returned by $\mathcal{A}$ in G; and (3) memory $m(Q, G, \mathcal{A})$, the memory needed to answer Q by $\mathcal{A}$ in G. For simplicity, we use t, q and m to denote the three metrics.

Response time and memory are rather familiar performance metrics. In contrast, the query answer quality metric is not straightforward. We next introduce a generic quality function $F(\cdot)$ for graph analytical queries.

Generic Quality Function. Given a query Q and its match $\phi(Q)$, we consider the following: (1) There is a node scoring function $S_V(u, \phi(u))$ that computes a similarity score (normalized to be in $(0, 1]$) between a query node u and its node matches $\phi(u)$ induced by $\phi(Q)$; and (2) similarly, there is an edge scoring function S_E that computes a score for each edge e in Q and its match $\phi(e)$.

A similarity function $S_V(\cdot)$ should consider both semantic constraint $L_V(\cdot)$ and topological constraint $T_V(\cdot)$. Each node match produces a similarity score $S_V(\cdot) = L_V(\cdot) * T_V(\cdot) \in (0, 1]$. In practice, $L_V(\cdot)$ supports various kinds of linguistic transformations such as synonym, abbreviation, and ontology _e.g._ "instructor" can be matched with "teacher" which allows a user to pose queries without having sophisticated knowledge about the vocabulary or schema of the graph [24]. Furthermore, query topological constraints such as node degrees are taken into account by $T_V(\cdot)$. Analogously, the similarity functions S_E and T_E are defined over edge matching. When the similarity functions are common for both nodes and edges, we do not write the subscript in the rest of the paper.

We consider a general quality function $F(\cdot)$ that aggregates the node and edge matching scores to produce a matching score defined as:

$$F(Q, \phi(Q)) = \frac{\sum_{v \in V_Q} S_V(v, \phi(v)) + \sum_{e \in E_Q} S_E(e, \phi(e))}{N}, \tag{1}$$

where N is a normalizer to get the score in $[0, 1]$. By default, a normalizer can be set to $|G|$, since $|\phi(Q)| \leq |G|$.

The quality of an answer depends on the query semantics. We will make use of the general function $F(\cdot)$ as a component to specialize the quality metric q for specific query classes.

Top-k Search Quality Function. Given a graph query Q, an approximate answer $\phi(Q)$, and an integer k, we define the quality function q_{topk} for top-k search as follows:

$$q_{topk}(Q, \phi(Q), k) = \frac{\sum_{i=1}^{k} F(Q, \phi_i(Q))}{k} \tag{2}$$

Here in its $F(\cdot)$, we set topological similarity function $T(\cdot) = 1$ since $\phi(Q)$ is isomorphic to Q. Note that the value of q_{topk} lies in $(0, 1]$ and is an average of qualities over all answers retrieved by top-k querying. That is, the closer the value is to 1, the more similar the labels of the matches are to that of the query.

Approximate Pattern Matching Quality Function. In the $F(\cdot)$ of this algorithm, since $\phi(Q)$ in simulation might be a topological approximation of Q, to compare the topology of the induced graph on $\phi(Q)$ to Q where the degrees of matches and the number of matched edges can be higher or lower than query nodes and edges, respectively, we set $T_V(V_Q) = min(\frac{deg(v)}{deg(V_Q)}, \frac{deg(V_Q)}{deg(v)})$, where $v \in \phi(V_Q)$ and $deg(v)$ (resp. $deg(V_Q)$) is the degree of node match v (resp. query node V_Q), in order to keep the quality metric in the range $(0, 1]$. Furthermore, we set $T_E(\cdot) = min(\frac{|E(\phi(Q))|}{|E_Q|}, \frac{|E_Q|}{|E(\phi(Q))|})$. The quality q_{Sim} is then defined as follows:

$$q_{Sim}(Q, \phi(Q)) = \frac{\sum_{\phi(Q)} F(Q, \phi(Q))}{|\phi(Q)|}, \tag{3}$$

where $|\phi(Q)|$ is the number of total matches retrieved by a simulation algorithm. We remark that in addition to considering the linguistic similarity by $L(\cdot)$, topological constraints are also assessed by $T(\cdot) \in (0, 1]$, affecting the overall quality of $F(\cdot)$. That is, the closer the structure of an answer is to the query, the closer the quality is to 1.

Regular Path Quality Function. Intuitively, the higher quality for reachability queries happens when query node pair (s, t) exactly matches the answer pair $(\phi(s), \phi(t))$ and also the length of shortest path between $\phi(s)$ and $\phi(t)$ is smaller (fewer number of hops between $\phi(s)$ and $\phi(t)$). Hence, given the data graph G and graph query Q, we set $S_E = 0$ and $T_V = \frac{1}{|E(\phi(Q))|}$. Therefore, the quality of the retrieved answers is defined as follows:

$$q_{reach}(Q, \phi(Q)) = \frac{\sum_{\phi(Q)} F(Q, \phi(Q))}{|V_Q||\phi(Q)|}, \tag{4}$$

where $|\phi(Q)|$ means the number of distinct pairs $(\phi(s), \phi(t))$ and $|V_Q| = 2$ for reachability queries.

2.3   Performance Prediction

In this subsection, our goal is to formulate QPP for graph analytical workloads. We consider a mixed workload $W = \{Q_1, \ldots, Q_n\}$ over a set of query classes $\mathcal{Q}$, where each query Q_i is an instance from a query class in $\mathcal{Q}$, and the vector $\mathcal{M}$, the multi-metrics performance to be predicted. We instantiate $\mathcal{M} = <t, q, m>$, where t, q, and m are the response time, answers quality, and memory usage as measure by the number of visited nodes, respectively. The problem of multi-metric graph query performance prediction, denoted as MGQPP, is to learn a prediction model $\mathcal{P}$ to predict the performance vector of each query instance in W with maximum accuracy measured by a specific metric.

A metric must measure how well the performance of future queries is likely to be predicted. We seek to minimize the error depending on the type of the queries and the scale of their performance values. Therefore, we use *R-Squared*, a widely used evaluation metric [8] to evaluate our framework. To empirically verify the robustness of our models, we also consider *mean absolute error* (MAE) besides *R-Squared*. MAE, defined as $\frac{1}{n}\sum_{i=1}^{n}|\hat{y} - y|$, is an absolute comparison of predictions and eventual outcomes [6].

GQPP as Regression. We approach MGQPP as a regression problem. We use the following construction:

Input: A data graph G, training workload W_T

Output: A prediction model $\mathcal{P}$ to predict a set of performance metrics $\mathcal{M}$ that maximizes the prediction accuracy of all metrics in the same time.

The problem is to learn, using a multi-label regression, a function $f(x) = y$ that maps a feature vector x of a query to a set of continuous values y corresponding to the exact response time, quality of the retrieved answer and number of visited nodes (as an indicator for memory usage) of the query.

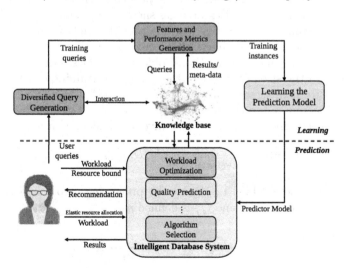

Fig. 2. Prediction framework for MGQPP

3   The Framework

Our multi-metric graph query performance prediction framework is illustrated in Fig. 2. Following statistical learning methodology, the framework derives a prediction model based on training sets (learning) and predicts query performance for test data points based on the derived model (prediction). Our goal is to adapt the framework to accomplish mixed graph analytical workload prediction.

3.1   Learning Phase

Besides the effect of feature selection, model choice, and loss function definition in any learning-based predictive frameworks, training data plays an important role in ensuring that a model is comprehensive enough to predict a wide variety of future inputs. Here, our training data is a set of queries, called training workload $\mathcal{W}_T$. In order to generate a good, small yet representative training data, MGQPP is armed with a diversified query generation module, which we describe below.

Diversified Query Generation. Given a query class $\mathcal{Q}$, a data graph G and a standard evaluation algorithm $\mathcal{A}$, the training workload $\mathcal{W}_T$ is a set of pairs $(\overrightarrow{Q_i}, \mathcal{M}(Q_i))$, where $\overrightarrow{Q_i}$ refers to a feature representation of query Q_i, and $\mathcal{M}(Q_i)$ is the actual performance metrics obtained by evaluating Q_i with algorithm $\mathcal{A}$.

An empirical study of over 3 million real-world SPARQL queries has shown that most queries are simple and include few triples patterns and joins [2]. Indeed, in order to simulate real-world queries, it is enough to generate small queries by a bounded random walk on the data graph. However, while a naive way of generating training queries is sampling the data graph using a random-walk [13], it does not guarantee the diversification of the training instances to provide additional information for the predictor. Hence, we adopt a batch mode active learning [23] to generate diversified queries. Research has shown that an intelligent selection of training data instances using active learning provides high accuracy with much fewer instances compared to a blind selection strategy [23].

We formulate the query generation problem as follows. Given a data graph G, a bound b on the size of the query, the number of training queries N, and a dissimilarity threshold σ, select a set of queries such that the diversity $d(Q_i)$ of each query Q_i compared to the current training set $\mathcal{W}_T$ is greater than σ. Given the query Q_i generated at the step i, the diversity is defined as:

$$d(Q_i) = \operatorname*{avg}_{\forall Q_j \in \mathcal{W}_T; j < i} CosDis(Q_i, Q_j), \qquad (5)$$

where Q_j is the query added to $\mathcal{W}_T$ at the processing step j and the Cosine Distance between queries Q_i and Q_j is defined as:

$$CosDis(Q_i, Q_j) = 1 - \frac{\overrightarrow{Q_i} \cdot \overrightarrow{Q_j}}{|\overrightarrow{Q_i}||\overrightarrow{Q_j}|}. \qquad (6)$$

Algorithm. The proposed query generation algorithm works as follows.

(1) Iteratively generate a connected subgraph as a query Q_i of query templates or a random walk, and use the query topology to build an unlabeled graph. Then, each query is assigned a type sampled from the top 20% most frequent types in the vocabularies of the knowledge base and $|Q_i|$ is set to be $\leq b$.

(2) Add the query Q_i to the training workload only if $d(Q_i) > \sigma$. The algorithm terminates when N queries are generated and added to the training workload.

Feature Generation and Learning Prediction Model. The learning framework generates the training workload $\mathcal{W}_T$. In particular, it generates queries (as discussed), evaluates the queries over the data graph G (stored and managed by the knowledge base) by invoking standard query evaluation algorithms, and collects the performance metrics and features for each query to construct $\mathcal{W}_T$. The predictive model is then derived by solving the multi-variable regression problem (details discussed in Sect. 3.2).

Features. As remarked earlier, graph analytical queries, unlike their relational counterparts, cannot be easily characterized by features from operators, algebra, and apriori query plans. Features from data alone may also be unreliable. We hence consider four classes of query-oriented features. The classes are called *Query, Sketch, Algorithm,* and *Quality* features since they characterize statistics from query instances, accessed data, search behavior, and similarity values, respectively. Later in the paper, we will use the shorthands Q, S, A and L to refer these four features, respectively.

(a) *Query features* encode the topological (*e.g.* query size, degree, cyclic) and semantic constraints (*e.g.* label, transformation functions [25]) from query terms.

(b) *Sketch features.* The idea is to exploit statistics that estimate the specificity and ambiguity of a query by "sketching" the data that will be accessed by the queries. These features may include the size of candidates (the nodes having the same or similar labels to some pattern query nodes), degree of sampled candidates, and statistics of sampled neighborhood of the candidates. By paying an affordable amount of time, these features significantly contribute to the prediction accuracy of graph queries (as verified in [19]).

(c) *Algorithm features* refer to the features that characterize the performance of graph querying algorithms. For example, top-k graph search typically decomposes a query to sub-queries, and assembles the complete results by aggregating partial matches as multi-way joins [24]. We found that features such as the number of decompositions and "joinable" candidates are very informative and critical to predict the cost of top-k search.

(d) *Quality features.* MGQPP also uses statistical features that directly affect the quality of the retrieved results. Such features include minimum, average, and maximum similarity values between a query and the candidates. Since the similarity computation is expensive, to aid efficient features computation at the prediction time, we follow an approach in which a one-time process calculates the features for any pair of nodes (resp. edges) in G and stores them in memory.

The proposed features can all be computed efficiently. Indeed, they can be extracted by fast linear scans and sampling over queries and data graphs, and are well supported by established database indexing techniques [24].

Feature Analysis. We studied the contribution of features in the framework by calculating their importance. For space considerations, we omit a complete description of the importance analysis we performed to select the features we use for each query algorithm, and instead, we discuss them only at a high level here.

Top-k Subgraph Queries. Algorithm and Sketch features were find to play the most important role in predicting the performance of top-k subgraph queries. Query features were found to be next in the importance ranking. Indeed, the performance of top-k subgraph queries may highly depend on the algorithm behavior (decomposition, n-way joins in the TA-style computation), which can be more critical than the number of joins (a plan-level feature) in a graph pattern [8].

Approx. Pattern Matching. Unlike top-k subgraph queries, we find Sketch features to be the most important for predicting the efficiency of the computation for dual-simulation queries. Next in importance for dual-simulation performance prediction were candidate size and degrees. Query size, in contrast, was found to be not as important.

Regular Path Queries. We found that quality of reachability queries is most determined by "average similarity values of candidates". Sketch features on source and target candidates come second in rank and query feature hop bound d comes third in rank. Indeed, the more candidates and more neighbors they have, and the more hops a query needs to visit, the more chance a reachability query have to find a match pair.

3.2 Prediction Phase

We use a multi-label learning framework as our primary predictive model. A multi-label or multi-output problem is a supervised learning problem with several outputs (potentially with correlations) to predict. In our case, we also observed that the output values related to the same input are themselves correlated. Hence, a better way is to build a single model capable of predicting all outputs at the same time. Such an approach will make the framework save training time, reduce complexity, and increase accuracy.

In order to build our multi-label model, we use XGBoost [3], an ensemble method, as our inner predictive model. XGBoost trains each subsequent model using residuals of current prediction and true values. Extensive studies have shown that XGBoost outperforms other methods on regression problems since it reduces the bias and variance at the same time. We note however that the GQPP problem has been addressed suitably by random forest regression models in the study [19].

In the next step, the prediction model is applied to predict the performance of new queries. Upon receiving a query workflow, the framework collects the queries, computes the query features, and predicts the query performance metrics vector. The predicted results can then be readily applied for resource allocation and workload optimization (see Sect. 4). Note that the proposed MGQPP framework can be specialized by "plugging in" other performance metrics to be applied for other type of graph queries.

4   Workload Optimization

We use resource bounded workload optimization as a practical application to illustrate one of the utilities of our MGQPP framework. We consider a mixed query workload $W = \{Q_1, \ldots, Q_n\}$ over a set of query classes Q, where each query Q_i is an instance of a query class in Q and is associated with a profit p_i. After execution of the query Q_i, a set of performance metrics $\mathcal{M}_i$ is associated with Q_i. Now using MGQPP, we can associate a predicted performance metrics $\hat{\mathcal{M}}_i$ to each query before its execution.

Resource-Bounded Query Selection. We formalize the multi-metric work-load optimization problem as the most profitable dominating skyline query selection problem. In a multi-dimensional dataset, a skyline contains the points that are not dominated by other points in any of the dimensions. A point dominates another point if it is as good or better in all dimensions and better in at least one dimension [11]. Using a modified version of a progressive skyline computation strategy, we propose an algorithm with performance guarantee.

Given workload W, integer k to retrieve top queries, a set of predicted performance metrics $\hat{\mathcal{M}}$, and resource bound $\mathcal{C} = \{c_1 \ldots c_m\}$ corresponding to the performance metrics $\mathcal{M}$, the problem is to find the most profitable dominating skyline queries $W' \subseteq W$ that maximize $\sum_{j=1}^{n} p_j x_j$, $x_j \in \{0, 1\}$, where $j = \{1, \ldots, n\}$ subject to the following two conditions.

(1) $\sum_{j=1}^{n} w_{ij} x_j \leq c_i$, $i = 1, \ldots, m$ where each query Q_j consumes an amount $w_{ij} > 0$ from each resource i (e.g. time, memory, 1-quality). The binary decision variables x_j indicate which queries are selected.
(2) Each query $Q_j \in W'$ is a skyline in W or by removing one or more queries in W', Q_j becomes a skyline in the updated W.

Progressive Skyline Query Selection. A skyline operator returns every query not dominated by any one of the rest of the queries in any of the performance metrics. Skyline operators have been found to be an important and popular technique in multidimensional environments for finding interesting and representative results. In practice, however, in order to solve the resource intensive

---

Algorithm skySel

Input: query workload $\mathcal{W} = \{\langle Q_0, \hat{M}_0, p_0 \rangle, \ldots\}$, integer k
resource bound $\mathcal{C} = \{c_1 \ldots c_m\}$.

Output: selected queries $\mathcal{W}'$.

1.   set $\mathcal{W}' \leftarrow \emptyset$; let S be a skyline operator;
        /* $\mathcal{L}$ in a decreasing order by profit p_i */
2.   let $\mathcal{L}$ be a priority queue;
3. $\mathcal{L} \leftarrow \mathcal{L} \cup S.\text{nextSkylineQueries}(\mathcal{W})$;
4.   **while** $|\mathcal{W}'| < k$ **and** $\mathcal{C}$ has enough resource **and** $\mathcal{L} \neq \emptyset$
        /* get the most profitable non-dominated query */
5.   query $Q_i \leftarrow \mathcal{L}.\text{pull}()$;
6. $\mathcal{W}.\text{remove}(Q_i)$;
7. $\mathcal{L} \leftarrow \mathcal{L} \cup S.\text{nextSkylineQueries}(\mathcal{W})$;
8. $\mathcal{W}' \leftarrow \mathcal{W}' \cup Q_i$; $\mathcal{C} \leftarrow \mathcal{C} - \hat{M}_i$;
9.   **return** $\mathcal{W}'$;

Fig. 3. skySel: Multi-metric query workload optimization algorithm

workload optimization problem, the domination constraint may be too restrictive to take the resource budget into account. In addition, profit maximization can be considered as an independent metric to be optimized since it is not an internal property of the query. Thus, we adapt progressive constrained skyline computation in order to guarantee selection of both resourced bounded and most profitable queries among the ones that are not dominated by the rest.

Our algorithm, denoted as skySel, is outlined in Fig. 3. The algorithm uses a priority queue $\mathcal{L}$ sorted in a decreasing order by profit (of queries). The queue $\mathcal{L}$ contains the queries that are not dominated by the rest of queries in $\mathcal{W}$ at any time. The algorithm skySel starts with an empty set of $\mathcal{W}'$ and uses a skyline operator S that progressively returns skyline queries (lines 1–2). It then populates the queue $\mathcal{L}$ with the first set of skyline queries (line 3). While the queue is not empty, there exists an available resource on all dimensions, and not enough queries are selected, it iteratively retrieves the most profitable non-dominated query Q_i from $\mathcal{L}$ (line 5), removes Q_i from the initial workload $\mathcal{W}$, updates the set of skyline queries in $\mathcal{L}$ with new queries and available resource vector $\mathcal{C}$, and adds Q_i to the set of selected queries (lines 6–8). When the algorithm terminates, $\mathcal{W}'$ is returned as the optimized query workload.

Correctness and Complexity. The algorithm skySel maintains two invariant at the beginning of each iteration: (I_1) the queries in $\mathcal{L}$ are not dominated by queries in the current $\mathcal{W} \cup \mathcal{L}$; and (I_2) the most profitable query in the set of non-dominated queries is selected as the top element in $\mathcal{L}$. The correctness of I_1 follows from the correctness of skyline computation and the correctness of I_2 follows from priority queue operations. Thus, the algorithm correctly finds the most profitable non-dominated queries.

A simple implementation of skyline takes $O(nlogn)$ to find the skyline queries [11]. The skyline operation is computed at most n times. Thus the overall complexity of algorithm skySel is $O(n^2 logn)$.

5   Experimental Evaluation

Using two real-world graphs, we conduct three sets of experiments to evaluate the following: (1) Performance of MGQPP over different metrics and a comparison to baselines; (2) Impact of diversified query workload generation vs. random generation on the accuracy of the predictors; and (3) Effectiveness of workload optimization using a case study.

Experimental Setting. We used the following setting.

Datasets. For the experiments, we use Pokec and DBpedia, two real-world graphs. Pokec[1] is a popular online social network in Slovakia. It has $1.6M$ users with 34 labels (*e.g.* region, language, hair color, etc.) and $30M$ edges among users. DBpedia[2] is a knowledge graph, consisting of $4.86M$ labeled entities (where each label is one of 1K labels such as "Place", "Person") and $15M$ edges.

Workload. We develop two query generators, a random generator using a random-walk with restart and a diversified query generator (see Sect. 3.1). We instantiate each generator for both graph pattern queries and reachability queries to construct training and test data sets over the two real-world networks.

Graph Pattern Queries. To generate graph pattern queries, we use the DBPSB benchmark [15], a DBpedia query benchmark. To achieve this, we use DBPSB query templates, and subsequently use the query topology to build an unlabeled graph. The graph is then assigned a type sampled from the top 20% most frequent types in the ontologies of Pokec and DBpedia. Furthermore, we set the maximum size of queries $b = 6$ (*i.e.* $\max |E_Q| = 6$) as it has been observed that most of the real-world SPARQL queries are small [2]. Although this process of query generation is a common practice [24], since we use these queries as a training data, it is important to consider the effect of each sample to the learning phase of the model. To address this, we employ the proposed diversified training query generation to make sure that the generated training data is informative to our learning model. (Details of the algorithm discussed in Sect. 3.1).

We draw the matching function $L(\cdot)$ from a library of similarity functions as in [25]. We then set an integer k drawn from $[10, 100]$.

Reachability Queries. For reachability queries $Q(s, t, d, G)$, we set $d \in [1, 4]$, and randomly select a pair of labels, from the top 20% most frequent labels in G. We sampled 4K queries, 1K for each d.

---

[1]   https://snap.stanford.edu/data/soc-pokec.html.
[2]   http://wiki.dbpedia.org/.

Algorithms. We implemented the following, all in Java.

(1) Standard query evaluation algorithms (Sect. 3), including:
    - STAR, the algorithm of [24] for top-k subgraph queries,
    - dual-simulation [14], for dual-sim queries, and
    - a variant of Breath-First Search, for reachability queries.
(2) Query workload optimization algorithms for top-k most profitable dominating queries, including
    - a progressive skyline computation algorithm (skySel) and
    - Rnd$_k$, a baseline algorithm that randomly selects k queries to be executed in the workload.

Predictive Model. We implemented the XGBoost model as our predictive model by leveraging the scikit-learn[3] library and APIs.

Metrics. We use two metrics as remarked in Sect. 2: R-Squared and MAE.

Test Platform. We ran all of our experiments on a Linux machine powered by an Intel 2.30 GHz CPU with 64 GB of memory. Each test is repeated 5 times and the averaged results are reported.

Result Overview. Here is a summary of our findings. Using the four classes of features and the XGBoost predictive model, we show the performance of analytical graph queries can be predicted quite accurately (**Exp-1**). Diversified training workload enables constructing a general model (**Exp-2**). Our case study verifies the effectiveness of our approach for query workload management (**Exp-3**). Furthermore, we found that using well-supported graph neighborhood and label indices [25], it takes on average 15.3 s to predict the performance of a workload of 433 queries, with total response time of 15 min. We next discuss our findings in details.

Exp-1: Performance of MGQPP. We estimate the accuracy of the XGBoost model using l-fold cross-validation [1]. We set $l = 5$.

Table 1. Performance evaluation measured in R^2

	Top-k		Dual-simulation		Reachability	
	DBPedia	Pokec	DBPedia	Pokec	DBPedia	Pokec
Time	0.819	0.985	0.818	0.926	0.928	0.869
Quality	0.978	0.991	0.827	0.906	0.973	0.985
Memory	0.981	0.998	0.995	0.995	0.938	0.993

Table 1 lists the *R-Squared* accuracy of XGBoost for the three query classes and the performance metrics response time, quality, and memory for both of the data sets Pokec and DBpedia. Over all datasets and performance metrics, we

---

[3] http://scikit-learn.org.

found that XGBoost attains an accuracy ranging between 61.64% and 99.84%. In addition, we found that the framework yields MAE of 420 ms on time, less than 0.0008 percent on quality, and 279.4 nodes on predicting the number of visited nodes as an indicator of memory usage in querying.

Actual vs. Predicted. Figure 4(a) shows a comparison between predicted and actual values of the XGBoost model for *top-k subgraph query* and the performance metric *time* for 1K queries. Figure 4(b) shows similar comparison for *dual simulation query* and performance metric *quality*, and Fig. 4(c) for *reachability query* and metric *memory*. In each of the cases, the results for the remaining two performance metrics are similar and are omitted for space considerations.

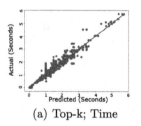
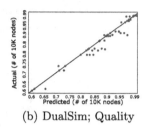
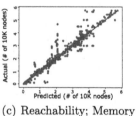

(a) Top-k; Time   (b) DualSim; Quality   (c) Reachability; Memory

Fig. 4. Actual vs. predicted values for different algorithms and metrics

Comparison with Related Works. As we mentioned in Sect. 1 related work, most of the related papers use SPARQL for querying semi-structured data. In fact, [19] is the first attempt that addressed the QPP in the context of general graph queries, although only on execution time. Anyhow, recent SPARQL query performance prediction approaches [7,26] used also DBPSB templates to generate queries and DBpedia as the underlying graph. Thus, we compare our results with their results (as reported in [7,26]) using the same metric denoted as "relative error" in Table 2. The results shows that our general framework outperforms recent approaches [7,26] in both accuracy and efficiency of training time.

Table 2. Performance comparison with the related works

	Model	Features	Relative err	1K Q's train (sec)
[7]	X-means+SVM+SVR	Algebra+GED	14.39%	1548.45
[26]	SVM+Weighted KNN	Algebra+BGP+Hybrid	9.81%	51.36
Ours	Multi-label XGBoost	Q+A+S+L	6.91%	35.33

Exp-2: Diversified Queries vs. Random Generation. Table 3 shows the accuracy of prediction using our diversified query generator compared with that of a simple random generation used as a baseline. It can be seen that diversified query generation outperforms the baseline by large margins consistently over all performance metrics.

Table 3. Diversified (Dvs) vs. Random (Rnd) query generation accuracy (%)

		Test					
		Time		Quality		Memory	
		Rnd	Dvs	Rnd	Dvs	Rnd	Dvs
Train	Rnd	40.59	25.78	74.51	66.83	62.06	36.2
	Dvs	**70.41**	**89.68**	**78.46**	**91.44**	**82.37**	**90.84**

Exp-3: Query Workload Optimization. We next conduct a case study to test the effectiveness of MGQPP for query workload optimization. The workloads are simulated as follows. (1) We generate a workload of $1K$ queries for each of the three query classes top-k subgraph, dual-sim, and reachability. (2) The queries are sent to each optimizer in batch. Given user input, the optimizer selects queries to be executed.

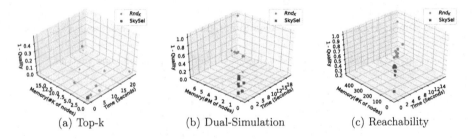

(a) Top-k   (b) Dual-Simulation   (c) Reachability

Fig. 5. Actual performance of selected queries, - skySel vs. Rnd$_k$ over DBpedia

Given workload W and $k = 10$, we invoke skySel to select k most profitable dominating queries and Rnd$_k$ as a random strategy. Figure 5(a), (b), and (c) demonstrate the selected queries by skySel and Rnd$_k$ for top-k, dual-sim, and reachability, respectively. The size of points shown in the figures are proportional to their profit. The closer the points are to the origin of coordinate system and the larger their size is, the better. The figures tell us that skySel outperforms Rnd$_k$ by selecting non-dominated queries with more profits. In addition, the results show that the query profit utilization of skySel algorithm is 54% more in comparison to Rnd$_k$ scenario on average of 10 different workloads.

6   Conclusion

We have presented a learning-based framework to predict performance of graph queries in terms of their response time, answer quality, and resource consumption. We introduced learning methods for both graph pattern queries, defined by subgraph isomorphism, dual-simulation, and reachability queries. We showed that by exploiting computationally efficient features from queries, sketches of the data to be accessed, and algorithm itself, multi-metric query performance can be accurately predicted using the proposed multi-label regression model. We also introduced a workload optimization strategy for selecting top-k best queries to be executed that maximizes the quality and minimizes the time and memory consumption. Our experimental study over real-world social networks and knowledge bases verifies the effectiveness of the learned predictors as well as the workload optimization strategy.

Acknowledgments. Sasani and Gebremedhin are supported in part by NSF CAREER award IIS-1553528. Namaki and Wu are supported in part by NSF IIS-1633629 and Huawei Innovation Research Program (HIRP).

References

1. Akdere, M., Çetintemel, U., Riondato, M., Upfal, E., Zdonik, S.B.: Learning-based query performance modeling and prediction. In: ICDE, pp. 390–401 (2012)
2. Arias, M., Fernández, J.D., Martínez-Prieto, M.A., de la Fuente, P.: An empirical study of real-world SPARQL queries. arXiv preprint arXiv:1103.5043 (2011)
3. Chen, T., Guestrin, C.: XGBoost: a scalable tree boosting system. In: KDD, pp. 785–794 (2016)
4. Fagin, R., Lotem, A., Naor, M.: Optimal aggregation algorithms for middleware. J. Comput. Syst. Sci. **66**(4), 614–656 (2003)
5. Fan, W., Li, J., Ma, S., Tang, N., Wu, Y.: Adding regular expressions to graph reachability and pattern queries. In: ICDE, pp. 39–50 (2011)
6. Guo, Q., White, R.W., Dumais, S.T., Wang, J., Anderson, B.: Predicting query performance using query, result, and user interaction features. In: RIAO (2010)
7. Hasan, R.: Predicting SPARQL query performance and explaining linked data. In: Presutti, V., d'Amato, C., Gandon, F., d'Aquin, M., Staab, S., Tordai, A. (eds.) ESWC 2014. LNCS, vol. 8465, pp. 795–805. Springer, Cham (2014). https://doi.org/10.1007/978-3-319-07443-6_53
8. Hasan, R., Gandon, F.: A machine learning approach to SPARQL query performance prediction. In: WI-IAT (2014)
9. Hauff, C., Hiemstra, D., de Jong, F.: A survey of pre-retrieval query performance predictors. In: Proceedings of the 17th ACM Conference on Information and Knowledge Management, pp. 1419–1420. ACM (2008)
10. Ilyas, I.F., Beskales, G., Soliman, M.A.: A survey of top-k query processing techniques in relational database systems. CSUR **40**, 11 (2008)
11. Kossmann, D., Ramsak, F., Rost, S.: Shooting stars in the sky: an online algorithm for skyline queries. In: VLDB, pp. 275–286 (2002)
12. Lu, J., Lin, C., Wang, W., Li, C., Wang, H.: String similarity measures and joins with synonyms. In: SIGMOD (2013)
13. Lu, X., Bressan, S.: Sampling connected induced subgraphs uniformly at random. In: Ailamaki, A., Bowers, S. (eds.) SSDBM 2012. LNCS, vol. 7338, pp. 195–212. Springer, Heidelberg (2012). https://doi.org/10.1007/978-3-642-31235-9_13
14. Ma, S., Cao, Y., Fan, W., Huai, J., Wo, T.: Capturing topology in graph pattern matching. VLDB **5**, 310–321 (2011)

15. Morsey, M., Lehmann, J., Auer, S., Ngonga Ngomo, A.-C.: DBpedia SPARQL benchmark – performance assessment with real queries on real data. In: Aroyo, L., Welty, C., Alani, H., Taylor, J., Bernstein, A., Kagal, L., Noy, N., Blomqvist, E. (eds.) ISWC 2011. LNCS, vol. 7031, pp. 454–469. Springer, Heidelberg (2011). https://doi.org/10.1007/978-3-642-25073-6_29
16. Namaki, M.H., Lin, P., Wu, Y.: Event pattern discovery by keywords in graph streams. In: IEEE Big Data (2017)
17. Namaki, M.H., Chowdhury, R.R., Islam, M.R., Doppa, J.R., Wu, Y.: Learning to speed up query planning in graph databases. In: ICAPS (2017)
18. Namaki, M.H., Sasani, K., Wu, Y., Ge, T.: BEAMS: bounded event detection in graph streams. In: ICDE, pp. 1387–1388 (2017)
19. Namaki, M.H., Sasani, K., Wu, Y., Gebremedhin, A.H.: Performance prediction for graph queries. In: NDA (2017)
20. Namaki, M.H., Wu, Y., Song, Q., Lin, P., Ge, T.: Discovering graph temporal association rules. In: CIKM, pp. 1697–1706 (2017)
21. Papadias, D., Tao, Y., Fu, G., Seeger, B.: Progressive skyline computation in database systems. TODS **30**, 41–82 (2005)
22. Wu, W., Chi, Y., Zhu, S., Tatemura, J., Hacigümüs, H., Naughton, J.F.: Predicting query execution time: Are optimizer cost models really unusable? In: ICDE, pp. 1081–1092 (2013)
23. Xu, Z., Hogan, C., Bauer, R.: Greedy is not enough: an efficient batch mode active learning algorithm. In: ICDMW, pp. 326–331 (2009)
24. Yang, S., Han, F., Wu, Y., Yan, X.: Fast top-k search in knowledge graphs. In: ICDE (2016)
25. Yang, S., Wu, Y., Sun, H., Yan, X.: Schemaless and structureless graph querying. VLDB **7**, 565–576 (2014)
26. Zhang, W.E., Sheng, Q.Z., Taylor, K., Qin, Y., Yao, L.: Learning-based SPARQL query performance prediction. In: Cellary, W., Mokbel, M.F., Wang, J., Wang, H., Zhou, R., Zhang, Y. (eds.) WISE 2016. LNCS, vol. 10041, pp. 313–327. Springer, Cham (2016). https://doi.org/10.1007/978-3-319-48740-3_23

A Privacy-Preserving Framework for Subgraph Pattern Matching in Cloud

Jiuru Gao[1], Jiajie Xu[1], Guanfeng Liu[1], Wei Chen[1], Hongzhi Yin[2],
and Lei Zhao[1($\boxtimes$)]

[1] School of Computer Science and Technology, Soochow University, Suzhou, China
jrgao@stu.suda.edu.cn, {xujj,gfliu,zhaol}@suda.edu.cn, wchzhg@gmail.com
[2] School of Information Technology and Electrical Engineering,
The University of Queensland, Brisbane, Australia
db.hongzhi@gmail.com

Abstract. The growing popularity of storing large data graphs in cloud has inspired the emergence of subgraph pattern matching on a remote cloud, which is usually defined in terms of subgraph isomorphism. However, it is an NP-complete problem and too strict to find useful matches in certain applications. In addition, there exists another important concern, i.e., how to protect the privacy of data graphs in subgraph pattern matching without undermining matching results. To tackle these problems, we propose a novel framework to achieve the privacy-preserving subgraph pattern matching via strong simulation in cloud. Firstly, we develop a k-automorphism model based method to protect structural privacy in data graphs. Additionally, we use a cost-model based label generalization method to protect label privacy in both data graphs and pattern graphs. Owing to the symmetry in a k-automorphic graph, the subgraph pattern matching can be answered using the outsourced graph, which is only a subset of a k-automorphic graph. The efficiency of subgraph pattern matching can be greatly improved by this way. Extensive experiments on real-world datasets demonstrate the high efficiency and effectiveness of our framework.

Keywords: Privacy-preserving · Subgraph pattern matching
Strong simulation · k-automorphism · Label generalization

1 Introduction

A graph can be a powerful model tool to represent objects and their relationships. The increasing number of applications that take use of graph data in recent years, such as disease transmission [1,2], communication patterns [3], and social networks [4–7], has promoted the development of graph data management, especially subgraph pattern matching. Typically, subgraph pattern matching is defined in terms of subgraph isomorphism [8,9], which is an NP-complete problem [10]. It is often too strict to catch sensitive matches, as it requires matches to have same topology with data graphs. The problem will hinder its applicability

© Springer International Publishing AG, part of Springer Nature 2018
J. Pei et al. (Eds.): DASFAA 2018, LNCS 10827, pp. 307–322, 2018.
https://doi.org/10.1007/978-3-319-91452-7_20

in some certain applications like social networks and crime detection. Our work focus on subgraph pattern matching via strong simulation [11]. Strong simulation is a revision of graph simulation, which imposes more flexible constraints on topology in data graphs, and it retains cubic-time complexity.

Example 1. *Consider a real-life social network shown in Fig. 1. Each vertex in graph G represents an entity, such as a human resources (HR_i) person, a development manager (DM_i), and a project manager (PM_i). Each directed edge in G indicates one recommendation relationship, e.g., edge $HR_1 \rightarrow PM_1$ represents HR_1 recommends PM_1. Each entity has some attributes like "name", "gender", "state", and "school".*

A headhunter wants to employ a DM to help a PM. A qualified candidate must live in Illinois and at the same time, he must recommend the PM and be recommended by the HR and PM. The headhunter issues a subgraph pattern matching of Q over G, as shown in Fig. 1. When subgraph isomorphism is taken, there is no match can be found, since there is no subgraph that has the same topology with the pattern Q in graph G. However, when it comes to strong simulation, we can find the subgraph G_1 is an appropriate match to pattern Q, since there exists a path (DM_1, PM_2, PM_1) from DM_1 to PM_1. Obviously, compared with the subgraph isomorphism that imposes a very strict constraint on the topology of the matched graphs, strong simulation provides a more flexible constraint.

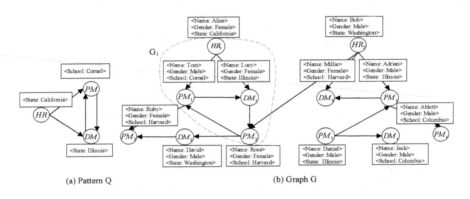

(a) Pattern Q   (b) Graph G

Fig. 1. An original data graph G and pattern graph Q.

Meanwhile, the popularity of storing the large number of data graphs in cloud to save storage brings another inevitable challenge, i.e., how to process users' queries without compromising sensitive information in cloud [12]. In many real scenarios, we can not make sure that the cloud platform is completely credible, as many adversaries may attack the cloud and cause serious privacy leakage. The main privacy leakage problem is the "identity disclosure" problem [13,14]. A naive anonymous approach is to remove all identifiable personal information before uploading the data graph to cloud. However, even though the data graph is uploaded without any sensitive information, it is still possible for an adversary to locate the target through structural attacks [13,15,16]. To protect privacy of data graphs from multiple structural attacks, many methods have been proposed [17–19]. One typical approach is k-automorphism, which uses the symmetry of the published data graph [19]. For each vertex v in a k-automorphic graph, there are at least $k-1$ structurally equivalent counterparts. An adversary can not distinguish v from the other $k-1$ symmetric vertices, because there is no structural difference between them.

Consider the Example 1 in Fig. 1, uploading the original graph G to cloud directly will cause privacy leakage. To address the problem, we propose the following solution. On one hand, we propose a k-automorphism model based method to protect structural privacy of data graphs. Firstly, we transform the original graph G to an "undirected" graph G^*. During the process, if an edge $u \to v$ is unidirectional, we will add an edge $v \to u$. For example, we add an edge $PM_2 \to DM_1$ for the edge $DM_1 \to PM_2$ in Fig. 1. Then, we can use the k-automorphism model to generate graph G^k, where $k = 2$ in Fig. 2. On the other hand, to protect the label privacy in both data graphs and pattern graphs, we apply a cost-model based label generalization technique [12], where each vertex label in G^k and Q is replaced by a label group. The mapping between label groups and vertex labels are given in the Label Correspondence Table (LCT), presented in Fig. 2(a).

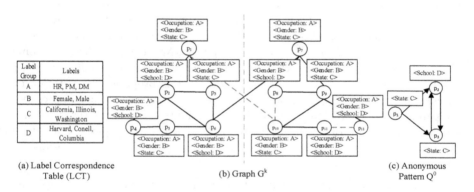

(a) Label Correspondence Table (LCT)

(b) Graph G^k

(c) Anonymous Pattern Q^0

Fig. 2. The k-automorphic graph G^k and anonymous pattern Q^0.

However, the solution suffers from the following limitation. During the generation of G^* and G^k, it may generate a large number of noise edges, which will result in more expensive storage cost and much larger communication overhead. Thus, we upload the outsourced graph G^0 (The definition of G^0 is given in Subsect. 4.3), which is only a subset of G^k, to cloud. Next, the cloud executes subgraph pattern matching via strong simulation of Q^0 over G^0 to obtain $r(Q^0, G^0)$, i.e., subgraph matches of Q^0 over G^0, and transmits it to the client side. On the basis of k-automorphic functions F_{k_i} $(i = 1, 2, \ldots, k-1)$, client can firstly compute $r(Q^0, G^k)$ according to $r(Q^0, G^0)$. Then, it filters out false positives based on the original data graph G and pattern Q to derive $r(Q,G)$. Note that we assume the client is the data owner who has access to the original graph G for the filtering step.

Contributions. The main contributions of our work are summarized as follows:

- To the best of our knowledge, we are the first to support privacy-preserving subgraph pattern matching via strong simulation in cloud.
- We re-design a cost-model based label generalization method to select effective vertex label combinations for anonymizing labels in both data graphs and pattern graphs.
- We conduct extensive experiments on several real-world datasets to study the efficiency and effectiveness of our framework.

The rest of the paper is organized as follows. Section 2 narrates the related work. Section 3 gives the problem formulation. Section 4 describes the main solution. Section 5 reports the experimental analysis. Section 6 concludes the paper.

2 Related Work

Strong Simulation. Subgraph pattern matching is typically defined in terms of subgraph isomorphism [8,9], an NP-complete problem [20]. It is often too restrictive to catch sensible matches. Subgraph simulation and its various extensions have been considered to lower the complexity [10,21,22]. Fan et al. [21] proposed bounded simulation which extended simulation by allowing bounds on the number of hops in pattern graphs and further, they extended it by incorporating regular expressions as edge constraints on pattern graphs [22]. Both the two extensions of simulation are in cubic-time. Nevertheless, the lower complexity comes with the price that they do not preserve the topology of data graphs and yield false matches. Thus, Ma et al. [11] proposed the notation of strong simulation by enforcing two additional conditions: the duality to preserve the parent relationships and the locality to eliminate excessive matches. Strong simulation is capable of capturing the topological structures of pattern and data graphs, and it retains the same cubic-time complexity of former extensions of graph simulation [10].

k-**Automorphism.** The question of how to publish information on graphs in a privacy-preserving way has been of interest for a number of years [13,23–25]. Most previous work focus on protecting data privacy from structural attacks [13,23,24]. Some of them assume that the adversary launches one type of structural attack only [13,23,24]. Liu and Terzi [13] studied how to protect privacy in published data from degree attack only. However, an attacker can launch multiple types of structural attacks to identify the target in practice. Thus, Zou et al. [19] proposed the k-automorphism framework. Each vertex in a k-automorphic graph has at least $k-1$ counterparts so that it is hard for an adversary to identify the vertex from others. The framework can protect privacy of data graph from multiple structural attacks and furthermore, the k-automorphism model does not need to delete any vertices or edges from data graph, which can significantly preserve the integrity of the data.

3   Problem Formulation

In this section, we first present the basic notations and definitions frequently used in this paper. Then we give a definition of our problem.

We model a social network as an attributed graph [12], $G = \{V(G), E(G), L_G(V(G))\}$, where $V(G)$ is the set of vertices, $E(G)$ is the set of edges, and $L_G(V(G))$ is the set of vertex labels. The notational convention of this paper are summarized in Table 1.

Table 1. Table of notations

Notations	Descriptions
d_Q	The diameter of pattern Q
G	The original data graph
G^*	The "undirected" graph by adding noise edges in G
G^k	The data graph released by k-automorphism model
G^o	The final outsourced graph uploaded to the cloud
$\widehat{G}[v, d_Q]$	The ball with center v and radius d_Q
Q	The original pattern graph
Q^o	The anonymous pattern graph of Q
$r(Q, G)$	The set of subgraph pattern matches of Q over G
$dis(u, v)$	The distance between vertex u and v

Definition 1 *Path*. *A directed path p is a sequence of nodes $(v_1, v_2, \ldots, v_n)$, where $i \in [1, n-1]$ and (v_i, v_{i+1}) is an edge in graph G. The number of edges in a path p is the length of p, denoted by $len(p)$.*

Definition 2 *Distance and diameter*. *Consider two nodes u, v in graph G, the distance from u to v is the length of the shortest undirected path from u to v, denoted by $dis(u, v)$. The diameter of the connected graph G is defined as the longest distance of all pairs of nodes in G, denoted by d_G. More specifically, $d_G = max\{dis(u, v)\}$ for all nodes u, v in graph G.*

Definition 3 *Ball* [11]. *For a node v in graph G, a ball is a subgraph of G, where v is the center node and r is the radius, denoted by $\widehat{G}[v, r]$. For all nodes u in $\widehat{G}[v, r]$, the shortest distance between u and v should satisfy $dis(u, v) \leq r$ and edges must exactly appear in graph G over the same node set.*

Consider the pattern graph Q and data graph G in Fig. 1, we can figure out that $d_Q = 1$ according to the Definition 2. If we take the vertex DM_1 as the center node and d_Q as the radius, then we can obtain the ball $\widehat{G}[DM_1, d_Q]$ (i.e., G_1), which is a subgraph of G based on Definition 3.

Definition 4 Subgraph Pattern Match. *Given a data graph $G = \{V(G),$ $E(G), L_G(V(G))\}$ and a pattern graph $Q = \{V(Q), E(Q), L_Q(V(Q))\}$, Q is a subgraph match to G via strong simulation, if there exists a node u in Q and a connected subgraph G_s of G such that:*

(1) There exists a match relation R, and for each pair (u, v) in R:
  (a) $L_Q(u) \subseteq L_{G_s}(v)$;
  (b) $\forall(u', u) \in E(Q)$, there exists a path $(v', \ldots, v)$ in $E(G_s)$;
  (c) $\forall(u, u') \in E(Q)$, there exists a path $(v, \ldots, v')$ in $E(G_s)$;

(2) G_s is contained in the ball $\widehat{G}[v, d_Q]$, where d_Q is the diameter of pattern Q. The set of subgraph pattern matches of Q over G via strong simulation is denoted as $r(Q, G)$.

Problem Definition. Given a data graph G and a pattern graph Q, our work is to find all subgraph pattern matches of Q over G via strong simulation in cloud, without compromising the privacy of both data graph G and pattern graph Q.

4   Privacy Preserving in Cloud

4.1   Structural Privacy

To protect the structural privacy in data graphs, we develop a novel approach based on k-automorphism model. When a directed data graph G is given, we firstly transform it to an "undirected" graph G^* by introducing noise edges. Then we convert G^* into graph G^k, where G^k satisfies the k-automorphic graph model.

Definition 5 k-automorphic Graph [12]. *A k-automorphic graph G^k is defined as $G^k = \{V(G^k), E(G^k)\}$, where $V(G^k)$ can be divided into k blocks and each block has $\left\lceil \frac{V(G^k)}{k} \right\rceil$ vertices.*

Intuitively, for any vertex v in a k-automorphic graph G^k, there are $k-1$ symmetric vertices. An adversary can hardly distinguish v from its structurally equivalent counterparts. Thus, the structural privacy in data graphs can be well preserved. According to Definition 5, we can transform G^* to a k-automorphic graph G^k as follows.

Firstly, we adopt the METIS algorithm [12,26] to partition the graph G^* into k blocks. In order to guarantee that each block has exactly $\left\lceil \frac{V(G^k)}{k} \right\rceil$ vertices, some noise vertices will be introduced if $V(G^*)$ can not be divided into k blocks. There is an efficient method to build Alignment Vertex Table (AVT) after the partition [19]. For example, we divide the graph G^k in Fig. 2(b) into two blocks and build the corresponding AVT, which is presented in Fig. 3(a). Each row in AVT denotes they are symmetric vertices, such as p_1 and p_7 in Fig. 2(b). Each column in AVT contains the vertices in one block, such as $(p_1, p_2, p_3, p_4, p_5, p_6)$ in the first block of G^k. According to the AVT, we define the k-automorphic function F_{k_1}, as shown in Fig. 3(b).

p_1	p_7
p_2	p_9
p_3	p_8
p_4	p_{12}
p_5	p_{11}
p_6	p_{10}

$$F_{k_1}(p_1) = p_7 \quad F_{k_1}(p_7) = p_1$$
$$F_{k_1}(p_2) = p_9 \quad F_{k_1}(p_9) = p_2$$
$$F_{k_1}(p_3) = p_8 \quad F_{k_1}(p_8) = p_3$$
$$F_{k_1}(p_4) = p_{12} \quad F_{k_1}(p_{12}) = p_4$$
$$F_{k_1}(p_5) = p_{11} \quad F_{k_1}(p_{11}) = p_5$$
$$F_{k_1}(p_6) = p_{10} \quad F_{k_1}(p_{10}) = p_6$$

(a) Alignment Vertex Table   (b) Automorphic Function

Fig. 3. The Alignment Vertex Table (AVT) and automorphic function.

Secondly, we perform block alignment and edge copy [19] to obtain the k-automorphic graph G^k. For example, we can obtain 2 isomorphic blocks: $B_0(p_1, p_2, p_3, p_4, p_5, p_6)$ and $B_1(p_7, p_8, p_9, p_{10}, p_{11}, p_{12})$ by adding noise edges (p_8, p_{10}) and (p_{11}, p_{12}) via block alignment in Fig. 2. According to the crossing edge (p_6, p_7) between 2 blocks, we add an edge (p_1, p_{10}) based on the edge copy techniques.

4.2 Label Privacy

Since the k-automorphism model based method can only protect structural privacy of the original graph G, we define a cost-model based label generalization method to protect label privacy of both data graph G and pattern graph Q. The method considers two factors: label matching and searching space, while estimating the number of candidates of a vertex u in Q^0, denoted as $sim(u)$.

According to the definition of strong simulation [11], when a vertex v in graph G^k matches the vertex u in pattern Q^0, it must firstly contain u's label groups. We let $|V_g(G^k, i)|$ and $|V_g(Q^0, i)|$ denote the set of vertices with the label group i in G^k and Q^0 that are obtained after the label generalization respectively. Then, we can define:

$$P^g_{G^k}(i) = \frac{|V_g(G^k, i)|}{|V(G^k)|}, P^g_{Q^0}(i) = \frac{|V_g(Q^0, i)|}{|V(Q^0)|} \tag{1}$$

$P^g_{G^k}(i)$ and $P^g_{Q^0}(i)$ estimate the probability of a vertex in G^k and Q^0 having an i-th label group after the label generalization, respectively. Then, the estimating number of vertices that can match vertex u in Q^0 while considering label matching can be defined as follows:

$$|V(G^k)| \sum_{i=1}^{\alpha} P^g_{G^k}(i) \cdot P^g_{Q^0}(i) \tag{2}$$

Next we need to consider the search space of checking whether each of u's parent vertices and child vertices can find matching vertices. We define the average in-degree $D_i(G^k)$ and average out-degree $D_o(G^k)$ to represent the in-degree and out-degree of vertex v (u's matching vertex) respectively. Similarly, $D_i(Q)$ and

$D_o(Q)$ represent the in-degree and out-degree of vertex u in Q^0 separately. Note that $D_o(Q^0) = D_o(Q)$, and $D_i(Q^0) = D_i(Q)$. Therefore, the maximum potential searching space of u's first child vertex is $D_o(G^k)^{2d_Q}$, and that of the second child vertex is $(D_o(G^k)-1)D_o(G^k)^{2d_Q-1}$. Thus, the total searching space of u's child vertices can be estimated as $D_o(G^k)^{2d_Q} \cdot (D_o(G^k)-1)D_o(G^k)^{2d_Q-1}\cdots(D_o(G^k)-D_o(Q)+1)D_o(G^k)^{2d_Q-1}$. We estimate it as $D_o(G^k)^{D_o(Q)} \cdot D_o(G^k)^{(2d_Q-1)^{D_o(Q)}}$ for simplicity, i.e., $D_o(G^k)^{D_o(Q)+(2d_Q-1)^{D_o(Q)}}$. Similarly, we can define the searching space of u's parent vertices as $D_i(G^k)^{D_i(Q)+(2d_Q-1)^{D_i(Q)}}$. Thus, the estimation of searching space can be defined as follows:

$$
\left\{ D_o(G^k) \left[\sum_{i=1}^{\alpha} P^g_{G^k}(i) P^g_{Q^0}(i) \right] \right\}^{D_o(Q)+(2d_Q-1)^{D_o(Q)}}
$$
$$
\cdot \left\{ D_i(G^k) \left[\sum_{i=1}^{\alpha} P^g_{G^k}(i) P^g_{Q^0}(i) \right] \right\}^{D_i(Q)+(2d_Q-1)^{D_i(Q)}}
$$

(3)

We assume the total labels in original graph G can be divided into α groups, each group contains θ different labels without loss of generality. We define $\langle p_1, p_2, p_3, \ldots, p_{\alpha\theta} \rangle$ to form a permutation of $\langle 1, 2, 3, \ldots, \alpha\theta \rangle$. According to [12], we can obtain that $P^g_{G^k}(i) \leq \sum_{j=1}^{\theta} P_G(p_{\theta(i-1)+j})$. Thus, we can define the *cost model*:

$$
|sim(u)| = |V(G^k)| \left[\sum_{i=1}^{\alpha} P^g_{G^k}(i) P^g_{Q^0}(i) \right] \cdot \left\{ D_o(G^k) \left[\sum_{i=1}^{\alpha} P^g_{G^k}(i) P^g_{Q^0}(i) \right] \right\}^{D_o(Q)+(2d_Q-1)^{D_o(Q)}} \cdot
$$
$$
\left\{ D_i(G^k) \left[\sum_{i=1}^{\alpha} P^g_{G^k}(i) P^g_{Q^0}(i) \right] \right\}^{D_i(Q)+(2d_Q-1)^{D_i(Q)}}
$$
$$
= |V(G^k)| D_o(G^k)^{D_o(Q)+(2d_Q-1)^{D_o(Q)}} \cdot D_i(G^k)^{D_i(Q)+(2d_Q-1)^{D_i(Q)}} \cdot
$$
$$
\left[\sum_{i=1}^{\alpha} P^g_{G^k}(i) P^g_{Q^0}(i) \right]^{D_o(Q)+D_i(Q)+(2d_Q-1)^{D_o(Q)}+(2d_Q-1)^{D_i(Q)}+1}
$$
$$
\leq |V(G^k)| D_o(G^k)^{D_o(Q)+(2d_Q-1)^{D_o(Q)}} \cdot D_i(G^k)^{D_i(Q)+(2d_Q-1)^{D_i(Q)}} \cdot
$$
$$
\left\{ \sum_{i=1}^{\alpha} \left[\sum_{j=1}^{\theta} P_G(p_{\theta(i-1)+j}) \right] \left[\sum_{j=1}^{\theta} P_Q(p_{\theta(i-1)+j}) \right] \right\}^{D_o(Q)+D_i(Q)+(2d_Q-1)^{D_o(Q)}+(2d_Q-1)^{D_i(Q)}+1}
$$

(4)

According to the *cost model* in Eq. (4), an effective permutation of $\langle 1, 2, 3, \ldots, \alpha\theta \rangle$, i.e., $\langle p_1, p_2, p_3, \ldots, p_{\alpha\theta} \rangle$, can decrease the cost of the searching space of pattern graph Q over G. We choose the component that concerns the label combination to define *Label Combination Cost*.

$$
cost(L) = \sum_{i=1}^{\alpha} \left[\sum_{j=1}^{\theta} P_G(p_{\theta(i-1)+j}) \right] \cdot \left[\sum_{j=1}^{\theta} P_Q(p_{\theta(i-1)+j}) \right]
$$

(5)

There is an iterative solution that can explore the optimal permutation to decrease $cost(L)$ according to Eq. (5). Firstly, a random label combination is generated; then, we try to swap two labels in two different label groups for each iteration. If the swap leads to smaller cost, we will keep the swap; otherwise, we will ignore that. When there is no swap that can lead to smaller cost, the iteration stops and we can obtain an effective permutation.

4.3   Outsourced Graph

After obtaining an anonymous k-automorphic graph G^k, a basic solution is to upload G^k to cloud directly. However, G^k is more larger than the original graph G since G^k contains large number of noise edges. Therefore, we only upload the outsourced graph, which is only a subset of G^k, denoted as G^0, to the cloud platform. The definition of G^0 is given below.

Definition 6 *Outsourced Graph.* *An outsourced graph is defined as $G^0 = \{V(G^0), E(G^0), L_{G^0}(V(G^0))\}$ where (1) $V(G^0)$ is the set of vertices in the first block of G^k (i.e. block B_0), denoted as $V(B_0)$, together with their neighbors within $2d_Q$-hops, denoted as $V(N_{2d_Q})$; (2) $E(G^0)$ is the set of edges that connected vertices within $V(B_0)$ and vertices between $V(B_0)$ and $V(N_{2d_Q})$; (3) $L_{G^0}(V(G^0))$ is the set of vertex labels in graph G^0.*

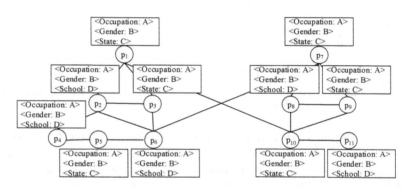

Outsourced Graph G^0

Fig. 4. The outsourced graph G^0 for G^k

According to Definition 6, we can generate an outsourced graph G^0 based on the graph G^k and upload it to cloud. For example, an outsourced graph G^0 (as shown in Fig. 4) can be generated based on graph G^k in Fig. 2. Although G^0 is a part of G^k, we can easily recover G^k based on G^0 together with k-automorphic functions $F_{k_i}(i = 1, 2, \ldots, k-1)$.

4.4   Result Processing

When G^0 and Q^0 are uploaded, the cloud executes subgraph pattern matching via strong simulation to obtain the matching result $r(Q^0, G^0)$ and transmits it to client. There are two steps for the client side to process the result.

Firstly, the client computes $r(Q^0, G^k)$ based on $r(Q^0, G^0)$ together with k-automorphic functions $F_{k_i}(i = 1, 2, \ldots, k-1)$ (Lines 1–3 in Algorithm 1). For each subgraph G_s in $r(Q^0, G^0)$, we can compute $F_{k_i}(G_s)$ $(i = 1, 2, \ldots, k-1)$ and add them to $r(Q^0, G^k)$. Then, we obtain the final $r(Q^0, G^k)$ by adding $r(Q^0, G^0)$ (Line 4 in Algorithm 1).

ALGORITHM 1. Result Processing Algorithm

Input: $r(Q^0, G^0)$ (The matching subgraphs of G^0 w.r.t. Q^0) and AVT
Output: $r(Q, G)$

1 $r(Q^0, G^k) := \emptyset$;
2 **for** $i := 1$ **to** $k - 1$ **do**
3 $\quad\lfloor\ r(Q^0, G^k) := r(Q^0, G^k) \cup F_{k_i}(r(Q^0, G^0))$;

4 $r(Q^0, G^k) := r(Q^0, G^k) \cup r(Q^0, G^0)$;
5 $r(Q, G) := \emptyset$;
6 **for** each subgraph $G_s \in r(Q^0, G^k)$ **do**
7 $\quad$ **for** each vertex $v \in V(G_s)$ **do**
8 $\quad\quad$ **if** $v \notin V(G)$ **then**
9 $\quad\quad\quad\lfloor$ remove node v from G_s;
10 $\quad\quad$ **else if** $L_G(v)$ do not match the corresponding vertex on pattern Q **then**
11 $\quad\quad\quad\lfloor$ remove node v from G_s;

12 $\quad$ **for** each edge $e \in E(G_s)$ **do**
13 $\quad\quad$ **if** $e \notin E(G)$ **then**
14 $\quad\quad\quad\lfloor$ remove edge e from G_s;

15 $\quad$ **if** G_s contains the connected component that matches to pattern Q **then**
16 $\quad\quad\lfloor\ r(Q, G) := r(Q, G) \cup G_s$;

17 **return** $r(Q, G)$

Secondly, the client needs to filter out the false matches in $r(Q^0, G^k)$ according to the original data graph G and pattern Q. For each matching subgraph G_s in $r(Q^0, G^k)$, if there exist vertices that are not contained in graph G or whose labels cannot match those of the corresponding vertices in the original pattern Q (We have anonymized the vertex labels in pattern Q via label generalization method), remove them from G_s (Lines 7–11 in Algorithm 1). Note that we have introduced noise edges when generating "undirected" graph G^* and k-automorphic graph G^k, if G_s contains edges that do not exist in the original graph G, remove them from G_s (Lines 12–14 in Algorithm 1). When all the noise vertices or edges and unmatch vertices are filtered out from G_s, we need to consider whether G_s is a candidate. We define that if there exists a subgraph which is a *match* (meets the requirements of strong simulation) to pattern Q in G_s, it is a right positive and we need to add it to $r(Q, G)$ (Lines 15–16 in Algorithm 1).

5   Experimental Study

5.1   Datasets and Setup

We evaluate our method in three real-world datasets in our experiments. The statistics on these datasets are given in Table 2.

Table 2. Real-world data graphs

Dataset	$|V|$	$|E|$	Number of labels
p2p-Gnutella08	6301	20777	62
Brightkite-edges	58228	428156	134
Web-NotreDame	325729	1090108	208

p2p-Gnutella08. p2p-Gnutella08 is a sequence of snapshots of the Gnutella peer-to-peer file sharing network collected in August 8, 2002. Nodes represent hosts in the Gnutella network topology and edges represent connections between the Gnutella hosts.

Brightkite-edges. Brightkite-edges is the friendship network collected using Brightkite's public API. Nodes correspond to users having checked-in Brightkite and directed edges correspond to relationships among them.

Web-NotreDame. Web-NotreDame is a web graph collected in 1999. Nodes represent pages from University of Notre Dame and directed edges represent hyperlinks between them.

SETUP. In our experiments, we compare four methods All_Ran, All_Eff, Part_Ran, and Part_Eff, where All_Ran applies the random label generalization method and upload G^k to cloud; All_Eff applies the cost-model based label generalization method introduced in Sect. 4.2 and upload G^k to cloud; Part_Ran applies the same label generalization approach with All_Ran but only upload G^0 to cloud; Part_Eff applies same label generalization method with All_Eff but upload G^0 to cloud.

All methods are implemented in C++. We use a Windows 10 PC with 2.30 GHz Intel Core i5 CPU and 8 GB of memory as the client side. The cloud server is on a virtualized Linux machine within Microsoft Azure Cloud with 4 CPU cores and 200 GB main memory.

5.2   Experiments Analysis

We evaluate the cost of our experiments from three aspects: time cost of generating G^k, time cost of pattern matching, and time cost of result processing in client.

Time Cost of Generating G^k. We first evaluate the performance of the proposed methods while generating graph G^k. In these experiments, we define that each label group contains two labels, i.e. the default value of θ is 2.

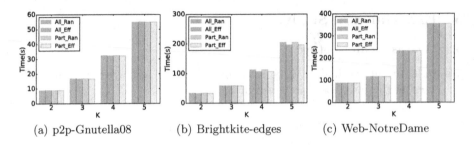

Fig. 5. Time cost in generating G^k

According to Fig. 5, the cost-model based label generalization method and the random label generalization method have similar performance while generating graph G^k, i.e., the four proposed methods have similar time cost. The reason is that all of them need to generate graph G^k firstly despite the ultimately uploaded graph is either G^k or G^0. We note that the time cost on the three datasets increases when k goes from 2 to 5. The reason is that when k increases, more and more noise edges are added to G^k, as shown in Table 3. Note that each "undirected" edge in G^k represents two directed edges. We can intuitively see that the number of noise edges has slightly difference when using different label generalization methods and increases with k.

Table 3. Number of noise edges in generating G^k

Dataset		$k = 2$	$k = 3$	$k = 4$	$k = 5$
p2p-Gnutella08	All_Ran	16417×2	34267×2	53195×2	71388×2
	Part_Ran				
	All_Eff	16309×2	34309×2	53195×2	71443×2
	Part_Eff				
Brightkite-edges	All_Ran	178278×2	367859×2	553810×2	753265×2
	Part_Ran				
	All_Eff	178674×2	368399×2	554794×2	752677×2
	Part_Eff				
Web-NotreDame	All_Ran	923266×2	1829324×2	2749760×2	3747812×2
	Part_Ran				
	All_Eff	923382×2	1846433×2	2745792×2	3767437×2
	Part_Eff				

Time Cost of Pattern Matching. Then we pay attention to the time cost of subgraph pattern matching via strong simulation in cloud. Firstly, we evaluate the time cost of the proposed methods while varying the number of edges in pattern Q, i.e., $|E(Q)|$. Pattern graphs are generated by randomly extracting subgraphs from the original data graph G. We use $|E(Q)|$ to control the size of pattern graphs. In these experiments, the value of k is set to 3. According to Fig. 6, we can clearly find out that Part_Eff performs better than the other three approaches on the three datasets. The one reason is that Part_Eff only uploads G^0 to cloud. Note that Part_Ran and Part_Eff are only different in label generalization. Thus, the results demonstrate the effectiveness of our cost-based label generalization method. The matching time increases with $|E(Q)|$ varying from 4 to 10, since the searching space will become larger for subgraph pattern matching when $|E(Q)|$ increases.

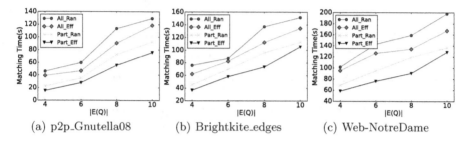

(a) p2p_Gnutella08   (b) Brightkite_edges   (c) Web-NotreDame

Fig. 6. Matching time vs. $|E(Q)|$. ($k = 3$)

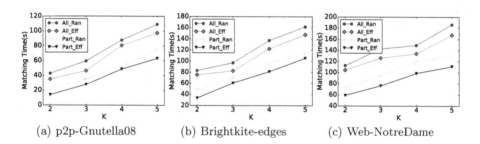

(a) p2p-Gnutella08   (b) Brightkite-edges   (c) Web-NotreDame

Fig. 7. Matching time vs. k. ($|E(Q)| = 6$)

Next, we evaluate the running time of subgraph pattern matching while varying the parameter k. The time cost increases with k varying from 2 to 5, as shown in Fig. 7. This is because $|E(G^0)|$ increases with k varying from 2 to 5, since more noise edges will be inserted when k becomes larger. The method Part_Eff, which uses the cost-model based label generalization method and uploads G^0 to cloud, has better performance than other methods in all three datasets. It demonstrates the superiority of our cost-model based generalization method as well.

Time Cost of Result Processing in Client. At last, we evaluate the performance of four methods involving result processing in the client side while varying parameters k and $|E(Q)|$ respectively. According to Fig. 8, the result processing time increases with k varying from 2 to 5 for all four methods, since client need to filter out more noise edges when k becomes larger. Note that both All_Ran and All_Eff upload G^k to cloud, the step to obtain $r(Q^0, G^k)$ on the basis of $r(Q^0, G^0)$ together with $F_{k_i}(i = 1, 2, \ldots, k-1)$ can be omitted. Thus, the time cost of result processing with All_Ran and All_Eff are smaller than the other two methods. However, the method Part_Eff that uses our cost-model based label generalization method still performs better than the Part_Ran.

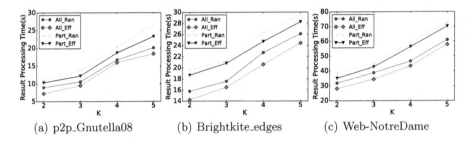

(a) p2p_Gnutella08   (b) Brightkite_edges   (c) Web-NotreDame

Fig. 8. Result processing time vs. k. ($|E(Q)| = 6$)

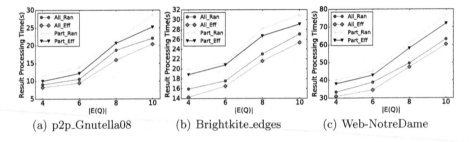

(a) p2p_Gnutella08   (b) Brightkite_edges   (c) Web-NotreDame

Fig. 9. Result processing time vs. $|E(Q)|$. ($k = 3$)

The result processing time increases with $|E(Q)|$ varying from 4 to 10, as shown in Fig. 9. The reason lies in that the search space will become larger for the filtering process when $|E(Q)|$ increases. Similarly, the time cost of All_Ran and All_Eff are smaller than other two methods. However, the gap is small compared to the time cost of pattern matching. As shown in Table 4, Part_Eff runs faster

Table 4. Overall running time (s) ($|E(Q)| = 6$ and $k = 3$)

Dataset	p2p-Gnutella08	Brightkite-edges	Web-NotreDame
Part_Eff	40	81.02	119.05
Part_Ran	49.1	91.55	140.22
All_Eff	56.02	98.93	160.89
All_Ran	70.12	114.42	181.56

than other three methods in terms of the overall running time. Note that the overall running time consists of the subgraph pattern matching time in cloud and the result processing time in the client side.

6 Conclusion

In this paper, we propose an effective framework to protect privacy of subgraph pattern matching via strong simulation in the cloud. Without losing utility, the framework protects structural and label privacy of both data graphs and pattern graphs. We introduce noise edges to transform a directed graph to an "undirected" graph so that the k-automorphism model based method can be applied to protect structural privacy of the data graph. We apply a cost-model based label generalization method to protect label privacy in both data graphs and pattern graphs additionally. In our framework, we only upload the outsourced graph to cloud so that the time cost of subgraph pattern matching can be decreased. Experiments on three real-world datasets illustrate the superior performance of our method.

Acknowledgements. This work was supported by the National Natural Science Foundation of China under Grant Nos. 61572335, 61572336, 61472263, 61402312 and 61402313, the Natural Science Foundation of Jiangsu Province of China under Grant No. BK20151223, and Collaborative Innovation Center of Novel Software Technology and Industrialization, Jiangsu, China.

References

1. Lu, H.-M., Chang, Y.-C.: Mining disease transmission networks from health insurance claims. In: Chen, H., Zeng, D.D., Karahanna, E., Bardhan, I. (eds.) ICSH 2017. LNCS, vol. 10347, pp. 268–273. Springer, Cham (2017). https://doi.org/10.1007/978-3-319-67964-8_26
2. Ray, B., Ghedin, E., Chunara, R.: Network inference from multimodal data: a review of approaches from infectious disease transmission. J. Biomed. Inform. **64**, 44–54 (2016)
3. Balsa, E., Pérez-Solà, C., Díaz, C.: Towards inferring communication patterns in online social networks. ACM Trans. Internet Technol. **17**(3), 32:1–32:21 (2017)
4. Yin, H., Zhou, X., Cui, B., Wang, H., Zheng, K., Hung, N.Q.V.: Adapting to user interest drift for POI recommendation. TKDE **28**(10), 2566–2581 (2016)
5. Yin, H., Hu, Z., Zhou, X., Wang, H., Zheng, K., Hung, N.Q.V., Sadiq, S.W.: Discovering interpretable geo-social communities for user behavior prediction. In: ICDE, pp. 942–953 (2016)
6. Xie, M., Yin, H., Wang, H., Xu, F., Chen, W., Wang, S.: Learning graph-based POI embedding for location-based recommendation. In: CIKM, pp. 15–24 (2016)
7. Yin, H., Wang, W., Wang, H., Chen, L., Zhou, X.: Spatial-aware hierarchical collaborative deep learning for POI recommendation. TKDE **29**(11), 2537–2551 (2017)
8. Aggarwal, C.C., Wang, H.: Managing and Mining Graph Data. Advances in Database Systems, pp. 11–52. Springer US, New York City (2010). https://doi.org/10.1007/978-1-4419-6045-0

9. Gallagher, B.: Matching structure and semantics: a survey on graph-based pattern matching. In: AAAI (2006)
10. Henzinger, M.R., Henzinger, T.A., Kopke, P.W.: Computing simulations on finite and infinite graphs. In: Annual Symposium on Foundations of Computer Science, pp. 453–462 (1995)
11. Ma, S., Cao, Y., Fan, W., Huai, J., Wo, T.: Strong simulation: capturing topology in graph pattern matching. ACM Trans. Database Syst. **39**(1), 4:1–4:46 (2014)
12. Chang, Z., Zou, L., Li, F.: Privacy preserving subgraph matching on large graphs in cloud. In: SIGMOD, pp. 199–213 (2016)
13. Liu, K., Terzi, E.: Towards identity anonymization on graphs. In: SIGMOD, pp. 93–106 (2008)
14. Tai, C., Tseng, P., Yu, P.S., Chen, M.: Identity protection in sequential releases of dynamic networks. TKDE **26**(3), 635–651 (2014)
15. Zhou, B., Pei, J.: Preserving privacy in social networks against neighborhood attacks. In: ICDE, pp. 506–515 (2008)
16. Li, J., Xiong, J., Wang, X.: The structure and evolution of large cascades in online social networks. In: Thai, M.T., Nguyen, N.P., Shen, H. (eds.) CSoNet 2015. LNCS, vol. 9197, pp. 273–284. Springer, Cham (2015). https://doi.org/10.1007/978-3-319-21786-4_24
17. Cheng, J., Fu, A.W., Liu, J.: K-isomorphism: privacy preserving network publication against structural attacks. In: SIGMOD, pp. 459–470 (2010)
18. Wu, W., Xiao, Y., Wang, W., He, Z., Wang, Z.: k-symmetry model for identity anonymization in social networks. In: EDBT, pp. 111–122 (2010)
19. Zou, L., Chen, L., Özsu, M.T.: K-automorphism: a general framework for privacy preserving network publication. PVLDB **2**(1), 946–957 (2009)
20. Ullmann, J.R.: An algorithm for subgraph isomorphism. J. ACM **23**(1), 31–42 (1976)
21. Fan, W., Li, J., Ma, S., Tang, N., Wu, Y., Wu, Y.: Graph pattern matching: from intractable to polynomial time. PVLDB **3**(1), 264–275 (2010)
22. Fan, W., Li, J., Ma, S., Tang, N., Wu, Y.: Adding regular expressions to graph reachability and pattern queries. In: ICDE, pp. 39–50 (2011)
23. Zhou, B., Pei, J.: Preserving privacy in social networks against neighborhood attacks. In: ICDE, pp. 506–515 (2008)
24. Tai, C.H., Yu, P.S., Yang, D.N., Chen, M.S.: Privacy-preserving social network publication against friendship attacks. In: SIGKDD, pp. 1262–1270 (2011)
25. Chen, S., Zhou, S.: Recursive mechanism: towards node differential privacy and unrestricted joins. In: SIGMOD, pp. 653–664 (2013)
26. Karypis, G., Kumar, V.: Analysis of multilevel graph partitioning. In: Supercomputing, p. 29 (1995)

Adaptive and Parallel Data Acquisition from Online Big Graphs

Zidu Yin, Kun Yue$^{(\boxtimes)}$, Hao Wu, and Yingjie Su

School of Information Science and Engineering, Yunnan University,
Kunming, China
kyue@ynu.edu.cn

Abstract. Acquisition of contents from online big graphs (OBGs) like linked Web pages, social networks and knowledge graphs, is critical as data infrastructure for Web applications and massive data analysis. However, effective data acquisition is challenging due to the massive, heterogeneous, dynamically evolving properties of OBGs with unknown global topological structures. In this paper, we give an adaptive and parallel approach for effective data acquisition from OBGs. We adopt the ideas of Quasi Monte Carlo (QMC) and branch & bound methods to propose an adaptive Web-scale sampling algorithm for parallel data collection implemented upon Spark. Experimental results show the effectiveness and efficiency of our method.

Keywords: Online big graph · Data acquisition · Adaptive collection
Parallel crawler · Spark

1 Introduction

In view of the infrastructure of knowledge engineering systems, data acquisition including collection and updating, has gained great attention accordingly for Web search, massive data analysis [6], data integration [1], knowledge extraction and fusion [4], etc. Regarding the organization of Internet contents, graphical structure is an essential feature represented by online big graph (OBG), such as linked Web pages, social network and knowledge base, shown in Fig. 1. An OBG consists of objects and connections, varying with respect to different kinds of specific contents.

Represented as OBGs without knowing the global topological structures before collected, data are massive, distributed, heterogeneous and fast changing [5]. Given the online property of OBG data acquisition, it begins with an empty set without prior knowledge, and meanwhile the OBGs could be explored gradually during acquisition. This is exactly our focus, where we discuss the effective and generic method for online data acquisition from OBGs to establish the basis for Web-scale data analysis.

Previous approaches for data collection from OBGs can be categorized into the following three categories: universal crawlers, hidden Web crawlers and preferential crawlers. First, universal crawlers primarily depend on local graph structures, such as the classic breadth-first [8] and deep-first methods [9]. That all areas are treated by the same importance degree is not consistent with the intuition in realistic situations. Second, hidden Web crawlers use keyword query, attribute and label extraction, and

© Springer International Publishing AG, part of Springer Nature 2018
J. Pei et al. (Eds.): DASFAA 2018, LNCS 10827, pp. 323–331, 2018.
https://doi.org/10.1007/978-3-319-91452-7_21

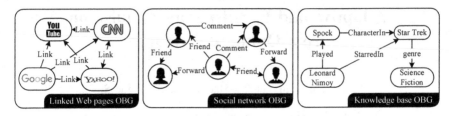

Fig. 1. Typical online big graphs

form-based approaches to find information hidden behind search engines. Third, preferential crawlers include topical crawler and focused crawler. Focused crawler [10] gives priority to those URLs during crawling and achieves high a probability that the found information satisfies user interest, which gives inspiration to our method.

In this paper we are to make use of graphical structures of the currently collected data to crawl more. OBG data with a high importance degree are data with much information or concerned contents, which are expected to be selected and collected.

Sampling has been widely used in statistic machine learning [7], by which the global outline shaping for a certain process could be done without much effort. Thus, using OBG's graphical structures, we adopt the ideas of Quasi-Monte Carlo (QMC) sampling [2] and branch & bound to propose the online algorithm, called BB-QMC. By this algorithm, we split the currently considered OBG recursively and hierarchically, and calculate the importance of its subsets, called areas. Then, we find the most important area adaptively for successive data collection. We further give the theoretic analysis of BB-QMC from effectiveness, complexity, standard error estimation, iteration bound, and conflict efficiency. By our BB-QMC, the importance of objects in an area could be measured by a scalable way, which makes it possible to parallelly acquire the data in that area, and the importance degree of areas could be ordered consequently. From the perspective of real applications upon OBGs, less important objects with low importance degrees could be discarded intuitively, which makes the total cost of data acquisition decreased correspondingly.

Current methodologies and tools on a single computer is hard to deal with very large volume of data [11]. Thus, focusing on the iterations during online data acquisition in response to the fast changing contents, our algorithms are implemented upon Spark.

Finally, we illustrate performance studies of our algorithm for online collection by experiments which results show the effectiveness and efficiency of our method.

2   Adaptive Data Collection Based on QMC Sampling

2.1   Basic Idea and Algorithm

Definition 1. An OBG is a directed graph $G_{ON} = \{O, E, R\}$, where $O = \{o_1, \ldots, o_s\}$ and $E = \{e_1, \ldots, e_l\}$ is the set of objects and connections respectively. Let $R = \{r_1, r_2, \ldots, r_v\}$ be the set of connection types in an OBG. Each connection in G_{ON} corresponds to an edge connecting two different objects with a type of connection, that is $e = (o_i, o_j, r_k), e \in E, o_j, o_j \in O, i \neq j$.

The most important OBG areas should be collected first, where the importance of an area is determined by the volume of contained information. For this purpose, we consider splitting G_{ON} into K areas according to the objects, denoted as $A = \{A_1, A_2, \ldots A_K\}$, and selecting the most important area by employing the idea of sampling. By repeating this process recursively and hierarchically, a K-fork tree will be built finally with a height of $\log_K |O|$, where $|O|$ is the number of objects in G_{ON}.

By modeling adaptive data collection as an optimization problem, we employ the branch & bound technique to obtain the importance degree of different scaled areas in every fork structure in the K-fork tree. A two-stage BB-QMC algorithm is proposed to denote the whole process.

In the first stage, we will gather each area's density as importance degree in parallel sampling which simultaneously obtains data from the OBG. the importance degree of each area is measured by gathering its density. Let $A = \{A_1, A_2, \ldots A_K\}$ be K areas, and the density of an area A_i is

$$\rho_{A_i} = \frac{\sum_{o \in A_i} D(o)}{|A_i|} \tag{1}$$

where $D(o)$ represents the connections of o pointing to other objects in an OBG.

Now we consider the idea to obtain ρ_{A_i}. For a given h dimensional space D, a subset of real space R_m, the integration (or mass) I of target function $f(X)$, where X is a set of points in D, could be calculated by $\int_D f(X) dX$. Given n randomly selected sampling points $\{x_1, x_2, \ldots, x_n\} \in D$, the Monte Carlo (MC) sampling [3] approximation of $I \approx MC_n \equiv \left(\int_D 1 dX\right) \frac{1}{n} \sum_{i=1}^{n} f(x_i)$. However, by MC sampling, randomly generated points may be too close to each other to be distinguished, while random point generation is also costly.

Frequently used with Halton sequence, QMC sampling [2] adopts deterministic approaches to generate low-discrepancy pseudo-random sequences as the sampling points, which has a higher coverage of the sampling area. Let b denote a prime number. Any number w ($w > 0$) could be represented as $d_j b^j + d_{j-1} b^{j-1} + \ldots + d_1 b + d_0$, in which $d_i \in \{0, 1, \ldots, b - 1\}$, and $i = 0, 1, \ldots, j$ The w-th element in a Halton sequence upon b is $\phi_b(w) = \frac{d_0}{b^1} + \frac{d_1}{b^2} + \ldots + \frac{d_j}{b^{j+1}}$. Note that $\phi_b(w) \in [0, 1]$ holds for any $w \geq 0$. This does make sense for OBG sampling, since OBG information could be gained comprehensively by QMC while avoiding the uneven sampling by random selection. By the QMC sampling of n objects, the approximation of ρ_{A_i} is calculated as

$$\rho_{A_i} \approx \frac{1}{n} \sum_{i=1}^{n} D(o_i) \tag{2}$$

In the second stage, we will choose the area to be collected in the next iteration. Density of areas visited in current and all previous iterations are recorded in a candidate pool $C = \{\rho_{A_1}, \rho_{A_2}, \ldots, \rho_{A_k}, \rho_x\}$. Area election, which makes BB-QMC adaptive, is fulfilled by the election criterion which always choose the area with the maximal density to be collected next. According to the fact that some parts without much information, like shattered isolated parts of an OBG, are unnecessary to collect, we give the following termination condition

$$\frac{\sum_{o\in\Delta o^c} D(o)}{|\Delta o^c|} < \rho_{\min} \tag{3}$$

where Δo^c is the set of acquired objects by the latest iteration, and $\rho_{\min}$ is the minimum acceptable density to make the collection process continued.

Now we give Algorithm 1 to summarize the above ideas to select fork structures in the K-fork tree of the original OBG.

Algorithm 1. BB-QMC

Input: area range A from A_1 to A_n, N_{split}, R_{samp}, P_{cand}
Output: $D_{obj}=\{ODC_1,\ ODC_2,\ ...,\ ODC_n\}$

```
IF A.start≠A.end AND Not (∑_{o∈Δo^c} D(o))/|Δo^c| < ρ_min THEN
    S[]←divide(A.start, A.end, N_split) // split into K ar-
    eas
    FOR EACH s IN S DO // s is the range of area
        mass←0; H[]←HaltonSeq(s, R_samp)
        FOR EACH e IN H DO
            IF D_obj.find(e) IS NULL THEN
                ODC←collect(e); D_obj.add(ODC)
                mass←mass+ODC.numConnection
            ELSE
                ODC←D_obj.find(e); mass←mass+ODC.numConnection
            END IF
        END FOR
        IF |s|>1 THEN
            density←mass/|H|; P_cand.add([s, density])
        END IF
        A←P_cand.findMax() // choose area with maximum density
        BB-QMC() // next iteration
    END FOR
END IF
```

2.2 Analysis

Effectiveness. The effectiveness of Algorithm 1 in a given area A is described as

$$S_A = \sum_{i=1}^{|A|} \sum_{j=1}^{i} D\left(o_j^c\right) \tag{4}$$

where $|A|$ is the number of object in area A. The more the early important objects are found and acquired, the larger the value S_A will be.

Complexity. In realistic situations, the most time consuming step when executing Algorithm 1 is data collection by visiting the Internet. By experiments, we found that all the collection time satisfies normal distribution, denoted as D_{CT}. If we suppose the expectation of D_{CT} is μ, then the average execution time of Algorithm 1 will be μn, which means that the complexity of Algorithm 1 is $O(n)$. Thus, the execution time of Algorithm 1 is linear, which will be verified by experiments in Sect. 3.

Standard Error Estimation. The standard error of sampling in BB-QMC is different in terms of the number of sampling points in each iteration. In view of the density of an area in Eq. (2), the standard error estimation can be inferred as $|A_i| \frac{\sigma_n}{\sqrt{n}}$, where $\sigma_n = \sqrt{\frac{1}{n-1} \sum_{i=1}^{n} \left(D(o_i) - \frac{1}{n} \sum_{i=1}^{n} D(o_i)\right)^2}$. This means that the standard error estimation of Algorithm 1 will nonlinearly decrease with the increase of sampling points.

Iteration Bound. The number of iterations of QMC sampling in Algorithm 1, denoted as N_{iter}, is determined by the number of areas (i.e., K) and that of objects (i.e., n). Following, we give the iteration bound by Theorem 1.

Theorem 1. Given n objects and K areas ($n > 0$, $K > 1$), the N_{iter} can be obtained as

$$N_{iter} = \begin{cases} 1 & 0 < n < K \\ n - K^{m-1} + \sum_{j=0}^{m-1} K^j & K^m \le n \le 2K^m \\ \sum_{j=0}^{m} K^j & 2K^m < n < K^{m+1} \end{cases} \tag{5}$$

where $m = \lfloor \pi \log_K n \rfloor$, $m \in Z^+$.

Conflict Efficiency. Sampling points in different iterations of the Algorithm 1 may be overlapped and cause conflicts, which are sampling points in the current iteration but already sampled in previous iterations. The conflict efficiency, denoted as E_{conf} is

$$E_{conf} = 1 - \frac{N_{conf}}{N_{iter} \cdot K \cdot |A_i| \cdot R_{samp}} \tag{6}$$

where $|A_i| \cdot R_{samp}$ is the number of sampling points in every sub-area A_i of the iteration, and N_{conf} is the total number of conflicts occurred in the entire process.

Following, we give the measurement of total conflicts in Theorem 2.

Theorem 2. Given an OBG with n objects, the total number of conflicts is

$$N_{conf} = \sum_{i=1}^{N_{iter}} \sum_{j=1}^{m_i} P_{ij} \tag{7}$$

where N_{iter} is obtained by Theorem 1; $m_i = \lfloor \log_K |A_i| \rfloor$; A_i is the sampling area of the i-th iteration; P_{ij} is the number of non-overlapping conflicts in the j-th level of children areas of A_i.

For space limitation, we ignore the proofs of Theorem 1 and Theorem 2.

3   Experimental Results

3.1   Experiment Setup

The performance tests of our proposed methods include effectiveness and efficiency. Our experiments were fulfilled upon a Spark cluster with 6 workers (each has a 4×3.6 GHz CPU and 128 GB RAM) sharing a Gigabit Ethernet switch. The version of Spark and HDFS is 1.6.1 and 2.5.2 respectively. Network speed is limited to 2 Mb/s.

3.2   Effectiveness

We selected Ber-Stan[1], Facebook[2], and Wikidata as typical linked Web page OBG, social network OBG and knowledge base OBG respectively to test the effectiveness of Algorithm 1. In our experiments, we used first 10000 entities in Wikidata on the wiki website, while simulated the environment of data crawling on Ber-Stan and Facebook benchmark datasets available in local disk.

To test the effectiveness of Algorithm 1, we used connection coverage which is calculated by $(\sum D(o^c))/|E|$, where $\sum D(o^c)$ is specified in Eq. (4). We compared the connection coverage and S_A (in Sect. 2) by Algorithm 1 (BB-QMC), sequence collection (Sequence), broad-first strategy (BFS), snowball sampling (Snowballing), random selection (Random), and original MC (BB-MC), shown in Fig. 2(a)–(c) respectively on Ber-Stan, Facebook and Wikidata.

It can be seen that, without considering the area importance, the connection coverage of Snowballing, Sequence and BFS is close to that of Random, and both that of BB-MC and BB-QMC are increased more considerably, while the latter is better than the former. Given a certain ρ_{min}, BB-QMC is terminated earlier than other methods, and achieves the smallest $|O^C|$. Upon the three types of OBGs, BB-QMC can make all connections in the original OBGs acquired efficiently.

We then tested the influence of R_{samp} (R) on connection coverage when using BB-QMC, shown in Fig. 3(a)–(c). It can be seen that better collection results have been achieved when R_{samp} is larger than 0.05 upon the three datasets. Meanwhile, we tested

---

[1] http://snap.stanford.edu/data/web-BerkStan.html.

[2] http://snap.stanford.edu/data/egonets-Facebook.html.

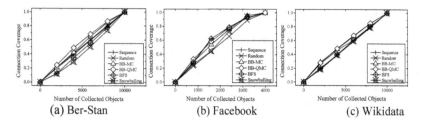

(a) Ber-Stan   (b) Facebook   (c) Wikidata

Fig. 2. Connection coverage of different methods for data collection

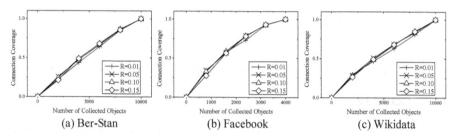

(a) Ber-Stan   (b) Facebook   (c) Wikidata

Fig. 3. Connection coverage of BB-QMC with different values of R_{samp}

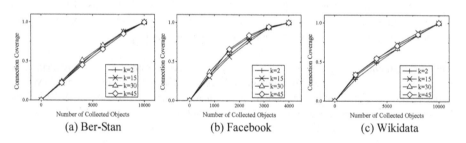

(a) Ber-Stan   (b) Facebook   (c) Wikidata

Fig. 4. Connection coverage of BB-QMC wiht different values of K

Table 1. S_A of different methods for data collection

Dataset	Sequence	Random	MCS	BFS	Snowballing	BB-QMCS
Wikidata	53.45	50.02	55.65	50.66	49.59	**56.45**
Ber-stan	49.99	51.06	53.46	46.27	43.05	**56.44**
Facebook	21.51	18.23	22.77	20.40	20.19	**22.83**

the influence of division number K on connection coverage, shown in Fig. 4(a)–(c), from which we can see that better results can be obtained when K is larger than 30.

We also calculated S_A, shown in Table 1. It is clear that the result obtained by BB-QMC is basically better than those by other methods.

3.3 Efficiency

We tested the efficiency of Algorithm 1 by execution time, speedup and parallel efficiency on Wikidata, shown in Figs. 5, 6 and 7 respectively. The execution time of data acquisition is linearly increased. The speedup and parallel efficiency are basically the ideal cases while the cost mainly depends on the network bandwidth, which ultimately verifies the efficiency of our method.

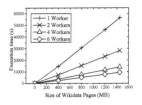
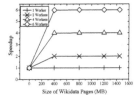
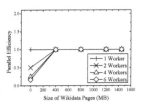

Fig. 5. Execution time   **Fig. 6.** Speedup   **Fig. 7.** Parallel efficiency

4 Conclusion

This paper explores parallel and adaptive data acquisition from OBGs. Our method adopts adaptive QMC sampling with importance degrees of different areas, and dominates data collection from OBGs in most situations compared with other classic ones. The method proposed in this paper could work for various kinds of OBGs by a scalable and efficient mechanism. Acquired online contents by our method could provide basis for big data analysis and knowledge engineering.

Acknowledgment. This paper was supported by the National Natural Science Foundation of China (Nos. 61472345, 61562090), Program for Excellent Young Talents of Yunnan University (No. WX173602), Research Foundation of Yunnan University (No. 2017YDJQ06), and Research Foundation of Educational Department of Yunnan Province (No. 2017ZZX228).

References

1. Yang, D., Xiao, Y., Tong, H., Zhang, J., Wang, W.: An integrated tag recommendation algorithm towards Weibo user profiling. In: Renz, M., Shahabi, C., Zhou, X., Cheema, M.A. (eds.) DASFAA 2015. LNCS, vol. 9049, pp. 353–373. Springer, Cham (2015). https://doi.org/10.1007/978-3-319-18120-2_21
2. Faure, H., Lemieux, C.: Improved Halton sequences and discrepancy bounds. Monte Carlo Methods Appl. **16**(3), 1–18 (2010)
3. Hammersley, J., Handscomb, D.: Monte Carlo methods. Appl. Stat. **14**(2/3), 347–385 (1964)
4. Sharma, A., Baral, C.: Automatic extraction of events-based conditional commonsense knowledge. In: Proceedings of Workshops at the 30th AAAI Conference on Artificial Intelligence, Phoenix, USA, pp. 527–531. AAAI (2016)
5. Surendran, S., Prasad, D., Kaimal, M.: A scalable geometric algorithm for community detection from social networks with incremental update. Soc. Netw. Anal. Min. **6**(1), 90:1–90:13 (2016)

6. Xi, S., Sun, F., Wang, J.: A cognitive crawler using structure pattern for incremental crawling and content extraction. In: IEEE International Conference on Cognitive Informatics, Beijing, China, pp. 238–244. IEEE (2010)

7. Wu, X., Chen, H., Wu, G., Liu, J., et al.: Knowledge engineering with big data. IEEE Intell. Syst. **30**(5), 46–55 (2015)

8. Stivala, A., Koskinen, J., Rolls, D., Wang, P., Robins, G.: Snowball sampling for estimating exponential random graph models for large networks. Soc. Netw. **47**, 167–188 (2016)

9. Urbani, J., Dutta, S., Gurajada, S., Weikum, G.: KOGNAC: efficient encoding of large knowledge graphs. In: International Joint Conference on Artificial Intelligence, New York, USA, pp. 3896–3902 (2016)

10. Wu, C., Hou, W., Shi, Y., Liu, T.: A Web search contextual crawler using ontology relation mining. In: International Conference on Computational Intelligence and Software Engineering, pp. 1–4. IEEE (2009)

11. Tsai, C., Lin, W., Ke, S.: Big data mining with parallel computing: a comparison of distributed and MapReduce methodologies. J. Syst. Softw. **122**, 83–92 (2016)

Answering the Why-Not Questions
of Graph Query Autocompletion

Guozhong Li$^{(\boxtimes)}$, Nathan Ng, Peipei Yi, Zhiwei Zhang, and Byron Choi

Department of Computer Science, Hong Kong Baptist University,
Kowloon Tong, Hong Kong
{csgzli,14222264,csppyi,cszwzhang,bchoi}@comp.hkbu.edu.hk

Abstract. Graph query autocompletion (GQAC) helps users formulate graph queries in a visual environment (a.k.a GUI). It takes a graph query that the user is formulating as input and generates a ranked list of query suggestions. Since it is impossible to accurately predict the user's target query, the current state-of-the-art of GQAC sometimes fails to produce useful suggestions. In such scenarios, it is natural for the user to ask why are useful suggestions not returned. In this paper, we address the why-not questions of GQAC. Specifically, given an intermediate query q, a target query q_t, and a GQAC system X, the why-not questions of GQAC seek for the minimal refinement of the configuration of X, with respect to a penalty model, such that at least one useful suggestion towards q_t appears in the returned suggestions. We propose a generic ranking function for existing GQAC systems. We propose a search algorithm for the why-not questions.

Keywords: Graph query · Query autocompletion · Why-not questions

1 Introduction

With the aid of graphical user interfaces GUIs, *e.g.*, Fig. 1, users can formulate graph queries by clicking and dragging visual constructs (*e.g.*, nodes and edges). However, clicking and dragging on a GUI can be cumbersome and error-prone. Hence, *graph query autocompletion* (GQAC) (*e.g.*, [3,8,9]) has been recently proposed to alleviate the burden of visual query formulation.

The conceptual GQAC procedure is shown in Fig. 2(a). Initially, a user has provided an initial query graph q. After invoking the GQAC, top-k (typically k is small) *query suggestions* (or simply *suggestions*) are returned to the user. The user may adopt one suggestion q' if useful ones present, or draw the desired increments manually. Users may repeat this procedure until they finished formulating the target query.

We have tested some publicly available prototype systems [9,10] and presented a scenario in Fig. 1, where existing systems fail to provide useful query suggestions. The current and the target query are shown in the middle of Fig. 1 and the right-hand side of Fig. 2(b). The top 3 suggestions are shown at the

© Springer International Publishing AG, part of Springer Nature 2018
J. Pei et al. (Eds.): DASFAA 2018, LNCS 10827, pp. 332–341, 2018.
https://doi.org/10.1007/978-3-319-91452-7_22

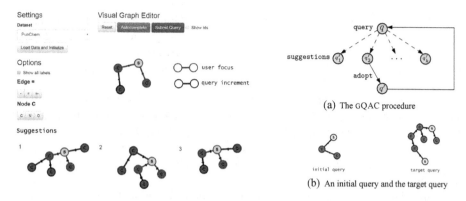

Fig. 1. GQAC without useful suggestions (the UI has been rearranged for presentation)

(a) The GQAC procedure

(b) An initial query and the target query

Fig. 2. The GQAC procedure and an initial query and the target query

bottom. It is easy to find that all 3 suggestions are useless for formulating the target query. In this scenario, a common question raised by users would be: *"why are the desired suggestions not returned?"* For instance, would the preference of suggestion ranking, that is data specific, be set improperly? In this paper, we formulate this as the *why-not questions of* GQAC. The answers to the questions give users insights to use GQAC further.

Related Work. The applications of QAC in search engines have been mature. In addition, there have been many studies on QAC for multi-word phrases, imprecise keywords or friendly GUI. As for GQAC, Mottin et al. [5] proposed graph query reformulation, which determines a reformulated query that maximally covers the *results* of the current query. Pienta et al. [8] and Li et al. [3] demonstrated interactive methods to produce *edge* or *node* suggestions for visual graph query construction. In contrast, this paper considers subgraph suggestions. Yi et al. [9] proposed a subgraph GQAC framework called AUTOG. Among these studies, they do not consider why-not questions. Moreover, our approach does not assume a specific GQAC system. Our algorithm for why-not questions is designed with a generic GQAC ranking function.

The concept of *why-not* proposed by Chapman and Jagadish [1] is devised to enhance the experience of using database systems. The first research about why-not questions on graph query is presented by Li et al. [2]. They refine the query graph directly because the problem they focus on was *similar graph matching*. While there have been many interesting studies on the why-not questions on query answering, to the best of our knowledge, there is no existing work on why-not questions on GQAC.

Background. We consider a graph database D containing a large collection of undirected graphs $\{g_1, g_2, g_3, ..., g_n\}$. Each graph is a 3-ary tuple $g = (V, E, l)$, where V and E are the vertex and edge sets of g, and l is the labeling function of g. The graph size is given as $|E|$. For illustration, the query formalism is subgraph isomorphism. That is, given a query graph q_t and a database D, the answers of q_t are the set of graphs R, where each graph in R contains q_t. This setting fits many applications including chemical compounds, co-purchase networks, etc.

In a GQAC system, the query that a user intends to formulate (a.k.a *target query*) is obviously not yet available. Query refinement on the query graph requires human intervention, *i.e.*, a manual construction of the query. This contradicts with query autocompletion. Therefore, this paper investigates an automatic way that only searches for changing the settings of a GQAC system, such that useful suggestions are returned.

While the why-not questions aim to be generally applicable to GQAC systems, we present this study with the current state-of-the-art called AUTOG [9] and a recent system GFOCUS [10] to present the questions in concrete terms. First, we present a generic ranking function (util) of a set of query suggestions Q by summarizing the current state-of-the-art GQAC systems and our experiments on them, as follows.

$$\text{util}(Q, \beta, \mu) = \beta \times R(Q) + (1 - \beta) \times S(Q)^{\mu}, \qquad (1)$$

where R is the normalized *result count* of Q; and S is the *diversity* of the structures among the suggestions in Q. We may omit Q when it is clear from the context. R and S are normalized, *i.e.*, $R, S \in [0,1]$. In particular, in [9], R is the normalized sum of the estimated result counts of Q, whereas S is the sum of the pairwise structural differences of Q. β is the user's preference to R and S. Large β implies that users may prefer to retrieve some results (*i.e.*, R), whereas small β implies users prefer different suggestion graphs (*i.e.*, S). We also include μ, the exponential factor of S, to tackle the discrepancies of the scales of R and S.

Second, we summarize the terminologies of these systems and their meanings. Given user's current query q_0 and target query q_t, a *useful suggestion* is considered to be a supergraph of q_0 and a proper subgraph of q_t. τ is a GQAC parameter for user's behavior. In [10], τ models the relative memory strength to remember his/her working subgraph f (a.k.a *user focus*) in q_0. Query suggestions are extending q_0 at f only. This paper models a configuration C of a GQAC system X as a 4-ary tuple: $C = (\mu, \tau, \beta, k)$.

Following the *query refinement* approach, we aim to compute a modification (except k) of the configuration of a GQAC system X, such that a useful suggestion appears in the top-k suggestions. We do not consider a modification of k because k is often bounded by the GUI, which displays the query suggestions, and a user has a fixed cognition resource to interpret suggestions. In addition, when devising the refined configuration, the modification of the parameters μ, τ and β are minimal, so that the original semantics of the settings are largely preserved. Hence, we propose the following *penalty model*.

$$\text{Penalty}(\mu^*, \tau^*, \beta^*) = \lambda_1 \times \frac{\Delta\mu}{\Delta\mu_{max}} + \lambda_2 \times \frac{\Delta\tau}{\Delta\tau_{max}} + \lambda_3 \times \Delta\beta, \qquad (2)$$

where $\lambda_1 + \lambda_2 + \lambda_3 = 1$ and $\lambda_i \in (0, 1)$ is a weight of the modifications on μ, τ, and β, denoted as $\Delta\mu$, $\Delta\tau$, and $\Delta\beta$. $\Delta\mu_{max}$ and $\Delta\tau_{max}$ are the maximum changes to the values of μ and τ. They are included for normalization. We then formalize the problem statement of this paper in Definition 1.

Definition 1. The Why-Not Questions of GQAC. *Given a GQAC system X, a graph database D, a target query q_t, a current query q_0 and a configuration $C = (\mu, \tau, \beta, k)$, and there is no useful suggestion, the* why-not questions *of graph query autocompletion is to search for the refined configuration $C^* = (\mu^*, \tau^*, \beta^*, k)$, such that*

- *X returns at least one useful suggestion for formulating q_t; and*
- *the penalty value, according to Formula (2) of C^*, is minimal.*   □

We remark that (i) it is possible that C^* does not exist. In that case, the why-not questions fail and users resort to manually compose their queries. (ii) The "target" query q_t can certainly be an *intermediate* query during query formulation.

2   Searching for the Answers to the Why-Not Questions

2.1   Illustrations of the Search for μ^* (search_μ^*) and τ^* (search_τ^*)

search_μ^*. Recall that we introduce μ in Formula (1) to tackle the possible discrepancies between the scale of R and S. The reason is that the result count R can sometimes be a large value (close to 1), whereas the structural diversity S can be very small. In addition, we observe from simulations that a simple change to the label of the suggestions can significantly change the structural diversity, which often has little relationship with q_t (Fig. 3).

Example 1. This example shows that the number of distinct labels in the query increments can significantly affect the diversity of query suggestions Q. Assume the diversity of Q is defined as the sum of the pairwise distance of suggestions in Q, where the distance is given by the fraction of subgraphs not in *maximum common edge subgraph* [9]. The sum of the pairwise distance of the 3 suggestions in Fig. 1 is $\frac{3}{5}$. Suppose we change only one label from C to O in Suggestion 3 (denoted as q_3), as shown in Fig. 3. The sum is increased to 1. The reason is that q_3 contains a label that only appears q_3 makes it different from the rest of suggestions in Q, which yields a high value of S. Hence, given q_t, search_μ^* search for μ^* that yields the correct number of distinct labels in Q.   □

search_τ^*. A major observation from [10] on GQAC is that query suggestions may sometimes be useless simply because the current query q is enlarged at wrong places. An example shown in Fig. 4. [10] predicts the red colored edge as the *user focus*, i.e., the subgraph the user is working at. Suggestions are formed by adding subgraph increments there. However, user focus cannot be perfectly predicted, such as an arbitrary focus shift. This is another scenario where the why-not questions are issued. Hence, we propose search_τ^* that enlarges the user focus until it covers the subgraph of q that a useful suggestion can form by adding a subgraph to it [4] (Fig. 5).

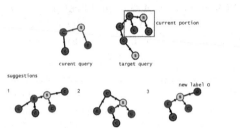

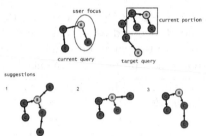

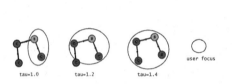

Fig. 3. GQAC without useful suggestions (with new distinct label 0)

Fig. 4. Incorrect user focus leads to useless suggestions (Color figure online)

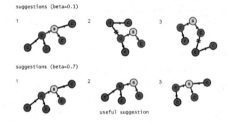

Fig. 5. An example of the search for τ

Fig. 6. An example of the search for β

2.2 Search for the Proper Suggestion Ranking Preference β (search_β^*)

Recall query suggestions are ranked according to Formula 1. A major reason why useful suggestions are not returned is that the preference β between R and S is not properly set. Further, the values of β are database specific, which makes it hard to manually set and interpret. It is also known that suggestion ranking is NP-hard to $|q_t|$ [9,10].

Example 2. Figure 6 shows two suggestion sets obtained using different β values. When increasing β from 0.1 to 0.7, one useful suggestion appears in the top-3 suggestions. The reason is that the target query retrieves a large answer set. When increasing β, similar suggestions having larger Rs are ranked higher. □

To search for a proper β^*, we propose search_β^*, shown in Algorithm 1. Algorithm 1 undertakes the simulated annealing approach, simply to avoid the search being trapped in local optima. A *search state* is the current β^* value. A *neighboring state* is $\beta^* \times d_\beta^\pm$, where d_β^+ and d_β^- are the rates for increasing and decreasing β^*, respectively.

In the search of β^*, we propose the following strategies. First, given a set of suggestions and a β value, increasing β may either increase or decrease the util value. Strategy 1 determines the next β value that increases (respectively, decreases) the util value of Q_0, denoted as the β^+ value (respectively, the β^- value). Strategy 2 states that if Q_1 has a util value higher than Q_0 but having

---

Algorithm 1. search_$\beta^*(\mu^*, \tau^*, \beta)$

Input : new parameters μ^*, τ^*, and current β
Output: refined β (β^*)

```
1  Initializing cr, T, Tmin ;                              /* cooling rate, temp., min. temp. */
2  Q0 ← executing gQAC with C' = (μ*, τ*, β, k);  β* ← β ;
3  while T > Tmin do
4  │   Q1 ← Q0 \ {qi} ∪ {qi'} for some i ∈ [1, k] ;        /* neighboring suggestions */
5  │   if util(Q1) ≥ util(Q0) then                         /* Strategy 1 & 2           */
6  │   │   if ∃ g ∈ Q1. g is a proper subgraph of qt then  /* a useful query           */
7  │   │   │   β*-Set ← β*-Set ∪ {β*};
8  │   │   else
9  │   │   │   β* ← β* × dβ⁻;  T ← T × (1 − cr); Q0 ← Q1 ;  /* cooling T */
10 │   else if P(Q0, Q1, T) > rand(0, 1) then              /* Random jump & Strategy1 & 2 */
11 │   │   if ∃ g ∈ Q1. g is a proper subgraph of qt then
12 │   │   │   β*-Set ← β*-Set ∪ {β*};
13 │   │   else
14 │   │   │   β* ← β* × dβ⁺;  T ← T × (1 − cr); Q0 ← Q1 ;  /* cooling T */
15 │   else
16 │   │   β* ← β* × dβ⁺;  T ← T × (1 − cr) ;               /* cooling T */
17 end
18 β* ← the β* value from β*-Set having the minimum penalty ;
19 return β* ;
```

---

no useful suggestions, then the search proceeds to next β attempts to "another direction", *i.e.*, to reduce util using β^-.

Strategy 1. *Denote Q_0 as the current query suggestions and β_0 as the current preference for the ranking, Eq. 1. β^+ and β^- are the next preferences that, respectively, increase and decrease util(Q_0), defined as follows.*

1. *If util(Q_0, $\beta \times d_\beta^+$) $\geq$ util(Q_0, β), then $\beta^+ = \beta \times d_\beta^+$ and $\beta^- = \beta \times d_\beta^-$.*
2. *Otherwise, $\beta^+ = \beta \times d_\beta^-$ and $\beta^- = \beta \times d_\beta^+$.* □

Strategy 2. *Consider another suggestions Q_1. Assume both Q_0 and Q_1 do not contain a useful suggestion. The search strategy for β is as follows:*

1. *If util(Q_1, β) $\geq$ util(Q_0, β), then we set β to β^-.*
2. *Otherwise, we set β to β^+.* □

The main ideas of search_$\beta^*(\mu^*, \tau^*, \beta)$ (Algorithm 1) are described as follows.

1. β is the current β^* value when a user asks the why-not question. μ^* and τ^* are obtained from Sec. 2.1 (detailed in [4]). In Line 1, Algorithm 1 initializes the parameters of a simulated annealing search. They are the cooling rate cr, the temperature T, and minimum temperature T_{min}. In Line 2, we obtain a set of suggestions Q_0 by using the current β, μ^*, and τ^*.
2. In Line 4, the algorithm determines query suggestions Q_1 that are slightly different from Q_0, by replacing one query q_i from Q_0 with a query q_i', which is a valid suggestion but was not ranked top k before.
3. In Line 5, it computes util(Q_1). In Lines 6–7, if Q_1 contains a useful suggestion, the search keeps the β^* value in a set. Otherwise, Line 9 essentially implements Strategy 1 and 2. Algorithm 1 proceeds the search to $\beta^* \times d_\beta^-$.

Algorithm 2. Answering the why-not questions of GQAC

Input : The target query q_t, current query q_0, current configuration $C = (\mu, \tau, \beta, k)$
Output: The refined configuration $C^* = (\mu^*, \tau^*, \beta^*, k)$
1 Let q_a be the subgraph added to q_0 to form q_t ;
2 Let c be the connection node of q_0 and q_a ;
3 $\mu^* = $ search_$\mu^*(q_t)$;   /* refine μ [4] */
4 $\tau^* = $ search_$\tau^*(c, \tau)$;   /* refine τ [4] */
5 $\beta^* = $ search_$\beta^*(\mu^*, \tau^*, \beta)$;   /* refine β */
6 **return** $C^* = (\mu^*, \tau^*, \beta^*, k)$;

---

4. If the util value of Q_1 is smaller than that of Q_0, we calculate the probability of $P(Q_0, Q_1, T) = e^{\frac{\text{util}(Q_1) - \text{util}(Q_0)}{T}}$, and compare it with rand$(0, 1)$, using the result to decide whether to jump to Q_1 (Line 10) or not. Lines 11–14 is a similar processing like the util$(Q_1) \geq$ util(Q_0) case, where the only difference is that it modifies β by a factor d_β^+ (Line 14), as presented in Strategy 2.

Putting These Together. We combine the proposed search algorithm for each parameter together. The pseudo-code for answering the why-not questions of GQAC is presented in Algorithm 2, which takes the target graph q_t, the current graph q_0, and the current configuration C as input. It returns the refined configuration C^* after refining each parameter.

3   Experimental Evaluation

This section presents some experimental results of the proposed algorithm. We investigate the suggestion quality (effectiveness of the algorithm) via user studies and simulations. Due to space limitations, we omit the detailed results and report them in [4].

Software and Hardware. We obtained the implementations of AUTOG [9] and GFOCUS [10] from their authors. We implemented the proposed algorithm on top of them. Our prototype is mainly implemented in C++. All the experiments were conducted on a machine with a 2.2 GHz Xeon E5-2630 processor and 256 GB memory, running Linux.

Datasets and Query Sets. We used the PUBCHEM [7] dataset, which is a real chemical compound dataset consisting of 1 million graphs. On average, there are 23.98 vertices and 25.76 edges of each graph. In the whole dataset, it has 81 different vertex labels and 3 different edge labels. The query sets are taken from [10], with the number of edges ranging from 4 to 24. Each query set contains 100 graphs.

Table 1. Default settings

Default settings	$|q|$	k	τ	α	β	δ_{max}	λ_1	λ_2
Value	20	10	1.2	0.1	0.1	5	0.2	0.4
Default settings (cont')	λ_3	μ_{max}	τ_{max}	β_{max}	d_τ	d_β^+	d_β^-
Value	0.4	5	3000	1.0	1.5	1.1	0.7

Table 2. Users' agreement level

	High	Medium	Low
AUTOG	4.12	3.07	2.45
YNOTG	4.73	3.40	2.27
GFOCUS	4.55	2.95	1.65
YNOTF	4.93	3.53	2.33

Default Settings. We ran extensive simulations on the parameters to determine proper default values (shown in Table 1). We will compare the results of 6 algorithms, including AUTOG, BaselineG, YNOTG, GFOCUS, BaselineF, and YNOTF. BaselineG and BaselineF are the baseline algorithms (enumerations of the parameters) applied to AUTOG and GFOCUS. YNOTG and YNOTF are the search algorithms (proposed in Sect. 2) applied to AUTOG and GFOCUS.

Quality Metrics. We report the number of useful and useless GQAC iterations of the simulations on AUTOG, denoted as #AUTO$_G$ and #N-AUTO$_G$, respectively. With the refinement, the ratio of the useless GQAC iterations that become useful is used to measure the effectiveness of the proposed baseline and advanced algorithms, denoted as useful$_B = \frac{\#\ of\ \#\text{N-AUTO}_G\ becomes\ useful\ by\ \text{BaselineG}}{\#\text{N-AUTO}_G} \times 100\%$ for BaselineG, and useful$_Y$ for YNOTG. The average penalty of BaselineG and YNOTG are reported, denote as Penalty$_B$ and Penalty$_Y$. The *total profit metric* [6,9,10], denoted as TPM, is used to quantify the percentage of mouse clicks saved by adopting suggestions in the visual formulation. It is calculated by $\text{TPM} = \frac{clicks\ saved\ by\ suggestions}{clicks\ without\ suggestions} \times 100\%$.

User Study. We invited 10 volunteers to test the usefulness of the suggestions after refinement by the proposed algorithms. Each volunteer was given six randomly shuffled 8-edge target queries from high to low TPM values obtained from the simulations. When there are no useful suggestions returned during one GQAC iteration with the default parameters, YNOTG or YNOTFwill be invoked to refine the configuration and obtained the refined suggestions. For each refinement, the volunteer reported the level of agreement to the statement "YNOTG (or YNOTF) is useful". A symmetric 5 level agree-disagree Likert scale is adopted, where 1 means "strongly disagree" and 5 means "strongly agree". The agreement level for the approaches are shown in Table 2.

After finishing the formulation of one query, the volunteers were asked to report their level of agreement to the usefulness of YNOTG and YNOTF for this query. The average agreement level for the queries with high, medium and low TPM values are 4.25, 3.43, 2.25 using YNOTG and 4.67, 3.59, 2.37 using YNOTF. Consistent with the findings of AUTOG (resp. GFOCUS), the correlation coefficient between the users' agreement levels and the TPM value is 0.914 (resp. 0.959) and the p-value is 0.0097 (resp. 0.0031) in YNOTG (resp. YNOTF).

Table 3. Quality metrics varying δ_{max} based on AutoG (PubChem)

δ_{max}	#AutoG	#N-AutoG	useful$_B$%	useful$_Y$%	TPM$_G$	TPM$_B$	TPM$_Y$	Penalty$_B$	Penalty$_Y$	TPM$_{\Delta_G}$%
1	1200	600	6	11	51	53	53	0.24	0.17	4
2	542	763	20	49	46	53	66	0.25	0.18	43
3	311	811	17	57	35	48	70	0.24	0.19	100
4	236	829	16	52	32	44	69	0.25	0.19	116
5	193	863	16	47	31	42	67	0.25	0.20	116

Table 4. Quality metrics varying δ_{max} based on GFocus (PubChem)

δ_{max}	#GFocus	#N-GFocus	useful$_B$%	useful$_Y$%	TPM$_F$	TPM$_B$	TPM$_Y$	Penalty$_B$	Penalty$_Y$	TPM$_{\Delta_F}$%
1	1366	434	37	39	58	64	65	0.35	0.17	12
2	458	412	43	62	54	67	74	0.28	0.17	37
3	341	704	33	65	47	62	77	0.30	0.17	64
4	278	1083	28	52	43	59	72	0.31	0.18	67
5	259	722	24	58	39	56	78	0.28	0.18	100

Table 5. Quality metrics varying $|q|$ based on AutoG (PubChem)

$|q|$	#AutoG	#N-AutoG	useful$_B$%	useful$_Y$%	TPM$_G$	TPM$_B$	TPM$_Y$	Penalty$_B$	Penalty$_Y$	TPM$_{\Delta_G}$%
8	108	298	17	30	37	49	58	0.23	0.20	57
12	213	434	21	34	43	57	64	0.23	0.19	49
16	327	592	14	29	44	53	62	0.22	0.20	41
20	419	745	15	34	44	54	64	0.23	0.18	45
24	500	943	12	31	44	51	63	0.24	0.18	43

Table 6. Quality metrics varying $|q|$ based on GFocus (PubChem)

$|q|$	#GFocus	#N-GFocus	useful$_B$%	useful$_Y$%	TPM$_F$	TPM$_B$	TPM$_Y$	Penalty$_B$	Penalty$_Y$	TPM$_{\Delta_F}$%
8	111	286	16	32	41	49	60	0.24	0.21	46
12	233	391	22	38	50	58	67	0.24	0.18	34
16	339	538	22	36	51	60	68	0.25	0.19	33
20	461	635	28	41	52	64	71	0.26	0.17	37
24	570	736	25	36	53	64	69	0.27	0.19	30

Effectiveness via Simulations. The results of large-scale simulations on different parameters are shown in Tables 3, 4, 5 and 6. We use useful$_B$%, useful$_Y$% to show the usefulness of BaselineG (or BaselineF) and YnotG (or YnotF); and TPM$_G$, TPM$_F$, TPM$_B$, and TPM$_Y$ to denote the TPM measures of AutoG, GFocus, BaselineG (or BaselineF), and YnotG (or YnotF), respectively. We use TPM$_{\Delta_G}$% (resp. TPM$_{\Delta_F}$%) to depict the TPM enhancement from AutoG (resp. GFocus) to YnotG (resp. YnotF). When compared to the baselines, AutoG and GFocus, the proposed algorithm performs clearly better in all quality metrics. Furthermore, the penalty of YnotG (resp. YnotF) is smaller than that of BaselineG (resp. BaselineF) in all the cases.

Efficiency via Simulations. Finally, we remark that the detailed evaluation of the efficiency of the six frameworks, *i.e.*, AUTOG, YNOTG, BaselineG, GFOCUS, YNOTF, and BaselineF is reported in [4]. The overall relationship among the average response times of the AUTOG-based and GFOCUS-based frameworks are: AUTOG< YNOTG< BaselineG; GFOCUS< YNOTF< BaselineF.

4   Conclusion

This paper studies the why-not questions of GQAC. A generic ranking function for GQAC systems is formulated. We propose a search algorithm to determine a minimal configuration modification of the GQAC system, w.r.t a penalty model. We conducted a user study and simulations to verify the performance of the algorithm.

References

1. Chapman, A., Jagadish, H.V.: Why not? In: SIGMOD, pp. 523–534 (2009)
2. Islam, M.S., Liu, C., Li, J.: Efficient answering of why-not questions in similar graph matching. TKDE **27**, 2672–2686 (2015)
3. Jayaram, N., Goyal, S., Li, C.: VIIQ: auto-suggestion enabled visual interface for interactive graph query formulation. PVLDB **8**, 1940–1951 (2015)
4. Li, G., Ng, N., Yi, P., Zhang, Z., Choi, B.: Answering the why-not questions of graph query autocompletion (2018). https://goo.gl/4Hpt5m
5. Mottin, D., Bonchi, F., Gullo, F.: Graph query reformulation with diversity. In: KDD, pp. 825–834 (2015)
6. Nandi, A., Jagadish, H.V.: Effective phrase prediction. In: PVLDB, pp. 219–230 (2007)
7. NLM: PubChem. ftp://ftp.ncbi.nlm.nih.gov/pubchem/
8. Pienta, R., Hohman, F., Tamersoy, A., Endert, A., Navathe, S., Tong, H., Chau, D.H.: Visual graph query construction and refinement. In: SIGMOD, pp. 1587–1590 (2017)
9. Yi, P., Choi, B., Bhowmick, S., Xu, J.: AutoG: a visual query autocompletion framework for graph databases. VLDB J. **26**, 347–372 (2017)
10. Yi, P., Choi, B., Zhang, Z., Bhowmick, S.S., Xu, J.: Gfocus: user focus-based graph query autocompletion (2018). https://goo.gl/MYYw94

Exploiting Reshaping Subgraphs
from Bilateral Propagation Graphs

Saeid Hosseini[2]([⊠]), Hongzhi Yin[1], Ngai-Man Cheung[2], Kan Pak Leng[2],
Yuval Elovici[2], and Xiaofang Zhou[1]

[1] The University of Queensland, Brisbane, Australia
`h.yin1@uq.edu.au`, `zxf@itee.uq.edu.au`
[2] Singapore University of Technology and Design, Singapore, Singapore
`saeid.hosseini@uq.net.au`, {`ngaiman_cheung,yuval_elovici`}`@sutd.edu.sg`,
`kanpl@stee.stengg.com`

Abstract. Given a graph over which defects, viruses, or contagions
spread, leveraging a set of highly correlated subgraphs is an appealing
research area with many applications. However, the challenges abound.
Firstly, an initial defect in one node can cause different defects in other
nodes. Second, while the time is the most significant medium to understand diffusion processes, it is not clear when the members of a subgraph
may change. Third, given a pair of nodes, a contagion can spread in both
directions. Previous works only consider the sequential time-window and
suppose that the contagion may spread from one node to the other during
a predefined time span. But the propagation can differ in various temporal dimensions (e.g. hours and days). Therefore, we propose a framework
that takes both sequential and multi-aspect attributes of the time into
consideration. Moreover, we devise an empirical model to estimate how
frequently the subgraphs may reshape. Experiment show that our framework can effectively leverage the reshaping subgraphs.

Keywords: Propagation graphs · Reshaping subgraphs
Diffusion networks

1 Introduction

Given a *bilateral propagation graph* over which the contaminations of the same
type spread in both directions, in this paper, we utilize miscellaneous temporal
attributes to reveal highly correlated subgraphs. We also study the effect of temporal dynamics on subgraph changes. Classical models in understanding of the
diffusion processes include Linear Threshold (LT), Independent Cascade (IC),
and Weighted Cascade(WC) [1]. As studied in the literature [2,3], group-based
immunization can perform better than classical models. Hence we have chosen
to exploit the groups of highly correlated nodes. Additionally, we propose an
evaluation metric which can infer how the nodes in each of the subgraphs may
change and when it is crucial to reinitiate the subgraphs via a *Trigger*. The
non-trivial task of subgraph mining can be applied to numerous scenarios [4–7].

© Springer International Publishing AG, part of Springer Nature 2018
J. Pei et al. (Eds.): DASFAA 2018, LNCS 10827, pp. 342–351, 2018.
https://doi.org/10.1007/978-3-319-91452-7_23

From an industry perspective, in our dataset, each node may raise one or more *alarms* indicating that a defect has happened in the node. As far as the sequential alarms occurred in a pair of nodes are of the same type, we can log an *alarm propagation* instance. The challenges are three-fold. Firstly, as alarms of the same type might have irrelevant semantics, one cannot employ the classic text understanding models [8,9] to trace propagation routes. Secondly, since the propagation may change by time, the number of nodes in a subgraph may change. Hence, it is not determined when we should reinitiate the subgraphs. Thirdly, the alarms among each pair of nodes may spread in both directions [10].

Given each pair of the nodes, the problem for exploiting the subgraphs initiates from learning of the edge weights. Subsequently, the subgraphs can be leveraged by getting the top k highest correlated pairs. Current sequential inference models (e.g. [11,12]) suppose that when two nodes raise an alarm one after the other in less than the temporal threshold, an instance of propagation can be recorded. In this paper, we employ miscellaneous temporal properties. Firstly, we utilize the time function with undirectional similarity metric to compute how the alarms of the same type can sequentially propagate between each pair of nodes. Secondly, subject to n temporal dimensions (i.e. hour, day, and week), we infer how correlated the pair of nodes raise the alarms of the same type. The propagation direction between two nodes of n_i and n_j might differ during the working days and weekends. Moreover, while n_i and n_j are tightly coupled during the day, they might be less correlated during after hours (9 pm to 6 am). To the best of our knowledge, this is the first work that, on the one hand, can extract the subgraphs from bidirectional propagation graphs, and on the other hand, can infer how the members of subgraphs may evolve. In this paper, we develop a generative model which employs multi-aspect temporal property to understand pairwise edge weights from bilateral propagation graphs. Also, we propose a unified framework that explores Multiple Attributes (Sequential and multi-dimension) of Time to Exploit Subgraphs (MATES) from bilateral propagation graphs. Our method can also reveal how frequently the subgraphs may reshape. The rest of the paper is as follows: Sect. 5 summarizes the related work; we present preliminary concepts in Sect. 2; our models and experiments are explained in Sects. 3 and 4 respectively; then Sect. 6 concludes the paper.

2   Problem Statement

2.1   Preliminary Concepts

Definition 1 *(alarm). Each alarm $a_i \in \mathbb{A}$ is a contagion, associated with a node, has a type and time-stamp.*

Problem 1 (sequential propagation). *Given A_i and A_j as the respective list of alarm times for n_i and n_j, our goal is to compute how a contagion $n_i.a_i$ raised in n_i at time t_i can cause another alarm $n_j.a_j$ in n_j at time $t_i + \zeta_t$.*

Given a node pair (n_i, n_j), propagation can be explained by a set of temporal latent dimensions (e.g. $\mathbb{T} = \{\boldsymbol{\tau}^h, \boldsymbol{\tau}^d, \boldsymbol{\tau}^w\}$ for hour $\boldsymbol{\tau}^h$, day $\boldsymbol{\tau}^d$, and week $\boldsymbol{\tau}^w$).

Given a dimension (e.g. τ^d) comprising l splits, we can create one or more *uni-dimensional temporal blocks* by merging similar splits (e.g. merging Tue. and Wed.). A *multi-dimensional temporal block* (e.g. $\tau = \{\tau^w : \{1,2\}, \tau^d : \{Sat., Sun.\}, \tau^h : \{9-12am\}\}$) is a set of uni-dimensional blocks in n aspects.

Definition 2 *(Correlation Weight). is the weight among each node pair that shows how correlated they can spread the contagions bidirectionally.*

Problem 2 (Exploiting correlation weights using multi-dimensional temporal blocks). *Given a set of latent temporal dimensions $\mathbb{T}$, and a set of multi-dimensional temporal blocks τ^b, our goal is to compute the correlation weight among node pairs (n_i, n_j) through $|\mathbb{T}|$ latent aspects. Given the solution for Problem 2, each group of joint nodes from top k edge weights can represent a subgraph.*

3   Methodology

Our model computes the correlation weight between a node pair using two modules of sequential temporal approach (Sect. 3.2) and multi-aspect temporal model (Sect. 3.1). The prob. of n_i to spread an alarm to n_j equates to the joint probability(Eq. 1).

$$Pr(n_j|n_i) \propto Pr(n_i, n_j) \tag{1}$$

3.1   Multi-aspect Temporal Property

Given multiple time dimensions denoted by τ, the multi-aspect temporal property can explain the bidirectional propagation among a pair of nodes (n_i, n_j). Based on the duration of the datasets, in this paper, we consider three dimensions of hour, day, and the week ($\mathbb{T} = \{\tau^h, \tau^d \tau^w\}$). To build the temporal blocks in each of three dimensions, we employ a similarity function like Pearson or Cosine to measure the level of correlation between split pairs (e.g. between 9 pm and 10 pm). Intuitively, if the alarm of the identical type is witnessed in both of the splits, it can support the similarity among them and increase the chance for them to merge and participate in a shared uni-aspect temporal block. Figures 1 and 2 illustrate the similarity weights between selected splits in three dimensions for stations 0 and 1 respectively. Given the similarity grids, a bottom-up hierarchical clustering model [13] can be used to construct the temporal blocks. Given three-dimension temporal blocks, we can propose a generative algorithm to exploit the correlation weights among the nodes - Problem 2. The joint probability for each pair of the nodes (n_i, n_j) (Eq. 1) to raise the same type alarm c_l denotes how harmonized they may raise the alarms in each of the three-fold temporal blocks (Eq. 2). Here $\eta_{w,d,h}^{iterate}$ represents the inner iterations for three dimensions where τ^b is the set of multi-aspect temporal blocks.

$$Pr(n_i, n_j) \propto \sum_{\eta_{w,d,h}^{iterate} \in \tau^b} Pr(n_i, n_j, \tau) \tag{2}$$

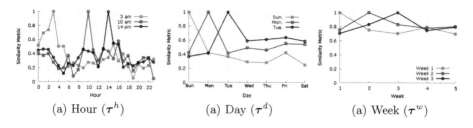

(a) Hour (τ^h)   (a) Day (τ^d)   (a) Week (τ^w)

Fig. 1. Temporal dimension similarity - station-0

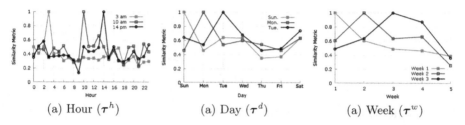

(a) Hour (τ^h)   (a) Day (τ^d)   (a) Week (τ^w)

Fig. 2. Temporal dimension similarity - station-1

Equation 3 formulates the joint probability for the pair of nodes alongside with the three-fold latent temporal factors. Also, Eq. 4 declares the non-temporal joint probability of the nodes ($Pr(n_i, n_j)$). To compute $Pr_\omega(n_j|n_i)$ we use Collaborative Filtering and consider the number of shared alarm types among n_i and n_j.

$$Pr(n_i, n_j, \tau) \propto Pr(n_i, n_j)Pr(\tau|n_i, n_j) \tag{3}$$

$$Pr(n_i, n_j) \propto Pr(n_i)Pr_\omega(n_j|n_i) \tag{4}$$

$$Pr(\tau|n_i, n_j) = Pr(\tau^h|\tau^{d,w}, n_i, n_j)\, Pr(\tau^d|\tau^w, n_i, n_j)\, Pr(\tau^w|n_i, n_j) \tag{5}$$

$$Pr(n_i, n_j, \tau) \propto Pr(n_i)Pr_\omega(n_j|n_i)\, Pr\left(\tau^h|\tau^{d,w}, n_i, n_j\right)$$
$$Pr\left(\tau^d|\tau^w, n_i, n_j\right)Pr(\tau^w|n_i, n_j) \tag{6}$$

We can place Eqs. 4 and 5 into Eq. 3 to get Eq. 6. Under three-fold temporal dimension, Eq. 6 formulates the scenario when each node pair raise the same alarm types.

$$\mathcal{C}(\Pi) = \sum_{<n_i, n_j> \in <\mathbb{N}, \mathbb{N}>} log(Pr(n_i, n_j; \Pi)) \tag{7}$$

We retrieve the set of missing parameters (Π) in Eqs. 6 and 7 using the Expectation-Maximization (EM) approach (Maximizing the log-prob. of $\mathcal{C}(\Pi)$).

E-step (Eq. 8): refreshes the joint expectation of the temporal variables for each node pair.

$$Pr(\tau|n_i, n_j) = \frac{Pr(n_i, n_j, \tau)}{\sum\limits_{\eta_{w,d,h}^{iterate}} Pr(n_i, n_j, \tau)} \tag{8}$$

M-Step (Eq. 9): maximizes the log-likelihood by the new values of each parameter in Π:

$$Pr(\tau^h|\tau^{d,w}, n_i, n_j) = \frac{Pr(\tau|n_i, n_j)}{\sum\limits_{\eta_{h'}^{iterate}} Pr(\tau^{h',d,w}|n_i, n_j)}$$

$$Pr(\tau^d|\tau^w, n_i, n_j) = \frac{\sum\limits_{\eta_h^{iterate}} Pr(\tau|n_i, n_j)}{\sum\limits_{\eta_{d',h}^{iterate}} Pr(\tau^{h,d',w}|n_i, n_j)} \tag{9}$$

$$Pr(\tau^w|n_i, n_j) = \frac{\sum\limits_{\eta_{d,h}^{iterate}} Pr(\tau|n_i, n_j)}{\sum\limits_{\eta_{w',d,h}^{iterate}} Pr(\tau^{h,d,w'}|n_i, n_j)}$$

3.2   Sequential Time Property

Given a pair of nodes (n_1, n_2), the sequential attribute of the time can explain how an alarm in one node can cause an alarm of the same type in the other. In an effective approach, we propose a method using the dissimilarity distance and the decay function.

$$b \in \mathbb{A}_j \wedge b.t > n_i.a.t \wedge b.t - n_i.a.t < \vartheta \tag{10}$$

Given ϑ, $\mathbb{A}_i$, and $\mathbb{A}_j$ as the sequential propagation threshold and the set of alarms for n_i and n_j (Eq. 10), to infer the sequential propagation, we can use *Euclidean* distance between each alarm in n_i $(n_i.a)$ with its closest possible propagation event in node n_j $(b \in \mathbb{A}_j)$. The unidirectional propagation from n_i to n_j $(d_{i,j} : n_i \rightarrow n_j)$ is computed via the average (Eq. 12) of the dissimilarities $(d_{i,j}^{n_i.a})$. Moreover, we normalize the unidirectional propagation $(d_{i,j})$ through dividing by maximum (Eq. 13). The final value is the average of the values in both directions (Eq. 14).

$$d_{i,j}^{n_i.a} = \sqrt{(n_j.b.t - n_i.a.t)^2} \tag{11}$$

$$d_{i,j} = \frac{\sum_{n_i.a \in \mathbb{A}_i} d_{i,j}^{n_i.a}}{|\mathbb{A}_i|} \tag{12}$$

$$d_{i,j} = \frac{d_{i,j}}{max_{(n_i,n_j) \in \mathbb{N}}\{d_{i,j}\}} \tag{13}$$

$$\overleftrightarrow{D}_{i,j} = avg(d_{i,j}, d_{j,i}) \tag{14}$$

Subsequently, the time-decay function can grant higher edge weights where the dissimilarity is lower.

$$Pr(n_i, n_j) = \{e^{-\lambda \overleftrightarrow{D}_{i,j}}|\tau_h = 1/\lambda\} \tag{15}$$

In Eq. 15, τ_h represents the half-time feature and λ is the decaying parameter.

4   Experiment

We compared the performance of the subgraphs mining models. The datasets were based on the propagated alarms in two metro stations in 2016.

4.1   Evaluation Metric

We used 80% of the data in each station for *parameter adjustments*. We then utilized the remaining 20% to compare the effectiveness of the methods. We examined how each method had correctly incorporated the nodes into the exploited subgraphs. Given a node n_a as a member of an exploited subgraph S_e, we define the hypothesis parameters as follows: *True Positive (TP)*: is where n_a raises an alarm type c_k at time t and one or more neighbors in S_e raise the same alarm type during temporal threshold $(t-\zeta_t, t+\zeta_t)$, while the competing model admits that n_a raises the alarm type c_k. *False Positive (FP)*: n_a does not raise the alarm type c_k and other neighbors in S_e raises the alarm type c_k during the threshold, and the model estimates that n_a raises the alarm type c_k. *True Negative (TN)*: n_a does not raise type c_k, and other neighbors in S_e raise the alarm type c_k during the threshold and the model estimates that n_a does not raise the alarm type c_k. *False Negative (FN)*: n_a raises alarm type c_k while the neighbors in S_e raise the alarm type c_k and the model estimates that n_a does raise type c_k. The F-Measure is used to compare the rivals. It can be computed using predefined hypothesis parameters.

The competitor methods are explained as follows: *Multi-Attribute Time trends to Exploit Subgraphs (MATES)* is our model. *Bidirectional Diffusion Inference* [11] represents the sequential temporal models. *Pair-wise Time Machine* (PTM) utilizes the dissimilarity distance and the decay (Sect. 3.2). *Concurrent temporal Three-layer Approach* (CTA) infers propagation using temporal layers. The MATES model computes the correlation weight using (Eq. 16) pair-wise time machine ($Pr_{ptm}(n_i, n_j)$) and multi-aspect model ($Pr_{cta}(n_i, n_j)$).

$$Pr_{mates}(n_i, n_j) \propto \alpha Pr_{ptm}(n_i, n_j) + (1 - \alpha) Pr_{cta}(n_i, n_j) \qquad (16)$$

The parameter α was set to 0.3 and 0.2 for Station-0 and Station-1 respectively. We obtained top k highest weights and retrieved the subgraphs to compare the methods using benchmark (Sect. 4.1). As depicted in Fig. 3, our devised unified framework is more effective than other methods.

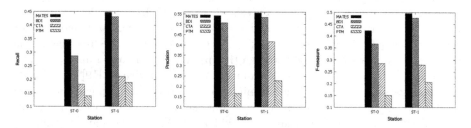

Fig. 3. Performance metrics

4.2   Estimating Reshaping Subgraphs via Trigger

The temporal behavior of the propagation process may change frequently and the changes in edge weights can accordingly update the subgraph to which a particular node belongs. We devise a simple but effective approach to estimate how often we need to trig the similarity grids, reapply the correlation weights between node pairs, and subsequently reinitialize the exploited subgraphs. Accordingly, we divide the dataset into two parts. We can also consider a portion of the recent data instead of the full dataset. We then use the first part to generate the similarity grids. Subsequently, we add the 50% portion by x% (e.g. 10%) and measure how the F-measure reduces. Figure 4 shows the empirical results. When the effectiveness reduces more than threshold ϕ, it is the time to reinitialize the subgraphs. e.g. if we set the ϕ to 0.02, we will need to reinitialize the subgraphs when the size of the dataset grows by 10%.

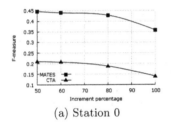

(a) Station 0

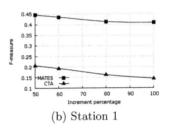

(b) Station 1

Fig. 4. Impact of evolving data on subgraph reshaping

5   Related Work

Spreading of the contagions (virus/message/alarm) over large graphs [11,14, 15] is interdisciplinary. Hence, we survey both the inference and intervention strategies in propagation networks.

Propagation Inference. Propagation inference algorithms are either cascade or threshold based. Cascade styles (e.g. SIS (no immunity), SIRS (temporary immunity), and SIR (permanent immunity)) consider two factors [16]. First, the current condition of the nodes (i.e. Susceptible, Infected, Removed, and Vigilant). Second, the way the contagion is being transmitted. However, the threshold based approaches (e.g. Linear Threshold (LT) [3,17,18]) consider a threshold for the behavior of the contagion. The weaker virus vanishes and the stronger [19] survives. [20] states that the contamination threshold is the first eigenvalue of the adjacency matrix. Nevertheless, other parameters are also important. The shape (e.g. star, chain, and clique) and the number of viruses [19] can change the propagation inference model. Similar to our work, the propagation can be bi-directional [10]. Previous works [11,12] only employ the sequential time property. We propose a unified framework which handles miscellaneous temporal attributes and estimates how frequently the subgraphs may reshape.

Immunization. Immunization strategies [21–23] intervene propagation processes to reduce the effects of the consequent damages. Random vaccination [24], non-spectral methods [18,25,26], game theory [27], and graph manipulation [18,28] focus on individual nodes. But the *group-based* approaches [15] can better control diffusion processes.

6   Conclusion

In this paper, given a propagation graph over which the alarms spread in both directions, we devise an effective framework (MATES) that employs Multi-Attributes Time trends to Exploit Subgraphs. We collectively compute the edge weights among node pairs using two components of the pair-wise time machine and the generative latent module. We also propose an empirical evaluation model to estimate how frequently we need to reinitialize the reshaping subgraphs. The comprehensive experiments verify that our unified framework outperforms current subgraph mining models. As a future work, it is appealing to employ the textual semantics to better exploit the reshaping subgraphs.

Acknowledgment. This work was supported by both ST Electronics and the National Research Foundation (NRF), Prime Minister's Office, Singapore under Corporate Laboratory @ University Scheme (Programme Title: STEE Infosec - SUTD Corporate Laboratory).

References

1. Domingos, P., Richardson, M.: Mining the network value of customers. In: Proceedings of the seventh ACM SIGKDD international conference on Knowledge discovery and data mining, pp. 57–66. ACM (2001)
2. Granovetter, M.: Threshold models of collective behavior. Am. J. Sociol. **83**(6), 1420–1443 (1978)
3. Kempe, D., Kleinberg, J., Tardos, E.: Maximizing the spread of influence through a social network. In: Proceedings of the Ninth ACM SIGKDD International Conference on Knowledge Discovery and Data Mining, pp. 137–146. ACM (2003)
4. Medlock, J., Galvani, A.P.: Optimizing influenza vaccine distribution. Science **325**(5948), 1705–1708 (2009)
5. Valente, T.W., Pitts, S.R.: An appraisal of social network theory and analysis as applied to public health: challenges and opportunities. Annu. Rev. Pub. Health **38**, 103–118 (2017)
6. Babishin, V., Taghipour, S.: Optimal maintenance policy for multicomponent systems with periodic and opportunistic inspections and preventive replacements. Appl. Math. Model. **40**(23), 10480–10505 (2016)
7. Cauchi, N., Macek, K., Abate, A.: Model-based predictive maintenance in building automation systems with user discomfort. Energy **138**, 306–315 (2017)
8. Hua, W., Wang, Z., Wang, H., Zheng, K., Zhou, X.: Short text understanding through lexical-semantic analysis. In: 2015 IEEE 31st International Conference on Data Engineering (ICDE), pp. 495–506. IEEE (2015)

9. Hosseini, S., Unankard, S., Zhou, X., Sadiq, S.: Location oriented phrase detection in microblogs. In: Bhowmick, S.S., Dyreson, C.E., Jensen, C.S., Lee, M.L., Muliantara, A., Thalheim, B. (eds.) DASFAA 2014. LNCS, vol. 8421, pp. 495–509. Springer, Cham (2014). https://doi.org/10.1007/978-3-319-05810-8_33

10. Peng, S., Wang, G., Zhou, Y., Wan, C., Wang, C., Yu, S.: An immunization framework for social networks through big data based influence modeling. IEEE Trans. Dependable Secure Comput (2017)

11. Gomez Rodriguez, M., Leskovec, J., Krause, A.: Inferring networks of diffusion and influence. In: Proceedings of the 16th ACM SIGKDD International Conference on Knowledge Discovery and Data Mining, pp. 1019–1028. ACM (2010)

12. Goyal, A., Bonchi, F., Lakshmanan, L.V.: Learning influence probabilities in social networks. In: Proceedings of the Third ACM International Conference on Web Search and Data Mining, pp. 241–250. ACM (2010)

13. Hosseini, S., Yin, H., Zhang, M., Zhou, X., Sadiq, S.: Jointly modeling heterogeneous temporal properties in location recommendation. In: Candan, S., Chen, L., Pedersen, T.B., Chang, L., Hua, W. (eds.) DASFAA 2017. LNCS, vol. 10177, pp. 490–506. Springer, Cham (2017). https://doi.org/10.1007/978-3-319-55753-3_31

14. Mathioudakis, M., Bonchi, F., Castillo, C., Gionis, A., Ukkonen, A.: Sparsification of influence networks. In: Proceedings of the 17th ACM SIGKDD International Conference on Knowledge Discovery and Data Mining, pp. 529–537. ACM (2011)

15. Zhang, Y., Adiga, A., Saha, S., Vullikanti, A., Prakash, B.A.: Near-optimal algorithms for controlling propagation at group scale on networks. IEEE Trans. Knowl. Data Eng. 28(12), 3339–3352 (2016)

16. Prakash, B.A., Chakrabarti, D., Valler, N.C., Faloutsos, M., Faloutsos, C.: Threshold conditions for arbitrary cascade models on arbitrary networks. Knowl. Inf. Syst. 33(3), 549–575 (2012)

17. Ghasemiesfeh, G., Ebrahimi, R., Gao, J.: Complex contagion and the weakness of long ties in social networks: revisited. In: Proceedings of the fourteenth ACM conference on Electronic Commerce, pp. 507–524. ACM (2013)

18. Khalil, E.B., Dilkina, B., Song, L.: Scalable diffusion-aware optimization of network topology. In: Proceedings of the 20th ACM SIGKDD International Conference on Knowledge Discovery and Data Mining, pp. 1226–1235. ACM (2014)

19. Prakash, B.A., Beutel, A., Rosenfeld, R., Faloutsos, C.: Winner takes all: competing viruses or ideas on fair-play networks. In: Proceedings of the 21st International Conference on World Wide Web, pp. 1037–1046. ACM (2012)

20. Ganesh, A., Massouli, L., Towsley, D.: The effect of network topology on the spread of epidemics. In: Proceedings of the IEEE INFOCOM 2005, 24th Annual Joint Conference of the IEEE Computer and Communications Societies, vol. 2, pp. 1455–1466. IEEE (2005)

21. Prakash, B.A., Adamic, L., Iwashyna, T., Tong, H., Faloutsos, C.: Fractional immunization in networks. In: Proceedings of the 2013 SIAM International Conference on Data Mining, SIAM 2013, pp. 659–667 (2013)

22. Roth, D.Z., Henry, B.: Social distancing as a pandemic influenza prevention measure: evidence review. National Collaborating Centre for Infectious Diseases (2011)

23. Shim, E.: Optimal strategies of social distancing and vaccination against seasonal influenza. Math. Biosci. Eng. 10, 1615–1634 (2013)

24. Cohen, R., Havlin, S., Ben-Avraham, D.: Efficient immunization strategies for computer networks and populations. Phys. Rev. Lett. 91(24), 247901 (2003)

25. Saha, S., Adiga, A., Prakash, B.A., Vullikanti, A.K.S.: Approximation algorithms for reducing the spectral radius to control epidemic spread. In: Proceedings of the 2015 SIAM International Conference on Data Mining, SIAM 2015, pp. 568–576 (2015)
26. Zhang, Y., Prakash, B.A.: Dava: distributing vaccines over networks under prior information. In: Proceedings of the 2014 SIAM International Conference on Data Mining, SIAM 2014, pp. 46–54 (2014)
27. Aspnes, J., Chang, K., Yampolskiy, A.: Inoculation strategies for victims of viruses and the sum-of-squares partition problem. In: Proceedings of the Sixteenth Annual ACM-SIAM Symposium on Discrete Algorithms. Society for Industrial and Applied Mathematics, pp. 43–52 (2005)
28. Chen, C., Tong, H., Prakash, B.A., Tsourakakis, C.E., Eliassi-Rad, T., Faloutsos, C., Chau, D.H.: Node immunization on large graphs: theory and algorithms. IEEE Trans. Knowl. Data Eng. **28**(1), 113–126 (2016)

Social Network Analytics

Sample Location Selection for Efficient Distance-Aware Influence Maximization in Geo-Social Networks

Ming Zhong[1]([⊠]), Qian Zeng[1], Yuanyuan Zhu[1], Jianxin Li[2], and Tieyun Qian[1]

[1] School of Computer, Wuhan University, Wuhan 430072, China
{clock,wennie,yyzhu,tyqian}@whu.edu.cn
[2] School of Computer Science and Software Engineering,
University of Western Australia, Crawley, WA 6009, Australia
jianxin.li@uwa.edu.au

Abstract. In geo-social networks, the distances of users to a location play an important role in populating the business or campaign at the location. Thereby, the problem of Distance-Aware Influence Maximization (DAIM) has been investigated recently. The efficiency of DAIM computation heavily relies on the sample location selection, because the online seeding performance is sensitive to the distance between sample location and promoted location, and the offline precomputation performance is sensitive to the number of samples. However, there is no work to fully study the problem of sample location selection w.r.t. DAIM in geo-social networks. To do this, we first formalize the problem under a reasonable assumption that a promoted location always adheres to the distribution of users. Then, we propose an efficient location sampling approach based on the heuristic anchor point selection and facility allocation techniques. Our experimental results on two real datasets demonstrate that our approach can improve the online and offline efficiency of DAIM approach like [9] by orders of magnitude.

1 Introduction

Motivation. The widely-used geo-position enabled devices (e.g., mobile phone, tablets, laptops, etc.) and services (e.g., geolocation, geocoding, geotagging, etc.) allow social networks to connect users with local places and events that match their interests. For example, there are currently a lot of popular geo-social network applications like Yelp, Gowalla, Facebook Places and Foursquare. Due to the obvious implication, many researches turn to focus on taking location information into account in the influence maximization problem of geo-social networks. Different from the traditional influence maximization, a typical scenario of influence maximization in geo-social networks is to promote a specific location like a newly opened restaurant or an upcoming sale activity, which is called

The original version of this chapter was revised: the acknowledgement section was updated. The correction to this chapter is available at https://doi.org/10.1007/978-3-319-91452-7_61

© Springer International Publishing AG, part of Springer Nature 2018
J. Pei et al. (Eds.): DASFAA 2018, LNCS 10827, pp. 355–371, 2018.
https://doi.org/10.1007/978-3-319-91452-7_24

query location. In that case, the users near the query location are more valuable to be influenced, because they are more likely to visit the location.

There are two typical problem definitions for the above scenario. The first one is called location-aware influence maximization (LAIM) [8]. The LAIM problem is to maximize the influence to only the users in a given query region, which is a rectangle containing the query location. As a shortcoming of LAIM problem, how to select an appropriate query region for a given query location is unclear. If the query region is too large, most users influenced by the selected seeds may distribute near the boundary of region, thereby being far away from the exact query location. If the query region is too small, many potential users near the query location but outside of the region will be neglected. To overcome the shortcoming of LAIM, the second one called distance-aware influence maximization (DAIM) [9] is proposed. For the DAIM problem, each user has a weight that is determined by the distance between it and the query location no matter whether it is in a query region, and the influence spread to users is adjusted according to their weight.

Typically, to address the DAIM problem, the existing approaches select a set of sample locations in the 2D space where the users are distributed, and precompute the influence spread with respect to the sample locations, which can be leveraged by the online seeding algorithms. Note that, the shorter the distance between sample locations and the given query location, the better the performance of online seeding algorithms. Moreover, the precomputation is very time-consuming for a sample location. Thus, given a budget of location sampling, we hope to minimize the objective distance between any possible query location and its nearest selected sample location.

However, the existing DAIM approaches focus on the seeding algorithms and only use naive sampling like random sampling [9] or equal cell sampling [10]. The naive sampling needs to sample a large number of locations to achieve a promising objective distance. Thus, it is unlikely to achieve a good online seeding performance without spending heavy precomputation overhead while using such naive sampling. Let us consider the following example.

Example 1. Figure 1(a) shows the geographical distribution of users in Brightkite, a real-world geo-social network. We can see that most users live in a few urban areas[1]. The naive sampling like equal cell sampling ignores this fact and try to reach an arbitrary point in the space within the minimum distance, as shown in Fig. 1(b). However, it is nonsense to promote a place that is far away from the users, since the users will hardly visit the place under the settings of DAIM. Instead, the possible query location in reality should adhere to the users, namely, be in a *"query zone"* around the users. Figure 1(c) shows an example of query zone comprised of circles centered at each user with an identical radius r. Thereby, any query location in the query zone is no farther from at least a user than r. Then, we can use more delicate sampling method to reduce the number of sample locations that is necessary for achieving the same objective distance.

---

[1] The 2D space in this example is the surface of earth. In reality, we only consider a city or even smaller district for location promotion. However, the situation of sparse user distribution remains the same.

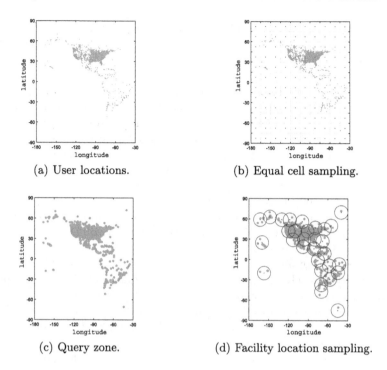

(a) User locations.   (b) Equal cell sampling.

(c) Query zone.   (d) Facility location sampling.

Fig. 1. A motivation example on a geo-social network named Brightkite. (Color figure online)

As shown in Fig. 1(d), the query zone can be covered by the red circles centered at only a few samples, and the radius of circles is equal to the half length of diagonal line of equal cells in Fig. 1(b), namely, both sampling methods have the same objective distance.

Therefore, we focus on the selection of sample locations in this paper. Our work is based on an important observation. That is, the users of real-world geo-social networks are usually distributed sparsely in a 2D space. It means, the real query location must not be an arbitrary point in the space, and should be close enough to some users. Otherwise, given a query location that has no user nearby, trying to maximize the influence to users with very low weights are actually meaningless. Consequently, we reasonably assume that the potential query location is not farther than a specific distance from the users. Under this assumption, we try to find a set of sample locations such that the maximum distance between any qualified query location and its nearest sample location is minimized. As a result, a DAIM approach can generally improve the online seeding performance with less offline precomputation overhead by using our approach for selecting sample locations.

Our Contributions. To address the above problem, we propose a novel sample location selection approach. Due to the NP-hardness of this problem, our approach deals with it heuristically as follows. Firstly, we select a set of *anchor points* from the query zone in various ways. Then, given a set of anchor points, we address an l-center problem, one subtype of classic facility location problem, which is to find l points in the space that can reach any anchor point with the minimum distance. Since the l-center problem is still NP-hard, we offer a heuristic algorithm to return l points as the final sample locations. Moreover, we estimate the upper bound of objective distance between a possible query location in the query zone and the selected sample locations, according to the anchor point selection strategy and the result of l-center problem.

Our contributions are generalized as follows.

- We formalize a novel and interesting sample location selection problem for distance-aware influence maximization in geo-social networks. In this problem, the query location must be in a particular query zone but not the whole 2D space under a reasonable assumption.
- We address the problem by simplifying it to an l-center problem for a set of anchor points selected from a given query zone. For that, we propose a flexible strategy for selecting anchor points that can improve the tightness of objective distance bound by investing more sampling time, and develop an efficient heuristic algorithm for dealing with l-center problem. Then, we can derive a safe and tight upper bound of the objective distance.
- We perform comprehensive experiments on two real-world datasets. Compared with [9], our approach can reduce the precomputation overhead significantly, and meanwhile, improve the efficiency of online seeding significantly.

The rest of this paper is organized as follows. We review the related work in Sect. 2. The formalized problem definition is given in Sect. 3. We present the sample location selection approach in Sect. 4. The experiment results are demonstrated in Sect. 5. Lastly, we conclude our work in Sect. 6.

2   Related Work

Influence Maximization in Social Network. Influence maximization problem is first defined by Kempe et al. [1], the authors also define the independent cascade model and the linear threshold model and prove the hardness of the problem in the paper. Since then, there are a large number of literatures on influence maximization, like [2–5] etc. In [2], the authors propose the CELF algorithm, which exploits the submodular property to significantly boost the traditional greedy approach. Chen et al. [3] propose the PMIA approach, the influence is considered to propagate only through the maximum influence path between users. While Cohen et al. [5] propose a bottom-k sketch based approach to reduce the cost influence estimation. The materialized sketch can be used as an oracle to evaluate the influence of any subset users.

Influence Maximization in Geo-Social Network. With the appearance of geo-position enabled devices and service, researchers begin to pay attention to the impact of geographical location on influence maximization. Zhu et al. attempt to measure the influence between users by considering social relation and location information in [6], while Cai et al. propose a novel network model and an influence propagation model in [7], they think the influence propagation should conduct both in online social networks and physical world. Li et al. [8] attempt to maximize the influence spread in a query region. However, it is non-trivial to determine an appropriate query region when conducting a location-aware promotion. The work which is most related to ours is by Wang et al. in [9,10], they propose the MIA-DA and RIS-DA approaches. The MIA-DA approach gives a priority based algorithm which compromises three pruning rules and a novel index structure. The RIS-DA comes up with an unbiased estimator for the distance-aware influence maximization, both of them need to estimate the necessary size of network samples for any potential query, but such process is very time-consuming.

Facility Location. Research in location theory formally started in 1909 by Weber [11], known as the father of modern location theory. He studies the problem of locating a single warehouse to minimize the total travel distance between the warehouse and a set of customers. Since then, many researchers have observed this problem in different areas, and there are some surveys about techniques for facility location problem, like [12,13]. Elzinga and Hearn give a geometric algorithm to solve the 1-center problem with Euclidean distances, and prove the correctness of the algorithm in [14]. Drezner and Wesolowsky [15] discusses the problem of locating a new facility among n given demand points by taking the l_p-norm distance into consideration, and proposes two heuristic and an optimal algorithms to solve the problem for a given l in time polynomial in n in [16]. Then Callaghan et al. [17] attempt to speed up the optimal method of Drezner in [16] by introducing neighbourhood reduction schemes and embedding an CPLEX policy.

Compared to these work, our approach focuses on combine facility allocation techniques into sample location selection, so that the objective distance derived by our approach can be much shorter. Thus, our approach can reduce the pre-computation overhead, and meanwhile, improve the efficiency of online seeding significantly.

3   Preliminary and Problem Definition

In this section, we first introduce the definition of DAIM problem and analyze the existing DAIM approaches of sample location selection, then give a formal definition of the problem proposed by us.

3.1   Distance-Aware Influence Maximization

We consider a geo-social network as a directed graph $G = (V, E)$, where V represents a set of users and $E = V \times V$ represents the relationships between users.

Each user $v \in V$ has a geographical location (x, y), where x and y represent the latitude and longitude respectively. We denote by $I(S, v)$ the probability that a node set $S \subseteq V$ can activate v under a specific propagation model. The traditional influence maximization problem is to find S with $|S| = k$ that maximizes $\sum_{v \in V} I(S, v)$. However, influence maximization in geo-social networks normally considers the promotion of a query location (like a restaurant). Intuitively, the users near the location are more likely to visit the location. We denote by $w(v, q)$ the weight of a user v with respect to a location q, and the weight depends on the distance between v and q. Thus, the definition of distance-aware influence maximization (DAIM) is given as follows.

Definition 1 *(Distance-Aware Influence Maximization). Given a geo-social network $G = (V, E)$, a query location q and a positive integer k, the problem of distance-aware influence maximization is to find a set S^* of k nodes in G which has the largest distance-aware influence spread, i.e.,*

$$S^* = \arg\max_{S \subseteq V}\{I_q(S) \| |S| = k\} \tag{1}$$

where $I_q(S) = \sum_{v \in V} I(S, v)w(v, q)$ is the distance-aware influence propagation of a node set S.

To address the DAIM problem, Wang et al. [9,10] propose two approaches, namely, MIA-DA and RIS-DA under the independent cascade model. MIA-DA extends the maximum influence arborescence model, and can achieve an approximation ratio of $1 - 1/e$. RIS-DA extends the reverse influence sampling model, and can achieve an approximate ratio of $1 - 1/e - \epsilon$ with at least $1 - \delta$ probability. According to the comparison in [9], RIS-DA is more precise but less efficient than MIA-DA.

Such DAIM approaches need to precompute the influence spread with respect to some sample locations. Then, based on the precomputed influence spread, they can derive the bounds of influence spread for any query location by investigating the relationship between the query location and the sample locations. Since the query location could be any point in the 2D space, they select sample locations distributed uniformly over the space. For example, MIA-DA partitions the space in to a number of equal cells, and selects the center of each cell as samples. While, RIS-DA selects sample locations randomly, and then partitions the space into Voronoi cells based on the set of samples. Therefore, there is surely a nearby sample location for an arbitrary promoted location, no matter which cell it is in.

3.2   Problem Definition

The above sample location selection methods result in heavy precomputation overheads and large index spaces in order to guarantee a good estimation of influence bounds. Let the number of user points be n, the number of seeds be k, the number of sample locations be l. The time complexity of precomputation for MIA-DA is $O(n^2)$ and for RIS-DA is $O(l^2 k^2 n \log n)$. Moreover, to derive tight bounds, the distance between the query location and its nearest sample location needs to be short enough. Since the sample locations are distributed uniformly over the space, the number of sample locations increases dramatically with the decrease of distance between sample locations and potential query locations.

In this paper, we argue that the query location in DAIM problem should consider the spatial distribution of users and should not be an arbitrary point in the 2D space. The possible query locations always follow the distribution of users in reality. For example, when a company needs to advertise for their products through the social network, they are more likely to select a query location which is in a densely populated location, but not far away from the crowd. Otherwise, there are no potential consumers with respect to the distance between them to the query location, and thereby addressing the DAIM problem is meaningless. So we have the following reasonable assumption of the query location distribution.

Assumption 1. *The given query location should follow the spatial distribution of users. Formally, given a positive real number r, there exists at least a user $v \in V$ for a query location q such that $dis(q, v) \leqslant r$.*

Intuitively, for a user, the area of activities is a circle centered at its location with a radius r, which is called *user circle*. Thus, only the query location in this circle can attract the user. All user circles compose a query zone Q, as shown in Fig. 1(c). We denote by $q \in Q$ that a point q located in the query zone Q. Under this assumption, the problem to be addressed in this paper can be formalized as follows.

Problem 1. Given a geo-social network $G = (V, E)$, a query zone Q defined by the locations of V and the radius r of user activities, and a location sampling budget l, find a set SL of l sample locations in the 2D space, such that the objective distance $D(SL, Q)$ is minimized. The objective distance is the maximum distance between any query location in Q and its nearest sample location in SL, namely, $D(SL, Q) = \max_{q \in Q} \min_{s \in SL} dis(q, s)$. For convenience, we denote by d_o the optimal objective distance.

For example, as shown in Fig. 2, there are two users v_1 and v_2, and the yellow circles comprise the corresponding query zone. The sample location s_1 is the middle point of line segment $v_1 v_2$. Thus, the farthest query location to s_1 is q_1 and q_3, and we have $D(\{s_1\}, Q) = \max_{q \in Q} dis(q, s_1) = dis(q_1, s_1)$ or $dis(q_3, s_1)$. For any other sample location s_2, suppose s_2 is closer to v_2, we have $D(\{s_2\}, Q) = dis(q_2, s_2) = dis(v_1, s_2) + r > dis(v_1, s_1) + r = dis(q_1, s_1)$. Suppose we only select one sample location, namely, $l = 1$. It is obviously that the minimum objective distance $d_o = dis(q_1, s_1)$, so that the optimal set of sample location is $\{s_1\}$. Given this set of sample locations, no matter which point in the query zone needs to be promoted, we can find a sample location within the optimal distance d_o.

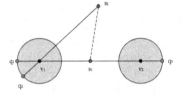

Fig. 2. A simple example of sample location selection problem. (Color figure online)

Problem Hardness. We briefly discuss the hardness of our problem as follows. Consider a query zone as an infinite set of points. If we only keep a fixed number of points in a query zone, the Problem 1 is reduced to the classic l-center problem [13]. It has been proved that the exact solution of l-center problem is NP-hard [18]. Approximation to the problem is also NP-hard when the error is small. Due to the hardness of l-center problem, the Problem 1 is also NP-hard where l is an infinite number.

4   Sample Location Selection

In this section, we firstly present a heuristic methodology to select sample locations for a given query zone in a 2D space, and develop efficient algorithms based on the studies of facility location problems.

4.1   Methodology

Due to the hardness of sample location selection problem defined above, we propose a heuristic approach to address it. The main idea is that, we select a set of discrete *anchor points* from the query zone, and find a given number of sample locations in the 2D space, such that each anchor point can reach the nearest sample location within the minimum distance d_a. Let d_z be the maximum distance between any point in the query zone and its nearest selected sample location. Although there could be some areas of query zone that can not be reached by selected sample locations within the distance d_a, namely, $d_a \leqslant d_z$, we can guarantee that $d_z - d_a$ is no more than $f(r)$ by selecting the anchor points with a particular strategy, where the function $f : r \mapsto (0, r]$ is determined by the strategy. It is certainly that any point in the query zone can reach its nearest selected sample location within a distance $d_a + f(r)$. Thus, we safely use the upper bound $d_a + f(r)$ of d_z (and of course d_o) as the final objective distance.

   In the followings, we introduce the strategy of selecting anchor points and the heuristics of selecting sample locations for a given set of anchor points.

Anchor Point Selection Strategy. We propose two strategies of anchor point selection, the baseline and the improved. The improved strategy can achieve a tighter bound of d_z than the baseline strategy.

The baseline strategy is to select the user points as anchor points. Let us consider a set of *result circles* with an identical radius $d_a + f(r)$ whose centers are the sample locations selected by our approach. The result circles of baseline strategy can cover the whole query zone when $f(r) = r$.

Lemma 1. *Given a set of sample locations selected by the baseline strategy, for any query location in the query zone, its distance to the nearest sample location is no more than $d_a + r$.*

Proof. We denote by u a user point, s the nearest sample location to u selected by the baseline strategy, and q a query location in the user circle of u, as shown in Fig. 3(a). We have $dis(s, q) \leqslant dis(s, u) + dis(u, q)$ according to the triangle inequality. Since $dis(s, u) \leqslant d_a$ and $dis(u, q) = r$, $dis(s, q) \leqslant d_a + r$. Thus, for any query location in the query zone, there exists at least a sample location like s such that the distance between them is no more than $d_a + r$.

Usually, the value of r is relatively very small, so that $d_a + r$ could be a tight bound of d_z. While, if the value of r is not that small, we can use an improved strategy to get a tighter bound, which selects more anchor points other than the user points. For each user circle, we divide its circumference into a number of equal arcs, and use the end points of these arcs as the additional anchor points. Here we only discuss about dividing the circumference into three equal arcs, as shown in Fig. 3(b). We have four anchor points, the black user point and the three green points on the circumference. In this case, the result circles of improved strategy can cover the whole query zone when $f(r) = \frac{r(2d_a+r)}{3d_a+r}$.

Lemma 2. *Given a set of sample locations selected by using the users points and three equal points on the circumference of user circles as anchor points, for any query location in the query zone, its distance to the nearest sample location is no more than $d_a + \frac{r(2d_a+r)}{3d_a+r}$.*

Proof. Consider the worst case that makes the value of $f(r)$ maximized (the trivial proof is omitted). As shown in Fig. 3(b), two sample circles are tangent to the query zone, the other one is intersect with the query zone in the original user point, so that the four anchor points are exactly covered by the sample circles. Since some area of the query zone are not reached, then in order to cover the whole query zone, these sample circles with a radius d_a must be extended to the result circles with a radius $d_a + f(r)$ and intersect in a same point, like the orange point shows. In Fig. 3(b), the blue points and the black point are all remove a same distance d and intersect in the orange point, consider the triangle, $dis(A, B) = r + d_a$, $dis(A, C) = d + d_a$, $dis(B, C) = d$, and $\theta = 60°$. According to the cosine theorem, $\cos 60° = \frac{d^2+(r+d_a)^2-(d+d_a)^2}{2d(d_a+r)}$. Then we can attain $d = \frac{r(2d_a+r)}{3d_a+r}$, due to $\frac{2d_a+r}{3d_a+r} < 1$, so $d < r$.

When r is varying in a certain region, the returned d_a of improved strategy is approximately the same with the returned d_a of baseline strategy. Since $f(r) = \frac{r(2d_a+r)}{3d_a+r}$ of improved strategy is smaller than $f(r) = r$ of baseline strategy, the

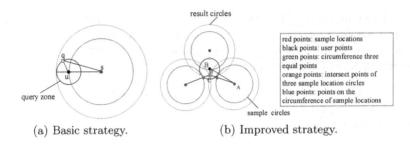

result circles

red points: sample locations
black points: user points
green points: circumference three
equal points
orange points: intersect points of
three sample location circles
blue points: points on the
circumference of sample locations

query zone

sample circles

(a) Basic strategy.   (b) Improved strategy.

Fig. 3. An example of anchor point selection. (Color figure online)

upper bound $d_a + f(r)$ is tighter in the improved strategy compared to the baseline strategy. Note that, selecting more anchor points will make the bound tighter, however the cost of sample location selection algorithm will increases, the detailed comparison of objective distance and efficiency will be presented in Sect. 5.

Sample Location Selection Heuristics. Given a set N of anchor points, we aim to find a set SL of l sample locations in the space to minimize the maximum distance between any anchor point and its nearest sample location, namely, $\max_{p \in N} \min_{s \in SL} dis(p, s)$, which is called l-center problem. In a nutshell, the heuristics of our solution to l-center problem is as follows. Let $\alpha = \{I_1, I_2, ..., I_l\}$ be a l-partition of N, namely, $\cup_{i=1}^{l} I_i = N$, where $I_i \subset N$. Given an optimal l-partition α, we find a center point for each $I_i \in \alpha$ by addressing a 1-center problem for I_i, and select the l center points as the final sample locations.

To get the optimal l-partition, we need to define an objective function. Let $F(I)$ be the optimal objective distance of 1-center problem for I. We have

$$F(I) = \min_{x \in X} \max_{p \in I} dis(p, x) \qquad (2)$$

where X is the set of all points in the space. For convenience, let $B(I)$ be the optimal point of 1-center problem for I. Then, let $F(\alpha)$ be the objective function for an l-partition. Thus, we have

$$F(\alpha) = \max_{i=1}^{l} F(I_i) \qquad (3)$$

Obviously, the optimal l-partition with respect to $F(\cdot)$ leads to the sample locations with the minimum objective distance. In particular, for $I_i \in \alpha$, if $F(I_i) = F(\alpha)$, then I_i is called *extremal subset*.

4.2   Algorithm

The pseudo code of sample location selection algorithm is given in Algorithm 1. Initially, we choose l anchor points as centers (line 1), and assign each other

---

Algorithm 1. sample location selection

Input: a set N of anchor points and a positive integer l
Output: l sample locations and F_α
1: choose l center points out of N;
2: assign each other anchor point to the subset of its nearest center by using Voronoi diagram;
3: **repeat**
4:     $i \leftarrow$ a point from $T(I_i)$, where I_i is the extremal set of α;
5:     choose a subset I_j other than the extremal subset I_i;
6:     **if** $F(I_j \cup \{i\}) < F(\alpha)$ **then**
7:         $I_j \leftarrow I_j \cup \{i\}$, $I_i \leftarrow I_i - \{i\}$;
8:     **end if**
9: **until** the value of $F(\alpha)$ does not change anymore
10: **return** the optimal point $B(I_i)$ of each subset $I_i \in \alpha$ and F_α;

Algorithm 2. 1-center problem algorithm

Input: a set I of anchor points
Output: the optimal center point $B(I)$, and $F(I)$
1: choose the initial center point (x_0, y_0), where $x_0 = \sum_{p \in I} x_p/|I|$, $y_0 = \sum_{p \in I} y_p/|I|$;
2: $I' \leftarrow$ the three points that are farthest from $(x^{(0)}, y^{(0)})$;
3: **while** there exists a point in $I - I'$ such that the distance between it and $B(I')$ is larger than $F(I')$ **do**
4:     $p' \leftarrow$ the farthest point from $B(I')$;
5:     $I' \leftarrow T(I' \cup \{p'\})$;
6: **end while**
7: **return** $B(I')$ and $F(I')$;

---

anchor point to the subset of its nearest center by leveraging the principle of Voronoi diagram (line 2). Then we refine the partition α of anchor points iteratively until the value of $F(\alpha)$ cannot be decreased (line 3–9). At each iteration, we try to move a point from a subset to another to get a better value of $F(\alpha)$. Straightforwardly, we can reallocate each anchor point to another subset, and choose the best plan. However, there are $N(l-1)$ possible plans, and not all of them can decrease the value of $F(\alpha)$. Thus, we give an efficient repartition method as follows. According to the study of minimum covering circle problem in [14], the value of $F(I)$ can be determined by no more than three points in I, the set of which is denoted by $T(I)$, namely, $F(I) = F(T(I))$. Given an extremal set I_i of α, we have $F(\alpha) = F(T(I_i))$, so that the value of $F(\alpha)$ will be changed if we remove a point $i \in T(I_i)$ from I_i. As a result, we only consider to reallocate the anchor points in $T(I_i)$ to achieve a better value of F_α. Lastly, the center points of the optimal partition of N is returned as the sample locations.

Algorithm 2 gives a solution to 1-center problem and the complexity is $O(n)$. Initially, for a subset I of anchor points, we choose a point (x_0, y_0) in the space as the center of I (line 1). Since $F(I) = F(T(I))$, we choose the three farthest points from (x_0, y_0) to compose a set I' as the possible $T(I)$ (line 2). Then we begin to update I' iteratively unless there is no point in $I - I'$ outside of the circle determined by I', namely, $I' = T(I)$ (line 3–6). At each iteration, we choose the farthest point p' from $B(I')$, and set the new I' as $T(I' \cup \{p'\})$. Lastly, we return $B(I')$ as the optimal center of I since $F(I) = F(T(I)) = F(I')$.

To get the center $B(I')$ of I' that has exact three points, the *three-point problem* is studied in [15]. The idea is that, first check if any two points p_1 and p_2 define the solution. If so, let $x = (x_1 + x_2)/2$ and $y = (y_1 + y_2)/2$, the distance between the other point p_3 and (x, y) is no more than $dis(p_1, p_2)/2$, and thus (x, y) is the center of these three points. Otherwise, we find a point inside the triangle of these three points as the center $B(I')$, which possesses equal distances to the three vertices of the triangle.

5   Experiments

Our experiments are conducted on a PC with Intel Core 3.2 GHz CPU and 16 GB memory. The algorithms are implemented in C++ with TDM-GCC 4.9.2 (Table 1).

Table 1. Experimental datasets.

Dataset	Node number	Edge number	Average in-degree	Average out-degree
Brightkite	58K	428K	7	7
Gowalla	100K	1.9M	13	13

5.1   Setup

Algorithms. There are four algorithms to be compared in our experiments. (1) RSQ extends RS to filter out the sample locations outside of the query zone, while RS is the original random sampling in [9]. The distance between query location and sample location is calculated by using Voronoi diagram. (2) K-means simply clusters the user points to a given number of groups with respect to distance, and select the cluster center as sample locations. (3) FLS is our facility-location-based sampling method with the baseline anchor point selection strategy. (4) FLS-3 is our facility location sampling method with an improved anchor point selection strategy which adds three points on each query zone circle into the anchor points.

Datasets. In our experiments, we use two real-world geo-social networks where users can share their check-ins. This check-ins represent users' locations, and the datasets are obtained from http://snap.stanford.edu/data/. Note that there are just 88.6% and 54.4% users have check-ins in the Brightkite and the Gowalla respectively, we need to pre-treat the datasets as follows: since there are a few users who don't have location information in Brightkite, so we randomly generate a location for them according to other users' location distribution. While in Gowalla, almost half of them have no check-ins, so we delete those users who don't have location information, in fact, there are 100K points in Gowalla used by us. The location information of added anchor points are calculated according to the original user points.

Parameters. The probability of edge (u, v) is set as $\frac{1}{N_{in}(v)}$, where $N_{in}(v)$ represents the number of incoming neighbours of v. The independent cascade model is used in influence spread, the size of seed set varies from 10 to 50, the interval is 10, and we run 10000 round for each returned seed set, we evaluate the average influence propagation of the returned seed set. The radius of each query zone is set to 10, for each query zone, the entire circumference is divided into 3 equal arcs, and add the end points of these equal arcs into the anchor points, the number of sample locations is set to 500, 1000, 1500 and 2000 relatively.

5.2 Effectiveness Analysis

We evaluate the effectiveness of sample location selection algorithms by four metrics. The first one is of course the objective distance, the other two, namely, the necessary size of network samples (simplified as sample size) and the response time of seed selection (simplified as seeding time), and the last one is the time consumption of offline index construction (simplified as indexing time), are used to demonstrate the impact of location sampling on a specific DAIM approach. Once we get the sample locations, we run the offline index construction and then the online seeding algorithm of RIS-DA [9]. RIS-DA needs to generate a set of network samples, the number of which is determined by the objective distance, in order to guarantee the $1 - 1/e - \epsilon$ approximate ratio of influence maximization. During online seeding, RIS-DA needs to deal with each network sample. Thus, the response time of seed selection also depends on the objective distance finally. Since the objective distance can be reduced by increasing the number of sample locations, but the efficiency of offline index construction will be influenced if the number of sample locations gets larger, so the indexing time indirectly depends on the objective distance.

As shown in Fig. 4(a) and (b), our facility-location-based sampling approach can achieve the best objective distance on both datasets with varying numbers of selected samples. Note that, the objective distance of RSQ shown in Fig. 4(a) and (b) is the average of results of 100 repeated tests, thereby avoiding the bias of random sampling. We can see that, the objective distance of naive sampling approaches such as RSQ is much worse than K-means, FLS and FLS-3. Although K-means can reduce the objective distance significantly, it is still not as effective as FLS and FLS-3. Compared with FLS, FLS-3 is even more effective when the number of selected samples is small, due to the improved anchor point selection strategy. Moreover, the objective distance decreases with the increase of the number of selected samples generally for all algorithms. While, the decrease of FLS and FLS-3 is not significant. Therefore, we can actually select a relatively

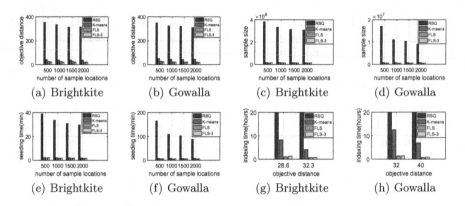

Fig. 4. Effectiveness evaluation.

small number of sample locations to get a good objective distance by using our facility-location-based sampling approach.

Due to the decrease of objective distance, FLS and FLS-3 improve the online performance of RIS-DA dramatically. As shown in Fig. 4(c) and (d), the necessary size of network samples of FLS and FLS-3 is orders of magnitude less than naive sampling. Further, the response time of seed selection is shown in Fig. 4(e) and (f). With the decrease of sample size, the response time is reduced to only a few minutes, so that we can quickly find a set of seed users for promotion of a given query location. In conclusion, given a fixed budget of location sampling, FLS and FLS-3 are quite effective to improve the online performance of DAIM approaches by reducing the objective distance.

We keep the objective distance same to evaluate the response time of offline index construction. As shown in Fig. 4(g) and (h), we can find the time consumption of K-means is longer than FLS and FLS-3, this is because for the same objective distance, the number of sample locations in K-means is greater than that in FLS and FLS-3. For example, to achieve the objective distance of 28.6, there are 1000 sample locations needed in K-means, while just 80 in FLS and 100 in FLS-3. Note that, the number of sample locations of RSQ will be greater than 2000 in order to achieve the same objective distance, then the response time will be several times larger than K-means, so we set a time limit of 20 h, once the running time goes beyond the time limit, the algorithm will be stopped. Due to the efficiency of offline index construction is sensitive to the number of sample locations, the offline time-consuming of RSQ and K-means will be enormous to achieve the same online performance compared with FLS and FLS-3. We can conclude that FLS and FLS-3 can well balance the overheads of offline precomputation and online performance.

5.3   Efficiency Analysis

We evaluate the efficiency of sample location selection algorithms by focusing on the response time of selecting certain number of sample locations. As shown

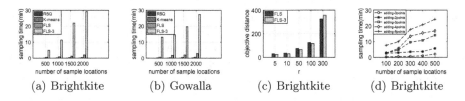

(a) Brightkite   (b) Gowalla   (c) Brightkite   (d) Brightkite

Fig. 5. Efficiency and anchor point selection evaluation.

in Fig. 5(a) and (b), RSQ runs the fastest among all the algorithms, K-means always outperforms FLS and FLS-3, and FLS-3 performs the worst. For RSQ, since it is just randomly selecting sample locations in the 2D space and filtering the sample locations outside of the query zone, there is little time consumption during the process. Compared with K-means, FLS and FLS-3 need to refine the partition and recalculate the center locations until the value of objective function is minimized. Since FLS and FLS-3 need to call the 1-center problem algorithm for all sample locations even there is a very tiny variety of the value of objective function in each iteration, such process will directly influence the response time of FLS and FLS-3. Furthermore, the number of anchor points in FLS-3 is three times more than FLS, when the number of sample locations is fixed, the number of points in each subset of α in FLS-3 will be larger than FLS, then the response time of 1-center problem algorithm in FLS-3 will be larger. Thus, we can find that the improved anchor point selection strategy may improve the objective distance, but the efficiency is decreased.

5.4 Anchor Point Selection Strategy Analysis

We evaluate the effectiveness of anchor point selection strategy by the objective distance. As shown in Fig. 5(c), we focus on the objective distance of baseline strategy and improved strategy when r is varying. Note that, considering the problem of efficiency, we choose 200 sample locations here to evaluate the objective distance. We can find the objective distance of FLS-3 is smaller than FLS when r is under a certain value like 100 in Fig. 5(c). When r is too large, the value of d_a in FLS-3 will be several times greater than FLS, and the value of $f(r)$ is increasing when r and d_a both increase. Thus, the objective distance of FLS-3 will be larger than FLS if r is too large. However, considering the users' regular activity area won't be too large in reality, so the objective distance of the improved strategy is better than the baseline strategy. As a result, we can achieve a tight upper bound by adding extra anchor points.

We evaluate the efficiency of anchor point selection strategy by comparing the response time of sampling when adding different number of anchor points. As shown in Fig. 5(d), except for the baseline strategy and improved strategy, we add extra 2, 4 and 6 anchor points. From the results, we can find the time consumption of each strategy is increasing when the number of sample locations increases, this is because the number of calling 1-center problem algorithm is increasing when the number of sample locations increases. For a fixed number of sample locations, since the number of points in each subset of α is increasing when we add more anchor points, the response time of sampling will be increasing. Thus, in order to balance the effectiveness and efficiency, it is important to determine the number of anchor points to add.

6 Conclusion

Sample location selection is crucial for the DAIM problem in geo-social network, but there is no work to fully study such a problem. The previous work mainly selects sample locations by naive methods such as random sampling or equal cell sampling, which can hardly achieve a good objective distance even when a large number of samples are selected. While, the online seeding performance is sensitive to the objective distance, and the precomputation overhead is sensitive to the sample number. In this paper, we propose the conception of query zone and reasonably formulate a novel problem of sample location selection for a given query zone. Due to the hardness of this problem, we solve our problem by selecting some anchor points from the query zone and finding a number of centers of the anchor points as the sample locations. Specifically, we propose a flexible strategy of anchor point selection and develop a heuristic partition refining algorithm to select centers. According to the experimental results, our approach can improve the efficiency of DAIM approach like [9] significantly. Since our approach can achieve a specific objective distance by selecting much less sample locations, we can balance the online performance and precomputation overhead effectively.

Acknowledgement. This paper was supported by National Natural Science Foundation of China under Grant No. 61202036, 61502349 and 61572376 and Natural Science Foundation of Hubei Province under Grant No. 2018CFB616.

References

1. Kempe, D., Kleinberg, J.M., Tardos, E.: Maximizing the spread of influence through a social network. In: SIGKDD, pp. 137–146 (2003)
2. Leskovec, J., Krause, A., Guestrin, C., Faloutsos, C., VanBriesen, J., Glance, N.: Cost-effective outbreak detection in networks. In: ACM KDD (2007)
3. Chen, W., Wang, C., Wang, Y.: Scalable influence maximization for prevalent viral marketing in large-scale social networks. In: SIGKDD, pp. 1029–1038 (2010)
4. Cohen, E., Delling, D., Pajor, T., Werneck, R.F.: Sketch-based influence maximization and computation: scaling up with guarantees. In: CIKM, pp. 629–638 (2014)
5. Chen, W., Yuan, Y., Zhang, L.: Scalable influence maximization in social networks under the linear threshold model. In: International Conference on Data Mining, pp. 88–97 (2010)
6. Zhu, W., Peng, W., Chen, L., Zheng, K., Zhou, X.: Modeling user mobility for location promotion in location-based social networks. In: SIGKDD, pp. 1573–1582 (2015)

7. Cai, J.L.Z., Yan, M., Li, Y.: Using crowdsourced data in location-based social networks to explore influence maximization. In: IEEE International Conference on Computer Communications, pp. 1–9 (2016)

8. Li, G., Chen, S., Feng, J., Tan, K., Li, W.: Efficient location-aware influence maximization. In: SIGMOD, pp. 87–98 (2014)

9. Wang, X., Zhang, Y., Zhang, W., Lin, X.: Efficient distance-aware influence maximization in geo-social network. IEEE Trans. Knowl. Data Eng. **29**(3), 599–612 (2017)

10. Wang, X., Zhang, Y., Zhang, W., Lin, X.: Distance-aware influence maximization in geo-social network. In: ICDE, pp. 1–12 (2016)

11. Weber, A.: Über den Standort der Industrien 1. Reine theorie des standordes, Tübingen, Germany, Teil (1909)

12. Arabani, A.B., Farahani, R.Z.: Facility location dynamics: an overview of classifications and applications. Comput. Ind. Eng. **62**(1), 408–420 (2012)

13. Irawan, C.A., Salhi, S.: Aggregation and non aggregation techniques for large facility location problems: a survey. Yugosl. J. Oper. Res. **25**(3), 313–341 (2015)

14. Elzinga, J., Hearn, D.W.: Geometrical solutions for some minimax location problems. Transp. Sci. **6**, 379–394 (1972)

15. Drezner, Z., Wesolowsky, G.O.: Single facility l_p-distance minimax location. SIAM J. Algebr. Discret. Methods **3**, 315–321 (1980)

16. Drezner, Z.: The p-centre problem-heuristic and optimal algorithm. J. Oper. Res. Soc. **35**(8), 741–748 (1984)

17. Callaghan, B., et al.: Speeding up the optimal method of Drezner for the p-centre problem in the plane. Eur. J. Oper. Res. **257**(3), 722–734 (2017)

18. Fowler, R.J., Paterson, M.S., Tanimoto, S.L.: Optimal packing and covering in the plane are NP-complete. Inf. Process. Lett. **12**(3), 133–137 (1981)

Identifying Topical Opinion Leaders in Social Community Question Answering

Tao Zhao[1], Hong Huang[2(✉)], and Xiaoming Fu[1]

[1] Institute of Computer Science, University of Goettingen, Goettingen, Germany
{Tao.Zhao,fu}@cs.uni-goettingen.de
[2] School of Computer Science, Huazhong University of Science and Technology,
Wuhan, China
honghuang@hust.edu.cn

Abstract. Social community question answering (SCQA) sites not only provide regular question answering (QA) service but also form a social network where users can follow each other. Identifying topical opinion leaders who are both expert and influential in SCQA becomes a hot research topic. However, existing works focus on either using knowledge expertise to find experts for improving the quality of answers, or measuring user influence to identify influential ones. In this paper, we propose QALeaderRank, a topical opinion leader identification framework, incorporating both the topic-sensitive influence and the topical knowledge expertise. To measure a user's topic-sensitive influence, we design a novel ranking algorithm that exploits both the social and QA features of SCQA, taking account of the network structure, topical similarity and knowledge authority. Besides, we incorporate three topic-relevant metrics to infer the topical expertise. Extensive experiments along with a user study demonstrate that QALeaderRank outperforms the compared state-of-the-art methods. QALeaderRank can also be used to identify multi-topic opinion leaders.

1 Introduction

Community question answering (CQA) site has become a popular platform for information needs [17], where users ask/answer questions and comment on posts. Compared to regular CQA platforms like Yahoo! Answers and Stack Overflow, an innovative type of social CQA (SCQA) sites has become popular, such as Quora and Zhihu, which provides social network function to connect users. As two most notable SCQA sites, Quora had around 100 million users and Zhihu ("Chinese Quora") had around 17 million users by the end of 2015. In these SCQA sites, users can follow each other to receive information updates from their followees. This built-in social function makes SCQA become a social platform [21]. Meanwhile, users usually publish posts involving diverse topics, resulting in different topic domains. For any specific topic(s), with the question answering (QA) and social functions of SCQA, some users tend to publish a large number of authoritative topic-related posts, which substantially affect other users' opinions, and

J. Pei et al. (Eds.): DASFAA 2018, LNCS 10827, pp. 372–387, 2018.
https://doi.org/10.1007/978-3-319-91452-7_25

even guide public opinion direction. In the light of the original concept of *opinion leader*[1] that is topic-irrelevant, we refer to these topic-relevant leaders as *topical opinion leaders*. They play a significant role in creating topic-related knowledge repositories, maintaining the activeness of the topic community, and even helping to control the development trend of public opinions.

Despite the important role that topical opinion leaders play in SCQA, the challenge of identifying topical opinion leaders is still intractable. According to the characteristics of topical opinion leaders in SCQA, a major challenge is how to find users who have both *strong topic-sensitive influence* and *high topic-relevant knowledge expertise* in given topic(s), as shown in Fig. 1. Most previous related works either focused on the knowledge expertise to find experts for improving the quality of answers in QA sites [18,19,25] (see Zone I+IV) or mainly measured the influence to identify influential users in social networks [3, 14,16,22] (see Zone I+II).

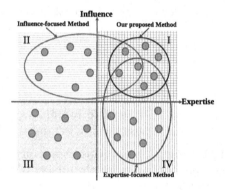

Fig. 1. User identification in terms of influence and expertise

To address this problem, we propose QALeaderRank, a topical opinion leader identification algorithm for SCQA that alleviates these shortcomings by incorporating the *topic-sensitive influence* and the *topic-relevant knowledge expertise*. To measure the true topic-sensitive influence of users, we propose a novel influence ranking algorithm called QARank which exploits both the *social* and *QA* features of SCQA. Two key problems are solved to build QARank: (i) inferring each user's topic interest and knowledge authority from published posts; (ii) confirming the existence of *homophily* in SCQA, which implies that a user follows one another on account of similar topic interest. QARank not only considers the network structure and the topical similarity between users like traditional influence measure methods (e.g., TwitterRank [22]), but also takes the topical knowledge authority into account. In addition, to measure the topical knowledge expertise, we incorporate three topic-relevant metrics that account for knowledge capacity,

---

[1] Opinion leaders give influential comments and opinions, put forward guiding ideas, agitate and guide the public to understand social problems [12].

satisfaction and contribution. Moreover, QALeaderRank can also be utilized to identify multi-topic opinion leaders with the popularity of multi-topic.

In this paper, employing a dataset crawled from a large SCQA site, Zhihu, as the basis of our study, we comprehensively study the QA and social features of SCQA. We also conduct an extensive evaluation for our proposed QALeaderRank with this dataset. Experimental results, along with an online user study, demonstrate that QALeaderRank greatly outperforms the compared state-of-the-art methods.

Our contributions are multi-folded:

- We analyze the social and QA features of SCQA and confirm the existence of *homophily* in the context of SCQA.
- To our knowledge, we are the first to propose an efficient algorithm (QALeaderRank) to tackle the issue of topical opinion leader identification in SCQA.
- To design QALeaderRank, we propose a novel topic-sensitive influence measure algorithm for SCQA, based on the QA and social features. Besides, we incorporate three topic-relevant metrics to measure topical expertise.
- We conduct extensive experiments to evaluate our proposed algorithm, along with an online user study.

2 Related Work

Previous related works can be classified into two main types: expertise-focused and influence-focused. We will discuss them in details as follows:

Expertise-Focused Methods. Most previous works on CQA sites study expert identification for the purpose of improving the quality of answers. For example, Bouguessa et al. [2] identified which experts would be answering open questions based on the number of best answers published by users. Riahi et al. [19] focused on finding experts for a newly posted question through investigating and comparing the suitability and performance of statistical topic models. Since the SCQA sites have gained increasing popularity recently, the need of identifying important users in SCQA sites has started to draw research interests. Song et al. [20] proposed a leading user detection model for Quora, taking into account authority, activity and influence. However, the user influence in the model is measured by its node in-degree in the social network (i.e., the number of followers), which cannot accurately capture the notion of influence in social networks [9,11]. Besides, all the factors in this model are topic-irrelevant.

Influence-Focused Methods. There are also a great number of works that study the issue of opinion leader or influential user identification in social medias. For the bulletin board systems (BBS), Zhai et al. [23] proposed interest-field based algorithms taking into account the network structure and user's interest to identify opinion leaders. In the microblogs, especially Twitter, there are amounts of works on identifying influential users [1,4,8,13,22]. One representative work is TwitterRank algorithm [22], an extension of PageRank [9]. TwitterRank identifies topic-sensitive influential users in Twitter considering both the topical similarity between users and the network structure. In general, most approaches mainly focus on measuring the user influence, which fail to identify topical opinion leaders in SCQA as SCQA users disseminate information by both the following relationship and the QA function.

3   Dataset

Before explaining our proposed algorithm, we begin in this section by describing our dataset and presenting initial analysis results on the QA and social features.

3.1   Dataset Collection

We gathered the Zhihu dataset through web-based crawls from March to June in 2016. We started our user crawls using a set of 10 popular Zhihu users. The crawls follow a BFS pattern through the following links of each user. In total, we crawled 1.41M+ individual users in Zhihu. Each user data contains the following fields: user ID, the lists of the user's followers and followees, the user's answers and questions posted. For each answer/question, we also crawled its topics (the topics of each question are edited by its author) and the number of received votes. As shown in Table 1, the user-based crawls produced 701K+ unique questions and 4.04M+ unique answers.

We crawled all the 160K+ unique topics in Zhihu utilizing the top-down tree-like topic structure provided by Zhihu. In the topic structure, there is only one root topic which has 6 child topics but no parent topic. The other topics except the lowest level topics have at least one parent topic and one child topic. For example, the topic "Fitness" has two parent topics "Sport" and "Health" while it has 31 child topics, such as "Muscle", "Bodybuilding", etc.

Table 1. Description of dataset

#users	#questions	#answers	#topics
1,411,669	701,982	4,047,183	160,664

3.2   Initial Analysis

In order to explore the QA and social features of SCQA, we first present the related statistical analysis of our crawled data. From this analysis, we can find that Zhihu has the similar QA and social features as Quora studied in [21].

Questions and Answers. As shown in Fig. 2a and b, the distribution of the number of questions/answers posted by each user follows power-law distribution. This means that a small portion of users posted a great number of questions/answers while most users posted a few ones. From Fig. 3a, we can also observe that 81% of the users did not ask any question and 72% of users did not publish any answer, which conforms to 80/20 rule.

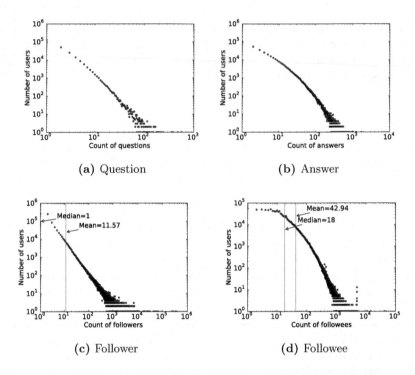

Fig. 2. Power law distribution of QA and following in Zhihu

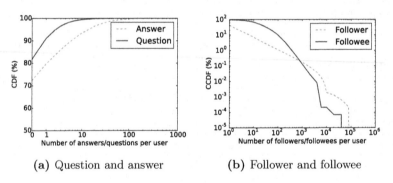

Fig. 3. Distribution of QA and following in Zhihu

Followers and Followees. In Zhihu, users can follow each other, which constructs a Twitter-like directed social network. We analyze the number of users' followers and followees to explore the social feature of Zhihu. Figure 2c and d respectively plots the distribution of the number of followers and followees per user, which follows power-law distribution. The exponential fitting parameter α for the follower count distribution is 1.84 with standard error 0.001, which is close to that of Twitter ($\alpha = 2.28$) [21]. The average numbers of followers and followees per user are around 12 and 43. As shown in Fig. 3b, about 38% of users have no follower and more than 99% of users have followees. This observation implies that Zhihu is a relatively dense social network like Twitter.

4 Topical Opinion Leader Identification in SCQA

Our study focuses on identifying topical opinion leaders, who have both strong topic-sensitive influence and high topic-relevant knowledge expertise in SCQA. Regarding measuring the topic-sensitive influence, we propose QARank algorithm in Sect. 4.1, and later employ three topic-relevant expertise metrics to measure the topical expertise in Sect. 4.2. Based on this, QALeaderRank, a topical opinion leader identification algorithm, can be developed. With the consideration of combining both topic-sensitive influence and topic-relevant knowledge expertise equally, users' ranking scores in topic T ($|T| \geq 1$), denoted as LR_T, can be calculated by $LR_T = Inf_T \times ES_T$. Inf_T denotes user's influence in topic T and ES_T means the topic-related expertise. Thus, the users who have high ranking scores are identified as topical opinion leaders.

4.1 Topic-Sensitive Influence Measure

In this section, before designing QARank, we first conduct topic preprocessing, and then confirm the existence of *homophily* in our dataset.

Topic Preprocessing. The topic preprocessing aims at identifying each user's topic interest. Every post can be related to one or (more typically) many unique topics. Directly leveraging these topics to represent the topic interest of posts/users is very intricate owing to their amount and diversity. To this end, we aggregate these topics into 7 major topics[2] which cover all the topic fields in Zhihu, using the tree-like topic structure of Zhihu. Using this topic aggregation method, each answer/question's topics of each user are transformed to the corresponding major topics in accordance to the topic relationship in the topic structure. Table 2 lists the descriptions of notations.

To identify each user's topic interest, we first compute the topic interest of each user's questions and answers over the major topics respectively. We can row normalize AM, QM into AM', QM' such that $||AM'_{i,.}||_1 = 1$ for each row $AM'_{i,.}$ and $||QM'_{i,.}||_1 = 1$ for each row $QM'_{i,.}$. Each row of these two matrices means the probability distribution of a user's interest in question/answer. Using a distance metric for probability distribution [6], the topic interest difference between questions and answers of user u_i can be calculated as:

$$QADiff(i) = \sqrt{D_{KL}(AM'_{i,.}||M) + D_{KL}(QM'_{i,.}||M)} \qquad (1)$$

---

[2] Besides 6 child topics of the root topic, we select another representative topic ("Science & Technology") that had not been edited into the topic structure due to some mistakes from Zhihu topic organization.

Table 2. Notation descriptions

Notation	Description
n	The total number of users
s	The total number of unique topics
A, Q	$n \times s$ matrix, where $A_{i,t}/Q_{i,t}$ contains the number of topic t in user u_i's answers/questions
V	$n \times s$ matrix, where $V_{i,t}$ contains the number of votes received by user u_i in topic t
AM, QM	$n \times 7$ matrix, where $AM_{i,t}/QM_{i,t}$ contains the number of major topic t in user u_i's answers/questions
CM	$n \times 7$ matrix, where $CM_{i,t}$ contains the number of major topic t in user u_i's posts (questions and answers), i.e., $CM_{i,t} = AM_{i,t} + QM_{i,t}$

where $M = \frac{1}{2}(AM'_{i,.} + QM'_{i,.})$, D_{KL} is the *Kullback-Leibler Divergence* which defines the divergence from distribution H to I as: $D_{KL}(H\|I) = \sum_i H(i) \log \frac{H(i)}{I(i)}$.

Figure 4 plots the distribution of topic interest difference between questions and answers of each user. The analysis is applied on a set of 181K+ users who posted at least one question and one answer. We can observe that the topic interests of their questions and answers for most users are similar, indicating each user's questions and answers are related to the similar topics. Hence, in this paper, the major topic probability distribution of posts published by each user is utilized to present each user's topic interest. Namely, after the row normalization, $CM'_{i,t}$ indicates the probability that user u_i is interested in topic t. Note that the topics are transformed to the corresponding major topics only in the user topic interest calculation process.

Homophily. To assist in measuring the true topical influence of each user, we examine whether *homophily* exists in the social network of our dataset, which has been observed in many social networks [15, 22]. The phenomenon shows that users follow each other because of similar topic interest, which means that the influence on each follower would depend on the topic interest. The question can assist in verifying whether *homophily* exists in Zhihu: *do users with "following" relationships have more similar topic interest than those without?*

The question can be formalized as a two-sample t-test: The null hypothesis is $H_0 : \mu_{follow} = \mu_{unfollow}$, and the alternative hypothesis is $H_1 : \mu_{follow} < \mu_{unfollow}$, where μ_{follow} is the mean topic interest difference between two users with "following" relationship, and $\mu_{unfollow}$ indicates the mean topic interest difference of those without. We design *homophily* testing and evaluation experiments based on a set of active Zhihu users who published at least 10 posts in total, denoted as U ($|U| = 124,445$). We conduct the two-sample t-test on the user congregation because about 92% of the users in our dataset have less than 30 followees. Sample 0 contains the topic interest difference of all the user

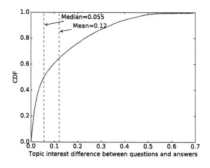

Fig. 4. Topic interest difference between Q&A

pairs with "following" relationships while Sample 1 contains the topic interest difference between each user and some randomly chosen users whom he/she does not follow. Note that the number of each user's chosen non-followees is identical to the number of each user's followees. The topic interest difference between two users is calculated as:

$$Diff(i,j) = \sqrt{D_{KL}(CM'_{i,.}||MM) + D_{KL}(CM'_{j,.}||MM)} \qquad (2)$$

where $MM = \frac{1}{2}(CM'_{i,.} + CM'_{j,.})$. The t-test result[3] shows that H_0 is rejected at significant level $\alpha = 0.01$ with a p-value of less than 1×10^{-17}. Therefore, we confirm that the *homophily* phenomenon does exist in Zhihu.

QARank Algorithm. In this part, we propose QARank, a novel topic-sensitive influence measure algorithm, taking into account the following three factors:

- *Network structure:* In SCQA, a user's influence is propagated to others through the following links between them. To this end, QARank considers the link structure, similar to the authority measure of a web page.
- *Topic interest:* Given *homophily*, a user's topical influence on his follower is stronger when their interests in this topic are more similar and vice versa. A user has different influence in different topic in the same social network.
- *Knowledge authority:* Generally a user's opinion is always accepted by his followers when his answers obtain many votes. Hence, the knowledge authority of a user plays an important role in his influence. Specifically, the more votes a user received, the more authoritative his followers think he is.

QARank is proposed as an extension of TwitterRank, which is modeled as a random surfer model. Let G be a directed graph where each node indicates a user and each directed edge denotes a "following" relationship between two users. A random surfer on the graph G visits each user with certain probability through

---

[3] The t-test result depends on the extent of the dataset normality. Skewness and kurtosis of these two samples are 1.19, 2.14 and 1.21, 2.09, which are considered acceptable in order to prove normal distribution [7].

following the corresponding edge. QARank differentiates itself from TwitterRank in that the topical knowledge authority is considered into the transition probability from one user to another and QARank can measure the multi-topic influence by leveraging Euclidean distance as the topic interest difference measure.

Therefore, each element of the transition matrix P_T for the topic set T ($|T| \geq 1$) is calculated as:

$$P_T(i,j) = \frac{|V_{j,T}|}{\sum_{k:u_i \ follows \ u_k} |V_{k,T}|} \times sim_T(i,j) \tag{3}$$

where

$$sim_T(i,j) = 1 - \sqrt{\sum_{t \in T}(CM'_{i,t} - CM'_{j,t})^2} \tag{4}$$

where $P_T(i,j)$ is the transition probability from follower u_i to followee u_j in the random surfer model. $|V_{j,T}| = \sum_{t \in T} V_{j,t}$ is the number of votes received by user u_j in topic T, and $\sum_{k:u_i \ follows \ u_k} |V_{k,T}|$ is the total number of votes received by all u_i's followees in topic set T. In the model, the number of topic-related votes received is regarded as the topical knowledge authority of a user.

In case of dangling nodes that do not have any out-degree and cyclic loops in the network, we apply random jump [9] by adding a teleportation vector $E_T = A''_{.,T}$, where $A_{.,T} = \sum_{t \in T} A_{.,t}$, and $A''_{.,T}$ is the column-normalized version of $A_{.,T}$ so that $||A''_{.,T}||_1 = 1$.

Given the transition probability matrix and the teleportation vector, using QARank, the topical influence scores of users in topic set T, known as Inf_T, can be calculated iteratively as follows:

$$Inf_T = \lambda P_T \times Inf_T + (1 - \lambda)E_T \tag{5}$$

4.2 Topic-Relevant Expertise Measure

Measuring topic-relevant expertise is of great importance in finding topical opinion leaders. To infer expertise, we employ three topic-relevant metrics accounting for the knowledge capacity, satisfaction and contribution in the paper.

Knowledge Capacity. In SCQA sites, replying to lots of questions in some topics means that one has rich topic-related knowledge while asking many topic-related questions is usually an indicator that one lacks knowledge about these topics. Thus, the *z-score* is adopted as an indicator to measure a user's knowledge capacity in specific topics [24]. Thus, the knowledge capacity of user u_i in topic set T is calculated by $KC_T(i) = (|A_{i,T}| - |Q_{i,T}|)/\sqrt{|A_{i,T}| + |Q_{i,T}|}$. Here $|A_{i,T}| = \sum_{t \in T} A_{i,t}$ is the number of answers published by user u_j in topic set T and $|Q_{i,T}| = \sum_{t \in T} Q_{i,t}$ sums up the number of questions asked by user u_i in topic set T. If answers are more than questions, KC is positive, otherwise it is negative.

Knowledge Satisfaction. Another key service in SCQA is to vote answers if a user agrees on them. The number of received votes means the satisfaction degree that an answer obtains. Hence, we use the average number of votes for the u_i's T-related answers as the knowledge satisfaction of user u_i in topic set T, which is defined as $KS_T(i) = |V_{i,T}|/|A_{i,T}|$.

Knowledge Contribution. Apart from the knowledge capacity and satisfaction, topical leaders should be active and make a great number of contributions to SCQA. Consequently, we choose the number of topic-related answers to measure the knowledge contribution of user u_i, i.e., $IC_T(i) = |A_{i,T}|$.

Before combining the above three factors, Min-Max normalization is adopted to rescale the range of factors to $[0, 1]$. Thus, KC_T, KS_T, and IC_T are transformed into the Min-Max normalized forms KC_T^*, KS_T^*, and IC_T^*. Given this, the expertise score ES_T of user u_i in topic set T ($|T| \geq 1$) is calculated by:

$$ES_T(i) = F(\beta KC_T^*(i), \gamma KS_T^*(i), (1 - \beta - \gamma)IC_T^*(i)) \tag{6}$$

where $F(x, y, z)$ means the expertise measure method, β and γ are two parameters tunning the weight. In order to compare with [20] equally in the evaluation section, in this paper, $F(x, y, z)$ is a weighted sum of three metrics. Note that the expertise measure can be replaced with other efficient methods [18,19].

5   Experiments and Evaluation

This section presents an extensive evaluation of QALeaderRank over 10 popular topics in Zhihu along with a user study, which shows our method greatly outperforms baseline methods in identifying topical opinion leaders.

5.1   Baseline Algorithms and Evaluation Metrics

We compare our **QALeaderRank (QALR)** with two baseline algorithms.

TwitterRank (TR): It measures users' topic-sensitive influence with the consideration of the topical similarity and the link structure [22]. Nevertheless, TwitterRank does not take any knowledge expertise into account.

InExRank (IR): Song et al. [20] proposed a topic-irrelevant method considering authority, activity and influence. In order to compare with this work, we extend it by incorporating topical expertise and following information (i.e., follower count) denoted as InExRank.

Two similarity metrics for comparing rankings are leveraged as follows:

OSim(r_1, r_2): It measures the overlap degree of two top k rankings r_1 and r_2 [10], which is defined as: $OSim(r_1, r_2) = \frac{|r_1 \cap r_2|}{k}$.

KSim(r_1, r_2): It considers the degree to which the relative ordering of two rankings r_1 and r_2 is in agreement [5]. Let $R = r_1 \cup r_2$, and $\theta_1 = R - r_1$. We extend r_1 by appending θ_1 to the tail of r_1 to yield r_1'. r_2' is analogously extended. Thus, the $KSim$ similarity can be calculated by:

$$KSim(r_1, r_2) = \frac{|\{(u,v)|r_1', r_2' \text{ agree on order of } (u,v)\}|}{|R| \times (|R| - 1)}$$

where $(u, v) \in R \times R$ ($u \neq v$) means u ranks in front of v.

Table 3. Ranking similarity among top 20 users identified by three algorithms

	OSim		KSim	
	Mean	Median	Mean	Median
QALR	0.24	0.19	0.42	0.42
IR	0.15	0.14	0.39	0.41
TR	0.96	0.95	0.96	0.96

5.2   Performance Evaluation

This section compares the performance of QALR and two baseline algorithms on our Zhihu dataset over 10 popular topics from different perspectives. These topics are "Movie" (T0), "Psychology" (T1), "Travel" (T2), "Food" (T3), "Fitness" (T4), "Internet" (T5), "Fashion" (T6), "Pioneer" (T7), "Design" (T8), "Finance" (T9). In the paper, for the simplicity we assume three expertise metrics are equally essential to the expertise measure, i.e., $\beta = \frac{1}{3}, \gamma = \frac{1}{3}$. Like TR, Teleportation parameter λ in QALR is also set as 0.85. As a result, we can get three user rankings identified by three methods.

Performance on Topic Correlation. We look at the ranking correlation between topic pairs for the three algorithms to compare their topic sensitivity. As shown in Table 3, TR identifies much more similar leaders (with high mean/median value) than IR and QALR, while QALR and IR can yield diversified top-ranked users in each topic. This is because TR considers the number of published tweets during computing transition probability, which makes one user who published many topic-irrelevant posts will also get high ranking score in the random surfer. Note that the ranking similarity of IR is a little less than that of QALR. This is because QALR considers more topical influence rather than primarily focusing on the topical expertise.

Performance on User Identification. Before comparing the performance, we first divide users into 4 types according to their influence and expertise. An illustration is given in Fig. 1. The 4 types of users are as follows:

- **Type I:** *Influential users with expertise* (Zone I in Fig. 1) have strong influence and high expertise in specific topic(s). They always have a great number of followers, publish many posts and receive a large number of votes.
- **Type II:** *Influential users without expertise* (Zone II in Fig. 1) have strong influence due to their popularity in other fields but publish very few posts and get few votes in specific topic(s).
- **Type III:** *Non-influential users without expertise* (Zone III in Fig. 1) seldom submit posts and do not influence others in given topic(s).
- **Type IV:** *Non-influential users with expertise* (Zone IV in Fig. 1) are not influential and have few followers. However, they like publishing posts.

Our work aims at identifying **type-I** users from all users. In the section we study the detailed information of opinion leaders identified by three algorithms to compare their identification accuracy.

Table 4. Statistic comparison of top 20 users identified by three algorithms

	Number of followers		Number of votes		Number of answers	
	Mean	Median	Mean	Median	Mean	Median
QALR	46922.59	6494.0	12245.54	4481.5	48.41	16.0
IR	43453.73	549.0	8389.55	1185.0	169.68	109.0
TR	56171.41	9261.5	4766.57	235.0	29.35	7.0

The results of QALR are conformant to our expectation. The top-ranked topical leaders identified by QALR mostly published lots of topic-related posts and received a great number of votes. They have many followers including some important followers, who are also top-ranked users identified by the algorithm. Obviously, they belong to *type-I*. Table 4 also shows that the top 20 users of QALR have much more votes than those of two baselines over 10 topics. Here we detail some of top 5 users of QALR due to the space limitation.

"wangxing" is identified as an opinion leader in topic "Pioneer". We find that he posted mainly about pioneer and has 61,268 followers including an important user "zhou-kui". Actually most of pioneer-related top 5 opinion leaders are successful company founders in real life. "wangxing" founded some popular websites such as meituan.com, fanfou.com and renren.com. "zhou-kui" is a partner of Sequoia Capital China. "dreamcog" founded a company named youxiamotors. Especially, "xiepanda", "liuniandate" and "WxzxzW" are identified as top 5 leaders in many topics because they are so-called *cewebrity*, who acquired fame via publishing a number of posts about various topics. For example, "xiepanda" posted mostly about movie, psychology, food, Internet and finance. He also often posted about fitness, fashion, pioneer and design. Besides, each of their answers always got 400+ votes in related topics.

However, TR identifies some *type-II* users. For example, "xiepanda", "liuniandate", "chuan-zhu", and "mazk" are identified as 4 out of top 5 users in topic "Travel". However, in fact, "xiepanda" did not post any content about travel, "liuniandate", "chuan-zhu", and "mazk" only posted one or two answers which received few votes. This is because the influence-focused TR ignores the topical knowledge expertise. Thus, from Table 4, although the mean/median follower count of top 20 users identified by TR is higher than that identified by QALR, TR is much less than QALeaderRank in terms of vote/answer count.

IR yields a number of *type-IV* users. For instance, in topic T3, "rou-si-23" only has 20 followers but published 192 related answers with 15 of maximal vote count and 0.58 of average vote count. "HuDP" posted 615 Internet-related answers that got 9 of maximal vote count and 0.34 of average vote count and only has 33 followers. One can image that these *type-IV* users may be paid posters, spammers or normal active but non-influential users, but cannot be indeed topical opinion leaders. This results from the accumulation of four factors in IR algorithm where one large factor (i.e., the number of answers) can greatly increase the final ranking score. Table 4 shows that the top 20 users identified

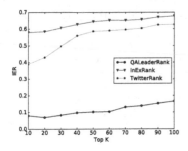
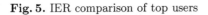
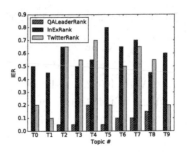

Fig. 5. IER comparison of top users   **Fig. 6.** IER comparison over topics

by IR got much less votes than those identified by QALR although the users of IR posted much more answers. Meanwhile, the top-ranked users of IR have much less followers than those of QALR. This is because IR measures influence using the number of followers while QALR measures the topical influence based on the link structure of the social network.

Performance on Identification Error Rate. As mentioned above, some users who are not leaders are wrongly identified by algorithms, such as *type-II* and *type-IV* users. To measure the fraction of apparently wrongly identified users, we propose a novel metric, ***Identification Error Rate*** (***IER***). For the top k users in topic t, the identification error rate can be calculated by:

$$IER(k,t) = \frac{|\{l_i|v_i^t \leq nv \ or \ f_i \leq nf, i \in [0,k)\}|}{k} \qquad (7)$$

where l_i is the i-th identified leaders. v_i^t is average vote count of l_i in topic t and f_i denotes l_i's follower count. nv and nf is average number of votes of all answers and average number of followers of all users respectively. We assume that l_i is a wrongly identified top-ranked user if his v_i^t or f_i is less than the mean value of all users. As shown in Table 1, we set $nv = 13.63$ and $nf = 11.57$.

Figure 5 shows the average *IER* of identified top k users over 10 topics for the three algorithms. We can observe that *IER* of QALR is always below 20% while IR and TR yield very high *IER*. As an example, Fig. 6 illustrates *IER* comparison of the top 20 users in each topic. Note that the rankings of QALR also lead to the lowest *IER* in each topic. In particular, the rankings yielded by QALR are of extremely high quality ($IER = 0$) in topics T0, T1, and T9. These observations further proves that our algorithm significantly outperforms the two baselines in SCQA.

Performance on Multi-topic Identification. QALR can identify multi-topic opinion leaders. For example, "big_caaat" is identified as a Internet-finance opinion leader. He posted frequently about Internet and Finance, who has 7939 followers including an important user "xiepanda". He published 83 Internet-related answers with 227 of average vote count and 48 finance-related answers with 163 of average vote count. It is worth noting that "liuniandate" is ranked as the No. 4 opinion leader across two topics "Movie" and "Psychology". However, the user is respectively ranked as No. 2 in these two topics. This is because QALR considers the general topical influence of the social network based on topical interest and knowledge expertise instead of the individual influence for each topic.

5.3   User Study

In order to further evaluate our algorithm, we conducted a user study to compare the proposed QALR with two baseline algorithms over 10 topics. By respectively selecting the top 20 users for each topic from the three algorithms, we got about 50 users in each topic because of some overlaps among the top 20 users of the three methods. Then we designed an online questionnaire that asks each participant to choose one topic that he/she focused on most frequently and rate each user's topical opinion influence using 5-point Likert scales. The questionnaire listed each user's name and three representative topic-related answers as tips. Before answering this questionnaire, each participant is required to realize 5 degrees of topical opinion influence as follows:

- *1 point: very weak, meaning you do not know the user or never view his posts.*
- *2 point: weak, meaning you browsed a few topic-related posts of the user.*
- *3 point: medium, meaning you agreed on some topic-related posts of the user.*
- *4 point: strong, indicating you browsed and agreed on most of his topical posts, and voted/commented on his answers.*
- *5 point: very strong, meaning you agreed on his topical posts absolutely, followed his update, and have invited or will invite him to answer questions.*

We spread the questionnaire to some professional online Zhihu discussion groups, the majority of whose members are active Zhihu users. Finally, we received 200 valid questionnaire responses (about 20 responses for each topic).

Average Rating Comparison. Using the ratings collected from those questionnaire responses, we calculate and give the comparison of average ratings of top 20 users identified by three algorithms as shown in Fig. 7. We find that the overall ratings of QALR over 10 topics are higher than those of two baseline methods. Note that average ratings of QALR seem not high, resulting from that the user study is in a cold-start situation with limited information from each participant.

Ranking Similarity Comparison. Figure 8 reports on the ranking similarity between real rankings and the top 20 rankings generated by the three algorithms over 10 topics. The real rankings for each topic is produced by ordering the average rating of each user in each topic. From Fig. 8a, we can observe that the rankings of QALR are much closer to the real rankings than those of two baselines over 10 topics in terms of the overlap similarity $OSim$. Especially, our algorithm yields much more prominent rankings in topics T0, T4, T8 and T9. Furthermore, from Fig. 8b, for nearly all the topics, the ordering accuracy of QALR is higher than those of two baseline algorithms. As a result, a majority of participants preferred the rankings of QALR.

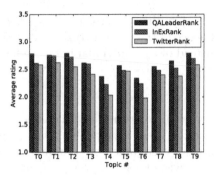

Fig. 7. Average rating comparison

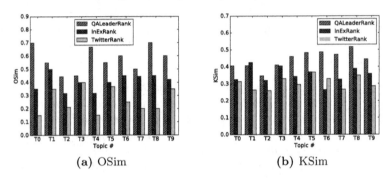

(a) OSim   (b) KSim

Fig. 8. Comparison of similarity between real rankings and identified rankings

6   Conclusion

This paper focuses on identifying topical opinion leaders in SCQA and proposes an efficient method called QALeaderRank, considering both the topic-sensitive influence and the topic-relevant expertise. To measure the true topical influence, by exploring the QA and social features, we propose a novel topic-sensitive influence measure algorithm named QARank for SCQA, incorporating the network structure, the topic interest similarity between users and the topical knowledge authority. In addition, we employ three topic-relevant expertise metrics for inferring the topical knowledge expertise of each user. Furthermore, QALeaderRank is able to identify multi-topic opinion leaders. Finally, the experimental results over 10 popular topics, along with the feedback from an online user study, show that QALeaderRank outperforms the compared state-of-the-art methods.

References

1. Bakshy, E., Hofman, J., Mason, W., et al.: Everyone's an influencer: quantifying influence on Twitter. In: WSDM, pp. 65–74 (2011)
2. Bouguessa, M., Dumoulin, B., Wang, S.: Identifying authoritative actors in question-answering forums: the case of Yahoo! answers. In: SIGKDD, pp. 866–874 (2008)

3. Bouguessa, M., Romdhane, L.B.: Identifying authorities in online communities. TIST **6**(3), 30 (2015)
4. Cha, M., Haddadi, H., Benevenuto, F., et al.: Measuring user influence in Twitter: the million follower fallacy. In: ICWSM, pp. 10–17 (2010)
5. Dwork, C., Kumar, R., Naor, M., et al.: Rank aggregation methods for the web. In: WWW, pp. 613–622 (2001)
6. Endres, D., Schindelin, J.: A new metric for probability distributions. IEEE TIT **49**, 1858–1860 (2003)
7. George, D.: SPSS for Windows Step by Step: A Simple Study Guide and Reference. Pearson Education India, Delhi (2011)
8. Ghosh, S., Sharma, N., Benevenuto, F., et al.: Cognos: crowdsourcing search for topic experts in microblogs. In: SIGIR, pp. 575–590 (2012)
9. Grin, S., Page, L.: The anatomy of a large-scale hypertextual web search engine. Comput. Netw. ISDN Syst. **30**, 107–117 (1998)
10. Haveliwala, T.: Topic-sensitive PageRank. In: WWW, pp. 517–526 (2002)
11. Kleinberg, J.: Authoritative sources in a hyperlinked environment. J. ACM **46**(5), 604–632 (1999)
12. Lazarsfeld, P.F., Berelson, B., en Gaudet, H.: The people's choice: how the voter makes up his mind in a presidential campaign. J. Consult. Psychol. **9**(5), 268 (1968)
13. Lee, C., Kwak, H., Park, H., et al.: Finding influentials based on the temporal order of information adoption in Twitter. In: WWW, pp. 1137–1138 (2010)
14. Li, F., Du, T.: Who is talking? An ontology-based opinion leader identification framework for word-of-mouth marketing in online social blogs. Dec. Support Syst. **51**(1), 190–197 (2011)
15. McPherson, M., Smith-Lovin, L., Cook, J.: Birds of a feather: homophily in social networks. Annu. Rev. Sociol. **27**(1), 415–444 (2001)
16. Miao, Q., Zhang, S., Meng, Y., et al.: Domain-sensitive opinion leader mining from online review communities. In: WWW, pp. 187–188 (2013)
17. Omari, A., Carmel, D., Rokhlenko, O., et al.: Novelty based ranking of human answers for community questions. In: SIGIR, pp. 215–224 (2016)
18. Pal, A., Konstan, J.: Expert identification in community question answering: exploring question selection bias. In: CIKM, pp. 1505–1508 (2010)
19. Riahi, F., Zolaktaf, Z., Shafiei, M., et al.: Finding expert users in community question answering. In: WWW, pp. 791–798 (2012)
20. Song, S., Tian, Y., Han, W., et al.: Leading users detecting model in professional community question answering services. In: GreenCom-iThings-CPSCom, pp. 1302–1307 (2013)
21. Wang, G., Gill, K., Mohanlal, M., et al.: Wisdom in the social crowd: an analysis of quora. In: WWW, pp. 1341–1352 (2013)
22. Weng, J., Lim, E., Jiang, J., et al.: TwitterRank: finding topic-sensitive influential twitterers. In: WSDM, pp. 261–270 (2010)
23. Zhai, Z., Xu, H., Jia, P.: Identifying opinion leaders in BBS. In: WI-IAT, pp. 398–401 (2008)
24. Zhang, J., Ackerman, M., Adamic, L.: Expertise networks in online communities: structure and algorithms. In: WWW, pp. 221–230 (2007)
25. Zhao, Z., Zhang, L., He, X., et al.: Expert finding for question answering via graph regularized matrix completion. TKDE **27**(4), 993–1004 (2015)

Personalized Geo-Social Group Queries in Location-Based Social Networks

Yuliang Ma[1](✉), Ye Yuan[1](✉), Guoren Wang[2], Xin Bi[3], and Yishu Wang[1]

[1] School of Computer Science and Engineering, Northeastern University,
Shenyang, China
ylma.neuer@gmail.com, yuanye@mail.neu.edu.cn
[2] School of Computer Science and Technology, Beijing Institute of Technology,
Beijing, China
[3] Sino-Dutch Biomedical and Information Engineering School,
Northeastern University, Shenyang, China

Abstract. Geo-social group query, one of the most important issues in LBSNs, combines both location and social factors to generate useful computational results, which is attracting increasing interests from both industrial and academic communities. In this paper, we propose a new type of queries, *personalized geo-social group* (PGSG) queries, which aim to retrieve both a user group and a venue. Specifically, a PGSG query intends to find a group-venue pattern (consisting of a venue and a group of users with size h), where each user in the group is socially connected with at least c other users in the group and the maximum distance of all the users in the group to the venue is minimized. To tackle the problem of the PGSG query, we propose GVPS, a novel search algorithm to find the optimal user group and venue simultaneously. Moreover, we extend the PGSG query to top-k personalized geo-social group (TkPGSG) query. Instead of finding the optimal solution in the PGSG query, the TkPGSG query is to return multiple feasibility solutions to guarantee the diversity. We propose an advanced search algorithm TkPH to address the TkPGSG query. Comprehensive experimental results demonstrate the efficiency and effectiveness of our proposed approaches in processing the PGSG query and the TkPGSG query on large real-world datasets.

1 Introduction

With the progress of location acquisition and wireless communication technology, people now are able to add location dimension into traditional social networks, which fosters a bunch of location-based social networks (LBSNs), such as, *Foursquare*, *Gowalla*, and *Yelp*. People can easily record and share their life experiences via their mobile devices in these service platforms. Therefore, individuals' location data and social data have been readily available from mobile devices. One of the most important applications in LBSNs, *geo-social group query*, combines both location and social factors to generate useful computational results, which is attracting increasing interests from both industrial and academic communities.

© Springer International Publishing AG, part of Springer Nature 2018
J. Pei et al. (Eds.): DASFAA 2018, LNCS 10827, pp. 388–405, 2018.
https://doi.org/10.1007/978-3-319-91452-7_26

In the literature of geo-social group queries, the authors in [11,23] aim to find a group of users close to a given rally point and to ensure that the selected users have a good social relationship. The authors in [22] aim to find the activity time and attendees with the minimum total social distance to the initiator. The authors in [6,7] explore a group of experts whose skills can cover all the requirements and the communication cost among group members is low. The authors in [9] retrieve a user group of size k where each user is interested in the query keywords and they are close to each other in the Euclidean space. Besides, some other types of geo-social group queries have been proposed, such as, geo-social k-cover group query [8], and geo-social group query with minimum acquaintance constraint [28]. While being useful in some applications, these queries mentioned above do not completely utilize new search potential brought by geo-social data.

The following scenario is very common in real life. Assume that user Alice wants to establish an activity (such as, a group of users P including Alice will have dinner together at a venue s). There are some constraints by satisfied: (1) the size of the group $|P|$ is 10. Each user in the group should know at least 3 other users, which can create good atmosphere in the activity; and (2) the maximum distance of all the users in the group to the venue is minimized. However, none of the existing works on geo-social group queries can be used to answer such a scenario. For example, the authors in [23] propose a novel query SSGQ to find a set of users close to a given venue. While, Alice (the query user) does not know such a input venue and wants to find a user group and a venue simultaneously. If we use the SSGQ query to model the above scenario, we need to address the SSGQ queries repeatedly, which is extremely expensive. Thus, the SSGQ query is unsuitable for modeling the above scenario.

Consequently, we propose a novel type of geo-social group queries, called Personalized Geo-Social Group (PGSG) queries. Specifically, a PGSG query intends to find a venue and a user group (including the query user) with size h, where each user in the group is socially connected with at least c other users in the group and the maximum distance of all the users in the group to the venue is minimized. We call such a pair of user group and venue as a *group-venue pattern*. Our proposed PGSG query can model the above mentioned scenario. The size constraint h is 10 and the social topology constraint c is 3. By modeling the scenario as a PGSG query, Alice can obtain a appropriate group-venue pattern (consisting of a user group with size 10 and a venue), which each user in the group is connected with at least 3 other users and the maximum distance of all the users in the group to the venue is minimized. Besides, the PGSG query can be used to model some other real applications, such as spatial task outsourcing [18–20], event planning [3,16,17].

Moreover, we extend the PGSG query to the top-k personalized geo-social group (TkPGSG) query. Instead of finding the optimal group-venue pattern defined in the PGSG query, the TkPGSG query is to return k group-venue patterns $\mathbb{X} = \{X_1, \cdots, X_k\}$ such that: (a) all the k group-venue patterns satisfy the social constraints (group size constraint and the social topology constraint); (b) any group-venue pattern $X_m \notin \mathbb{X}$ satisfying the social constraints has a cost (the maximum distance of all the users in the group to the venue) that exceeds that of any group-venue pattern $X_i \in \mathbb{X}$.

Challenges. The PGSG query is a hard problem to be tackled. The challenge is threefold.

Firstly, the PGSG query aims to find a user group (including the query user) with size h, where each user is socially connected with at least c other users in the group. If we directly extract all such groups by simply enumerating all possible combinations, the search space is large and redundant. Therefore, the first challenge is how to extract possible user groups efficiently.

Secondly, there are infinitely many user group combinations and venues, which make infeasibility to examine all group-venue patterns to the PGSG query. Therefore, the second challenge is how to efficiently find the optimal group-venue pattern for the PGSG query.

Thirdly, in order to guarantee the diversity of query processing, one common and effective way is to return multiple query results. We extend the PGSG query to the TkPGSG query. It is inefficient to tackle the TkPGSG query by invoking multiple PGSG queries. Thus, the third challenge is how to efficiently return top-k group-venue patterns for the query user.

Our Proposed Methods. In order to tackle the PGSG query efficiently, we propose a novel search algorithm, called *group-venue pattern search* (GVPS). The intuition of GVPS is that we expand a user group from the query user by a breadth search strategy. With the group expanding processing, GVPS reduces the venue search space by a derived lower bound and upper bound of spatial distance. Moreover, based on the GVPS algorithm, we propose an advanced top-k personalized geo-social group query algorithm, namely *top-k group-venue patterns hunter* (TkPH), to tackle the TkPGSG query efficiently.

Contributions. To summarize, we make following contributions in this paper.

- We propose a new type of geo-social group queries called Personalized Geo-Social Group (PGSG). Specifically, a PGSG query aims to find a venue and a group with size h, where each user in the group is socially connected with at least c other users in the group and the sum of the distance from every user in the group to the venue is minimized.
- We extend the PGSG query to the top-k personalized geo-social group (TkPGSG) query. Instead of finding the optimal group-venue pattern defined in PGSG, the TkPGSG query is to return multiple query results to guarantee the diversity of query processing.
- To tackle the problem of PGSG query, we propose a search algorithm, called *group-venue pattern search* (GVPS). Besides, we propose an advanced top-k group-venue patterns search algorithm, namely TkPH, for processing the TkPGSG query.
- Extensive experiments are conducted to demonstrate the efficiency and effectiveness of the proposed approaches on real-world datasets.

The rest of this paper is organized as follows. We formally define the PGSG query and the TkPGSG query in Sect. 2. We present the details of PGSG query algorithm in Sect. 3. In Sect. 4, we present the proposed algorithm to process the TkPGSG query. We show an extensive experimental evaluation in Sect. 5, and overview the related works in Sect. 6. In Sect. 7, we conclude this paper.

2   Problem Formulation

In this section, we describe the terms and notations that we use throughout the paper, and formally define the Personalized Geo-social Group (PGSG) queries.

LBSNs allow users to search location-tagged contents within their social graphs, and consist of the new social structures made up of individuals. We call the integration of location data and social data, generated by individuals in LBSNs, as *geo-social data*.

We model a LBSN as two main components. As shown in Fig. 1(a), the first component is the underlying social network $G = (V, E)$, where each vertex $u \in V$ is a user. Each edge $e \in E$ denotes an acquainted relation (e.g., friendship) between the two users it connects. Moreover, each user u maintains a pair of coordinates indicating the user's location, which can be extracted from the user's profiles (such as home address) or discovered by the existing work [27]. We take the user's location as one part of the input, since user location extraction is not the focus of our work. In this paper, we take the venues in LBSNs as spatial objects. Thus, the second component of a LBSN is a set of spatial venues S. Each venue $s \in S$ is a spatial object associated with a pair of coordinates indicating its geographical position. Figure 1(b) shows an example of the spatial locations of users and venues in a LBSN.

A PGSG query aims to find a user group (including the query user) with size h, where each user in the group is socially connected with at least c other users in the group.

Definition 1 *(c-core). For a graph* $G = (V, E)$*, a maximal connected subgraph* $G' = (V', E')$ *of* G *is a c-core, if each vertex* $u \in V'$ *has degree at least* c.

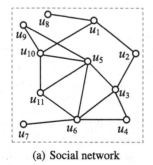

(a) Social network

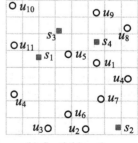

(b) Spatial location

Fig. 1. An example of a location-based social network.

The concept of c-core was first proposed by Seidman in [15], which can be widely applied to describe the complex topologies of social networks and reveal the hierarchical structures of networks.

Definition 2 *(Induced Subgraph).* *Given a graph $G = (V, E)$, for any subset V' of V, and edge set E':*

$$E' = \{(u, v)|u, v \in V' \text{ and } (u, v) \in E\},$$

we call $G' = (V', E')$ is an induced subgraph of V' in G, denoted as $G[V']$.

Definition 3 *(Connected c-core Component).* *Given a graph $G = (V, E)$, a subset V' of V, $G[V']$ is a connected c-core component, if $G[V']$ is a connected component and $\min_{u \in V'} deg_{G[V']}(u) \geq c$.*

Definition 4 *(Group-Venue Pattern).* *Given a user group $P = \{u_1, u_2, \cdots, u_h\}$, a venue s, we call such a pair of user group and venue a group-venue pattern, denoted as $X = (P, s)$. The distance of a group-venue pattern X is the maximum distance of all the users in $X.P$ to $X.s$. That is, we have:*

$$Dist(X) = Dist(X.P, X.s) = Max_{u_i \in P} dist(u_i, s) \tag{1}$$

where $dist(u_i, s)$ is the Euclidean distance.

Remember that a *personalized geo-social group* (PGSG) query aims to find a venue and a user group (including the query user) with size h, where the group is a connected c-core component and the maximum distance of all the users in the group to the venue is minimized. Formally, we define the PGSG query as follows:

Definition 5 *(Personalized Geo-Social Group (PGSG) Query).* *Given an underlying social network $G = (V, E)$ and a spatial venue set $S = \{s_1, s_2, \cdots, s_m\}$, the personalized geo-social group query $q = \langle u_q, h, c \rangle$, where u_q is the query user who initiate such a query, h is the group size constraint, and c indicates a social constraint c-core, aims to retrieve a user group $P \subseteq V$ and a venue $s \in S$ such that:*

(1) P includes u_q and $|P| = h$;
(2) $G[P]$ is a connected c-core component;
(3) $Dist(P, s)$ is minimized.

Example 1. Take Fig. 1 as an example, assume a PGSG query $q = \langle u_5, 4, 2 \rangle$, which means the query user is u_5, the group size constraint is 4. The user group $P = \{u_5, u_9, u_{10}, u_{11}\}$ contains 4 users including u_5, and the induced subgraph P in G is a 2-core. In this condition, the group-venue pattern $X = (P, s_3)$ is the result of q, since $Dist(P, s_3)$ is minimized.

In order to guarantee the diversity of query processing, we extend the PGSG query to the top-k personalized geo-social group (TkPGSG) query. Instead of finding the optimal group-venue pattern defined in the PGSG query, the TkPGSG query aims to find k group-venue patterns. Formally, we define the TkPGSG query as follows:

Definition 6 *(Top-k   Personalized   Geo-Social   Group   (TkPGSG)* *Query).* *Given an underlying social network $G = (V, E)$ and a spatial venue set $S = \{s_1, s_2, \cdots, s_m\}$, the top-k personalized geo-social group query $q = \langle u_q, h, c, k \rangle$, where u_q is the query user who initiate such a query, h is the group size constraint, c indicates a social constraint c-core, and k the number of group-venue patterns needed to return, aims to retrieve k group-venue patterns $\mathbb{X} = \{X_1, X_2, \cdots, X_k\}$ such that:*

(1) for any $X_i \in \mathbb{X}$, X_i includes u_q and $|X_i.P| = h$;
(2) for any $X_i \in \mathbb{X}$, $G[X_i.P]$ is a connected c-core component;
(3) for any $X_i \in \mathbb{X}$, and any $X_j \notin \mathbb{X}$ such that $X_j.P$ satisfies (1) and (2), $Dist(X_i) \leq Dist(X_j)$.

If the number of group-venue patterns satisfying the conditions (1) and (2) in Definition 6, the TkPGSG query returns all these group-venue patterns. It's worth noting that the TkPGSG query can also be extended to other metrics, such as average distance and minimal distance.

Example 2. Continue to use Fig. 1 as an example. Assume a TkPGSG query $q = \langle u_5, 4, 2, 3 \rangle$, which means the query user is u_5, the group size constraint is 4, the core constraint is 2, q aims to find 3 group-venue patterns. There are four user groups: $P_1 = \{u_5, u_3, u_4, u_6\}$, $P_2 = \{u_5, u_3, u_6, u_{11}\}$, $P_3 = \{u_5, u_6, u_{10}, u_{11}\}$, and $P_4 = \{u_5, u_9, u_{10}, u_{11}\}$, where each group satisfies the condition (1) and (2) presented in Definition 6. The results of q are: $X_1 = (P_1, s_1)$, $X_2 = (P_2, s_1)$, and $X_3 = (P_4, s_3)$. The user group P_3 can not be included in the results, since the distance of any group-venue pattern X_j containing P_3 exceeds the third-largest distance in the results.

3   PGSG Query Processing

In this section, we discuss how to process the PGSG query. We firstly introduce a baseline algorithm, and then propose a novel search algorithm.

3.1   Baseline Algorithm

According to our problem statement, a PGSG query aims to find a venue s and a user group P (including the query user u_q) with size h, where the induced subgraph $G[P]$ is a connected c-core component and the maximum distance of all the users in P to s is minimized. To process the PGSG query, a baseline solution can be readily described as follows. Firstly, we enumerate all the c-core groups including the query user with size h. We regard these c-core groups as group candidates (called GCS). To find a feasible venue s_i for each group $P_i \in GCS$, we utilize the method proposed in [12], which aims to find the aggregate nearest neighbor for a given query point set in spatial databases. We take the group-venue pattern $X_i = (P_i, s_i)$ as a candidate solution. Finally, the group-venue pattern with the minimum distance is returned as the result.

The methods proposed in [12] to address the aggregate nearest neighbor query are based on a spatial access method, R-tree [5]. In order to index social relations in LBSNs, a new concept, core bounding rectangles (CBRs), has been proposed in [8,28]. Based on this concept, SaR-tree structure [28] and enhanced SaR-tree [8] are proposed to facilitate social relations processing. But they both relies on inputs of corresponding geo-social group queries, which are unsuitable for our PGSG queries. In this paper, we utilize the R-tree in all our proposed algorithms to speed up spatial distance computation.

3.2   Group-Venue Pattern Search

Apparently, the baseline algorithm is time-consuming and impractical. It is inefficient to enumerate all c-core groups including the query user with size h, since most of these groups are not the final result. Besides, it is expensive and time consuming to find a feasible venue for each above enumerated group. Thus, we propose a novel search algorithm, called *group-venue pattern search* (GVPS).

The intuition of the GVPS algorithm can be present as follows. GVPS expands a user group from the query user by a breadth search strategy. With the group expansion processing, GVPS reduces the venue search space by a derived lower bound and upper bound of spatial distance. And then, a user group and a venue can be retrieved simultaneously by a novel group-venue pattern search method.

Group Expansion. When a PGSG query $q = \langle u_q, h, c \rangle$ is initiated, we firstly adopt a core decomposition algorithm to obtain the c-core G_c. And then, we take the query user u_q as a center to do breadth search in G_c. The size of a user group increases with the breadth expansion. In order to accelerate processing, we select multiple vertices in one breadth expansion. In each expansion, we select the vertex with maximum degree in the current group as the new center for next expansion. When the size of a group P_i reaches h, we check whether the induced subgraph $G[P_i]$ is a connected c-core component. If yes, P_i must be a connected c-core component and contain the query user, since we start to expand a user group from the query user and select the neighbors of the center in each expansion.

Take the social graph G in Fig. 1 as an example. Assume a PGSG query $q = \langle u_{11}, 4, 2 \rangle$. Figure 2 shows an example of group expansion. Figure 2(a) shows the 2-core of G. We start to do breadth search from the query user u_{11}. We select at least 2 neighbors of u_{11}. The expansions from v_{11} are shown in Fig. 2(b). As the size constraint h is 4 in q, b_1 has already finished expansion. In a user group (such as b_2) needed to be extended, we selected the node with maximum degree as new center to do next group expansion. The size of b_2 is 3, we can add only *one* user into the current user group. The user groups extended from b_2 are shown in Fig. 2(c).

Group-Venue Pattern Search. Before illustrating the produce of group-venue pattern search method, we derive the upper bound and the lower bound of the distance of a group-venue pattern X. Note that X consists of a user group

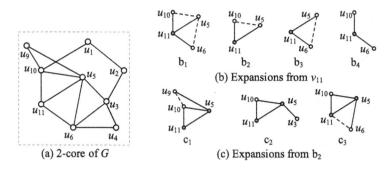

(a) 2-core of G

(b) Expansions from v_{11}

(c) Expansions from b_2

Fig. 2. Example of group expansion.

and a venue. According to the problem definition of the PGSG query, one of the query constraint is that the returned user group in a pattern should include the query user. Thus, in the rest of this section, discussions are restricted to a user group including the query user, unless otherwise stated.

Theorem 1. *Given a user group P including the query user u_q, and a venue s, the* ***upper bound*** *of the maximum distance of a user in P to s is $dist(u_q, s) + \max_{u_i \in P} dist(u_i, u_q)$.*

Proof. Suppose the maximum distance of any user in P to s is the distance of a user u_0 to s, denoted as $dist(u_0, s)$. As the user group P includes the query user u_q, we have $dist(u_0, s) < dist(u_q, s) + dist(u_0, u_q)$ by the triangle inequality. This theorem is proved, since $dist(u_0, u_q) \leq \max_{u_i \in P} dist(u_i, u_q)$.

Theorem 2. *Given a user group P including the query user u_q, and a venue s, the* ***lower bound*** *of the maximum distance of a user in P to s, $Dist(P, s)$, is $\max\{dist(u_q, s), \max_{u_i \in P} dist(u_i, u_q) - dist(u_q, s)\}$.*

Proof. As the user group P includes the query user u_q, the maximum distance of a user in P to a venue s is at least the distance of u_q to s, i.e., $dist(u_q, s) \leq Dist(P, s)$. Suppose the maximum distance of any user in P to s is the distance of a user u_0 to s, by the triangle inequality, we have $|dist(u_q, s) - dist(u_q, u_0)| < Dist(P, s)$. If $dist(u_q, s) \geq dist(u_q, u_0)$, we have $|dist(u_q, s) - dist(u_q, u_0)| < dist(u_q, s) < Dist(P, s)$. If $dist(u_q, s) < dist(u_q, u_0)$, we have $dist(u_q, u_0) - dist(u_q, s) < Dist(P, s)$. And than, $\max_{u_i \in P} dist(u_i, u_q) - dist(u_q, s)\} < Dist(P, s)$ can be derived by the triangle inequality. Thus, we take $\max\{dist(u_q, s), \max_{u_i \in P} dist(u_i, u_q) - dist(u_q, s)\}$ as the lower bound of $Dist(P, s)$.

Now, the GVPS algorithm is ready to be presented. GVPS maintains the current minimum distance of the current optimal group-venue pattern X^*. By leveraging the lower bound (denoted as $LDist(P, s)$) and the upper bound (denoted as $UDist(P, s)$) of the distance of a group-venue pattern $X = (P, s)$, the search space can be reduced. In order to retrieve the user group and the

---

Algorithm 1. Group-Venue Pattern Search (GVPS)

Input: A graph $G = (V, E)$, $S = \{s_1, s_2, \cdots, s_m\}$, a PGSG query $q = \langle u_q, h, c \rangle$

Output: A group-venue pattern $X^* = (P^*, s^*)$

1 Initialize $X^* \leftarrow \emptyset$;

2 Initialize $Dist(X^*) \leftarrow \infty$;

3 Initialize priority queue $UQ \leftarrow u_q$, $DQ \leftarrow \emptyset$, $DV \leftarrow \emptyset$;

4 Find the c-core G_c of G;

5 **while** $UQ \neq \emptyset$ **do**

6 $\quad$ $P_i \leftarrow UQ.dequeue()$;

7 $\quad$ **if** $|P_i| < h$ **then**

8 $\quad\quad$ Select the node $v \in P_i$ with highest degree and $v \notin DV$;

9 $\quad\quad$ $DV \leftarrow DV \cup v$;

10 $\quad\quad$ **for** each set $Vp \subseteq N(v)/P_i$, $c \leq |Vp| + d_{G[P_i]}(v)$ **and** $|Vp| + |P_i| \leq h$ **do**

11 $\quad\quad\quad$ induce a subgraph G_i of $P_i \cup Vp$ in G_c;

12 $\quad\quad\quad$ **if** $G_i \not\subset UQ \cup DQ$ **then**

13 $\quad\quad\quad\quad$ $UQ \leftarrow UQ \cup G_i$;

14 $\quad$ **if** $|P_i| = h$ and $P_i \not\subset DQ$ **then**

15 $\quad\quad$ $DQ \leftarrow DQ \cup P_i$;

16 $\quad\quad$ **if** $\{v \in P_i : c > |deg_{G[P_i]}(v)|\} \neq \emptyset$ **then**

17 $\quad\quad\quad$ Continue;

18 $\quad\quad$ Generate a venue candidate set S_i for P_i;

19 $\quad\quad$ $Dist(X^*) \leftarrow Dist(P_i, s_0)$;

20 $\quad\quad$ **for** each venue $s_j \in S_i$ **do**

21 $\quad\quad\quad$ **if** $LDist(P_i, s_j) \geq Dist(X^*)$ **then**

22 $\quad\quad\quad\quad$ Continue;

23 $\quad\quad\quad$ **if** $UDist(P_i, s_j) < Dist(X^*)$ **or** $Dist(P_i, s_j) < Dist(X^*)$ **then**

24 $\quad\quad\quad\quad$ Update $P^* \leftarrow P_i$, $s^* \leftarrow s_j$;

25 Return X^*;

---

venue of X^* simultaneously, GVPS retrieves the venues along with the user group expansion processing. That is, when the size of a user group P_i reaches h, GVPS utilizes the lower bound to prune venues. GVPS prunes the pattern $X_i = (P_i, s_j)$, if $LDist(X_i) \geq Dist(X^*)$. On the other hand, GVPS leverages the upper bound to check whether the group can emerge in the final query result. If $UDist(X_i) < Dist(X^*)$, GVPS updates the current optimal solution without more distance computation. Let s_0 be the nearest venue of u_q. Instead of taking the whole venues in a LBSN as the search space for a user group P_i, we generate a venue candidate set S_i. That is, we adopt $Dist(P_i, s_0)$ to do a range query at each user in P_i. The venues in these query ranges are regarded as venue candidates. By this way, we can largely reduce the venue search space.

Algorithm 1 details the procedure of the GVPS algorithm. At the beginning, GVPS initially sets the final optimal group-venue pattern X^* to $\emptyset$ and its distance $Dist(X^*)$ to ∞ (lines 1–2). GVPS maintains two priority queues UQ

and DQ. UQ stores user groups that have not been extended. DQ manages the user groups that have already been extended. Besides, the vertices in G are stored in DV if they are impossible to be included in any other user groups except these already lie in UQ and DQ. GVPS exploits c-core to prune users in the original social graph (line 4). And then, GVPS leverages the above-mentioned group expansion to search user groups (lines 6–13).

When the size of a user group P_i reaches h, GVPS checks whether the induced subgraph of P_i is a connected c-core component (lines 14–17). If yes, GVPS generates a venue candidate set S_i for P_i and update $Dist(X^*)$ (lines 18–19). Here, we only update the value of $Dist(X^*)$ rather than X^*, since an earlier given $Dist(X^*)$ can bring a better pruning effectiveness. And then, GVPS utilizes the lower bound to reduce the search space (lines 21–22). That is, a group-venue pattern (P_i, s_j) can be pruned directly, if $LDist(P_i, s_j) \geq Dist(X^*)$. Next, the upper bound can be used to speed up the update of current optimal solution (lines 23–24). Final, the optimal group-venue pattern X^* is returned as the result of the PGSG query.

The following example illustrates how Algorithm 1 works.

Example 3. Continue to use Fig. 1 as an example. Assume a PGSG query $q = \langle u_{11}, 4, 2 \rangle$. GVPS starts to do group expansions from u_{11} (as shown in Fig. 2). When the size of a user group P reaches 4 (suppose $P = \{u_5, u_9, u_{10}, u_{11}\}$), GVPS executes multiple range queries to generate a venue candidate set. Referring to Fig. 1 again, s_1 is the nearest venue of u_{11}. GVPS utilizes $Dist(P, s_1)$ (i.e., $dist(u_9, s_1)$) to do range query at each user in P. We can see that s_2 does not lie in any query range. Thus s_2 can pruned directly about P. Finally, the group-venue pattern (P, s_3) is returned as the optimal result of q.

Complexity Analysis. According to Algorithm 1, we can analyze the time complexity of GVPS algorithm from the following three aspects. Let a given graph $G = (V, E)$ and a PGSG query $q = \langle u_q, h, c \rangle$. Firstly, the time complexity of *core decomposition* $O(|V| + |E|)$ by following the existing work [2]. Secondly, the time complexity of *group expansion* is $O(\bar{d}^h \cdot h^2)$, where $\bar{d}$ is the average degree of the vetices in G, $O(\bar{d}^h)$ is the time complexity of packing a user group with size h from u_q, and $O(h^2)$ is the time of estimating whether the induced subgraph of a user group with size h constitutes a c-core. Thirdly, the time complexity of once group-venue pattern search is $O(h^2 \cdot |S|)$, where $|S|$ is the number of venues in a LBSN. Overall the time complexity of the GVPS algorithm is $O(|V| + |E| + \bar{d}^h \cdot (h^2 + h^2 \cdot |S|))$.

4 TkPGSG Query Processing

As mentioned above, we extend the PGSG query to the top-k personalized geo-social group (TkPGSG) query. Instead of finding the optimal group-venue pattern defined in the PGSG query, the TkPGSG query is to return k group-venue patterns $\mathbb{X} = \{X_1, \cdots, X_k\}$. Although the parameter k is a small positive integer, it is time-consuming and inefficient to invoke multiple Algorithm 1 for TkPGSG query processing. Thus, we propose an advanced search algorithm (top-k group-venue patterns hunter, TkPH) to tackle the TkPGSG query.

The TkPH algorithm is designed based on the GVPS algorithm presented in Subsect. 3.2. Thus, we only elaborate the group-venue pattern search phase, and the group expansion is the same with the GVPS algorithm.

TkPH maintains $\mathbb{X}$ to store the current best k group-venue patterns. In $\mathbb{X}$, the patterns are sorted in ascending order of their group-venue distance. Given a user group P including the query user u_q, and a venue s, let $Dist(X_k^*)$ be the distance of the k-th group-venue pattern, the group-venue pattern (P, s) can be pruned directly if $LDist(P, s) \geq Dist(X_k^*)$. On the other hand, the group-venue pattern (P, s) can be updated into the $\mathbb{X}$, if $UDist(P, s) < Dist(X_k^*)$. That is, we update $\mathbb{X}$ by replacing the k-th group-venue pattern with (P, s).

Once a user group P_i includes u_q with size h, and $G[P_i]$ is a connected c-core component, TkPH retrieves k nearest venues for P_i to constitute k group-venue patterns. By this way, an initial solution can be obtained quickly. An earlier viable solution can bring greater pruning performance. Consequently, the distance of the k-th group-venue pattern can be leveraged to reduce the search space.

The following example illustrates how this algorithm works. As the group expansion phase of TkPH algorithm is similar to that in the GVPS algorithm, thus we simply show how TkPH keeps tracking of the current best k results.

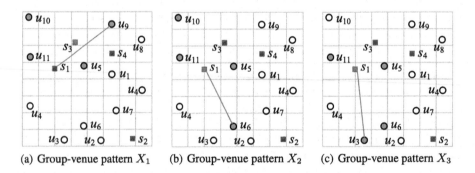

(a) Group-venue pattern X_1 (b) Group-venue pattern X_2 (c) Group-venue pattern X_3

Fig. 3. An example of the TkPH algorithm. (Color figure online)

Example 4. Take the social network in Fig. 1 as an example. Assume a TkPGSG query $q = \langle u_{11}, 4, 2, 2 \rangle$. The group expansion procedure is shown in Fig. 2. As shown in Fig. 3(a), suppose $P_1 = \{u_5, u_9, u_{10}, u_{11}\}$ is the first user group that satisfies the size constraint and social core constraint, we take $\mathbb{X} = \{(P_1, s_3), (P_1, s_1)\}$ as an initial solution. And then, we utilize $Dist(P_1, s_1)$ to reduce the search space. As shown in Fig. 3(a), (b), and (c), we can find three group-venue patterns (consisting of the grey user nodes and the red venue in each figure). Finally, $\mathbb{X} = \{X_1, X_2\}$ is returned as the result of q, since $Dist(X_3) > Dist(X_2)$.

5   Experiments

In this section, we experimentally study the performance of the proposed approaches. We perform a series of sensitivity tests to study the impact of query parameters with real-world datasets. In the following, we first present the experimental settings, and then analyze the experimental results.

5.1   Experimental Settings

Algorithms. We implement four various algorithms: BL, BL+GE, GVPS, and TkPH. Specifically, BL denotes the baseline algorithm presented in Subsect. 3.1. Note that, to find all the connected c-core components with size h, BL utilizes brute-force strategy to do user group expansion. Then, BL adopts the methods proposed in [12] to find a optimal venue for each extracted user group. BL+GE denotes the baseline algorithm with the group expansion strategy presented in Subsect. 3.2. GVPS denotes the group-venue search algorithm presented in Subsect. 3.2. TkPH is the top-k group-venue patterns hunter algorithm presented in Sect. 4. All algorithms are implemented in C++. All experiments are conducted on a computer with 3.20 GHz Intel i5-6500 CPU and 16 GB memory.

Dataset. We use three real world datasets, i.e., Gowalla, Brightkite, and Foursquare. Table 1 gives the details. The first two datasets are downloaded from (http://snap.stanford.edu/data/index.html). We crawled the Foursquare datasets via Foursquare API[1] from November 2014 to January 2016. This dataset has 76,503 users and 1,531,357 social edges. We obtained 299,995 venues located in Singapore.

Table 1. Some statistics of datasets

Datasets	Gowalla	Brightkite	Foursquare
# of venues	1,280,969	772,789	299,995
# of users	196,591	58,228	76,503
# of social edges	950,327	214,078	1,531,357

5.2   Experimental Results

Efficiency. The objective of this set of experiments is to study the efficiency of the proposed algorithms with different real datasets.

We first test the efficiency of BL, BL+GE, and GVPS algorithms to address the PGSG query. Note that the PGSG query can be model as $\langle u_q, h, c \rangle$, where u_q is the query user, h is the user group size constraint, and c is the core number. We set the value of size constraint h to be 8, c to be 3, and the degree of query user to be 9. Figure 4(a) shows the runtime (average time of 50 queries) of the three

---

[1] https://developer.foursquare.com/.

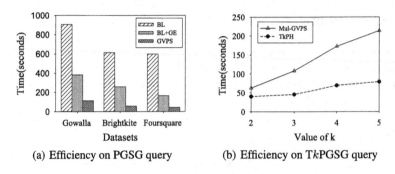

Fig. 4. Efficiency

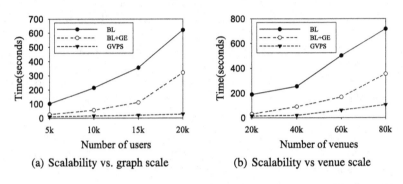

Fig. 5. Scalability

algorithms on the three real datasets. It can be seen that the BL+GE algorithm
runs faster that the BL algorithm, which indicates that the group expansion
strategy presented in Subsect. 3.2 can efficiently reduce the search space. The
runtime is related to the number of venues and the size of social networks. The
reason is that our aim is to find a group-venue pattern. The group extraction
processing relies on the size of social network and the venue search depends on
the number of venues.

To study the efficiency of TkPH algorithm to tackle the TkPGSG query, we
construct a comparative algorithm by invoking multiple the GVPS algorithm,
denoted as Mul-GVPS. With the same parameter setting of h, c, and degree of u_q
in Fig. 4(a), the experimental results of the TkPH algorithm and the Mul-GVPS
algorithm on Gowalla datasets are shown in Fig. 4(b). We observe that the TkPH
algorithm runs faster than the Mul-GVPS algorithm. The reason is that TkPH
algorithm keeps track of the current best k results rather than tracking only
the current best result. The TkPH algorithm is proposed based on the GVPS
algorithm. In the following experiments, we just show the performance of the
algorithms on tackling the PGSG query.

Scalability. We study the effect of the scale of graph (vertex number) and the number of venues on the performance of our proposed algorithms, which can evaluate the scalability of our proposed algorithms.

By extracting the subgraph of the social graph in Gowalla datasets, we obtain 4 datasets containing 5,000, 10,000, 15,000, and 20,000 users, respectively. The corresponding induced graphs can be obtained. Figure 5(a) shows the runtime of the proposed algorithms (the size constraint h is 6, $c = 2$, and the degree of the query user is 6) with the number of users increasing. Analogously, by extracting the subset of the venues in Gowalla datasets, we obtain 4 datasets containing 20K, 40K, 60K, and 80K venues, respectively. Figure 5(b) shows the runtime of the proposed algorithms (the size constraint h is 6, $c = 2$, and the degree of the query user is 6) with the number of venues increasing. It can be seen that the relative ratio of GVPS algorithm changes only slightly, and the other two are opposite. Thus, the scalability of GVPS algorithm outperforms the other two algorithms.

Varying the Query Parameters. In this set of experiments, we study the effect of the three parameters on the performance of our proposed algorithms. Particularly, we regulate u_q for fixed coreness and size constraint pairs to estimate the effect of the degree of query user. Similarly, we evaluate the impact of core number c (or size constraint h) by fixing the other two parameters.

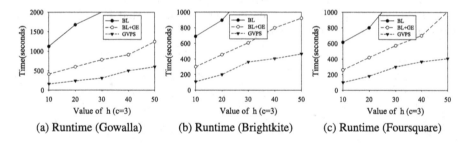

(a) Runtime (Gowalla)   (b) Runtime (Brightkite)   (c) Runtime (Foursquare)

Fig. 6. Varying size constraint h

Varying h. In this set of experiments, we study the effect of h in the PGSG query. Let $c = 3$, we alter the size constraint h from 10 to 50. We regard the average runtime of 50 queries as the performance measure of our proposed algorithms, where the degree of query users is 9. Figure 6 shows the results on the three datasets.

For a fixed coreness and degree of query user, the three algorithms run slower with the size constraint h increasing. The reason is that a query with a larger size constraint h is more likely to have a greater search space, that is, all the algorithms are required to settle more group expansions.

Varying c. In this set of experiments, we study the effect of c in the PGSG query. Let the size constraint $h = 20$, we vary c from 2 to 10. We regard the average

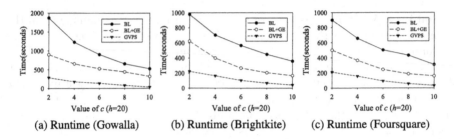

Fig. 7. Varying c

runtime of 50 queries as the performance measure of our proposed algorithms, where the degree of query users is 9. The results on the three real datasets is shown in Fig. 7. The three algorithms run faster as the coreness c increases. The reason is that a query with a larger coreness c is more likely to have a stronger social constraint, which indicates that the query has fewer group expansions to process.

Varying the Query User (Degree). In this set of experiments, we study the effect of the degree with respect to the query user. Let the size constraint $h = 20$ and $c = 4$, we vary the query users with different degree from 10 to 50. Figure 8 shows the running time of the three algorithms over the three real datasets. For a fixed size constraint h and core constraint c pair, the three algorithms run slower with the degree of query user increasing. The reason is that a larger degree of query user leads to a larger search space.

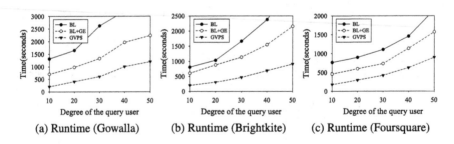

Fig. 8. Varying query user (degree)

6   Related Works

Our work is related to three main research areas: (1) group recommendation, (2) activity-partner recommendations, and (3) geo-social group query.

Group Recommendations. A group recommendation system suggests items to a group of users engaged in a group activity. By taking the different preferences of different group members into consideration, most of the studies on group recommendations aims to recommend items for a given user group [1,10,13,14, 24,26]. Whereas, in this paper, we focus on retrieving a user group rather than research on a given user group.

Activity-Partner Recommendations. Tu *et al.* [21] propose and study the problem of recommending activity partners to Web uses for activity items suggested to them. Besides, they study the use of partner factor for improving activity recommendation. She *et al.* [16,17] propose a novel problem, utility-aware social event-participant planning (USEP), that provides personalized event planning for each participant. By taking the minimum-participant requirement constraint for each evnet, Cheng *et al.* [3] formalize the global event planning with constraints (GEPC) problem, and its incremental variant. Our work is different from this research area. Our work is not only to find a group of users (partners), but also to find a venue (activity item), which yields specified social constraint and spatial constraint in LBSNs.

Geo-Social Query Processing. With the increasing prevalence of location-based services, geo-social queries considering both spatial and social relations are attracting increasing attention [4,6–9,11,22,23,25,28]. The most related work to ours is [23], which study the Socio-Spatial Group Query (SSGQ): given a query venue s, the SSGQ returns a group users, such that (a) each user in the group is socially connected with at least a number of other members, and (b) the sum of distances of all members in the group to s is minimized. Our work is different from [23] in that, (a) the PGSG query aims to find both a venue and a user group simultaneously, (b) the user group is a connected c-core component with a specified size, and (c) the maximum distance of all the users in the group to the venue is minimized.

7   Conclusions

In this paper, we propose a new type of geo-social group query to find both a venue and a user group. We formally define the problem as *Personalized Geo-Social Group* (PGSG) query, which aims to find a group-venue pattern (consisting of a venue and a group of users with size h), where each user in the group is socially connected with at least c other users in the group and the maximum distance of all the users in the group to the venue is minimized. We study the problem of PGSG query and propose a novel search algorithm to find the optimal user group and venue simultaneously. Moreover, we extend the PGSG query to top-k personalized geo-social group (TkPGSG) query. Instead of finding the optimal solution in PGSG query, the TkPGSG query is to return multiple feasibility solutions to guarantee the diversity. We propose an advanced search algorithm TkPH to address the TkPGSG query. Extensive experimental results demonstrate the effectiveness and efficiency of the proposed approaches. A direction for future work is to investigate the issue of social trust and how to integrate social trust into geo-social group query.

Acknowledgments. This research is partially funded by the National Natural Science Foundation of China (No. 61572119, 61622202, U1401256, 61732003, 61729201, 61702086) and the Fundamental Research Funds for the Central Universities (No. N150402005).

References

1. Amer-Yahia, S., Roy, S.B., Chawlat, A., Das, G., Yu, C.: Group recommendation: semantics and efficiency. Proc. VLDB Endow. **2**(1), 754–765 (2009)
2. Batagelj, V., Zaversnik, M.: An O(m) algorithm for cores decomposition of networks. Comput. Sci. **1**(6), 34–37 (2003)
3. Cheng, Y., Yuan, Y., Chen, L., Giraud-Carrier, C., Wang, G.: Complex event-participant planning and its incremental variant. In: 2017 IEEE 33rd International Conference on Data Engineering, ICDE, pp. 859–870. IEEE (2017)
4. Fang, Y., Cheng, R., Li, X., Luo, S., Hu, J.: Effective community search over large spatial graphs. Proc. VLDB Endow. **10**(6), 709–720 (2017)
5. Guttman, A.: R-Trees: A Dynamic Index Structure for Spatial Searching, vol. 14. ACM, New York (1984)
6. Lappas, T., Liu, K., Terzi, E.: Finding a team of experts in social networks. In: Proceedings of the 15th ACM SIGKDD International Conference on Knowledge Discovery and Data Mining, pp. 467–476. ACM (2009)
7. Li, C.T., Shan, M.K.: Team formation for generalized tasks in expertise social networks. In: IEEE Second International Conference on Social Computing, pp. 9–16 (2010)
8. Li, Y., Chen, R., Xu, J., Huang, Q., Hu, H., Choi, B.: Geo-social k-cover group queries for collaborative spatial computing. IEEE Trans. Knowl. Data Eng. **27**(10), 2729–2742 (2015)
9. Li, Y., Wu, D., Xu, J., Choi, B., Su, W.: Spatial-aware interest group queries in location-based social networks. Data Knowl. Eng. **92**, 20–38 (2014)
10. Li, Y.M., Chou, C.L., Lin, L.F.: A social recommender mechanism for location-based group commerce. Inf. Sci. **274**, 125–142 (2014)
11. Liu, W., Sun, W., Chen, C., Huang, Y., Jing, Y., Chen, K.: Circle of friend query in geo-social networks. In: Lee, S., Peng, Z., Zhou, X., Moon, Y.-S., Unland, R., Yoo, J. (eds.) DASFAA 2012. LNCS, vol. 7239, pp. 126–137. Springer, Heidelberg (2012). https://doi.org/10.1007/978-3-642-29035-0_9
12. Papadias, D., Tao, Y., Mouratidis, K., Hui, C.K.: Aggregate nearest neighbor queries in spatial databases. ACM Trans. Database Syst. (TODS) **30**(2), 529–576 (2005)
13. Quijano-Sanchez, L., Recio-Garcia, J.A., Diaz-Agudo, B., Jimenez-Diaz, G.: Social factors in group recommender systems. ACM Trans. Intell. Syst. Technol. (TIST) **4**(1), 8 (2013)
14. Quijano-Sanchez, L., Sauer, C., Recio-Garcia, J.A., Diaz-Agudo, B.: Make it personal: a social explanation system applied to group recommendations. Expert Syst. Appl. **76**, 36–48 (2017)
15. Seidman, S.B.: Network structure and minimum degree. Soc. Netw. **5**(3), 269–287 (1983)
16. She, J., Tong, Y., Chen, L.: Utility-aware social event-participant planning. In: Proceedings of the 2015 ACM SIGMOD International Conference on Management of Data, pp. 1629–1643. ACM (2015)

17. She, J., Tong, Y., Chen, L., Cao, C.C.: Conflict-aware event-participant arrangement and its variant for online setting. IEEE Trans. Knowl. Data Eng. **28**(9), 2281–2295 (2016)
18. Tong, Y., Chen, L., Zhou, Z., Jagadish, H.V., Shou, L., Lv, W.: SLADE: a smart large-scale task decomposer in crowdsourcing. IEEE Trans. Knowl. Data Eng. (2018). https://doi.org/10.1109/TKDE.2018.2797962
19. Tong, Y., She, J., Ding, B., Wang, L., Chen, L.: Online mobile micro-task allocation in spatial crowdsourcing. In: 2016 IEEE 32nd International Conference on Data Engineering, ICDE, pp. 49–60. IEEE (2016)
20. Tong, Y., Wang, L., Zhou, Z., Ding, B., Chen, L., Ye, J., Xu, K.: Flexible online task assignment in real-time spatial data. Proc. VLDB Endow. **10**(11), 1334–1345 (2017)
21. Tu, W., Cheung, D.W., Mamoulis, N., Yang, M., Lu, Z.: Activity recommendation with partners. ACM Trans. Web (TWEB) **12**(1), 4 (2017)
22. Yang, D.N., Chen, Y.L., Lee, W.C., Chen, M.S.: On social-temporal group query with acquaintance constraint. Proc. VLDB Endow. **4**(6), 397–408 (2011)
23. Yang, D.N., Shen, C.Y., Lee, W.C., Chen, M.S.: On socio-spatial group query for location-based social networks. In: ACM SIGKDD International Conference on Knowledge Discovery and Data Mining, pp. 949–957 (2012)
24. Yuan, Q., Cong, G., Lin, C.Y.: COM: a generative model for group recommendation. In: Proceedings of the 20th ACM SIGKDD International Conference on Knowledge Discovery and Data Mining, pp. 163–172. ACM (2014)
25. Yuan, Y., Lian, X., Chen, L., Sun, Y., Wang, G.: RSkNN: kNN search on road networks by incorporating social influence. IEEE Trans. Knowl. Data Eng. **28**(6), 1575–1588 (2016)
26. Zhang, C., Gartrell, M., Minka, T., Zaykov, Y., Guiver, J., et al.: GroupBox: a generative model for group recommendation (2015)
27. Zheng, Y., Zhang, L., Ma, Z., Xie, X., Ma, W.Y.: Recommending friends and locations based on individual location history. Acm Trans. Web **5**(1), 1–44 (2011)
28. Zhu, Q., Hu, H., Xu, C., Xu, J., Lee, W.C.: Geo-social group queries with minimum acquaintance constraints. VLDB J. **26**(5), 709–727 (2017)

Tracking Dynamic Magnet Communities: Insights from a Network Perspective

Chang Liao[1,2], Yun Xiong[1,2(✉)], Xiangnan Kong[3], and Yangyong Zhu[1,2]

[1] Shanghai Key Laboratory of Data Science, School of Computer Science, Fudan University, Shanghai, China
{cliao15,yunx}@fudan.edu.cn
[2] Shanghai Institute for Advanced Communication and Data Science, Fudan University, Shanghai, China
[3] Worcester Polytechnic Institute, Worcester, MA, USA

Abstract. Communities, such as user groups, companies and countries, are important objects in social systems. Recently, researchers have proposed numerous quantitative indicators to measure the attractiveness of a community. However, most of these indicators are mainly under static settings and lack the predictive power of future impact/attractiveness. Meanwhile, in many real-world applications, especially in finance, it is of great interest for the stakeholders to identify the communities, not necessarily the most influential ones at the moment, but the future leaders for years to come. Given the increasing availability of entity-community interaction evolution records, it's natural to exploit them to model the network changes of communities. We refer the change of community interaction as attention flow and define communities that will sustainably attract more entities' attentions than others in a future time interval as dynamic magnet communities. We study the problem of dynamic magnet community identification based on entity-community interaction evolution records. Two major challenges are identified as follows: (1) temporal dynamics, it's difficult to model the rising-declining trend of interactions; (2) sustainability constraints, the effect of attention flow on community prosperity is complex, where too rapid attention growth increases the corruption risks. In response, we propose to model the interaction network evolution of different communities over time by lasso based growth curve fitting. Taking sustainable attention flow into account, we measure attention flow utility from benefit and risk perspectives, and further present a hybrid approach of local and global ranking to track dynamic magnet communities. Due to the lack of dataset for testing, we collected a dataset of international business merger and acquisition network among different countries in the world. The experimental results demonstrate the effectiveness of our proposed model.

1 Introduction

Communities, such as user groups, companies and countries, are important terms in social systems and are defined as bounded sets of patterned relations among

© Springer International Publishing AG, part of Springer Nature 2018
J. Pei et al. (Eds.): DASFAA 2018, LNCS 10827, pp. 406–424, 2018.
https://doi.org/10.1007/978-3-319-91452-7_27

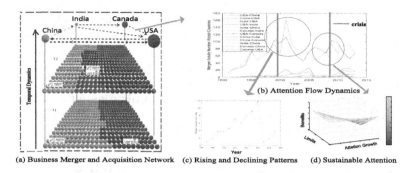

(a) Business Merger and Acquisition Network   (c) Rising and Declining Patterns   (d) Sustainable Attention

Fig. 1. An illustration of cross-border business merger and acquisition network (Cross border business merger and acquisition (M&A) are corporate finance strategies dealing with the buying, selling, dividing and combining of different companies across countries): (1) Communities (countries) draw entities' attention (companies merger activities) from each other, shown from micro to macro perspective; (2) the evolution of attention flow (Interaction refers to aggregated community interaction between $[t_o, t_c)$, while attention flow refers to the change of interaction between $[t_c - 1, t_c)$; where t_o is the beginning time and t_c is the current time) obeys "rising and declining" process, while too rapid attention (merger) growth precedes with high increase of community (financial) recession.

social entities. In recent years, remarkable efforts have been devoted to characterize and rank communities [24,29]. By successfully capturing these top ranked communities, it facilitates the establishment of people's decision on selecting these communities. For example, newcomers can quickly recognize the popularity/importance of each community, which guides them to join the promising communities.

Despite the frequent use of numerous quantitative indicators [21] to measure the impact of a community, they generally lack the predictive power of future impacts, such as detecting future leading communities within the expected time interval in the future. While almost everyone knows the top communities in an industry, finding the smarter crowds of high potential to prosper (not necessarily influential at the moment) in the near future is more important. In other words, the investment value of a community depends on its future performance. By successfully identifying these potential communities, it facilitates the establishment of investment strategies and brings in tremendous rewards.

With the increasing availability of entity-community interaction records among communities, it's natural to exploit them to reveal and model the network changes of communities. We provide an example of such community interaction network as in Fig. 1(a), where the companies migrate from one country to another through cross-border business merger and acquisition activities (For example, when *Alibaba Group* in China bought *Vendio* in USA, there's an attention flow from China to USA). Figure 1(b) shows the entity-community interaction evolution records, which can potentially reveal the impact and potential impact of communities. In other words, the trend of attention flow can indicate the communities of investment potential to some degree.

To fill this gap of the lack of research on potential communities identification, as well as motivated by the available entity-community interaction information, we propose to identify the communities which will prosper in the given future time interval. We define dynamic magnet community as the community which will have the strong abilities/potentials to attract others sustainably in the given future time interval, and formalize the task as **dynamic magnet communities tracking** based on entity-community interaction information. We summarize the unique challenges as follows:

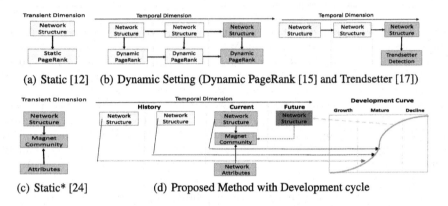

(a) Static [12]   (b) **Dynamic Setting (Dynamic PageRank [15] and Trendsetter [17])**

(c) Static* [24]   (d) **Proposed Method with Development cycle**

Fig. 2. Comparison of different methods for dynamic magnet communities detection.

Temporal Dynamics: To identify dynamic magnet communities, the straightforward method is PageRank [12], which is limited by its lack of temporal information modeling, as in Fig. 2(a). When combined with temporal dynamics modeling, dynamic system based method [15] or exponential time decay based method [17] are applied, as in Fig. 2(b). However, such work focuses on predicting the future change mainly through using a monotonically decreasing/increasing function (linear or exponential). However, as is shown in Fig. 1(c), there are two distinct conditions (1) flow evolves differently: slowly or quickly; (2) flow evolution has distinct life stage (growth and declining) patterns. And thus, a simple function cannot effectively describe such process. While growth curve [22] can statistically model this development cycle, it's still very hard to estimate the parameters with limited data samples. For instance, while some community interactions begin very recently, it's very difficult to estimate corresponding growth curves.

Sustainability Constraints: Suppose we can successfully capture the evolution of interaction network, it is still not clear how to detect dynamic magnet communities. Diffusion based method can only exploit influence information of communities, which is not equaled to what dynamic magnet communities stands for. [24] proposes to detect magnet communities by exploiting both network and attribute information in Fig. 2(c), but it focuses on static settings as well as extensive attributes (which is hard to collect). Meanwhile, the effect of attention flow on community prosperity is complex, where only sustainable attention flow counts for the prosperity of a community. As depicted in Fig. 1(d), too small quantity of attention flow is a reflection of insufficient vitality, while too big of it may increase the risk of community recession. How to formally define and integrate such constraints with diffusion based ranking method is non-trivial. For example, the financial recession is often followed by large quantity of foreign merger/acquisition events, also shown in Fig. 1(b).

To alleviate the aforementioned issues, we introduce DMC (Dynamic Magnet Community Tracking) framework to detect dynamic magnet communities together with predicting attention flow, as illustrated in Fig. 2(d). Compared with the conventional settings, DMC can unfold more sophisticated and microscopic evolution process of attention flow. The major contributions are summarized as follows:

- We motivate and propose a new problem called tracking dynamic magnet communities, with particular emphasis on integrating entity-community interaction network. It is immensely helpful, especially in investment market. The problem of influence analysis has not been studied in this context, to the best of our knowledge.
- The proposed model integrates magnet score quantification with attention flow evolution prediction. Lasso based growth curve model is adopted to model the "growth-declining" trend of attention flow from a mechanistic perspective. We further extract features of network dynamics from gain/risk perspective to encode community attention flow sustainability and integrate them in a joint constrained optimization framework.
- Since there is no available dataset to justify our model, we are the first time to introduce cross-border merger and acquisition dataset into computer science community. We believe such data is of great research and practical value, for cross-border business merger and acquisition is a major form of global investment.

2 Preliminaries

In this section, we introduce several concepts and then formulate the problem. Several important mathematical notations are listed in Table 1.

2.1 Community Projected Flow Network Construction

In this subsection, we organize the input interaction records as dynamic Entity Community Network and transform it to dynamic community network, as well as the community projected flow network by predicting future community interactions.

Dynamic Entity-Community Interaction Network: Communities are group of entities that gather together. Combined with the time information,

Table 1. Several important mathematical notations

Notations	Description	Notations	Description
$t/\triangledown t$	Time/future time interval	n/ne	Community/entity Size
g_G, g_P	Growth curve function	$\alpha, \gamma, \lambda, \varphi$	Weighting parameters
Γ	Entity-community interaction network	Γ^C	Community interaction network
Ee	Community-entity tensor $\mathbb{R}^{n \times ne \times t}$	L, O, R, S, Q	Loss function
$\tilde{\mathbf{G}}$	Projected flow network $\mathbb{R}^{n \times n \times \triangledown t}$	**G**	Attention flow tensor $\mathbb{R}^{n \times n \times t}$
E	Community interaction tensor $\mathbb{R}^{n \times n \times t}$	$\tilde{\mathbf{E}}$	Predicted interaction $\mathbb{R}^{n \times n \times \triangledown t}$
β	Interaction trend parameters $\mathbb{R}^{n \times 3}$	**d**	Community size $\mathbb{R}^{n \times 1}$
f	GDP time series $\mathbb{R}^{n \times t}$	**I**	Identity matrix $\mathbb{R}^{n \times n}$
$\tilde{\mathbf{m}}$	Dynamic magnet score $\mathbb{R}^{n \times \triangledown t}$	**W**	Similarity matrices $\mathbb{R}^{n \times n}$

a dynamic entity-community interaction network $\Gamma_t(Ve, \mathbf{Ee}_t)$, where Ve are two types of nodes (entities and communities), and $\mathbf{Ee}_t$ are the interactions between entities and communities at time t, as is shown in Fig. 3(a). When an entity migrates from a community to another (say merger activity), there is a directed link (attention) between the two communities.

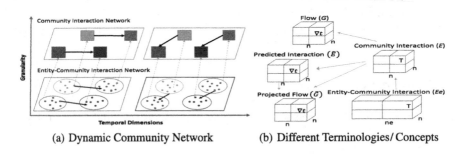

(a) Dynamic Community Network   (b) Different Terminologies/Concepts

Fig. 3. Illustration of community interaction network extraction

Dynamic Community (Interaction) Network: Given dynamic entity-community interaction network, a dynamic community network can be transformed from it and is represented as $\Gamma_t^C(V, \mathbf{E}_t, \mathbf{d})$, where V are the communities and $\mathbf{d}$ is the quantity of entities for each community. $\mathbf{E}_t$ are interactions observed among communities at time t, also shown in Fig. 3(a).

Community Projected Flow Network: The community Projected flow network $\tilde{\mathbf{G}}_{\triangledown t}$ with future $\triangledown t$ interval is defined as the overall attention flow[1] predicted in the future time interval $\triangledown t$ by averaging. It is determined as

$$\tilde{\mathbf{G}}_{\triangledown t} = 1/|\triangledown t| \sum_{t'=t+1}^{t+\triangledown t} \tilde{\mathbf{E}}_{t'-t} - \mathbf{E}_t, \tag{1}$$

The future community interaction refers to $\tilde{\mathbf{E}}_{\triangledown t}$, the change of which is predicted attention flow in the future time interval $\triangledown t$. We further explain the aforementioned notations with Fig. 3(b): (1) the input is entity-community interaction tensor $\mathbf{Ee}$; (2) by transforming from entity-community interaction tensor, community interaction is $\mathbf{E}$ and predicted interaction is $\tilde{\mathbf{E}}$; (3) flow refers to overall change of interactions among communities, that is, attention flow at current time is $\mathbf{G}_t = (\mathbf{E}_t - \mathbf{E}_{t-1})$, while predicted attention flow around future time interval $\triangledown t$ is as Eq. (1).

2.2   Dynamic Magnet Community Detection

Dynamic Magnet Community: Dynamic Magnet Communities are defined as communities that have high-ranked dynamic magnet score. The dynamic magnet score of a community $V(i)$ is defined as $\tilde{\mathbf{m}}_i$, which is to evaluate its potential ability to attract attention flow from others steadily. As in Eq. (2), R is the objective function that ensures the score $\tilde{\mathbf{m}}$ consistent with community Projected Flow Network $\tilde{\mathbf{G}}$ globally, while S are the constraints which guarantee the scores $\tilde{\mathbf{m}}$ are consistent with network information and community size $\mathbf{d}$ locally as much as possible:

$$\tilde{\mathbf{m}}^* = \arg \min_{\tilde{\mathbf{m}}} R(\tilde{\mathbf{m}}; \tilde{\mathbf{G}}) \qquad s.t. \quad S(\tilde{\mathbf{m}}; \mathbf{d}, \tilde{\mathbf{G}}, \mathbf{G}) \tag{2}$$

where parameter $\tilde{\mathbf{m}}$ is time-sensitive, and we represent $\tilde{\mathbf{m}}_{\triangledown t}$ as dynamic magnet score for future $\triangledown t$ time intervals[2]. By measuring the dynamic magnet score for each community, we can identify the dynamic magnet communities by ranking and selecting the communities with high dynamic magnet score[3]. We regard a community as a dynamic magnet community if the expected attention flow for the future time interval $\triangledown t$ meets the following two requirements: (1) better attention flow from other high ranked communities; (2) sustainable attention flow growth compared with internal community size, which will be discussed later.

Similar definitions can be classified into three categories, and the differences are as follows: (1) in static influence ranks [12,24], temporal information is lost in these methods; (2) among dynamic influence metrics [15,17], temporal infor-

---

[1] The start of it is based on the current time t.

[2] $\tilde{\mathbf{m}}$ is taken dynamic magnet score matrix for future time and $\tilde{\mathbf{m}}_{\triangledown t}$ is taken as dynamic magnet score vector at future time $\triangledown t$.

[3] While it is a ranking problem, threshold or top k can be defined randomly.

mation has not been utilized fully; (3) for domain specific methods [9, 18, 25], they cannot be easily generalized to other applications. In contrast to them, we attempt to detect dynamic magnet communities by predicting all the future changes and the proposed method is domain independent.

3 Dynamic Magnet Community Model

In this section, we develop a model, called DMC (Dynamic Magnet Community), to detect dynamic magnet communities. The main framework is shown in Fig. 4(a). There are two procedures, as in Fig. 4(b): (1) community Projected Flow Network constructed by evolution prediction; (2) dynamic magnet communities detection based on community projected flow network.

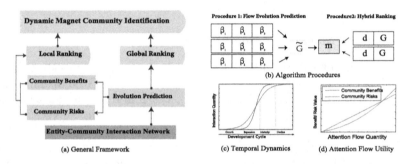

Fig. 4. DMC Framework: (1) community Projected Flow Network construction by lasso based growth curve as in Eq. (5); (2) measure dynamic magnet score by combining global ranking with sustainable attention flow based local ranking as in Eq. (8).

3.1 Community Projected Flow Network Construction

The community Projected flow is defined as the difference between the predicted interactions and current interactions by averaging. And the community projceted flow network can be built by predicting the future overall interactions given the current interactions, as in Eq. (1). To predict the future overall interactions $\tilde{\mathbf{E}}_{\nabla t}$, we need to model the evolution of community interaction network $\mathbf{E}$. As is shown in Eq. (3), L is the objective function that minimizes the error between estimated value based on β and real value of $\mathbf{E}$, while O are the constraints which guarantee the community interaction evolutions are correlated together with community size d.

$$\beta^* = \arg\min_{\beta} L(\beta; \mathbf{E}) \qquad s.t. \quad O(\beta; \mathbf{d}) \tag{3}$$

To model the rising and declining trend of the change of community interactions, as in Fig. 4(c), growth model is adopted. Growth model [22] is a technique that statistically assesses change and correlates of change, which is used in statistics to determine the type of growth pattern of the quantity.

Two growth curve models are applied here. *The Pearl Curve*: The function $g_P(t) := g_P(t; \beta_1, \beta_2, \beta_3) := \frac{\beta_3}{1 + \beta_2 \exp(-\beta_1 t)}$ is a widely used logistic function. The curve converges to the initial value $(y = 0)$ at $t = -\infty$ and reaches its upper limit β_3 at $t = \infty$. *The Gompertz Curve*: Another function to model the increase or decrease of a system is $g_G(t) := g_G(t; \beta_1, \beta_2, \beta_3) := \exp(\beta_3 + \beta_1 \beta_2{}^t), t \geq 0$, where $\beta_1, \beta_3 \in R$ and $\beta_2 \in (0,1)$. The Gompertz model has an initial value of 0 at $t = -\infty$ and an upper limit of $\exp(\beta_3)$ at $t = \infty$. There are three parameters in both types of growth curve: the theoretical upper limit of growth curve determined by parameter β_3, as well as deviation and rate of growth related with parameters β_1 and β_2. The estimations of these parameters are in Eq. (4) by the least square method with T time data points.

$$L(\mathbf{E}_t(i,j); \beta_1, \beta_2, \beta_3) = \sum_{t=1}^{T} (\mathbf{E}_t(i,j) - g_P(t, \beta_1, \beta_2, \beta_3))^2 \qquad (4)$$

The parameters β_1 and β_2 can be estimated by minimizing the cost function L. Meanwhile, β_3 can be learnt through interpolation method.

Lasso based Evolution Estimation: However, with little training data, it cannot provide enough information for the estimation of parameters $\{\beta_{i,j,1}, \beta_{i,j,2}, \beta_{i,j,3}\}$ in growth curve model. Specially, the trend of new interactions between communities cannot be estimated for **the lack of history data**.

Fortunately, there are some similar evolution patterns between different attention interactions. Intuitively, it is reasonable to believe that those evolution patterns can aid in parameter estimations for new interaction growth curve. That is, the more similar two communities are, the less differences their growth curves parameters are. For instance, recalling that in Fig. 1(b), similar cross-border mergers have similar evolution patterns. The similarities between the communities as $\mathbf{W}_{i,i'} = \exp\{-\frac{\|\mathbf{d}(i) - \mathbf{d}(i')\|^2}{\sigma^2}\}$, where $\mathbf{d}$ is community size vector. And then, we can incorporate their similarity based lasso as constraints $O(\beta; \mathbf{d})$ into the problem of estimating interaction growth curve parameters. The **optimization function** is displayed in Eq. (5), where the former part is on the error of interaction growth curve fitting and the latter part is to penalize differences between the variables at similar interactions.

$$\min \sum_{ij} \sum_{t} (\mathbf{E}_t(i,j) - g_P(t, \beta_{i,j,1}, \beta_{i,j,2}, \beta_{i,j,3}))^2 + \lambda \sum_{ii'jj'} W_{i,i'} W_{j,j'} \|\beta_{i,j} - \beta_{i',j'}\|_2$$
$$s.t. \quad \beta_{i,j,1} > 0, \beta_{i,j,2} > 0, \beta_{i,j,3} > 0$$
$$(5)$$

The parameter $\beta_{i,j}$ refers to parameters $\{\beta_{i,j,1}, \beta_{i,j,2}, \beta_{i,j,3}\}$ of the growth curve for corresponding interactions from community i to j. Parameter λ is a weighting parameter, and the larger it is, the more emphasis the optimization function puts on the similarity requirement of the corresponding growth curves. The estimation of $\beta_{i,j}$ can be taken as a weighted median of other similar interaction solutions. Alternating Direction Method of Multipliers (ADMM) [1] can be a good solution.

3.2   Measuring Dynamic Magnet Score

With the estimated community Projected Flow Network $\tilde{\mathbf{G}}_{\nabla t}$, a straightforward way to measure the dynamic magnet score of a community is to use the random walk with restart. Mathematically, to ensure the estimated dynamic magnet score $\tilde{\mathbf{m}}^*_{\nabla t}$ consistent with the real dynamic magnet score $\tilde{\mathbf{m}}_{\nabla t}$, the **objective function** R is as follows:

$$R = ||\tilde{\mathbf{m}}^*_{\nabla t} - \tilde{\mathbf{m}}_{\nabla t}||_2 = ||(\alpha\tilde{\mathbf{G}}_{\nabla t} - \mathbf{I})\tilde{\mathbf{m}}_{\nabla t} + (1 - \alpha)\frac{\mathbf{e}}{|V|}||_2 \qquad (6)$$

In particular, the ranking scores should be as close to the stationary distribution of Markov process as possible, iteratively updated as $\tilde{\mathbf{m}}^*_{\nabla t} = \mathbf{m}^{(s+1)}_{\nabla t} = \alpha\tilde{\mathbf{G}}_{\nabla t}\tilde{\mathbf{m}}^{(s)}_{\nabla t} + (1 - \alpha)\frac{\mathbf{e}}{|V|}$, where s is the iteration number and α is the damping factor, with flow between communities as $\tilde{\mathbf{G}}_{\nabla t}$ and average reset probability as $\frac{\mathbf{e}}{|V|}$.

Sustainable Attention Flow Based Local Ranking: Just accounting for network diffusion model still cannot achieve a remarkable result. In other words, a community of good investment is not a community of just good influence, but of better attention flow growth and weaker corruption risks, as shown in Fig. 4(d). To be more concrete, (1) it is the difference between the predicted attention flow (Projected Flow Network) and current attention flow that accounts for the investment gain potentials; (2) the strong external attention flow also brings in high risks, where the attention flow should meet the needs of the present without compromising the community stability. Motivated by this, we extract investment value features from gain and risk perspectives: (1) the gain as attention flow growth $\tilde{\mathbf{G}}_{\nabla t} - \mathbf{G}_t$; (2) the risk as external attention flow number compared with the community size $-\tilde{\mathbf{G}}^2_{\nabla t}/\mathbf{d}$ (d is the community size). We refine the combination of these two parts as the **utility of attention flow** and integrate it to sustainable attention flow requirements. The attention flow utility based local ranking can be formalized as $S(\tilde{\mathbf{m}}; \mathbf{d}, \mathbf{G}, \tilde{\mathbf{G}})$, and the **objective function** S with regard to $\tilde{\mathbf{m}}$ is

$$\begin{aligned} S = \sum_{i,j} (\tilde{\mathbf{m}}_{\nabla t}(i) - \tilde{\mathbf{m}}_{\nabla t}(j))((\tilde{\mathbf{G}}_{\nabla t}(j,i) - \mathbf{G}_t(j,i) - \gamma\frac{\tilde{\mathbf{G}}^2_{\nabla t}(j,i)}{\mathbf{d}(i)}) \\ -(\tilde{\mathbf{G}}_{\nabla t}(i,j) - \mathbf{G}_t(i,j) - \gamma\frac{\tilde{\mathbf{G}}^2_{\nabla t}(i,j)}{\mathbf{d}(j)})), \end{aligned} \qquad (7)$$

where γ is the weighting parameter between the benefit and the risk, accounting for both community development growth and community stability cost. In particular, the local ranking that maximizes the sum of the weight of unviolated ranking community pairs, which can be also taken to subtract the weight of all rank-violating community pairs from the utility of attention flow perspectives.

Hybrid of Global and Local Ranking: Here, we jointly model diffusion based ranking and local (pairwise) ranking in a unified objective function. Formally, our objective function at high level is to minimize Q, the Frobenius norm of the predicted dynamic magnet score and the real value. Sustainable flow based pairwise ranking constraints to fulfil requirements is displayed when

$\tilde{m}_{\nabla t}(i) - \tilde{m}_{\nabla t}(j) \geq 0$, and φ is casted as the weighting parameter between global ranking and local ranking constraints. Aiming to estimate $\arg\min_{\tilde{m}_{\nabla t}} Q$, we represent the **generalized objective function** Q as:

$$Q = ||(\alpha\tilde{\mathbf{G}}_{\nabla t} - \mathbf{I})\tilde{\mathbf{m}}_{\nabla t} + (1 - \alpha)\frac{\mathbf{e}}{|V|}||_2 - \varphi\tilde{\mathbf{m}}_{\nabla t}^T\mathbf{c}$$

$$s.t. \quad \mathbf{c}(i) = \gamma(\sum_{u \in in}(\frac{\tilde{G}_{\nabla t}^2(i,u)}{\mathbf{d}_u} - \frac{\tilde{G}_{\nabla t}^2(u,i)}{\mathbf{d}_j}) - \sum_{v \in out}(\frac{\tilde{G}_{\nabla t}^2(i,v)}{\mathbf{d}_v} - \frac{\tilde{G}_{\nabla t}^2(v,i)}{\mathbf{d}_i}))$$
$$+(\sum_{u \in in}(\tilde{\mathbf{G}}_{\nabla t}(u,i) - \tilde{\mathbf{G}}_{\nabla t}(i,u)) - \sum_{v \in out}(\tilde{\mathbf{G}}_{\nabla t}(v,i) - \tilde{\mathbf{G}}_{\nabla t}(i,v))) \qquad (8)$$
$$-(\sum_{u \in in}(\mathbf{G}_t(u,i) - \mathbf{G}_t(i,u)) - \sum_{v \in out}(\mathbf{G}_t(v,i) - \mathbf{G}_t(i,v)))$$

To solve the optimization problem stated in Eq. (8), it proves to be a convex optimization problem, and many strategies can be brought in.

Theorem 1. *The objective function Q stated is a convex function.*

The objective function can be expanded via a mathematical transformation as

$$\min_{\tilde{\mathbf{m}}_{\nabla t}} \tilde{\mathbf{m}}_{\nabla t}^T(\alpha\tilde{\mathbf{G}}_{\nabla t} - \mathbf{I})^T(\alpha\tilde{\mathbf{G}}_{\nabla t} - \mathbf{I})\tilde{\mathbf{m}}_{\nabla t} - \varphi\tilde{\mathbf{m}}_{\nabla t}^T\bar{\mathbf{c}}, \qquad (9)$$

where $\bar{\mathbf{c}} = \mathbf{c} + ((\mathbf{I} - \alpha\tilde{\mathbf{G}}_{\nabla t})\frac{\mathbf{e}}{|V|})/\varphi$. Straightforwardly, while $(\alpha\tilde{\mathbf{G}}_{\nabla t} - \mathbf{I})^T(\alpha\tilde{\mathbf{G}}_{\nabla t} - \mathbf{I})$ is symmetric and positive semi-definite, the objective function is a convex function. The convexity of objective function and the existence of global minimum ensures there are polynomial time algorithms that can find the minimum solutions. In such ways, we can measure the dynamic magnet score $\tilde{m}$ for each community efficiently.

4   Experiments

Experiments are conducted on machines with Intel(R) Xeon(R) CPU of 2.5 GHz and 192 G memory. Lasso tools [4] in Python are adopted for attention flow evolution prediction. And the dynamic magnet community detection task is written in Matlab.

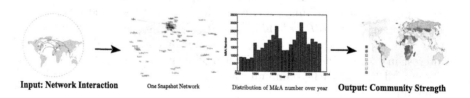

Input: Network Interaction   One Snapshot Network   Distribution of M&A number over year   **Output: Community Strength**

Fig. 5. Cross-border merger and acquisition data description

4.1   Experimental Settings

Dataset Description the International M&A (Merger and Acquisition) Dataset: As for entity-community interaction records, we collect data for all cross-border mergers from the SDC Thomson Platinum database, with announcement dates are between January 1, 1990 and December 31, 2012, comprising a sample of 39,754 observations. Figure 5 shows the basic statistics: (1) one year snapshot of merger and acquisition events among countries; (2) the distribution of M&A number over years. Meanwhile, we collect the listed companies number for each country. Finally, the dynamic entity-community interaction network is derived as the input data, comprising of 208 communities and their related entity migration records within the 23 years.

Ground Truth: Despite many kinds of country rankings, there is no one that directly captures the property of the dynamic magnet score. Nevertheless, we select two related indexes: (1) the Gross Domestic Product (GDP)[4], (2) the number of $fortune500$[5] for each country. Both indexes measure the country's prosperity: the former one for the value of the market, while the latter one for the commercial atmosphere (affection strength of investors). To account for dynamic magnet communities, these two metrics are modified as $gGDP@\nabla t$ (next ∇t years GDP growth Ranking) and $fortune500@\nabla t$ (the average number ranking of $fortune500$ within the following ∇t years).

The Synthetic Dataset: To demonstrate the scalability of the model, we generate synthetic data similar to the real data we are interested in modeling. That is, we generate the synthetic dataset with the same settings of international merger and acquisition dataset. We vary and amplify the real dataset by expanding its size. The number of interactions ranges from 852 to 6,840, and the number of communities ranges from 208 to 13,332.

Baseline Method. We compare the proposed model DMC with four baselines:

- *PageRank Algorithm:* [12] is the classic method for influence quantification. We choose PageRank with one simple but important variation as our baseline. We replace its original binary adjacent matrix with real-valued matrix.
- *Dynamic PageRank (dPageRank) Algorithm:* [15] proposes a dynamical system that captures changes to the network centrality of nodes as external interest in those nodes varies. By slight modification of our settings, we treat the overall flow of each community as external event, and it can rank each community.
- *Trendsetter Detection:* [17] proposes a ranking strategy that focuses on the ability to push new ideas that will be successful in the future. When assuming recent relations are more important, exponential weights are given to recent interactions.

---

[4] It is the monetary measure of the market value in a period of time.
[5] http://fortune.com/fortune500/2016.

– *Magnet Community (MC) Detection*: [24] formalizes attention flow, attention quality, and attention persistence with combination of community feature into a graph ranking formulation based on constraint quadratic programming. We take community size **d** as community features.

We evaluate our method from three perspectives: (1) evaluation on dynamic magnet community detection; (2) model analysis; (3) dynamic magnet score based applications.

4.2   Evaluation for Detection Results

In this part, we evaluate the effectiveness of dynamic magnet community detection. As for detection accuracy measurement, we compare the community dynamic magnet score based ranking result. Two metrics are used: (1) *Discounted Cumulative Gain (DCG)* [6], a judgement of ranking which emphasizes the correctness of high ranked objects; (2) *Weighted Pairwise Distance (wDist)* [24], another judgement with emphasizing on ranking order and correctness.

To evaluate the derived score's ability to identify magnet communities in the next ∇t years, $gGDP@\nabla t$ ranking result is used as ground truth. We perform prediction with ∇t varying from 1 to 5, starting from the year 2007. Figure 6(a) compares the normalized marginal gain on DCG values for different methods. As it shows, DMC outperforms the other methods significantly. Figure 6(b) also shows that DMC result has smaller $wDist$ (better performance). Fairly speaking, both *trendsetter algorithm* and *dPageRank algorithm* do have the ability to capture the future potential communities, but they cannot utilize temporal information to the fullest. Both exponential decay and linear dynamic system cannot model the "rising-decreasing" trend characteristics. As with the other two methods, they generally lack the prediction power and perform even worse. What's more, while the results of other baselines remain the same varying with time interval ∇t, DMC displays different ranking results. This is of great significance for both long-term and short-term investors, as it can detect future potential communities and the time to prosper.

To reduce the variance of model performance, we compare the normalized marginal gain on DCG values along the 13 years from year 2000 to year 2012. It aims to evaluate the predictive ability of the derived score in next year magnet community identification. The reason why we choose $\nabla t = 1$ lies in the limited data samples (time intervals). Figure 6(c) compares the normalized marginal gain on DCG values along the 13 years from year 2000 to year 2012 on $gGDP@1$ metrics. As it shows, DMC outperforms the other methods. Figure 6(d) also shows that DMC result has smaller normalized weighted distance. The performance curve is non-smooth, possibly due to violating ranking results on less developed countries. Nevertheless, our proposed model consistently beats others and shows its priority.

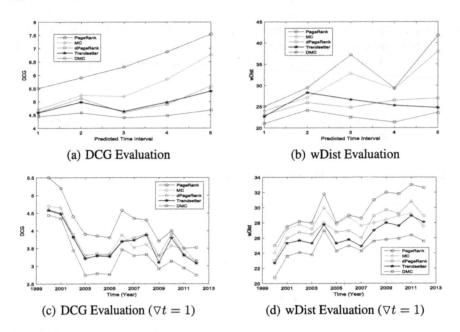

(a) DCG Evaluation

(b) wDist Evaluation

(c) DCG Evaluation ($\nabla t = 1$)

(d) wDist Evaluation ($\nabla t = 1$)

Fig. 6. Detection accuracy results

4.3   Model Analysis

We evaluate the effectiveness and efficiency of the proposed model based on inter-action network evolution prediction and sustainable flow based hybrid ranking.

Interaction Network Evolution Prediction: We evaluate the community interaction network evolution prediction results with root mean squared error(RMSE). Two growth curves and their extensions are compared in modeling temporal dynamics: (1) Gompertz Curve Fitting; (2) Pearl Curve Fitting; (3) Gompertz Curve with lasso; (4) Pearl Curve with lasso. We compare the evolution prediction accuracy for the next 5 years after the year 2007. Figure 7(a) indicates that pearl curve with lasso has better performance. As is shown in Fig. 7(a), it outperforms Pearl Curve by almost 30%, despite that the RMSE of the latter is already low. The result confirms the superiority of our model in capturing the trend of merger flow across countries.

As for lasso weight λ in Fig. 7(b), the performance remains good between 10 and 15. Too large a value of λ is not good as the objective function would put too much weight on similarity between community interactions and ignores their individual characteristics and vice versa. To test the model scalability, we evaluate it by using the synthetic data. The results are shown in Fig. 7(c). As is shown, the running time changes smoothly with the growing number of communities, and that the algorithm can be run in acceptable time.

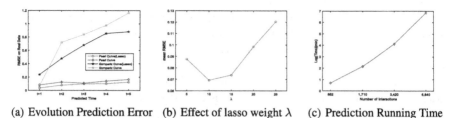

(a) Evolution Prediction Error   (b) Effect of lasso weight λ   (c) Prediction Running Time

Fig. 7. Interaction network evolution prediction

Sustainable Flow Based Hybrid Ranking: We now evaluate the dynamic magnet score in the next year. Using a real dataset along the 13 years from 2000 to 2012, we take the ranking accuracy DCG as the evaluation metric. We compare the benefits of modeling attention flow utility, and the results in Fig. 8(a) suggest the utility modeling is effective in identifying dynamic magnet communities. We also compare the proposed model with only global based ranking and only local based ranking. The results in Fig. 8(b) indicate the necessity of the hybrid ranking. To test the model scalability, we evaluate it by using the synthetic data. The results are shown in Fig. 8(c). As is shown, the running time changes smoothly with the growing number of communities, and the algorithm can yield acceptable running time.

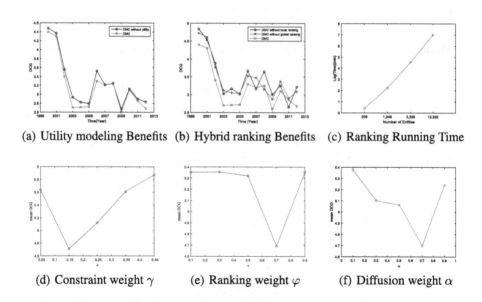

(a) Utility modeling Benefits   (b) Hybrid ranking Benefits   (c) Ranking Running Time

(d) Constraint weight γ (e) Ranking weight φ (f) Diffusion weight α

Fig. 8. Benefits on sustainable flow based hybrid ranking

For constraint weight γ, since it determines the relative weight about the benefits and costs the mergers bring in, it is more application dependent.

The choice of the parameter γ can be determined by cross-validation, or one can use some prior knowledge. We start from the year 2007 and vary the value of ∇t from 1 to 5 to obtain the average DCG value on ranking accuracy evaluation. In Fig. 8(d), we can see the result is consistent around the value from 0.15 to 0.25. As for ranking weight φ and α in Fig. 8(e) and (f), though the performance varies with parameter values, the change is bounded by a small range.

4.4   Dynamic Magnet Score Application Case Study

Table 2 lists the top 10 countries identified by DMC and baseline methods in the year 2009 for forecasting the following future time interval $\nabla t = 3$ (others are fine). Both $gGDP@3$ and $fortune500@3$ are served as ground truth. Intuitively, the more agreement a ranking list has with $gGDP@3$ and $fortune500@3$, the better the performance is. We take the hit number of the detected top 10 communities with ground truth for evaluation. From the table, we can see DMC has the best matching results with single $gGDP@3$ metrics and combined $gGDP@3 + fortune500@3$ metrics. $Trendsetter$ and $dPageRank$ perform slightly better than MC and $PageRank$, as temporal information is integrated. Although it seems the detected communities are quite trivial and redundant compared with traditional indicator based evaluation, the main contribution is measuring community potentials by utilizing publicly available information. While traditional indicators are collected with extensive efforts, our model is based on information available online that costs far less.

Table 2. Top 10 Dynamic magnet communities on merger/acquisition network

gGDP@3	Fortune 500@3	DMC(8/9)	MC(5/7)	Trend setter(6/7)	Page Rank(4/6)	dPage Rank(5/7)
China	United States	United States	United States	United States	United States	United States
United States	China	Canada	Australia	Canada	Canada	United Kingdom
Russia	Japan	Australia	Sweden	Australia	Mexico	Canada
Japan	France	United Kingdom	United Kingdom	United Kingdom	United Kingdom	Australia
Brazil	Germany	Japan	India	France	India	France
Australia	United Kingdom	China	China	China	China	Turkey
Canada	Switzerland	France	France	India	France	Hong Kong
South Korea	South Korea	Brazil	Brazil	Bosnia	Spain	China
United Kingdom	Netherlands	Germany	Germany	Abrabia	Germany	Spain
Arabia	Canada	India	Serbia	Singapore	Turkey	Italy

To further demonstrate the practical value of the proposed dynamic magnet score, we apply it to GDP forecasting tasks. To begin, a linear regression model

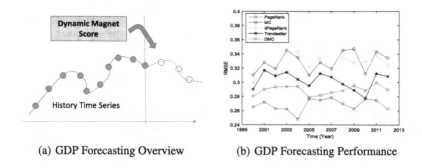

(a) GDP Forecasting Overview   (b) GDP Forecasting Performance

Fig. 9. The benefits of dynamic magnet score

is set up, where the function is $\mathbf{f}(t) \approx [\mathbf{f}(t-1)\mathbf{f}(t-2)...\mathbf{f}(t-w)]\,\mathbf{b}$, with $\mathbf{b}$ as parameter vector and $\mathbf{f}(t)$ referring to the GDP value at time t. In order to predict the GDP at time $t+1$ (denoted as $\mathbf{f}(t+1)$), the data before time t (the year 2000) is taken for training with $w = 10$, as shown in Fig. 9(a). Meanwhile, the dynamic magnet score $\tilde{m}_1$ for the next year is added as a new feature and incorporated into the linear regression model. The RMSE(root mean squared error) performance on GDP prediction is shown in Fig. 9(b), where the dynamic magnet score based prediction is much better than others. To explain, the growth curve fitting can model the evolution of attention flow more precisely, and the sustainable attention flow based hybrid ranking can encode the complex relationship between attention flow and community prosperity to some degree.

5 Related Work

5.1 Influence Analysis

[19] surveys social influence analysis research. [26] focuses on how to effectively exploit rich heterogeneous information to improve the ranking performance. To compare, these methods are supervised, while our task is to gauge a new type of influence (magnetic force) with unsupervised settings.

From application perspectives, this paper belongs to country evaluation category. [20] defines a new metric for the country fitness, while [2] investigates on measuring the competitiveness of countries. [5] attempts to measure internet gravitation between countries by website link information. In this work, we utilize cross-border business merger and acquisition interaction data to measure attractiveness potential of countries.

5.2 Time Series Modeling

When considering influence analysis in time series, [8] proposes to compute the intensities and bursts of some hidden communities over time. [16] applys the technology growth curve model to evaluate and predict scientific technology capacity within a country. Hot topic prediction is introduced in [9]. Although these models succeed in using temporal information, the essential 'rising and declining' characteristics with limited data samples in our settings is lacking modeling.

For temporal information modeling, [23] derives a mechanistic model for the citation dynamics of individual papers. Meanwhile, a survey of growth model is in [22]. [7] presents a methodology for modeling time drifting of user preferences in the context of recommender systems. [27] adapts the inverted-U-curve to model the dynamic evolution process of user interests for a better recommendation. However, limited work has combined temporal dynamics modeling with influence analysis sophisticatedly.

5.3   Community Characterization

Our paper is a contribution to the growing literatures on the study of community characterization in the recent years. [21] proposes to characterize the communities detected in heterogeneous complex systems. When interacted with influence analysis, [11] introduces to analyze social influence at the granularity of communities. [3] attempts to track the progression of the community strength. [10] proposes 'community gravity' that measures how strongly a user might be attracted to, while the authors in [24] propose to identify magnet community in social network. One step further, our work aims at tracking dynamic magnet communities bound to prosper in near future, together with uncovering their evolution process.

With regard to community dynamics, [14] investigates the group growth dynamics, while [28] proposes to model group evolution. [13] discovers a strong dichotomy among groups in terms of life-cycle: social messaging groups come together, grow new members, and evolve over time. Above all, our work aims to further integrate community dynamics into influence evaluation.

6   Conclusion

In this paper, we discuss a practical but challenging problem: tracking dynamic magnet communities. This problem is new and brings in many benefits, especially in financial investment domains. DMC (Dynamic Magnet Community) model has been presented, which can track dynamic magnet communities by predicting all the future changes with lasso based growth curve model effectively. To account for sustainable attention flow, local ranking to account for both gains and risks is considered and integrated by a hybrid approach. Experimental results on real dataset show that our method outperforms the baselines and makes much better predictions. To the best of knowledge, this is the first paper to incorporate growth curve and utility modeling into influence analysis and also the first to introduce merger and acquisition dataset into data mining society. Future work can be classified into two aspects: firstly, the competitive relation can be extracted to further enhance the performance of the detection task; secondly, the proposed model will be explored to other domains, such as social media.

Acknowledgment. This work is supported in part by the National Natural Science Foundation of China Projects No. 91546105, No. U1636207, the Shanghai Science and Technology Development Fund No. 16JC1400801, No.17511105502.

References

1. Boyd, S., Parikh, N., Chu, E., Peleato, B., Eckstein, J.: Distributed optimization and statistical learning via the alternating direction method of multipliers. Found. Trends Mach. Learn. **3**(1), 1–122 (2011)
2. Cristelli, M., Gabrielli, A., Tacchella, A., Caldarelli, G., Pietronero, L.: Measuring the intangibles: a metrics for the economic complexity of countries and products. PLoS One **8**(8), e70726 (2013)
3. Du, N., Jia, X., Gao, J., Gopalakrishnan, V., Zhang, A.: Tracking temporal community strength in dynamic networks. IEEE Trans. Knowl. Data Eng. **27**(11), 3125–3137 (2015)
4. Hallac, D., Leskovec, J., Boyd, S.: Network lasso: clustering and optimization in large graphs. In: Proceedings of the 21th ACM SIGKDD International Conference on Knowledge Discovery and Data Mining, pp. 387–396 (2015)
5. Internetmap (2016). http://internet-map.net/
6. Järvelin, K., Kekäläinen, J.: Cumulated gain-based evaluation of ir techniques. ACM Trans. Inf. Syst. (TOIS) **20**(4), 422–446 (2002)
7. Koren, Y.: Collaborative filtering with temporal dynamics. In: ACM SIGKDD International Conference on Knowledge Discovery and Data Mining, pp. 447–456 (2009)
8. Li, J., Cheung, W.K., Liu, J., Li, C.: On discovering community trends in social networks. In: IEEE/WIC/ACM International Joint Conferences on Web Intelligence and Intelligent Agent Technologies 2009, WI-IAT 2009, vol. 1, pp. 230–237 (2009)
9. Liu, W., Deng, Z.H., Gong, X., Jiang, F., Tsang, I.W.: Effectively predicting whether and when a topic will become prevalent in a social network. In: Proceedings of the Twenty-Ninth AAAI Conference on Artificial Intelligence, pp. 210–216 (2015)
10. Matsuo, Y., Yamamoto, H.: Community gravity: measuring bidirectional effects by trust and rating on online social networks. In: Proceedings of the 18th International Conference on World wide web, pp. 751–760 (2009)
11. Mehmood, Y., Barbieri, N., Bonchi, F., Ukkonen, A.: CSI: community-level social influence analysis. In: Blockeel, H., Kersting, K., Nijssen, S., Železný, F. (eds.) ECML PKDD 2013. LNCS (LNAI), vol. 8189, pp. 48–63. Springer, Heidelberg (2013). https://doi.org/10.1007/978-3-642-40991-2_4
12. Page, L.: The PageRank citation ranking : bringing order to the web, vol. 9, no. 1, pp. 1–14 (1998). http://www-db.stanford.edu/?backrub/pageranksub.ps
13. Qiu, J., Li, Y., Tang, J., Lu, Z., Ye, H., Chen, B., Yang, Q., Hopcroft, J.E.: The lifecycle and cascade of wechat social messaging groups. In: International Conference on World Wide Web, pp. 311–320 (2016)
14. Ribeiro, B.: Modeling and predicting the growth and death of membership-based websites. In: Proceedings of the 23rd International Conference on World Wide Web, pp. 653–664 (2014)
15. Gleich, D.F., Rossi, R.A.: A dynamical system for PageRank with time-dependent teleportation. Internet Math. **10**(1–2), 188–217 (2012)

16. Ryu, J., Byeon, S.C.: Technology level evaluation methodology based on the technology growth curve. Technol. Forecast. Soc. Chang. **78**(6), 1049–1059 (2011)
17. Saez-Trumper, D., Comarela, G., Almeida, V., Baeza-Yates, R., Benevenuto, F.: Finding trendsetters in information networks. In: ACM SIGKDD International Conference on Knowledge Discovery and Data Mining, pp. 1014–1022 (2012)
18. Sinatra, R., Wang, D., Deville, P., Song, C., Barabási, A.L.: Quantifying the evolution of individual scientific impact. Science **354**(6312), aaf5239 (2016)
19. Sun, J., Tang, J.: A survey of models and algorithms for social influence analysis. In: Aggarwal, C. (ed.) Social Network Data Analytics, pp. 177–214. Springer, Boston (2011). https://doi.org/10.1007/978-1-4419-8462-3_7
20. Tacchella, A., Cristelli, M., Caldarelli, G., Gabrielli, A., Pietronero, L.: A new metrics for countries' fitness and products' complexity. Sci. Rep. **2**, 723 (2012)
21. Tumminello, M., Miccichè, S., Lillo, F., Varho, J., Piilo, J., Mantegna, R.N.: Community characterization of heterogeneous complex systems. J. Stat. Mech Theory Exp. **2011**(01), P01019 (2011)
22. Von Rosen, D.: The growth curve model: a review. Commun. Stat. Theory Methods **20**(9), 2791–2822 (1991)
23. Wang, D., Barabsi, A.L.: Quantifying long-term scientific impact. Science **342**(6154), 127–32 (2013)
24. Wang, G., Zhao, Y., Shi, X., Yu, P.S.: Magnet community identification on social networks. In: Proceedings of the 18th ACM Sigkdd International Conference on Knowledge Discovery and Data Mining, pp. 588–596 (2012)
25. Wang, S., Xie, S., Zhang, X., Li, Z., Yu, P.S., Shu, X.: Future influence ranking of scientific literature. In: Proceedings of the 2014 SIAM International Conference on Data Mining, pp. 749–757. SIAM (2014)
26. Wei, W., Gao, B., Liu, T.Y., Wang, T.: A ranking approach on large-scale graph with multidimensional heterogeneous information. IEEE Trans. Cybern. **46**(4), 930–944 (2016)
27. Xu, Y., Hong, X., Peng, Z., Yang, G., Yu, P.S.: Temporal recommendation via modeling dynamic interests with inverted-U-curves. In: Navathe, S.B., Wu, W., Shekhar, S., Du, X., Wang, X.S., Xiong, H. (eds.) DASFAA 2016. LNCS, vol. 9642, pp. 313–329. Springer, Cham (2016). https://doi.org/10.1007/978-3-319-32025-0_20
28. Zhang, T., Cui, P., Faloutsos, C., Lu, Y., Ye, H., Zhu, W., Yang, S.: Come-and-go patterns of group evolution: a dynamic model. In: The ACM SIGKDD International Conference, pp. 1355–1364 (2016)
29. Zheng, V.W., Zheng, V.W., Zhu, F., Chang, C.C., Huang, Z.: From community detection to community profiling. Proc. VLDB Endow. **10**(7), 817–828 (2017)

Discovering Strong Communities with User Engagement and Tie Strength

Fan Zhang[1], Long Yuan[1(✉)], Ying Zhang[2], Lu Qin[2], Xuemin Lin[1],
and Alexander Zhou[3]

[1] University of New South Wales, Sydney, Australia
fan.zhang3@unsw.edu.au, {longyuan,lxue}@cse.unsw.edu.au
[2] Centre for AI, University of Technology Sydney, Sydney, Australia
{ying.zhang,lu.qin}@uts.edu.au
[3] University of Queensland, Brisbane, Australia
alexander.zhou@uqconnect.edu.au

Abstract. In this paper, we propose and study a novel cohesive subgraph model, named (k,s)-core, which requires each user to have at least k familiars or friends (not just acquaintances) in the subgraph. The model considers both user engagement and tie strength to discover strong communities. We compare the (k,s)-core model with k-core and k-truss theoretically and experimentally. We propose efficient algorithms to compute the (k,s)-core and decompose the graph by a particular sub-model k-fami. Extensive experiments show (1) our (k,s)-core and k-fami are effective cohesive subgraph models and (2) the (k,s)-core computation and k-fami decomposition are efficient on various real-life social networks.

1 Introduction

Graphs are widely used to represent the abundant interactions in social networks, where each vertex represents a user and each edge represents a relationship between two users. A variety of cohesive subgraph models have been proposed to find social communities, while most of which suffer from computational intractability and other drawbacks. Clique [12] is the most cohesive subgraph model where each vertex is adjacent to every other vertex in the clique. Due to the exponential number of maximal cliques in most social networks and the NP-completeness of the clique decision problem [5], a lot of clique-relaxation models are proposed.

The increasing volume of real-life social networks requires outstanding computation efficiency on cohesive subgraph models, which leads us to the k-core [14] and k-truss [6], the popular and well-studied models with polynomial computation time. The k-core is defined as a maximal subgraph where each vertex has a *degree* of at least k (i.e., at least k neighbors) in the k-core. The k-core is computed by deleting every vertex with degree less than k, which is efficient. However, the simple definition leads to promiscuous subgraphs and thus the k-core is considered as "seedbeds, within which cohesive subsets can precipitate

© Springer International Publishing AG, part of Springer Nature 2018
J. Pei et al. (Eds.): DASFAA 2018, LNCS 10827, pp. 425–441, 2018.
https://doi.org/10.1007/978-3-319-91452-7_28

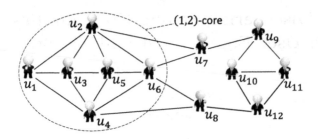

Fig. 1. Motivation example

out" [14]. Another concern of the k-core is that the definition on the neighbor number of a vertex treats each incident edge equally (i.e., the strength of every edge is always same). However, the ties (i.e., edges) in social networks have quite different strength, e.g., two users can be acquaintances who only met once, or close friends who meet every day. So the k-truss model is proposed and defined as a maximal subgraph where each edge has a *support* of at least is k (i.e., is contained in at least k triangles) in the k-truss. The support of an edge has been shown to be effective on dynamically estimating the strength of the edge [6,8,16]. The k-truss is computed by deleting every edge with support less than k, and then delete isolated vertices. In Fig. 1, we show a social network G, consisting of 12 users who form the 3-core. The 4-core of G is empty. The 1-truss of G is G minus the edges $e(u_7, u_9)$ and $e(u_8, u_{12})$. The 2-truss is empty.

There are two concerns of the k-truss model. (1) All the edges with support less than k are deleted including those between the vertices in the k-truss, which is not consistent with reality. Like in Fig. 1, the edges $e(u_7, u_9)$ and $e(u_8, u_{12})$ are not a part of the 1-truss. The friendship between two users, even if weak, always exists in a community as long as the two users are in the community. Besides, the enforced deletion of weak ties in k-truss makes the tie strength estimation inaccurate, as some triangles are unnecessarily deleted. (2) In social communities, the existence of a relationship between two users is dependent on the existence of the two users (i.e., the communities are vertex-oriented). Since the computation of k-truss is based on the removal of weak ties in the network, which makes the k-truss edge-oriented (i.e., the existence of a vertex in k-truss is decided by the existence of its incident edges). Like in Fig. 1, there are many edges with support 2 initially, while the 2-truss computation recursively deletes the edges with support less than 2, which leads to an empty 2-truss.

To address the above concerns, we introduce the (k,s)-core model which is a maximal subgraph where each vertex has an *engagement* of at least k (i.e., at least k strong ties) in the (k,s)-core. Towards tie strength, like in k-truss, an edge is a strong tie if it has a support of at least s (i.e., is contained in at least s triangles) in the (k,s)-core. In Fig. 1, the $(1,2)$-core consists of $u_1, u_2, ..., u_6$ where each user has at least a strong tie. We can see the $(1,2)$-core is tightly connected which cannot be found by k-core or k-truss for any k. The definition of (k,s)-core ensures there is sufficient number of close friends for each user in the (k,s)-core, which strongly encourages the user to keep engaged in the (k,s)-core. Besides, this definition preserves all the weak ties as long as the incident vertices exist in the (k,s)-core, which is more consistent with reality and allows for a more accurate estimation of tie strength. The definition of vertex engagement ensures that the (k,s)-core is vertex-oriented and possesses more potential on computation efficiency than k-truss. An efficient algorithm is proposed to compute the (k,s)-core.

The two parameters in (k,s)-core enable the model to have high flexibility in regards to adjusting different requirements for user engagement and tie strength. However, this makes the decomposition more complex as it needs to compute all the (k,s)-cores for any given k and s. To make the decomposition more afford-able, we introduce a representative sub-model k-fami which is a $(k, k - 1)$-core. We propose an efficient algorithm to decompose a graph into hierarchical struc-tures by the k-fami. Extensive experiments show our (k,s)-core computation and k-fami decomposition are more efficient than k-truss computation and its decom-position, respectively. With the definitions based on characteristics of social com-munities, our (k,s)-core and k-fami produce more convincing cohesive subgraphs for finding strong communities.

2   Problem Definition

In this section, we give some notations and formally define the cohesive subgraph models including our novel (k,s)-core. The notations are summarized in Table 1.

We consider an unweighted and undirected graph $G = (V, E)$, where V (resp. E) represents the set of vertices (resp. edges) in G. We denote $n = |V|$, $m = |E|$ and assume $m > n$. $N(u, G)$ is the set of adjacent vertices of u in G. We say a vertex u is incident to an edge e, or e is incident to u, if u is one of the endpoints of e. Let S denote a subgraph of G. We use $deg(u, S)$, the *degree* of u in S, to represent the number of adjacent vertices of u in S. When the context is clear, we omit the input graph in notations, such as $deg(e)$ for $deg(e, G)$.

Definition 1 k-core. *Given a graph G, a subgraph S is the k-core of G, denoted by $C_k(G)$, if (i) S satisfies the degree constraint, i.e., $deg(u, S) \geq k$ for every $u \in S$; and (ii) S is maximal, i.e., any subgraph $S' \supset S$ is not a k-core.*

User Engagement. For each user (vertex), the k-core model uses the number of acquaintances (neighbors) in the k-core to measure the engagement of this user. In our (k,s)-core model, we consider the number of friends or familiars to better represent the engagement of a user.

Towards the k-core model, one straightforward concern is that the relation-ships (edges, i.e., ties) between users are enforced to have equal strength, which is not consistent with reality. Consequently, the model of k-truss is proposed where each tie has different strength. We define a triangle as a cycle of length 3 in the graph. A containing-e-triangle is a triangle which contains e. The *support* of e in S, i.e., $sup(e, S)$, represents the number of containing-e-triangles in S.

Table 1. Summary of notations

Notation	Definition
G	An unweighted and undirected graph
u, v	A vertex in the graph
$e; e(u, v)$	An edge in the graph; the edge with u and v as endpoints
n, m	The number of vertices and edges in G, respectively
$N(u, G)$	The set of adjacent vertices of u in G
$deg(u, G)$	The number of adjacent vertices of u in G
$sup(e, G)$	The number of triangles each containing e in G
k, s	The thresholds
$eng(u, G)$	The number of edges where each edge e has $sup(e, G) \geq s$ and e is incident to u in G
$C_k(G); T_k(G)$	the k-core of G; the k-truss of G
$C_{k,s}(G); F_k(G)$	the (k,s)-core of G; the k-fami of G
$fn(u)$	Fami number of the vertex u
$E(u, G)$	The edge set where each edge is incident to u and is in G

Definition 2 k-truss. *Given a graph G, a subgraph S is the k-truss of G, denoted by $T_k(G)$, if (i) $sup(e, S) \geq k$ for every edge $e \in S$; (ii) S is maximal, i.e., any subgraph $S' \supset S$ is not a k-truss; and (iii) S is non-trivial, i.e., no isolated vertex in S.*

Tie Strength. For each tie (edge), the k-truss model uses the number of triangles containing it (common neighbors of two endpoints) in the k-truss to estimate the strength of this tie. All weak ties are deleted in k-truss. In our (k,s)-core model, we preserve the weak ties between community members to better estimate the strength of a tie.

In real-life social network, the relationship between two users is concurrent with the existence of the users. This means the weak ties between the users in k-truss should not be deleted. Furthermore, the enforced removing of the weak ties leads to the incompletion of some triangles and thus the inaccuracy on the estimation of tie strength. To overcome these concerns, we firstly define strong tie and strong engagement as the following.

Definition 3 *strong tie.* *Given a graph G and an integer s, an edge e is called a strong tie in G if $sup(e, G) \geq s$; or a weak tie if $sup(e, G) < s$.*

We use $eng(u, S)$, the *engagement* of u in S, to represent the number of strong ties where each edge e has $sup(e, S) \geq s$ and e is incident to u.

Definition 4 *strong engagement.* *Given a graph G and an integer k, a vertex u is strongly engaged in G if u is incident to at least k strong ties in G, i.e., $eng(u, G) \geq k$; or weakly engaged if $eng(u, G) < k$.*

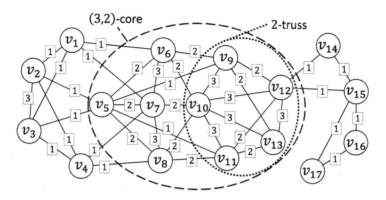

Fig. 2. Running example

If a user has at least k close friends or familiars in a community, he/she is strongly encouraged to stay engaged in the community, which naturally leads us to the following definition for modeling strong communities.

Definition 5 *(k,s)-core. Given a graph G, a subgraph S is the (k,s)-core of G, denoted by $C_{k,s}(G)$, if (i) every vertex in S is strongly engaged in S, i.e., $eng(u,S) \geq k$ for each $u \in S$; and (ii) S is maximal, i.e., any subgraph $S' \supset S$ is not a (k,s)-core.*

The (k,s)-core model ensures that each inside user has at least k strong ties in the (k,s)-core, and each tie is preserved if the two corresponding users (endpoints) exist in the (k,s)-core. Consequently, the (k,s)-core overcomes the above mentioned concerns in k-core and k-truss, and can be more consistent with real-life scenarios. Note that we have $C_{k,0}(G) = C_k(G)$, $C_{k,s}(G) \subseteq C_k(G)$ and $T_k(G) \subseteq C_{k+1,k}(G)$ according to above definitions.

Example 1. In Fig. 2, the social network G consists of 17 vertices where each edge is labeled by its support in G. The G itself is a 2-core, 1-truss and (2,1)-core, respectively. The $C_{3,2}(G)$ is induced by $\{v_5, v_6, v_7, v_8, v_9, v_{10}, v_{11}, v_{12}, v_{13}\}$ which is a tightly connected vertex set. Note that although there are some edges between $\{v_1, v_2, v_3, v_4\}$ and $C_{3,2}(G)$, each of the edges has a support of only 1, which means the connection is weak. The $C_{3,2}(G)$ cannot be found by k-core or k-truss because (i) the $C_3(G) = C_4(G)$ is induced by $G \setminus \{v_{14}, v_{15}, v_{16}, v_{17}\}$; (ii) the $C_5(G) = \emptyset$; (iii) the $T_2(G)$ is induced by $v_9, v_{10},..., v_{13}$; and (iv) the $T_3(G) = \emptyset$. Note that in the computation of $T_2(G)$, the edge supports decrease accordingly when removing each edge with a support of less than 2.

In real-life applications, the value of k (resp. s) can be determined by users based on their requirement for engagement level (resp. tie strength), or learned according to ground-truth communities. The parameters k and s provide more flexibility on adjusting the resulting communities from (k,s)-core.

3   (k,s)-Core Computation

In this section, we introduce an efficient algorithm for finding the (k,s)-core. The following theorem shows we can correctly compute the (k,s)-core on a small k'-core.

Theorem 1. *When $k > 0$, the (k,s)-core of G is a subgraph of k'-core of G (i.e., $C_{k,s}(G) \subseteq C_{k'}(G)$) where $k' = max(k, s+1)$.*

Proof. The (k,s)-core of G, $C_{k,s}(G)$, is a subgraph of the k-core of G because each vertex in $C_{k,s}(G)$ has at least k neighbors in $C_{k,s}(G)$. When $k > 0$, each vertex in $C_{k,s}(G)$ is incident to at least a strong tie which is contained in at least s triangles in $C_{k,s}(G)$, which means the vertex has a degree of at least $s+1$. Since $k' = max(k, s+1)$, every vertex in $C_{k,s}(G)$ has a degree of at least k', i.e., the (k,s)-core is a subgraph of the k'-core.

Theorem 1 allows us to compute the k'-core first as a base for (k,s)-core computation. The algorithm for computing k-core is shown in Algorithm 1 with a time complexity of $O(m+n)$.

Algorithm 2 gives the algorithm for computing the (k,s)-core. In Line 1, we compute the k'-core. Then we do triangle counting on the k'-core to generalize the support and engagement values in Line 2. For each vertex with insufficient engagement (Line 3), we delete the vertex and its incident edges (Line 12 and 5) where the support and engagement values are updated accordingly. Specifically, in Line 4, we can delete the incident edge of u one by one. Then we update the affected engagement value (Line 7, and 11) and the edge support in affected triangles (Line 8 and 9). Note that we do not need to update the vertex engagements which are already less than k and the edge supports which are already less than s.

Example 2. In Fig. 2, the social network G consists of 17 vertices where each edge is labeled by its support in G. In the computation of (3,2)-core, we firstly compute the 3-core of G, which deletes $\{v_{14}, v_{15}, v_{16}, v_{17}\}$ from G. Then we compute the support for each edge in current G and count the engagement for each vertex in G. We push v_1, v_2, v_3 and v_4 in queue for deletion since their engagement is less than 3. Note that when we delete a vertex and its incident edges, we do not need to update the corresponding supports and engagements which are already insufficient. Like when v_1 is deleted, we do not need to update

---

Algorithm 1. ComputeCore(G, k)

   Input   : G : a social network, k : degree constraint
   Output : $C_k(G)$
1 **while** exists $u \in G$ with $deg(u, G) < k$ **do**
2    $\lfloor$ $G := G \setminus \{u \cup E(u, G)\}$;
3 **return** G

Algorithm 2. ComputeKSCore(G, k, s)

Input : G : a social network, k : engagement constraint, s : tie strength
constraint

Output : $C_{k,s}(G)$

1 $k' := max(k, s+1)$; $G :=$ **ComputeCore**(G, k');

2 $s(e) := sup(e, G)$ for each $e \in G$; $d(u) := eng(u, G)$ for each $u \in G$;

3 **while** exists $u \in G$ with $d(u, G) < k$ **do**

4 $\quad$ **for each** $v \in N(u)$ **and** $d(v, G) \geq k$ **do**

5 $\quad\quad$ $G := G \setminus e(u, v)$;

6 $\quad\quad$ **if** $s(e(u, v)) \geq s$ **then**

7 $\quad\quad\quad$ $d(v) := d(v) - 1$;

8 $\quad\quad$ **for each** $w \in N(u) \cap N(v)$ **and** $d(w) \geq k$ **and** $s(e(v, w)) \geq s$ **do**

9 $\quad\quad\quad$ $s(e(v, w)) := s(e(v, w)) - 1$;

10 $\quad\quad\quad$ **if** $s(e(v, w)) = s - 1$ **then**

11 $\quad\quad\quad\quad$ $d(w) := d(w) - 1$;

12 $\quad$ $G := G \setminus \{u \cup E(u, G)\}$;

13 **return** G

---

the engagement of v_2 and the support of $e(v_2, v_3)$. Once the engagement of a
vertex drops from 3 to 2, it is pushed into the queue. We get the (3,2)-core after
deleting every vertex and its incident edges in the queue.

Complexity. The most time-consuming steps are computing $sup(e, G)$ for each
e (Line 2) and its update (Line 8) which both take $O(m^{1.5})$ [18]. The vertex
deletion and edge deletion take $O(n)$ and $O(m)$ respectively. So the time complexity is $O(m^{1.5})$. We need $O(n)$ space to store engagement set and $O(m)$ space
to store the neighbor set and edge support set in G. So the space complexity is
$O(m)$.

Correctness. The correctness is straightforward if no (1) $d(u, G) < k$ in Line 4
and no (2) $d(w) \geq k$ and $s(e(v, w)) \geq s$ in Line 8. The reason for (1) and
$d(w) \geq k$ in (2) is that all the vertices with already less than k engagements will
be deleted with their incident edges, the update for their engagements and edge
supports is not necessary. The reason for $s(e(v, w)) \geq s$ in (2) is that all the
edges with less than s supports are already weak ties which cannot be affected
by the deletion of other edges. Note that the existence of edges is concurrent
with the existence of their incident vertices.

4   Fami Decomposition

In this section, we propose the model of k-fami and its decomposition algorithm. Firstly, we introduce the following theorem which reveals the hierarchical
structure from the (k,s)-core.

Theorem 2. *Given k and s, the (k,s)-core of G is a subgraph of (k', s')-core of G (i.e., $C_{k,s}(G) \subseteq C_{k',s'}(G)$) if $k \geq k'$ and $s \geq s'$.*

Proof. When $s \geq s'$, we have $C_{k,s}(G) \subseteq C_{k,s'}(G)$ because (i) every strong tie e in $C_{k,s}(G)$ is also a strong tie on $sup(e) \geq s'$; and (ii) each vertex in $C_{k,s}(G)$ has at least k strong ties to fulfill the requirement for existing in $C_{k,s'}(G)$. When $k \geq k'$, we have $C_{k,s'}(G) \subseteq C_{k',s'}(G)$ because $eng(u) \geq k'$ for each $u \in C_{k,s'}(G)$. Consequently, $C_{k,s}(G) \subseteq C_{k',s'}(G)$ if $k \geq k'$ and $s \geq s'$.

For a given k and s, Theorem 2 shows that we can find a (k', s')-core which contains the (k,s)-core ($k' \leq k$ and $s' \leq s$) and a (k'', s'')-core which is contained in the (k,s)-core ($k'' \geq k$ and $s'' \geq s$). It motivates us to introduce a particular sub-model k-fami as a representation for the (k,s)-core to show the hierarchical structure of a graph. The computation of all the (k,s)-core for any k and s is time-consuming due to the large number of combinations of k and s. Our k-fami decomposition can produce the hierarchical structure of a graph in $O(m^{1.5})$, which runs faster than k-truss decomposition in our experiments.

Definition 6 k-fami. *Given a graph G, a subgraph S is the k-fami of G (k-familiar in full), denoted by $F_k(G)$, if S is a $(k, k-1)$-core, i.e., $S = C_{k,k-1}(G)$.*

With the k-fami model, every vertex in the graph can have a fami number.

Definition 7 *fami number*. *Given a graph G, the fami number of a vertex u is k^*, denoted by $fn(u, G)$, if (i) there is a k^*-fami which contains u, i.e., $u \in F_{k^*}(G)$; and (ii) there is no other $k' > k^*$ such that $u \in F_{k'}(G)$.*

Fami decomposition is to compute the fami number for every vertex in the graph. Algorithm 3 presents an algorithm for fami decomposition. Line 1 to 3 initialize the arguments including $d(u)$ (engagement) for each vertex u and $s(e)$ (support) for each edge e. Line 4 orders the vertices with increasing order of their engagements. Note that the order can be updated in $O(1)$ time in Line 12, 16, 19 and 20 by using bin sort. Line 5 orders the edges with increasing order of supports. The order can also be updated in $O(1)$ time in Line 14 by using bin sort. A good implementation can be found in [10]. Then we compute the k-fami from $k = 1$ which computes the $(k-1)$-fami. The algorithm terminates and produces all fami numbers when G becomes empty in Line 6. In Line 7 and 8, the fami number of every vertex u with less than k engagement is recorded as $k - 1$. For each neighbor of u with at least k engagement, we delete the edge $e(u, v)$ and update its engagement if necessary (Line 9 to 12). We also update the supports and engagements affected by the deletion of $e(u, v)$ (Line 13 to 16). Note that we do not need to update the engagements of vertices with less than k engagements and the supports of their incident edges, because these vertices are already in the waiting list to be deleted (Line 7). After deleting all vertices with less than k engagements, we increase k by 1 in Line 21 and update the engagements since some edges becomes weak ties with the increase of k in 18 to 20.

Algorithm 3. FamiDecomp(G)

Input : G : a social network
Output : $fn(u)$ for every $u \in G$
1 $k := 1$; $G' := G$;
2 $d(u) := deg(u, G)$ for each $u \in G$;
3 $s(e) := sup(e, G)$ for each $e \in G$;
4 order the vertices in G with increasing order of $d(u)$ for each u;
5 order the edges in G with increasing order of $s(e)$ for each e;
6 **while** G is not empty **do**
7   **while** exists $u \in G$ with $d(u) < k$ in the order **do**
8       $fn(u) := k - 1$;
9   **for each** vertex $v \in N(u)$ and $d(v) \geq k$ **do**
10          $G := G \setminus e(u, v)$;
11   **if** $s(e(u, v)) \geq k - 1$ **then**
12            $d(v) := d(v) - 1$ and reorder the vertices in G;
13   **for each** $w \in N(u) \cap N(v)$ and $d(w) \geq k$ and $s(e(v, w)) \geq k - 1$ **do**
14               $s(e(v, w)) := s(e(v, w)) - 1$ and reorder the edges in G;
15   **if** $s(e(v, w)) = k - 2$ **then**
16                  $d(w) := d(w) - 1$ and reorder the vertices in G;
17       $G := G \setminus \{u \cup E(u, G)\}$;
18   **for each** edge $e(u, v) \in G$ with $s(e(u, v)) = k - 1$ in the order **do**
19       $d(u) := d(u) - 1$ and reorder the vertices in G;
20       $d(v) := d(v) - 1$ and reorder the vertices in G;
21    $k := k + 1$;
22 **return** $fn(u)$ for every $u \in G'$

Example 3. In Fig. 2, the social network G consists of 17 vertices where each edge is labeled by its support in G. In the k-fami decomposition, we firstly compute the support for each edge in G and count the engagement for each vertex in G. Then the edges are ordered with increasing supports and the vertices are ordered with increasing engagements. Note that the orders are implemented in integer buckets with $O(1)$ update time as in [10]. Then we compute the k-fami from $k = 1$ to k_{max}. When $k = 1$, there is no $u \in G$ with $d(u) < k$. Then we lift the strong tie threshold by 1, which decrease the $d(u)$ and $d(v)$ by 1 for each $e(u, v)$ with support $k - 1$, and reorder the vertices. Then we lift k by 1 and find the vertices with $d(u) < k$ in the order. The computation of k-fami is the same as in Algorithm 2 except that $s = k + 1$ and the reorder for vertices and edges. When $k = 2$, there is still no $u \in G$ with $d(u) < k$. When $k = 3$, 8 vertices with $d(u) < k$ are deleted (v_1, ..., v_4, v_{14}, ..., v_{17}) and marked with fami number 2. Recursively, the algorithm terminates when all the vertices are deleted and marked. We thus have the k-fami for k from 1 to 3.

Complexity. In Algorithm 3, the most time-consuming steps are computing $sup(e)$ for every $e \in G$ (Line 3) and updating the edge supports (Line 13 and 14), which takes $O(m^{1.5})$ time [18]. The removal of all vertices and edges takes $O(m)$ time. The orders, engagement reorders and support reorders take $O(m)$ time. So the time complexity of Algorithm 3 is $O(m^{1.5})$. Towards the space complexity, the neighbor set for every vertex, the edge support set and the support order dominate the complexity, where each takes $O(m)$ space. So the space complexity of Algorithm 3 is $O(m)$.

Correctness. We show that for each k in Algorithm 3, it computes a correct $(k-1)$-fami. When $k = 1$, the isolated vertices are removed from G (Line 17) with fami number 0 (Line 8). When $k = 2$, the engagement numbers are correctly updated (Line 18 to 20). Then every vertex with less than k engagement is deleted, where the incident edges are deleted one by one (Line 10 and 17). Note that the engagements of vertices with less than k engagement need not to be updated according to edge deletions because these vertices are already in waiting list for deletion. The supports of their incident edges also need not to be updated. Besides, the supports less than $k-1$ need not to be updated because they are already weak ties and cannot affect the vertex engagements. At Line 21, all vertices with less than k engagement are deleted with correct updates of all supports and engagements. Current G is a $(k-1)$-fami. For $k > 2$, the correctness can be ensured by recursion and Theorem 2.

5   Experimental Evaluation

5.1   Experimental Setting

Datasets. Eight real-life networks were deployed in our experiments and we assume all vertices in each network are initially engaged. The original data of DBLP was downloaded from http://dblp.uni-trier.de/ and the others from http://snap.stanford.edu/. In DBLP, we consider each author as a vertex and there is an edge for a pair of authors if they have at least one co-authored paper. There are existing vertices and edges in other datasets. Table 2 shows the statistics of the 8 datasets, listed in increasing order of their edge numbers.

Algorithms. To the best of our knowledge, no existing work investigates the (k,s)-core and k-fami. We tested 4 algorithms (k-core, k-truss, k-fami and ks-core) to produce and compare different resulting subgraphs. We also implemented and evaluated the decomposition algorithms including core decomposition (coreDecomp), truss decomposition (trussDecomp) and our fami decomposition (famiDecomp). Table 3 shows the summary of the algorithms.

Parameters. We conducted experiments under different settings by varying the engagement constraint k from 3 to 80 and the support constraint s from 10 to 50. We also report the result of 3 graph decompositions.

All programs were implemented in standard C++ and compiled with G++ in Linux. All experiments were performed on a machine with Intel Xeon 2.8 GHz CPU and Redhat Linux System.

Table 2. Statistics of datasets

Dataset	Nodes	Edges	d_{avg}	k_{max}^{core}	k_{max}^{truss}	k_{max}^{fami}	$|\triangle|$
Facebook	4,039	88,234	43.7	115	95	102	1,612,010
Brightkite	58,228	194,090	6.7	52	40	43	449,717
Gowalla	196,591	456,830	4.7	43	21	25	1,061,143
YouTube	1,134,890	2,987,624	5.3	51	17	24	3,056,386
DBLP	1,566,919	6,461,300	8.3	118	117	118	15,389,320
Pokec	1,632,803	8,320,605	10.2	27	18	19	6,971,538
LiveJournal	3,997,962	34,681,189	17.4	360	350	353	177,820,130
Orkut	3,072,441	117,185,083	76.3	253	76	83	627,584,181

Table 3. Summary of algorithms

Algorithm	Description
k-core	Computing the k-core [3], i.e., Algorithm 1
k-truss	Computing the k-truss [6]
k-fami	Computing the k-fami, i.e., Algorithm 2
ks-core	Computing the (k,s)-core, i.e., Algorithm 2
coreDecomp	Core decomposition in [10], i.e., computing the largest k for every vertex $u \in G$ such that the k-core contains u
trussDecomp	Truss decomposition in [18], i.e., computing the largest k for every vertex $u \in G$ such that the k-truss contains u
famiDecomp	Fami decomposition, i.e., Algorithm 3

5.2 Effectiveness

Statistics. We report the maximum core number (k_{max}^{core}), truss number (k_{max}^{truss}) and fami number (k_{max}^{fami}) on each dataset in Table 2. The k_{max}^{fami} is usually between the values of k_{max}^{core} and k_{max}^{fami}, which shows our k-fami model captures unique hierarchical structures of the graphs. We show the number of vertices in k-core, k-truss and k-fami in Fig. 3. When $k = 15$, Fig. 3(a) shows the size of k-fami is always between k-core and k-truss where the difference varies on all datasets due to the different natures of the datasets. Figure 3(b) and (c) show the decrease of the size in 3 models with the growth of k. When $k = 25$, the k-truss in Gowalla is empty. The margin between the sizes of k-fami and the other 2 models varies with different k.

In Fig. 4, we report the size of (k,s)-core with different k and s. Figure 4(a) shows the trend of the (k,s)-core size when we fix s and vary k. Figure 4(b) shows the trend of the (k,s)-core size when we fix k and vary s. In both Fig. 4(a) and (b), for a given s, We can see that the (k,s)-core sizes are almost the same when $k \leq s$ because they all belong to $(s + 1)$-core according to Theorem 1. When

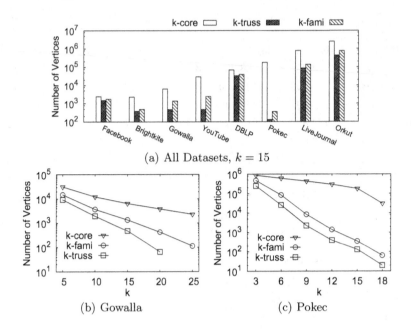

(a) All Datasets, $k = 15$

(b) Gowalla

(c) Pokec

Fig. 3. Vertex number in k-core, k-truss and k-fami

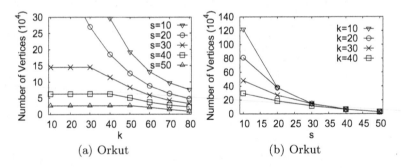

(a) Orkut

(b) Orkut

Fig. 4. Vertex number in (k,s)-core with different k and s

$k > s$, the (k,s)-core becomes smaller with the increase of k. Figure 4 reveals the hierarchical structure of a graph with the changing of k and s in the (k,s)-core.

Case Study on DBLP. Figure 5 depicts the k-core, k-truss and k-fami on the DBLP-30 dataset with $k = 15$. In DBLP-30, to make this case study visible, each edge between two authors represents that there are at least 30 co-authored papers between the two authors. The whole graph in Fig. 5 is the k-core of DBLP-30. The (k,s)-core excludes the sparse group at the bottom right corner with 5 authors. The k-truss is formed by all the square vertices (in blue) which excludes all the sphere vertices (in red) and their incident edges from (k,s)-core. We can see all the square and sphere vertices connect tightly, which shows

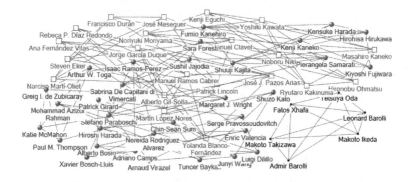

Fig. 5. Case study of k-core, k-truss and k-fami on DBLP

the advantages of our (k,s)-core model. Specifically, the (k,s)-core is superior than the other 2 models in the sense that (1) the k-core is relatively large since it tolerates some vertices with low engagement and (2) the k-truss enforcedly excludes all the weak ties which makes the tie strength estimation (number of triangles) inaccurate and the non-concurrence of the vertices and its incident edges in resulting communities.

5.3   Efficiency

Decompositions. In Fig. 6, we report the decomposition time for k-core, k-truss and k-fami. The coreDecomp and trussDecomp are the state-of-the-art algorithms for in-memory core decomposition and truss decomposition, respectively. Figure 6 shows that coreDecomp is faster than trussDecomp and famiDecomp because it does not need triangle listing and support updates on the graph. However, coreDecomp treats each edge equally and ignores the difference in tie strength. Our famiDecomp algorithm outperforms trussDecomp in running time by up to 2 times, because famiDecomp is a vertex-oriented algorithm where the existence of edges depends on the incident vertices, while trussDecomp is an edge-oriented algorithm. Besides, our famiDecomp can avoid unnecessary updating of some supports and engagements as Algorithm 3 shows.

Effect of k and s. We show the effect of k in k-core, k-truss and k-fami computation in Fig. 7. As discussed above, k-core is faster than k-truss or k-fami but it ignores the strength of ties in discovering cohesive subgraphs. The running time of k-truss and k-fami becomes smaller as k increases because both of them compute a k'-core first which reduces the candidate set for their computation. So performance of 3 algorithms tends to be closer when k becomes larger. In Fig. 8, we show the running time of (k,s)-core on different k and s. The runtime of ks-core becomes smaller when $max(k, s + 1)$ becomes larger, because ks-core computes a $max(k, s + 1)$-core first to reduce the candidate set. For the same reason, when we fix s and $k \leq s$, the runtime of ks-core does not change much. It also explains the consistent runtime of ks-core when we fix k and $s < k$.

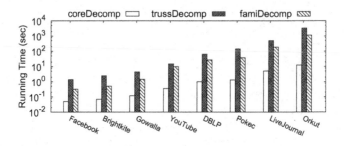

Fig. 6. Running time for graph decompositions

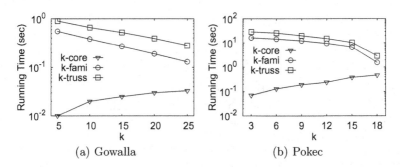

(a) Gowalla   (b) Pokec

Fig. 7. Running time of k-core, k-truss and k-fami computation

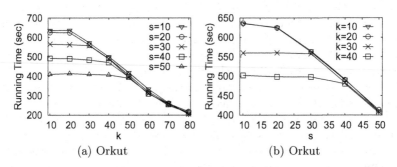

(a) Orkut   (b) Orkut

Fig. 8. Running time of (k,s)-core with different k and s

Different Datasets. Figure 9 reports the running time of k-core, k-truss and k-fami on all datasets with $k = 15$. k-core still outperforms k-truss and k-fami by ignoring the tie strength in computation. Our k-fami is faster than k-truss on all datasets because k-fami is a vertex-oriented algorithm and avoids updating unnecessary support and engagement values in Algorithm 2. It further confirms the effectiveness of our (k,s)-core model in producing cohesive subgraphs with the consideration of tie strength.

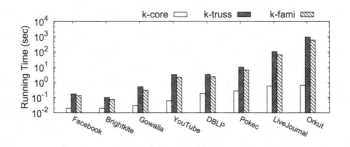

Fig. 9. Running time of k-core, k-truss and k-fami computation

6   Related Work

There are various cohesive subgraph models to accommodate different scenarios in the literature. Clique [12] is an extremely cohesive subgraph where every vertex is adjacent to every other vertex in the clique. Because the definition of clique is usually too restrictive, some clique relaxation models have been proposed, such as k-plex [15], k-core [14] and k-truss [6], and so on. Among these cohesive subgraph models, k-core and k-truss are the widely studied models with polynomial computation time.

Seidman [14] proposes the k-core where each vertex has at least k neighbors in the k-core. The k-core has a wide spectrum of applications such as social contagion [17], community detection [21], user engagement [20] and so on. Batagelj and Zaversnik [2] present an algorithm for core decomposition of a graph with time complexity of $O(m + n)$. Zhang et al. [22] propose a fast order-based algorithm to maintain k-core in dynamic graphs. Bhawalkar et al. [4] propose the problem of anchored k-core to prevent network unraveling. Zhang et al. [19] present an efficient algorithm to solve the anchored k-core problem. However, the k-core uses vertex degree to determine the user engagement which treats each edge equally. In real-life social networks, the strength of user relationships (edges) varies a lot and cannot be always identical [7].

Further considering the strength of ties, Cohen [6] proposes the model of k-truss where every edge exists in at least k triangles in the k-truss with its decomposition algorithm in $O(\sum_{v \in V(G)} (\deg(v)^2))$ time. Rotabi et al. [13] show that most strong tie detection methods are based on structural information, especially on triangles. Wang and Cheng [18] reduce the time complexity of truss decomposition to $O(m^{1.5})$ and study the I/O efficient truss decomposition. Shao et al. [16] study the k-truss detection problem on distributed systems and propose an efficient parallel algorithm. Zhao and Tung [23] use the k-truss to capture the cohesion in social interactions and propose a visualization system based on k-truss. Huang et al. [8] study the k-truss based community model, which further requires edge connectivity inside the community. Akbas and Zhao [1] propose a truss-equivalence based index to speed up the search of the truss based community. Huang and Lakshmanan [9] study the attributed k-truss community search where the largest attribute relevance score is satisfied.

However, k-truss deletes all the weak ties even if the corresponding users exists in the k-truss, which makes the tie strength estimation inaccurate and that some users are excluded from k-truss unreasonably. Besides, k-truss is edge-oriented while social communities are user(vertex)-oriented.

To the best of our knowledge, we for the first time propose the novel (k,s)-core to overcome above concerns in k-core and k-truss. Lee et al. [11] propose the (k,d)-core where each vertex has at least k neighbors in the subgraph and each edge is contained in at least d triangles in the subgraph. The (k,d)-core is essentially a subgraph of d-truss with additional requirement for vertex degree of at least k, which can be regarded as a strengthened d-truss. Thus, the model still has the same concerns as in k-truss and is different than our (k,s)-core model.

7 Conclusion

In this paper, we propose a novel cohesive subgraph, (k,s)-core, which requires each user to have at least k familiars or friends in the subgraph. The (k,s)-core addresses the concerns in k-core and k-truss including (1) k-core enforces the strength of each tie to be equal; (2) k-truss deletes all the weak ties; and (3) k-truss is edge-oriented while social communities are user(vertex)-oriented. We propose an efficient algorithm to compute the (k,s)-core. A particular k-fami is introduced to efficiently decompose a graph. Extensive experiments validate the effectiveness of our models and the efficiency of our algorithms.

Acknowledgments. Fan Zhang and Long Yuan are supported by Huawei YBN2017100007. Ying Zhang is supported by ARC FT170100128 and DP180103096. Lu Qin is supported by ARC DP160101513. Xuemin Lin is supported by NSFC 61672235, ARC DP170101628, DP180103096 and Huawei YBN2017100007.

References

1. Akbas, E., Zhao, P.: Truss-based community search: a truss-equivalence based indexing approach. PVLDB **10**(11), 1298–1309 (2017)
2. Batagelj, V., Zaversnik, M.: An O(m) algorithm for cores decomposition of networks. CoRR, cs.DS/0310049 (2003)
3. Batagelj, V., Zaversnik, M.: Fast algorithms for determining (generalized) core groups in social networks. Adv. Data Anal. Classif. **5**(2), 129–145 (2011)
4. Bhawalkar, K., Kleinberg, J., Lewi, K., Roughgarden, T., Sharma, A.: Preventing unraveling in social networks: the anchored k-core problem. SIAM J. Discrete Math. **29**(3), 1452–1475 (2015)
5. Bron, C., Kerbosch, J.: Finding all cliques of an undirected graph (algorithm 457). Commun. ACM **16**(9), 575–576 (1973)
6. Cohen, J.: Trusses: cohesive subgraphs for social network analysis. National Security Agency Technical Report, p. 16 (2008)
7. Granovetter, M.S.: The strength of weak ties. Am. J. Sociol. **78**(6), 1360–1380 (1973)
8. Huang, X., Cheng, H., Qin, L., Tian, W., Yu, J.X.: Querying k-truss community in large and dynamic graphs. In: SIGMOD, pp. 1311–1322 (2014)

9. Huang, X., Lakshmanan, L.V.S.: Attribute-driven community search. PVLDB **10**(9), 949–960 (2017)
10. Khaouid, W., Barsky, M., Venkatesh, S., Thomo, A.: K-core decomposition of large networks on a single PC. PVLDB **9**(1), 13–23 (2015)
11. Lee, P., Lakshmanan, L.V.S., Milios, E.E.: CAST: a context-aware story-teller for streaming social content. In: CIKM, pp. 789–798 (2014)
12. Luce, R.D., Perry, A.D.: A method of matrix analysis of group structure. Psychometrika **14**(2), 95–116 (1949)
13. Rotabi, R., Kamath, K., Kleinberg, J.M., Sharma, A.: Detecting strong ties using network motifs. In: WWW, pp. 983–992 (2017)
14. Seidman, S.B.: Network structure and minimum degree. Soc. Netw. **5**(3), 269–287 (1983)
15. Seidman, S.B., Foster, B.L.: A graph-theoretic generalization of the clique concept. J. Math. Sociol. **6**(1), 139–154 (1978)
16. Shao, Y., Chen, L., Cui, B.: Efficient cohesive subgraphs detection in parallel. In: SIGMOD, pp. 613–624 (2014)
17. Ugander, J., Backstrom, L., Marlow, C., Kleinberg, J.: Structural diversity in social contagion. PNAS **109**(16), 5962–5966 (2012)
18. Wang, J., Cheng, J.: Truss decomposition in massive networks. PVLDB **5**(9), 812–823 (2012)
19. Zhang, F., Zhang, W., Zhang, Y., Qin, L., Lin, X.: OLAK: an efficient algorithm to prevent unraveling in social networks. PVLDB **10**(6), 649–660 (2017)
20. Zhang, F., Zhang, Y., Qin, L., Zhang, W., Lin, X.: Finding critical users for social network engagement: the collapsed k-core problem. In: AAAI, pp. 245–251 (2017)
21. Zhang, F., Zhang, Y., Qin, L., Zhang, W., Lin, X.: When engagement meets similarity: efficient (k, r)-core computation on social networks. PVLDB **10**(10), 998–1009 (2017)
22. Zhang, Y., Yu, J.X., Zhang, Y., Qin, L.: A fast order-based approach for core maintenance. In: ICDE, pp. 337–348 (2017)
23. Zhao, F., Tung, A.K.H.: Large scale cohesive subgraphs discovery for social network visual analysis. PVLDB **6**(2), 85–96 (2012)

Functional-Oriented Relationship Strength Estimation: From Online Events to Offline Interactions

Chang Liao[1,2], Yun Xiong[1,2(✉)], Xiangnan Kong[3], Yangyong Zhu[1,2], Shimin Zhao[4], and Shanshan Li[5]

[1] Shanghai Key Laboratory of Data Science, School of Computer Science,
Fudan University, Shanghai, China
`{cliao15,yunx}@fudan.edu.cn`
[2] Shanghai Institute for Advanced Communication and Data Science,
Fudan University, Shanghai, China
[3] Worcester Polytechnic Institute, Worcester, MA, USA
[4] Technical Center of Shanghai Shengtong Metro Group Co. Ltd., Shanghai, China
[5] School of Computer, National University of Defense Technology, Changsha, China

Abstract. Link mining/analysis over network has received widespread attention from researchers. Recently, there has been growing interest in measuring relationship strength between entities based on attribute similarity. However, limited work has assessed the competitive advantage of functional elements in relationship strength quantification. The functional elements embody the growth/development nature of the relationship. Motivated by the availability of large volumes of online event records that can potentially reveal underlying functional socio-economic characteristics, we study the problem of offline relationship strength estimation with functional elements awareness from online events. Two major challenges are identified as follows: (1) informal information, online events are of high dimensions, and not all the learnt functions of online events are predictive to offline interactions; (2) heterogeneous dependency, it's hard to measure the relationship strength by modeling functional elements with network effects jointly. To handle these challenges, we propose generalized relationship strength estimation model (**gStrength**), a novel approach for relationship strength estimation. First, we define the combination of latent roles and observed groups as generalized roles, and present generalized role constrained latent topic model to make the extracted latent functions compatible with offline interactions. Second, we model the functional elements and further extend them to structural dependency settings to quantify relationship strength. We apply this approach to the political and economic application scenario of measuring international investment relations. The experimental results demonstrate the effectiveness of the proposed method.

1 Introduction

Link mining/analysis, *e.g.*, link existence prediction between entities, is an important task that has been well studied by the social network mining

© Springer International Publishing AG, part of Springer Nature 2018
J. Pei et al. (Eds.): DASFAA 2018, LNCS 10827, pp. 442–459, 2018.
https://doi.org/10.1007/978-3-319-91452-7_29

community. However, most existing researches only address conventional link prediction tasks, where links only have one of two possible states: active and inactive. In this work, we take on the task of relationship strength estimation. Instead of characterizing relations with binary states, it designates a continuous-valued strength [23,27]. Having a good estimation of the relationship strength between entities can yield huge benefits. Investors and governments can use this information to identify critical relations (for instance, the fragile relations that are bound to dissolve), enabling precise strike against potential risk factors. For example, in the 2008 financial crisis, many banks were reported to dissolve their relations with poorly interacted firms, possibly leading to risk cascading effect. Capturing fragile (risky) relations in advance is crucial to such risk management.

In real world scenarios, relationships with strong strength are not necessarily ones with the most attribute similarity, but with function complementarities. For example, an individual's access to opportunities and resources can only be exploited if it is linked with others with diverse functions, as shown in [11]. The functional elements indicate the competitive advantages of growth/development of the relationship, which is of great significance, especially in economic and political fields. In contrast, limited work has integrated functional elements between entities in measuring relationship strength. Unfortunately, offline attributes representing functional effects are expensive or unavailable to collect. For example, although the quantity and quality of economic data available in developing countries have improved in recent years, certain key indicators of economic development remain unavailable or expensive to collect. Meanwhile, large volumes of online event information are being generated and can potentially reveal such underlying socio-economic characteristics. Rather than pre-determined attributes, communities might be better characterized by probabilistic models of event patterns. Such latent recurring patterns might best reflect the complex nature and diverse functions of relationship. To fill the gap of the lack of functional elements consideration in quantifying relationship strength and motivated by the available online event data, as is shown in Fig. 1, we combine online events and traditional data (group, offline attributes, auxiliary information) to recover relationship strength from offline interaction graph, addressing the following challenges:

Informal Information: Online information is of high dimensions, and directly leveraging it for latent function extraction is troublesome. To be more concrete, the online events and offline interactions are of different modalities. Not all online events can be used to predict offline interactions, as in Fig. 1. For example, online events like subtle non-commercial or non-political communications are not causally related to offline economic interactions. Thus, the proposed model needs to automatically extract latent functions from online information that is compatible with offline information.

Structural Dependency: Suppose we have derived the latent function of each relationship, it is still difficult to measure the functional-aware relationship strength under unsupervised settings. That is, the relations associated with the same entity or the same group are likely to be dependent, given the resource

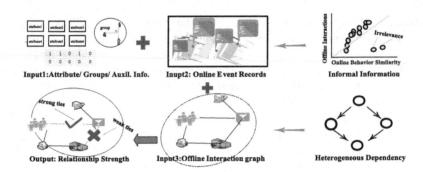

Fig. 1. A toy example of functional relationship strength estimation between five entities. As is shown, we quantify the relationship strength from offline interaction graph by integrating online event information and traditional data, addressing informal information and heterogeneous dependency problems

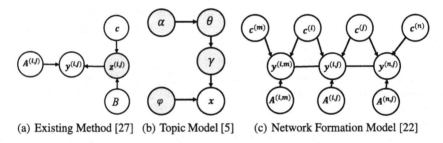

(a) Existing Method [27]   (b) Topic Model [5]   (c) Network Formation Model [22]

Fig. 2. Methods on relationship strength estimation. x represents online event between each pair of entities and z denotes the relationship strength. y is the interaction and c is the group. B is the attributes and A is the interaction specific information. Further, in (b), functional elements are extracted by latent topic model, where α is the hyperparameters, γ is the topics, θ is the document-topic distributions and φ is the topic-word distributions.

constraints on relationship formation and maintenance, as is shown in Fig. 1. Ignoring such structural dependency would lead to the overfitting of the latent function effects. Thus, it is necessary to model the value of functional elements and extend it to network dependency circumstances jointly, which is very difficult.

To address these challenges, we attempt to make full use of online event information and traditional dataset, in contrast to attribute based models shown in Fig. 2(a). We propose to combine online latent role with offline group as generalized role, which is to ensure that latent functions extracted from online information conform to offline information by subjecting it to generalized role constraint, compared with latent topic model in Fig. 2(b). Moreover, we integrate structural dependency to encode structural dependency in quantifying relationship strength, based on network formation model in Fig. 2(c). The main contributions are as follows:

- We introduce a new task of measuring functional-oriented relationship strength between entities from online events. To the best of our knowledge, limited work has considered to measure relationship strength with function awareness.
- We formulate Generalized Relationship Strength Model (**gStrength**). More specifically, we extract latent topics from online events and utilize them to measure relationship strength. Moreover, both informal information and structural dependency challenges are addressed.
- With no direct evaluation metrics, we evaluate our approach indirectly. We also apply the method to international investment relationship strength quantification. The results demonstrate that our method can potentially serve as a complementary tool for international relationship study.

2   Problem Formulation

In this section, we first introduce several key terms and then present the formal problem definition. Table 1 shows the important notations used in this paper.

Table 1. Several important mathematical notations

Notations	Meaning	Notations	Meaning
$\mathbf{Y}$	Interaction graph	$\mathbf{Z}$	Relationship Strength
$\mathbf{C}$	Entity group representation	$\mathbf{X}$	Entity event representation
$\mathbf{Cr}$	Relationship group representation	$\mathbf{Xr}$	Relationship event representation
$\rho, w, \mathbf{d}$	Coefficient parameters	$\mathbf{\Phi}$	Function-event distribution
$\mathbf{R}$	Generalized role	$\mathbf{\Upsilon}$	Topics/function
$\mathbf{\Theta}$	Relationship-topic distribution	α	Dirichlet prior vector
$\mathbf{A}$	Auxiliary information	K	Number of topics
$\mathbf{B}, \bar{\mathbf{B}}$	Attributes/generalized attributes	u/v	Entity/event indicator
$\hbar$	Structure statistical	s	Similarity values

Definition 1 (Event Network). *Given n entities, we denote an attribute information matrix with b attributes as $\mathbf{B} \in \mathbb{R}^{n \times b}$, group set as $\mathbf{C} \in \mathbb{R}^{n \times g}$, an online event tensor with p types of events as $\mathbf{X} \in \mathbb{R}^{n \times n \times p}$ and auxiliary interaction specific information $\mathbf{A} \in \mathbb{R}^{n \times n \times t}$ with t is auxiliary information dimension for interactions (e.g., total number of interactions). An event network consists of these four parts and is defined as $G = \langle \mathbf{X}, \mathbf{B}, \mathbf{C}, \mathbf{A} \rangle$.*

As an example, attribute and auxiliary information is displayed in Fig. 3(a). We transform both event and group information from entity representation to relationship representation with $m = n^2$. The relationship based event $\mathbf{Xr} \in \mathbb{R}^{m \times p}$ is transformed from entity based event $\mathbf{X}$ as in Fig. 3(b), where p is the event type. Meanwhile, the relationship based group $\mathbf{Cr} \in \mathbb{R}^{m \times q}$ is shown in

446   C. Liao et al.

Fig. 3(c), where q is the squared size to entity group size g. For example, suppose there are three observed groups $\mathbf{C} = \{6,7,8\}$ in the interacted entities. After transforming group information from entity to relationship representation, it can be indicated as $\mathbf{Cr} = \{(6,7),(6,8),(7,6),(7,8),(8,6),(8,7),(6,6),(7,7),(8,8)\}^1$.

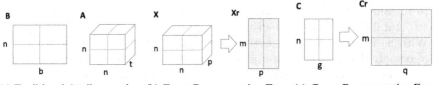

(a) Traditional Attribute and Auxiliary Information   (b) Event Representation From Entity to Relationship   (c) Group Representation From Entity to Relationship

Fig. 3. Event network illustration (the shaded are intermediate results)

Definition 2 (Generalized Role/Function Distribution). *In an interconnected system, relationship between entities may play multiple roles[2], denoted as* $\mathbf{R}$*, where* $r \in \mathbf{R}$ *is the role indicator. Note that* $\mathbf{R}$ *is latent and unobservable. To integrate entity group information* $\mathbf{Cr}$ *with role information* $\mathbf{R}$*,* **we consider the group** $\mathbf{C}$ **as a special type of roles** $\mathbf{R}$*. We define the combination of the latent roles* $\mathbf{R}$ *and group information* $\mathbf{C}$ *as generalized roles* $\mathbf{R}$*, where* $r \in \mathbf{R}$ *is the generalized role indicator. Each relationship between entities is assumed to have a multinomial distribution over generalized roles* $\mathbf{R}$*, denoted as* Θ*.*

For example, assuming there are 5 shared latent alternative groups and the observed groups $\mathbf{Cr}$ in the different relationships. A relationship between entity i with observed group $\mathbf{C}(i) = 6$ and entity j with observed group $\mathbf{C}(j) = 7$ is represented as $\mathbf{Cr}(i,j) = (6,7)$. Its corresponding generalized role can be represented as $\mathbf{R} = \{1,2,3,4,5,(6,7)\}$. $\Theta_r^{(v)}$ denotes the probability for relationship v to play role r, and it is subject to $\sum_r \Theta_r^{(v)} = 1$. This is the key concept of the paper. By modeling online topic extraction with generalized role constraints, we can make the extracted topics more compatible, achieving more flexible online-to-offline modality transformation, as will be shown in Eq. (4). Furthermore, the proposed model can extend to network dependency settings, relieving the overfitting problem of just considering functional elements, as will be shown in Eq. (8).

Definition 3 (Relationship Strength Estimation). *The relationship strength* $\mathbf{Z} \in \mathbb{R}^{n \times n}$ *represents the full spectrum of relationship of continuous value, in contrast to coarse representation in interaction graph* $\mathbf{Y} \in \{0,1\}^{n \times n}$*. Given an event network* $G = \langle \mathbf{X}, \mathbf{B}, \mathbf{C}, \mathbf{A} \rangle$*, the goal of relationship strength estimation is to quantify relationship strength* $\mathbf{Z}$ *from interaction graph* $\mathbf{Y}$*.*

---

[1] In the rest of the paper, both $\mathbf{Cr}$ and $\mathbf{Xr}$ are used directly for simplicity and convenience.

[2] Role is exchangeable with function element and latent topic.

This is the problem definition of this paper. We assume the observed inter-action graph **Y** is generated from underlying continuous-valued relationship strength **Z** with auxiliary interaction specific information **A**. As with relation-ship strength **Z**, it is measured by network structures and attributes, together with generalized role distribution (function distribution) extracted from event records and group information.

3 Proposed Method

The problem is to quantify relationship strength, as in Fig. 4(a). We first present the simplified model, then the generalized relationship strength model (**gStrength**) by addressing the informal information and structural dependency challenges. The main framework of **gStrength** consists of two generative pro-cesses and a discriminative process: (1) online event generation from latent func-tions; (2) interaction formation on relationship strength and auxiliary informa-tion; (3) relationship strength measured by latent functions, network structures and attributes.

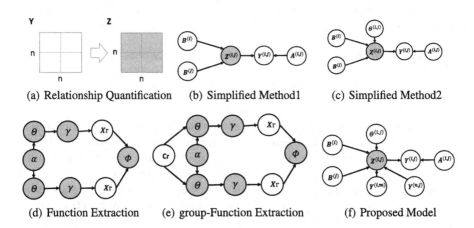

(a) Relationship Quantification   (b) Simplified Method1   (c) Simplified Method2

(d) Function Extraction   (e) group-Function Extraction   (f) Proposed Model

Fig. 4. Overall framework of gStrength: (1) the problem studied is to recover con-tinuous value of relationship strength **Z** from coarse representation **Y**; (2) combined with latent dirichlet allocation model, it can model the generation of event **Xr** with generalized role by constraining hyperparameter α with group **Cr**; (3) by modeling network dependency, it can model the interaction **Y** formation process more precisely with learnt functions; (4) relationship strength **Z** is measured by latent functions and structural dependency, as well as attribute similarity. Meanwhile, the blank circle rep-resents observed values, and the yellow ones as latent parameters.

3.1 Simplified Method

For the simplest version 1, the intuition is that relationship strength is quantified by attribute values and interaction is generated based on this hidden strength.

Mathematically, the relationship strength $\mathbf{Z}$ is measured by attribute values $\bar{\mathbf{B}}^3$. It assumes that interaction graph $\mathbf{Y}$ is generated from relationship strength $\mathbf{Z}$ and auxiliary information $\mathbf{A}$. Each interaction is independent of others. As in Fig. 4(b), the joint distribution of $\mathbf{Z}$ and $\mathbf{Y}$ is derived as

$$
\begin{aligned}
\Pr\left(\mathbf{Y}, \mathbf{Z} | \mathbf{w}, \mathbf{d}, \bar{\mathbf{B}}, \mathbf{A}\right) &= \prod_{(i,j)} \Pr\left(\mathbf{Z}^{(i,j)} | \bar{\mathbf{B}}^{(i)}, \bar{\mathbf{B}}^{(j)}, \mathbf{d}\right) \cdot \Pr\left(\mathbf{Y}^{(i,j)} | \mathbf{Z}^{(i,j)}, \mathbf{A}^{(i,j)}, \mathbf{w}\right) \\
&\propto \prod_{(i,j)} \frac{\exp(-(\mathbf{d}^\top s(\bar{\mathbf{B}}^{(i)}, \bar{\mathbf{B}}^{(j)}) - \mathbf{Z}^{(i,j)})^2 / 2\sigma^2)}{(2\pi)^{1/2}\sigma} \cdot \frac{\exp\left(\sum_{(i,j)} (\mathbf{Z}^{(i,j)} + \mathbf{w}^\top \mathbf{A}^{(i,j)})\right)}{\prod_{(i,j)} \left(1 + \exp\left(\mathbf{Z}^{(i,j)} + \mathbf{w}^\top \mathbf{A}^{(i,j)}\right)\right)},
\end{aligned}
\tag{1}
$$

where $\mathbf{w}$ and $\mathbf{d}$ are weighting parameters for corresponding interaction specific information $\mathbf{A}$ and the similarity of generalized attributes $\bar{\mathbf{B}}$. This model is not good in real applications, as the functional elements of characterizing relationship strength are not considered.

For the simplest version 2, we extract the function elements from online events. Formally, we learn the latent function[4] and model relationship strength by combining it. To learn the latent function distribution $\boldsymbol{\Theta}$ from online events $\mathbf{X}$ ($\mathbf{Xr}$), we adapt the idea of topic modeling [5], where events and relationships are analogized as words and documents respectively. As in Fig. 4(d), the generation of online event $\mathbf{X}$ ($\mathbf{Xr}$) based on latent topic distribution can be specified as

$$
\begin{aligned}
\Pr(\mathbf{Xr}, \boldsymbol{\Theta}, \boldsymbol{\Phi}, \boldsymbol{\Upsilon}, \alpha) &= \prod_v \Pr(\alpha) \Pr(\boldsymbol{\Theta}^{(v)} | \alpha) \Pr(\mathbf{Xr}^{(v)} | \boldsymbol{\Theta}^{(v)}, \boldsymbol{\Phi}, \boldsymbol{\Upsilon}^{(v)}) \\
&\propto \prod_v \frac{\Gamma(\sum_{k=1}^K \alpha_k)}{\prod_{k=1}^K \Gamma(\alpha_k)} \prod_{k=1}^K \left(\boldsymbol{\Theta}_k^{(v)}\right)^{\alpha_k - 1} \prod_u \boldsymbol{\Theta}_{\boldsymbol{\Upsilon}_u^{(v)}}^{(v)} \boldsymbol{\Phi}_u^{(\boldsymbol{\Upsilon}_u^{(v)})},
\end{aligned}
\tag{2}
$$

where hyperparameter α follows beta distribution, $\boldsymbol{\Theta}$ belongs to dirichelt distribution based on α, $\mathbf{Xr}$ is then generated from function distribution $\boldsymbol{\Theta}$ and event distribution $\boldsymbol{\Phi}$. As for others, $\Gamma()$ is the Gamma function, v is the relationship indicator, K is the topic number and u is the event dimension.

By exploiting the learnt function distribution, we assume the relationship strength is measured by both the attribute similarity and latent functions. As in Fig. 4(c), together with auxiliary information $\mathbf{A}$, we can derive the probability of network formation based on relationship strength as

$$
\begin{aligned}
\Pr(\mathbf{Y}, \mathbf{Z} | \mathbf{A}, \bar{\mathbf{B}}, \mathbf{w}, \mathbf{d}, \rho, \boldsymbol{\Theta}) &= \prod_{(i,j)} \Pr\left(\mathbf{Z}^{(i,j)} | \bar{\mathbf{B}}^{(i)}, \bar{\mathbf{B}}^{(j)}, \boldsymbol{\Theta}, \rho, \mathbf{d}\right) \cdot \Pr\left(\mathbf{Y}^{(i,j)} | \mathbf{Z}^{(i,j)}, \mathbf{A}^{(i,j)}, \mathbf{w}\right) \\
&\propto \prod_{(i,j)} \frac{\exp(-(\rho_c^\top \boldsymbol{\Theta}^{(i,j)} + \mathbf{d}^\top s(\bar{\mathbf{B}}^{(i)}, \bar{\mathbf{B}}^{(j)}) - \mathbf{Z}^{(i,j)})^2 / 2\sigma^2)}{(2\pi)^{1/2}\sigma} \cdot \frac{\exp\left(\sum_{(i,j)} (\mathbf{Z}^{(i,j)} + \mathbf{w}^\top \mathbf{A}^{(i,j)})\right)}{\prod_{(i,j)} \left(1 + \exp\left(\mathbf{Z}^{(i,j)} + \mathbf{w}^\top \mathbf{A}^{(i,j)}\right)\right)},
\end{aligned}
\tag{3}
$$

where $\mathbf{w}$, $\mathbf{d}$ and ρ_c are weighting parameters for corresponding interaction specific information $\mathbf{A}$, similarity of generalized attributes $\bar{\mathbf{B}}$ and latent functions $\boldsymbol{\Theta}$.

---

[3] We refer both attribute $\mathbf{B}$ and group $\mathbf{C}$ as generalized attribute $\bar{\mathbf{B}}$.

[4] Function/Topic is assigned to every relationship and indicated by $\boldsymbol{\Theta}^{(i,j)}$ from relationship representation, here v is the order number for interaction $\mathbf{Y}^{(i,j)}$.

The model parameters can be learnt easily via Newton method in an EM algorithm style.

Although the simplified model with online events would be applicable for relationship strength estimation under our settings, it still has two limitations: (1) online events are informal and not all of them are predictive to offline interactions; (2) the relations associated with the same entity or the same group are likely to be dependent.

3.2 Generalized Role-Aware Latent Function Extraction

The method in Eq. (2) doesn't distinguish online and offline information, thus cannot perform transformation from online events to offline interactions flexibly. To address the informal information challenge, we integrate generalized role $\mathbf{R}$ to constrain it with hyperparameters, as in Fig. 4(e). We specify it as

$$
\Pr(\mathbf{Xr}, \mathbf{\Theta}, \mathbf{\Phi}, \mathbf{\Upsilon}, \alpha | \mathbf{R}) = \prod_{v=1} \Pr(\alpha | \mathbf{R}) \Pr(\mathbf{\Theta}^{(v)} | \alpha) \Pr(\mathbf{Xr}^{(v)} | \mathbf{\Theta}^{(v)}, \mathbf{\Phi}, \mathbf{\Upsilon}^{(v)})
$$
$$
\propto \prod_{cr=1}^{Ncr} \prod_{v=1}^{N_v^{cr}} \frac{\Gamma(\sum_{k=1}^{K^{cr}} \alpha_k^{cr})}{\prod_{k=1}^{K^{cr}} \Gamma(\alpha_k^{cr})} \prod_{k=1}^{K^{cr}} (\Theta_k^{(v)})^{\alpha_k^{cr}-1} \prod_{u=1}^{(v)} \Theta_{\Upsilon^{(v)}}^{(v)} \Phi_u^{(\Upsilon_u^{(v)})} \tag{4}
$$

where N_{cr} is relationship group number and K^{cr} is topic number in relationship group [5] cr, while others are the same with Eq. (2). To make the extracted functions consistent with offline information, generalized role information is utilized. That is, the relationships $\mathbf{Xr} \in \mathbb{R}^{m \times p}$ are divided into N_c observed groups: $\mathbf{Xr} = \{\mathbf{Xr}^{cr}\}_{cr=1}^{N_{cr}}$ with N_v^{cr} relationship number per group cr. In contrast to standard LDA that uses a fixed hyperparameter α as in (2), we set $\alpha = [\alpha(0), \alpha(1), ..., \alpha(N_{cr})]$, where $\alpha(0)$ is shared latent roles. Then the hyperparameter $\alpha^{cr} = [\alpha(0), \alpha(cr)]$ conforms to generalized role $\mathbf{R}$, mathematically represented as $\Pr(\alpha | \mathbf{R})$ as in Eq. (4). Finally, the latent function distribution $\mathbf{\Theta}$ is represented as generalized role based topic distribution. It consists of group constrained function distribution $\mathbf{\Theta}_{cr}$ and shared latent function distribution $\mathbf{\Theta}_{-cr}$. By reconciling the observed group information and event records for discovering latent functions between entities, it can benefit functional elements extraction in a more interpretable and effectiveness way.

For the parameter estimation, the Gibbs-EM algorithm [3] is adopted, where the iteration procedure to update the parameters is carried on until converges:
The Gibbs sampling for updating topic[6] $\mathbf{\Upsilon}_u^{(v)}$ is:

$$
p(\mathbf{\Upsilon}_u^{(v)} | \mathbf{\Upsilon}_u^{(-v)}, \mathbf{Xr}, c, \alpha, \beta) = \frac{n_{\Upsilon,u}^{-v,u} + \beta}{\sum_w (n_{\Upsilon,w}^{-v,u} + \beta)} \frac{n_{v,\Upsilon}^{-v,u} + \alpha_\Upsilon}{\sum_k (n_{v,k}^{-v,u} + \alpha_k)} \tag{5}
$$

---

[5] Group is assigned to every relationship and indicated by $\mathbf{C}^{(cr)}$ from relationship representation, here cr is the order number for group information $\mathbf{Cr}^{(i,j)}$.

[6] $n_{\Upsilon,u}$ and $n_{v,\Upsilon}$ denotes the count number of events and topics correspondingly, others are of the same styles and omitted for explanation due to limited space.

As each topic/function $\mathbf{\Upsilon}$ entails dirichlet distributions over function-event parameter $\mathbf{\Phi}$ and relationship-function parameter $\mathbf{\Theta}$, parameters $\mathbf{\Phi}$ and $\mathbf{\Theta}$ can be point-estimated as:

$$\mathbf{\Phi}_u^{(k)} = \frac{n_{k,u}+\beta}{\sum_u (n_{k,u}+\beta)}; \qquad \mathbf{\Theta}_k^{(v)} = \frac{n_{v,k}+\alpha_k}{\sum_k (n_{v,k}+\alpha_k)} \tag{6}$$

In addition to the estimation of hyperparameter α, it can be transformed as

$$\frac{\partial L}{\partial \alpha_k} = \sum_{s=1}^{N_s} \left(\sum_{v=1}^{N_v^c} (\Psi(\alpha^{0,c}) - \Psi(n_{v,.}^{0,c} + \alpha^{0,c})) \right) + \sum_{v=1}^{N_v^c} (\Psi(n_{v,k} + \alpha_k) - \Psi(\alpha_k)), \tag{7}$$

where N_s is the sampling counts. The parameter can be then learnt by fixed point iterations as (7), where Ψ is digamma function [20].

3.3   Functional-Aware Relationship Strength Estimation

In addition, the relationship strength is also determined by network structures, which corresponds to the structural dependency challenge. To combine functional elements with network effects, we assume the relationship strength $\mathbf{Z}$ is also determined by its structural dependency $\mathbf{Y}$ with ρ_r as weighting parameter and specify it as:

$$\Pr\left(\mathbf{Z}^{(i,j)} | \mathbf{Y}^{-(i,j)}, \bar{\mathbf{B}}, \mathbf{\Theta}, \rho, \mathbf{d}\right) = \frac{\exp(-(\rho_c^\top \mathbf{\Theta}^{(i,j)} + \rho_r^\top \hbar(\mathbf{Y}^{-(i,j)}) + \mathbf{d}^\top s(\bar{\mathbf{B}}^{(i)}, \bar{\mathbf{B}}^{(j)}) - \mathbf{Z}^{(i,j)})^2 / 2\sigma^2)}{(2\pi)^{1/2}\sigma} \tag{8}$$

As in Fig. 4(f), an offline interaction $\mathbf{Y}^{(i,j)}$ is generated partially according to network structure $\mathbf{Y}^{-(i,j)}$. And $\hbar(\mathbf{Y}^{-(i,j)})$ indicates network statistics vector, and is specified as reciprocality $\mathbf{Y}^{(j,i)}$, neighborhood $\sum_{n=\backslash j} \mathbf{Y}^{(i,n)} + \sum_{m=\backslash j} \mathbf{Y}^{(m,j)}$ as well as triangle $\sum_{p=\backslash i,j} \mathbf{Y}^{(j,p)} \mathbf{Y}^{(p,i)}$. In other words, the model assumes interaction between entities is reciprocal and dependent, given resource constraints as with the first and second part of $\hbar()$. As for triangle feature, it is unlikely to encounter a triple of entities that two of the ties are strong and the third is missing.

As for relationship strength estimation with structural dependency, we add auxiliary parameter ε for every relationship pair. More specifically, we put $\varepsilon^{(i,j)} = -(\rho_c^\top \mathbf{\Theta}^{(i,j)} + \rho_r^\top \hbar(\mathbf{Y}^{-(i,j)}) + \mathbf{d}^\top s(\bar{\mathbf{B}}^{(i)}, \bar{\mathbf{B}}^{(j)})) + \mathbf{Z}^{(i,j)}$ and the probability of $\mathbf{Z}$ given ε is 1. Then Eq. (1) can be transformed to Eq. (9) as:

$$\Pr(\mathbf{Y}, \mathbf{Z} | \mathbf{A}, \bar{\mathbf{B}}, \mathbf{w}, \mathbf{d}, \rho, \mathbf{\Theta}) = \prod_{(i,j)} \Pr(\mathbf{Y}^{(i,j)} | \mathbf{Y}^{-(i,j)}, \mathbf{A}, \bar{\mathbf{B}}, \varepsilon^{(i,j)}, \mathbf{w}, \mathbf{d}, \rho, \mathbf{\Theta}) \cdot \Pr(\varepsilon^{(i,j)})$$

$$= \frac{\exp\left(\sum_{(i,j)} (\rho_c^\top \mathbf{\Theta}^{(i,j)} + \rho_r^\top \hbar(\mathbf{Y}^{-(i,j)}) + \mathbf{d}^\top s(\bar{\mathbf{B}}^{(i)}, \bar{\mathbf{B}}^{(j)}) + \varepsilon^{(i,j)} + \mathbf{w}^\top \mathbf{A}^{(i,j)})\right)}{\prod_{(i,j)} \left(1 + \exp(\rho_c^\top \mathbf{\Theta}^{(i,j)} + \rho_r^\top \hbar(\mathbf{Y}^{-(i,j)}) + \mathbf{d}^\top s(\bar{\mathbf{B}}^{(i)}, \bar{\mathbf{B}}^{(j)}) + \varepsilon^{(i,j)} + \mathbf{w}^\top \mathbf{A}^{(i,j)})\right)} \cdot \prod_{(i,j)} \frac{\exp(-\varepsilon^{(i,j)2}/2\sigma^2)}{(2\pi)^{1/2}\sigma} \tag{9}$$

In such setting, auxiliary parameter ε occurs in both relationship strength quantification process and network formation process. The coefficient parameters

ρ_c, ρ_r, $\mathbf{d}$, $\mathbf{w}$ and auxiliary parameter ε can be learnt by Newton-Raphson method, where σ is set as 0.2. In the end, as for relationship strength $\mathbf{Z}$, it can be measured by $\rho_c^\top \mathbf{\Theta}^{(i,j)} + \rho_r^\top \hbar(\mathbf{Y}^{-(i,j)}) + \mathbf{d}^\top s(\bar{\mathbf{B}}^{(i)}, \bar{\mathbf{B}}^{(j)}) + \varepsilon^{(i,j)} = \mathbf{Z}^{(i,j)}$.

4 Experiments

We conduct various experiments to evaluate the proposed method. All experiments are conducted on machines with Intel(R) Xeon(R) CPU of 2.5 GHz and 192G memory.

4.1 Data Description

We conduct experiments on both real-world data and synthetic data[7] to demonstrate the effectiveness of the proposed model.

International Relation Related Dataset. Fig. 5 presents an overview of international relationship dataset. It consists of the following:

– **Cross-Border Investment Dataset:** It is from the SDC Thomson Platinum database, which comprises of the cross-border merger and acquisition records. We regard the interaction between two countries as 1 (0 otherwise) if they meet the following conditions: (1) announcement dates are between January 1, 1990 and December 31, 2012; (2) the merger is completed. And then, we obtain an interaction graph of 208×208 entries with binary values.
– **Global Event Dataset:** It consists of over 200 million records of real-world events from 1979 to the present [16], also called **GDELT**. From this event record data, we collect online event records between every two countries by accumulating the total number of articles involved from the year 1990 to the year 2012, as the same time interval with cross-border investment dataset. Finally, we have a collection of 1,624,860 events with 263 event types. One thing to note here is that **GDELT** data also describes temporal interaction between countries, we are here to accumulate them together along temporal dimension to treat as events, in order to meet up with the experimental design of the proposed model.
– **National Attributes Survey Dataset:** It is downloaded from "https:// data.worldbank.org/". After removing missing values, we derive 245 countries and 4 basic attributes (*GDP, Area, FixedIncome, FixedInvestment*). In this dataset, the political and economic status of each country is evaluated and assigned to one of three levels. The level assignment to each country is set as the group information.

---

[7] Auxiliary information $\mathbf{A}$ is not specified around the whole paper, but we take it in the framework whenever interaction auxiliary information is available for generality.

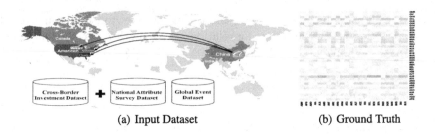

(a) Input Dataset   (b) Ground Truth

Fig. 5. Dataset description

To integrate and align them together, it ends up with an offline interaction graph with 164 × 164 entries from cross-border investment dataset, traditional group information (3 groups) and 4 attribute information from National Attributes Survey Dataset, as well as online event records with 263 dimensions from global event dataset, as in Fig. 5(a). The merger number between countries is served as ground truth for relationship strength, as in Fig. 5(b).

Synthetic Dataset . We generate synthetic data similar to the real data. That is, we generated the synthetic dataset with the same settings of international relationship dataset. We vary and amplify the dataset by expanding its size based on the real dataset. The number of entities ranges from 164 to 137,760, and the number of events ranges from 852 to 100,160, as well as event dimension from 815 to 205,792.

4.2   Comparative Methods

As with competing methods, existing methods on relationship estimation are classified into attribute based methods and network based methods. Other methods with different purposes that can provide intermediate results as relationship strengths are also considered. From each category, we select several representative methods as baselines: (1) attribute based (**Strength** and **Denoise**), (2) network based (**Smooth** and **Neighbor**), (3) other method (**RandomWalk**):

- **Strength:** [27] is a latent variable model to infer relationship strength based on profile similarity and interactions. It can be seen as a nearly relaxed version of **gStrength**, which lacks network effects modeling. The attribute and group information[8] is utilized.
- **Denoise:** [9] integrates multiple social interactions for denoising social networks. Though with different purposes, the relationship strength estimation results along the intermediate process can be brought directly with the four attribute information.

---

[8] The similarity function is specified as $[\mathbf{B}^{(i)} \bullet \mathbf{B}^{(j)}, C(i) \bullet C(j)]$, where $\bullet$ is the dot product-based score operator.

- **Smooth:** [28] is a neighborhood smoothing based method to estimate the expectation of the adjacency matrix. The interaction network structure expectation/probability is taken as relationship strength for comparison.
- **Neighbor:** [21] argues that social networks can be modeled as the outcome of processes that result from local social neighborhoods. It is another relaxed version of the proposed model that lacks online event information.
- **RandomWalk:** [4] combines network structure with attributes, interactions, groups and event records of nodes and edges into a unified link prediction algorithm, where the relationship strength along the learning process can be obtained.

4.3   Performance Evaluation

We evaluate the proposed model *gStrength* together with other baselines from three perspectives: (1) link classification/ranking based evaluation; (2) network denoising based evaluation; (3) network statistics recovery based evaluation.

Relationship Estimation Result Analysis. For our first evaluation, we evaluate the learnt relationship strength from both classification and ranking dissimilarity perspectives. In the task, we measured how well the estimated relationship strengths conform with the ground truth.

Recall that in the original dataset, we divide the real number of mergers for each relationship into two categories: the strength larger than mean value as positive and negative vice versa. To evaluate the effectiveness of link classification, we rank the pairs of entities by the learnt relationship strength and measure the area under the ROC curve (AUC) for the ranked pair. The results are shown in Fig. 6(a). The clear gain in AUC indicates that the model can automatically identify and distinguish relationship strength. As is shown, *gStrength* method performs slightly better than *RandomWalk*, as the network structural dependency is not directly emphasized in *gStrength*. As for performance gain from other baselines, it demonstrates the benefits of modeling all these features properly.

The intersection dissimilarity measure (IDM) [6] is adopted to measure the dissimilarity ranking between real relationship strength $\bar{\mathbf{Z}}$ and predicted relation tion strength $\mathbf{Z}$ as $IDM = 1/2\iota \sum_\iota |\mathbf{Z}_\iota \Delta \bar{\mathbf{Z}}_\iota|/2\iota$, where Δ denotes the symmetric set difference. The larger the value of IDM is, the less similar $\bar{\mathbf{Z}}$ and $\mathbf{Z}$ are. The gain of the results on *gStrength* demonstrates the effectiveness of the proposed model, as is shown in Fig. 6(b). While in this task, the advantage of our method over *RandomWalk* appears limited, as we will show later, our method can significantly outperform it in the other two tasks.

With Application to Denoising. For our second evaluation, we apply it to network denoising task of removing noisy links. Since there is no indicators about links being noisy or not, we verify it indirectly - comparing the performance of denoising by identifying the injected noises. That is, we first randomly inject 10 noisy links and measure the strengths of both the real ones (852 links) and injected ones (10 links). And then, we rank the links according to the relationship

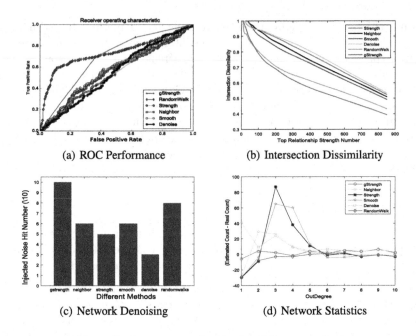

(a) ROC Performance   (b) Intersection Dissimilarity

(c) Network Denoising   (d) Network Statistics

Fig. 6. Performance evaluation for different methods

strength, select the low-ranked 100 ones, and plot the hit number (10 noisy links) in Fig. 6(c). As is shown, we can successfully achieve better network denoising performance, significantly outperforming *RandomWalk*.

With Application to Network Statistics Recovery. For our third evaluation of estimating relationship strength, we set the number of edges the same with original network and take the top ranked estimated relation strength to generate synthetic network. To demonstrate our model's superiority, we use the *OutDegree* to measure the difference of structure between the real network and the generated synthetic network. We plot the difference between the degree of nodes of real network and that of the corresponding nodes in the generated network in Fig. 6(d). From it, we can see that the network generated by our model is more similar to the ground truth than the baselines. Especially, our model significantly outperforms *RandomWalk*.

4.4   Model Analysis

To further study how the proposed model works, we conduct the following experiments: (1) performance comparison on functional elements encoding; (2) performance comparison on structural dependency encoding; (3) model scalability and parameter sensitivity.

Functional Elements Encoding. We compare different strategies on modeling functional elements: (a) proposed model as in Eq. (8); (b) without group constrained latent function as in Eq. (4); (c) without latent function as in Eq. (1). Figure 7(a) shows the proposed method outperforms the others on the same task of link classification as described in Subsect. 4.3. The AUC result suggests that the generalized role aware settings can successfully improve the model performance by extracting topics from both offline group and online event information jointly.

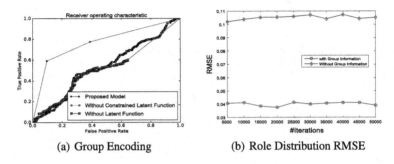

(a) Group Encoding   (b) Role Distribution RMSE

Fig. 7. Benefits on functional elements encoding

To further demonstrate the benefit of extracting latent functions by utilizing group information, we also run our model over the synthetic dataset and plot the RMSE in the inferred latent variables within or without group information modeling. The results in Fig. 7(b) show that our model is capable of recovering latent parameter (function distribution), demonstrating the benefits of explicitly exploiting group information.

Structural Dependency Encoding. We compare different strategies on modeling functional elements: (a) proposed model as in Eq. (8); (b) proposed model without modeling structural dependency as in Eq. (3). Figure 8(a) shows the proposed method outperforms the others on the same task of link classification as described in subsection 4.3. The results suggest that the structural dependency encoding can successfully improve the expressiveness and performance of the model.

We show the result of structural dependency coefficient parameters in Fig. 8(b). The values of reciprocity, neighborhood and triangle structures are positive, which is consistent with our intuitions. These results further demonstrate the necessity of modeling structural dependency jointly when estimating relationship strength.

Model Scalability and Parameter Sensitivity. To test the model scalability, we evaluate it by using the synthetic data. The default settings are: event number(852), event dimension(815), entity number(208), function number(34). The results are shown in Fig. 9(a), (b) and (c). From them, the running time changes smoothly with the growing number of events, the dimensions of events and the number of entities. The results show our model is scalable and can be flexibly

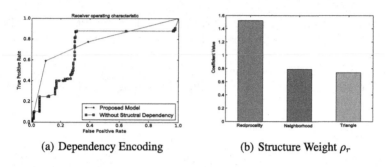

(a) Dependency Encoding   (b) Structure Weight ρ_r

Fig. 8. Benefits on structural dependency exploitation

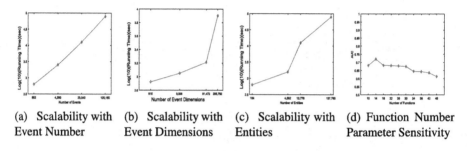

(a) Scalability with   (b) Scalability with   (c) Scalability with   (d) Function Number
Event Number   Event Dimensions   Entities   Parameter Sensitivity

Fig. 9. Model scalability and parameter sensitivity

adapted to large-scale interaction graph with rich online events. We also discuss how parameter K (the number of functions) influences the performance of our method in Fig. 9(d). It displays the AUC performance with different dimensions of parameter K on the task of link classification as in Subsect. 4.3. We find it changes very smoothly, highlighting the robustness of the proposed model ($gStrength$).

5   Related Work

Relationship Strength Estimation: A popular research direction of network analysis is link prediction. [1] surveys the state of the art of link prediction method. The key difference is that we focus on modeling relationship strength rather than relationship existence. When measuring relationship strength, [10] has leveraged social media data to quantify tie strength, while [27] estimates the relationship strength by integrating interaction information and multiple similarity features. By utilizing network information, [28] proposes to estimate probabilities (strength) of network edges from the observed adjacency matrix. However, existing work has not utilized online event information with functional elements awareness. From application perspectives, unlike previous work focusing on employing online event records to support online predictions, this paper utilizes online country event dataset and offline third-party merger/acquisition records to quantify offline national investment relationship strength conversely.

Latent Topic Model: Faced with high-dimensional data issues, latent topic model [5] has gained popularity for modeling a large amount of high dimensional text documents. When integrated with attribute and network constraints, it can be transformed to supervised settings. Along with this category, [14] introduces semi-supervised latent topic model to identify rare and subtle anomalies in computer vision domain. [12] attempts to capture all the information of a social network such as links, communities, attributes, roles and activities in a unified manner for modeling social network. [26] proposes a novel targeted topic model to enable focused analysis on any specific aspect of interest. [19] proposes a novel social-aware semi-supervised topic model by reconciling the observed user characteristics and social network structure for discovering the latent reasons behind social connections and further extracting users' potential profiles. To the best of our knowledge, none of the existing work has addressed the difference between offline and online information on relationship strength estimation. A similar work is [18], which attempts to make use of the online behaviors and offline actions jointly to promote recommendation for offline retailing. The main difference lies in that there are also large quantities of offline action records compared with our limited offline attribute settings.

Network Formation Modeling: [17] studies how links are formed in social networks and formulates the problem of predicting reciprocity and triadic closure into a graphical model. [25] uses the principle of strong triadic closure to characterize the strength of relationships in social networks and studies the problem from running time perspectives. [22] presents network models for social selection processes, where individuals form social relationships on the basis of certain characteristics they possess. However, to the best of our knowledge, there's no work to combine structural features based network formation model with relationship strength quantification task. A similar work [23] that considers the problem of inferring the strength of social ties from network structure perspectives is proposed recently. However, it has not modelled the attribute information, let alone functional-aware features. To compare, our paper is the first to track relationship strength estimation problem by utilizing functional element, together with structural dependency awareness.

Online Public Data Analysis: As for online public data utilization sceneries, [2] studies the effect that social networking has on user's offline behavior, [8] has surveyed the usage of it in economic research, while [7] has discussed opportunities associated with online administrative data. When applied to country/region evaluation, [15] combines publicly available satellite imagery to predict poverty. [13] proposes to infer socio-economic maps from large-scale spatio-temporal data. Focusing on international relations measurement, traditional way to measure international relations is based on survey data. While survey data can generally describe economic characteristics, it possibly suffers from sparsity problems and even with incomplete values. Aiming at this problem, [24] tries to infer international relations from online event data. Along this line, this paper is the first to exploit online event data and offline investment interaction information for offline international relationship strength estimation.

6   Concluding Remarks

In this paper, we propose a new problem of functional-oriented relationship strength estimation by integrating online events. Addressing both informal information and structural dependency challenges, a generalized relationship strength model (**gStrength**) is proposed. In particular, we conduct the task of international relationship strength estimation from online global event data and offline cross-border investment records. The experimental results show that our work has some potential use for international relationship study. As future work, it can be applied to relationship dissolution among communities, such as bank tie strength quantification.

Acknowledgment. This work is supported in part by the National Natural Science Foundation of China Projects No. 91546105, No. U1636207, the National High Technology Research and Development Program of China No. 2015AA020105, the Shanghai Science and Technology Development Fund No. 16JC1400801, No. 17511105502, No. 16511102204.

References

1. Hasan, M.A., Zaki, M.J.: A survey of link prediction in social networks. In: Aggarwal, C. (ed.) Social Network Data Analytics, pp. 243–275. Springer, Boston (2011)
2. Althoff, T., Jindal, P., Leskovec, J.: Online actions with offline impact: how online social networks influence online and offline user behavior, pp. 537–546 (2017)
3. Andrieu, C., Freitas, N.D., Doucet, A., Jordan, M.I.: An introduction to MCMC for machine learn. Mach. Learning **50**(1–2), 5–43 (2003)
4. Backstrom, L., Leskovec, J.: Supervised random walks: predicting and recommending links in social networks. In: Proceedings of the Fourth ACM International Conference on Web Search and Data Mining, pp. 635–644. ACM (2011)
5. Blei, D.M., Ng, A.Y., Jordan, M.I.: Latent Dirichlet allocation. J. Mach. Learn. Res. **3**, 993–1022 (2003)
6. Boldi, P.: TotalRank: ranking without damping. In: Special Interest Tracks and Posters of the International Conference on World Wide Web, pp. 898–899 (2005)
7. Connelly, R., Playford, C.J., Gayle, V., Dibben, C.: The role of administrative data in the big data revolution in social science research. Soc. Sci. Res. **59**, 1–12 (2016)
8. Einav, L., Levin, J.: The data revolution and economic analysis. Innov. Policy Econ. **14**(1), 1–24 (2014)
9. Gao, H., Wang, X., Tang, J., Liu, H.: Network denoising in social media. In: 2013 IEEE/ACM International Conference on Advances in Social Networks Analysis and Mining (ASONAM), pp. 564–571. IEEE (2013)
10. Gilbert, E., Karahalios, K.: Predicting tie strength with social media. In: SIGCHI Conference on Human Factors in Computing Systems, pp. 211–220 (2009)

11. Granovetter, M.: The strength of weak ties: a network theory revisited. Sociol. Theor. **1**(6), 201–233 (1983)
12. Han, Y., Tang, J.: Probabilistic community and role model for social networks. In: Proceedings of the 21th ACM SIGKDD International Conference on Knowledge Discovery and Data Mining, pp. 407–416. ACM (2015)
13. Hong, L., Frias-Martinez, E., Frias-Martinez, V.: Topic models to infer socio-economic maps. In: Thirtieth AAAI Conference on Artificial Intelligence, pp. 3835–3841 (2016)
14. Hospedales, T.M., Li, J., Gong, S., Xiang, T.: Identifying rare and subtle behaviors: a weakly supervised joint topic model. IEEE Trans. Pattern Anal. Mach. Intell. **33**(12), 2451–2464 (2011)
15. Jean, N., Burke, M., Xie, M., Davis, W.M., Lobell, D.B., Ermon, S.: Combining satellite imagery and machine learning to predict poverty. Science **353**(6301), 790–794 (2016)
16. Leetaru, K., Schrodt, P.A.: Gdelt: global data on events, location, and tone, 1979–2012. In: ISA Annual Convention, vol. 2, pp. 1–49 (2013)
17. Lou, T., Tang, J., Hopcroft, J., Fang, Z., Ding, X.: Learning to predict reciprocity and triadic closure in social networks. ACM Trans. Knowl. Discov. Data **7**(2), 5 (2013)
18. Luo, P., Yang, S., Yang, S., Yang, S., Yang, S., He, Q.: From online behaviors to offline retailing. In: ACM SIGKDD International Conference on Knowledge Discovery and Data Mining, pp. 175–184 (2016)
19. Ma, C., Zhu, C., Fu, Y., Zhu, H., Liu, G., Chen, E.: Social user profiling: a social-aware topic modeling perspective. In: Candan, S., Chen, L., Pedersen, T.B., Chang, L., Hua, W. (eds.) DASFAA 2017. LNCS, vol. 10178, pp. 610–622. Springer, Cham (2017). https://doi.org/10.1007/978-3-319-55699-4_38
20. Minka, T.: Estimating a Dirichlet distribution. In: UAI-2002, vol. 39, no. 3273, p. 115 (2013)
21. Pattison, P., Robins, G.: Neighborhood-based models for social networks. Sociol. Method. **32**(1), 301–337 (2010)
22. Robins, G., Elliott, P., Pattison, P.: Network models for social selection processes. Soc. Netw. **23**(1), 1–30 (2001)
23. Rozenshtein, P., Tatti, N., Gionis, A.: Inferring the strength of social ties: a community-driven approach. In: Proceedings of the 23rd ACM SIGKDD International Conference on Knowledge Discovery and Data Mining, pp. 1017–1025. ACM (2017)
24. Schein, A., Paisley, J., Blei, D.M., Wallach, H.: Bayesian poisson tensor factor-ization for inferring multilateral relations from sparse dyadic event counts. In: Proceedings of the 21st ACM SIGKDD International Conference on Knowledge Discovery and Data Mining, pp. 1045–1054. ACM (2015)
25. Sintos, S., Tsaparas, P.: Using strong triadic closure to characterize ties in social networks. In: Proceedings of the 20th ACM SIGKDD International Conference on Knowledge Discovery and Data Mining, pp. 1466–1475. ACM (2014)
26. Wang, S., Chen, Z., Fei, G., Liu, B., Emery, S.: Targeted topic modeling for focused analysis. In: Proceedings of the 22nd ACM SIGKDD International Conference on Knowledge Discovery and Data Mining, pp. 1235–1244. ACM (2016)
27. Xiang, R., Neville, J., Rogati, M.: Modeling relationship strength in online social networks. In: Proceedings of the 19th International Conference on World Wide Web, pp. 981–990. ACM (2010)
28. Zhang, Y., Levina, E., Zhu, J.: Estimating network edge probabilities by neighbor-hood smoothing. Biometrika **104**(4) (2015)

Incremental and Adaptive Topic Detection over Social Media

Konstantinos Giannakopoulos[(✉)] and Lei Chen[(✉)]

Department of Computer Science and Engineering,
Hong Kong University of Science and Technology,
Clear Water Bay, Kowloon, Hong Kong
{kgaa,leichen}@cse.ust.hk

Abstract. Social media like Twitter and Facebook are very popular nowadays for sharing users' interests. However, the existing solutions on topic detection over social media overlook time and location factors, which are quite important and useful. Moreover, social media are frequently updated. Thus, the proposed detection model should handle the dynamic updates. In this paper, we introduce a topic model for topic detection that combines time and location. Our model is equipped with incremental estimation of the parameters of the topic model and adaptive window length according to the correlation of consecutive windows and their density. We have conducted extensive experiments to verify the effectiveness and efficiency of our proposed Incremental Adaptive Time Location (IncrAdapTL) model.

1 Introduction

In recent years, the use of online social networks like Twitter have been spread. Hundreds of thousands of short messages are exchanged between users. Research has been done on detection of topics on messages that users publish. Each tweet, consists of the main text message, and additional useful information like timestamp and location coordinates. All this information is used by researchers in order to incorporate time and location in the proposed topic detection models.

In this paper, we propose a generative, LDA-based topic model for topic detection in tweets. Our model incorporates time-zones and location regions. We process input data with sliding windows with incremental re-evaluation of the topic model parameters and adaptive window lengths for faster processing.

Most of the previous existing works on this field that propose generative topic models use either only location or only time separately. However, we combine both. In addition, previous works do not handle incremental updates of model's parameters. We propose incremental updates where we do not need to process all the tweets in the sliding time windows. It is not necessary to process the same tweets in consecutive windows. Moreover, we propose adaptive window lengths. There are time periods where more tweets are posted and time periods where less tweets appear. We take advantage of sparse windows by increasing the window

© Springer International Publishing AG, part of Springer Nature 2018
J. Pei et al. (Eds.): DASFAA 2018, LNCS 10827, pp. 460–473, 2018.
https://doi.org/10.1007/978-3-319-91452-7_30

length. This improves accuracy of detected topics in sparse windows. As far as we are concerned, previous research works do not use any mechanism to handle this situation.

The main contributions of our proposed model are the following:

- Firstly, we introduce incremental update of the model parameters between consecutive windows. Our proposed model used sliding windows for processing messages. It does not need to process the old messages of each window. It processes only the new tweets and we do not need to re-evaluate from scratch the model parameters in each window.
- Secondly, we introduce adaptive window lengths for processing data. We observe that more tweets and different topics are posted during day-time than during night-time. So, we adapt the window length according to the correlation of consecutive windows and according to their density for faster processing.

The rest of the paper is organized as follows. In Sect. 2 we review the related work, in Sect. 3 we present our topic model, in Sect. 4 we evaluate our approach.

2   Related Work

In this section we review some previous research papers that have proposed topic models for topic detection. Topic models are based on the original LDA that is introduced in [2].

Firstly, we present *temporal topic models* that were proposed in previous works. A nonparametric mixture model for topic modeling over time is introduced in [5]. TOT [11] is a non-Markov continuous-time model of topical trends. In this model, words and continuous time are generated by a topic associated with a user. Dynamic Topic Model (DTM) [1] captures the evolution of topics over time. It shows topic distribution in various time intervals.

Secondly, we discuss *spatial topic models* that were introduced in previous works. Geographical topic discovery and comparison in presented in [13]. It presents three models: a location-driven model where GPS documents are clustered into topics based on their locations, a text-driven model where geographical topics are detected based on topic modeling with regularization by spatial information, a location-text joint model, a.k.a. LGTA (Latent Geographical Topic Analysis), which combines geographical clustering and topic modeling into one framework. GLDA (Geo Latent Dirichlet Allocation) [9] extends LDA for location recommendation. Paper [7] addresses the problem of modeling geo-graphical topical patterns on Twitter by introducing a sparse generative model.

Thirdly, we show few research works that combine time and location in *Spatio-Temporal topic models*. Paper [10] processes users' check-in. It detects topics and proposed a POI recommendation system with spatial and temporal information of user movements and interests. It proposes two models: USTTM and MSTTM for local (within a city) and global area (between cities) respectively. A Spatio-Temporal Topic (STT) model for location recommendation is presented in [8]. It processes users' check-ins to combine geographical influence and temporal activity patterns.

In addition, topic detection has been achieved through *wavelet analysis*. A lightweight event detection using wavelet signal analysis of hashtag occurrences in the twitter public stream is presented in [4].

Moreover, LDA-based methods for topic detection are SparseLDA [12] and O-LDA [3]. These methods describe real-time approaches to detect latent topics in data streams. In addition, topic mixtures estimated from an LDA model [6] are used to identify hot and cold topics.

3 Approach

We propose an LDA-based generative model for topic detection that incorporates time and location, that we call 'IncrAdapTL'. We identify two time-zones according to tweet time-stamps: day-time [6am–6pm] and night-time [6pm–6am]. The collected locations are the districts from the city of Hong Kong.

Our proposed model processes input data with sliding windows. We introduce incremental update of model's parameters between consecutive windows, and adaptive window lengths. We call our model Incremental Adaptive Time Location (IncrAdapTL) model and we present it in Algorithm 2.

3.1 Generative Process

In Table 1 we list the notations of parameters that we use. In Fig. 1 we present the topic model of IncrAdapTL and in Algorithm 1 its generative process. Our model consists of four distributions: word multinomial distribution per topic ϕ, topic multinomial distribution per tweet θ, timezone multinomial distribution per tweet ω, and location multinomial distribution per tweet ψ.

For each word w of each tweet message m, first we draw a timezone t from a multinomial distribution ω of timezones per tweet message, then we draw a location l from a multinomial distribution ψ of locations per tweet message, and finally we draw a topic z using the sampling process described in Sect. 3.2.

Algorithm 1. Generative Process

---

1 **for** *each tweet m* **do**
2   **for** *each word w of the tweet m* **do**
3   Draw a timezone $t \sim Mult(\omega)$;
4   Draw a location $l \sim Mult(\psi)$;
5   Draw a topic $z \sim p(k|t,l)$;
6   **end**
7 **end**

Table 1. Notation of parameters

Variable	Notation
ϕ	Word distribution per topic
θ	Topic distribution per tweet message
ω	Timezone distribution per tweet message
ψ	Location distribution per tweet message
t	A chosen timezone
l	A chosen location
z	A chosen topic
m	A tweet message
M	Total number of tweets
N	Total number of words for each tweet
V	Vocabulary size
K	Total number of topics
T	Total number of timezones
L	Total number of locations
$n_{w,k}$	Occurrences of a word w given a topic k
$\sum\limits_{w=1}^{V} n_{w,k}$	Total number of words assigned to topic k
$n_{m,k}$	Occurrences of a topic k given a tweet m
$\sum\limits_{k=1}^{K} n_{m,k}$	Total number of topics assigned to tweet m
$n_{t,m}$	Occurrences of a timezone t given a tweet m
$\sum\limits_{t=1}^{T} n_{t,m}$	Total number of timezones assigned to a tweet m
$n_{l,m}$	Occurrences of a location l given a tweet m
$\sum\limits_{l=1}^{L} n_{l,m}$	Total number of locations assigned to tweet m

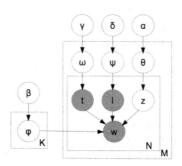

Fig. 1. Topic model

3.2   Sampling

Each drawn topic depends on the sampled timezone and on the sampled location by estimating the following probability:

$$p(k|t,l) \sim \frac{n_{w,k} + \beta}{\sum\limits_{w=1}^{V} n_{w,k} + V\beta} \times \frac{n_{m,k} + \alpha}{\sum\limits_{k=1}^{K} n_{m,k} + K\alpha} \times \frac{n_{t,m} + \gamma}{\sum\limits_{t=1}^{T} n_{t,m} + T\gamma} \times \frac{n_{l,m} + \delta}{\sum\limits_{l=1}^{L} n_{l,m} + L\delta} \quad (1)$$

3.3   Incremental

The IncrAdapTL model uses incremental re-estimation of topic model parameters. In the following algorithms we notate 'increm' mode when we estimate parameters from the previous window incrementally without processing all the tweets of each window. We notate 'estim' mode when we estimate parameters non-incrementally. In the latter, we re-estimate the parameters from scratch, by processing all the tweets of every window.

At this part, we explain the IncrAdapTL Algorithm 2. Our algorithm processes a stream of tweet data using sliding windows. In the first window of the stream of data [lines: 3–7] we run our model in the 'estim' mode (initialization and sampling). There are no previous parameters saved on the model, we run in non-incremental model because we need to process all the tweets of the first window.

In the rest windows of the stream [lines: 8–15]. Firstly, we load intermediate results from the previous window for incremental update of the model parameters. Secondly, we make decision of adaptive window length (we describe this in detail in Algorithm 5 of Sect. 3.4). Thirdly, we run our model in the 'increm' mode (initialization and sampling). Finally, in both modes, we save the window intermediate results [lines: 16–17].

Both modes, 'estim' and 'increm', have two steps: initialization and sampling. The sampling step is the same in both modes. During sampling, for each word of every tweet document a timezone, a location, and a topic are assigned. The initialization step of each mode differs.

Initialization of the 'estim' mode is presented in Algorithm 3. First we initialize the counters that are used in the estimation of the probabilities: $n_{w,k}, n_{m,k}, \sum\limits_{w=1}^{V} n_{w,k}, \sum\limits_{k=1}^{K} n_{m,k}$, with 0. We pass through all the tweets of the current window (old and new). We cannot benefit from the tweets that already existed in the previous window. Then, for every word of each tweet message we randomly choose a topic k and we increment the proper counters above by 1.

Initialization of the 'increm' mode is presented in Algorithm 4. Our model processes tweet datasets with sliding windows. In order to avoid passing through the tweets that existing in the previous window, we keep an tweet index in the stream of tweets, i.e. tweetIndex, from the previous window [lines: 3, 4]. So, we load the counters, $n_{m,k}$ and $\sum\limits_{k=1}^{K} n_{m,k}$, that are related with the topic-tweet

---

Algorithm 2. Window Process of Incremental Adaptive Time Location (IncrAdapTL) model

---

1 Global Initialization - Collection of global dictionary, locations, timezones;
2 **for** *each window* **do**
3   **if** *windowCounter == 1* **then**
4   /* Run in 'estim' mode. */
5   Initialization of 'estim' mode;
6   Sampling;
7   **end**
8   **else**
9   Load intermediate results from previous window;
10   /* Adaptive window length. */
11   Adaptive window length decision;
12   /* Incremental parameter estimation. Run in 'increm' mode. */
13   Initialization of 'increm' mode;
14   Sampling;
15   **end**
16   /* For incremental update. */
17   Save window intermediate results;
18 **end**

Algorithm 3. Initialization of 'estim' mode

---

1 /* Initialize counters with zero. */
2 $n_{w,k} = 0, \sum_{w=1}^{V} n_{w,k} = 0, n_{m,k} = 0, \sum_{k=1}^{K} n_{m,k} = 0;$
3 **for** *each tweet m* **do**
4   **for** *each word w* **do**
5   /* Draw a topic k randomly. */
6   k = Random(K);
7   /* Increment proper counters by one. */
8        $n_{w,k} \mathrel{+}= 1; \sum_{w=1}^{V} n_{w,k} \mathrel{+}= 1; n_{m,k} \mathrel{+}= 1; \sum_{k=1}^{K} n_{m,k} \mathrel{+}= 1;$
9   **end**
10 **end**

---

distribution θ. In addition, when we have slided the window we have updated the counters, $n_{w,k}$ and $\sum_{w=1}^{V} n_{w,k}$ that are related with the word-topic distribution ϕ. So, in [line: 2] of Algorithm 4, we load the updated values of $n_{m,k} \sum_{k=1}^{K} n_{m,k}, n_{w,k}$ and $\sum_{w=1}^{V} n_{w,k}$. These counters contain the information of the overlap between consecutive windows.

Algorithm 4. Initialization of 'increm' mode

---

1 /* Load counters from previous window. */

2 Load previous $n_{m,k}, \sum_{k=1}^{K} n_{m,k}, n_{w,k}, \sum_{w=1}^{V} n_{w,k}$;

3 /* Load the index in tweet stream. */

4 Load tweetIndex;

5 **for** *each tweet m after tweetIndex* **do**

6   **for** *each word w* **do**

7   /* Draw a topic k randomly. */

8   k = Random(K);

9   /* Increment proper counters by one. */

10        $n_{w,k}\ +\!= 1;\ \sum_{w=1}^{V} n_{w,k}\ +\!= 1;\ n_{m,k}\ +\!= 1;\ \sum_{k=1}^{K} n_{m,k}\ +\!= 1;$

11   **end**

12 **end**

Then, in [lines: 5–12] we process only the new tweets of the current window. For each word of every tweet after the tweetIndex, we update $n_{w,k}, n_{m,k}, \sum_{w=1}^{V} n_{w,k}, \sum_{k=1}^{K} n_{m,k}$ as before.

After initialization we perform sampling as we have mentioned above in Algorithm 2. We have described the sampling method in Sect. 3.2. The sampling process remains the same in both 'estim' and 'increm' modes.

3.4 Adaptive Window

Our second contribution is that the IncrAdapTL model uses adaptive window lengths. We have observed that the number of posted tweets varies between night-time and day-time in particular districts and in total. Throughout a day, there are sparse and dense windows. The tweet density of windows affects the performance of a topic model. Thus, in sparse windows we increase the window length in order to process more tweets. On the other hand, in dense windows we decrease the window lengths, so that we can focus on smaller time period.

So, we introduce different window lengths for more efficient processing of input stream in terms of time and accuracy. We start with window of 2 h length and we double it until it reaches the length of 8 h. Hence, we have three window lengths: windows of 2 h, 4 h, 8 h. In each case, the overlap with the previous window has length of 1 h.

As we have shown above in Algorithm 2, during processing of each window our model decides adaptively the length of the next window $i+1$ [lines: 10–11]. This decision is made as follows: First, we sample $r\%$ of the current window i. $r = \frac{\#\text{tweets in window}}{\text{window length in hours}} * 0.001$. We observe that the number of tweets per hour, i.e. $\frac{\#\text{tweets in window}}{\text{window length in hours}}$, ranges between 100 and 300. So, we transform this number into a percentage between 10% and 30%. We use high sampling ratio for dense windows and low sampling ratio for sparse windows. This is how our sampling rate is estimated in every window.

After we have collected the samples of the current window i, we compare the topic distribution of the samples $sample_i$ with the topic distribution of the previous window $increm_{i-1}$ by estimating the $\chi^2 - test$. We present this in Algorithm 5.

Algorithm 5. Adaptive window length decision

---
1 /* sample $r\%$ $sample_i$ mode */
2 Run the topic model in 'sample' mode;
3 /* $sample_i \sim increm_{i-1}$ */
4 $\chi^2 - test$ for topic-tweets distributions comparison of $sample_i$ and $increm_{i-1}$;

In Algorithm 6 we explain the steps for applying the $\chi^2 - test$. First [line: 1], we map similar topics between the 'sample' mode in current window i and the 'increm' mode of the previous window, $i-1$. We use the Jaccard distance for this topic similarity. We detect 15 topics in every mode and every topic consists of 10 words. Then, in [lines: 2, 3], we collect the tweet-topic distributions in $sample_i$ and in $increm_{i-1}$.

Algorithm 6. $\chi^2 - test$ for topic-tweets-distribution in each window i

---
1 Map similar topics between $sample_i$ and $increm_{i-1}$;
2 Collect tweets-per-topic distribution in $sample_i$;
3 Collect tweets-per-topic distribution in $increm_{i-1}$;
4 Estimate the $\chi^2 - test$ between $sample_i$ and $increm_{i-1}$;
5 **if** $\chi^2 > critical\ value$ **then**
6   /* Reject H_0 */
7   **if** $current\ window\ i\ is\ more\ dense\ than\ window\ i-1$ **then**
8   /* more dense, smaller window */
9   make next window $(i+1)$ length half;
10   **end**
11   **else**
12   /* more sparse, larger window */
13   make next window $(i+1)$ length double;
14   **end**
15 **end**
16 **else**
17   /* Insufficient evidence to reject H_0 */
18   keep same window length;
19 **end**

Then, in [line: 4], we use the $\chi^2 - test$ in order to test if tweet-topic distributions of $sample_i$ and $increm_{i-1}$ are similar. We consider null hypothesis H_0 that they come from same distribution, with significance level: $\alpha = 0.05$.

H_0: tweet-topic distributions of $sample_i$ and $increm_{i-1}$ are similar.
H_1: not H_0.

Then, in [lines: 5–15], if the χ^2 is larger than the critical value, then we reject the H_0. In this case, the distributions are not similar and we change the length of the window. If the current window i is more dense than the previous window $(i-1)$, then we make next window $(i+1)$ length half [lines: 7–10]. Otherwise, if the current window i is more sparse, then we make next window $(i+1)$ length double [lines: 11–14]. The window length ranges between 2 and 8 h. The overlap between consecutive windows is fixed to 1 h. Density metric is the comparison of tweets per hour $\frac{\#\text{tweets in window}}{\text{window length in hours}}$ between current and previous window.

When we have insufficient evidence to reject H_0 [lines: 16–19], we consider that the distributions are similar and we keep the same window length for next window.

4   Evaluation

In this section we present the experiments for the evaluation of our Incremental Adaptive Time Location (IncrAdapTL) model. We perform two sets of experiments. In the first set we compare the running times between IncrAdapTL and its non-incremental and non-adaptive version (TL). We show that IncrAdapTL processes the same dataset faster. In the second set of experiments, we show how the accuracy of IncrAdapTL changes in relationship with window lengths.

4.1   Characteristics of Datasets

Firstly, we present details on the datasets we use. We have crawled tweets from Hong Kong. We identify 22 districts, and two time-zones: day-time [6am–6pm], night-time [6pm–6am]. We use three datasets. As we present in Table 2, dataset A consists of 73K tweets crawled from the 21st December, 2015 to the 3rd January, 2016; dataset B includes 47K tweets from the 15th January to the 25th January; dataset C contains 77K from the 28th January to the 14th February.

We crawl tweets from the internet Twitter4j API[1] and Snowball[2]. We collect data from the area of Hong Kong. The goal of our work is the detection of discussed topics in different districts of the city, in different time-zones. We separate a day period into two time-zones: day-time [6am–6pm], night-time [6pm–6am].

---

[1] http://twitter4j.org.
[2] http://snowball.tartarus.org.

Table 2. Datasets

Dataset	Dates	Number of tweets
A	21/12/2015–03/01/2016	73,192
B	15/01/2016–25/01/2016	47,585
C	28/01/2016–14/02/2016	77,974

4.2 Execution Time

In the first set of experiments, we compare the execution times in milliseconds of our Incremental Adaptive Time Location (IncrAdapTL) model, as we presented in Algorithms 2 and 6 with the non-incremental and non-adaptive version of our model (TL). In the TL model, every window has a fixed length of two hours (non-adaptive) and in every window we run the 'estim' mode, i.e. estimation of the model parameters from scratch by processing all the tweets of each window, (non-incremental), as we described in Sect. 3.3.

We show that our proposed model, IncrAdapTL, can process the same datasets in less total execution time. We present the results for each dataset in Fig. 2 and in Table 3.

We observe that dataset A is processed by IncrAdapTL in 987 s, and in 1,214 s by TL. IncrAdapTL needs 81% of the TL's time. Similarly, IncrAdapTL processes dataset B in 629 s, and TL in 782 s. The difference is 80%. Also,

Execution Time (ms)

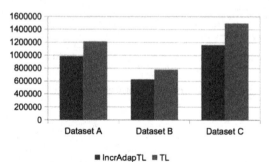

Fig. 2. Execution times in (ms)

Table 3. Execution times in (ms)

	'IncrAdapTL'	'TL'
Dataset A	987,219	1,214,082
Dataset B	629,736	782,205
Dataset C	1,161,124	1,496,003

IncrAdapTL processes dataset C in 1,161 s, whereas TL in 1,496 s. This is the 77% of TL's time. The experiments show that IncrAdapTL is better. The trend also shows that our method can scale well to very large data sets.

4.3   Accuracy

In the second set of experiments, we estimate the accuracy of our model. In every window, we compare our Incremental Adaptive Time Location (IncrAdapTL) model, as we presented in Algorithms 2 and 6, with the 'estim' mode, i.e. estimation of the model parameters from scratch (non-incremental). The result of the 'estim' mode is our ground truth, because in this mode processes all the tweets of every window and estimate the parameters from scratch. In these experiments, each window length of 'estim' mode (non-incremental) and 'increm' mode (incremental) are the same.

Results for dataset A are presented in Fig. 3; for dataset B in Fig. 4; and for dataset C in Fig. 5. In every graph we observe how our model's window length changes during the processing of the stream of data (adaptive). We see the sparse windows with 8 h length and the dense windows with 2 h length. Also, we see

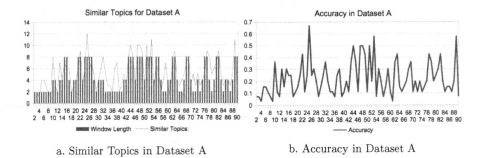

a. Similar Topics in Dataset A   b. Accuracy in Dataset A

Fig. 3. Dataset A

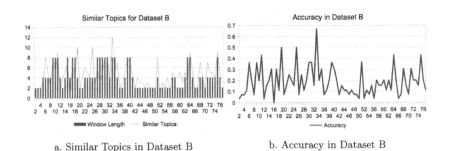

a. Similar Topics in Dataset B   b. Accuracy in Dataset B

Fig. 4. Dataset B

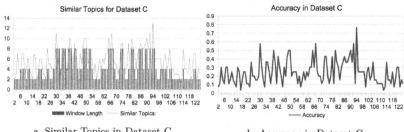

a. Similar Topics in Dataset C   b. Accuracy in Dataset C

Fig. 5. Dataset C

consecutive windows with the same length when the χ^2 value is small, and there is high correlation with the previous window, as we described in Algorithm 6.

Also, we observe that the number of similar topics is improved in the case of sparse windows, then the window length grows to 8 h. The number of common topics follows the length of windows in all datasets.

4.4 Qualitative Analysis

In addition, we show some concrete examples in each dataset that our model can detect some interesting topics. Dataset A presented in Table 4 contains tweet related with travel, Christmas and New Year. Dataset B in Table 5 includes few fashion events and entertainment trends. Dataset C in Table 6 contains topics related with travel, Chinese New Year, entertainment.

Table 4. Similar topics in dataset A

Mode	Topic	Keywords
estim	Christmas	christmas, eve, dinner, everyone, restaurant, kowloon, #HongKong, first
increm	Christmas	christmas, eve, #christmas, restaurant, hongkong, going, #HongKong, city
estim	Travel	#HongKong, #Asia, #Travel, #2015, #Exploration, #Christmas, disney, #Adventure
increm	Travel	#Asia, #Travel, #2015, #Exploration, #Adventure, #Holiday, super
estim	New Year	hong, kong, #2016, #happynewyear, love, #HongKong, #HappyNewYear, posted
increm	New Year	#2016, #happynewyear, #HongKong, #HappyNewYear, #hongkong, #HK, park, #newyear, #hk, peak
estim	New Year	good, first, countdown, morning, #2016HK, fireworks#, hi, hope, people, guys
increm	New Year	love, countdown, posted, #2016HK, fireworks#, hi, bye, life
estim	New Year	year, new, happy, first, #2016, best, start, everyone
increm	New Year	year, new, happy, hk, 2016, photo, posted, see, day, first

Table 5. Similar topics in dataset B

Mode	Topic	Keywords
estim	Fashion	out, new, life, fox, fur, #furry, doing, design, vest
increm	Fashion	central, new, fox, fur, #furry, year, design, vest, #furs, #furvest
estim	Ocean Park	posted, photo, kong, park, hong, ocean, adventure, kok, hotel,
increm	Ocean Park	posted, photo, park, ocean, adventure, away, sure, please, #travel,
estim	Entertainment	devonseron, day, bemylady, #BMLAngSimula, devon, central, seron, china, onitsshowtime
increm	Entertainment	devonseron, bemylady, #BMLAngSimula, devon, seron, onitsshowtime, itsshowtime, tweet, guangzhou, ever
estim	Fashion	sha, tsim, tsui, people, collection, fashionally, #6, womenswear, #fashion, #fashionally
increm	Fashion	time, collection, fashionally, #6, womenswear, #fashion, two, #fashionally, class, side
estim	Sports	#NBAVote, kobe, bryant
increm	Sports	#NBAVote, kobe, bryant, big
estim	Career	#Hiring, #CareerArc, our, #Jobs, #job, see, #HongKong, team, #Zhuhai, latest
increm	Career	#Hiring, #CareerArc, our, #Jobs, #job, #HongKong, team, #Zhuhai, latest, opening

Table 6. Similar topics in dataset C

Mode	Topic	Keywords
estim	Location	hong, kong, airport, international, hkg, islands, district, station, disneyland,
increm	Location	hong, kong, airport, international, hkg, islands, district, disneyland, ocean,
estim	Entertainment	#MrAndMrsSotto, ako, best, wishes, congrats, ang, first
increm	Entertainment	#MrAndMrsSotto, ako, wishes, bossing, congrats, forever, #Shenzhen
estim	Entertainment	#HBLPSL, #AbKhelKeDikha, #PSLT20, runs, overs, wright, new, russell, balls, batsman
increm	Entertainment	#HBLPSL, #AbKhelKeDikha, #PSLT20, runs, overs, gone, bowling, new, batsman, imran
estim	Travel	#travelling, #travelgram, #wanderlust, #travel, #wanderer, #explore, furniture, world
increm	Travel	#travelling, #travelgram, #wanderlust, #travel, #wanderer, #explore, last, miss
estim	Chinese New Year	new, year, happy, chinese, eve, lunar, 2016, year's, coming
increm	Chinese New Year	year, happy, one, lunar, home, spring, market, first
estim	Chinese New Year	new, year, happy, chinese, monkey, eve, batsman, night, lunar, everyone
increm	Chinese New Year	new, year, happy, chinese, monkey, eve, everyone, family, year's, friends
estim	Chinese New Year	year, new, happy, chinese, monkey, lunar, eve, wish, central, #gathering
increm	Chinese New Year	year, new, chinese, lunar, eve, time, wish, hotel, #familydinner, #qualitytime

5   Conclusion

In this paper we propose an Incremental Adaptive Time Location (IncrAdapTL) topic model for topic detection in tweets. This is an LDA-style generative topic model that incorporates time-zones (taken from time-stamps) and locations extracted from tweet stream API. We propose an incremental way of updating the parameters between consecutive windows and an adaptive window length in relationship with the correlation of consecutive windows and density, for faster processing. We evaluate IncrAdapTL by comparing total execution time and accuracy using three tweet datasets.

Acknowledgment. The work is partially supported by the Hong Kong RGC GRF Project 16207617, National Grand Fundamental Research 973 Program of China under Grant 2014CB340303, the National Science Foundation of China (NSFC) under Grant No. 61729201, Science and Technology Planning Project of Guangdong Province, China, No. 2015B010110006, Webank Collaboration Research Project, and Microsoft Research Asia Collaborative Research Grant.

References

1. Blei, D.M., Lafferty, J.D.: Dynamic topic models. In: Proceedings of the 23rd International Conference on Machine Learning, pp. 113–120. ACM (2006)
2. Blei, D.M., Ng, A.Y., Jordan, M.I.: Latent Dirichlet allocation. J. Mach. Learn. Res. **3**(Jan), 993–1022 (2003)
3. Canini, K., Shi, L., Griffiths, T.: Online inference of topics with latent Dirichlet allocation. In: Artificial Intelligence and Statistics, pp. 65–72 (2009)
4. Cordeiro, M.: Twitter event detection: combining wavelet analysis and topic inference summarization. In: Doctoral Symposium on Informatics Engineering, pp. 11–16 (2012)
5. Dubey, A., Hefny, A., Williamson, S., Xing, E.P.: A nonparametric mixture model for topic modeling over time. In: Proceedings of the 2013 SIAM International Conference on Data Mining, pp. 530–538. SIAM (2013)
6. Griffiths, T.L., Steyvers, M.: Finding scientific topics. Proc. Natl. Acad. Sci. **101**(suppl 1), 5228–5235 (2004)
7. Hong, L., Ahmed, A., Gurumurthy, S., Smola, A.J., Tsioutsiouliklis, K.: Discovering geographical topics in the twitter stream. In: Proceedings of the 21st International Conference on World Wide Web, pp. 769–778. ACM (2012)
8. Hu, B., Jamali, M., Ester, M.: Spatio-temporal topic modeling in mobile social media for location recommendation. In: 2013 IEEE 13th International Conference on Data Mining (ICDM), pp. 1073–1078. IEEE (2013)
9. Kurashima, T., Iwata, T., Hoshide, T., Takaya, N., Fujimura, K.: Geo topic model: joint modeling of user's activity area and interests for location recommendation. In: Proceedings of the sixth ACM International Conference on Web Search and Data Mining, pp. 375–384. ACM (2013)
10. Liu, Y., Ester, M., Qian, Y., Hu, B., Cheung, D.W.: Microscopic and macroscopic spatio-temporal topic models for check-in data. IEEE Trans. Knowl. Data Eng. **29**, 1957–1970 (2017)
11. Wang, X., McCallum, A.: Topics over time: a non-Markov continuous-time model of topical trends. In: Proceedings of the 12th ACM SIGKDD International Conference on Knowledge Discovery and Data Mining, pp. 424–433. ACM (2006)
12. Yao, L., Mimno, D., McCallum, A.: Efficient methods for topic model inference on streaming document collections. In: Proceedings of the 15th ACM SIGKDD International Conference on Knowledge Discovery and Data Mining, pp. 937–946. ACM (2009)
13. Yin, Z., Cao, L., Han, J., Zhai, C., Huang, T.: Geographical topic discovery and comparison. In: Proceedings of the 20th International Conference on World Wide Web, WWW 2011, pp. 247–256. ACM, New York (2011)

Incorporating User Grouping
into Retweeting Behavior Modeling

Jinhai Zhu[1,2], Shuai Ma[1,2(✉)], Hui Zhang[1,2], Chunming Hu[1,2], and Xiong Li[3]

[1] SKLSDE Lab, Beihang University, Beijing, China
[2] Beijing Advanced Innovation Center for Big Data and Brain Computing,
Beijing, China
{zhujh,mashuai,zhangh,hucm}@act.buaa.edu.cn
[3] National Computer Network Emergency Response Technical Team/Coordination
Center of China, Beijing, China
li.xiong@foxmail.com

Abstract. The variety among massive users makes it difficult to model
their retweeting activities. Obviously, it is not suitable to cover the overall users by a single model. Meanwhile, building one model per user is not
practical. To this end, this paper presents a novel solution, of which the
principle is to model the retweeting behavior over user groups. Our system, GruBa, consists of three key components for extracting user based
features, clustering users into groups, and modeling upon each group.
Particularly, we look into the user interest from different perspectives
including long-term/short-term interests and explicit/implicit interests.
We have evaluated the performance of GruBa using datasets of real-world
social networking applications, showcasing its benefits.

Keywords: User grouping · Social networks · Behavior modeling

1 Introduction

Social media is overwhelming nowadays, and popular social networks, e.g., Facebook, Twitter and Weibo, have attracted massive users [6,10,16,23,34]. These
users behave variously, knowledge of whom is significant for various applications
such as recommendation system and activity analysis. Hence there is an emer-
gent demand of developing systems and algorithms that could properly model
user behaviors, which has already attracted the attention from both academia
and industry [7,10,12,18,19,34,35].

Central to user behavior modeling is the need to choose the right granularity
of model (i.e., how many users share one model), as well as the variety of features
to be utilized for differentiating users. Already, there exist works of building a
single model for all the users [10,34]. Apparently, such model bears the limitation
of being coarse. On the other hand, modeling each user is not practical, due to
the tremendous number of users.

The key driver of our work is the observation that in social media applications, users could fall into groups and each group shares representative behaviors. As one example, consider the film *Brave Heart*, fans of which are probably addicted to highland, bagpipe and war films, and thus likely to retweet blogs of these topics. Particularly, we study the retweeting behavior of users and our work can be readily generalized to other behaviors of "likes" and "comments" as well. In the realm of social network behavior modeling, few work has been done over grouping, which however has been proved to be effective in other fields such as economic behavior analysis. This motivates us to incorporate user grouping into the retweeting behavior modeling, filling the gap of existing studies that build a single model for all users.

The contributions of our work are as follows:

(1) We present a system named GruBa with the novel perspective to model user behaviors over groups instead of a single model for all users.
(2) We leverage user interests to facilitate the modeling of retweeting behavior and look into interests with various dimensions, including long-term/short-term interests and explicit/implicit interests.
(3) We offer a clustering method K-Gru to deal with complex vectors, serving as an extension for standard K-Prototype algorithm.
(4) We evaluate the performance of GruBa using real-world datasets, showcasing its benefits against competitive state of the art approaches.

Our paper is organized as follows. Section 2 gives the problem formulation and system overview, followed by detailed explanations in Sects. 3, 4 and 5. Section 6 is performance evaluation, followed by related work in Sect. 7 and conclusions in Sect. 8.

2   Overview of System GruBa

2.1   Problem Formulation

We consider the retweeting behavior of users in social media. For simplicity, assuming the microblogs that a user can retweet come from those owned by her/his followees.

Definition 1. *A microblog $M_b = (O, T, M, flag)$ has its owner O (a.k.a. user in this paper) to whom M_b belongs (either tweeted or retweeted), the generated time T of M_b, the message content M, and flag denoting M_b is retweeted or originally tweeted by O. Here we use 1 and 0 to denote retweeted and tweet, respectively.*

Definition 2. *Given a user u, we adopt B_u, R_u and E_u to represent her/his microblogs B_u, followers R_u and followees E_u, respectively, in which a follower/followee is a user.*

Note that $M_b.O$ is a user, and B_u is a set of microblogs.

Providing a set of users $\mathbb{U}$ and their associated microblogs $\mathbb{B}$, system GruBa builds a retweeting model for each group of $\mathbb{U}$, such that given a microblog b and a follower f of its onwer $b.O$, i.e., $f \in R_{b.O}$, 1 or 0 is returned regarding whether f retweets b or not.

2.2   GruBa Framework

System GruBa is designed from the ground up as a system for modeling users' retweeting behavior in social media, and Fig. 1 shows the architectural components of GruBa.

Sina Microblog Data. It is the data crawled to be processed by GruBa, i.e., data of microblogs and users.

Key Modules. GruBa consists of three key modules.

(1) Feature Extraction: By coalescing the microblog data, each user is depicted by a bunch of features, which are grouped into three categories. They are features of *Basics* (e.g., the number of followers and followees), *Behavior* (e.g., the frequency and the popular slots of retweeting) and *Interest* (e.g., long and short term interests, as well as explicit and implicit interests). These features are extracted from the stored Sina Weibo data by Feature Extraction module, and serve as the input of the User Clustering module.
(2) User Clustering: Providing the user-based features, User Clustering takes charge of the clustering task such that each user falls into a proper group.
(3) Behavior Modeling: For each group obtained by User Clustering, Behavior Modeling builds a model by employing both positive and negative samples (i.e., microblogs labeled with retweeted and not retweeted), on which the user retweeting behaviors are also tested.

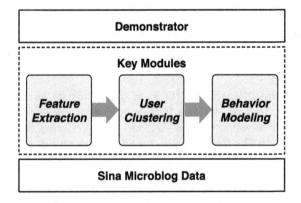

Fig. 1. GruBa architecture

Demonstrator. At the top layer of GruBa, it is the Demonstrator for visualizing all aspects of the system, e.g., profiling of user groups.

The distinctive feature of System GruBa models user retweeting behaviors over groups instead of a single model for all the users [10,34].

3 Feature Extraction

With the underlying Sina Microblog Data, the Feature Extraction module is responsible for mining the user characteristics, and produces three classes of features for each user: *Basic Feature*, *Behavior Feature* and *Interest Feature*, referred to as *Feature Data* in GruBa.

3.1 Basic Feature

The *Basic Feature* employs a vector I to depict the basic characteristics of a user u.

$$I_u = (G_u, P_u, \#R_u, \#E_u, R_{ee,u}, U_{t,u}),\tag{1}$$

in which the variable details are illustrated in Table 1.

Table 1. Illustration of variables in basic feature

Variables	Illustration
G_u	Gender of user u
P_u	Province of user u
$\#R_u$	Number of followers
$\#E_u$	Number of followees
$R_{ee,u}$	A ratio defined as the number of followers over that of followees, i.e., $\frac{\#R_u}{\#E_u}$
$U_{t,u}$	User type (as illustrated in Table 2)

Table 2. Category of user type

Types	Illustration
0	$\#E_u \le 50$ and $\#R_u \le 50$
1	$\frac{\#E_u}{\#R_u} \ge 5$
2	$\frac{\#R_u}{\#E_u} \ge 5$
3	Other cases

3.2 Behavior Feature

Unlike *Basic Feature*, the *Behavior Feature* of a user u is certain statistics regarding the retweeting behavior of u, shown below:

(a) the number of owned microblogs $\#B_u$,
(b) the ratio $R_{oc,u}$ that is the number of retweeted microblogs over that of originally tweeted, i.e., $\frac{\#(B_u|flag==1)}{\#(B_u|flag==0)}$,
(c) the average number of retweeted and tweeted microblogs per week: $\#W_{r,u}$ and $\#W_{t,u}$,
(d) the normalized vectors regarding the time distribution of a user's retweeting/tweeting behavior: $P_{rt,u} = (p'_{r0}, p'_{r1}, \dots, p'_{r11})$, $P_{tt,u} = (p'_{t0}, p'_{t1}, \dots, p'_{t11})$, where p'_{r0}/p'_{t0} is the probability that the retweeting/tweeting activity happens from 0am to 2am, p'_{r1}/p'_{t1} is the probability that the retweeting/tweeting activity happens from 2am to 4am, and so on, and

(e) the normalized vectors with respect to the gap distribution of a user's retweeting/tweeting behavior: $P_{rg,u} = (p''_{r0}, p''_{r1}, \ldots, p''_{r5})$, $P_{tg,u} = (p''_{t0}, p''_{t1}, \ldots, p''_{t5})$, in which p''_{r0}/p''_{t0} is the probability that the gap between two retweeted/tweeted microblogs is within 1 min. Ditto for p''_{r1}/p''_{t1} (1 min to 1 h), p''_{r2}/p''_{t2} (1 to 12 h), p''_{r3}/p''_{t3} (12 to 24 h), p''_{r4}/p''_{t4} (24 to 48 h) and p''_{r5}/p''_{t5} (more than 48 h).

In summary, the *Behavior Feature* H_u of user u consists of the following:

$$(\#B_u, R_{oc,u}, \#W_{r,u}, P_{rt,u}, P_{rg,u}, \#W_{t,u}, P_{tt,u}, P_{tg,u}). \tag{2}$$

3.3   Interest Feature

Different from the slightly straightforward *Basic Feature* and *Behavior Feature*, *Interest Feature* involves a process of labeling users with their interested topics based on their tweeted and retweeted microblogs. In short, with a given lexicon (made by some professionals) consisting of several *topics*, the interest feature of a user u is a normalized vector, in which each entry refers to the degree that u is interested in the corresponding *topic*.

Definition 3. *A lexicon L consists of a set of topics γ such that each topic is associated with a set of cell words c, in which each cell word depicts an aspect of the topic.*

Definition 4. *For a user u, each microblog $b \in B_u$ is represented by a set of words w.*

Definition 5. *The interest feature P_u of a user u is a normalized vector*

$$(p_0, p_1, \ldots, p_{x-1}), \tag{3}$$

in which user u matches x topics in lexicon L, and p_i refers to the degree that u is interested in topic i.

The similarity p_i ($i \in [0, x-1]$) will be detailed in each scenario (explicit/implicit interest analysis, towards words/topics, etc).

In GruBa, a word, either in the form of c or w, acts as the minimum unit for analysis. Hence, the similarity $sim(w, c)$ of a word pair (w, c) could be generalized to the similarity of a microblog against one topic $sim(b, \gamma)$, and finally to a user u versus each topic γ in lexicon $sim(u, \gamma)$; topics with similarity satisfying certain thresholds are allocated to user u and constitute the interests of u.

System GruBa employs a well established lexicon to discover the explicit interests of users. When no proper explicit interests are found, two algorithms are leveraged to identify implicit interests: (1) TF-IDF (Term-Frequency and Inverse Document-Frequency), and (2) Twitter-LDA [37] (a method to discover topics from Twitter, by applying the standard Latent Dirichlet Allocation (LDA) model [3]), both of which adopt word2vector [25] to measure word similarity.

We now explain the detailed process for mining the interest features of a user.

Step 1: Each microblog b in B_u of user u is decomposed into a word set WS.

Step 2: Explicit interests are explored. Specifically, every word w in WS is sent to match each cell word c of lexicon topics. If w and c are identical, $sim(w, c) = 1$, and $sim(w, c) = 0$, otherwise.

The similarity of b with a lexicon topic γ is defined as

$$sim(b, \gamma) = \sum_{i,j} sim(w_i, c_j). \tag{4}$$

If $sim(b, \gamma)$ satisfies a certain threshold (3 by default), topic γ is labeled to microblog b; the user u is then discovered having an explicit interest (topic γ). Thus, by looking into the similarity of b against all topics in lexicon L, the explicit interests of u are derived, in the form of interest feature (see Definition 5).

Step 3: If the $sim(b, \gamma)$ in **Step 2** cannot meet the threshold, i.e., explicit interest discovery over user u fails, the implicit interests of user u are further mined, by running the following steps 3.1 and 3.2 in parallel.

Step 3.1: A metric *TF-IDF weight* W_f is computed by employing TF-IDF to calculate the weight distribution of words in microblog b:

$$W_f = \{(w_i, p_i)\}, \tag{5}$$

where w_i refers to a single word, of which the weight is p_i, with $\sum_i p_i = 1$.

To compute such weight p_i for word w_i, a metric p_i'' is first calculated as:

$$p_i'' = \frac{|b_i|}{|b|} * \log(\frac{|D|}{|D_i| + 1}), \tag{6}$$

in which we use the operator $|\ |$ to measure the cardinality, such that $|b_i|$ is the occurrences of word w_i in microblog b, and $|b|$ the total occurrences of all words in b. $|D|$ is the total number of microblogs in the dataset and $|D_i|$ is the number of microblogs where w_i appears. Hence, each word w_i shall get an initial weight p_i'', upon which the normalization is performed and p_i is obtained, resulting the *TF-IDF weight*.

TF-IDF based similarity is then calculated. For example, the similarity of W_f over a single topic γ in lexicon, written as $sim(W_f, \gamma)$, is defined as formula 7. Here $W_f = \{(w_i, p_i)\}$, $\gamma = \{c_j\}$, VEC_{W_f} is defined as formula 8, VEC_t is defined as formula 9, where N_γ is the number of words in topic γ and $vec[w]$ is the word vector of word w returned by word2vector [25].

$$sim(W_f, \gamma) = VEC_{W_f} \cdot VEC_t, where \tag{7}$$

$$VEC_{W_f} = \sum_i p_i * vec[w_i], and \tag{8}$$

$$VEC_t = \sum_j \frac{1}{N_\gamma} * vec[c_j].$$

(9)

Step 3.2: Similarly, another metric *Twitter-LDA weight* W_w is obtained by using Twitter-LDA to result the word weight distribution of microblog b. Unlike TF-IDF, Twitter-LDA first trains the overall microblogs, allocating each microblog with a *tag*. The structure of *tag* is as follows:

$$W_t = \{(w_i', p_i')\},$$

(10)

where w_i' refers to a word in *tag* W_t, and p_i' is the probability that w_i' appears in microblogs with the said *tag*, with $\sum_i p_i' = 1$ ($|W_t| = 30$ in this work by default). Subsequently, W_t is leveraged to conclude W_w, i.e., $W_w = W_t$, which shares the format with that of W_f.

Step 4: Hence, the similarity of a microblog b against a lexicon topic γ is given by:

$$sim(b, \gamma) = \alpha * sim(W_f, \gamma) + (1 - \alpha) * sim(W_t, \gamma),$$

(11)

where the α is a parameter by which GruBa could set flexible priorities between TF-IDF and Twitter-LDA.

Step 5: Repeat Steps 1 to 4 for the microblog b over every topic in lexicon, i.e., $\forall \gamma_k \in L$ results one similarity value of $sim(b, \gamma_k)$. Such computation further extends to all the microblogs owned by user u, such that: $\forall b_m \in B_u$, $\forall \gamma_k \in L$, there exists a similarity of $sim(b_m, \gamma_k)$. Hence, the overall similarity of user u over lexicon topics $\{\gamma\}$ (i.e., L), written as $S(u, L)$, could be denoted by a vector:

$$S(u, L) = (s_0, s_1, \ldots, s_{n-1}),$$

(12)

where n refers to the cardinality of L (i.e., number of topics in L) and s_k is the overall similarity of user u over topic γ_k, which is given by:

$$s_k = \sum_m sim(b_m, \gamma_k).$$

(13)

Among the n dimensions of $S(u, L)$, those with top x (3 in GruBa) similarity values are selected to label the implicit interests of user u, which results an x dimensional vector P_u as described in Definition 5. Similarly, interest features of all users are returned.

As a result, the *Feature Data* F_u for a user u is

$$F_u = (I_u, H_u, P_u),$$

(14)

where I_u, H_u and P_u are the *Basic Feature*, *Behavior Feature* and *Interest Feature* of u, respectively.

4   User Clustering

Providing the *Feature Data*, the User Clustering module takes the charge of grouping each user concerned into a proper cluster, as illustrated in Algorithm 1. The idea is to enumerate a number of clustering trials (line 3) and select the optimal solution with the best Silhouette coefficient value (v in line 13). In principle, each trial (referred by t in line 3) first performs a clustering task (line 4 to be detailed in Sect. 4.1), resulting a cluster (by $l(u)$) for each user u (line 5); then, each user obtains a Silhouette coefficient value $v(u)$ stemmed from the in/out-cluster distances (lines 7–9; shall be illustrated in Sect. 4.2); finally, the averaged Silhouette coefficient value of all users serves as the Silhouette coefficient value of the current trial, written as $v(t)$ (line 11), by which the said selection process is conducted (line 13).

Next, we shall now first detail how GruBa performs the clustering task and subsequently illustrate the computation for the metric of Silhouette coefficient value.

Algorithm 1. User Clustering in GruBa

1: Input: *Feature Data* F of a set of users, the minimum/maximum number of clusters N_i and N_a
2: Output: Optimal user clustering result R

3: **for all** $t \in [N_i, N_a]$ **do**
4:      perform K-Gru over F to get t clusters;
5:      clustering result $R'(t) = \{(u, l(u))\}$ with cluster info $l(u)$ for each user u;
6:      **for all** $u \in \{u\}$ **do**
7:          in-cluster distance $d_i(u)$;
8:          out-cluster distance $d_o(u)$;
9:          Silhouette coefficient value $v(u) := \frac{(d_o - d_i)}{max(d_o, d_i)}$;
10:     **end for**
11:     $v(t) := Avg\{v(u)\}$;
12: **end for**
13: **if** $v(a) = Max\{v(t)\}$ **then**
14:     $R := R'(a)$;
15: **end if**
16: **return** R.

4.1   K-Gru: Clustering in GruBa

In GruBa, the clustering rests on an optimized K-Prototype [17] algorithm, named K-Gru in this work. Similar as K-Prototype, K-Gru randomly selects the cluster kernels among samples and employs the minimum distance between them to determine an initial result, upon which the clustering tasks are iterated until the results are stable.

Unlike K-Prototype that supports vector samples in which each dimension is of numerical/categorical, K-Gru could also handle the case where a dimension is

Table 3. Dimension types in *Feature Data* vector

Types	Data dimensions
Numerical data	$\#R_u, \#E_u, R_{ee,u}, \#B_u, R_{oc,u}, \#W_{r,u}, \#W_{t,u}$
Categorical data	$G_u, P_u, U_{t,u}$
Normalized vectors	$P_{rt,u}, P_{rg,u}, P_{tt,u}, P_{tg,u}, P_u$

one normalized vector. Recall the sample data for User Clustering, i.e., *Feature Data* in form of vectors (see formula 14), of which the data type regarding each dimension is shown as Table 3.

As aforementioned, the clustering of K-Gru rests on the distance between vector samples, where the dimensions are combined with numbers (normalized to 0–1 range), categories and normalized vectors. For simplicity, we shall first illustrate the distance calculation of the simple vectors with a mono data type on each dimension, and then demonstrate that of complex vectors.

For numerical vectors $Y' = (y_0', y_1', \ldots)$ and $Z' = (z_0', z_1', \ldots)$, the Euclidean distance [8] between Y' and Z' is given by:

$$D_n(Y', Z') = \sum_e (y_e' - z_e')^2. \tag{15}$$

For categorical vectors $Y'' = (y_0'', y_1'', \ldots)$ and $Z'' = (z_0'', z_1'', \ldots)$, the Hamiltonian distance [17] of Y'' and Z'' is:

$$D_h(Y'', Z'') = \sum_e H_e, \tag{16}$$

where H_e refers to the Hamiltonian distance over each dimension, with $H_e = 0$ if y_e'' and z_e'' share the identical value, and $H_e = 1$, otherwise.

Regarding two vectors where each dimension is a normalized vector per se, Cosine Similarity is leveraged to compute the distance. Then, the distance between such two vectors $Y^* = (Y_0^*, Y_1^*, \ldots)$ and $Z^* = (Z_0^*, Z_1^*, \ldots)$ is:

$$D_v(Y^*, Z^*) = \sum_e (1 - Y_e^* \cdot Z_e^*), \tag{17}$$

where $\cdot$ refers to the dot product operation between two normalized vectors Y_e^* and Z_e^*.

Putting these together, the distance regarding the complex vectors $Y = (Y_0, Y_1, \ldots)$ and $Z = (Z_0, Z_1, \ldots)$ in K-Gru, referred to as GruBa Distance, is defined as:

$$D_g(Y, Z) = \sum_e G_e, \tag{18}$$

where the distance on each dimension G_e is given by:

$$G_e = \begin{cases} (Y_e - Z_e)^2 & \text{if } Y_e/Z_e \text{ is numerical} \\ H_e \ (1 \ or \ 0) & \text{if } Y_e/Z_e \text{ is categorical} \\ 1 - Y_e \cdot Z_e & \text{if } Y_e/Z_e \text{ is of normalized vector} \end{cases} \tag{19}$$

4.2   Silhouette Coefficient Metric Computation

In system GruBa, the Silhouette coefficient metric serves as the fundamental criteria for deriving an optimal clustering result. Providing a clustering result, each user is associated with a cluster.

Definition 6. *The in-cluster distance $d_i(u)$ is the average distance to all the other users in the same cluster:*

$$d_i(u) = Avg\{D_g(F_u, F_{u''}) \mid u'' \in l \wedge u \neq u''\}. \tag{20}$$

Definition 7. *The out-cluster distance $d_o(u)$ is measured as the minimum of the distances $\{d^*\}$ between u and other clusters:*

$$d_o(u) = Min\{d^*(u, l') \mid l' \neq l\}, \tag{21}$$

where d^* is given by:

$$d^*(u, l') = Avg\{D_g(Y_u, Y_{u'}) \mid u' \in l'\}. \tag{22}$$

Definition 8. *The Silhouette coefficient value $v(u)$ is defined as:*

$$v(u) = \frac{(d_o - d_i)}{max(d_o, d_i)}. \tag{23}$$

Intuitively, a good clustering solution should have a bigger d_o and a smaller d_i, such that samples with obvious differences go to various clusters and vice versa. When d_o is far more than d_i, Silhouette coefficient value approaches to 1. Hence, the larger Silhouette coefficient value is, the better clustering performs, by which the optimal solution is selected.

5   Group Based Behavior Modeling

Recall the central problem of GruBa, where the retweeting behaviors of users are modeled. Specifically, such model is built by the Group Modeling module for each user group and thus named as group model. To avoid ambiguity, we shall use the term of *items* to denote the data for training the group model. A given *item* is either positive or negative.

Definition 9. *An item E involves a microblog b and a user f such that $f \in R_{b,O}$, i.e., f is a follower of the owner of microblog b.*

$$E \in \begin{cases} positive\ items & if\ f\ retweeted\ b \\ negative\ items & if\ f\ did\ not\ retweet\ b \end{cases} \tag{24}$$

And the data of item E consists of three parts.

(1) **User Info** contains a list of aforementioned metrics $\{G_u, P_u, \#R_u, \#E_u, R_{ee,u}\}$.

(2) **Microblog Info** refers to metrics related to the microblog b. The number that b has been retweeted, commented, liked and the length of $b.M$ (microblog message) are considered. What is more, we consider the correlation between b and recent event, where the latter is expressed as several core words returned by Ring [1]. Here we compute *TF-IDF weight* W_f of $b.M$, and get correlation metric C_h of b and event by formula 7.

(3) **Interaction Info** includes seven correlation metrics: $\#B_u$, $R_{oc,u}$, $\#W_{r,u}$, $\#W_{t,u}$, microblog b versus the user u's *Interest Feature* P_u (a.k.a. long-term/stable interest in this work), and b's timestamp versus the time distribution of u's retweeting behavior $P_{rt,u}$. In addition, we consider u's short-term interest, which is mined from u's recent microblogs (e.g., within 30 days) and calculated by *TF-IDF*, namely W_s. The correlation between microblog b and u's short-term interest is computed by W_f and W_s (using formula 7).

The modeling of retweeting behavior of groups is treated as a classification problem, and we utilize the random forest classifier to address it. Details for random forest [15] are omitted here for space reason. The advantage of this classification model lies in that it could integrate different features conveniently, and the obtained group behavior model could learn what a positive/negative item looks like over each metric mentioned above.

Here we use accuracy to evaluate our model. To define accuracy, we set four variables: E_{tp}, E_{fp}, E_{tn} and E_{fn}. For a given item E, if E is a positive item and our model also determines it a positive item, then we set E_{tp} to 1, else we set E_{tp} to 0. If E is a negative item and our model determines it a positive item, then we set E_{fp} to 1, else we set E_{fp} to 0. If E is a negative item and our model determines it a negative item, then we set E_{tn} to 1, else we set E_{tn} to 0. If E is a positive item and our model determines it a negative item, then we set E_{fn} to 1, else we set E_{fn} to 0. So the accuracy can be defined as:

$$accuracy = \frac{\sum_E (E_{tp} + E_{tn})}{\sum_E (E_{tp} + E_{tn} + E_{fp} + E_{fn})} \tag{25}$$

6   Performance Evaluation

In this section, we shall first detail the experimental setting, and we then present the evaluation results and analyses, showing the benefit of GruBa against state of the art approaches.

6.1   Experimental Setting

Experiments were run on a machine with two Intel Xeon E5C2630 2.4 GHz CPUs and 64 GB of Memory, running 64 bit Windows 7 professional system. We have employed a real-world dataset Sina Weibo that consists of 24 million microblogs that are associated with 43.5K users.

With respect to the parameters of GruBa, we use the default values as mentioned in previous sections. Particularly, for the Feature Extraction module, for practical reasons, we employed a smaller testing dataset (with manually labeled topics for yardstick) to obtain the proper value of α for extracting *Interest Feature*; For the User Cluster module, we studied the clustering solutions with the minimum/maximum number of clusters 2 and 10; For the Behavior Modeling module, the recent 30 days microblogs of users are used for their short-term interest analysis, and popular words in the latest 24 h are returned by Ring as the Hot Event keywords [1].

6.2 Results and Analyses

Next, we shall report the performance of system GruBa over each component.

Exp-1: Feature Extraction. Figure 2 shows the testing results of using various α values. We report the interest accuracy with various α values. Suppose the label set manually labeled for each microblog is A, the label set our method labeled is B. The interest accuracy is defined as $\frac{A \cap B}{A \cup B}$.

System GruBa reaches the optimal results when α is 0.7, upon which the interest feature extracting is performed for the overall dataset with 43.5K users and 24 million microblogs. In general, it performs well when α falls into $[0, 0.8]$.

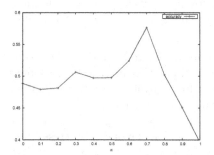
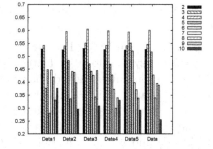

Fig. 2. Interest extraction: varying α.   **Fig. 3.** Silhouette coefficient tests.

Exp-2: User Clustering. Figure 3 depicts the Silhouette coefficient values of multiple clustering solutions, with the cluster number varied from 2 to 10. Specially, we used different testing datasets, with *Data* containing the overall 43.5K users, and each of {*Data1,..., Data5*} contains 10K randomly selected users. Except for *Data1*, solutions for {*Data2,..., Data5*} are the best for 4 clusters.

Exp-3: Behavior Modeling. Figure 4(a) shows the performance of GruBa against the state of the art approach LRC-BQ [34]. We evaluate the performance using the metrics of accuracy. LRC-BQ does not deal with user grouping. Hence, we not only study the modeling effectiveness per group (i.e., "Group-One/Two/Three/Four" with user clustering), but also examine GruBa versus

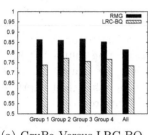

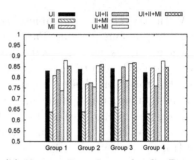

(a) GruBa Versus LRC-BQ.   (b) Various Data Items for GruBa.

Fig. 4. Performance of GruBa

LRC-BQ in the case that all users are in a single group (i.e., "All-Users"). The results show that:

(1) With user clustering, GruBa performs better than LRC-BQ in most cases.
(2) For GruBa, having user clustering is better than the alternative single group. Ditto for LRC-BQ.

Figure 4(b) explores the performance of GruBa when using alternative data items for modeling. By default, GruBa uses "UI+II+MI", i.e., items of users (UI), microblogs (MI) and interactions (II). How about using other combinations of the above item(s)? As shown in Fig. 4(b), the default setting wins in most cases.

Exp-4: Case Study for Feature Extraction. In this test, we show the results of our demo system for user features extraction. Considering the huge amount of users, we carefully selected one typical user for analysis. Here we chose Mary (a famous drama and movie actress in China) as an example.

The feature extraction result for Mary is depicted in Fig. 5. Then we can see the basic information of Mary in Fig. 5(a): her nickname is Actress Mary (演员马丽) and she is from Beijing (北京). Mary has more followers than followees. The Sina Microblog tag she made for herself is "actress" (演员). As to the long-term interest, she is interested with stage performance (舞台表演), drama (话剧表演), film (电影光影) and so on as shown in Fig. 5(b), which is consistent with her tag. The probability distribution of tweeting and retweeting indicates that she is more active at night than daytime as shown in Fig. 5(c) and (d). According to Fig. 5(e) and (f), the interval between her two tweeted/retweeted microblogs is mostly within 48 h, showing she is an active user. Mary's short-term interest, e.g. from 09/01/2016 to 09/30/2016, is shown in Fig. 5(g), which indicates she had been busy with promoting the drama "Earl of Oolong Mountain" (乌龙山伯爵). So the results are in line with expectations, as drama (话剧) and stage (舞台) in Fig. 5(g).

To conclude, by modeling user behaviors over groups instead of a single model for all users such as LRC-BQ, we improve the average accuracy over LRC-BQ by 6%. What deserves to be mentioned is that the performance of LRC-BQ is also improved significantly by user clustering.

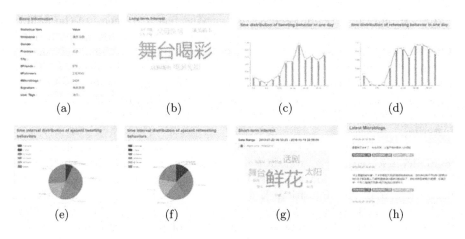

(a)   (b)   (c)   (d)

(e)   (f)   (g)   (h)

Fig. 5. Visualization for user features.

7 Related Work

In this section, we review related work in literature from the aspects of analyzing features, mining groups and modeling behaviors within the realm of social network modeling. As aforementioned, GruBa leverages the user features of basics, behavior and interest.

For feature analysis, there has been existing work of mining user features, such as race [27], gender [4], age [26], political preference [27,30] and occupation [9]. Our work, however, does not focus on the mining process, but uses the mined features as the input for user clustering and group based behavior modeling.

Studies of behavior analysis put emphasis on exploring the characteristics. For example, [20] proposed a model that can properly explain various time distributions of user behaviors by theoretical analysis; [13] studied the user activity distribution of one day/week; [5] provided the PowerWall distribution of Facebook users, identifying a number of surprising behaviors and anomalies. Considering the behavior characteristics, GruBa makes use of them to feed the modeling process.

There have been established work of extracting user interests. [22] mined the user interests by exploring keywords of microblogs with the aid of word frequency and machine translation. [32] proposed a method of extending the topic model to obtain user interests. Also, [24] used a knowledge base and [11] provided a solution of using hashtags for interest analysis. [21] summarized user interest by exploring the mentioned celebrities; Similarly, [2] leveraged the followed experts to result interest characteristics. Different from these existing solutions, GruBa employs a cell lexicon to properly express user interests, in which Twitter-LDA [37] and TF-IDF are employed.

Approaches of grouping users in social networks could fall into a variety of categories. [27] grouped users by the info of race, political view and etc. [36] studied the social groups on Facebook and Wechat, resulting various patterns of group evolution. [31] proposed a time-varying factor to measure the affinity between users and groups such that proper group proposals are recommended. More recent studies also look into mining user communities. [33] employed matrix decomposition to mine user communities; [14,28] considered followees info for user clustering; [29] proposed an incremental algorithm to mine user communities using modular degree as the clustering yardstick. Providing the diversity of user features, GruBa employs basic, behavior and interest features into user clustering.

The main problem of current approaches for behavior modeling lies in that the model is for either the overall users or a single user. [10,34] discovered that users' retweeting behavior is largely influenced by their followees, whereas [19] employed matrix decomposition, [18] used collaborative filtering methods and [35] leveraged statistical models for retweeting analysis. [7,12] focused on identifying whether the retweeting is fraudulent or of protest. Whereas our work GruBa builds the retweeting model for each user group, instead of the mono model for all users or one model per user.

8   Conclusions

In this work, we have presented GruBa, a system to model the retweeting behavior of users in social media. GruBa departures from existing work by grouping users into clusters, during which features of basics, behavior and interests are extracted. Specially, we have studied interest features from various perspectives, such as long-term/short-term interests and explicit/implicit interests. Finally, we have provided a performance evaluation of GruBa by using real-world datasets to demonstrate the benefits of our system GruBa.

Acknowledgments. Ma is supported in part by NSFC U1636210, 973 Program 2014CB340300, NSFC 61421003, and MSRA Collaborative Research Program. Li is supported in part by NSFC U1636123 & 61403090. For any correspondence, please refer to Shuai Ma.

References

1. http://ring.act.buaa.edu.cn/#/
2. Bhattacharya, P., et al.: Inferring user interests in the Twitter social network. In: RecSys (2014)
3. Blei, D.M., Ng, A.Y., Jordan, M.I.: Latent dirichlet allocation. JMLR **3**, 993–1022 (2003)
4. Ciot, M., Sonderegger, M., Ruths, D.: Gender inference of Twitter users in Non-English contexts. In: EMNLP (2013)
5. Devineni, P., Koutra, D., Faloutsos, M., Faloutsos, C.: Facebook wall posts: a model of user behaviors. Soc. Netw. Anal. Min. **7**(1), 6:1–6:15 (2017)
6. Duan, L., Ma, S., Aggarwal, C.C., Ma, T., Huai, J.: An ensemble approach to link prediction. IEEE Trans. Knowl. Data Eng. **29**(11), 2402–2416 (2017)

7. Ranganath, S., et al.: Predicting online protest participation of social media users. CoRR, abs/1512.02968 (2015)
8. Everitt, B.: Cluster Analysis. Heinemann Educational Books Ltd., Portsmouth (1974)
9. Fan, Y., Chen, Y., Tung, K., Wu, K., Chen, A.L.P.: A framework for enabling user preference profiling through Wi-Fi logs. In: ICDE (2016)
10. Feng, W., et al.: Retweet or not?: personalized Tweet re-ranking. In: WSDM (2013)
11. Feng, W., Wang, J.: We can learn your #hashtags: connecting Tweets to explicit topics. In: ICDE (2014)
12. Giatsoglou, M., Chatzakou, D., Shah, N., Faloutsos, C., Vakali, A.: Retweeting activity on Twitter: signs of deception. In: Cao, T., Lim, E.-P., Zhou, Z.-H., Ho, T.-B., Cheung, D., Motoda, H. (eds.) PAKDD 2015. LNCS (LNAI), vol. 9077, pp. 122–134. Springer, Cham (2015). https://doi.org/10.1007/978-3-319-18038-0_10
13. Guo, Z., et al.: Characterizing user behavior in weibo. In: MUSIC (2012)
14. He, C., Ma, H., Kang, S., Cui, R.: An overlapping community detection algorithm based on link clustering in complex networks. In: MILCOM (2014)
15. Ho, T.K.: Random decision forests. In: Proceedings of the 3rd International Conference on Document Analysis and Recognition (1995)
16. Hu, R., Aggarwal, C.C., Ma, S., Huai, J.: An embedding approach to anomaly detection. In: 32nd IEEE International Conference on Data Engineering, ICDE 2016, Helsinki, Finland, 16–20 May 2016, pp. 385–396 (2016)
17. Huang, Z.: Clustering large data sets with mixed numeric and categorical values. In: PAKDD (1997)
18. Jiang, B., et al:. Retweeting behavior prediction based on one-class collaborative filtering in social networks. In: SIGIR (2016)
19. Jiang, B., Liang, J., Sha, Y., Wang, L.: Message clustering based matrix factorization model for retweeting behavior prediction. In: CIKM (2015)
20. Jiang, Z., et al.: Understanding human dynamics in microblog posting activities. J. Stat. Mech: Theory Exp. **2013**(02), P02006 (2013)
21. Lim, K.H., Datta, A.: Interest classification of Twitter users using Wikipedia. In: OpenSym (2013)
22. Liu, Z., Chen, X., Sun, M.: Mining the interests of Chinese microbloggers via keyword extraction. Front. Comput. Sci. China **6**(1), 76–87 (2012)
23. Ma, S., Li, J., Hu, C., Lin, X., Huai, J.: Big graph search: challenges and techniques. Front. Comput. Sci. **10**(3), 387–398 (2016)
24. Michelson, M., Macskassy, S.A.: Discovering users' topics of interest on Twitter: a first look. In: Workshop on Analytics for Noisy Unstructured Text Data, (in conjunction with CIKM) (2010)
25. Mikolov, T., Sutskever, I., Chen, K., Corrado, G.S., Dean, J.: Distributed representations of words and phrases and their compositionality. In: NIPS (2013)
26. Park, S., Han, S.P., Huh, S., Lee, H.: Preprocessing uncertain user profile data: inferring user's actual age from ages of the user's neighbors. In: ICDE (2009)
27. Pennacchiotti, M., Popescu, A.: A machine learning approach to Twitter user classification. In: ICWSM (2011)
28. Ruan, Y., Fuhry, D., Parthasarathy, S.: Efficient community detection in large networks using content and links. In: WWW (2013)
29. Shiokawa, H., Fujiwara, Y., Onizuka, M.: Fast algorithm for modularity-based graph clustering. In: AAAI (2013)
30. Volkova, S., Coppersmith, G., Durme, B.V.: Inferring user political preferences from streaming communications. In: ACL (2014)

31. Wang, X., et al.: Recommending groups to users using user-group engagement and time-dependent matrix factorization. In: AAAI (2016)
32. Xu, Z., Lu, R., Xiang, L., Yang, Q.: Discovering user interest on Twitter with a modified author-topic model. In: WI (2011)
33. Yang, J., Leskovec, J.: Overlapping community detection at scale: a nonnegative matrix factorization approach. In: WSDM (2013)
34. Zhang, J., Liu, B., Tang, J., Chen, T., Li, J.: Social influence locality for modeling retweeting behaviors. In: IJCAI (2013)
35. Zhang, Q., Gong, Y., Guo, Y., Huang, X.: Retweet behavior prediction using hierarchical dirichlet process. In: AAAI (2015)
36. Zhang, T., Cui, P., Faloutsos, C., Lu, Y., Ye, H., Zhu, W., Yang, S.: come N go: a dynamic model for social group evolution. TKDD **11**(4), 41:1–41:22 (2017)
37. Zhao, W.X., Jiang, J., Weng, J., He, J., Lim, E.-P., Yan, H., Li, X.: Comparing Twitter and traditional media using topic models. In: Clough, P., Foley, C., Gurrin, C., Jones, G.J.F., Kraaij, W., Lee, H., Mudoch, V. (eds.) ECIR 2011. LNCS, vol. 6611, pp. 338–349. Springer, Heidelberg (2011). https://doi.org/10.1007/978-3-642-20161-5_34

Maximizing Social Influence
for the Awareness Threshold Model

Haiqi Sun[1]([✉]), Reynold Cheng[1], Xiaokui Xiao[2], Jing Yan[1], Yudian Zheng[1],
and Yuqiu Qian[1]

[1] The University of Hong Kong, Pok Fu Lam, Hong Kong
{hqsun,ckcheng,jyan,ydzheng2,yqqian}@cs.hku.hk
[2] Nanyang Technological University, Singapore, Singapore
xkxiao@ntu.edu.sg

Abstract. Given a social network G, the *Influence Maximization* (IM) problem aims to find a *seed set* $S \subseteq G$ of k users. These users are advertised, or *activated*, through marketing campaigns, with the hope that they will continue to influence others in G (e.g., by spreading messages about a new book). The goal of IM is to find the set S that achieves an optimal advertising effect or *expected spread* (e.g., make the largest number of users in G know about the book).

Existing IM solutions make extensive use of *propagation models*, such as Linear Threshold (LT) or the *Independent Cascade* (IC). These models define the *activation probability*, or the chance that a user successfully gets activated by his/her neighbors in G. Although these models are well-studied, they overlook the fact that a user's influence on others decreases with time. This can lead to an over-estimation of activation probabilities, as well as the expected spread.

To address the drawbacks of LT and IC, we develop a new propagation model, called *Awareness Threshold* (or AT), which considers the fact that a user's influence decays with time. We further study the *Scheduled Influence Maximization* (or SIM), to find out the set S of users to activate, as well as *when* they should be activated. The SIM problem considers the time-decaying nature of influence based on the AT model. We show that the problem is NP-hard, and we develop three approximation solutions with accuracy guarantees. Extensive experiments on real social networks show that (1) AT yields a more accurate estimation of activation probability; and (2) Solutions to the SIM gives a better expected spread than IM algorithms on the AT model.

1 Introduction

The *Influence maximization (IM)* problem, first proposed by Kempe et al. [22], has been extensively studied in recent years (e.g., [6,9,18,26–29,36,37]). The IM plays a fundamental role in *viral marketing*, a business promotion strategy that employs the *word-of-mouth* effect, where the advertisement is based on customers spreading news about something (e.g., a new electronic product) in social networks. The main goal of IM is to find, given an *influence graph G*, a

© Springer International Publishing AG, part of Springer Nature 2018
J. Pei et al. (Eds.): DASFAA 2018, LNCS 10827, pp. 491–510, 2018.
https://doi.org/10.1007/978-3-319-91452-7_32

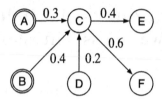

Fig. 1. Illustrating the IM problem

seed set S of k users such that the number of users in G influenced by S (or *expected spread*) is maximized.

Figure 1 shows an influence graph, which is derived from the social relationship among users. Each node in the graph represents a social network user. The value of each edge is the *influence probability*, i.e., the chance that a user is affected or *activated* by another one. Suppose that $k = 2$, and a seed set $S = \{A, B\}$ is chosen by an IM solution. A company, which wants to advertise a product, can promote it to A and B (e.g., by giving them discounts), hoping that the expected spread (i.e., the number of people influenced by A and B to buy its product) is maximal.

To enable IM, *propagation models* have been used to govern the flow of influence in G. For instance, in *Linear Threshold (LT)* [23], a node $v \in G$ is activated, if and only if the sum of the influence probabilities induced by v's active neighbors on v exceed some threshold T. In Fig. 1, suppose that $T = 0.5$, and A and B have been previously activated (e.g., through a company's marketing campaign). Node C then receives an activation probability of $0.3 + 0.4 = 0.7$ from A and B and gets activated. In another well-studied model, *Independent Cascade (IC)*, C is activated by A and B with a probability of $1 - (1 - 0.3) \times (1 - 0.4)$, or 0.58. Once C becomes active, we can then use the propagation model again to activate C's neighbors (i.e., E and F). Based on these models, efficient IM algorithms have been designed to find the best set S of seed nodes that yield the highest expected spread (or the number of nodes activated) [12,19,27,36,37].

The main problem of IC and LT is that they overlook the fact that the influence of a user received from other nodes *decreases* with time. For example, in Fig. 1, the influence probabilities on C due to A is not always 0.3 as claimed by IC and LT; rather, it gradually drops as time goes by. Once A has influenced C, A's effect on C becomes less profound, possibly because C's memory or awareness of what A told him/her fades. In our experiments performed on social networks (*Twitter* and *Digg*), the average activation probability of a node, which is a function of influence probabilities, drops quickly with time. However, IC and LT do not consider this factor, and subsequently over-estimate the expected spread significantly. This calls for a better propagation model.

Awareness Threshold. To deal with the issues of decaying influence probabilities, we study the *awareness threshold* (or AT in short). This propagation model is inspired by *brand awareness*, suggesting that a person u's impression about the brand of a product is affected by two factors: (1) *build*, which means

Table 1. Frequently used notations

Notation	Description
$w_{u,v}$	The propagation probability defined on edge $e_{u,v}$
$N_v^{(t)}$	Active in-degree neighbor set of user v at step t
θ_v	Activation threshold of user v
$A_v^{(t)}$	Awareness level of user v at step t
S/U	Seed set/scheduled seed set
k	Number of seeds
$t_{u,v}$	The time at which u attempts to activate v
$t^{(v)}$	The activation time of user v
$t^{(u,v)}$	Time delay defined on the edge $e_{u,v}$
$\sigma_m(S)/\sigma_m(U)$	The expected spread of S/U under model m

u's awareness about a product is increased through exposure to more marketing activities [8,13]; and (2) *decay*, which says that u's awareness drops exponentially because he/she receives no further advertisement about the product [4,7,13,16]. The AT model integrates brand awareness into the influence propagation process. Compared to IC and LT, AT attains a more accurate estimation of spread in our experiments.

Influence Maximization Under AT. We further study the IM problem under the AT model. We show that this problem is NP-hard. We thus develop an efficient solution with accuracy guarantees on the expected spread. We observe that the expected spread can be further improved through a *scheduled* activation of seed nodes. To explain, in the IM problem, the set S of k nodes is activated at the same time. However, it may be wiser by activating these nodes at different time instants, taking into account their brand awareness. Hence, in addition to deciding the seed nodes, we also want to determine the exact time that they are activated. We term this variant of the IM problem *scheduled influence maximization* (or *SIM*). Because SIM is NP-hard, we design a fast approximate solution, called the *Two-Phase Search* (TPS). Under the TPS framework, we study three scheduling policies, namely the *Breadth First Schedule*, the *Depth First Schedule*, and the *Bucket Schedule*. Our results on real datasets show that SIM solutions yield higher spread than their IM algorithms (Table 1).

The rest of the paper is as follows. In Sect. 2, we discuss the research background. Section 3 discusses the limitations of existing propagation models. In Sect. 4, we describe the AT model. We then present our IM and SIM solutions for AT in Sects. 5 and 6 respectively. We discuss the experiment results in Sect. 7, and related works in Sect. 8. Section 9 concludes.

2   Preliminaries

2.1   Problem Definition

Given a directed graph $G = \{V, E, W\}$, V is the user set, and E is the set of edges (e.g., the user influence relation). W is the set of probabilities associated on E, where for each $e_{u,v} \in E$, the weight $w_{u,v} \in [0, 1]$ defined on the edge captures the probability that u will influence v. Under a fixed budget k and a propagation model m (e.g. LT), the *IM* problem looks for a seed set S consisting of no more than k users, whose *expected spread* is the largest. Its definition is $S^* = \text{argmax}_{S \subseteq V \& |S| \leq k}\{\sigma_m(S)\}$, where influence function $\sigma_m(S)$ outputs the number of active users under model m, when each node in S is activated. To address the *IM* problem, it is important to model how the information is propagated, as will be shown next.

2.2   Propagation Models

There are two major classic propagation models, namely *LT* and *IC* model [23]. In both models, information (or *influence*) spreads from the seed set to other nodes as discrete time steps unfold.

LT Model. Initially, all seed users are initialized as active and others stay inactive. At each discrete time step, each inactive user will be activated if the weighted sum contributed by her active neighbors is larger than a pre-defined threshold. Once a user is activated, the user will stay active. The influence propagation terminates when no more activations are possible.

IC Model. Here the propagation process is also triggered by a set of active seed users. At each time step, the newly-activated user (i.e., activated at the previous step), influences each of her inactive neighbors with a pre-defined probability. If the activation is successful, then the inactive neighbor(s) will be activated and remain active in further steps. The influence propagation terminates when no more users can be activated.

2.3   Activation Probability

Under a specific propagation model, at time step t, given an inactive user v and her active neighbor set $N_v^{(t)}$, activation probability of v is defined as the likelihood that v can be activated by $N_v^{(t)}$. Specifically, in *LT* and *IC*, the activation probability values of v are denoted by $\sum_{u \in N_v^{(t)}} w_{u,v}$ and $1 - \prod_{u \in N_v^{(t)}} (1 - w_{u,v})$ respectively.

3   Limitations of IC and LT

As discussed in Sect. 2.3, *IC* and *LT* both adopt the following assumption: *At time t, for a given inactive user v and her active neighbor set $N_v^{(t)}$, the activation*

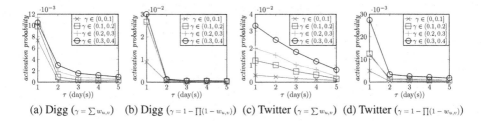

(a) Digg $(\gamma = \sum w_{u,v})$ (b) Digg $(\gamma = 1 - \prod(1 - w_{u,v}))$ (c) Twitter $(\gamma = \sum w_{u,v})$ (d) Twitter $(\gamma = 1 - \prod(1 - w_{u,v}))$

Fig. 2. Activation probability

probability of v estimated by LT (IC) *model is computed based on the influence weights associated with his/her incoming edges.*

However, as will be shown in the following case studies on real datasets, the assumption is inconsistent with the reality. Suppose that each active neighbor $u \in N_v^{(t)}$ always attempts to activate v. Then at step t, we define the average timespan, denoted by τ, as $\tau = \frac{1}{|N_v^{(t)}|} \sum_{u \in N_v^{(t)}} (t - t_{u,v})$. This is the average time that v is reactivated by u.

Next we conduct experiments to show the relationship between activation probability and the average timespan τ.

Datasets. Two real-world datasets, *Digg* and *Twitter* [1], are used in this experiment. *Digg* and *Twitter* are both social network applications, where *Digg* serves as a platform for sharing and voting stories, and *Twitter* provides users with chances of sharing news, photos, etc. The data from these two sources both consists of two parts, namely graph and *actionlog*. The graph contains social links among users, and *actionlog* records voting/tweeting history of users in the form of <user_id, action_id, action_time>, e.g. in *Twitter*, <001, 002, 1250772872> means that user *001* re-tweeted an URL (numbered as 002) on the date *1250772872* (Unix Timestamps). In [18], the propagation procedure is defined by action follow. For instance, in *Digg*, if user u votes "Cinderella", and later on her follower v also votes it, we consider the action of voting "Cinderella" has propagated from u to v. Table 2 shows the statistics.

Parameter Setting. We set the parameters of Twitter and Digg as follows:

(1) All probabilistic weights assigned on social links are learned from *actionlog* data by the methods proposed in [17] and settings are referred to [18].
(2) Due to the sparsity of our datasets, not all action seeds, i.e. first voters of stories or initial posters of tweets, are included in the social graph. Adopting the strategy in [18], we take the users who first vote stories (post the tweets) among their friends as seeds.

Observations. To assess *LT* and *IC* model, we define the *accumulated influence*, denoted by γ, as respectively $\gamma = \sum w_{u,v}$ and $\gamma = 1 - \prod(1 - w_{u,v})$. Then conditioned on fixed γ and increasing τ, values of *activation probability* are shown in Fig. 2. Specifically, accumulated influence γ is fixed within an interval $(x, x+0.1]$,

Table 2. Statistics of datasets

	Digg	Twitter	NetHEPT	DBLP
# Nodes	71K	736K	15K	914K
# Edges	1.7M	36M	62K	6.6M
Avg. Degree	24	50	4.1	7.2
# Actionlogs	3M	2.8M	-	-

and for each γ, average timespan τ is varied from 1 day to 5 days. In experiments, x is set from 0 to 0.9. Due to space constraints, only some representative results are shown here.

Observe that in both of *Digg* and *Twitter*, for each settled τ, with greater γ issued, the activation probability reaches a higher level, which is consistent with the theoretical result concluded by classic models (Sect. 2.3). However, for each γ, the activation probability decreases with an increasing τ, which demonstrates that even if the accumulated influence is fixed, activation probability decays when the average timespan grows, however, this is not observed by classic models.

In particular, the activation probability can be overestimated by *LT* and *IC* models. During a specific marketing campaign, these classic models will give an over-prediction on the expected spread.

We next present the awareness threshold (AT) model, which addresses the drawback of the LT and IC.

4   Awareness Threshold (AT)

Brand awareness is a term in the field of marketing science, which quantifies the extent to which a brand is recognized by potential users, and is the primary goal of advertising. Awareness level of user v at time t, denoted by $A_v^{(t)}$, reflects the likelihood of her adoption behaviors [15,25]. By measuring user awareness level, business agents are able to predict their marketing achievement, i.e., the number of users who will adopt specific products.

The awareness has two dimensions, *build* and *decay*, namely as below.

Awareness Build. Viral marketing attempts to spread brand content over the whole social network. When marketing campaign is launched in social networks, advertising messages are broadcast through the social connection between user pairs. Some individuals are in the circles which are full of brand information, they are thus exposed to advertising messages adequately. However, other social groups may only touch these information to a limited extent. User awareness level grows as increasing copies of advertising are exposed to them. With higher awareness level in minds, users tend to adopt the brand (or get activated in the setting of propagation models) [4,7].

Decay Effect. In the absence of further advertising exposures, customer awareness will decline and eventually decay to negligible levels [4,7,13,16]. One common function to model the decay effect is assuming the awareness will decrease exponentially w.r.t time.

Formally, for a specific user v, her awareness decay can be mathematically modeled as follows:

$$A_v^{(t)} = \Delta A_v^{(t)} + \lambda_v \cdot A_v^{(t-1)}, \qquad \lambda_v \in (0,1) \tag{1}$$

where $A_v^{(t)}$ and $A_v^{(t-1)}$ represents her awareness level at step t and $t-1$, and $\Delta A_v^{(t)}$ is the awareness increment contributed by new advertising exposure.

Our Model. To incorporate the above two effects, we build our Awareness Threshold (AT) model. Like IC and LT models, AT model simulates social network as a directed graph $G = \{V, E, W, T, \Lambda\}$, and each user is in the status of active (product adopter) or inactive. Meanwhile, different from IC and LT models, in AT model, there are two additional sets ($T \subseteq \mathbb{Z}^+$ and Λ) assigned to the graph G. T records time delays attached to social links. Since information diffusion progress among friend network will not be finished in a moment, $t^{(u,v)} \in T$ implies that it takes $t^{(u,v)}$ time steps for messages to propagate from u to v by the edge $e_{u,v}$. Many researches [10,30] have been conducted on how to calculate T. Moreover, for each user v, Λ contains her decay factor $\lambda_v \in (0,1)$.

We denote $t^{(v)}$ as the time at which v gets activated. (If v stays inactive, $t^{(v)} = +\infty$.) Let $\mathbb{1}_{\{.\}}$ denote an indicator function which returns 1 if its argument is true; 0, otherwise. For example, $\mathbb{1}_{\{5=2\}} = 0$ and $\mathbb{1}_{\{5=5\}} = 1$. When time $t \geq 1$, awareness score of inactive user v, denoted by $A_v^{(t)}$, is given by Eq. 1. Specifically, in AT model, $\Delta A_v^{(t)}$ is defined as follows:

$$\Delta A_v^{(t)} = \sum_{u \in N_v^{(t)}} w_{u,v} \cdot \mathbb{1}_{\{t^{(u)}+t^{(u,v)}=t\}}.$$

The term $t^{(u)} + t^{(u,v)} = t$ indicates that v receives the influence forwarded by u at t. The threshold of v, notated by θ_v, controls her activation condition, and v will be activated only if her awareness score reaches θ_v.

Intuitively, the update of $A_v^{(t)}$ contains two parts: (1) $\Delta A_v^{(t)}$: the awareness increments brought by active neighbors of v; (2) $\lambda_v \cdot A_v^{(t-1)}$: the awareness remained from last step.

Formally speaking, we define the awareness score of inactive user v at time t as follows:

Definition 1 (Awareness Score). *For the time t, awareness score of inactive user $v \in V$, i.e. $A_v^{(t)}$, is defined as*

$$\lambda_v \cdot A_v^{(t-1)} + \sum_{u \in N_v^{(t)}} w_{u,v} \cdot \mathbb{1}_{\{t^{(u)}+t^{(u,v)}=t\}} \tag{2}$$

if $t \geq 1$; and $A_v^{(t)} = 0$ if $t = 0$.

In AT model, if all seeds ($\forall s \in S$) are activated initially, starting from them, information spreads as time goes in discrete steps:

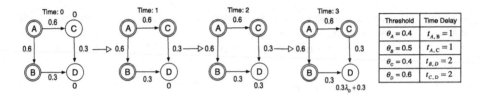

Fig. 3. Example of SIM.

(1) $t = 0$, $\forall s \in S$, s is active. $\forall v \in V - S$, v is inactive.
(2) $t \geq 1$, the information is broadcast by every newly-activated user u, i.e. $t^{(u)} = t - 1$, and propagated to each inactive neighbor v:
   - At $t^{(u)} + t^{(u,v)}$, v will capture an awareness increment $w_{u,v}$.
   - If $A_v^{(t)} \geq \theta_v$, v is activated.
   - Meanwhile, awareness score of every inactive individual j is weakened by her decay factor λ_j, as formalized by Eq. 2.
(3) It terminates when no further activation is possible.

Figure 3 illustrates this process with an example. Noting that in Fig. 3, the double circle in a node indicates that the user is active, and the value next to an inactive user represents her awareness score at a specific time.

5   Influence Maximization

In this section, we will first prove that under AT model, the NP-hardness of IM problem still holds, and then provide solution towards IM problem based on the AT model.

Theorem 1. *Under awareness threshold model, the influence maximization problem is NP-hard.*

Proof. In AT model, when there is no decay inside user awareness, i.e. for each user v, $\lambda_v = 1$, then solving influence maximization problem is exactly equivalent to addressing it under LT model whose **NP**-hardness is proved in [23]. Thus AT-based influence maximization can be proved as **NP**-hard.

We attempt to solve this NP-hard problem approximately. In the literature of IM, if influence function $\sigma_m(\cdot)$ is *monotone*, i.e. $\sigma_m(S) \leq \sigma_m(T)$ whenever $S \subseteq T$, and *submodular* which means if $S \subseteq T$, then $\forall v \notin T$ $\sigma_m(S \cup \{v\}) - \sigma_m(S) \geq \sigma_m(T \cup \{v\}) - \sigma_m(T)$, then the greedy algorithm [23] provides a result which approximates the optimal value within a factor of $(1 - \frac{1}{e})$ [33]. However, under AT model, influence function $\sigma_{AT}(\cdot)$ is not submodular (refer to Example 1). Inspired by [31], we exploit the *Sandwich Approximation* technique to solve IM problem with solution-dependent lower bound.

Algorithm 1. Sandwich Approach

Input: $G, k, \sigma_{AT}, \sigma_L, \sigma_U$
Output: seed set S
1 $S_L \leftarrow ALG(G,k, \sigma_L)$;
2 $S_U \leftarrow ALG(G,k, \sigma_U)$;
3 $S_{AT} \leftarrow ALG(G,k, \sigma_{AT})$;
4 **return** $\text{argmax}_{S \in \{S_L, S_U, S_{AT}\}} \sigma_{AT}(S)$;

Example 1. *Here a propagation example is taken to show the non-submodularity of σ_{AT}. According to the social graph presented in Fig. 3, when seed set is assigned as $\{A\}$, $\{A,C\}$, $\{A,B\}$ and $\{A, B, C\}$ respectively, then $\sigma_{AT}(\{A,B\}) - \sigma_{AT}(\{A\}) < \sigma_{AT}(\{A,B,C\}) - \sigma_{AT}(\{A,C\})$. As mentioned above, influence function $\sigma_{AT}(\cdot)$ is submodular iff $\forall S \subseteq T$ and $v \notin T$, the equation $\sigma_{AT}(S \cup v) - \sigma_{AT}(S) \geq \sigma_{AT}(T \cup v) - \sigma_{AT}(T)$ holds. So $\sigma_{AT}(\cdot)$ is not submodular.*

Let σ_L and σ_U be non-negative, monotone and submodular set functions defined on user set V, i.e. $\sigma_L : 2^V \to \mathbb{R}_{\geq 0}$ and $\sigma_U : 2^V \to \mathbb{R}_{\geq 0}$, such that $\forall S \subseteq V$, $\sigma_L(S) \leq \sigma_{AT}(S) \leq \sigma_U(S)$. The sandwich approach is described in Algorithm 1, where ALG represents a greedy-manner approximation algorithm.

Theorem 2 [31]. *The sandwich algorithm is able to generate a seed set S_{sand} such that $\sigma_{AT}(S_{sand})$ is not lower than*

$$max\{\frac{\sigma_{AT}(S_U)}{\sigma_U(S_U)}, \frac{\sigma_L(S^{opt})}{\sigma_{AT}(S^{opt})}\} \cdot (1 - \frac{1}{e}) \cdot \sigma_{AT}(S^{opt})$$

where S^{opt} is the optimal solution.

In accordance with Theorem 2, sandwich approach approximates the optimal result within a solution-dependent factor. Under this algorithmic framework, construction of bound functions, namely σ_L and σ_U, is the next critical issue.

One instance of σ_L can be built by a modification on AT model, that replacing Eq. 2 with the equation below:

$$max_{u \in N_v^{(t)}} \{w_{u,v} \cdot \mathbb{1}_{\{t^{(u)} + t^{(u,v)} = t\}}\}, \tag{3}$$

where the decay factor is set as 0, and $\sum$ is replaced with max, i.e., at each step, only the weight which contributes most is picked to update awareness score. Similarly, to generate σ_U, we alter Eq. 2 to the equation:

$$A_v^{(t-1)} + \sum_{u \in N_v^{(t)}} w_{u,v} \cdot \mathbb{1}_{\{t^{(u)} + t^{(u,v)} = t\}}, \tag{4}$$

where the decay factor is 1. Intuitively, by adoption of above modifications, for specified seed set S, we have $\sigma_L(S) \leq \sigma_{AT}(S) \leq \sigma_U(S)$.

Theorem 3. *Influence function σ_L (σ_U), which is built by modifying AT model in the way that altering Eq. 2 to Eq. 3 (Eq. 4), is non-negative, monotone and submodular.*

Algorithm 2. Two-Phase Search

Input: $G, k, \sigma_{AT}, \sigma_{LT}$
Output: scheduled seed set U^*
1 $S^g \leftarrow Selection(G, k, \sigma_{LT})$;
2 $U^0 \leftarrow \emptyset$;
3 **foreach** $s_i \in S^g$ **do**
4 | Add $(s_i, 0)$ into U^0
5 $U^* \leftarrow Schedule(G, \sigma_{AT}, U^0)$;
6 **return** U^*;

Proof. Clearly, both σ_L and σ_U output non-negative numbers. We first prove the Monotonicity and submodularity of σ_L.

Suppose the propagation model corresponding to σ_L is called AT_L. Given a social graph G, for each $e_{u,v} \in E$, if $w_{u,v} < \theta_v$, $e_{u,v}$ is cut from G, then the generated graph is denoted by g. Under AT_L model, given a seed set S, $\forall v \in G - S$, v can be activated by S in G iff there is a path from S to v in g. So an intuitive result is that for any seed set $T \subseteq S$, $\sigma_L(T) \leq \sigma_L(S)$, i.e. σ_L is monotone. Consider the quantity $\sigma_L(S \cup v) - \sigma_L(S)$ $(v \notin S)$, it is the number of nodes which are reachable from v but unreachable from S in g. Obviously, if $T \subseteq S$, we have $\sigma_L(S \cup v) - \sigma_L(S) \leq \sigma_L(T \cup v) - \sigma_L(T)$. So σ_L is submodular.

Similarly, propagation model associated with σ_U is named by AT_U. Conditioned on AT_U model, for any inactive user v, $A_v^{(t)}$ is actually equal to the sum of all $w_{u,v}$ received. So for issued seed set S, its expected influence under AT_U equals the value under LT, i.e. $\sigma_U(S) = \sigma_{LT}(S)$. Because of the monotonicity and submodularity of σ_{LT}, it is trivial that σ_U is monotone and submodular.

Obviously, Algorithm 1 is an approximation solution of IM problem, with the approximate ratio of $max\{\frac{\sigma_{AT}(S_U)}{\sigma_U(S_U)}, \frac{\sigma_L(S^{opt})}{\sigma_{AT}(S^{opt})}\} \cdot (1 - \frac{1}{e}) \cdot \sigma_{AT}(S^{opt})$.

6   Scheduled Influence Maximization

Viral marketing is to maximize brand awareness or product adoption over social networks, where the level of 'brand awareness or product adoption' is equivalent as user activation by influence propagation study. However, we observed that In AT model, the reach of information is also highly related to the schedule of seeds, i.e. the time step at which each seed is initially activated. Intuitively, the schedule of activating seeds is also essential in Viral marketing. Motivated by this issue, Sect. 6 proposed a new IM problem: *Scheduled Influence Maximization (SIM) problem* and addressed it with an approximation algorithm.

6.1   Scheduled Influence Maximization

Let $S = \{s_i | 1 \leq i \leq k\} \subseteq V$ be a k-element seed set, and $\Gamma = \{t_i | 1 \leq i \leq k\} (t_i \in N)$ be the corresponding schedule set, i.e. during a specific marketing campaign, each seed s_i is artificially activated at t_i. A *scheduled seed set* $U = \{(s_i, t_i) | s_i \in S, t_i \in \Gamma, 1 \leq i \leq k\}$ contains every selected seed and its activation time. In this paper, taking a scheduled seed set U as input, influence function $\sigma_{AT}(U)$ outputs the number of active users finally generated. Under AT model, formal definition of *Scheduled Influence Maximization (SIM)* problem is given as follows.

Definition 2 *(SIM Problem). Given a directed social graph $G = \{V, E, W, T, \Lambda\}$, and an integer budget $k \leq |V|$, search a scheduled seed set U, where $|U| \leq k$, that maximizes $\sigma_{AT}(U)$.*

Intuitively, for each tuple $(s_i, t_i) \in U$, if $t_i = 0$, SIM is trivially reduced to IM problem. As proved by Theorem 4, the hardness result on SIM still holds.

Theorem 4. *SIM Problem under AT model is **NP**-hard.*

Proof. In AT model, if $\forall v \in V$, decay factor $\lambda_v = 1$. Then for any possible seed user set, arranging their activation time cannot change final output. So scheduled influence maximization problem is actually identical to the conventional influence maximization problem. As proved before, under AT model, influence maximization problem is **NP**-hard. Thus addressing scheduled influence maximization is also **NP**-hard.

6.2 Two-Phase Search

An intuitive observation is that, compared to IM problem, the search space of SIM is enlarged dramatically. To reduce the time cost of searching solution result, we design the *Two-Phase Search (TPS)* algorithm which divides the result exploration procedure into two phases, namely seed selection and seed schedule. The algorithmic details are described in Algorithm 2. To be specific, seed set S^g is obtained by running greedy-manner approximation approach (*Selection*), like *IMM* [36], and their activation steps are managed by the *Schedule* module. In this work, three schedule algorithms, *Breadth First Schedule (BFS)*, *Depth First Schedule (DFS)*, and *Bucket Schedule (BS)* are proposed to implement the module.

Algorithm 3. Breadth First Schedule

    Input: G, σ_{At}, U
    Output: scheduled seed set U
  1 $loop \leftarrow true$;
  2 **while** $loop$ **do**
  3      $loop \leftarrow false$;
  4   **foreach** $(s_i, t_i) \in U$ **do**
  5          $\varphi \leftarrow \sigma_{AT}(U), t_i \leftarrow t_i + 1$;
  6   **if** $\sigma_{AT}(U) \leq \varphi$ **then**
  7              $t_i \leftarrow t_i - 1$;
  8   **else**
  9              $loop \leftarrow true$;
10 **return** U;

BFS and DFS. As detailed in Algorithms 3 and 4, taking U^0 (as mentioned in Algorithm 2, in which every seed is activated at step 0) as input, *BFS* and *DFS* operate the schedule by deferring activation time heuristically. Specifically, in both of these two methods, seed-step pairs (s_i, t_i) are picked in a round-robin fashion for deferment test, and each test will increase t_i by one step if expected spread is raised accordingly.

Algorithm 4. Depth First Schedule

Input: G, σ_{AT}, U
Output: scheduled seed set U
1 $loop \leftarrow true$;
2 **while** $loop$ **do**
3   **foreach** $(s_i, t_i) \in U$ **do**
4       $loop \leftarrow true$;
5   **while** $loop$ **do**
6          $loop \leftarrow false$;
7          $\varphi \leftarrow \sigma_{AT}(U), t_i \leftarrow t_i + 1$;
8   **if** $\sigma_{AT}(U) \leq \varphi$ **then**
9             $t_i \leftarrow t_i - 1$;
10   **else**
11             $loop \leftarrow true$;
12 **return** U;

Algorithm 5. Bucket Schedule

Input: G, σ_{AT}, U, B
Output: scheduled seed set U
1 $loop \leftarrow true$;
2 **while** $loop$ **do**
3    $loop \leftarrow false$;
4    $\varphi \leftarrow \sigma_{AT}(U)$;
5   **foreach** $(s_i, t_i) \in U$ **do**
6       $t_i \leftarrow \text{argmax}_{t_i \in [0,B]}\{\sigma_{At}(U)\}$;
7   **if** $\sigma_{AT}(U) > \varphi$ **then**
8          $loop \leftarrow true$;
9 **return** U;

The difference is that as shown in *BFS* (Algorithm 3: lines 4 to 9), when one deferment test on (s_i, t_i) is finished, the next seed-step pair will be chosen. Differently, for specified tuple (s_i, t_i), *DFS* keeps deferring t_i, namely setting $t_i \leftarrow t_i + 1$, until no improvement on $\sigma_{AT}(U)$ is achieved, then next tuple will be loaded, shown in Algorithm 4: lines 3 to 11. Both of these two methods terminate if no further promotion on $\sigma_{AT}(U)$ is possible.

BS. Another heuristic is that assuming U^+ is generated from U^0 by a decent schedule, and $\forall (s_i, t_i) \in U^+$, $t_i \in [0, B]$, where interval $[0, B]$ is called a *bucket*.

BS targets to return a result which approaches to U^+ as much as possible (pseudo-code is listed in Algorithm 5). Similarly, BS also schedules seeds in rotation. When pair (s_i, t_i) is picked, t_i will be assigned as the value which maximizes $\sigma_{AT}(U)$ among all integer elements in $[0, B]$. If no spread improvement is achieved in the whole round, schedule iteration exits and final result is returned.

6.3   Theoretical Analysis

To evaluate the performance of U^* outputted by TPS, we first give Lemma 1 by which Theorem 5 next guarantees that TPS achieves an approximation ratio of $\alpha(1 - 1/e)$, where $\alpha = \sigma_{AT}(U^0)/\sigma_{LT}(S^g)$.

Lemma 1. *Given arbitrary scheduled seed set U and the corresponding seed set S, i.e. $|U| = |S|$ and $\forall(s_i, t_i) \in U$, $s_i \in S$, the expected spread of U is bounded by $\sigma_{LT}(S)$, formally $\sigma_{AT}(U) \leq \sigma_{LT}(S)$.*

Proof. Under AT model, for an issued scheduled seed set U, its expected spread $\sigma_{AT}(U)$ is computed by counting active users finally generated. An observation is that, if each decay factor $\lambda_v = 1$, since awareness score $A_v^{(t)}$ never declines over time, the expected spread of U reaches its peak value, formally $\sigma_{AT}(U) \leq \sigma_{AT}^{(\lambda=1)}(U)$. Meanwhile, conditioned on $\lambda_v = 1$, an intuitive result is that for specified seed set, its expected spread $\sigma_{AT}^{(\lambda=1)}$ is free of seeds schedule. Suppose U^0 is the scheduled seed set that $|U^0| = |U|$, and $\forall(s_i, t_i) \in U, (s_i, 0) \in U^0$. Thus we have, $\sigma_{AT}^{(\lambda=1)}(U) = \sigma_{AT}^{(\lambda=1)}(U^0)$. Moreover, with cutting of decay factor, user awareness score equals sum of the probabilistic weights received, which is actually equivalent to the pattern defined in LT model. So $\sigma_{AT}^{(\lambda=1)}(U^0) = \sigma_{LT}(S)$. Based on the above analysis, we have $\sigma_{AT}(U) \leq \sigma_{LT}(S)$. Thus the Lemma is proved.

Theorem 5. *Suppose $\alpha = \sigma_{AT}(U^0)/\sigma_{LT}(S^g)$. Algorithm 2 approximates the optimum to within a factor of $\alpha(1 - 1/e)$. Formally $\sigma_{AT}(U^*) \geq \alpha(1 - 1/e)\sigma_{AT}(U^{opt})$.*

Proof. Let $S' = \{s_i | (s_i, t_i) \in U^{opt}\}$. For the submodularity, monotonicity and non-negativity of σ_{LT} [23], thus

$$\sigma_{LT}(S^g) \geq (1 - \frac{1}{e})\sigma_{LT}(S^{opt}) \geq (1 - \frac{1}{e})\sigma_{LT}(S')$$

By Lemma 1, we have $\sigma_{LT}(S') \geq \sigma_{AT}(U^{opt})$, thus $\sigma_{LT}(S^g) \geq (1 - \frac{1}{e})\sigma_{AT}(U^{opt})$. Since $\sigma_{LT}(S^g) = \frac{\sigma_{AT}(U^0)}{\alpha}$, and $\sigma_{AT}(U^*) \geq \sigma_{AT}(U^0)$. Therefore, we can derive $\sigma_{AT}(U^*) \geq \alpha(1 - \frac{1}{e})\sigma_{AT}(U^{opt})$.

7   Experiments

In this paper, we have tested the proposed solution of the SIM problem based on *AT* model. The general experiments are divided into two parts, the spread prediction and spread promotion (For space limitation, spread computation details are mentioned in our full technical report [2]). We have conducted intensive experiments on public datasets to demonstrate the effectiveness and efficiency of our solution.

7.1   Dataset

In addition to the dataset utilized in Sect. 3, other real social graphs [3] are also used in our experimental evaluation, and their details are illustrated in Table 2. All the graphs are constructed with directed edges and commonly-used for influence maximization research [20]. Specifically, *NetHEPT* and *DBLP* contains co-author(s) data in High Energy Physics (Theory) section of arXiv and DBLP Computer Science Bibliography respectively. Since the datasets of *NetHEPT* and *DBLP* do not include *actionlogs*, which means actual propagation size is unavailable, Experiment 7.3 is only conducted on *Digg* and *Twitter*. For other experiments, all 4 datasets are used.

7.2   Parameter Setting and Implementation

All the weights associated on social edges are learned from user historical data by the method mentioned in Sect. 3. Specifically, for *NetHEPT* and *DBLP*, the weights are provided by the author of [20]. Refer to [30], for each edge $e_{u,v}$, the attached time delay $t^{(u,v)}$ follows a Poisson distribution where the mean is randomly sampled from $\{1, 2, \cdots, 20\}$. In reality, decay factor varies among different individuals. The best way to decide each decay factor is to learn from the historical data of each user. Unfortunately we do not have access to sufficient such real data sets for the purpose of experiments. Thus, we use synthesized setting. For each user v, her decay factor λ_v is set as a randomly-sampled value from $[1 - \theta_v, 1)$, which means lower threshold is more likely to generate a higher decay factor. Since easily-activated individuals usually show more interest on the propagated information, we thus assign them a slower speed of awareness-decay. For the computation of $\sigma_{AT}(U)$ and $\sigma_m(S)$, we adopt the Monte Carlo approach [23], instead of using a fixed number of samples, we determine the sample number dynamically, which is detailed in our technical report.

7.3   Evaluation Results

The evaluation can be generally grouped into two parts. In the first part, we measured the spread estimation of propagation models (including classic models, and *AT* model (Sect. 7.3)). Second, we study the improvement of influence in the social network, and the efficiency of each scheduling method (Sect. 7.3). All

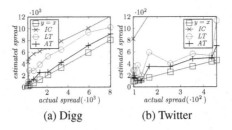

(a) Digg   (b) Twitter

Fig. 4. Spread estimation

experiments are performed on an Ubuntu-installed machine equipped with an Intel Xeon(R) CPU E5-1620 v2 @ 3.70 GHz ×8 and 16GB memory. Results are obtained by running each experiment 10 times and taking the average.

Model Evaluation. Since the aforementioned decay effect is not adopted in model construction, classic propagation models usually over-estimates user activation probability, which eventually causes over-prediction in influence spread. This section targets to evaluate propagation models in the way that comparing the real influence spread with the spread values estimated by models. By observing the estimation gap, the effectiveness of propagation models can be evaluated. Experimentally, for specified seed set S, its *expected spread* is compared with the *actual spread*. For ground truth, the seeds of each action are set according to Sect. 3, and actual spread is the number of users who performed the same action, namely propagation size.

As shown in Fig. 4, in contrast to actual propagation size, the expected spread given by IC and LT models severely suffers from over-prediction. In *Digg*, the model-caused extra prediction reaches to 60% of the actual spread. While in *Twitter*, expected spread even exceeds the actual value by numerous times.

In AT model, activation steps of seeds are set according to their action timestamps. Specifically, timestamp of the earliest action is recorded as $t = 0$, then $t = 1$ denotes the next timestamp. (in our dataset, there are 12 h between two timestamps.) As demonstrated by empirical results, expected spread given by AT model approaches the actual value closer. Since AT model captures decay nature inside user awareness, when it is utilized for spread estimation purpose, over-prediction problem can be eased much.

Spread Promotion. As addressed above, decent schedule on seeds (activation time) can contribute to enlarge influence spread. Our second set of experiments compare spread promotion created by different schedule algorithms. To measure the spread promotion, *spread increase rate* is defined first. Formally, with fixed budget issued, $spread\,increase\,rate = \frac{\sigma_{AT}(U) - \sigma_{AT}(S)}{\sigma_{AT}(S)}$, where U denotes scheduled seed set returned by TPS (Algorithm 2), and S represents seed set outputted by non-scheduling *sandwich approach* (Algorithm 1). Figure 5(a) to (d) report spread increase rate achieved, varying k from 5 to 30. Each line is labeled by the schedule algorithm invoked in TPS.

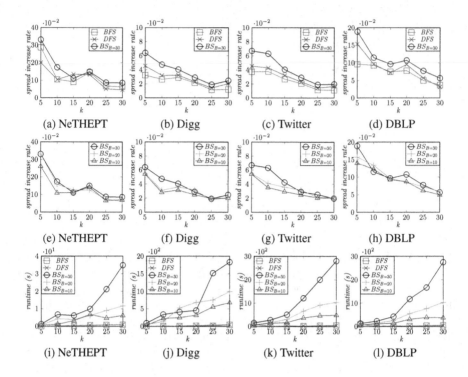

Fig. 5. Spread increase rate and efficiency comparison

Overall, when $B = 30$, BS, notated as $BS_{B=30}$, outperforms BFS and DFS over all budget values. Moreover, under $k = 5$, invocation of $BS_{B=30}$ raises spread increase rate to its maximal value. In *NetHEPT*, the maximum reaches to 33%, while in *DBLP*, it is 18%. Another observation is that as growing budget issued, spread increase rate tends to go downward. Experiment results show that when incremental seeds are allocated on a social graph, $\sigma_{AT}(S)$ increases faster than $\sigma_{AT}(U) - \sigma_{AT}(S)$ which denotes the additional spread created by *TPS*. Hence issuing k incrementally causes the rate decreases.

We are also interested in investigating how assignment of B affects performance of BS. Theoretically, with a greater B inputted, BS explores solutions in a larger search space, so better results may be returned. In practice, as shown in Fig. 5(e) to (h), a bigger B generally boosts spread increase rate, but merely in a limited extent. By contrast, bucket size addition results in much extra computation time (discussed in Sect. 7.3).

Efficiency Analysis. This section evaluates computation cost of running schedule algorithms. As a function of seed number, runtime of schedule algorithms is reported in Fig. 5(i) to (l). It can be seen that *BFS* and *DFS* scale well with respect to k. For instance, in the largest dataset *Twitter* and *DBLP*, *DFS* only takes around 40 s to schedule 30 seeds, and *BFS* performs equally. Comparatively, as seeds are incrementally imported, *BS* requires significantly more time to handle the schedule task, especially when k is greater than 20. In addition to seed number, if bucket size B is enlarged, BS also heavily suffers from consequent inefficiency issue. Figure 5(i) and (l) show that conditioned on $k = 30$, time cost of BS_{20} is even tripled when B is increased to 30.

8 Related Work

8.1 Influence Propagation Model

Classic models, LT and IC model have been introduced in Sect. 2. Apart from these two models, there are also extensive research on other models simulating the influence propagation.

Time-Dependent Model. In *Latency Aware Independent Cascade (LAIC)* model [30], for each piece of propagation (e.g. at time t_a user u sends a specific message to her friend v, then at t_b, v receives it), the time delay, i.e. $t_b - t_a$, is expressed as a delay weight associated with each social link. Similarly, this time delay is also incorporated by *Independent Cascade model with Meeting events (IC-M)* [10]. In the context, at each time step newly-activated users perform activation on their inactive neighbors by some pre-determined probabilities. If it fails to occur, active users will continue activation attempt in subsequent steps until success. The time span from initial attempt to the final reflects this time delay. Other time-dependent models like *Delayed Linear Threshold* and *Delayed Independent Cascade* [32] are constructed in similar manners. All these models integrate the aforementioned time delay into influence propagation mechanism, so they can track information propagation progress over time. But the decay effect inside user awareness is omitted, which means the fundamental problem in classic models is unsolved.

Decay Model. Motivated by the intuition that fresh news exhibits more attractiveness than out-of-date information, *Independent Cascade with Novelty Decay (IC_{ND})* [14] is proposed to model information diffusion conditioned on that novelty inside the diffused message decays as its exposure frequency increasing. For instance, the initial weight assigned on edge $e_{u,v}$ is w, where v is inactive. If n neighbors (except u) have attempted to activated v (and failed), i.e. a specific message M has been exposed to v n times, then w will keep decreasing as n grows, which indicates that M becomes less attractive to v. In IC_{ND},edge weights decay is dependent on the times of information exposure, which are actually related to the network structure and diffusion trace. So the time-dependent nature of decay effect is not revealed.

Other models like *Time-varying Independent Cascade model (TV-IC)* [34] and *Cascade Model with Diffusion Decay (CMDD)* [38] all build edge weights as time-decay functions, i.e. as time lapse, the weights vary non-increasingly. These models are constructed under the hypothesis that information freshness decays as a function of time, i.e. it becomes less attractive over time. For instance, at step t, the weight associated on $e_{u,v}$ is w, at subsequent steps, even if the propagated message M has never been exposed to v, value of w still decreases. However, above hypothesis is challenged by experimental results given in [35], it turns out that edge weights diminishing only occurs if messages are not fresh to the social circles. So back to the last instance, weight of $e_{u,v}$ should keep invariant if v has not touched M. Previous time-decay models only simply relate influence relationship decay and time lapse together, and omit the rationale behind this decay effect, which causes that the fundamental defined by these models is inconsistent with influence propagation nature. Thus existing decay mechanisms are unable to depict propagation dynamics.

8.2   Scalable Influence Maximization

By exploiting submodularity of influence functions, Leskovec *et al.* propose *CELF* technique [27] which accelerates the classic Greedy dramatically without compromise on accuracy guarantee. Then an improved version, called *CELF++* [19], incrementally contributes to efficiency enhancement. However, running of *CELF* (and *CELF++*) triggers thousands of sampling operation, which causes heavy time cost in practice. Then several algorithms aim to solve IM heuristically, e.g. PMIA [11], IRIE [24] and Simpath [20]. Subsequently, the technique breakthrough, namely reverse reachable subgraph sampling, is achieved by Borgs *et al.*, theoretically, the novel approach *RIS* [6] is able to provide elegant performance guarantee in both of efficiency and effectiveness. Inspired by *RIS*, Tang *et al.* develop *TIM* [37] and *IMM* [36]. Also, there are many problems motivated by IM problem. Competitive Influence Maximization (CIM) [5], a widely discussed problem, considers multiple agents attempting to maximize their influence over the social network competitively. Another instance is Influence Blocking Maximization (IBM) [21] which targets to minimize the influence triggered by the competing seeds.

However, all these problems assume that seeds are activated initially, and seeds scheduling has not been well studied.

9   Conclusion

In this paper, based on the *decaying* observation and verified experiments, we propose a new model *Awareness Threshold (AT)*. This model enables a more accurate estimation of the propagation process. Conditioned on *AT* model, a tailored algorithm is introduced to address influence maximization (IM) problem approximately. Further, under the *AT* model, selection and schedule of seeds collaboratively impact final information spread. Structured on this investigation, *scheduled influence maximization (SIM)* problem is proposed next. To tackle this challenge, *Two-Phase Search* method is developed to approximate the optimal value with a lower bound guarantee. Finally, the approach utility is evaluated by intensive experiments.

Acknowledgment. We would like to thank the reviewers for the insightful comments. Haiqi Sun, Jing Yan, Yudian zheng, and Reynold Cheng were supported by the Research Grants Council of Hong Kong (RGC Projects HKU 17229116 and 17205115) and the University of Hong Kong (Projects 104004572, 102009508, 104004129).

References

1. http://www.isi.edu/integration/people/lerman/downloads.html
2. https://www.dropbox.com/s/g8bfafun0n055ou/main.pdf?dl=0
3. https://www.dropbox.com/s/zhqnvsy5a5xocfa/simpath_icdm11_datasets.zip?dl=0
4. Almon, S.: The distributed lag between capital appropriations and expenditures. Econom.: J. Econom. Soc. **33**, 178–196 (1965)
5. Bharathi, S., Kempe, D., Salek, M.: Competitive influence maximization in social networks. In: Deng, X., Graham, F.C. (eds.) WINE 2007. LNCS, vol. 4858, pp. 306–311. Springer, Heidelberg (2007). https://doi.org/10.1007/978-3-540-77105-0_31
6. Borgs, C., Brautbar, M., Chayes, J., Lucier, B.: Maximizing social influence in nearly optimal time. In: SODA (2014)
7. Broadbent, S.: Accountable Advertising: A Handbook for Managers and Analysts. Admap Publications, Henley-on-Thames (1997)
8. Brown, G.: Modelling advertising awareness. Statistician **35**, 289–299 (1986)
9. Chen, S., Fan, J., Li, G., Feng, J., Tan, K.-L., Tang, J.: Online topic-aware influence maximization. PVLDB **8**, 666–677 (2015)
10. Chen, W., Lu, W., Zhang, N.: Time-critical influence maximization in social networks with time-delayed diffusion process. In: AAAI (2012)
11. Chen, W., Wang, C., Wang, Y.: Scalable influence maximization for prevalent viral marketing in large-scale social networks. In: KDD (2010)
12. Chen, W., Wang, Y., Yang, S.: Efficient influence maximization in social networks. In: KDD (2009)
13. Dubé, J.-P., Hitsch, G.J., Manchanda, P.: An empirical model of advertising dynamics. Quant. Mark. Econ. **3**, 107–144 (2005)
14. Feng, S., Chen, X., Cong, G., Zeng, Y., Chee, Y.M., Xiang, Y.: Influence maximization with novelty decay in social networks. In: AAAI (2014)
15. Ferguson, R.: Word of mouth and viral marketing: taking the temperature of the hottest trends in marketing. J. Consum. Mark. **25**, 179–182 (2008)
16. Fry, T.R., Broadbent, S., Dixon, J.M., et al.: Estimating advertising half-life and the data interval bias. Technical report, Monash University, Department of Econometrics and Business Statistics (1999)
17. Goyal, A., Bonchi, F., Lakshmanan, L.V.: Learning influence probabilities in social networks. In: WSDM (2010)
18. Goyal, A., Bonchi, F., Lakshmanan, L.V.: A data-based approach to social influence maximization. PVLDB **5**, 73–84 (2011)
19. Goyal, A., Lu, W., Lakshmanan, L.V.: CELF++: optimizing the greedy algorithm for influence maximization in social networks. In: WWW (2011)
20. Goyal, A., Lu, W., Lakshmanan, L.V.: Simpath: an efficient algorithm for influence maximization under the linear threshold model. In: ICDM (2011)
21. He, X., Song, G., Chen, W., Jiang, Q.: Influence blocking maximization in social networks under the competitive linear threshold model. In: Proceedings of the 2012 SIAM International Conference on Data Mining, pp. 463–474. SIAM (2012)

22. Kempe, D., Kleinberg, J., Tardos, E.: Maximizing the spread of influence through a social network. In: KDD (2003)
23. Kempe, D., Kleinberg, J., Tardos, É.: Influential nodes in a diffusion model for social networks. In: Caires, L., Italiano, G.F., Monteiro, L., Palamidessi, C., Yung, M. (eds.) ICALP 2005. LNCS, vol. 3580, pp. 1127–1138. Springer, Heidelberg (2005). https://doi.org/10.1007/11523468_91
24. Kim, J., Kim, S.-K., Yu, H.: Scalable and parallelizable processing of influence maximization for large-scale social networks? In: ICDE (2013)
25. Kirby, J.: Viral marketing. In: Connected marketing (2012)
26. Lei, S., Maniu, S., Mo, L., Cheng, R., Senellart, P.: Online influence maximization. In: KDD (2015)
27. Leskovec, J., Krause, A., Guestrin, C., Faloutsos, C., VanBriesen, J., Glance, N.: Cost-effective outbreak detection in networks. In: KDD (2007)
28. Li, G., Chen, S., Feng, J., Tan, K.-L., Li, W.: Efficient location-aware influence maximization. In: SIGMOD (2014)
29. Lin, S.-C., Lin, S.-D., Chen, M.-S.: A learning-based framework to handle multi-round multi-party influence maximization on social networks. In: KDD (2015)
30. Liu, B., Cong, G., Xu, D., Zeng, Y.: Time constrained influence maximization in social networks. In: ICDM (2012)
31. Lu, W., Chen, W., Lakshmanan, L.V.: From competition to complementarity: comparative influence diffusion and maximization. PVLDB 9, 60–71 (2015)
32. Mohammadi, A., Saraee, M., Mirzaei, A.: Time-sensitive influence maximization in social networks. J. Inf. Sci. 41, 765–778 (2015)
33. Nemhauser, G.L., Wolsey, L.A., Fisher, M.L.: An analysis of approximations for maximizing submodular set functions. Math. Program. 14, 265–294 (1978)
34. Ohsaka, N., Yamaguchi, Y., Kakimura, N., Kawarabayashi, K.: Maximizing time-decaying influence in social networks. In: Frasconi, P., Landwehr, N., Manco, G., Vreeken, J. (eds.) ECML PKDD 2016. LNCS (LNAI), vol. 9851, pp. 132–147. Springer, Cham (2016). https://doi.org/10.1007/978-3-319-46128-1_9
35. Steeg, G.V., Ghosh, R., Lerman, K.: What stops social epidemics? In: ICWSM (2011)
36. Tang, Y., Shi, Y., Xiao, X.: Influence maximization in near-linear time: a martingale approach. In: SIGMOD (2015)
37. Tang, Y., Xiao, X., Shi, Y.: Influence maximization: near-optimal time complexity meets practical efficiency. In: SIGMOD (2014)
38. Zhang, Z., Wu, H., Yue, K., Li, J., Liu, W.: Influence maximization for cascade model with diffusion decay in social networks. In: Che, W., et al. (eds.) ICYCSEE 2016. CCIS, vol. 623, pp. 418–427. Springer, Singapore (2016). https://doi.org/10.1007/978-981-10-2053-7_37

A Time-Aware Path-Based Publish/Subscribe Framework

Mengdi Jia[1], Yan Zhao[1], Bolong Zheng[2,3], Guanfeng Liu[1], and Kai Zheng[4(✉)]

[1] School of Computer Science and Technology, Soochow University, Suzhou, China
mdjia@stu.suda.edu.cn, {zhaoyan,gfliu}@suda.edu.cn
[2] School of Data and Computer Science, Sun Yat-sen University, Guangzhou, China
zhengblong@mail.sysu.edu.cn
[3] Aalborg University, Aalborg, Denmark
[4] Big Data Research Center, University of Electronic Science and Technology
of China, Chengdu, China
zhengkai@uestc.edu.cn

Abstract. Nowadays, massive geo-tagged records are generated on the social media. These records are useful when the users intend to plan a trip and are interested in some specific topics along the trip. With such redundant records, a publish/subscribe system has been designed to allow the users who are interested in certain information (i.e., the subscribers) to receive messages from some message generators (i.e., the publishers). Existing efforts on publish/subscribe mainly focus on the textual content or the spatial location of the subscribers, while leaving the consideration of incorporating the subscribers' moving behaviors and temporal information. Therefore, in this paper, we propose a Time-aware Path-based Publish/Subscribe (TPPS) model, where we propose a filtering-verification framework that contains two kinds of filters, i.e., time-aware location-based filter and time-aware region-based filter, with considering both temporal information and moving behaviors, and filtering unrelated subscriptions for each message. We evaluate the efficiency of our approach on a real-world dataset and the experimental results demonstrate the superiority of our method in both efficiency and effectiveness.

Keywords: Time-aware · Path-based · Publish/subscribe

1 Introduction

Traditional publish/subscribe systems are used to deliver the messages from publishers (message producers) to subscribers (message consumers), where subscribers can submit their interests in the form of subscription. Once a new message is generated and published to the system, the publish/subscribe system is responsible for assigning it efficiently to all the subscribers whose subscriptions are similar to the message [1–3].

© Springer International Publishing AG, part of Springer Nature 2018
J. Pei et al. (Eds.): DASFAA 2018, LNCS 10827, pp. 511–528, 2018.
https://doi.org/10.1007/978-3-319-91452-7_33

Generally, content-based subscriptions specify the subscription only based on the properties of the messages without the consideration of the spatial and temporal information [4–6]. However, with the rapid development of mobile web computing and on-line social networks, there are more requirements given by subscribers, where they are interested in not only the textual contents but also the associated spatial and temporal information. For example, when a subscriber registers for a tour route that contains some interested locations and corresponding planned arrival time. Along this route, she may want to receive useful information around these locations, which helps the subscribers to get better travel experiences. For example, from the perspective of a businessman, the information that when her potential customers will show up around is extremely useful, so that she knows when to push the advertisements to attract more customers. A commuter goes to work at 8:00 and returns home at 17:00 everyday. She probably wants to buy breakfast on the way to work or go shopping on the way back. Specifically, if she receives a message that "You can enjoy half price discount if you buy any products from Starbucks today" at 7:30 and there is just a Starbucks by the side of her way to work, there will be a high chance that she will go and buy a cup of coffee.

The existing location-aware publish/subscribe systems deal with the spatial and content features of subscriptions mainly by using boolean filters, delivering the up-to-date messages associated with geo-tagged information to subscribers if both the location and content match with the corresponding subscriptions [7–10]. However, these works fail to take temporal matching into consideration, which cannot accurately represent users' interests and may result in a delayed message delivering. For instance, in a user-specified region, the system may continue delivering the messages to the user even if she has already left this region. Moreover, by using boolean filters, the system can only get exact matchings between subscriptions and messages. In other words, the boolean filters fail to measure the extent how similar a message is to a subscription, which is not able to flexibly satisfy users' requirements.

In this paper, we propose a Time-aware Path-based Publish/Subscribe (TPPS) model. Specifically, in TPPS model, subscribers register path-based subscriptions to capture their interests while publishers post messages associated with their publish time in the publish/subscribe system. A subscription includes a set of textual keywords, indicating the subscriber's interests and a planned path, which consists of a finite sequence of interested locations associated with corresponding planned arrival time. Instead of boolean expressions, TPPS considers the similarity between a subscription and a message, where not only textual relevance and spatial proximity, but also temporal approximation are taken into account. Once a message is posted, TPPS aims to find all the subscriptions which are similar to the message. Note that it actually consists of two sub-problems: (a) for each message, we need to deliver it to the subscribers whose subscriptions match this message; and (b) for each subscriber, we need to find the suitable messages for each location along her planned path. Compared to the traditional methods, the challenge of our problem lies in that, once each user's path is associated with moving features and the planned time is taken into account, pruning infeasible messages gets more difficult than the conventional settings that just specify a valid range for each user without considering temporal information [7–10].

We propose a filtering-verification framework, which comprises two parts: the filtering stage and the verification stage, to deal with this challenge. In the filtering stage, we devise two effective filters, including time-aware location-based filter and time-aware region-based filter, to prune a large number of irrelevant subscriptions for each message and obtain some promising subscription candidates. In the verification stage, we verify these candidates to generate the final suitable subscriptions for each message.

Our contributions in this work can be summarized as follows:

(1) We formulate the TPPS problem on path-based subscriptions which supports flexible requirements from different subscribers.
(2) We devise a filtering-verification framework to deliver messages to relevant subscribers efficiently. Especially in the filtering stage, we propose two filters to prune irrelevant subscriptions, i.e., time-aware location-based filter and time-aware region-based filter. In the time-aware location-based filtering part, we have integrated the inverted index, spatial pruning technique and prefix filtering to implement pruning. In the time-aware region-based filtering part, we partition the locations into regions and use R-tree as an index structure, achieving more effective pruning.
(3) We conduct an experimental study on a real-world dataset (Geolife Trajectories 1.3) and the experimental results demonstrate the superiority of our method in both efficiency and effectiveness.

The remainder of this paper is organized as follows. Section 2 introduces the preliminary concepts and gives an overview of the filtering-verification framework. Two kinds of filtering algorithms are presented in Sect. 3, followed by the verification process introduced in Sect. 4. We show the experimental results in Sect. 5 and discuss the related work in Sect. 6. Finally, we conclude this paper in Sect. 7.

2 Problem Statement

In this section, we present some preliminary concepts and give an overview of the proposed filtering-verification framework. Table 1 summarizes the major notations used throughout the paper.

2.1 Preliminary Concepts

Definition 1 (Path-based Subscription). *A path-based subscription, denoted as $s = (K, P, \alpha, \beta, \tau)$, contains a set of textual keywords K, a path P, two parameters: α, β, and a threshold τ. The set of textual keywords is denoted as $K = \{k_1, k_2, ..., k_{|K|}\}$, where each keyword is associated with a*

Table 1. Summary of notations

Notation	Definition
s	A subscription
$s.K$	A set of keywords contained in s
$s.P$	Path contained in s
$s.l$	Location contained in $s.P$
$t(s.l)$	Time when s passes by location l
m	A message
$m.K$	A set of keywords contained in m
$m.l$	Location of m
$m.t$	Creation time of m
$\mathcal{S}$	A subscription set
$\mathcal{C}$	A candidate set
$\mathcal{LL}(l)$	Time-aware location-based list of location l
R	A region
$\mathcal{RL}(R)$	Time-aware region-based list of $\mathcal{R}$

weight $w(k_i)$ calculated by the inverted document frequency (IDF) [11]. The path $P = \{(l_1, t_1), (l_2, t_2), ..., (l_{|P|}, t_{|P|})\}$ is a finite sequence of spatio-temporal points (l_i, t_i), in which t_i is a user-specified time scheduled to arrive at location l_i. α, β and τ that represent users' different needs are set by users themselves. α, β, and $(1 - \alpha - \beta)$ are the parameters that balance the importance between textual relevance, spatial proximity and temporal approximation. α and β are directly proportional to the importance of textual relevance and spatial proximity, respectively, and $(1 - \alpha - \beta)$ is positively linked to the weight of the temporal approximation. τ is a subscriber-specified minimum similarity threshold that determines whether a subscription is similar to a message.

Note that as each keyword is unique for each person, which means that the importance of each keyword is the same for a particular user, we use IDF rather than TF-IDF to generate keywords weight.

Definition 2 (Massage). *A message, posted by a publisher, is denoted as $m = (K, l, t)$, where $K = \{k_1, ..., k_{|K|}\}$ is a set of textual keywords, which has the same meaning as the keywords of subscriptions, l is a location point with latitude and longitude, and t is the creation time of m.*

For instance, Fig. 1 shows an example of a subscription set $\mathcal{S} = \{s_1, s_2, s_3\}$ and a messages m. Each subscription contains a set of keywords, a path, two parameters and a threshold, such as $s_1 = (\{k_1, k_2, k_3\}, P_1, 0.5, 0.4, 0.6)$. The message is associated with a set of keywords, a location and a creation time, i.e., m located at $(7, 3)$ contains a set of keywords $\{k_1, k_2, k_4\}$ and is published at 8:15.

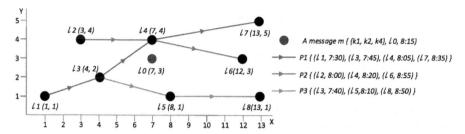

The subscription set

id	subscription
s_1	$(\{k_1, k_2, k_5\}, P_1, 0.5, 0.4, 0.6)$
s_2	$(\{k_3, k_4, k_5\}, P_2, 0.4, 0.4, 0.5)$
s_3	$(\{k_4, k_5\}, P_3, 0.4, 0.3, 0.8)$

The weight of keywords

id	keyword	weight
k_1	discount	0.5
k_2	Starbucks	0.4
k_3	Costa	0.3
k_4	coffee	0.2
k_5	drinks	0.1

Fig. 1. Running example ($MAXDIST = 15, D = 0.9$)

Next we define the similarity between a subscription and a message in three dimensions, i.e., text, space and time.

Definition 3 (Textual Similarity). *The textual similarity between s and m is defined as the ratio between the total weight of keywords shared by s and m and the total weight of keywords of s, which can be computed as follows:*

$$KSIM(s, m) = \frac{\sum_{k \in s.K \cap m.K} w(k)}{\sum_{k \in s.K} w(k)} \qquad (1)$$

where $w(k)$ is the weight of k.

Definition 4 (Spatial Similarity). *The spatial similarity between s and m at location $l \in s.P$ is defined as follows:*

$$SSIM_l(s, m) = \max\{0, 1 - \frac{d(l, m.l)}{MAXDIST}\} \qquad (2)$$

where $d(l, m.l)$ is Euclidian distance between l and $m.l$. $MAXDIST$ is the maximum Euclidian distance between subscriptions and messages that the user can tolerate.

Definition 5 (Temporal Similarity). *The temporal similarity between s and m with respect to location $l \in s.P$ is defined as follows:*

$$TSIM_l(s, m) = \begin{cases} 0 & \Delta < 0 \\ D^\Delta & otherwise. \end{cases} \qquad (3)$$

$$\Delta = t(s.l) - m.t$$

where $t(s.l)$ is the time when s arrives at location l, and $D(0 < D < 1)$ is base number that determines the rate of recency decay. The function is monotonically decreasing with Δ.

Then we combine the textual, spatial and temporal similarity to quantify the similarity between s and m, called Textual and Spatial-Temporal (TST) similarity. More formally,

Definition 6 (TST Similarity). *TST similarity between s and m with respect to the location $l \in s.P$ is denoted as follows:*

$$SIM_l(s,m) = \alpha \cdot KSIM(s,m) + \beta \cdot SSIM_l(s,m) + (1-\alpha-\beta) \cdot TSIM_l(s,m) \quad (4)$$

where α and β are preference parameters defined by a subscriber.

Definition 7 (Similar Subscription). *Given a message m and a subscription s, we denote s as the similar subscription for m, which should satisfy the following two conditions:*

(1) the message has been already published at location l when the subscriber arrives at l, i.e., $m.t \leqslant t(s.l)$, and
(2) the similarity between m and s does not violate the textual and spatio-temporal constraints, i.e., $SIM(s.l, m) \geqslant s.\tau$.

Problem Statement. Given a set of subscriptions, i.e., $\mathcal{S} = \{s_1, s_2, ..., s_{|\mathcal{S}|}\}$ and a message m, the TPPS problem aims to find all similar subscriptions for m.

2.2   Framework Overview

Figure 2 shows the overview of the solution to TPPS problem, i.e., the filtering-verification framework, which comprises two parts, i.e., filtering and verification.

Fig. 2. Framework overview

Taking a message and a subscription set as input, the filtering process, including the location-based filtering and the region-based filtering, generates a set of promising subscription candidates by pruning a massive number of irrelevant subscriptions for the message. The verification process calculates the similarity between each subscription candidate and the given message to obtain the matching subscriptions for the massage. We will detailed discuss each part in the following sections.

3   Filtering Process

In this section, we discuss in detail about two filters, i.e., Time-aware Location-based Filter (TLF) and Time-aware Region-based Filter (TRF), used in the filtering process.

3.1   Time-Aware Location-Based Filter

The main challenge of TPPS problem lies in huge search space when enumerating all locations for each subscription in $\mathcal{S}$, which increases exponentially with respect to the number of subscriptions. To cope with the vast amounts of subscriber enumerating, in this part, we apply the inverted index, a highly efficient data structure, to all the location points. In specific, we create a time-aware location-based list for each location l. More formally,

Definition 8 (Time-aware Location-based List). *The time-aware location-based list of a location l, denoted as $\mathcal{LL}(l)$, is a list of triples $\langle s, q, t(s.l)\rangle$, where s is one of the subscriptions containing the location l, q is the location order of l in $s.P$, and $t(s.l)$ is the time when s will arrive at l. The elements of $\mathcal{LL}(l)$ are sorted by the time $t(s.l)$ in descending order.*

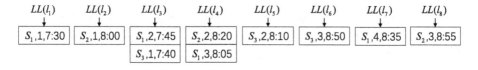

Fig. 3. Time-aware location-based lists

Figure 3 shows the time-aware location-based lists for the previous example in Fig. 1. Consider $s_1 = (\{k_1, k_2, k_3\}, P_1, 0.5, 0.4, 0.6)$ whose path is $P_1 = \{(l_1, 7\!:\!30), (l_3, 7\!:\!45), (l_4, 8\!:\!05), (l_7, 8\!:\!35)\}$, we insert $\langle s_1, 1, 7\!:\!30\rangle, \langle s_1, 2, 7\!:\!45\rangle, \langle s_1, 3, 8\!:\!05\rangle, \langle s_1, 4, 8\!:\!35\rangle$ into inverted list of l_1, l_3, l_4, l_7 respectively, where 1,2,3,4 denote location orders of l_1, l_3, l_4, l_7 in P_1. It is worth noting that, after each inserting, each list is still arranged in descending order according to time. For instance, $\mathcal{LL}(l_3) = \{\langle s_1, 2, 7\!:\!45\rangle, \langle s_3, 2, 7\!:\!40\rangle\}$ and $\mathcal{LL}(l_4) = \{\langle s_2, 2, 8\!:\!20\rangle, \langle s_1, 3, 8\!:\!05\rangle\}$.

Apart from the time-aware location-based list, we extend the prefix filtering technique to TPPS problem, inspired by the prefix filtering method in data cleaning [12].

We first fix a global ordering on the keywords of all subscriptions and messages according to their keyword weight and then sort the keywords of each subscription and message based on the global ordering. For instance, in Fig. 1, keywords of all subscriptions and the message, i.e., $\{k_1, k_2, k_3, k_4, k_5\}$, are sorted based on their weight. In addition, keywords of s_1, s_2, s_3 and m are sorted based on the ordering of $\{k_1, k_2, k_3, k_4, k_5\}$.

Given a subscription $s = (K, P, \alpha, \beta, \tau)$ and a message m, as neither the spatial similarity nor temporal similarity can exceed 1, we can deduce a lower textual similarity bound between s and m according to the Eq. 4, i.e.,

$$K_{LB}(s) = \frac{\tau + \alpha - 1}{\alpha} \tag{5}$$

Based on this lower bound, we can derive the prefix length p for s, stated as below.

(1) p is the minimum number satisfying $\frac{\sum_{i=p}^{|s.K|} w(k_i)}{\sum_{i=1}^{|s.K|} w(k_j)} < K_{LB}(s)$ if $K_{LB}(s) > 0$.

(2) p is the length of $s.K$ plus two, i.e., $p = |s.K| + 2$ if $K_{LB}(s) \leq 0$.

Then we can define the prefix of a subscription as following.

Definition 9 (Prefix of subscription). *Given a subscription s and its prefix number p, the prefix of s is*

$$Pre(s) = \{k_1, k_2, ..., k_{p-1}\} \tag{6}$$

where $k_i \in s.K$ if $i \leq |s.K|$ and $k_i = *$ $(w(*) = 0)$ if $i = |s.K|+1$ $(1 \leq i \leq p-1)$.

The following lemma states the relationship of the keywords between a message and its similar subscriptions.

Lemma 1. *Given a message m, a subscription s and its prefix $Pre(s)$, if $K_{LB}(s) > 0$, we have: if s is similar to m, $Pre(s)$ and $m.K$ must have an intersection* [12].

According to this lemma, finding the similar subscriptions for a message m can be reduced to finding the subscriptions whose keyword prefixes have overlap with $m.K$. In other words, we can prune a subscription if the prefix of it has no overlap with $m.K$.

Continuing with the previous example in Fig. 1, as for s_1, we can calculate $K_{LB}(s_1) = \frac{0.6+0.5-1}{0.5} = 0.2 > 0$. As $\frac{w(k_2)+w(k_5)}{w(k_1)+w(k_2)+w(k_5)} = \frac{0.4+0.1}{0.5+0.4+0.1} = 0.5 > K_{LB}(s_1) = 0.2$ and $\frac{w(k_5)}{w(k_1)+w(k_2)+w(k_5)} = \frac{0.1}{0.5+0.4+0.1} = 0.1 < 0.2$, we have $p = 3$ and $Pre(s_1) = \{k_1, k_2\}$. Considering s_2, $K_{LB}(s_2) = \frac{0.4+0.5-1}{0.4} = -0.25 < 0$. Therefore $p = |s_2.K| + 2 = 5$ and $Pre(s_2) = \{k_3, k_4, k_5, *\}$. Similarly, we can compute the prefix of s_3: $Pre(s_3) = \{k_4\}$. For the message m whose keywords are $m.K = \{k_1, k_2, k_4\}$, s_1, s_2 may be similar to m as m contains common keywords of both $Pre(s_1)$ and $Pre(s_2)$ while s_3 can be pruned safely as m does not include any common keyword of $Pre(s_3)$.

Moreover, we have the following observations, which can be utilized to design TLF. Given a message m and a subscription s, if s is a candidate of m, s has to satisfy the following two conditions:

(1) the time $t(s.l)$ when s arrive at l must be not earlier than $m.t$ based on Definition 5, and

(2) the prefix of s, i.e., $Pre(s)$, must share at least one common keyword with $m.K$ based on Lemma 1.

Algorithm 1. Time-aware Location-based Filter

Input: m: A message; $\mathcal{S}$: Subscription set
Output: $\mathcal{C}$: Candidate set of m

1 $\mathcal{C} = \emptyset$;
2 sort $m.K$ and add $*$ to $m.K$;
3 **for** *each l in $\mathcal{S}$* **do**
4   **if** $d(l, m.l) \leqslant MAXDIST$ **then**
5   **for** $\langle s, q, t(s.l) \rangle \in \mathcal{LL}(l)$ **do**
6   **if** $t(s.l) < m.t$ **then**
7                  $\lfloor$ judge the next location in $\mathcal{S}$;
8   **else if** $Pre(s) \cap m.K = \emptyset$ **then**
9                  $\lfloor$ judge the next element in $\mathcal{LL}(l)$;
10   **else**
11                 $\lfloor$ $\mathcal{C}$.add($\langle s, q \rangle$);

12 **return** $\mathcal{C}$;

TLF creates the time-aware location-based lists for all locations and calculates the prefix for each subscription In the off-line stage and then generates a candidate set in the on-line stage. Algorithm 1 shows the on-line stage of TLF. The main idea is that when we get a new message m, for each location l whose distance to m, i.e., $d(l, m.l)$, is no larger than $MAXDIST$, we check each element of the time-aware location-based list of l, i.e., $\langle s, q, t(s.l) \rangle \in \mathcal{LL}(l)$ in sequence (line 2–5). If the current element satisfies $t(s.l) < m.t$, the loop will terminate in advance for pruning current subscription and all the subscriptions behind it since the elements have been sorted by $t(s.l)$ in descending order (line 6–7). Otherwise, we check if there are any common keywords between $Pre(s)$ and $m.K$ (line 8–11). Specifically, if $Pre(s)$ and $m.K$ have no intersection, we just prune the current subscription; otherwise, we add $\langle s, q \rangle$ into the candidate set $\mathcal{C}$.

3.2 Time-Aware Region-Based Filtering

TLF can filter out a large number of irrelevant subscriptions and reduce the running time. However, it still needs to traverse all locations in $\mathcal{S}$ for a message, which is time-consuming if the locations are very dense. Therefore, we use TRF, in which locations are divided into regions based on their locality, to prune a region if all locations in this region have low similarity to the message.

First, we propose a time-aware region-based list for each region in the index, stated as below.

Definition 10 (Time-aware Region-based List). *The time-aware region-based list for a region R, denoted as $\mathcal{RL}(R)$, is a list of pairs $\langle c, t_{max} \rangle$, where c is the children of R and t_{max} is the maximal value of the time among the children of R. Similar to the time-aware location-based list, $\mathcal{RL}(R)$ is also sorted in descending order according to the maximal time t_{max}.*

Due to the fact that children of R could be a set of regions or locations, there will be two kinds of time-aware region-based list, as described below.

(1) R is a non-leaf node. In such condition, the children of R are some regions. Therefore, each $\langle c, t_{max} \rangle \in \mathcal{RL}(R)$ denotes a subregion of R and the corresponding maximal time of c.
(2) R is a leaf node. In this case, the children of R are some locations. Thus, each $\langle c, t_{max} \rangle \in \mathcal{RL}(R)$ denotes the child locations of R and the corresponding maximum time of c.

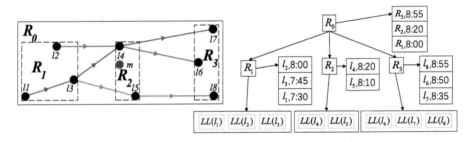

Fig. 4. Illustration of indexing structure

Figure 4 shows an illustration of indexing structure for $\mathcal{S} = \{s_1, s_2, s_3\}$, where each region is associated with a time-aware location-based list. For example, the non-leaf node R_0 whose children are three regions, i.e., R_1, R_2 and R_3 is associated with $\mathcal{RL}(R_0) = \{\langle R_3, 8 : 55 \rangle, \langle R_2, 8 : 20 \rangle, \langle R_1, 8 : 00 \rangle\}$. The leaf node R_1 whose children are three locations, i.e., l_1, l_2 and l_3 is associated with $\mathcal{RL}(R_1) = \{\langle l_2, 8:00 \rangle, \langle l_3, 7:45 \rangle, \langle l_1, 7:30 \rangle\}$.

After the partition, we can calculate the minimum spatial distance between the message m and each region R, denoted by $MINDIST(m, R)$. If $m \in R$, then we have $MINDIST(m, R) = 0$; otherwise, $MINDIST(m, R)$ is the minimum value among the distances from m to the four boundaries of region R, which can be easily computed in $(O)(1)$ time. Obviously, if $MINDIST(m, R) > MAXDIST$, for each location $l \in R$, as we have $d(m, l) \leqslant MINDIST(m, R)$, we can deduce that $d(m, l) > MAXDIST$ which means subscriptions located at l cannot be similar to m. Therefore, we directly prune entire region R if $MINDIST(m, R) > MAXDIST$.

Algorithm 2. Time-aware Region-based Filtering

Input: m: A message; $\mathcal{T}$: indexing structure for $\mathcal{S}$
Output: $\mathcal{C}$: Candidate set of m

1 sort $m.K$, add $*$ to $m.K$, $\mathcal{C} = \emptyset$, $\mathcal{Q}.Push(\mathcal{T}.root)$;
2 **while** $\mathcal{Q}$ *is not empty* **do**
3      $R = \mathcal{Q}.Pop()$;
4   **if** $R.t_{max} \geqslant m.t$ *and* $MINDIST(m, R) \leqslant MAXDIST$ **then**
5   **if** R *is a non-leaf node* **then**
6   **for** each $\langle R_c, t_{max} \rangle \in \mathcal{RL}(R)$ **do** $\mathcal{Q}.Push(R_c)$ **if** $t_{max} \geqslant m.t$;
7   **if** R *is a leaf node* **then**
8   **for** *each* $\langle l, t_{max} \rangle \in \mathcal{RL}(R)$ **do**
9   **if** $t_{max} < m.t$ **then** judge the next region in $\mathcal{Q}$;
10   **else**
11   **for** $\langle s, i, t(s.l) \rangle \in \mathcal{LL}(l)$ **do**
12   **if** $t(s.l) < m.t$ **then**
13   judge the next region in $\mathcal{RL}(R)$;
14   **else if** $Pre(s) \cap m.K \neq \emptyset$ **then**
15   Find the first matching keyword k_f of s and m;
16   **if** $K_{LB}^{+}(l_i, m.l) \leqslant K_{UB}(s|k_f)$ **then**
17                          $\mathcal{C}.add(\langle s, i \rangle)$;
18   **else**
19   judge the next location in $\mathcal{LL}(l)$;

20 return $\mathcal{C}$;

Having calculated the prefix $Pre(s)$ for each subscription and supposing k_f is the first matching keyword between $Pre(s)$ and $m.K$, i.e., $m.K$ does not contain any keyword before k_f in $Pre(s)$, we can estimate an upper textual similarity bound for s, i.e,

$$K_{UB}(s|k_f) = \frac{\sum_{i=f}^{|s.K|} w(t_i)}{\sum_{i=1}^{|s.K|} w(t_i)} \tag{7}$$

However, we cannot know which keyword will be the first matching keyword before m is given. Therefore, we calculate the upper textual similarity bound for each $k_i \in Pre(s)$ in the off-line phase, preparing for TRF in the on-line stage.

After calculating $K_{UB}(s|k_f)$, we can calculate a tighter lower keyword similarity bound for s with respect to a location $l \in s.P$ based on Eq. 4 as followed.

$$K_{LB}^{+}(l, m) = \frac{\tau - (1 - \alpha) \cdot SSIM_l(s, m) - (1 - \alpha - \beta) \cdot TSIM_l(s, m)}{\alpha} \tag{8}$$

For any subscription s, if its upper textual similarity bound to the message m with respect to location l is smaller than the lower textual similarity bound, i.e., $K_{UB}(s|k_i) < K_{LB}^{+}(l, m)$, s can be pruned with respect to l.

In off-line stage of TRF, we first calculate $Pre(s)$ for each $s \in \mathcal{S}$ and initialize $\mathcal{LL}(l)$ for each location. Then we divide locations into hierarchical regions and initialize $\mathcal{RL}(R)$ for each region. Finally we build an R-tree, denoted as $\mathcal{T}$), as the indexing structure.

With the help of these pre-processing, TRF filters unrelated subscriptions on-line, as shown in Algorithm 2. When we get a new message m, we first push the root of $\mathcal{T}$ into a queue $\mathcal{Q}$ and then pop regions from $\mathcal{Q}$ in sequence (line 1–3). For each popped region R, if its maximum time $R.t_{max}$ satisfies $R.t_{max} \geqslant m.t$ and the minimum distance from m to R is smaller than $MAXDIST$ (line 4), we consider the following two cases: (a) R is a non-leaf node (line 5–6) in which the subscriptions of R will be accessed and checked in sequence if $t_{max} \geqslant m.t$. (b) R is a leaf node (line 7–19) where we directly check elements in the time-aware region-based list $\mathcal{RL}(R)$. For each element $\langle l, t_{max} \rangle \in \mathcal{RL}(R)$, we prune the current subscription and all subscriptions behind $\langle l, t_{max} \rangle$ and then judge the next region (line 9) if $t_{max} < m.t$. Otherwise, we check the time-aware location-based list of $l \in \langle l, t_{max} \rangle$ (line 11–19).

4   Verification Process

After the filtering procedure generated a small number of subscriptions, i.e., a candidate set, we calculate accurate TST similarity to verify these candidates.

Algorithm 3 shows the process of verification. For the candidate set $\mathcal{C}$ generated by TLF or TRF, we verify each survived candidate $s \in \mathcal{C}$ one by one by computing the TST similarity between s and m to generate final answers. Specifically, we check if a candidate satisfies the requirements of temporal similarity (line 4) and if the similarity is larger than the threshold of the subscription

---

Algorithm 3. Verification Process

Input: m: A message; $\mathcal{C}$: Candidate set
Output: $\mathcal{A}$: Answer set of m

```
1   A = ∅;
2   for ⟨s, i⟩ ∈ C do
3       if t(s.lᵢ) ≥ m.t then
4           calculate the TST similarity SIMₗᵢ(s, m);
5           if SIMₗᵢ(s, m) ≥ s.τ then
6               if there exists any ⟨s, j⟩ ∈ C then
7                   if SIMₗᵢ(s, m) > SIMₗⱼ(s, m) then
8                       A.delete(⟨s, j⟩);
9                       A.add(⟨s, i⟩);

10              else
11                  A.add(⟨s, i⟩);

12  return A;
```

(line 6). If a candidate satisfies these two conditions, we check if this candidate has already been one of the answers (line 7): if yes, we choose the location where we can get a larger similarity as the answer (line 8–10); otherwise, we just need to add this candidate to the answer set (line 11).

5 Experiments

In this section, extensive experiments are performed to evaluate our proposed methods in efficiency and effectiveness.

5.1 Experimental Setup

Dataset. We use a real dataset, i.e., GeoLife 1.3 [13–15]. GeoLife is a real-life dataset collected in (Microsoft Research Asia) GeoLife project by 182 users. GeoLife 1.3 contains 17621 trajectories with a total distance of 129295 km and a total duration of 50176 h. We collect about 150 words from different users and use these words to randomly generate 1–10 keywords for each user. Each message contains 1–50 keywords collected from tweets which collected from Twitter randomly, their locations and creation time are randomly generated based on the subscriptions.

Table 2. Parameter settings

Parameter	Default value
Preference for keywords α	$[0, 1]$(randomly)
Preference for space β	$[0, 1 - \alpha]$(randomly)
Threshold of the similarity τ	$[0.5, 1]$(randomly)
Maximum user-tolerated Euclidian distance $MAXDIST$	1.5 km
Base number of temporal similarity function D	0.9
Number of locations contained in all subscriptions' path N	10^6

Parameters. The default values of parameters used in our experiments are summarized in Table 2. Notice that as the length of each subscription's path is inconsistent, we use the number of locations contained in paths of all subscriptions in $\mathcal{S}$, denoted as N, to measure the size of the subscription set. In addition, we use IDF to calculate the weight of keywords. When we varied a parameter, others will be in default range.

Baseline. We extend the spatial-keyword search method IR-tree [16,17] to support the TPPS problem.

Experimental Setting. All the algorithms are implemented in Python and run on a PC with Intel(R) Core(TM) i7-6500U @2.50 GHZ and 8 GB RAM.

5.2   Experimental Results

In this part, we evaluate the performance of our methods, i.e., TLF and TRF, and IR-tree to TPPS problem. We use candidate percentage, which is the ratio between N of the candidate set and N of the original subscription set as a metric of pruning ability.

First, we evaluate the effect of α, β and τ. The weighting parameter α, β in the similarity function between a message and a subscription is to let users specify their preference between textual similarity, spatial proximity and temporal approximation. A large α means textual similarity is more important while a large β means temporal similarity is more important, and vice versa. τ is a threshold that determines whether a subscription is similar to a message. α, β changes from 0.1 to 0.9 while τ changes from 0.5 to 1. The results is shown in Figs. 5 and 6, where we can see that TRF performs best and TLF performs better than IR-tree.

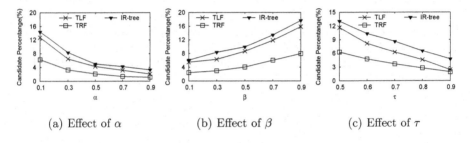

(a) Effect of α (b) Effect of β (c) Effect of τ

Fig. 5. Effect of parameters on pruning ability

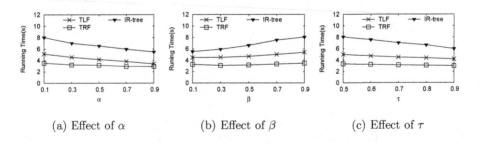

(a) Effect of α (b) Effect of β (c) Effect of τ

Fig. 6. Effect of parameters on running time

Effect of α. As shown in 5(a) and 6(a), both the candidate percentage and running time of these three methods, i.e., TLF, TRF and IR-tree, are decreasing with the increase of α. This is because, the number of subscriptions that can fully cover the subscribers' keywords is reduced. Note that the increase of α leads to that the candidates which have lower spatial similarity or lower temporal similarity can be included, however, they will be filtering out again as subscriptions is require a high textual similarity.

Effect of β **.** As shown in Fig. 5(b) and 6(b), with the increase of β, both the candidate percentage and running time of these three methods are decreasing. Although the increase of β will lead to a higher demand for spatial similarity, the requirements on textual similarity and temporal similarity are decreasing. Thus subscribers, which have lower spatial similarity but higher temporal similarity and higher textual similarity, will be included again. So both the number of candidates and the running time are increasing as β increases.

Effect of τ. As shown in Fig. 5(c) and 6(c), the greater the value of τ, the less candidate percentage and the less running time. The reason is that the larger value of τ can filter out more subscriptions which have small similarity. Therefore, with the increase of τ, the running time reduces.

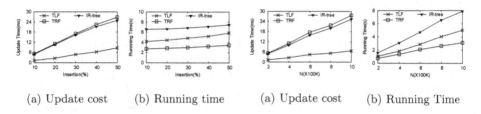

(a) Update cost   (b) Running time   (a) Update cost   (b) Running Time

Fig. 7. Update on the same $\mathcal{S}$ **Fig. 8.** Update on different $\mathcal{S}$

In addition, we evaluate the cost of updating the subscription set and the running time after updates. We first create indexes for a subscription set then insert subscriptions into it. We perform this evaluation under two conditions: (a) insert different number of subscriptions into a same subscription set; and (b) insert the same percentage of subscriptions into different subscription sets.

Update Cost. Figure 7 shows the results of inserting subscriptions containing different number of locations (vary from 10% to 50%) into the indexes of a subscription set ($N = 400K$) while Fig. 8 shows the results of inserting subscriptions containing the same percentage (20%) of locations into each index of subscription sets with different N (changes from $400K$ to $1M$). We can see that both the update time and the running time after updates are increasing with the increase of updates. However, the update cost of TLF is smallest, but its performance are decreasing with the increasing of the size of subscription set. The running time of IR-tree is largest. This is because TLF and TRF have early termination strategy while IR-tree have not.

Finally, in order to study the scalability of IR-tree and the proposed algorithms, we generate five subscription sets containing from two hundred thousand to one million by random selection from the original dataset. Then we run experiments on these five subscription sets, the results of which are shown in Fig. 9.

Scalability. As shown in Fig. 9, TLF and TRF scaled much better than IR-tree, because TLF and TRF use tighter bounds to generate candidates. In addition, TRF performs best as IR-tree have largest running time and the performance of

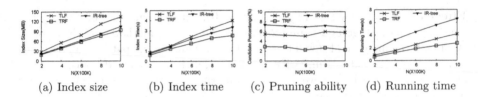

| (a) Index size | (b) Index time | (c) Pruning ability | (d) Running time |

Fig. 9. Effect of N on scalability

TLF decreases with the increasing of N. This is because TRF not only utilizes a hierarchical structure but also prune dissimilar subscriptions.

6   Related Work

Content-Based Publish/Subscribe Framework. Existing work on content-based top-k publish/subscribe systems [18–26] match the published objects to subscriptions only if these subscriptions rank among the top-k in the subscription set. Most of these systems set that, during a pre-specified time interval, the relevance of the objects remains constant. And these objects are expired and replaced by the most relevant unexpired objects once its lifetime exceeds the specified time interval. This setting is different from us, because we do not just take the top-k subscriptions as results, because some of these top-k results may not be similar to the published objects.

Location-Aware Publish/Subscribe Framework. Wang et al. [8] and Jiang et al. [10] design a location-aware publish/subscribe system which utilizes spatial overlap to evaluate spatial similarity and utilizes the "AND" and "OR" semantics to evaluate textual relevancy. Chen et al. [7] propose an efficient index to match a stream of boolean range continuous queries over a stream of geo-textual objects. However, since they focus on the "AND" and "OR" semantics to evaluate textual relevancy and do not consider the moving behavior of each user, their techniques cannot be applied to our proposed TPPS problem.

7   Conclusion

In this paper we study a novel problem in publish/subscribe system, namely TPPS, to deliver the messages associated with spatio-temporal information to the suitable subscribers with planned paths by comparing the similarity between messages and subscriptions. In order to achieve high effectiveness and efficiency, we addressed a few challenges by proposing different strategies to prune plenty of irrelevant subscriptions and generate a set of promising subscription candidates, and designing a verification algorithm to obtain the similar subscriptions for each message. To the best of our knowledge, it is the first work that utilizes the subscribers' planned path and time information to delivery messages to subscribers. Extensive experiments have been conducted using a real dataset, i.e.,

GeoLife 1.3, to demonstrate the powerful pruning ability and high efficiency of our approaches. One of our future work is to parallelize our solution so that it can be used in more data-intensive applications.

Acknowledgement. This research is partially supported by the Natural Science Foundation of China (Grant No. 61502324, 61532018).

References

1. Gupta, A., Sahin, O.D., Agrawal, D., El Abbadi, A.: Meghdoot: content-based publish/subscribe over P2P networks. In: Jacobsen, H.-A. (ed.) Middleware 2004. LNCS, vol. 3231, pp. 254–273. Springer, Heidelberg (2004). https://doi.org/10.1007/978-3-540-30229-2_14
2. Vom Fachbereich Informatik: Large-scale content-based publish/subscribe systems. Technische Universitat, vol. 60, no. 3, p. 435C450 (2002)
3. Eugster, P.T., Felber, P.A., Guerraoui, R., Kermarrec, A.M.: The many faces of publish/subscribe. ACM Comput. Surv. **35**(2), 114–131 (2003)
4. Shang, S., Chen, L., Wei, Z., Jensen, C.S., Zheng, K., Kalnis, P.: Trajectory similarity join in spatial networks. PVLDB **10**(11), 1178–1189 (2017)
5. Shang, S., Zheng, K., Jensen, C.S., Yang, B., Kalnis, P., Li, G., Wen, J.: Discovery of path nearby clusters in spatial networks. IEEE Trans. Knowl. Data Eng. **27**(6), 1505–1518 (2015)
6. Shang, S., Ding, R., Zheng, K., Jensen, C.S., Kalnis, P., Zhou, X.: Personalized trajectory matching in spatial networks. VLDB J. **23**(3), 449–468 (2014)
7. Chen, L., Cong, G., Cao, X.: An efficient query indexing mechanism for filtering geo-textual data. In: SIGMOD, pp. 749–760 (2013)
8. Wang, Y., Wang, Y., Wang, T., Feng, J.: Location-aware publish/subscribe. In: SIGKDD, pp. 802–810 (2013)
9. Wang, X., Zhang, Y., Zhang, W., Lin, X., Wang, W.: AP-tree: efficiently support continuous spatial-keyword queries over stream. In: ICDE, pp. 1107–1118 (2015)
10. Jiang, H., Zhao, P., Sheng, V.S., Liu, G., Liu, A., Wu, J., Cui, Z.: An efficient location-aware publish/subscribe index with Boolean expressions. In: Wang, J., Cellary, W., Wang, D., Wang, H., Chen, S.-C., Li, T., Zhang, Y. (eds.) WISE 2015. LNCS, vol. 9418, pp. 216–231. Springer, Cham (2015). https://doi.org/10.1007/978-3-319-26190-4_15
11. Church, K., Gale, W.: Inverse Document Frequency (IDF): A Measure of Deviations from Poisson, pp. 283–295. Springer, Netherlands (1999)
12. Chaudhuri, S., Ganti, V., Kaushik, R.: A primitive operator for similarity joins in data cleaning. In: ICDE, p. 5 (2006)
13. Zheng, Y., Zhang, L., Xie, X., Ma, W.Y.: Mining interesting locations and travel sequences from GPS trajectories. In: WWW, pp. 791–800 (2009)
14. Zheng, Y., Li, Q., Chen, Y., Xie, X., Ma, W.Y.: Understanding mobility based on GPS data. In: UbiComp, pp. 312–321 (2008)
15. Zheng, Y., Xie, X., Ma, W.Y.: GeoLife: a collaborative social networking service among user, location and trajectory. IEEE Data Eng. Bull. **33**(2), 32–39 (2010)
16. Li, Z., Lee, K.C.K., Zheng, B., Lee, W.C., Lee, D., Wang, X.: IR-tree: an efficient index for geographic document search. IEEE TKDE **23**(4), 585–599 (2011)
17. Felipe, I.D., Hristidis, V., Rishe, N.: Keyword search on spatial databases. In: ICDE, pp. 656–665 (2008)

18. Haghani, P., Michel, S., Aberer, K.: Evaluating top-k queries over incomplete data streams. In: CIKM, pp. 877–886 (2009)
19. Haghani, P., Aberer, K., Michel, S.: The gist of everything new: personalized top-k processing over web 2.0 streams. In: CIKM, pp. 489–498 (2010)
20. Aberer, K.: Top-k/w publish/subscribe: finding k most relevant publications in sliding time window w. In: DEBS, pp. 127–138 (2008)
21. Shraer, A., Gurevich, M., Fontoura, M., Josifovski, V.: Top-k publish-subscribe for social annotation of news. PVLDB 6(6), 385–396 (2013)
22. Chen, L., Cong, G., Cao, X., Tan, K.L.: Temporal spatial-keyword top-k publish/subscribe. In: ICDE, pp. 255–266 (2015)
23. Zheng, B., Su, H., Hua, W., Zheng, K., Zhou, X., Li, G.: Efficient clue-based route search on road networks. TKDE 29(9), 1846–1859 (2017)
24. Zheng, K., Zheng, B., Xu, J., Liu, G., Liu, A., Li, Z.: Popularity-aware spatial keyword search on activity trajectories. WWWJ 20(4), 749–773 (2017)
25. Zheng, B., Zheng, K., Xiao, X., Su, H., Yin, H., Zhou, X., Li, G.: Keyword-aware continuous knn query on road networks. In: ICDE, pp. 871–882 (2016)
26. Zheng, B., Yuan, N.J., Zheng, K., Xie, X., Sadiq, S.W., Zhou, X.: Approximate keyword search in semantic trajectory database. In: ICDE 2015, pp. 975–986 (2015)

Direction Recovery in Undirected Social Networks Based on Community Structure and Popularity

Yi-Ming Wen, Chang-Dong Wang$^{(\boxtimes)}$, and Kun-Yu Lin

School of Data and Computer Science, Sun Yat-sen University, Guangzhou, China
Yi-Ming.Wen@outlook.com, changdongwang@hotmail.com, kunyulin14@outlook.com

Abstract. Directionality is a significant property of social networks, which enables us to improve our analytical tasks and have a deeper understanding about social networks. Unfortunately, the potential directionality is hidden in undirected social networks. The previous studies on recovering directionality in undirected social networks mostly focus on the microscopic patterns discovered in the existing directed social networks. In this paper, we attempt to recover the directionality based on the macroscopic community structure. To this end, a variant of the existing modularity model, called behavioural modularity, is designed for discovering community membership of nodes. Assuming that members in the same community have higher behavioural similarity, we introduce the concept of the intra-community popularity, and then estimate directionality of undirected ties based on the community structure and the intra-community popularity. Accordingly, we propose a novel Community and Popularity based Direction Recovering (CPDR) approach to recover the directionality of undirected social networks. Experimental results conducted on three real-world social networks have confirmed the effectiveness of the proposed approach on direction recovery.

Keywords: Direction recovery · Popularity · Community detection
Behavioural similarity

1 Introduction

With the rapid development of modern information technology, a large number of networks are generated, such as e-mail networks [1], file sharing networks [2], voting networks [3,4], etc. In these networks, directionality is a significant property, since it is beneficial in many valuable tasks, including link prediction [5], community detection [6,7], product recommendation [8], etc. However, the directionality is hidden in undirected networks. Our task is to recover the directions of undirected social networks.

The previous studies about direction recovery [9,10] mostly focus on microscopic patterns. In [9], the authors find out four patterns by analyzing existing

---

© Springer International Publishing AG, part of Springer Nature 2018
J. Pei et al. (Eds.): DASFAA 2018, LNCS 10827, pp. 529–537, 2018.
https://doi.org/10.1007/978-3-319-91452-7_34

directed networks, i.e. degree consistency pattern, triad status consistency pattern, similarity consistency pattern and collaborative consistency pattern, and then recover the directions of undirected networks by minimizing the global inconsistency to these patterns. However, these patterns do not always exist in all networks, which results in a bad performance in some networks. In this paper, to recover the directions of networks, instead of microscopic patterns, we focus on the macroscopic community structure.

Modularity is a popular model for measuring the community structure in both directed [7] and undirected networks [11]. In the previous study about modularity in directed networks [7], directionality is used to measure the connection strength between nodes. However, more information is hidden in directionality, such as, behaviour of individuals. Behavioural similarity should be high between nodes in the same community. For instance, in a citation network, physicists are in the same community. They may cite a number of famous papers about physics, and therefore, they have higher similarity in behaviours. In undirected networks, a tie between two individuals indicates higher possibility for them to be in the same community. However, it does not work in directed networks, since in a directed relationship, two individuals have different behaviours, one as the source while the other as the target. Therefore, we need to measure modularity with regard to behavioural similarity.

In a community, some individuals can be regarded popular when a large number of other individuals in this community propose a relationship to them. For example, in the communication networks of classes, teachers or those who are helpful in solving problems are popular since many other individuals send them messages to ask questions. Chances are high that a relationship is proposed towards those with high popularity. The concept of popularity has been studies in community detection and link analysis [12].

From the perspective of the macroscopic community structure, we propose a novel Community and Popularity based Directions Recovering (CPDR) approach to recover the directions in undirected social networks. In particular, a variant of the existing modularity model, called behavioural modularity, is designed for recovering community membership of nodes, which is suitable for the task of recovering directions. Moreover, we introduce the concept of the intra-community popularity and then estimate the direction probability based on the resulting community membership of nodes as well as the intra-community popularity of nodes. Finally, the directions can be recovered based on the direction probability. Experiments are conducted on three real-world datasets, the results of which have confirmed that the proposed method has significantly outperformed the baselines.

2   The Proposed Model

Consider an undirected graph $\mathcal{G} = (\mathcal{V}, \mathcal{E})$ where $\mathcal{V}$ denotes the node set and $\mathcal{E}$ denotes the edge set, the binary adjacency matrix is denoted by $A \in \mathbb{R}^{|\mathcal{V}| \times |\mathcal{V}|}$. For simplicity, we set that when there is an edge between node i and j, both (i, j)

and (j, i) belong to the edge set $\mathcal{E}$. Our task is to recover the direction of each edge based on the community structure and the intra-community popularity of node.

Assume that the nodes $\mathcal{V}$ are divided into c communities, two matrices $C \in \mathbb{R}^{|\mathcal{V}| \times c}$ and $P \in \mathbb{R}^{|\mathcal{V}| \times c}$ are introduced to characterize the community structure and the intra-community popularity of node respectively, where C_{ik} represents the community membership probability of node i belonging to community k, and P_{ik} represents the popularity of node i in community k.

Besides we introduce the direction probability matrix $S \in \mathbb{R}^{|\mathcal{V}| \times |\mathcal{V}|}$, where S_{ij} represents the probability for node i pointing to node j and $S_{ij} = 0$ when edge between node i and node j does not exist in $\mathcal{G}$. In our approach, we first initialize S as a potential direction probability matrix derived from the node degree. In social network, chances are high that an edge starts from a high-degree node and points to a low-degree node. Therefore, $\forall(i, j) \in \mathcal{E}$, we initialize the direction probability S_{ij} as

$$S_{ij} = \sigma(\deg(i) - \deg(j)) \qquad (1)$$

where $\sigma(x) = \frac{1}{1+e^{-x}}$, and $\deg(i)$ is the degree of node i in $\mathcal{G}$.

After initializing S, the first phase of our approach is to generate the community structure C by leveraging the modularity model with regard to behavioural similarity, which is called behavioural modularity. Then, in the second phase, the direction probability matrix S can be derived from the resulting community structure (i.e. matrix C) as well as the intra-community popularity of node (i.e. matrix P), where S and P are updated in an interplay manner. Finally, in the third phase, we recover the directions in $\mathcal{G}$ with S.

2.1 Modeling Community Membership with Behavioural Modularity

In undirected networks, an edge between node i and node j indicates a higher similarity in their community membership. However, in a directed network, an edge from node i to node j does not indicate a higher propensity for them to have similar community membership, because for a directed edge, node i and node j have different behaviours, i.e., node i is the source while node j is the target. On the contrary, a higher behavioural similarity between two nodes indicates their similarity in community membership. To this end, a behavioural modularity is designed, which is a variant of the existing modularity.

In the existing modularity model for direct network [7], the strength of bond between node i and j is modeled as $A_{ij} - E_{ij}$, where A is the adjacency matrix of the directed network and $E_{ij} = \frac{\sum_{j'} A_{ij'} \sum_{i'} A_{i'j}}{|\mathcal{E}|}$ is the expected number of edges between node i and node j in a randomly rewired graph preserving the degree distribution. However, as mentioned in the previous paragraph, such model is not suitable for characterizing the behavioural similarity, since $A_{ij} - E_{ij}$ in such model only models the strength of bond between node i and j rather than the behaviour of node i proposed to j. Therefore, in this paper, we propose to model the behaviour of node i proposed to j, denoted as H_{ij}, by means of the direction probability matrix S

$$H_{ij} = S_{ij} - E'_{ij} \qquad (2)$$

where $E'_{ij} = \frac{\sum_{j'} S_{ij'} \sum_{i'} S_{i'j}}{|\mathcal{E}|}$. The i-th row of H, i.e. H_i, is the behavioural vector of node i. The behavioural similarity between i and j can be obtained as $H_i H_j^T$. Based on the idea of the existing modularity, we propose the behavioural modularity Q_{ij} in directed networks as $Q_{ij} = H_i H_j^T C_i C_j^T$. Summing up over every pair of nodes, we can obtain the behavioural modularity Q of the graph:

$$Q = \sum_{i \in \mathcal{V}} \sum_{j \in \mathcal{V}} H_i H_j^T C_i C_j^T \qquad (3)$$

which can be rewritten as

$$Q(C) = tr(C^T H H^T C) \quad \text{s.t.} \quad \sum_{k'=1}^{c} C_{ik'} = 1, C_{ik} \geq 0, \forall i, k \qquad (4)$$

by introducing the constraint of the community membership. By maximizing the modularity (4), the community membership matrix C can be generated.

2.2   Estimating Direction Probability

In the second phase, the direction probability matrix S can be derived from the resulting community structure (i.e. matrix C) as well as the intra-community popularity of node (i.e. matrix P), where S and P are updated in an interplay manner.

In directed networks, it is assumed that nodes with a larger number of the other nodes pointing to (i.e. larger in-degree) are of higher popularity, such as the teachers in classes, a well-known professor in a department, etc. In particular, for modeling the intra-community popularity of node j in community k, P_{jk} is defined by summing up the edges starting from community k and pointing to node j

$$P_{jk} = \sum_{i \in \mathcal{N}(j)} C_{ik} S_{ij} \qquad (5)$$

where $\mathcal{N}(j)$ is the set of neighbours of j.

Based on the computed community membership and popularity of each node, we can estimate the direction probability. Given the community membership of each node, the edges of $\mathcal{G}$ can be divided into two types: (i) the two nodes of an edge are in different communities, and (ii) the two nodes of an edge are in the same community. Provided that node i is in community k, to measure the probability of node i pointing to node j, denoted as $Pr(j|i;k)$, we consider the following two cases.

Case (i): Node j is Not in Community k: Since nodes in the same community have higher behavioural similarity, node i tends to propose an edge to node j

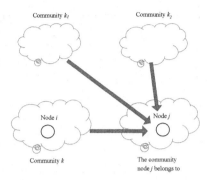

Fig. 1. In case (i), node i has a higher propensity to propose an edge to node j if j has higher popularity in community k. Intuitively, node j can be regarded as popular in community k when P_{jk} is higher than j's popularity in the other communities.

if the majority of the other nodes in community k are pointing to node j, that is, when node j has a higher popularity in communities k. As shown in Fig. 1, node i has higher propensity to propose an edge to j if j is more popular in community k than in the other communities. Therefore, we estimate $Pr(j|i;k)$ by comparing P_{jk} and the average popularity of node j among communities, denoted as $\bar{P}_j$. Since the average popularity differs greatly among communities, the difference between P_{jk} and $\bar{P}_j$ is unsuitable for measuring the degree in which P_{jk} is higher than $\bar{P}_j$, because, for example, for j whose $\bar{P}_j$ is relatively large, even a P_{jk} is slightly larger, the absolute difference between them can be large, even larger than the one when $\bar{P}_j$ is relatively small and P_{jk} is several times larger than $\bar{P}_j$. On the contrary, $\frac{P_{jk}}{\bar{P}_j}$ better shows such degree since the fraction can better eliminate the difference among communities. With such measurement, we can estimate the probability in this case with the following formula.

$$Pr_1(j|i;k) = \frac{P_{jk}}{P_{jk} + \bar{P}_j} \quad \text{where} \quad \bar{P}_j = \frac{\sum_{k'=1}^{c}(1 - C_{jk'})P_{jk'}}{\sum_{k'=1}^{c}(1 - C_{jk'})} \quad (6)$$

Note that we only compare P_{jk} with the popularity of j among all the other communities except the community node j belongs to, since in this case we discuss the type of edges whose two nodes are in different communities. Moreover, a node usually has higher popularity in the community it belongs to than in the others. As a consequence, the popularity of node j in the community it belongs to is larger than the one in other communities, and therefore even when P_{jk} is relatively high, the comparison between them may hide the relatively high P_{jk}. Hence, here $\bar{P}_j$ is calculated as the average popularity of j among all communities except the community node j belongs to. Since C is a community membership probability matrix, it is infeasible to exactly point out the community node j belongs to. Therefore, we use $1 - C_{jk}$ as the non-membership of node j to community k, based on which $\bar{P}_j$ can be derived as the average popularity of j among all communities weighted by its non-membership.

Case (ii): Node j is in Community k: In this case, the higher the popularity of node j is in community k, the higher probability it is for node i to point to node j. Therefore, we can estimate $Pr(j|i;k)$ by comparing P_{jk} and the average popularity of nodes in community k, denoted as $\bar{P}_k$. The comparison uses the fraction between them for the same reason in case (i). The estimation is as follows,

$$Pr_2(j|i;k) = \frac{P_{jk}}{P_{jk} + \bar{P}_k}$$

where $\bar{P}_k$ is the average popularity of all nodes weighted by their membership to community k, i.e. $\bar{P}_k = \frac{\sum_{j'=1}^{|\mathcal{V}|} C_{j'k} P_{j'k}}{\sum_{j'=1}^{|\mathcal{V}|} C_{j'k}}$.

Summing up the two cases mentioned above, we can estimate $Pr(j|i;k)$ by:

$$Pr(j|i;k) = (1 - C_{jk}) Pr_1(j|i;k) + C_{jk} Pr_2(j|i;k) \tag{7}$$

Based on $Pr(j|i;k)$, by integrating all communities [12], we can estimate S_{ij} by

$$\hat{S}_{ij} = \sum_{k=1}^{c} C_{ik} Pr(j|i;k) \tag{8}$$

The estimated S_{ij} should preserve the structure of the original undirected network. In the previous study, the constraint $S_{ij} + S_{ji} = 1$ is used. Under such constraint, a large pointing probability of a direction means a weak one in the opposite direction, which neglects the existence of communications in both directions. To preserve the structure of the original undirected network and take bilateral communications into account simultaneously, we can adapt the following constraint, $\forall (i,j) \in \mathcal{E}, S_{ij} + S_{ji} > 1$ and $0 < S_{ij} < 1$.

Therefore, we can minimize the following objective function to find out the direction probability matrix S that minimizes the error between S and $\hat{S}$ with the above constraint.

$$\min_{S} f(S) = \sum_{(i,j) \in \mathcal{E}} (S_{ij} - \hat{S}_{ij})^2 \quad \text{s.t. } \forall (i,j) \in \mathcal{E}, \ S_{ij} + S_{ji} > 1, \ 0 < S_{ij} < 1 \tag{9}$$

where $\hat{S}_{ij}$ can be obtained by (8). Notice that, $\hat{S}_{ij}$ is not a constant, since $\hat{S}_{ij}$ changes with P and P changes with S. That is, S and P will be updated in an interplay manner. Gradient descent is used to maximize the objective function (4) and minimize (9). With the computed direction probability matrix S, we can recover the directions by labeling every edge (i,j) satisfying that $S_{ij} \geq \theta$ as proposed by node i towards node j, where θ is a probability threshold.

3   Experimental Evaluation

In this section, some experiments are conducted on three real-world datasets to confirm the effectiveness of the proposed method[1].

Three datasets are used in our experiments.

- email-Eu-core[2]: The network was generated using email data from a large European research institution. The e-mails represent communication between institution members (the core). It contains 1005 nodes and 25571 edges.
- Gnutella peer-to-peer[3]: A sequence of snapshots of the Gnutella peer-to-peer file sharing network from August 2002. Nodes represent hosts in the Gnutella network topology and edges represent connections between the Gnutella hosts. It contains 8846 nodes and 31839 edges.
- wiki-Vote[4]: The network contains all the Wikipedia voting data from the inception of Wikipedia till January 2008. Nodes in the network represent wikipedia users and a directed edge from node i to node j represents that user i voted on user j. It contains 7115 nodes and 103689 edges.

Two baseline methods are compared in our experiments.

- ReDirect-T/SF [9]: This algorithm iteratively updates the direction of every edge to minimize the local inconsistency to four consistency patterns. In each iteration, when updating the direction of an edge, it assumes that the directions of the other edges are known. For this method, the parameters are set as default values or tuned as suggested in [9].
- Random: We randomly assign direction to each edge in undirected networks and generate random directed networks. Here we set the probability for a direction to exist as 0.5.

In our method, the parameters are set as: $\theta = 0.3$ on all the datasets, $c = 50$ on email-Eu-core and wiki-Vote, and $c = 100$ on Gnutella peer-to-peer.

The comparison results in terms of Accuracy and Precision are shown in Table 1. According to Table 1, our approach outperforms the ReDirect-T/SF method and the random method on all the datasets. In particular, the proposed CPDR method has achieved about 19.41% and 9.27% improvements in terms of accuracy and precision on all the three datasets over the two compared methods. Meanwhile, the performance of ReDirect-T/SF method is even worse than the random method, suggesting that our CPDR method still works well when the microscopic patterns are inapplicable. In conclusion, the comparison results confirm the effectiveness of the proposed CPDR method leveraging the community structure and the intra-community popularity of nodes.

---

[1] The Python code and the benchmark datasets are available at https://www.drop box.com/sh/dhvhosmzxko0jk2/AACMzPhrm0XdrYZSHI6db_Cda?dl=0   (Extracting Password: DASFAA2018).
[2] http://snap.stanford.edu/data/email-Eu-core.html.
[3] http://snap.stanford.edu/data/p2p-Gnutella05.html.
[4] http://snap.stanford.edu/data/wiki-Vote.html.

Table 1. Comparison results. The best results are highlighted in bold.

Method	email-Eu-core		Gnutella peer-to-peer		wiki-Vote	
	Accuracy	Precision	Accuracy	Precision	Accuracy	Precision
ReDirect-T/SF	0.524	0.730	0.455	0.471	0.1915	0.263
Random	0.567	0.779	0.498	0.498	0.502	0.514
CPDR	**0.711**	**0.812**	**0.622**	**0.592**	**0.541**	**0.537**
Improvements	25.42%	4.33%	25.06%	18.94%	7.76%	4.53%

4   Conclusion

In this paper, we study how to recover the directions in social networks based on the community membership and the intra-community popularity of nodes. We introduce a new method to measure the modularity in directed networks with regard to behavioural similarity of nodes. Besides, we propose an approach to estimate the direction probability based on the intra-community popularity and community membership of nodes. Experiments conducted on three real-world social networks have confirms the effectiveness of our approach.

Acknowledgments. This work was supported by NSFC (61502543) and Tip-top Scientific and Technical Innovative Youth Talents of Guangdong special support program (2016TQ03X542).

References

1. Yin, H., Benson, A.R., Leskovec, J., Gleich, D.F.: Local higher-order graph clustering. In: KDD, pp. 555–564 (2017)
2. Ripeanu, M., Foster, I.T., Iamnitchi, A.: Mapping the Gnutella network: properties of large-scale peer-to-peer systems and implications for system design. CoRR cs.DC/0209028 (2002)
3. Leskovec, J., Huttenlocher, D., Kleinberg, J.: Predicting positive and negative links in online social networks. In: WWW, pp. 641–650 (2010)
4. Leskovec, J., Huttenlocher, D., Kleinberg, J.: Signed networks in social media. In: Proceedings of the SIGCHI Conference on Human Factors in Computing Systems, pp. 1361–1370 (2010)
5. Liben-Nowell, D., Kleinberg, J.: The link-prediction problem for social networks. J. Assoc. Inf. Sci. Technol. **58**(7), 1019–1031 (2007)
6. Liu, L., Xu, L., Wangy, Z., Chen, E.: Community detection based on structure and content: a content propagation perspective. In: ICDM, pp. 271–280 (2015)
7. Leicht, E.A., Newman, M.E.: Community structure in directed networks. Phys. Rev. Lett. **100**(11), 118703 (2008)
8. Ma, H., Zhou, T.C., Lyu, M.R., King, I.: Improving recommender systems by incorporating social contextual information. ACM Trans. Inf. Syst. **29**(2), 9 (2011)
9. Zhang, J., Wang, C., Wang, J., Yu, J.X., Chen, J., Wang, C.: Inferring directions of undirected social ties. IEEE Trans. Knowl. Data Eng. **28**(12), 3276–3292 (2016)

10. Peng, X.-R., Huang, L., Wang, C.-D.: A hybrid approach for recovering information propagational direction. In: Liu, D., Xie, S., Li, Y., Zhao, D., El-Alfy, E.-S.M. (eds.) ICONIP 2017. LNCS, vol. 10638, pp. 357–367. Springer, Cham (2017). https://doi.org/10.1007/978-3-319-70139-4_36

11. Newman, M.E.: Modularity and community structure in networks. Proc. Nat. Acad. Sci. **103**(23), 8577–8582 (2006)

12. Yang, T., Jin, R., Chi, Y., Zhu, S.: Combining link and content for community detection: a discriminative approach. In: KDD, pp. 927–936. ACM (2009)

Detecting Top-k Active Inter-Community Jumpers in Dynamic Information Networks

Xinrui Wang[1], Hong Gao[1], Jinbao Wang[1(✉)], Tianbai Yue[2],
and Jianzhong Li[1]

[1] Harbin Institute of Technology, Harbin, China
wangxinrui@stu.hit.edu.cn, {honggao,wangjinbao,lijzh}@hit.edu.cn
[2] Harbin Institute of Petroleum, Harbin, China
letianbai1005@qq.com

Abstract. Dynamic information networks, containing evolving objects and links, exist in various applications. Mining such networks is more challenging than mining static ones. In this paper, we propose a novel concept of *Active Inter-Community Jumpers* (*AICJumpers*) for dynamic information networks, which are objects changing communities frequently over time. Given communities of several snapshots in a dynamic network, we devise a time-efficiency top-k *AICJumpers* detection algorithm with a sliding window model. After denoting the jump score which captures how frequently an object changes communities over time, we encode the community changing trajectory of each object as bit vectors and transform jump scores computation into bitwise *and*, *or* and *xor* operations between bit vectors. We further propose a slide-based strategy for space and time saving. Experiments on both real and synthetic datasets show high effectiveness and efficiency of our methods as well as the significance of the *AICJumper* concept.

Keywords: Active inter-community jumpers detection
Dynamic information networks · A sliding window model · Bit vectors

1 Introduction

Information networks, containing a large number of objects interacting with each other, are ubiquitous in many applications, e.g. online bibliographic databases like DBLP, movie websites like IMDB and social websites like Facebook. There has been numerous research work in the last decade on static information network mining, including clustering [10,11], outlier detection [2,4], etc.

In reality, information networks are usually dynamic where objects and links among them change drastically over time. Recently, some progress has been made in dynamic information network mining, such as community detection [9], link prediction [8], etc. Given communities of several snapshots in a dynamic network, we observe that there exist some objects changing communities frequently

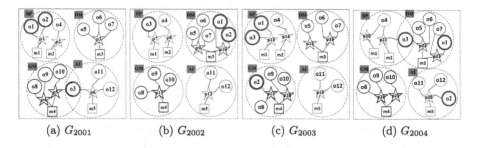

(a) G_{2001} (b) G_{2002} (c) G_{2003} (d) G_{2004}

Fig. 1. Four snapshots of a DBLP dynamic network (Color figure online)

over time. In the following, we present examples and discuss the importance of identifying such objects in practical applications.

Example Application. A DBLP network consists of three types of objects: papers, authors and conferences. There exist connections between papers/authors and papers/conferences with the relationship of "write/written by" and "publish/published by", respectively. Figure 1 shows an example of a DBLP network with four snapshots for 2001–2004. Each sub-figure is a snapshot containing the data of one year, since conferences are often held annually. There are four communities marked in different colors in Fig. 1 which represent four research areas: security and privacy (SP), data management (DM), graphics and multimedia (GM) and artificial intelligence (AI). $\bigcirc$, $\star$ and $\square$ represent the author, paper and conference respectively. An author o_1 moves between two communities SP and DM for 2001–2004. Other two authors, o_2 and o_3, travel among four communities for 2001–2004. o_1, o_2 and o_3 might be multi-field researchers or topic-oriented authors who are always catching current hotspots.

To the best of our knowledge, previous research in dynamic network mining has paid little attention to the objects which change communities frequently over time. In this paper, we call such special objects *AICJumpers* and propose an algorithm to continuously detect top-k *AICJumpers* in a dynamic information network with a sliding window model. We firstly define the jump sore considering both the number of community changes and the changing range for each object to measure how frequently it changes communities over time. Then we transform the process of computing jump scores into building several bit vectors and conducting bitwise *and, or, xor* operations between them so that the computation time are greatly reduced. We also propose a slide-based strategy to further save the memory and time cost by storing only two bit vectors for each object in the memory and computing intermediate results of jump scores slide by slide. Finally, we conduct experiments on both real and synthetic dynamic information networks to demonstrate the effectiveness and efficiency of our methods.

Related Work. Recently, there has been growing interest in detecting community outliers whose behaviors deviate significantly from the majority of community members for both static [2,4] and dynamic networks [5,6]. The definitions of *AICJumpers* in this paper and community outliers in [5,6] are essentially different although they all consider both the temporal and community dimensions. Either a normal object or a community outlier in [5,6] can be an *AICJumper*, the determining factor is whether it changes communities frequently over time. So extracting the patterns which reflects the normal community evolutionary trend of most community members to discover outliers in [5,6] is not adapted to the detection of *AICJumpers* in this paper.

Trajectory outlier detection has been extensively studied in recent years [1,3, 7]. The existing efforts on trajectory outlier detection can not detect *AICJumpers* because: (1) the community changing trajectory of each object in a dynamic network is discrete and contains no location information so that distance-based methods for trajectory outlier detection [1,7] are inapplicable; (2) *AICJumpers* discovery can not utilize additional features such as the speed or direction of motions of objects in trajectory outlier detection [3].

2   Problem Definition

An *information network*, denoted as G, is a 4-tuple $G = (V, E, L, \Phi)$, where V and E represent the vertex set and edge set of G, L is the set of type labels on vertices and Φ is a function mapping each vertex to its corresponding type label in L. The subset of vertices with the same type label $l \in L$ is denoted as $V^l \subseteq V$. In the network $G = (V, E, L, \Phi)$, a vertex subset $C \subseteq V$ is called a *community* if vertices inside C are cohesively connected by internal edges within C. In this paper, we assume that all communities on G have already been previously found by some pre-defined algorithms.

A *dynamic information network*, denoted as $\mathcal{G} = (G_1, G_2, \ldots, G_t, \ldots)$, is an infinite sequence of information networks, where G_t is a snapshot of the network associated with timestamp t. We use $\mathcal{C}_t = \{C_t^1, C_t^2, \ldots, C_t^{|\mathcal{C}_t|}\}$ to denote the set of communities on G_t, where $|\mathcal{C}_t|$ is the number of communities on G_t. Note that C_t^i and $C_{t'}^i$ represent the same community C^i at different timestamps t and t'. An object could belong to several communities at the same timestamp. W.l.o.g., we assume that each community always exists on each snapshot, that is, no new community appears or old community vanishes at any timestamp. Therefore, the number of communities, denoted as $M = |\mathcal{C}_t|$, is the same on each G_t.

Given a starting time t and a fixed *window size* τ, for a dynamic network $\mathcal{G} = (G_1, G_2, \ldots, G_t, \ldots)$, a *sliding window* W_q is the set of τ snapshots: $G_t, G_{t+1}, \ldots, G_{t+\tau-1}$, where q is the ID of the window (q is an integer greater than 0). When new snapshots arrive, the window slides to incorporate μ new snapshots of the dynamic network, and then the oldest μ snapshots will be discarded from the current window. μ denotes the *slide size* which characterizes the evolving speed of the dynamic network. W_q's next window is defined as W_{q+1}, which consists of τ snapshots: $G_{t+\mu}, G_{t+1+\mu}, \ldots, G_{t+\tau-1+\mu}$.

An object o moving from one community C_t^i to another community C_{t+1}^j ($i \neq j$) indicates that it *changes communities* from timestamp t to timestamp $t + 1$. If o doesn't belong to any community at $t + 1$, the frequency that o

changes communities will be measured from t to $t + x$ $(x \geq 1)$ where $t + x$ is the latest timestamp when o belongs to at least one community after t.

Given a window W_q, an object o is an *AICJumper* if its jump score ranks within the top-k among its same type of objects. The *JumpScore* of an object o, capturing how frequently it changes communities in W_q, is defined as follows:

$$JumpScore(o) = \varepsilon F(o) + (1 - \varepsilon)R(o) \tag{1}$$

where $F(o)$ is the number of community changes of o in W_q, $R(o)$ is the changing range of o in W_q, ε is a parameter to decide the importance of $F(o)$ and $R(o)$ in Eq. (1).

The top-k *AICJumpers* detection problem is summarized as follows.

• **Input:** A dynamic network $\mathcal{G} = (G_1, G_2, \ldots, G_t, \ldots)$, the set of communities $\mathcal{C}_t$ for each G_t, a sliding window with the window size τ and the slide size μ, the parameters ε, an integer k and the vertex type $l \in L$.
• **Output:** Top-k *AICJumpers* among objects with the type l in the current window W_q when the window slides.

3   Top-k AICJumpers Detection

JumpScore Definition and Computing. For each object o, we maintain a bit vector $\overrightarrow{A(o)_t} = \{(a(o)_t^1, a(o)_t^2, \ldots, a(o)_t^i, \ldots, a(o)_t^M) | a(o)_t^i \in \{0, 1\}\}$ at timestamp t. $a(o)_t^i$ equals 1 if o belongs to C_t^i at t, otherwise $a(o)_t^i$ equals 0.

Given a window W_q, its starting time t and its window size τ, the *community changes* of o in W_q is calculated as follows:

$$F(o) = \sum_{i=t}^{t+\tau-2} \delta(\overrightarrow{A(o)_i} \oplus \overrightarrow{A(o)_{i+1}}) \tag{2}$$

where $\oplus$ is the bitwise *xor* operation and $\delta(\overrightarrow{B}) = \sum_{r=1}^{|\overrightarrow{B}|} b^r$ (b^r is the r_{th} element of a bit vector $\overrightarrow{B}$, $|\overrightarrow{B}|$ is the number of elements of $\overrightarrow{B}$).

The *changing range* of o in W_q is calculated as follows:

$$R(o) = \sum_{i=t}^{t+\tau-2} \frac{\delta(\bigvee_{j=t}^{t+\tau-1} \overrightarrow{A(o)_j})}{\delta(\overrightarrow{A(o)_i} \wedge \overrightarrow{A(o)_{i+1}}) + 1} \tag{3}$$

where $\vee$ ($\wedge$ resp.) is the bitwise *or* (*and* resp.) operation and the function $\delta(\cdot)$ has the same definition as Eq. (2).

When $t = 1$, $\tau = 4$, we can get $F(o_1) = F(o_2) = 6$, $F(o_3) = 4$, $R(o_1) = 6$, $R(o_2) = 12$, $R(o_3) = 8$ in Fig. 1.

Algorithm Description. Algorithm 1 illustrates how to continuously detect top-k *AICJumpers* in a dynamic network. The algorithm maintains a min heap

---

Algorithm 1. Top-k *AICJumpers* Detection Algorithm

Input : A dynamic network $\mathcal{G} = (G_1, G_2, \ldots, G_t, \ldots)$, the set of communities $\mathcal{C}_t$ for each G_t, a sliding window with the window size τ and the slide size μ, the parameters ε, an integer k and the vertex type $l \in L$

Output: The set of top-k *AICJumpers* in the current window

1:   $t \leftarrow 1$; $W_q.start \leftarrow t$; $W_q.end \leftarrow W_q.start + \tau - 1$;
2:   **for** *each snapshot G_t* **do**
3:      **for** *each object $o \in V^l$* **do**
4:        construct a bit vector $\overrightarrow{A(o)_t}$;
5:      **if** $t == W_q.end$ **then**
6:        $\mathcal{H} \leftarrow \emptyset$;
7:        **for** *each object $o \in V^l$* **do**
8:          compute $F(o)$ using Eq. (2);
9:          compute $R(o)$ using Eq. (3);
10:         $JumpScore(o) \leftarrow$ compute the $JumpScore$ of o using Eq. (1);
11:         **if** $\mathcal{H}$ *has less than k distinct elements* **then**
12:           insert $(o, JumpScore(o))$ into $\mathcal{H}$;
13:         **else**
14:           $\theta \leftarrow$ the smallest $JumpScore$ in $\mathcal{H}$;
15:           **if** $(JumpScore(o) > \theta)$ **then**
16:             pop the element with the smallest $JumpScore$ in $\mathcal{H}$;
17:             insert $(o, JumpScore(o))$ into $\mathcal{H}$;
18:        output $\mathcal{H}$;
19:      $W_q.start \leftarrow W_q.start + \mu$; $W_q.end \leftarrow W_q.start + \tau - 1$;

$\mathcal{H}$ which contains at most k entries to compute the top-k *AICJumpers*. Each entry in $\mathcal{H}$ has the form $(o, JumpScore(o))$.

For each snapshot G_t, the algorithm constructs the bit vector $\overrightarrow{A(o)_t}$ for each object o in V^l (lines 3–4). Then, if the current timestamp equals the ending time of the current window W_q, the min heap $\mathcal{H}$ is emptied (line 6). For each object o in V^l, it uses τ bit vectors, $\overrightarrow{A(o)_{W_q.start}}, \overrightarrow{A(o)_{W_q.start+1}}, \ldots, \overrightarrow{A(o)_{W_q.end}}$, to compute $F(o)$ and $R(o)$ according to Eqs. (2) and (3) (lines 8–9). Subsequently, $JumpScore(o)$ is acquired using Eq. (1) (line 10). If $\mathcal{H}$ has less than k entries, $(o, JumpScore(o))$ is inserted into $\mathcal{H}$ (lines 11–12). Otherwise, the algorithm checks whether $JumpScore(o)$ is lager than the smallest JumpSocre in $\mathcal{H}$. If so, the entry with the smallest $JumpScore$ in $\mathcal{H}$ is replaced by $(o, JumpScore(o))$ (lines 13–17). After all objects in V^l being considered, the entries in $\mathcal{H}$ are output as top-k *AICJumpers* in W_q (line 18). Finally, the window slides (line 19) and the new top-k results of the next window will be acquired (going back to line 2).

Slide-Based Improvement. As mentioned above, in each window, for each object o in V^l, we store τ bit vectors in the memory. There may be insufficient memory when τ is large. Observing Eqs. (2) and (3), every time bitwise *and, or* and *xor* operations are conducted, it only needs two bit vectors corresponding to two latest contiguous timestamps. Therefore, we can simply maintain two latest bit vectors for each object in the memory.

When the window slides, $\tau - \mu$ bit vectors built in W_q keep the same in W_{q+1}. It's time-wasting to conduct bitwise *and, or* and *xor* operations again among these $\tau - \mu$ bit vectors in W_{q+1}. Motivated by this, we divide each window into

$\lceil \frac{\tau}{\mu} \rceil$ slides. At the end of each slide, we compute the slide's $F(o)$ and $R(o)$ for each object o in V^l using bitwise *and, or* and *xor* operation results. Then, at the end of each window, we combine each slide's $F(o)$ and $R(o)$ in the window to acquire the final $F(o)$ and $R(o)$. Thus, when the window slides, we only need to conduct bitwise operations among μ bit vectors in the latest slide.

By storing only two bit vectors for each object in the memory and computing $F(o)$ and $R(o)$ slide by slide, both the memory and time cost is saved compared to Algorithm 1. The pseudo-code of the slide-based top-k *AICJumpers* detection algorithm is omitted for space limitation.

Complexity Analysis. When τ, μ and M are not large, which is reasonable in reality, the time and space complexity for both algorithms are linear in the number of objects in V^l. More details are omitted for space limitation.

4   Experiment

Synthetic Datasets. We generate synthetic datasets to show the efficiency of our algorithms: the *Top-k AICJumpers Detection algorithm* (TAD) and the *Slide-based Top-k AICJumpers Detection algorithm* (STAD) by varying three major cost factors: the window size τ, the slide size μ and the snapshot size $|V^l|$ which is the number of objects in V^l. As the number of different communities is usually small in reality, we set $M = 20$. ε and k can be set as different values to require different groups of top-k *AICJumpers*. We fix $\varepsilon = 0.5$ and $k = 500$ only for illustration and it would not influence the efficiency evaluation.

• Varying Snapshot Size. As shown in Fig. 2, when the snapshot size increases, the CPU time and peak memory for both algorithms increase as well because it needs to build bit vectors and compute *JumpScores* for a larger number of objects in each window. STAD is a little superior than TAD with respect to the averaged CPU time in Fig. 2(b) by applying the slide-based strategy to reduce the repeated computation. STAD outperforms TAD in terms of memory consumption in Fig. 2(c) as the former only stores two bit vectors for each object.

• Varying Window Size. In Fig. 3(a) and (b), when the window size increases, the CPU time of both algorithms increase as more bit vectors are built and processed for *JumpScore* computation. In Fig. 3(c), the memory consumption for TAD increases more significantly than STAD when the window size increases as the latter only stores two bit vectors in the memory for each object. Note that it runs out of memory for TAD when the window size is 1000.

• Varying Slide Size. In Fig. 4(a), the CPU time of the first window for both algorithms is nearly unchanged as the window size is unchanged. When the slide size increases, more new bit vectors need to be built, so the averaged CPU time of other subsequent windows for both algorithms increases in Fig. 4(b). When $\frac{\mu}{\tau}$ is large enough like 50%, the averaged CPU time of both algorithms is nearly the same as the effect of the slide-based strategy degrades. In Fig. 4(c), the memory cost for TAD is almost unchanged as it always stores τ bit vectors for each object. For STAD, when $\frac{\mu}{\tau}$ increases from 2% to 25%, the memory cost decreases rapidly

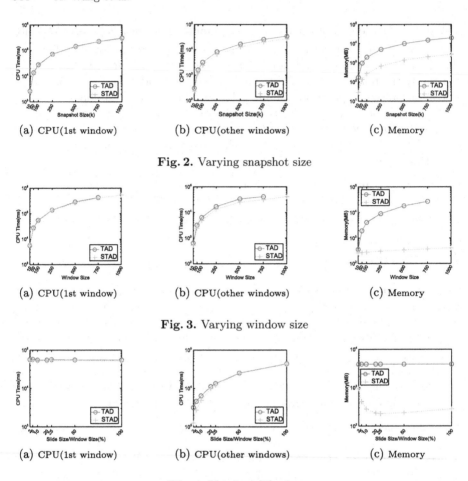

Fig. 2. Varying snapshot size

Fig. 3. Varying window size

Fig. 4. Varying slide size

since the number of slides for each window declines so that fewer intermediate $F(o)$ and $R(o)$ results need to be kept in the memory for final combination. When $\frac{\mu}{\tau}$ increases from 25% to 100%, the memory cost increases slowly as the slide size becomes larger and more bitwise operation results need to be stored.

Case Study. We perform experiments with a real dataset to justify that the top-k *AICJumpers* returned by our algorithms are significant. The dataset is a subset of DBLP for 7 research areas: architecture and systems (AS), computer networks (CN), security and privacy (SP), data management (DM), graphics and multimedia (GM), artificial intelligence (AI) and Ubiquitous Computing (UC) for 2000–2015. The number of communities is 7 and each community consists of several top tier conferences with papers and authors. The data of each year is modeled as a snapshot. An author belongs to a research area if he or she wrote a paper published in a conference classified into the area. We can obtain which communities each author belongs to by simply preprocessing.

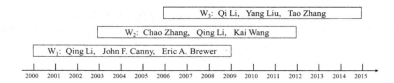

Fig. 5. Top-3 *AICJumpers* for each window

Setting $\tau = 10$, $\mu = 3$ and $\varepsilon = 0.5$, it only takes hundreds of milliseconds for each window to return top-3 *AICJumpers* from 99232 authors using our algorithms. All the results shown in Fig. 5 are manually verified to be multi-field researchers or topic-oriented authors.

5   Conclusion

In this paper, we propose the concept of *AICJumpers* in dynamic information networks. Such objects frequently change communities over time. By encoding the community changing trajectory of each object as several bit vectors and transforming the complex *JumpScore* computation process into bitwise *and*, *or*, *xor* operations, we propose a high time-efficiency top-k *AICJumpers* detection algorithm with a sliding window model. To reduce the memory overhead and repeated computation time, a slide-based strategy is also introduced. Experiments on several synthetic datasets show the high performance of the proposed algorithms. Case studies on the DBLP dataset reveal some interesting and meaningful *AICJumpers*.

Acknowledgements. This work is partially supported by the Key Research and Development Plan of National Ministry of Science and Technology under grant No. 2016YFB1000703.

References

1. Bu, Y., Chen, L., Fu, A.W.C., Liu, D.: Efficient anomaly monitoring over moving object trajectory streams. In: Proceedings of the 15th ACM SIGKDD International Conference on Knowledge Discovery and Data Mining, pp. 159–168. ACM (2009)
2. Gao, J., Liang, F., Fan, W., Wang, C., Sun, Y., Han, J.: On community outliers and their efficient detection in information networks. In: Proceedings of the 16th ACM SIGKDD International Conference on Knowledge Discovery and Data Mining, pp. 813–822. ACM (2010)
3. Ge, Y., Xiong, H., Zhou, Z.H., Ozdemir, H., Yu, J., Lee, K.C.: TOP-EYE: top-k evolving trajectory outlier detection. In: Proceedings of the 19th ACM International Conference on Information and Knowledge Management, pp. 1733–1736. ACM (2010)
4. Gupta, M., Gao, J., Han, J.: Community distribution outlier detection in heterogeneous information networks. In: Blockeel, H., Kersting, K., Nijssen, S., Železný, F. (eds.) ECML PKDD 2013. LNCS (LNAI), vol. 8188, pp. 557–573. Springer, Heidelberg (2013). https://doi.org/10.1007/978-3-642-40988-2_36

5. Gupta, M., Gao, J., Sun, Y., Han, J.: Community trend outlier detection using soft temporal pattern mining. In: Flach, P.A., De Bie, T., Cristianini, N. (eds.) ECML PKDD 2012. LNCS (LNAI), vol. 7524, pp. 692–708. Springer, Heidelberg (2012). https://doi.org/10.1007/978-3-642-33486-3_44
6. Gupta, M., Gao, J., Sun, Y., Han, J.: Integrating community matching and outlier detection for mining evolutionary community outliers. In: Proceedings of the 18th ACM SIGKDD International Conference on Knowledge Discovery and Data Mining, pp. 859–867. ACM (2012)
7. Lee, J.G., Han, J., Li, X.: Trajectory outlier detection: a partition-and-detect framework. In: IEEE 24th International Conference on Data Engineering, 2008, ICDE 2008, pp. 140–149. IEEE (2008)
8. Sun, Y., Han, J., Aggarwal, C.C., Chawla, N.V.: When will it happen?: relationship prediction in heterogeneous information networks. In: Proceedings of the Fifth ACM International Conference on Web Search and Data Mining, pp. 663–672. ACM (2012)
9. Sun, Y., Tang, J., Han, J., Gupta, M., Zhao, B.: Community evolution detection in dynamic heterogeneous information networks. In: Proceedings of the Eighth Workshop on Mining and Learning with Graphs, pp. 137–146. ACM (2010)
10. Sun, Y., Yu, Y., Han, J.: Ranking-based clustering of heterogeneous information networks with star network schema. In: Proceedings of the 15th ACM SIGKDD International Conference on Knowledge Discovery and Data Mining, pp. 797–806. ACM (2009)
11. Xu, X., Yuruk, N., Feng, Z., Schweiger, T.A.: SCAN: a structural clustering algorithm for networks. In: Proceedings of the 13th ACM SIGKDD International Conference on Knowledge Discovery and Data Mining, pp. 824–833. ACM (2007)

Sequence and Temporal Data Processing

Distributed In-Memory Analytics for Big Temporal Data

Bin Yao[1,2], Wei Zhang[1,3], Zhi-Jie Wang[4], Zhongpu Chen[1], Shuo Shang[5(✉)],
Kai Zheng[6(✉)], and Minyi Guo[1]

[1] Department of Computer Science and Engineering, Shanghai Jiao Tong University,
Shanghai, China
[2] Guangdong Province Key Laboratory of Big Data Analysis and Processing,
Guangzhou, China
[3] Guangdong Province Key Laboratory of Popular High Performance Computers,
Guangzhou, China
[4] School of Data and Computer Science, Sun Yat-sen University, Guangzhou, China
[5] Extreme Computing Research Center,
King Abdullah University of Science and Technology, Mecca, Saudi Arabia
shuo.shang@kaust.edu.sa
[6] School of Computer Science and Engineering,
University of Electronic Science and Technology of China, Chengdu, China
zhengkai@uestc.edu.cn

Abstract. The temporal data is ubiquitous, and massive amount of temporal data is generated nowadays. Management of big temporal data is important yet challenging. Processing big temporal data using a distributed system is a desired choice. However, existing distributed systems/methods either cannot support native queries, or are disk-based solutions, which could not well satisfy the requirements of high throughput and low latency. To alleviate this issue, this paper proposes an In-memory based Two-level Index Solution in Spark (ITISS) for processing big temporal data. The framework of our system is easy to understand and implement, but without loss of efficiency. We conduct extensive experiments to verify the performance of our solution. Experimental results based on both real and synthetic datasets consistently demonstrate that our solution is efficient and competitive.

Keywords: Big temporal data · Distributed in-memory analytics
Apache Spark · Temporal queries

1 Introduction

Temporal data management has been studied tens of years and has gained increasingly interest recently [17,26], due to its wide applications. For example, users may wish to investigate the demographic information of an administrative region (e.g., California) at a specific time (e.g., five years ago). Querying a historical version of the database (like above) is usually referred to as time

© Springer International Publishing AG, part of Springer Nature 2018
J. Pei et al. (Eds.): DASFAA 2018, LNCS 10827, pp. 549–565, 2018.
https://doi.org/10.1007/978-3-319-91452-7_36

travel [5,11,28]. As another example, in the quality assurance department users may wish to analyze how many orders are delayed as a function of time, thereby querying all historical versions of the database over a certain time period. Queries like mentioned above are usually called temporal aggregation [10,19,20].

In the literature, there are already a large bulk of papers addressing the problems of time travel and temporal aggregation queries (see e.g., [5,11,20,21, 25,28]). Yet, most of prior works focused on developing single-machine-based solutions, and few attention has been made on developing distributed solutions for handling big temporal data. Nowadays, various *apps*, e.g., web apps and Internet of things (IoT) apps, generate massive amount of temporal data. It is urgently needed to efficiently process big temporal data. In particular, it is challenging to handle such a large volume of temporal data in traditional database systems, since the limited computing ability of a single-machine based system. Clearly, processing such a large volume of temporal data using a distributed system should be a good choice. Recently, distributed temporal analytics for big data have been also investigated (e.g., [9,39]). These works share at least two common features: (i) they are distributed *disk-based* temporal analytics; and (ii) time travel and temporal aggregation queries are not covered in their papers. With the surging data size, these solutions could not well meet the demand of high throughput and low latency.

Spark SQL [37] is such an engine, which extends Spark (a fast distributed *in-memory* computing engine) to enable us to query the data with SQL interface inside Spark programs. To support distributed in-memory analytics for big temporal data with high throughput and low latency, this paper proposes an In-memory based Two-level Index Solution in Spark (ITISS). To the best of our knowledge, none of existing big data systems (e.g., Apache Hadoop, Apache Spark) provide native support for temporal data queries, and none of prior works develop distributed in-memory based solution for processing time travel and temporal aggregation over big temporal data. To summarize, the main contributions of our work are as follows:

- We propose a distributed in-memory analytics framework for big temporal data. Our framework is easy to understand and implement, but without loss of efficiency.
- We present targeted algorithms for answering time travel and temporal aggregation queries, by fully utilizing the proposed framework that adopts a two-level index structure.
- We implement our framework in Apache Spark, and extend the Apache Spark SQL to support declarative SQL interface that enables users to perform temporal queries with a few lines of SQL statements.
- We conduct a comprehensive experimental evaluation for our proposed solution, using both real and synthetic temporal data. The experimental results consistently demonstrate the efficiency and competitiveness of our proposal.

The rest of this paper is organized as follows. Section 2 formulates our problem. Section 3 presents our proposed framework for big temporal data, including a distributed indexing structure, the query procedures, and the implementation details based on Apache Spark. We present the experimental evaluation in Sect. 4. Section 5 reviews prior works most related to ours, and Sect. 6 concludes this paper.

2 Problem Definition

Specifically, this paper attempts to achieve two representative operations (i.e., *time travel* and *temporal aggregation*) over *temporal data* in distributed environments. Nevertheless, our framework and algorithms described later can be easily extended to support other operations (e.g., *temporal join*) and other data (e.g., *bitemporal data* [7]). In what follows, we formally define our problems. (For ease of reference, Table 1 lists the frequently used notations.)

Let D be a temporal dataset containing $|D|$ temporal records $\{t_1, t_2, \ldots, t_{|D|}\}$. Each record t_i ($i \in [1, |D|]$) is a quadruple in the form of ($key, value, start, end$), where key corresponds to the id of the record, $start$ and end are the starting and ending timestamps of a time interval during which the record is alive. Further, given a version (or timestamp) v and a record t_i, we say that record t_i exists in version v (i.e., record t_i is alive in version v), if and only if $v \in [t_i.start, t_i.end)$.

Time travel establishes a consistent view for the history of a database, and it is one of the most significant temporal operations in temporal databases. Here we address two widely used time travel operations, i.e., *time travel exact-match query* and *time travel range query*. Both of the operations can support querying the past version of a database. Their major difference is that the input of *exact-match query* uses a specific value, while the input of *range query* uses a given range [5]. Specifically, their formal definitions are as formulated below.

Definition 1 (Time travel exact-match query). *Given a time travel exact-match query $Q_e = \{key, v\}$, we are asked to retrieve the record (denoted as θ) from D such that,*

$$\theta = \{t_i \in D \mid t_i.key = key \wedge t_i.start \leq v \wedge v < t_i.end\}.$$

Table 1. Frequently used notations

Notation	Description
D	A temporal dataset
t_i	The i-th temporal record of D
I_p	A partition interval
Q_e	Time travel exact-match query
Q_r	Time travel range query
Q_a	Temporal aggregation query
g	A temporal aggregation operator, e.g. *SUM*, *MAX*

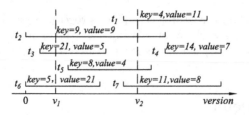

Fig. 1. Temporal aggregation

As an example, consider a simple temporal database with 7 temporal records as shown in Fig. 1. When $Q_e = \{21, v_1\}$, the query returns t_3; in contrast, when $Q_e = \{21, v_2\}$, the query returns $\varnothing$.

Definition 2 (Time travel range query). *Given a time travel range query $Q_r = \{start\_key, end\_key, v\}$, we are asked to retrieve a set θ of records from D such that,*

$$\theta = \{t_i \in D \mid start\_key \le t_i.key \wedge t_i.key \le end\_key \wedge t_i.start \le v \wedge v < t_i.end\}.$$

As an example (see also Fig. 1), when $Q_r = \{7, 22, v_1\}$, the query returns $\{t_2, t_3\}$; in contrast, when $Q_r = \{7, 22, v_2\}$, the query returns $\{t_2, t_5, t_7\}$.

Temporal aggregation is a common operation in temporal database, and usually is challenging and expensive. Since temporal aggregation was introduced by [21], it has been heavily studied. In this paper we focus on aggregation (e.g., MAX, SUM) conducted at a specific timestamp. Formally, the temporal aggregation operation is defined as follows.

Definition 3 (Temporal aggregation query). *Given a temporal aggregation query $Q_a = \{g, v\}$ where g is an aggregation operator such as MAX, we are asked to return an aggregate value (denoted as θ) based on D such that,*

$$\theta = g\{t_i \in D \mid t_i.start \le v \wedge v < t_i.end\}.$$

Consider also the example shown in Fig. 1. When $Q_a = \{MAX, v_1\}$, the query returns 21 (since $max\{9, 21, 5\} = 21$); in contrast, when $Q_a = \{SUM, v_1\}$, the query returns 32 (since $4 + 9 + 8 + 11 = 32$).

Note that, compared with prior works, in this paper our focus is on big temporal data in distributed environments. As discussed in Sect. 1, a straightforward implementation based on existing distributed systems is inefficient and ineffective; in the next section we present our solution in detail.

3 Our Solution

In this section, we first describe the distributed processing framework. Then, we show how to achieve time travel and temporal aggregation queries based on the proposed framework. Finally, we discuss the implementation details of deploying the framework on the classic distributed computing engine — Apache Spark.

3.1 System Framework

At a high level, our framework consists of three parts: (i) Partition unit. It is responsible for partitioning all data into distributed (slave) nodes. Usually, we should guarantee that each node has roughly the same size of data, in order to keep the *load balance*. (ii) Local index unit. Within each partition, the local indexes are maintained to avoid a "full" scanning, and so may help us boost the query efficiency. In addition, each partition also maintains a *partition interval* (explained later) for the global index construction. And (iii) global index unit. In the master node a global index is designed to prune "unpromising" partitions in advance. This can avoid checking each (individual) partition, and thus may help us reduce the CPU cost and/or network transmission cost. In our design, the master node collects all partition intervals from each partition in slave nodes, and then builds the global index based on the collected partition intervals. The architecture of our framework is shown in Fig. 2. It is easy to understand that our framework adopts a two-level indexing structure, which can avoid visiting irrelevant candidates (e.g., partitions and local records) as much as possible. Although the rational behind the framework is simple, it is definitely efficient as demonstrated later. In what follows, we discuss important issues in each unit.

▶ *Partition Method.* Typically, load balance is a desired goal when partitioning the general data. As to the temporal data, another desired goal is to minimize the overlap of partition intervals. To achieve these goals, in our design we partition the temporal data by interval (known as *range partition*). As an example, assume one wants to partition six temporal records, shown in Fig. 3(a), into two partitions P_1 and P_2. He/she can first sort these temporal records by their intervals, obtaining the sorted records $(t_3, t_2, t_6, t_4, t_5, t_1)$. To balance the size of each partition, he/she can evenly split the sorted records into two. As a result, P_1 contains

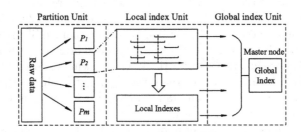

Fig. 2. The architecture of our system framework

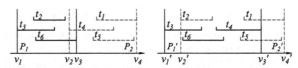

(a) Range partition method (b) Hash partition method

Fig. 3. Different partition methods

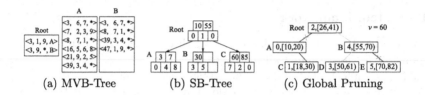

Fig. 4. Indexes used in our system

first three records (t_3, t_2, t_6), and correspondingly P_2 contains (t_4, t_5, t_1). This way, the partition interval of P_1 is $[v_1, v_3)$, and that of P_2 is $[v_2, v_4)$. In particular, the interval overlap of P_1 and P_2 is $v_3 - v_2$, which is the minimum overlap.

Note that, although using *hash* to partition the data is widely used for other data domain such as streaming data (since the data can be evenly allocated via this manner), it could be not appropriate for the context of our concern. The major reason is that partitioning in such a way could cause many overlaps (among partition intervals). For example, consider the temporal data shown in Fig. 3(b). After finishing hash partition, P'_1 contains (t_3, t_4, t_6) and P'_2 consists of (t_1, t_2, t_5). One can easily see that the interval overlap of P'_1 and P'_2 is $v'_3 - v'_2$, which is much larger than that of P_1 and P_2.

▶ *Local Index Method.* As mentioned earlier, the local index is used to manage the temporal data in each partition. In the literature, there are already on-shelf index structures to support time travel queries such as *multiversion B-tree* [5] and *time-index* [11]. In our paper, we use multiversion B-tree (shorted as MVB-Tree) as an example. For ease of understanding, Fig. 4(a) shows this index structure. The first entry of the root points to its leaf child A, which contains all the records that are alive from version 1 to 9 (excluded). In the leaf nodes, each entry represents a record, where ∗ means that this record is still alive now.

Also, there are already existing index structures (e.g., [29,35]) to support temporal aggregation queries. Here we use the index (named SB-Tree) developed in [35] as an example. The SB-Tree node is composed of two arrays, as illustrated in Fig. 4(b). One of the arrays stores the intervals, which is used for pointing to the children nodes, and another stores the aggregate values. To calculate an aggregation using the SB-Tree, one can search the tree from the root to the leaf, and aggregate the values in its path.

Note that, although this paper adopts the MVB-Tree and SB-Tree, it is not compulsory to use these indexes. In other words, other on-shelf indexes, or more powerful indexes developed in the future can be also used in our framework.

▶ *Global Index Method.* As discussed previously, the global index manages the partition intervals. Since each partition interval is a pair of version numbers, and is comparable by starting value and length of the interval, naturally we can use the binary search tree to maintain partitions' interval information. Note that, for each partition in slave nodes, there are many *time intervals (of records)*. Nevertheless, we only use one *partition interval* for a partition. To understand the partition interval, consider a simple example with three time intervals $\{[u_1, u_2), [u_3, u_4), [u_5, u_6)\}$ in a partition. Then, the partition interval is $[min\{u_1, u_3, u_5\}, max\{u_2, u_4, u_6\})$. This way, each partition interval in the global index essentially corresponds to a partition in slave nodes. This implies that, in the query processing, if a partition interval can be pruned, then the corresponding partition can be pruned safely. Based on this intuition, in our design each node in the global tree maintains a key-value pair $<I_p, id>$, where I_p and id refer to the partition interval and its corresponding partition, respectively.

3.2   Query Processing

The query evaluation in our framework consists of two phases: (i) global pruning, and (ii) local look-up.

The first phase essentially is to fully utilize the global index and the version v (in the query input) to prune "unrelated" partitions. To understand, consider an example shown in Fig. 4(c). Assume one wants to prune partitions that does not belong to version 60, he/she can traverse the global index to examine the partition interval. As a result, only two partitions ($id = 3$ and $id = 4$) can be regarded as the candidates. In contrast, the second phase mainly retrieves, in each candidate partition, the "qualified" records, based on the local indexes and part of query inputs. As an example, consider Fig. 4(a) and assume a time travel exact-match query $Q_e = \{key = 8, v = 6\}$; the local look-up first finds the entry that belongs to version 6 at the root node. Then, it checks the child A, in which we can find an entry with $key = 8$, and its valid time interval is $[1, *)$ containing 6. This completes the local look-up. In what follows, we cover detailed query algorithms for time travel and temporal aggregation queries.

▶ *Time Travel Queries.* We first discuss the time travel exact-match query, followed by the time travel range query. Algorithm 1 shows the pseudo-codes of the *time travel exact-match query*. Note that, Line 2 is used to perform global pruning, detailed in Algorithm 2. After finishing the global pruning at the master node, we obtain the ids of candidate partitions, which are stored in P. Then, the

---

Algorithm 1. ExactMatchQuery (key, v)

```
 1  R ← ∅
 2  P ← GlobalPruning(v, r_g)        // r_g is the root of global index
 3  foreach p in P do
 4  │   root ← r_l                   // r_l is the root of local index
 5  │   while root is not leaf do
 6  │   │   root ← child of root whose route directs to key and v
 7  │   end while
 8  │   if key exists in root then
 9  │   │   add record containing key to R
10  │   end if
11  end foreach
12  return R
```

Algorithm 2. GlobalPruning $(v,\ root)$

---

1 $R \leftarrow \varnothing$
2 **if** $root \neq null$ **then**
3   **if** $v \in root.I_p$ **then**
4   add $root.id$ to R
5   **end if**
6   GlobalPruning(v, $root.left$)
7   GlobalPruning(v, $root.right$)
8 **end if**
9 **return** R

Algorithm 3. RangeQuery $(start\_key, end\_key, v, root)$

---

1 $R \leftarrow \varnothing$
2 $P \leftarrow$ GlobalPruning(v, r_g)   // r_g is the root of global index
3 **foreach** p *in* P **do**
4   **if** $root$ *is not leaf* **then**
5          $start\_c \leftarrow$ *child* of $root$ whose route directs to $start\_key$ and v
6          $end\_c \leftarrow$ *child* of $root$ whose route directs to end_key and v
7          $children \leftarrow$ all children between $start\_c$ and end_c
8   **foreach** *node in* $children$ **do**
9   RangeQuery($start\_key, end\_key, v, node$)
10   **end foreach**
11   **else if** *key exists in* $root$ **then**
12   add *record* containing *key* to R
13   **end if**
14 **end foreach**
15 **return** R

---

local look-up (Lines 3–11) retrieves the results in each partition; here local look-ups for all these candidate partitions are distributed to the cluster and executed in parallel. Note that, the algorithm for *time travel range query* is similar to Algorithm 1. The difference is that, we do not need to find the *child* for the given key (Line 6). Instead, we maintain an array named *children* that can direct to [$start\_key, end\_key$], and then examine each node in *children*. More details are shown in Algorithm 3.

▶ *Temporal Aggregation Queries.* When processing the temporal aggregation queries, the global pruning process is the same with that for the time travel queries. Yet, the local look-up phase works in a different way. In brief, in each candidate partition, it first finds the *child* of the *root* so that the interval contains version v. If *child* is a leaf node, we just return the aggregate value (denoted as r) in it. Otherwise, we recursively find the aggregate value (denoted as s) of v in *child*, and return the aggregate value of r and s. The pseudo-codes are shown in Algorithm 4.

Algorithm 4. TemporalAggregation $(g, v, root)$

---

1 $P \leftarrow \text{GlobalPruning}(v, r_g)$
2 **foreach** p *in* P **do**
3     $child \leftarrow child$ of $root$ which satisfies $v \in child.interval$
4   **if** *child is leaf* **then**
5 | return $child.value$
6   **else**
7 | return $g(child.value, TemporalAggregation(g, v, child)$
8   **end if**
9 **end foreach**

3.3 Implementation on Apache Spark

In Apache Spark the resilient distributed dataset (RDD) is fault-tolerant and can be stored in memory to support fast data reusing without accessing disk. In this section, we elaborate how to implement our framework in Apache Spark.

To support partition method suggested in Sect. 3.1, we extend Spark's **RangePartitioner**. Note that, Spark's **RangePartitioner** is developed for the general purpose data partition; it cannot effectively support *partition by interval*. To achieve this function, we implement the comparison procedure for interval data format, and integrate it to Spark **RangePartitioner**.

As to the implementation of global index in Spark, we first collect all the partition intervals distributed in the slaves, and then we build a binary search tree as the global index in the master node. The implementation of local indexes in Spark is basically different from the above. One can easily know that RDD is the basic abstraction in Spark, and it represents a partitioned collection of elements that can be operated in parallel. Meanwhile, a partition wraps its dataset records according to its partitioner. Particularly, we observe that RDD is designed for sequential access. This incurs that one cannot build indexes over RDDs *directly*. To deploy the local indexes over RRDs, we use a method suggested in [34]. In brief, we first load all the temporal records (in a partition) into memory, and then construct the local index structure; afterwards, the memory (used to store the original temporal data) is released, and we persist the local index in memory to support subsequent queries.

In addition, it would be nice to enable users to write concise SQL statements to support analytics for big temporal data. Yet, in Apache Spark there is no corresponding SQL commands. To this end, we develop new Spark SQL operations/commands to support temporal data analytics. Several major changes are as follows.

• We design a new keyword "**VERSION**" to support temporal operations with SQL statements. This new keyword can help us reinterpret the **AS OF** subclause inherited from SQL Server, endowing it with the new meaning by modifying the SQL plan in the Spark SQL engine. Specifically, **FOR VERSION AS OF** *version_number* means specifying a *version_number*, where **VERSION** is just the newly introduced keyword. For instance, users can use the following SQL statements to execute a time travel exact-match query mentioned in Sect. 2.

SELECT * FROM D **WHERE** key = '9'
FOR VERSION AS OF v_2.

- In order to manage indexes for temporal data, we also develop index management SQL statement. Users can specify the index structure by using **USE** *index_type*, where *index_type* is the keyword for a specific index name (e.g., MVB-TREE, SBTREE). For example, to create a SB-tree index called "sbt" for table D, one can use the following SQL commands:

CREATE INDEX sbt **ON** D **USE** SBTREE.

---

4   Experiments

4.1   Experiment Setup

In our experiments, we use both real and synthetic datasets described as follows. The real dataset **SX-ST** is extracted from a temporal network on the website Stack Overflow [24]. The network has 2.6 million nodes representing users, and 63 million edges in form of (u, v, t), where u and v are the ids of source and target users respectively, and t is the interaction time between these two users. Specifically, we extract users who interacted with others more than once. And we treat each of these users as a record, in which two consecutive interaction timestamps of the user are regarded as the interval of the record, and the value of the record is the total number of interactions related to the users. This gives us about 0.4 million records. Following the schema of SX-ST, we also generate the synthetic dataset, shorted as **SYN**. Specifically, in SYN the starting timestamp of a record is generated randomly, and the length of the interval is uniformly distributed between the minimum and maximum length of that in SX-ST. The size of SYN ranges from 1 million to 4 billion (i.e., $[10^6, 4 \times 10^9]$) records, taking from 32 MB up to 166 GB disk space. The default setting is 5×10^8 records.

To measure the performance of our system, we adopt two widely-used evaluation metrics: (i) runtime (i.e., query latency) and (ii) throughput. To obtain the runtime, we repeatedly perform 10 queries for each test case, and calculate the average value. On the other hand, the throughput is evaluated as the number of queries performed *per* minute. Additionally, we also examine the performance of indexes used in our system.

We compare our system with two baselines: (i) a Naive In-memory based Solution on Spark (**NISS**). It partitions all temporal records randomly using the default method in Spark, and stores the data in memory of the distributed system. These partitions are collected and managed via RDD, which allows us to manipulate the data in parallel. To achieve temporal queries, NISS uses predicates (e.g., *WHERE* predicate) provided by Spark SQL, to launch a scanning on the data. By checking each record according to the condition presented in the query input, NISS can obtain the query result. For example, when an aggregation query with MAX operator is detected, NISS checks each partition in parallel. For each partition, it scans the whole partition and determines the "max" value of all the records which are alive in version v. Finally, it collects all "local" max values from partitions and finds the "global" max value. And (ii) a distributed disk-based solution named **OcRT**, which is extended from OceanRT [39]. Note that, OceanRT employs a hashing of temporal data blocks according to the temporal attributes of records; this behaviour essentially serves as a global index. In our baseline, we implement this hashing process by grouping the starting value of intervals to form a partition. In addition, OceanRT runs multiple computing units on one physical node and connects these units using Remote Direct Memory Access (RDMA); this behaviour is roughly the same with the executors in Apache Spark. More importantly, our adapted solution OcRT stores the data on disks, which is the same with that in OceanRT.

All experiments are conducted on a cluster containing 5 nodes with dual 10-core Intel Xeon E5-2630 v4 processors @ 2.20 GHz and 256 GB DDR4 RAM. All these nodes are connected to a Gigabit Ethernet switch, running Linux operating system (Kernel 4.4.0-97) with Hadoop 2.6.5 and Spark 1.6.3. One of these 5 nodes is selected as the *master* and the remaining 4 machines are *slaves*. The configuration is totally 960 GB main memory and 144 virtual cores in our cluster, which is deployed in standalone mode. In our experiments, the size of HDFS block is 128 MB. The default partition size (a.k.a., the size of each partition) contains 10^5 records. The fanout of local index(es) is set to 100.

4.2 Experimental Results

Figure 5 investigates the index cost of our system. For the local indexes, the construction time of SB-Tree (SBT) is much faster than that of MVB-Tree (MVBT),

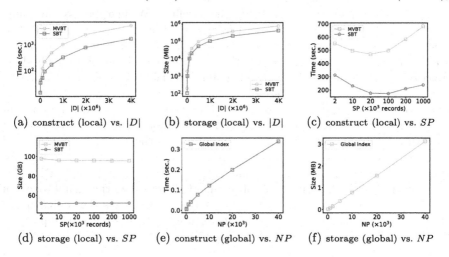

(a) construct (local) vs. $|D|$ (b) storage (local) vs. $|D|$ (c) construct (local) vs. SP

(d) storage (local) vs. SP (e) construct (global) vs. NP (f) storage (global) vs. NP

Fig. 5. Index construction time and storage overhead vs. $|D|$, SP and NP

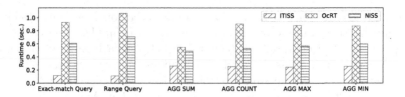

Fig. 6. Time travel and temporal aggregation queries on the SX-ST dataset

as shown in Fig. 5(a). This is mainly because MVBT requires *node copy* and has about 2 times of operations (e.g., insertion and deletion) than SB-Tree. Even so, the indexing time is acceptable. For example, indexing 4 billion records using MVBT takes only 1.54 h. As we expected, Fig. 5(b) shows that indexing storage overhead increases with the size of dataset. Besides, we also show the results by varying the size of partition (SP); see Fig. 5(c) and (d). It can be seen that there is a non-linear relationship between SP and the index construction time (cf., Fig. 5(c)). This is mainly because the index construction time is influenced by not only the size of each partition but also the total number of partitions. In our experiments, the "good" partition size falls in the range from $20K$ to $200K$ records. This is essentially why we choose $SP = 100K$ as the default setting (recall Sect. 4.1). Note that, an appropriate choice on the number of partitions and the size of each partition can both improve system throughput and query latency. Meanwhile, we can see that SP makes less impact on the index size (cf., Fig. 5(d)). This further shows that the index size is mainly related to the dataset size $|D|$. On the other hand, one can see that the construction of the global index is very fast; about 330 ms even if NP is set to the largest value (cf., Fig. 5(e)). This is mainly because the global index size is very small, e.g., only about 3 MB even when $NP = 40K$ (cf., Fig. 5(f)). In addition, as we expected, the global index size is strictly proportional to NP.

Next, we compare our method with the baselines. We first discuss the results on the SX-ST dataset. It can be seen from Fig. 6 that the execution of NISS is slow, although it also stores the data in-memory. This is mainly because the full scan over the dataset in partitions is time-consuming. As to OcRT, the hashing process can perform partition pruning, but the lack of local index makes it slow, since it needs in-partition full scanning. The reason why OcRT is slower than NISS could be due to two points: (i) OcRT is disk-based solution; and (ii) the partition pruning effect of OcRT is weak when it is confronted with relatively small dataset like SX-ST. Compared to the baselines, our method takes only about 0.3 s for temporal aggregation queries, and less than 0.2 s for time travel. It is about 3× faster than NISS, and 4× faster than OcRT. This demonstrates the competitiveness of our method. On the other hand, one can see that different aggregation queries (e.g, SUM, MAX) have the similar query cost. In what follows, when we discuss aggregation queries, we mainly report the SUM aggregation query results for saving space.

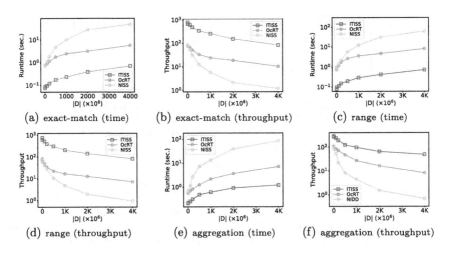

Fig. 7. Time travel and temporal aggregation queies on the SYN dataset

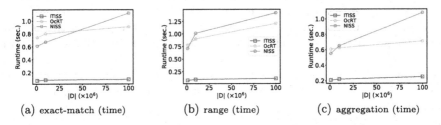

Fig. 8. An enlarged drawing. Here $|D|$ ranges from 1×10^6 to 100×10^6 (Color figure online)

Figure 7 covers the comparison results on the synthetic (SYN) data, which is much larger than the SX-ST dataset. For time travel exact-match queries, one can easily see from Fig. 7(a) that our solution is 3–7 times faster than OcRT. Our solution outperforms NISS about one order of magnitude on both *runtime and throughput* (cf., Fig. 7(a) and (b)) when dataset size $|D|$ ranges from 10^6 to 4×10^9 records; especially, it outperforms NISS near to two orders of magnitude when $|D| = 4 \times 10^9$. This essentially demonstrates the superiorities of our solution. Also, we can see that the performance of our framework drops much slower than that of others, which essentially shows us that our framework has much better scalability. This is mainly because the partition pruning in our framework is much more powerful on larger datasets. Another interesting phenomenon is that, OcRT here is obviously better than NISS (cf., Fig. 7(a), (c), and (e)), while it is inferior to NISS in the previous test (cf., Fig. 6). This is mainly because SX-ST is relatively small, compared to SYN. Figure 8 well explains this phenomenon (see the crossing point between the red and blue lines).

As we expected, when we execute the time travel range queries (cf., Fig. 7(c) and (d)), our solution presents the similar performance, compared against the

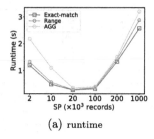
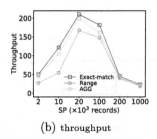

(a) runtime   (b) throughput

Fig. 9. Temporal operations vs. SP

exact-match queries. For example, the running time for both queries is close and has the similar growth tendency. On the other hand, for temporal aggregation queries, one can see from Fig. 7(e) that, the runtime of aggregation query is a little longer than that of time travel operations. This is mainly because it needs to check many more records. Similarly, in Fig. 7(f), the throughput of the aggregation query has the similar characteristics.

Figure 9 shows the impact of partition size SP on the performance of temporal queries. We can see from Fig. 9(a) that, the "good" partition size for both time travel and temporal aggregation queries is between 20K and 100K records. Meanwhile, it can be seen from Fig. 9(b) that the throughput is even more sensitive to partition size. This shows the significance of number of partitions in distributed systems.

5   Related Work

In the field of temporal databases, prior works addressed various issues related to temporal data (see several representative surveys [17,18,30]).

In the literature, most of early works concentrated on semantics of time [6], logical modelling [33] and query languages [4] for temporal data. Recently, some researchers addressed the problem of discovering/mining interesting information [27] from temporal data, such as trend analysis [15] and data clustering [36]. Other works addressed query or search issues for temporal data, such as top-k queries [26] and membership queries [22]. Some optimal problems related to temporal data are also investigated, such as finding optimal splitters for large temporal data [23]. Similar to general databases, in temporal databases join operation is also a common operation; researches on this topic can be found in [13]. Since temporal data is involved with an evolving process, researchers have attempted to model evolutionary traces [32], and to trace various elements in temporal databases, such as tracing evolving subspace clusters [16]. The aforementioned works are related to ours (since these works also handle temporal data). Yet, it is not hard to see that they are clearly different from ours, since our work focuses on time travel and temporal aggregation queries, instead of the above problems such as trend analysis and logical modeling.

Nevertheless, there are already existing works addressing the problems of time travel [1,3,5,11,20,28,31] and temporal aggregation [10,11,20,21,25,38] queries. For example, Kaufmann *et al.* [20] proposed a unified data structure called timeline index for processing queries on temporal data, in which they use column storage to mange temporal data. General-purpose temporal index structures can be found in [5,11]. Furthermore, SAP HANA [12] provides a basic form of time travel queries based on restoring a snapshot of a past transaction. ImmortalDB [28] is another system that supports time travel queries. From industry perspective, database vendors, such as Oracle [3], IBM [31], Postgres [1] and SQL Server [2], also integrate time travel queries into theirs systems. On the other hand, Kline and Snodgrass [21] introduced the first algorithm for computing temporal aggregation on constant intervals. Later, algorithm for temporal aggregation based on AVL Trees was proposed [8]. Furthermore, temporal aggregation with range predicates [38], or over extreme cases such as null time intervals [10], are also investigated. Attempts for temporal aggregation with a multiprocessor machine can be found in [20,25]. Efficient indexing structures supporting temporal aggregation are discussed in [11,29,35]. A major feature of the aforementioned proposals or systems is that, they focused on single-machine-based solutions, while few attention has been made on developing distributed solutions for handling big temporal data.

Essentially, we also realize that, distributed temporal analytics for big data have been also investigated in recent years [9,39]. And they are different from the early work [14] (in which the data being processed is relatively small). Nevertheless, these works share at least two common features: (i) they are distributed disk-based temporal analytics instead of distributed in-memory based temporal analytics; and (ii) time travel and temporal aggregation queries are not covered in their papers. Thus, they are different from our work.

6   Conclusion

In this paper we suggested a distributed in-memory analytics framework for big temporal data and implemented it on Spark. Our framework used a two-level index structure to enhance the pruning power. It also provided declarative SQL query interface that enables users to perform typical temporal operations with a few lines of SQL statements. We conducted extensive experiments to demonstrate the superiorities of our solution. In the future, we plan to extend this framework to support more temporal queries.

Acknowledgments. This work was supported by the National Basic Research Program (973 Program, No. 2015CB352403), the NSFC (U1636210, 61729202, 91438121, 61672351, 61472453, U1401256, U1501252, U1611264, U1711261 and U1711262), the National Key Research and Development Program of China (2016YFB0700502), the Scientific Innovation Act of STCSM (15JC1402400), the Opening Projects of Guangdong Key Laboratory of Big Data Analysis and Processing (201808), Guangdong Province Key Laboratory of Popular High Performance Computers of Shenzhen University (SZU-GDPHPCL2017), and the Microsoft Research Asia.

References

1. Postgres 9.2 highlight - range types. http://paquier.xyz/postgresql-2/postgres-9-2-highlight-range-types
2. Temporal Tables. https://docs.microsoft.com/en-us/sql/relational-databases/tables/temporal-tables
3. Workspace Manager Valid Time Support. https://docs.oracle.com/cd/B28359_01/appdev.111/b28396/long_vt.htm#g1014747
4. Ahn, I., Snodgrass, R.: Performance evaluation of a temporal database management system. In: SIGMOD (1986)
5. Becker, B., Gschwind, S., Ohler, T., Seeger, B., Widmayer, B.: An asymptotically optimal multiversion B-tree. VLDBJ (1996)
6. Bettini, C., Wang, X.S., Bertino, E., Jajodia, S.: Semantic assumptions and query evaluation in temporal databases. In: SIGMOD (1995)
7. Bliujute, R., Jensen, C.S., Saltenis, S., Slivinskas, G.: R-tree based indexing of now-relative bitemporal data. In: VLDB (1998)
8. Böhlen, M., Gamper, J., Jensen, C.S.: Multi-dimensional aggregation for temporal data. In: Ioannidis, Y., Scholl, M.H., Schmidt, J.W., Matthes, F., Hatzopoulos, M., Boehm, K., Kemper, A., Grust, T., Boehm, C. (eds.) EDBT 2006. LNCS, vol. 3896, pp. 257–275. Springer, Heidelberg (2006). https://doi.org/10.1007/11687238_18
9. Chandramouli, B., Goldstein, J., Duan, S.: Temporal analytics on big data for web advertising. In: ICDE (2012)
10. Cheng, K.: On computing temporal aggregates over null time intervals. In: Benslimane, D., Damiani, E., Grosky, W.I., Hameurlain, A., Sheth, A., Wagner, R.R. (eds.) DEXA 2017. LNCS, vol. 10439, pp. 67–79. Springer, Cham (2017). https://doi.org/10.1007/978-3-319-64471-4_7
11. Elmasri, R., Wuu, G.T., Kim, Y.J.: The time index: an access structure for temporal data. In: VLDB (1990)
12. Färber, F., et al.: The SAP HANA database-an architecture overview. IEEE Data Eng. Bull. (2012)
13. Gao, D., Jensen, S., Snodgrass, R.T., Soo, D.: Join operations in temporal databases. VLDBJ (2005)
14. Gendrano, J.A.G., Huang, B.C., Rodrigue, J.M., Moon, B., Snodgrass, R.T., Parallel algorithms for computing temporal aggregates. In: ICDE (1999)
15. Gollapudi, S., Sivakumar, D.: Framework and algorithms for trend analysis in massive temporal data sets. In: CIKM (2004)
16. Günnemann, S., Kremer, H., Laufkötter, C., Seidl, T.: Tracing evolving subspace clusters in temporal climate data. DMKD **24**, 387–410 (2012)
17. Gupta, M., Gao, J., Aggarwal, C.C., Han, J.: Outlier detection for temporal data: a survey. TKDE (2014)
18. Jensen, C.S., Snodgrass, R.T.: Temporal data management. TKDE (1999)
19. Kaufmann, M., Fischer, P.M., May, N., Ge, C., Goel, A.K., Kossmann, D.: Bitemporal timeline index: a data structure for processing queries on bi-temporal data. In: ICDE (2015)
20. Kaufmann, M., Manjili, A.A., Vagenas, P., Fischer, P.M., Kossmann, D., Färber, F., May, N.: Timeline index: a unified data structure for processing queries on temporal data in SAP HANA. In: SIGMOD (2013)
21. Kline, N., Snodgrass, R.T.: Computing temporal aggregates. In: ICDE (1995)
22. Kollios, G., Tsotras, V.J.: Hashing methods for temporal data. TKDE (2002)

23. Le, W., Li, F., Tao, Y., Christensen, R.: Optimal splitters for temporal and multi-version databases. In: SIGMOD (2013)
24. Leskovec, J., Krevl, A.: SNAP datasets: stanford large network dataset collection (2014). http://snap.stanford.edu/data
25. Leung, T.C., Muntz, R.R.: Temporal query processing and optimization in multi-processor database machines. In: VLDB (1992)
26. Li, F., Yi, K., Le, W.: Top-k queries on temporal data. VLDBJ (2010)
27. Loglisci, C., Ceci, M., Malerba, D.: A temporal data mining framework for ana-lyzing longitudinal data. In: Hameurlain, A., Liddle, S.W., Schewe, K.-D., Zhou, X. (eds.) DEXA 2011. LNCS, vol. 6861, pp. 97–106. Springer, Heidelberg (2011). https://doi.org/10.1007/978-3-642-23091-2_9
28. Lomet, D., et al.: Transaction time support inside a database engine. In: ICDE (2006)
29. Ramaswamy, S.: Efficient indexing for constraint and temporal databases. In: Afrati, F., Kolaitis, P. (eds.) ICDT 1997. LNCS, vol. 1186, pp. 419–431. Springer, Heidelberg (1997). https://doi.org/10.1007/3-540-62222-5_61
30. Roddick, J.F., Spiliopoulou, M.: A survey of temporal knowledge discovery paradigms and methods. TKDE (2002)
31. Saracco, C.M., et al.: A matter of time: temporal data management in DB2 10. Technical report, IBM (2012)
32. Wang, P., Zhang, P., Zhou, C., Li, Z., Yang, H.: Hierarchical evolving Dirichlet processes for modeling nonlinear evolutionary traces in temporal data. DMKD **31**, 32–64 (2017)
33. Wang, X.S., Jajodia, S., Subrahmanian, V.: Temporal modules: an approach toward federated temporal databases. In: SIGMOD (1993)
34. Xie, D., Li, F., Yao, B., Li, G., Zhou, L., Guo, M.: Simba: efficient in-memory spatial analytics. In: SIGMOD (2016)
35. Yang, J., Widom, J.: Incremental computation and maintenance of temporal aggre-gates. In: ICDE (2001)
36. Yang, Y., Chen, K.: Temporal data clustering via weighted clustering ensemble with different representations. TKDE (2011)
37. Zaharia, M., Chowdhury, M., Das, T., Dave, A., Ma, J., McCauley, M., Stoica, I.: Resilient distributed datasets: a fault-tolerant abstraction for in-memory cluster computing. In: NSDI (2012)
38. Zhang, D., Markowetz, A., Tsotras, V.J., Gunopulos, D., Seeger, B.: On computing temporal aggregates with range predicates. TODS (2008)
39. Zhang, S., Yang, Y., Fan, W., Lan, L., Yuan, M.: OceanRT: real-time analytics over large temporal data. In: SIGMOD (2014)

Scalable Active Constrained Clustering for Temporal Data

Son T. Mai[1(✉)], Sihem Amer-Yahia[1], Ahlame Douzal Chouakria[1],
Ky T. Nguyen[1], and Anh-Duong Nguyen[2]

[1] CNRS, Univ. Grenoble Alpes, Grenoble, France
{mtson,sihem.amer-yahia,ahlame.douzal,
trung-ky.nguyen}@univ-grenoble-alpes.fr
[2] University of Rennes 1, Rennes, France
duongnguyenhumg@gmail.com

Abstract. In this paper, we introduce a novel interactive framework to handle both instance-level and temporal smoothness constraints for clustering large temporal data. It consists of a constrained clustering algorithm, called $CVQE+$, which optimizes the clustering quality, constraint violation and the historical cost between consecutive data snapshots. At the center of our framework is a simple yet effective active learning technique, named *Border*, for iteratively selecting the most informative pairs of objects to query users about, and updating the clustering with new constraints. Those constraints are then propagated inside each data snapshot and between snapshots via two schemes, called *constraint inheritance* and *constraint propagation*, to further enhance the results. Experiments show better or comparable clustering results than state-of-the-art techniques as well as high scalability for large datasets.

Keywords: Semi-supervised clustering · Active learning
Interactive clustering · Incremental clustering · Temporal clustering

1 Introduction

In semi-supervised clustering, domain knowledge is typically encoded in the form of instance-level *must-link* and *cannot-link* constraints [9] for aiding the clustering process, thus enhancing the quality of results. Such constraints specify that two objects must be placed or must not be placed in the same clusters, respectively. Constraints have been successfully applied to improve clustering quality in real-world applications, e.g., identifying people from surveillance cameras [9] and aiding robot navigation [8]. However, current research on constrained clustering still faces several major issues described below.

Most existing approaches assume that we have a set of constraints beforehand, and an algorithm will use this set to produce clusters [2,8]. Davidson et al. [6] show that the clustering quality varies significantly using different equi-size

© Springer International Publishing AG, part of Springer Nature 2018
J. Pei et al. (Eds.): DASFAA 2018, LNCS 10827, pp. 566–582, 2018.
https://doi.org/10.1007/978-3-319-91452-7_37

sets of constraints. Moreover, annotating constraints requires human intervention, an expensive and time consuming task that should be minimized as much as possible given the same expected clustering quality. Therefore, how to choose a *good* and *compact* set of constraints rather than randomly selecting them from the data has been the focus of many research efforts, e.g., [1,15,19].

Many approaches employ different *active learning* schemes to select the most meaningful pairs of objects and then query experts for constraint annotation [1,15]. By allowing the algorithms to choose constraints themselves, we can avoid insignificant ones, and expect to have high quality and compact constraint sets compared to the randomized scheme. These constraints are then used as input for constrained clustering algorithms to operate. However, if users are not satisfied with the results, they are asked to provide another constraint set and start the clustering again, which is obviously time consuming and expensive.

Other algorithms follow a *feedback* schema which does not require a full set of constraints in the beginning [5]. They iteratively produce clusters with their available constraints, show results to users, and get feedback in the form of new constraints. By iteratively refining clusters according to user feedback, the acquired results fit users' expectations better [5]. Constraints are also easier to select with an underlying cluster structure as a guideline, thus reducing the overall number of constraints and human annotation effort for the same quality level. However, exploring the whole data space for finding meaningful constraints is also a non-trivial task for users.

To reduce human effort, several methods incorporate *active learning* into the feedback process, e.g., [13–15,19]. At each iteration, the algorithm automatically chooses pairs of objects and queries users for their feedback in terms of *must-link* and *cannot-link* constraints instead of leaving the whole clustering results for users to examine. Though these active feedback techniques are proven to be very useful in real-world tasks such as document clustering [13], they suffer from very high runtime since they have to repeatedly perform clustering as well as exploring all $O(n^2)$ pairs of objects to generate queries to users each time.

In this paper and its preliminary version [18], we develop an efficient framework to cope with the above problems following the iterative active learning approach as in [13,19]. However, instead of examining all pairs of objects, our technique, called *Border*, selects a small set of objects around cluster borders and queries users about the most uncertain pairs of objects. We also introduce a constraint inheritance approach based on the notion of μ-nearest neighbors for inferring additional constraints, thus further boosting performance. Finally, we revisit our approach in the context of evolutionary clustering [4]. Evolutionary clustering aims to produce high quality clusters while ensuring that the clustering does not change dramatically between consecutive timestamps. This scheme is very useful in many application scenarios. For example, doctors want to track groups patients based on their treatment progresses each year. They may expect that existing groups do not change much over time if there are minor changes in the data. However, the clustering process should be able to reflex the changes if there are significant differences in the new data. Therefore, we propose to formulate a temporal smoothness constraint into our framework and add a time-fading factor to our constraint propagation.

Contributions. Our contributions are summarized as follows:

- We introduce a new algorithm CVQE+ that extends CVQE [8] with weighted must-link and cannot-link constraints and a new object assignment scheme.
- We propose a new algorithm, Border, that relies on active clustering and constraint inheritance to choose a small number of objects to solicit user feedback for. Beside the active selection scheme for pairs of objects, Border employs a constraint inheritance method for inferring more constraints, thus further enhancing the performance.
- We present an evolutionary clustering framework which incorporates instance-level and temporal smoothness constraints for temporal data. To the best of our knowledge, our algorithm is the first framework that combines active learning, instance-level and temporal smoothness constraints.
- Experiments are conducted for six real datasets to demonstrated the performance of our algorithms over state-of-the-art ones.

Outline. The rest of the paper is organized as follows. We formulate the problem in Sect. 2. Our framework is described in Sect. 3. Experiments are presented in Sect. 4. Section 5 discusses related works. Section 6 concludes the paper.

2   Problem Formulation

Let $D = \{(d, t)\}$ be a set of $|D|$ vectors $d \in \mathbb{R}^p$ observed at time t. Let $S = \{(S_s, D_s, ts_s, te_s)\}$ be a set of preselected $|S|$ data snapshots. Each S_s starts at time ts_s, ends at time te_s and contains a set of objects $D_s = \{(d, t) \in D \mid ts_s \leq t < te_s\}$. Two snapshots S_s and S_{s+1} may overlap but must satisfy the time order, i.e., $ts_s \leq ts_{s+1}$ and $te_s \leq te_{s+1}$. For each snapshot S_s, let $ML_s = \{(x, y, w_{xy}) \mid (x, y) \in D_s^2\}$ and $CL_s = \{(x, y, w_{xy}) \mid (x, y) \in D_s^2\}$ be the set of *must-link* and *cannot-link* constraints of S_s with a degree of belief of $w_{xy} \in [0, 1]$. Initially, ML_s and CL_s can be empty.

In this paper, we focus on the problem of grouping objects in all snapshots into clusters. Our goals are (1) reduce the number of constraints thus reducing the constraint annotation costs (2) make the algorithm scale well with large datasets and (3) smooth the gap between clustering results of two consecutive snapshots, i.e., ensure temporal smoothness.

3   Our Proposed Framework

Figure 1 illustrates our framework which relies on two algorithms, Border and CVQE+. Our framework starts with a small (or empty) set of constraints in each snapshot. Then, it iteratively produces clustering results and receives refined

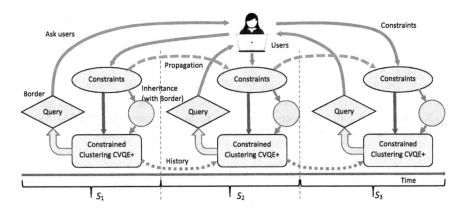

Fig. 1. Our active temporal clustering framework

constraints from users in the next iterations. This process is akin to feedback-driven algorithms for enhancing clustering quality and reducing human annotation effort [5]. However, instead of passively waiting for user feedback as in [5], our algorithm, Border, *actively* examines the current cluster structure, selects β pairs of objects whose clustering labels are the least certain, and asks users for their feedback in terms of instance-level constraints. Examining all possible pairs of objects to select queries is time consuming due the quadratic number of candidates. To ensure scalability, Border limits its selection to a small set of most promising objects. When there are new constraints, instead of reclustering from scratch as in [13,19], our algorithm, CVQE+, incrementally updates the cluster structures for saving computation times. We also aim to ensure a smooth transition between consecutive clusterings [4]. We additionally introduce two novel concepts: (1) the *constraint inheritance* scheme for automatically inferring more constraints inside each snapshot and (2) the *constraint propagation* scheme for propagating constraints between different snapshots. These schemes help significantly reduce the number of constraints that users must enter into the systems for acquiring a desired level of clustering quality by automatically adding more constrained based on the annotated ones. To the best of our knowledge, Border is the first framework that combines active learning, instance-level and temporal smoothness constraints.

3.1 Constrained Clustering Algorithm

For each snapshot S_s, we use constrained kMeans for grouping objects. Generally, any existing techniques such as MPCK-Means [2], CVQE [8] or LCVQE [17] can be used. Here we introduce CVQE+, an extension of CVQE [8] to cope with weighted constraints, to do the task.

The New Algorithm CVQE+. Let $C = \{C_i\}$ be a set of clusters. The cost of C_i is defined as its vector quantization cost VQE_i and the constraint violation costs ML_i and CL_i (where $ML_i \subseteq ML$ and $CL_i \subseteq CL$ are the sets of *must-link*

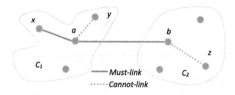

Fig. 2. Assigning a pair of constrained object (a, b) to clusters in CVQE+. All constraints starting and ending at a and b will be considered

and *cannot-link* constraints related to C_i) as follows. Note that, our ML_i cost is symmetric compared to [8].

$$Cost_{C_i} = Cost_{VQE_i} + Cost_{ML_i} + Cost_{CL_i} \qquad (1)$$

$$Cost_{VQE_i} = \sum_{x \in C_i} (c_i - x)^2$$

$$Cost_{ML_i} = \sum_{(a,b) \in ML_i \wedge vl(a,b)} w_{ij}(c_i - c_{\pi(a,b,i)})^2$$

$$Cost_{CL_i} = \sum_{(a,b) \in CL_i \wedge vl(a,b)} w_{ij}(c_i - c_{\varphi(i)})^2$$

where, $vl(a, b)$ is true for (a, b) that violates *must-link* or *cannot link* constraints, c_i is the center of cluster C_i, $\pi(a, b, i)$ returns the center of clusters of a or b (not including cluster C_i), and $\varphi(i)$ returns the nearest cluster center of C_i. Note that, $Cost_{ML_i}$ is symmetric compared to [8].

Taking the derivative of $Cost_{C_i}$, the new center of C_i is updated as:

$$\frac{dCost_{C_i}}{dc_i} = \frac{dCost_{VQE_i}}{dc_i} + \frac{dCost_{ML_i}}{dc_i} + \frac{dCost_{CL_i}}{dc_i} \qquad (2)$$

$$c_i = \frac{\displaystyle\sum_{x \in C_i} x + \sum_{(a,b) \in ML_i \wedge vl(a,b)} w_{ij}C_{\pi(a,b,i)} + \sum_{(a,b) \in CL_i \wedge vl(a,b)} w_{ij}C_{\varphi(i)}}{\displaystyle |C_i| + \sum_{(a,b) \in ML_i \wedge vl(a,b)} w_{ij} + \sum_{(a,b) \in CL_i \wedge vl(a,b)} w_{ij}}$$

For each constraint (a, b), CVQE+ assigns objects to clusters by examining all k^2 cluster combinations for a and b like CVQE. The major difference is that when we calculate the violation cost, we consider all constraints starting and ending at a and b instead of only the constraint (a, b) as in CVQE [7] or LCVQE [7], which is very sensitive to the cost change when some constraints share the same objects (changing these objects affects all their constraints) as illustrated in Fig. 2. The assigning cost for (a, b) will include the violation costs for $(a, x), (a, y)$ and (b, z) as well. Thus, this scheme is expected to improve the clustering quality of CVQE+ compared to CVQE and LCVQE.

Complexity Analysis. Let n be the number of objects, m be the number of constraints, k be the number of clusters. CVQE+ has time complexity

$O(rkn + rk^2m^2)$ which is higher than $O(rkn + rk^2m)$ of CVQE due to the fact that all related constraints must be examined while assigning a constraint, where r is the number of iterations of the algorithm. Since k and m are constants, CVQE+ is thus has linear time complexity to the number of objects n. It also require $O(n)$ space for storing objects and constraints.

3.2   Active Constraint Selection

We introduce an active learning method called Border for selecting pairs of objects and query users for constraint types. The general idea is examining objects lying around borders of clusters since they are the most uncertain ones and choosing a block of β pairs of objects to query users until the query budget δ is reached. Here, β and δ are predefined constants.

Active Learning with Border. To avoid examining all pairs of objects, Border chooses a subset of $m = min(O(\sqrt{n}), M)$ objects located at the boundary of the clusters as the main targets since they are the most uncertain ones, where M is a predefined constant (default as 100). This bound limits the number of pairs as a constant, thus reducing the number of pairs needed for examining in the subsequence steps. For each object a in cluster C_i, the border score of a is defined as:

$$bor(a) = \frac{(a - c_i)^2}{(a - c_{\varphi(i)})^2(1 + ml(a))(1 + cl(a))} \tag{3}$$

where $ml(a)$ and $cl(a)$ are the sums of weights of must and cannot-link constraints of a. Here, we favor objects that have fewer constraints for increasing constraint diversity. This also fits well with our constraint inheritance scheme. Moreover, by considering the distance to the second nearest cluster center $c_{\varphi(i)}$, we focus more on objects that are close to the boundaries of two clusters rather than ones that are far away from other clusters, which may not bring much benefit to clarify the groups. For each cluster C_i, we select $m|C_i|/n$ top objects based on their border score distribution in C_i. This can be done by building a histogram with $O(\sqrt{|C_i|})$ bins (a well-known rule of thumb for the optimal histogram bin) [3]. Then, objects are taken sequentially from the outermost bins. This scheme ensures that all clusters are considered based on their current sizes. Bigger clusters contribute more objects than smaller ones since their changes will more likely affect the final clustering result. Moreover, by using histogram bins, we give equal changes to objects within a bin since these objects might have the same importances for clarifying the clusters.

For each selected object a, we estimate the uncertainty of a w.r.t. the current clustering result as:

$$sco(a) = ent(\mu nn(a)) + \frac{vl(ml(a)) + vl(cl(a))}{ml(a) + cl(a) + 1} \tag{4}$$

where $ent(\mu nn(a))$ is the entropy of class labels of μ nearest neighbors of a and $vl(ml(a))$ and $vl(cl(a))$ are the sums of violated must-link and cannot-link

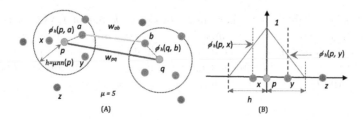

Fig. 3. (A) Constraint inheritance from (p, q) to (a, b). (B) The effect of the object b on its neighbors

constraints of a. A high $score(a)$ means that a is in high uncertain areas with different mixed class labels and a high number of constraint violations. And thus, it should be focused on.

We divide $m^2 = O(n)$ pairs of selected objects into two sets: the set of inside cluster pairs X and between cluster pairs Y, i.e., for all $(x, y) \in X : label(x) = label(y)$ and for all $(x, y) \in Y : label(x) \neq label(y)$. For a pair $(x, y) \in X$, it is sorted by $val(x, y) = \frac{(x-y)^2(1+sco(x))(1+sco(y))}{(1+ml(x)+cl(x))(1+ml(y)+cl(y))}$. For $(x, y) \in Y$, $val(x, y) = \frac{(x-y)^2(1+ml(x)+cl(x))(1+ml(y)+cl(y))}{(1+sco(x))(1+sco(y))}$. The larger val is, the more likely x and y belong to different clusters and vice versa. Moreover, in Y, we tend to select pairs with more related constraints to strengthen the current clusters, while we try to separate clusters in X by considering pairs with fewer related constraints. We choose top $\beta/2$ non-overlapped largest val pairs of X and top $\beta/2$ non-overlapped smallest pairs of Y in order to maximize the changes in clustering results (inside and between clusters). To be concrete, if a pair (a, b) was chosen, all pairs starting and ending with a or b will not be considered for enhancing the constraint diversity, which can help to bring up better performance. If all pairs are excluded, we select the remainder randomly.

We show β pairs to users to ask for the constraint type and add their feedback to the constraints set and update clusters until the total number of queries exceeds a predefined budget δ as illustrated in Fig. 1.

Constraint Inheritance in Border. For further reducing the number of queries to users, the general idea is to infer new constraints automatically based on annotated ones. Our inheritance scheme is based on the concept of μ nearest neighbors below.

Let h be the distance between an object p and its μ nearest neighbors. The influence of p on its neighbor x is formulated by a triangular kernel function $\phi_h(p, x)$ centered at p as in Fig. 3. Given a constraint (p, q, w_{pq}), for all $a \in \mu nn(p)$ and $b \in \mu nn(q)$, we add (a, b, w_{ab}) to the constraints set, where w_{ab} is defined as:

$$w_{ab} = w_{pq}\phi_h(p, a)\phi_h(q, b) \tag{5}$$

The general intuition is that the label of an object a tends to be consistent with its closest neighbors (which is commonly used in data classification such as nearest neighbor classification [12]). This scheme is expected to increase the clustering quality, especially when combined with the active learning approach of the algorithm Border described above.

During the inheritance scheme, if a pair of objects (a, b) is inherited from two constraints (c, d) and (e, f) with inherited weights w_1 and w_2, respectively, its weight and type are determined as follows:

$$w_{ab} = \begin{cases} max(w_1, w_2) \ if \ type(c, d) = type(e, f) \\ |w_1 - w_2| \ otherwise \end{cases} \tag{6}$$

where $type(c, d)$ is the constraint type of (c, d) (either *must-link* or *cannot-link*). And $type(a, b)$ is determined by $type(c, d)$ if $w_1 > w_2$ and vice versa. The general idea here is that if (a, b) is influenced by two constraints with different kinds, it will follow the one with the highest influence. Note that, if (a, b) belongs to the main constraint set, we exclude it from the constraint inheritance scheme since we consider it as annotated by users and thus it is confident.

Updating Clusters. At each iteration, instead of performing clustering again for updating the clustering result with the new set of constraints, we propose to update it incrementally for saving runtime. To do so, we only need to take the old cluster centers and update them following Eq. 1 with the updated constraints set. The intuition behind this is that new constraints is more likely to change clusters locally. Thus, starting from the current state might make the algorithm converges faster. In Sect. 4, we show that this updating scheme acquire the same quality but converge much faster than re-clustering from scratch.

Complexity Analysis. Similarly to CVQE+, Border has $O(n)$ time and space complexity at each iteration and thus has $O(\delta n/\beta)$ time complexity overall, where δ is the budget limitation and β is the number of selected objects at each iteration described above.

3.3 Temporal Smoothness Constraints

The general idea of temporal smoothness [4] is that clusters not only have high quality in each snapshot but also do not change much between sequential time frames. It is useful in many applications where the transition between different snapshots is smoothed for consistency.

Temporal Smoothness. To ensure the smoothness, we re-define the cost of cluster C_i of snapshot S_s in Eq. 1 by enforcing a historical cost from its previous snapshot as follows:

$$TCost_{VQE_i} = (1 - \alpha)Cost_{VQE_i} + \alpha Hist(C_i, S_{s-1}) \tag{7}$$

where $Hist(C_i, S_{s-1})$ is the historical cost of cluster C_i between two snapshots S_s and S_{s-1} and α is a regulation factor to balance the current clustering quality and the historical cost. This cost keeps the new clusters do not deviate too much from clusters from previous snapshot while performing clustering. We define the historical cost as follows:

$$Hist(C_i, S_{s-1}) = (c_i - \psi(C_i, S_{s-1}))^2 \qquad (8)$$

where $\psi(C_i, S_{s-1})$ returns the closest cluster center to C_i in snapshot S_{s-1}. Obviously if two clustering results are too different, indicated by high historical cost, the penalty will be higher thus focing the algorithm to lower down the overall cost by creating clusters closer to those of the previous snapshot.

Taking the derivation of Eq. (7) as in Eq. (1), we can update the cluster centers as follows:

$$c_i = \frac{(1 - \alpha)A + \alpha\psi(C_i, S_{s-1})}{(1 - \alpha)B + \alpha} \qquad (9)$$

where A and B are respectively the numerator and the denominator given in Eq. 2 for updating clusters.

Constraint Propagation. Whenever we have a new constraint (x, y, w_{xy}) in snapshot S_s, we propagate it to snapshots $S_{s'}$ where $s' > s$ if $x, y \in S'$. The intuition is that if x and y are linked (either by must or cannot-link) in S_s, they are more likely to be linked in $S_{s'}$. Thus we add the constraint (x, y, w'_{xy}) to $S_{s'}$ where:

$$w'_{xy} = w_{xy}\frac{te_s - ts_{s'}}{te_{s'} - ts_s} \qquad (10)$$

where $(te_s - ts_{s'})/(te_{s'} - ts_s)$ is a time fading factor. This scheme helps to increase the clustering quality by putting more constraints into the clustering algorithm like the inheritance scheme. Since propagated constraints are not considered as user annotated ones, we treat it as non-confident constraints in our model and will not build offspring for them like those in the main constraint set described in the inheritance scheme above.

4 Experiments

Experiments are conducted on a workstation with 4.0 Ghz CPU and 32 GB RAM using Java. We use 6 datasets Iris, Ecoli, Seeds, Libras, Optdigits, and Wdbc acquired from the UCI archives[1]. The numbers of clusters k are acquired from the ground truths. Constraint queries are also simulated from the ground truths by adding a *must-link* if two objects have the same labels or a *cannot-link* if they have different labels. We use Normalized Mutual Information (NMI) [16] for assessing the clustering quality. NMI score is in $[0, 1]$ where 1 means a perfect clustering result compared to the ground truth and vice versa. All results are averaged over 10 runs.

---

[1] http://archive.ics.uci.edu/ml/.

4.1   Constrained Clustering

Performance of CVQE+. Figure 4 shows comparisons among CVQE+ and existing techniques including kMeans, MPCK-Means [2], CVQE [8] and LCVQE [17] over different sets of randomly selected constraints. CVQE+ consistently outperforms or acquires comparable results to CVQE and others for most datasets (except the Libras dataset where it is outperformed by LCVQE), especially when the number of constraints is large. This can be explained by the way CVQE+ assigns objects to clusters. By considering all related constraints while assigning cluster labels for objects, it can better optimize the overall cost function, thus leading to better clustering quality. Compared to its predecessor algorithm CVQE or LCVQE, it deals well with constraint overlap (constraints that share the same objects), which increases with the number of constraints. Note that, when the constraint set is empty, CVQE+ produces clustering in the similar way with k-Means. Thus, the clustering quality does not start from 0.

Noise Robustness. For studying the effect of noisy constraints on CVQE+, we randomly choose some constraints and change them from *must-link* to *cannot-link* and vice versa. Figure 5 shows the clustering quality of different algorithms w.r.t. the percentages of noisy constraints from 2% to 8% for real datasets. As we can see, for all algorithms, when the number of noisy constraints increases, the clustering quality decreases accordingly. However, CVQE+ tends to be more affected by noise than its related techniques CVQE and LCVQE. Though its point assignment scheme helps to increase the clustering quality as discussed above, it makes CVQE+ more sensitive to noise since one noisy constraint will affect the assignment cost for all of its related constraints. Nevertheless, in our experimented data, CVQE+ still acquires better (or equivalent) clustering results than CVQE and LCVQE under the same noisy conditions in most

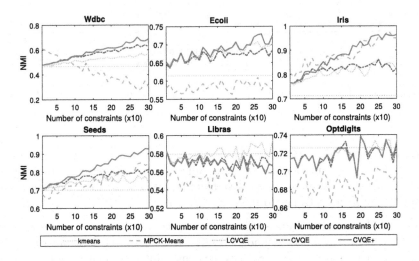

Fig. 4. Performance of CVQE+ compared to others

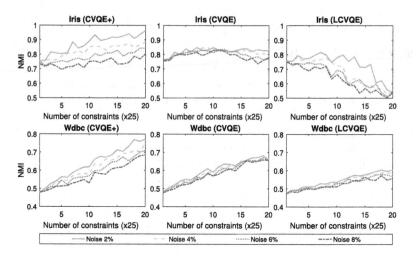

Fig. 5. Effect of noisy constraints on CVQE+ (left), CVQE (middle) and LCVQE (right) for the Iris and Wdbc datasets

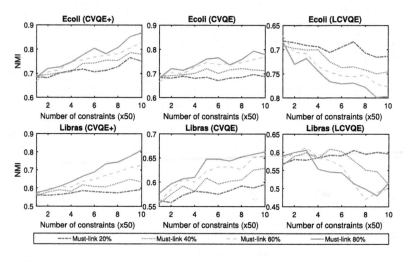

Fig. 6. Effect of constraint types on CVQE+ (left), CVQE (middle) and LCVQE (right) for the Ecoli and Libras datasets

cases as seen in Fig. 5. However, we only use maximum 500 constraints in our experiments. If the number of noisy constraints become bigger, CVQE+ may not be the winner. Developing an effective algorithm to cope with noisy constraints is thus an interesting target to pursue.

Effect of Constraint Types. Figure 6 shows the performance of CVQE+ and its related techniques CVQE and LCVQE when the number of *must-link* constraints increases from 20% to 80% of the constraint sets. The clustering quality of CVQE+ and CVQE increases with the number of *must-link* constraints, while that of LCVQE decreases. This can be explained by the ways they calculate the constraint violation costs for the *must-link* and especially the *cannot-link* constraints. LCVQE treats violated *cannot-link* constraints more properly than CVQE and CVQE+. Thus, it deals well with higher number of those constraints.

4.2 Active Constraint Selection

We study the performance of Border in comparison with other state-of-the-art active learning techniques. Unless otherwise stated, the budget limitation δ is set to 200, the query size $\beta = 10$ and the neighborhood size $\mu = 4$.

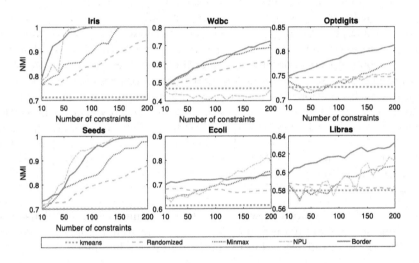

Fig. 7. Comparison among different active learning techniques

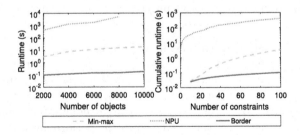

Fig. 8. Runtimes of different techniques

Active Constraint Selection. Figure 7 shows comparisons between Border, NPU [19], Huang [19] (a modified version of [13] for working with non-document data), Min-max [15], Explorer-Consolidate [1], and a randomized method (Huang and Consolidate are removed from Fig. 7 for readability). Border acquires better results than others on Libras, Wdbc and Optidigits, comparable results on Iris and Ecoli. For the Seeds dataset, it is outperformed by NPU. The difference is because Border tends to strengthen existing clusters by fortifying both the cluster borders and inter connectivity for groups of objects rather than connecting a single object to existing components like NPU and Huang. Moreover, since it iteratively studies the clustering results for selecting constraints, it has better performance than non-iterative methods like Consolidate and Min-max.

Runtime Comparison. For studying the runtime of Border on large-scale datasets, we create five synthetics datasets of sizes 2000 to 10000 consisting of 5 Gaussian clusters and measure the time for acquiring 100 constraints. The results are shown in Fig. 8. Border is orders of magnitude faster than other methods in selecting pairs to query. For 1000 objects, it takes Border 0.1 s while NPU and Min-max need 439.4 and 3.0 s, which is 4394 and 30 times slower than Border. For 10000 objects, Border, NPU and Min-max consumes 0.18, 5216.3 and 18.2 s, respectively. It is due to the fact that Border does not evaluate all pairs of objects at each iteration. Thus, it does much less works than others and faster. Besides, NPU and Min-max are implemented in Matlab which is slower than Border in Java. Nevertheless, the higher the number of objects and constraints, the higher the runtime differences. For 10000 objects, Border is around 28979.4 and 101.1 times faster than NPU and Min-max, respectively. Hence, its runtime performance makes Border an effective technique to cope with very large datasets.

Cluster Update. Figure 9 shows the NMI and the number of iterations of our algorithm for the Ecoli dataset. The NMI scores are comparable, while it takes fewer iterations for our algorithm to converge in its update mode.

Effect of the Block Size β. Figure 10 shows the performance of Border when the query block size β varies from 10 to 30. As we can see, the smaller the value of β is, the better the performance of Border since the cluster structure is assessed more frequently, thus leading to better constraints to be selected at each iteration.

Effect of the Constraint Inheritance Scheme. Figure 11 shows the effect of the parameter μ on our algorithm Border via the inheritance scheme. Typically, its performance will increase with μ until it reaches the peak and then decreases as shown for the dataset Iris. This can be explained by the neighborhood influence scheme of Border. When μ is large enough, the number of wrong constraints will be increased, thus lower down the performance of Border. However, the peak value of μ is actually dataset dependence and thus is very hard to predict. Taking the dataset Optdigits as an example, the performance of Border still increases when $\mu = 5$. However, with $\mu = 3$, Border starts perform worse on the dataset Seeds. Unfortunately, the value of μ is highly data dependent and is hard to select. In our experiments, we observe that the value of μ around 2 to 4 is overall good for most datasets. Thus, we choose $\mu = 4$ as a default value.

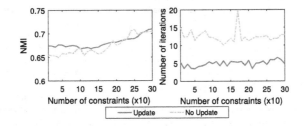

Fig. 9. Update vs. fully reclustering for the Ecoli dataset

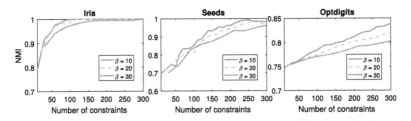

Fig. 10. The effect of the query block size β on the performance of Border

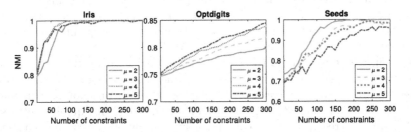

Fig. 11. The effect of the neighborhood size μ on the performance of Border

4.3   Temporal Clustering

For studying the temporal clustering result, we divide the datasets into different snapshots and measuring the clustering quality using the ground truths provided for the full datasets.

Temporal Clustering. Figure 12 shows the active temporal clustering results for three snapshots of the Optdigits dataset (we set $\alpha = 0.5$). As we see, our active learning scheme can help boost clustering quality inside each snapshot compared to the original kMeans or a randomized constraint selection method. With the constraint propagation scheme (Border-Propagation), the clustering results are further boosted compared to Border. For example, in Snapshot 2 and 3, Border-propagation performs much better than Border without the constraint propagation scheme. Since we only consider forward propagation, the clustering result in Snapshot 3 will be more affected than Snapshot 2 and Snapshot 1.

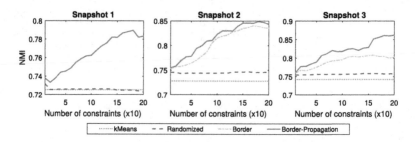

Fig. 12. Temporal clustering on the dataset Optdigits

For example, in Snapshot 3 the difference between Border and Border-Propagation is much higher than in Snapshot 2. In case of interest, we can easily extend the algorithm for a backward propagation scheme.

5   Related Work

Constraint Clustering. There are many proposed constrained clustering algorithms such as MPC-kMeans [2], CVQE [8] and LCVQE [17]. These techniques optimize an objective function consisting of the clustering quality and the constraint violation cost like our algorithm CVQE+. CVQE+ is an extension of CVQE [8], where we extend the cost model to deal with weighted constraints, make the must-link violation cost symmetric and change the way each constraint is assigned to clusters by considering all of its related constraints. This makes cluster assignment more stable, thus enhancing the clustering quality. Interested readers are referred to [7] for a comprehensive survey on constrained clustering methods.

Active Learning. Most existing techniques employ *active learning* for acquiring a desired constraints set before or during clustering. In [1], the authors introduce the Explorer-Consolidating algorithm to select constraints by exploiting the connected-components of must-link ones. Min-max [15] extends the Consolidation phase of [1] by querying most uncertain objects rather than randomly selecting them. These techniques produce constraints sets before clustering. Thus, they cannot exploit the cluster labels for further enhancing performance. Huang et al. [13] introduce a framework that iteratively generates constraints and updates clustering results until a query budget is reached. However, it is limited to a probabilistic document clustering algorithm. NPU [19] also uses connected-components of must-link constraints as a guideline for finding most uncertain objects. Constraints are then collected by querying these objects again existing connected components like the Consolidate phase of [1]. Though more effective than pre-selection ones, these techniques typically have a quadratic runtime which makes them infeasible to cope with large datasets like Border. Moreover, Border relies on border objects around clusters to build constraints rather than must-link graphs [1,19]. The inheritance approach is closely related to the constraint propagation in the multi-view clustering algorithm [10,11] for transferring constraints among different views. The major difference is that we use the μ-nearest neighbors rather than the ϵ-neighborhoods which is limited to Gaussian clusters and can lead to an excessive number of constraints.

Temporal Clustering. Temporal smoothness has been introduced in the evolution framework [4] for making clustering results stable w.r.t. the time. We significantly extend this framework by incorporating instance-level constraints, active query selections and constraint propagation for further improving clustering quality while minimizing constraint annotation effort.

6   Conclusion

We introduce a scalable novel framework which incorporates an iterative active learning scheme, instance-level and temporal smoothness constraints for coping with large temporal data. Experiments show that our constrained clustering algorithm, CVQE+, performs better than existing techniques such as CVQE [8], LCVQE [17] and MPC-kMeans [1]. By exploring border objects and propagating constraints via nearest neighbors, our active learning algorithm, Border, results in good clustering results with much smaller constraint sets compared to other methods such as NPU [19] and Min-max [15]. Moreover, it is orders of magnitude faster making it possible to cope with large datasets. Finally, we revisit our approach in the context of evolutionary clustering adding a temporal smoothness constraint and a time-fading factor to our constraint propagation among different data snapshots. Our future work aims at providing more expressive support for user feedback. We are currently using our framework to track group evolution of our patient data with sleeping disorder symptoms.

Acknowledgment. This work is supported by the CDP Life Project.

References

1. Basu, S., Banerjee, A., Mooney, R.J.: Active semi-supervision for pairwise constrained clustering. In: SDM, pp. 333–344 (2004)
2. Bilenko, M., Basu, S., Mooney, R.J.: Integrating constraints and metric learning in semi-supervised clustering. In: ICML (2004)
3. Birgé, L., Rozenholc, Y.: How many bins should be put in a regular histogram. ESAIM: Probab. Stat. **10**, 24–45 (2006)
4. Chakrabarti, D., Kumar, R., Tomkins, A.: Evolutionary clustering. In: SIGKDD, pp. 554–560 (2006)
5. Cohn, D., Caruana, R., Mccallum, A.: Semi-supervised clustering with user feedback. Technical report (2003)
6. Davidson, I.: Two approaches to understanding when constraints help clustering. In: KDD, pp. 1312–1320 (2012)
7. Davidson, I., Basu, S.: A survey of clustering with instance level constraints. TKDD (2007)

8. Davidson, I., Ravi, S.S.: Clustering with constraints: feasibility issues and the k-means algorithm. In: SDM, pp. 138–149 (2005)
9. Davidson, I., Ravi, S.S., Ester, M.: Efficient incremental constrained clustering. In: KDD, pp. 240–249 (2007)
10. Eaton, E., desJardins, M., Jacob, S.: Multi-view clustering with constraint propagation for learning with an incomplete mapping between views. In: CIKM, pp. 389–398 (2010)
11. Eaton, E., desJardins, M., Jacob, S.: Multi-view constrained clustering with an incomplete mapping between views. Knowl. Inf. Syst. **38**(1), 231–257 (2014)
12. Han, J.: Data Mining: Concepts and Techniques. Morgan Kaufmann Publishers Inc., San Francisco (2005)
13. Huang, R., Lam, W.: Semi-supervised document clustering via active learning with pairwise constraints. In: ICDM, pp. 517–522 (2007)
14. Huang, Y., Mitchell, T.M.: Text clustering with extended user feedback. In: SIGIR, pp. 413–420 (2006)
15. Mallapragada, P.K., Jin, R., Jain, A.K.: Active query selection for semi-supervised clustering. In: ICPR, pp. 1–4 (2008)
16. Nguyen, X.V., Epps, J., Bailey, J.: Information theoretic measures for clusterings comparison: is a correction for chance necessary? In: ICML, pp. 1073–1080 (2009)
17. Pelleg, D., Baras, D.: K-means with large and noisy constraint sets. In: Kok, J.N., Koronacki, J., Mantaras, R.L., Matwin, S., Mladenič, D., Skowron, A. (eds.) ECML 2007. LNCS (LNAI), vol. 4701, pp. 674–682. Springer, Heidelberg (2007). https://doi.org/10.1007/978-3-540-74958-5_67
18. Chouakria, A.D., Mai, S.T., Amer-Yahia, S.: Scalable active temporal constrained clustering. In: EDBT (2018)
19. Xiong, S., Azimi, J., Fern, X.Z.: Active learning of constraints for semi-supervised clustering. IEEE Trans. Knowl. Data Eng. **26**(1), 43–54 (2014)

Nearest Subspace with Discriminative Regularization for Time Series Classification

Zhenguo Zhang[1,2], Yanlong Wen[2], Ying Zhang[2], and Xiaojie Yuan[2(✉)]

[1] Department of Computer Science and Technology, Yanbian University,
977 Gongyuan Road, Yanji 133002, People's Republic of China
[2] College of Computer and Control Engineering, Nankai University,
38 Tongyan Road, Tianjin 300350, People's Republic of China
{zhangzhenguo,wenyanlong,zhangying,yuanxiaojie}@dbis.nankai.edu.cn

Abstract. For time series classification (TSC) problem, many studies focus on elastic distance measures for comparing time series and complete the task with the help of Nearest Neighbour (NN) classifier. This is mainly due to the fact that the order of variables is a crucial factor for time series. Unlike the NN classifier only considers one training sample, in this paper, we propose an improved Nearest Subspace (NS) classifier to classify new time series. By adding a discriminative regularization item, the improved NS classifier takes full advantage of all training time series of one class. Two kinds of discriminative regularization items are employed in our method. One is directly calculated based on Euclidean distance of time series. For the other, we obtain the regularization items from a lower-dimensional subspace. Two well-known dimensional reduction methods, Generalized Eigenvector Method (GEM) and Local Fisher Discriminant Analysis (LFDA), are employed to complete this task. Furthermore, we combine these improved NS classifiers through ensemble schemes to accommodate different time series datasets. Through extensive experiments on all UCR and UEA datasets, we demonstrate that the proposed method can gain better performance than NN classifiers with different elastic distance measures and other classifiers.

Keywords: Time series · Nearest subspace
Discriminative regularization · Ensemble

1 Introduction

Time series data have attracted significant interest in data mining community due to the fact that the series data are present in our daily life and have been studied in many application domains. As the most prominent task, time series classification (TSC) problems are most studied and many methods are proposed in recent years [2]. Unlike other data, the order of variables is a crucial factor for time series. The empirical studies have shown that the simple 1-NN classifier

© Springer International Publishing AG, part of Springer Nature 2018
J. Pei et al. (Eds.): DASFAA 2018, LNCS 10827, pp. 583–599, 2018.
https://doi.org/10.1007/978-3-319-91452-7_38

can get high accuracy and is very hard to beat [4]. So numerous TSC studies focus on how to calculate the distance between two time series. The standard benchmark distance measure is Euclidean distance (ED), but it cannot handle the slight phase shift of time series. To overcome this drawback, Ratanamahatana and Keogh [26] propose to utilize dynamic time warping (DTW) distance to mitigate against distortions in the time axis. By extensive experiments, Ding et al. [9] validate that DTW with a proper warping window size set is commonly accepted as the standard measures. Another way to deal with this phase shift is based on edit distance to measure the similarity.

Due to the characteristic of DTW and edit distance, lots of studies try to improve elastic distance measures based on these two techniques for TSC task. Jeong *et al.* [14] propose a modified DTW that weights against large warping (WDTW) by adding a multiplicative weight penalty. Another improvement of DTW, called DDTW, which transforms the time series into a series of first-order differences [17]. By this transformation, DDTW can avoid the scenarios that a single point on one time series may map onto a large number of points of another time series. Meanwhile, it also can be used in conjunction with standard DTW to calculate similarity between time series simultaneously [11]. Several approaches based on the edit distance are also proposed, such as edit distance with real penalty (ERP) [6], time warp edit (TWE) [22] and move-split-merge (MSM) [27] distance. A recent work [20] combines most of elastic distance measures through ensemble schemes to improve the accuracy.

Although the NN classifier with elastic distance measure has the advantages of simplicity, the label prediction of new instance relies on the special training samples. In this paper, we propose an improved nearest subspace (NS) method to deal with the TSC problem, which considers all training samples of one class when predicting the new instance's label. This is a linear combination way and the similar idea has been used in other research field [7,19]. In this method, the most important factor is the combination coefficients of training samples. We get the coefficients based on the differences between the new instance and each training sample. Concretely, we calculate the distances between them and take these distances as weighted items to regularize the coefficients. In this way, the training time series which has a large distance with the new instance will be assigned a small coefficient. It's the biggest difference between our method and NN classifier. By the discriminative regularization item, the improved NS method not only considers the most similar training samples with a new instance but also takes other training time series into account.

To achieve higher performance, we map the time series into the low-dimensional space where the samples with the same label are closer and the different classes are more separate. Thus, the weighted regularization items obtained in the low-dimensional space are more powerful for the improved NS method. We employ two well-known dimensionality reduction methods, GEM [15] and LFDA [28], for this task. Furthermore, ensemble schemes which combine different weighted regularization items are proposed to suit different kinds of time series dataset. The extensive experiments demonstrate that the proposed method can gain better performance than NN classifier with different elastic distance measures.

2 Background and Related Work

2.1 NN-Based Time Series Classification

As mentioned in Sect. 1, most studies have been directed at finding techniques that can compensate for small misalignments between time series. Two main elastic distance measures, DTW and edit distance, have been widely studied.

DTW-Based Elastic Distance Measures. DTW is considered as the standard benchmark elastic distance measure to find an optimal alignment between two given sequences [16]. The standard DTW utilizes a pointwise distance matrix to record the cumulative distance from the start point pair to current point pair and employs the dynamic programming method to complete the process. Two aspects of DTW have been studied in recent years. One is speedup technique because the standard DTW has a quadratic time and space complexity. Some works have reduced it to nearly linear time complexity [25]. Another improvement is to change the calculation way of cumulative distance. A weighted form of DTW (WDTW) [14] is proposed to reduce warping by adding a multiplicative weight item to penalize points with higher phase difference between a test point and a reference point. In standard DTW, there is a scenario where a single point on one time series may map onto a large number of points on the other time series, which lead to pathological results. To avoid these singularities, a modification of DTW, called DDTW, is proposed by transforming the time series into a series of first-order difference [17]. On the basis of this idea, Górecki et al. [11] use a weighted combination of DTW on raw time series and DDTW on first order differences for NN classification. An extension of DDTW that uses DTW in conjunction with transforms and derivatives is proposed by Górecki and Łuczak [12]. They propose a new distance function by combining three distances: DTW distance between time series, DTW distance between derivatives of time series, and DTW distance between transforms of time series.

Edit Distance-Based Elastic Distance Measures. The initial edit distance technique is longest common subsequence (LCSS) distance which is extended to handle the real-valued time series from discrete series by using a distance threshold. A point pair from two time series can be considered as a match if their distance is less than the predefined threshold. Like LCSS, edit distance on real sequences (EDR) [5] also use a distance threshold to define a series match, but the difference is EDR employs a constant penalty to deal with the scenario of non-matching point pair. The drawback of EDR is it does not satisfy triangular inequality. Chen et al. [6] revise the weakness of EDR by utilizing the distance between point pairs when there is no gap and a constant when gaps occur. TWE [22] and MSM [27] are two effective edit distance-based approaches proposed in recent years. TWE makes full use of the characteristics of LCSS and DTW, which allows warping in the time axis and combines the edit distance with Lp-*norms*. In MSM the similarity of two different time series is calculated by using a series of operations to transform a given time series into the target time series.

Other Elastic Distance Measures. Batista *et al.* consider the complexity invariance problem for time series similarity measures and propose a parameter-free method, complexity invariant distance (CID) [13], to solve this problem. They describe a method for weighting a distance measure to compensate for the differences in the complexity when two time series are compared. The sum of squares of the first differences is used to measure the complexity. Except using individual elastic distance measure to calculate the similarity of two time series, Lines and Bagnall [20] combine 11 elastic distance measures through simple ensemble schemes and get significantly better classification accuracy.

2.2 Basic Nearest Subspace Algorithm

The NN classifier may be the simplest supervised method to predict the label of a test instance. Essentially, it seeks the best representation of a test instance in term of one training sample. Unlike NN algorithm, the nearest subspace (NS) classifier (e.g. [18,21]) takes all training samples of each class into consideration and tries to find the best representation by fitting the test instance. Formally, we assign a test instance $\boldsymbol{y}$ to class i if the distance from $\boldsymbol{y}$ to the subspace spanned by all samples $\boldsymbol{X}_i = [\boldsymbol{x}_{i,1}, \ldots, \boldsymbol{x}_{i,n_i}]$ of class i is the smallest one among all classes, i.e.,

$$r_i(\boldsymbol{y}) = \min_{\boldsymbol{\alpha}_i \in \mathbb{R}^{n_i}, i \in \{1, \ldots, K\}} \|\boldsymbol{y} - \boldsymbol{X}_i \boldsymbol{\alpha}_i\|_2 \tag{1}$$

where $\boldsymbol{\alpha}_i$ is a fitting coefficient vector.

One may notice that the Eq. (1) is easily overfitting which makes the problem ill-posed when we attempt to get the best solution. In general, we can introduce an additional regularization item to prevent overfitting. An alternative is to restrict the variation of $\boldsymbol{\alpha}$ by adding an L_2-regularization term:

$$\widetilde{\boldsymbol{\alpha}}_i = \arg\min_{\boldsymbol{\alpha}_i \in \mathbb{R}^{n_i}} \|\boldsymbol{y} - \boldsymbol{X}_i \boldsymbol{\alpha}_i\|_2^2 + \lambda \|\boldsymbol{\alpha}_i\|_2^2 \tag{2}$$

3   Improved Nearest Subspace Classifier

In Eq. (2), L_2-norm is employed to overcome the ill-posed problem of NS algorithm. It's optional to restrict parameter $\boldsymbol{\alpha}_i$, but for classification, the uniform weight for each element $\alpha_{i,j}(j \in \{1, \ldots, n_i\})$ does not consider the differences between training samples. In this paper, we want to utilize a non-uniform regularization to improve nearest subspace classifier. Like NS with L_2-regularization, we still calculate the residual for each class and assign the test instance to the class with the minimum residual. The difference is that we utilize Tikhonov regularization [10] with non-uniform weight to replace the simple L_2-regularization. The non-uniform regularization can penalize the dissimilarity training samples with a specific test instance $\boldsymbol{y}$ from being assigned large contributions when constructing $\widetilde{\boldsymbol{y}}_i$. Thus, the approximation coefficients can be formulated as

$$\widetilde{\boldsymbol{\alpha}}_i = \arg\min_{\boldsymbol{\alpha}_i \in \mathbb{R}^{n_i}} \|\boldsymbol{y} - \boldsymbol{X}_i \boldsymbol{\alpha}_i\|_2^2 + \lambda \|\boldsymbol{\Gamma}_{i,y} \boldsymbol{\alpha}_i\|_2^2 \tag{3}$$

where $\boldsymbol{\Gamma}_{i,y}$ is the weight matrix to measure the similarity between $\boldsymbol{y}$ and each sample of class i. In the following, we utilize Euclidean distance to describe $\boldsymbol{\Gamma}_{i,y}$ and discuss the effect of $\boldsymbol{\Gamma}_{i,y}$ for approximation coefficients $\boldsymbol{\alpha}_i$.

3.1   ED-Based Regularization

For TSC, the widely studied NN-based classifiers have demonstrated that the distance is a proper alternative to measure the similarity of time series. So we first use Euclidean distance as the weight to restrict regularization term. Concretely, for each element of $\boldsymbol{\alpha}_i$ ($\alpha_{i,j}$), we calculate the distance between the test instance $\boldsymbol{y}$ and each training sample $\boldsymbol{X}_{i,j}$ and use this distance to restrict $\alpha_{i,j}$, i.e.,

$$\boldsymbol{\Gamma}_{i,y} = diag(\|\boldsymbol{y} - \boldsymbol{X}_{i,1}\|_2, \ldots, \|\boldsymbol{y} - \boldsymbol{X}_{i,n_i}\|_2) \tag{4}$$

By $\boldsymbol{\Gamma}_{i,y}$, the training samples that is the most similar to $\boldsymbol{y}$ in terms of Euclidean distance can get more contribution to construct the approximation instance than those which are dissimilar. After weight matrix is determined, Eq. (3) can be rewritten as

$$f(\boldsymbol{\alpha}_i) = \frac{1}{2}\|\boldsymbol{y} - \boldsymbol{X}_i\boldsymbol{\alpha}_i\|^2 + \lambda\|\boldsymbol{\Gamma}_{i,y}\boldsymbol{\alpha}_i\|^2 \tag{5}$$

This is convex in $\boldsymbol{\alpha}_i$, so we can find its the optimal coefficients $\widetilde{\boldsymbol{\alpha}}_i$ by setting the derivative of $f(\boldsymbol{\alpha}_i)$ to 0:

$$\widetilde{\boldsymbol{\alpha}}_i = (\boldsymbol{X}_i^T\boldsymbol{X}_i + \lambda\boldsymbol{\Gamma}_{i,y}^T\boldsymbol{\Gamma}_{i,y})^{-1}\boldsymbol{X}_i^T\boldsymbol{y} \tag{6}$$

Thus, $\widetilde{\boldsymbol{y}}_i$ is

$$\widetilde{\boldsymbol{y}}_i = \boldsymbol{X}_i\widetilde{\boldsymbol{\alpha}}_i = \boldsymbol{X}_i(\boldsymbol{X}_i^T\boldsymbol{X}_i + \lambda\boldsymbol{\Gamma}_{i,y}^T\boldsymbol{\Gamma}_{i,y})^{-1}\boldsymbol{X}_i^T\boldsymbol{y} = \boldsymbol{H}_i\boldsymbol{y} \tag{7}$$

So $\boldsymbol{H}_i$ can be considered as a projection matrix.

To further investigate the effect of $\boldsymbol{\Gamma}_{i,y}$ when constructing $\widetilde{\boldsymbol{y}}_i$, we analyze the properties of $\boldsymbol{H}_i$ in terms of eigen-decomposition. Because $\boldsymbol{H}_i$ contains two matrices, $\boldsymbol{X}_i$ and $\boldsymbol{\Gamma}_{i,y}$, we employ the generalized singular value decomposition (GSVD) [1] between them to do this task.

For matrices $\boldsymbol{X}_i$ and $\boldsymbol{\Gamma}_{i,y}$, their GSVD is given by

$$\boldsymbol{X}_i = \boldsymbol{U}\boldsymbol{\Sigma}_1[\boldsymbol{0}, \boldsymbol{\Omega}]\boldsymbol{Q}^T, \quad \boldsymbol{\Gamma}_{i,y} = \boldsymbol{V}\boldsymbol{\Sigma}_2[\boldsymbol{0}, \boldsymbol{\Omega}]\boldsymbol{Q}^T \tag{8}$$

where $\boldsymbol{U}$, $\boldsymbol{V}$ and $\boldsymbol{Q}$ are unitary matrices, $\boldsymbol{\Omega}$ is upper triangular and non-singular matrix. Since $\boldsymbol{\Gamma}_{i,y}$ is a diagonal square matrix, $\boldsymbol{U}$, $\boldsymbol{V}$ and $\boldsymbol{Q}$ are orthogonal and matrices $\boldsymbol{\Sigma}_1$ and $\boldsymbol{\Sigma}_2$ are non-negative diagonal, which hold that $\boldsymbol{\Sigma}_1^T\boldsymbol{\Sigma}_1 = \lceil\sigma_{X,1}^2, \sigma_{X,2}^2, \ldots, \sigma_{X,r}^2\rfloor$ and $\boldsymbol{\Sigma}_2^T\boldsymbol{\Sigma}_2 = \lceil\sigma_{\Gamma,1}^2, \sigma_{\Gamma,2}^2, \ldots, \sigma_{\Gamma,r}^2\rfloor$, where $r = rank([\boldsymbol{X}_i^T, \boldsymbol{\Gamma}_{i,y}^T])$. Two properties of GSVD is $0 \leq \sigma_{X,i}, \sigma_{\Gamma,i} \leq 1$ and $\boldsymbol{\Sigma}_1^T\boldsymbol{\Sigma}_1 + \boldsymbol{\Sigma}_2^T\boldsymbol{\Sigma}_2 = \boldsymbol{I}_r$ which implies $\sigma_{X,i}^2 + \sigma_{\Gamma,i}^2 = 1$.

By these decompositions, $\boldsymbol{H}_i$ of Eq. (7) is formulated as

$$\boldsymbol{H}_i = \boldsymbol{U}\boldsymbol{\Sigma}_1(\boldsymbol{\Sigma}_1\boldsymbol{\Sigma}_1 + \lambda\boldsymbol{\Sigma}_2\boldsymbol{\Sigma}_2)\boldsymbol{\Sigma}_1\boldsymbol{U}^T = \boldsymbol{U}\boldsymbol{\Sigma}\boldsymbol{U}^T \tag{9}$$

As can be seen, $\boldsymbol{\Sigma}$ is still a diagonal matrix with the values σ_k:

$$\sigma_k = \frac{\sigma_{X,i}^2}{\sigma_{X,i}^2 + \lambda\sigma_{\Gamma,i}^2} = \frac{1 - \sigma_{\Gamma,i}^2}{1 + (\lambda - 1)\sigma_{\Gamma,i}^2} \tag{10}$$

Apparently, $\sigma_k \in [0, 1]$. Since $\boldsymbol{U}$ is an orthogonal matrix, the decomposition of Eq. (9) can represent the eigen decomposition of $\boldsymbol{H}_i$. It is clear that the values of $\sigma_k(k = 1, \ldots, r)$ depend on three parts: the structure of the training samples of class i (i.e., $\sigma_{X,i}$), the regularization parameter (λ) and the distances between $\boldsymbol{y}$ and each sample of class i (i.e., $\sigma_{\Gamma,i}$). This means that different test instances have different amount of shrinkage when constructing the approximation instance even using the same training samples.

Due to $\lambda > 0$ and $\sigma_{\Gamma,i} \in [0, 1]$, σ_k and $\sigma_{\Gamma,i}$ have an inverse relationship. To be more specific, when $\boldsymbol{y}$ has large distances with the training samples of $\boldsymbol{X}_i$ (that is, the non-zero entries of $\boldsymbol{\Gamma}_{i,y}$ have large values), the values of σ_k are large and the eigenvalues of $\boldsymbol{H}_i$ are small. This suggests that the classes whose training samples are distant from test instance lead to a stiffer shrinkage penalty. The large penalty will make the obtained $\widetilde{\boldsymbol{y}}_i$ dissimilar to $\boldsymbol{y}$. Back to Eq. (3), large distances in $\boldsymbol{\Gamma}_{i,y}$ make $\boldsymbol{\alpha}_i$ become small, which results in a large residual. One thing should be noticed that $\boldsymbol{\Gamma}_{i,y}$ is an ensemble of distances and only one or several small values have little effect for the final $\boldsymbol{\alpha}_i$. This is the primary difference with NN-based approaches.

3.2   GEM-Based Regularization

Through the analysis of $\boldsymbol{H}_i$, we find the training sample that is distant from test instance will lead to a small coefficient which makes little effect for calculating the residual. Therefore, if we map the time series into a lower-dimensional subspace where the samples of each class become more closed while the different classes are more separated, the prediction in terms of $\|\boldsymbol{y} - \widetilde{\boldsymbol{y}}_i\|_2$ will be more accurate. In this section, we employ the state-of-the-art supervised feature extraction technique, generalized eigenvector method (GEM) [15], to complete this task.

GEM exploits the simple second-order structure of the training samples and extracts the discriminative features from the generalized eigenvectors of the class conditional second moments. In [15], the author suggests that an alternative would be to use the covariance matrix instead of the second moment. GEM gets the direction by maximizing the ratio of projected data variances between different classes. Thus the direction v is calculated as

$$v = \arg\max_v \frac{v^T C_i v}{v^T (C_j + \frac{\gamma}{d} trace(C_j))v} \tag{11}$$

where C_i is the covariance matrix of class i, d is the dimensionality of samples, γ is a small multiple and $trace()$ denotes the trace of a matrix. The item $\frac{\gamma}{d} trace(C_j)$ is used to solve the problem that C_j is rank deficient when there are few training samples [24]. This objection function is solved by the generalized eigenvectors of $C_i v = \mu(C_j + \frac{\gamma}{d} trace(C_j))v$. Generally, there are many eligible eigenvalues, so we can get a transformation matrix V by merging all vs.

When V is obtained, we can calculate the weight matrix $\Gamma_{i,y}^G$ (distinguished from Eq. (4)). Recall that each element in the diagonal of $\Gamma_{i,y}$ is the distance between test instance y and each training sample of class i. We calculate the diagonal values of $\Gamma_{i,y}^G$ by

$$D_G(V, y, X_{i,j}) = \|Vy - VX_{i,j}\|_2 = \sqrt{(y - X_{i,j})^T \Phi (y - X_{i,j})} \tag{12}$$

where $\Phi = V^T V$, $X_{i,j}$ denotes the j-th training sample of class i.

The above procedure is suited for the scenario of two class. For multi-class problem involving $K(K > 2)$ classes, we use two strategies to get the projection matrix: *one-vs-rest* and *multiple one-vs-one*. The transformation matrix and weight matrix are denoted as $V^{(1:r)}$, $\Gamma_{i,y}^{G(1:r)}$, $V^{(1:1)}$ and $\Gamma_{i,y}^{G(1:1)}$, respectively.

3.3   LFDA-Based Regularization

Local Fisher discriminant analysis (LFDA) is another effective method which embeds the high-dimensional data into a lower-dimensional subspace. In this section, we employ LFDA to help the construction of $\Gamma_{i,y}$. To deal with the multi-modal problem of sample distribution, LFDA employs a concept of affinity to represent the neighbour relationship of two samples. Let W be the affinity matrix, and an elaborate and proper definition of the affinity between x_i and x_j is provided by [29]: $W_{i,j} = exp(-d^2(x_i, x_j)/\eta_i \eta_j)$, where $d(x_i, x_j)$ is a distance measure, $\eta_i = d(x_i, x_i^{knn})$ is a local scaling parameter in terms of the neighbourhood of x_i and x_i^{knn} is the knn-nearest neighbour of x_i. By this definition, a large distance between two samples will lead to a small affinity. If the values of local within-class scatter matrix (S_{lw}) and local between-class scatter matrix (S_{lb}) used in LFDA are weighted by the affinity, the far-apart samples have less influence, which can preserve the local structure of the data. Based on affinity matrix, the S_{lw} and S_{lb} are defined as

$$S_{lw} = \frac{1}{2} \sum_{i,j=1}^{n} W_{ij}^w (x_i - x_j)(x_i - x_j)^T \tag{13}$$

$$S_{lb} = \frac{1}{2} \sum_{i,j=1}^{n} W_{ij}^b (x_i - x_j)(x_i - x_j)^T \tag{14}$$

where

$$W_{ij}^w = \begin{cases} W_{ij}/n_c, & l_i = l_j = c \\ 0, & l_i \neq l_j \end{cases}, W_{ij}^b = \begin{cases} W_{ij}(1/n - 1/n_c), & l_i = l_j = c \\ 1/n, & l_i \neq l_j \end{cases} \tag{15}$$

where l_i, l_j are class labels, n_c is the number of training samples of c-th class. As usual, we still use [24]'s solution to regularize S_{lw}. The reason is that S_{lw}

will be the denominator matrix in the objection function of LFDA. Using S_{lw} and S_{lb}, the LFDA transformation matrix T is defined as

$$T = \arg \max_{T} \frac{T^T S_{lb} T}{T^T (S_{lw} + \frac{\gamma}{d} trace(S_{lw})) T} \qquad (16)$$

Now, we can project the samples into a low-dimensional subspace by T and recalculate the weight matrix $\Gamma_{i,y}$ of Eq. (4). For the sake of distinction, we denote it as $\Gamma_{i,y}^L$ and each element is obtained by calculating

$$D_L(T, y, X_{i,j}) = \|Ty - TX_{i,j}\|_2 = \sqrt{(y - X_{i,j})^T \Psi (y - X_{i,j})} \qquad (17)$$

where $\Psi = T^T T$.

Except the transformation matrix is obtained by Eq. (16), for $K(K > 2)$ classes, we also use *one-vs-rest* strategy to find the discriminative projection directions and merge them to form transformation matrix $(T^{(1:r)})$. We use $\Gamma_{i,y}^{L(1:r)}$ to denote the weight matrix of this case.

3.4   Effect of GEM and LFDA-Based Regulation for Improved NS

In Sects. 3.2 and 3.3, we respectively employ GEM and LFDA to transform the samples into a low-dimensional space where the obtained weight matrix is expected to have a proper shrinkage penalty for approximation coefficients α_i. No matter GEM or LFDA, the purpose is to make the samples with the same label more closed in the new space while the samples from different classes are apart. So by comparing the distance relationships calculated in original sample space, the distances obtained in new low-dimensional space become more meaningful for the classification task. What is different is that GEM excepts more information contained in data distribution while LFDA tries to make the inter-class more separated. Therefore, GEM uses the covariance matrix as the basis while LFDA employs the collection of class-conditional mean feature vectors. In the low-dimensional space constructed by LFDA, the inter-class separability is increased, which penalizes the class whose memberships are most distant from y. Meanwhile, the samples that are truly neighbours of y are also seen as neighbours, which gives better information on within-class distance relationships with y and offers more benefit for classification.

Empirically, LFDA seems to be a more suitable choice for classification task to restrict α_i. But there are various kinds of time series datasets in our daily life and no one technique is a panacea which can deal with all situations. Bagnall and Lines's works [3, 20] have demonstrated this point and they suggest that an ensemble scheme of different approaches is an alternative. In the next section, we will utilize all these weight matrices to propose an ensemble NS classifier.

4   Ensemble NS Classifier

An ensemble of classifiers is a set of base classifiers, which has been widely studied in machine learning domain. When classifying new samples, the decisions of all base classifiers are combined through a fusion way. In Sect. 3, we design 5 different weight matrices: $\boldsymbol{\Gamma}_{i,y}$, $\boldsymbol{\Gamma}_{i,y}^{G(1:r)}$, $\boldsymbol{\Gamma}_{i,y}^{G(1:1)}$, $\boldsymbol{\Gamma}_{i,y}^{L}$ and $\boldsymbol{\Gamma}_{i,y}^{L(1:r)}$, where $\boldsymbol{\Gamma}_{i,y}$ is directly constructed by calculating the distance between test instance $\boldsymbol{y}$ and each training sample of class i while the other 4 weight matrices are constructed in different low-dimensional spaces. Our ensemble approach is transparent, i.e., we first construct the five different forms of NS classifiers by these weight matrices and then take the simple voting schemes to make a decision for each test instance. For convenience, we denote these five different forms of NS classifiers as NS_ED, NS_G(1:r), NS_G(1:1), NS_L and NS_L(1:r).

We separately test three weighting schemes for the ensemble classifier: *equal*, *proportional* and *best*. The *equal* scheme, as the name implies, sets the equal weight for each classifier when classifying new instance. Clearly, it is the fastest scheme, but it does not consider the difference of base classifiers. The *proportional* and *best* schemes exploit unequal voting weight. A widely used technique is to calculate the cross-validation accuracy of each base classifier on the training set and then the weight is set based on the accuracy [20]. The *proportional* scheme assigns the wights as the cross-validation accuracy directly while the *best* scheme takes a binary strategy which assigns a weight of 1 to the base classifier with the highest cross-validation accuracy and 0 to the others. In the next Section, these schemes are all employed for TSC task. Here, we propose an ensemble algorithm of improved NS classification with five weight matrices by the *best* scheme.

Algorithm 1. Ensemble procedure of improved NS algorithm

Input: training set $\boldsymbol{X} = [\boldsymbol{X}_1, \boldsymbol{X}_2, \ldots, \boldsymbol{X}_K]$, test instance $\boldsymbol{y}$,
    cross-validation accuracy of different NS classifier $\boldsymbol{cv\_acc}$ and λ
Output: the label of $\boldsymbol{y}$

1 $[\sim, maxIndex] = max(\boldsymbol{cv\_acc})$;
2 Calculate $\boldsymbol{V}^{(1:r)}$ and $\boldsymbol{V}^{(1:1)}$ based on Eq.(11);
3 Calculate $\boldsymbol{T}$ and $\boldsymbol{T}^{(1:r)}$ based on Eq.(16);
4 $res = zeros(K, 5)$;
5 **for** $i = 1 : K$ **do**
6   Calculate $\|\boldsymbol{y} - \boldsymbol{X}_{i,j}\|_2$, $\|\boldsymbol{V}^{(1:r)}\boldsymbol{y} - \boldsymbol{V}^{(1:r)}\boldsymbol{X}_{i,j}\|_2$, $\|\boldsymbol{T}\boldsymbol{y} - \boldsymbol{T}\boldsymbol{X}_{i,j}\|_2$,
        $\|\boldsymbol{V}^{(1:1)}\boldsymbol{y} - \boldsymbol{V}^{(1:1)}\boldsymbol{X}_{i,j}\|_2$, $\|\boldsymbol{T}^{(1:r)}\boldsymbol{y} - \boldsymbol{T}^{(1:r)}\boldsymbol{X}_{i,j}\|_2$ $(j \in [1 : n_i])$;
7   Construct $\boldsymbol{\Gamma}_{i,y}$, $\boldsymbol{\Gamma}_{i,y}^{G(1:r)}$, $\boldsymbol{\Gamma}_{i,y}^{G(1:1)}$, $\boldsymbol{\Gamma}_{i,y}^{L}$ and $\boldsymbol{\Gamma}_{i,y}^{L(1:r)}$;
8   Get the optimal coefficients $\boldsymbol{\Theta} = (\widetilde{\boldsymbol{\alpha}}_i, \widetilde{\boldsymbol{\alpha}}_i^{G(1:r)}, \widetilde{\boldsymbol{\alpha}}_i^{G(1:1)}, \widetilde{\boldsymbol{\alpha}}_i^{L}, \widetilde{\boldsymbol{\alpha}}_i^{L(1:r)})$;
9   Calculate the residuals $\boldsymbol{r}_i = \|\boldsymbol{y} - \boldsymbol{X}_i\boldsymbol{\Theta}\|$;
10      $res(i, :) = \boldsymbol{r}_i$;
11 **end**
12 $[\sim, labelVector] = min(res)$;
13 label($\boldsymbol{y}$)$=labelVector(maxIndex)$;

The time consuming of Algorithm 1 mainly depends on two parts: one is the calculation of weight matrices, the other is the solution of optimal coefficients.

For optimal coefficient $\widetilde{\alpha}_i$ of Eq. (6), it takes $O(n_i^2 m)$ where n_i is the number of i-th class and m is the length of time series. For weight matrix Γ, the GEM and LFDA consume most of running time but they only need to be computed once. When using GEM, the projection vector of Eq. (11) requires $O(m^2)$, thus, the *one-vs-one* way takes $O(K^2 m^2)$ and *one-vs-rest* strategy needs $O(Km^2)$. When using LFDA, Eq. (16) needs $O(n^2 k)$ where k is the number of selected projection vectors and $k = 1$ for *one-vs-one* strategy. So the total time complexity is $O(max(Kn_i^2 m, max(K^2 m^2, n^2 k)))$.

5   Experiments

We use 93 diverse time series datasets from UCR[1] and UEA[2] repository to perform our experiments. These datasets are common used for time series data mining task, which cover a wide range of domains, such as image outlines, motions and sensor readings [3]. All these datasets have default *train/test* splits and we don't change them as other methods. In experiments, we use the non-parametric *Friedman test* [8] to evaluate the performance of our method. All experiments are performed on Windows 10 platform with Intel i7 CPU and 16 GB memory, and the programming language is Matlab.

In Eq. (3), the parameter λ needs to be estimated and we get it through cross-validation on the training dataset. When using LFDA to construct weight matrix, the parameter knn is the key factor to deal with the multi-modal problem. In other works using LFDA, like [23], a fixed value 7 has been widely used. We take the same strategy to set a fixed value for each dataset in experiments. But for some time series datasets in UCR archive, there are only several training samples in each class, so we set $knn = 3$ for small datasets and $knn = 5$ for others.

5.1   Results of NS Classifier with Different Regularization

We first compare the classification accuracy of NS classifiers with the uniform $L2$-norm (denoted as NS_L2) and other 5 non-uniform regularization items. Figure 1 gives the critical difference diagram of *Friedman test* where NS_ED, NS_G(1:r), NS_G(1:1), NS_L and NS_L(1:r) stand for the improved NS classifier with different non-uniform regularization, representatively. Three ensemble schemes are represented as BEST, PROP and EQUAL.

From Fig. 1, we can find that all NS classifiers with non-uniform regularization are clearly superior to NS_L2. The *Friedman test* shows that LFDA-based regularization has better results than GEM and ED-based regularization. It's because the low-dimensional space obtained by LFDA is more suitable for classification. To further exhibit the improvement of non-uniform regularization for

---

[1] http://www.cs.ucr.edu/~eamonn/time_series_data/.
[2] http://www.uea.ac.uk/computing/machine-learning/shapelets/shapelet-data.

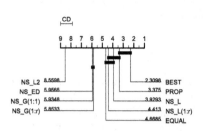

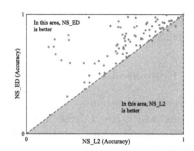

Fig. 1. Critical difference diagram for NS classifiers with different regularization and three ensemble NS classifiers on 93 datasets.

Fig. 2. Scatter plot of test accuracy of NS_L2 against NS_ED on 93 datasets.

NS when classifying time series, we draw the scatter plot of NS_L2 against NS_ED and Fig. 2 shows the result. In this comparison, NS_ED is better on 79 datasets, NS_L2 only wins 5 datasets and they tie on 9 datasets. This result fully illustrates that the non-uniform regularization based on distance is an effective shrinkage penalty for approximation coefficient.

As to one-vs-one and one-vs-rest strategy used in GEM and LFDA, there are little differences in average predictive accuracy based on *Friedman test* (Fig. 1). But on almost all datasets with $K(K > 3)$ classes, the classification accuracy is different. Table 1 shows the results of part datasets. It's easy to understand that the low-dimensional spaces constructed by these strategies are slightly different even though the projected time series are all class separative. Just as this diversity, our ensemble NS classifier can get higher accuracy. Although BEST scheme has the highest average accuracy (see Fig. 1), it doesn't mean that the other 2 ensemble schemes are useless. Theoretically, BEST scheme can gain the best result in all datasets. But in practice, the priori accuracy of each classifier obtained by cross-validation approach may be inaccurate, which lead to EQUAL and PROP schemes can achieve higher classification accuracies on some datasets. Table 2 presents some examples.

Table 1. Some results (%) of NS with different regularization

Dataset	NS_G(1:r)	NS_G(1:1)	NS_L	NS_L(1:r)
Adiac	75.7	53.2	73.7	75.7
Symbols	87.1	82.6	91.9	91.5
PP_Thumb	72.1	74	73.8	73.2
MALLAT	93	95.3	93.9	95.5

Table 2. Accuracy (%) of 3 ensemble NS classifiers of part datasets

Dataset	BEST	EQUAL	PROP
FordA	89.6	89.8	89.2
SwedishLeaf	93.1	93	93.6
MPOAG	79.5	79.8	79.8

In the following parts of this paper, we denote the ensemble NS classifier with BEST scheme as ENS and use it to compare with other methods.

5.2 Compared with Baselines

In this section, we compare ENS with the widely used classic classifiers in other domains, which cover decision tree algorithm C4.5, Naïve Bayes, Bayesian Network, Random Forest, Rotation Forest and SVM and 1-NN classifier with elastic distance measures including basic DTW [26], WDTW [14], CID [13], DDTW [17], DTD_C [12], MSM [27] and TWE [22]. Figure 3 shows the critical difference diagram. For more details, one can refer to the Appendix (Table 3).

As can be seen, compared with these baselines, ENS has significant advantage according to the average rank (4.31) while the best classic classifier, Rotation Forest, gets 5.07 and the best NN-based method, MSM, has the average rank of 6.07 on all 93 datasets. The experiment indicates that, even though the elastic distance measures can solve the slight phase shift contained in time series to some extent, they don't make full use of the information of the other time series in the same class. That's why, even using Euclidean distance in the objective function, ENS still outperforms them. One thing to be noticed that these classifiers and ENS all take the whole time series as the basis to classify the new instance. This result illustrates that ENS is more suitable for TSC task.

The Elastic Ensemble classifier (EE) is a combination of NN classifiers with 11 primary elastic distance measures [20]. This technique is similar to our method. Unfortunately, we can not get the results of EE for some datasets on our computer due to the time and space consuming. So we collect the results from other literatures and compare them with our method. Figure 4 shows the comparison of EE and ENS on UCR archives. As can be seen, even EE contains 11 different elastic distance measures, ENS is still slightly better. To be specific, ENS wins 45 times, loses 38 times and draws 2 times on all 85 datasets. The result illustrates that some characteristics of time series are usually dispersed in several samples or even the entire class instead of a single time series. Therefore, the overall consideration can get the general information of each class.

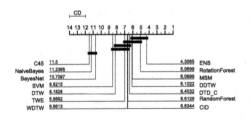

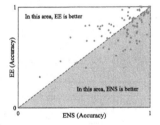

Fig. 3. Critical difference diagram for NS classifiers with other baselines on 93 datasets.

Fig. 4. Scatter plot of ENS against EE on 85 datasets.

5.3   Scalability

To test the scalability of ENS, we use the largest time series dataset *Stra-LightCurves* of UCR archives for the task. It contains 9,236 starlight time series of length 1,024. There are 3 types of star objects: *Eclipsed Binaries, Cepheids* and *RR Lyrae Variables*. The training set has 1,000 time series in default partition. Two key factors for testing the scalability of TSC methods are the number of training samples and the length of each sample. Here, we select NN classifiers with the different elastic distance measures as the comparison based on the following consideration: (1) most traditional classifiers have poor accuracy; (2) EE [20] is too slow to be run in our computer. The DTD_C [12] method is also dropped from the comparison because of the time-consuming.

First, we vary the number of training samples from 100 to 1,000 while the length is fixed to 1,024. To be fair, we use uniform sampling to select samples from each class. In this case, DD_DTW takes more than 5 h even when the number is 100, so we omit it. Meanwhile, we get only 8 results for NNDTW and 3 results for MSM and TWE due to the same reason. The total running time of different methods and the corresponding classification accuracy are depicted in Fig. 5. The running time (Fig. 5(a)) clearly show that our method ENS is more scalable when training data size increases. For classification accuracy, Fig. 5(b) demonstrates that ENS can gain higher accuracy when we have more training time series while other methods have little improvement. This trend also proves that the overall consideration of each time series class is benefit for predicting the label of a test instance.

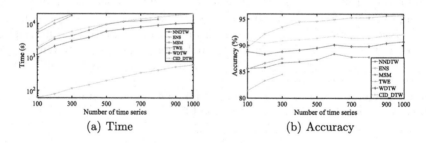

(a) Time   (b) Accuracy

Fig. 5. Results when varying the number of training samples.

Then, we fix the number of training samples to 1,000 and vary the length from 100 to 1,024. Again, some methods cannot finish in a reasonable time, so we only have 3 results for DD_DTW, 5 results for TWE, 6 results for MSM and 9 results for NNDTW. Figure 6 shows the results. In this case, same as the trends when varying the number of training samples, i.e., ENS is clearly faster (Fig. 6(a)) and, with the length increasing, ENS is advantageous. As for classification accuracy, there are a little fluctuate for all methods when time series length increases (Fig. 6(b)). The reason is that the training samples are intercepted from the

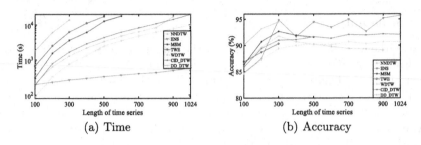

(a) Time   (b) Accuracy

Fig. 6. Results when varying the length of time series

original time series and each sample doesn't contain the entire information. But basically, the accuracy is increased when the length becomes larger and ENS is better than other methods.

6   Conclusions

In this paper, we propose a method, nearest subspace with discriminative regularization, for time series classification problem. In our method, a weight matrix is utilized to incur a shrinkage penalty for linear combination coefficients. We demonstrate these weight matrices are beneficial for constructing an approximation time series. To enhance the performance, two well-known dimensional reduction methods, GEM and LFDA, are employed to get four more class separability matrices. Furthermore, an ensemble scheme is used to combine these techniques, called ENS. We verify the performance of our method by comparing with classic classifiers, NN-based methods with elastic distance measure, ensemble methods on UCR and UEA time series datasets. The results show our method gains satisfactory results in terms of classification accuracy and scalability.

Acknowledgements. This work is supported by the National Natural Science Foundation of China [grant numbers 61772289 and 61702285].

Appendix

Table 3. Accuracy (%) of different methods in UCR and UEA repository.

dataset	BN	C45	NB	RandF	RotF	SVM	DTW	WDTW	CID	DDTW	DTD_C	MSM	TWE	EE	ENS
50words	16	41.8	56.3	68.1	66.2	64.6	69	77.1	78	75.4	75.4	81.3	79.3	82	70.5
Adiac	50.1	54.2	56.3	63.7	77.5	44.2	60.4	60.6	62.4	70.1	70.1	62.7	63.4	66.5	75.7
ArrowHead	57.1	60.6	54.3	71.4	73.7	73.1	70.3	81.7	82.9	78.9	72	80.6	79.4	81.1	86.9
Beef	60	53.3	66.7	73.3	86.7	90	63.3	70	63.3	66.7	66.7	46.7	60	63.3	90
BeetleFly	85	90	75	80	90	80	70	70	75	65	65	60	70	75	90
BirdChicken	60	80	55	75	85	65	75	75	75	80	85	85	85	80	90
Car	56.7	56.7	60	66.7	80	85	73.3	78.3	76.7	80	78.3	90	91.7	83.3	90
CBF	85.4	67.3	89.6	89	92.9	87.8	99.7	99.7	99.9	99.7	98	96.9	99.1	99.8	93.4
Chlorine.	59.9	68.8	34.6	71.3	84.7	58.4	64.8	64.9	64.9	70.8	71.3	62.8	64.8	65.6	91.8
CinC.	83.1	59	84.6	73.8	80.6	46.9	65.1	85.9	94.6	72.5	85.2	96.2	84.6	94.2	97.3
Coffee	96.4	92.9	92.9	100	100	100	100	100	100	100	100	89.3	100	100	100
Computers	66.4	56.8	58	60.8	70	49.2	70	70	62	71.6	71.6	57.6	63.2	70.8	63.2
Cricket_X	31.3	27.4	37.9	58.7	63.3	39	75.4	79	77.7	75.4	75.4	76.4	74.9	81.3	63.8
Cricket_Y	28.7	32.8	39.5	61	61	45.9	74.4	75.9	71.3	77.7	77.4	77.2	75.4	80.5	62.3
Cricket_Z	32.8	32.1	39.5	60.8	65.6	37.9	75.4	76.2	73.1	77.4	77.4	76.4	73.6	78.2	61.5
Diatom.	94.1	71.6	87.9	87.3	87.3	95.4	96.7	96.7	93.5	96.7	91.5	94.8	94.8	94.4	95.4
DPOA	73.4	69.1	72.7	77	74.8	71.2	77	69.8	63.3	70.5	66.2	74.1	71.2	69.1	85.3
DPOC	72.8	75.7	67.4	78.3	75.7	66.3	71.7	71.7	73.6	73.2	72.5	73.6	73.9	72.8	83
DistalP.	66.2	62.6	64	70.5	70.5	70.5	59	60.4	62.6	61.2	57.6	64	63.3	64.7	84.5
DP_Little	44.7	43.9	46.2	60.5	69.9	67.8	50.5	50.5	50.9	60.2	50.5	49.1	50.2	-	58.9
DP_Middle	37.2	49.1	42.5	57.2	71.2	70.7	54.6	54.6	52	60.9	54.6	52.9	54.3	-	75.7
DP_Thumb	45.1	51.9	42.3	64	71.5	72.7	52.6	52.6	51.8	59.4	51	52.9	52.4	-	74.4
Earthquakes	56.8	69.8	66.9	74.8	74.8	64	71.9	72.7	71.9	70.5	70.5	69.1	67.6	74.1	82.3
ECG200	75	80	77	79	85	81	77	88	89	83	84	91	89	88	92
ECG5000	85.3	89.9	88.4	93.5	94.6	93.8	92.4	92.5	92.7	92.4	92.4	93.2	92.7	93.9	94.4
ECGFiveD.	78	72.1	79.7	72.1	90.8	97.6	76.8	79.6	78.2	76.9	82.2	89.1	82.9	82	99.7
ElectricDevices	46.3	55.9	51.3	65.5	78.6	49.8	60.2	61.3	61.6	59.2	59.4	65.9	59.3	66.3	57.7
FaceAll	67.3	54.9	69.3	73.3	91.1	72	80.8	79.4	86.4	90.2	89.9	80.9	78.6	84.9	86.2
FaceFour	89.8	71.6	84.1	85.2	81.8	88.6	83	87.5	87.5	83	81.8	94.3	85.2	90.9	83
FacesUCR	61.2	47.8	72.7	78.2	80.3	75.8	90.5	92	89.5	90.4	90.8	97.1	92	94.5	83.2
FISH	65.7	56.6	65.1	76	82.9	85.1	82.3	85.1	83.4	94.3	92.6	93.1	93.1	96.6	90.3
FordA	51.6	56.2	49.2	77.2	84.5	49.6	55.5	65.3	77.2	72.3	76.5	71.9	61.8	73.8	89.6
FordB	50.7	52.6	53.5	65.4	77.2	52.7	62	59.9	62.7	66.7	65.3	64.2	60.9	66.2	89.6
Gun_Point	85.3	77.3	78.7	94	92	80	90.7	98	92.7	98	98.7	97.3	95.3	99.3	93.3
Ham	70.5	53.3	75.2	71.4	71.4	60	46.7	58.1	51.4	47.6	55.2	51.4	51.4	57.1	81
HandO.	84.6	88.1	81.1	90.5	91.1	89.2	88.1	87	86.2	86.8	86.5	87.6	87	88.9	87.3
Haptics	36.4	35.1	43.8	45.5	43.8	40.6	37.7	37	42.5	39.9	39.9	44.2	41.6	39.3	51.3
Herring	53.1	57.8	57.8	59.4	65.6	60.9	53.1	53.1	51.6	54.7	54.7	56.3	51.6	57.8	68.8
InlineSkate	17.6	26.9	21.6	35.1	37.1	30.9	38.4	40.4	41.5	56.2	50.9	45.3	42.2	46	32.2
InsectW.	61.4	50.7	62.8	65.6	63.6	64.3	35.5	57.4	57.8	35.5	47.3	57.5	53.4	59.5	64.6
ItalyP.	93.2	94.7	90.1	96.6	97.3	97.2	95	95	95.6	95	95.1	94.4	94.8	96.2	97.3
LargeK.	54.9	48.5	45.9	56.8	60.8	42.7	79.5	79.5	79.2	79.5	79.5	66.9	75.2	81.1	50.9
Lighting2	73.8	62.3	67.2	73.8	68.9	72.1	86.9	90.2	86.9	86.9	86.9	82	83.6	88.5	78.7
Lighting7	68.5	54.8	64.4	69.9	72.6	71.2	72.6	76.7	69.9	67.1	65.8	75.3	75.3	76.7	68.5
MALLAT	95.5	75.1	84.6	86.5	94.9	86.7	93.4	93.8	92.5	94.9	92.7	93.2	91.2	94	95.8
Meat	91.7	86.7	93.3	93.3	96.7	96.7	93.3	93.3	93.3	93.3	93.3	93.3	93.3	93.3	98.3
Medical1.	40.1	62.5	44.9	72.1	77.2	61.6	73.7	73.7	74.2	73.7	74.5	74.1	71.1	74.2	69.9
MPOAG	60.4	55.2	59.1	60.4	57.1	61.7	50	51.9	51.3	53.9	50	49.4	51.9	55.8	79.5
MPOC	70.4	69.8	55.3	81.4	80.1	63.6	69.8	75.3	76.3	73.2	74.2	75.3	76.3	78.4	80.5
MiddleP.	50.6	49.4	51.3	55.8	63	57.8	50.6	51.3	51.3	48.7	50	49.4	48.7	51.3	74.2
MoteStrain	85.6	78.7	84.3	88.9	88	86.7	83.5	85.9	79.6	83.3	76.8	85.5	79.7	88.3	88.7
MP_Little	44.7	45.4	44	59.1	67.9	69.1	55.3	55.3	57.2	60.5	55.3	60.5	56.9	-	73.6
MP_Middle	41.1	50.2	41.4	61.2	71.2	72.9	52.2	52.2	53.2	64.8	51	53.8	51.8	-	78.4
NonI.1	81.2	71.9	81.1	87.9	90.5	92.3	79	81.6	83.7	80.6	84.1	81.6	82	84.6	93.7
NonI.2	83.9	78.5	82.7	91.2	92.5	94.2	86.5	88.4	87.9	89.3	89	88.3	88.8	91.3	95
OliveOil	90	83.3	90	90	86.7	86.7	83.3	83.3	86.7	83.3	86.7	83.3	86.7	86.7	93.3
OSULeaf	31	34.3	38	51.2	57	44.2	59.1	62.4	62	88	88.4	77.3	77.7	80.6	60.3
Phalanges	73.1	73.4	65.4	81.7	86	64.7	72.8	74.7	76.2	73.9	76.1	75.2	76	77.3	82.9
Phoneme	11.3	6.5	13.7	14.7	13	9.4	22.8	16.1	22.1	26.9	26.8	29.2	27	30.5	16.5
Plane	98.1	96.2	94.3	99	99	98.1	100	100	100	100	100	100	100	100	99
PP_Little	35.3	49.9	40.2	62.5	68.4	69.9	49.9	49.9	48.4	53	50.1	50.5	50.9	-	73.5
PP_Middle	41.9	52.1	38.3	62.3	69.8	75.2	48.8	48.8	47.4	54.4	49.8	48.5	46.5	-	77.5
PP_Thumb	46.4	51	41.6	65	70.2	70.2	51	51	51	52.6	51.6	52.7	50.4	-	74
PPOAG	85.4	83.4	82.4	86.8	85.4	85.4	80.5	80.5	79	80	79.5	79	80.5	80.5	88.3
PPOC	72.2	79.7	64.6	86.6	86.3	82.5	78.4	78.4	79	79.4	79.4	78.7	80.8	80.8	90
PPTW	74.6	73.7	73.7	81	82.4	82.4	76.1	75.1	75.1	76.6	77.1	75.6	74.6	76.6	82.5
RD	50.4	46.4	38.4	51.5	56.5	35.2	46.4	42.7	44.5	44.5	44.5	48.3	51.2	43.7	43.5
ScreenType	44.8	37.3	43.5	43.5	44	38.1	39.7	41.1	40.5	42.9	43.7	45.9	42.7	44.5	44.8
ShapeletSim	52.2	48.9	46.7	48.9	41.1	48.9	65	75.6	74.4	61.1	60	86.7	57.8	81.7	53.9
ShapesAll	58.3	47.2	62.3	77	74.3	71.3	76.8	81.2	80.8	85	83.8	87.2	86	86.7	80.3
SmallK.	65.3	57.3	51.5	66.9	72.8	46.1	64.3	67.5	67.5	64	64.8	69.9	64.5	69.6	39.7
SonyAIBOR.	74	65.6	93	63.7	80.9	70.4	72.5	73.7	81.5	74.2	71	73	68.1	70.4	82.2
SonyAIBOR.II	79	68.6	78.7	79	80.8	81.8	83.1	83.1	87.7	89.2	89.2	87.1	85.3	87.8	86.4
StarLightC.	79.1	90.7	79.1	95	96.9	91.9	90.7	89.5	91.8	96.2	96.2	86.8	88.3	92.6	95.7
Strawberry	81.9	93.8	76.5	95.7	97.3	91.9	94.1	94.3	94.3	95.4	95.7	94.3	94.3	94.6	98.9
SwedishLeaf	83.4	65.8	85.8	87	88.2	84.2	79.2	87.4	88.2	90.1	89.6	89.6	89.1	91.5	93.1
Symbols	89.5	62.7	79.8	90.2	79.3	87	95	95	94.1	95.3	96.3	94.9	96	96	91.9
SyntheticC.	92.7	81	96	94.3	97.3	92.3	99.3	99.3	97.3	99.3	99.7	97.3	98.7	99	98.7
ToeS.1	57.9	59.2	55.3	64.9	53.1	54.4	77.2	79.4	73.7	80.7	80.7	81.6	82	82.9	68.9
ToeS.2	56.2	50	63.8	76.9	58.5	54.6	83.8	89.2	87.7	74.6	71.5	75.4	78.5	89.2	84.6
Trace	82	79	80	78	93	73	100	100	99	100	99	93	99	99	84
Two_Patterns	73.3	71.8	69.9	72.4	97	94.1	90.5	90.5	88.1	97.8	98.5	94.7	97.4	97.1	91.8
TwoLeadECG	46.3	65.1	45.7	85.8	92.8	82.2	100	100	99.8	100	100	99.9	99.9	100	92.2
uW._X	89.1	78.4	89.1	93.9	94.4	89.4	89.2	96.6	96.3	93.5	93.8	96.3	95.3	96.8	78.8
uW._Y	66.9	60.1	66.5	76.2	78.3	65.9	72.8	77.4	79	77.9	77.5	76.9	77.1	80.5	72.4
uW._Z	58.5	56.3	55.6	69.4	71.4	62.8	63.4	69.3	72.3	71.6	69.8	70.2	68.5	72.6	72.7
UW.All	57.3	57.3	56.3	72	72.3	61.4	65.8	67.6	70.6	69.6	67.9	70	68.4	72.4	97.5
wafer	95.8	98.2	70.8	99.4	99.4	96	98	99.7	99.4	98	99.3	99.7	99.6	99.7	99.7
Wine	61.1	68.5	57.4	74.1	94.4	83.3	57.4	57.4	61.1	57.4	61.1	59.3	57.4	57.4	85.2
WordsS.	25.1	38.1	47.6	57.1	59.9	51.3	64.9	74.5	75.7	73	73	76.3	74.9	77.9	61.9
Worms	44.2	40.3	39	54.5	61	39	58.4	53.2	61	58.4	64.9	57.1	57.1	66.2	51.9
Worms.	57.1	55.8	51.9	63.6	68.8	49.4	62.3	57.1	67.5	64.9	62.3	67.5	61	68.8	67.4
yoga	60.6	68.4	54.2	81.5	82.4	57.2	85.3	84.4	85.6	85.6	86.5	87.2	0.83	87.7	84.1

References

1. Abdi, H.: Singular value decomposition (SVD) and generalized singular value decomposition. In: Encyclopedia of Measurement and Statistics, pp. 907–912. Sage, Thousand Oaks (2007)
2. Bagnall, A., Lines, J., Bostrom, A., Large, J., Keogh, E.: The great time series classification bake off: a review and experimental evaluation of recent algorithmic advances. Data Min. Knowl. Disc. **31**(3), 606–660 (2017)
3. Bagnall, A., Lines, J., Hills, J., Bostrom, A.: Time-series classification with COTE: the collective of transformation-based ensembles. IEEE Trans. Knowl. Data Eng. **27**(9), 2522–2535 (2015)
4. Batista, G.E., Wang, X., Keogh, E.J.: A complexity-invariant distance measure for time series. In: Proceedings of the 2011 SIAM International Conference on Data Mining, pp. 699–710. SIAM (2011)
5. Chen, L., Ng, R.: On the marriage of Lp-norms and edit distance. In: Proceedings of the Thirtieth International Conference on Very Large Data Bases, vol. 30, pp. 792–803. VLDB Endowment (2004)
6. Chen, L., Özsu, M.T., Oria, V.: Robust and fast similarity search for moving object trajectories. In: Proceedings of the 2005 ACM SIGMOD International Conference on Management of Data, pp. 491–502. ACM (2005)
7. Chi, Y., Porikli, F.: Connecting the dots in multi-class classification: from nearest subspace to collaborative representation. In: 2012 IEEE Conference on Computer Vision and Pattern Recognition (CVPR), pp. 3602–3609. IEEE (2012)
8. Demšar, J.: Statistical comparisons of classifiers over multiple data sets. J. Mach. Learn. Res. **7**(Jan), 1–30 (2006)
9. Ding, H., Trajcevski, G., Scheuermann, P., Wang, X., Keogh, E.: Querying and mining of time series data: experimental comparison of representations and distance measures. Proc. VLDB Endow. **1**(2), 1542–1552 (2008)
10. Fuhry, M., Reichel, L.: A new Tikhonov regularization method. Numerical Algorithms **59**(3), 433–445 (2012)
11. Górecki, T., Łuczak, M.: Using derivatives in time series classification. Data Min. Knowl. Disc. **26**, 1–22 (2013)
12. Górecki, T., Łuczak, M.: Non-isometric transforms in time series classification using DTW. Knowl.-Based Syst. **61**, 98–108 (2014)
13. Gustavo, E., Batista, A., Keogh, E.J., Tataw, O.M., Vinícius, M., de Souza, A., et al.: CID: an efficient complexity-invariant distance for time series. Data Min. Knowl. Disc. **28**(3), 634 (2014)
14. Jeong, Y.S., Jeong, M.K., Omitaomu, O.A.: Weighted dynamic time warping for time series classification. Pattern Recogn. **44**(9), 2231–2240 (2011)
15. Karampatziakis, N., Mineiro, P.: Discriminative features via generalized eigenvectors. In: Proceedings of the 31st International Conference on Machine Learning, ICML 2014, pp. 494–502. JMLR.org (2014)
16. Keogh, E.: Exact indexing of dynamic time warping. In: Proceedings of the 28th International Conference on Very Large Data Bases, pp. 406–417. VLDB Endowment (2002)
17. Keogh, E.J., Pazzani, M.J.: Derivative dynamic time warping. In: Proceedings of the 2001 SIAM International Conference on Data Mining, pp. 1–11. SIAM (2001)
18. Lee, K.C., Ho, J., Kriegman, D.J.: Acquiring linear subspaces for face recognition under variable lighting. IEEE Trans. Pattern Anal. Mach. Intell. **27**(5), 684–698 (2005)

19. Li, W., Tramel, E.W., Prasad, S., Fowler, J.E.: Nearest regularized subspace for hyperspectral classification. IEEE Trans. Geosci. Remote Sens. **52**(1), 477–489 (2014)
20. Lines, J., Bagnall, A.: Time series classification with ensembles of elastic distance measures. Data Min. Knowl. Disc. **29**(3), 565–592 (2015)
21. Liu, Y., Ge, S.S., Li, C., You, Z.: k-NS: a classifier by the distance to the nearest subspace. IEEE Trans. Neural Netw. **22**(8), 1256–1268 (2011)
22. Marteau, P.F.: Time warp edit distance with stiffness adjustment for time series matching. IEEE Trans. Pattern Anal. Mach. Intell. **31**(2), 306–318 (2009)
23. Pedagadi, S., Orwell, J., Velastin, S., Boghossian, B.: Local fisher discriminant analysis for pedestrian re-identification. In: Proceedings of the IEEE Conference on Computer Vision and Pattern Recognition, pp. 3318–3325 (2013)
24. Platt, J.C., Toutanova, K., Yih, W.: Translingual document representations from discriminative projections. In: Proceedings of the 2010 Conference on Empirical Methods in Natural Language Processing, pp. 251–261. Association for Computational Linguistics (2010)
25. Rakthanmanon, T., Campana, B., Mueen, A., Batista, G., Westover, B., Zhu, Q., Zakaria, J., Keogh, E.: Addressing big data time series: mining trillions of time series subsequences under dynamic time warping. ACM Trans. Knowl. Disc. Data (TKDD) **7**(3), 10 (2013)
26. Ratanamahatana, C.A., Keogh, E.: Three myths about dynamic time warping data mining. In: Proceedings of the 2005 SIAM International Conference on Data Mining, pp. 506–510. SIAM (2005)
27. Stefan, A., Athitsos, V., Das, G.: The move-split-merge metric for time series. IEEE Trans. Knowl. Data Eng. **25**(6), 1425–1438 (2013)
28. Sugiyama, M.: Local fisher discriminant analysis for supervised dimensionality reduction. In: Proceedings of the 23rd International Conference on Machine Learning, pp. 905–912. ACM (2006)
29. Zelnik-Manor, L., Perona, P.: Self-tuning spectral clustering. In: Advances in Neural Information Processing Systems, pp. 1601–1608 (2005)

Efficient Approximate Subsequence Matching Using Hybrid Signatures

Tao Qiu$^{(\boxtimes)}$, Xiaochun Yang, Bin Wang, Yutong Han, and Siyao Wang

School of Computer Science and Engineering, Northeastern University,
Shenyang 110819, Liaoning, China
{qiutao,hanyutong,wangsiyao}@stumail.neu.edu.cn,
{yangxc,binwang}@mail.neu.edu.cn

Abstract. In this paper, we focus on the problem of approximate subsequence matching, also called the read mapping problem in genomics, which is finding similar subsequences (A subsequence refers to a substring which has consecutive characters) of a query (DNA subsequence) from a reference genome under a user-specified similarity threshold k. Existing methods first extract subsequences from a query to generate signatures, then produce candidate positions using the generated signatures, and finally verify these candidate positions to obtain the true mapping positions. However, there exist two main issues in these works: (1) producing many candidate positions; and (2) generating large numbers of signatures, among which many signatures are redundant. To address the above two issues, we propose a novel filtering technique, called *hybrid signatures*, which can achieve a better balance between the filtering ability of signatures and the overhead of producing candidate positions. Accordingly, we devise an adaptive algorithm to produce candidate positions using hybrid signatures. Finally, the experimental results on real-world genomic sequences show that our method outperforms state-of-the-art methods in query efficiency.

Keywords: Read mapping · Approximate subsequence matching
Hybrid signatures

1 Introduction

Similar to web applications, another area that has recently witnessed a rapid surge in the amount of data being produced is the similar biosequence search. In this paper, we focus on the problem of finding similar subsequences of a query (DNA subsequence) in a large genome reference sequence, which is known as the read mapping problem in genomics. This problem shares core technical challenges with the similar substring matching problem studied in the database research literature.

The work is partially supported by the National Natural Science Foundation of China (Nos. 61572122, U1736104, 61532021).

© Springer International Publishing AG, part of Springer Nature 2018
J. Pei et al. (Eds.): DASFAA 2018, LNCS 10827, pp. 600–609, 2018.
https://doi.org/10.1007/978-3-319-91452-7_39

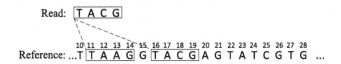

Fig. 1. An example of a read mapping with similarity threshold $k = 1$.

Given a query that is a DNA subsequence called a *read*, the read mapping problem is to find all subsequences in a reference sequence that are similar to the query under hamming distance. Figure 1 shows an example of mapping a read TACG to a reference under hamming distance threshold $k = 1$. There are two mapping positions, in which the read occurs approximately in position 11 and occurs exactly in position 16.

To solve this problem, a variety of algorithms have been proposed under a filtering-and-verification framework [1–3,7]. They first utilize the subsequences of a read as signatures and identify candidate positions by the matching positions of signatures, then verify these candidate positions to find true mapping results. The pigeonhole principle [6] is utilized to generate signatures from a read, that is *at least c subsequences must be exactly matched on a mapping result among $k+c$ non-overlapping subsequences of the read* ($c \geq 1$). Hence, given the $k + c$ non-overlapping subsequences of a read, among which any c subsequences compose a signature.

The filtering scheme with $k+1$ subsequences can avoid verifying a lot of candidate positions in most of cases, but (1) it still produces too many candidate positions when some subsequences are frequently occurred in the reference. In general, a signature with c subsequences occurs less frequently than the signature with only one subsequence, accordingly the filtering scheme with $k + c$ subsequences ($c > 1$) could produce fewer candidate positions than the $k + 1$ filtering scheme. However, (2) there has a high computational cost for producing candidates since $\binom{k+c}{c}$ signatures[1] must be obtained, among which some signatures could be redundant.

To solve these problems, we propose a flexible way to generate signatures by extending individual subsequence to multiple subsequences signature, and guarantee that all mapping positions are found. Unlike existing methods which require any signature contains c subsequences, our method can generate signatures that contain a different number of subsequences, which are called *hybrid signatures*. Hybrid signatures can achieve a better balance between the filtering ability of signatures and the overhead of producing candidate positions. The following are the main contributions of the paper.

- We propose a novel filtering technique, called hybrid signatures, for the read mapping problem, which can utilize the signatures with variable number of segments.
- We devise an adaptive algorithm to generate hybrid signatures and use them to produce candidate positions.

---

[1] $\binom{n}{r}$ means the number of r-combinations for a set with size n.

– We conduct experiments using real genomic datasets, and demonstrate that our method outperforms state-of-the-art methods in query efficiency.

The rest of the paper is organized as follows. Section 2 gives the problem formulation and related work. Section 3 proposes the filtering scheme using hybrid signatures and the algorithm of producing candidate positions using hybrid signatures. Experimental results are presented and analyzed in Sect. 4. Section 5 concludes the paper.

2   Preliminaries

(1) Problem Formulation

A genomic sequence is a concatenation of multiple characters in the alphabet $\Sigma = \{A, C, G, T, N\}$. For a sequence S, we use $|S|$ to denote its length, $S[i]$ to denote its i-th character (starting from 1), and $S[i, j]$ to denote the subsequence from its i-th character to its j-th character. A subsequence with consecutive characters is called a **segment**. The following two similarity measures are widely used in the read mapping problem [2,3].

Definition 1. *(Hamming distance) For two sequences S_1 and S_2 that have the same length (i.e. $|S_1| = |S_2|$), the hamming distance between S_1 and S_2 is the number of positions i such that $S_1[i] \neq S_2[i]$ where $1 \leq i \leq |S_1|$, denoted by $hd(S_1, S_2)$.*

For example, given two strings $S_1 = $ GAACC and $S_2 = $ AACCT, the hamming distance $hd(S_1, S_2)$ is 3 since $S_1[1] \neq S_2[1]$, $S_1[3] \neq S_2[3]$ and $S_1[5] \neq S_2[5]$. Given a similarity measure, two sequences are deemed similar if their distance is within a user-specified similarity threshold k.

Problem Definition. Given a read Q, a reference T and a similarity threshold k for a similarity measure, the *read mapping problem* is to find all positions p of T such that the similarity between Q and each subsequence of T starting at p is within k.

Positional Inverted Index Based on q-grams. A *q-gram* is a subsequence of S with length q. All q-grams of S can be obtained by sliding a window of length q on S, e.g., there are three 2-grams (TA, AC and CG) for the read in Fig. 1. For a q-gram g, we use $I(g)$ to denote the inverted list of g which records the starting positions of matches of g in T, and $|I(g)|$ is the list size. The *positional q-gram inverted index* for a reference T includes the inverted lists of all q-grams in T. We use 2-gram for all running examples.

(2) Related Work

There are a large number of techniques for supporting approximate subsequence matching. In the database research literature, many methods have been proposed for the approximate substring matching problem [4,5,7]. These methods generally follow a filtering and verification framework and the pigeonhole principle is used to identify candidate positions in the filtering phase. Yang *et al.* utilize an even partition scheme to do approximate substring matching on compressed data [4]. Wang *et al.* also use the even partition scheme to identify candidates for the problem of finding approximate substring pairs from two string collections [7].

In genomics, there are a number of techniques proposed for solving read mapping problem. Most of them focus on reducing the number of candidates. Hobbes [1] is a pioneer work that identifies candidates using inverted lists of non-overlapping q-grams. Hobbes2 [2] and Hobbes3 [3] further reduce the number of candidates by utilizing additional prefix q-grams. Masai [8] focus on reducing candidates for multiple reads by building an index on the input reads. A recent work Bitmapper [9] reduces the mapping time by verifying multiple candidate positions at the same time. Other popular tools (e.g., Bowtie [10], Bowtie2 [11], BWA [12]) achieve a significant improvement in mapping time by identifying a few best mapping positions for each read. However, they cannot guarantee to find all mapping positions for the reads, which is often more desirable for many applications [13,14].

3 A Filtering Scheme Using Hybrid Signatures

In this section, we first propose the filtering scheme using hybrid signatures, then give the candidate generation algorithm that uses hybrid signatures.

3.1 Hybrid Signatures

According to the pigeonhole principle, we know k *errors cannot modify all $k + 1$ non-overlapping segments of a sequence*, since an error can only affect at most 1 segment. This property can be generalized by selecting $k + c$ non-overlapping segments.

Lemma 1. *Given two sequences S_1, S_2 and the similarity threshold k, if S_1 and S_2 are similar, then at least c segments in a set of $k + c$ non-overlapping segments of S_1 are exactly matched on S_2.*

Given a query Q, let $\mathcal{S}_{k+c}$ be a set of $k + c$ non-overlapping segments of Q. According to Lemma 1, at least c segments in $\mathcal{S}_{k+c}$ must be exactly matched on a mapping result of Q. That is, a *signature* is composed of c segments in $\mathcal{S}_{k+c}$, and we denote it by $t = (s_1, s_2, \cdots, s_c)$. If $c > 1$, we say t is a *multi-segments signature*.

Given a sequence S and a segment s_i of S, the subsequences of S except s_i is called the *complementary subsequence* of s_i, denoted by $S_{\bar{s}_i}$. Actually, when s_i can exist in the middle of S, $S_{\bar{s}_i}$ could consists of two subsequences.

Extended Multi-segments Signatures. For the $k + 1$ filtering scheme with segments set $\mathcal{S}_{k+1}$, let s_i be a segment in $\mathcal{S}_{k+1}$. If s_i is regarded as a signature,

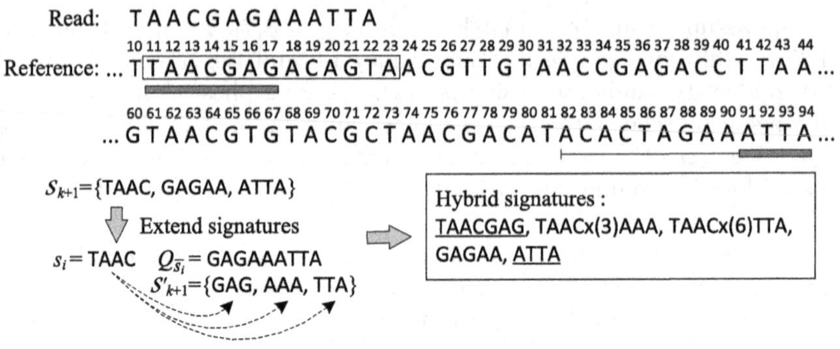

Read: T A A C G A G A A A T T A

10 11 12 13 14 15 16 17 18 19 20 21 22 23 24 25 26 27 28 29 30 31 32 33 34 35 36 37 38 39 40 41 42 43 44
Reference: ... T T A A C G A G A C A G T A A C G T T G T A A C C G A G A C C T T A A ...

60 61 62 63 64 65 66 67 68 69 70 71 72 73 74 75 76 77 78 79 80 81 82 83 84 85 86 87 88 89 90 91 92 93 94
... G T A A C G T G T A C G C T A A C G A C A T A C A C T A G A A A T T A ...

S_{k+1}={TAAC, GAGAA, ATTA}

⬇ Extend signatures

s_i = TAAC $Q_{\bar{s}_i}$ = GAGAAATTA
S'_{k+1}={GAG, AAA, TTA}

Hybrid signatures :
TAACGAG, TAACx(3)AAA, TAACx(6)TTA,
GAGAA, ATTA

Fig. 2. An example of using hybrid signatures to produce candidate positions.

this means k errors can only modify the complementary subsequence of s_i (i.e., $Q_{\bar{s}_i}$). Next, we continue to apply the $k+1$ filtering scheme for the complementary subsequence $Q_{\bar{s}_i}$ and compute $k + 1$ non-overlapping segments S'_{k+1} for $Q_{\bar{s}_i}$, so we get at least one segment in S'_{k+1} must be exactly matched on a mapping result. In this case, s_i and one segment s'_j in S'_{k+1} are required to be exactly matched on a mapping result simultaneously. In other words, s_i and s'_j compose a signature (s_i, s'_j), called *extended multi-segments signature* (a.k.a. extended signatures) of s_i and denoted by t_e.

In this way, the signatures are not required to contain certain number of segments while the non-extended signature is still an individual segment from S_{k+1}. We call these signatures **hybrid signatures**, and use S_h to denote them. In this work, we only show computing hybrid signatures based on $k + 1$ filtering scheme.

Example 1. Consider the running example in Fig. 2, $S_{k+1} = \{\text{TAAC}, \text{GAGAA}, \text{ATTA}\}$ is a non-overlapping segment set of the read Q, 6 candidate positions are produced using S_{k+1} as signatures. Next, we extend the segment $s_i = \text{TAAC}$, another set $S'_{k+1} = \{\text{GAG}, \text{AAA}, \text{TTA}\}$ is obtained from $Q_{\bar{s}_i} = \text{GAGAAATTA}$, so we get the extended signatures (TAAC,GAG), (TAAC,AAA) and (TAAC,TTA). Since GAGAA and ATTA are not extended, the final hybrid signatures are {(TAAC,GAG),(TAAC,AAA),(TAAC,TTA), GAGAA, ATTA}. By the extended multi-segments signatures, 4 candidates produced by TAAC are reduced, and only 2 candidate positions are produced.

Note that an extended signature t_e can be further extended in the same way by computing $k + 1$ non-overlapping segments for the complementary subsequence of t_e.

3.2   Candidate Generation Using Hybrid Signatures

We first review the method of producing candidate positions using inverted index [3], this lays the foundation for the adaptive algorithm which utilizes hybrid signatures.

(1) *Producing Candidate Positions for Signatures Using q-gram Inverted Index*

Let s_i be a segment of a query Q with starting position π_i (i.e., $Q[\pi_i, \pi_i + |s_i| - 1] = s_i$). For a matching position p_{s_i} of s_i on the reference, a relative position $\bar{p}_{s_i} = p_{s_i} - \pi_i + 1$ is called a *normalized position* of s_i. Obviously, if s_i is a signature, then the normalized positions of s_i are the corresponding candidate positions. Given the inverted list $I(s_i)$ of s_i, $I^N(s_i) = \{p_{s_i} - \pi_i + 1 | p_{s_i} \in I(s_i)\}$ is the normalized positions of s_i.

First, we consider a signature t is a q-gram of Q. In this case, we can directly obtain the list $I(t)$ of matching positions of t from the inverted index, and $I^N(t)$ can also be obtained since the starting position of t on Q is easily computed. For the signatures which are variable-length segments or multi-segments signatures, the following properties that initially proposed in [3] are used to compute their normalized position lists.

- If a signature t is a variable-length segment, let $\{g_1, \cdots, g_n\}$ be the overlapping or consecutive q-grams that composed of t, then $I^N(t) = \cap_{i=1}^{n}(I^N(g_i))$.
- If a signature t is a multi-segments signature, then t can be represented by $t = (s_1, \cdots, s_n)$. The list of normalized positions $I^N(t)$ is computed by $\cap_{i=1}^{n}(I^N(s_i))$.

(2) *Adaptively Producing Candidates Using Hybrid Signatures*

Next, we introduce how to use hybrid signatures to produce candidate positions. The algorithm is described in Algorithm 1, called CandGenerator, which is a recursive algorithm and each recursion is used to determine if a signature is needed to be extended and produce the candidate positions of extended signatures.

Initially, we use the read Q and all positions of T as the inputs S and $I^N(t)$ of CandGenerator. In each recursion, we first check if S can be divided into $k+1$ non-overlapping segments, each of which has length $\geq q$, by checking if $|S| \geq (k+1) \cdot q$ (line 1). The reason is that only the length of each segment is not less than q, the q-gram inverted index can be used. Next, $\mathcal{S}_{k+1}$ is computed from S by a given strategy (line 3), (e.g., even partition [4,5,7] or optimized partition strategy [3]). If t is extended, each segment s_i in $\mathcal{S}_{k+1}$ is used to compose a new extended signature t' with t (line 7).

The algorithm checks if the new signature t' is further extended by recursively invoking CandGenerator (lines 8–9). Basically, we can extend each t if $\mathcal{S}_{k+1}$ is available. However, there is a high computational cost to extend all signatures. Hence, we extend a signature t only if $|I^N(t)|$ is greater than an user-specific threshold α (line 4).

Suppose there are m signatures are extended and C is the cost of computing $k+1$ non-overlapping segments from a sequence. In each recursion, the candidate positions can be computed in time $\mathcal{O}((k+2)|T|)$ since the size of any normalized list is bounded by $|T|$. To summarize, CandGenerator has time complexity $\mathcal{O}(m(C + (k+2)|T|))$.

Algorithm 1. CandGenerator - Producing Candidate Positions

Input : Normalized list $I^N(t)$ of a signature t, a subsequence S of query read Q
Output : The list of candidate positions

1 **if** $|S| \geq (k+1) \cdot q$ **then**
2 $\quad$ **return** $I^N(t)$
3 Compute $k+1$ non-overlapping segments $\mathcal{S}_{k+1}$ from S;
4 **if** $|I^N(t)| > \alpha$ **then**
5 $\quad$ $I^N_{temp} \leftarrow \emptyset$; // initialize a temp list
6 $\quad$ **for** *each segment s_i in $\mathcal{S}_{k+1}$* **do**
7 $\quad\quad$ $I^N(t') \leftarrow I^N(t) \cap I^N(s_i)$;
8 $\quad\quad$ $S_{\overline{t_e}} \leftarrow S - s_i$; // get the complementary subsequence
9 $\quad\quad$ $I^N_{temp} \leftarrow I^N_{temp} \cup$ CandGenerator$(I^N(t'), S_{\overline{t_e}})$;
10 $\quad$ **return** I^N_{temp}
11 **else**
12 $\quad$ **return** $I^N(t)$

---

4   Experiments

In this section, we present experimental results on three real-world datasets.

Experimental Setup. The experiments were carried out on a PC with an Intel i7-6700 3.4 GHz Processor and 8 GB RAM, running Ubuntu 14.04 OS. All algorithms were implemented in C++. The following three datasets are used as reference sequences.

- *Human genome.* The Human genome (HG18)[2] with length 1 billion.
- *Mouse genome.* The Mouse genome (MGSCv37 chr1[3]) with length 198 million.
- *Drosophila genome.* Drosophila genome[4] extracted from FlyBase 5.42 has 165 million characters.

Positional 9-gram inverted index is built for each genome sequence. We picked 10k 100-bp single-end reads (i.e., each read contains 100 characters) from the "1000 genome project[5]" as queries for each genome. For our algorithm, we use an even partition strategy to compute $k+1$ non-overlapping segments $\mathcal{S}_{k+1}$ from a sequence.

Evaluating Different Filtering Schemes. The first experiment compares the filtering ability of different filtering schemes. Figure 3(a)–(c) show the number

---

[2] http://hobbes.ics.uci.edu/.
[3] http://hgdownload.cse.ucsc.edu/goldenPath/mm9/chromosomes/.
[4] http://fruitfly.org/sequence/.
[5] ftp://ftp-trace.ncbi.nih.gov/1000genomes/.

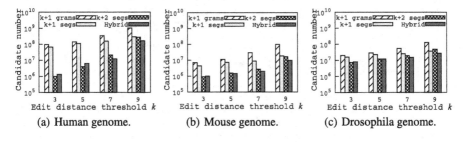

Fig. 3. Candidate number of different filtering schemes.

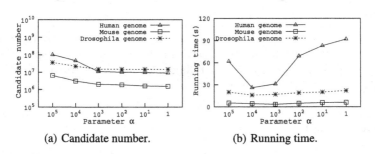

Fig. 4. Impact of the parameter α for the adaptive method under $k = 7$.

of candidates produced by different filtering schemes on three references. More candidates are produced as the increasing of similarity threshold k since more errors can occur in a mapping result. Among four filtering schemes, Hybrid produces the least candidates under most of thresholds, such as $k = 5$, 7 and 9 in Mouse and Drosophila genomes, $k = 7$ and 9 in Human genome. On the contrary, $k + 1$ grams always gets the most candidates.

Evaluating Adaptive Method. Recall the adaptive algorithm in Sect. 3.2, an user-specific threshold α is used to determine if a signature is needed to be extended. In this experiment, we test the impact of α for the adaptive method under $k = 7$. As α decreases, fewer candidates are produced since more extended signatures are utilized with smaller α, as shown in Fig. 4(a). In Fig. 4(b), the running time decreases when varying α from 10^5 to 10^4, but it increases after 10^4 since there has a high computational cost for producing candidates if too many extended signatures are used. This phenomenon is most obvious in Human genome, the reason is that more candidates can be reduced by a few extended signatures in the large reference.

Comparison with Alternative Algorithms. In the last experiment, we compare the performance of the adaptive method with the state-of-the-art methods Hobbes3 [3], BWA [12], BWA* [12] (approximate algorithm, only parts of results are found) and Bitmapper [9]. The running time of methods is shown in Fig. 5. We can see Hybrid achieves the best performance when $k \geq 5$ in three genomes, Hobbes3 and Bitmapper are the followed methods. BWA* is faster than other

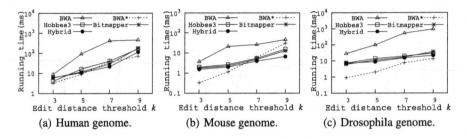

Fig. 5. Performance comparison with alternative algorithms.

methods in some cases, such as $k = 1$ and 2 in Human and Mouse genomes. However, BWA is much slower than other methods, e.g., 452 s is spent by BWA when $k = 9$ in Human genome, while Hybrid, Hobbes3 and Bitmapper only spend 113 s, 175 s and 163 s, respectively.

5   Conclusion

In this paper, we study the problem of efficient read mapping. We propose a novel filtering technique, named hybrid signatures, to improve the candidate generation for a read query. Hybrid signatures achieve a better balance between the filtering ability of signatures and the overhead of producing candidates. We design an adaptive algorithm to produce candidates using hybrid signatures. Experiments are conducted on the real genomes and show that our proposed algorithm outperforms the existing algorithms.

References

1. Ahmadi, A., Behm, A., Honnalli, N., Li, C., Xie, X.: Hobbes: optimized gram-based methods for efficient read alignment. Nucleic Acids Res. **40**, e41 (2012)
2. Kim, J., Li, C., Xie, X.: Improving read mapping using additional prefix grams. BMC Bioinf. **15**(1), 42 (2014)
3. Kim, J., Li, C., Xie, X.: Hobbes3: dynamic generation of variable-length signatures for efficient approximate subsequence mappings. In: ICDE 2016. IEEE (2016)
4. Yang, X., Wang, B., Li, C., Wang, J., Xie, X.: Efficient direct search on compressed genomic data. In: ICDE 2013, Brisbane, Australia, 8–12 April 2013, pp. 961–972 (2013)
5. Yang, X., Wang, Y., Wang, B., Wang, W.: Local filtering: improving the performance of approximate queries on string collections. In: SIGMOD 2015, pp. 377–392 (2015)
6. Qin, J., Wang, W., Xiao, C., Lu, Y., Lin, X., Wang, H.: Asymmetric signature schemes for efficient exact edit similarity query processing. ACM Trans. Database Syst. **38**(3), 16 (2013)
7. Wang, J., Yang, X., Wang, B., Liu, C.: LS-Join: local similarity join on string collections. IEEE Trans. Knowl. Data Eng. **29**(9), 1928–1942 (2017)

8. Siragusa, E., Weese, D., Reinert, K.: Fast and accurate read mapping with approximate seeds and multiple backtracking. Nucleic Acids Res. **41**, e78 (2013)

9. Cheng, H., Jiang, H., Yang, J., Xu, Y., Shang, Y.: BitMapper: an efficient all-mapper based on bit-vector computing. BMC Bioinf. **16**, 192 (2016)

10. Langmead, B., Trapnell, C., Pop, M., Salzberg, S.: Ultrafast and memory-efficient alignment of short dna sequences to the human genome. Genome Biol. **10**, r25 (2009)

11. Langmead, B., Salzberg, S.: Fast gapped-read alignment with Bowtie 2. Nat. Methods **9**, 357–359 (2012)

12. Li, H., Durbin, R.: Fast and accurate short read alignment with Burrows-Wheeler transform. Bioinformatics **25**, 1754–1760 (2009)

13. Newkirk, D., Biesinger, J., Chon, A., Yokomori, K.: AREM: aligning short reads from ChIP-sequencing by expectation maximization. J. Comput. Biol. **18**, 1495–1505 (2011)

14. Roberts, A., Pachter, L.: Streaming fragment assignment for realtime analysis of sequencing experiments. Nat. Methods **10**, 71–73 (2013)

Trajectory and Streaming Data

MDTK: Bandwidth-Saving Framework for Distributed Top-k Similar Trajectory Query

Zhigang Zhang, Jiali Mao, Cheqing Jin[(✉)], and Aoying Zhou

School of Data Science and Engineering, East China Normal University,
Shanghai, China
{zgzhang,jlmao1231}@stu.ecnu.edu.cn, {cqjin,ayzhou}@sei.ecnu.edu.cn

Abstract. During the past decade, with the popularity of smartphones and other mobile devices, big trajectory data is generated and stored in a distributed way. In this work, we focus on the DTW distance based top-k query over the distributed trajectory data. Processing such a query is challenging due to the limited network bandwidth and the computation overhead. To overcome these challenges, we propose a communication-saving framework MDTK (<u>M</u>ulti-resolution based <u>D</u>istributed <u>T</u>op-<u>K</u>). MDTK sends the bounding envelopes of the reference trajectory from coarse to finer-grained resolutions and devises a level-increasing communication strategy to gradually tighten the proposed upper and lower bound. Then, distance bound based pruning strategies are imported to reduce both the computation and communication cost. Besides, we embed techniques including: indexing, early-stopping and cascade pruning, to improve the query efficiency. Extensive experiments on real datasets show that MDTK outperforms the state-of-the-art method.

Keywords: Top-k query · Communication cost · DTW distance
Trajectory data

1 Introduction

In this paper, we aim at processing such a top-k query: "given a reference trajectory Q and a trajectory dataset D which is stored on distributed nodes, compare Q against all trajectories in D to find the k most similar ones". The system model is given in Fig. 1. Here, we have a coordinator site that communicates to all M remote sites, each of which stores a huge collection of trajectories. The coordinator acts as a search engine by accepting query reference Q and searching the k most similar trajectories across M remote sites. This query is practical in real world scenarios. For example, video cameras are set up above many roads to capture the traces of moving vehicles continuously. The transport department is interested in finding trajectories similar to a given driving pattern such as "waving" or "swerving" to detect potential drunk drivers [8].

© Springer International Publishing AG, part of Springer Nature 2018
J. Pei et al. (Eds.): DASFAA 2018, LNCS 10827, pp. 613–629, 2018.
https://doi.org/10.1007/978-3-319-91452-7_40

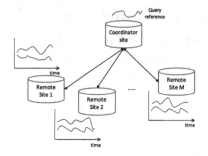

Fig. 1. Distributed processing model

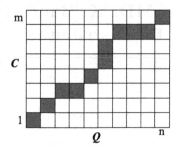

Fig. 2. Warping path for DTW

Given the system model and the query problem, there are two processing strategies: (i) Collect all distributed data into one site, and then use centralize-based solutions to process the query. (ii) Send the query reference $\mathcal{Q}$ to all remote sites, in which the pairwise similarities are computed precisely. The first strategy is unacceptable due to limited storage, communication resources and privacy issues. While, the second one suffers the problem of communication overhead especially when the amount of remote sites is tremendous, or the size of reference is significant.

In our prior work [21], we have considered the top-k query under the same system model, by adopting the second strategy and presenting a communication-efficient approach for Euclidean distance similarity measure. In that work, the reference trajectory is decomposed into multi-resolution Haar wavelet coefficients, and the coefficients are sent from the coarsest level to the finer ones. As finer-grained coefficients are transmitted, tighter similarity bounds are derived to prune candidates. This Haar wavelet-based solution is good, but it is much tied to Euclidean distance and cannot extend to other measures, such as DTW (Dynamic Time Warping) distance which is believed a better distance measure for trajectory data especially when trajectories are sampled at different paces.

In this paper, we research the DTW distance based top-k query. Although the DTW distance is powerful, the computation complexity is quadratic in the length of trajectories. To accelerate the top-k query, a common way is to do early pruning of unqualified candidate using computationally cheaper lower bounds. [14] discussed existing lower bounds and showed the trade-off between their tightness and computational complexity. BandWidth [7] is the only work that introduced how to apply such lower bounds to reduce the communication cost. However, none of existing work simultaneously take the time efficiency and communication into consideration.

To reduce the communication cost in the proposed query, we apply the multi-resolution based BE (Bounding Envelope, a series of MBRs) structure to decompose the trajectory data. At first, the reference is divided into a series of disjoint segments, each of which is represented by its MBR. These MBRs consist the coarsest level BE. Then, each segment is further divided into shorter ones to compose a finer resolution which discloses more detailed information about the reference. Based on the multi-resolution property, the coordinator sends BEs of $\mathcal{Q}$ in a level-wise manner. Then, we propose new upper and lower bounds for DTW distance to achieve a better pruning result than the state-of-the-art method which uses only lower bounds. Finally, we devise our communication and computation efficient algorithm MDTK to process the query. Extensive experiments show that our algorithm outperforms state-of-the-art method. The main contributions of this work are summarized as follows:

- We propose upper and lower bounds for DTW by using BE of the reference, and show that these bounds are tightened when finer-grained BEs are used.
- We devise MDTK framework to save communication cost for the proposed query. MDTK prunes candidates levelly by utilizing the gradually tighten bounds and guarantees no false dismissal. Besides, MDTK embeds three computation speed-up techniques to improve the time efficiency.
- We conduct extensive experiments to evaluate the performance of MDTK and show its superiority over existing methods.

The rest of this paper is organized as follows. Section 2 discusses the related work. In Sect. 3, we define the problem formally and review background knowledge. Furthermore, the multi-resolution based DTW distance bounds are introduced in Sect. 4. We introduce MDTK framework in Sect. 5 and evaluate its performance in Sect. 6. Finally, a brief conclusion is provided in Sect. 7.

2   Related Work

Distributed Trajectory Similarity Queries. [12] proposes four schemes to improve the time efficiency of the top-k query. These frameworks executes at different parallelism to balance the communication cost and time efficiency. SmartTrace [4] and SmartTrace[+] [5] take communication cost into consideration from the perspective of crowd-sourcing. They only consider a special case where each remote site (represented by a smart phone) only maintains one trajectory. Whereas, we consider a more general scenario where each site maintains multiple trajectories. So, the proposed methods in these work [4,5,12] cannot solve the proposed query. Besides, [20] targets at the scenario where each trajectory is divided into a few subsequences that are stored in different remote sites. Its goal is to solve the top-k query quickly. Moreover, [16] treats trajectory as a sequence of segments and proposes a segment based framework to process the query on top of Spark. Both bitmap index and a dual index are used to improve the query efficiency. However, communication cost is not considered in the above two work.

Data Reduction for Time Series. To efficiently perform the similarity query, data/dimension reduction techniques have been widely used. [6] utilizes Discrete Fourier Transform (DFT) to reduce the number of dimensions and the transformed result is indexed by an R-tree. Other techniques include Singular Value Decomposition (SVD) [9], Discrete Wavelet Transform (DWT) [13], Piecewise

Aggregate Approximation (PAA) [18], Adaptive Piecewise Constant Approximation (APCA) [2] and many other variants. [1] focuses on spatio-temporal data.

Bandwidth Optimizations for the Query. To reduce the communication cost in the distributed query, [17,21] proposed level-wise approaches by leveraging the multi-resolution property of Haar wavelets to decompose the query reference, in which only small fraction of Haar coefficients need to be sent to remote sites for pruning candidates. In these work, [21] is specially designed for multi-dimensional time series data. However, these work is Haar wavelet-based and only applies for Euclidean distance based cases. Although Chan [3] showed that an approximation of DTW distance can be computed by using Haar wavelet, this method leads to the problem of false dismissal which is usually unacceptable. BandWidth [7] is the only work that investigated the case when DTW is adopted as the distance measure and saves communication cost simultaneously. It builds BEs of multi-resolution for the reference and sends them from the coarsest level to the finest one to prune candidates gradually by using lower bounds. However, only a small fraction of remote sites are joined to prune candidates at one time, and other sites should be waiting until they are informed. So, it usually waits a long time until all sites are traversed. Besides, it prunes candidates by only using lower bound and converges slower than the upper and lower bound based prune-strategy.

3   Preliminary

3.1   Problem Statement

The trajectory of a moving object can be represented as a time series in which each point is a d-dimensional data that contains temporal, spatial and other information such as speed, angle. So, given a reference trajectory $Q = \{q_1, q_2, \cdots, q_n\}$ of length n and a candidate $C = \{c_1, c_2, \cdots, c_m\}$ of length m, where $q_i, c_j \in R^d$ ($1 \leq i \leq n, 1 \leq j \leq m$). The DTW distance between them is defined as:

$$DTW(Q,C) = f(n,m) \tag{1}$$

$$f(i,j) = \begin{cases} 0 & if \quad i = 0 \wedge j = 0, \\ \infty & if \quad i = 0 \otimes j = 0, \\ d(q_i, c_j) + min\{f(i-1, \\ \quad j-1), f(i-1,j), f(i,j-1)\} & otherwise \end{cases}$$

In the above definition, $d(q_i, c_j)$ represents the squared Euclidean distance between q_i and c_j. Due to different scales of the attributes, data in each dimension should be normalized. The whole computation procedure can be viewed as searching the optimal path from the warping matrix in which each $cell[i,j]$

Table 1. Notations

Symbol	Definition
Q	A query reference of length n denoted as $\{q_1, \cdots q_n\}$
C	A candidate trajectory of length m denoted as $\{c_1, \cdots c_m\}$
s_l^i	The MBR of i-th segment in the l-th level resolution
Q_l	The l-th level BE of Q
QL_l	The level-l lower bound of Q, denoted as $\{qL_{l,1}, \cdots, qL_{l,n}\}$
QU_l	The level-l upper bound of Q, denoted as $\{qU_{l,1}, \cdots, qU_{l,n}\}$
k	The number of requested results
$q_{l,i}$	The level-l MBR for point q_i, $q_{l,i} = \langle qL_{l,i}, qU_{l,i} \rangle$

denotes the cost to math q_i and c_j (shown in Fig. 2). The optimal path has the following constraints: (i) it starts from $cell[i,j]$ and ends in $cell[n,m]$; (ii) it only can move one step in each cell and only can move in the following three possible directions in each cell: up, right and up-right; (iii) the summarization of cost in the optimal path should be the minimal of all possible paths. It needs to mention that although DTW measures a distance-like quantity between two given sequences, it doesn't guarantee the triangle inequality to hold. So, indexing based methods cannot be applied as they don't guarantee no false dismissal.

Next, we list important notations in Table 1 and formalize the problem of Distributed Top-k (DTK) trajectory query below:

Definition 1 ($DTK(Q, D, distance, k)$). *Given reference Q, trajectory dataset D and result set size k, this query returns a result set S such that: (i) $|S| = k$, $S \subseteq D$, and (ii) $\forall C \in S$, $C' \in D - S$, $DTW(Q,C) \leq DTW(Q,C')$.*

Our goal is to minimize the communication time cost and improve the time efficiency of the query.

3.2 Multi-resolution Bounding Envelopes

To reduce the communication cost, our solution only sends some sketch data of the reference to compute lower and upper bounds for each candidate. Existing lower bounds usually use the maximum and minimum values to represent original time series. In this work, we divide the trajectory into a few disjoint segments and compute an MBR for each segment. The sequence of such MBRs consists a BE for the original trajectory. Finally, multiple resolutions of BEs are constructed to obtain sketch data of different granularities.

$$ub^k = \max(q_i^k : q_j^k), \quad lb^k = \min(q_i^k : q_j^k); \ (1 \leq i < j \leq n; 1 \leq k \leq d) \quad (2)$$

Bounding Envelope. There have been many techniques proposed for segmenting time series data [7,11]. Assume that the reference Q is segmented into n'

Table 2. The multi-resolution representation of $\mathcal{Q}$

$\mathcal{Q}_0$	$s_0^0 = [2, 5, 4]$			
$\mathcal{Q}_1$	$s_1^0 = [2, 5, 2]$		$s_1^1 = [3, 4, 2]$	
$\mathcal{Q}_2$	$s_2^0 = [5, 5, 1]$	$s_2^1 = [2, 2, 1]$	$s_2^2 = [4, 4, 1]$	$s_2^3 = [3, 3, 1]$
$\mathcal{Q}$	5	2	4	3

disjoint but time-adjacent segments. As we will segment the trajectory into different granularities, we use $s_l^p = [lb, ub, len]$ to denote the MBR data of a trajectory segment $\{\boldsymbol{q}_i, \boldsymbol{q}_{i+1}, \cdots, \boldsymbol{q}_j\}$, where l represents the level number, p represents the position number in that level. lb and ub of s_l^p are computed according to Eq. 2, and len is length of the segment. The list of $\{s_l^1, \cdots s_l^{n'}\}$ consists the level-l BE of $\mathcal{Q}$, which is denoted as $\mathcal{Q}_l$. At level-0, the whole reference is viewed as a segment. Then, we cut each segment in level-l into R disjoint ones to get the level-$(l+1)$ segments. This procedure stops until no segments can be divided. An example of one-dimensional time series is given in Table 2, where $R = 2$, shows the multi-resolution BEs of $\mathcal{Q} = \{5, 2, 4, 3\}$. Without loss of generality, in the following examples and our experiment we also use $R = 2$ for ease of exposition. It needs to mention that we may also apply unequal-sized segmentation technique proposed in [7] to get trajectory segments.

Multi-resolution Property. Given BE $\mathcal{Q}_l$, we can construct a boundary for $\mathcal{Q}$ which is composed of a lower bound sequence $QL_l = \{qL_{l,0}, \cdots, qL_{l,n-1}\}$ and a upper bound sequence $QU_l = \{qU_{l,0}, \cdots, qU_{l,n-1}\}$. Basically, QL_l and QU_l are constructed by repeating the lb and ub values by len times of the level-l respectively. For example, given $\mathcal{Q}_1$ of Table 2, $QL_1 = \{2, 2, 3, 3\}$ and $QU_1 = \{5, 5, 4, 4\}$. Obviously, QL_l and QU_l have the same length as the original query reference $\mathcal{Q}$. Let $\boldsymbol{q}_{l,i} = \langle qL_{l,i}, qU_{l,i} \rangle$, and we have the following conclusion: $qL_{l+1,i}^k \leq qL_{l,i}^k \leq q_i^k \leq qU_{l,i}^k \leq qU_{l+1,i}^k$, where $1 \leq i \leq n, 1 \leq k \leq d$. Besides, we observed that half data of $\mathcal{Q}_{l+1}$ comes from $\mathcal{Q}_l$ when $R = 2$. So, when moving from level l to $l+1$, we only need to transfer the other half data.

4 Multi-resolution Based Bounds

In the above section, we show that the BE can give an approximation of original trajectory with fewer data. In this section, our goal is to give a similarity range for each candidate after receiving the BE of the reference.

4.1 Distance Bounds for Points

Given two points $\boldsymbol{q}_i$ and $\boldsymbol{c}_j$ which are from trajectory $\mathcal{Q}$ and $\mathcal{C}$ respectively, and $\boldsymbol{q}_{l,i}$ is constructed from $\mathcal{Q}_l$. Then, we get Euclidean distance range for $\boldsymbol{q}_i$ and $\boldsymbol{c}_j$ by using $\boldsymbol{q}_{l,i}$.

Lemma 1. $d_{lb}(\boldsymbol{q}_{l,i}, \boldsymbol{c}_j) \leq d(\boldsymbol{q}_i, \boldsymbol{c}_j) \leq d_{ub}(\boldsymbol{q}_{l,i}, \boldsymbol{c}_j)$, where $\boldsymbol{q}_{l,i} = \langle qL_{l,i}, qU_{l,i} \rangle$, $d_{lb}(\boldsymbol{q}_{l,i}, \boldsymbol{c}_j)$ and $d_{ub}(\boldsymbol{q}_{l,i}, \boldsymbol{c}_j)$ are computed below:

$$d_{lb}(\boldsymbol{q}_{l,i}, \boldsymbol{c}_j) = \sum_{k=1}^{d} \begin{cases} (c_j^k - qU_{l,i}^k)^2 & if \ c_j^k > qU_{l,i}^k, \\ 0 & if \ qL_{l,i}^k \leq c_j^k \leq qU_{l,i}^k, \\ (qL_{l,i}^k - c_j{}^k)^2 & if \ c_j^k < qL_{l,i}^k. \end{cases} \tag{3}$$

$$d_{ub}(\boldsymbol{q}_{l,i}, \boldsymbol{c}_j) = \sum_{k=1}^{d} \begin{cases} (c_j^k - qL_{l,i}^k)^2 & if \ c_j^k > qU_{l,i}^k, \\ (qU_{l,i}^k - qL_{l,i}^k)^2 & if \ qL_{l,i}^k \leq c_j^k \leq qU_{l,i}^k, \\ (qU_{l,i}^k - c_j{}^k)^2 & if \ c_j^k < qL_{l,i}^k. \end{cases} \tag{4}$$

Lemma 1 is easy to prove according to the definition of Euclidean distance. In Lemma 2, we further show that these similarity bounds get tighter when finer grained information are used.

Lemma 2. *The similarity bounds between two points increase at higher resolutions, i.e.,* $d_{ub}(\boldsymbol{q}_{l+1,i}, \boldsymbol{c}_j) \leq d_{ub}(\boldsymbol{q}_{l,i}, \boldsymbol{c}_j)$ *and* $d_{lb}(\boldsymbol{q}_{l+1,i}, \boldsymbol{c}_j) \geq d_{lb}(\boldsymbol{q}_{l,i}, \boldsymbol{c}_j)$.

Proof. Since $qL_{l,i}^k \leq qL_{l+1,i}^k \leq \boldsymbol{q}_i^k \leq qU_{l+1,i}^k \leq qU_{l,i}^k$, $k \in [1, d]$, we have:

$$\begin{cases} (c_j^k - qL_{l+1,i}^k)^2 \leq (c_j^k - qL_{l,i}^k)^2 & if \ c_j^k > qU_{l,i}^k \\ (c_j^k - qL_{l+1,i}^k)^2 \leq (qU_{l,i}^k - qL_{l,i}^k)^2 & if \ qU_{l+1,i}^k < c_j^k \leq qU_{l,i}^k, \\ 0 = 0 & if \ qL_{l+1,i}^k < c_j^k \leq qU_{l+1,i}^k, \\ (qU_{l+1,i}^k - c_j^k)^2 \leq (qU_{l,i}^k - qL_{l,i}^k)^2 & if \ qL_{l,k}^j \leq c_j^k \leq qL_{l+1,i}^k, \\ (qU_{l+1,i}^k - c_j^k)^2 \leq (qU_{l,i}^k - c_j^k)^2 & otherwise \end{cases} \tag{5}$$

After summation of all dimensions we have $d_{ub}(\boldsymbol{q}_{l+1,i}, \boldsymbol{c}_j) \leq d_{ub}(\boldsymbol{q}_{l,i}, \boldsymbol{c}_j)$. Similarly, for the lower bound we have:

$$\begin{cases} (c_j^k - qU_{l+1,i}^k)^2 \geq (c_j^k - qU_{l,i}^k)^2 & if \ c_j^k > qU_{l,i}^k \\ (c_j^k - qU_{l+1,i}^k)^2 \geq 0 & if \ qU_{l+1,i}^k < c_j^k \leq qU_{l,i}^k, \\ 0 = 0 & if \ qL_{l+1,i}^k < c_j^k \leq qU_{l+1,i}^k, \\ (qL_{l+1,i}^k - c_j^k)^2 \geq 0 & if \ qL_{l,k}^j \leq c_j^k \leq qL_{l+1,i}^k, \\ (qL_{l+1,i}^k - c_j^k)^2 \geq (qL_{l,i}^k - c_j^k)^2 & otherwise \end{cases} \tag{6}$$

In summarization of all dimensions, $d_{lb}(\boldsymbol{q}_{l+1,i}, \boldsymbol{c}_j) \geq d_{lb}(\boldsymbol{q}_{l,i}, \boldsymbol{c}_j)$ holds.

4.2   Multi-resolution Based Bounds for Trajectories

In this section, we introduce upper and lower bounds for trajectories which are denoted as $M_{ub}(\mathcal{Q}_l, \mathcal{C})$ and $M_{lb}(\mathcal{Q}_l, \mathcal{C})$ respectively, where $\mathcal{Q}_l$ is the level-l BE of $\mathcal{Q}$. The definition of $M_{ub}(\mathcal{Q}_l, \mathcal{C})$ is given below:

$$M_{ub}(\mathcal{Q}_l, \mathcal{C}) = f'(n, m) \tag{7}$$

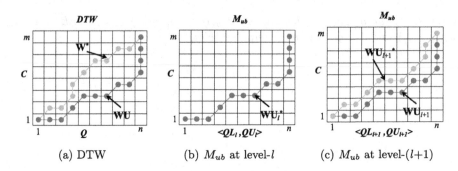

(a) DTW   (b) M_{ub} at level-l (c) M_{ub} at level-$(l+1)$

Fig. 3. M_{ub} upper bound

$$f'(i,j) = \begin{cases} 0 & if \quad i = 0 \wedge j = 0, \\ \infty & if \quad i = 0 \otimes j = 0, \\ d_{ub}(\boldsymbol{q}_{l,i}, \boldsymbol{c}_j) + \min\{f'(i-1, \\ j-1), f'(i-1,j), f'(i,j-1)\} & otherwise \end{cases}$$

Similarly, $M_{lb}(\mathcal{Q}_l, \mathcal{C})$ is defined by replacing the $d_{ub}(\boldsymbol{q}_{l,i}, \boldsymbol{c}_j)$ with $d_{lb}(\boldsymbol{q}_{l,i}, \boldsymbol{c}_j)$. Now, we give the following theorems to show that M_{ub} and M_{lb} are the upper and lower bounds of DTW distance.

Theorem 1. $DTW(\mathcal{Q}, \mathcal{C}) \leq M_{ub}(\mathcal{Q}_l, \mathcal{C})$.

Proof. Suppose W^* and WU_l^* are the optimal warping path for computing $DTW(\mathcal{Q}, \mathcal{C})$ and $M_{ub}(\mathcal{Q}_l, \mathcal{C})$ separately (shown in Fig. 3(a) and (b)). We copy WU_l^* to the warping matrix of $DTW(\mathcal{Q}, \mathcal{C})$ and call the copy one as WU. Let $DTW(W^*)$ and $M_{ub}(WU_l^*)$ denote the summation of $d(\boldsymbol{q}_i, \boldsymbol{c}_j)$ and $d_{ub}(\boldsymbol{q}_{l,i}, \boldsymbol{c}_j)$ of all cells on the warping path W^* and WU_l^* separately. Apparently, $DTW(WU) \leq M_{ub}(WU_l^*)$ because for each pair $(\boldsymbol{q}_i, \boldsymbol{c}_j)$, we have $d_{ub}(\boldsymbol{q}_{l,i}, \boldsymbol{c}_j) \geq d(\boldsymbol{q}_i, \boldsymbol{c}_j)$ according to Lemma 1. Moreover, as W^* is the optimal warping path, we have $DTW(W^*) \leq DTW(WU)$. Finally, we get that

$$DTW(\mathcal{Q}, \mathcal{C}) = DTW(W^*) \leq DTW(WU) \leq M_{ub}(WU_l^*) = M_{ub}(\mathcal{Q}_l, \mathcal{C}) \quad (8)$$

Theorem 2. $M_{lb}(\mathcal{Q}_l, \mathcal{C}) \leq DTW(\mathcal{Q}, \mathcal{C})$.

The proof of Theorem 2 is similar to that of Theorem 1. The above two theorems show that a distance range for can be computed by using less data and computation than the exact DTW distance.

Theorem 3. $M_{ub}(\mathcal{Q}_l, \mathcal{C})$ decreases as l increases, i.e., $M_{ub}(\mathcal{Q}_{l+1}, \mathcal{C}) \leq M_{ub}(\mathcal{Q}_l, \mathcal{C})$.

Proof. Let WU_l^* be the optimal warping path for $M_{ub}(\mathcal{Q}_l, \mathcal{C})$ (in Fig. 3(b)), and WU_{l+1}^* for $M_{ub}(\mathcal{Q}_{l+1}, \mathcal{C})$ (in Fig. 3(c)). In addition, let WU_{l+1} be the same path copied from WU_l^* to the warping matrix of $M_{ub}(\mathcal{Q}_{l+1}, \mathcal{C})$. Then we have

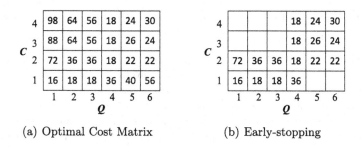

(a) Optimal Cost Matrix   (b) Early-stopping

Fig. 4. DTW distance computing with/without early-stopping

$M_{ub}(WU_{l+1}) \leq M_{ub}(WU_l^*)$ because $d_{ub}(q_{l+1,i}, c_j) \leq d_{ub}(q_{l,i}, c_j)$ for each corresponding pair of (q_i, c_j) according to Lemma 2. Moreover, It is apparent that $M_{ub}(WU_{l+1}^*) \leq M_{ub}(WU_{l+1})$. Finally, we get that:

$$M_{ub}(Q_{l+1}, C) = M_{ub}(WU_{l+1}^*) \leq M_{ub}(WU_{l+1}) \leq M_{ub}(WU_l^*) = M_{ub}(Q_l, C) \quad (9)$$

Theorem 4. $M_{lb}(Q_l, C)$ increases as l increases, i.e., $M_{lb}(Q_l, C) \leq M_{lb}(Q_{l+1}, C)$.

The proof of Theorem 4 is similar to that of Theorem 3. These two theorems show that the proposed similarity bounds are tightened when more information are given. Besides, it is obvious that the BE of the reference overlaps with the original trajectory. And at that condition, both the upper and lower bounds are equal to the DTW distance.

5   Implementation of MDTK

In this section, we introduce techniques to improve query efficiency first and then detail our MDTK framework.

5.1   Query Speed Up

The similarity bounds can prune candidates with less computation, but the cost for such bounds is still quadratic. Therefore, comparing Q_l to all trajectories becomes intractable for large distributed trajectory dataset. We are seeking ways to avoid examining the trajectories that are distant to our queries and reduce the quadratic computation. In this paper, following three techniques are adopted: indexing, early-stopping and cascade pruning.

Trajectory Indexing. Trajectory indexing techniques have been widely researched [19]. In summary of these methods, our indexing and query executing strategy in each remote site contain the following steps: (i) The local trajectories in a remote site r are segmented into MBRs, which are stored in an R-tree like index T_r. (ii) After receiving Q_l, we probe each MBR of the reference into T_r and find the leaf nodes that are intersected with it. (iii) Based on the intersections, we find qualified candidates.

Early-Stopping. This kind of technique has been used for reducing computation cost [14,15]. It works like that: During search processing, we maintain the best-so-far distance value as the pruning threshold. Lower bounding distances greater than this value does not need to be computed. Figure 4 shows how the early-stopping technique works for the DTW-based top-k query. The number in $cell[i,j]$ represents the cost of optimal path from $cell[1,1]$ to $cell[i,j]$. Figure 4(a) shows the procedure of exact DTW distance computation. While Fig. 4(b) shows the result when we have known the best-so-far distance is "28". The values in white cells are omitted since they are certainly greater than 28, and we do not need to compute them. Early-stopping can be applied to the exact DTW distance calculation as well as to our M_{lb} lower bound.

Cascade Pruning. The idea of cascade pruning is to leverage lower bounds that are loose but have low computation costs to do the early pruning, and then to use those that are tight but have high computation cost to do subsequent pruning. We adopt LB_KimFL and LB_Keogh, two widely used lower bounds, to do early pruning. LB_KimFL is computed as follows: $LB\_KimFL(\mathcal{Q}_l, \mathcal{C}) = \max\{d_{lb}(\boldsymbol{q}_{l,1}, \boldsymbol{c}_1), d_{lb}(\boldsymbol{q}_{l,n}, \boldsymbol{c}_m)\}$, which uses the first and last point of two trajectories. The computation of LB_Keogh is defined in Eq. 10, which requires building two assistant sequences firstly. The parameter r in Eq. 10 is a global constraint to confine the searching area which is detailed in [10]. The computation cost of these lower bounds are $O(1)$ and $O(\min\{n,m\})$ respectively, which are smaller than the quadratic cost of M_{lb}. It's easy to prove that these two lower bounds also increase as l increases and we omit such proofs.

$$upper_{l,i}^j = \max\{qU_{l,i-r}^j : qU_{l,i+r}^j\}, \quad lower_{l,i}^j = \min\{qL_{l,i-r}^j : qL_{l,i+r}^j\}; j \in [1, d]$$

$$LB\_Keogh = \sum_{i=1}^{\min\{n,m\}} d_{lb}(\langle lower_{l,i}, upper_{l,i}\rangle, \boldsymbol{c}_i) \quad (10)$$

5.2 MDTK Processing Framework

Now, we describe the procedure of our multi-resolution based MDTK framework. MDTK consists of two parts, one runs in the coordinator, while the other runs in the remote sites. These two parts communicate interactively, and the whole procedure can be divided into following three phases:

Initial. In this phase, the coordinator site sends $\mathcal{Q}_l$ ($l = 1$ in default) to remote sites in the set $\mathcal{R}$ (lines 1–2 in Algorithm 1). After receiving $\mathcal{Q}_l$, Each remote site retrieves its local index tree, to find trajectories that are intersected with $\mathcal{Q}_l$. Each remote site reports the number of candidates back to the coordinator and terminates if no candidates are left (lines 1–5 in Algorithm 2). Then, the coordinator removes sites which contain no candidates from set $\mathcal{R}$ (lines 3–4 in Algorithm 1). We use θ to denote the pruning threshold, S_r to maintain the M_{ub} and M_{lb} bounds for those local candidates in TS_r.

Algorithm 1. MDTK in the coordinator site

Input: reference trajectory $\mathcal{Q}$, k;
Output: the k most similar trajectories to $\mathcal{Q}$;

1 $\mathcal{R} \leftarrow$ the set of all remote sites;
2 sendToRemoteSites($\mathcal{R}$, $\mathcal{Q}_l$);
3 $CSS \leftarrow$ getFromRemoteSites($\mathcal{R}$)
4 $\mathcal{R} \leftarrow \{x.r | x \in CSS, x.|S_r| > 0\}$;   /*initial pruning remote sites*/
5 **for** $i=l+1;\ ;i++$ **do**
   /* global pruning threshold update*/
6 $\quad$ sendToRemoteSites($\mathcal{R}$, $\mathcal{Q}_i$);
7 $\quad$ $GUBS \leftarrow$ getFromRemoteSites($\mathcal{R}$);
8 $\quad$ $gkub \leftarrow argmin_\tau(|x \in GUBS, x < \tau| \geq k)$;
9 $\quad$ sendToRemoteSites($\mathcal{R}$, $gkub$);
   /*get the number of candidates from all sites in $\mathcal{R}$*/
10 $\quad$ $CSS \leftarrow$ getFromRemoteSites($\mathcal{R}$);
11 $\quad$ $\mathcal{R} \leftarrow \{x.r | x \in CSS, x.|S_r| > 0\}$;   /*pruning remote sites*/
12 $\quad$ $sum \leftarrow \sum_{x \in CSS} x.|S_r|$;   /*get the number of candidates*/
13 $\quad$ **if** $sum == k$ **then**
14 $\quad\quad$ sendToRemoteSites($\mathcal{R}$, *finish*);
15 $\quad\quad$ **break**;

16 $IDs \leftarrow$ getFromRemoteSites($\mathcal{R}$);
17 **return** IDs;

Iterative Pruning. In this phase, the coordinator site levelly sends the BEs and remote sites prune until only k candidates left. Firstly, the coordinator site sends level-i BE $\mathcal{Q}_i$ to all candidate remote sites (line 6 in Algorithm 1). After receiving $\mathcal{Q}_i$, each remote site prunes candidates by using LB_KimFL, LB_Keogh and M_{lb}. While computing M_{lb}, early-stopping strategy is adopted. If the M_{lb} bound of a candidate is smaller than θ, then c_lb is set to M_{lb}. Otherwise, c_lb is set to "∞". Then, we update the M_{ub} upper bounds for left candidates and return the top-k smallest ones to the coordinator (lines 8–18 in Algorithm 2). Secondly, the coordinator site collects local upper bound values and sends k-th smallest one to candidate remote sites. Then, these sites update θ with this value and prune candidates whose lower bounds are greater than θ. Next, each of them sends the number of local candidates to the coordinator, and stops running if all trajectories are pruned. Thirdly, the coordinator receives the number of candidates from remote sites in $\mathcal{R}$ and removes those who contain no candidates from $\mathcal{R}$. If the total number of candidates is greater than k, the iteration continues. If this value equals to k which means the results have been found, the coordinator will stop the iteration and sends the *finish* signal to remote sites (lines 10–15 in Algorithm 1). Note that the iteration will terminate since the similarity bounds are equal to the exact DTW distance when the finest resolution BE is used.

Algorithm 2. MDTK in remote site r

Input: local trajecotry set TS_r, local trajectory indexing tree T_r;

1   $\mathcal{Q}_l \leftarrow$ getFromCoordinator();

2   $TS_r \leftarrow$ Intersection(T_r, $\mathcal{Q}_l$) ;   /* pruning with index, filter operation*/

3   SendToCoordinator($|TS_r|$);

4   **if** $|TS_r| == 0$ **then**

5 | return;

6   $\theta \leftarrow \infty$; $S_r \leftarrow$ bounds set for trajectories in TS_r;

7   **while** $m \leftarrow$ getFromCoordinator() **do**

8 | **if** $m == \mathcal{Q}_i$ **then**

9 | | **foreach** $c \in TS_r$ **do**

10 | | | **if** $LB\_KimFL(\mathcal{Q}_i, c) > \theta \,||\, LB\_Keogh(\mathcal{Q}_i, c) > \theta$ **then**

11 | | | | TS_r.remove(c) ;   /* filter operation */

12 | | | $c\_lb \leftarrow$ lowerBoundEarlyStopping($M_{lb}(\mathcal{Q}_i, c), \theta$);

13 | | | **if** $c\_lb == \infty$ **then**

14 | | | | TS_r.remove(c);   /* filter operation */

15 | | | **else**

16 | | | | updateBounds(S_r,c,c_lb,$M_{ub}(\mathcal{Q}_i, c)$);

17 | | $lkub \leftarrow \{\alpha.ub \,|\, \alpha \in S_r, \alpha.ub \le argmin_\tau(|x \in S_r, x.ub < \tau| \ge k)\}$;

18 | | SendToCoordinator($lkub$);

19 | **else if** $m == gkub$ **then**

20 | | $\theta \leftarrow m$; $S_r \leftarrow \{\beta \,|\, \beta \in S_r,\, \beta.lb \le \theta\}$;   /* pruning with θ */

21 | | SendToCoordinator($\langle r, |S_r| \rangle$);

22 | | **if** $|S_r| = 0$ **then**

23 | | | return ;

24 | **else**

25 | | SendToCoordinator($\{a.ID \,|\, a \in S_r\}$);

26 | | return;

Results Retrieving. In this step, all remaining remote sites in set $\mathcal{R}$ send the IDs of local candidates back to the coordinator after receiving the *finish* signal. The coordinator site collects these IDs from candidate sites and sends them back to the user.

5.3   Performance Analysis

Before the analysis of MDTK, we first detail our baseline method BandWidth. BandWidth consists of followings phases: (i) Pruning threshold retrieving: It initially sends level-1 bounding envelope to all remote sites to compute lower bounds (the one-dimensional case of M_{lb}) for all candidates and retrieves the sites that contain the top-k smallest lower bounds. Then, these sites get the reference from the coordinator site, compute the exact DTW distance values for all trajectories in them and get the global k-th smallest one τ as the pruning

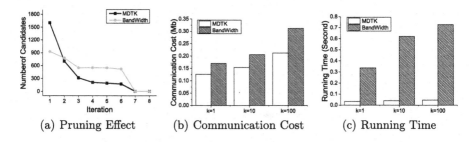

(a) Pruning Effect   (b) Communication Cost   (c) Running Time

Fig. 5. Performance comparison

threshold. (ii) Group-by-group traversing: BandWidth divides all sites into a few groups, each containing m sites. It levelly transports BEs to sites in one group. After receiving a BE, these sites compute tighter lower bounds and prune candidates with τ. Besides, they will calculate the exact DTW distance for un-pruned candidates to tighten the pruning threshold τ. Finally, BandWidth will repeat the procedure for other groups until all sites are traversed.

In comparison with BandWidth, there are two significant advantages of MDTK. Firstly, the pruning threshold θ in MDTK is initialized with the global k-th smallest upper bound, while the corresponding threshold τ in BandWidth is chosen from the exact DTW distance values trajectories who have the top-k smallest lower bounds. We can see that τ is more complex in computation as it needs to compute the exact DTW distance. Besides, τ cannot guarantee a tight pruning threshold since trajectories that with small lower bounds do not mean their DTW distance values are also small. Secondly, the group-by-group traversing strategy in BandWidth is really time consuming as other sites should be waiting until a group of sites are finished, and when traversing a group of sites the exact DTW distance also need to be computed. On the contrary, all remote sites are pruned simultaneously in MDTK, which leads to a higher running efficiency.

6 Experiments

6.1 Experimental Setup

We evaluate the performance of MDTK in this section and adopt the BandWidth algorithm as the baseline. All codes, written in Java, were evaluated on a 10-node clustering running Spark 1.5.2 over Ubuntu 12.0.4. Each node is equipped with an 8 cores Intel E5335 2.0 GHz processor and 16 GB memory.

We use two real-world datasets, *T-Small* and *T-Big*, both of which are generated from Beijing Taxis [21] and the length of trajectories in these datasets are greater than 4,096. *T-Small* contains 10,000 trajectories from October 1st to 7th, and *T-Big* contains 1,000,000 trajectories from November 1st to December 31st. Trajectories in the datasets are 0–1 normalized for each dimension.

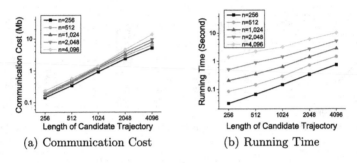

(a) Communication Cost   (b) Running Time

Fig. 6. Performance in respect of trajectory length

6.2   Efficiency and Scalability of MDTK

We first compare the performance of MDTK with BandWidth on *T-Small*. In this experiment, we set k to 1 and the length of reference to 256. We also assume that each remote site contains only one trajectory. From Fig. 5(a), we can see that more than 83% candidates are pruned at the first iteration by using the proposed bounds. The cascade pruning does not work for the first iteration since the pruning threshold θ in Algorithm 2 is initialized with "∞". After that, the number of candidates drops quickly when finer-grained BEs are used to prune candidates. Besides, we see that BandWidth prunes more candidates as it uses an exact DTW distance as the pruning threshold which is tighter than our global top-k upper bound. While, in latter iterations, our upper bound gets tighter and achieves a better pruning effect than the distance threshold in BandWidth. Moreover, as the sizes of BEs at last few steps dominate the total communication cost. So, BandWidth requires more communication consumption than MDTK which is shown in Fig. 5(b). Finally, Fig. 5(c) shows that BandWidth is far slower than MDTK. This is because BandWidth divides the remote sites into groups, and traverses these groups in a one after another way.

We next study the performance of MDTK in respect of the length of trajectories on *T-Small*. In this experiment, we set k to 1, and vary the length of reference (n) and candidates (X-axis) from 256 to 4,096. From Fig. 6(a), we can see that the communication cost increase linearly with the length of candidate trajectories. This result is intuitive, as longer candidates require more information of the reference to get an equivalent distance approximation of shorter candidates. Besides, we find that given a trajectory dataset, longer references consumes more communication cost to find the result. The explanation of this phenomena is similar to that of the above. But, it needs to note that when the length of reference increases two times, the communication cost increases less than two times, which means that MDTK performs well for long references. Next, we show the efficiency of MDTK in Fig. 6(b) by varying the length of trajectories. We can see that the running time increases linearly to the length of both the reference and candidates. This is because the computation complexity of our bounds is linear to these lengths.

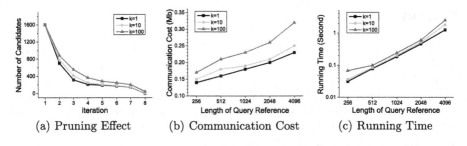

(a) Pruning Effect   (b) Communication Cost   (c) Running Time

Fig. 7. Performance of MDTK in respect of k

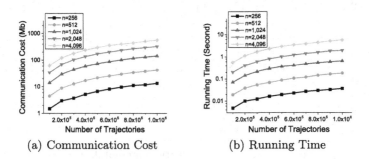

(a) Communication Cost   (b) Running Time

Fig. 8. Scalability of MDTK

We then report the impacts of k on the performance of MDTK since the tightness of our bound is closely related to the value of k. We evaluate the impacts from three aspects: the number of candidates left after each iteration, the communication and time cost. We set the length of reference to 256 and vary the value of k from 1 to 100. Figure 7(a) shows the pruning effect of MDTK on *T-Small* when the length of candidates is set to 256. At the first iteration, the spatial trajectory index prunes almost 84% candidates, which is irrelevant to the value of k. At latter iterations, we find that smaller values of k prune more candidates than greater ones. This is because smaller k leads to a tighter upper bound for pruning candidates and more remote sites are pruned. So, we get that a smaller value of k for the query leads to less communication cost which is shown in Fig. 7(b). Besides, as more candidates are pruned at each iteration for a smaller k, we need to compute fewer candidate bounds which leads to quicker query results for smaller ks. So, we get that from Fig. 7(c) when k is set to 1, MDTK runs faster than the cases when k is set to 10 and 100.

Finally, we research the scalability of MDTK on *T-Big* regarding communication cost and time efficiency. We divide the dataset into 10,000 partitions to simulate the case when 10,000 remote sites are joined in the query and set the length of trajectory data set to 256 and the value of k to 1. Figure 8 shows the communication and time consumption when we vary the number of trajectories from 100,000 to 1,000,000. We can see that the increase of both consumptions grow slower than the increment of trajectory dataset size, which means MDTK can scale well for large trajectory datasets.

7   Conclusion

In this paper, we present MDTK, a bandwidth saving framework, to process DTW distance based top-k query over distributed trajectories. MDTK distributes the bounding envelopes of query reference to remote sites in a level-wise fashion. Coordinator sites tighten the proposed similarity bounds when receiving finer-grained bonding envelopes. The upper and lower bound based pruning strategy in MDTK achieves a better prune efficiency than existing only lower bound based method. Besides, MDTK imports three optimization techniques to improve the query efficiency. Extensive experiments on real trajectory data show that MDTK is efficient, scalable and outperforms state-of-the-art method.

Acknowledgement. Our research is supported by the National Key Research and Development Program of China (2016YFB1000905), NSFC (61370101, 61532021, U1501252, U1401256 and 61402180), Shanghai Knowledge Service Platform Project (No. ZF1213).

References

1. Cao, H., Wolfson, O., Trajcevski, G.: Spatio-temporal data reduction with deterministic error bounds. VLDB J. **15**(3), 211–228 (2006)
2. Chakrabarti, K., Keogh, E., Mehrotra, S., Pazzani, M.: Locally adaptive dimensionality reduction for indexing large time series databases. ACM Trans. Database Syst. (TODS) **27**(2), 188–228 (2002)
3. Chan, F.P., Fu, A.C., Yu, C.: Haar wavelets for efficient similarity search of time-series: with and without time warping. TKDE **15**(3), 686–705 (2003)
4. Costa, C., Laoudias, C., Zeinalipour-Yazti, D., Gunopulos, D.: SmartTrace: finding similar trajectories in smartphone networks without disclosing the traces. In: Proceedings of the 27th ICDE, pp. 1288–1291 (2011)
5. Demetrios, Z.Y., Christos, L., Constandinos, C.: Crowdsourced trace similarity with smartphones. TKDE **25**(6), 1240–1253 (2013)
6. Faloutsos, C., Ranganathan, M., Manolopoulos, Y.: Fast subsequence matching in time-series databases. In: Proceedings of the 1994 ACM SIGMOD, pp. 419–429 (1994)
7. Hsu, C.C., Kung, P.H., Yeh, M.Y., Lin, S.D., Gibbons, P.B.: Bandwidth-efficient distributed k-nearest-neighbor search with dynamic time warping. In: Proceedings of the 2015 ICBD, pp. 551–560. IEEE (2015)
8. Jiangpeng, D., Jin, T., Xiaole, B., Zhaohui, S., Dong, X.: Mobile phone based drunk driving detection. In: Proceedings of the 2010 ICPCTH, pp. 1–8. IEEE (2010)
9. Kanth, K.V.R., Agrawal, D., Singh, A.K.: Dimensionality reduction for similarity searching in dynamic databases. In: Proceedings of the 1998 ACM SIGMOD, pp. 166–176 (1998)
10. Keogh, E.: Exact indexing of dynamic time warping. In: Proceedings of the 28th VLDB, pp. 406–417 (2002)
11. Keogh, E.J., Chu, S., Hart, D.M., Pazzani, M.J.: An online algorithm for segmenting time series. In: Proceedings of the 2001 ICDM, pp. 289–296 (2001)
12. Papadopoulos, A.N., Manolopoulos, Y.: Distributed processing of similarity queries. Distrib. Parallel Databases **9**(1), 67–92 (2001)

13. Popivanov, I., Miller, R.J.: Similarity search over time-series data using wavelets. In: Proceedings of the 18th ICDE, pp. 212–221 (2002)
14. Rakthanmanon, T., Campana, B.J.L., Mueen, A.: Searching and mining trillions of time series subsequences under dynamic time warping. In: The 18th ACM SIGKDD, pp. 262–270 (2012)
15. Sakurai, Y., Yoshikawa, M., Faloutsos, C.: FTW: fast similarity search under the time warping distance. In: Proceedings of the 24th ACM PODS, pp. 326–337 (2005)
16. Xie, D., Li, F., Phillips, J.M.: Distributed trajectory similarity search. PVLDB **10**(11), 1478–1489 (2017)
17. Yeh, M.Y., Wu, K.L., Yu, P.S., Chen, M.S.: LeeWave: level-wise distribution of wavelet coefficients for processing kNN queries over distributed streams. PVLDB **1**(1), 586–597 (2008)
18. Yi, B., Faloutsos, C.: Fast time sequence indexing for arbitrary Lp norms. In: Proceedings of 26th VLDB, pp. 385–394 (2000)
19. Zheng, Y., Zhou, X. (eds.): Computing with Spatial Trajectories. Springer, New York (2011). https://doi.org/10.1007/978-1-4614-1629-6
20. Zeinalipour-Yazti, D., Lin, S., Gunopulos, D.: Distributed spatio-temporal similarity search. In: Proceedings of the 2006 CIKM, pp. 14–23 (2006)
21. Zhang, Z., Wang, Y., Mao, J., Qiao, S., Jin, C., Zhou, A.: DT-KST: distributed top-k similarity query on big trajectory streams. In: Proceedings of the 22nd DASFAA, Part I, pp. 199–214 (2017)

Modeling Travel Behavior Similarity
with Trajectory Embedding

Wenyan Yang[1], Yan Zhao[1], Bolong Zheng[2,3(✉)], Guanfeng Liu[1],
and Kai Zheng[4]

[1] School of Computer Science and Technology, Soochow University, Suzhou, China
20164227005@stu.suda.edu.cn, {zhaoyan,gfliu}@suda.edu.cn
[2] School of Data and Computer Science, Sun Yat-sen University, Guangzhou, China
zhengblong@mail.sysu.edu.cn
[3] Aalborg University, Aalborg, Denmark
[4] Big Data Research Center, University of Electronic Science and Technology
of China, Chengdu, China
zhengkai@uestc.edu.cn

Abstract. The prevalence of GPS-enabled devices and wireless communication technologies has led to myriads of spatial trajectories describing
the movement history of moving objects. While a substantial research
effort has been undertaken on the spatio-temporal features of trajectory data, recent years have witnessed the flourish of location-based web
applications (i.e., Foursquare, Facebook), enriching the traditional trajectory data by associating locations with activity information, called
activity trajectory. These trajectory data contain a wealth of activity
information and offer unprecedented opportunities for heightening our
understanding about human behaviors. In this paper, we propose a novel
framework, called TEH (Trajectory Embedding and Hashing), to mine
the similarity among users based on their activity trajectories. Such user
similarity is of great importance for individuals to effectively retrieve
the information with high relevance. With the time being separated into
several slots according to the activity-based temporal distribution, we
utilize trajectory embedding technique to mine the sequence property of
the activity trajectories by treating them as paragraphs. Then a hash-
based method is presented to reduce the dimensions for improving the
efficiency of users' similarity calculation. Finally, extensive experiments
on a real activity trajectory dataset demonstrate the effectiveness and
efficiency of the proposed methods.

Keywords: Activity trajectory · User similarity
Trajectory embedding

1 Introduction

The increased popularity of various GPS-equipped smart devices, such as smart
phones, personal navigation devices, and on-board diagnostics, has resulted in

© Springer International Publishing AG, part of Springer Nature 2018
J. Pei et al. (Eds.): DASFAA 2018, LNCS 10827, pp. 630–646, 2018.
https://doi.org/10.1007/978-3-319-91452-7_41

a huge volume of spatial trajectories (e.g., the GPS trajectories of vehicles), each of which consists of a sequence of time-ordered spatial points. This inspires massive efforts on analyzing the huge scale trajectory data from various aspects in the literature, including trajectory clustering [1,2], and knowledges/patterns mining from trajectory data [3–5], to name a few.

These studies mainly put the focus on the spatio-temporal features on the trajectory data. However, recent years have witnessed a revolution in location-based social network (LBSN) services, such as Foursquare[1], Facebook[2], and Flicker[3], in which large amounts of users' trajectories associated with activity information have been accumulated. These trajectories, namely activity trajectories [6,7], contain footprints (or check-ins) which offers the data about where and when a user has been as well as what she has done. For instance, at Foursquare, users can check in the places where they are visiting and leave tips about what they are doing. Intuitively, an activity trajectory of a user can be formed, which is a finite sequence of time-stamped locations. Each of the activity trajectory is associated with a keyword to describe a specific activity performed at this location. For example, T_{r1}, T_{r2} and T_{r3} in Fig. 1 exemplify the activity trajectories, where each point is attached with a keyword to illustrate an activity performed by the users. Note that how to extract and classify the activities is orthogonal to the approaches in our work, so each activity is just regarded as a unique entry of a pre-defined activity vocabulary. With these trajectory data, people intend to know more information about behavior preferences and daily habits [2,8].

In order to solve this problem, some existing studies have been proposed to learn user-specific activities [9,10] and mine knowledge [11–13] from the individual spatial data and trajectory information. However, these works do not explore the correlation between users, which is of great significance to effectively retrieve information matching user' tastes and recommend new friends sharing similar behaviors with them. Then in the literature, Lv et al. [8] mine the similarity between users based on the routine activity, which is extracted from the raw trajectory data. Taking the semantics of visited positions into account, Chen et al. [14] propose some basic principles to measure user similarity based on trajectory patterns. Li et al. [15] extract latent topics of users' check-ins by using topic modelling and then adopt these topics to measure user similarity. However, these existing works concentrate on analyzing the geographic and semantic feature of locations to learn the users' similarity, but neglect temporal features, which is significant in the similarity measurement.

In this paper, we focus on mining similar users based on their time-ordered activity trajectories regardless of geographic feature, and proposing a framework, referred to as Trajectory Embedding and Hashing (TEH), to model users' activity trajectory histories and to explore the similarity between users. The main idea of our approach is that, observing that performing the same activity in different time may refer to different lifestyles, we utilize Gaussian Mixture

---

[1] http://www.foursquare.com.
[2] http://www.facebook.com.
[3] http://www.flickr.com.

Model (GMM) [16] widely used in density estimation to model the temporal distribution of activities and apply a time-partition procedure to divide time into independent segments. A trajectory embedding algorithm is then developed to mine the sequence property of the activity trajectories by treating them as paragraphs. Finally, a hashing method is presented for the purpose of dimensionality reduction.

Our main contributions are summarized as follows:

1. We mine the similarity between users based on both their activity information and the temporal distribution feature of these activities embedded in the trajectories.
2. We utilize Gaussian Mixture Model to fit the temporal distribution of activities, which can achieve good effect, and then divide time into segments based on this distribution.
3. To the best of our knowledge, this is the first work that applies paragraph vector model, a widely used natural language processing technique, to map the activity trajectories and users into feature vectors, and then adopt a hash-based similarity calculation method to find similar users.
4. We conduct extensive experiments on real trajectory dataset, which empirically demonstrate the good efficiency and effectiveness of our proposed model.

The rest of this paper is organized as follows. Section 2 surveys the related work under different problem settings. Section 3 introduces the preliminary concepts and overviews our model. The proposed algorithms and related techniques are presented in Sect. 4, followed by the empirical study in Sect. 5. Finally, Sect. 6 concludes this paper and outlines the directions to further extend our work.

2   Related Work

In this section, we discuss prior work related to user similarity mining approaches and Neural Embedding Representation, which is the core technology in our work.

2.1   User Similarity

With the rapid development of positioning technologies and GPS devices, it is possible to record the movement trajectories of users in time. This attracts increasing attention of researchers, followed by a large number of trajectory research works [17–20]. Zheng et al. [21] propose spatial query processing for fuzzy objects. Shang et al. [22] study a novel problem of planning unobstructed paths in traffic-aware spatial networks. Zheng et al. [23] studies approximate keyword search in semantic trajectory.

Most of the above work tend to attach great importance to trajectory data itself by ignoring the similarity of moving objects. Recently, taking both the sequence property of user behaviors and the hierarchy property of geographic locations into account, Li et al. [24] develop a framework to model users' location histories and measure the similarity among them. A two-stage approach is proposed by Lv [8] to mine users' long-term activity similarity based on their trajectories. They first introduce routine activities to capture users' long-term activity regularities and then calculate user similarity hierarchically based on the routine activities. Lee and Chung [25] propose a method to calculate the user similarity on the basis of the semantics of the location. However, these approaches mine the user similarity only based on the spatial and semantical features without considering the time distribution of activities in locations. Our work is to find behaviorally similar users based on the temporal distribution feature of activities embedded in all check-in points of the trajectories.

2.2 Neural Embedding Representation

Neural embedding representations are widely used in Natural Language Processing (NLP) [26,27]. Mikolov et al. [28] propose two novel model architecture - continuous Bags-of-Words and Skip-gram for learning continuous vector representations of words from large datasets. These distributed representations can capture word similarities in a semantic level and have good compositionality. Paragraph vector is introduced on the basis of word vector by Le and Mikolov [29], which projects words and documents into a single semantic space. It compensates for some key weaknesses of bag-of-words models, which lose the ordering of words and ignore their semantic information.

Meanwhile, these distributed representation methods are applied to the work in other fields successfully. Tajima et al. [30] model user activities on the Web with paragraph vectors. They consider users and web activities as paragraphs and words respectively. The distributed representations are used among the user-related prediction tasks. Zhao et al. [31] utilize paragraph vectors on the traditional medical system for similar cases recommendation. They try to utilize the semantic representation of sentences in a continuous space to understand the cases. Zhang et al. [32] use multi-modal embedding to achieve online local event detection in geo-tagged tweet streams. They capture short-text semantics by learning embedding of the location, time and text, and then perform online clustering by using a novel Bayesian mixture model. In this paper, we will capture the potential behavior features of users via trajectory embedding technology.

3 Preliminaries

In this section, we present some preliminary concepts and give an overview of the proposed model. Table 1 lists the major notations used throughout the paper.

3.1 Preliminary Concepts

Definition 1 (Activity). *An activity, denoted by α, is a type of action that a user can take at some place of interest (i.e., dining, sport and entertaining). We use $\mathbb{A}$ to denote the union of all the activities, which can be performed by the users.*

Table 1. Summary of notations

Notation	Definition
α	Activity
$\mathbb{A}$	Activity set
p	Check-in point
$p.l$	Location of check-in point p
$p.t$	Time stamp of check-in point p
$p.\alpha$	Activity of check-in point p
T_r	Activity trajectory
θ	Time partition ratio
U_i	User vector
W_i	Check-in vector
b_i	A binary code
k	Binary code length

Definition 2 (Check-in Point). *A check-in point, denoted by* $p = <p.l, p.t, p.\alpha>$, *is a time-stamped location point associated with an activity p.α, where p.l : (x, y) stands for the longitude and latitude information of the location at time stamp p.t.*

Definition 3 (Activity Trajectory). *An activity trajectory, denoted by T_r, is a finite sequence of check-in points, i.e., $T_r = (p_1, p_2, \ldots, p_n)$.*

An activity trajectory T_r is a sequence of historical records representing not only where and when a user has been but also what she has done. Different users may exhibit similarity with respect to their traversed activity trajectories, and therefore two users can be correlated based on their activity trajectory similarity. In the rest of the paper, we will use *trajectory* and *activity trajectory* interchangeably when the context is clear.

For instance, Fig. 1 shows an example of several users $U = \{u_1, u_2, u_3\}$ and all activities $\mathbb{A}$ along the routes of the users. Each user has an activity trajectory, in which the spatio-temporal data and the information about her activities are embedded.

3.2 The Framework of TEH

Figure 2 shows the overview of the proposed framework, which basically comprised three parts: time partition, trajectory embedding and hash-based similarity calculation. In this work, we first employ the Gaussian mixture model to obtain the temporal distribution of activities. Then we divide the whole day into multiple segments based on the temporal distribution. In the second step, a trajectory embedding algorithm, which can capture the sequence property of

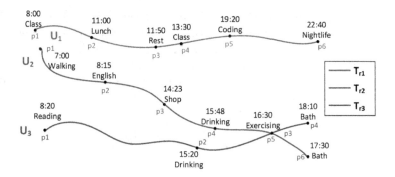

Fig. 1. Running example

the activity trajectories, is designed to map users' historical activity trajectories and users into vectors for quantifying these data. Due to the high computational cost of the high dimensional vectors, we introduce a hash-based user similarity calculation method for dimensionality reduction by mapping these vectors into compact binary codes, which can improve computational efficiency. We will detail the discussion for each part in the next section.

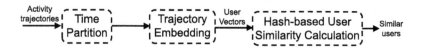

Fig. 2. TEH framework overview.

4   TEH Model

In this section, we detail the process of the proposed TEH model, including distribution-based time partition, trajectory embedding and hash-based user similarity calculation.

4.1   Distribution-Based Time Partition

Temporal Distribution of Activity. We observe that people performing the same activity in different time slots may have different behavior characteristics. In other words, two persons who perform a same activity in the same time slot can be considered as similar users, which play an important role in recommendation applications (i.e., potential friend recommendation, product recommendation). In this subsection, we design a method of time slot partition in terms of temporal distribution of all activities. We first introduce the Gaussian mixture model to model the temporal distribution of all activities, and then divide the whole day into time slots.

A Gaussian mixture model, widely used in density estimation, is employed to fit the temporal distribution in terms of the number of users performing the activity. Specifically, given an activity α, we denote the number of users performing this activity α at the time unit t as y_t. The Gaussian mixture model can be expressed as Eq. (1), assuming that the component $N(y_t, \mu_k, \sigma_k^2)$ obeys the single-Gaussian distribution.

$$p(y_t) = \sum_{k=1}^{K} \omega_k N(y_t, \mu_k, \sigma_k^2) \tag{1}$$

$$s.t. \sum_{i=1}^{n} \omega_k = 1, 0 \leq \omega_k \leq 1.$$

where K is the number of mixture components, and ω_k is the corresponding weight of the k-th component. μ_k and σ_k^2 represent the mean and covariance of the k-th component.

The parameters of $\omega_k, \mu_k, \sigma_k^2$ can be estimated by Expectation-Maximization (EM) algorithm, which is implemented by the repeating iterations between Expectation step and Maximization step as follows:

1. Initialize all parameters randomly. One day will be split into 24 units and each hour represents a unit.
2. In the E-step, the posterior probability of observation t belonging to k-th component is given as follow:

$$\gamma_{tk} = \frac{w_k N(y_t, \mu_k, \sigma_k^2)}{\sum_{k=1}^{K} w_k N(y_t, \mu_k, \sigma_k^2)}, t = 1, 2, \ldots, T; k = 1, 2, \ldots, K \tag{2}$$

where t is the time unit when user conducts the activity.
3. In the M-step, the parameters will be derived under new posterior probabilities. The updated parameters are computed by Eq. 3.

$$\mu_k = \frac{\sum_{t=1}^{T} \gamma_{tk} y_t}{\sum_{t=1}^{T} \gamma_{tk}}, k = 1, 2, \ldots K$$

$$\sigma_k^2 = \frac{\sum_{t=1}^{T} \gamma_{tk}(y_t - \mu_k)^2}{\sum_{t=1}^{T} \gamma_{tk}}, k = 1, 2, \ldots, K \tag{3}$$

$$\omega_k = \frac{\sum_{t=1}^{T} \gamma_{tk}}{T}, k = 1, 2, \ldots K$$

4. Repeat steps 2 and 3 until converge.

As a result, we can get the distribution of all the activities. Figure 3 depicts the probability density distribution of the activity - Nightlife. It can be clustered as a mix of two Gaussian distributions. For each activity, by observing the change of number of people performing this activity with time, we will choose the K value based on the number of peak points and obtain the distribution of the

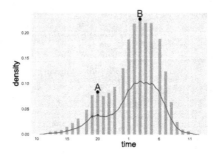

Fig. 3. Activity time distribution of Nightlife.

activity. For instance, Fig. 3 shows the time distribution of nightlife, in which the number of peak points (i.e. A and B) determines the K value (i.e., $K = 2$).

Time Partition. Based on the distribution of the activity, the comparative ratio of density in adjacent time unit, which is usually defined as the ratio between the small density and large density of two adjoining time units, is utilized to partition time. A threshold θ is set to partition time. Specifically, if the ratio of density is greater than or equal to the threshold, the two time units can be merged into a same time slot. Take Fig. 3 as an example, the comparative density ratio between unit 19 and 20 is 0.872. In the same time slot, users participating in the same activity can be considered to have similar behavior. Time partition will cluster crowds with similar density together for each activity.

4.2   Trajectory Embedding

As mentioned earlier, an activity trajectory is a finite sequence of check-in points, each of which is associated with a specific activity. To measure the similarity between users, we take into account not only the activities involved in the trajectories, but also the sequence of the activities. Hence, we employ a trajectory embedding algorithm that maps users' trajectories and users into vectors in order to capture the sequence property of these check-ins.

The trajectory embedding algorithm summarizes the sequence of each user's activities using the Paragraph Vector [29], which is an unsupervised approach that learns continuous distributed vector representations from pieces of text. In specific, we consider activities with corresponding time as words, and users as paragraphs, each of which has a main idea that represents the topic of the whole paragraph. In our work, this topic refers to behavior preference of each user. Thus, people having the similar topics would share similar preferences. The Paragraph vector model adds a vector to extract the topic based on the word embedding, which is the user vector what we need to compute the similarity between users.

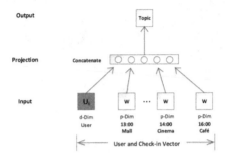

Fig. 4. The architecture of embedding.

Figure 4 demonstrates the architecture of trajectory embedding. In this model, the concatenation of user vector with a context of some check-in points is used to conclude the topic. The user vector represents the missing information from the current context and can act as a memory of the topic of the activity trajectory.

After mapping each user into a unique vector that is represented by a column in matrix U and each check-in into a unique vector that is represented by a column in matrix W, we apply the Paragraph Vector model to sequences of user trajectories. Given a sequence of check-ins $(c_{i,1}, c_{i,2}, \ldots, c_{i,N_i})$ generated by a user u_i and N_i is the size of this sequence, the objective of the vector model for this sequence is to maximize the average log probability:

$$\frac{1}{N_i} \sum_{n=j}^{N_i-j} logp(c_{i,k}|c_{i,k-j}, \ldots, c_{i,k+j}, u_i). \tag{4}$$

where $(c_{i,k-j}, \ldots, c_{i,k+j})$ is a sliding window, $c_{i,k}$ is the target check-in and other check-ins are contexts. The probability can be calculated using the softmax function according to PV_DM [29] as follows:

$$p(c_{i,k}|c_{i,k-j}, \ldots, c_{i,k+j}, u_i) = \frac{exp(\omega_{c_{i,k}}^{\mathrm{T}} v_i)}{\sum_{c \in |C|} exp(\omega_c^{\mathrm{T}} v_i)}. \tag{5}$$

where C is the union of all the possible check-ins, $\omega_{c_{i,k}}$ is the output vector of check-in $c_{i,k}$, and v_i is the concatenated input vector of the context check-in vectors $v_{c_{i,k-j}}, \ldots, v_{c_{i,k+j}}$ and user vector U_i, that is $v_i = [v_{c_{i,k-j}}^{\mathrm{T}}, \ldots, v_{c_{i,k+j}}^{\mathrm{T}}, U_i]^{\mathrm{T}}$. The user vector U_i can be viewed as a feature extraction of user behavior intentions and is what we need to compare the similarity between users.

The user vectors and check-in vectors are usually trained with Stochastic Gradient Descent (SGD), in which the gradient is obtained via backpropagation. In [29], Le and Mikolov develop hierarchical softmax based on a binary Huffman tree structure for fast training. In this paper, we employ negative sampling method [33] for optimization instead of hierarchical softmax. We randomly choose a check-in point and K negative instances at each time segment. Then we minimize the following objective to log probability for the sampled check-in points:

$$J_c = -log\sigma(\omega_{c_{i,k}}^{\mathrm{T}} v_i) - \sum_{j=1}^{J} log\sigma(-\omega_j^{\mathrm{T}} v_i). \tag{6}$$

where $\sigma(x) = 1/(1 + exp(-x))$ is a sigmoid function. Then this objective can be trained using SGD. As the updating rules for parameters is easily derived, we omit the detail in this paper.

Take Fig. 1 as an example, we can obtain the vector representations of the three users through trajectory embedding model. Specifically, if we set the size of vectors and sliding windows to 5 and 3 respectively, we can get three 5-dimensional user vectors.

4.3   Hash-Based User Similarity Calculation

The main computational challenge lies in highly time-consuming similarity calculation for all the user vectors with high dimensions. Hence, we employ a dimensionality reduction method, spherical hashing algorithm, to transform the user vectors into the corresponding binary codes, and then compute the similarity between users based on spherical Hamming distance.

Binary Code Learning. In this section, we utilize spherical hashing algorithm based on hyperspheres to learn the compact binary codes from user vectors.

Given a set of user vectors in a D-dimensional space, $\mathbb{U} = \{u_1, u_2, \ldots, u_n\}$, $u_i \in \mathbb{R}^d$, a binary code corresponding to a user vector u_i is expressed by $b_i = \{0, 1\}^k$, where k is the length of the code. Since each binary code has k bits, we need k different hash functions to map user vectors. Each hash function, $h_m(u)(1 \le m \le k)$, is defined with a pivot $p_m \in \mathbb{R}^d$ and a radius $t_m \in \mathbb{R}^+$ as following:

$$h_m(u) = \begin{cases} 0, & \text{if} \quad d(p_m, u) > t_m \\ 1, & \text{if} \quad d(p_m, u) \le t_m \end{cases} \tag{7}$$

where $d(x, y)$ denotes the Euclidean distance between vector x and y. The hash function determines whether the vector u is inside the hypersphere whose center is p_m and radius is t_m, where p_m and t_m for the k different hyperspheres can be obtained by an iterative optimization process [34]. To save space, the procedure of calculating p_m and t_m is omitted here. Through multiple hash functions, we can obtain the binary code for each user vector.

Note that the multiple hash function construction must follow two criteria, balanced partitioning of data space for each hash function and the independence between hash functions. Balanced partitioning indicates that the user vectors have the equal probability to be divided inside and outside the given hypersphere, so we define each hash function h_m to have the equal probability for +1 and −1 bits respectively as following:

$$Pr[h_m(u) = 1] = \frac{1}{2}, \quad u \in \mathbb{U}, 1 \le m \le c \tag{8}$$

A probabilistic event E_m is defined to represent the case of $h_m(u) = 1$. As we all know, two events E_i and E_j are independent if and only if $Pr[E_i, E_j] = Pr[E_i] \cdot [E_j]$. After the balanced partitioning process of user vectors for each bit (Eq. (8)), the independence between two bits should satisfy the following equation:

$$Pr[h_i(u) = 1], h_j(u) = 1]$$
$$= Pr[h_i(u) = 1] \cdot Pr[h_j(u) = 1] \qquad (9)$$
$$= \frac{1}{2} \cdot \frac{1}{2} = \frac{1}{4}$$

where $u \in \mathbb{U}, 1 \leq i, j \leq c$.

User Similarity Calculation. Next we explain how to measure the similarity between two users based on their binary codes. As mentioned before, the user vectors, reflecting the users' behavior characteristics, are mapped into the corresponding binary codes. The two users are more similar with each other if they have more similar binary codes. In this part, we apply the spherical Hamming distance [34], $d_{shd}(b_i, b_j)$, to measure the similarity between two binary codes to find the similar users. $d_{shd}(b_i, b_j)$ is defined as follows:

$$d_{shd}(b_i, b_j) = \frac{|b_i \oplus b_j|}{|b_i \wedge b_j|}. \qquad (10)$$

where $b_i \oplus b_j$ is the XOR bit operation and $b_i \wedge b_j$ is the AND bit operation. $|\cdot|$ denotes the number of +1 bits in the binary code. Two users are more similar in behaviors with the smaller distance between them.

5 Experiment

In this section, we conduct extensive experiments to validate the effectiveness and efficiency of our model. All the algorithms are implemented on an Intel Core i5-6200U CPU @ 2.40 GHZ with 8 GB RAM.

5.1 Experimental Setup

Datasets. We use a real trajectory dataset generated in Foursquare by 2070 users who live in California, USA. Each record is associated with a check-in point of a user, which contains user id, venue name and category, locations and timestamp. There are 483813 check-in records in this dataset.

Parameters. Table 2 lists all the parameters we used throughout the experiments, that all the parameters are assigned the default values unless specified explicitly.

Table 2. Default values of parameters

Parameter	Default value	Description
θ	0.8	Time partition ratio
d	300	User vector dimension
c	64	Binary code length

Baselines. We compare our approach with the two baselines. The first is Jaccard index [35], which is a popular method to measure the similarity. The Jaccard similarity between two users, (i.e., u and v), can be computed as follows:

$$sim_j(u, v) = \frac{T_{r,u} \bigcap T_{r,v}}{T_{r,u} \bigcup T_{r,v}} \tag{11}$$

where $T_{r,u}$ is the activity trajectory of user u.

The second baseline is to compute the Pearson coefficient between users to measure the users similarity. In specific, each user is denoted by a matrix $M_{m \times n}$, where each element $u_{i,j}$ represents the probability that the user conducts the activity i at time unit j. The Pearson coefficient between user u and user v can be computed as follows:

$$sim_p(u, v) = \frac{\sum_m \sum_n (U_{mn} - \overline{U})(V_{mn} - \overline{V})}{\sqrt{(\sum_m \sum_n (U_{mn} - \overline{U})^2)(\sum_m \sum_n (V_{mn} - \overline{V})^2)}} \tag{12}$$

Evaluation Method. A user study method is employed to evaluate the performance of our approaches and the two baselines, in which 100 volunteers are required to vote for the most similar user for each user. For each experiment, we run 200 test cases, which are randomly chosen from the dataset. We use the voting results to evaluate the effectiveness of all the methods.

5.2 Experimental Results

In this subsection, we will show the effectiveness and efficiency of our proposed model based on user study results. We declare some notations shown in the following figures. TEH represents the similarity measure of our proposed model. Trajectory embedding and cosine similarity (TEC) is our model without hashing step and use cosine similarity to compare users. Jaccard and Pearson denote the two baseline methods.

Effectiveness of TEH. Through user study, we obtain the score of three methods. According to the score, we could judge the superiority of our method. The user voting results are listed below.

Table 3. Effect of θ

θ	0.7	0.75	0.8	0.85	0.9
TEH	110	118	124	120	115
Jaccard	52	45	41	46	48
Pearson	38	37	35	34	37

Effect of θ. The votes of three methods are shown in Table 3. It can be seen from the table that the performance increases originally. With lower ratio, more people can be mistaken as similar users who are not in fact. Then, as θ continues to increases, the votes of our model will decline. A higher ratio means more subtle time partition, which will split the relationship between users. Besides, the votes of our model are far more than that of baselines, which directly shows that our model is better than the baselines.

Table 4. Effect of d

d	200	300	400	500
TEH	115	124	126	127
Jaccard	46	41	40	39
Pearson	39	35	34	34

Effect of d. The voting result of all the methods are presented with the change of user vector dimension d in Table 4. We can observe, for our model, the votes increase as d grows. However, we also notice that the increase becomes slower when $d \geq 300$ since with higher vector dimensions there is increasing chance that subsequently added dimensions are redundant.

Table 5. Effect of k

k	32	64	128	256
TEH	116	124	125	126
Jaccard	48	41	40	38
Pearson	36	35	35	36

Effect of k. In Table 5, it depicts the votes of three methods changing over binary code length. As we can see, at the beginning the performance improves as k increases. Moreover, we observe that there is little improvement on votes after k exceeds 64. Because binary code length reaches a certain value, the relationship between users can be well preserved.

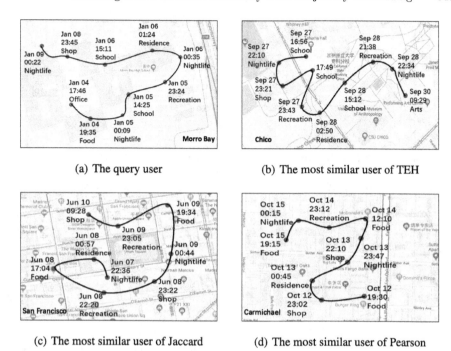

(a) The query user

(b) The most similar user of TEH

(c) The most similar user of Jaccard

(d) The most similar user of Pearson

Fig. 5. The query user and the most similar user of three methods.

Figure 5 shows that, given a query user, our model can find out more similar users than baselines. From the picture, we find that the query user often stay at school in the afternoon and have colorful nightlife. Besides, she sleeps late. The two baselines discover the most similar users for the query user, just considering the users' nightlife activity, while ignoring the learning at school in the afternoon. Our model can find a user, whose behavior is more similar to that of the query user, i.e., learning at school in the afternoon and staying up late. Our model can exactly capture the sequence property of user trajectory, while baselines only take the proportion of user activities into account. The baselines usually just consider the major activity and leave the consideration of the remaining activity.

Efficiency of TEH. In this part, we vary the values of parameters in Table 2 to compare our model with baselines and observe the effect of these parameters in efficiency. We use the total running time of finding the top-n similar user for each user to represent the efficiency.

Effect of n. Figure 6(a) shows the running time of different methods. We can perceive that TEH improves a lot than TEC and baselines in efficiency. When the number of users is large, calculating cosine similarity of user vectors is highly time-consuming. Therefore, it is particularly vital to operate hashing procedure. Besides, the running time increases a little as n increases. With n value increasing, we need to search more users who are similar to the query user.

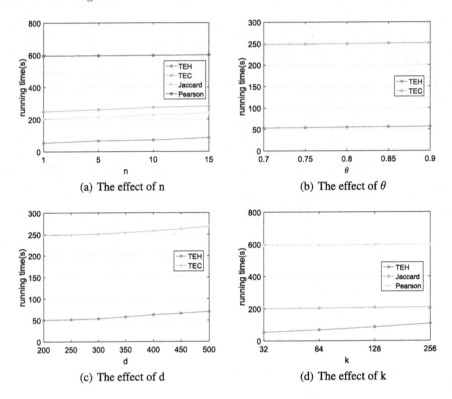

Fig. 6. The effect of parameters.

Effect of θ. In Fig. 6(b), we find that θ has little effect on the running time of our methods. This is because different θ values result in different time partition, however, it has little impact on the running time of trajectory embedding.

Effect of d. From Fig. 6(c), we can find that the efficiency will cut down as the user vector size increases. This is mainly because high vector dimension will lead to more time for calculating the similarity between users.

Effect of k. In Fig. 6(d), the overall time of TEH will increase drastically with k increasing, since it takes more time to train binary code. When the binary code length increases, TEH need to construct more hash functions, which will lead to a longer training time.

6   Conclusion

In this paper, we propose a novel framework, referred to as TEH, to model the users' activity trajectories and effectively mine the similarity between users. The core of our framework is the trajectory embedding technique and hash-based similarity measurement, which employ paragraph vector model to map the trajectories into vectors by treating them as paragraphs and then transform the vectors into compact binary codes for efficient user similarity calculation. To our best knowledge, this is the first work applying the paragraph vector model, a widely used Natural Language Processing technique, to model the trajectory data as vectors. Through a series of experiments, we validate our proposal and demonstrate that the proposed TEH framework has excellent performance under various conditions. As for future work, we plan to design more sophisticated user similarity mining algorithm by taking the spatial information and users' social relationships into account.

Acknowledgement. This research is partially supported by the Natural Science Foundation of China (Grant Nos. 61502324, 61532018).

References

1. Shang, S., Zheng, K., Jensen, C.S., Yang, B., Kalnis, P., Li, G., Wen, J.R.: Discovery of path nearby clusters in spatial networks. TKDE **27**(6), 1505–1518 (2015)
2. Hung, C.-C., Peng, W.-C., Lee, W.-C.: Clustering and aggregating clues of trajectories for mining trajectory patterns and routes. VLDB J. **24**(2), 169–192 (2015)
3. Shang, S., Chen, L., Wei, Z., Jensen, C.S., Zheng, K., Kalnis, P.: Trajectory similarity join in spatial networks. PVLDB **10**(11), 1178–1189 (2017)
4. Shang, S., Ding, R., Zheng, K., Jensen, C.S., Kalnis, P., Zhou, X.: Personalized trajectory matching in spatial networks. VLDB J. **23**(3), 449–468 (2014)
5. Su, H., Zheng, K., Zeng, K., Huang, J., Sadiq, S., Yuan, N.J., Zhou, X.: Making sense of trajectory data: a partition-and-summarization approach. In: 2015 IEEE 31st International Conference on Data Engineering (ICDE), pp. 963–974. IEEE (2015)
6. Kai, Z., Zheng, B., Xu, J., Liu, G., An, L., Li, Z.: Popularity-aware spatial keyword search on activity trajectories. WWWJ **20**(4), 1–25 (2016)
7. Zheng, K., Shang, S., Yuan, N.J., Yang, Y.: Towards efficient search for activity trajectories. In: ICDE, pp. 230–241 (2013)
8. Lv, M., Chen, L., Chen, G.: Mining user similarity based on routine activities. Inf. Sci. **236**(1), 17–32 (2013)
9. Liao, L., Patterson, D.J., Fox, D., Kautz, H.: Building personal maps from GPS data. Ann. N. Y. Acad. Sci. **1093**(1), 249 (2006)
10. Liao, L., Fox, D., Kautz, H.: Learning and inferring transportation routines. In: AAAI, pp. 348–353 (2004)
11. Shang, S., Ding, R., Yuan, B., Xie, K., Zheng, K., Kalnis, P.: User oriented trajectory search for trip recommendation. In: EDBT, pp. 156–167 (2012)
12. Su, H., Zheng, K., Huang, J., Wang, H., Zhou, X.: Calibrating trajectory data for spatio-temporal similarity analysis. VLDB J. **24**(1), 93–116 (2015)
13. Zheng, K., Zheng, Y., Yuan, N.J., Shang, S., Zhou, X.: Online discovery of gathering patterns over trajectories. TKDE **26**(8), 1974–1988 (2014)
14. Chen, X., Lu, R., Ma, X., Pang, J.: Measuring user similarity with trajectory patterns: principles and new metrics. In: Chen, L., Jia, Y., Sellis, T., Liu, G. (eds.) APWeb 2014. LNCS, vol. 8709, pp. 437–448. Springer, Cham (2014). https://doi.org/10.1007/978-3-319-11116-2_38
15. Li, W., Jiang, J., Li, G.: Mining user similarity based on users trajectories. In: ICCCBDA, pp. 557–562 (2014)

16. Stauffer, C., Grimson, W.E.L.: Adaptive background mixture models for real-time tracking. In: CVPR, p. 2246 (1999)
17. Zheng, K., Su, H., Zheng, B., Shang, S., Xu, J., Liu, J., Zhou, X.: Interactive top-k spatial keyword queries. In: IEEE, pp. 423–434, May 2015
18. Zheng, B., Zheng, K., Xiao, X., Su, H., Yin, H., Zhou, X., Li, G.: Keyword-aware continuous KNN query on road networks. In: ICDE, pp. 871–882 (2016)
19. Zheng, B., Su, H., Hua, W., Zheng, K., Zhou, X., Li, G.: Efficient clue-based route search on road networks. TKDE **29**(9), 1846–1859 (2017)
20. Wang, H., Zheng, K., Xu, J., Zheng, B., Zhou, X., Sadiq, S.: SharkDB: an in-memory column-oriented trajectory storage. In: Proceedings of the 23rd ACM International Conference on Information and Knowledge Management, pp. 1409–1418. ACM (2014)
21. Zheng, K., Zhou, X., Fung, P.C., Xie, K.: Spatial query processing for fuzzy objects. VLDB J. **21**, 1–23 (2012)
22. Shang, S., Liu, J., Zheng, K., Lu, H., Pedersen, T.B., Wen, J.: Planning unobstructed paths in traffic-aware spatial networks. GeoInformatica **19**(4), 723–746 (2015)
23. Zheng, B., Yuan, N.J., Zheng, K., Xie, X., Sadiq, S., Zhou, X.: Approximate keyword search in semantic trajectory database. In: ICDE, pp. 975–986 (2015)
24. Li, Q., Zheng, Y., Xie, X., Chen, Y., Liu, W., Ma, W.Y.: Mining user similarity based on location history. In: ACM SIGSPATIAL, p. 34 (2008)
25. Lee, M.-J., Chung, C.-W.: A user similarity calculation based on the location for social network services. In: Yu, J.X., Kim, M.H., Unland, R. (eds.) DASFAA 2011. LNCS, vol. 6587, pp. 38–52. Springer, Heidelberg (2011). https://doi.org/10.1007/978-3-642-20149-3_5
26. Mikolov, T., Le, Q.V., Sutskever, I.: Exploiting similarities among languages for machine translation. Computer Science (2013)
27. Collobert, R., Weston, J., Karlen, M., Kavukcuoglu, K., Kuksa, P.: Natural language processing (almost) from scratch. J. Mach. Learn. Res. **12**(1), 2493–2537 (2011)
28. Mikolov, T., Chen, K., Corrado, G., Dean, J.: Efficient estimation of word representations in vector space. Computer Science (2013)
29. Le, Q.V., Mikolov, T.: Distributed representations of sentences and documents. In: ICML 2014, vol. 32, pp. II-1188–II-1196 (2014)
30. Tagami, Y., Kobayashi, H., Ono, S., Tajima, A.: Modeling user activities on the web using paragraph vector. In: International Conference on World Wide Web, pp. 125–126 (2015)
31. Zhao, Y., Wang, J., Wang, F.Y., Shi, X., Lv, Y.: Paragraph vector based retrieval model for similar cases recommendation. In: Intelligent Control and Automation, pp. 2220–2225 (2016)
32. Zhang, C., Liu, L., Lei, D., Yuan, Q., Zhuang, H., Hanratty, T., Han, J.: TrioVecEvent: embedding-based online local event detection in geo-tagged tweet streams. In: ACM SIGKDD, pp. 595–604 (2017)
33. Mikolov, T., Sutskever, I., Chen, K., Corrado, G., Dean, J.: Distributed representations of words and phrases and their compositionality. In: NIPS, pp. 3111–3119 (2013)
34. Heo, J.P., Lee, Y., He, J., Chang, S.F.: Spherical hashing. In: CVPR, pp. 2957–2964 (2012)
35. Jaccard, P.: The distribution of the flora in the alpine zone. New Phytol. **11**(2), 37–50 (2010)

MaxBRkNN Queries for Streaming Geo-Data

Hui Luo[1]([✉]), Farhana M. Choudhury[1], Zhifeng Bao[1], J. Shane Culpepper[1], and Bang Zhang[2]

[1] School of Science, RMIT University, Melbourne, Australia
{hui.luo,farhana.choudhury,zhifeng.bao,shane.culpepper}@rmit.edu.au
[2] CSIRO, Canberra, Australia
Mattbang.Zhang@data61.csiro.au

Abstract. The problem of maximizing bichromatic reverse k nearest neighbor queries (MaxBRkNN) has been extensively studied in spatial databases, where given a set of facilities and a set of customers, a MaxBRkNN query returns a region to establish a new facility p such that p is a kNN of the maximum number of customers. In the literature, current solutions for MaxBRkNN queries are predominantly static. However, there are numerous applications for dynamic variations of these queries, including advertisements and resource reallocation based on streaming customer locations via social media check-ins, or GPS location updates from mobile devices. In this paper, we address the problem of continuous MaxBRkNN queries for streaming objects (customers). As customer data can arrive at a very high rate, we adopt two different models for recency information (sliding windows and micro-batching). We propose an efficient solution where results are incrementally updated by reusing computations from the previous result. We present a *safe interval* to reduce the number of computations for the new objects, and prune the objects that cannot affect the result. We perform extensive experiments on datasets integrated from four different real-life data sources, and demonstrate the efficiency of our solution by rigorously comparing how different properties of the datasets can affect the performance.

1 Introduction

Given two distinct types of objects, a set P of facilities and a set O of customers, if a facility p ($p \in P$) is a kNN of a customer o ($o \in O$), then o is one of the *Bichromatic Reverse k Nearest Neighbor* (BRkNN) of p. Given a set of facilities and a set of customers, a *Maximizing Bichromatic Reverse k Nearest Neighbor* (MaxBRkNN) query returns a region to establish a new facility p such that p is a kNN of the maximum number of customers [6,12,19]. In this study, we explore the problem of MaxBRkNN queries over streaming geo-data in spatial databases. This problem is critical in many real-time resource supply scenarios. For example, when a disaster happens, how can supplies be allocated dynamically to different rescue stations? The optimal location p with the greatest need for supplies should be updated based on patient arrivals in near real-time.

© Springer International Publishing AG, part of Springer Nature 2018
J. Pei et al. (Eds.): DASFAA 2018, LNCS 10827, pp. 647–664, 2018.
https://doi.org/10.1007/978-3-319-91452-7_42

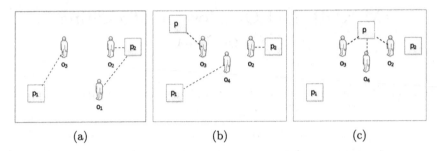

Fig. 1. Motivation example

However, existing MaxBRkNN studies neglect the fact that the cardinality of the objects in the current spatial region can change continuously as new objects are arriving and expiring. Below is an example of MaxBRkNN on streaming objects. For ease of illustration, we will consider $k = 1$ in the following example, which can be easily extended to the case $k > 1$.

Example 1. Figure 1a shows the location of two facilities p_1 and p_2, and the location of three customers o_1, o_2, o_3 for the time instance t_0. As the set of customers changes, due to the arrival of new customers and the departure of other customers, Fig. 1b and c present two alternatives for new facility p placements at time t_1. The reverse kNN of each facility is shown with a connecting dotted line in the figures.

In Fig. 1a, $RNN(p_1, P) = \{o_3\}$, $RNN(p_2, P) = \{o_1, o_2\}$, where $RNN(p_i, P)$ denotes the set of RNN customers for p_i in P. After the arrival of a new customer o_4 and the departure of a customer o_1 at time t_1, Fig. 1b and c show two alternative location choices for placing a new p that can serve the maximum number of customers. The location shown in Fig. 1b is not a suitable choice as $RNN(p, P \cup p) = \{o_3\}$, where the location shown in Fig. 1c is the optimal choice as $RNN(p, P \cup p) = \{o_2, o_3, o_4\}$. As the set of customers changes, the optimal location needs to be continuously updated.

In this paper, we address the problem of continuous updates in MaxBRkNN queries for streaming objects, where, given a set P of static facility locations, a stream of locations of customers O, a positive integer k, the problem is to update the optimal region in space to place a new facility p such that p has the maximum number of BRkNNs in O. To maintain recency information and minimize memory costs, a sliding window model is imposed on the stream, and a customer is valid only while it remains in the window.

We consider two commonly used sliding window models: *count-based* and *time-based* windows [8]. Both are represented by a window size $|W|$ and a slide size Δw, which are either a fixed number of objects for *count-based* windows, or time intervals for *time-based* windows. We also consider two variants of count-based windows: (i) real-time updates where Δw is '1', i.e., the result needs to be updated each time a new customer arrives; and (ii) micro-batching of customers for $\Delta w > 1$, i.e., the Δw customer changes are processed together to update the result. We propose solutions robust to both windows, however, the choice of an appropriate sliding window setting will depend on the intended application.

To the best of our knowledge, there is no prior work on supporting MaxBRkNN queries over streaming objects. Existing approaches present static-only solutions [6,12,19]. In large datasets, static MaxBRkNN solutions are incapable of efficiently supporting sliding window models since the number of redundant computations are incurred as a function of the window size. Therefore, we propose new approaches to incrementally update the current result set based on the previous computations.

In our proposed approach, we reuse the computations for the customers shared between two consecutive windows to update the optimal result in order to avoid redundant computations. Specifically, every time the sliding window shifts a distance of Δw, the results of the previous window and the overlapping objects are used to update the information of the new objects locally. We present the notion of a *"safe interval"* to reduce the number of computations for updated objects. Then only the objects that can influence the optimal region will be evaluated to update the result.

The rest of the paper is organized as follows. Section 2 reviews the related work. Section 3 formalises the MaxBRkNN problem. Section 4 proposes our baseline algorithm and optimization method. Section 5 evaluates the proposed methods through extensive experiments on real dataset, and Sect. 6 concludes the paper.

2 Related Work

Several studies have investigated the problem of finding a location or a region in space to establish a new facility such that the facility can serve the maximum number of customers based on different optimization criteria. The related body of work includes query processing on (i) static objects and (ii) dynamic objects.

2.1 Facility Location Selection Queries on Static Objects

Based on the optimization criteria, the facility location selection problem can be categorized mainly as Maximizing Bichromatic Reverse k Nearest Neighbor (MaxBRkNN) queries and distance aggregation queries.

MaxBRkNN Queries. Wong et al. [12] introduced the MAXOVERLAP algorithm to solve the MaxBRkNN problem. Here, a circle is first defined for each object $o \in O$, denoted by the Nearest Location Circle (NLC), where the center of the circle is o and the radius is the distance between o and its kNN. They reduced the problem of finding a region in space to the problem of finding the intersection point of the NLCs that is covered by the largest number of NLCs. The optimal region is obtained from the overlap of such NLCs. The MAXOVERLAP algorithm was extended to support the L_p-norm and three-dimensional space in a later work by the same authors [13].

Zhou et al. [19] presented the MAXFIRST algorithm. In contrast to the other studies, they consider the probability of a customer o visiting each of the ith nearest facilities while finding the result. They construct a Quadtree by iteratively partitioning the space into quadrants until each quadrant is fully covered by an NLC. For each quadrant, an *upper* and a *lower bound* of a number of NLCs that intersect with that quadrant are computed. In each iteration, a quadrant with the highest *upper bound* (which is more likely to become a part of the optimal region) is further partitioned into four quadrants. The process continues until the upper and the lower bound of the quadrants converge.

Liu et al. [7] presented an approach called MAXSEGMENT to reduce the search space by transforming the optimal region search problem to the optimal interval search problem in a one-dimensional space. The authors use a plane sweep-like method to find the optimal interval. Finally, the optimal interval is transformed back to the optimal region in the original two-dimensional space and returned as the result.

Lin et al. [6] presented the OPTREGION algorithm to solve the MaxBRkNN problem. The key idea is to index the set of facilities with a kd-tree to find the k nearest facilities of each object $o \in O$, and obtain an NLR (region enclosed by NLC) for each o. A *sweepline* algorithm is employed to generate the intersection lists for every NLR by traversing a "line" along the y-coordinates. The optimal region is obtained from the NLRs containing the maximal intersection point.

The existing solutions rely on the fact that the set of objects is static, and most of the solutions construct an index over the objects (or the NLCs). Therefore, these solutions are not easily extendable to the streaming object scenario.

Distance Aggregation Queries. Qi et al. [11] have explored the optimal location selection query, which finds a location for a new facility that minimizes the average distance from each customer to its closest facility. An *influence set* to manage a potential location p that includes customers for whom the nearest facility distance is reduced if a new facility is established at p. A similar problem was explored in other work [4, 10, 14, 15, 18] which try to find a location for a new facility such that the maximum distance between the facility and any customer is minimized. Papadias et al. [9] found a location that minimizes the sum of the distances of a facility placed in that location from the customers. These queries focus on an aggregation (such as average or summation) over the distances of the optimal location from the objects. All of these approaches only consider static objects and do not directly address the streaming MaxBRkNN problem explored in this paper.

2.2 Facility Location Selection Queries on Dynamic Objects

Ghaemi et al. [3] studied the MaxBRkNN problem for moving objects and facilities in road networks, but the solution can only work when $k = 1$ (the nearest facility). Their approach relies heavily on pre-computation, and uses multiple lookup tables to answer the queries online. Specifically, for each o, they store all edges (or parts of edges) with a distance less than or equal to its nearest

Table 1. Related work on MaxBRkNN

Study	Input		Distance
	O	P	
[6, 7, 12, 13, 19]	Static	Static	Euclidean
[3]	Moving	Moving	Network
Our work	Streaming	Static	Euclidean

facility. The set of these edges is denoted as the "local network" of o. The information of whether an edge or a part of an edge belongs to the local network of multiple objects is also stored. When the location of an object is updated, the most promising edges that could be the optimal location for a new facility are obtained using the pre-computed information. Additionally, three atomic operations to support complex movement operations are proposed. An assumption for these pre-computed local networks is that, an object can move only to a neighboring location. In contrast, as a streaming object can arrive anywhere in the space (or along any edge), the local network cannot be pre-computed for that object. Thus it is not easy to extend this method to solve our problem.

Table 1 summarizes the existing related work and their problem settings. Our work can also be used for other distance functions. Although the contributions of the existing work are important, there exists a research gap between these approaches and some real-life applications. The MaxBRkNN problem has not been explored previously in the streaming object setting, and the existing methods are not easily extensible for our problem.

3 Problem Formulation

Let P be a set of static facilities, where each $p \in P$ is defined as a pair $(p.lat, p.lng)$, representing its geo-spatial location. Let O be a stream of tuples $\langle o.lat, o.lng, o.t \rangle$ in the order of their arrival time $o.t$, where each item represents a customer $o \in O$ and $(o.lat, o.lng)$ represents its geo-spatial location.

We adopt the sliding window model where an object (customer) is valid while it belongs to the current sliding window W. The window size can be specified by time, count, and the update size as the number of insertions and deletions between two consecutive windows [5]. A time-based window contains the objects whose arrival time is within $|W|$ most recent time-slots, where Δw can be different. In a count-based window, $|W|$ and the update size (slide size) Δw are constant. The window contains the $|W|$ most recent data objects where the window updates for each new Δw ($1 \leq \Delta w \leq |W|$) object arrivals. A small value of Δw represents real-time updates, where a larger Δw depicts micro-batching of objects. When the context is clear, we use the terms 'customer' and 'object' interchangeably. Let O_n be the set of customers inserted, and O_o be the set of customers expired from the current window W. Before defining our problem, we first present the necessary preliminaries.

k **Nearest Neighbor Circle (c).** Let kNN (o) be the k nearest neighbor facility of a customer o. The k nearest neighbor circle c_o is a circle with the location of o as the center and the distance between o and kNN (o) as the radius. Let C be the set of kNN circles of all of the customers in the current window.

Intersection Circles Set (IS). Given a kNN circle c_o of a customer o, the intersection circle set IS_o is the set of circles that contain or intersect with c_o.

Maximal Intersection Point $(s^\top)$. As there is at least one intersection point when any two circles overlap, let s_o be the intersection point in circle c_o with the largest number of overlapping circles from C, and η_s be the number of the circles overlapping with s_o. Let $s^\top$ be the intersection point with the largest η_s, and $\eta^\top$ is its corresponding number of overlapping circles.

Definition 1 (Maximal Intersection Region, R). Given C, we define the maximal intersection region R such that (i) For $\forall\ r \in R$, $|RkNN(r, P)|$ is maximal, where r is a point location; (2) For $\forall r, r' \in R$, $RkNN(r, P) = RkNN(r', P)$.

Problem Statement. Given a set P of static facility locations, a stream of customers O, a positive integer k, and a sliding window model on O, the continuous MaxBRkNN problem on a stream is to continuously update the Maximal Intersection Region, R for the customers valid in the updated window. There may exist multiple Maximal Intersection Regions.

4   Algorithm

In this section we propose the following different solutions to address the MaxBRkNN problem on streaming objects: (i) As there are multiple studies that address the MaxBRkNN query for static objects, first we apply one of the approaches directly to solve the problem on streaming objects (customers) as our baseline. (ii) The baseline is originally designed for static objects and does not reuse any computation for the streaming objects. Instead, we propose an optimized solution where computations are shared among the consecutive windows as much as possible. (iii) We further propose two more optimizations: (a) *safe interval*, and (b) pruning of objects that cannot update the result from the previous window, on top of our proposed solution to improve overall efficiency.

4.1   Baseline Algorithm

We adopt the solution, OPTREGION, proposed by Lin et al. [6] for our baseline, as they have shown that their solution consistently outperforms two other state-of-the-art solutions (MAXOVERLAP [12] and MAXFIRST [19]). The OPTREGION solution applies the principle *region-to-point transformation* [12] to find the intersection point overlapping with the maximum number of circles. If such a maximal intersection point $s^\top$ is found, then the maximal intersection region R (the result of the MaxBRkNN query) can be easily obtained from the circles overlapping with $s^\top$.

Algorithm 1. Baseline (W, kd-tree over P)

1: $\eta^\uparrow \leftarrow 1$
2: **for** o in W **do**
3:    compute c_o of o using the kd-tree
4: **for** o in W **do**
5:    $IS_o \leftarrow$ Set of circles overlapping with c_o by a sweepline algorithm
6:    **for** c_i in IS_o **do**
7:       update IS_i
8: Sort $|IS|$ in descending order of $|IS_o|$
9: **for** IS_o in IS **do**
10:    **if** $|IS_o| > \eta^\uparrow$ **then**
11:       compute the exact η of c_o
12:       **if** $\eta > \eta^\uparrow$ **then**
13:          update $\eta^\uparrow$ and $s^\uparrow$
14:    **else** Break
15: find the intersection of all circles containing $s^\uparrow$ and update R
16: **return** R

As the OPTREGION algorithm is proposed for static objects, every time the sliding window updates, we invoke the algorithm in our baseline. Algorithm 1 shows the pseudocode of the baseline. Here, the set of the facilities P is indexed using a kd-tree. The index is initially built before processing any queries. The input of the algorithm is the set of all objects in the initial window W and the kd-tree over P. Whenever the window updates, the steps of the OPTREGION algorithm are executed. There are three main steps in the algorithm:

1. **Find the kNN circle c_o:** For each object o in the window, the kd-tree over P is used to find the k nearest neighbors of o. Then the c_o for each o (Lines 2-3) is constructed, where the center is the location of o, and the radius is the distance from o to its kNN.
2. **Find the set of intersecting circles:** For each circle c_o, the sweepline algorithm outlined in Algorithm 1 of Lin et al. [6] is used to determine the set IS_o of the circles intersecting with c_o. This sweepline algorithm scans along the y-coordinate of each circle that is valid in the current window from top to bottom to find IS_o. Let the highest and the lowest y-coordinate of a c_o be $y_o^\uparrow$ and $y_o^\downarrow$, respectively. When the sweepline reaches the $y_o^\uparrow$ of c_o, c_o is inserted in a status tree (AVL tree). However, when the bottom point $y_o^\downarrow$ is encountered, c_o is deleted from the status tree. This status tree, which is updated dynamically, saves the candidate IS_o of the current circle.
3. **Find the maximal intersection point $s^\uparrow$:** Here, the maximum number of circles overlapping with an intersecting point s of a c_o can be at most IS_o (the number of circles intersecting with c_o). The set IS_o of the objects are sorted and considered in descending order of cardinality. We maintain the maximum value $\eta^\uparrow$ of intersecting circles for any intersection point found so far. If the IS_o of any c_o is greater than $\eta^\uparrow$, the actual number of intersections for c_o is computed by sweeping around the perimeter of c_o and finding the intersection point s with IS_o. The value of $s^\uparrow$ and $\eta^\uparrow$ are updated if necessary.

4. Finally, the optimal region R is constructed as the intersecting region of all of the circles overlapping with the point $s^{\uparrow}$, and R is returned as the result. The process is repeated when the window updates again.

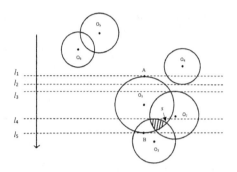 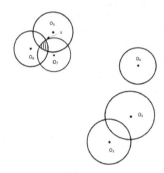

Fig. 2. An example of the initial process.   **Fig. 3.** An example of the first slide.

Example 2. Figure 2 illustrates an example of the initial process of the baseline. For ease of presentation, we omit the set P from the figure. Let A and B be the top point and bottom point of c_1, respectively. The five states of the sweepline of c_1 are l_1, l_2, l_3, l_4 and l_5. In the l_1 state, the AVL tree has two nodes: c_1 and c_4. In l_2, c_4 is removed from the AVL tree. For l_3, the top point of c_2 is encountered, and c_2 is inserted into the AVL tree. In l_4, c_3 is inserted in the tree. And finally for l_5, the sweepline reaches the bottom point B, and c_1 is deleted. We then stop updating IS_1. We choose the union of circles (except itself) under these states as the upper bound of IS_1, which is $\{c_2, c_3, c_4\}$. Correspondingly, we also need to update the IS of each circle c_i in IS_1, for example, c_1 should be added to IS_4. Note that $IS_1 = \{c_2, c_3, c_4\}$, c_4 does not intersect with c_1 since they do not have any common intersection point. We sort $|IS_o|$ in descending order to get: $|IS_1| = 3$, $|IS_2| = 2$, $|IS_3| = 2$, $|IS_4| = 1$, $|IS_5| = 1$, $|IS_6| = 1$. We traverse c_1 with the largest $|IS_o|$, and s (shown in the figure) is returned because it has the largest η, which equals to 3. Then we update $\eta^{\uparrow}$ to 3 and $s^{\uparrow}$ to s. The next iteration results in early termination as $|IS_2| < 3$. Finally, R is found (shown as the shaded area) as the area overlapped by c_1, c_2 and c_3, which overlaps with $s^{\uparrow}$.

An example of the update phase for the first slide is illustrated in Fig. 3 for $\Delta w = 1$. From the example in Fig. 2, o_1 is expired and o_7 arrives. We repeat a new computation from scratch to find the new R (shown as a shaded area in the example).

Drawbacks of Baseline Algorithm. When the sliding window updates, the baseline algorithm repeats the process for all of the objects in W. However, for a count-based window, only Δw objects are inserted, where $W - \Delta w$ objects are common in the two consecutive windows. For a time-based window, $W - O_o$ objects are common in the two consecutive windows.

Algorithm 2. Optimized Algorithm (W', W, kd-tree over P)

1: $O_n \leftarrow W - W'$
2: $O_o \leftarrow W' - W$
3: $\eta \leftarrow 1$
4: **for** each o in O_n **do**
5:      compute c_o of o using the kd-tree
6: **for** each o in O_n **do**
7:      $IS_o \leftarrow$ Set of circles overlapping with c_o by a sweepline algorithm
8:      **for** c_i in IS_o **do**
9:          update IS_i
10: **for** each o in O_o **do**
11:      **for** c_i in IS_o **do**
12:          update IS_i
13: $C \leftarrow$ Set of c_o for all $o \in W$
14: Lines 9 - 13 of Algorithm 1
15: Find the intersection of all circles containing $s^\dagger$ and update R
16: **return** R

Thus, the baseline algorithm requires a substantial number of repeated computations. Therefore, we propose the following refinements to the algorithm that only consider the inserted and expired objects (that are not common between two consecutive windows) when updating the optimal region.

4.2 Algorithmic Improvements

The key idea of our proposed algorithm is to share the computations between two consecutive windows whenever possible. After initializing the sliding window, the optimal region R is first obtained by any of the existing MaxBRkNN algorithms on static objects for the objects in the initial W. Then each time the window updates, our proposed optimization algorithm outlined in Algorithm 2 is called to update the result R. Here, the input of the algorithm is the previous window W', the current window W, and the kd-tree over P. Algorithm 2 consists of the following steps:

- The set of the newly inserted objects O_n and the expired objects O_o are obtained from the previous and the current window.
- **Updating only for the required objects:** In contrast to the baseline where the c_o of each object in the current window is computed, we compute the kNN circle c_o of only the objects in O_n using a sweepline algorithm, and compute the set IS_o of circles intersecting with each $o \in O_n$. The set of intersecting circles IS_o is also updated for each expired object $o \in O_o$ (Lines 10–12). Thus, instead of the steps in Lines 2–7 of Algorithm 2 where the computations are done for each object in W, the computations are now done only for the objects in O_n and O_o that are not common between W and W'.

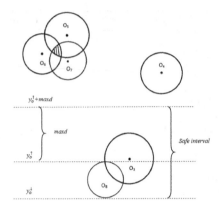

Fig. 4. Illustration of pruning rule 1.

- **Finding the maximal intersecting point, $s^\uparrow$:** We execute Lines 9–13 of Algorithm 1 to find $s^\uparrow$. In contrast to the baseline, we do not sort IS_o based on their cardinality again, as only a subset of them are likely to change.
- Finally, similar to the baseline, we compute the optimal region R from the point $s^\uparrow$ and return R.

Example 3. From the example in Figs. 2 and 3, let $\Delta w = 1$, o_1 expires, and o_7 arrives as a new object. First, we construct c_7 and compute $IS_7 = \{c_5, c_6\}$ for the new object using the sweepline algorithm. Then we update $IS_2 = \{c_3\}$ and $IS_3 = \{c_2\}$ because c_1 is expired, and then update $IS_5 = \{c_6, c_7\}$ and $IS_6 = \{c_5, c_7\}$ for the arriving object c_7. As $|IS_5|$ has the maximal value, then we sweep around c_5 to find the maximal intersection point s (shown in Fig. 3). Finally, the previous $s^\uparrow$ is replaced by the new s, and the result region R (shown as the shaded area) is also updated. It is worth noting that we only need to compute the new circles (c_7 in this example) in our algorithm, but all of the valid circles require recomputations in the baseline algorithm.

Drawbacks of the Optimization Algorithm. Although Algorithm 2 improves the performance by reusing computations, it has the following drawbacks:

1. In Line 7, the sweepline algorithm outlined in Algorithm 1 by Lin et al. [6] is employed to determine the set IS_o of the intersecting circles for each object $o \in O_n$. This sweepline algorithm scans along the y-coordinate of each valid circle in the current window from top to bottom to find IS_o. However, when only a subset of the circles change (inserted or expired), sweeping along the y-coordinate of each circle results in many unnecessary circle scans.
2. The value $\eta^\uparrow$ denotes the maximum number of circles overlapping with the intersecting point found so far. Each time $\eta^\uparrow$ is initialized as '1'. However, the optimal region R of the previous window, and thus the $\eta^\uparrow$ for the previous window can be decreased by at most Δw due to the expired circles.

In order to overcome these drawbacks, we propose two different pruning rules.

Pruning Rule 1: Safe Interval. The sweepline algorithm presented in Algorithm 1 by Lin et al. [6] scans the circles from top to bottom along the y-coordinate. In contrast, we want to determine a safe interval around the circle c_o such that any circle outside the safe interval cannot overlap with c_o, thus avoid scanning the unnecessary circles for c_o in the sweepline algorithm.

Let, the maximum diameter among the circles valid in the current window be $maxd$, and the highest and the lowest value in the y-coordinate of a c_o are $y_o^\uparrow$ and $y_o^\downarrow$, respectively. As the sweep is performed in a top-down manner, only the circles whose highest y-coordinate are within $y_o^\uparrow + maxd$ and $y_o^\downarrow$ can possibly overlap with the circle c_o. Thus we only need to scan the safe interval $y_o^\uparrow + maxd$ to $y_o^\downarrow$ in the sweepline algorithm for a circle c_o.

Example 4. In Fig. 4, c_8 is generated after the second slide. The circles c_4, c_5, c_6, c_7 are pruned as their top points are not within safe interval, while c_3 is added into IS_8.

Pruning Rule 2: Using the Result of the Previous Window. As shown in Line 10 of Algorithm 1 (which is also executed in Line 14 of Algorithm 2), we only need to compute the exact η of an object o if the corresponding $|IS_o|$ is greater than $\eta^\uparrow$. As only O_o circles can expire, the $\eta^\uparrow$ of the previous window (the maximum number of overlapping circles) can decrease by at most O_o. Thus, we can update the algorithm where additional input is the $\eta^\uparrow$ of the previous window, and initialize $\eta^\uparrow$ of the current window as $\eta^\uparrow - |O_o|$. If $\eta^\uparrow - |O_o|$ is less than '1', we initialize $\eta^\uparrow$ as '1'. For a count-based window, $|O_o| = \Delta w$.

5 Experimental Evaluation

In this section, we present the experimental evaluation for our solutions when continuously updating results for MaxBRkNN on sliding windows. In particular, we compare our proposed optimization solution (OP) with the baseline (BA), and further evaluate the benefit of applying each of the proposed pruning rules – pruning rule 1 (P1) and pruning rule 2 (P2) as proposed in Sect. 4.2.

5.1 Experimental Settings

All of our experiments were conducted on a Intel(R) Core(TM) i5-7200U CPU@2.50 GHz processor and 8 GB memory, running a Ubuntu 17.10 operating system. All algorithms were implemented in C++. The GCC version was 7.2.0.

Datasets. We conduct all experiments on an integrated real dataset, which is a combination of five different real check-in datasets. Table 2 shows a summary of all the datasets before integration. Each dataset contains the location of the points of interest (POIs) and the check-in locations of the users. The Foursq_NYC dataset collected from Foursquare contains POIs and check-in information of only New York City. The rest of the datasets consist of check-ins and

Fig. 5. Check-ins in NYC

Table 2. Dataset: check-in information.

Data source	Check-ins	Proportion
Foursq_NYC [16]	227,428	15.29%
Foursq_Global [17]	289,727	19.48%
LSSD [1]	710,827	47.79%
Gowalla [2]	139,171	9.36%
BrightKite [2]	120,359	8.18%

POIs, where the locations are distributed all around the world. These datasets are collected from Foursquare [16,17], the Location Sharing Services Dataset (LSSD) [1], Gowalla [2], and BrightKite [2].

We combine these datasets and take a subset, denoted as 'NYC', where the locations of the POIs and check-ins are all within New York City. We get 1,471,074 check-ins from mid April 2008 to mid September 2013 after duplicate removal, and 299,698 unique POIs in NYC. A graphical view of the check-in locations of this dataset is shown in Fig. 5 as a heatmap.

To conduct the experiments, an area where the check-ins are located is chosen as a pre-defined percentage of the total area of the dataset. Note that, area size is a parameter of our experiments where the default area is 1%. We issued a range query with the area size in a random location. If the number of check-ins in that region is greater than 10% of the total check-ins, we denote that as a high density area. Similarly, we find mid and low density areas with check-in numbers – around 5% and 1% of the total check-ins. We report the average performance for 50 shifts of the sliding window for each setting.

Table 3. Parameters

Parameter	Symbol	Range
Window size	$|W|$	3000, 3500, **4000**, 4500, 5000
Slide size	Δw	1, 100, 200, **400**, 600, 800
No. of k nearest neighbors	k	2, 5, **10**, 20, 50
Density of check-ins	d	1%, 5%, **10%**
Area of check-ins	a	0.5%, **1%**, 2%, 4%

Evaluation and Parameterization. We study the efficiency of the baseline (BA), our proposed solution (OP), pruning rule 1 on top of OP (OPF), and both pruning rule 1 and pruning rule 2 together on top of OP (OPFS) as a function of varying parameters. The parameters and their ranges are listed in Table 3, where the default values are in bold. For all experiments, a single parameter is varied while the rest are fixed as their default.

Table 4. Runtime comparison of BA and OPFS (ms).

Model	BA	OPFS
Micro-batch	1,429	189
Real-time	1,348	54

Among the parameters, Δw denotes the number of objects updated in the window. Since Δw of a general count-based window is a fixed integer between 1 to $|W|$, in contrast to a time-based window where Δw varies, we present our experimental results for count-based window to better demonstrate the effect of varying a single parameter. However, all of our proposed approaches are applicable for both types of windows.

For each Δw-sized update of the window, we study the impact of each parameter on: (i) the total runtime, and (ii) the percentage of iterations pruned by P1 and P2 with respect to the OP. As a small value of Δw represents real-time updates, and a larger Δw depicts micro-batching of objects, we present our experimental results in two parts:

- Section 5.2.1 presents the real-time update experiments, where Δw is 1, and a single parameter other than Δw is varied while keeping the rest as default.
- Section 5.2.2 presents the experiment for micro-batching, where Δw is varied in a set of experiments from 1 to 800, and the default Δw is set as 400.

5.2   Performance Evaluation

We first compare the baseline (BA) and our proposed OPFS (which applies both pruning rules on top of our OP method) for the default parameter settings and show the runtime of both methods in Table 4. As BA applies an existing algorithm originally designed for static objects, the runtime of BA is about one to two orders of magnitude higher than OPFS for the streaming query. Due to this huge difference in performance, we exclude BA from the rest of the experimental evaluation as this is representative of the best case for the baseline algorithm, and compare the performance among OP, OPF, and OPFS only.

5.2.1   Real-Time Processing ($\Delta w = 1$)

Varying $|W|$. Figure 6a shows the effect of varying the size $|W|$ of a window. As more kNN circles are likely to intersect with each other as $|W|$ increases, the runtime increases for all of the three methods as $|W|$ increases. As P1 reduces the number of circles to be checked while finding the intersection of the newly arrived objects, the percentage of pruning by P1 decreases for a higher $|W|$ (where there are more circles), as shown in Fig. 6b. In contrast, the percentage of pruning from P2 increases with $|W|$. The reason is that, although more circles may intersect with each other, there are not many intersection points that can become

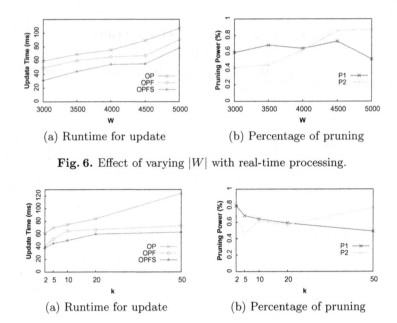

(a) Runtime for update   (b) Percentage of pruning

Fig. 6. Effect of varying $|W|$ with real-time processing.

(a) Runtime for update   (b) Percentage of pruning

Fig. 7. Effect of varying k with real-time processing.

a candidate better than the previous window's result, thus more intersection points can be pruned from consideration.

Varying k. The radius of each kNN circle increases w.r.t. k. As the radius increases, more circles intersect, and thus the runtime increases with k in all of the approaches (Fig. 7a). The runtime of the OP approach increases rapidly for $k > 20$, where the other two methods do not vary much. As more circles intersect, the percentage of pruning by $P1$ decreases with the increase of k, as shown in Fig. 7b.

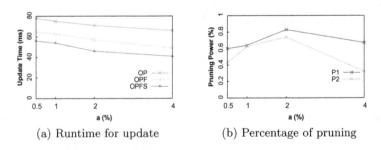

(a) Runtime for update   (b) Percentage of pruning

Fig. 8. Effect of varying a with real-time processing.

Varying a. We vary the size of the area where the check-ins are located as a percentage of the total area of the dataset. A higher percentage of the area (a)

Table 5. Varying d with real-time processing.

Density	Update time (ms)				Pruning power	
	BA	OP	OPF	OPFS	$P1$	$P2$
1%	904	61	52	45	0.60	0.29
5%	1,225	74	61	54	0.67	0.58
10%	1,348	76	65	57	0.64	0.63

denotes that locations of the objects in a window are sparser. As the chances of circles intersecting with each other decrease as their density becomes sparser, the runtime decreases (Fig. 8a).

Varying d. Table 5 shows the performance of the four approaches for varying d. As more circles are likely to intersect with each other for a higher percentage of density, the runtime increases for each method. The runtime of BA is around one to two orders of magnitude higher than the other methods. As the result region is likely to have a higher number of intersecting circles for a higher density, $P2$ shows a greater benefit by pruning the intersecting points that cannot be better than the previous window's result.

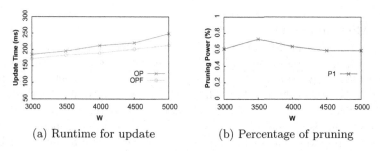

(a) Runtime for update   (b) Percentage of pruning

Fig. 9. Effect of varying $|W|$ with micro-batching.

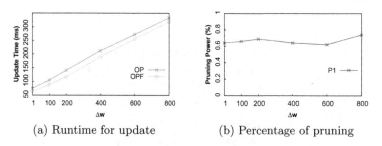

(a) Runtime for update   (b) Percentage of pruning

Fig. 10. Effect of varying Δw with micro-batching.

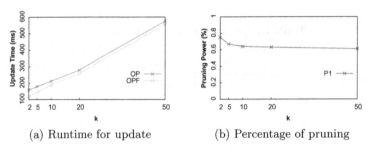

(a) Runtime for update   (b) Percentage of pruning

Fig. 11. Effect of varying k with micro-batching.

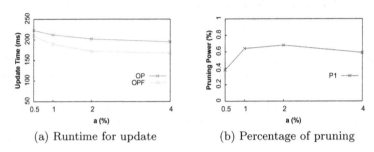

(a) Runtime for update   (b) Percentage of pruning

Fig. 12. Effect of varying a with micro-batching.

5.2.2 Micro-batching ($\Delta w \geq 1$)

Here, we show the performance when varying different parameters in a more general sliding window setting, where the Δw can be greater than or equal to 1. The runtime and the percentage of pruning by varying each parameter are shown in Figs. 9a, 10, 11 and 12b. The trends are mostly similar to the previous set of experimental evaluations, except that $P2$ is more effective for real-time processing. This is because $P2$ relies on the difference that can happen to the result of the previous window, i.e., $\eta^\uparrow - \Delta w$, but this value becomes smaller for a larger value of Δw, thus the pruning effect is reduced in larger windows. As the rest performance results are similar, we focus on the effect of varying Δw in this section.

Table 6. Varying d with micro-batching.

Density	Update time (ms)				Pruning power
	BA	OP	OPF	OPFS	$P1$
1%	1,408	186	157	157	0.64
5%	1,429	198	164	164	0.69
10%	1,489	212	189	189	0.72

Varying Δw. A larger Δw denotes that a smaller number of objects are shared between two consecutive windows. As shown in Fig. 10a, the runtime is proportional to the increase of Δw, as more objects that are not common among the windows (updated objects) need to be processed (Table 6).

6 Conclusion

In this paper, the problem of maximizing bichromatic reverse k nearest neighbor queries on streaming geo-data is introduced for the first time. We proposed an efficient solution where results are incrementally updated by reusing computations from the previous result. Our solution can work on both the count-based and the time-based sliding window models, thereby supporting both real-time processing and micro-batch processing. Extensive experiments based on real datasets have been conducted to verify the efficiency of our solution.

Acknowledgements. This work was partially supported by ARC DP170102726, DP180102050, and NSFC 61728204, 91646204. Zhifeng Bao is supported by a Google Faculty Award.

References

1. Cheng, Z., Caverlee, J., Lee, K., Sui, D.Z.: Exploring millions of footprints in location sharing services. In: ICWSM 2011, pp. 81–88 (2011)
2. Cho, E., Myers, S.A., Leskovec, J.: Friendship and mobility: user movement in location-based social networks. In: SIGKDD, pp. 1082–1090 (2011)
3. Ghaemi, P., Shahabi, K., Wilson, J.P., Banaei-Kashani, F.: Continuous maximal reverse nearest neighbor query on spatial networks. In: GIS, pp. 61–70 (2012)
4. Cardinal, J.J., Langerman, S.: Min-max-min geometric facility location problems. In: EWCG, pp. 149–152 (2006)
5. Li, J., Maier, D., Tufte, K., Papadimos, V., Tucker, P.A.: No pane, no gain: efficient evaluation of sliding-window aggregates over data streams. SIGMOD Rec. **34**(1), 39–44 (2005)
6. Lin, H., Chen, F., Gao, Y., Lu, D.: OptRegion: finding optimal region for bichromatic reverse nearest neighbors. In: Meng, W., Feng, L., Bressan, S., Winiwarter, W., Song, W. (eds.) DASFAA 2013. LNCS, vol. 7825, pp. 146–160. Springer, Heidelberg (2013). https://doi.org/10.1007/978-3-642-37487-6_13
7. Liu, Y., Wong, R.-W., Wang, K., Li, Z., Chen, C., Chen, Z.: A new approach for maximizing bichromatic reverse nearest neighbor search. Knowl. Inf. Syst. **36**(1), 23–58 (2013)
8. Mouratidis, K., Bakiras, S., Papadias, D.: Continuous monitoring of top-k queries over sliding windows. In: SIGMOD, pp. 635–646 (2006)
9. Papadias, D., Shen, Q., Tao, Y., Mouratidis, K.: Group nearest neighbor queries. In: ICDE, pp. 301–312 (2004)
10. Qi, J., Xu, Z., Xue, Y., Wen, Z.: A branch and bound method for min-dist location selection queries. In: ADC, pp. 51–60 (2012)
11. Qi, J., Zhang, R., Kulik, L., Lin, D., Xue, Y.: The min-dist location selection query. In: ICDE, pp. 366–377 (2012)

12. Wong, R.C.-W., Özsu, M.T., Yu, P.S., Fu, A.W.-C., Liu, L.: Efficient method for maximizing bichromatic reverse nearest neighbor. PVLDB **2**(1), 1126–1137 (2009)
13. Wong, R.C.-W., Özsu, M.T., Fu, A.W.-C., Yu, P.S., Liu, L., Liu, Y.: Maximizing bichromatic reverse nearest neighbor for lp-norm in two and three-dimensional spaces. PVLDB **20**(6), 893–919 (2011)
14. Yan, D., Zhao, Z., Ng, W.: Efficient algorithms for finding optimal meeting point on road networks. PVLDB **4**(11), 968–979 (2011)
15. Yan, D., Zhao, Z., Ng, W.: Efficient processing of optimal meeting point queries in Euclidean space and road networks. Knowl. Inf. Syst. **42**(2), 319–351 (2015)
16. Yang, D., Zhang, D., Zheng, V.W., Yu, Z.: Modeling user activity preference by leveraging user spatial temporal characteristics in LBSNs. Trans. SMC **45**(1), 129–142 (2015)
17. Yang, D., Zhang, D., Qu, B.: Participatory cultural mapping based on collective behavior data in location-based social networks. TIST **7**(3), 30:1–30:23 (2016)
18. Zhang, P., Lin, H., Gao, Y., Lu, D.: Aggregate keyword nearest neighbor queries on road networks. GeoInformatica **22**, 1–32 (2017)
19. Zhou, Z., Wu, W., Li, X., Lee, M.L., Hsu, W.: MaxFirst for MaxBRkNN. In: ICDE, pp. 828–839 (2011)

Free-Rider Episode Screening via Dual Partition Model

Xiang Ao[1,2]([✉]), Yang Liu[1,2], Zhen Huang[3], Luo Zuo[1,2], and Qing He[1,2]

[1] Key Lab of Intelligent Information Processing of Chinese Academy of Sciences
(CAS), Institute of Computing Technology, CAS, Beijing 100190, China
{aoxiang,liuyang17z,zuol,heqing}@ict.ac.cn
[2] University of Chinese Academy of Sciences, Beijing 100049, China
[3] Tsinghua University, Beijing, China
huangzh2.13@sem.tsinghua.edu.cn

Abstract. One of the drawbacks of frequent episode mining is that over-
whelmingly many of the discovered patterns are redundant. Free-rider
episode, as a typical example, consists of a real pattern doped with some
additional noise events. Because of the possible high support of the inside
noise events, such free-rider episodes may have abnormally high support
that they cannot be filtered by frequency based framework. An effective technique for filtering free-rider episodes is using a partition model
to divide an episode into two consecutive subepisodes and comparing
the observed support of such episode with its expected support under
the assumption that these two subepisodes occur independently. In this
paper, we take more complex subepisodes into consideration and develop
a novel partition model named EDP for free-rider episode filtering from
a given set of episodes. It combines (1) a dual partition strategy which
divides an episode to an underlying real pattern and potential noises; (2)
a novel definition of the expected support of a free-rider episode based on
the proposed partition strategy. We can deem the episode interesting if
the observed support is substantially higher than the expected support
estimated by our model. The experiments on synthetic and real-world
datasets demonstrate EDP can effectively filter free-rider episodes compared with existing state-of-the-arts.

Keywords: Episode dual partition · Interesting pattern discovery
Episode mining · Sequence mining

1 Introduction

One of the defects in frequent pattern mining is that there are abundant redundant patterns in the very large number of output patterns [22]. As a result, how

Z. Huang—This work was done when Zhen was visiting Institute of Computing
Technology, CAS.

The original version of this chapter was revised: For detailed information please see the
erratum to this chapter available at https://doi.org/10.1007/978-3-319-91452-7_61

© Springer International Publishing AG, part of Springer Nature 2018
J. Pei et al. (Eds.): DASFAA 2018, LNCS 10827, pp. 665–683, 2018.
https://doi.org/10.1007/978-3-319-91452-7_43

to effectively reduce redundancy of the output becomes an essential problem of current research [6, 8, 10, 11, 13, 16, 18, 23]. Frequent episode mining [14] (FEM for short), as one of the sub-topics of frequent pattern mining, which aims at discovering frequently appeared ordered sets of events from a single symbol (event) sequence, is facing the similar problem as well. Moreover, FEM algorithms usually suffer from inefficient processing since checking whether a sequence covers a general episode is an **NP**-hard problem [20]. It makes redundancy reduction in FEM particularly crucial because users will be reluctant to obtain redundant and useless patterns after a long time waiting.

In this paper, our purpose is to reduce redundancy of frequent episodes. Specifically, we focus on screening *free-rider episodes* and identifying underlying patterns from a given set of frequent episodes. Here free-rider episode, as a prototypical example of redundancy in episodes, consists of a real pattern doped with additional noise events.

While filtering patterns to reduce redundancy is well-studied for itemsets, it is still underexplored for episodes. The most straightforward approach to solve such problem is to compare observations against expectations predicted by independence model [9, 12, 17]. Interestingly enough, episodes occurrences in a single sequence may be dependent to each other, and hence support is no longer a sum of independent variables [18]. A more effective manner is to utilize partition models in which an episode is divided into subepisodes and the observed support is compared with the expected support produced by the model under specialized assumptions [16, 18]. However, constructing a good partition model has significant challenges that existing approaches present some limitations, and we summarize them as follows.

- Noise events could be doped *before*, *after*, or *inside* a real pattern. For example, $a \rightarrow x \rightarrow b$ could be a free-rider episode, where $a \rightarrow b$ is the real pattern and x is the noise. Here $a \rightarrow b$ is a non-prefix subepisode of $a \rightarrow x \rightarrow b$. However, most existing partition models fail to take non-prefix subepisodes into account. For example, in [18], $a \rightarrow x \rightarrow b$ can only be divided into two consecutive subepisodes, e.g. $\{a \rightarrow x, b\}$ or $\{a, x \rightarrow b\}$.
- It is challenging to define the expected support of an episode in one long sequence. The major reason is that the support of an episode is defined as its number of occurrences in a sequence, and the events in such sequence may be dependent to each other. Hence we have to carefully use independence assumptions in this scenario. However, independence assumptions are widely used in redundancy reduction in database of itemsets or sequences [18, 23]. In that problem, the support of pattern is usually defined as the number of the transactions/sequences covering the pattern, and each individual transaction/sequence is naturally regarded independent with each other. As a result, it is difficult to adapt traditional ideas to our problem.
- The search space of partitions suffers from "combination explosion" as it increases exponentially to the number of events in an episode. Hence pruning strategies are needed to perform an efficient filtering.

To address the challenges and overcome the limitations of the existing methods, we propose a novel partition model named EDP (**E**pisode **D**ual **P**artition).

In EDP, we partition the events of an episode into two distinct categories, called *informative* and *random* respectively, and we make the assumption that informative events form the real pattern while random events are noises. Here we do not restrict the position of either informative or random events and thus non-prefix subepisodes are taken into consideration. Then we build a *generative model* for random sequences which are considered producing free-rider episodes by the following ways. First, we fix the occurrences of informative events according to their positions in the original sequence. Second, we generate random events with an assumption that they are independent of potential real patterns. Next, based on these random sequences, we define the expected support of a free-rider episode and adopt an automaton based algorithm to compute. Finally we deem an episode redundant if its observed support can be explained by the estimated expected support. Otherwise, it will be an informative pattern. The experiments on both synthetic and real-world dataset about finance and text demonstrate the effectiveness of our model.

2   Related Work

We categorize related papers in the literature as follows.

Mining Frequent Episodes. Mining frequent episodes from event sequence was first introduced by Mannila et al. [14] where they defined episodes as directed acyclic graphs and considered two kinds of methods counting support, i.e. sliding window and minimal occurrence. Mining general episodes can be intricate and computing-intensive because discovering whether a sequence covers a general episode is **NP**-hard [20]. As a result, efforts have been focused on mining subclasses of episodes, such as serial episodes [3–5], closed episodes [19,20], episodes with unique labels [2,15].

Filtering Episodes Based on Expectation. Our EDP model belongs to such category. That is to define the expected support for an episode and compare with its observed support for filtering or ranking tasks. Some existing efforts compare the observed support against the independence model [9,12,17]. In such models, they use a null hypothesis that the data are generated by a stationary and independent stochastic process. Recent work estimated the expected support by considering different partitions of episodes. Tatti [18] divided an episode into two consecutive subepisodes and downplayed the importance of such episode if its observed support could be explained by the two subepisodes. Such model failed to take non-prefix subepisodes into account when perform partitioning. Therefore it may be difficult to detect free-rider episodes when noise events are interweaved with a real pattern. Some idea for filtering sequential patterns can be adapted to episodes as well. For example, Petitjean et al. proposed SkOPUS [16] in which reordering candidates were made from two subpattern partitions, and the expected support was defined based on the mean of support of all possible candidates in the original sequences. However, such work may distort the expected support as some re-orderings may be surprisingly rare in the original sequence.

Episode Set Mining. There is also another kind of efforts to score a set of episodes which best explain the data. Most of them are Minimum Description Length (MDL) based approaches for scoring a set of patterns relative to a dataset as well as a heuristic search method to construct the set [7,8,10,11,21]. Our work is different with such episode set mining since episode set mining methods select episodes based on how well we can describe the data using the episodes while our purpose is to filter free-rider episodes from real patterns based on how well their support can be explained by random sequences.

3   Definitions and Problem Statement

In this section, we introduce preliminaries, definitions and problem statement. Let $\Omega = \{e_1, e_2, \ldots, e_m\}$ be a finite event alphabet. We will use $\Omega = \{\mathsf{a}, \mathsf{b}, \mathsf{c}, \mathsf{d}\}$ for all examples in this paper. An event sequence $S = \langle (E_1, t_1), (E_2, t_2), \cdots, (E_n, t_n) \rangle$, is an ordered sequence of events, where each $E_i \subseteq \Omega$ consists of all events associated with time stamp t_i, $1 \leqslant i \leqslant n$, and $t_j < t_k$ for any $1 \leqslant j < k \leqslant n$. In addition, n is the *length* of event sequence S, denoted by $\mathrm{len}(S) = n$.

For example, Fig. 1 shows an event sequence S with $\mathrm{len}(S) = 10$. S will be used as the running example in this paper.

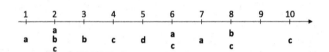

Fig. 1. The example event sequence.

An episode (refer to serial episode in this paper) α is a totally ordered events in the form of $e_{\alpha_1} \to e_{\alpha_2} \to \cdots \to e_{\alpha_k}$ where $\alpha_i \in [1, m]$ and thus $e_{\alpha_i} \in \Omega$ for all $1 \leqslant i \leqslant k$. We can abbreviate $\alpha = e_{\alpha_1} \to e_{\alpha_2} \to \cdots \to e_{\alpha_k}$ as $e_{\alpha_1} e_{\alpha_2} \cdots e_{\alpha_k}$. We will use such abbreviation to represent episode in this paper unless otherwise specified. The *length* of an episode is defined as the number of events in the episode. An episode of length k is called a k-episode. We call an episode with length 0 as an *empty episode*, and denote it by $\emptyset$. A minimal occurrence window of α is a time-window $[t_s, t_e]$ which contains an occurrence of α, such that no proper sub-window of it contains another occurrence of α. t_s and t_e are called *start time* and *end time*, respectively. Usually there is an additional threshold of maximal window size δ such that $t_e - t_s < \delta$. The set of all distinct minimal occurrence window of an episode α is denoted by $\mathrm{moSet}(\alpha)$.

For example, $\mathrm{moSet}(\mathsf{abc}) = \{[2, 4], [7, 10]\}$ if $\delta = 4$ in Fig. 1. Though $[1, 4]$ contains an occurrence of episode abc, it is not a minimal occurrence window of abc because it subsumes the time window $[2, 4]$ which contains another occurrence of abc.

The support of an episode α, denoted by $\mathrm{sp}(\alpha)$, is defined as the number of distinct minimal occurrence windows of α in sequence $\boldsymbol{S}$, i.e. $\mathrm{sp}(\alpha) = |\mathrm{moSet}(\alpha)|$. We call an episode α *frequent* if its support passes a user-specific *minimum support threshold* $min\_sup > 0$, i.e. $\mathrm{sp}(\alpha) \geqslant min\_sup$. Otherwise, it is infrequent. Consider two episodes $\alpha = e_{\alpha_1} \cdots e_{\alpha_k}$ and $\beta = e'_{\alpha_1} \cdots e'_{\alpha_m}$ where $m \leqslant k$, the episode β is a *subepisode* of α, denoted by $\beta \preceq \alpha$, if and only if there exist m integers $1 \leqslant i_1 \leqslant i_2 \cdots \leqslant i_m \leqslant k$ such that $e'_{\alpha_j} = e_{\alpha_{i_j}}$ where $1 \leqslant j \leqslant m$. We call an episode $H_j^\alpha = e_{\alpha_1} \cdots e_{\alpha_j}$ a *j-prefix* episode of α if $0 \leqslant j \leqslant k$. Specifically, 0-prefix episode of any non-empty episode α is the empty episode $\emptyset$.

For example, all possible subepisodes of $\alpha =$ abc include $\emptyset$, a, b, c, ab, ac, bc and abc. All prefix episodes of $\alpha =$ abc are $H_0^{abc} = \emptyset, H_1^{abc} =$ a, $H_2^{abc} =$ ab, $H_3^{abc} =$ abc.

Next, we define the concept of free-rider episode and give a formal statement of the problem. We assume the knowledge of generative rules behind real patterns in advance and give the following assumption: A *generative rule* $\alpha \Rightarrow e$ states that an event e will be more likely to occur after an occurrences of an episode α.

For example, ab $\Rightarrow$ c is the generative rule indicates that the probability of occurrence of event c will increase in some time stamps after episode ab occurs. With the defined generative rule, free-rider events in an episode can be seen as those events that no generative rule is associated with.

Free-Rider Event. A free-rider event e_{α_i} in an episode $\alpha = e_{\alpha_1} \cdots e_{\alpha_k}$, where $1 \leqslant i \leqslant k$, is an event that there is no generative rule associated with in the form as follows.

1. $\beta \Rightarrow e_{\alpha_i}$, where β is a non-empty subepisode of $(i-1)$-prefix episode of α, i.e., $\beta \preceq H_{i-1}^\alpha$ or
2. $\beta' \Rightarrow e_{\alpha_j}$, where $j > i$ and β' containing e_{α_i} is a non-empty subepisode of $(j-1)$-prefix episode of α, i.e., $e_{\alpha_i} \in \beta' \preceq H_{j-1}^\alpha$.

For instance, consider episode $\alpha =$ abc, b will not be a free-rider event if a $\Rightarrow$ b holds or if ab $\Rightarrow$ c holds. However, generative rule ac $\Rightarrow$ b has no impact on whether b is a free-rider event of the episode abc or not. In other words, the event order in α should be respected. Then we can describe free-rider episodes as follows.

Free-Rider Episode. An episode $\alpha = e_{\alpha_1} \cdots e_{\alpha_k}$ is a *free-rider episode* if it contains at least one free-rider event e_{α_i} where $1 \leqslant i \leqslant k$.

We argue that filtering free-rider episodes by definition is intricate since it may be hard to identify free-rider events by checking these implicit generative rules. To close the gap, however, we devise a novel expected support of a free-rider episode and take *Lift*, which is the ratio of observed support to the expectation, as the measure to determine whether an episode is a free-rider episode or not. We can then filter the episodes whose Lift values fail to exceed the minimum lift threshold min_lift and rank the rest by Lift with a descending order as their interestingness scores. Hence, the problem of free-rider episode screening in this paper is formulated as follows.

Problem Statement of Free-rider Episode Screening. Given an event sequence S and a set of frequent episodes on S, the free-rider episode screening problem is to rank the episodes with the Lift produced by the EDP model and filter the episodes whose Lift values are less than *min_lift*.

4   The EDP Model

The key ingredient to our problem is to compute the expected support of a free-rider episode, and we thus propose the EDP model. It involves an episode partition strategy, a generative model for random sequences based on episode partitions, and an expected support definition of a free-rider episode based on the generated random sequences. The inherent idea behind our model is that if the support of an episode can be simulated by that from a group of random sequences, it cannot be a real pattern.

4.1   Episode Partition Strategy

We first detail the episode partition strategy in EDP. Since pattern and noise events in an episode may interweave in complex ways we first make the following assumption to simplify the problem. Specifically, we assume that an event of an episode will not simultaneously be a part of real pattern and a noise event. Such hard assignment could simplify the processes of sequence generation and expectation calculation that will be introduced in the following parts. Formally, given an episode $\alpha = e_{\alpha_1} \ldots e_{\alpha_k}$, there is no $i, j \in [1, k]$ and $i \neq j$ such that $e_{\alpha_i} = e_{\alpha_j}$ while e_{α_i} is a free-rider event and e_{α_j} is not.

Take the episode abb as an example. There are two event b in such episode. We assume it is not possible that the first b is a noise event while the second b is not. The practical usefulness of our model will not be limited by this assumption, as shown in our experiments. Under such assumption, we have the following partition strategy.

Definition 1 (Dual Partition). *Let Ω_α be the set of event alphabet of an episode α. We divide Ω_α into two distinct parts called* informative *and* random *events, which are denoted by $\mathcal{I}_\alpha$ and $\overline{\mathcal{I}_\alpha}$, respectively. We assume the events in $\mathcal{I}_\alpha$ form the real pattern, and the events in $\overline{\mathcal{I}_\alpha}$ are regarded as noises. Then the episode α is partitioned as a subepisode of α which consists of events in $\mathcal{I}_\alpha$, and noise events in $\overline{\mathcal{I}_\alpha}$.*

Such partition indeed takes non-prefix subepisodes of α into account when performing partitioning. For example, consider an episode $\alpha = $ abcd, we have $\Omega_\alpha = \{$a, b, c, d$\}$. Ω_α can be partitioned as $\mathcal{I}_\alpha = \{$a, c$\}$ and $\overline{\mathcal{I}_\alpha} = \{b, d\}$ or $\mathcal{I}_\alpha = \{$a, b, d$\}$ and $\overline{\mathcal{I}_\alpha} = \{c\}$. Under these two partitions, α can be divided into a subepisode ac and a noise event set $\{$b, d$\}$ or abd and $\{$c$\}$. Both ac and abd are non-prefix subepisodes of abcd.

4.2   Generate Random Sequences

Next we devise a generative model (denoted by $\mathcal{M}_{\mathcal{I}_\alpha}$), for a given episode α, generating random sequences based on the aforementioned episode dual partition $\mathcal{I}_\alpha$ and $\overline{\mathcal{I}_\alpha}$.

The basic idea for $\mathcal{M}_{\mathcal{I}_\alpha}$ is that we consider the potential real pattern, which consists of informative events, follows implicit generative rules. Their occurrences in the original event sequence are thus meaningful. While the random events, which are independent of the real patterns, could occur at any time stamp. Hence the model generates random sequences as follows. First, the length of every random sequence is the same as the original event sequence. Second, the model produces informative events according to their occurrences in the original sequence. Third, the random events are generated independently over all possible time stamps.

Formally, we first define the probability of generating an event e at a specific time stamps t_j under $\mathcal{M}_{\mathcal{I}_\alpha}$ as follows.

$$P(e|\mathcal{M}_{\mathcal{I}_\alpha}, t_j) = \begin{cases} 1 & e \in \mathcal{I}_\alpha \ \& \ e \in E_j \\ 0 & e \in \mathcal{I}_\alpha \ \& \ e \notin E_j \\ p_{ind}(e) & e \in \overline{\mathcal{I}_\alpha} \end{cases} \tag{1}$$

In Formula 1, E_j is the event set associated with time stamp t_j in the original event sequence S, and $p_{ind}(e)$ is the occurring probability[1] of event e in S.

Next we define the probability for generating an event set E_j' based on $\mathcal{M}_{\mathcal{I}_\alpha}$ at a specific time stamp t_j as follows.

$$P(E_j'|\mathcal{M}_{\mathcal{I}_\alpha}, t_j) = \prod_{e \in E_j'} P(e|\mathcal{M}_{\mathcal{I}_\alpha}, t_j) \prod_{e \in \Omega_\alpha \backslash E_j'} (1 - P(e|\mathcal{M}_{\mathcal{I}_\alpha}, t_j)) \tag{2}$$

We can generate event set for every time stamp with Formula 2 and eventually obtain a random sequence by assuming every event set is independent to each other. Here we can safely use the independent assumption here since we attempt to generate random sequences without specialized rules. Remember that the length of a random sequence is the same as the length of S. Therefore, the probability to generate a specific random sequence $\hat{S}$ can be given as Eq. 3, where n denotes the length of S.

$$P(\hat{S}|\mathcal{M}_{\mathcal{I}_\alpha}, n) = \prod_{j=1}^{n} P(E_j'|\mathcal{M}_{\mathcal{I}_\alpha}, t_j) \tag{3}$$

For example, consider the event sequence shown in Fig. 1 and the episode $\alpha = \text{abc}$. Suppose $\mathcal{I}_\alpha = \{\text{a}, \text{b}\}$ and $\overline{\mathcal{I}_\alpha} = \{\text{c}\}$. The generative model $\mathcal{M}_{\{\text{a},\text{b}\}}$ fixes all the occurrences of event a and b as in the sequence shown in Fig. 1 and generates event c by $p_{ind}(\text{c}) = 0.5$ ($\text{sp}(\text{c}) = 5$ and $\text{len}(S) = 10$). The illustration of $\mathcal{M}_{\{\text{a},\text{b}\}}$ is shown as Fig. 2 in which the fixed informative events are shown

---

[1] $p_{ind}(e)$ can be calculated by $\frac{\text{sp}(e)}{\text{len}(S)}$, where $\text{sp}(e)$ is the support of event e in S.

under the time stamps while the random event c is shown above with the grey font. For probability of the candidate event set at each time stamp, it is the multiplication of all possible events in it, e.g. $P(E_7' = \{a, c\} | \mathcal{M}_{\{a,b\}}, 7) = P(E_7' = \{a\} | \mathcal{M}_{\{a,b\}}, 7) = 0.5$ and $P(E_9' = \{c\} | \mathcal{M}_{\{a,b\}}, 9) = P(E_9' = \{\emptyset\} | \mathcal{M}_{\{a,b\}}, 9) = 0.5$. Though E_9 has no event in the original sequence, the model may generate events based on the dual partition.

Fig. 2. The generative model $\mathcal{M}_{\{a,b\}}$ for episode abc. Occurrences of event a and b are fixed according to their occurrences in the original sequence. Event c is generated with its occurring probability in the original sequence.

4.3   Expected Support Definition

Next we define the expected support of a free-rider episode on these random sequences. Recall that given an episode α, EDP generates random sequences with events in Ω_α based on the model $\mathcal{M}_{\mathcal{I}_\alpha}$, and we assume, in $\mathcal{M}_{\mathcal{I}_\alpha}$, events belonging to $\mathcal{I}_\alpha$ have implicit generative rules but events in $\overline{\mathcal{I}_\alpha}$ do not. Hence based on our definition every event e' in $\overline{\mathcal{I}_\alpha}$ will be a free-rider event to α in each generated sequence and α in these sequences will become a free-rider since $e' \in \alpha$. As a consequence, denoted the set of random sequences generated by a given generative model $\mathcal{M}_{\mathcal{I}_\alpha}$ as $\mathrm{Seq}(\mathcal{M}_{\mathcal{I}_\alpha})$, we match the minimal occurrences of α on each generated sequence $\hat{S} \in \mathrm{Seq}(\mathcal{M}_{\mathcal{I}_\alpha})$ and define the expected support of α, which is regarded as a free-rider episode, as follows.

$$\mathbb{E}_{\hat{S} \sim P(\hat{S} | \mathcal{M}_{\mathcal{I}_\alpha}, n)}[\mathrm{sp}(\alpha | \hat{S})] = \sum_{\hat{S} \in \mathrm{Seq}(\mathcal{M}_{\mathcal{I}_\alpha})} \mathrm{sp}(\alpha | \hat{S}) \cdot P(\hat{S} | \mathcal{M}_{\mathcal{I}_\alpha}, n) \qquad (4)$$

where $\mathrm{sp}(\alpha | \hat{S})$ denotes the support of the episode α in a generated sequence $\hat{S}$.

It is no doubt different generative models will derive different expected support for an episode α. To minimize the false positive, we define the formal expected support as the maximum expected support value produced by a specific generative model. Hence we define the expected support of an episode α, denoted as $\mathrm{ExpSup}(\alpha)$, as follows.

$$\mathrm{ExpSup}(\alpha) = \max_{\mathcal{M}_{\mathcal{I}_\alpha}} \mathbb{E}_{\hat{S} \sim P(\hat{S} | \mathcal{M}_{\mathcal{I}_\alpha}, n)}[\mathrm{sp}(\alpha | \hat{S})] \qquad (5)$$

With such definition, we adopt the following $Lift(\alpha)$ to measure the deviation between the observed support of an episode α and its expectation and regard it as a free-rider episode if $Lift(\alpha)$ does not exceed the minimum lift threshold min_lift.

$$Lift(\alpha) = \frac{\mathrm{sp}(\alpha)}{\mathrm{ExpSup}(\alpha)} \qquad (6)$$

Under this formula, we actually make the following assumption: For an episode α in an event sequence $\boldsymbol{S}$, if we could find a generative model producing free-rider episodes in the same form of α by EDP which can generate α more frequently or close to the observed support of α in $\boldsymbol{S}$, then α in $\boldsymbol{S}$ will be redundant.

4.4 Time Complexity of EDP

In this subsection, we analyze the time complexity of the proposed EDP. Though the search space of possible partitions of an episode α is $2^{|\Omega_\alpha|}$, where Ω_α is the alphabet of α, we could stop calculation and screen α if we find one $\mathcal{M}_{\mathcal{I}_\alpha}$ such that the value of $\dfrac{\mathrm{sp}(\alpha)}{\mathbb{E}_{\hat{\boldsymbol{S}} \sim P(\hat{\boldsymbol{S}}|\mathcal{M}_{\mathcal{I}_\alpha},n)}[\mathrm{sp}(\alpha|\hat{\boldsymbol{S}})]}$ is clearly less than min_lift. It indicates there exists one group of random sequences that provides free-rider episodes of α can explain α's observed support in the input sequence. As a result, we can explore pruning techniques based on such property.

For a real pattern, on the other hand, we need to enumerate all partitions to ascertain its interestingness. Hence the time complexity will be $O(2^{|\Omega_\alpha|})$. However, in our EDP each partition has no impact on others, the process can thus be highly parallelized without hurting the performance. The experimental results will show EDP can achieve significant speedup in running time if we distribute the checking into multi-processes.

5   Expected Support Calculation

So far our problem is how to compute the expected support of an episode α given a generative model $\mathcal{M}_{\mathcal{I}_\alpha}$. In this section, we devise an algorithm computing the expected support defined in Eq. 4 with the help of automaton.

5.1 Tracking Minimal Occurrences with Automaton

Automaton is an effective tool to track minimal occurrence of episode in sequences. Given an episode $\alpha = e_{\alpha_1} e_{\alpha_2} \cdots e_{\alpha_k}$, we can obtain a unique automaton $\mathcal{A}_\alpha$ in which every node (also known as state) denotes a prefix subepisode of α, namely H_i^α where $0 \leqslant i \leqslant k$. The label on each edge denotes the event that can trigger state transition. We denote the transition function by $T(H, E)$ where H is the current state of an automaton and E is an event set. The first state and the last state of an automaton are called the source state and the sink state, respectively. Since minimal occurrences may overlap to each other on the event sequence, we need to simultaneously manage multiple automatons which do not reach the sink state during scanning the sequence. We call them active automatons.

5.2   The Algorithm

The pseudo code of expected support calculation is shown as Algorithm 1.

In our algorithm, we do not need to generate any specific random sequence. Instead, we adopt the probability of event set generation and the probability of the appearance of active automatons to estimate the expected support defined in Eq. 4. Since $\mathcal{M}_{\mathcal{I}_\alpha}$ may generate multiple different event sets for each time stamp, we need to simultaneously manage more than one possible active automaton lists and consider all possible event sets for every time stamp. In more detail, we use a variable L_i to store active automatons at time stamp t_i and first initialize $L_0 = \{H_0^\alpha\}$. Since L_0 is determined, its appearance probability is one, and the list of active automatons, denoted as $\mathcal{L}_0$, contains the only element L_0 (Line 1–2).

Then we sequentially consider every pair of possible active finite automaton list L_{i-1} and possible event set E_i' produced by the generative model $\mathcal{M}_{\mathcal{I}_\alpha}$ for every time stamp t_i where $1 \leqslant i \leqslant \text{len}(S)$ (Line 3–7). For each active $\mathcal{A}(\alpha)$ in L_{i-1}, we invoke the state transitive function by consuming its current state H_c

---

Algorithm 1. Expected support calculation given $\mathcal{M}_{\mathcal{I}_\alpha}$

Input: α: a k-episode
$\mathcal{M}_{\mathcal{I}_\alpha}$: a generative model constructed by episode partition $\mathcal{I}_\alpha$
S: the original event sequence
Output: $\mathbb{E}_{\hat{S} \sim P(\hat{S}|\mathcal{M}_{\mathcal{I}_\alpha}, n)}(\text{sp}(\alpha|\hat{S}))$: the expected support of an episode α given a generative model $\mathcal{M}_{\mathcal{I}_\alpha}$

1 initialize $L_0 = \{H_0^\alpha\}$, $\mathcal{L}_0 = \emptyset$, $P(L_0) = 1.0$, $\mathbb{E}_{\hat{S} \sim P(\hat{S}|\mathcal{M}_{\mathcal{I}_\alpha}, n)}(\text{sp}(\alpha|\hat{S})) = 0$
2 $\mathcal{L}_0 \leftarrow \mathcal{L}_0 \cup \{L_0\}$
3 **for** $i = 1$ *to* $\text{len}(S)$ **do**
4      $\mathcal{L}_i = \emptyset$
5   **foreach** *possible active automaton list* $L_{i-1} \in \mathcal{L}_{i-1}$ **do**
6          $P(L_i) = 0$
7   **foreach** *possible event set* E_i' *produced by* $\mathcal{M}_{\mathcal{I}_\alpha}$ **do**
8   **foreach** *active* $\mathcal{A}(\alpha) \in L_{i-1}$ **do**
9                  $H_c \leftarrow$ current state of $\mathcal{A}(\alpha)$
10                 $H_t = T(H_c, E_i')$
11   **if** *TRANSIT i.e.* $H_t \neq H_c$ **then**
12   **if** *SINK i.e.* $H_t = H_k^\alpha$ **then**
13                         $\lfloor$ increase $\mathbb{E}_{\hat{S} \sim P(\hat{S}|\mathcal{M}_{\mathcal{I}_\alpha}, n)}(\text{sp}(\alpha|\hat{S}))$ by $P(L_{i-1}) \cdot P(E_i'|\mathcal{M}_{\mathcal{I}_\alpha}, t_i)$
14   **else**
15                         $\lfloor$ add a copy of $\mathcal{A}(\alpha)$ whose current state is H_t to L_i
16   **if** *SOURCE i.e.* $H_c = H_0^\alpha$ **then**
17                         $\lfloor$ add an initialized $\mathcal{A}(\alpha)$ whose current state is H_0^α to L_i
18   **forall the** $\mathcal{A}(\alpha) \in L_{i-1}$ *not TRANSIT* **do**
19                  $H_c \leftarrow$ current state of $\mathcal{A}(\alpha)$
20   **if** *there is no* $\mathcal{A}(\alpha)' \in L_i$ *whose current state is* H_c **then**
21                      $\lfloor$ add a copy of $\mathcal{A}(\alpha)$ to L_i
22   increase $P(L_i)$ by $P(L_{i-1}) \cdot P(E_i'|\mathcal{M}_{\mathcal{I}_\alpha}, t_i)$, update $\mathcal{L}_i$
23 **return** $\mathbb{E}_{\hat{S} \sim P(\hat{S}|\mathcal{M}_{\mathcal{I}_\alpha}, n)}(\text{sp}(\alpha|\hat{S}))$

---

and the event set E_i' and acquire the output state H_t (Line 8–10). If a transition occurs, we check both H_c and H_t for automaton management. In particular, if H_t reaches the sink state H_k^α, we obtain a minimal occurrence of α and increase the expected support of α by $P(L_{i-1}) \cdot P(E_i'|\mathcal{M}_{\mathcal{I}_\alpha}, t_i)$ (Line 11–13). Otherwise, we add the updated $\mathcal{A}(\alpha)$ to L_i. We meanwhile check whether H_c is a source state and add an initialized automaton to L_i if it is (Line 16–17). For the rest automatons without state transitions, we will determine whether they can be added to L_i since we are interested in minimal occurrence of episode (Line 18–21). Finally we update the probability of the appearance of active automatons L_i for further computations (Line 22).

For example, consider the episode abc and the generative model $\mathcal{M}_{ab}$ as shown in Fig. 2. When $i = 3$, the only possible active automaton list $L_2 \in \mathcal{L}_2$, having three active automatons, contains $\{H_0^{abc}, H_1^{abc}, H_2^{abc}\}$. Here H_i^{abc} denotes the current state of the corresponding automaton. For possible event sets, there are two possible event sets, i.e. $E_3' = \{b\}$ and $E_3'' = \{b, c\}$, with the probability of 0.5 of each, respectively. Considering every pair of the active automaton and the event set, we detect $T(H_2^{abc}, E_3'')$ derives the SINK state. We thus add the expected support of the episode abc under the generative model $\mathcal{M}_{ab}$ by $P(L_2) \cdot P(E_3''|\mathcal{M}_{ab}, t_3) = 0.5$. Then we update the probability of active automatons and derive $\mathcal{L}_3 = \{L_3\} = \{H_0^{abc}, H_2^{abc}\}$ with $P(L_3) = 1$.

5.3 Time Complexity Analysis

Here we analyze the time complexity of the algorithm, i.e., Algorithm 1, for expected support calculation given a specific generative model. We only discuss the worst case for simplicity. Given a specific generative model $\mathcal{M}_{\mathcal{I}_\alpha}$ of an episode α, for each time stamp t_i, there are at most $|\alpha|$ active automatons in each active automatons list L_i, $2^{|\overline{\mathcal{I}_\alpha}|}$ possible event set E_j and $2^{|\alpha|}$ possible lists of active automatons, respectively. Hence the time complexity for precessing a time stamp t_i in the wort case is $O(|\alpha| \cdot 2^{|\overline{\mathcal{I}_\alpha}|+|\alpha|})$. Since there are all together n time stamps on the event sequence $\boldsymbol{S}$, the overall time complexity of Algorithm 1 will be $O(n \cdot |\alpha| \cdot 2^{|\overline{\mathcal{I}_\alpha}|+|\alpha|})$ where n denotes the length of the event sequence. Though it is exponential in $|\overline{\mathcal{I}_\alpha}|$ and $|\alpha|$, this number might not be very large as usually the episode to be checked has limited length and so is $|\overline{\mathcal{I}_\alpha}|$.

6 Experiments

In this section, we present the results of our experimental studies.

6.1 Datasets and Experiment Settings

We conducted the experiments on both synthetic and real-world datasets, and Table 1 illustrates their statistics.

The **SYN** dataset refers to a synthetic dataset which is generated as follows. We use an alphabet of 52 events, i.e., $a \cdots z$ and $A \cdots Z$ for generating a sequence whose length is set to 10,000. In such sequence, we planted two episodes and a high frequency noise event. The first episode abc with no gaps was embedded 300 times into randomly selected time stamps. The second episode, a 4-episode defg, can have a gap between any two consecutive events. The gap is a positive integer drawn from a truncated normal distribution $N(2, 2)$. The episode defg was randomly planted 300 times in the sequence as well. These two patterns might overlap but there are no generative rules between them. Third, we planted a high frequency event X with $p_{ind}(X) = 0.3$ into randomly selected time stamps of the sequence. Such event might become a free-rider event that doped with the embedded patterns. Finally, we filled the sequence using the rest events which were generated from a uniform distribution. They are considered as random noises. As a consequence, all the episodes rather than the non-empty subepisodes of abc and defg are considered as free-riders in such dataset.

Table 1. Statistical information of datasets

Dataset	Sequence len.	Events in alphabet	Avg. events per timestamp	Avg. occurrences per event
SYN	10,000	52	1.4	267.3
STK	9,334	50	5.55	1,037
JMLR	75,645	3,846	1.0	19.67

The **STK** dataset consists of daily prices of stocks from Chinese stock market. It includes 50 blue chip stocks with their price information over 26 years from December 19th, 1990 to July 14th, 2016. The events for a stock are generated by the increase ratio of price on each trading day. If the ratio is positive, we generate an increase event for such stock on the corresponding day. For example, if the stock of Bank of China increases on December 19th, 1990, we will generate an event as "Bank of China+" for that day. Otherwise, we will not produce event for the stock of Bank of China on that day. Similar operations were performed for all the 50 stocks in such dataset.

The **JMLR** dataset consists of the abstracts of the papers published on the Journal of Machine Learning Research which was adopted in [18]. We treat each word as an event and built an event sequence by connecting every sentence together. Such dataset was preprocessed by stemming the words and removing the stop words. The details can be found in [18].

We compare our EDP model against the following state-of-the-art methods:

1. **PRT** [18]: A partition model that divides an episode into two consecutive sub-episodes and detects whether its frequency can be explained by the two sub-episodes.
2. **SkOPUS** [16]: A partition model that compares the frequency of a pattern with its re-ordering candidates consisting of two sub-pattern partitions.

3. **EGH** [12]: An independence baseline that connects the significance of an episode with a stationary hidden markov model.
4. **IND:** The degraded version of the proposed EDP model which considers every event is a random event.

Since all the compared methods require a set of candidate patterns, we start from a set of frequent episodes, denoted as $\mathcal{F}$, in which we require $|\Omega_\alpha| > 1$ for every $\alpha \in \mathcal{F}$.[2] PRT and SkOPUS are designed for multiple sequences rather than a single event sequence, we thus transfer our event sequence into database of short sequences by segmenting it with fixed length sliding windows to fit the methods. For mining frequent episodes and deriving $\mathcal{F}$, we adopted the DFS algorithm [1], which is a state-of-the-art minimal occurrence based frequent episode mining algorithm.

6.2   Results on Data with Known Patterns

We first demonstrate our results on the SYN dataset having known patterns. In particular, we mined $\mathcal{F}$ by setting $min\_sup = 200$ and $\delta = 12$ and finally obtained $|\mathcal{F}| = 325$. Such setting ensures the embedded real patterns are contained in $\mathcal{F}$. Our purpose is to check whether the compared methods can unearth these planted patterns and rank them as high as possible.

Table 2. Precision@k on SYN dataset.

Top k	Precision@k				
	IND	EGH	SkOPUS	PRT	EDP
1	100.0%	0.0%	100.0%	100.0%	100.0%
2	100.0%	50.0%	100.0%	100.0%	100.0%
3	100.0%	66.7%	100.0%	100.0%	100.0%
4	100.0%	75.0%	100.0%	100.0%	100.0%
5	80.0%	60.0%	100.0%	100.0%	100.0%
6	66.7%	50.0%	100.0%	83.3%	100.0%
7	71.4%	57.1%	85.7%	71.4%	100.0%
8	75.0%	62.5%	75.0%	62.5%	100.0%
9	66.7%	66.7%	66.7%	66.7%	100.0%
10	60.0%	60.0%	60.0%	60.0%	100.0%
11	54.5%	54.5%	54.5%	63.6%	100.0%
12	50.0%	58.3%	50.0%	66.7%	100.0%
13	46.2%	53.8%	46.2%	69.2%	100.0%
14	42.9%	57.1%	50.0%	71.4%	100.0%
15	40.0%	60.0%	53.3%	73.3%	100.0%

Table 3. Details of top 15 episodes.

Top k	SkOPUS	PRT	EDP
1	d → e → f → g	d → e → f → g	e → f → g
2	a → b → c	d → e → g	f → g
3	d → e → f	d → f → g	e → f
4	e → f → g	e → f → g	e → g
5	d → e → g	d → e → f	d → e → f
6	c → X → d	a → c → d	d → f
7	b → c → X → d	a → b → c → d	d → e
8	b → c → d	b → c → d	a → b
9	b → X → d	a → b → c	b → c
10	a → b → c → d	a → b → d	a → c
11	a → b → X → d	f → g	a → b → c
12	a → c → X → d	e → f	d → e → f → g
13	f → g	e → g	d → f → g
14	e → f	d → e	d → e → g
15	e → g	d → f	d → g

We explored every compared method on the $\mathcal{F}$ and ranked the episodes by the deviation between the observed support and the expectation. For our

---

[2] Here we do not take any frequent event or the episode consisting of single event into account.

EDP and IND model, we set $min\_lift = 1$ and ranked the episodes whose Lift exceeds min_lift with descending order. For other methods we directly ranked by the deviation used in their papers. We report precision at k in Table 2, that is, proportions of embedded patterns that are included in the top-k patterns returned by each approach. Remember that non-empty subepisodes of abc and defg except the single events are real patterns based on our definition as there exists underlying generative rules to explain it. As a result, there could be 15 embedded patterns in the dataset. What we concern about is the ability of each approach to recognize these embedded patterns.

From the table, we observe the independence models, i.e. EGH and IND, do not perform well as expected, while the partition based models are more effective. Our EDP model, with the highest precision, ranks all the embedded patterns at top 15 rankings and significantly outperforms other methods. PRT and SkOUPS begin to report false positive patterns after 5th and 6th, respectively. In addition, we observe EDP provides 20 outputs whose Lift is greater than the threshold which was 1 in our setting. However we observe a large gap between the lift measure of the 15th and 16th in the results of EDP, which doubles the gap between the 1st and 15th. Though 16th-20th are false positives, we argue that we can filter the additional noises by finer tuning the threshold based on the gaps in the lift and our method already returns the targets on top 15.

Next we look closer to the results returned by the three partition models. Table 3 demonstrates the top 15 episodes produced by the three partition based models. Among them, we mark the false positive patterns in blue font. From the table we observe that SkOPUS cannot handle the high frequency noise event, i.e. X, very well, and X could be doped inside patterns. On the other hand, SkOPUS may underestimate the expectation of patterns with longer length as their re-ordering candidates may be rare in the original sequence. Thus, the top 5 of SkOPUS are 3-episode or longer, and it can return defg and abc very early. The ineffectiveness of PRT comes from its incorrectness in identifying the event from other patterns. For example, it fails to recognize d is a free-rider event to the pattern abc. However, our EDP can correctly filter such episodes. As the fact that EDP can recognize more complicated redundant patterns, we conjecture the reason is that our partition model takes non-prefix episodes into account during the partition.

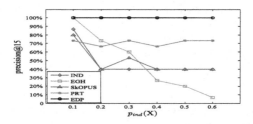

Fig. 3. The effect of $p_{ind}(\mathsf{X})$.

We next tune the probability of the embedded high frequency noise event, i.e. $p_{ind}(\mathsf{X})$, from 0.1 to 0.6 and visualize the Precision@15 of all compared methods as shown in Fig. 3. From the figure, we observe that the probability of X has few effects on EDP. It always achieves the highest precisions with 100%. Second, the measure of IND, SkOPUS and PRT may vary around 40% to 75% as $p_{ind}(\mathsf{X})$ varies. The only special case happens when $p_{ind}(\mathsf{X}) = 0.1$, both IND and SkOPUS achieve more than 80% in precision. EGH, on the other hand, is much more sensitive to the $p_{ind}(\mathsf{X})$, and its performance drops significantly as the $p_{ind}(\mathsf{X})$ increases. However, it performs well when the appearance probability of X is low.

As the synthetic experiments clearly show that IND and EGH overall are less effective than the others, we do not include them in the following experiments.

6.3   Results on Real-World Dataset

Results on STK Dataset. We start from the most frequent 1,000 episodes mined from the STK dataset by setting $min\_sup = 200$ and $\delta = 5$. We also ranked each episode by the deviation between the observed support and the expectation, and meanwhile we set min_lift to 1 to screen redundant episodes for our method. We were surprised to find that each approach returned very different results on such dataset. Since the episodes related to stocks are not easy to understand, we design the following measure to demonstrate the results.

Table 4. The percentage of the most frequent 10 events in the top k episodes in STK.

Top k	SkOPUS	PRT	EDP
1	33.3%	100.0%	0.0%
2	50.0%	100.0%	0.0%
3	44.4%	100.0%	0.0%
4	33.3%	100.0%	0.0%
5	33.3%	100.0%	20.0%
6	33.3%	100.0%	33.3%
7	33.3%	100.0%	28.6%
8	33.3%	100.0%	37.5%
9	33.3%	100.0%	33.3%
10	33.3%	100.0%	30.0%

For the top k episodes given by each approach, we exhibit the percentage of the top 10 frequent events are included in. The reason we choose top 10 frequent events is that they have relatively high occurrence probabilities ranging from 0.2 to 0.26, which is twice to the average occurrence probability of events in such dataset (see Table 1). From the results on the synthetic dataset, we have known that real patterns may dope with some highly frequent but not related events, which makes the episode become a free-rider. As a result, we argue that such ratio is higher may indicate that there might be more possibility to have free-rider episodes in the top k episodes.

Table 4 demonstrates the results, and we observe PRT achieves surprisingly high percentages. EDP outperforms SkOPUS especially when k is less than 5. To validate further, we check the top episodes produced by every method. We find that, as an example, the highest ranking episode returned by PRT indicates that the increase of a software company (SH.600100+) is followed by the increase of a security corporation (SH.600109+) and then leads to the increase of a metallurgical enterprise (SH.600111+). All the three events belong to different industry sectors and are included in top ten frequent events. In fact, from our common sense, we can hardly believe these three stocks have convincing correlations with each other. The results of SkOPUS is similar with that in the SYN dataset that the underlying patterns are always doped with the most frequent event, while the results provided by EDP are clearly different. For example, the episode "**Industrial and Commrcl Bank of China+**" → "**Bank of China+**" and "**Minsheng Bank+**" → "**China Merchants Bank+**" rank at 1st and 2rd, respectively. Such patterns are more convincing and easy to understand from the stock names. The stocks in the same episode may have implicit correlations as they come from the same industry sector. We also find more examples from the results returned by EDP model but further discussions on possible related stocks are out of the scope of this paper.

Table 5. Top 10 output episodes on JMLR dataset.

Top k	SkOPUS	PRT	EDP
1	support → vector → machin (138)	support → vector (168)	support → vector (168)
2	support → vector (168)	support → vector → machin (138)	real → world (78)
3	vector → machin (151)	support → machin (142)	support → vector → machin (138)
4	support → machin (142)	vector → machin (151)	support → machin (142)
5	data → set → model (79)	real → world (78)	vector → machin (151)
6	base → algorithm → base (78)	featur → select (101)	bayesian → network (95)
7	data → set → data → set (76)	bayesian → network (95)	solv → problem (84)
8	paper → learn → algorithm (75)	problem → solv (79)	data → set (298)*
9	set → learn → set (75)	bayesian → model (92)	machin → learn (77)*
10	real → data (97)	solv → problem (84)	real → data (97)

Results on JMLR Dataset. For the JMLR dataset, we are given $1,023$ frequent episodes in $\mathcal{F}$ by setting $min\_sup = 50$ and $\delta = 12$. Then we explored the three partition models on such dataset. We visualize the top 10 episodes ranked by each method in Table 5. In such table, the number in the bracket is the support of corresponding episode. Since there are no ground truth patterns in such data, we can hardly compare which method is better in a quantitative manner. However, we find the following interesting observations. First, confirm the ones in the previous comparisons, SkOPUS outputs free-rider episodes like "**data → set → model**" and "**base → algorithm → base**" at very early. In addition, SkOPUS seems to have negative correlations with support since it prefers longer patterns. We know the JMLR dataset is a sequence without simultaneous events, hence support based on minimal occurrence holds anti-monotonicity. PRT performs well on such data. Every pattern demonstrated in the table is meaningful, and all of them have lower frequency which is less than 200. Our EDP also provides meaningful results and has few correlations with pattern support. It can rank both high frequency as well as low frequency episode high, e.g. "**data**" → "**set**" (298) and "**machin**" → "**learn**" (77). We mark them by asterisk in the table. Since the basic idea of EDP is checking whether the observed support can be simulated by the random sequences generated by specific partitions. Episodes with high or low support may have possibility to pass such test. We conjecture it may be the reason why EDP can return both general and specific patterns.

6.4 Scalability

Finally we evaluate the efficiency of EDP. Since EDP can be highly distributed, we mainly focus on the scalability of EDP by varying the number of processes. To this end, we implemented EDP with Python 3.5, distributed each partition model of an episode into different processes and verified its scalability on a shared memory computing system. The system equipped with two Intel Xeon E5-2620 CPUs and 128 GB RAM running on Linux CentOS 6.5. Figure 4 demonstrates the results. From the figure we can see EDP can achieve significant speedup as the number of processes increases on all datasets we investigated. Among them, EDP holds the most advanced speedup on the STK dataset. It is because we adopt small δ when performing mining such that the episodes on STK have less length and smaller alphabet.

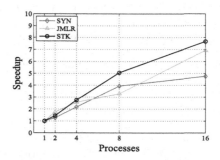

Fig. 4. The scalability of EDP.

7 Conclusion

In this paper, we proposed EDP model for reducing redundancy of frequent episodes in long single sequence. EDP first separates an episode into informative and random events, then builds a generative model for random sequences based on the partitions. Next we define the expected support of an episode if it is considered as a free-rider with these random sequences and devise an efficient algorithm to compute. Finally, the redundancy of an episode is determined by the deviation between the observation and the estimated expectation. Experimental results on synthetic and real-world datasets clearly exhibit the effectiveness of the proposed method.

Acknowledgement. This research is supported by the National Natural Science Foundation of China (No. 61602438, 91546122, 61573335), National key R&D program of China (No. 2017YFB1002104), Guangdong provincial science and technology plan projects (No. 2015B010109005).

References

1. Achar, A., Ibrahim, A., Sastry, P.S.: Pattern-growth based frequent serial episode discovery. DKE **87**, 91–108 (2013)
2. Achar, A., Laxman, S., Viswanathan, R., Sastry, P.S.: Discovering injective episodes with general partial orders. DMKD **25**, 67–108 (2012)
3. Ao, X., Luo, P., Li, C., Zhuang, F., He, Q.: Online frequent episode mining. In: ICDE (2015)
4. Ao, X., Luo, P., Li, C., Zhuang, F., He, Q., Shi, Z.: Discovering and learning sensational episodes of news events. In: WWW (2014)
5. Ao, X., Luo, P., Wang, J., Zhuang, F., He, Q.: Mining precise-positioning episode rules from event sequences. IEEE TKDE **30**, 530–543 (2018)
6. Bertens, R., Vreeken, J., Siebes, A.: Keeping it short and simple: summarising complex event sequences with multivariate patterns. In: KDD (2016)
7. Bhattacharyya, A., Vreeken, J.: Efficiently summarising event sequences with rich interleaving patterns. In: SDM (2017)
8. Fowkes, J., Sutton, C.: A subsequence interleaving model for sequential pattern mining. In: KDD (2016)
9. Robert, G., Atallah, M.J., Szpankowski, W.: Reliable detection of episodes in event sequences. In: KAIS (2005)
10. Ibrahim, A., Sastry, S., Sastry, P.S.: Discovering compressing serial episodes from event sequences. KAIS **47**, 405–432 (2016)
11. Lam, H.T., Mörchen, F., Fradkin, D., Calders, T.: Mining compressing sequential patterns. Stat. Anal. Data Mining **7**, 34–52 (2014)
12. Laxman, S., Sastry, P.S., Unnikrishnan, K.P.: A formal connection discovering frequent episodes and learning hidden Markov models. IEEE TKDE **17**, 1505–1517 (2005)
13. Mampaey, M., Vreeken, J., Tatti, N.: Summarizing data succinctly with the most informative itemsets. ACM TKDD **6**, 16 (2012)
14. Heikki, M., Toivonen, H., Inkeri Verkamo, A.: Discovery of frequent episodes in event sequences. DMKD **1**, 259–289 (1997)

15. Pei, J., Wang, H., Liu, J., Wang, K., Wang, J., Yu, P.S.: Discovering frequent closed partial orders from strings. IEEE TKDE **18**, 1467–1481 (2006)
16. Petitjean, F., Li, T., Tatti, N., Webb, G.I.: Skopus: mining top-k sequential patterns under leverage. DMKD **30**, 1086–1111 (2016)
17. Tatti, N.: Discovering episodes with compact minimal windows. DMKD **28**, 1046–1077 (2014)
18. Tatti, N.: Ranking episodes using a partition model. DMKD **29**, 1312–1342 (2015)
19. Tatti, N., Cule, B.: Mining closed strict episodes. In: ICDM (2010)
20. Tatti, N., Cule, B.: Mining closed episodes with simultaneous events. In: KDD (2011)
21. Tatti, N., Vreeken, J.: The long and the short of it: summarising event sequences with serial episodes. In: KDD (2012)
22. Vreeken, J., Tatti, N.: Interesting patterns. In: Aggarwal, C.C., Han, J. (eds.) Frequent Pattern Mining, pp. 105–134. Springer, Cham (2014). https://doi.org/10.1007/978-3-319-07821-2_5
23. Webb, G.I.: Self-sufficient itemsets: an approach to screening potentially interesting associations between items. In: ACM TKDD (2010)

Maximize Spatial Influence of Facility Bundle Considering Reverse k Nearest Neighbors

Shenlu Wang[1]($\boxtimes$), Ying Zhang[2], Xuemin Lin[1],
and Muhammad Aamir Cheema[3]

[1] The University of New South Wales, Sydney, Australia
{swan398,lxue}@cse.unsw.edu.au
[2] QCIS, The University of Technology, Sydney, Australia
ying.zhang@uts.edu.au
[3] Monash University, Melbourne, Australia
aamir.cheema@monash.edu

Abstract. Consider a two dimensional Euclidean space, let F be a set of points representing facilities and U be a set of points representing users. Spatial influence of a facility is the number of users who have this facility as one of their k nearest facilities. This is because that users normally prefer to go to their nearby facilities and are naturally influenced the most by their nearest facilities. Given a facility bundle of size t, spatial influence of the bundle is the number of distinct users influenced by any one of them. Existing works on facility selection problem find out top-t facilities with the highest spatial influence. However, the literature lacks study on this problem when the t facilities have the highest spatial influence as a bundle. We are the first to study the problem of Maximizing Bundled Reverse k Nearest Neighbors (MB-RkNN), where the spatial influence of a facility bundle of size t is maximized. We prove its NP-hardness, and propose a branch-and-bound best first search algorithm that greedily select the currently best facility until we get t facilities. We introduce the concept of kNN region such that a group of users have their kNN facilities all belong to the same kNN region. This sharing property of kNN region allows us to avoid redundant calculation with dynamic programming technique. We conduct experiments on real data sets and show that our algorithm is orders of magnitudes better than our baseline algorithm both in terms of CPU time and IO cost.

1 Introduction

Consider a two dimensional Euclidean space, let F be a set of points representing facilities (e.g., gas stations, supermarkets) and U be a set of points representing users (e.g., cars, residents). Given a facility $f \in F$, the **spatial influence** of a facility is the number of users who have this facility as one of their k nearest facilities. These users are said to be influenced by the facility. The set of all such users is referred to as the influence set of the facility. Given a facility bundle of

© Springer International Publishing AG, part of Springer Nature 2018
J. Pei et al. (Eds.): DASFAA 2018, LNCS 10827, pp. 684–700, 2018.
https://doi.org/10.1007/978-3-319-91452-7_44

size t (e.g., a set of t facilities), the **spatial influence** of the facility bundle is the number of **distinct** users influenced by any one of them.

In the literature, a user's k nearest facilities are also referred to as its bichromatic k nearest neighbors (kNN), and a facility's reverse k nearest users are also referred to as its bichromatic reverse k nearest neighbors (RkNN). The concept of influence is defined based on reverse nearest neighbors (RNN) queries [6], reverse k nearest neighbors (RkNN) queries [1], and reverse top-k queries [11]. A great number of existing works study facility allocation problem [2,4,13,14,19] and facility/product selection problem [3,5,7,11,12,15,17,18] based on the concept of influence. In this paper, we focus on facility selection problem, and we want to maximize spatial influence.

Existing works on facility selection problem find out t facilities such that no other facility has higher spatial influence than any one of the t facilities, either based on RNN queries [15,17,18] or RkNN queries [7]. However, the literature lacks study on this problem when the t facilities are selected as a facility bundle. The distinctive feature of facility bundle is that the combined spatial influence of the t facilities is considered other than the spatial influence of the t facilities individually. The scenario of facility bundling finds a lot of real world applications.

Consider an **example** of facility bundling, where the government wants to promote or do a survey on a new policy by campaign at a couple of supermarkets, so that the residents shopping at those supermarkets (e.g., influence sets) will be covered (e.g., influenced). For consideration of budget, only a limited number of supermarkets, let's say t supermarkets, can be selected, but the government wants to cover as much residents as possible (e.g., maximize spatial influence). Notice that in this example, a user may go shopping at two selected supermarkets, but should not be counted as two heads. As a result, we can not simply select the top-t facilities with the largest spatial influence. Instead, we need to select a facility bundle with the highest spatial influence among all possible facility bundles of size t.

Consider another **example**, where we have a corporate group manages a large number of facilities. The corporate group wants to create a facility bundle by selecting t facilities and promote a joint membership so that users of any one of them can also get discount at all the other selected facilities. The joint membership aims to increase the number of potential users for all the selected facilities without extra customer acquisition costs. In this example, to maximize the number of potential users of every selected facility, we need to maximize the spatial influence of the selected facility bundle.

The example in Fig. 1 illustrates the reason why we cannot adapt existing works on facility selection problem to facility bundle selection problem. f_1, f_2 and f_3 are 3 facilities. Assume the green triangles inside the 3 circles, C_1, C_2 and C_3, are their RkNN users, then their spatial influence are 5, 4 and 3, respectively. When $t = 2$, existing works would select f_1 and f_2 because they have the largest spatial influence individually. However, the spatial influence of $\{f_1, f_2\}$ is 6, which is smaller than the spatial influence of $\{f_1, f_3\}$, which is 8. As a facility bundle of size 2, $\{f_1, f_3\}$ is better than $\{f_1, f_2\}$. Motivated by this, we study the

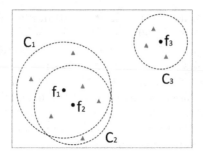

Fig. 1. MB-RkNN (Color figure online)

problem of maximizing bundled reverse k nearest neighbors (**MB-RkNN**), so that the spatial influence of a facility bundle of size t is maximized.

It can be shown that the MB-RkNN problem is NP-hard by reducing an existing NP-complete problem called the Maximum Coverage Problem to MB-RkNN problem in polynomial time. Hence, we propose a greedy algorithm that iteratively select the currently best facility that increases spatial influence of the current facility bundle the most. We assume that the data sets, namely the facilities and users are indexed by R-tree, respectively. The greedy algorithm is a branch-and-bound best first search algorithm on R-tree. Moreover, we introduce the concept of kNN region, which is an area shared by a group of users, such that the kNN facilities of them are all contained in the same area. kNN region shared between users allows us to avoid redundant calculation with dynamic programming technique.

Our contribution can be summarized as follows.

- We introduce the concept of spatial influence of a facility and a facility bundle. We are the first to study the problem of Maximizing Bundled Reverse k Nearest Neighbor (MB-RkNN).
- We prove that MB-RkNN problem is NP-hard and propose a branch-and-bound best first search algorithm that greedily select the currently best facility. We introduce the concept of kNN region that is shared by a group of users. This allows us to avoid redundant calculation with dynamic programming technique.
- We conduct experiments on real data sets. Experiments show that our algorithm is orders of magnitudes better than our baseline algorithm both in terms of CPU time and IO cost.

The rest of the paper is organized as follows. Section 2 review the literature. Section 3 present a formal definition of MB-RkNN problem, prove its NP-hardness, and present a baseline algorithm. Section 4 present our algorithm. Section 5 present experimental results. Section 6 conclude our work.

2   Related Work

Reverse k Nearest Neighbors Queries. Reverse Nearest Neighbor (RNN) query is first introduced by Korn and Muthukrishnan in [6]. Given a set F of facilities and a set U of users, the reverse nearest neighbors of a facility $f \in F$ are the users whose nearest neighbors in f. The set of such users is the *influence set* of f, and the *influence* of f can be measured by the number of users whose nearest neighbors is f. A generalization of RNN query is reverse k nearest neighbor (RkNN) query, where users are assumed to prefer not just their nearest facility but their k nearest facilities. Most of the existing techniques adopt a pruning and verification framework to solve RkNN queries. The two most notable pruning techniques are *regions-based pruning* [8,16] and *half-space pruning* [1,10]. The best known algorithm in terms of IO cost is *influence zone* [1] and the-state-of-the-art algorithm in terms of overall running time is *SLICE* [16].

Facility Allocation Problem. Wong *et al.* [14] introduce the MaxOverlap problem and proposed the MaxOverlap algorithm in the two-dimensional case when the metric space is the L_2-norm space. They then extend the problem definition to any Minkowski metric of order 1 or above for two- and three-dimensional spaces and propose appropriate algorithmic solutions [13]. These two works maximize spatial influence based on RNN queries. Zhou *et al.* [19] find an optimal region that maximizes spatial influence based on RkNN queries. More generally, Huang *et al.* [4] select a set of t optimal locations from a given set of candidate locations such that no other candidate location is more influential than any one of them. Alternatively, Chen *et al.* [2] finds out a set of t regions such that setting up t new facilities, one in each region, will collectively attract the maximum number of RNN users. The distinctive feature of this work is that it optimizes the combined influence of t selected regions rather than the influence of any single region. However, facility allocation problem is intrinsically different from facility selection problem. Locations are represented by either points or regions, and are either given or not given, but can never be collocated with existing facilities. In other words, what is being selected can never be existing facilities.

Product Selection Problem. In product selection problem, top-t most influential products are selected based on the influence, which is defined as the cardinality of a product's reverse top-k user preference [11]. The selected products have significant impact in the market individually. Koh *et al.* [5] considers coverage, and select a set of products that collectively covers the maximum possible number of reverse top-k. Gkorgkas *et al.* [3] considers diversity, and select a set of products that are most diverse to each other based on centroid of reverse top-k. Wang *et al.* [12] considers a combination of coverage and diversity. Product selection problem is intrinsically different from facility selection problem as the influence of a product is defined based on reverse top-k queries, but not RkNN queries.

Facility Selection Problem. Previous works tried to find out t facilities such that no other facility is more spatially influential than any one of the t selected facilities, either based on RNN queries [15,17,18] or RkNN queries [7]. Sun *et al.* [9] study the problem of constructing a heat (e.g., spatial influence) map, where points share the same set of RNN are grouped together. However, the literature lacks study on facility selection problem when the t facilities have the highest spatial influence as a bundle.

3   Preliminaries

In this section, we present a formal definition of our problem, prove it is NP-hard, and present our baseline algorithm.

3.1   Problem Definition

Consider a two dimensional Euclidean space, let F be a set of points representing facilities and U be a set of points representing users.

Definition 1 *(kNN, RkNN, Facility Bundle and Spatial Influence).* *Given a query user $u \in U$ and a number k, a kNN query of u retrieves every facility $f \in F$ such that f is one of the k nearest facilities from u. A facility like this is a kNN (facility) of u. The set of all such facilities, denoted as $kNN(u)$, is kNN set of u.*

   Given a query facility $f \in F$ and a number k, an RkNN query of f retrieves every user $u \in U$ such that $f \in kNN(u)$. A user like this is a RkNN (user) of f. The set of all such users, denoted as $RkNN(f)$, is RkNN set of f, and is also called the influence set of f.

   Let S be a facility bundle, and $|S|$ be the size of S. Given a number t, a facility bundle of size t is a set of t facilities. The influence set of S, denoted as $RkNN(S)$ is the union of the influence sets of all included facilities.

   Spatial influence of a facility f or a facility bundle S is the number of users in its influence set, denoted as $|RkNN(f)|$ and $|RkNN(S)|$, respectively.

Definition 2 *(Influence Zone* [1]*).* *Influence zone of f, denoted as $Z(f)$, is a region such that for every user u, if $u \in Z(f)$ we have $u \in RkNN(f)$, and if $u \notin Z(f)$ we have $u \notin RkNN(f)$.*

   Influence zone is a star-shaped polygon such that for every point $p \in Z(f)$, the line segment $\overline{fp}$ lies entirely in the polygon.

Definition 3 *(MB-RkNN).* *Given a number t, Maximizing Bundled Reverse k Nearest Neighbors (MB-RkNN) query retrieves a facility bundle S such that $|S| = t$, and $|RkNN(S)|$ is maximized.*

3.2   NP-Hardness

Next, we show that the MB-RkNN problem is NP-hard by reducing an existing NP-complete problem called the Maximum Coverage Problem to MB-RkNN in polynomial time.

Maximum Coverage Problem: Given a number t and a collection of sets $S = \{S_1, S_2, S_3, \cdots, S_m\}$, the objective is to find a subset $S' \subset S$ such that $|S'| \leq t$ and the number of covered elements $|\bigcup_{S_i \in S'} S_i|$ is maximum.

We reduce the Maximum Coverage Problem to MB-RkNN as follows. Sets $S_1, S_2, S_3, \cdots, S_m$ correspond to the influence sets of the facilities. The collection of sets S is equal to the collection of the influence sets of all facilities. It is easy to see that this transformation can be constructed in polynomial time, and it can be easily verified that when the problem is solved in the transformed MB-RkNN problem, the original Maximum Coverage Problem is also solved. Since the Maximum Coverage Problem is an NP-complete problem, MB-RkNN is NP-hard.

3.3   Baseline Algorithm

Since MB-RkNN problem is NP-hard, it is straight forward to employ a greedy algorithm. Iteratively, the currently best facility is selected until we have got a facility bundle of desired size. To determine the currently best facility, we first define the contribution of a facility.

Definition 4 (Contribution). *Let S be the current facility bundle. The contribution of a facility f, denoted as $N(f)$, is $|RkNN(S \bigcup f)| - |RkNN(S)|$, or in other words $|RkNN(f) \setminus RkNN(S)|$.*

A facility f is said to be the currently best facility if $f \notin S$ and for every other facility $f' \in F$ we have $N(f) \geq N(f')$.

To determine the currently best facility in each iteration, it is necessary to calculate influence set for every facility. As widely adopted, we assume that facilities and users are each indexed by an R-tree, and we employ the state-of-the-art algorithm *SLICE* [16] to do this. Then, we can calculate contribution of all facilities, and the currently best facility can be identified by doing set minus as described in Definition 4.

Notice that, since S changes after each iteration, the contribution of the same facility might also change after each iteration. Therefore, the contribution of all facilities need to be calculated again after each iteration. We noticed that there are duplicated calculations. Let $S_1 = \{f_1\}$ and $S_2 = \{f_1, f_2\}$ be the current facility bundle after the 1st and the 2nd iteration, respectively. To calculate $N(f)$ with S_1, we do $RkNN(f) \setminus RkNN(f_1)$. And then, to calculate $N(f)$ with S_2, we do $RkNN(f) \setminus (RkNN(f_1) \bigcup RkNN(f_2))$, which is equivalent to $RkNN(f) \setminus RkNN(f_1) \setminus RkNN(f_2)$. That is, we can exclude $RkNN(f_1)$ from $RkNN(f)$ after we have got S_1, and then exclude $RkNN(f_2)$ from $RkNN(f)$ after we have got S_2.

Inspired by this, we store kNN sets for every user in addition to influence sets for every facility. Every time when a facility f is to be inserted into S, we can for each user $u \in RkNN(f) \setminus RkNN(S)$, easily retrieve the facilities $f' \in kNN(u)$ and remove u from $RkNN(f')$. Note that this can actually be done by maintaining a counter for each influence set. As a result, the currently best facility is simply the facility with the highest counter for its influence set.

4   Techniques

4.1   Solution Overview

Similar as the baseline, we employ a greedy algorithm that iteratively selects the currently best facility and inserts it into S until $|S| = t$. We give you an overview of our algorithm in this section.

Let F^R be the R-tree indexing facilities, and U^R be the R-tree indexing users. Let F_e and U_e be a node or an entry on F^R and U^R, respectively. We first give you a brief introduction to the concept of maximum possible contribution and candidate RkNN set, to assist your understanding of the overview of our algorithm. More detailed information will be given in Sects. 4.3 and 4.4.

Definition 5 (Maximum Possible Contribution). *The Maximum Possible Contribution (MPC) of F_e, denoted as $N_e(F_e)$, is an upper bound of the contribution of F_e. Thus, $N_e(F_e) \geq N(F_e)$.*

Since MPC is no smaller than the contribution of F_e, higher value of MPC indicates higher chance of containing or being the next currently best facility. Therefore, we can access F_es in decreasing order of their MPC. And this allows us to design a branch-and-bound best first search algorithm. At this stage, we initialize one's MPC by the number of facilities belong to its candidate RkNN set, which is a super set of its RkNN set. However, $N_e(F_e)$ might take a tighter estimation, and we will elaborate on this further in Sect. 4.4.

Definition 6 (Candidate RkNN Set). *The Candidate RkNN Set of F_e, denoted as $RkNN_e(F_e)$, is a set of U_e such that for every facility $f \in F_e$ we have $RkNN(f) \subseteq RkNN_e(F_e)$. The cardinality of the candidate RkNN set of F_e, denoted as $|RkNN_e(F_e)|$ is the number of facilities indexed by $U_e \in RkNN_e(F_e)$.*

Recall the concept of influence zone, RkNN users of a facility occupies an area around it. Facilities that are geographically close to each other have their influence zone largely overlapped. This is the intuition behind this concept of candidate RkNN set. That is to group nearby facilities together, so that they can share the same candidate RkNN set without including too much false positives, and we can calculate one candidate RkNN set once for a few facilities.

Algorithm 1 illustrates the overview of our algorithm. S is initialized as an empty set in line 1. Let $F^R.root$ and $U^R.root$ be the root node of F^R and U^R, respectively. $RkNN_e(F^R.root)$ is initialized to $U^R.root$. $N_e(F^R.root)$ is initialized to $|RkNN_e(F^R.root)|$, which is equal to $|U|$ (line 2). Let H be a max heap, it is initialized by $<N(F^R.root), F^R.root>$. Line 3 iterate through F_e in descending order of $N_e(F_e)$. As long as H is not empty and $|S| < t$, a heap entry F_e is dequeued from H in line 4. We check if F_e is a node or a facility. Let F_e^c be a child of F_e. If F_e is a node (line 5), then for every F_e^c, we calculate $RkNN_e(F_e^c)$, initialize $N_e(F_e^c)$ and insert $<N_e(F_e^c), F_e^c>$ into H (If F_e^c is a facility, then we calculate $RkNN(F_e^c)$, and initialize $N_e(F_e^c)$ to $|RkNN(F_e^c)|$). If F_e is a facility (line 9), we check if $N_e(F_e)$ is set to $N(F_e)$. Note that when $N(F_e)$ is not calculated, $N_e(F_e)$ is considered as not set to $N(F_e)$. If it is not (line 10), we calculate

---

Algorithm 1. *MB-RkNN*(F^R, U^R, t)

Input : F^R : an R-tree indexing facilities, U^R : an R-tree indexing users
$\quad\quad\quad$ t : number of facilities to be selected, H : a max heap
Output: $S : S \subseteq F$, $|S| = t$

1 $S := \emptyset$; $RkNN_e(F^R.root) := U^R.root$;

2 $N_e(F^R.root) := |U|$; $H \leftarrow\; < N_e(F^R.root), F^R.root >$;

3 **while** $H \neq \emptyset$ and $|S| < t$ **do**

4 $\quad$ dequeue an entry F_e from H;

5 $\quad$ **if** F_e *is a node* **then**

6 $\quad\quad$ **for** *child F_e^c of F_e* **do**

7 $\quad\quad\quad$ calculate $RkNN_e(F_e^c)$;

8 $\quad\quad\quad$ $N_e(F_e^c) := |RkNN_e(F_e^c)|$; $H \leftarrow\; < N_e(F_e^c), F_e^c >$;

9 $\quad$ **else** $\hfill$ /* F_e is a facility */

10 $\quad\quad$ **if** $N_e(F_e) \neq N(F_e)$ **then** $\hfill$ /* $N(F_e)$ is not calculated */

11 $\quad\quad\quad$ $N(F_e) := |RkNN(F_e) \setminus RkNN(S)|$;

12 $\quad\quad\quad$ $N_e(F_e) := N(F_e)$; $H \leftarrow\; < N(F_e), F_e >$;

13 $\quad\quad$ **else**

14 $\quad\quad\quad$ $S \leftarrow F_e$; clear $N(F_e)$ for all F_e;

15 **return** S

$N(F_e)$, update $N_e(F_e)$, and insert $<N_e(F_e), F_e>$ into H. If it is (line 13), we add it to S, and clear all calculated values of $N(F_e)$ as they are invalidated by the facility newly added to S.

4.2 kNN Region

Before we can explain how to calculate candidate RkNN set, we need the concept of kNN region. Recall that kNN facilities of a user occupies an area around it. Users that are geographically close to each other have their kNN sets largely overlapped. This is the intuition behind the concept of kNN region. That is to group nearby users together, so that they can share the same kNN region without including too much false positives, and we can calculate one kNN region once for a few users. And this is also the reason why we use U_e instead of individual users in candidate RkNN set.

Definition 7 *(kNN Region of a Point).* *Let p be a point, the kNN region of p, denoted as $R(p)$, is the circle centered at p with radius equal to the distance from p to its kth nearest facility.*

Lemma 1. *Let p be a point, $kNN(p) \in R(p)$.*

Proof. We omit the proof of Lemma 1 as it is obvious from Definition 7.

Definition 8 *(kNN Region of a Line Segment).* *(As illustrated in Fig. 2) Let v_1 and v_2 be two points that defines a line segment $\overline{v_1 v_2}$. The kNN region of $\overline{v_1 v_2}$, denoted as $R(v_1 v_2)$, is the minimum circle that contains $R(v_1)$ and $R(v_2)$.*

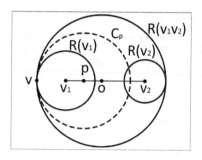

Fig. 2. kNN region of a line segment

Lemma 2. *Let v_1 and v_2 be two points that defines a line segment $\overline{v_1 v_2}$. Let p be a point on $\overline{v_1 v_2}$. $kNN(p) \in R(p) \subseteq R(v_1 v_2)$.*

Proof. As shown in Fig. 2. $kNN(v_1) \in R(v_1)$ and $kNN(v_2) \in R(v_2)$. Since $R(v_1 v_2)$ is the smallest circle that contains both $R(v_1)$ and $R(v_2)$, let o be the center of $R(v_1 v_2)$, o must lie on $\overline{v_1 v_2}$, and divides $\overline{v_1 v_2}$ into two segments, $\overline{v_1 o}$ and $\overline{o v_2}$. Without lose of generality, we assume that p lies on $\overline{v_1 o}$. This implies that we can find a circle C_p that contains $R(v_1)$ and is contained by $R(v_1 v_2)$. Since $R(v_1)$ has at least k facilities, C_p must have at least k facilities. Thus, $kNN(p) \in R(p) \subseteq C_p \subseteq R(v_1 v_2)$.

Definition 9 (*kNN Region of a Rectangle*). *(As illustrated in Fig. 3) Let r be a rectangle, v_1, v_2, v_3, v_4 be the four vertex of r. The kNN region of r, denoted as $R(r)$, is the union of $R(v_1 v_2)$, $R(v_2 v_3)$, $R(v_3 v_4)$, $R(v_1 v_4)$ and r.*

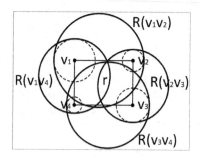

Fig. 3. kNN region of rectangle

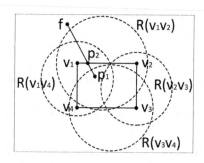

Fig. 4. Lemma 3

Lemma 3. *Given a rectangle r, and a point $p \in r$. $kNN(p) \in R(p) \subseteq R(r)$.*

Proof. As shown in Fig. 4. A point $p \in r$ either lies strictly inside r (e.g., p_1) or on the boundary (e.g., p_2). Cases like p_2 is clear from Lemma 2, we prove cases like p_1 by contradiction.

Let f be a facility such that $f \notin R(r)$ and $f \in kNN(p_1)$. Since $f \notin R(r)$, f must be outside of r, and the line segment $\overline{p_1 f}$ intersects the boundary of r. Without lose of generality, let p_2 be the intersection point. Then we have $kNN(p_2) \in R(p_2) \subseteq R(r)$.

Since $f \in kNN(p_1)$, we have $p_1 \in RkNN(f)$, and $p_1 \in Z(f)$. Since influence zone is a star-shaped polygon, and p_2 lies on $\overline{p_1 f}$, we have $p_2 \in Z(f)$, and $p_2 \in RkNN(f)$. Thus, $f \in kNN(p_2) \in R(p_2) \subseteq R(r)$. This contradicts with the assumption $f \notin R(r)$.

4.3   Inherit Candidate RkNN Set

From the definition of candidate RkNN set, it is valid to set it as coarse as simply the root node of U^R, or as fine as exactly the RkNN set. The former would cover too much false positives. Recall that the cardinality of candidate RkNN set is used as the key to access F_es in a best first manner, with too much false positives, the ordering will lose effectiveness. The later cannot effectively group users together.

To strike a balance here, we can initialize one's candidate RkNN set as $U^R.root$, gradually open up $U_e \in RkNN_e(F_e)$ until all of them are of the right size, in other words, the right level on the tree. When we open up U_e, we need to know whether U_e^c, a child of U_e should be pruned from $RkNN_e(F_e)$. This can be determined with the help of $R(U_e)$.

Lemma 4 (Pruning). U_e can be pruned from $RkNN_e(F_e)$ when $R(U_e)$ and F_e do not overlap with each other.

Proof. Let r be the Minimum Bounding Rectangle (MBR) of U_e. From Lemma 3, we have $kNN(u) \in R(U_e)$ for every user $u \in U_e$. Since F_e do not overlap with $R(U_e)$, we have $f \notin R(U_e)$ and $f \notin kNN(u)$ for every facility $f \in F_e$. In other words, we have $u \notin RkNN(f)$ for every pair of $u \in U_e$ and $f \in F_e$. Thus, U_e can be pruned from $RkNN_e(F_e)$.

Next, we show that calculation of candidate RkNN set do not always have to start from $U^R.root$.

Lemma 5 (Inheritance Property). $RkNN_e(F_e^c) \subseteq RkNN_e(F_e)$.

Proof. Since F_e^c is a child of F_e, F_e contains F_e^c. Thus, if F_e^c overlaps with $R(U_e)$, F_e must also overlaps with $R(U_e)$. In other words, if $U_e \in RkNN_e(F_e^c)$, we must have $U_e \in RkNN_e(F_e)$ (e.g., $RkNN_e(F_e^c) \subseteq RkNN_e(F_e)$).

Lemma 5 implies that $RkNN_e(F_e^c)$ can be obtained by filtering on $U_e \in RkNN_e(F_e)$. This is referred to as the inheritance property of candidate RkNN set.

Algorithm 2 illustrates the process of inheritance. Line 1 loop through all $U_e \in RkNN_e(F_e)$. The ones who have their kNN region overlap with F_e^c are selected in line 2. If U_e is an user (line 3), U_e is added to $RkNN_e(F_e^c)$. If U_e is a node (line 5), all children U_e^c of U_e is examined. If $R(U_e^c)$ is not determined

---

Algorithm 2. *inherit*(F_e, F_e^c)

Input : F_e : a node on facility R-tree, F_e^c : a child node/facility of F_e
Output: $RkNN_e(F_e^c)$

1 **for** U_e *in* $RkNN_e(F_e)$ **do**
2   **if** F_e^c *overlaps* $R(U_e)$ **then**
3   **if** U_e *is a user* **then**
4               $RkNN_e(F_e^c) \leftarrow U_e$;
5   **else**   /* U_e is a node */
6   **for** *child* U_e^c *of* U_e **do**
7   **if** $R(U_e^c)$ *is not determined* **then**
8   calculate $R(U_e^c)$;
9   **if** F_e^c *overlaps* $R(U_e^c)$ **then**
10                      $RkNN_e(F_e^c) \leftarrow U_e^c$;

(line 7), then $R(U_e^c)$ is calculated. Note that we only calculate the same kNN region once, although it may be used several times. If F_e^c overlaps $R(U_e^c)$ (line 9), it is added to $RkNN_e(F_e^c)$.

Note that in this algorithm, when we descend one level down on F^R, namely from F_e to F_e^c, we also descend one level down on U^R, namely from U_e to U_e^c. This is to say, whenever it is applicable, the distance from $U_e \in RkNN_e(F_e)$ to $U^R.root$ is no shorter than the distance from F_e to $F^R.root$.

4.4   Maximum Possible Contribution and Contribution

Recall that in Algorithm 1, maximum possible contribution is initialized in line 8. However, it might be initialized to a tighter value.

Lemma 6. *Let f be a facility such that $f \in S$, and F_e' be a common ancestor of F_e and f, then we have $N(F_e) \leq |RkNN_e(F_e')| - N(f)$.*

Proof. We omit the proof of Lemma 6 as it is obvious.

Let f^a and F_e^a be an ancestor of f and F_e, respectively. Inspired by Lemma 6, when we add a facility f to S (Algorithm 1 line 14), we decrease $N_e(f^a)$ by $N(f)$ for all f^as. And when we initialize $N_e(F_e)$ (Algorithm 1 line 8), we assign it the minimum value among $|RkNN_e(F_e)|$ and all $N_e(F_e^a)$s.

Recall that in Algorithm 1 we have $N(F_e) := |RkNN(F_e) \setminus RkNN(S)|$ in line 11. However, there are cases when $|RkNN(F_e) \setminus RkNN(S)| = |RkNN(F_e)|$. This happens when the influence zone of F_e is disjoint with every influence zone of facilities in S. When this happens, we set $N(F_e)$ to $|RkNN(F_e)|$ directly. The following Lemmas specifies the criteria when two influence zones are disjoint.

Lemma 7. *Let f_1 and f_2 be two facilities. If there is a point p_1 that has both $f_1 \in kNN(p_1)$ and $f_2 \in kNN(p_1)$, then there is a point p_2 on the perpendicular bisector of f_1 and f_2 that has both $f_1 \in kNN(p_2)$ and $f_2 \in kNN(p_2)$.*

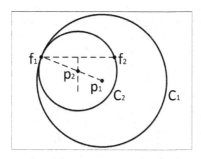

Fig. 5. Lemma 7.

Proof. As shown in Fig. 5. Without lose of generality, we assume that the line segment $\overline{f_1 p_1}$ intersects the perpendicular bisector of f_1 and f_2 at p_2. Let C_1 be a circle that centered at p_1 and passes through f_1. Since we have $f_1 \in kNN(p_1)$ and $f_2 \in kNN(p_1)$, there are at most k facilities in C_1. Let C_2 be a circle that centered at p_2 and passes through f_1 and f_2. It is clear that C_1 contains C_2. Therefore, there are at most k facilities in C_2. Thus, we have $f_1 \in kNN(p_2)$ and $f_2 \in kNN(p_2)$.

Lemma 8. *Let f_1 and f_2 be two facilities. If $Z(f_1)$ do not intersect with the perpendicular bisector between f_1 and f_2, then $Z(f_1)$ and $Z(f_2)$ are disjoint.*

Proof. We prove this by contradiction. Assume $Z(f_1)$ and $Z(f_2)$ are not disjoint, and there is a point p_1 such that $f_1 \in kNN(p_1)$ and $f_2 \in kNN(p_1)$. From Lemma 7, we can find a point p_2 on the perpendicular bisector between f_1 and f_2 that has both $f_1 \in kNN(p_2)$ and $f_2 \in kNN(p_2)$. Thus $p_2 \in Z(f_1)$. This conflict with the assumption that $Z(f_1)$ do not intersect with the perpendicular bisector between f_1 and f_2.

Lemma 9. *Let f_1 and f_2 be two facilities, $d(f_1, v)$ be the distance from f_1 to the furthest point v on $Z(f_1)$. If f_2 is more than twice the distance $d(f_1, v)$ away from f_1, then $Z(f_1)$ and $Z(f_2)$ are disjoint.*

Proof. Since f_2 is more than twice the distance of $d(f_1, v)$ away from f_1, $Z(f_1)$ do not intersect with the perpendicular bisector between f_1 and f_2. From Lemma 8, $Z(f_1)$ and $Z(f_2)$ are disjoint.

Inspired by Lemma 9, when we add a facility to S in Algorithm 1 line 14, we calculate its influence zone [1], and find out the distance from the facility to the furthest point on its influence zone. Then, in line 11, we check the distance from F_e to all facilities in S against Lemma 9. If Lemma 9 always holds, then we can set $N(F_e)$ to $|RkNN(F_e)|$ directly.

5   Experiments

In this section, we present our experimental evaluation. All algorithms are implemented in C++, compiled by g++ with flag $-O3$. The experiments are run on

a 32-bit PC with Intel Xeon 2.40 GHz dual CPU and 4 GB memory running
Debian Linux. As moderate data sets may be completely loaded in main memory, we use main memory R*-tree with 4kB pagesize, and we evaluate CPU time
mainly. To cooperate cases where disk-based index is used, we also report the
number of page requests.

Algorithms. We compare our algorithm (**DP**) with our baseline algorithm
(**RKNN**). Both algorithms calculate exact RkNN sets, and are approximate
algorithms using the same criteria to select the currently best facility. Their
results are expected to be the same with an assumption that no two facilities
have equal spatial influence (e.g., same number of RkNN users). Hence, we compare their performance only on CPU time and I/O cost.

Data Sets. We use the real data set obtained from Yelp Dataset Challenge[1]. We
extract 144072 points as facilities from business file[2]. A small random variable is
added so that no two facilities share the same latitude and longitude. We extract
946600 points as users from tip file[3]. We treat each record in this file as a user.
The location of a user is the latitude and longitude of the associated facility. A
small random variable is added so that no two users share the same latitude and
longitude. We than pick the required number of facilities and users out of the
extracted points uniformly at random.

Parameters. Parameters and values used are specified in Table 1, default settings are in bold.

Table 1. Experimental settings

Parameter	Values
$|F|$	1k, 5k, 10k, 20k, **50k**, 80k, 100k
$|U|$	5k, 10k, 20k, 50k, **100k**, 200k, 500k
k	5, **10**, 15, 20, 30, 50
t	1, **3**, 5, 8, 10

Effect of $|F|$. Figure 6 show the effect of the size of the data sets. Specifically,
Fig. 6(a) show the CPU time and Fig. 6(b) show the I/O cost. Since in real world
scenario, it does not make sense to have $|U| < |F|$, when we vary the value of
$|F|$, we also vary the value of $|U|$, and we keep the ratio $\frac{|U|}{|F|} = 2$ unchanged.
Note that, log scale are used for y-axis in both Fig. 6(a) and (b). We can see
that DP is orders of magnitudes better than RKNN both in terms of CPU time
and I/O cost, and also scales better.

---

[1] https://www.yelp.com/dataset/challenge.
[2] *yelp_academic_dataset_business.json*.
[3] *yelp_academic_dataset_tip.json*.

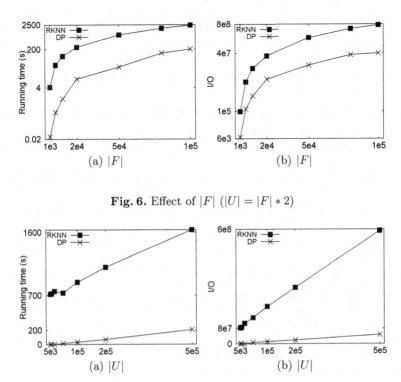

Fig. 6. Effect of $|F|$ $(|U| = |F| * 2)$

Fig. 7. Effect of $|U|$

Effect of $|U|$. Figure 7 show the effect of the size of the data sets. Different from Fig. 6, we do not keep the ratio of $\frac{|U|}{|F|}$ unchanged. Instead, we keep the value of $|F|$ fixed, and vary only $|U|$. Without log scale for y-axis, it is clear that DP is orders of magnitudes better than RKNN both in terms of CPU time and I/O cost, and also scales much better.

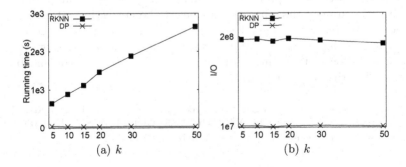

Fig. 8. Effect of k

698   S. Wang et al.

Effect of k. Figure 8 show the effect of k. It is clear that DP is orders of magnitudes better than RKNN both in terms of CPU time and I/O cost. Specifically, in Fig. 8(a), RKNN demonstrates linear growth in terms of CPU time while DP demonstrates no visible growth. In Fig. 8(b), they both demonstrate no visible growth in terms of I/O cost.

Effect of t. Figure 9 show the effect of t. It is clear that DP is orders of magnitudes better than RKNN both in terms of CPU time and I/O cost. None of them demonstrate visible growth.

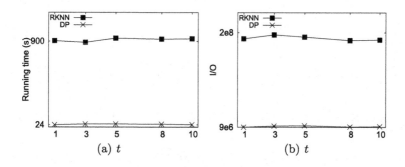

Fig. 9. Effect of t

6   Conclusion

Consider a two dimensional Euclidean space, let F be a set of points representing facilities and U be a set of points representing users. The spatial influence of a facility is the number of users who have this facility as one of their k nearest facilities. Given a facility bundle of size t, the spatial influence of the bundle is the number of distinct users influenced by any one of them. We are the first to study the problem of Maximizing Bundled Reverse k Nearest Neighbors (MB-RkNN), such that the spatial influence of a facility bundle of size t is maximized. We prove that MB-RkNN problem is NP-hard, and propose a branch-and-bound best first search algorithm that greedily select the currently best facility until we get t facilities. We also employ dynamic programming technique to avoid redundant calculation. Extensive experiments on real data sets show that our algorithm is orders of magnitudes better than our baseline algorithm both in terms of CPU time and IO cost.

Acknowledgment. Ying Zhang is supported by ARC FT170100128 and DP18010 3096.Xuemin Lin is supported by NSFC61672235, DP170101628 and DP180103096. Muhammad Aamir Cheema is supported by ARC DP180103411.

References

1. Cheema, M.A., Lin, X., Zhang, W., Zhang, Y.: Influence zone: efficiently processing reverse k nearest neighbors queries. In: Proceedings of the 27th International Conference on Data Engineering, ICDE 2011, Hannover, Germany, 11–16 April 2011, pp. 577–588 (2011)
2. Chen, F., Lin, H., Qi, J., Li, P., Gao, Y.: Collective-k optimal location selection. In: Gertz, M., Renz, M., Zhou, X., Hoel, E., Ku, W.-S., Voisard, A., Zhang, C., Chen, H., Tang, L., Huang, Y., Lu, C.-T., Ravada, S. (eds.) SSTD 2017. LNCS, vol. 10411, pp. 339–356. Springer, Cham (2017). https://doi.org/10.1007/978-3-319-64367-0_18
3. Gkorgkas, O., Vlachou, A., Doulkeridis, C., Nørvåg, K.: Finding the most diverse products using preference queries. In: Proceedings of the 18th International Conference on Extending Database Technology, EDBT 2015, Brussels, Belgium, 23–27 March 2015, pp. 205–216 (2015)
4. Huang, J., Wen, Z., Qi, J., Zhang, R., Chen, J., He, Z.: Top-k most influential locations selection. In: Proceedings of the 20th ACM Conference on Information and Knowledge Management, CIKM 2011, Glasgow, United Kingdom, 24–28 October 2011, pp. 2377–2380 (2011)
5. Koh, J., Lin, C., Chen, A.L.P.: Finding k most favorite products based on reverse top-t queries. VLDB J. **23**(4), 541–564 (2014)
6. Korn, F., Muthukrishnan, S.: Influence sets based on reverse nearest neighbor queries. In: Proceedings of the 2000 ACM SIGMOD International Conference on Management of Data, Dallas, Texas, USA, 16–18 May 2000, pp. 201–212 (2000)
7. Li, C., Wang, E.T., Huang, G., Chen, A.L.P.: Top-n query processing in spatial databases considering bi-chromatic reverse k-nearest neighbors. Inf. Syst. **42**, 123–138 (2014)
8. Stanoi, I., Agrawal, D., El Abbadi, A.: Reverse nearest neighbor queries for dynamic databases. In: 2000 ACM SIGMOD Workshop on Research Issues in Data Mining and Knowledge Discovery, Dallas, Texas, USA, 14 May 2000, pp. 44–53 (2000)
9. Sun, Y., Zhang, R., Xue, A.Y., Qi, J., Du, X.: Reverse nearest neighbor heat maps: a tool for influence exploration. In: 32nd IEEE International Conference on Data Engineering, ICDE 2016, Helsinki, Finland, 16–20 May 2016, pp. 966–977 (2016)
10. Tao, Y., Papadias, D., Lian, X.: Reverse kNN search in arbitrary dimensionality. In: Proceedings of the Thirtieth International Conference on Very Large Data Bases, Toronto, Canada, 31 August - 3 September 2004, pp. 744–755 (2004)
11. Vlachou, A., Doulkeridis, C., Nørvåg, K., Kotidis, Y.: Identifying the most influential data objects with reverse top-k queries. PVLDB **3**(1), 364–372 (2010)
12. Wang, S., Cheema, M.A., Zhang, Y., Lin, X.: Selecting representative objects considering coverage and diversity. In: Second International ACM Workshop on Managing and Mining Enriched Geo-Spatial Data, GeoRich@SIGMOD 2015, Melbourne, VIC, Australia, 31 May 2015, pp. 31–38 (2015)
13. Wong, R.C., Özsu, M.T., Fu, A.W., Yu, P.S., Liu, L., Liu, Y.: Maximizing bichromatic reverse nearest neighbor for lp-norm in two- and three-dimensional spaces. VLDB J. **20**(6), 893–919 (2011)
14. Wong, R.C., Özsu, M.T., Yu, P.S., Fu, A.W., Liu, L.: Efficient method for maximizing bichromatic reverse nearest neighbor. PVLDB **2**(1), 1126–1137 (2009)
15. Xia, T., Zhang, D., Kanoulas, E., Du, Y.: On computing top-t most influential spatial sites. In: Proceedings of the 31st International Conference on Very Large Data Bases, Trondheim, Norway, 30 August - 2 September 2005, pp. 946–957 (2005)

16. Yang, S., Cheema, M.A., Lin, X., Zhang, Y.: SLICE: reviving regions-based pruning for reverse k nearest neighbors queries. In: IEEE 30th International Conference on Data Engineering, Chicago, ICDE 2014, IL, USA, 31 March - 4 April 2014, pp. 760–771 (2014)
17. Zhan, L., Zhang, Y., Zhang, W., Lin, X.: Finding top k most influential spatial facilities over uncertain objects. In: 21st ACM International Conference on Information and Knowledge Management, CIKM 2012, Maui, HI, USA, 29 October - 02 November 2012, pp. 922–931 (2012)
18. Zhan, L., Zhang, Y., Zhang, W., Lin, X.: Finding top k most influential spatial facilities over uncertain objects. IEEE Trans. Knowl. Data Eng. **27**(12), 3289–3303 (2015)
19. Zhou, Z., Wu, W., Li, X., Lee, M., Hsu, W.: MaxFirst for MaxBRkNN. In: Proceedings of the 27th International Conference on Data Engineering, ICDE 2011, Hannover, Germany, 11–16 April 2011, pp. 828–839 (2011)

A Road-Aware Neural Network for Multi-step Vehicle Trajectory Prediction

Jingze Cui, Xian Zhou, Yanmin Zhu$^{(\boxtimes)}$, and Yanyan Shen

Department of Computer Science and Engineering,
Shanghai Jiao Tong University, Shanghai, China
{misaki1,zhouxian,yzhu,shenyy}@sjtu.edu.cn

Abstract. Multi-step vehicle trajectory prediction has been of great significance for location-based services, e.g., actionable advertising. Prior works focused on adopting pattern-matching techniques or HMM-based models, where the ability of accurate prediction is limited since patterns and features are mostly extracted from historical trajectories. However, these methods may become weak to multi-step trajectory prediction when new patterns appear or the previous trajectory is incomplete.

In this paper, we propose a neural network model combining road-aware features to solve multi-step vehicle trajectory prediction task. We introduce a novel way of extracting road-aware features for vehicle trajectory, which consist of intra-road feature and inter-road feature extracted from road networks. The utilization of road-aware features helps to draw the latent patterns more accurately and enhances the prediction performances. Then we leverage LSTM units to build temporal dependencies on previous trajectory path and generate future trajectory. We conducted extensive experiments on two real-world datasets and demonstrated that our model achieved higher prediction accuracy compared with competitive trajectory prediction methods.

Keywords: Multi-step trajectory prediction
Road-aware features · LSTM

1 Introduction

Recent years have witnessed the rapid developments of wireless communication techniques, which gives rise to the availability of a huge number of GPS-enabled devices such as locators on taxies. And lots of location-based services such as navigation have been well developed, which brings great conveniences in various aspects of people's life, especially in transportation systems.

Typically, predicting vehicle future locations is the fundamental basis of most existing location-based services, (e.g., navigation route planning, location-based advertising, traffic management and so on), which has raised great attention to trajectory prediction problem. Furthermore, instead of predicting only for the

© Springer International Publishing AG, part of Springer Nature 2018
J. Pei et al. (Eds.): DASFAA 2018, LNCS 10827, pp. 701–716, 2018.
https://doi.org/10.1007/978-3-319-91452-7_45

next step (e.g., next one minute), the multi-step trajectory prediction is more favorable to most location-based services. The reason is that, results of multi-step prediction offer a relatively long-term trend information [16], which is crucial for applications like traffic management.

In this paper, we focus on solving the multi-step vehicle trajectory prediction problem, i.e., predicting vehicle locations in next few steps. The key challenges lie to (1) accurately extract the latent patterns, and (2) effectively build temporal dependencies on previous trajectory path. Specifically, GPS data usually suffers from noise interference and partly missing problems. Meanwhile, the road networks in reality are mostly complicated, which makes it difficult to extract the real latent patterns accurately. Moreover, it is complex to capture the tendency of varying trajectories when applying traditional machine learning approaches due to the intricate road networks.

Previous methods mostly employed pattern-matching techniques [3,7,14] or HMM-based models [12,15]. In such methods, the ability of trajectory prediction is limited when new patterns appear, or previous trajectory information is incomplete. Besides, it is inferior to build a prediction model under assumptions like vehicles proceed following the shortest route [1] or frequent trajectory patterns [6]. Such assumptions are unreasonable in some scenarios, e.g., travellers may want to take another longer route with multiple visiting places.

As for these problems, we have several key observations about road network. Firstly, the intrinsic property of road itself plays an important role in vehicle trajectories. For example, there are a large amount of trajectories going through the primary road, and the path through one-way road is always directional. With the information of inner propensity, the intra-road features can be used to help predict future trajectory with higher accuracy and confidence. Secondly, the connection between roads presents regularity, which can be learned from large quantity of trajectory data. We find that the connections are more determined by extrinsic factors, e.g., urban district structure. Usually, the connection between roads in different districts changes when the functionality of these districts is changed. Failure to utilize the inter-road features may compromise overall prediction performance.

Based on the above observations, we propose a neural network model combining road-aware features to solve multi-step trajectory prediction problem. First of all, we partition the road network into uniform road segments and project trajectories into sequences of road segments. Secondly, we extract the road-aware features and encode them into intra-road feature and inter-road feature embeddings for each road segment. The intra-road feature involves main intrinsic properties (e.g., road type). The inter-road feature is extracted based on social interactions with neighboring road segments, which is embedded in a continuous vector space. And we learn the inter-road feature embedding for each road segment via network embedding techniques by constructing a graph which reflects the correlations between roads. Thirdly, we propose **Road-Aware-LSTM** model (**RA-LSTM**) based on long short-term memory network, which was proposed by [2] to specifically learn long-term dependencies in a sequence. Given a query trajectory, after projecting into sequence of road segments, RA-LSTM takes as inputting the road-aware feature vectors of corresponding road segments, which concatenate intra-road and inter-road feature embeddings. Then we leverage LSTM to effectively capture temporal dependencies in the input sequence (i.e., sequence of road-aware feature vectors), and generate future trajectory.

The main contributions of this paper are as follows:

- We introduce a novel way of extracting road-aware features for vehicle trajectories. The road-aware features consist of intra-road feature and inter-road feature, which are able to help extract latent patterns accurately from trajectories and lead to a substantial increase in prediction accuracy according to our experiments.
- We propose RA-LSTM to achieve multi-step trajectory prediction. The model leverages LSTM units to capture temporal features from previous trajectory path.
- We conduct extensive experiments to evaluate the performance of proposed model on two real-world trajectory data sets. The experimental results show that our model outperforms three baseline methods by 28%–65% on two datasets.

The rest of the paper is organized as follows. Section 2 gives problem definitions. In Sect. 3, we provide an overview of the proposed model and preliminaries. Section 4 describes the details of our prediction model. The experimental results are presented in Sect. 5 and related work are given in Sect. 6. Finally, we conclude this paper in Sect. 7.

2   Problem Definition

To exploit road network information, we use road records downloaded from OSM[1]. Originally, roads are represented as variable-length sequences. We partition these roads into segments of identical length for fine-grained prediction.

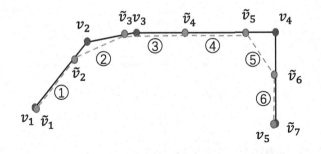

Fig. 1. Illustration of constructing road segments.

---

[1] http://www.openstreetmap.org/.

Definition 1 (Road Segment Set). *Let $\mathcal{A} = \{r_i | i = 1, 2, \cdots, M\}$ be road network records downloaded from OSM for a city (e.g., Shanghai) where r_i indicates a road. We propose a road network partition algorithm(details in Sect. 3.2) to partition variable-length roads in $\mathcal{A}$ into road segments of equal length. And we denote the road segment set as $\mathcal{RS} = \{s_i | i = 1, 2, \cdots, N\}$. We use length g to control the granularity of road segments. Intuitively, for a constant $\mathcal{A}$, smaller value of g creates a larger number of road segments and fine-grained trajectory predictions.*

For example, given a road $r = v_1 v_2 \cdots v_5$ as illustrated in Fig. 1 where v_i means geographical vertex of road r. After partition, we get $\tilde{r} = \tilde{v}_1 \tilde{v}_2 \cdots \tilde{v}_7$ and create 6 segments of same length: $\tilde{v}_1\tilde{v}_2, \tilde{v}_2\tilde{v}_3, \cdots, \tilde{v}_6\tilde{v}_7$ marked as red dashed lines in Fig. 1.

Definition 2 (Trajectory Sequence). *We consider a set of historical trajectories occurred in a city. Let $\mathcal{T}$ be a set of trajectories over one city. For some trajectory $T = p_1 p_2 \cdots p_l$, $p_i = (lon_i, lat_i)$ indicates longitude and latitude coordinates at timestamp i. Through map matching, each p_i corresponds to a unique road segment s_{t_i}. And, we denote trajectory sequence as $S = s_{t_1} s_{t_2} \cdots s_{t_l}$, where each geographical point p_i is contained in road segment s_{t_i}.*

Definition 3 (Problem Definition). *Consider a road network $\mathcal{A}$ in a city, given a query trajectory $T = p_1 p_2 \cdots p_l$, we focus on multi-step trajectory prediction problem which is to predict next k trajectory points $p_{l+1} p_{l+2} \cdots p_{l+k}$.*

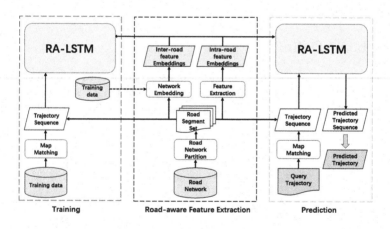

Fig. 2. Road-aware trajectory prediction framework

3 Overview and Preliminaries

3.1 Overview

Figure 2 provides an overview of our approach to the trajectory prediction task. The whole framework contains three parts: road-aware feature extraction, training and prediction. Road-aware feature extraction is applied to extract intrinsic and social features for each road segment, which are encoded as intra-road and inter-road feature embeddings. Training phase uses training data and road information to train a RA-LSTM. Given a query trajectory, we use the well-trained RA-LSTM and road feature embeddings to generate future trajectory. Road-aware feature extraction and training can be processed offline in advance.

Road-Aware Feature Extraction: Given a road network, road network partition is firstly applied to obtain road segment set. For each road segment, we extract road-aware features, which contain intra-road features and inter-road features. We learn inter-road feature embeddings via network embedding (details in Sect. 4.2), which is based on social connections between road segments. The intra-road features (details in Sect. 4.1) are mainly extracted from intrinsic properties (e.g., road types).

Training: Training data is created from historical vehicle trajectories occurred in a city. Through map matching, we obtain trajectory sequences based on road segment set and training data. RA-LSTM (details in Sect. 4.3) takes feature vectors of trajectory sequence as input and recognizes latent temporal features via LSTM network to generate future trajectory. Model parameters are updated using gradient descent during training phase. Training data is also useful for network embedding (details in Sect. 4.2).

Prediction: Given a query trajectory, we firstly convert query trajectory into trajectory sequence via map matching. Then the feature vectors of trajectory sequence are taken as input in RA-LSTM. Finally, the output of RA-LSTM is converted back to geographical points.

3.2 Preliminaries

Road Network Partition. Road partition is applied to convert road network $\mathcal{A}$ to road segment set $\mathcal{RS}$. Assume that a road r in $\mathcal{A}$ is $v_1 v_2 \cdots v_m$ and after partition we get $\tilde{r} = \tilde{v}_1 \tilde{v}_2 \cdots \tilde{v}_{\tilde{m}}$. Define $D(v_1, v_2)$ as the distance between geographical vertices v_1 and v_2. We implement Algorithm 1 for each road r in road network $\mathcal{A}$ and add corresponding segments to $\mathcal{RS}$. For each road r in $\mathcal{A}$ (line 1), first we use an array to store accumulative distance from each geographical vertex to start vertex v_1 in road r (lines 2–5), and calculate the number of vertices in $\tilde{r}$ (line 6). Then, we generate new vertices by solving linear and distance equations and add each segment into $\mathcal{RS}$ (lines 7–12). Finally, the last new vertex is the last origin vertex and we get the last segment of road r (line 13).

Trajectory Projection. We use trajectory projection to convert original trajectory T into trajectory sequence S. The widely used map matching approach based on HMM [12] is utilized to achieve trajectory projection. The core idea of the method is that a trajectory T is modeled to move in the light of a Markov process between road segments. Road segments are considered as hidden states

---

Algorithm 1. Road Partition Algorithm

Input: road network $\mathcal{A}$, equal length g
Output: road segment set $\mathcal{RS}$
1: **for** each road r in network $\mathcal{A}$ **do**
2:   cd =new array$[m]$, $cd[1] = 0$
3:   **for** $i = 2$ to m **do**
4:     $cd[i] = cd[i-1] + D(v_{i-1}, v_i)$
5:   **end for**
6:   $\tilde{m} = \lceil cd[m]/g \rceil + 1$, $\tilde{v}_1 = v_1$
7:   **for** $i = 2$ to $\tilde{m} - 1$ **do**
8:     $seglen = (i - 1) \cdot g$.
9:     find j s.t. $cd[j] \leq seglen < cd[j+1]$
10:    find $\tilde{v}_i$ s.t.:
        (1) $\tilde{v}_i$ on the line defined by v_j and v_{j+1}
        (2) $D(\tilde{v}_i, v_j) = seglen - cd[j]$
11:    $\mathcal{RS} \Leftarrow \tilde{v}_{i-1}\tilde{v}_i$
12:   **end for**
13:   $\tilde{v}_{\tilde{m}} = v_m$, $\mathcal{RS} \Leftarrow \tilde{v}_{\tilde{m}-1}\tilde{v}_{\tilde{m}}$
14: **end for**
15: **return** $\mathcal{RS}$

---

and geographical points are considered as observable states. Emission Probability describes how close the geographical point and road segment are. And state transition probability describes the condition probability from a road segment to another. Then Viterbi algorithm is applied to find the trajectory sequence S with the highest probability corresponding to T.

In prediction phase, we convert predicted trajectory sequence S back to trajectory geographical points. To simplify our model and calculation, we use midpoint of each segment as the corresponding geographical point. The reason is that we know the equal length of segments is g and assume that the predicted points are following uniform distribution in a segment. Suppose a predicted point on the segment is in location bg which $0 \leq b \leq 1$. Then we have the expected error $E(b)$ of the predicted point by Eq. 1:

$$E(b) = \int_0^{bg} (bg - x)\frac{1}{g}dx + \int_{bg}^{g} (x - bg)\frac{1}{g}dx = (\frac{1}{2} - b + b^2)g \qquad (1)$$

The minimum value of $E(b)$ is $g/4$ when $b = 1/2$. So we set the midpoint of the segment as the predicted point for prediction. Then we can get predicted trajectory T from predicted trajectory sequence S.

4   Road-Aware Trajectory Prediction

4.1   Intra-road Feature Extraction

In this part, we present how to extract the intra-road features for each road segment and construct an intra-road feature matrix denoted as M_{intra}. We extract

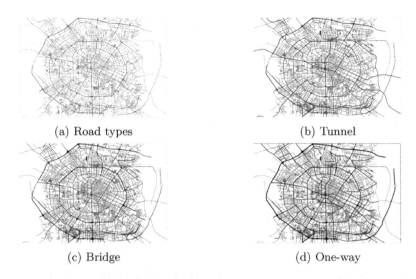

(a) Road types   (b) Tunnel

(c) Bridge   (d) One-way

Fig. 3. Illustration of four intra-road features. (Color figure online)

four intra-road features: "road type", "tunnel", "bridge" and "one-way". Figure 3 gives an example of these features, using the data downloaded from OSM. The data describes the road network in the center of Chengdu in 2017.

We explain these four features respectively as follows:

RoadType. This feature describes the main traits of road which has 27 types in Fig. 3(a). Different colors of the line mean different types. Such as "primary","residential" and "track", they have been built for different application. Roads in different types also have different lengths and widths which affect the driver's choice of route.

Tunnel. The feature has only two value 0 and 1 in the data structure. 0 means the road is not a tunnel while 1 means it is. Red lines in Fig. 3(b) mean tunnels and blue lines mean other roads. Tunnels are scattered throughout the area and usually extend very long. Traveling in tunnels will limit vehicle's speed and sometimes lose GPS signals.

Bridge. Similar as **Tunnel**, red lines represent bridges and blue lines represent others in Fig. 3(c). Bridges concentrate in the edge area and have many intersections. Bridges have height information and may guide the travel trajectory of vehicles, such as a circle.

One-way. Similar as **Tunnel** and **Bridge**, it is shown in Fig. 3(d). One-way means the road is unidirectional for drivers and guides vehicle's driving tendency.

Since these features contain only category information, intra-road feature vectors concatenate the representations of each feature in one-hot encoding. Assume that the length of intra-road feature vector is f and number of segments in segment set $\mathcal{RS}$ is n in total. We take the intra-road feature vectors as row vectors to form intra-road feature matrix M_{intra} of size $(n \times f)$.

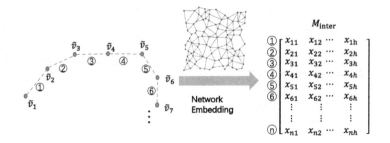

Fig. 4. Illustration of inter-road feature extraction

4.2   Inter-road Feature Extraction

In this section, we extract inter-road features based on social connections between road segments.

We use DeepWalk [9] which is proposed to learn latent representations of vertices in network, to learn inter-road feature embedding matrix denoted as M_{inter}. DeepWalk deploys truncated random walks to generate sequences of vertices in the network, and trains sequences with Skip-gram model [5] like sentences. After training, each vertex in network has an embedding representation.

If we only consider the location information of roads, road segment set $\mathcal{RS}$ can be regarded as an undirected graph with no weight. Since what we want is the inter-road feature vector representations of road segments not road vertices, in our context, road segments can be considered as "vertices" and their intersection vertices can be considered as "edges" in our network graph. To simplify road network graph and prominent its effect on trajectories, cause the road segments vehicles passed are connected, we take each two adjacent segments in a trajectory sequence S as a pair of adjacent points in the graph. So all training sequences can construct a relatively simple graph.

After training, represented vectors of every segment according to order of road segments form a matrix M_{inter}(Fig. 4). There are in total n segments in $\mathcal{RS}$ and assume each one is represented by a $(h \times 1)$ vector. Similar as M_{intra}, the size of M_{inter} is $(n \times h)$.

4.3   Road-Aware Neural Network Prediction Model

In this part we first show the process of constructing RA-LSTM based on LSTM and two road feature matrices M_{inter} and M_{intra}. Then, we present details of training and prediction.

RA-LSTM. Figure 5 provides detailed structure of RA-LSTM. Given input trajectory sequence $s_{t_1}, s_{t_2}, ..., s_{t_l}$, we firstly encode each input road segment s_i as one-hot representation vector d_i, $d_i \in R^N$, where N is the number of road segments. For each input road segment s_i, we get corresponding intra-road

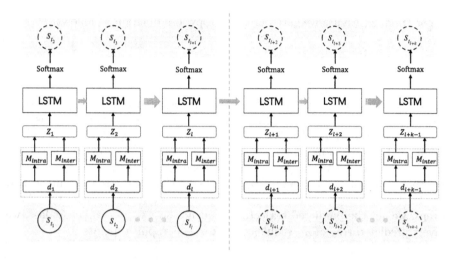

Fig. 5. Road-aware neural network prediction model

feature vector $M \cdot d_i$ and inter-road feature vector $M \cdot d_i$. And the concatenation of $M \cdot d_i$ and $M \cdot d_i$ is taken as final input feature vector denoted as Z_i, in Eq. 2.

$$Z_i = [M_{intra}^T \cdot d_i; M_{inter}^T \cdot d_i] \tag{2}$$

where $[;]$ indicates the operation of concatenating two vectors. $M_{intra}^T \cdot d_i \in R^{f \times 1}$, $M_{inter}^T \cdot d_i \in R^{h \times 1}$ and $Z_i \in R^{(f+h) \times 1}$.

Then LSTM network takes feature vector Z_i as input one by one. The key equations of generating hidden states in LSTM are the following.

$$
\begin{aligned}
i_t &= \sigma(W_{xi} \cdot Z_t + W_{hi} \cdot H_{t-1} + b_i) \\
f_t &= \sigma(W_{xf} \cdot Z_t + W_{hf} \cdot H_{t-1} + b_f) \\
o_t &= \sigma(W_{xo} \cdot Z_t + W_{ho} \cdot H_{t-1} + b_o) \\
C_t &= f_t \circ C_{t-1} + i_t \circ tanh(W_{xc} \cdot Z_t + W_{hc} \cdot H_{t-1} + b_c) \\
H_t &= o_t \circ tanh(C_t)
\end{aligned}
\tag{3}
$$

where Z_t is the input vector at t-th step; i_t, f_t and o_t are input, forget and output gates; C_t and H_t are internal memory and hidden state respectively; 'o' is the Hadamard product and $\sigma(\cdot)$ is sigmoid function; different kernel matrices W and biases b are parameters to be learned.

Afterwards, we use a softmax layer to get final output vector y, $y \in R^N$, where each value represents the probability of road segment at next step. The road segment with highest probability is the predicted $\hat{s}_i$ at i-th step.

Training and Prediction. We train and test our RA-LSTM model for multi-step trajectory prediction.

For training, we use weighted cross-entropy loss function as follows.

$$loss(\hat{s}, s) = -\frac{1}{l}\sum_{i=1}^{l}(s_i \log \hat{s}_i) \tag{4}$$

where $\hat{s}_i$ is our prediction, s_i is the real segment label and l is the length of trajectory sequence. And the mini-batch gradient descent algorithm is used to update parameters in RA-LSTM.

For prediction, assume that $s_{t_1}s_{t_2}\cdots s_{t_l}$ is the query trajectory sequence. We input it into RA-LSTM and get future road segments $s_{t_{l+1}}, s_{t_{l+2}}\cdots s_{t_{l+k}}$ one by one. Specifically, first we generate $s_{t_{l+1}}$ as the first predicted road segment, and we take it into neuron network as the input road segment for next step. After k steps, we get the predicted trajectory sequence $s_{t_{l+1}}s_{t_{l+2}}\cdots s_{t_{l+k}}$. At last, we convert predicted trajectory sequence back to geographical points.

5   Experiments

5.1   Experimental Setup

Datasets. We use two real-world trajectory datasets to evaluate our model. The corresponding road network datasets are acquired from OSM. Table 1 summaries the details:

Table 1. Trajectory dataset table

Dataset	Trajectories				Road network		
	Coverage	# of vehicles	# of trajectories	Time interval	# of roads	# of vertices	# of features
Shanghai	10 km × 8 km	74	4859	10s	193	1616	36
Chengdu	5 km × 5 km	2428	18991	6s	942	6094	31

Shanghai. The data was collected from device OBD of private cars in Fengxian district, Shanghai from July 2014 to January 2015. The covered area is the suburbs of Shanghai with coverage of 10 km×8 km. The unit of time is set as 10 s, which means the time interval between successive points is 10 s.

Chengdu. We use the collected taxi data from DiDi company in November 2016. The data contains 18991 orders and each order records a taxi trajectory of geographical coordinates. The covering area is downtown of Chengdu, where the number of roads are much more than that in Shanghai with narrower coverage. Due to the completeness of this dataset, we set the time interval as 6 s.

Baseline Methods. We compare our RA-LSTM model with both basic and advanced methods as follows:

- RMF [13] It is a descriptive model-based path prediction method, which computes motion function to capture movements.
- R2-D2 [15] R2-D2 is a HMM-based probabilistic method which predicts future trajectory path based on a few reference trajectories.
- LSTM [2] It is the vanilla version of LSTM model without road-aware features, where we use one-hot encoding of road segments as input.

Experimental Settings. We interpolate the trajectories to get synchronized timestamp 10 s and 6 s on Shanghai and Chengdu respectively, using Moving Object Cache, which is also used in R2-D2. We divide 80% dataset for training and 20% dataset for testing. In training dataset, we randomly select 20% for validation. We set the length of road segment $g = 50$ m The dimension of inter-road feature embedding and the number of hidden states in LSTM are both set as 128. The parameter of comparison methods are selected using validation set.

The experiments were conducted on a PC with Intel Core i5-7300 2.5 GHz, Nvidia GTX 1050ti and 16 GB RAM. RMF, LSTM and RA-LSTM were implemented in python with toolkit tensorflow 1.4.0, and R2-D2 was implemented in Java with JDK 1.7.

Measurement. We consider the input query trajectory with 5 previous geographical points and predict future locations for the next 5 steps. We evaluate all methods using *Average distance error* as follows.

$$Average\ distance\ error = \frac{1}{N} \sum_i \sum_t \|p_i^t - \hat{p}_i^t\|_2$$

where $\hat{p}_i^t$ and p_i^t are predicted coordinates and ground truth of query trajectory i at t-th step; N is the number of all the predicted value; and the errors are measured by Euclidean distance.

5.2   Results

Comparison of Different Methods. Table 2 shows the comparison results of 4 approaches over two datasets. All methods predict future trajectory for no more than 0.02 s and can be used in real-world applications. As we can see, all the methods get lower distance error in Chengdu than that in Shanghai. The main reason is that time interval dataset in Shanghai is longer than Chengdu, which makes it harder to capture latent patterns. Overall, RA-LSTM achieves the best results, which give 41% and 47% lower error than R2-D2. The reasons are two-fold. First, the input of R2-D2 prediction filter is restricted to a small set of reference trajectories, which cannot accurately reflect the latent patterns while RA-LSTM uses road-aware features in a global perspective. Second, probability

Table 2. Average distance error

Dataset	RMF	R2-D2	LSTM	RA-LSTM
Shanghai	247.9 8m	149.25 m	122.53 m	**87.78 m**
Chengdu	100.22 m	88.63 m	66.18 m	**46.18 m**

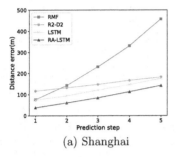 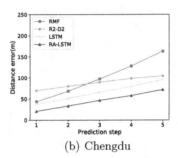

(a) Shanghai   (b) Chengdu

Fig. 6. Step-wise average distance error

of predicted path is computed based on HMM model, where the states are set as the cluster of reference points, which occurs inevitable error. It is also the main reason that R2-D2 performs worse than LSTM. For both datasets, RMF obtains the worst results. This is because the moving function does not take environment constraints into account and it lacks necessary historical experience. It is worth mentioning that RA-LSTM outperforms significantly than vanilla LSTM, giving the improvements by 28% and 30% as showed in the Table 2. The improvements confirms that the utilization of road-aware features enhances performances for trajectory prediction.

Results on Step-Wise Prediction. For further analysis, we present the step-wise prediction results over two datasets in Fig. 6. Generally, RA-LSTM performs the best for all steps, achieving 22%–67% and 25%–65% lower distance error than other methods for dataset Shanghai and Chengdu, respectively. The two figures show that RMF is better than R2-D2 for the first step, but worse in the further steps. RMF is designed for predicting next step by constructing motion function. But in multi-step prediction, it takes the predicted location as input for further step, which accumulates distance error. The distance error of LSTM and RA-LSTM grow a little faster than R2-D2, and RMF behaves worst. One of the reason is, not all query trajectories are predictable in R2-D2, which is explained in the next part. The other important reason is that, different from R2-D2 which predicts the whole path, RMF and RA-LSTM make trajectory prediction step by step. The predicted point is taken as input for next step, which causes accumulated error. And we see that the distance error of LSTM and RA-LSTM increase more slowly than RMF, demonstrating that memory cells in LSTM are able to capture long-term features effectively.

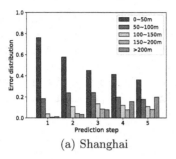 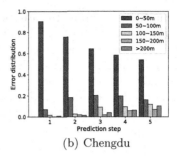

(a) Shanghai   (b) Chengdu

Fig. 7. Distance error distribution

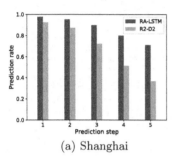 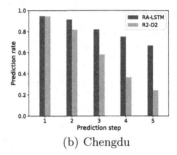

(a) Shanghai   (b) Chengdu

Fig. 8. Comparison of prediction rate

Comparison of Prediction Rate. R2-D2 selects the reference trajectories of high similarity (i.e., higher than a given threshold θ) with query trajectory. Those query trajectories with no reference trajectories are regarded as unpredictable. Therefore, R2-D2 adopts a metric called *prediction rate*, which is the fraction of the number of predictable query trajectories over the total number of query trajectories. The prediction rate reflects the robustness of a prediction model. Obviously, in R2-D2, those unpredictable query trajectories are still 'predictable' if we set a weakened threshold, although the distance error is probably large.

To compare with R2-D2 on the metric of prediction rate, we list error distribution for each step in Fig. 7. We set a threshold $\hat{\theta}$ and consider the query trajectories with distance error higher than $\hat{\theta}$ as unpredictable. The thresholds $\hat{\theta}$ are set as 200 and 100 for Dataset Shanghai and Chengdu, respectively. The thresholds are approximately twice of average distance error reported by RA-LSTM and no more than twice of that in R2-D2. Then we compare the prediction rate with R2-D2 in each step as showed in Fig. 8. Prediction rate of RA-LSTM at each step is higher than R2-D2 and the superiority is more obvious in the further step, which validates the effectiveness of RA-LSTM on multi-step trajectory prediction.

Results of Ablation Test. In Fig. 9, we present ablation results for RA-LSTM. Specifically, 'IntraFea-LSTM' refers to LSTM model with only intra-road feature embedding and 'InterFea-LSTM' represents LSTM model with only inter-road feature embeddings. From the Fig. 9(a) and (b), RA-LSTM with road-aware features that contains both intra and inter features achieves the best or comparable performance for all steps on both datasets. Either IntraFea-LSTM or InterFea-LSTM performs better than vanilla LSTM, which demonstrates both intra and inter features of roads assist in recognizing latent patterns. InterFea-LSTM gives better results than IntraFea-LSTM on both datasets. We see that inter-road feature learned by exploiting connections between roads is more helpful than intra-road feature.

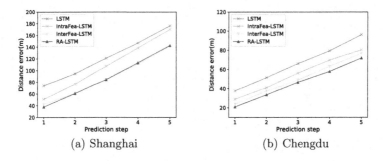

(a) Shanghai   (b) Chengdu

Fig. 9. Results of ablation test

6   Related Work

There are many researches about vehicle trajectory prediction, which can be divided into three types: trajectory pattern mining, prediction based on moving function and Markov model and using road networks.

Pattern-Matching Prediction. Pattern-based prediction methods are dedicated to find frequent trajectories and discover their movement rules. In [7,8], Morzy et al. proposed an improved Apriori algorithm and improved PrefixSpan algorithm. A hybrid prediction model which combines motion function and trajectory patterns was utilized in [3] using object's recent movement to predict. For next location prediction without involving the path to the destination, the method proposed in [6] built and trained a decision tree named T-pattern tree to find best matching result in prediction. In addition, besides using only geographical patterns, Ying et al. [14] proposed an approach also using both geographical and semantic patterns of trajectory. Pattern-based methods consider the association with moving objects which loses the generality of the prediction. If you query for moving objects that do not fit any pattern, their trajectories are unpredictable.

HMM-Based Prediction. This part mainly introduces motion function and HMM. RMF [13], a most accuracy motion function which constructs a state matrix by query trajectory and uses Singular Value Decomposition(SVD) to calculate the predicted point. The drawback of moving function is that it has no communication with environment and it can not learn information from history trajectories. Methods based on HMM are suitable for estimating future locations in discrete space. In [15], Zhou et al. proposed a "semi-lazy" approach to make prediction in dynamic environment which is used as baseline in our experiments. This method finds grid-based reference trajectories, and the longest reference path whose probability of retrieval is greater than a preset threshold is retrieved as the predicted path by HMM. Authors in [11] improved the HMM for with self-adaptive parameter selection. These methods focus on prediction algorithm design regardless of external environment information. Our approach adds road-aware features for vehicle trajectory prediction which has been proven more effective than not adding.

Prediction with Road Network. Before our works, there are several related works using road network for prediction. A maximum likelihood model and a greedy algorithm were built for long-term travel path prediction in [4]. But the method only considers the last location of query trajectories which may lose history information. Qiao et al. [10] proposed a three-in-one Trajectory-Prediction model in road-constrained transportation networks called TraPlan. Predictive Tree has also been proposed to maintain the reachability of road segments using additional information such as road length [1]. Most of them use large-scale road network dasebase to match trajectories with high similarity for prediction, while our model adopts network embeddings to encapsulate road-aware features for accurate trajectory predictions.

7   Conclusion

In this paper, we study the problem of multi-step vehicle trajectory prediction and propose a road-aware neural network solution. Unlike previous approaches relying on historical trajectories, we introduce a novel way of extracting road-aware features for vehicle trajectories, which consist of intra-road feature and inter-road feature exploited from road networks. And we leverage LSTM units to capture temporal dependencies on previous trajectory path to predict future trajectory. The experiments on two real-world datasets demonstrated that (1) our proposed model RA-LSTM outperforms other competitive methods; (2) the road-aware feature extracted from road networks can effectively improve the multi-step prediction performance.

Acknowledgment. This research is supported in part by 973 Program (No. 2014CB340303), NSFC (No. 61772341, 61472254, 61170238, 61602297 and 61472241), and Singapore NRF (CREATE E2S2). This work is also supported by the Program for Changjiang Young Scholars in University of China, the Program for China Top Young Talents, and the Program for Shanghai Top Young Talents.

716   J. Cui et al.

References

1. Hendawi, A.M., Bao, J., Mokbel, M.F., Ali, M.: Predictive tree: an efficient index for predictive queries on road networks. In: 2015 IEEE 31st International Conference on Data Engineering (ICDE), pp. 1215–1226. IEEE (2015)
2. Hochreiter, S., Schmidhuber, J.: Long short-term memory. Neural Comput. **9**(8), 1735–1780 (1997)
3. Jeung, H., Liu, Q., Shen, H.T., Zhou, X.: A hybrid prediction model for moving objects. In: IEEE 24th International Conference on Data Engineering, ICDE 2008, pp. 70–79. IEEE (2008)
4. Jeung, H., Yiu, M.L., Zhou, X., Jensen, C.S.: Path prediction and predictive range querying in road network databases. VLDB J. **19**(4), 585–602 (2010)
5. Mikolov, T., Chen, K., Corrado, G., Dean, J.: Efficient estimation of word representations in vector space. arXiv preprint arXiv:1301.3781 (2013)
6. Monreale, A., Pinelli, F., Trasarti, R., Giannotti, F.: WhereNext: a location predictor on trajectory pattern mining. In: Proceedings of the 15th ACM SIGKDD International Conference on Knowledge Discovery and Data Mining, pp. 637–646. ACM (2009)
7. Morzy, M.: Prediction of moving object location based on frequent trajectories. In: Levi, A., Savaş, E., Yenigün, H., Balcısoy, S., Saygın, Y. (eds.) ISCIS 2006. LNCS, vol. 4263, pp. 583–592. Springer, Heidelberg (2006). https://doi.org/10.1007/11902140_62
8. Morzy, M.: Mining frequent trajectories of moving objects for location prediction. In: Perner, P. (ed.) MLDM 2007. LNCS (LNAI), vol. 4571, pp. 667–680. Springer, Heidelberg (2007). https://doi.org/10.1007/978-3-540-73499-4_50
9. Perozzi, B., Al-Rfou, R., Skiena, S.: DeepWalk: online learning of social representations. In: Proceedings of the 20th ACM SIGKDD International Conference on Knowledge Discovery and Data Mining, pp. 701–710. ACM (2014)
10. Qiao, S., Han, N., Zhu, W., Gutierrez, L.A.: TraPlan: an effective three-in-one trajectory-prediction model in transportation networks. IEEE Trans. Intell. Transp. Syst. **16**(3), 1188–1198 (2015)
11. Qiao, S., Shen, D., Wang, X., Han, N., Zhu, W.: A self-adaptive parameter selection trajectory prediction approach via hidden Markov models. IEEE Trans. Intell. Transp. Syst. **16**(1), 284–296 (2015)
12. Raymond, R., Morimura, T., Osogami, T., Hirosue, N.: Map matching with hidden Markov model on sampled road network. In: 2012 21st International Conference on Pattern Recognition (ICPR), pp. 2242–2245. IEEE (2012)
13. Tao, Y., Faloutsos, C., Papadias, D., Liu, B.: Prediction and indexing of moving objects with unknown motion patterns. In: Proceedings of the 2004 ACM SIGMOD International Conference on Management of Data, pp. 611–622. ACM (2004)
14. Ying, J.J.C., Lee, W.C., Weng, T.C., Tseng, V.S.: Semantic trajectory mining for location prediction. In: Proceedings of the 19th ACM SIGSPATIAL International Conference on Advances in Geographic Information Systems, pp. 34–43. ACM (2011)
15. Zhou, J., Tung, A.K., Wu, W., Ng, W.S.: A semi-lazy approach to probabilistic path prediction in dynamic environments. In: Proceedings of the 19th ACM SIGKDD International Conference on Knowledge Discovery and Data Mining, pp. 748–756. ACM (2013)
16. Zhou, X., Shen, Y., Zhu, Y., Huang, L.: Predicting multi-step citywide passenger demands using attention-based neural networks. In: Proceedings of the Eleventh ACM International Conference on Web Search and Data Mining, pp. 736–744. ACM (2018)

Secure Data Aggregation with Integrity Verification in Wireless Sensor Networks

Ying Liu[1,2], Hui Peng[3], Yuncheng Wu[1,2], Juru Zeng[1,2], Hong Chen[1,2(✉)],
Ke Wang[4], Weiling Lai[1,2], and Cuiping Li[1,2]

[1] Key Laboratory of Data Engineering and Knowledge Engineering
of Ministry of Education, Beijing, China
[2] School of Information, Renmin University of China, Beijing, China
chong@ruc.edu.cn
[3] The Fifth Electronic Research Institute of MIIT, Guangzhou, China
[4] School of Computing Science, Simon Fraser University, Burnaby, BC, Canada

Abstract. In recent years, wireless sensor networks (WSNs) have become a useful tool for environmental monitoring and information collection due to their strong sensory ability. Whereas WSNs utilize wireless communication and is usually deployed in an outdoors environment, which make them vulnerable to be attacked and then lead to the privacy disclosure of the monitored environment. SUM, as one common query among the queries of WSNs, is important to acquire a high-level understanding of the monitored environment and establish the basis for other advanced queries. In this paper, we present a secure hash-based privacy preservation mechanism called HP2M, which not only preserves the privacy of the monitored environment during SUM aggregation query, but also could achieve exact SUM aggregation. Furthermore, an integrity verification mechanism is proposed to verify the integrity of SUM aggregation result, which could alarm the system once data packets transmitted through the networks are modified. One main characteristic of HP2M and the proposed integrity verification mechanism is that they are lightweight with a small bandwidth consumption. Finally, some numerical experiments are performed to demonstrate the efficiency of our proposed approach.

1 Introduction

In recent decades, wireless sensor networks (WSNs) have been widely used in the field of environmental monitoring and information collection, such as smart-home [1], battlefield surveillance [2], wild animal tracking [3,4] and so on, because sensor devices are powerful to perceive environmental information (eg. temperature, humidity). What's more, WSNs are easily extensible and have flexible self-organization. Most operation modes of existing large-scale WSNs follow the architecture as shown in Fig. 1 to acquire the information about the monitored environment. For example, the summation of water consumed in distinct zones, the average value of toxic gas density in the battlefield the count of sensor nodes

© Springer International Publishing AG, part of Springer Nature 2018
J. Pei et al. (Eds.): DASFAA 2018, LNCS 10827, pp. 717–733, 2018.
https://doi.org/10.1007/978-3-319-91452-7_46

which have perceived wild animal tracking. However, with the outdoors environment and wireless communication, WSNs are vulnerable to be attacked by adversaries, leading to the privacy disclosure of the monitored environment [5]. For instance, the data packets containing water consumption can be eavesdropped by adversary to infer whether a family member is home or not. Poachers compromise infrared sensor nodes recording the trail of wild animals to detect animals' location, and then hunt illegally. So it is important and draws more and more attentions how to operate WSNs while resisting illegal attacks and preserving the privacy of monitored environment [6,7].

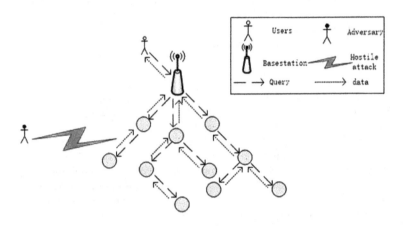

Fig. 1. Operation mode of WSNs

Aggregation queries to the WSNs include exemplary ones including MAX/MIN, MEDIAN and summary ones including COUNT, SUM, AVERAGE [8]. With these queries, users can acquire a comprehensive understanding of the monitored environment. All the summary queries could be processed as SUM query. For example, COUNT query could be seen as SUM where the sensor data is set as 1 or 0. AVERAGE query can be drawn from SUM by $AVERAGE = \frac{SUM}{N}$, where N is the COUNT of sensor nodes in the network. As a result, in this paper we focus on solving secure SUM aggregation query in WSNs. That is to collect SUM aggregation result when resisting illegal attacks and keeping sensor data, partial and final aggregation results secret from any other third party. It is necessary to point out that the accuracy of SUM aggregation result and energy consumption are two important factors to evaluate the secure SUM aggregation schemes. What's more, integrity verification is usually tied firmly to the secure SUM aggregation. Integrity verification in secure SUM aggregation aims to verify whether there is tamper, deletion, or injection behaviors happens to the data packets during aggregation. If these behaviors happens, it will violate the accuracy of aggregation result.

Roughly speaking, current solutions for secure SUM aggregation are divided into hop-by-hop encryption mechanism, end-to-end encryption mechanism, and unencrypted mechanism [9]. In hop-by-hop encryption mechanism, data are decrypted, aggregated and re-encrypted in intermediate nodes [10–12], which could resist eavesdropping on the wireless link effectively. However, intermediate nodes become the foremost targets, the compromising of which results in the leak of partial aggregation result. The end-to-end encryption mechanism [13,14] could address the weakness in hop-by-hop encryption by decrypting data only at the basestation with aggregation performed on encrypted data directly. But they are both computationally intensive which is less practical for sensor nodes with limited resource. Some other solutions not based on encryption mechanism include data camouflage [15,16], secure router finding [17], etc. Data camouflage mechanism [15,16] hides sensitive data among a set of camouflage values, bringing great redundant communication cost and an inexact result. With a long observation and enough background knowledge about the monitored environment, the effect of the data camouflage mechanism [15] is decreased to useless. The secure router finding mechanism and periodical updating in [17] consume large quantities of energy in the network.

When it comes to integrity verification in secure SUM aggregation, methods [11,13,16] use redundant information and comparison in the basestation to perform integrity verification, which introduce different levels of additional energy cost depending on the selection of parameters. Methods [18,19] concentrate on the detection of nodes' program integrity, based on the idea that nodes' data must change once the program of nodes is modified manually. Obviously, it is useless to the modification behavior on wireless links, and they all need continuous interaction to verify integrity, increasing communication load.

Motivated by the shortcomings of the related works, we present a secure hash-based privacy preservation mechanism called HP2M and an integrity verification method for SUM aggregation result. The HP2M and the integrity verification worked synergistically could achieve the best performance of SUM aggregation. First, HP2M has higher security. HP2M can resist eavesdropping and compromising attacks effectively. Second, the SUM aggregation result is exact because the hash value introduced for perturbation can be removed. The integrity verification method is based on HP2M, redundancy and polynomial algebraic properties. It is competent to detect injection/tamper/deletion behaviors to data packets transmitted through the networks, verifying the integrity of SUM aggregation result. Finally, the energy cost of our method is the lowest when compared with other schemes which could achieve secure aggregation and integrity verification at the same time.

The rest of the paper is organized as follows. Section 2 surveys the related work, Sect. 3 contains problem statement, Sect. 4 proposes our secure SUM aggregation and integrity verification methods. In Sect. 5, we evaluate the efficiency of the proposed schemes compared with other schemes. Section 6 concludes this paper.

2   Related Work

The open nature of wireless communication network makes it easy for adversaries to launch attacks to the WSNs, leading to private data of the monitored environment leaked. Many schemes have been proposed for secure SUM aggregation in WSNs [9]. Some methods are based on encryption mechanism, which can be divided into hop-by-hop encryption and end-to-end encryption. Some other methods are based on unencryption mechanisms, like data camouflage, secure router finding. There are also many related studies about integrity verification.

2.1   Secure Aggregation Based on Encryption Mechanism

Data is decrypted, aggregated and re-encrypted in intermediate nodes in hop-by-hop encryption mechanism [10–12], which can resist eavesdropping effectively. CPDA [10] applies hop-by-hop encryption and data perturbation mechanism to protect data privacy. It conceals true sensor readings by adding random seeds and private random numbers and cluster head nodes calculate the desired aggregate value based on the algebraic properties of matrix operation. iPDA [11] and ESMART [12] slice raw readings of sensor nodes into pieces and encrypt each part of the pieces. iPDA [11] sends encrypted data slices to different aggregators within its vicinity. ESMART [12] utilizes secret times of data slicing and dynamic numbers of data pieces to strength the privacy preservation. Nevertheless, all the hop-by-hop encrypted mechanism in secure SUM data aggregation are under the threat of aggregator compromised attack. The compromising of aggregator leads to the danger of partial aggregation results leakage. In addition, the above three methods introduce additional communication cost required for privacy preservation. The accuracy of ESMART [12] is determined by the degree to which the data is aggregated in basestation.

The end-to-end encryption mechanism [13] decrypts data only at basestation and performs aggregation on encrypted data directly, addressing the weakness in hop-by-hop encryption. SIES [13] achieves end-to-end encryption through homomorphic functions. But the encryption and homomorphic functions increase the request for node computation ability. What is worse is that the communication cost also increases. So it's not suitable for large-scale wireless sensor networks with resource-limited sensor nodes.

2.2   Secure Aggregation Based on Unencryption Mechanism

Method [15,16] present data camouflage mechanism for secure sum aggregation. KIPDA [15] obfuscates private data by hiding them among a set of camouflage values, enabling k-indistinguishability for data aggregation. But it increases the communication cost. iPPHA [16] introduces a generalization histogram method for secure data aggregation where sensor nodes map their readings to a histogram and aggregation is performed on histograms. But the bandwidth consumed by the scheme and the computation complexity experienced by the aggregators will increase as the increase of the intervals of the histogram and the number of sensor nodes. Meanwhile, data generalization results in an approximate aggregation result.

The method [17] presents a trust value computation mechanism to find secure routers in case of private data leaked. The trust values of nodes are assessed from historical communication behavior and neighbor nodes suggestions, and the influence of the suspect ones will be decreased in the network. The work of route-finding should be updated periodically, consuming the energy of networks.

2.3 Related Work on Integrity Verification

Methods [11,13,16] utilize redundant information and comparison in the basestation to perform integrity verification, introducing additional energy cost. iPDA [11] constructs two disjoint aggregation trees in network. The sensor data is sliced in two different forms and aggregated in the two trees independently after encryption. The comparison of two aggregation results in basestation reflects the integrity of aggregation result. SIES [13] incorporates secret share into sensor readings, and the basestation compares the complete secret shares extracted from aggregation result and a vested value. If they are equal, there is no modifying behavior during data transmission. iPPHA [16] constructs another aggregation tree beside sensor readings aggregation tree to aggregate redundant information which the basestation employ to verify the integrity of aggregation result. Those methods introduce different levels of additional energy cost depending on the selection of parameters, for example, the number of sliced pieces in iPDA, the length of redundancy information in iPPHA.

Methods [18,19] concentrate on detecting the integrity of nodes' program, based on the idea that nodes data must change once the program of nodes is modified manually. In paper [18], the clustered heads are equipped with trusted platform module and serve as verifiers to verify the integrity of the program of a node based on hardware. In paper [19], the protocol can be concluded as distributed monitoring of neighboring nodes aliveness based on software. Methods [18,19] all need continuous interaction to verify integrity, which means increasingly cost of communication.

3   Problem Statement

We describe the data SUM aggregation model, the attack model, and system goals to be studied in this section.

3.1   Data SUM Aggregation Model

The data SUM aggregation model in WSNs is shown in Fig. 1. Users pose queries to the WSNs to acquire some statistic information about the monitored environment, for example, the summation of water consumed in distinct zones, the average value of toxic gas density in battlefield, the count of sensor nodes which have perceived wild animal tracking. The basestation with adequate energy works as the interface of WSNs by receiving those queries and returning query results. Meanwhile, it publishes these queries to the network and becomes the destination of aggregated sensor data. We assume the basestation is absolutely secure and will never be compromised by adversaries. N sensor nodes with limited energy are mainly responsible for sensing data of certain environment indicators (e.g. water consumption, gas density, monitoring wildlife-tracking state) periodically in every time epoch (for instance, the 0 second, the 2th second, the 4th second,...). According to some specific query, they process and transmit sensor data to parent nodes layer by layer, achieving data aggregation.

When users pose SUM aggregation query for time epoch t, which means to acquire the summation of all the sensor readings generated at time t, the basestation publishes the query to the network level by level. The sensor node s_i $(i = 1, 2 \ldots N)$ submits its sensor readings $x_i(t)$ generated at time epoch t. We assume that $x_i(t)$ is a positive integer with 32 bits (other data types can be encoded as positive integers via simple translation and scaling operations [20]). The sensor readings are aggregated following the tree routing structure like TAG. TAG [21] is a classic in-network SUM aggregation scheme without any privacy preservation and integrity verification.

3.2   Attack Model

The adversary attempts to acquire privacy data like sensor readings, partial or final aggregation results through eavesdropping, compromising and other means, and then learns the conditions of the monitored environment. What's worse, the adversary would like to modify the data packets, destroying aggregation results. For better expression, We classify the mainstream attacks and normalize the attack behaviors as followed. We assume adversaries are equipped with laptops, the computational and storage abilities of which are civilian-grade. These devices can capture and analyze data packets, even modify or falsify these data packets through technical software. They also can invade the operation system of compromised sensor nodes and acquire its secure aggregation scheme. What's more, those devices own enough storage space to store data packets.

- **Eavesdropping attack.** The adversary eavesdrops on the wireless communication links and captures data packets. **Simple eavesdropping** is that adversaries figure out the private data merely depending on captured data packets. Besides, adversaries can launch **deducing-eavesdropping** by utilizing known background knowledge about the monitored environment to deduce privacy data from captured data packets. The background knowledge can be learned by adversaries through field survey. Typical deducing-eavesdropping are: (1) **temporal-deducing attack**: the sensor readings in a node are relatively stable at two or more time epoch. It can be used by the adversary to infer the real data after an enough long-term recording data packets. (2) **correlation-deducing attack**: the reading data of two or more sensor nodes are known to be similar. Adversaries can infer sensitive readings by analyzing the data packets of these sensors together.

- **Compromising attack.** Single node compromised leads to the secure aggregation scheme of this node leaked, which adversaries utilize to acquire larger-scale privacy disclosure. Adversaries launch **backward attack** in time dimension to acquire real data generated before node compromised by utilizing obtained secure scheme and stored data packets. They can also launch **transverse attack** to acquire the secure aggregation scheme of other uncompromised nodes. Multiple node compromised can be used for adversaries to launch **node collusion**. The compromised sensor nodes and its parent node collude with each other to acquire the secure aggregation scheme of other child nodes which is not compromised yet.
- **Tampering attack.** The adversary can capture data packets and modify these data packets or just delete them. They can also falsify data packets and inject them into the network, destroying the aggregation result.

3.3 System Goals

Based on the data aggregation model and attack model, we design a secure aggregation mechanism to prevent privacy data from disclosure and an integrity verification mechanism to detect tampering attack in time, guaranteeing the integrity of aggregation result with four criteria:

- **Privacy.** The privacy of data aggregation in WSNs means to collect aggregation result while keeping sensor data, partial and final aggregation results secret from any other third party. The privacy-preserving mechanism we designed should resist these attacks mentioned above.
- **Accuracy.** The SUM aggregation query requests the result at time t, $sum(t)$, should be equal to the summation of all the sensor readings at time t namely, $sum(t) = \sum_1^N x_i(t)$.
- **Integrity verification.** The integrity verification mechanism should detect illegal behaviors like injecting/replacing/deleting to data packets, verifying the integrity of aggregation result.
- **Efficiency.** The efficiency of privacy-preservation aggregation scheme and integrity verification scheme in WSNs is that achieving secure and exact data aggregation with least energy consumption when compared with other remaining schemes.

4 Our Approach

In this section, We first introduce our approach, HP2M, which can achieve secure SUM aggregation with exact aggregation result, and discuss its privacy performance when it deals with attack behaviors. Then integrity verification scheme and its utility are followed.

Algorithm 1. HP2M

Pre-processing:
   preload each sensor node s_i with the code of $hash_{iter}(\cdot)$, $H(\cdot)$ and initial value $k_{i,0}$. For time epoch t, every sensor node s_i $(i = 1, 2 \ldots N)$ generates data $x_i(t)$.

1. Perturbation-add Phase(in sensor node)
   (a) key parameter generation: $k_{i,t} = hash_{iter}(k_{i,t-1})$
   (b) perturbing value generation: $\Delta x = H(k_{i,t}, t)$
   (c) adding perturbing data: $x_i'(t) = x_i(t) + \Delta x$
2. Merging Phase(in intermediate node).
   (a) merge on perturbed value: $Res_{partial}(t) = \sum x_i'(t)$
3. Perturbation-remove Phase(in basestation).
   (a) calcalte perturbing data $H(k_{i,t}, t)$ for every node.
   (b) remove perturbing data from aggregation result:

$$SUM(t) = Res_{total}(t) - \sum_{i=1}^{N} H(k_{i,t}, t)$$

4.1   Hash-Based Privacy Preserving Mechanism

The procedure of HP2M is described in Algorithm 1. We preload each sensor node s_i with the code of the function generating k which is presented as $hash_{iter}(\cdot)$, function generating perturbing value which is presented as $H(\cdot)$ and initial value $k_{i,0}$ before settle it outdoors. The functions of $hash_{iter}(\cdot)$ and $H(\cdot)$ have the properties that (1) they perform like random functions with the output independent with each other. (2) it is infeasible to infer the input from the output. The function of $hash_{iter}(\cdot)$ generates the important parameter k utilizing $k_{i,0}$ in the form of iteration. The process of iteration is: $k_{i,t} = hash_{iter}(k_{i,t-1})$.

   $k_{i,0}$ preloaded in sensor node is different among sensor nodes. With the mapping of $hash_{iter}(\cdot)$, the $k_{i,t}$ varies among different sensor nodes and different time epoch. Note that during the processing of iteration, once the $k_{i,t}$ is generated, the $k_{i,t-1}$ is deleted.

   The parameter $k_{i,t}$ is one of the input of $H(\cdot)$ and the other is time epoch t, making the perturbing value various in sensor nodes and time. Then, the perturbing data is added with sensor readings, generating perturbed data $x'(t)$. The basestation keeps the codes of $hash_{iter}(\cdot)$ and $H(\cdot)$ and $k_{i,0}$ for each sensor node s_i. So it can calculate the perturbing value for every node correctly and remove the summation of perturbing values from aggregation result exactly.

   Now we discuss the security performance of HP2M against eavesdropping and compromising attacks. We assume the size of Δx is M bit.

Theorem 1. *The probability of breaking HP2M and acquire privacy sensor data is 2^{-M}.*

Proof: Δx is a random number with the range of 0–$(2^M - 1)$ because $H(\cdot)$ is a random function. The value of Δx can be vastly different and independent for different sensor nodes, camouflaging the real data $x_i(t)$ with the probability of

2^{-M} that adversary can guess Δx from $x_i'(t)$ correctly. So it has the ability to resist simple-eavesdropping attack once M is big enough. Even on the condition that $x_i(t)$ have temporal or correlation relationships, the random-number perturbing mechanism can break the relationships. The difficulty for adversaries to launch deducing-eavesdropping to crack the privacy is the same as that in simple-eavesdropping attack.

As to compromising attack, backward attack is invalid because $k_{i,t-1}$ is deleted timely once $k_{i,t}$ is generated and $hash_{iter}(\cdot)$ is infeasible to reverse. Even the adversary obtains secure scheme of compromised node and stores data packets generated before compromising, it cannot acquire the real privacy data, owing to the fact that the privacy preservation scheme in one node can be seen as different and independent in each time epoch. What's more, the parameter $k_{i,t}$ is a random function, which makes the privacy preservation mechanism of different nodes different and independent from each other. So the transverse attack and node collusion attack is useless.

Then, we should find out a proper function to implement $hash_{iter}(\cdot)$ and $H(\cdot)$. The SHA-256 function is a preferable selection, based on the facts that SHA-256 is a common cryptographic hash function with the variable length input which is less than $2^{64} - 1$ bits and fixed length output of 256 bits. Without knowing the input, the output of SHA-256 is distinguished from that of a truly random function with negligible probability [13], which satisfy the first property of $hash_{iter}(\cdot)$ and $H(\cdot)$. In addition, it is infeasible to generate a message from its hash value except by trying all possible messages [22], which consumes tremendous energy and time that the adversary cannot bear, satisfying the second property of $hash_{iter}(\cdot)$ and $H(\cdot)$. So SHA-256 is qualified as the implementation of $H(\cdot)$ and $hash_{iter}(\cdot)$. But if we use the output of SHA-256 as Δx directly, the result of the addition operation, namely $x_i'(t)$, will be 256 bits. The communication load for transmitting data during aggregation is increased greatly. Based on the consideration of saving communication cost, we choose some sequential bits of the output of SHA-256 as Δx, which is M we mentioned above. Obviously, if M is too small, with a smaller value range of Δx it will violate the perturbation effect. A superior selection of M is 32. That is, Δx and $x_i'(t)$ are both 32 bits (we ignore the overflow of result in addition operation), which can make $x_i'(t)$ generate perturbation effect as much as possible on the condition of introducing no additional communication cost because the original data, $x_i(t)$, is 32 bits, too. The maximum of the probability that attacks can succeed under HP2M is 2^{-32} which is little enough.

4.2 Integrity Verification Mechanism

The integrity verification mechanism in aggregation scheme is important to resist tampering attack. In this paper, we propose an integrity verification mechanism to verify whether the data is modified during transmission. The steps of integrity verification mechanism we proposed are summarized in Algorithm 2.

Algorithm 2. Integrity verification mechanism

Pre-processing:

preload each sensor node s_i with the codes of $hash_{iter}(\cdot)$, $H(\cdot)$, d initial values: $k_{i,0}^1\ k_{i,0}^2\dots k_{i,0}^d$, d zoom factors $Z_1\ Z_2\dots Z_d$. For time epoch t, every sensor node $s_i\ (i=1,2\dots N)$ generates data $x_i(t)$.

1. Performing d circulations to achieve multiple perturbation to sensor data $x_i(t)$ (in sensor node).

$j \leftarrow 1$; While $j <= d$ do

$$k_{i,t}^j = hash_{iter}(k_{i,t-1}^j);\quad x_i^j(t) = Z_j * x_i(t) + H(k_{i,t}^j,t);\quad j++;$$

End While

2. Merging data in multiways (in intermediate node).

$j \leftarrow 1$; While $j <= d$ do

$$Res_{partial}^j(t) = \sum x_i^j(t);\quad j++;$$

End While

3. Integrity verification Phase (in basestation).

$j \leftarrow 1$; While $j <= d$ do

$$Verify\_z_j = \left\{ Res_{total}^j(t) - \sum_{i=1}^{n} H(k_{i,t}^j,t)\right\} * \frac{1}{Z_j};\quad j++;$$

End While

If all the $Verify\_z_j$ are equal, we can say data is not modified during transmission.

The integrity verification mechanism is based on the algorithm of HP2M. The perturbing data has the same source and property as that in HP2M, but there are d initial values $(k_{i,0}^1, k_{i,0}^2 \dots k_{i,0}^d)$, d zoom factors $(Z_1, Z_2 \dots Z_d)$ to achieve the multiway perturbation. The number of d is set artificially with the minimum value of 2. We will explain later that the selection of d reflects the balance between the effect of the integrity verification mechanism and energy consumption. $Z_j\ (=1,2\dots d)$ in every sensor nodes are the same. When a sensor node transmits sensor data to the base station, it transmits d copies of perturbed data.

The basestation keeps values of initial values: $k_{i,0}^1\ k_{i,0}^2 \dots k_{i,0}^d$ for every sensor node, d zoom factors $Z_1\ Z_2 \dots Z_d$ and codes of $hash_{iter}(\cdot)$, $H(\cdot)$.

The basestation can calculate the perturbing values because it keeps values of initial values: $k_{i,0}^1\ k_{i,0}^2 \dots k_{i,0}^d$ for every sensor node, d zoom factors $Z_1\ Z_2 \dots Z_d$ and codes of $hash_{iter}(\cdot)$, $H(\cdot)$. It can verify integrity through comparing $verify\_z_j$ with each other. If the result of the comparison is unequal, we can tell the data is modified during transmission. Otherwise, the integrity of aggregation result is guaranteed, and $verify\_z_j$ can be accepted as the final aggregation result. What is more, even if the node was compromised, and the values of hash functions, zoom factors were modified, the serious condition would be detected through this verification mechanism.

Now we analyze the probability of breaking this mechanism, which means the adversaries can tamper, inject or delete data transmitted through the network illegally without being detected under this mechanism.

Theorem 2. *The probability of breaking the integrity verification mechanism we proposed is $\frac{1}{(2^M-1)^{d-1}}$, where M is the size of Δx and d is the times of perturbations.*

Proof: In this paper, we verify the integrity of aggregation result by comparing the values of $verify\_Z_j$ $(j = 1, 2 \ldots d)$ to detect illegal behavior from adversaries. If the result of the comparison is unequal, we can tell the data is modified during transmission, and the tampering attack is resisted successfully. Otherwise, we say there is no modifying behavior to data packets. If adversaries attempt to tamper, inject or delete data transmitted through the network illegally without being detected, the modified values of the set of SUMs should keep the corresponding values for verifying equal with each other.

Specifically, let R be the legitimate set of final SUMs transmitted to the basestation.

$$R = (Res_{total}^1(t), Res_{total}^2(t), \ldots Res_{total}^d(t))$$

Let V be the set of the corresponding values for verifying.

$$V = (verify\_Z_1, verify\_Z_2 \ldots verify\_Z_d)$$

The legitimate set, R, makes the elements in V equal with each other.

Note that there is a one to one relationship between $verify\_Z_j$ and $Res_{total}^j(t)$ in the same formula (the same j), which means for any selection of $Res_{total}^j(t)$, there exists only one value for $verify\_Z_j$ to satisfy this formula, and the reverse also holds.

If we assume the $Res_{total}^1(t)$ was modified into $(Res_{total}^1(t))'$, the corresponding value for verifying, $verify\_Z_1$, would be changed to $(verify\_Z_1)'$. For concealing the change of $verify\_Z_1$ and the inequality among $verify\_Z_j$, the adversaries should modify the rest of R to ensure the values in the rest of V are equal to $(verify\_Z_1)'$. And there exists only one set of modified values, $(Res_{total}^2(t))' \ldots (Res_{total}^d(t))'$, to make the corresponding values $(verify\_Z_2)' \ldots (verify\_Z_d)'$ to be equal to $(verify\_Z_1)'$ because of the one to one relationship between $verify\_Z_j$ and $Res_{total}^j(t)$.

For adversaries, the another $d-1$ data packets they captured are M-bit data without any background knowledge. Every data packet has $2^M - 1$ values as alternates for adversaries to select. So the number of possible modified values for $(Res_{total}^2(t))' \ldots (Res_{total}^d(t))'$ is $(2^M - 1)^{d-1}$.

So the probability for adversaries to modified $(Res_{total}^2(t))' \ldots (Res_{total}^d(t))'$ correctly is $\frac{1}{(2^M-1)^{d-1}}$. As we discuss before, the proper selection of M is 32 in consideration of saving communication cost. The minimal value of d is 2.

We can conclude that the integrity verification we proposed can resist tampering attack and verify the integrity of aggregation result with the breaking probability of $\frac{1}{2^{32}-1}$ at most. And there is a balance between the strength of integrity verification and energy consumption. With a larger selection of d, the probability that adversaries can modify $(Res_{total}^2(t))' \ldots (Res_{total}^d(t))'$ correctly is smaller.

5   Evaluation

In this section, we evaluate the efficiency of our methods. First, we utilize a table to demonstrate the performance of our methods and many other mechanisms when they deal with various attacks. Then, we compare the energy consumption of our method with other mechanisms by experiments simulation.

5.1   Resistibility Comparison

Different mechanisms have different performances when they deal with attacks, which can be summarized as a comprehensive Resistibility. We present the comparison of comprehensive resistibility in Table 1. As the table demonstrates, compared with other privacy-preservation schemes, our methods have a more comprehensive resistibility, which can resist all these attacks perfectly. The reason and concrete mechanism are illustrated Sect. 4.

Table 1. Comparison of Different Privacy-Preserving Schemes Performance

Schemes	Simple eaves-dropping	Temporal-deducing attack	Correlation-deducing attack	Backward attack	Transverse attack	Node collusion	Tamper attack
CPDA [10]	√	√	√	×	√	×	×
iPDA [11]	√	√	√	×	×	×	√
ESMART [12]	√	×	×	×	√	×	×
SIES [13]	√	√	√	×	√	√	√
KIPDA [15]	√	×	×	×	√	√	×
iPPHA [16]	√	√	√	×	√	√	√
Our method	√	√	√	√	√	√	√

(1) √ means the scheme can resist the attack
(2) × means the scheme cannot resist the attack

5.2   Energy Consumption

The energy consumption of privacy preservation and integrity verification consists in computation energy consumption and communication energy consumption. We compare the energy consumption of our method with iPDA [11], SIES [13], iPPHA [16] through experiments. Because the latter three methods achieve privacy preservation and integrity verification in SUM aggregation together. Moreover, TAG [21] scheme is added into the comparison as the baseline. These protocols are implemented on OMNet++4.1 [23], and our sensor data set comes from Grand-St-Bernard Deployment [24], a small SensorScope network which deployed at the Great St. Bernard Pass between Switzerland and Italy to monitor meteorology.

Table 2 concludes the relative parameters and typical values used in experiments and analysis. The symmetric encryption algorithm in iPDA which is not indicated in paper [11] is implemented with typical DES algorithm. The hash function used in iPPHA for generating redundancy information is SHA-256. All the final data demonstrated in figures are the averages derived from 50 repeated trials.

Table 2. Summary of parameters and values

Parameter	Meaning	Typical value settings
N	Number of nodes	1024
F	Aggregator fanout	4
S	Number of sliced pieces in iPDA	2
K	Number of intervals of histogram in iPPHA	10
L	Length of redundancy information in iPPHA	32 bits
d	Times of perturbation in our method	2

Computation Energy Consumption: There exists a relationship between computation energy consumption($Energy_{compu}$), and operation time($Time_{op}$): $Energy_{compu} = Time_{op} * Freq_{clock} * Energy_{perClock}$ where $Freq_{clock}$ is the machine's dominant frequency where the procedure is carried on, and $Energy_{perClock}$ is the energy consumed per tick. $Energy_{compu}$ rises in direct proportion to $Time_{op}$ based on the fact that $Freq_{clock}$ and $Energy_{perClock}$ are static in the same type of machine, for example, sensor nodes TelosB, or the computer where the OMNet++ simulator software is carried on. Admittedly, the computer is much powerful than that of a sensor node and, thus, it solely facilitates the comparison of the schemes.

So we can utilize the operation time($Time_{op}$) on OMNet++ to demonstrate the comparison of computation energy consumption.

As it is presented in Fig. 2, the operation time of our method retains a comparable performance to iPDA, SIES, iPPHA. The aggregators refer in particular to those sensor nodes with child nodes. Specifically, the operation time in sensor node of our method is higher than that in TAG because of the generation of $k_{i,t}$ and perturbing value Δx but is much lower than iPDA with the minimum value of S and SIES. iPPHA has a little better performance than our method because the times of using SHA-256 for generating redundancy in iPPHA is fewer than our method. However, the advantage will diminish owing to the fact that it is affected by the increase of K and L. The operation time in aggregators of our method is basically flat with TAG and lower than other schemes. The operation time in basestation of our method is a little high, because the intensive computing like removing perturbing data, verifying the integrity of aggregation result

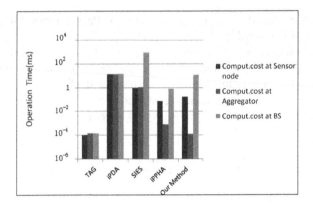

Fig. 2. Comparison of operation time of different methods

is performed in basestation. The energy-intensive computation in basestation will not affect the practical application of our method since the basestation is resource-rich with great computation ability.

Next, we focus on the effect of fanout numbers on operation time in aggregators. More fanouts mean aggregators need to increase the time of addition operation, which results in more operation time and more computation energy consumed. With the varying of F, ranging from 2 to 6, the operation time in aggregators is changing, the result of which is shown in Fig. 3(a). The effect of fanout numbers on operation time in aggregators of iPDA [11] is neglected because the aggregation structure of iPDA [11] is improvised during aggregation, we cannot set its fanout manually.

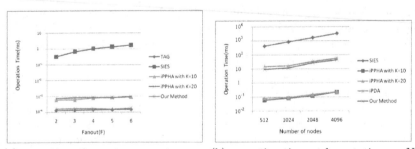

(a) operation time at aggregators vs. (b) operation time at basestation vs. N fanout

Fig. 3. Effect of different factors on operation time

The operation time on aggregators of our method with different fanout is similar to TAG. The procedure performed in aggregators is a simple algebraic addition on integers in our method and TAG, the operation time of which remains relatively stable, not influenced by the selection of F greatly. In comparison, the operation time of SIES and iPPHA in aggregators grow apparently with the enlargement of F, which is caused by the increased times of modular addition operation in aggregators. Then, we study the affection of N on operation time in basestation. As Fig. 3(b) demonstrated, the operation time in basestation of all the three methods increases along with the enlargement of network scale. SIES is 2 orders of magnitude larger than our method in terms of operation time. The reason that operation time in basestation of iPPHA is lower than our method is that the times of using SHA-256 for generating corresponding redundancy in iPPHA is fewer than our method.

Communication Energy Consumption: We apply the algorithms of TAG, iPDA with diverse S, SIES, iPPHA and our method on simulated WSNs with different scales. The algorithm of TAG is to be the baseline. The result is shown in Fig. 4.

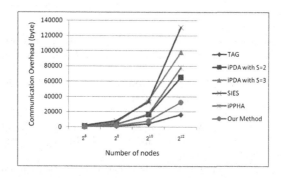

Fig. 4. Communication cost vs. N

Obviously, our method can achieve privacy preservation and integrity verification with the lowest communication overhead. With the enlargement of network scale, communication overhead in our method will not increase as quickly as other schemes.

Theoretically, the sensor data are positive integers with 32 bits (4 bytes). We assume the number of nodes deployed in the network is N. The total communication overhead of TAG is 4N bytes. The size of a data packet in our method transmitted in the network is no longer than that in TAG schemes, and there is no additional data transmission process during aggregation compared to TAG. We just transmit data for d copies, where $d = 2$, so the total communication overhead is 8N bytes.

We can conclude that our method is lightweight with a small bandwidth consumption. What is more, the energy consumption will not be greatly influenced by the factors of fanout value and network scale.

6 Conclusion

In this paper, we present a mechanism which can address both privacy preservation of exact data aggregation and integrity verification in wireless sensor networks synergistically. Through analyzing the comprehensive resistibility and performing relative experiments about energy consumption, our method has a comprehensive ability to resist attacks and verify the integrity of aggregation result, and the mechanism is lightweight with a small bandwidth consumption with the comparison of other privacy and integrity preservation schemes. In the future work, we will focus on addressing the problem where some nodes in the network miss its contribution which will lead to the incorrect sum result. A primary idea is to apply prime number to mark every node.

Acknowledgment. This work is supported by National Science Foundation of China (No. 61532021, 61772537, 61772536, 61702522), and National Key R & D program of China (No. 2016YFB1000702).

References

1. Chen, C., Lin, M., Lin, W.: Designing and implementing a lightweight WSN MAC protocol for smart home networking applications. J. Circ. Syst. Comput. **26**(03), 1–20 (2017)
2. Yang, Z.P., Huang, Q.Z., Wang, M.: Study on application of wireless sensor networks in military highway transportation battlefield environment. Equip. Environ. Eng. **11**(03), 78–82 (2014)
3. Ramadan, R., Elfouly, F., Mahmoud, M., Dessouky, M.: Efficient data reporting in a single and multi-object tracking using WSNs. In: IEEE Symposium Series on Computational Intelligence (2015)
4. Badescu, A.M., Cotofana, L.: A wireless sensor network to monitor and protect tigers in the wild. Ecol. Ind. **57**, 447–451 (2015)
5. Lupu, T.-G., Rudas, I., Demiralp, M., Mastorakis, N.: Main types of attacks in wireless sensor networks. In: WSEAS Proceedings International Conference on Recent Advances in Computer Engineering, no. 9 (2009)
6. Ali, A.W., et al.: Energy efficieny in routing protocol and data collection approaches for WSN: a survey. In: 2015 International Conference on Computing, Communication and Automation (ICCCA), pp. 540–545 (2015)
7. Sang, Y., Shen, H., Inoguchi, Y., Tan, Y., Xiong, N.: Secure data aggregation in wireless sensor networks: a survey. In: 2006 Seventh International Conference on Parallel and Distributed Computing, Applications and Technologies, PDCAT 2006, pp. 315–320 (2006)
8. Morell, A., Correa, A., Barcelo, M., Vicario, J.L.: Data aggregation and principal component analysis in WSNs. IEEE Trans. Wirel. Commun. **15**(6), 3908–3919 (2016)
9. YongJian, F., Hong, C., Ying, Z.X.: Data privacy preservation in wireless sensor networks. Chin. J. Comput. **35**(6), 1131–1146 (2012)
10. He, W., Liu, X., Nguyen, H., Nahrstedt, K., Abdelzaher, T.: PDA: privacy-preserving data aggregation in wireless sensor networks. In: 26th IEEE International Conference on Computer Communications, INFOCOM 2007, pp. 2045–2053. IEEE (2007)

11. He, W., Nguyen, H., Liu, X., Nahrstedt, K.: iPDA: an integrity-protecting private data aggregation scheme for wireless sensor networks. In: Military Communications Conference, MILCOM 2008, pp. 1–7 (2008)
12. Li, C., Liu, Y.: ESMART: energy-efficient slice-mix-aggregate for wireless sensor network. Int. J. Distrib. Sens. Netw. **2013**(2), 1–9 (2013)
13. Papadopoulos, S., Kiayias, A., Papadias, D.: Secure and efficient in-network processing of exact SUM queries. In: IEEE International Conference on Data Engineering, pp. 517–528 (2011)
14. Desai, H.K., Jinwala, D.C.: Secure aggregation of exact sum queries with integrity protection for wireless sensor networks. Int. J. Comput. Appl. **73**(15), 9–17 (2013)
15. Groat, M.M., He, W., Forrest, S.: KIPDA: k-indistinguishable privacy-preserving data aggregation in wireless sensor networks. In: 2011 Proceedings of IEEE INFOCOM, pp. 2024–2032 (2011)
16. Chen, W., Yu, L., Gao, D.: A privacy preserving histogram aggregation algorithm with integrity verification support. Chin. J. Electron. **42**(11), 2268–2272 (2014)
17. Pavithra, M.N., Chinnaswamy, C.N., Sreenivas, T.H.: Privacy preservation scheme for WSNs using signature and trust value computation. Int. J. Eng. Tech. Res. **5**(6), 50–53 (2016)
18. Agrawal, S., Das, M.L., Mathuria, A., Srivastava, S.: Program integrity verification for detecting node capture attack in wireless sensor network. In: Jajodia, S., Mazumdar, C. (eds.) ICISS 2015. LNCS, vol. 9478, pp. 419–440. Springer, Cham (2015). https://doi.org/10.1007/978-3-319-26961-0_25
19. Al-Riyami, A., Zhang, N., Keane, J.: An adaptive early node compromise detection scheme for hierarchical WSNs. IEEE Access **4**, 4183–4206 (2016)
20. Nath, S., Yu, H., Chan, H.: Secure outsourced aggregation via one-way chains. In: ACM SIGMOD International Conference on Management of Data, SIGMOD 2009, Providence, Rhode Island, USA, 29 June–July 2009, pp. 31–44 (2009)
21. Madden, S., Franklin, M.J., Hellerstein, J.M., Hong, W.: TAG: a tiny aggregation service for ad-hoc sensor networks. In: Symposium on Operating Systems Design and Implementation Copyright Restrictions Prevent ACM From Being Able To Make the Pdfs for This Conference Available for Downloading, pp. 131–146 (2002)
22. Cryptographic hash function. https://en.wikipedia.org/wiki/Cryptographic_hash_function
23. Omnet++4.1. https://omnetpp.org/
24. "Grand-st-bernarddeployment," http://lcav.epfl.ch/cms/lang/en/pid/86035

A Parallel Spatial Co-location Pattern Mining Approach Based on Ordered Clique Growth

Peizhong Yang, Lizhen Wang$^{(\boxtimes)}$, and Xiaoxuan Wang

School of Information Science and Engineering, Yunnan University,
Kunming 650091, China
pzyang0924@163.com, lzhwang@ynu.edu.cn

Abstract. Co-location patterns or subsets of spatial features, whose instances are frequently located together, are particularly valuable for discovering spatial dependencies. Although lots of spatial co-location pattern mining approaches have been proposed, the computational cost is still expensive. In this paper, we propose an iterative mining framework based on MapReduce to mine co-location patterns efficiently from massive spatial data. Our approach searches for co-location patterns in parallel through expanding ordered cliques and there is no candidate set generated. A large number of experimental results on synthetic and real-world datasets show that the proposed method is efficient and scalable for massive spatial data, and is faster than other parallel methods.

Keywords: Spatial data mining · Co-location patterns · Ordered clique
Parallel algorithm · MapReduce

1 Introduction

The spatial co-location pattern mining is one of the spatial knowledge discovery technologies, and it is intended to discover a subset of spatial features whose instances are frequently located together. The spatial co-location pattern mining has many applications [1], and various co-location pattern mining methods have been proposed. But most methods are serial processing and inefficient when handling massive spatial data. As a solution, the methods of parallel co-location pattern mining are imperative. However, little research pays attention to the parallel co-location pattern mining. In this work, we propose a parallel spatial co-location pattern mining approach based on ordered clique growth, and it needs not to generate candidate sets and check clique instances. The algorithm is implemented on Apache Spark, and extensive experiments are conducted to evaluate the efficiency. Experimental results demonstrate that our method is efficient and scalable for mining co-location patterns from massive spatial data.

The main contributions of this work can be summarized as follows: (1) The ordered clique expanding method in a level-wise manner is proposed. (2) An iterative framework for spatial co-location pattern mining based on ordered clique growth is provided. (3) We suggest a pruning strategy to cut out some ordered cliques early to speed up the mining process. (4) A parallel spatial co-location pattern mining algorithm based on MapReduce is proposed.

J. Pei et al. (Eds.): DASFAA 2018, LNCS 10827, pp. 734–742, 2018.
https://doi.org/10.1007/978-3-319-91452-7_47

The rest of this paper is organized as follows: Sect. 2 reviews related work. Section 3 presents the basic concept of spatial co-location mining and the MapReduce. In Sect. 4, a novel parallel spatial co-location pattern mining method is provided. Section 5 shows experimental evaluations and the paper will conclude on Sect. 6.

2 Related Work

A large number of methods have been developed to discover co-location patterns. Huang et al. [1] defined the participation index to measure the prevalence of co-location patterns and proposed the *Apriori-like* method join-based. Based on the participation index, a partial-join approach [3] and a join-less approach [2] proposed by Yoo and Shekhar to improve mining efficiency. However, above methods are difficult to avoid generating huge candidate patterns and storing massive clique instances. To address this problem, various optimization algorithms have been proposed [4–6, 8], but these methods are serial processing and they are inefficient when processing massive spatial data. Yoo et al. [7] proposed a parallel spatial co-location pattern mining algorithm based on MapReduce. It can handle massive spatial data, but the candidate clique instances generation and the clique checking operation are still time-consuming. In this work, a novel parallel spatial co-location pattern mining approach based on ordered clique growth is proposed. There is no candidate set generated and the clique testing operations can be avoided in our method.

3 Basic Concepts

In a spatial database, let $F = \{f_1, f_2, ..., f_m\}$ be a set of spatial features and $O = \{o_1, o_2, ..., o_n\}$ be a set of instances of F. Two instances have the spatial neighbor relationship R, if the distance between two instances is less than a threshold d. A co-location pattern $Cl = \{f_1...f_k\}$ is a subset of spatial features, whose instances frequently form clique under R. A set of instances I is a row instance of Cl, if (1) I contains all features of Cl and no proper subset of I does so, and (2) all instances of I form a clique. The set of all row instances of Cl is called table instance, denoted as $T(Cl)$. The *Participation Index* (PI) is defined [1] to evaluate the prevalence of co-location pattern. The participation index of Cl is the minimum *Participation Ratio* (PR) for all spatial features in Cl. The participation ratio of f_i in Cl can be computed as:

$$PR(Cl, f_i) = \frac{\text{number of distinct instances of } f_i \text{ in } T(Cl)}{\text{number of global instances of } f_i} \quad (1)$$

Cl is called a prevalent co-location pattern, if the participation index of Cl is not less than a given prevalence threshold *min_prev*. The participation index is anti-monotone [1], which means the *PI* of a pattern is not bigger than the *PI* of its sub-patterns.

MapReduce is a programming model which provides a highly scalable and flexible framework for data-oriented parallel computing. A MapReduce job is executed in two main phases of user defined data transformation functions, *map* and *reduce*. In the first phase, the *key-value* pairs are processed by Mapper instances and the output of the *map* function is another set of intermediate *key-value* pairs. The intermediate process of moving the intermediate *key-value* pairs from the map tasks to the assigned Reducer is called *shuffle* phase. At the completion of the *shuffle*, all *values* associated with the same *key* are fed to a Reducer and processed by the reduce function. Each Reducer generates a set of new *key-value* pairs as the output of the job.

4   A Parallel Approach

The generation of table instance set is the most time-consuming operation in co-location pattern mining. Generating row instance in parallel is an effective way to improve the efficiency of the mining process. This section presents a method to discover co-location patterns through expanding ordered clique in parallel.

4.1   Ordered Clique Growth

Definition 1 (Ordered Clique). Given an instance set $I_k = \{o_1, o_2,..., o_k\}$, $o_i \in O$, $1 \leq i \leq k$, if $f(o_i) \leq f(o_j)$ and $R(o_i, o_j)$ holds for every $1 \leq i \leq j \leq k$, I_k is called an ordered clique. The set I_k contains k instances, thus I_k is a size-k ordered clique.

$f(o_i)$ is the feature type of instance o_i. $f(o_i) \leq f(o_j)$ represents that the feature type of o_i is not greater than o_j in alphabetical order.

Definition 2 (Neighbor Set of Instance). Given an instance $o_i \in O$, the neighbor set of instance o_i is defined as:

$$NSI(o_i) = \{o_k | o_k \in O \wedge R(o_i, o_k) \wedge (f(o_i) < f(o_k))\}$$

That is, the neighbor set of o_i consists of some instances who has spatial neighbor relationship with o_i and whose feature type is bigger than o_i in alphabetical order.

Definition 3 (Neighbor Set of Clique). Given a size-k ordered clique $I_k = \{o_1, o_2,..., o_k\}$, $o_i \in O$, $1 \leq i \leq k$. The neighbor set of clique I_k is defined as:

$$NSC(I_k) = NSI(o_1) \cap \cdots \cap NSI(o_k)$$

Lemma 1. Given a size-k ordered clique I_k, if $o \in NSC(I_k)$, appending o to I_k constitutes a new clique $I_{k+1} = \{o_1, o_2,..., o_k, o\}$, and I_{k+1} is a size-$(k + 1)$ ordered clique.

Proof. If $o_i \in I_k$, $1 \leq i \leq k$, and $o \in NSC(I_k)$. Thus, $o \in NSI(o_i)$ by Definition 3, that is, the instance o has spatial neighbor relationship with all instances in I_k, and the feature type of o is bigger than all instances in I_k lexicographically according to Definition 2. Therefore, appending o to I_k to form a size-$(k + 1)$ clique I_{k+1}, and it is ordered clique by definition.

4.2 An Iterative Parallel Mining Framework

According to Lemma 1, given an ordered clique I_k, the size-$(k + 1)$ ($k \geq 1$) ordered cliques prefixed with I_k can be produced by appending an element in $NSC(I_k)$ to I_k. Obviously, the operation of expanding ordered cliques can be performed in parallel. A size-k ordered clique corresponds to a row instance of a size-k ordered co-location pattern. When all size-k ordered cliques are generated, table instances for all size-k ordered co-location patterns can be collected easily. Starting from the size-2 ordered cliques, we can search for all ordered co-location patterns level by level. Naturally, an iterative parallel co-location pattern mining framework based on ordered clique growth is suggested. The mining framework is given in Fig. 1. In addition, a pruning strategy is proposed to narrow searching space.

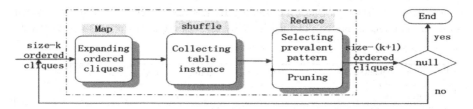

Fig. 1. The mining framework

Pruning Strategy. Given a size-k ordered clique I_k, and I_k is a row instance of the pattern Cl. If Cl is not prevalent, the size-$(k + 1)$ ordered cliques prefixed with I_k can be pruned.

Depending on the anti-monotone property of the participation index, if the pattern Cl is not prevalent, the super set of Cl must be not prevalent. A size-$(k + 1)$ ordered clique prefixed with I_k corresponds to a size-$(k + 1)$ co-location pattern Cl', and Cl' must be the super set of Cl. We do not need to search for the pattern Cl' because the pattern Cl is not prevalent, thus the size-$(k + 1)$ ordered cliques who are prefixed with I_k can be pruned.

4.3 Parallel Algorithm

In this subsection, the parallel co-location pattern mining algorithm based on ordered clique growth (PCPM_OC) is presented. The algorithm is described by MapReduce procedure presented in Fig. 2.

Firstly, we count and store the number of instances per spatial feature utilizing Job1 presented in Fig. 2(a). It is preparation for future participation index calculation.

The task of generating the neighbor set of instance is accomplished by Job2 presented in Fig. 2(b). The pair of instances who have spatial neighbor relationship is the input for Mapper. Spatial neighbor relationships can be obtained in advance by the parallel neighbor searching method proposed in [7]. Then, the pair $<o_i, o_j>$ is the output of Mapper and the feature type of o_i is smaller than o_j lexicographically. At *shuffle* phase, the neighbor of per instance will be collected in a set. The pair, $<o, NSI(o)>$, be fed to Reducer and then the pair will be stored for the operation of expanding ordered clique. In order to construct the initial input of Job3, an instance is considered as a size-1 ordered clique. An ordered clique I consists of 3 parts, the pattern corresponding to I, the instance set of I and the neighbor set of clique. The new *key-value* pair on behalf of a size-1 ordered clique is emitted as the output of Reducer.

The procedure of searching for prevalent co-location patterns is shown in Fig. 2(c). The Mapper is fed with size-k ordered cliques, and performs the operation of expanding ordered clique according to Lemma 1. For a size-k ordered clique I_k, the size-$(k + 1)$ ordered cliques prefixed with I_k are emitted. At the completion of the shuffle, the ordered cliques associated with the same pattern will be gathered. In Reducer, computing the participation index of the pattern, and then storing prevalent pattern and pruning the searching space for the next iteration. Depending on the pruning strategy, just emitting the size-$(k + 1)$ ordered cliques corresponding to prevalent patterns and whose neighbor set is not null. Proceeding from size-1 ordered cliques, all prevalent co-location patterns can be obtained level-by-level through executing Job3 iteratively.

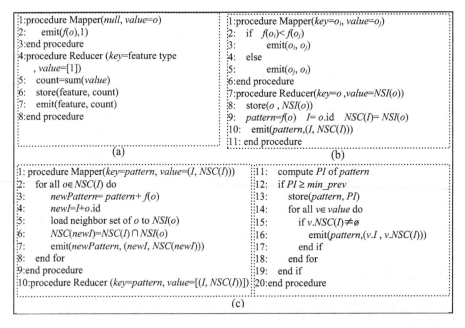

Fig. 2. The parallel spatial co-location pattern mining approach based on ordered clique growth, (a) Job1: procedure for counting the number of instances per spatial feature, (b) Job2: procedure for generating the neighbor set of instance, (c) Job3: procedure of searching for prevalent co-location patterns

Our algorithm has a few advantages: (1) High degree of parallelism. The expanding operation for each size-k ordered clique can be carried out independently, and it can resolve the problem that generating table instances is time-consuming in co-location pattern mining. (2) Iterative execution. Because of the iterative mining framework, we can re-use the previously processed information. (3) Effective pruning. A pruning strategy is proposed to narrow the searching space. (4) No candidate sets. There is not any candidate pattern and candidate clique instance to be generated. (5) No clique checking. Our method ensures that each expanding operation generates ordered cliques.

5 Experimental Evaluation

We implement our algorithm in Spark library functions. The performance evaluation is conducted on the cluster that deployed Hadoop and Spark. The cluster consists of one master node and six worker nodes with the following characteristics: (1) CPU per node: Intel Core i7-6700, @3.40 GHz; (2) Memory per node: 8 GB. Experiments are conducted on real and synthetic data sets. Two real data sets are used. The first is a plant dataset of the "Three Parallel Rivers of Yunnan Protected Areas" that contains 25 spatial features and 13,348 spatial instances. The second is the POI of Beijing that contains 63 spatial features and 303,895 spatial instances. The synthetic data sets are generated based on the spatial data generator described in [1].

5.1 Compared with Serial Mining Methods

In the first experiment, we evaluate the efficiency of the PCPM_OC compared with two state-of-the-art serial co-location mining algorithms, the JoinBase [2] and the JoinLess [3]. We use a small data set, the plant dataset, because the serial algorithms easily overflow on large data sets. In Fig. 3(a), we set the participation index threshold *min_prev* to 0.1. The running time of the JoinBase increases dramatically with the increasing of the distance threshold d and the efficiency of the JoinLess is slightly better than the PCPM_OC when d is smaller, owing to the PCPM_OC requires extra cost. When $d > 3,000$ m, the advantage of the PCPM_OC can be reflected. In Fig. 3(b), we set d to 3,000 m. The running time of three algorithms decreases with the increasing of *min_prev*, because fewer patterns need to be searched when *min_prev* is larger. Similarly, the JoinBase is the least efficient. When *min_prev* is smaller, more patterns are searched, the PCPM_OC is better than two serial algorithms, because the large number of calculation is executed in parallel.

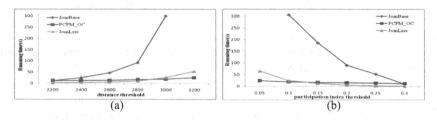

(a)   (b)

Fig. 3. Running time over the plant dataset: (a) by the distance thresholds, (b) by the participation index thresholds

5.2   Compared with PCPM_SN

In the second experiment, we compare the efficiency of the PCPM_OC with the parallel algorithm (PCPM_SN) proposed in [7]. We implement the PCPM_SN algorithm in Spark and use the data set of the POI of Beijing. First, the effect of the distance threshold is evaluated. In Fig. 4(a), we set min_prev to 0.4. The running time of two methods is increased with the increasing of d, because a larger value of d means more instances could form cliques. The performance of the PCPM_OC is better than the PCPM_SN, as there is no candidate clique generation and clique checking operation in the PCPM_OC. Next, we assess the effect of the participation index thresholds. In Fig. 4(b), we set d to 300 m. As min_prev becomes higher, more patterns dissatisfy the condition of prevalent co-location patterns. Naturally, the performance of two algorithms improves. When min_prev is smaller, the running time of the PCPM_OC is less than the PCPM_SN obviously. Because more patterns are searched means that more candidate clique instances need to be generated and more clique checking operations are required to be undertaken in the PCPM_SN, but these are not required in the PCPM_OC.

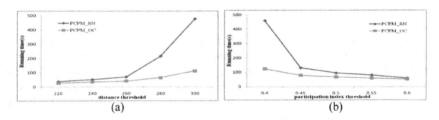

(a)   (b)

Fig. 4. Running time over the POI of Beijing: (a) by the distance thresholds, (b) by the participation index thresholds.

5.3   Scalability Evaluation

In order to assess the scalability, some experiments are conducted on synthetic data sets. We generate spatial instances and randomly distribute them into a 10,000 × 10,000 space, and $min\_prev = 0.1$, $d = 20$.

In Fig. 5(a), we assess the effect of the number of spatial features. The average of instances per spatial feature is 10,000. The running time is increased with the increasing of the number of spatial features, because more potential patterns are required to be searched when increasing the number of spatial features. The performance of the PCPM_OC is preferable to the PCPM_SN, especially when the number of spatial features is larger.

In Fig. 5(b), we assess the impact of the number of spatial instances and set the number of spatial features to 100. As the number of spatial instances becomes larger, the running time of two algorithms is raised. The distribution of spatial instances is more densely and more spatial instances could form cliques when the number of total spatial instances is larger. The performance of the PCPM_OC outperforms the PCPM_SN, especially when the number of spatial instances is larger.

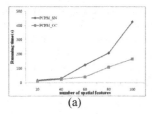 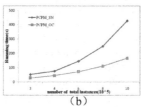 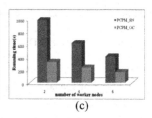

(a)   (b)   (c)

Fig. 5. Running time over synthetic data sets: (a) by the number of spatial features, (b) by the number of spatial instances, (c) by the number of worker nodes

In Fig. 5(c), the performance of speedup is evaluated. The number of spatial features is 100 and the number of spatial instances is 1,000,000. Increasing worker nodes, the running time of two algorithms is reduced obviously, because the mining tasks are allocated to more nodes to execute. In the case of the same number of worker nodes, the performance of the PCPM_OC is better than the PCPM_SN.

6 Conclusions

In this work, we propose a parallel approach for mining co-location patterns from massive spatial data. Each worker node conducts the co-location pattern mining process through expanding ordered cliques level by level. In our method, there is no candidate set generating and clique checking. The experiments show that our approach has a significant improvement in efficiency and has better scalability. Our method is effective for handling massive spatial data, but collecting table instance sets are still time-consuming. Moreover, the issue of load balance is to be explored also. The above questions will be focused on our future researches.

Acknowledgement. This work is supported by the National Natural Science Foundation of China (61472346, 61662086, 61762090), the Natural Science Foundation of Yunnan Province (2015FB114, 2016FA026), and the Project of Innovative Research Team of Yunnan Province.

References

1. Huang, Y., Shekhar, S., Xiong, H.: Discovering colocation patterns from spatial data sets: a general approach. IEEE Trans. Knowl. Data Eng. **16**(12), 1472–1485 (2004)
2. Yoo, J.S., Shekhar, S.: A joinless approach for mining spatial colocation patterns. IEEE Trans. Knowl. Data Eng. **18**(10), 1323–1337 (2006)
3. Yoo, J.S., Shekhar, S.: A partial join approach for mining co-location patterns. In: The 12th Annual ACM International Workshop on Geographic Information Systems, pp. 241–249 (2004)
4. Wang, L., Bao, X., Zhou, L.: Redundancy reduction for prevalent co-location patterns. IEEE Trans. Knowl. Data Eng. **30**(1), 142–155 (2018)
5. Xiao, X., Xie, X., Luo, Q., Ma, W.: Density based co-location pattern discovery. In: 16th ACM SIGSPATIAL, pp. 1–10 (2008)

6. Lin, Z., Lim, S.J.: Fast spatial co-location mining without cliqueness checking. In: International Conference on Information and Knowledge Management, pp. 1461–1462 (2008)
7. Yoo, J.S., Boulware, D., Kimmey, D.: A parallel spatial co-location mining algorithm based on MapReduce. In: IEEE International Congress on Big Data, pp. 25–31 (2014)
8. Wang, L., Bao, X., Chen, H., Cao, L.: Effective lossless condensed representation and discovery of spatial co-location patterns. Inf. Sci. **436–437**(2018), 197–213 (2018)

RDF and Knowledge Graphs

Multi-query Optimization in Federated RDF Systems

Peng Peng[1(✉)], Lei Zou[2,3], M. Tamer Özsu[4], and Dongyan Zhao[2]

[1] Hunan University, Changsha, China
hnu16pp@hnu.edu.cn
[2] Peking University, Beijing, China
{zoulei,zhaodongyan}@pku.edu.cn
[3] Beijing Institute of Big Data Research, Beijing, China
[4] University of Waterloo, Waterloo, Canada
tamer.ozsu@uwaterloo.ca

Abstract. This paper revisits the classical problem of multiple query optimization in federated RDF systems. We propose a heuristic query rewriting-based approach to share the common computation during evaluation of multiple queries while considering the cost of both query evaluation and data shipment. Furthermore, we propose an efficient method to use the interconnection topology between RDF sources to filter out irrelevant sources and share the common computation of intermediate results joining. The experiments over both real and synthetic RDF datasets show that our techniques are efficient.

1 Introduction

Now, many data providers publish, share and interlink their datasets using open standards such as RDF and SPARQL [3]. In RDF, data is represented as triples of the form ⟨subject, property, object⟩. By deeming subjects and objects as the vertices, and properties as the directed edges from the corresponding subjects to objects, an RDF dataset can be viewed as a directed labeled graph G. On the other hand, SPARQL is a query language to retrieve and manipulate data stored in RDF format. When many data providers publish their RDF data, they often store the actual triple files at their own *autonomous* sites some of which are *SPARQL endpoints* that can execute SPARQL queries. An autonomous site with a SPARQL endpoint is called an *RDF source* in this paper.

To integrate and provide transparent access over many RDF sources, federated RDF systems have been proposed [1,5,14,15,18], in which, a control site is introduced to provide a common interface for users to issue SPARQL queries. However, existing federated RDF systems only consider query evaluation for a single query and miss the opportunity for multiple query optimization. Real SPARQL query workloads reveal that many SPARQL queries are often posed simultaneously. For example, in a real SPARQL query workload over DBPedia[1],

---

[1] http://aksw.org/Projects/DBPSB.html.

© Springer International Publishing AG, part of Springer Nature 2018
J. Pei et al. (Eds.): DASFAA 2018, LNCS 10827, pp. 745–765, 2018.
https://doi.org/10.1007/978-3-319-91452-7_48

there are in average more than six SPARQL queries per second. Furthermore, 97% queries in the workload are isomorphic to one of the 163 frequent patterns. There is room for sharing computation when executing these queries. Therefore, it is desirable to design a multiple SPARQL query optimization strategy.

Consider a batch of queries (e.g., Q_1, Q_2 and Q_3 in Fig. 1) that are posed simultaneously over the federated RDF system in Fig. 1. Specifically, Q_1 is to find out all news about Canada; Q_2 retrieves all people who graduated from Canadian universities; and Q_3 is to retrieve all semantic web-related workshops held in Canada. Obviously, there are some common substructures over these three queries, which suggests the possibility of some sharing computation. This motivates us to revisit the classical problem of multiple query optimization in the context of federated RDF systems.

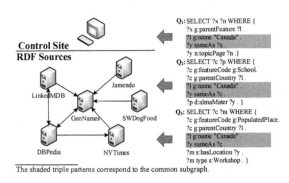

Fig. 1. Multiple federated SPARQL queries

Although multiple query optimization has been well studied in distributed relational databases [12], some techniques commonly referred to as data movement and data/query shipping [10] are not easily applicable in federated RDF systems. In federated RDF systems, we cannot require one source to send intermediate results directly to another source for join processing [7]. Meanwhile, there is only one proposal about multiple SPARQL query optimization in literature [11], but *only in the centralized environment*, where all RDF datasets are collected in one physical database. It is a method based on query rewriting. However, there is no data shipment in the centralized environment, and rewriting multiple queries as [11] in federated RDF systems may generate many remote requests for data shipment as shown in the experiments.

To the best of our knowledge, this is the first study of multiple SPARQL query optimization over federated RDF systems, with the objective to reduce the query response time and the number of remote requests. In this paper, we propose a cost model-driven greedy approach based on query rewriting for multiple query optimization in federated RDF systems. We use both "OPTIONAL" and "FILTER" clauses of SPARQL to rewrite multiple queries with commonalities while considering the cost for both query evaluation and data shipment.

In addition, we also study relevant source selection and partial match joins in federated RDF systems, which do not arise in the centralized counterpart. We propose a topology structure-based source selection and study how to share common computation in joining partial matches in federated RDF systems.

2 Background

In this section, we review the terminology that we use throughout this paper. First, in the context of federated RDF systems, an RDF graph G is a combination of many RDF graphs located at different source sites.

Definition 1 *(Federated RDF System).* *A federated RDF system is defined as $W = (S, g, d)$, where (1) S is a set of source sites that can be obtained by looking up URIs in an implementation of W; (2) $g : S \rightarrow 2^{E(G)}$ is a mapping that associates each source with a subgraph of RDF graph G; and (3) $d : V(G) \rightarrow S$ is a partial, surjective mapping where looking up URI of vertex u results in the retrieval of the source represented by $d(u) \in S$. $d(u)$ is called the* host source of *u, and is unique for a given URL of vertex u.*

Obviously, RDF graph G is formed by collecting all subgraphs at different sources, i.e., $\bigcup_{s \in S} g(s) = G$. Note that, although there may be multiple RDF sources that describe an entity identified by vertex u, u can be only dereferenced by the host source $d(u)$.

On the other hand, SPARQL is a structured query language over RDF where primary building block is the basic graph pattern (BGP).

Definition 2 *(Basic Graph Pattern).* *A basic graph pattern is denoted as $Q = \{V(Q), E(Q), L\}$, where $V(Q) \subseteq V(G) \cup V_{Var}$ is a set of vertices, where $V(G)$ denotes vertices in RDF graph G and V_{Var} is a set of variables; $E(Q) \subseteq V(Q) \times V(Q)$ is a set of edges in Q; each edge e in $E(Q)$ either has a property in L or the property is a variable.*

In federated RDF systems, a BGP match may span over different sources as follows.

Definition 3 *(BGP Match over Federated RDF System).* *Consider an RDF graph G, a federated RDF system $W = (S, g, d)$ and a BGP Q that has n vertices $\{v_1, \ldots, v_n\}$. For $S' \subseteq S$, a subgraph M of $\bigcup_{s \in S'} g(s)$ with n vertices $\{u_1, \ldots, u_n\}$ is said to be a* match *of Q if and only if there exists a function μ from $\{v_1, \ldots, v_n\}$ to $\{u_1, \ldots, u_n\}$, where the following conditions hold: (1) if v_i is not a variable, $\mu(v_i)$ and v_i have the same URI or literal value ($1 \leq i \leq n$); (2) if v_i is a variable, there is no constraint over $\mu(v_i)$ except that $\mu(v_i) \in \{u_1, \ldots, u_n\}$; (3) if there exists an edge $\overrightarrow{v_i v_j}$ in Q, there also exists an edge $\overrightarrow{\mu(v_i)\mu(v_j)}$ in $\bigcup_{s \in S'} g(s)$; furthermore, $\overrightarrow{\mu(v_i)\mu(v_j)}$ has the same property as $\overrightarrow{v_i v_j}$ unless the label of $\overrightarrow{v_i v_j}$ is a variable.*

The set of matches for Q over S' is denoted as $[\![Q]\!]_{S'}$.

Our notion of a SPARQL query can be defined recursively as follows by combining BGPs using the following standard SPARQL algebra operations.

Definition 4 (SPARQL Query). *Any BGP is a SPARQL query. If Q_1 and Q_2 are SPARQL queries, then expressions $(Q_1 \text{ AND } Q_2)$, $(Q_1 \text{ UNION } Q_2)$, $(Q_1 \text{ OPT } Q_2)$ and $(Q_1 \text{ FILTER } F)$ are also SPARQL queries.*

The results of a query Q over sources S' are defined as follows.

Definition 5 (SPARQL Result over Federated RDF System). *Given a federated RDF system $W = (S, g, d)$, the result of a SPARQL query Q over a set of sources $S' \subseteq S$, denoted as $[\![Q]\!]$, is defined recursively as follows:*

1. *If Q is a BGP, $[\![Q]\!]_{S'}$ is defined in Definition 3.*
2. *If $Q = Q_1 \text{ AND } Q_2$, then $[\![Q]\!]_{S'} = [\![Q_1]\!]_{S'} \bowtie [\![Q_2]\!]_{S'}$*
3. *If $Q = Q_1 \text{ UNION } Q_2$, then $[\![Q]\!]_{S'} = [\![Q_1]\!]_{S'} \cup [\![Q_2]\!]_{S'}$*
4. *If $Q = Q_1 \text{ OPT } Q_2$, then $[\![Q]\!]_{S'} = ([\![Q_1]\!]_{S'} \bowtie [\![Q_2]\!]_{S'}) \cup ([\![Q_1]\!]_{S'} \setminus [\![Q_2]\!]_{S'})$*
5. *If $Q = Q_1 \text{ FILTER } F$, then $[\![Q]\!]_{S'} = \Theta_F([\![Q_1]\!]_{S'})$*

If $S' = S$, i.e., the whole federated RDF system W, we call $[\![Q]\!]_S$ the results of Q over federated RDF system W.

The problem to be studied in this paper is defined as follows:
Given a set of SPARQL queries $\mathcal{Q}$ and a federated RDF system $W = (S, g, d)$, our problem is to find the results of each query in $\mathcal{Q}$ over W.

3   Framework

A federated RDF system consists of a control site as well as some RDF sources. The control site is amenable to receive the SPARQL queries, decompose them into several subqueries on their relevant sources and do some global optimizations, while the RDF sources actually store the RDF graphs. Generally speaking, our approach consists of five steps: *query decomposition and source selection, query rewriting, local evaluation, postprocessing* and *partial match join* (see Fig. 2). We briefly review the five steps before we discuss them in details in upcoming sections. Note that only *local evaluation* is conducted over the remote sources and the other four steps work at the control site.

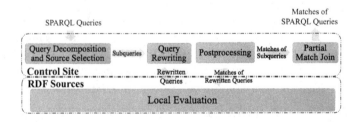

Fig. 2. Scheme for federated SPARQL query processing

First, in the step of query decomposition and source selection, we decompose the input queries into a set of subqueries expressed over relevant sources. Given a batch of SPARQL queries $\{Q_1, \ldots, Q_n\}$ posed simultaneously, we decompose each query Q_i into $\{q_i^1@S(q_i^1), \ldots, q_i^{m_i}@S(q_i^{m_i})\}$ where $S(q_i^j)$ is the set of relevant sources for subquery q_i^j (see Sect. 4). Then, in the step of query rewriting, we use FILTER and OPTIONAL operators to rewrite these subqueries as fewer queries to reduce the number of remote requests and improve the query performance. After query rewriting, we obtain a set of rewritten queries $\hat{Q}_s$ that will be sent to source s (see Sect. 5). In the step of local evaluation, we send the set of rewritten queries to their relevant sources and evaluate them there, and the results will be returned back to the control site (see Sect. 6). In the step of post-processing, the control site checks each local evaluation result against each query in Q_s (see Sect. 6). Last, in the step of partial match join, for each subquery q_i^j, we collect and join all the matches at each relevant source in $S(q_i^j)$. Considering the context of multiple SPARQL queries over a federated RDF system, we propose an optimized solution to avoid duplicate computation in join processing (see Sect. 7).

4   Query Decomposition and Source Selection

Most existing solutions [1,5,6,13–15,18] decompose the input query based on the triple patterns. Given a query, each triple pattern maps to a subquery, and they select the relevant sources based on the values of its subject, property and object. If a source is exclusively selected for some connected triple patterns, they can be combined together to form a larger subquery. Note that, if a group of triple patterns shares exactly the same set of multiple RDF sources, they cannot be combined, because combining them together may miss some matches crossing different RDF sources.

For example, the existing solutions decompose the query Q_1 in Fig. 1 into three subqueries q_1^1, q_1^2 and q_1^3, as shown in Fig. 3. q_1^1 has a relevant source "GeoNames"; q_1^2 has five sources except for "Jamendo" and q_1^3 has a relevant source "NYTimes".

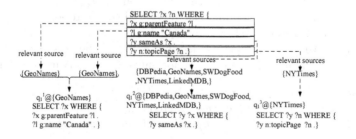

Fig. 3. Existing query decomposition and source selection result for Q_1

However, the existing solutions may overestimate the set of relevant sources. To filter out more irrelevant sources, we employ the interconnection structures among sources. A crawler proposed in [16] can be used for the federated RDF system to figure out crossing edges between different sources, and then we define the *source topology graph* (maintained in the control site) as follows:

Definition 6 (Source Topology Graph). *Given a federated RDF system* $W = (S, g, d)$, *the corresponding* source topology graph $T = (V(T), E(T))$ *is an undirected graph, where (1) each vertex in* $V(T)$ *corresponds to a source* $s_i \in S$; *(2) there is an edge between vertices* s_i *and* s_j *in* T, *if and only if there is at least one edge* $\overrightarrow{u_i u_j} \in g(s_i)$ *(or* $\overrightarrow{u_j u_i} \in g(s_i)$ *or* $\overrightarrow{u_i u_j} \in g(s_j)$ *or* $\overrightarrow{u_j u_i} \in g(s_j)$*), where* $d(u_i) = s_i$ *and* $d(u_j) = s_j$.

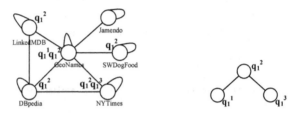

(a) Annotated Source Topology Graph for Q_1 (b) Join Graph $JG(Q_1)$ of Q_1

Fig. 4. Example join graph and annotated source topology graph for Q_1

We propose a source topology graph (STG)-based pruning rule to filter out irrelevant sources. We firstly annotate each source in STG with its relevant decomposed subqueries. For example, "NYtimes" is a relevant source to q_1^3, we annotate "NYTimes" in STG with q_1^3. Then, each query leads to an annotated source topology graph. Figure 4(a) shows the annotated STG T^* for query Q_1. Meanwhile, for a BGP Q, we build a *join graph* (denoted as $JG(Q)$), where each vertex indicates a subquery of Q and an edge between two vertices in the join graph if and only if the corresponding subqueries are connected in the original SPARQL query. Figure 4(b) shows the join graph $JG(Q_1)$.

Given a join graph $JG(Q)$ and the annotated source topology graph T^*, we find all homomorphism matches of $JG(Q)$ over T^* (SPARQL semantic is based on homomorphism). If a subquery q does not map to a source s in any match, s is not a relevant source of subquery q. We formalize this observation in Theorem 1.

Theorem 1. *Given a join graph* $JG(Q)$ *and its corresponding annotated source topology graph* T^*, *for a subquery* q, *if there exists a homomorphism match* m *of* Q^* *over* T^* *containing* q, *then* $m(q)$ *is the relevant source of* q.

Proof. Given a source s pruned by the theorem for subquery q, there do not exist any homomorphism matches of q over T^* that contains s. Then, there exists another subquery q' of which relevant sources do not contain triples that can join with results of q in s through $JG(Q)$. Hence, s cannot contribute any final results and can be pruned.   □

Based on the above pruning rule, for example, sources "LinkedMDB" or "SWDogFood" do not match q_1^2, so both of them can be pruned from the relevant sources of q_1^2.

5   Query Rewriting

As mentioned in Sect. 1, when multiple SPARQL queries are posed simultaneously, there is room for sharing computation when executing these queries. This section proposes a cost-driven query rewriting scheme to rewrite them (at the control site) into fewer SPARQL queries. To simplify presentation, we assume that the SPARQL queries originally issued at the control site are BGPs, since BGPs are the building block of SPARQL queries. Our solution is easily extensible to handle general SPARQL queries.

5.1   Intuition

We first discuss how to rewrite multiple queries with the common substructure into a single SPARQL query. Specifically, we utilize "OPTIONAL" and "FILTER" operators to make use of common structures among different queries for rewriting.

FILTER-Based Rewriting. Let us consider two subqueries $q_2^1@\{GeoNames\}$ and $q_3^1@\{GeoNames\}$ in Fig. 5, which are decomposed from Q_2 and Q_3. Although they distinguish from each other at the first triple pattern, the only difference is constant bounded to objects in the first triple pattern. We can rewrite the two queries using FILTER, as shown in Fig. 5. In other words, if some subqueries issued at the same source have the common structure except the constants on some vertices (subject or object positions), they can be rewritten as a single query with FILTER.

Fig. 5. Rewriting queries using OPTIONAL and FILTER operators

Formally, if a set of subqueries $\{q_1@\{s\}, q_2@\{s\}, \ldots, q_n@\{s\}\}$ has the same structure p except for some vertex labels (i.e., constants on vertices), we rewrite them as follows.

$$\hat{q}@\{s\} = p \; FILTER(\bigvee_{1 \leq i \leq n} (\bigwedge_{v \in V(q_i)} f_i(v) = v))@\{s\}$$

where f_i is a bijective isomorphism function from q_i to p.

For the FILTER operators, we can perform selection over main pattern results. Thus, the result cardinality of SPARQL with FILTER operator is upper bounded by result cardinality of its main graph pattern. In addition, unlike OPTIONAL operators, FILTER operators do not introduce extra joins and only generate a little more partial matches. Hence, the cost of data shipment will increase little.

OPTIONAL-Based Rewriting. Let us consider two subqueries over Geo-Names, $q_1^1@\{GeoNames\}$ and $\hat{q}_1@\{GeoNames\}$, as shown in Fig. 5. They share a common substructure, i.e., triple pattern "?l g:name "Canada"". Therefore, they can be rewritten to a single query, where "?l g:name "Canada"" maps to the main pattern. The subgraphs that q_1^1 and q_2^1 minus "?l g:name "Canada"" map to two OPTIONAL clauses, respectively. The query rewriting is illustrated in Fig. 5. The rewritten query can avoid one remote request for GeoNames.

Formally, given a set of subqueries $\{q_1@\{s\}, q_2@\{s\}, \ldots, q_n@\{s\}\}$ over a source s, if p is the common subgraph among $q_1, \ldots, q_n$, we rewrite these sub-queries as a query with OPTIONAL operators as follows.

$$\hat{q}@\{s\} = p \; OPT \; (q_1 - p) \; OPT \; (q_2 - p) \; \ldots \; OPT \; (q_n - p)@\{s\}$$

Because existing RDF stores implement OPTIONAL operators using left-joins, the result cardinality of a SPARQL query with OPTIONAL operator is also upper bounded by result cardinality of its main graph pattern. Thus, the cardinality of the rewritten query does not increase much. Meanwhile, using OPTIONAL operators adds extra left-joins for each optional clause which results in larger local evaluation cost. This needs a trade off between many queries being rewritten together and many optional clauses. We propose a cost model to measure the rewriting benefits in Sect. 5.2.

5.2   Cost Model

The cost of a rewriting strategy can be expressed with respect to the total time. The total time is the sum of all time (also referred to as cost) components. There are two time components for evaluating a rewriting strategy: time for local evaluation and time for data shipment. We discuss the two components respectively in the following sections.

Cost Model for BGPs. As mentioned in [11], selective triple patterns in BGP have higher priorities in evaluation. We verify the principle in two real and synthetic RDF repositories, DBpedia and WatDiv. For DBpedia, we randomly sample 10000 queries from the real SPARQL workload; for WatDiv, we generate 12500 queries from 125 templates provided in [2]. Given a triple pattern e, its selectivity is defined as $sel(e) = \frac{\|[e]\|}{|E(G)|}$, where $\|[e]\|$ denotes the number of matches of e and $|E(G)|$ denotes the number of edges in RDF graph G. The experimental results are shown in Fig. 6. As the experimental results show, the cardinality of a query is positively associated with the selectivity of the most selective triple pattern for both real and synthetic datasets.

Based on the above observation, we define the cardinality of evaluating a basic graph pattern Q as follows.

$$card(Q) = min_{e \in E(Q)}\{sel(e)\}$$

where $sel(e)$ is the selectivity of triple pattern e in Q.

In real applications, for estimating the selectivity of triple pattern, we can employ the heuristics introduced by [19]. Meanwhile, the relative coefficients, T_{CPU} and T_{MSG}, are greatly influenced by the resources of each RDF sources and the network topology of the federated RDF system, which can be estimated offline.

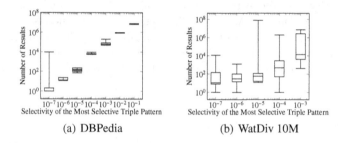

(a) DBPedia

(b) WatDiv 10M

Fig. 6. Relationship between the cardinality and the most selective triple pattern

Cost Model for General SPARQLs. Then, we extend the cost model to handle general SPARQLs. The design of our cost model is motivated by the way in which a SPARQL query is evaluated on popular RDF stores. This includes a well-justified principle that the graph patterns in OPTIONAL clauses and the expressions in FILTER operators are evaluated on the results of the main pattern (for the fact that the graph pattern in the OPTIONAL clause is a left-join and the FILTER operator is selection) [11]. Hence, given a SPARQL Q, its cardinality $card(Q)$ is defined as follows.

$$card(Q) = \begin{cases} min_{e \in E(Q)}\{sel(e)\} & \textit{if } Q \textit{ is a BGP}; \\ min\{card(Q_1), card(Q_2)\} & \textit{if } Q = Q_1 \textit{ AND } Q_2; \\ card(Q_1) + card(Q_2) & \textit{if } Q = Q_1 \textit{ UNION } Q_2; \\ card(Q_1) + \Delta_1 & \textit{if } Q = Q_1 \textit{ OPT } Q_2; \\ card(Q_1) + \Delta_2 & \textit{if } Q = Q_1 \textit{ FILTER } F; \end{cases} \quad (1)$$

where Δ_1 and Δ_2 are empirically trivial values [11].

Then, given a set of subqueries $\mathcal{Q} = \{q_1@S, q_2@S, \ldots, q_n@S\}$ over a source S using OPTIONAL and FILTER, if p is their common subgraph among $q_1, \ldots, q_n$, we rewrite them into a SPARQL query $\hat{q}$. The cost of evaluating $\hat{q}$ is defined as follows.

$$cost_{DS}(\hat{q}) = card(\hat{q}) \times T_{MSG} = (\min_{e \in p}\{sel(e)\} + \Delta_1 + \Delta_2) \times T_{MSG}$$

Because Δ_1 and Δ_2 are trivial values, we ignore the trivial variables Δ_1 and Δ_2, and have the following cost function of data shipment.

$$cost_{DS}(\hat{q}) = \min_{e \in p}\{sel(e)\} \times T_{MSG}$$

In addition, since the OPTIONAL and FILTER operators are evaluated on the result of main pattern on popular RDF stores, the time for local evaluation is also based on the cardinality of the main pattern. Furthermore, since each OPTIONAL operator adds simply a left-join with the results of the main pattern on popular RDF stores, the time for local evaluation of multiple OPTIONAL operators is approximately equal to the time of multiple execution of main pattern. We verify the above principle by using WatDiv on three popular RDF stores, Jena, Sesame and Virtuoso. We generate some queries with multiple OPTIONAL operators by using the query rewriting algorithm proposed in Sect. 5.3. The results in Fig. 7 show that the query response time of a query with multiple OPTIONALs is proportional to the number of OPTIONALs on all RDF stores.

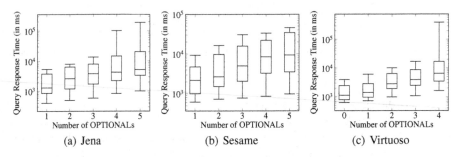

(a) Jena   (b) Sesame   (c) Virtuoso

Fig. 7. Experiment results of the relationship between the local evaluation cost and the number of OPTIONALs

Thus, given a SPARQL Q, its time for local evaluation can be estimated as follows.

$$cost_{LE}(Q) = \begin{cases} \min_{e \in E(Q)}\{sel(e)\} \times T_{CPU} & \text{if } Q \text{ is a BGP;} \\ \min\{cost_{LE}(Q_1), cost_{LE}(Q_2)\} & \text{if } Q = Q_1 \text{ AND } Q_2; \\ cost_{LE}(Q_1) + cost_{LE}(Q_2) & \text{if } Q = Q_1 \text{ UNION } Q_2; \\ cost_{LE}(Q_1) \times (|OPT_{Q_2}| + 1) + \Delta_3 & \text{if } Q = Q_1 \text{ OPT } Q_2; \\ cost_{LE}(Q_1) + \Delta_4 & \text{if } Q = Q_1 \text{ FILTER } F; \end{cases} \quad (2)$$

where $|OPT_{Q_2}|$ is the number of OPTIONALs in Q_2, and Δ_3 and Δ_4 are also empirically trivial values. Here, the definition of the local evaluation cost of a query containing OPTIONAL operators is recursive, which means that the local evaluation cost of a query is proportional to the number of OPTIONAL operators.

Thus, the local evaluation cost of evaluating a rewritten query, $\hat{q}$, is as follows.

$$cost_{LE}(\hat{q}) = (|OPT_{\hat{q}}| \times \min_{e \in p}\{sel(e)\} + \Delta_3 + \Delta_4) \times T_{CPU}$$

where $|OPT_{\hat{q}}|$ is the number of OPTIONAL operators in $\hat{q}$.

Similar to the cost function of data shipment, since Δ_3 and Δ_4 are trivial values can also be ignored, the cost function of local evaluation is as follows.

$$cost_{LE}(\hat{q}) = |OPT_{\hat{q}}| \times \min_{e \in p}\{sel(e)\} \times T_{CPU}$$

In summary, given a set of subqueries $\mathcal{Q}$ over a source S, we define the cost of a specific query rewriting as follows.

Definition 7 (Rewriting Cost). *Given a set of subqueries $\mathcal{Q}$ on a source s using OPTIONAL and FILTER, if p is their common subgraph among queries in $\mathcal{Q}$, we rewrite them into a SPARQL query $\hat{q}$. The cost of the rewriting is the cost of the rewritten query $\hat{q}$ with main basic graph pattern p as shown in the following formula:*

$$cost(\mathcal{Q}, \hat{q}) = cost(\hat{q}) = cost_{LE}(\hat{q}) + cost_{DS}(\hat{q}) = \min_{e \in p}\{sel(e)\} \times (|OPT_{\hat{q}}| \times T_{CPU} + T_{MSG})$$

5.3   Query Rewriting Algorithm

The problem of query rewriting is that given a set $\mathcal{Q}$ of subqueries $\{q_1, \ldots, q_n\}$, we compute a set $\hat{\mathcal{Q}}$ of rewritten queries $\{\hat{q}_1, \ldots, \hat{q_m}\}$ $(m \leq n)$ with the smallest cost. Note that each rewritten query $\hat{q}_i$ $(i = 1, \ldots, m)$ comes from rewriting a set of original subqueries in $\mathcal{Q}$, where these subqueries share the same main pattern p_i.

Generally speaking, we find the optimal set of common patterns P, where each subquery contains at least one patterns in P. According to Sect. 5.1, if a set of subqueries can be rewritten as a rewritten query $\hat{q}$, they must share one common main pattern p. Therefore, we have the following equation.

$$cost(\mathcal{Q}, \hat{\mathcal{Q}}) = \sum_{\hat{q} \in \hat{\mathcal{Q}}} cost(\hat{q}) = \sum_{p \in P} \min_{e \in p}\{sel(e)\} \times (|OPT_{\hat{q}}| \times T_{CPU} + T_{MSG}) \quad (3)$$

where $|OPT_p|$ is the number of OPTIONAL operators that the rewritten query contains.

Given a set of original queries $\mathcal{Q}$, Eq. 3 is a set-function with respect to set P, i.e., a set of main patterns. Unfortunately, finding the optimal rewriting is a NP-complete problem as discussed in the following theorem.

Theorem 2. *Given a set of subqueries $\mathcal{Q}$, finding an optimal rewriting $\hat{\mathcal{Q}}$ to minimize the cost function in Eq. 3 is a NP-complete problem.*

Proof. We prove that by reducing the weighted set cover problem into the problem of selecting the optimal set of patterns. We map each set in S to a pattern and each element in the universe U to a subquery. A set s in S containing an element e in U maps to the pattern corresponding to s hitting the subquery corresponding to e. The weight of a set s in S is the cost of its corresponding pattern. Hence, finding the smallest weight collection of sets from S whose union covers all elements in U is equivalent to the problem of selecting the optimal set of patterns. Since the weighted set cover problem is NP-complete [4], selecting the optimal set of patterns is also NP-complete. □

Then, we propose a greedy algorithm that iteratively selects the locally optimal triple pattern in Algorithm 1. Let Q denote all original subqueries. At each iteration, we select a triple pattern e_{max} with the largest value $\frac{|Q'|}{sel(e_{max}) \times (|OPT_{\hat{q}}| \times T_{CPU} + T_{MSG})}$, where Q' denote all subqueries containing e_{max}. Then, we extract the largest common pattern p of queries in Q'. It should be noted that the common pattern contains the triple pattern e_{max}, so we can find the largest common pattern by exploring from e_{max}, which is much cheaper. We divide Q' into several equivalence classes, where each class contains subqueries with the same structure except for some constants on subject or object positions. Subqueries in the same equivalence class can be rewritten to a query pattern with FILTER operators as discussed in Sect. 5.1, and all queries in Q' can be rewritten into SPARQL $\hat{q}$ with OPTIONAL operator using e_{max} as the main pattern. We remove queries in Q' from Q and iterate until Q is empty.

Algorithm 1. Query Rewriting Algorithm

Input: A set of subqueries Q.
Output: A set of rewritten queries sets Q_{OPT}.
1 **while** $Q \neq \emptyset$ **do**
2   Select the triple pattern e_{max} with the largest value $\frac{|Q'|}{sel(e_{max}) \times (|OPT_{\hat{q}}| \times T_{CPU} + T_{MSG})}$, where Q' is the set of subqueries containing e_{max};
3   Extract the largest common pattern p of queries in Q' by exploring from e_{max}, since all queries in Q' contain e_{max};
4   Initialize a rewritten query $\hat{q}$, where p is its main pattern;
5   Divide Q' into a collection of equivalence classes C, where each class contains subqueries isomorphic to each other;
6   **for** *each class $C \in C$* **do**
7   Generalize a pattern p' isomorphic to all patterns in C;
8   Build a query pattern with p';
9   Add FILTER operators by mapping p' to patterns in C;
10   Add the pattern into $\hat{q}$ as an OPTIONAL pattern;
11   Add $\hat{q}$ into Q_{OPT};
12     $Q = Q - Q'$;
13 Return Q_{OPT};

For example, given subqueries $q_1^1@\{GeoNames\}$, $q_2^1@\{GeoNames\}$ and $q_3^1@$ $\{GeoNames\}$ in Fig. 5, we select the triple pattern "?1 g:name "Canada"" in the first step. It is contained by the three subqueries. We divide them into two equivalence classes $\{q_1^1\}$, $\{q_2^1, q_3^1\}$ according to the query structure. Then, we rewrite $\{q_2^1, q_3^1\}$ using FILTER operator. Finally, we rewrite the three queries using OPTIONAL operator using "?1 g:name "Canada"" and obtain the result rewritten query as shown in Fig. 5.

Theorem 3. *The total cost of patterns selected by Algorithm 1 is no more than* $(1 + ln| \cup_{q \in \mathcal{Q}} E(q)|) \times cost_{opt}$, *where* $\cup_{q \in \mathcal{Q}} E(q)$ *is the set of triple patterns of all subqueries in* $\mathcal{Q}$ *and* $cost_{opt}$ *denotes the smallest cost of patterns that are contained by all subqueries.*

Proof. According to Eq. 3, selecting patterns to hit subqueries is equivalent to selecting triple patterns to hit subqueries. Thus, although we only select the most beneficial triple pattern in Algorithm 1 (Line 2), it is equivalent to selecting the most beneficial pattern graph to hit subqueries. A result in [4] shows that the approximation ratio of the greedy algorithm for the weighted set-cover problem is $(1 + ln| \cup_{q \in \mathcal{Q}} E(q)|)$. □

6   Local Evaluation and Postprocessing

A set of subqueries $\mathcal{Q}$ that will be sent to source s are rewritten as queries $\hat{\mathcal{Q}}$ and evaluated at source s. Let $[\![\hat{q}]\!]_{\{s\}}$ denote the result set of $\hat{q}$ ($\in \hat{\mathcal{Q}}$) at source s, and $\hat{q}$ is obtained by rewriting a set of original subqueries in $\mathcal{Q}$. Thus, $[\![\hat{q}]\!]_{\{s\}}$ is always the union of the results of the subqueries that are rewritten, and we track the mappings between the variables in the rewritten query and the variables in the original subqueries. The result of a rewritten query might have empty (null) columns corresponding to the variables from the OPTIONAL operators. Therefore, a result in $[\![\hat{q}]\!]_{\{s\}}$ may not conform the description of every subquery in $\mathcal{Q}$.

We should identify the valid overlap between each result in $[\![\hat{q}]\!]_{\{s\}}$ and each subquery in $\mathcal{Q}$, and check whether a result in $[\![\hat{q}]\!]_{\{s\}}$ belongs to the relevant sources of a subquery. We return to each query the result it is supposed to get. Specifically, we perform an intersection between each result in $[\![\hat{q}]\!]_{\{s\}}$ and each subquery. The algorithm distributes the corresponding part of this result to $q@\{s\}$ as one of its query results, if the result meet the following two conditions: (1) the columns of this result corresponding to those columns of a subquery $q@s \in \mathcal{Q}$ are not null; and (2) the columns of the result meet the constraints in the FILTER operators rewritten from q. This step iterates over each row and each subquery in $\mathcal{Q}$. The checking on $[\![\hat{q}]\!]_{\{s\}}$ only requires a linear scan on $[\![\hat{q}]\!]_{\{s\}}$, and can be done on-the-fly as the results of $\hat{q}@\{s\}$ are streamed out from the evaluation.

7   Joining Partial Matches

After obtaining the matches of subqueries, we need to join them together to form complete results. Assume that an original query Q_i $(i = 1, \ldots, n)$ is decomposed into a set of subqueries $\{q_i^1@S(q_i^1), \ldots, q_i^{m_i}@S(q_i^{m_i})\}$. We need to obtain query result $[\![Q_i]\!]$ by joining $[\![q_i^1]\!]_{S(q_i^1)}, \ldots, [\![q_i^{m_i}]\!]_{S(q_i^{m_i})}$ together. In the following, we abbreviate $[\![q_i^{m_i}]\!]_{S(q_i^{m_i})}$ to $[\![q_i^{m_i}]\!]$.

The straightforward method is to join subquery matches for each original SPARQL query independently. However, considering multiple queries, there may exist some common computation in joining partial matches. Taking advantage of these common join structures, we can speed up the query response time for multiple queries.

Formally, given subqueries $q_i^{i_1}$ and $q_i^{i_2}$ for query Q_i and subqueries $q_j^{j_1}$ and $q_j^{j_2}$ for query Q_j, if $q_i^{i_1}$ has the same structure to $q_i^{i_2}$, $q_j^{j_1}$ has the same structure to $q_j^{j_2}$ and the join variables between $q_i^{i_1}$ and $q_i^{i_2}$ are the same to the join variables between $q_j^{j_1}$ and $q_j^{j_2}$, then we can merge $[\![q_i^{i_1}]\!] \bowtie [\![q_i^{i_2}]\!]$ and $[\![q_j^{j_1}]\!] \bowtie [\![q_j^{j_2}]\!]$ into $([\![q_i^{i_1}]\!] \cup [\![q_j^{j_1}]\!]) \bowtie ([\![q_i^{i_2}]\!] \cup [\![q_j^{j_2}]\!])$.

The use of the above optimization technique is beneficial if the cost to merge the same two joins is less than the cost of executing two joins separately. To illustrate the potential benefit of the above optimization technique, let us compare the costs of the two alternatives: $[\![q_i^{i_1}]\!] \bowtie [\![q_i^{i_2}]\!]$ and $[\![q_j^{j_1}]\!] \bowtie [\![q_j^{j_2}]\!]$ versus $([\![q_i^{i_1}]\!] \cup [\![q_j^{j_1}]\!]) \bowtie ([\![q_i^{i_2}]\!] \cup [\![q_j^{j_2}]\!])$.

The cost of executing $[\![q_i^{i_1}]\!] \bowtie [\![q_i^{i_2}]\!]$ and $[\![q_j^{j_1}]\!] \bowtie [\![q_j^{j_2}]\!]$ separately is the sum of the costs of two joins, $min\{card([\![q_i^{i_1}]\!]), card([\![q_i^{i_2}]\!])\} + min\{card([\![q_j^{j_1}]\!]), card([\![q_j^{j_2}]\!])\}$. On the other hand, the cost of executing $([\![q_i^{i_1}]\!] \cup [\![q_j^{j_1}]\!]) \bowtie ([\![q_i^{i_2}]\!] \cup [\![q_j^{j_2}]\!])$ is $min\{card([\![q_i^{i_1}]\!] \cup [\![q_j^{j_1}]\!]), card([\![q_i^{i_2}]\!] \cup [\![q_j^{j_2}]\!])\}$. Then, our optimization technique is better if it acts as a sufficient reducer, that is, if $[\![q_i^{i_1}]\!]$ and $[\![q_j^{j_1}]\!]$ overlap a lot and $[\![q_i^{i_2}]\!]$ and $[\![q_j^{j_2}]\!]$ overlap a lot. Otherwise, we do two joins separately. It is important to note that neither approach is systematically the best; they should be considered as complementary.

We can find common join substructures by using frequent subgraph mining technique [20]. Specifically, we can first find a common substructure among the join graphs of all queries, where vertices (i.e., the subqueries) in the common substructure have the largest benefit. We perform the join for this common substructure; and iterate the above process. Obviously, we can do this part only once to avoid duplicate computation.

8   Experimental Evaluation

In this section, we evaluate our federated multiple query optimization method (FMQO) over both real (FedBench) and synthetic datasets (WatDiv). We compare our system with two existing federated SPARQL query engines: FedX [18] and SPLENDID [5].

8.1   Setting

WatDiv. WatDiv [2] is a benchmark that enables diversified stress testing of RDF data management systems. We generate a WatDiv dataset of 10 million triples. WatDiv provides its own workload generator, so we directly use WatDiv's own workload generator of WatDiv to generate different workloads for testing.

FedBench. FedBench [17] is a comprehensive benchmark suite for federated RDF systems. It includes 6 real cross-domain RDF datasets and 4 real life science domain RDF datasets with 7 federated queries for each RDF dataset. To enable multiple query evaluation, we use these 14 queries as seeds and generate the workload of queries isomorphic to the benchmark queries in our experiments.

We conduct all experiments on a cluster of machines running Linux, each of which has one CPU with four cores of 3.06 GHz, 16 GB memory and 150 GB disk storage. The prototype is implemented in Java. At each site, we install Sesame 2.7 to build up an RDF source. Each source can only communicate with the control site through HTTP requests and cannot communicate with each other. For FedBench, we assume that each dataset is resident at a source site. For WatDiv, we first use METIS [8] to divide the schema graph of the collection into 4 connected subgraphs. Then, we place all instances of the same type in a source that subgraphs of the schema graph.

8.2   Evaluation of Proposed Techniques

In this section, we use WatDiv and a query workload of 150 queries of 10 templates to evaluate each proposed technique in this paper. In other words, 150 queries are posed simultaneously to the federated RDF systems storing WatDiv.

Effect of the Query Decomposition and Source Selection Technique. First, we evaluate the effectiveness of our source topology-based technique proposed in Sect. 4. In Fig. 8, we compare our technique with the baseline without any optimizations during source selection (denoted as FMQO-B). We also compare the source selection method proposed in [6,13], which is denoted as QTree and uses the neighborhood information in the source topology to prune some irrelevant sources for each triple patterns.

As shown in Fig. 8, FMQO-Basic does not prune any sources, so it leads to the most number of remote requests and the largest query response time. QTree only uses the neighborhood information and does not consider the whole topology of relevant sources, so the effectiveness of its pruning rule is limited. In contrast, many queries contain triple patterns containing constants with high selectivity, so these triple patterns can be localized to a few sources. Then, for other triple patterns, if the relevant sources are far from relevant sources of the selective triple patterns in the source topology graph, they can be filtered out by our method. Thus, our method leads to the smallest numbers of remote requests and the least query response time.

Effect of the Rewriting Strategies. In this experiment, we compare our SPARQL query rewriting techniques using only OPTIONAL operators (denoted

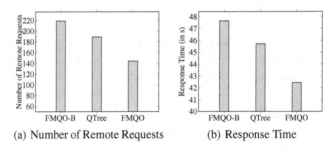

(a) Number of Remote Requests   (b) Response Time

Fig. 8. Evaluating source topology-based source selection technique

as OPT-only) and only FILTER operators (denoted as FIL-only). We also re-implement the rewriting strategies proposed in [11] (denoted as Le et al.) to rewrite subqueries. Our query rewiring technique is denoted as FMQO. Figure 9 shows the experimental results.

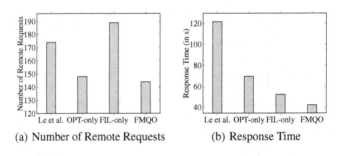

(a) Number of Remote Requests   (b) Response Time

Fig. 9. Evaluating different rewriting strategies

Given a workload, since the number of subqueries sharing common substructures is more than the number of subqueries having the same structure, FIL-only leads to the largest number of rewritten queries, which indicates most remote requests. Le et al. first cluster all subqueries into some groups, and then find the maximal common edge subgraphs of the group of subqueries for query rewriting. In contrast, OPT-only and FMQO use some triple patterns to hit subqueries. In real applications, most maximal common edge subgraphs found by Le et al. also contain our selected triple patterns. Hence, Le et al. generate more rewritten queries which means more remote requests. Finally, FMQO obtains the smallest number of rewritten queries.

Since OPT-only generates smaller number of rewritten queries and share more computation than Le et al., OPT-only can result in faster query response time. If two queries have the same main pattern, the OPTIONAL operator is slower than the FILTER operator, since the former is based on left-join and the latter is based on selection. Hence, although FIL-only generate more rewritten queries, it takes about half time of Le et al., as shown in Fig. 13(b) and two thirds of OPT-only. Furthermore, our rewriting technique using both OPTIONAL and FILTER operators has the best performance, because it takes advantages of both OPTIONAL and FILTER operators.

Evaluation of the Cost Model. In this section, we evaluate the effectiveness of our cost model and cost-aware rewritten strategy in Sect. 5.2. We design a baseline (FMQO-R) that randomly select triple patterns to rewrite subqueries. As shown in Fig. 10(a), because the patterns with lower cost are shared by more subqueries and can result in fewer rewritten queries, our cost-based selection causes fewer remote requests. In addition, in our cost-based rewriting strategy, we prefer selective query patterns, resulting in lower query response times, as shown in Fig. 10(b). Generally speaking, the cost model-based approach can provide speed up of twice.

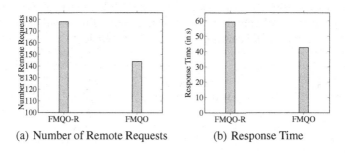

(a) Number of Remote Requests   (b) Response Time

Fig. 10. Evaluating cost model

Effect of Optimization Techniques for Joins. We evaluate our optimized join strategy proposed in Sect. 7. We design a baseline that runs multiple federated queries with only rewriting strategies but not our optimization techniques for joins (denoted as FMQO-QR). Although this technique does not affect the number of remote requests, it reduces the join cost by making use of common join structures. In general, it reduces join processing time by 10%, as shown in Fig. 11.

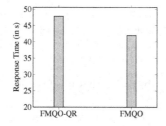

Fig. 11. Effect of optimization techniques for joins

8.3   Performance Comparison

In this experiment, we using both WatDiv and FedBench to test the performance of our method in Figs. 12 and 13. We design a baseline that runs multiple federated queries sequentially (denoted as No-FMQO). This baseline does not employ any optimizations proposed in this paper. We also compare our method with FedX and SPLENDID.

Due to query rewriting, FMQO can merge many subqueries into fewer rewritten queries, which results in smaller number of remote requests, as shown in Fig. 12. FMQO can reduce the number of remote accesses by 1/2–2/3, compared with No-FMQO. Since FedX and SPLENDID do not provide their numbers of remote requests, we do not compare FMQO with FedX and SPLENDID. In terms of evaluation times, the cost-driven rewriting strategy in our method guarantees that rewritten queries are always faster than evaluating them sequentially, as shown in Fig. 13. Note that, FedX and SPLENDID always employ the semijoin to join partial matches. In WatDiv, since almost all partial matches participate in the join, the semijoin is not always efficient.

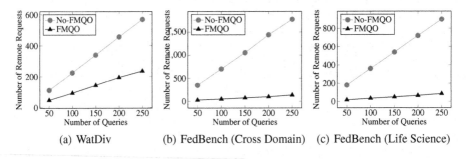

(a) WatDiv   (b) FedBench (Cross Domain)   (c) FedBench (Life Science)

Fig. 12. Number of remote requests

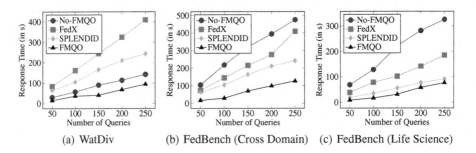

(a) WatDiv   (b) FedBench (Cross Domain)   (c) FedBench (Life Science)

Fig. 13. Response time

9   Related Work

There are two threads of related work: SPARQL query processing in federated RDF systems and multiple SAPRQL queries optimization.

Federated Query Processing. Many approaches [1,5,6,13–15,18] have been proposed for federated SPARQL query processing. Since RDF sources in federated RDF systems are autonomous, the major differences among existing approaches are the query decomposition and source selection approaches.

First, most papers find the relevant RDF sources for all triple patterns based on the metadata. In particular, the metadata in DARQ [14] is named service descriptions, which describes the data available from a data source in form of capabilities. SPLENDID [5] uses Vocabulary of Interlinked Datasets (VOID) as the metadata. QTree [6,13] is a variant of RTree, where its leaf stores a set of source identifiers, including one for each source of a triple approximated by the node. HiBISCuS [15] relies on capabilities to compute the metadata. For each source, HiBISCuS defines a set of capabilities which map the properties to their subject and object authorities. ANAPSID [1] dynamically generates the query plan while considering data availability and run-time conditions. Moreover, ANAPSID implement a memory-based caching mechanism.

Besides the metadata-assisted methods, FedX [18] performs the source selection by using ASK queries. FedX sends ASK queries for each triple pattern to the RDF sources. Based on the results, It annotates each pattern in the query with its relevant sources.

Multiple SPARQL Queries Optimization. Le et al. [11] first discuss how to optimize multiple SPARQL queries evaluation, but only in a centralized environment. It first finds out all maximal common edge subgraphs (MCES) among a group of query graphs, and then rewrite the queries into queries with OPTIONAL operators. In the rewritten queries, the MCES constitutes the main pattern, while the remaining subquery of each individual query generates an OPTIONAL clause. Konstantinidis et al. [9] discuss how to optimize multiple SPARQL queries evaluation over multiple views. They find out some atomic join operations among multiple queries and compute them just once.

10   Conclusion

In this paper, we study multiple query optimization over federated RDF systems. Our optimization identifies common subqueries with a cost model and rewrites queries into equivalent queries. We also discuss how to efficiently select relevant sources and join intermediate results. Experiments show that our optimizations are effective.

Acknowledgement. This work was supported by The National Key Research and Development Program of China under grant 2016YFB1000603, NSFC under grant 61702171, 61622201 and 61532010, and the Fundamental Research Funds for the Central Universities. Özsu's work was supported in part by Natural Sciences and Research Council (NSERC) of Canada.

References

1. Acosta, M., Vidal, M.-E., Lampo, T., Castillo, J., Ruckhaus, E.: ANAPSID: an adaptive query processing engine for SPARQL endpoints. In: Aroyo, L., Welty, C., Alani, H., Taylor, J., Bernstein, A., Kagal, L., Noy, N., Blomqvist, E. (eds.) ISWC 2011. LNCS, vol. 7031, pp. 18–34. Springer, Heidelberg (2011). https://doi.org/10.1007/978-3-642-25073-6_2
2. Aluç, G., Hartig, O., Özsu, M.T., Daudjee, K.: Diversified stress testing of RDF data management systems. In: Mika, P., Tudorache, T., Bernstein, A., Welty, C., Knoblock, C., Vrandečić, D., Groth, P., Noy, N., Janowicz, K., Goble, C. (eds.) ISWC 2014. LNCS, vol. 8796, pp. 197–212. Springer, Cham (2014). https://doi.org/10.1007/978-3-319-11964-9_13
3. Berners-Lee, T.: Linked Data - Design Issues. W3C (2010)
4. Chvatal, V.: A greedy heuristic for the set-covering problem. Math. Oper. Res. 4(3), 233–235 (1979)
5. Görlitz, O., Staab, S.: SPLENDID: SPARQL endpoint federation exploiting VOID descriptions. In: COLD (2011)
6. Harth, A., Hose, K., Karnstedt, M., Polleres, A., Sattler, K., Umbrich, J.: Data summaries for on-demand queries over linked data. In: WWW, pp. 411–420 (2010)
7. Hose, K., Schenkel, R., Theobald, M., Weikum, G.: Database foundations for scalable RDF processing. In: Polleres, A., d'Amato, C., Arenas, M., Handschuh, S., Kroner, P., Ossowski, S., Patel-Schneider, P. (eds.) Reasoning Web 2011. LNCS, vol. 6848, pp. 202–249. Springer, Heidelberg (2011). https://doi.org/10.1007/978-3-642-23032-5_4
8. Karypis, G., Kumar, V.: Multilevel graph partitioning schemes. In: ICPP, pp. 113–122 (1995)
9. Konstantinidis, G., Ambite, J.L.: Optimizing query rewriting for multiple queries. In: IIWeb, pp. 7:1–7:6 (2012)
10. Kossmann, D.: The state of the art in distributed query processing. ACM Comput. Surv. 32(4), 422–469 (2000)
11. Le, W., Kementsietsidis, A., Duan, S., Li, F.: Scalable multi-query optimization for SPARQL. In: ICDE, pp. 666–677 (2012)
12. Li, J., Deshpande, A., Khuller, S.: Minimizing communication cost in distributed multi-query processing. In: ICDE, pp. 772–783 (2009)
13. Prasser, F., Kemper, A., Kuhn, K.A.: Efficient distributed query processing for autonomous RDF databases. In: EDBT, pp. 372–383 (2012)
14. Quilitz, B., Leser, U.: Querying distributed RDF data sources with SPARQL. In: Bechhofer, S., Hauswirth, M., Hoffmann, J., Koubarakis, M. (eds.) ESWC 2008. LNCS, vol. 5021, pp. 524–538. Springer, Heidelberg (2008). https://doi.org/10.1007/978-3-540-68234-9_39
15. Saleem, M., Ngonga Ngomo, A.-C.: HiBISCuS: hypergraph-based source selection for SPARQL endpoint federation. In: Presutti, V., d'Amato, C., Gandon, F., d'Aquin, M., Staab, S., Tordai, A. (eds.) ESWC 2014. LNCS, vol. 8465, pp. 176–191. Springer, Cham (2014). https://doi.org/10.1007/978-3-319-07443-6_13
16. Schmachtenberg, M., Bizer, C., Paulheim, H.: Adoption of the linked data best practices in different topical domains. In: Mika, P., Tudorache, T., Bernstein, A., Welty, C., Knoblock, C., Vrandečić, D., Groth, P., Noy, N., Janowicz, K., Goble, C. (eds.) ISWC 2014. LNCS, vol. 8796, pp. 245–260. Springer, Cham (2014). https://doi.org/10.1007/978-3-319-11964-9_16

17. Schmidt, M., Görlitz, O., Haase, P., Ladwig, G., Schwarte, A., Tran, T.: FedBench: a benchmark suite for federated semantic data query processing. In: Aroyo, L., Welty, C., Alani, H., Taylor, J., Bernstein, A., Kagal, L., Noy, N., Blomqvist, E. (eds.) ISWC 2011. LNCS, vol. 7031, pp. 585–600. Springer, Heidelberg (2011). https://doi.org/10.1007/978-3-642-25073-6_37
18. Schwarte, A., Haase, P., Hose, K., Schenkel, R., Schmidt, M.: FedX: optimization techniques for federated query processing on linked data. In: Aroyo, L., Welty, C., Alani, H., Taylor, J., Bernstein, A., Kagal, L., Noy, N., Blomqvist, E. (eds.) ISWC 2011. LNCS, vol. 7031, pp. 601–616. Springer, Heidelberg (2011). https://doi.org/10.1007/978-3-642-25073-6_38
19. Stocker, M., Seaborne, A., Bernstein, A., Kiefer, C., Reynolds, D.: SPARQL basic graph pattern optimization using selectivity estimation. In: WWW, pp. 595–604 (2008)
20. Yan, X., Han, J.: gSpan: graph-based substructure pattern mining. In: ICDM, pp. 721–724 (2002)

Distributed Efficient Provenance-Aware Regular Path Queries on Large RDF Graphs

Yueqi Xin[1,2], Xin Wang[1,2(✉)], Di Jin[1,2], and Simiao Wang[1,2]

[1] School of Computer Science and Technology, Tianjin University, Tianjin, China
{xinyueqi,wangx,jindi,wangsimiao}@tju.edu.cn
[2] Tianjin Key Laboratory of Cognitive Computing and Application, Tianjin, China

Abstract. With the proliferation of knowledge graphs, massive RDF graphs have been published on the Web. As an essential type of queries for RDF graphs, Regular Path Queries (RPQs) have been attracting increasing research efforts. However, the existing query processing approaches mainly focus on the standard semantics of RPQs, which cannot provide provenance of the answer sets. We propose dProvRPQ that is a distributed approach to evaluating provenance-aware RPQs over big RDF graphs. Our Pregel-based method employs Glushkov automata to keep track of matching processes of RPQs *in parallel*. Meanwhile, four optimization strategies are devised, including edge filtering, candidate states, message compression, and message selection, which can reduce the intermediate results of the basic dProvRPQ algorithm dramatically and overcome the counting-paths problem to some extent. The proposed algorithms are verified by extensive experiments on both synthetic and real-world datasets, which show that our approach can efficiently answer the provenance-aware RPQs over large RDF graphs.

Keywords: Regular path query · Provenance-aware · RDF graph
Pregel

1 Introduction

With the increasing popularity of knowledge graphs, *Resource Description Framework* (RDF) has been widely recognized as a flexible graph-like data model to represent large-scale knowledge bases. It has become essential to realize efficient and scalable query processing for big RDF graphs in various domains, such as social networking [12] and bioinformatics [7], stored in distributed clusters. As one of the fundamental operations for querying graph data [2], regular path queries (RPQs) can explore RDF graphs in a navigational manner, which is an indispensable building block in most graph query languages. The latest version of the standard query language of RDF, SPARQL 1.1 [6], has provided the *property path* [9] feature which is actually an implementation of RPQ semantics. In particular, answering an RPQ $Q = (x, r, y)$ over an RDF graph T is to find a set

© Springer International Publishing AG, part of Springer Nature 2018
J. Pei et al. (Eds.): DASFAA 2018, LNCS 10827, pp. 766–782, 2018.
https://doi.org/10.1007/978-3-319-91452-7_49

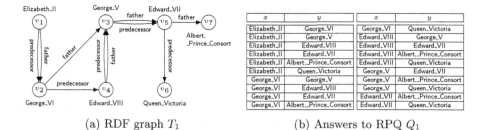

(a) RDF graph T_1 (b) Answers to RPQ Q_1

Fig. 1. An example RDF graph T_1 and answers to RPQ Q_1

of pairs of resources (v_0, v_n) such that there exists a path ρ in T from v_0 to v_n, where the label of ρ, denoted by $\lambda(\rho)$, satisfies the regular expression r in Q.

However, from the above standard semantics of RPQs, we cannot tell what such a path ρ from v_0 to v_n looks like. To provide the *provenance* why a pair of resources in an RDF graph satisfies Q, we focus on the *provenance-aware* semantics of RPQs which actually returns a *subgraph* of the RDF graph consisting of all the "witness triples". For example, Fig. 1(a) depicts an RDF graph T_1 excerpted from DBpedia, which shows predecessor and father relationships among seven British monarchs [14]. The RPQ $Q_1 = (x, (\text{predecessor}|\text{father})^+, y)$ asks to find pairs of monarchs (v_0, v_n) such that v_0 can navigate to v_n via one or more predecessor or father edges. The answers under the standard semantics to Q_1 are shown in Fig. 1(b). In contrast, the provenance-aware answer to Q_1 is a subgraph that contains all the paths whose labels satisfy Q_1. In this example, the subgraph (i.e., answer) is exactly T_1, which can efficiently encode the conventional answers to Q_1 in Fig. 1(b).

Currently, there have been some research works on RPQs over RDF graphs under both standard and provenance-aware semantics. To answer RPQs under the standard semantics, some approaches leverage views [4] or other auxiliary structures, such as "rare labels" [8]. The RPQ evaluation system Vertigo [11] is implemented based on *Brzozowski's derivatives* using the Giraph parallel framework. Wang et al. [15] employ the partial evaluation to obtain partial answers to RPQs in parallel and assemble the partial answers using an automata-based algorithm. However, the above methods may lead to potential large intermediate results and suffer from performance bottleneck when evaluating RPQs on large-scale RDF graphs. Although Dey et al. [5] have done the first work to investigate provenance-aware RPQs, they translate RPQs into standard Datalog queries, which is hardly scalable when evaluating on large RDF graph data. Another representative work [13] is based on product automata to evaluate provenance-aware RPQs, which may incur the costly construction process of product automata and excessive communications when handling large-scale RDF graphs.

To this end, in this paper, we propose a Pregel-based parallel approach to answering provenance-aware RPQs using *Glushkov automata*, which consists of a series of supersteps. The query processing starts with the vertices in an RDF graph to match against the states in the corresponding automaton of the RPQ;

in each superstep, one hop of edges in the paths of the RDF graph are matched forward to obtain the intermediate partial answer to the provenance-aware RPQ. In addition, we design four optimization strategies to reduce vertex computation and message passing cost: (1) the *edge-filtering* technique filters out those edges whose labels not occurring in r; (2) the *candidate-states* technique implies all the possible matched states of a vertex, which can avoid traversals via outgoing edges of the vertex; (3) to address the *counting-paths* problem [1], we further propose another two techniques, which combine multiple equivalent messages into a single message and compress the messages that are sent via different outgoing edges.

Our main contributions include: (1) we propose an automata-based distributed algorithm, called dProvRPQ, for RPQs under the provenance-aware semantics using the Pregel graph parallel computing framework; (2) four optimization strategies are presented to reduce the overhead of the basic dProvRPQ algorithm and address the counting-paths problem to some extent; and (3) the extensive experiments were conducted to verify the efficiency and scalability of the proposed method on both synthetic and real-world datasets.

The rest of this paper is organized as follows. Section 2 reviews related work. In Sect. 3, we introduce preliminary definitions of RPQs. In Sect. 4, we describe in detail the dProvRPQ algorithm for answering provenance-aware RPQs. We then present the optimization techniques in Sect. 5. Section 6 shows experiment results, and we conclude in Sect. 7.

2   Related Work

Most of the existing approaches aim to evaluate RPQs under the standard semantics, but relatively fewer works focus on RPQs under the provenance-aware semantics. Currently, we are not aware any distributed approach to provenance-aware RPQs. We classify the existing approaches into the following two categories.

2.1   Standard Semantics of RPQs

(1) Standalone RPQs. The approach proposed in [4] answers RPQs using views, which can be interpreted as checking whether a pair of nodes is one of the answers. The view-based approach for RPQs has been extensively investigated, while the types of data and queries that this approach can handle are restricted under certain assumptions. Koschmieder and Leser [8] propose a rare-labels-based approach that decomposes RPQs into a series of smaller RPQs. The rare labels denote the elements in RPQs that have few matches by utilizing the labels and their frequencies in data graph, which can reduce the search space. However, the performance of the method highly depends on a specific query decomposition and selectivity of rare labels.

(2) Parallel and Distributed RPQs. A distributed algorithm for evaluating RPQs on large-scale RDF graphs is proposed in [15], which is the first work

to investigate RPQs using partial evaluation. It employs a dynamic programming method to compute partial answers in parallel, which are then assembled to obtain the final results using an automata-based algorithm. Nevertheless, the experiments on the real-world datasets are not shown in the paper. Nolé and Sartiani [11] exploit Brzozowski's derivatives of regular expressions to evaluate RPQs in a vertex-centric and message-passing-based manner, which is implemented on top of the Giraph framework. However, the experimental results are only evaluated on the Erdös-Rényi models and the power-law graphs, lacking experiments on synthetic and real-world RDF graphs to verify the algorithm.

2.2 Provenance-Aware Semantics of RPQs

(1) Standalone RPQs. Dey et al. [5] first translate the provenance-aware RPQs into standard Datalog queries or SQL queries, in which auxiliary predicates are introduced to evaluate queries represented by Datalog. In this work, two evaluators for RPQs and provenance-aware RPQs are both implemented on the relational DBMS. However, from the experimental results, we can observe that the approach is hardly scalable for large-scale RDF graphs.

(2) Parallel and Distributed RPQs. Wang et al. [13] propose an automata-based approach, which employs product automata for evaluating RPQs under the provenance-aware semantics in parallel. The product automaton is constructed using two NFA converted from the regular expression of an RPQ and the RDF data graph, respectively. Then the answer paths are extracted by running the product automaton recursively. Nevertheless, the product automata construction in this method may incur high overhead and excessive communication cost when dealing with large-scale RDF graphs.

Unlike the above previous works, we propose a distributed algorithm dProvRPQ for evaluating provenance-aware RPQs on big RDF graphs. To the best of our knowledge, it is the first work to implement an efficient and scalable evaluation of provenance-aware RPQs using the Pregel parallel graph computing model.

3 Preliminaries

In this section, we introduce the definitions of relevant background knowledge.

Definition 1 (RDF graph). *Let U and L be the disjoint infinite sets of URIs and literals, respectively. A tuple $(s, p, o) \in U \times U \times (U \cup L)$ is called an RDF triple, where s is the subject, p is the predicate (a.k.a. property), and o is the object. A finite set of RDF triples is called an RDF graph.*

Given an RDF graph $T = (V, E, \Sigma)$, where V, E, and Σ denote the set of vertices, edges, and edge labels in T, respectively. Formally, $V = \{s \mid (s, p, o) \in T\} \cup \{o \mid (s, p, o) \in T\}$, $E = \{(s, o) \mid (s, p, o) \in T\}$, and $\Sigma = \{p \mid (s, p, o) \in T\}$. In

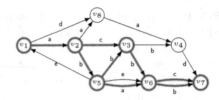

Fig. 2. RDF graph T_2 and the provenance-aware answer to RPQ Q_2 (Color figure online)

addition, we define an infinite set Var of variables that is disjoint from U and L. An example *RDF graph* T_2 is shown in Fig. 2, which consists of 15 triples (i.e., edges). For instance, (v_1, a, v_2) is an RDF triple as well as an edge with label a in T_2, and $V_{T_2} = \{v_i \mid 1 \leq i \leq 8\}$, $\Sigma_{T_2} = \{\mathsf{a,b,c,d,e}\}$.

Definition 2 (Regular path queries). *Let* $Q = (x, r, y)$ *be a regular path query over an RDF graph* $T = (V, E, \Sigma)$, *where* $x, y \in Var$ *are variables, and* r *is a regular expression over the alphabet* Σ. *Regular expression* r *is recursively defined as* $r ::= \varepsilon \mid p \mid r/r \mid r|r \mid r^*$, *where* $p \in \Sigma$ *and* $/$, $|$, *and* $*$ *are concatenation, alternation, and the Kleene closure, respectively. The shorthands* r^+ *for* r/r^* *and* $r?$ *for* $\varepsilon|r$ *are also allowed.* $L(r)$ *denotes the language expressed by* r *and* $\lambda(\rho)$ *is the label of path* ρ. *The answer set of* Q *under the standard semantics, denoted by* $[\![Q]\!]_T$, *is defined as* $\{(x, y) \mid \exists$ *a path* ρ *in* T *from* x *to* y *s.t.* $\lambda(\rho) \in L(r)\}$.

Given a regular expression r, let $Pos(r) = \{1, 2, \dots, |r|\}$ be the set of positions in r, where $|r|$ is the length of r. Thus, the symbols in r can be denoted as $r[1], r[2], \dots, r[|r|]$.

Definition 3 (Automata of RPQs). *Given an RDF graph* T *and an RPQ* $Q = (x, r, y)$ *over* T, *the* automaton *of RPQ* Q *is the Glushkov automaton* A_Q *converted from the regular expression* r *by using the Glushkov's construction algorithm* [3]. *The function* first(r) *(resp.* last$(r))$ *is the set of positions in* r *that can match the first (resp. last) symbol of some string in* $L(r)$, *and the function* follow(r, i) *is the set of positions in* r *that can follow position* i *when matching some string in* $L(r)$. A_Q *is defined as a 5-tuple* $(St, \Sigma, \delta, q_0, F)$, *where (1)* $St = \{0\} \cup Pos(r)$ *is a finite set of states, (2)* Σ *is the alphabet of* r, *(3)* $\delta : St \times \Sigma \rightarrow \mathcal{P}(St)$ *is the transition function, (4)* $q_0 = 0$ *is the initial state, (5) and* F *is the set of final states. Here,* δ *and* F *are further defined as follows:*

$$\delta(q, a) = \begin{cases} \{i \mid i \in \mathsf{first}(r) \wedge r[i] = a\} & \text{if } q = q_0 \\ \{i \mid i \in \mathsf{follow}(r, q) \wedge r[i] = a\} & \text{if } q \in Pos(r) \end{cases}$$

$$F = \begin{cases} \{q_0\} \cup \mathsf{last}(r) & \text{if } \varepsilon \in L(r) \\ \mathsf{last}(r) & \text{otherwise} \end{cases}$$

Example 1. Given an RPQ $Q_2 = (x, r_2, y)$ and $r_2 = (\mathsf{a/b^*/c})^+|(\mathsf{a/b})^+$, we build $A_{Q_2} = \{St, \Sigma, \delta, q_0, F\}$ based on r_2, where $St = \{0, 1, 2, 3, 4, 5\}$, $\Sigma = \{\mathsf{a,b,c}\}$, $q_0 = \{0\}$, $F = \{3, 5\}$, and the transition function δ is represented in the form of the transition graph shown in Fig. 3(a).

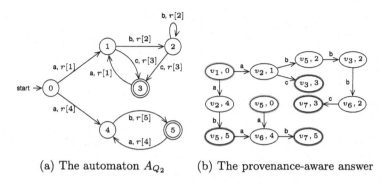

(a) The automaton A_{Q_2} (b) The provenance-aware answer

Fig. 3. The automaton of RPQ Q_2 and the provenance-aware answer (Color figure online)

Definition 4 (Provenance-aware answers to RPQs). *Given an RDF graph T and an automaton $A_Q = (St, \Sigma, \delta, q_0, F)$ of an RPQ Q, the provenance-aware answer to Q over T, denoted by $[\![Q]\!]_T^{\mathfrak{p}}$, is defined as a 5-tuple $(V_{\mathfrak{p}}, E_{\mathfrak{p}}, L_{\mathfrak{p}}, I_{\mathfrak{p}}, F_{\mathfrak{p}})$, where (1) $V_{\mathfrak{p}} \subseteq V \times St$ is a set of vertices, (2) $E_{\mathfrak{p}} \subseteq V_{\mathfrak{p}} \times V_{\mathfrak{p}}$ is a set of edges, (3) $L_{\mathfrak{p}}$ is a function that assigns each edge a label in Σ, and (4) $I_{\mathfrak{p}} = \{(v, q_0) \mid v \in V\} \subseteq V_{\mathfrak{p}}$ and $F_{\mathfrak{p}} = \{(v, q_f) \mid v \in V \wedge q_f \in F\} \subseteq V_{\mathfrak{p}}$ are the sets of start and final vertices, respectively. Here, $[\![Q]\!]_T^{\mathfrak{p}}$ is constructed by the following process: for each path $v_0 a_0 v_1 \cdots v_{n-1} a_{n-1} v_n$ in T such that there exists a sequence of states $st_0, st_1, \ldots, st_n$ in St satisfying $st_0 = q_0$, $st_{i+1} \in \delta(st_i, a_i)$ for $i \in \{0, \ldots, n-1\}$, and $st_n \in F$, a path $\rho = v_{\mathfrak{p}_0} a_0 v_{\mathfrak{p}_1} \cdots v_{\mathfrak{p}_{n-1}} a_{n-1} v_{\mathfrak{p}_n}$ in $[\![Q]\!]_T^{\mathfrak{p}}$ is constructed, where $v_{\mathfrak{p}_i} = (v_i, st_i)$ for $i \in \{0, \ldots, n\}$ and by definition $v_{\mathfrak{p}_0} \in I_{\mathfrak{p}} \wedge v_{\mathfrak{p}_n} \in F_{\mathfrak{p}}$.*

Example 2. The provenance-aware answer to Q_2 over T_2 as defined in Definition 4 is shown in Fig. 3(b), which can be considered as the extended version of the provenance-aware to Q_2 (as a subgraph of T_2 marked in green in Fig. 2) that attaches the matched states of A_{Q_2} to the corresponding vertices.

Pregel is a vertex-centric parallel model for graph computation [10]. The computation in Pregel is composed of a sequence of iterations, i.e., *supersteps*, conforming to the *Bulk Synchronous Parallel* (*BSP*) model. For each vertex v, we define $Val(v)$ as the set of values associated with v. Within a superstep, each vertex executes the user-defined computation to get and update $Val(v)$ in *parallel*.

Definition 5 (Pregel framework). *Given an RDF graph T as the input data, each vertex $v \in V$ is in one of the two states, i.e., active and inactive. The function* superStep *gets the number of the current superstep. In the first superstep, all the vertices are active. The function* voteToHalt*: $V \to \emptyset$ is called by a vertex v to deactivate itself. The entire computation terminates when all vertices are inactive. Let M be the set of messages. Within a superstep, the user-defined function* compute*(T,M) is executed on each active vertex in parallel. The function* sendMsg*: $V \times V \times M \to \emptyset$ is to send messages M from one vertex v to another vertex v'. An inactive vertex will be reactivated by incoming messages* *sent to it. When* compute*(T, M) is invoked on each active vertex v, it (1) first receives messages (i.e., each m ∈ M) sent to v in the previous superstep, (2) gets and/or updates Val(v), (3) modifies M to generate the set of new messages M', and (4) invokes* sendMsg*(v, v', M') to send M' to the adjacent vertex v'.*

4   The dProvRPQ Algorithm

In this section, we propose the Pregel-based algorithm for answering provenance-aware RPQs, which employs the automata introduced in Sect. 3. First, we describe the overall evaluation, then elaborate the computation in each vertex of each superstep. Finally, we discuss the cycle detection mechanism to avoid infinite matching when evaluating on cyclic RDF graphs.

4.1   Provenance-Aware RPQs Based on Pregel

The overall evaluation process dProvRPQ is shown in Algorithm 1, in which we construct an automaton $A_Q = \{St, \Sigma, \delta, q_0, F\}$ of an RPQ $Q = (x, r, y)$ (line 2). In each superstep, every active vertex v invokes compute *in parallel* (line 4) to match against a state $q \in St$, while the updated partial answers are maintained in $Val(v)$. The matching process is executed repeatedly in the following supersteps until the computation terminates. When the entire computation completes, we combine $Val(v)$ of each vertex v to obtain the final provenance-aware answer set $[\![Q]\!]_T^p$ (line 5).

Algorithm 1. dProvRPQ

   Input : An RDF graph $T = (V, E, \Sigma)$ and an RPQ $Q = (x, r, y)$
   Output: The provenance-aware answer set $[\![Q]\!]_T^p$
1 Compute $r[i]$, first(r), last(r), and follow(r, i) based on r of Q;   /* $0 < i \leq |r|$ */
2 Build the automaton $A_Q = \{St, \Sigma, \delta, q_0, F\}$ of Q;
3 $M_r \leftarrow \emptyset$;   /* initialize the message set */
4 compute(T, M_r);   /* compute in every vertex v in parallel */
5 $[\![Q]\!]_T^p \leftarrow \bigcup_{v \in V} Val(v)$;
6 **return** $[\![Q]\!]_T^p$;

The vertex computation compute is shown in Algorithm 2. It is executed at every vertex $v \in V$ *in parallel*, which has the following three phases:

(1) In the first superstep (lines 2–7), v is considered to be matched against $q \in$ first(r) only if there exists an outgoing edge (v, v') such that the label of (v, v') is the same as $r[q]$ (lines 3–6). Then, the matched message m is generated, which formally is a 3-tuple $(v, q, r[q])$. The message set M_s is a set of m, which is sent to the adjacent vertices by invoking sendMsg (line 7).

(2) As to the remaining supersteps (lines 8–22), if the set M_r of the receiving messages from the adjacent vertices in the previous superstep is empty, v is deactivated by voteToHalt (line 22), otherwise each active vertex is matched

---

Algorithm 2. compute(T, M_r)   /* compute in every vertex v in parallel */

Input : An RDF graph T and a set M_r of receiving messages
1 $V_n \leftarrow$ all the adjacent vertices of v;
2 **if** superStep $= 1$ **then**
3   **foreach** *outgoing edge* (v, v') *of* v **do**
4   **if** $\exists\ q \in$ first(r) $\wedge r[q] = \lambda((v, v'))$ **then**
5              $m \leftarrow ((v, q, r[q]))$;
6              $M_s \leftarrow M_s \cup \{m\}$;

7   **foreach** $v' \in V_n$ **do** sendMsg(v, v', M_s) ;

8 **else**
9   **if** $M_r \neq \emptyset$ **then**
10   **foreach** $m' \in M_r$ **do**
11             $(v_t, q, r[q]) \leftarrow$ the last element of m';
12   **if** $q \in$ last(r) **then**
13                 $m' \leftarrow$ append $(v, q_f, r[q])$ to m' ;   /* $q_f \in F$ */
14   ρ_f is the equivalent path of m' ;   /* $\lambda(\rho_f) \in L(r)$ */
15                 $Val(v) \leftarrow Val(v) \cup \{\rho_f\}$; /* $Val(v)$ is the set of partial answers */
16             $R_{q'} \leftarrow$ follow(r, q);   /* $R_{q'}$ is the set of next possible states of q */
17   **foreach** *outgoing edge* (v, v') *of* v **do**
18   **if** $\exists\ q' \in R_{q'} \wedge r[q'] = \lambda((v, v'))$ **then**
19                     $m \leftarrow$ append $(v, q', r[q'])$ to m';
20                     $M_s \leftarrow M_s \cup \{m\}$;

21   **foreach** $v' \in V_n$ **do** sendMsg(v, v', M_s) ;

22   **else** voteToHalt;

---

forward based on the messages in M_r (lines 10–21). First, the set $R_{q'}$ of the next possible states is computed by follow w.r.t. the current matched state q (line 16). Next, if v has an outgoing edge labeled with the same symbol as $r[q']$ ($q' \in R_{q'}$), a new message m is built by appending $(v, q', r[q'])$ to m' and then added to the message set M_s to be sent (lines 17–20). Finally, sendMsg(v, v', M_s) is invoked to send M_s from v to v' (line 21). Meanwhile, v checks whether the current matched state q is a final state (i.e., $q \in$ last(r)). If it is, then $(v, q_f, r[q])$ is appended to m' to form the answer path ρ_f, which is added to $Val(v)$ (lines 12–15).

(3) In sendMsg(v, v', M_s), the condition of sending a message $m \in M_s$ from v to v' is that the current matched state q' satisfies $r[q'] = \lambda((v, v'))$. When v' receives messages from different vertices, it merges all the messages into M_r.

The correctness of the dProvRPQ algorithm is guaranteed by the following theorem.

Theorem 1. *Given an RPQ $Q = (x, r, y)$ over an RDF graph T, $(v_0, v_n) \in$ $[\![Q]\!]_T$ iff $\exists\ \{(v_0, q_0), (v_1, q_1), \ldots, (v_n, q_n)\} \in [\![Q]\!]_T^p$ in dProvRPQ.*

Proof. (Sketch) (i) **"If"** direction: for $\{(v_0, q_0), (v_1, q_1), \ldots, (v_n, q_n)\} \in [\![Q]\!]_T^p$, $\exists$ a path ρ_1 in T from v_0 to v_n and a path ρ_2 in A_Q from q_0 to q_n. It can be observed that $q_i \in \delta(q_{i-1}, \lambda((v_{i-1}, v_i)))$, for $1 \in \{1, \ldots, n\}$, holds in dProvRPQ. The label of ρ_1 is the same as that of ρ_2, i.e., $\lambda(\rho_1) \in L(r)$. Therefore, $(v_0, v_n) \in [\![Q]\!]_T$. (ii) **"Only if"** direction: for $(v_0, v_n) \in [\![Q]\!]_T$, assume a path $v_0 a_0 \cdots v_{n-1} a_{n-1} v_n$ in

T such that there exists a sequence of states $st_0 \ldots st_n$ in St of A_Q in dProvRPQ satisfying $st_0 = q_0$, $st_{i+1} \in \delta(st_i, a_i)$ for $i \in \{0, \ldots, n-1\}$, and $st_n \in F$. The path $\rho = (v_0, q_0)a_0(v_1, q_1)\cdots(v_{n-1}, q_{n-1})a_{n-1}(v_n, q_n)$ is constructed in dProvRPQ, i.e., $\{(v_0, q_0), (v_1, q_1), \ldots, (v_n, q_n)\} \in [\![Q]\!]_T^p$.   □

Theorem 2. *The complexity of the* dProvRPQ *algorithm is bounded by* $O(|deg_m^+|^k \cdot |r|^k \cdot |deg_m^-|^{k-1})$, *where* $|r|$ *is the length of the regular expression* r *in* Q, k *is the total number of supersteps, and* $|deg_m^+|$ *(resp.* $|deg_m^-|$*) is the maximum outdegree (resp. indegree) of the vertices in* T.

Proof. (Sketch) (i) **Basis:** When $k = 1$, for only one superstep, there exists a vertex that is matched $O(|deg_m^+| \cdot |r|)$ times since at most $|deg_m^+|$ outgoing edges of the vertex are matched against the states in $first(r)$. Thus, the complexity is $O(|deg_m^+| \cdot |r|)$. (ii) **Induction step:** For k ($k \geq 1$) supersteps, the complexity is $O(|deg_m^+|^k \cdot |r|^k \cdot |deg_m^-|^{k-1})$ as the induction hypothesis. Thus, there exists a vertex that is matched $O(|deg_m^+|^k \cdot |r|^k \cdot |deg_m^-|^{k-1})$ times, and all these matches are sent as messages via the outgoing edges. Then, for $(k+1)$ supersteps, since the maximum number of incoming edges of a vertex may be $|deg_m^-|$, the receiving message set of the vertex includes $O(|deg_m^+|^k \cdot |r|^k \cdot |deg_m^-|^k)$ messages. Next, at most $|deg_m^+|$ outgoing edges of the vertex are matched against at most $|r|$ states in the automaton based on the receiving messages. Thus, the vertex is matched $O(|deg_m^+|^{k+1} \cdot |r|^{k+1} \cdot |deg_m^-|^k)$ times, which is the complexity for $(k+1)$ supersteps.   □

In particular, $|deg_m^+|$ and $|deg_m^-|$ can be further reduced to $|r|$ by our optimization techniques in Sect. 5. Thus, the complexity of the optimized dProvRPQ algorithm is $O(|r|^{3k-1})$.

4.2   Cycle Detection

For an RPQ $Q = (x, r, y)$, when closure operations * and/or + occur in r, it may lead to infinite matching when evaluating on cyclic RDF graphs. Therefore, we introduce a cycle detection mechanism in dProvRPQ to ensure that a vertex is not matched with the same state twice in intermediate partial answers. For a message m' in a set of receiving messages, if a 3-tuple message $(v, q, r[q]) \in m'$, then it cannot be added to m' again for generating a new message in the following supersteps.

5   Optimization Strategies

Cost Estimation. The cost of the Pregel computation is determined as the sum of the cost of all supersteps. The cost of each superstep consists of the following terms: (1) w_i is the maximum cost of vertex computation among all vertices in a superstep; (2) h_i is the maximum number of messages sent or received by each vertex; and (3) l is the cost of the barrier synchronization at the end of a superstep. Thus, the cost of a Pregel-based algorithm is $\sum_{i=1}^{k} w_i + g \sum_{i=1}^{k} h_i + kl$, where g is the cost to deliver a message and k is the number of supersteps. Thus, the total cost of the dProvRPQ algorithm can be estimated by the following formula, where c is the cost to do a single match operation in Algorithm 2

$$c \sum_{i=1}^{k} |deg_m^+|^i \cdot |r|^i \cdot |deg_m^-|^{i-1} + g \sum_{i=1}^{k} |deg_m^+|^{i+1} \cdot |r|^i \cdot |deg_m^-|^{i-1} + kl$$

5.1 Edge Filtering

In order to improve the scalability of the dProvRPQ algorithm, we design the *edge-filtering* technique, which only loads the edges labeled with the symbols that occur in r of $Q = (x, r, y)$. We use Σ_r to denote the subset of the alphabet that appears in r. Then, an edge (v, a, u) in the input RDF graph T is loaded if and only if $a \in \Sigma_r$. In Example 2, with edge filtering, only the edges labeled with the symbols in $\Sigma_{r_2} = \{a, b, c\}$ are involved in the processing.

5.2 Candidate States

In Algorithm 2, the traversal operations (lines 3 and 17) are not efficient when evaluating RPQs on large-scale RDF graphs. Thus, we leverage the *priori knowledge* in a given query RPQ Q over an RDF graph T to construct an auxiliary structure called *candidate states*, denoted by R_c, which keeps the states of the RPQ automaton that v can match against. We compare the states in R_c with that in $R_{q'}$ instead of the costly iteration of all adjacent vertices. Algorithm 3 is an optimized version of lines 17–20 in Algorithm 2 by using the candidate-states technique, in which the modified matching process is: (1) when v receives a message $m' \in M_r$, $R_{q'}$ will be derived; (2) if $q' \in R_c \cap R_{q'}$, a new message m will be built by appending $(v, q', r[q'])$ to m', otherwise no new message will be generated for m'.

Algorithm 3. candidateStates()

---
1 $R_c \leftarrow$ the states of A_Q that v can match against;   /* R_c is calculated only once */
2 **foreach** $q' \in R_c \cap R_{q'}$ **do**
3 | $m \leftarrow$ append $(v, q', r[q'])$ to m';
4 |_ $M_s \leftarrow M_s \cup \{m\}$;

Example 3. Consider the RDF graph T_2, the RPQ Q_2, and the automaton $A_{Q_2} = \{St, \Sigma, \delta, q_0, F\}$ in Example 2. Since the outgoing edges of v_5 are labeled with the symbols in $\{a, b, e\}$ and $(r[1] = r[4] = a) \wedge (r[2] = r[5] = b)$, the candidate states for v_5 is $R_c = \{1, 2, 4, 5\}$. If v_5 receives a message $m' = ((v_1, 1, a), (v_2, 2, b))$ from v_2, then $R_{q'} = \{2, 3\}$. Thus, $(v_5, 2, r[2])$ is appended to m' to generate a new matched message m.

5.3 Message Compression

In Algorithm 2, a message $m \in M_s$ may be sent several times when there exist some outgoing edges labeled with the same symbol. To reduce the cost of message passing, in Algorithm 4, we compress the duplicate messages to be sent. Thus, Algorithm 4 is an optimized version of line 20 in Algorithm 2. We employ a sequence S_m of messages, which is attached to the vertex v, to keep the original uncompressed messages. Then m is compressed into a sequence $((v, q', i))$ consisting of only one element, where q' is the matched state and i denotes the index of the original message in S_m.

Algorithm 4. messageCompression()

---

1 $S_m \leftarrow$ the sequence of the messages attached to the vertex v;
2 **if** $m \notin M_s$ **then**
3 $\quad \lfloor \ M_s \leftarrow M_s \cup \{m\}$;

4 **else if** $m \notin S_m$ **then**
5 $\quad \mid \quad M_s \leftarrow M_s \setminus \{m\}$;
6 $\quad \mid \quad S_m \leftarrow$ append m to S_m;
7 $\quad \mid \quad i \leftarrow$ the index of m in S_m;
8 $\quad \mid \quad m \leftarrow ((v, q', i))$;   /* the compressed message to be sent */
9 $\quad \lfloor \quad M_s \leftarrow M_s \cup \{m\}$;

Example 4. For RDF graph T_2 in Example 2, when computing in v_3, there exist messages that can be compressed. When v_3 receives a message $m' = ((v_1, 1, \mathsf{a}), (v_2, 2, \mathsf{b}), (v_5, 2, \mathsf{b}))$, a new message $m = ((v_1, 1, \mathsf{a}), (v_2, 2, \mathsf{b}), (v_5, 2, \mathsf{b}), (v_3, 2, \mathsf{b}))$ will be built and then sent via two outgoing edges (v_3, b, v_6) and (v_3, b, v_4). Next, the original message m is appended to the sequence S_m, and m is compressed to become $((v_3, 2, 0))$, where the index $i = 0$ since m is the first element of S_m.

5.4 Message Selection

It turns out that it is inevitable to cause the counting-paths problem for a distributed Pregel-based algorithm to generate provenance-aware answers to RPQs, which may incur the prohibitively expensive overhead [1]. Given an RDF graph T_3 in Fig. 4, the vertex x has k incoming edges labeled with a and n outgoing edges labeled with b. If x receives a set M of k messages via the incoming edges $(v_1, x), (v_2, x), \ldots, (v_k, x)$, for each message $m' \in M$, n new messages may be built by appending $(v, q_x, r[q_x])$ to m' for each state q_x satisfying $r[q_x] = \mathsf{b}$. In the worst case, $k \times |r| \times n$ messages can be generated, which is actually the Cartesian product of the k receiving messages and n outgoing edges. It is known that the Cartesian product is the key factor in causing the counting-paths problem.

To partly address it, we reduce the number of matches between a vertex and a state, which is the dominant cost in vertex computation of each superstep. Let a message set $M = \{m'_1, \ldots, m'_k\}$ be a subset of M_r. If the matched states of the

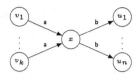

Fig. 4. RDF graph T_3 with the Cartesian product

last elements of all the messages in M are the same, we just select any message m' from M and append the element $(v, q_x, r[q_x])$ to m' to built the sending message. This optimization technique is referred to as *message selection*, which can avoid the Cartesian product to reduce the number of matches from $O(k \cdot |r| \cdot n)$ to $O(|r| \cdot n)$.

In addition, we also employ the message compression technique in Sect. 5.3 to reduce the cost of message passing to help alleviate the counting-paths problem even further.

6 Experimental Evaluation

In this section, we evaluate the performance of our method. We conducted extensive experiments to verify the efficiency and scalability of the proposed algorithms on both synthetic and real-world datasets.

6.1 Experimental Settings

The proposed algorithms were implemented in Scala using Spark GraphX, which were deployed on a 10-site cluster in the *Tencent Cloud*[1]. Each site in this cluster installs a 64-bit CentOS 7.3 Linux operating system, with a 4-core CPU and 16 GB memory. Our algorithm was executed on Java 1.8, Scala 2.11, Hadoop 2.7.4, and Spark 2.2.0.

Table 1. Datasets

Datasets	$|V|$	$|E|$
LUBM10	314,853	1,316,700
LUBM100	3,301,718	1,387,997
LUBM200	6,574,860	27,643,644
WatDiv10	158,118	1,109,678
WatDiv100	1,526,677	10,958,704
WatDiv200	3,228,213	24,098,747
DBpedia	5,526,330	18,295,010

Table 2. Regular path queries

Simple RPQs	
$Q_1 = (a/b)|(c/d)$	$Q_3 = a/b/(c|d|e)$
$Q_2 = a/b/c/d$	$Q_4 = (a|b)/(a|c)$
Complex RPQs	
$Q_5 = (a/a)^+$	$Q_9 = (a/b)^+|(c/d)^+$
$Q_6 = (a|b|c)^+$	$Q_{10} = (a/(b/c)^*)^+|(d/e)^+$
$Q_7 = a^+$	$Q_{11} = ((a/b)/(c|d)^*)^+/(e|f)^*$
$Q_8 = a/(a|b|c)^*$	$Q_{12} = (a|b)^+/(c|d)^+$

---

[1] https://cloud.tencent.com/.

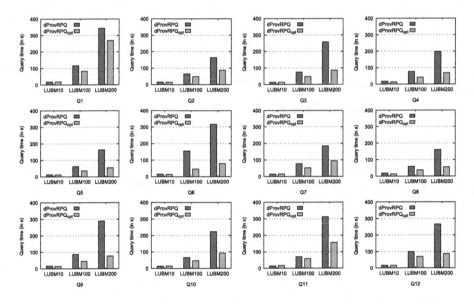

Fig. 5. The experimental results of efficiency on LUBM datasets

For the verification of our algorithms, we use three datasets, including two benchmark datasets (LUBM[2] and WatDiv[3]) and a real-world dataset (DBpedia[4]), which are listed in Table 1. For an RPQ Q, if it contains the closure operators $*$ and/or $+$, it is a complex query, otherwise it is a simple one. We designed twelve RPQs, denoted by Q_1 to Q_{12} in Table 2, which cover most typical patterns of RPQs.

6.2   Experimental Results

We compared the query time of the basic algorithm and the optimized algorithm with the edge-filtering and candidate-states strategies, which are denoted as dProvRPQ and dProvRPQ$_{opt}$, respectively, using 12 queries over 3 datasets.

Exp 1. Efficiency of the Algorithms. (1) *LUBM dataset.* Both dProvRPQ and dProvRPQ$_{opt}$ are executed on LUBM datasets, whose results are illustrated in Fig. 5. It can be observed that the algorithms scale linearly with the size of the data. However, the time of Q_1, Q_3, and Q_6 increases rapidly along with the increasing size of the data because the query results have reached a relatively large scale (i.e., millions of paths). In most cases, the query time of dProvRPQ$_{opt}$ is reduced significantly compared with dProvRPQ, which verifies the effectiveness of our optimization strategies. However, when Q_3 and Q_{11} are evaluated on

---

[2] http://swat.cse.lehigh.edu/projects/lubm/.
[3] http://dsg.uwaterloo.ca/watdiv/.
[4] http://wiki.dbpedia.org/.

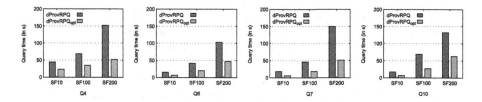

Fig. 6. The experimental results of efficiency on WatDiv datasets

LUBM10, dProvRPQ$_{opt}$ takes slightly longer time than dProvRPQ. The reason is that Σ_r of these queries involve more various symbols than other queries and LUBM10 is relatively small in size, which result in filtering out fewer useless edges than other queries.

(2) *WatDiv dataset.* Four representative queries (Q_4, Q_6, Q_7, and Q_{10}) are selected from Table 2 and evaluated on the WatDiv datasets of varying scale factors (SF), i.e., 10, 100, and 200, respectively. In Fig. 6, it can be observed that dProvRPQ$_{opt}$ outperforms dProvRPQ for all queries. The query time of dProvRPQ$_{opt}$ is on average 41.59% of that of dProvRPQ. Due to the diversified predicates in WatDiv, dProvRPQ$_{opt}$ can reduce the times of traversals and filter out more useless edges in comparison with the evaluation on the LUBM datasets.

(3) *DBpedia dataset.* The experimental results on the DBpedia dataset, as shown in Fig. 7(a), indicate that dProvRPQ$_{opt}$ performs better than dProvRPQ in all cases. We notice that the query time of Q_3 is more than other queries. In fact, the number of the intermediate partial results has reached millions of paths in the query processing of Q_3. For dProvRPQ$_{opt}$, the most significant improvement is for the most complex query Q_{11}, which takes 52.44% of the query time of dProvRPQ.

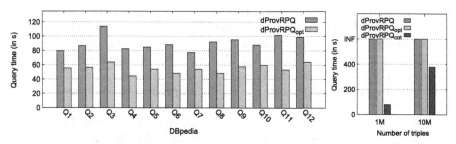

(a) The experimental results of efficiency on DBpedia datasets   (b) Counting-paths

Fig. 7. Queries on DBpedia datasets and counting-paths alleviation

Exp 2. Scalability of the Algorithms. In order to evaluate the scalability of dProvRPQ and dProvRPQ$_{opt}$, we use the LUBM100 and DBpedia datasets, with four representative queries (Q_2, Q_4, Q_5, and Q_{11}) selected from Table 2. The query time on the different number of sites, varying from 4 to 10, is shown in Fig. 8(a) and (b). The query time of dProvRPQ and dProvRPQ$_{opt}$ decreases with the number of sites increasing, which confirms that our algorithms can take full advantage of the vertex-centric Pregel framework for graph parallel computing. Moreover, the average speedup ratio of dProvRPQ is 1.21 times of dProvRPQ$_{opt}$.

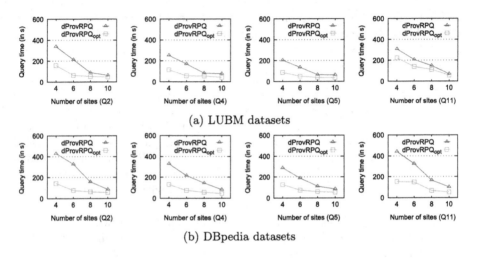

(a) LUBM datasets

(b) DBpedia datasets

Fig. 8. Scalability by varying the number of sites

Exp 3. Effectiveness of the Counting-Paths Alleviation Strategy. The optimized algorithm dProvRPQ$_{cnt}$ was implemented with the message-compression and message-selection techniques to partly address the counting-paths problem. Due to the limited lengths of the answers to the queries in Table 2, dProvRPQ$_{cnt}$ cannot reach its full potential. To this end, we generate RDF graphs w.r.t. the data model of WatDiv[5] by constructing structures like T_3 in Fig. 4. Meanwhile, we design an RPQ $Q_c = x_1/x_2/\ldots/x_{10}$ by covering the predicates that may generate the Cartesian product. It is obvious that the lengths of the answer paths to Q_c are much longer than that of the previous queries. In Fig. 7(b), it can be observed that dProvRPQ and dProvRPQ$_{opt}$ cannot finish within the time limit (10^4 s), denoted by INF, while dProvRPQ$_{cnt}$ can return the answers in 78.39 s and 377.56 s over the RDF graphs that contain 1 million and 10 million triples, respectively. Thus, dProvRPQ$_{cnt}$ can effectively alleviate the counting-paths problem.

---

[5] http://dsg.uwaterloo.ca/watdiv/watdiv-data-model.txt.

7   Conclusion

In this paper, we propose a novel method for answering provenance-aware RPQs over large RDF graphs by using the Pregel parallel graph computing framework. We also devise four optimization techniques, among which the edge-filtering and candidate-states techniques can significantly improve the performance of RPQs, and the message-compression and message-selection strategies are employed to alleviate the counting-paths problem. The extensive experiments were conducted on both synthetic and real-world datasets, which have verified the effectiveness, efficiency, and scalability of our method.

Acknowledgments. This work is supported by the National Natural Science Foundation of China (61572353, 61772361), the National High-tech R&D Program of China (863 Program) (2013AA013204), and the Natural Science Foundation of Tianjin (17JCYBJC15400).

References

1. Arenas, M., Conca, S., Pérez, J.: Counting beyond a Yottabyte, or how SPARQL 1.1 property paths will prevent adoption of the standard. In: Proceedings of the 21st International Conference on World Wide Web, pp. 629–638. ACM (2012)
2. Barceló, P., Libkin, L., Lin, A.W., Wood, P.T.: Expressive languages for path queries over graph-structured data. ACM Trans. Database Syst. (TODS) **37**(4), 31 (2012)
3. Brüggemann-Klein, A.: Regular expressions into finite automata. Theoret. Comput. Sci. **120**(2), 197–213 (1993)
4. Calvanese, D., De Giacomo, G., Lenzerini, M., Vardi, M.Y.: Answering regular path queries using views. In: 16th International Conference on Data Engineering, Proceedings, pp. 389–398. IEEE (2000)
5. Dey, S., Cuevas-Vicenttín, V., Köhler, S., Gribkoff, E., Wang, M., Ludäscher, B.: On implementing provenance-aware regular path queries with relational query engines. In: Proceedings of the Joint EDBT/ICDT 2013 Workshops, pp. 214–223. ACM (2013)
6. Harris, S., Seaborne, A., Prudhommeaux, E.: SPARQL 1.1 query language. W3C Recomm. **21**(10) (2013). https://www.w3.org/TR/sparql11-query/
7. Jupp, S., Malone, J., Bolleman, J., Brandizi, M., Davies, M., Garcia, L., Gaulton, A., Gehant, S., Laibe, C., Redaschi, N., et al.: The EBI RDF platform: linked open data for the life sciences. Bioinformatics **30**(9), 1338–1339 (2014)
8. Koschmieder, A., Leser, U.: Regular path queries on large graphs. In: Ailamaki, A., Bowers, S. (eds.) SSDBM 2012. LNCS, vol. 7338, pp. 177–194. Springer, Heidelberg (2012). https://doi.org/10.1007/978-3-642-31235-9_12
9. Kostylev, E.V., Reutter, J.L., Romero, M., Vrgoč, D.: SPARQL with property paths. In: Arenas, M., et al. (eds.) ISWC 2015. LNCS, vol. 9366, pp. 3–18. Springer, Cham (2015). https://doi.org/10.1007/978-3-319-25007-6_1
10. Malewicz, G., Austern, M.H., Bik, A.J., Dehnert, J.C., Horn, I., Leiser, N., Czajkowski, G.: Pregel: a system for large-scale graph processing. In: Proceedings of the 2010 ACM SIGMOD International Conference on Management of data, pp. 135–146. ACM (2010)

11. Nolé, M., Sartiani, C.: Regular path queries on massive graphs. In: Proceedings of the 28th International Conference on Scientific and Statistical Database Management, p. 13. ACM (2016)
12. Tong, Y., She, J., Meng, R.: Bottleneck-aware arrangement over event-based social networks: the max-min approach. World Wide Web **19**(6), 1151–1177 (2016)
13. Wang, X., Ling, J., Wang, J., Wang, K., Feng, Z.: Answering provenance-aware regular path queries on RDF graphs using an automata-based algorithm. In: Proceedings of the 23rd International Conference on World Wide Web, pp. 395–396. ACM (2014)
14. Wang, X., Wang, J.: ProvRPQ: an interactive tool for provenance-aware regular path queries on RDF graphs. In: Cheema, M.A., Zhang, W., Chang, L. (eds.) ADC 2016. LNCS, vol. 9877, pp. 480–484. Springer, Cham (2016). https://doi.org/10.1007/978-3-319-46922-5_44
15. Wang, X., Wang, J., Zhang, X.: Efficient distributed regular path queries on RDF graphs using partial evaluation. In: Proceedings of the 25th ACM International on Conference on Information and Knowledge Management, pp. 1933–1936. ACM (2016)

Discovering Graph Patterns for Fact Checking in Knowledge Graphs

Peng Lin[1], Qi Song[1], Jialiang Shen[3], and Yinghui Wu[1,2(✉)]

[1] Washington State University, Pullman, USA
{plin1,qsong,yinghui}@eecs.wsu.edu
[2] Pacific Northwest National Laboratory, Richland, USA
[3] Beijing University of Posts and Telecommunications, Beijing, China
shenjialiang@bupt.edu.cn

Abstract. Given a knowledge graph and a fact (a triple statement), fact checking is to decide whether the fact belongs to the missing part of the graph. This paper proposes a new fact checking method based on supervised graph pattern mining. Our method discovers discriminant graph patterns associated with the training facts. These patterns can then be used to construct classifiers based on either rules or latent features. (1) We propose a class of *graph fact checking rules* (GFCs). A GFC incorporates graph patterns that best distinguish true and false facts of generalized fact statements. We provide quality measures to characterize useful patterns that are both discriminant and diversified. (2) We show that it is feasible to discover GFCs in large graphs, by developing a supervised pattern discovery algorithm. To find useful GFCs as early as possible, it generates graph patterns relevant to training facts, and dynamically selects patterns from a pattern stream with small update cost per pattern. We further construct two GFC-based models, which make use of ordered GFCs as predictive rules and latent features from the pattern matches of GFCs, respectively. Using real-world knowledge bases, we experimentally verify the efficiency and the effectiveness of GFC-based techniques for fact checking.

1 Introduction

Knowledge graphs have been adopted to support emerging applications, *e.g.,* web search [6], recommendation [26], and decision making [14]. A knowledge graph consists of a set of facts. Each fact is a triple statement $<v_x, r, v_y>$, where v_x and v_y denote a *subject* entity and an *object* entity, respectively, and r refers to a *predicate* (a relation) between v_x and v_y. One of the cornerstone tasks for knowledge base management is *fact checking*. Given a knowledge graph G, and a fact t, it is to decide whether t belongs to the missing part of G. Fact checking can be used to (1) directly refine incomplete knowledge bases [2,6,19,25], (2) provide cleaned evidence for error detection in dirty knowledge bases [3,13,21,36], and (3) improve the quality of knowledge search.

© Springer International Publishing AG, part of Springer Nature 2018
J. Pei et al. (Eds.): DASFAA 2018, LNCS 10827, pp. 783–801, 2018.
https://doi.org/10.1007/978-3-319-91452-7_50

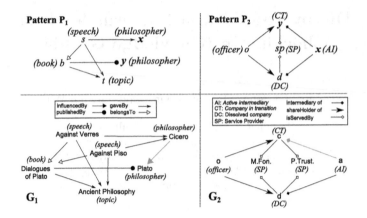

Fig. 1. Facts and associated graph patterns

Real-world facts in knowledge graphs are often associated with nontrivial regularities that involve both topological and semantic constraints beyond paths. Given observed facts, such regularities can be usually captured by graph patterns and their matches associated with the facts. Consider the following example.

Example 1. The graph G_1 in Fig. 1 illustrates a fraction of DBpedia [19] that depicts the facts about philosophers. A user is interested in finding "whether a philosopher v_x is *influenced by* another philosopher v_y". Given a relation influencedBy between philosophers (influencedBy($philosopher, philosopher$)) and an instance <Cicero, influencedBy, Plato> verified to be true in G_1, a graph pattern P_1 can be extracted to define influencedBy by stating that "*if a philosopher v_x (e.g., "Cicero") gave one or more speeches (e.g., "Against Piso") that cited a book of v_y (e.g., "Dialogues of Plato") with the same topic, then "v_x is likely to be influenced by v_y*".

Graph patterns with "weaker" constraints may not explain facts well. Consider a graph pattern P_1' obtained by removing the edge belongsTo($speech, book$) from P_1. Although "Cicero" and "Plato" also matches P_1', a false fact <Cicero, influencedBy, John Stuart Mill> also matches P_1' (not shown). Therefore, discriminant patterns that well distinguish observed true and false facts should be discovered.

As another example, consider graph G_2, a fraction of a real-world offshore activity network [16] in Fig. 1. To find whether an active intermediary (AI) a is likely to serve a company in transition (CT) c, a pattern P_2 that explains such an action may identify G_2 by stating "*a is likely an intermediary of c if it served for a dissolved company d which has the same shareholder o and one or more providers with c.*"

These graph patterns can be easily interpreted as rules, and the matches of the graph patterns readily provide instance-level evidence to "explain" the facts. These matches also indicate more accurate predictive models for various facts. We ask the following question: *How to characterize useful graph patterns and efficiently discover useful graph patterns to support fact checking in large knowledge graphs?*

Contribution. We propose models and pattern discovery algorithms that explicitly incorporate discriminant graph patterns to support fact checking in knowledge graphs.

(1) We introduce *graph fact checking rules* (GFCs) (Sect. 2). GFCs incorporate discriminant graph patterns as the antecedent and generalized triple patterns as the consequent, and characterize fact checking with graph pattern matching. We adopt computationally efficient pattern models to ensure tractable fact checking via pattern matching.

We also develop statistical measures (*e.g.,* support, confidence, significance, and diversity) to characterize useful GFCs (Sect. 3). Based on these measures, we formulate the top-k GFC discovery problem to mine useful GFCs for fact checking.

(2) We develop a feasible supervised pattern discovery algorithm to compute GFCs over a set of training facts (Sect. 4). In contrast to conventional pattern mining, the algorithm solves a submodular optimization problem with provable optimality guarantees, by a single scan of a stream of graph patterns, and incurs a small cost for each pattern.

(3) To evaluate the applications of GFCs, we apply GFCs to enhance rule-based and learning-based models, by developing two such classifiers. The first model directly uses GFCs as rules. The second model extracts instance-level features from the pattern matches induced by GFCs to learn a classifier (Sect. 4.2).

(4) Using real-world knowledge bases, we experimentally verify the efficiency of GFC-based techniques (Sect. 5). We found that the discovery of GFCs is feasible over large graphs. GFC-based fact checking also achieves high accuracy and outperforms its counterparts using Horn clause rules and path-based learning. We also show that the models are highly interpretable by providing case studies.

Related Work. We categorize the related work as follows.

Fact Checking. Fact checking has been studied for both unstructured data [10,28] and structured (relational) data [15,37]. These work rely on text analysis and crowd sourcing. Automatic fact checking in knowledge graphs is not addressed in these work. Beyond relational data, several methods have been studied to predict triples in graphs.

(1) Rule-based models extract association rules to predict facts. AMIE (or its improved version AMIE+) discovers rules with conjunctive Horn clauses [11, 12] for knowledge base enhancement. Beyond Horn-rules, GPARs [8] discover association rules in the form of $Q \Rightarrow p$, with a subgraph pattern Q and a single edge p. It recommends users via co-occurred frequent subgraphs.

(2) Supervised link prediction has been applied to train predictive models with latent features extracted from entities [6,18]. Recent work make use of path features [4,5,13,29,34]. The paths involving targeted entities are sampled from 1-hop neighbors [5] or via random walks [13], or constrained to be shortest paths [4]. Discriminant paths with the same ontology are grouped to generate positive and negative examples in [29].

Rule-based models are easy to interpret but usually cover only a subset of useful patterns [24]. It is also expensive to discover useful rules (*e.g.*, via subgraph isomorphism) [8]. On the other hand, latent feature models are more difficult to be interpreted [12,24] compared with rule models. Our work aims to balance the interpretability and model construction cost. (a) In contrast to AMIE [12], we use more expressive rules enhanced with graph patterns to express both constant and topological context of facts. Unlike [8], we use approximate pattern matching for GFCs instead of subgraph isomorphism, since the latter may produce redundant examples and is computationally hard in general. (b) GFCs can induce useful and discriminant features from patterns and subgraphs, beyond path features [5,13, 34]. (c) GFCs can be used as a standalone rule-based method. They also provide context-dependent features to support supervised link prediction to learn highly interpretable models. These are not addressed in [8,12].

Graph Pattern Mining. Frequent pattern mining defined by subgraph isomorphism has been studied for a single graph. GRAMI [7] discovers frequent subgraph patterns without edge labels. Parallel algorithms are also developed for association rules with subgraph patterns [8]. In contrast, (1) we adopt approximate graph pattern matching for feasible fact checking, rather than subgraph isomorphism as in [7,8]. (2) we develop a more feasible stream mining algorithm with optimality guarantees on rule quality, which incurs a small cost to process each pattern. (3) Supervised graph pattern mining over observed ground truth is not discussed in [7,8]. In contrast, we develop algorithms that discover discriminant patterns that best distinguish observed true and false facts.

Graph Dependency. Data dependencies have been extended to capture errors in graph data. Functional dependencies for graphs (GFDs) [9] enforce topological and value constraints by incorporating graph patterns with variables and subgraph isomorphism. These hard constraints (*e.g.*, subgraph isomorphism) are useful for detecting data inconsistencies but are often violated by incomplete knowledge graphs for fact checking tasks. We focus on "soft rules" to infer new facts towards data completion rather than identifying errors with hard constraints [27].

2   Fact Checking with Graph Patterns

We review the notions of knowledge graphs and fact checking. We then introduce a class of rules that incorporate graph patterns for fact checking.

Knowledge Graphs. A knowledge graph [6] is a directed graph $G = (V, E, L)$, which consists of a finite set of nodes V, a set of edges $E \subseteq V \times V$, and for each node $v \in V$ (resp. edge $e \in E$), $L(v)$ (resp. $L(e)$) is a label from a finite alphabet, which encodes the content of v (resp. e) such as properties, relations or names.

For the example in Fig. 1, a subject $v_x =$ "Cicero" carries a *type* $x =$ "philosopher" and an object $v_y =$ "Plato" carries a *type* $y =$ "philosopher". A fact $<v_x, r, v_y> = <\text{Cicero}, \texttt{influencedBy}, \text{Plato}>$ is an edge e in G encoded with label "influencedBy" between the subject v_x and the object v_y.

Fact Checking in Knowledge Graphs. Given a knowledge graph G and a new fact $<v_x, r, v_y>$, where v_x and v_y are in G, the task of fact checking is to learn and use a model M to decide whether the relation r exists between v_x and v_y [24]. This task can be represented by a binary query in the form of $<v_x, r?, v_y>$, where the model M outputs "true" or "false" for the query. A variant of fact checking is a class of queries in the form of $<v_x, r, v_y?>$, where the model M outputs a set of possible values of type y for specified v_x and r that are likely to be true in G.

Graph Pattern. A graph pattern $P(x, y) = (V_P, E_P, L_P)$ (or simply P) is a directed graph that contains a set of pattern nodes V_P and pattern edges E_P, respectively. Each pattern node $u_p \in V_P$ (resp. edge $e_p \in E_P$) has a label $L_P(u_p)$ (resp. $L_P(e_p)$). Moreover, it contains two designated *anchored nodes* u_x and u_y in V_P of type x and y, respectively. Specifically, when it contains a single pattern edge with label r between u_x and u_y, P is called a *triple pattern*, denoted as $r(x, y)$.

Pattern Matches. Given a graph pattern $P(x, y)$ and a knowledge graph G, a node match $v \in V$ of a pattern node u_p has the same label of node u_p, and an edge match $e = (v, v')$ of a pattern edge $e_p = (u, u')$ is induced by the matches v and v' of nodes u and u', respectively. We say a pattern P *covers* a fact $<v_x, r, v_y>$ in G if v_x and v_y match its anchored nodes u_x and u_y, respectively. Specifically, a pattern P can cover a fact by enforcing two established semantics.

(1) Subgraph patterns [7] define pattern matching in terms of subgraph isomorphism, induced by bijective functions.
(2) *Approximate patterns* [22,32] specify a matching relation R with constraints below to preserve both parent and child relations of P. For each pair $(u, v) \in R$,
   – for every edge $e = (u, u') \in E_P$, there exists an edge match $e' = (v, v') \in E$; and
   – for every edge $e = (u'', u) \in E_P$, there exists an edge match $e' = (v'', v) \in E$.

To ensure feasible fact checking in large knowledge graphs, we adopt approximate patterns for our model (see "Semantics").

We now introduce our rule model that incorporates graph patterns.

Rule Model. A *graph fact checking rule* (denoted as GFC) is in the form of $\varphi : P(x, y) \rightarrow r(x, y)$, where (1) $P(x, y)$ and $r(x, y)$ are two graph patterns carrying the same pair of anchored nodes (u_x, u_y), and (2) $r(x, y)$ is a triple pattern and is not in $P(x, y)$.

Semantics. A GFC $\varphi : P(x,y) \rightarrow r(x,y)$ states that "*a fact $<v_x, r, v_y>$ holds between v_x and v_y in G, if (v_x, v_y) is covered by P*". To ensure computationally efficient fact checking, we prefer approximate patterns instead of subgraph patterns. Subgraph isomorphism may be an overkill in capturing meaningful patterns [22,31,32] and is expensive (NP-hard). Moreover, it generates (exponentially) many isomorphic subgraphs, and thus introduces redundant features for model learning. In contrast, it is in $O(|V_P|(|V_P| + |V|)(|E_P| + |E|))$ time to find whether a fact is covered by an approximate pattern [22]. The tractability carries over to the validation of GFCs (Sect. 4).

Example 2. Consider the patterns and graphs in Fig. 1. To verify the influence between philosophers, a GFC $\varphi_1 : P_1(x,y) \rightarrow$ `influencedBy`(x,y). Pattern P_1 has two anchored nodes x and y, both with type `philosopher`, and covers the pair (`Cicero`, `Plato`) in G_1. Another GFC $\varphi_2 : P_2(x,y) \rightarrow$`intermediaryOf`$(x,y)$ verifies the service between a pair of matched entities (`a`, `c`). Note that with subgraph isomorphism, P_1 induces two subgraphs of G_1 that only differ by entities with label `speech`. It is not practical for users to inspect such highly overlapped subgraphs.

Remarks. We compare GFCs with two models below. (1) Horn rules are adopted by AMIE+ [11], in the form of $\bigwedge B_i \rightarrow r(x,y)$, where each B_i is an atom (fact) carrying variables. It mines only closed (each variable appears at least twice) and connected (atoms transitively share variables/entities to all others) rules. We allow general approximate graph patterns in GFCs to mitigate missing data and capture richer context features for supervised models (Sect. 4). (2) The association rules with graph patterns [8] have similar syntax with GFCs but adopt strict subgraph isomorphism for social recommendation. In contrast, we define GFCs with semantics and quality measures (Sect. 3) specified for observed true and false facts to support fact checking.

3   Supervised GFC Discovery

To characterize useful GFCs, we introduce a set of measures that extend their counterparts from established rule-based models [12] and discriminant patterns [38], specialized for a set of training facts. We then formalize supervised GFC discovery problem.

Statistical Measures. Our statistical measures are defined for a given graph $G = (V, E, L)$ and a set of *training facts* Γ. The training facts Γ consists of two classes: true facts Γ^+, which contains validated facts that hold in G; and false facts Γ^-, which contains triples not in G, respectively. Following common practice in knowledge base completion [24], we adopt the silver standard to construct Γ from G, where (1) $\Gamma^+ \subseteq E$, and (2) Γ^- are created following partial closed world assumption (see "Confidence").
       We use the following notations. Given a GFC $\varphi : P(x,y) \rightarrow r(x,y)$, graph G, facts Γ^+ and Γ^-, (1) $P(\Gamma^+)$ (resp. $P(\Gamma^-)$) refers to the set of training facts in

Γ^+ (resp. Γ^-) that are covered by $P(x,y)$ in Γ^+ (resp. Γ^-). $P(\Gamma)$ is defined as $P(\Gamma^+) \cup P(\Gamma^-)$, *i.e.*, all the facts in Γ covered by P. (2) $r(\Gamma^+), r(\Gamma^-)$, and $r(\Gamma)$ are defined similarly.

Support and Confidence. The support of a GFC $\varphi : P(x,y) \to r(x,y)$, denoted by $\mathsf{supp}(\varphi, G, \Gamma)$ (or simply $\mathsf{supp}(\varphi)$), is defined as

$$\mathsf{supp}(\varphi) = \frac{|P(\Gamma^+) \cap r(\Gamma^+)|}{|r(\Gamma^+)|}$$

Intuitively, the support is the fraction of the true facts as instances of $r(x,y)$ that satisfy the constraints of a graph pattern $P(x,y)$. It extends head coverage, a practical version for rule support [12] to address triple patterns $r(x,y)$ with not many matches due to the incompleteness of knowledge bases.

Given two patterns $P_1(x,y)$ and $P_2(x,y)$, we say $P_1(x,y)$ *refines* $P_2(x,y)$ (denoted by $P_1(x,y) \preceq P_2(x,y)$, if P_1 is a subgraph of P_2 and they pertain to the same pair of anchored nodes (u_x, u_y). We show that GFC support preserves anti-monotonicity in terms of pattern refinement.

Lemma 1. *For graph G, given any two* GFCs $\varphi_1 : P_1(x,y) \to r(x,y)$ *and* $\varphi_2 : P_2(x,y) \to r(x,y)$, *if* $P_1(x,y) \preceq P_2(x,y)$, $\mathsf{supp}(\varphi_2) \leq \mathsf{supp}(\varphi_1)$.

Proof Sketch: It suffices to show that any pair (v_{x_2}, v_{y_2}) covered by P_2 in G is also covered by $P_1(x,y)$. Assume there exists a pair (v_{x_2}, v_{y_2}) covered by P_2 but not by P_1, and assume *w.l.o.g.* v_{x_2} does not match the anchored node u_x in P_1. Then there exists either (a) an edge (u_x, u) (or (u, u_x)) in P_1 such that no edge (v_{x_2}, v) (or (v, v_{x_2})) is a match, or (b) a node u as an ancestor or descendant of u_x in P_1, such that no ancestor or descendant of v_{x_2} in G is a match. As P_2 refines P_1, both (a) and (b) lead to that v_{x_2} is not covered by P_2, which contradicts the definition of approximate patterns. □

Following rule discovery in incomplete knowledge base [12], we adopt *partial closed world assumption* (PCA) to characterize the confidence of GFCs. Given triple pattern $r(x,y)$ and a true instance $<v_x, r, v_y> \in r(\Gamma^+)$, PCA assumes that a missing instance $<v_x, r, v_y'>$ of $r(x,y)$ is a false fact if v_y and v_y' have different values. We define a normalizer set $P(\Gamma^+)_N$, which contains the entity pairs (v_x, v_y) that are in $P(\Gamma^+)$, and each pair has at least a false counterpart under PCA. The confidence of φ in G, denoted as $\mathsf{conf}(\varphi, G, \Gamma)$ (or simply $\mathsf{conf}(\varphi)$), is defined as

$$\mathsf{conf}(\varphi) = \frac{|P(\Gamma^+) \cap r(\Gamma^+)|}{|P(\Gamma^+)_N|}$$

The confidence measures the probability that a GFC holds over the entity pairs that satisfy $P(x,y)$, normalized by the facts that are assumed to be false under PCA. We follow PCA to construct false facts in our experimental study.

Significance. A third measure quantifies how significant a GFC is in "distinguishing" the true and false facts. To this end, we extend the G-test score [38]. The

G-test tests the null hypothesis of whether the number of true facts "covered" by $P(x, y)$ fits its distribution in the false facts. If not, $P(x, y)$ is considered to be statistically significant. Specifically, the score (denoted as $\text{sig}(\varphi, p, n)$, or simply $\text{sig}(\varphi)$) is defined as

$$\text{sig}(\varphi) = 2|\Gamma^+|(p \ln \frac{p}{n} + (1 - p) \ln \frac{1-p}{1-n})$$

where p (resp. n) is the frequency of the facts covered by pattern P of φ in Γ^+ (resp. Γ^-), i.e., $p = \frac{|P(\Gamma^+)|}{|\Gamma^+|}$ (resp. $n = \frac{|P(\Gamma^-)|}{|\Gamma^-|}$). As $\text{sig}(\cdot)$ is not anti-monotonic, a common practice is to use a "rounded up" score to find significant patterns [38]. We adopt an upper bound of $\text{sig}(\cdot)$, denoted as $\hat{\text{sig}}(\varphi, p, n)$ (or $\hat{\text{sig}}(\varphi)$ for simplicity), which is defined as $\max\{\text{sig}(\varphi, p, \delta), \text{sig}(\varphi, \delta, n)\}$, where $\delta > 0$ is a small constant (to prevent the case that $\text{sig}(\cdot) = \infty$). It is not hard to show that $\hat{\text{sig}}(\cdot)$ is anti-monotonic in terms of pattern refinement, following a proof similar to Lemma 1. Besides, we normalize $\hat{\text{sig}}(\varphi)$ to $\hat{\text{nsig}}(\varphi)$ in range $[0, 1]$ by a sigmoid function, i.e., $\hat{\text{nsig}}(\varphi) = \tanh(\hat{\text{sig}}(\varphi))$.

Redundancy-Aware Selection. In practice, one wants to find GFCs with both high significance and low redundancy. Indeed, a set of GFCs can be less useful if they "cover" the same set of true facts in Γ^+. We introduce a bi-criteria function that favors significant GFCs that cover more diversified true facts. Given a set of GFCs $\mathcal{S}$, when the set of true facts Γ^+ is known, the *coverage score* of $\mathcal{S}$, denoted as $\text{cov}(\mathcal{S})$, is defined as

$$\text{cov}(\mathcal{S}) = \text{sig}(\mathcal{S}) + \text{div}(\mathcal{S})$$

The first term, defined as $\text{sig}(\mathcal{S}) = \sqrt{\sum_{\varphi \in \mathcal{S}} \hat{\text{nsig}}(\varphi)}$, aggregates the total significance of GFCs in $\mathcal{S}$. The second term, defined as $\text{div}(\mathcal{S}) = \left(\sum_{t \in \Gamma^+} \sqrt{\sum_{\varphi \in \Phi_t(\mathcal{S})} \text{supp}(\varphi)} \right) / |\Gamma^+|$, where $\Phi_t(\mathcal{S})$ refers to the GFCs in $\mathcal{S}$ that cover a true fact $t \in \Gamma^+$, quantifies the diversity of $\mathcal{S}$, following a diversity reward function [20]. Intuitively, it rewards the diversity in that there is more benefit in selecting a GFC that covers new facts, which are not covered by other GFCs in $\mathcal{S}$ yet. Both terms are normalized to $(0, \sqrt{|\mathcal{S}|}]$.

The coverage score favors GFCs that cover more distinct true facts with more discriminant patterns. Moreover, adding a new GFC φ to a set $\mathcal{S}$ improves its significance and coverage at least as much as adding it to any superset of $\mathcal{S}$ (diminishing gain to $\mathcal{S}$). That is, $\text{cov}(\cdot)$ is well defined in terms of submodularity [23], a property widely used to justify goodness measures for set mining. Define the marginal gain $\text{mg}(\varphi, \mathcal{S})$ of a GFC φ to a set $\mathcal{S}$ ($\varphi \notin \mathcal{S}$) as $\text{cov}(\mathcal{S} \cup \{\varphi\}) - \text{cov}(\mathcal{S})$. We show the following result.

Lemma 2. *The function* $\text{cov}(\cdot)$ *is a monotone submodular function for* GFCs, *that is, (1) for any two sets* $\mathcal{S}_1$ *and* $\mathcal{S}_2$, *if* $\mathcal{S}_1 \subseteq \mathcal{S}_2$, *then* $\text{cov}(\mathcal{S}_1) \leq \text{cov}(\mathcal{S}_2)$, *and (2) for any two sets* $\mathcal{S}_1$ *and* $\mathcal{S}_2$, *if* $\mathcal{S}_1 \subseteq \mathcal{S}_2$ *and for any* GFC $\varphi \notin \mathcal{S}_2$, $\text{mg}(\varphi, \mathcal{S}_2) \leq \text{mg}(\varphi, \mathcal{S}_1)$.

It is easy to verify that $\mathsf{cov}(\cdot)$ is a monotone submodular function by the definition of monotone submodularity and that both $\mathsf{sig}(\mathcal{S})$ and $\mathsf{div}(\mathcal{S})$ are defined in terms of square root functions. Due to space limit, we omit the detailed proof.

We now formulate the supervised top-k GFC discovery problem over observed facts.

Top-k Supervised GFC Discovery. Given graph G, support threshold σ and confidence threshold θ, and training facts Γ^+ and Γ^- as instances of a triple pattern $r(x, y)$, and integer k, the problem is to identify a set $\mathcal{S}$ of top-k GFCs that pertain to $r(x, y)$, such that (a) for each GFC $\varphi \in \mathcal{S}$, $\mathsf{supp}(\varphi) \geq \sigma$, $\mathsf{conf}(\varphi) \geq \theta$, and (b) $\mathsf{cov}(\mathcal{S})$ is maximized.

4 Discovery Algorithm

4.1 Top-k GFC Mining

The supervised discovery problem for GFCs is not surprisingly intractable. A naive "enumeration-and-verify" algorithm that generates and verifies all k-subsets of GFC candidates that cover some examples in Γ is clearly not practical for large G and Γ. We consider more efficient algorithms.

"Batch + Greedy". We start with an algorithm (denoted as GFC_batch) that takes a batch pattern discovery and a greedy selection as follows. (1) Apply graph pattern mining (*e.g.*, Apriori [17]) to generate and verify all the graph patterns $\mathcal{P}$. The verification is specialized by an operator Verify, which invokes the pattern matching algorithm in, *e.g.*, [22] to compute the support and confidence for each pattern. (2) Invoke a greedy algorithm to do k passes of $\mathcal{P}$. In each iteration i, it selects the pattern P_i, such that the corresponding GFC $\varphi_i : P_i(x, y) \rightarrow r(x, y)$ maximizes the marginal gain $\mathsf{cov}(\mathcal{S}_{i-1} \cup \{\varphi_i\}) - \mathsf{cov}(\mathcal{S}_{i-1})$, and then it updates $\mathcal{S}_i$ as $\mathcal{S}_{i-1} \cup \{\varphi_i\}$.

GFC_batch guarantees a $(1 - \frac{1}{e})$ approximation, following Lemma 2 and the seminal result in [23]. Nevertheless, it requires the verification of all patterns before the construction of GFCs. The selection further requires k passes of all the verified patterns. This can be expensive for large G and Γ.

We can do better by capitalizing on stream-based optimization [1,32]. In contrast to "batch" style mining, we organize newly generated patterns in a stream, and assemble new patterns to top-k GFCs with small update costs. This requires a single scan of all patterns, without waiting for all patterns to be verified. We develop such an algorithm to discover GFCs with optimality guarantees, as verified by the result below.

Theorem 1. *Given a constant $\epsilon > 0$, there exists a stream algorithm that computes top-k GFCs with the following guarantees:*

(1) It achieves an approximation ratio $(\frac{1}{2} - \epsilon)$;
(2) It performs a single pass of all processed patterns $\mathcal{P}$, with update cost in $O((b + |\Gamma_b|)^2 + \frac{\log k}{\epsilon})$, where b is the largest edge number of the patterns, and Γ_b is the b hop neighbors of the entities in Γ.

Algorithm GFC_stream

Input: Graph G, training facts Γ, support threshold σ,
              confidence threshold θ, integer k, triple pattern $r(x, y)$.
Output: Top-k GFCs $\mathcal{S}$ pertaining to $r(x, y)$.

1.   set $\mathcal{S} := \emptyset$; set $\mathcal{P} := \emptyset$; maxpcov = 0;
2.   graph pattern $P := \mathsf{PGen}(G, \Gamma, \sigma, \theta).\mathrm{next}()$;
3.   **while** $P \neq$ null **do**
4.   **if** P is a size-1 pattern **and** maxpcov$<$cov(P) **then**
5.   maxpcov $:=$ cov(P);
6.   **if** P contains more than one edge **then**
7. $\mathcal{S} := \mathsf{PSel}(P, \mathcal{S}, \mathcal{P}, r(x, y), \mathsf{maxpcov})$;
8.   pattern $P := \mathsf{PGen}(G, \Gamma, \sigma, \theta).\mathrm{next}()$;
9.   **return** $\mathcal{S}$;

Fig. 2. Algorithm GFC_stream

As a proof of Theorem 1, we next introduce such a stream discovery algorithm.

"Stream + Sieve". Our supervised discovery algorithm, denoted as GFC_stream (illustrated in Fig. 2), interleaves pattern generation and GFC selection as follows.

(1) *Pattern stream generation.* The algorithm GFC_stream invokes a procedure PGen to produce a pattern stream (lines 2, 8). In contrast to GFC_batch that verifies patterns against entire graph G, it partitions facts Γ to blocks, and iteratively spawns and verifies patterns by visiting local neighbors of the facts in each block. This progressively finds patterns that better "purify" the labels of only those facts they cover, and thus reduces unnecessary enumeration and verification.

(2) *Selection On-the-fly.* GFC_stream invokes a procedure PSel (line 7) to select patterns and construct GFCs on-the-fly. To achieve the optimality guarantee, it applies the stream-sieving strategy in stream data summarization [1]. In a nutshell, it estimates the optimal value of a monotonic submodular function $F(\cdot)$ with multiple "sieve values", initialized by the maximum coverage score of single patterns (Sect. 3) maxpcov $= \max_{P \in \mathcal{P}}(\mathsf{cov}(P))$ (lines 4–5), and eagerly constructs GFCs with high marginal benefit that refines sieve values progressively.

The above two procedures interact with each other: each pattern verified by PGen is sent to PSel for selection. The algorithm terminates when no new pattern can be verified by PGen or the set $\mathcal{S}$ can no longer be improved by PSel (as will be discussed). We next introduce the details of procedures PGen and PSel.

Procedure PGen. Procedure PGen improves its "batch" counterpart in GFC_batch by locally generating patterns that cover particular sets of facts, following a manner of decision tree construction. It maintains the following structures in each iteration i: (1) a pattern set $\mathcal{P}_i$, which contains graph patterns of size (number of pattern edges) i, and is initialized as a size-0 pattern that contains anchored nodes u_x and u_y only; (2) a partition set $\Gamma_i(P)$, which records the sets of facts $P(\Gamma^+)$ and $P(\Gamma^+)$, is initialized as $\{\Gamma^+, \Gamma^-\}$, for each pattern $P \in \mathcal{P}_i$. At iteration i, it performs the following.

(1) For each block $B \in \Gamma_{i-1}$, PGen generates a set of graph patterns $\mathcal{P}_i$ with size i. A size-i pattern P is constructed by adding a triple pattern $e(u, u')$ to its size-$(i-1)$ counterpart P' in $\mathcal{P}_{i-1}$. Moreover, it only inserts $e(u, u')$ with instances from the neighbors of the matches of P', bounded by $P'(\Gamma)$.
(2) For each pattern $P \in \mathcal{P}_i$, PGen computes its support, confidence and significance (G-test) as in procedure Verify of algorithm GFC_batch, and prunes $\mathcal{P}_i$ by removing unsatisfied patterns. It refines $P'(\Gamma^+)$ and $P'(\Gamma^-)$ to $P(\Gamma^+)$ and $P(\Gamma^-)$ accordingly. Note that $P(\Gamma^+) \subseteq P'(\Gamma^+)$, and $P(\Gamma^-) \subseteq P'(\Gamma^-)$. Once a promising pattern P is verified, PGen returns P to procedure PSel for the construction of top-k GFCs $\mathcal{S}$.

Procedure PSel. To compute the set of GFCs $\mathcal{S}$ that maximizes $\mathsf{cov}(\mathcal{S})$ for a given $r(x, y)$, it suffices for procedure PSel to compute top k graph patterns that maximize $\mathsf{cov}(\mathcal{S})$ accordingly. It solves a submodular optimization problem over the pattern stream that specializes the sieve-streaming technique [1] to GFCs.

Sieve-Streaming [1]. Given a monotone submodular function $F(\cdot)$, a constant $\epsilon > 0$ and element set $\mathcal{D}$, sieve-streaming finds top-k elements $\mathcal{S}$ that maximizes $F(\mathcal{S})$ as follows. It first finds the largest value of singleton sets $m = \max_{e \in \mathcal{D}} F(\{e\})$, and then uses a set of sieve values $(1+\epsilon)^j$ (j is an integer) to discretize the range $[m, k * m]$. As the optimal value, denoted as $F(\mathcal{S}^*)$, is in $[m, k * m]$, there exists a value $(1 + \epsilon)^j$ that "best" approximates $F(\mathcal{S}^*)$. For each sieve value v, a set of top patterns $\mathcal{S}_v$ is maintained, by adding patterns with a marginal gain at least $(\frac{v}{2} - F(\mathcal{S}_v))/(k - |\mathcal{S}_v|)$. It is shown that selecting the sieve of best k elements produces a set $\mathcal{S}$ with $F(\mathcal{S}) \geq (\frac{1}{2} - \epsilon)F(\mathcal{S}^*)$ [1].

A direct application of the above sieve-streaming for GFCs seems infeasible: one needs to find the maximum $\mathsf{cov}(\varphi)$ (or $\mathsf{cov}(P)$ for fixed $r(x, y)$), which requires to verify the entire pattern set. Capitalizing on data locality of graph pattern matching, Lemma 2, and Lemma 1, we have good news.

Lemma 3. *It is in $O(|\Gamma_1|)$ time to compute the maximum* $\mathsf{cov}(P)$.

This can be verified by observing that $\mathsf{cov}(\cdot)$ also preserves anti-monotonicity in terms of pattern refinement. That is, $\mathsf{cov}(P') \leq \mathsf{cov}(P)$ if $P \preceq P'$. Thus, $\max_{P \in \mathcal{P}} \mathsf{cov}(P)$ is contributed by a single-edge pattern. That is, procedure PSel only needs to cache at most $|\Gamma_1|$ size-1 patterns from PGen to find the global maximum $\mathsf{cov}(P)$ (lines 4–5 of GFC_stream). The rest of PSel follows the sieve-streaming strategy, as illustrated in Fig. 3. The GFCs are constructed with the top-k graph patterns (line 8).

Procedure PSel$(P, \mathcal{S}, \mathcal{P}, r(x,y), \mathsf{maxpcov})$

1.   **if** $\mathcal{P} := \emptyset$ **then**
2.   set $\mathcal{S}_V := \{(1+\epsilon)^j | (1+\epsilon)^j \in [\mathsf{maxpcov}, k * \mathsf{maxpcov}]\}$;
3.   **for each** $v_j \in \mathcal{S}_V$ **do** $\mathcal{S}_{v_j} := \emptyset$;
4.   **for each** $v_j \in \mathcal{S}_V$ **do**
5.   **if** $\mathsf{mg}(P, \mathcal{S}_{v_j}) \geq (\frac{v_j}{2} - \mathsf{cov}(\mathcal{S}_{v_j}))/(k - |\mathcal{S}_{v_j}|)$ **and** $|\mathcal{S}_{v_j}| < k$ **then**
6. $\mathcal{S}_{v_j} := \mathcal{S}_{v_j} \cup \{P\}$;
7.   **if** all sets $\mathcal{S}_{v_j}$ have size k **then**
8. $\mathcal{S} := \mathsf{constructGFCs}(\mathcal{S}_{v_j})$ $(j \in [1, \log_{(1+\epsilon)}(k * \mathsf{maxpcov})])$
9.   **return** $\mathcal{S}$;

Fig. 3. Procedure PSel

Optimization. To further prune unpromising patterns, procedure PGen estimates an upperbound $\hat{\mathsf{mg}}(P, \mathcal{S}_{v_j})$ (line 5 of PSel) without verifying a new size-i pattern P. If $\hat{\mathsf{mg}}(P, \mathcal{S}_{v_j}) < (\frac{v_j}{2} - \mathsf{cov}(\mathcal{S}_{v_j}))/(k - |\mathcal{S}_{v_j}|)$, P is skipped without further verification.

To this end, PGen first traces to a GFC $\varphi' : P'(x,y) \rightarrow r(x,y)$, where P' is a verified sub-pattern of P, and P is obtained by adding a triple pattern r' to P'. It estimates an upper bound of the support of the GFC $\varphi : P(x,y) \rightarrow r(x,y)$ as $\hat{\mathsf{supp}}(\varphi) = \mathsf{supp}(\varphi') - \frac{l}{|r(\Gamma^+)|}$, where l is the number of the facts in $r(\Gamma^+)$ that have no instance of r' in their i hop neighbors (thus cannot be covered by P). Similarly, one can estimate an upper bound for p and n in the formula of $\mathsf{sig}(\cdot)$, and thus get an upper bound for $\hat{\mathsf{nsig}}(\varphi)$. For each t in Γ^+, denote term $\sqrt{\sum_{\varphi \in \Phi_t(\mathcal{S})} \mathsf{supp}(\varphi)}$ in $\mathsf{div}(\mathcal{S})$ as T_t, it then computes $\hat{\mathsf{mg}}(P, \mathcal{S})$ as $\frac{\hat{\mathsf{nsig}}(\varphi)}{2\sqrt{\mathsf{sig}(\mathcal{S})}} + \left(\sum_{t \in P(\Gamma^+)} \frac{\hat{\mathsf{supp}}(\varphi)}{2\sqrt{T_t}}\right)/|\Gamma^+|$. One may prove that this is an upper bound for $\hat{\mathsf{mg}}(P, \mathcal{S})$, by applying the inequality $\sqrt{\alpha + \beta} - \sqrt{\alpha} \leq \beta/(2\sqrt{\alpha})$ to each square root term in $\mathsf{sig}(\cdot)$ and $\mathsf{div}(\cdot)$. We found that this effectively reduces redundant verifications (see Sect. 5).

Performance Analysis. Denote the total patterns verified by the algorithm GFC_stream as $\mathcal{P}$, it takes $O(|\mathcal{P}|(b + |\Gamma_b|)^2)$ time to compute the pattern matches and verify the patterns. Each time a pattern is verified, it takes $O(\frac{\log k}{\epsilon})$ time to update the set $\mathcal{S}_v$. Thus the update time for each pattern is in $O((b + |\Gamma|_b)^2 + \frac{\log k}{\epsilon})$.

The approximation ratio follows the analysis of sieve stream summarization in [1]. Specifically, (1) there exists a sieve value $v_j = (1+\epsilon)^j \in [\mathsf{maxpcov}, k * \mathsf{maxpcov}]$ that is closest to $F(\mathcal{S}^*)$, say, $(1 - 2\epsilon)F(\mathcal{S}^*) \leq v_j \leq F(\mathcal{S}^*)$; and (2) the set $\mathcal{S}_{v_j}$ is a $(\frac{1}{2} - \epsilon)$ answer for an estimation of $F(\mathcal{S}^*)$ with sieve value v_j. Indeed, if $\mathsf{mg}(P, \mathcal{S}_{v_j})$ satisfies the test in PSel (line 5), then $\mathsf{cov}(\mathcal{S}_{v_j})$ is at least $\frac{v_j|\mathcal{S}|}{2k} = \frac{v_j}{2}$ (when $|\mathcal{S}| = k$). Following [1], there exists at least a value $v_j \in \mathcal{S}_V$ that best estimates the optimal $\mathsf{cov}(\cdot)$, and thus achieves approximation ratio $(\frac{1}{2} - \epsilon)$. Thus, selecting the GFCs with patterns from the sieve set with the largest coverage value guarantees approximation ratio $(\frac{1}{2} - \epsilon)$.

The above analysis completes the proof of Theorem 1.

4.2   GFC-Based Fact Checking

The GFCs can be applied to enhance fact checking as rule models or via supervised link prediction. We introduce two GFC-based models.

Using GFCs as Rules. Given facts Γ, a rule-based model, denoted as GFact_R, invokes algorithm GFC_stream to discover top-k GFCs S as fact checking rules. Given a new fact $e = <v_x, r, v_y>$, it follows "hit and miss" convention [12] and checks if there exists a GFC φ in S that covers e (*i.e.*, both its consequent and antecedent cover e). If so, GFact_R accepts e, otherwise, it rejects e.

Using GFCs in Supervised Link Prediction. Useful instance-level features can be extracted from the patterns and their matches induced by GFCs to train classifiers. We develop a second model (denoted as GFact) that adopts the following specifications.

Features. For each example $e = <v_x, r, v_y> \in \Gamma$, GFact constructs a feature vector of size k, where each entry encodes the presence of the ith GFC φ_i in the top-k GFCs S. The class label of the example e is *true* (resp. *false*) if $e \in \Gamma^+$ (resp. Γ^-).

By default, GFact adopts Logistic Regression, which is experimentally verified to achieve slightly better performance than others (*e.g.*, Naive Bayes and SVM). We find that GFact outperforms GFact_R over real-world graphs (See Sect. 5).

5   Experimental Study

Using real-world knowledge bases, we empirically evaluate the efficiency of GFCs discovery and the effectiveness of GFC-based fact checking.

Datasets. We used five real-world knowledge graphs, including (1) YAGO [33] (version 2.5), a knowledge base that contains $2.1M$ entities with 2273 distinct labels, $4.0M$ edges with 33 distinct labels, and $15.5K$ triple patterns; (2) DBpedia [19] (version 3.8), a knowledge base that contains $2.2M$ entities with 73 distinct labels, $7.4M$ edges with 584 distinct labels, and $8.2K$ triple patterns; (3) Wikidata [35] (RDF dumps 20160801), a knowledge base that contains $10.8M$ entities with 18383 labels, $41.4M$ edges of 693 relations, and $209K$ triple patterns; (4) MAG [30], a fraction of an academic graph with $0.6M$ entities (*e.g.*, papers, authors, venues, affiliations) of 8565 labels and $1.71M$ edges of 6 relationships (cite, coauthorship), and (5) Offshore [16], a social network of offshore entities and financial activities, which contains $1M$ entities (*e.g.*, companies, countries, person) with 357 labels, $3.3M$ relationships (*e.g.*, establish, close) with 274 labels, and 633 triple patterns. We use Offshore mostly for case studies.

Methods. We implemented the following methods in Java: (1) algorithm GFC_stream, compared with (a) its "Batch + Greedy" counterpart GFC_batch (Sect. 4), (b) algorithm AMIE+ [11] that discovers AMIE rules, (c) PRA [6,18], the Path Ranking Algorithm that trains classifiers with path features from random walks, and (d) KGMiner [29], a variant of PRA that makes use of features from discriminant paths; (2) fact checking models GFact_R and GFact, compared with learned models (and also denoted) by $\mathsf{AMIE+}$, PRA, and $\mathsf{KGMiner}$, respectively. For practical comparison, we set a pattern size (the number of pattern edges) bound $b = 4$ for GFC discovery.

Model Configuration. For fair comparison, we made effort to calibrate the models and training/testing sets under consistent settings. (1) For the supervised link prediction methods (GFact, PRA, and $\mathsf{KGMiner}$), we sample 80% of the facts in a knowledge graph as the training facts Γ, with instances of in total 107 triple patterns, and 20% edges as testing set $\mathcal{T}$. Each triple pattern has $5K$-$50K$ instances. In Γ (resp. $\mathcal{T}$), 20% are true examples Γ^+ (resp. $\mathcal{T}^+$), and 80% are false examples Γ^- (resp. $\mathcal{T}^-$). We generate Γ^- and $\mathcal{T}^-$ under PCA (Sect. 3) for all the models. For all methods, we use Logistic Regression to train the classifiers, same as the default settings of PRA and $\mathsf{KGMiner}$. (2) For rule-based methods GFact_R and $\mathsf{AMIE+}$, we discover rules that cover the same set of Γ^+. We set the size of $\mathsf{AMIE+}$ rule body to be 3, comparable to the number of pattern edges in our work.

Overview of Results. We find the following. (1) It is feasible to discover GFCs in large graphs (**Exp-1**). For example, it takes 211 s for $\mathsf{GFC\_stream}$ to discover GFCs over YAGO with 4 million edges and 3000 training facts. On average, it outperforms $\mathsf{AMIE+}$ by 3.4 times. (2) GFCs can improve the accuracy of fact checking models (**Exp-2**). For example, it achieves additional 30%, 20% and 5% gain of precision over DBpedia, and 20%, 15% and 16% gain of F_1 score over Wikidata when compared with $\mathsf{AMIE+}$, PRA, and $\mathsf{KGMiner}$, respectively. (3) Our case study shows that GFact yields interpretable models (**Exp-3**).

We next report the details of our findings.

Exp-1: Efficiency. We report the efficiency of $\mathsf{GFC\_stream}$, compared with $\mathsf{AMIE+}$, $\mathsf{GFC\_batch}$ and PRA over DBpedia. As $\mathsf{KGMiner}$ has unstable learning time and is not comparable, the result is omitted.

Varying $|E|$. Fixing $|\Gamma^+| = 15K$, support threshold $\sigma = 0.1$, confidence threshold $\theta = 0.005$, $k = 200$, we sampled 5 graphs from DBpedia, with size (number of edges) varied from $0.6M$ to $1.8M$. Figure 4(a) shows that all methods take longer time over larger $|E|$, as expected. (1) $\mathsf{GFC\_stream}$ is on average 3.2 (resp. 4.1) times faster than $\mathsf{AMIE+}$ (resp. $\mathsf{GFC\_batch}$) due to its approximate matching scheme and top-k selection strategy. (2) Although $\mathsf{AMIE+}$ is faster than $\mathsf{GFC\_stream}$ over smaller graphs, we find that it returns few rules due to low support. Enlarging rule size (*e.g.*, to 5), $\mathsf{AMIE+}$ does not run to completion. (3) The cost of PRA is less sensitive due to that it samples a (predefined) fixed number of paths.

Varying $|\Gamma^+|$. Fixing $|E| = 1.5M$, $\sigma = 0.1$, $\theta = 0.005$, $k = 200$, we varied $|\Gamma^+|$ from $3K$ to $15K$. As shown in Fig. 4(b), while all the methods take longer time for larger $|\Gamma^+|$, $\mathsf{GFC\_stream}$ scales best with $|\Gamma^+|$ due to its stream selection strategy. $\mathsf{GFC\_stream}$ achieves comparable efficiency with PRA, and outperforms $\mathsf{GFC\_batch}$ and $\mathsf{AMIE+}$ by 3.54 and 5.1 times on average, respectively.

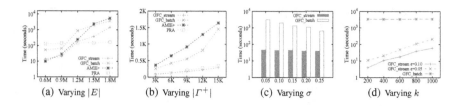

(a) Varying $|E|$ (b) Varying $|\Gamma^+|$ (c) Varying σ (d) Varying k

Fig. 4. Efficiency of GFC_stream

Table 1. Effectiveness: average accuracy.

Model	YAGO				DBpedia				Wikidata				MAG			
	Pred	Prec	Rec	F_1	Pred	Prec	Rec	F_1	Pred	Prec	Rec	F_1	Pred	Prec	Rec	F_1
GFact	**0.89**	**0.81**	0.60	**0.66**	**0.91**	**0.80**	0.55	**0.63**	**0.92**	**0.82**	0.63	**0.68**	**0.90**	0.86	**0.62**	**0.71**
GFact$_R$	0.73	0.40	0.75	0.50	0.70	0.43	0.72	0.52	0.85	0.55	0.64	0.55	0.86	0.78	0.55	0.64
AMIE+	0.71	0.44	0.76	0.51	0.69	0.50	0.85	0.58	0.64	0.42	0.78	0.48	0.70	0.53	0.62	0.52
PRA	0.87	0.69	0.34	0.37	0.88	0.60	0.41	0.45	0.90	0.65	0.51	0.53	0.77	0.88	0.21	0.32
KGMiner	0.87	0.62	0.36	0.40	0.88	0.75	0.60	0.63	0.90	0.63	0.49	0.52	0.76	0.74	0.17	0.27

Varying σ. Fixing $|E| = 1M, \theta = 0.005, k = 200$, we varied σ from 0.05 to 0.25. As shown in Fig. 4(c), GFC_batch takes longer time over smaller σ, due to more patterns and GFC candidates need to be verified. On the other hand, GFC_stream is much less sensitive. This is because it terminates early without verifying all patterns.

Varying k. Fixing $G = 1.5M$, $\sigma = 0.1, \theta = 0.005$, we varied k from 200 to 1000. Figure 4(d) shows that GFC_stream is more sensitive to k due to it takes longer to find k best patterns for each sieve value. Although GFC_batch is less sensitive, the major bottleneck is its verification cost. In addition, we found that with larger ϵ, less number of patterns are needed, thus GFC_stream takes less time.

Exp-2: Accuracy. We report the accuracy of all the models in Table 1.

Rule-Based Models. We apply the same support threshold $\sigma = 0.1$ for AMIE+ and GFact$_R$. We set $\theta = 0.005$ for GFact$_R$, and set $k = 200$. We sample 20 triple patterns $r(x,y)$ and report the average accuracy. As shown in Table 1, GFact$_R$ constantly improves AMIE+ with up to 21% gain in prediction rate, and with comparable performance for other cases. We found that AMIE+ reports rules with high support but not necessarily meaningful, while GFCs capture more meaningful context (see Exp-3). Both models have relatively high recall but low precision, due to that they have better chance to cover missing facts but may introduce errors when hitting false facts.

Supervised Models. We next compare GFact with supervised link prediction models (Table 1). GFact achieves the highest prediction rates and F_1 scores. It outperforms PRA with 12% gain on precision and 23% gain on recall on average, and outperforms KGMiner with 16% gain on precision and 19% recall. Indeed, GFact extracts useful features from GFCs with both high significance and diversity, beyond path features.

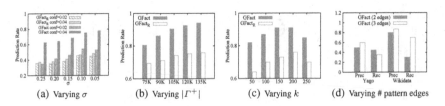

(a) Varying σ (b) Varying $|\Gamma^+|$ (c) Varying k (d) Varying # pattern edges

Fig. 5. Impact factors to accuracy

We next evaluate the impact of factors to the model accuracy using Wikidata.

Varying σ and θ. Fixing $|\Gamma^+| = 135K$, we varied σ from 0.05 to 0.25 for both GFact and $GFact_R$. We select 20 patterns with 0.02 confidence and 20 patterns with 0.04 confidence, respectively. Figure 5(a) shows that both GFact and $GFact_R$ has lower prediction rate when support threshold (resp. confidence) is higher (resp. lower). This is because fewer patterns can be discovered with higher support, leading to more "misses" in facts; while higher confidence lead to stronger association of patterns and more accurate predictions. In general, GFact achieves higher prediction rate than $GFact_R$.

Varying $|\Gamma^+|$. Fixing $\sigma = 0.01, \theta = 0.005, k = 200$, we vary $|\Gamma^+|$ from $75K$ to $135K$ as shown in Fig. 5(b). It tells us that GFact and $GFact_R$ have higher prediction rate with more positive examples. Their precisions (not shown) follow the similar trend.

Varying k. Fixing $\sigma = 0.01, \theta = 0.005, |\Gamma^+| = 2500$, we varied k from 50 to 250. Figure 5(c) shows the prediction rate first increases, and then decreases. For rule-based model, more rules increase the accuracy by covering more true facts, while increasing the risk of hitting false facts. For supervised link prediction, the model will be under-fitting with few features for small k, and will be over-fitting with too many features due to large k. We observe that $k = 200$ is a best setting for high prediction rate.

Varying b. Fixing $|E| = 4M, \sigma = 0.01, k = 1000$ and $\theta = 0.005$, we select 200 size-2 patterns and 200 size-3 patterns to train the models. Figure 5(d) verifies an interesting observation: smaller patterns contribute more to recall and larger patterns contribute more to precision. This is because smaller patterns are more likely to "hit" new facts, while larger patterns have stricter constraints for correct prediction of true fact.

Exp-3: Case Study. We perform case studies to evaluate the application of GFCs.

Accuracy. We show 3 relations and report accuracy in Fig. 6. These relations are non-functional, and may contain incomplete subjects and objects, where PCA may not hold [11]. We found that GFact complements AMIE+ and mitigate such disruptions with context enforced by graph patterns.

$r(x,y)$	GFact		GFact$_R$		AMIE+	
	Pred	Prec	Pred	Prec	Pred	Prec
r_1 (YAGO): *created*(director, movie)	0.81	0.56	0.33	0.13	0.39	0.12
r_2 (DBpedia): *derived*(music, music)	0.81	0.53	0.75	0.40	0.32	0.22
r_3 (Offshore): *serves*(person, company)	0.90	0.96	0.92	0.91	0.77	0.46

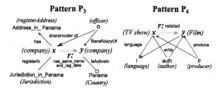

Fig. 6. Accuracy: case study.

Fig. 7. Real-world GFCs discovered by GFact

Interpretability. We further illustrate two top GFCs in Fig. 7, which contribute to highly important features in GFact with high confidence and significance over a real-world financial network Offshore and DBpedia, respectively.

(1) GFC φ_3 : $P_3(x,y) \rightarrow$ has_same_name_and_reg_date(x,y) states that two (anonymous) companies are likely to have the same name and registration date if they share shareholder and beneficiary, and one is registered and within jurisdiction in Panama, and the other is active in Panama. This GFC has support 0.12 and confidence 0.0086, and is quite significant. For the same $r(x,y)$, AMIE+ discovers a top rule as registerIn$(x$, Jurisdiction_in_Panama$) \wedge$ registerIn$(y$, Jurisdiction_in_Panama$)$ implies x and y has the same name and registration date. This rule has a low prediction rate.

(2) GFC φ_4 : $P_4(x,y) \rightarrow$ relevant(x,y) states that a TV show and a film have relevant content if they have the common language, authors and producers. This GFC has support 0.15 and a high confidence and significant score. Within bound 3, AMIE+ reports a top rule as Starring$(x,z) \wedge$ Starring$(y,z) \rightarrow$ relevant(x,y), which has low accuracy.

6   Conclusion

We have introduced GFCs, a class of rules that incorporate graph patterns to predict facts in knowledge graphs. We developed a stream-based rule discovery algorithm to find useful GFCs for observed true and false facts. We have shown that GFCs can be readily applied as rule models or provide useful instance-level features in supervised link prediction. Our experimental study has verified the effectiveness and efficiency of GFC-based techniques. We are evaluating GFCs with real-world graphs and pattern models. One future topic is to develop scalable GFCs-based models and methods with parallel graph mining and distributed rule learning.

Acknowledgments. This work is supported in part by NSF IIS-1633629 and Huawei Innovation Research Program (HIRP).

References

1. Badanidiyuru, A., Mirzasoleiman, B., Karbasi, A., Krause, A.: Streaming submodular maximization: massive data summarization on the fly. In: SIGKDD (2014)
2. Carlson, A., Betteridge, J., Kisiel, B., Settles, B., Hruschka Jr., E.R., Mitchell, T.M.: Toward an architecture for never-ending language learning. In: AAAI (2010)
3. Chen, Y., Wang, D.Z.: Knowledge expansion over probabilistic knowledge bases. In: SIGMOD (2014)
4. Ciampaglia, G.L., Shiralkar, P., Rocha, L.M., Bollen, J., Menczer, F., Flammini, A.: Computational fact checking from knowledge networks. PLoS One **10**, e0141938 (2015)
5. Cukierski, W., Hamner, B., Yang, B.: Graph-based features for supervised link prediction. In: IJCNN (2011)
6. Dong, X., Gabrilovich, E., Heitz, G., Horn, W., Lao, N., Murphy, K., Strohmann, T., Sun, S., Zhang, W.: Knowledge vault: a web-scale approach to probabilistic knowledge fusion. In: KDD (2014)
7. Elseidy, M., Abdelhamid, E., Skiadopoulos, S., Kalnis, P.: GraMI: frequent subgraph and pattern mining in a single large graph. PVLDB **7**, 517–528 (2014)
8. Fan, W., Wang, X., Wu, Y., Xu, J.: Association rules with graph patterns. PVLDB **8**, 1502–1513 (2015)
9. Fan, W., Wu, Y., Xu, J.: Functional dependencies for graphs. In: SIGMOD (2016)
10. Finn, S., Metaxas, P.T., Mustafaraj, E., O'Keefe, M., Tang, L., Tang, S., Zeng, L.: TRAILS: a system for monitoring the propagation of rumors on Twitter. In: Computation and Journalism Symposium, New York City, NY (2014)
11. Galárraga, L., Teflioudi, C., Hose, K., Suchanek, F.M.: Fast rule mining in ontological knowledge bases with AMIE+. VLDB J. **24**, 707–730 (2015)
12. Galárraga, L.A., Teflioudi, C., Hose, K., Suchanek, F.: AMIE: association rule mining under incomplete evidence in ontological knowledge bases. In: WWW (2013)
13. Gardner, M., Mitchell, T.M.: Efficient and expressive knowledge base completion using subgraph feature extraction. In: EMNLP (2015)
14. Goodwin, T.R., Harabagiu, S.M.: Medical question answering for clinical decision support. In: CIKM (2016)
15. Hassan, N., Sultana, A., Wu, Y., Zhang, G., Li, C., Yang, J., Yu, C.: Data in, fact out: automated monitoring of facts by FactWatcher. VLDB **7**, 1557–1560 (2014)
16. ICIJ: Offshore dataset. https://offshoreleaks.icij.org/pages/database
17. Jiang, C., Coenen, F., Zito, M.: A survey of frequent subgraph mining algorithms. Knowl. Eng. Rev. **28**, 75–105 (2013)
18. Lao, N., Mitchell, T., Cohen, W.W.: Random walk inference and learning in a large scale knowledge base. In: EMNLP (2011)
19. Lehmann, J., Isele, R., Jakob, M., Jentzsch, A., Kontokostas, D., Mendes, P.N., Hellmann, S., Morsey, M., Van Kleef, P., Auer, S., et al.: DBpedia-a large-scale, multilingual knowledge base extracted from Wikipedia. Semant. Web **6**, 167–195 (2015)
20. Lin, H., Bilmes, J.: A class of submodular functions for document summarization. In: ACL/HLT (2011)
21. Lin, Y., Liu, Z., Sun, M., Liu, Y., Zhu, X.: Learning entity and relation embeddings for knowledge graph completion. In: AAAI (2015)
22. Ma, S., Cao, Y., Fan, W., Huai, J., Wo, T.: Capturing topology in graph pattern matching. VLDB **5**, 310–321 (2011)

23. Nemhauser, G.L., Wolsey, L.A., Fisher, M.L.: An analysis of approximations for maximizing submodular set functions-I. Math. Program. **14**, 265–294 (1978)
24. Nickel, M., Murphy, K., Tresp, V., Gabrilovich, E.: A review of relational machine learning for knowledge graphs. Proc. IEEE **104**, 11–33 (2016)
25. Niu, F., Zhang, C., Ré, C., Shavlik, J.W.: Deepdive: web-scale knowledge-base construction using statistical learning and inference. VLDS **12**, 25–28 (2012)
26. Passant, A.: dbrec—music recommendations using DBpedia. In: Patel-Schneider, P.F., Pan, Y., Hitzler, P., Mika, P., Zhang, L., Pan, J.Z., Horrocks, I., Glimm, B. (eds.) ISWC 2010. LNCS, vol. 6497, pp. 209–224. Springer, Heidelberg (2010). https://doi.org/10.1007/978-3-642-17749-1_14
27. Paulheim, H.: Knowledge graph refinement: a survey of approaches and evaluation methods. Semant. Web **8**, 489–508 (2017)
28. Shao, C., Ciampaglia, G.L., Flammini, A., Menczer, F.: Hoaxy: a platform for tracking online misinformation. In: WWW Companion (2016)
29. Shi, B., Weninger, T.: Discriminative predicate path mining for fact checking in knowledge graphs. Knowl.-Based Syst. **104**, 123–133 (2016)
30. Sinha, A., Shen, Z., Song, Y., Ma, H., Eide, D., Hsu, B.j.P., Wang, K.: An overview of microsoft academic service (MAS) and applications. In: WWW (2015)
31. Song, C., Ge, T., Chen, C., Wang, J.: Event pattern matching over graph streams. VLDB **8**, 413–424 (2014)
32. Song, Q., Wu, Y.: Discovering summaries for knowledge graph search. In: ICDM (2016)
33. Suchanek, F.M., Kasneci, G., Weikum, G.: YAGO: a core of semantic knowledge. In: WWW (2007)
34. Thor, A., Anderson, P., Raschid, L., Navlakha, S., Saha, B., Khuller, S., Zhang, X.-N.: Link prediction for annotation graphs using graph summarization. In: Aroyo, L., Welty, C., Alani, H., Taylor, J., Bernstein, A., Kagal, L., Noy, N., Blomqvist, E. (eds.) ISWC 2011. LNCS, vol. 7031, pp. 714–729. Springer, Heidelberg (2011). https://doi.org/10.1007/978-3-642-25073-6_45
35. Vrandečić, D., Krötzsch, M.: Wikidata: a free collaborative knowledgebase. Commun. ACM **57**, 78–85 (2014)
36. Wang, Q., Liu, J., Luo, Y., Wang, B., Lin, C.Y.: Knowledge base completion via coupled path ranking. In: ACL (2016)
37. Wu, Y., Agarwal, P.K., Li, C., Yang, J., Yu, C.: Toward computational fact-checking. PVLDB **7**, 589–600 (2014)
38. Yan, X., Cheng, H., Han, J., Yu, P.S.: Mining significant graph patterns by leap search. In: SIGMOD (2008)

KAT: Keywords-to-SPARQL Translation Over RDF Graphs

Yanlong Wen, Yudong Jin, and Xiaojie Yuan[(✉)]

College of Computer and Control Engineering,
Nankai University, Nankai, People's Republic of China
{wenyanlong,jinyudong,yuanxiaojie}@dbis.nankai.edu.cn

Abstract. In this paper, we focus on the problem of translating key-words into SPARQL query effectively and propose a novel approach called KAT. KAT takes into account the context of each input keyword and reduces the ambiguity of input keywords by building a keyword index which contains the class information of keywords in RDF data. To explore RDF data graph efficiently, KAT builds a graph index as well. Moreover, a context aware ranking method is proposed to find the most relevant SPARQL query. Extensive experiments are conducted to show that KAT is both effective and efficient.

Keywords: Keywords-to-SPARQL · Two-facet index · Context aware

1 Introduction

RDF is a framework for Web resource description standardized by W3C, which provides a flexible way of representing knowledge. SPARQL is a structured query language for RDF, but composing SPARQL queries could be extremely difficult for users who are not familiar with the schema of underlying RDF data or the syntax of SPARQL. Keywords-to-SPARQL [3,9] provides a more simple way of exploring RDF data by translating keywords into SPARQL query automatically. However, the existing solutions may not answer keyword queries effectively and efficiently due to the following reasons. First, context information among input keywords is not taken into account. Second, schema information of RDF graph, such as RDF classes related to a certain keyword, is not exploited.

In this paper, we propose KAT for keywords-to-SPARQL translation, which takes into account both context and schema information. The main contributions are summarized as follows:

1. A novel keywords-to-SPARQL approach called KAT is proposed to translate input keywords into accurate SPARQL queries. KAT takes advantage of context information among the keywords to improve the relevance of results.
2. A two-facet index is presented. The keyword index contains RDF class information and helps solve keyword disambiguation, while the graph index is constructed to speed up the graph exploration.

© Springer International Publishing AG, part of Springer Nature 2018
J. Pei et al. (Eds.): DASFAA 2018, LNCS 10827, pp. 802–810, 2018.
https://doi.org/10.1007/978-3-319-91452-7_51

3. Effectiveness and efficiency of our approach are evaluated on both real and benchmark RDF datasets. Experimental results show that KAT performs better than previous approaches in both effectiveness and efficiency.

The rest of the paper is organized as follows. Section 2 formulates our problem. Section 3 provides an overview and details of KAT. Section 4 presents evaluation results. Section 5 surveys related work. Section 6 makes a conclusion.

2   Preliminaries

An RDF dataset is a collection of (*subject, predicate, object*) triples, where *subject* is an *entity* or a *class*, *object* is an *entity*, a *class* or a *literal value* and *predicate* expresses the attribute associated with the *subject*. An RDF dataset can be treated as a graph, where vertices represent *subjects* and *objects* and edges represent *predicates*, as shown in Fig. 1. Formally, an RDF data graph is defined as follows.

Definition 1 *An RDF data graph is a tuple $G = (V, L, E)$, where*

1. $V = V_E \cup V_C \cup V_L$ *is a collection of vertices corresponding to subjects and objects in RDF data, where V_E, V_C and V_L are collections of entity vertices, class vertices and literal vertices, respectively.*
2. $L = L_R \cup L_A \cup \{type, subClassOf\}$ *is a collection of edge labels corresponding to predicates in RDF data, where L_R and L_A are collections of inter-entity edge labels and entity-literal edge labels, respectively.*
3. $E = \{e\,(\overrightarrow{v_1, v_2})\,|v_1, v_2 \in V, e \in L\}$ *is a collection of directed edges.*

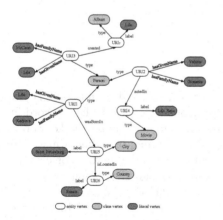

Fig. 1. RDF data graph

SELECT ?p WHERE {
?p rdf:type y:person.
?p y:actedIn ?m.
?m rdf:type y:movie.
?m rdfs:label "Lila_Says" }

Fig. 2. SPARQL query

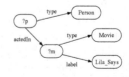

Fig. 3. SPARQL query graph

A SPARQL query is a collection of triples with parameters. For example, the SPARQL query in Fig. 2 would find persons who acted in the movie "Lila Says" from a YAGO dataset. A SPARQL query can be modeled as a graph as well. Figure 3 shows the query graph corresponding to the SPARQL query in Fig. 2. A SPARQL query graph can be viewed as a subgraph of the RDF data graph, replacing several vertices with parameters. Accordingly, the key of keywords-to-SPARQL method is to find the subgraph that contains the input keywords and meets the requirement of the user.

3   Keywords-to-SPARQL Translation

3.1   Overview

An overview of KAT is depicted in Fig. 4. During the offline process, a two-facet index, including a keyword index and a graph index, is created. When the user submits a keyword query, the keywords are mapped to relevant vertices and edges on the RDF graph called keyword elements by the keyword index. An exploration is then carried out on the graph index for finding subgraphs containing the keywords. The subgraphs are ranked by a context aware method and the top ones are translated into SPARQL queries.

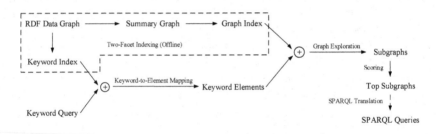

Fig. 4. Overview of KAT

3.2   Two-Facet Indexing

Keyword Indexing. The keyword index is constructed for mapping keywords to corresponding keyword elements. In previous approaches, the keyword index is simply an inverted index. Thus all the elements containing the keywords are regarded as candidates when conducting the mapping, including a mass of useless ones. This is mainly caused by the ambiguity of keywords. For example, a keyword query {person, movie, lila} is given to ask for the persons who acted in the movie "Lila Says". However, the keyword "lila" is quite an ambiguous word which could also mean a person named Lila, an album named Lila or even a metro named Lila.

To reduce the ambiguity of words, a hybrid index is proposed as the keyword index, in which class information of the terms in the RDF data is indexed. As shown in Fig. 5, each term points to a B-tree built on all possible RDF classes related to the term. The elements that contains the term are denoted in the

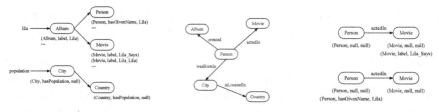

Fig. 5. Hybrid index   **Fig. 6.** Summary graph   **Fig. 7.** Subgraphs

form of (v_C, l_A, v_L) and assigned to the corresponding nodes on the B-tree, where $v_C \in V_C$, $l_A \in L_A$ and $v_L \in V_L$. Accordingly, keyword mapping for a keyword w is denoted by the function $f : (w, C) \rightarrow K$, where C is a set of RDF classes that w may belong to and K is a set of keyword elements. Specifically, the set C consists of the classes mentioned by unambiguous keywords such as "person" and "movie" and the classes that reach them within n hops on the summary graph, based on the common assumption that the information need of the user can be modeled in terms of entities which are closely related [9]. In this way keywords are mapped to a small and valuable set of keyword elements. For example, the metro named Lila would not be mapped in the case of the exemplary keyword query.

Graph Indexing. The graph index is built for exploration on graph. Since it is the structure of data graph that we are interested in, a small-scale summary graph that captures relations among classes of entities is adopted as the graph index. The summary graph of the RDF graph in Fig. 1 is shown in Fig. 6. When conducting graph exploration, the widely used backward search algorithm [4,7,9] is adopted to find subgraphs on the graph index. The algorithm starts from keyword elements and then performs an iterative traversal along edges until connecting vertices are found. Given the exemplary keyword query, subgraphs in Fig. 7 would be obtained after the exploration.

3.3 Context Aware Ranking

Ranking is an important work since it has a great influence on effectiveness of results. Standard methods are based on the structures of the subgraphs, i.e., smaller subgraphs tend to be correct answers. To overcome the limitations, a context aware ranking method is proposed to compute the relevance of subgraphs to the user's search intention. We investigated numbers of keyword queries given by different users and found that close keywords tend to indicate similar information. For example, a user tends to put a keyword "movie" next to "lila" to indicate the meaning of "lila". Thus the context among keywords can help find out the true requirement. Specifically, three kinds of context information are taken into account. First, the relevance of a keyword element (v_C, l_A, v_L) referring to a literal vertex increases if there is a keyword referring to a class semantically similar to v_C. Second, the relevance of a keyword element $(v_C, l_A, null)$

referring to an entity-literal edge increases if there is a keyword referring to a class semantically similar to v_C. Third, the relevance of a keyword element (v_C, l_A, v_L) referring to a literal vertex increases if there is a keyword referring to an entity-literal edge l_A whose related class is semantically similar to v_C.

The algorithm for keyword elements scoring is shown in Algorithm 1. The scores are initialized to 0 and increased based on the above rules. Line 3–7 follows the first two rules and line 8–12 follows the third one. The increment depends on semantic similarity between RDF classes and how close are the corresponding keywords in the sequence. Since classes tend to be similar if they share more super classes, the semantic similarity between classes c_1 and c_2 is denoted by

$$\text{sim}(c_1, c_2) = \frac{|S(c_1, O) \cap S(c_2, O)|}{|S(c_1, O) \cup S(c_2, O)|} \tag{1}$$

where $S(c, O)$ is the set of super classes of c in ontology O. Given two keyword elements k_1 and k_2, the relevance increment of k_2 according to k_1 is calculated by

$$\text{rel}(k_1, k_2) = \frac{\text{sim}(k_1.v_C, k_2.v_C)}{|\text{pos}(k_1) - \text{pos}(k_2)|} \tag{2}$$

where $\text{pos}(k)$ stands for the position of the keyword corresponding to keyword element k in the sequence of the keyword query. Finally, the score of a subgraph is calculated by dividing the sum of scores of involved keyword elements by the path length of the subgraph, i.e., the number of edges. The subgraphs are ranked by the scores and the top ones are translated into SPARQL queries.

Algorithm 1. Keyword Elements Scoring

Input: $K = (K_1, \ldots, K_n)$.
Output: Scores of elements in K.

1: **for** each $K_i \in K$ **do**
2:     **for** each $k \in K_i$ **do**
3:         **if** k is a class vertex **then**
4:             $\text{score}(k) \leftarrow 1$;
5:             **for** each entity-literal edge and literal vertex $k' \in K - K_i$ **do**
6:                 $\text{score}(k') \leftarrow \text{score}(k') + \text{rel}(k, k')$;
7:             **end for**
8:         **else if** k is an entity-literal edge **then**
9:             **for** each literal vertex $k' \in K - K_i$ **do**
10:                $\text{score}(k') \leftarrow \text{score}(k') + \text{rel}(k, k')$;
11:            **end for**
12:        **end if**
13:    **end for**
14: **end for**

4   Evaluation

4.1   Datasets and Keyword Queries

Real datasets YAGO with 946 classes and 2331482 entities, DBLP with 9 classes and 5934486 entities and benchmark dataset LUBM with 14 classes and 1355801 entities are used. Since there are not any benchmark keyword queries, ten different queries and corresponding intentions are given by the users who participate in the experiments for each data set and representative ones can be found in Table 1. QYi, QDi and QLi denote the queries for YAGO, DBLP and LUBM, respectively. Both effectiveness and efficiency experiments are conducted to compare KAT with Popularity-Cost [9] and Shortest-Path [3]. It is noteworthy that the experiments are conducted on a PC with 3.41 GHz Intel Core i7, 16 GB memory, 1TB SATA hard disk and Windows 10 OS.

Table 1. Keyword queries

#	Keyword query	Query intention
QY1	Person movie Lila	Persons, played a role in movie "Lila Says"
QY2	Artist Grammy award city Moscow	Artists, born in Moscow and won Grammy award
QY3	Westminster University Nobel prize	Nobel prize winners, graduated from Westminster University
QY4	Computer game company country Japan	Computer games, produced by companies of Japan
QY5	Mountain country China	Mountains, located in China
QY6	Acted movie Paris	Persons, played a role in movie "Paris"
QY7	Airport BFN country South Africa	Airports, airport code "BFN", located in South Africa
QY8	Created album Bad Girl	Artists, created the album "Bad Girl"
QY9	Artist Grammy award country England	Artists, born in England and won the Grammy award
QD1	Article machine learning journal artificial intelligence	Articles, about machine learning, published in journals of artificial intelligence
QD2	Journal data mining date 2011	Journals of data mining, published in 2011
QD3	Person a hybrid collaborative filtering recommendation mechanism for P2P networks	Persons, published "a hybrid collaborative filtering recommendation mechanism for P2P networks"
QL1	Graduate student University 20	Graduate students of University 20
QL2	Full professor University 15	Full professors of University 15
QL3	Undergraduate student department 1	Undergraduate students of department 1
QL4	Teacher course 5	Teachers of course 5
QL5	Teaching assistant course 10	Teaching assistants of course 10
QL6	Department University 15	Departments of University 15

4.2 Experimental Results

In order to assess the effectiveness of the generated SPARQL queries and their rankings, a standard IR metric *Reciprocal Rank* (RR) defined as $RR = 1/r$ is adopted, where r denotes the rank of the correct result, i.e., the SPARQL query that meets the search intention. Popularity-Cost introduces three different scoring functions and the best one is chosen. The RR scores are shown in Fig. 8. Since ambiguity of keywords is reduced and subgraphs with high relevance are found by the strategies, KAT performs better than others. Popularity-Cost focuses on the popularity of vertices and edges, so the correct but unpopular results are always ignored. Shortest-Path tries to identify the correct result by calculating the path length but the SPARQL needed is not always in a simple structure.

For evaluating the efficiency, average time needed for result generation is measured as the metric. In consideration of the top-k algorithm in Popularity-Cost, the k is set to ten in our experiments, which has been proved performing best in their paper. The experimental result is shown in Fig. 9. Popularity-Cost adopts a top-k algorithm but still needs to explore massive useless vertices on the RDF graph. KAT achieves better performance by filtering irrelevant information with the index. Shortest-Path performs poorly since shortest paths between pairs of vertices on the graph need to be computed.

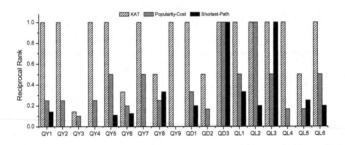

Fig. 8. Reciprocal ranks of different approaches

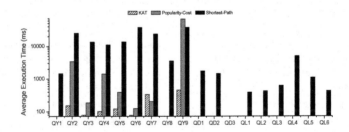

Fig. 9. Average execution time of different approaches

A hybrid keyword index is constructed in KAT. The novel index is compared with traditional inverted index adopted by previous work in both size and indexing time, as shown in Fig. 10. Since class information of terms in the RDF data is indexed, it requires a bit larger disk storage and it needs more time for index construction. However, the slight increase makes a great contribution to both effectiveness and efficiency of online execution.

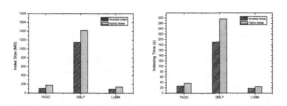

Fig. 10. Comparison of different keyword index

5   Related Work

Compared with approaches for keyword search on graph [4–6] and RDF [1,2,7], there is not much work on keywords-to-SPARQL [3,9]. Backward search is a typical algorithm for graph exploration [4,7,9] and bidirectional search [5] makes an improvement. A cluster-based exploration approach [1] generates results with path templates and *Kargar and An* [6] focus on finding r-cliques. Keywords-to-SPARQL solutions for RDF appear recently. *Tran et al.* [9] adopt backward search for finding subgraphs, while *Gkirtzou et al.* [3] utilize an algorithm based on shortest path computation. To achieve better efficiency, different strategies have been adopted. *He et al.* [4] store connectivity information offline to speed up the backward search algorithm. *Mass and Sagiv* [8] propose a pruning algorithm based on virtual documents to select top-n keyword elements. In terms of effectiveness, most of the ranking methods combine structural features with IR technique and *Tran et al.* [9] introduce the popularity.

6   Conclusion

The goal of this paper is to generate accurate SPARQL queries from a few keywords. To this end, a keyword index that contains RDF class information is built for disambiguation and a graph index for exploration. A context aware method is used for ranking. In the future, more semantical information among keywords would be considered for further improving and there is potential to speed up graph exploration by top-k algorithms.

Acknowledgement. This work is supported by National Natural Science Foundation of China (grant No. 61772289) and National 863 Program of China (grant No. 2015AA015401).

References

1. De Virgilio, R., Cappellari, P., Miscione, M.: Cluster-based exploration for effective keyword search over semantic datasets. In: Laender, A.H.F., Castano, S., Dayal, U., Casati, F., de Oliveira, J.P.M. (eds.) ER 2009. LNCS, vol. 5829, pp. 205–218. Springer, Heidelberg (2009). https://doi.org/10.1007/978-3-642-04840-1_17
2. Elbassuoni, S., Blanco, R.: Keyword search over RDF graphs. In: Proceedings of the 20th ACM International Conference on Information and Knowledge Management, pp. 237–242. ACM (2011)
3. Gkirtzou, K., Papastefanatos, G., Dalamagas, T.: RDF keyword search based on keywords-to-SPARQL translation. In: Proceedings of the First International Workshop on Novel Web Search Interfaces and Systems, pp. 3–5. ACM (2015)
4. He, H., Wang, H., Yang, J., Yu, P.S.: BLINKS: ranked keyword searches on graphs. In: Proceedings of the 2007 ACM SIGMOD International Conference on Management of Data, pp. 305–316. ACM (2007)
5. Kacholia, V., Pandit, S., Chakrabarti, S., Sudarshan, S., Desai, R., Karambelkar, H.: Bidirectional expansion for keyword search on graph databases. In: Proceedings of the 31st International Conference on Very Large Data Bases, pp. 505–516. VLDB Endowment (2005)
6. Kargar, M., An, A.: Keyword search in graphs: finding r-cliques. Proc. VLDB Endowment 4(10), 681–692 (2011)
7. Le, W., Li, F., Kementsietsidis, A., Duan, S.: Scalable keyword search on large RDF data. IEEE Trans. Knowl. Data Eng. 26(11), 2774–2788 (2014)
8. Mass, Y., Sagiv, Y.: Virtual documents and answer priors in keyword search over data graphs. In: EDBT/ICDT Workshops (2016)
9. Tran, T., Wang, H., Rudolph, S., Cimiano, P.: Top-k exploration of query candidates for efficient keyword search on graph-shaped (RDF) data. In: 2009 IEEE 25th International Conference on Data Engineering, ICDE 2009, pp. 405–416. IEEE (2009)

Text and Data Mining

A Scalable Framework for Stylometric Analysis of Multi-author Documents

Raheem Sarwar[1(✉)], Chenyun Yu[1], Sarana Nutanong[1],
Norawit Urailertprasert[2], Nattapol Vannaboot[2],
and Thanawin Rakthanmanon[2,3]

[1] Department of Computer Science, City University of Hong Kong,
Kowloon Tong, Hong Kong SAR, China
{rsarwar2-c,chenyunyu4-c}@my.cityu.edu.hk, s.nutanon@cityu.edu.hk
[2] Department of Computer Engineering, Kasetsart University, Bangkok, Thailand
{norawit.u,nattapol.v}@ku.th, thanawin.r@ku.ac.th
[3] Vidyasirimedhi Institute of Science and Technology, Rayong, Thailand

Abstract. *Stylometry* is a statistical technique used to analyze the variations in the author's writing styles and is typically applied to authorship attribution problems. In this investigation, we apply stylometry to authorship identification of multi-author documents (AIMD) task. We propose an AIMD technique called *Co-Authorship Graph (CAG)* which can be used to collaboratively attribute different portions of documents to different authors belonging to the same community. Based on CAG, we propose a novel AIMD solution which (i) significantly outperforms the existing state-of-the-art solution; (ii) can effectively handle a larger number of co-authors; and (iii) is capable of handling the case when some of the listed co-authors have not contributed to the document as a writer. We conducted an extensive experimental study to compare the proposed solution and the best existing AIMD method using real and synthetic datasets. We show that the proposed solution significantly outperforms existing state-of-the-art method.

Keywords: Stylometry · Authorship identification
Co-Authorship Graph · Multi-author documents

1 Introduction

Authorship attribution (AA) aims to infer authorship information from documents [6]. Authorship attribution has several variations depending upon the type of information to be inferred. One of the extensively investigated variation is *authorship identification* [5]. "*Authorship identification* aims at identifying the true author of a disputed document from a set of candidate authors" [15]. The main idea of authorship identification is that, by computing stylometric feature from documents and building a classification model on them, we can distinguish between documents written by different authors [15].

© Springer International Publishing AG, part of Springer Nature 2018
J. Pei et al. (Eds.): DASFAA 2018, LNCS 10827, pp. 813–829, 2018.
https://doi.org/10.1007/978-3-319-91452-7_52

One useful generalization of authorship identification problem that has received relatively little attention is *authorship identification of multi-author documents (AIMD)* [5]. The AIMD problem can be defined as follows. *Given a corpus of multi-author documents labeled with their authors, identify the authors of an anonymous multi-author document from a set of authors of a given corpus* [16].

Existing authorship identification techniques designed to handle single-author documents are not applicable to multi-author documents [5]. This is because, single-author authorship attribution techniques rely on the assumption that every text sample (document) has only one single label (author). However, the AIMD problem requires the ability to (i) infer the writing style of each individual author from a corpus of multi-author documents; and (ii) make a multi-label prediction for each document.

One prominent application domain of AIMD is *bibliometrics*, in which AIMD can help improve the processes of measuring and analyzing the collaborative natures among a community of researchers [18]. Instead of attributing the entire paper to all the listed authors, one can use AIMD techniques to perform a more fine-grained analysis. Specifically, different parts of the same document can be attributed to different authors on the author list. Such an authorship identification capability can help the information retrieval system in the following ways: (i) scholarly search engines may implement an author specific search in which the researchers can look for text sample written by a particular author; and (ii) a researcher may wish to construct individual author profiles reflecting the contributions of each author in different scientific fields. In addition, AIMD techniques can also be used to identify researchers who had been involved actively in writing and mentors who are giving feedback and providing ideas. Another aspect of the AIMD is the peer-review system of the academic conferences where both the reviewers and the authors of the paper stay anonymous. This notion can be challenged by showing that it is possible for a reviewer to reveal the identity of the authors of scientific papers by using the AIMD framework.

Several existing studies [4,9] on *authorship identification of multi-author documents (AIMD)* have shown some success on corpora consisting of scientific papers using the *citations* included in each paper. However, their success was achieved mostly in constrained scenarios, e.g., identifying the authors of papers sharing the self-citations. Along with the citation information, Payer et al. [16] have also made use of topic-information and some common stylometric features such as the frequencies of most common words.

The main difference between our work and a great majority of existing studies is that we make use of only the stylometric features. (see Sect. 2.2 for more details). Specifically, our features are topic-independent [7,19]. Hence, unlike most of the existing studies, our solution is also applicable to corpora where citation information is not available and the documents have different topics.

In summary, existing AIMD studies have the following limitations. (i) The accuracy levels of existing AIMD techniques can still be greatly improved. For example, the state-of-the-art stylometry based technique [5] reports an accuracy level less than 30% on a corpus containing over 360 candidate authors.

(ii) Existing techniques are adversely affected by an increase in the number of co-authors. For example, Dauber et al. [5] reported a drop in accuracy level from 25% to 16% as the number of authors had increased from 2 to 7. (iii) To the best of our knowledge, existing AIMD techniques do not tackle the issue of *non-writing authors (NWA)* [5,16]. However, NWAs do exist in real world scenarios. For example, in a scientific/engineering article, it is not necessary that all listed co-authors had contributed as *writers*.

In this investigation, we propose a solution to overcome the aforementioned limitations. The main challenge of AIMD is the lack of *"ground truth"* information. That is, most documents in the training set are associated with multiple authors. Hence, we need the ability to attribute different parts of the same document to different authors on the author list. In order to address this challenge, we propose a method which collaboratively learns individual writing styles from multiple co-authored documents called *Co-Authorship Graph (CAG)*.

Figure 1 illustrates the basic concept behind our CAG method. It shows four documents where each document contains three fragments. Each edge linking two fragments denotes that they are stylistically similar to each other. We initially assume that each fragment is associated with all listed authors. For example, the author list of three fragments $D_{1.1}$, $D_{1.2}$ and $D_{1.3}$ is $[A, B, C]$. That is, $D_{1.1}$, $D_{1.2}$, and $D_{1.3}$ could have been written by A, B, or C. The figure also shows that $D_{1.1}$ is stylistically similar to $D_{3.3}$ and $D_{4.2}$, which could have been written by $[C, D, A]$ and $[D, A, B]$, respectively. Since $D_{3.3}$ must have been written by one of the authors in $[A, B, C]$, we can see that A is the only author common to the three author lists. As a result, we can deduce that $D_{1.1}$ must have been written by A. Following the same principle, we can also deduce that the author of $D_{1.2}$ is B and the author of $D_{1.3}$ is C. The full result is given in the table on the right side of Fig. 1. In order to adopt the basic concept illustrated in Fig. 1 to a real-world corpus we have to address the following issues. First, the intersection between

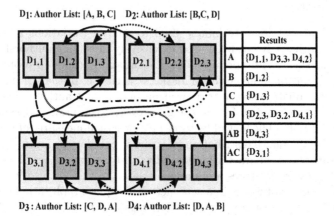

D_1: Author List: [A, B, C] D_2: Author List: [B,C, D]

	Results
A	$\{D_{1.1}, D_{3.3}, D_{4.2}\}$
B	$\{D_{1.2}\}$
C	$\{D_{1.3}\}$
D	$\{D_{2.3}, D_{3.2}, D_{4.1}\}$
AB	$\{D_{4.3}\}$
AC	$\{D_{3.1}\}$

D_3 : Author List: [C, D, A] D_4: Author List: [D, A, B]

Fig. 1. Co-Authorship Graph

multiple author lists (obtained from stylistically similar fragments) may not result with exactly 1 author. Second, the intersection between multiple author lists may result with no author at all being identified. Third, a number of listed authors may *not* have contributed as writers. In this paper, we formulate an AIMD solution that can handled these stated issues in a real world corpus.

In order to demonstrate the effectiveness of our method, we apply it to one synthetic dataset and two real datasets. We also compare our method against the best-existing AIMD solution [5]. Results from our experimental studies show that our method outperforms the best existing technique in all three datasets. The contributions of this investigation can be summarized as follows.

- We propose an AIMD technique called *Co-Authorship Graph (CAG)* which can be used to collaboratively attribute text fragments in a set of documents to distinguish authors in the same community.
- Based on the CAG technique, we propose a novel AIMD solution which (i) significantly outperforms the existing state-of-the-art solution; (ii) can effectively handle a larger number of co-authors; and (iii) is capable of handling NWAs.
- We conducted an extensive experimental study comparing the proposed method and the best existing AIMD method [5] using real and synthetic datasets.

The rest of the paper is organized as follows. Section 2 reviews previous studies on authorship attribution for single- and multi-author documents. Section 3 presents the proposed solution. Section 4 reports results from our extensive experimental studies. Section 5 contains our concluding remarks.

2   Literature Review

2.1   Stylometry

Stylometry is a statistical technique used to analyze variations in the writing styles of the authors. It has been used extensively in solving authorship attribution problems such as authorship identification and authorship profiling [15].

Stylometric features are stylistic markers/attributes of the writing style that can help discriminate between texts written by different authors. There are different types of stylometric features, e.g., lexical, structural, and syntactic features [5,7,11,14,16]. The lexical features are statistical measures of lexical variations such as word length distributions [11] and vocabulary richness [7]. Examples of lexical features are character-based and word-based measures of lexical variations [16]. Structural features are markers related to the layout of the text, e.g., the average number of words in a sentence or in a paragraph [11]. The examples of the syntactic features are part-of-speech tags and function words [14].

Payer et al. [16] proposed a solution for AIMD and applied it on a corpus of academic papers. They calculated a set of 10,727 features from the academic papers out of which 399 were stylometric and 2,374 content based, while the rest of the features 7,954 were based on citations. Later on, Dauber [5] proposed a solution for AIMD using the "Writeprints Limited features set" [1]. It includes content-specific, lexical, structural, syntactic and idiosyncratic features.

Comparison to Our Work. In this investigation, we use a set of 56 stylometric features which can be categorized into three types, namely, *syntactic, lexical* and *structural* features [7,11,14]. Specifically, we use 27 lexical [7,11], 2 structural [11], and 27 syntactic features [14]. These features are explained in Appendix A. The features used in this investigation differ those adopted in existing studies in several ways. Unlike the existing feature sets [4,5,9,16], our set of features contains only stylometric features. Specifically, these features are topic-independent [7,19]. As a result, our solution is also applicable to corpora in which citations information is not available and the documents address different topics. Moreover, we use a set of 56 features which is smaller than feature sets used in existing AIMD studies [4,5,9,16]. As a result, in comparison to existing studies, the proposed solution requires less storage and is computationally less expensive.

2.2 Authorship Identification

From the viewpoint of the context of this investigation, existing studies on authorship identification can be categorized into two types (i) authorship identification of single-author documents (AISD); and (ii) authorship identification of multi-author documents (AIMD). The main idea behind AISD is to identify the true author of a disputed document from a set of candidate authors. Existing studies of AISD have reported good results [3,17]. However, as already explained in the introduction section, existing authorship identification techniques designed to handle single-author documents are inapplicable to multi-author documents [5]. Since this investigation focuses on AIMD, we limit the discussion on AISD in interest of brevity.

AIMD. The AIMD problem can be defined as follows. *Given a corpus of multi-author documents labeled with their co-authors, identify the co-authors of an anonymous multi-author document from the authors in the given corpus* [16]. Several existing studies [4,9] on AIMD have shown some success for corpora that consist of scientific papers using only the *citations* made in each paper. However, their success was achieved mostly in a constrained scenario, e.g., identifying the authors of papers sharing the self-citations or in a specific domain such as Physics [9] or Machine Learning [4]. Specifically, Bradley et al. [4] reported less than 71% accuracy while Hill et al. [9] reported less than 50% accuracy. Later, Payer et al. [16] proposed a solution for AIMD and applied it on a corpus that consists of academic papers. Their method made use of citations-based features, stylometric features, and topic-based features. Hence, most of the existing solutions for AIMD are inapplicable to a corpus where the documents do not have citations such as novels or harassment letters, in addition their performance may turn worse when the corpus contains documents on multiple topics or may be performing topic classification. Our proposed solution is based purely on stylometric features. We do not make use of any other information such as topic information or citation information for the AIMD task. As a result, the feature set used in this investigation is topic-independent [7, 19].

There are several other variations of AIMD which are comparatively easier to implement than the aforementioned variation and have shown promising results. For example, one of the AIMD variations used a training set of single-author documents which makes this variation easy to tackle. However, this requirement may not be realistic in real-world scenarios in which the training sample themselves are also multi-author documents [8]. In addition, the study reported a drastic accuracy drop as the number of co-authors in one document increases, i.e., from 50% to 30% after increasing the number of co-authors from 2 to 3 [5]. Another variation of AIMD assumed that each co-author group had a sufficient number of writing samples for training [5]. Due to the combinatoric nature of collaborative patterns of researchers in a community, we consider this assumption to be unrealistic.

Since in the AIMD task, each document is associated with more than one author, where each author can be considered as a label, one can also consider this problem as a multi-label (ML) classification task. One of the popular ML classifiers is the *multi-label k-nearest neighbor (MLkNN) classifier* [22]. As with the regular kNN method, MLkNN identifies the k nearest neighbors with respect to a given test instance. To make a multi-label prediction, MLkNN derives statistical information from the label sets of identified kNNs, e.g., the number of neighbors for each label. Finally, it applies the *Maximum A Posteriori (MAP)* principle to determine the label set of the given test instance [22]. It can be seen that this multi-label classification task naturally fits the AIMD problem definition. However, existing AIMD studies have reported that transforming each multi-label sample into multiple single-label samples improves the classification accuracy [5, 16].

3   Proposed Solution

In this section, we show how the *collaborative authorship prediction* concept introduced in Sect. 1 can be realized. Our proposed solution consists of two preprocessing steps: *feature extraction* and *co-authorship graph training*. After the preprocessing steps, the trained data is used to make a multi-authorship prediction for any query document. Our design principle is based on the concept of *probabilistic multi-class, multi-label classification*. That is, each training sample is associated with multiple labels where each label is associated with a probability. Given a query sample Q, probabilistic labels of stylistically similar samples with respect to Q are used to derive a probabilistic prediction. In this way, we can accurately capture the multi-author nature in both training samples and test samples.

3.1   Preprocessing: Feature Extraction

In this subsection, we discuss the feature extraction process. Each document is represented as a collection of fragments where each fragment is represented as a set of points. Each point is calculated from 1,000 tokens (sequences of characters separated by white spaces) using the stylometric feature described in Appendix A. In this way, authorship predictions are made per fragment and the prediction for an entire document is an aggregation over multiple fragments associated with the same document.

The main motivation of this "collection of point sets" representation is two-fold. First, a point set can capture how one's writing style varies within the same document. Second, different parts of the same document can be associated with different authors. Note that in order to obtain reliable stylometric information for each data point, the number of tokens for each data point should be set to at least 1,000 [20]. However, with this number of tokens, we can have only 12 data points for each 12,000-token document, which is insufficient for our analysis. Hence, we apply the sliding window method to generated data points from overlapping token sequences. This process is illustrated in Fig. 2(a) with a sliding window increment of 100 tokens and the window size of 1,000, which are the value we use in this paper. In this way, we can generate, 111 data points from a 12,000-token document. The same principle is also applied to fragments in order for us to obtain a sufficient number of fragments for our analysis as shown in Fig. 2(b) with a fragment sliding window increment of 2 data points and a fragment size of 6 data points. In this way, we can generate 53 fragments from a 12,000-token document.

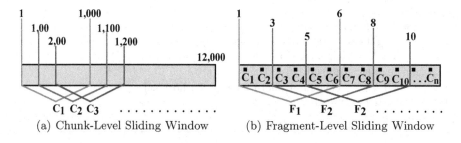

(a) Chunk-Level Sliding Window   (b) Fragment-Level Sliding Window

Fig. 2. Feature extraction

3.2   Preprocessing: Co-Authorship Graph Training

As stated in the introduction, the main challenge of the AIMD problem is that each document in the corpus at hand can be associated with multiple authors. Due to its combinatoric nature, the same list of authors may not be repeated in the corpus. In addition, some of the authors on the author list may not have contributed as writers to the document, making the AIMD problem more complicated. Hence, an AIMD predictive method must be able to infer the authorships of each document without relying on the absolute ground truth information.

In this investigation, we propose a novel AIMD solution based on the observation that *stylistically similar fragments should have been written by a similar group of authors.* As a result, we propose a data structure called *Co-authorship Graph (CAG)* to capture the stylistic similarity between these fragments. We also propose an iterative algorithm which attempts to identify the true writer of each fragment.

The structure of the CAG construction process is given in Algorithm 1. Recall that after the feature extraction process, each document is represented as a collection of point sets (fragments), where each data point corresponds to one feature vector. The algorithm iterates through all fragments from all documents (Lines 4 to 9). CAG edges can be constructed by identifying k stylistically similar fragments for each document fragment. We use *modified Hausdorff distance (MHD)* [12] as the distance between two fragments. Specifically, the procedure GetKNN(F, Fragments) finds k fragments in "Fragments" with the smallest MHDs from F (Line 5). These neighbors are the graph's edges, while the distances (MHDs) are edge weights. We assume that each fragment F is associated with the list of document authors F.AuthorList which may include one or more non-writing authors (NWA). The *probability mass function* (PMF) over the author list is initialized by giving each author on the list the same probability (Lines 8 to 9). After iterating through *all fragments from all documents*, the CAG is returned (Line 10).

Algorithm 1. CAGConstruction

1: **procedure** CAG CONSTRUCTION
2:     Vertices ← []
3:     Edges ← []
4:     **for** F in Fragments **do**
5:         Neighbors ← GetKNN(F, Fragments)
6:         **for** N in Neighbors **do**
7:             Edges.Append((F, N))
8:         F.PMF ← GenerateUniformPMF(F.AuthorList)
9:         Vertices.Append(F)
10:     **return** G(Vertices, Edges)

We illustrate now, how a CAG can be constructed using the example given in Fig. 3. The example contains 4 documents and each document is associated with 4 listed authors as shown in the figure. In this example, we set the ground truth as follows. First, only the first three authors contributed as writers to the respective document, e.g., only authors A, B, and C wrote different parts of D_1, while author W is an NWA. Similarly, authors X, Y, and Z are NWAs of D_2, D_3, and D_4, respectively. Note that this ground truth information is hidden from the model.

Figure 3 also illustrates the initial PMF of each document fragment. Since *the ground truth regarding the non-writer authors is hidden from the model.* All fragments of all documents are associated initially with all listed authors with the equal probability. For example, the author PMFs of $D1.1$, $D1.2$, $D1.3$ are uniform, e.g., $\{A : 0.25, B : 0.25, C : 0.25, W : 0.25\}$. The initial PMFs of the other fragments in the figures are derived in the same fashion.

After executing the CAG construction algorithm (Algorithm 1) we obtain the edges connecting stylistically similar fragments together. For example, according to the edges identified, we can see that the fragment $D_{1.1}$ is stylistically similar to $D_{3.3}$ and $D_{4.2}$. Similarly, $D_{1.2}$ is stylistically similar to $D_{2.1}$ and $D_{4.3}$. As can be seen, although all fragments in the same document are associated initially with the same authors with equal probability, they are connected to different sets of stylistically similar fragments with different author lists. Next, we will show that these differences can be used to collaboratively identify the author who had contributed as a writer of each fragment through Algorithm 2.

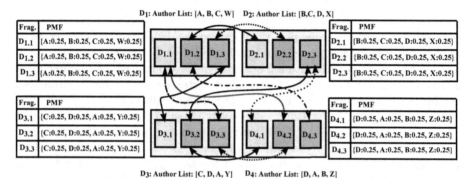

Fig. 3. Co-Authorship Graph: each vertex represents a document fragment and each edge represents the 2 nearest neighbors of each fragments. The dotted and dashed patterns are used to only help distinguish overlapping crossing edges.Initial PMFs of all fragments are given in the corresponding tables.

The purpose of *Co-Authorship Graph (CAG)* training is to alter the PMF of each fragment in order to better reflect the true writer(s) of that fragment. Algorithm 2 shows how the PMF of each CAG vertex can be updated. The same algorithm is executed at each vertex in multiple iterations (called *supersteps*). In the algorithm, each vertex corresponds to a document fragment and each edge denotes the stylistic similarity between two fragments. Each vertex (fragment) keeps track of the top$-k$ most similar fragments as neighbors. The algorithm contains three main parts: *Receive, Compute* and *Send*.

- *Receive (Lines 5 to 10).* The vertex receives the PMFs from its neighbors.
- *Compute (Line 11).* The vertex PMF is updated as the weighted average of all neighbors' PMFs. These weights are obtained from the distances of the neighbors through the *Probabilistic k Nearest Neighbor* method with the *radial basis function (Gaussian) kernel* [10]. The total weight is assumed to have been normalized to 1.
- *Send (Lines 12 to 13).* The updated PMF is sent to the neighbors.

Algorithm 2. CAG Training

1: **procedure** UPDATECAGVERTEX
2:     NeighborPMFs ← []
3:     NeighborDistances ← []
4:     V ← ThisVertex
5:     **for** N in V.GetNeighbors() **do**
6:         PMF ← ReceivePMF(N)
7:         PMF ← RemoveNonAuthors(PMF, V.AuthorList)
8:         PMF ← Renormalize(PMF)
9:         NeighborPMFs.Append(PMF)
10:        NeighborDistances.Append(Distance(V, N))
11:    V.PMF ← ComputeWeightedAvg(NeighborPMFs, NeighborDistances)
12:    **for** N in V.GetNeighbors() **do**
13:        SendPMF(N, V.PMF)

At each superstep, the same process described in Algorithm 2 is repeated in all vertices and supersteps are repeated until all PMFs converges.

Consider now how Algorithm 2 operates in the context of the example given in Fig. 3. Consider the fragment $D_{1.1}$. The vertex receives 2 PMFs from its 2 neighbors $D_{3.3}$ and $D_{4.2}$ as $\{C : 0.25, D : 0.25, A : 0.25, Y : 0.25\}$ and $\{D : 0.25, A : 0.25, B : 0.25, Z : 0.25\}$, respectively (Line 4). Each PMF is compared against the author list $[A, B, C, W]$ to remove the authors that do not appear in the author list of D_1 (Line 5). In this case, D and Y are disregarded for $D_{3.3}$. Similarly, D and Z are disregarded for $D_{4.2}$. After re-normalization, we obtain $\{C : 0.5, A : 0.5\}$ and $\{A : 0.5, B : 0.5\}$ as the PMFs for $D_{3.3}$ and $D_{4.2}$, respectively. For ease of exposition, we assume that all 2 NNs have the same distance to its respective fragment and hence contributes to the fragment's PMF equally. As a result, the weighted average of the two PMFs is $\{A : 0.5, B : 0.25, C : 0.25\}$ after the first superstep.

Following the same process we obtain $\{A : 0.25, B : 0.5, C : 0.25\}$ for $D_{1.2}$, $\{A : 0.25, B : 0.25, C : 0.5\}$ for $D_{1.3}$, $\{B : 0.5, C : 0.25, D : 0.25\}$ for $D_{2.1}$, $\{B : 0.25, C : 0.5, D : 0.25\}$ for $D_{2.2}$, $\{B : 0.25, C : 0.25, D : 0.5\}$ for $D_{2.3}$, $\{C : 0.5, D : 0.25, A : 0.25\}$ for $D_{3.1}$, $\{C : 0.25, D : 0.5, A : 0.25\}$ for $D_{3.2}$, $\{C : 0.25, D : 0.25, A : 0.5\}$ for $D_{3.3}$, $\{D : 0.5, A : 0.25, B : 0.25\}$ for $D_{4.1}$, $\{D : 0.25, A : 0.5, B : 0.25\}$ for $D_{4.2}$, and $\{D : 0.25, A : 0.25, B : 0.5\}$ for $D_{4.3}$. We can see that all PMFs are becoming less uniform after only the first superstep.

For each document, the PMFs will converge to the following values.

- Document D_1: $\{A : 1\}$ for $D_{1.1}$, $\{B : 1\}$ for $D_{1.2}$, and $\{C : 1\}$ for $D_{1.3}$.
- Document D_2: $\{B : 1\}$ for $D_{2.1}$, $\{C : 1\}$ for $D_{2.2}$, and $\{D : 1\}$ for $D_{2.3}$.
- Document D_3: $\{C : 1\}$ for $D_{3.1}$, $\{D : 1\}$ for $D_{3.2}$, and $\{A : 1\}$ for $D_{3.3}$.
- Document D_4: $\{D : 1\}$ for $D_{4.1}$, $\{A : 1\}$ for $D_{4.2}$, and $\{B : 1\}$ for $D_{4.3}$.

As can be seen, the NWAs of each document are not included in the PMFs and the author lists of D_1, D_2, and D_3 are correctly identified as $[A, B, C]$, $[B, C, D]$, $[C, D, A]$, and $[D, A, B]$, respectively.

3.3 Multi-authorship Prediction

In this subsection, we explain how we can make a multi-authorship prediction for a query document $\mathcal{Q}$ using the trained document fragments obtained from the two preprocessing steps. Algorithm 3 provides the structure of this process. The query document $\mathcal{Q}$ is decomposed into multiple query fragments. For each query fragment Q (Lines 4 to 11), we find the k nearest neighbors using the same GetKNN() function introduced in the CAG construction step (cf. Algorithm 1). In a fashion similar to that in the CAG training process (cf. Algorithm 2), the PMFs of the neighboring fragments and their distances with respect to Q are used to compute the weighted average to make a single prediction. After obtaining the PMFs of all query fragments (Line 12), we compute the average PMF to make a final prediction for the entire document.

Algorithm 3. Authorship Identification

```
 1: procedure MULTI-AUTHORSHIPPREDICTION
 2:     FragmentPMFs ← []
 3:     QueryFragments ← GetDocumentFragments(𝒬)
 4:     for Q in QueryFragments do
 5:         Neighbors ← GetKNN(Q, Fragments)
 6:         NeighborPMFs ← []
 7:         for N in Neighbors do
 8:             NeighborPMFs.Append(PMF)
 9:             NeighborDistances.Append(Distance(Q, N))
10:         Q.PMF ← ComputeWeightedAvg(NeighborPMFs, NeighborDistances)
11:         FragmentPMFs.Append(Q.PMF)
12:     return GetDocumentPMF(FragmentPMFs)
```

According to Fig. 3, given that there is a query document Q_1 with $Q_{1.1}$ and $Q_{1.2}$ as its fragments. We assume that $D_{1.1}$ and $D_{3.3}$ are identified as the 2 NNs of $Q_{1.1}$, $D_{1.3}$ and $D_{3.1}$ are identified as the 2 NNs of $Q_{1.2}$. We can then obtain $Q_{1.1}$ and $Q_{1.2}$ predictions as the following PMFs: $\{A : 1.0\}$ and $\{C : 1.0\}$, respectively. As a result, the document prediction for Q_1 is $\{A : 0.5, C : 0.5\}$, i.e., A and C are the authors of Q_1.

4 Performance Evaluation

In this section, we report results from our experimental studies. We compare the performance of the proposed solution against the best existing stylometry-based method for *authorship identification in multi-author documents (AIMD)* [5] and its improved version.

Competitive Methods. The competitive method presented in *stylometric authorship attribution of collaborative documents (AICD)* [5] is based on a linear *support vector machine (SVM)* classifier. For the training documents, AICD makes use of copy transformation in which m single-label samples is created from each training sample associated with m labels. In this way, we can associate each single-label sample to one label at a time [21]. As for the features, AICD extracts the *"Writeprints Limited Features Set"* [1] from multi-author documents using the JStylo tool [13]. The output from the linear SVM classifier is converted into a probabilistic distribution. AICD uses the most probable m authors as their result, where m is the given number of co-authors.

Furthermore, we formulate an improved variant of the AICD, named as I-AICD in this paper. In I-AICD, we use the sliding window method to generate chunks of 1,000 tokens and follow the same procedure as used in AICD. To this end, we aggregate our chunk level predictions by having each chunk vote for its most likely author. As for both of the techniques mentioned above, the 5-fold cross-validation is used for evaluation.

4.1   Experimental Setup

In this subsection, we describe the datasets used in this investigation along with the performance measures and parameters settings. One synthetic dataset and two real datasets are used to evaluate the three methods.

Synthetic Dataset. To generate a corpus of multi-author documents, we retrieved a collection of 23,096 single-author documents written by a set of 8,698 authors from online Project Gutenberg[1]. We first found a set A of authors such that each author $a_i \in A$ had 15 or more single-author documents, where each document had at least 6,000 tokens. Assume that D_a is a document written by m authors in A. The document D_a is generated by randomly selecting m authors $\{a_1, ..., a_m\}$ from A. For each author a_i, we obtained a text sample of $\mathcal{L}/m$ tokens where $\mathcal{L}$ is the synthetic document length. In this way, each author in the same document has the same number of tokens. Note that once a single-author document had been used in a multi-author document, it was never used again in any other document to avoid any possible training-testing sample contamination. Furthermore, each co-author set $\{a_1, ..., a_m\}$ was unique.

Real Datasets. As for real datasets, we retrieved two sets of research papers from arXiv.org: (i) Computer Sciences; and (ii) Social Sciences. Specifically, we sampled a set of papers from the real word datasets such that each author had his/her name appear in 5 papers. As can be seen from Table 1, for the Computer Science papers, we got a resulting dataset of 1,957 papers from a set of 707 authors. As for Social Sciences papers, we got a dataset of 616 papers from a set of 300 authors.

Parameter Settings. We tested different values for each parameter in order to find the most appropriate value. In the interest of conciseness, we display only the final results of this test. For synthetic dataset, the size of each synthetic document($\mathcal{L}$) was fixed at 12,000 tokens. The chunk size was set at 1,000 tokens

---

[1] https://www.gutenberg.org.

Table 1. Statistics of the datasets

	Synthetic dataset	Real dataset (computer science)	Real dataset (social sciences)
#Authors	1,360	707	300
#Documents	3,600	1,957	616
#Tokens	43,200,000	22,139,274	15,613,718

and the fragment size at 6,000. Chunk-level and fragment-level sliding window increments were set to 100 and 2,000 tokens, respectively. The k value of 10 was used for the top-k retrieval.

Evaluation Measures. Two types of measures were used in this experiment. (i) **Accuracy (A):** The accuracy indicates the discrepancy of a prediction with respect to the ground truth, which was defined as the number of correctly predicted authors divided by the size of the true co-author set. (ii) **Guess-one (G):** A document was considered correct if the prediction contains at least one of the true authors.

4.2 Experimental Results

Our experimental studies were designed to verify whether our proposed method can handle (i) a larger number of co-authors (than those used in existing studies); and (ii) non-writing authors (NWAs). To control the number m of co-authors and the number ϕ of NWAs, these studies were conducted on our synthetic dataset. In addition, we also conducted experimental studies on two real datasets to show that our method can handle real-world corpora. Results from the studies are reported as follows.

Effect of Number m of Authors. We studied the effect of number m of co-authors on accuracy by varying m between 2 and 5. The number ϕ of NWAs was set to the default value of 0. These m values were chosen because they conformed with the numbers of co-authors in the real datasets used in these experimental studies. Moreover, the bibliometric analysis of different disciplines shows that mostly the average number of authors per paper are 5 or less [2]. Figure 4(a) shows that our method was the best performer, while the improved variant of the competitive method *I-AICD* performed slightly better than *AICD*. Furthermore, the performance gap between our proposed method and *I-AICD* increases as the number m of co-authors increases. Our method can handle a larger number of authors better than the two competitive methods. We can also see that our method had maintained the perfect *guess-one* accuracy in all cases.

Effect of Number ϕ of Non-writing Authors. We study the effect of number ϕ of non-writing authors (NWA) on the accuracy as we vary ϕ from 0 to 2. The number m of co-authors is set to the default value of 3. As can be seen from

Fig. 4(b), including non-writing (NWA) authors into the list of actual authors negatively affects the prediction accuracy. Specifically, the accuracy level drops from 83.24 to 74.05 as we increase the value of ϕ from 0 to 1, while further increasing ϕ to 2 has no significant effect on the accuracy. The figure also shows that our method continues to be the best performer in this study, while *I-AICD* performs substantially better than *AICD*. Since *AICD* and *I-AICD* are not designed to handle NWAs, the accuracy levels of the two methods drastically drop as the ϕ is increased from 0 to 2. We can also see that our method had maintained the perfect *guess-one* accuracy in all cases.

Real Datasets. We evaluated the proposed method on real datasets. Note that unlike the synthetic dataset, the real datasets do not contain the ground truth regarding the number of NWAs of each document. As a result, for accuracy measurements, we assumed that all listed authors are assumed to be the writing authors. This assumption makes the measured accuracies of all methods lower than their actual values. However, it allows us to compare the three methods using real-world data. As can be seen from Table 2, the proposed method significantly outperformed the two competitive methods. Note that due to the unknown NWAs in the corpora, the accuracy level of our method reported here was lower than those of the synthetic datasets. We can also see that our method maintained the perfect *guess-one* accuracy in all cases.

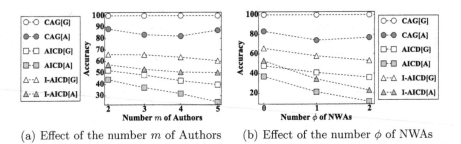

(a) Effect of the number m of Authors   (b) Effect of the number ϕ of NWAs

Fig. 4. Comparison of CAG performance against competitors: method[G] denotes guess-one accuracy and method[A] denotes the accuracy.

Table 2. Real dataset results

Method	Computer science		Social science	
	Accuracy	Guess-one	Accuracy	Guess-one
CAG	72.17	100	42.46	100
I-AICD	26.02	48.21	29.41	54.49
AICD	16.46	31.26	21.31	40.15

5   Conclusions

We have presented a solution for authorship identification of multi-author documents. The crux of our solution lies in the ability to probabilistically attribute different parts (fragments) of the same documents to different subsets of co-authors. Specifically, we have proposed a data structure called the *Co-Authorship Graph (CAG)* to capture stylistic similarity between pairs of fragments across the entire document corpus. We have also formulated a CAG training algorithm to learn the true writer(s) of each fragment. We evaluated the proposed solution using one synthetic dataset and two real datasets. Our experimental results have shown that our method had (i) significantly outperformed the best existing solution; (ii) could effectively handle a larger number of co-authors; and (iii) could handle non-writer authors (NWAs).

Appendix A Stylometric Features

The stylometric features used in this investigation are shown in Table 3. For features 5 to 12, N represents count of words and V represents count of distinct words. For Features 6 and 9, V_i represents the count of words that occur i times.

Table 3. List of stylometric features

Lexical features		
1. N: Total #words	2. V: Total #distinct words	3. Average word length
4. S.D. of word lengths	5. $\frac{V}{N}$	6. $VR(K) = \frac{10^4(\sum i^2 V_i - N)}{N^2}$
7. $VR(R) = \frac{V}{\sqrt{N}}$	8. $VR(C) = \frac{\log V}{\log N}$	9. $VR(H) = \frac{(100 \log N)}{(1-V_1)/V}$
10. $VR(S) = \frac{V_2}{V}$	11. $VR(k) = \frac{\log V}{\log(\log N)}$	12. $VR(LN) = \frac{(1-V^2)}{V^2(\log N)}$
13. Entropy of word freq. ditri.	14. Total number of chars	15. Freq. of alpha chars
16. Freq. of uppercase chars	17. Freq. of lowercase chars	18. Freq. of numeric chars
19. Freq. of special chars	20. Freq. of white spaces	21. Freq. of punctuations
22. Alpha char ratio	23. Uppercase char ratio	24. Lowercase char ration
25. Numeric char ratio	26. Special char ratio	27. White spaces ratio
Syntactic features		
28. Freq. of nouns	29. Freq. of proper nouns	30. Freq. of pronouns
31. Freq. of ordinal adjs.	32. Freq. of comparative adjs.	33. Freq. of superlative adjs.
34. Freq. of advs.	35. Freq. of comparative advs.	36. Freq. of superlative advbs.
37. Freq. of modal auxiliaries	38. Freq. of bases form verbs	39. Freq. of past verbs
40. Freq. of present part. verbs	41. Freq. of past part. verbs	42. Freq. of particles
43. Freq. of wh-words	44. Freq. of conjunctions	45. Freq. of numerical words
46. Freq. of determiners	47. Freq. of existential theres	48. Freq. of existential to
49. Freq. of prepositions	50. Freq. of genitive markers	51. Freq. of quotations
52. Freq. of commas	53. Freq. of terminators	54. Freq. of symbols
Structural Features		
55. Total number of sentence	56. Avg. #words per sentence	

References

1. Abbasi, A., Chen, H.: Writeprints: a stylometric approach to identity-level identification and similarity detection in cyberspace. ACM Trans. Inf. Syst. **26**(2), 7:1–7:29 (2008)
2. Akhavan, P., Ebrahim, N.A., Fetrati, M.A., Pezeshkan, A.: Major trends in knowledge management research: a bibliometric study. Scientometrics **107**(3), 1249–1264 (2016)
3. Baron, G.: Influence of data discretization on efficiency of Bayesian classifier for authorship attribution. Procedia Comput. Sci. **35**, 1112–1121 (2014)
4. Bradley, J.K., Kelley, P.G., Roth, A.: Author identification from citations. Technical report, Department of Computer Science, Carnegie Mellon University, Pittsburgh, PA, USA (2008)
5. Dauber, E., Overdorf, R., Greenstadt, R.: Stylometric authorship attribution of collaborative documents. In: Dolev, S., Lodha, S. (eds.) CSCML 2017. LNCS, vol. 10332, pp. 115–135. Springer, Cham (2017). https://doi.org/10.1007/978-3-319-60080-2_9
6. Giannella, C.: An improved algorithm for unsupervised decomposition of a multi-author document. JASIST **67**(2), 400–411 (2016)
7. Grieve, J.: Quantitative authorship attribution: an evaluation of techniques. LLC **22**(3), 251–270 (2007)
8. Hassan, S.U., Sarwar, R., Muazzam, A.: Tapping into intra- and international collaborations of the organization of Islamic cooperation states across science and technology disciplines. Sci. Public Policy **43**(5), 690–701 (2016)
9. Hill, S., Provost, F.: The myth of the double-blind review? Author identification using only citations. ACM SIGKDD Explor. Newsl. **5**(2), 179–184 (2003)
10. Holmes, C., Adams, N.: A probabilistic nearest neighbour method for statistical pattern recognition. J. R. Stat. Soc. Ser. B Stat. Methodol. **64**(2), 295–306 (2002)
11. Li, J., Zheng, R., Chen, H.: From fingerprint to writeprint. Commun. ACM **49**(4), 76–82 (2006)
12. Lipikorn, R., Shimizu, A., Kobatake, H.: A modified Hausdorff distance for object matching. Pattern Recogn. **1**, 566–568 (1994)
13. McDonald, A.W.E., Afroz, S., Caliskan, A., Stolerman, A., Greenstadt, R.: Use fewer instances of the letter "i": toward writing style anonymization. In: Fischer-Hübner, S., Wright, M. (eds.) PETS 2012. LNCS, vol. 7384, pp. 299–318. Springer, Heidelberg (2012). https://doi.org/10.1007/978-3-642-31680-7_16
14. Mosteller, F., Wallace, D.L.: Inference and Disputed Authorship: The Federalist. Addison-Wesley, Reading (1964)
15. Nutanong, S., Yu, C., Sarwar, R., Xu, P., Chow, D.: A scalable framework for stylometric analysis query processing. In: ICDM (2016)
16. Payer, M., Huang, L., Gong, N.Z., Borgolte, K., Frank, M.: What you submit is who you are: a multimodal approach for deanonymizing scientific publications. IEEE Trans. Inf. Forensics Secur. **10**(1), 200–212 (2015)
17. Ramnial, H., Panchoo, S., Pudaruth, S.: Authorship attribution using stylometry and machine learning techniques. In: Berretti, S., Thampi, S.M., Srivastava, P.R. (eds.) Intelligent Systems Technologies and Applications. AISC, vol. 384, pp. 113–125. Springer, Cham (2016). https://doi.org/10.1007/978-3-319-23036-8_10
18. Rexha, A., Klampfl, S., Kröll, M., Kern, R.: Towards a more fine grained analysis of scientific authorship: predicting the number of authors using stylometric features. In: Proceedings of the Third Workshop on BIR Co-located with the 38th (ECIR 2016), Padova, Italy, 20 March 2016, pp. 26–31 (2016)

19. Sboev, A., Litvinova, T., Gudovskikh, D., Rybka, R., Moloshnikov, I.: Machine learning models of text categorization by author gender using topic-independent features. Procedia Comput. Sci. **101**, 135–142 (2016)
20. Stamatatos, E.: A survey of modern authorship attribution methods. JASIST **60**(3), 538–556 (2009)
21. Tsoumakas, G., Katakis, I.: Multi-label classification: an overview. IJDWM **3**(3), 1–13 (2007)
22. Zhang, M., Zhou, Z.: ML-KNN: a lazy learning approach to multi-label learning. Pattern Recogn. **40**(7), 2038–2048 (2007)

Is a Common Phrase an Entity Mention or Not? Dual Representations for Domain-Specific Named Entity Recognition

Jiangtao Zhang[1]($\boxtimes$), Juanzi Li[1], Xiao-Li Li[2], Yixin Cao[1], Lei Hou[1], and Shuai Wang[1]

[1] Department of Computer Science and Technology, Tsinghua University, Beijing 100084, China
zhang-jt13@mails.tsinghua.edu.cn, lijuanzi@tsinghua.edu.cn, caoyixin2011@gmail.com, greener2009@gmail.com, 18813129752@163.com
[2] Institute for Infocomm Research, A*STAR, Singapore 138632, Singapore
xlli@i2r.a-star.edu.sg

Abstract. Named Entity Recognition (NER) for specific domains is critical for building and managing domain-specific knowledge bases, but conventional NER methods cannot be applied to specific domains effectively. We found that one of reasons is the problem of *common-phrase-like entity mention* prevalent in many domains. That is, many *common phrases* frequently occurring in general corpora may or may not be treated as named entities in specific domains. Therefore, determining whether a common phrase is an entity mention or not is a challenge. To address this issue, we present a novel BLSTM based NER model tailored for specific domains by learning *dual representations* for each word. It learns not only *general* domain knowledge derived from an external large scale general corpus via a word embedding model, but also the *specific* domain knowledge by training a *stacked deep neural network* (SDNN) integrating the results of a low-cost *pre-entity-linking* process. Extensive experiments on a real-world dataset of movie comments demonstrate the superiority of our model over existing state-of-the-art methods.

1 Introduction

Named Entity Recognition (NER), which seeks to locate named entities in text and classifies them into pre-defined categories, is a fundamental task in natural language processing. Especially, domain-specific named entities are critical for building and managing specific domain knowledge bases that are very useful for many applications. For example, identifying the names of movies, actors and directors in public reviews is helpful for producers to better understand optimal actor-director combinations and the audience preference, so that they can produce more popular works such as *"House of Cards"* released

© Springer International Publishing AG, part of Springer Nature 2018
J. Pei et al. (Eds.): DASFAA 2018, LNCS 10827, pp. 830–846, 2018.
https://doi.org/10.1007/978-3-319-91452-7_53

by Netflix.[1] Recognizing the names of drugs and diseases in biomedical texts facilitates the analysis of drug-drug interaction, drug adverse reaction and drug development [13]. In addition, discovering product names from social media sites and Twitters/Microblogs, is beneficial for companies to monitor user opinions on their products and to recommend suitable products to their customers subsequently [26].

Most existing NER methods [1,6,8,19,21] focus on identifying *general named entities* such as names of person, location and organization in *English* (or *western language*) texts and evaluate their recognition performance on some well-established datasets such as CoNLL-2003 [22] and OntoNotes 5.0 [15]. Particularly, some methods [1,8,21] typically use HMM, SVM, and CRF models by leveraging multi-level linguistic features, lexicon features and corpus statistics features. Recently neural network models, such as FNN and RNN, have been widely used in NER task [3,6,19] by training word embeddings on a large corpus such as Wikipedia and Gigaword.[2] Although these NER methods have achieved good performance on recognizing general named entities in English texts, e.g., 90.80% F1 measure on CoNLL-2003 [21] and 91.25% on CoNLL-2003 [19], they cannot easily be adapted to specific domains and different language datasets due to the following reasons: (1) most linguistic features used in English or western languages could not work on non-English languages such as Chinese; (2) different domains typically have different kinds of entity categories e.g., disease and drug names in medical domain, film and actor names in movie domain so that the features leveraged by general NER methods are not suitable for specific domains; (3) a unique challenge, *common-phrase-like entity mention issue*, is prevalent in many specific domains regardless of languages.

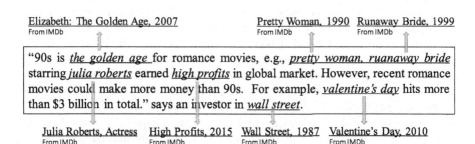

Fig. 1. An example of domain-specific NER. The red mentions denote CFEMs and the blue ones represent CREMs. (Color figure online)

We observe that many common phrases frequently occurring in general corpora should potentially be treated as named entities in some specific domains. We define these *common phrases* denoting real named entities in a certain domain as *Common-phrase-like Real Entity Mentions* (CREMs). Figure 1 shows a real

---

[1] https://www.netflix.com/jp-en/title/70178217.
[2] https://catalog.ldc.upenn.edu/ldc2011t07.

example of NER task in movie domain. Entity mentions *"pretty woman"*, *"runaway bride"* and *"valentine's day"* are three films in IMDb,[3] an online database related to movies. These entities occur in general corpora (e.g., Wikipedia) more frequently referring to their general meanings: *a beautiful woman, escaped bride* and *a holiday* respectively. For example, based on statistics on Wikipedia, the mention *"valentine's day"* refers to a common phrase having its general meaning, i.e., *a holiday*, with a dominant probability of 97.4%, compared to a specific film entity: *"Valentine's Day, 2010, USA"* with a small probability of 2.6%. Clearly, such kind of mentions tend to be treated as common phrases with their general meaning rather than domain-specific named entities by existing methods which rely on corpus statistics information, leading to a low recall of NER—miss detection of real entity mentions.

To discover more real entities, some NER methods [12,29,31,32] [?] leverage external lexicons to include potential entities. However, these methods suffer from another issue of introducing too much noise, *Common-phrase-like Fake Entity Mentions* (CFEMs) since many common phrases listed in these lexicons should not be recognized as real named entities according to their context. In Fig. 1, *"the golden age"*, *"high profits"* and *"wall street"* are three names of entities/films in IMDb, but none of them refers to a movie in the context. As such, these common phrases will be recognized as named entities *incorrectly* by lexicon-based NER methods, as they are listed in movie lexicon and have similar context with other real movie names, e.g., *"pretty woman"* and *"valentine's day"*, hurting recognition precision dramatically.

In summary, determining whether a common phrase is a *real* or *fake* named entity, i.e., distinguishing CREM and CFEM accurately in specific domains becomes a challenge. To address this problem, instead of representing each word with only one single vector used in conventional NER methods, we propose to learn *dual representations* of each word: (1) *general domain knowledge representation*—facilitating CFEM detection—which captures the general meaning of words via their textual context from a large scale *general* corpus; (2) *specific domain knowledge representation*—assisting CREM detection—which encodes the specific domain knowledge by training a *Stacked Deep Neural Network* (SDNN) leveraging the results of a *pre-entity-linking* process, i.e., linking entity mentions to their reference entities in a given domain-specific knowledge base. Then we propose a novel BLSTM-based model integrating **D**ual **R**epresentations i.e., *general domain* vs *specific domain* knowledge for domain-specific **NER**, called DRNER in this paper, to discover CREMs and filter out CFEMs accurately.

Contributions. Our main contributions are as followings:

1. To the best of our knowledge, this is the first work to distinguish two types of common phrases: CFEMs and CREMs, which are critical for domain-specific NER task.

---

[3] http://www.imdb.com.

2. Instead of learning one single embedding for each word in conventional NER methods, we learn dual representations to distinguish CFEM and CREM. The method is more generalized than traditional methods—it neither uses language dependent information, nor feature engineering—and thus applicable to different languages and different datasets.
3. We conducted extensive experiments on a manually annotated dataset of real-world movie comments, and the results demonstrate that our DRNER model outperforms existing state-of-the-art methods significantly.
4. We released a new dataset publicly (including manually labeled data and a movie domain knowledge base) that could be very useful for researchers to benchmark their technologies to improve the NER research in specific domains.

2   Preliminaries and Problem Definition

In this section, we first provide the definitions of some basic concepts and then define our task.

A *domain-specific knowledge base* (DSKB) defines a set of representational primitives to model domain knowledge from different perspectives, which can be defined as DSKB $= \{\mathcal{C}, \mathcal{E}, \mathcal{R}\}$. $\mathcal{C} = \{c_1, \cdots, c_{|\mathcal{C}|}\}$ represents a set of *categories* or *concepts*. $\mathcal{E} = \{e_1, \cdots, e_{|\mathcal{E}|}\}$ denotes *entities*: concrete objects or instances of concepts and each $e_i \in \mathcal{E}$ belongs to one or more categories in $\mathcal{C}$. The relationship $r_i \in \mathcal{R} = \{r_1, \cdots, r_{|\mathcal{R}|}\}$ between entities and categories can be formalized as a *subject-property-object* triple (s, p, o), where $s \in \mathcal{E} \cup \mathcal{C}$; $o \in \mathcal{E} \cup \mathcal{C} \cup \mathcal{L}$; p denotes a *property* to describe the attribute of an entity or a concept such as actor's name, movie's director; $\mathcal{L}$ is a set of literals or strings [30].

Let us define a phrase/words $m = \langle w_1, \cdots, w_{|m|} \rangle$ occurring in a textual document which is denoted to an entity in DSKB as an *entity mention* where $w_i \in m$ is a word.

Finally, we formulate our domain-specific NER task as a sequence labeling problem. Given a text from a specific domain $\mathbf{s} = \langle w_1, w_2, \cdots, w_{|\mathbf{s}|} \rangle$ and predefined categories $\mathcal{C} = \{c_1, \cdots, c_{|\mathcal{C}|}\}$, the output of our task is to generate a list of tags t_i for each word $w_i \in \mathbf{s}$. $t_i \in \mathcal{T} = \{cp | \forall c \in \mathcal{C}, \forall p \in \mathcal{P} - \{O\}\} \cup \{O\}$ is a *category-position* combinatorial tag for w_i, where $\mathcal{P} = \{B, I, E, S, O\}$ is a set of *position* tags indicating the position information of a word located in an entity mention. Specifically, B, I and E stand for beginning, intermediate, ending positions of a *multi-word* entity, S represents a *single-word* entity and O denotes outside of any entity.

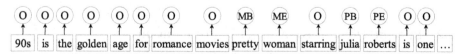

Fig. 2. An example of sequence labeling for domain-specific NER.

According to this definition, for a CREM $m_r = \langle w_{r1}, \cdots, w_{r|m_r|} \rangle$, $t_{r1} = cB$, $t_{r|m_r|} = cE$ and $t_{ri} = cI$, where $1 < i < |m_r|$. For a CFEM $m_f = \langle w_{f1}, \cdots,$

$w_{f|m_f}\rangle$, $t_{ri} = O$, where $i = 1, \cdots, |m_f|$. Figure 2 gives an example of sequence labeling for NER in which $\mathcal{C} = \{P, M\}$, P represents persons in movie domain such as actors, directors, producers and M denotes films, TV series and shows. The common phrase *"the golden age"* is a CFEM (label O) and *"pretty woman"* is a CREM (label MB, ME).

3 Our Proposed Methodology

3.1 The Proposed Overall Framework

Figure 3 shows the overall framework of our proposed model. The input of our model is a word sequence $\langle \ldots, w_i, w_{i+1}, w_{i+2}, w_{i+3}, \ldots \rangle$ and output is a tag sequence $\langle \ldots, t_i, t_{i+1}, t_{i+2}, t_{i+3} \ldots \rangle$ where $t_i \in \mathcal{T}$. Our model consists of three parts. The first part is to learn the specific domain knowledge representation (red rectangle vector) by training a SDNN based on the results of a pre-entity-linking process. The second part is to learn the general domain knowledge representation (blue rounded rectangle vector) by training a distributed word embedding model on a large scale corpus: Baidu Baike,[4] a large scale Chinese, collaborative, web-based encyclopedia. The last part is a BLSTM sequence labeling model. Particularly, the learned dual representations are concatenated to feed into the main BLSTM network and a tag constraint layer (details in Sect. 3.4) is stacked on the output of the BLSTM to produce the best tag sequence by introducing a transition matrix.

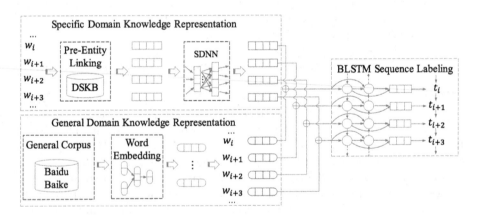

Fig. 3. Overview of our DRNER model. (Color figure online)

3.2 Specific Domain Knowledge Representation

In order to integrate the external specific domain knowledge into our model, we should bridge the input texts to a DSKB. Therefore, we perform a low cost pre-entity-linking process which links the entity mentions occurring in texts to their

[4] https://baike.baidu.com/.

reference entities in DSKB. Then we leverage some entity embedding models trained on DSKB to capture the *entity relationships* in DSKB. Nevertheless, this idea seems not perfect since we perform an entity linking (EL) before NER, i.e., linking mentions prior to recognizing mentions and the pre-entity-linking results could be noisy. As such, we propose a SDNN trained on a labeled EL training set to revise the results of the pre-entity-linking. The detailed process is described as follows.

Pre-entity-Linking. We first construct a mapping lexicon $\mathcal{D} = \{\langle key_i, value_i \rangle |$ $i = 1, \cdots, |\mathcal{D}|\}$ where key_i is a surface name of an entity, such as abbreviation, variation, and nickname and $value_i$ is its corresponding reference entity derived from the given DSKB and Wikipedia [23,30]. Given a word sequence of inputs $\mathbf{s} = \langle w_1, \cdots, w_{|\mathbf{s}|} \rangle$, we generate a *Candidate Mention set* $M_c = \{m_j | m_j \in \mathcal{D}.key, j = 1, \cdots, |M_c|\}$ by performing a longest match to the lexicon $\mathcal{D}$ where $m_j = \langle w_{jk} | w_{jk} \in \mathbf{s}, k = 1, \cdots, |m_j| \rangle$ is a possible entity mention and $\mathcal{D}.key = \{key_i | i = 1, \cdots, |\mathcal{D}|\}$. As many candidate entity mentions in M_c are also common phrases that may not always be real entities, such as *"the golden age"* and *"high profits"* in Fig. 1, it will inevitably introduce many CFEMs.

Then, for each candidate entity mention $m_j \in M_c$, we build its *Candidate Entity set* $E(m_j) = \{e_k | e_k \in \mathcal{E}, k = 1, \cdots, |E(m_j)|\}$ based on the $\langle key_i, value_i \rangle$ mappings in $\mathcal{D}$ [23]. Next, we perform a simple EL method to calculate the linking confidence $el_{e_k}(m_j)$ between m_j and $\forall e_k \in E(m_j)$. Specifically, we extract the context words around m_j in a certain window and content words of the abstract (or the first paragraph of description) of e_k in DSKB to compose two bag-of-words, and subsequently calculate Jaccard distance between them denoted as $el_{e_k}(m_j)$. Note that $el_{e_k}(m_j)$ indicates the strength of a mention m_j being linked to an entity e_k. Finally, for mention m_j, we choose two special entities: (1) the entity with the highest score as the most possible entity, denoted as $e^h(m_j) = \arg\max_{e_k \in E(m_j)}(el_{e_k}(m_j))$; (2) the entity with the lowest score as the least possible entity $e^l(m_j) = \arg\min_{e_k \in E(m_j)}(el_{e_k}(m_j))$.

We observe that each linked entity e_k has rich semantic information in DSKB, such as its properties and relationships with other entities. To leverage such information in our model for NER, we represent e_k as a real-value vector preserving such knowledge information. There are several knowledge embedding models which learn vector embeddings for both entities and relationships such as TransE [2], TransH [28] and TransR [17]. In this paper, we choose TransR model as our *entity embedding* model as it is more suitable for modeling 1-to-N, N- to-1 and N-to-N relationships—in a specific domain, most of relationships between entities are 1-to-N or N-to-1. For example, in movie domain, a movie is related to several actors/actresses and an actor/actress usually stars in many films. We also pre-train TransR model on DSKB in advance and then get its vector representation $\mathbf{e}_k = [v_{k_1}, \cdots, v_{k_n}]$, embedding specific domain knowledge for every linked entity e_k.

In short, through our pre-entity-linking process, given a word sequence $\mathbf{s}$, we get a set of candidate entity mentions M_c and their corresponding entity vectors. That is, $\forall m_j \in M_c$, we get its most possible entity vector $\mathbf{e}^h(m_j)$ and least possible entity vector $\mathbf{e}^l(m_j)$.

Stacked Deep Neural Network. Since we have the entity vector for each mention, an intuitive idea is that we can directly feed the highest entity $\mathbf{e}^h(m_j)$ for m_j into BLSTM. But it has two problems. Firstly, through our pre-entity-linking process, $\forall m_j \in M_c$ will be linked to a real entity e_k in DSKB. However, this will also link those CFEMs, e.g., "*the golden age*", to entities (wrong entity linking). Secondly, even for a real entity mention m_j, the result of pre-entity-linking $e^h(m_j)$ is not always correct (linking wrong entity) due to the ambiguity of the mention. For example, "*Valentine's Day*" can be represented as two different films: an American film and an Australian film, and the result of pre-entity-linking could be possibly wrong. To address these problems, we train a SDNN model (a fully connected two-layer neural network) on a labeled EL dataset (adding EL labels to our NER dataset) to revise the noisy pre-entity-linking results before stacking them into the BLSTM network. Specifically, For $\forall w_i \in \mathbf{s}$, if $w_i \in m_j$, we get:

$$
\begin{aligned}
\mathbf{x}_i =& \mathbf{e}^h(m_j) \oplus \mathbf{e}^l(m_j) \oplus [el_{e^h(m_j)}(m_j)] \\
& \oplus [el_{e^l(m_j)}(m_j)] \oplus [pos_{m_j}(w_i)] \oplus [len(m_j)], \\
\mathbf{h}_i =& \sigma(\mathbf{W}_{hx}\mathbf{x}_i + \mathbf{b}_h), \\
\mathbf{g}_i =& \sigma(\mathbf{W}_{gh}\mathbf{h}_i + \mathbf{b}_g), \\
\mathbf{o}_i =& \mathbf{W}_{og}\mathbf{g}_i + \mathbf{b}_o
\end{aligned}
\tag{1}
$$

where $\oplus$ concatenates two vectors, $pos_{m_j}(w_i)$ denotes the l-th position of the current word w_i located in mention m_j, and $len(m_j)$ is the length of m_j. Our model parameters $\mathbf{W}_*$ and $\mathbf{b}_*$ are learned by minimizing the *mean squared error* (MSE) between output of the model $\mathbf{o}_i$ and the ground truth entities in training set. We denoted Eq. 1 as $sdnn(.)$ function.

If $w_i \notin \forall m_j \in M_c$, we simply set $\mathbf{x}_i$ to a zero vector.

Through SDNN, we obtain a *specific* domain knowledge representation for each word w_i. Obviously, if a candidate mention m_j is a CFEM, such as "*the golden age*" in Fig. 1, the output of the SDNN for every word in "*the golden age*" will be like a near-zero vector because its context and description of the TV series "*the golden age (2012)*" in DSKB share few common information (few overlapping words) which leads to lower the highest linking confidence score $el_{e^h(m_j)}(m_j)$, i.e., they do not like the movie entity in DSKB. Instead, for a CREM m_j, such as "*pretty woman*" in Fig. 1, the output of each component word is similar to the entity embedding vector of the American movie "*Pretty Woman (2010)*" in DSKB, because (1) the highest EL score $el_{e^h(m_j)}(m_j)$ tends to be high due to similarity between context and description, and (2) the movie "*pretty woman*" and other entities such as "*Runaway Bride (1999)*" and "*Valentine's Day (2010)*" are tightly related—starring the same actress, belonging to the same type of films (romance movie) and produced in the same country (USA) etc.

3.3 General Domain Knowledge Representation

For each word w_i, in addition to embedding its specific domain knowledge, we represent its *general* domain knowledge to capture its semantic relationship and context information among words. Here, we choose GloVe [20] which is an unsupervised learning algorithm for obtaining vector representations for words. Training is performed on aggregated global word-word co-occurrence statistics from a corpus, and the resulting representations showcase interesting linear substructures of the word vector space. We aim to leverage the results of GloVe trained on Baidu Baike to capture the general meanings of CFEMs, facilitating us to recognize them as common phrases instead of entity mentions. As such, we get a real-value vector $we(w_i) = [u_1, \cdots, u_n]$ which embeds *general domain knowledge* (general context information) for every word w_i.

Finally, we obtain dual representations for each word w_i as shown in Fig. 3: red vector for specific domain representation and blue vector for general domain representation.

$$\mathbf{w}_i = we(w_i) \oplus sdnn(w_i) \tag{2}$$

3.4 Sequence Labeling Model

LSTM [14] is a special kind of RNN, capable of learning long-term dependencies. It has been proved to be state-of-the-art sequence labeling model in language model, speech recognition and POS tagging etc [16]. It is defined as follows:

$$
\begin{aligned}
\mathbf{i}_i &= \sigma(\mathbf{W}_{iw}\mathbf{w}_i + \mathbf{W}_{ih}\mathbf{h}_{i-1} + \mathbf{b}_p), \\
\mathbf{f}_i &= \sigma(\mathbf{W}_{fw}\mathbf{w}_i + \mathbf{W}_{fh}\mathbf{h}_{i-1} + \mathbf{b}_f), \\
\mathbf{o}_i &= \sigma(\mathbf{W}_{ow}\mathbf{w}_i + \mathbf{W}_{oh}\mathbf{h}_{i-1} + \mathbf{b}_o), \\
\mathbf{g}_i &= tanh(\mathbf{W}_{gw}\mathbf{w}_i + \mathbf{W}_{gh}\mathbf{h}_{i-1} + \mathbf{b}_g), \\
\mathbf{c}_i &= \mathbf{f}_i \odot \mathbf{c}_{i-1} + \mathbf{i}_i \odot \mathbf{g}_i, \\
\mathbf{h}_i &= \mathbf{o}_i \odot tanh(\mathbf{c}_i),
\end{aligned}
\tag{3}
$$

where σ is the logistic sigmoid function, and $\mathbf{i}, \mathbf{f}, \mathbf{o}, \mathbf{c}$ and $\mathbf{h}$ are input gate, forget gate, output gate, cell vector and hidden vector respectively. $\mathbf{W}_{*w}$ is the transformation matrix from the input to LSTM states and $\mathbf{W}_{*h}$ is the recurrent transformation matrix between the recurrent states h_i.

For NER task, as we have access to both left and right context of the sentence at every word t, we compute a left context representation $\overrightarrow{\mathbf{h}_i}$ and right context representation $\overleftarrow{\mathbf{h}_i}$ respectively (BLSTM [11]) to achieve better performance. Then the output of a word using this model is obtained by concatenating its left and right context representations, $\mathbf{h}_i = [\overrightarrow{\mathbf{h}_i}; \overleftarrow{\mathbf{h}_i}]$. In our model, the vector of each input word $\mathbf{w}_i$ has been represented by Eq. 2. Note that there are strong dependencies across output labels. For example, tag I could not directly follow O, S cannot occur directly in front of E, and E could only follow B or I. Therefore, instead of modeling tagging decisions independently, we model them jointly using a labeling constraint layer [16,19]. We first define a tag-transition matrix A between each word's output. Each element $A_{i,j} \in A$ represents the possibility of transition from tag t_i to t_j in successive words ($A_{0,i}$ is the possibility for starting with tag t_i). We consider $blstm(\cdot)$ to be the output of the BLSTM network. $blstm(w_i)_{t_i}$ corresponds to the score of the word w_i labeled as t_i in a sentence. Given a word sequence $\mathbf{s} = \langle w_1, \cdots, w_{|\mathbf{s}|} \rangle$ with corresponding tag list $\mathbf{t} = \langle t_1, \cdots, t_{|\mathbf{s}|} \rangle$, the score of it can be given as follows:

$$f(\mathbf{s}, \mathbf{t}) = \sum_{i=1}^{|\mathbf{s}|} (A_{t_{i-1}, t_i} + blstm(w_i)_{t_i}) \tag{4}$$

A softmax over all possible tag sequences yields a probability for the tag list $\mathbf{t}$:

$$p(\mathbf{t}|\mathbf{s}) = \frac{e^{f(\mathbf{s}, \mathbf{t})}}{\sum_{\hat{\mathbf{t}} \in \mathbf{T_s}} e^{f(\mathbf{s}, \hat{\mathbf{t}})}} \tag{5}$$

where $\mathbf{T_s}$ represents all possible tag sequences for a word sequence $\mathbf{s}$.

Training and Testing. During training, we use the maximum conditional likelihood estimation. For a training set $\{(\mathbf{s}_k, \mathbf{t}_k)|k = 1, \cdots, m\}$, the logarithm of the likelihood (a.k.a. the log-likelihood) is given by:

$$\mathcal{L} = \sum_k \log p(\mathbf{t}|\mathbf{s}) = \sum_k f(\mathbf{s}, \mathbf{t}) - \sum_k \log \sum_{\hat{\mathbf{t}} \in \mathbf{T_s}} e^{f(\mathbf{s}, \hat{\mathbf{t}})} \tag{6}$$

Our model parameters and transition matrix are learned by maximizing $\mathcal{L}$.

Decoding is to find the optimal tag sequence with highest conditional probability:

$$\mathbf{t}^* = \arg\max_{\mathbf{t}} p(\mathbf{t}|\mathbf{s}) \tag{7}$$

This is a typical dynamic programming problem and can be solved with Viterbi algorithm [25].

4   Experiments

We implement BLSTM and SDNN using Keras library[5] on a server with 12 3.5GHz CPU cores, 16GB memory, 2 33MHz GPU processors with 12G built-in memory and Ubuntu 14.04. The basic setups of Keras are set as follows: the activation function is RELU, using *sgd* optimizer (stochastic gradient descent optimizer) with constant learning rate *0.01*.

---

[5] https://keras.io.

4.1 Dataset

To the best of our knowledge, there is no publicly available benchmark dataset for the task of domain-specific NER. Thus, we create a gold standard dataset for our task. We first establish a DSKB from a well-known Chinese website Douban containing 23 concepts, 91 properties, 197,940 entities and more than 3 million triples (RDFs). We then crawl the user generated comments covering movies and TV shows from Douban, including 524 reviews, 480 short comments, 298 group discussions and 7 synthetic reviews.

For our experiments, the set of pre-defined categories is $C = \{person, movie\}$ where *person* denotes the name of an actor/actress, a director or a producer, etc and *movie* indicates the name of a film or a TV show.

Finally, we spend considerable time to manually label the dataset including NER and EL task. Note that the annotation task is very time-consuming. Both DSKB and the annotated dataset have been *released publicly*.[6]

Table 1 shows the statistics of our dataset. There are 7972 *real entity mentions* (REMs) in total, containing 4278 *persons* and 3694 *movies*. $\overline{|M|}$ and $\overline{|E(m)|}$ denote the average number of mentions in one comment and the average number of candidate entities per mention respectively. Since it is difficult to clearly define the common phrase, we can simply treat each n-gram occurring in the dataset which is also listed in $\mathcal{D}$ *and* occurs over a certain times (200 times used in our experiments) in Baidu Baike as the common phrase entity mention, including CREM and CFEM. Notice the number of the CFEMs is huge (too much noise) which is more than 14 times than that of REMs, and the CREMs is about 16.6% of the REMs. Furthermore, the average number of candidate entities per mention ($\overline{|E(m)|}$) is less than two, which indicates that it is not challenging to disambiguate the extracted entity mentions in movie domain.

Table 1. Statistics of the datasets

Comment#	Token#	REM#	P#	M#	CFEM#	CREM#	$\overline{|M|}$	$\overline{|E(m)|}$
1309	864661	7972	4278	3694	112004	1320	6.09	1.89

We employ the standard NER evaluation metrics, namely, *precision, recall* and *f1-measure* to measure the performance of various methods. Additionally, we apply 10-fold cross validation on our dataset: 8 folds for training, 1 fold for validation and 1 fold for test.

4.2 Experiment Results and Analysis

Experiment Settings. The configurations of our model are listed in Table 2. We employ BLSTM network and SDNN with 2 hidden layers respectively. Both dimension size are 200, following the paper [27]. Dropout layers are applied on hidden layers of BLSTM and SDNN in order to reduce overfitting.

---

[6] https://github.com/naxier/MovieEL.

Table 2. Parameter settings of BLSTM and SDNN

	Input		Hidden 1 & 2	Output size	Softmax layer	Dropout	Batch size	Epochs
	WE	SDNN						
BLSTM	100	100	200	200	9	0.5	64	128
SDNN	204		200	100	—	0.5	64	128

Network Structure. We evaluate deep structure which uses multiple BLSTM layers to choose the best network structure for our experiments. Table 3 compares the performance of BLSTM with one and two hidden layers. As expected, it shows that the two-layer BLSTM achieves better performance than one-layer BLSTM, which indicates that training multiple-layer neural networks can improve NER performance while increasing training time and complexity. Based on this observation, in following experiments, the hidden layers of our BLSTM network is set to 2. As [27] observes that model performance is not sensitive to the dimension size of hidden layer, we fix it to 200.

Table 3. Comparison of hidden layers

Method	Precision	Recall	F1	T_time
One-layer	0.782	0.723	0.751	0.74 h
Two-layers	**0.817**	**0.748**	**0.781**	1.77 h

Table 4. Different configurations.

Methods	Precision	Recall	F1
$DRNER_{we}$	0.664	0.423	0.517
$DRNER_{ee}$	0.758	0.620	0.682
$DRNER_{sdnn}$	0.792	**0.753**	0.772
DRNER	**0.817**	0.748	**0.781**

Methods for Comparison. We implement four methods, including one baseline and three state-of-the-art methods, to compare with our proposed model.

1. *Baseline*: In this method, we only use BLSTM network to conduct the NER task, without capturing general domain knowledge representation or specific domain knowledge representation, where each word is randomly embedded into dense vector of fixed size,[7] and then fed into the BLSTM network.
2. *Conditional Random Fields (CRF)*: It is the most popular method in traditional NER tasks which has achieved satisfactory performance in general domain. Here we employed CRF++ toolkit[8] using a bundle of traditional features such as morphological, POS and word segmentation features combing with lexicon features (derived from $\mathcal{D}$) on our constructed dataset.
3. *CNN+BLSTM+CRF (CBC)*: This method proposed in [19] is the state-of-the-art method, which integrates CNN, BLSTM and CRF to conduct end-to-end sequence labeling by capturing word-level and character-level representations. Almost all recent works conducting NER use this model and achieve

---

[7] Using Embedding Layer in Keras.
[8] https://taku910.github.io/crfpp/.

good performance in general domain including systems participating in EDL (Entity Discovery and Linking) track at NIST TAC-KBP[9]. In our experiments, we implement this method on our dataset by training GloVe on Baidu Baike (the same as our method for fair comparison) to capture the word-level representations, and stacking a CRF layer on output of the BLSTM to capture the mutual interdependency between output tags, while discarding the CNN model that aims to capture character-level representations, which is not suitable for non-English task (e.g., our task).

4. *Iterative Joint Model* (*IJM*): Recently [30] proposed an iterative joint model which conducts EL and NER processes in an iterative manner by constructing complicated features to mutually enhance the performance of both tasks. We also implement this method on our dataset for fair comparison.

Results and Discussion

Comparison with State-of-the-Art Methods. Table 5 shows the detailed comparison results. The precision and recall of the *Baseline* are both low because it doesn't include the general and specific domain knowledge but only leverage the capability of learning long-term dependencies of the main BLSTM network. As the most popular method in traditional NER method, *CRF* combining with lexicon features doesn't work well (especially in terms of precision) on our dataset due to the fact that there are a lot of common phrases falsely recognized as entity mentions, i.e., CFEMs, although it has been shown high performance in general area. In contrast, the precision of *CBC* increases significantly as it includes the word-level representation trained on a large scale general corpus, which captures the general domain knowledge of the word. Nevertheless, its recall is still low because the word embedding only considers the general context information, while the meanings of some common phrases in specific domain have been converted to CREMs. Furthermore, from the table we notice *IJM* achieves much higher precision as it aims to detect and filter out fake named entity through iterating EL and NER process by introducing complicated features. However, it doesn't take CREMs into consideration, that is, a lot of CREMs are also filtered out. Thus its recall is still low. Finally, by capturing both general domain knowledge and specific domain knowledge, our proposed DRNER method achieves the best performance without manually constructing any features. It is 4.4% higher than the second best *IJM*, which relies heavily on costly constructed features.

Besides, since the *movie* category contains more common-phrase-like entity mentions than *person* category, more significant improvement of our method comes from *movie* category than *person* category. Note that there still exist some common-phrase-like *person names*, such as "文章(the article)" referring to a famous Chinese actor.

An example of experimental results is shown in Fig. 4. We observe that the output of *baseline* is almost wrong. While *CBC* and *IJM* can achieve better results, our *DRNER* model generates almost the same output with the ground truth except wrong recognition of "文章(the article)". The reason that our model

---

[9] http://nlp.cs.rpi.edu/kbp/.

Table 5. Comparison with other systems.

Approach	*person*			*movie*			overall		
	Precision	Recall	F1	Precision	Recall	F1	Precision	Recall	F1
Baseline	0.488	0.301	0.372	0.412	0.289	0.340	0.450	0.295	0.357
CRF	0.708	0.645	0.675	0.622	0.601	0.611	0.674	0.625	0.649
CBC	0.785	0.669	0.722	0.740	0.612	0.632	0.767	0.642	0.699
IJM	**0.829**	0.719	0.770	0.793	0.621	0.697	0.814	0.674	0.737
DRNER	0.827	**0.768**	**0.796**	**0.806**	**0.725**	**0.763**	**0.817**	**0.748**	**0.781**

Text & Ground Truth	Baseline	CRF	CBC	IJM	DRNER
文章最后想表达的是，中国人再努力奋斗，也不会创造出[钢琴师]$_M$，更不会拍出[美丽人生]，最多也只是[许鞍华]，还有[葛大爷]$_M$未成名时的[活着]$_M$··· At the end of the article, the author expressed that Chinese will never create movies like [**The Pianist**]$_M$ or even [Life Is Beautiful]$_M$, no matter how much they struggle. What they could create most are [The Golden Age]$_M$ directed by [Xu Anhua]$_M$ which had a very low rating and [Uncle Ge]$_M$'s [Lifetimes]$_M$ when he had not become a celebrity ...	文章最后想表达的是，[中国人(Chinese)]$_M$再努力奋斗，更不会拍出美丽人生，最多只是[许鞍华(Xu Anhua)]$_P$无人问津的黄金时代，还有葛大爷未成名时的活着···	[文章(the article)]$_P$最后想表达的是，[中国人(Chinese)]$_M$再努力奋斗，也不会创造出钢琴师，更不会拍出[美丽人生(Life Is Beautiful)]$_M$，最多也只是[许鞍华(Xu Anhua)]$_P$无人问津的黄金时代，还有葛大爷未成名时的活着···	文章最后想表达的是，中国人再努力奋斗，也不会创造出钢琴师，更不会拍出[美丽人生([Life Is Beautiful)]$_M$，最多也只是[许鞍华(Xu Anhua)]$_P$无人问津的黄金时代，还有[葛大爷(Uncle Ge)]$_P$未成名时的活着···	[文章(the article)]$_P$最后想表达的是，中国人再努力奋斗[奋斗(Struggle)]$_M$，也不会创造出[钢琴师(The pianist)]$_M$，更不会拍出[美丽人生(Life Is Beautiful)]$_M$，最多也只是[许鞍华(Xu Anhua)]$_P$无人问津的[黄金时代(The Golden Age)]$_M$，还有[葛大爷(Uncle Ge)]$_P$未成名时的活着···	[文章(the article)]$_P$最后想表达的是，中国人再努力奋斗，也不会创造出[钢琴师(The pianist)]$_M$，更不会拍出[美丽人生(Life Is Beautiful)]$_M$，最多也只是[许鞍华(Xu Anhua)]$_P$无人问津的[黄金时代(The Golden Age)]$_M$，还有[葛大爷(Uncle Ge)]$_P$未成名时的[活着(Lifetimes)]$_M$···

Fig. 4. An example of our experimental results. The red ones are CFEMs and blue ones denote CREMs. (Color figure online)

fails to filter out the CFEM "文章" is that there exists a famous Chinese actor named "文章" starring a well-known TV series "奋斗(struggle)"—occurring by accident in the same context—which misleads the result of SDNN through pre-entity-linking (damaging *specific domain knowledge representation*). However, for CFEM "奋斗", it usually cooccurs with the word "努力(endeavor)" in large Chinese corpora so that it is easily filtered out by our *general domain knowledge representation*.

Comparison with Different Components. We further study the effectiveness of *each component* in our proposed model, which is configured into the following 4 different settings:

1. *DRNER$_{we}$*: Only put the word embedding (GloVe trained on Baidu Baike) into BLSTM network. That is, only use the *general* domain knowledge (one single representation).
2. *DRNER$_{ee}$*: Directly put the result of the pre-entity-linking, i.e., the entity embedding with the highest score into the BLSTM network combining the word embedding without training SDNN (dual representations without SDNN).
3. *DRNER$_{sdnn}$*: Stack SDNN into BLSTM combining the word embedding (dual representations with SDNN).

4. *DRNER*: Our final model which stacks dual representations of each word into BLSTM and adds a labeling constraint layer on the output of the BLSTM (dual representations with SDNN combining with labeling constraint).

From Table 4, we can see that the performance of $DRNER_{ee}$ is better than $DRNER_{we}$ due to the pre-entity-linking process which brings the specific domain knowledge. However, the precision of $DRNER_{ee}$ is not good because a lot of CFEMs are introduced by directly feeding the pre-entity-linking result into BLSTM network. At the same time, because this method heavily relies on the quality of pre-entity-linking result, lots of CREMs are also filtered out if highest linking confidence is weak or they are linked to wrong entities, which leads to lower recall. As comparison, $DRNER_{sdnn}$ can produce a significant improvement of both precision and recall because the pre-trained SDNN is introduced to revise the result of pre-entity-linking by leveraging the EL training set. When incorporating the labeling constraint layer, the precision is further increased (2.5%) because the output tags have been optimized by capturing the constraint between sequential tags through the introduced transition matrix while the recall is decreased slightly (0.5%) as some illegal tags tend to be labeled as O (Others).

5   Related Work

NER with Domain Adaptation. While most existing approaches focus on recognizing the named entities in general domain, some studies have been proposed to deal with NER problem adapted to certain specific domains which mainly focus on biomedical domain. Specifically, early researches [9,10] have exploited domain-specific rules and knowledge resources, such as gazetteers and dictionaries to refine and identify the entity boundaries. More recent work [4,12,29,32] apply the traditional approaches used in general domain to the specific domain which have centered around HMMs, SVMs, and CRFs, relying on a rich set of hand-crafted features. However, their performance depends heavily on the quality and quantity of the selected features. Very recently, [24] proposed a hybrid model, which adds a stacked auto-encoder to a text-based deep neural network for NER specialized for a domain, i.e., Japanese chess. However, all of them did not consider the CFEMs and CREMs issues and could not achieve satisfactory performance.

NER with Joint Model. Another thread of related work uses a joint model between NER and EL to enhance the performance of both tasks, e.g., [18] achieved improvement of 0.4% F1 on CoNLL03 dataset by training a joint model that is able to capture the mutual dependency between NER and EL. [7] combined coreference resolution, entity linking, and NER into a single CRF model and designed cross-task interaction factors to facilitate the integration. Their system achieved state-of-the-art results on the OntoNotes dataset. While most of joint models aim to deal with the NER in general domain, recently [30] proposed a joint model between NER and EL tasks in an interactive manner to solve the fake named entity issue and achieved good performance in a specific domain. However, all these existing joint models need to manually define a wide variety of domain-specific features and they cannot be easily adapted to a new domain.

NER with Deep Learning. While there has been a long history of research applying CRF model to NER, recently some attempts have focused on applying deep neural network to NER. [6] presented SENNA, a unified tagging solution based on deep feed-forward neural network which uses task-independent features and word embeddings learnt from unlabeled text to achieve near state-of-the-art results on POS tagging, chunking, and NER. Based on this work, [16,27] also proposed a unified tagging solution based on BLSTM instead of feed-forward neural network which incorporates CRF layer to enhance the performance of chunking and NER. [5,19] presented a hybrid architecture combining BLSTM and CNN, which automatically detects word- and character-level features to achieve good performance. However, all these work are only suitable for English dataset as they all rely on rich language-dependent features or character-level features. Additionally, all these works aim to provide a unified solution in general domain which conduct experiments on CoNLL-2003 dataset, without considering the characteristics in specific domains. Therefore, these methods cannot be applied to specific domains effectively.

6   Conclusion and Future Work

The traditional NER systems aim to recognize the names of persons, locations and organizations, etc in general domain. However, they cannot achieve good results for domain-specific tasks. While recently some methods have been proposed to address the domain-specific issues, they heavily rely on hand-crafted language dependent features and do not consider the CREMs and CFEMs issues which are not effective to be used in many real-world NER applications. In this paper, we proposed a novel BLSTM-based model with dual representations to perform NER in specific domains. The proposed model is highly effective compared with existing approaches without constructing any hand-crafted language-dependent features. In the future, we will evaluate our method in other domains, such as biomedical domain. In addition, we will explore some new unsupervised methods that is very useful when training dataset is not available.

Acknowledgments. The work is supported by major national research and development projects (2017YFB1002101), NSFC key project (U1736204, 61661146007), Fund of Online Education Research Center, Ministry of Education (No. 2016ZD102), and THU-NUS NExT Co-Lab.

References

1. Ando, R.K., Zhang, T.: A framework for learning predictive structures from multiple tasks and unlabeled data. J. Mach. Learn. Res. **6**, 1817–1853 (2005)
2. Bordes, A., Usunier, N., Garcia-Duran, A., Weston, J., Yakhnenko, O.: Translating embeddings for modeling multi-relational data. In: Advances in Neural Information Processing Systems, vol. 26, pp. 2787–2795 (2013)
3. Cao, Y., Huang, L., Ji, H., Chen, X., Li, J.: Bridge text and knowledge by learning multi-prototype entity mention embedding. In: ACL (2017)
4. Cao, Y., Li, J., Guo, X., Bai, S., Ji, H., Tang, J.: Name list only? Target entity disambiguation in short texts. In: EMNLP (2015)
5. Chiu, J.P.C., Nichols, E.: Named entity recognition with bidirectional LSTM-CNNs. TACL **4**, 357–370 (2016)
6. Collobert, R., Weston, J., Bottou, L., Karlen, M., Kavukcuoglu, K., Kuksa, P.P.: Natural language processing (almost) from scratch. J. Mach. Learn. Res. **12**, 2493–2537 (2011)
7. Durrett, G., Klein, D.: A joint model for entity analysis: coreference, typing, and linking. In: TACL (2014)
8. Florian, R., Ittycheriah, A., Jing, H., Zhang, T.: Named entity recognition through classifier combination. In: Proceedings of the Seventh Conference on Natural Language Learning at HLT-NAACL 2003, CONLL 2003, vol. 4, pp. 168–171 (2003)
9. Fukuda, K., Tsunoda, T., Tamura, A., Takagi, T.: Information extraction: identifying protein names from biological papers. In: PSB, pp. 707–718 (1998)
10. Gaizauskas, R., Demetriou, G., Humphreys, K.: Term recognition and classification in biological science journal articles. In: Proceedings of the Computational Terminology for Medical and Biological Applications Workshop of the 2nd International Conference on NLP, pp. 37–44 (2000)
11. Graves, A., Schmidhuber, J.: Framewise phoneme classification with bidirectional LSTM and other neural network architectures. Neural Netw. **18**(5–6), 602–610 (2005)
12. Gu, B.: Recognizing nested named entities in GENIA corpus. In: Proceedings of the BioNLP Workshop on Linking Natural Language Processing and Biology at HLT-NAACL 2006, pp. 112–113 (2006)
13. Henriksson, A., Dalianis, H., Kowalski, S.: Generating features for named entity recognition by learning prototypes in semantic space: the case of de-identifying health records. In: 2014 IEEE International Conference on Bioinformatics and Biomedicine (BIBM), pp. 450–457 (2014)
14. Hochreiter, S., Schmidhuber, J.: Long short-term memory. Neural Comput. **9**(8), 1735–1780 (1997)
15. Hovy, E., Marcus, M., Palmer, M., Ramshaw, L., Weischedel, R.: Ontonotes: the 90% solution. In: NAACL-Short 2006, pp. 57–60 (2006)
16. Huang, Z., Xu, W., Yu, K.: Bidirectional LSTM-CRF models for sequence tagging. CoRR abs/1508.01991 (2015)
17. Lin, Y., Liu, Z., Sun, M., Liu, Y., Zhu, X.: Learning entity and relation embeddings for knowledge graph completion. In: Proceedings of the Twenty-Ninth AAAI Conference on Artificial Intelligence, pp. 2181–2187. AAAI (2015)
18. Luo, G., Huang, X., Nie, Z., Lin, C.-Y.: Joint named entity recognition and disambiguation. In: EMNLP, pp. 879–888 (2015)
19. Ma, X., Hovy, E.H.: End-to-end sequence labeling via bi-directional LSTM-CNNs-CRF. CoRR abs/1603.01354 (2016)

20. Pennington, J., Socher, R., Manning, C.D.: Glove: global vectors for word representation. In: EMNLP, vol. 14, pp. 1532–1543 (2014)
21. Ratinov, L., Roth, D.: Design challenges and misconceptions in named entity recognition. In: CoNLL (2009)
22. Tjong Kim Sang, E.F., De Meulder, F.: Introduction to the CoNLL-2003 shared task: language-independent named entity recognition. In: HLT-NAACL 2003, vol. 4, pp. 142–147 (2003)
23. Shen, W., Wang, J., Han, J.: Entity linking with a knowledge base: issues, techniques, and solutions. Trans. Knowl. Data Eng. **27**, 443–460 (2015)
24. Tomori, S., Ninomiya, T., Mori, S.: Domain specific named entity recognition referring to the real world by deep neural networks. In: ACL, vol. 2, Short Papers (2016)
25. Viterbi, A.: Error bounds for convolutional codes and an asymptotically optimum decoding algorithm. IEEE Trans. Inf. Theor. **13**, 260–269 (2006)
26. Wang, J., Zhao, W.X., Wei, H., Yan, H., Li, X.: Mining new business opportunities: identifying trend related products by leveraging commercial intents from microblogs. In: EMNLP, pp. 1337–1347 (2013)
27. Wang, P., Qian, Y., Soong, F.K., He, L., Zhao, H.: A unified tagging solution: bidirectional LSTM recurrent neural network with word embedding. CoRR abs/1511.00215 (2015)
28. Wang, Z., Zhang, J., Feng, J., Chen, Z.: Knowledge graph embedding by translating on hyperplanes. In: AAAI, pp. 1112–1119 (2014)
29. Yang, Z., Lin, H., Li, Y.: Exploiting the contextual cues for bio-entity name recognition in biomedical literature. J. Biomed. Inform. **41**, 580–587 (2008)
30. Zhang, J., Li, J., Li, X.-L., Shi, Y., Li, J., Wang, Z.: Domain-specific entity linking via fake named entity detection. In: Navathe, S.B., Wu, W., Shekhar, S., Du, X., Wang, X.S., Xiong, H. (eds.) DASFAA 2016. LNCS, vol. 9642, pp. 101–116. Springer, Cham (2016). https://doi.org/10.1007/978-3-319-32025-0_7
31. Zhang, J., Cao, Y., Hou, L., Li, J., Zheng, H.-T.: XLink: an unsupervised bilingual entity linking system. In: Sun, M., Wang, X., Chang, B., Xiong, D. (eds.) CCL/NLP-NABD -2017. LNCS (LNAI), vol. 10565, pp. 172–183. Springer, Cham (2017). https://doi.org/10.1007/978-3-319-69005-6_15
32. Zhao, S.: Named entity recognition in biomedical texts using an hmm model. In: Proceedings of the International Joint Workshop on Natural Language Processing in Biomedicine and Its Applications, JNLPBA 2004, pp. 84–87 (2004)

Recognizing Textual Entailment with Attentive Reading and Writing Operations

Liang Liu[1], Huan Huo[1(✉)], Xiufeng Liu[2], Vasile Palade[3], Dunlu Peng[1],
and Qingkui Chen[1]

[1] Department of Computer Science,
University of Shanghai for Science and Technology, Shanghai, China
huo_huan@yahoo.com
[2] Technical University of Denmark, Kongens Lyngby, Denmark
[3] Faculty of Engineering, Environment and Computing,
Coventry University, Coventry, UK

Abstract. Inferencing the entailment relations between natural language sentence pairs is fundamental to artificial intelligence. Recently, there is a rising interest in modeling the task with neural attentive models. However, those existing models have a major limitation to keep track of the attention history because usually only one single vector is utilized to memorize the past attention information. We argue its importance based on our observation that the potential alignment clues are not always centralized. Instead, they may diverge substantially, which could cause the problem of long-range dependency. In this paper, we propose to facilitate the conventional attentive reading operations with two sophisticated writing operations - *forget* and *update*. Instead of utilizing a single vector that accommodates the attention history, we write the past attention information directly into the sentence representations. Therefore, higher memory capacity of attention history could be achieved. Experiments on Stanford Natural Language Inference corpus (SNLI) demonstrate the superior efficacy of our proposed architecture.

1 Introduction

Recognizing Textual Entailment (RTE) refers to the task of determining whether the meaning of one sentence (denoted as *hypothesis*) can be inferred from the other (denoted as *premise*) [1]. It covers three basic semantic relationships in human languages - *Entailment, Contradiction, Neutral* (a.k.a containment, exclusion, implicitivity), which lays the foundation for applications that are highly associated with natural logic [2] (e.g. semantic search and question answering).

Early attempts on RTE [3,4] had long been restricted to rather small datasets (e.g. FracCaS[1] and SICK[2], etc.). This situation alleviated after researchers

---

[1] https://nlp.stanford.edu/~wcmac/downloads/.
[2] http://clic.cimec.unitn.it/composes/sick.html.

© Springer International Publishing AG, part of Springer Nature 2018
J. Pei et al. (Eds.): DASFAA 2018, LNCS 10827, pp. 847–860, 2018.
https://doi.org/10.1007/978-3-319-91452-7_54

verified the competitive performance acquired by applying recursive neural networks [5,6] to the problem and decided to build an open-domain and large-scale dataset in hope of getting more out of neural models. The release of Stanford Natural Language Inference (SNLI) corpus [7] is the tipping point of many neural models afterwards. However, people soon realize that separately encoding two sentences and feeding the concatenation into a softmax classifier is insufficient to model the complex interactions between them, which facilitates the utilization of attention mechanism [8] that establishes the word or phrase-level alignments between *hypothesis* and *premise*.

However, most neural attention models mainly utilize a memory of only the most recent history and fail to exploit long-range dependencies. The reason lies in the fact that there is only one single vector that keeps track of the past attention information, which is insufficient in terms of parameters. Consider the following example:

- *a man and a young boy are eating on a city sidewalk next to a cardboard box and a red shopping cart.*
- *father is eating with his son on a city sidewalk.*

It is fairly easy to conclude that the second sentence is neutral to the first one, because the first sentence does not indicate any parent-child relationship. However, when an existing attentive model is trying to identify the same issue, even if it separately aligns "father" vs "man" as well as "son" vs "boy", it still needs to link the information of both alignments to make the right decision, because "boy" vs "man" is not as close as "son" vs "father". However, these two alignments are five spans apart, longer than the four-spans limitation observed by [9].

To address this problem, we propose a novel architecture with a sophisticated writing mechanism to empower the conventional attentive reading operations that only reads from *premise*. Inspired by recent advances on Neural Turing Machines [10], two types of writing operations (i.e. *forget* and *update*) are introduced to incrementally act on the representations of *premise*. Therefore, rather than using only one single vector, we resort to a whole chunk of memory to ensure the storage of the past attention information. We refer to this architecture as RW-Attention.

Empirical results on the SNLI corpus show that while the proposed architecture achieves superior performance than the current state-of-the-art [11] by 0.4%, it outperforms *word-by-word* attention [12] by almost 4% only via introducing the writing mechanism.

The rest of the paper is structured as follows: We introduce our method in Sect. 2, present the experiments in Sect. 3, survey related work in Sect. 4 and conclude our work in Sect. 5.

2   RTE Architecture

In this section, we will elaborate on our RTE Architecture with both attentive reading and writing. We refer to it as RW-Attention. Figure 1 gives an overview of

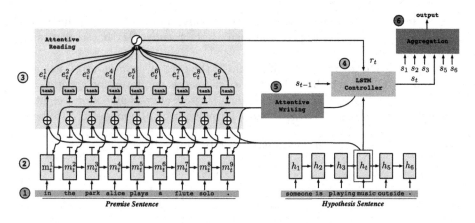

Fig. 1. The overview of our proposed RTE Architecture RW-Attention.

this architecture. It is composed of six major components (each one is annotated with a number referring to which stage it is executed). Let us first detail 1 & 2 & 3 that make up the reading module of the architecture.

2.1   Refined Reading Module

To our best knowledge, *word-by-word* attention proposed by [12] first applies the attention mechanism [8] to the RTE task, in order to give more interaction when encoding a sentence pair. We refine the attentive reading operation as a part of our RW-Attention architecture, which includes the following components:

1. **Embedding.** We formalize each RTE task as a triple (P, H, y). Therefore, given a *premise* $P = (x_1^p, \cdots, x_{\ell_P}^p)$ with the length ℓ_P and a *hypothesis* $H = (x_1^h, \cdots, x_{\ell_H}^h)$ with the length ℓ_H, our goal is to predict the label:

$$y* = argmax_{y \in \gamma} Pr(y|P, H) \qquad (1)$$

where $\gamma = \{Entailment, Neutral, Contradiction\}$, $Pr(y|P, H)$ is the conditional probability that we will discuss later. Here, each x is a fixed-sized embedding vector of the corresponding word, which can be initialized either by the pre-trained word embeddings like GloVe [13] and word2vec [14] or random vectors.

2. **Encoding.** In order to incorporate more contextual information, the embedding representations of *premise* and *hypothesis* are separately passed through a LSTM to yield the new contextual representations $\bar{P}$ and $\bar{H}$. Take the *hypothesis* for example, at each time-step we have its contextual representation:

$$h_t = LSTM(h_{t-1}, x_t) \qquad (2)$$

where $t = 1, \cdots, \ell_H$. Due to the space limitation, we skip the description of LSTM architecture and readers can refer to [15] for more details. Note that in Fig. 1, we denote the representations of *premise* as M instead of $\bar{P}$, because $\bar{P}$ only initializes the memory of *premise* M that is constantly changing afterwards.

3. **Attentive Reading.** This component is the core of *word-by-word* attention proposed by [12], which as the authors claimed is employed for *hypothesis* to do *word-by-word* reasoning via assigning the soft-aligned attention weights to the representations of *premise*. More formally, it can be modeled as follows:

$$e_{ti} = \nu^T \cdot \tanh(W^h h_t + W^m m_t^i + W^r r_{t-1}) \tag{3}$$

$$\alpha_{ti} = \frac{\exp(e_{ti})}{\sum_{j=1}^{\ell_P} \exp(e_{tj})} \tag{4}$$

$$r_t = \sum_{i=1}^{\ell_P} \alpha_{ti} m_t^i + \tanh(W^e r_{t-1}) \tag{5}$$

where the vector $\nu \in R^d$ and all matrices $W^* \in R^{d \times d}$ are the training parameters. Here, $\alpha_{ti} \in [0,1]$ is the attention weight that indicates the degree to which h_t in *hypothesis* is relevant to m_t^i in *premise*. Namely, the higher the attention weight, the more possible the information of m_t^i will be retained by the *so-called* attention-weighted representation of *premise* r_t.

If we take a close look at how this reading process works, it can be considered as a plain RNN: $r_t = RNN(r_{t-1}, h_t)$, with only one single vector r_t keeping track of the attention history, which is insufficient to face the challenge of long-range dependencies. Besides, *premise* is often much longer than *hypothesis*, many phrases or clauses that have no counterparts in *hypothesis* may disturb the recording of attention history by taking over long-range spans. Since later the role of r_t as tracing attention history will be replaced by the whole memory of *premise*, in order to avoid redundancy, r_{t-1} will no longer be utilized. Therefore, Eqs. 3 to 5 are refined as follows:

$$e_{ti}^R = \nu^{RT} \cdot \tanh(W^{Rh} h_t + W^{Rm} m_t^i) \tag{6}$$

$$\alpha_{ti}^R = \frac{\exp(e_{ti}^R)}{\sum_{j=1}^{\ell_P} \exp(e_{tj}^R)} \tag{7}$$

$$r_t = \sum_{i=1}^{\ell_P} \alpha_{ti}^R m_t^i \tag{8}$$

where superscript R denotes the Reading process.

2.2   Improved Writing Module

Attention mechanism, to some extent, is an act of memory retrieval. When most existing models take the encoded representations of *premise* as a fixed chunk of memory, here we argue the opposite, that according to [16], memory retrieval is not simply pulling a book from a shelf and returning it back as it was before. Instead, the memory is constantly changing every time being retrieved. Therefore, we can turn to this memory to do more than just memorizing the contextual information of *premise*, but the intermediate alignment information as well. We leverage this intuition to propose here a novel writing module upon the basic reading module, which simulates the way how visited memory in our mind is being modified:

4. **LSTM Controller.** For writing operations, we first need to figure out what exactly should be written into the memory of *premise* as well as how to update it. The simplest idea is to use r_t calculated in the previous layer. However, it is only a compressed vector representation that helps us distinguish which parts have been aligned and which parts are not. Besides, since *premise* may diverge substantially from *hypothesis* in terms of their meanings, semantic equivalence will not always be a given. In this work, we believe semantic matching is more critical than semantic equivalence, which requires the attention history to accommodate the information not only from correct alignments but also from valid matchings. Therefore, we design an intermediate structure called LSTM controller to bridge the gap between attentive reading and writing. Distinct from the recurrent controller in Neural Turing Machines (NTM) [10], its primary task is to do local inference by matching r_t with h_t:

$$s_t = LSTM(s_{t-1}, [r_t, h_t]) \tag{9}$$

The generated s_t will be used in the following layer to modify the current memory of *premise* as well as sent to the aggregation layer for inference composition.

5. **Attentive Writing.** This layer is the core of our architecture. In order to write the attention information (i.e. s_t) to the memory of *premise*, we design two types of writing operation - *forget* and *update*, which is inspired by the writing operation in Neural Turing Machines [10]. *Forget* is adopted to specify the values to be forgotten or removed on each dimension of M_t. This is implemented by assigning a *forget* vector F_t to each memory cell m_t^i at time-step t:

$$\tilde{m}_{t+1}^i = m_t^i [1 - \alpha_{ti}^W F_t] \tag{10}$$

where $F_t = f \cdot (W^{forget} s_t)$, $W^{forget} \in R^{d \times d}$. f can be any nonlinearity function, here we adopt typical sigmoid σ, in consistency with the forget gate in LSTM. *Forget* operation shares a very similar idea with forget gate in LSTM: the elements of a memory cell are reset to 1 if both the assigned weight and *forget* vector are 1; they are left unchanged if either the weight or *forget* vector is 0.

Besides the *forget* operation, the other writing operation *update* is of the same importance, which helps to encourage the storage of new extracted or aligned information. It works upon the memory that has just been modified by *forget* operation:

$$m_{t+1}^i = \tilde{m}_{t+1}^i + \alpha_{ti}^W U_t \tag{11}$$

Where $U_t = \sigma \cdot (W^{update} s_t)$, $W^{update} \in R^{d \times d}$.

In terms of α_{ti}^W, it has been proven by [17,18] that using the same normalized attention weights with attentive reading can help boost the performance in NMT, we test the same strategy in our experiment and obtain the positive result in our task.

6. **Aggregation.** In this final layer, we aggregate all the outputs from the LSTM controller at each time-step into a fix-length vector by sum-pooling:

$$s = \sum_{t=1}^{\ell_H} s_t \tag{12}$$

where $s \in R^d$ is the final representation of the sentence-pair. Therefore, all the word or phrase-level inferences are composed to make the final decision. More specifically, we feed s into a multi-layer perceptron (MLP) classifier, which contains two fully-connected layers with tanh activation and a softmax output layer:

$$Pr(y|P,H) = softmax(W^p s + b^p) \tag{13}$$

where $W^p \in R^{3 \times d}$, $b^p \in R^3$, $Pr(y|P,H) \in R^3$ is a normalized vector with each element being the probability of choosing a class in $\{Entailment, Neutral, Contradiction\}$. According to the Eq. 1, the predicted label $y*$ is the one with the highest probability.

3   Experimental Results

In this section, we evaluate the performance of our proposed architecture, which is conducted in two parts: quantitative evaluation and qualitative analysis. First, let us introduce some implementation details.

3.1   Settings

– **Data.** We conduct our experiment on the Stanford Natural Language Inference (SNLI) corpus [7]. This dataset is widely used in testing the effectiveness of neural models for RTE task, which contains 570,152 sentence pairs, each annotated with one of the following labels: *Entailment, Contradiction, Neutral* and -, where - indicates a lack of consensus from human annotators. In our experiment, we discard the sentence-pairs labeled with -, which leads to a total number of 549,367 sentence-pairs left for training, 9,842 for evaluation and 9,824 for testing.
– **Training Parameters.** We collect 34,877 unique words from train/dev/test data as our vocabulary, among which 30,626 can be found with corresponding pre-trained word embeddings in 840B-GloVe [13]. For OOV words, we use a uniform distribution between -0.05 and 0.05 as initialization. The length for word embeddings and the hidden states are both 300. Note that the embeddings matrix is not being updated in our settings. Moreover, for mini-batch training, we add an extra token NULL as the padding word in each sentence

with a vector of 0s. We use mini-batch of a relatively-large size 256 in the training. ADAM [19] is adopted for optimization with the first momentum coefficient β_1 set to 0.9 and the second one β_2 set to 0.999. The learning rate lr is initialized to 0.001 with a decay ratio of 0.95 for every 1000 global training steps. To avoid overfitting, dropout is introduced to LSTM encoders, namely setting some dimensions of the hidden states to 0 with a probability (0.8 in our experiment) randomly. We set the total number of training epochs to 30 with an early stopping of 3, during which the model with highest validation score (a.k.a accuracy) will be stored and treated as the optimal model that will be further used to report the corresponding testing score.

– **Training Objective.** RTE task is essentially a sentence-pair classification problem, therefore in our settings we adopt cross entropy loss as our objective function for training. More formally, given the ground-truth label y_i for each sentence-pair in training data and the parameter set θ for training, the objective function can be conducted as follows:

$$J(\theta) = -\sum_i^N y_i \log(y_i^*) + \frac{\lambda}{2}||\theta||^2 \tag{14}$$

where N is the number of sentence-pairs in training data, y_i^* is calculated in Eqs. 1 and 13, λ is ℓ_2-regularization strength. After trying with a small grid search in the range of [0.0, 1E−4, 3E−4, 1E−3], we found that using ℓ_2-regularization will slow down the training process with a certain amount of performance degradation. Therefore, we do not enforce ℓ_2 constraints on the training parameters in this work.

3.2 Quantitative Results

In Table 1, we present the performance comparisons of our proposed architectures on SNLI corpus with the previously reported models. All the listed models are attention-based and will be further discussed in Sect. 4, except for the first one with two separate encoders that we demonstrate to give readers a general idea of how a naive neural architecture performs on the RTE task.

According to the table, we have the following observations:

– We implement the *word-by-word* attention [12] in this work since it is the basis of RW-Attention. We believe it is an important baseline for knowing the efficacy of our architecture. However, even with the dimension of vector representations lifted from 100 to 300, its performance (83.2%) is still worse than the reported one (83.5%).
– RW-Attention achieves an accuracy of 87.2% of test accuracy, which is better than the current state-of-the-art [11] (86.8%). Meanwhile, it outperforms *word-by-word* attention (our implementation) by a large margin (4.0%). As can be seen from the table, the listed methods, even some with fancy techniques, could only achieve limited increments. On the contrast, our proposed

Table 1. Comparison between our attention architecture with other existing models on SNLI corpus. Column d specifies the dimension of the sentence representations. #Parameters is the number of parameters without the word embeddings. Note that we do not update the word embeddings during training.

Model	d	#Parameters	Train	Test
LSTM encoder [7]	300	3.0M	83.9	80.6
word-by-word attention [12]	100	250k	85.3	83.5
DF-LSTM [21]	600	2.8M	85.9	85.0
Stacked TC-LSTMs [22]	50	190k	86.7	85.1
mLSTM [20]	300	1.9M	92.0	86.1
LSTMN with deep attention fusion [23]	300	1.7M	87.3	85.7
LSTMN with deep attention fusion [23]	450	3.4M	88.5	86.3
Decomposable attention model [11]	200	382k	89.5	86.3
[11] + with intra-sentence attention	200	582k	90.5	86.8
word-by-word attention [12] (our implementation)	300	2.5M	88.4	83.2
mLSTM [20] (our implementation)	300	2.8M	90.4	85.9
RW-Attention	300	3.5M	90.3	**87.2**
RW-Attention with Bi-LSTM	300	9.8M	91.8	87.3

model is now capable of performing textual entailment recognition close to the human testers (87.5%). Furthermore, in order to minimize the influence of the potential statistical and data separation issues, we take almost the same settings of the basic *word-by-word* attention model to train our model. Therefore, this improvement is observed mostly due to the proposed structure itself. It further verifies the existence of long-range dependency of *word-by-word* attention and the introduced writing mechanism is effective in solving this problem.

- We also implement *mLSTM* [20] because it is currently the only model that works on improving *word-by-word* attention. However, even though it adopts an extra LSTM which seems to be analogy to our LSTM controller (but with different purposes, we will discuss about it latter), RW-Attention is still able to outperform it by 1.3%, further showing the effectiveness of the writing mechanism.
- We also test RW-Attention with more sophisticated Bi-LSTM encoders in order to get more contextual information from the sentence pairs. However, no matter how we try with the different settings, the result is only 0.1% better than the original RW-Attention. From our observations, when Bi-LSTM is utilized, the rising number of training parameters might dawdle the training process. However, training a neural network model on SNLI corpus is already very much likely to overfit. As a result, the overfitting could happen earlier, making the potential optimal even harder to get to.

In Fig. 2, we show how accuracy changes on the training and testing data for the three models implemented in this paper. First thing we should notice

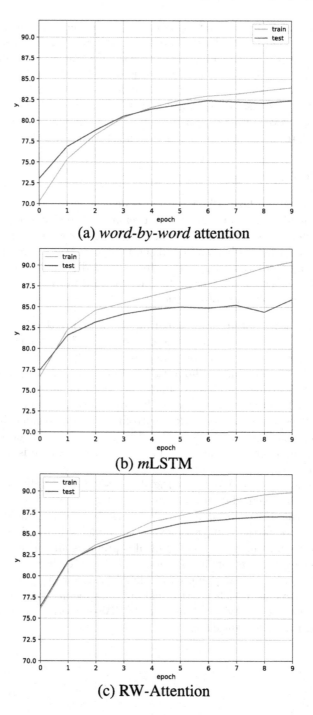

(a) *word-by-word* attention

(b) *m*LSTM

(c) RW-Attention

Fig. 2. Each subplot illustrates the first 10 epochs of the training process for the corresponding model.

is that both $mLSTM$ and our proposed RW-Attention converge in the first 10 training epochs. On the contrast, *word-by-word* attention (Fig. 2(a)) converges after almost 30 epochs, which happens to be the hardest one to train. Therefore, due to the space limitation, only 10 epochs of training results are included in this paper. Another thing we observe is that $mLSTM$ (Fig. 2(b)) converges very fast. Interestingly, it rises even higher than RW-Attention (Fig. 2(c)) in the first few epochs. However, after only 5 to 6 epochs, the model starts to overfit and no further improvement has been observed afterwards.

3.3   Qualitative Analysis

To better understand how RW-Attention is dealing with the long-range dependency problem of conventional attention mechanism, we visualize three attention matrices of the same example ("a man and a young boy are eating on a city sidewalk next to a cardboard box and a red shopping cart." vs "father is eating with his son on a city sidewalk.") in Fig. 3, from three distinct models. Recall that a darker color indicates a higher probability of word in *premise* being attended over. We have the following observations:

- Note that both *word-by-word* attention (Fig. 3(a)) and $mLSTM$ (Fig. 3(b)) perfectly resolve the semantic coherences: "father" vs "man" and "son" vs "boy". However, also recall from the previous analysis in Sect. 1, that merely aligning these synonyms without taking into considerations the differences between "father" vs "son" and "man" vs "boy" in terms of their closeness tends to confuse the model to make the wrong decision. Since neither one of them involves a mechanism to deal with long-range dependencies, the alignment from "son" in *hypothesis* to "boy" in *premise* is not conditioned on the previous alignment with "father". Therefore, this example is wrongly classified as *Entailment*.
- On the contrary, RW-Attention (Fig. 3(c)) is able to establish the semantic relatedness between "father" vs "man" and "son" vs "boy". It should be considered that the color intensity of "son" vs "boy" is much lighter than those in the first two matrices, suggesting that the model is doing alignment not only based on the current word "son", but also previous alignment information.
- Also note that *premise* includes a long subordinate clause "next to a cardboard box and a red shopping cart" that has no counterpart in *hypothesis*. While the previous models are still making some meaningless alignments (e.g. "on" vs "cart" or "city" vs "shopping"), RW-Attention is capable of fully neglecting this irrelevant clause.

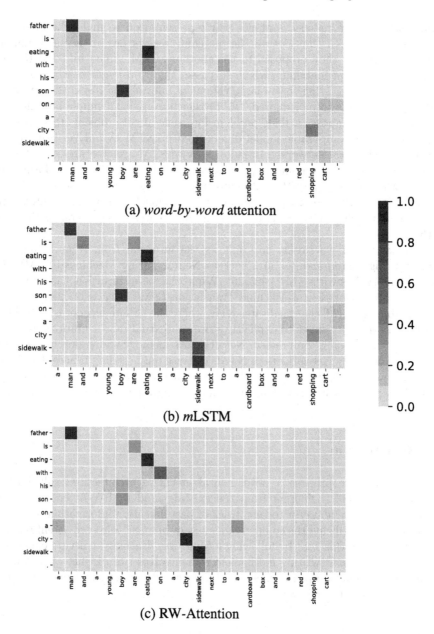

(a) *word-by-word* attention

(b) *m*LSTM

(c) RW-Attention

Fig. 3. An example that illustrates the superior performance of RW-Attention over other two models with conventional attention mechanism. (Color figure online)

4   Related Works

Attention mechanism has been frequently sought by a variety of recent works on RTE. [12] first introduced a *word-by-word* attention model, which however only picks up the reduced sentence-pair representation to make the final prediction. [20] argued that not all the matching results are of the same importance. Therefore, they proposed to use another LSTM, called $mLSTM$ to perform *word-by-word* matching, which surpasses the *word-by-word* attention by a large margin, suggesting that a large volume of information lies in the attention history that needs to be explored. [11] further introduced a simple but effective model that decomposes a RTE problem into several subproblems to solve separately. This decomposable model differs from $mLSTM$ by only utilizing naive feed-forward neural networks instead of heavy LSTM architectures, which substantially reduces the amount of parameters without sacrificing its performance.

Another line of works [21,22,24] emphasizes the significance of sentence interactions. The basic idea is to ensure the information flow from one sentence to the other, which intuitively is like reading one sentence while considering the other.

Unlike the above works, ours follows the intuition from recently popular memory-augmented architectures such as Neural Turing Machines (NTM) [10] and Memory Networks (MN) [25,26]. While the same intuition has benefited most the realm of Neural Machine Translation (NMT) [17,18,27,28], we only observe one from [23] that facilitates RTE, where the LSTM encoder is embedded with extra memory tapes to explicitly store contextual representations of input tokens without being compressed. Note that its memory addressing tends to be more like Memory Networks than NTM, because only reading is accepted. This paper, however, further explores the significance of writing and demonstrates one of its potential contributions on handling long-range dependencies in terms of the attention history.

5   Conclusion

In this paper we propose a novel neural attentive architecture, called RW-Attention for RTE. Different from the existing models, RW-Attention is not restricted to a single vector to keep track of past attention information. Instead, it is capable of utilizing the complete memory of *premise* to record attention history. It achieves this goal by following a joint deep learning structure enhanced with a sophisticated writing mechanism that involves two distinct operations - *forget* and *update*. Empirical analysis shows the efficacy of the proposed architecture on SNLI benchmark.

References

1. Dagan, I., Glickman, O., Magnini, B.: The PASCAL recognising textual entailment challenge. In: Quiñonero-Candela, J., Dagan, I., Magnini, B., d'Alché-Buc, F. (eds.) MLCW 2005. LNCS (LNAI), vol. 3944, pp. 177–190. Springer, Heidelberg (2006). https://doi.org/10.1007/11736790_9
2. Lakoff, G.: Linguistics and natural logic. Synthese 22(1), 151–271 (1970)
3. MacCartney, B.: Natural Language Inference. Stanford University, Stanford (2009)

4. Pavlick, E.: Compositional lexical semantics in natural language inference. Ph.D. dissertation, University of Pennsylvania (2017)
5. Bowman, S.R., Potts, C., Manning, C.D.: Recursive neural networks for learning logical semantics. CoRR, abs/1406.1827 (2014). http://arxiv.org/abs/1406.1827
6. Bowman, S.R., Potts, C., Manning, C.D.: Learning distributed word representations for natural logic reasoning. In: Proceedings of the Association for the Advancement of Artificial Intelligence Spring Symposium, AAAI, pp. 10–13 (2015)
7. Bowman, S.R., Angeli, G., Potts, C., Manning, C.D.: A large annotated corpus for learning natural language inference. CoRR, abs/1508.05326 (2015). http://arxiv.org/abs/1508.05326
8. Bahdanau, D., Cho, K., Bengio, Y.: Neural machine translation by jointly learning to align and translate. CoRR, abs/1409.0473 (2014). http://arxiv.org/abs/1409.0473
9. Daniluk, M., Rocktäschel, T., Welbl, J., Riedel, S.: Frustratingly short attention spans in neural language modeling. CoRR, abs/1702.04521 (2017). http://arxiv.org/abs/1702.04521
10. Graves, A., Wayne, G., Danihelka, I.: Neural turing machines. CoRR, abs/1410.5401 (2014). http://arxiv.org/abs/1410.5401
11. Parikh, A.P., Täckström, O., Das, D., Uszkoreit, J.: A decomposable attention model for natural language inference. CoRR, abs/1606.01933 (2016). http://arxiv.org/abs/1606.01933
12. Rocktäschel, T., Grefenstette, E., Hermann, K.M., Kociský, T., Blunsom, P.: Reasoning about entailment with neural attention. CoRR, abs/1509.06664 (2015). http://arxiv.org/abs/1509.06664
13. Pennington, J., Socher, R., Manning, C.: GloVe: global vectors for word representation. In: Proceedings of the 2014 Conference on Empirical Methods in Natural Language Processing, EMNLP, pp. 1532–1543 (2014)
14. Mikolov, T., Sutskever, I., Chen, K., Corrado, G.S., Dean, J.: Distributed representations of words and phrases and their compositionality. In: Advances in Neural Information Processing Systems, pp. 3111–3119 (2013)
15. Hochreiter, S., Schmidhuber, J.: Long short-term memory. Neural Comput. 9(8), 1735–1780 (1997). https://doi.org/10.1162/neco.1997.9.8.1735
16. McDermott, K.B., Roediger, H.L.: Memory (Encoding, Storage, Retrieval). Noba Textbook Series: Psychology. DEF Publishers, Champaign (2016). https://doi.org/nobaproject.com
17. Meng, F., Lu, Z., Li, H., Liu, Q.: Interactive attention for neural machine translation. CoRR, abs/1610.05011 (2016). http://arxiv.org/abs/1610.05011
18. Wang, M., Lu, Z., Li, H., Liu, Q.: Memory-enhanced decoder for neural machine translation. CoRR, abs/1606.02003 (2016). http://arxiv.org/abs/1606.02003
19. Kingma, D.P., Ba, J.: Adam: a method for stochastic optimization. CoRR, abs/1412.6980 (2014). http://arxiv.org/abs/1412.6980
20. Wang, S., Jiang, J.: Learning natural language inference with LSTM. CoRR, abs/1512.08849 (2015). http://arxiv.org/abs/1512.08849
21. Liu, P., Qiu, X., Chen, J., Huang, X.: Deep fusion LSTMs for text semantic matching. In: Proceedings of the 54th Annual Meeting of the Association for Computational Linguistics, ACL 2016, Berlin, Germany, 7–12 August 2016, vol. 1, Long Papers. The Association for Computer Linguistics (2016). http://aclweb.org/anthology/P/P16/P16-1098.pdf
22. Liu, P., Qiu, X., Huang, X.: Modelling interaction of sentence pair with coupled-LSTMs. CoRR, abs/1605.05573 (2016). http://arxiv.org/abs/1605.05573

23. Cheng, J., Dong, L., Lapata, M.: Long short-term memory-networks for machine reading. CoRR, abs/1601.06733 (2016). http://arxiv.org/abs/1601.06733
24. Sha, L., Chang, B., Sui, Z., Li, S.: Reading and thinking: re-read LSTM unit for textual entailment recognition. In: Calzolari, N., Matsumoto, Y., Prasad, R. (eds.) 26th International Conference on Computational Linguistics, COLING 2016. Proceedings of the Conference, Technical Papers, Osaka, Japan, 11–16 December 2016, pp. 2870–2879. ACL (2016). http://aclweb.org/anthology/C/C16/C16-1270.pdf
25. Weston, J., Chopra, S., Bordes, A.: Memory networks. CoRR, abs/1410.3916 (2014). http://arxiv.org/abs/1410.3916
26. Sukhbaatar, S., Szlam, A., Weston, J., Fergus, R.: End-to-end memory networks. In: Cortes, C., Lawrence, N.D., Lee, D.D., Sugiyama, M., Garnett, R. (eds.) Annual Conference on Neural Information Processing Systems. Advances in Neural Information Processing Systems, Montreal, Quebec, Canada, 7–12 December 2015, vol. 28, pp. 2440–2448 (2015). http://papers.nips.cc/paper/5846-end-to-end-memory-networks
27. Meng, F., Lu, Z., Tu, Z., Li, H., Liu, Q.: Neural transformation machine: a new architecture for sequence-to-sequence learning. CoRR, abs/1506.06442 (2015). http://arxiv.org/abs/1506.06442
28. Feng, Y., Zhang, S., Zhang, A., Wang, D., Abel, A.: Memory-augmented neural machine translation. CoRR, abs/1708.02005 (2017). http://arxiv.org/abs/1708.02005

Interpreting Fine-Grained Categories from Natural Language Queries of Entity Search

Denghao Ma[1], Yueguo Chen[1(✉)], Xiaoyong Du[1], and Yuanzhe Hao[2]

[1] Renmin University of China, Beijing, China
{madenghao,chenyueguo,duyong}@ruc.edu.cn
[2] Shandong University, Jinan, China
haoyuanzhe@mail.sdu.edu.cn

Abstract. The fine-grained target categories/types are very critical for improving the performance of entity search because they can be used for retrieving relevant entities by filtering irrelevant entities with a high confidence. However, most solutions of entity search face an urgent problem, i.e., the lack of fine-grained target categories of queries, which are hard for users to explicitly specify. In this paper, we try to interpret fine-grained categories from natural language based queries of entity search. We observe that entity search queries often contain terms specifying the contexts of the desired entities, as well as a topic of the desired entities. Accordingly, we propose to interpret fine-grained categories of entity search queries from the context perspective and the topic perspective. Therefore, we propose an approach by formalizing both context-based category model and topic-based category model, to tackle the category interpreting task. Extensive experiments on two widely-used test sets: INEX-XER 2009 and SemSearch-LS, indicate significant performance improvement achieved by our proposed method over the state-of-the-art baselines.

Keywords: Entity search · Fine-grained category
Type interpretation

1 Introduction

Entity search is to return a ranked list of entities in response to a query. It has received considerable attention in the past few years. Previous work [1–5,7] has shown that exploiting the category information of entities can significantly improve the performance of entity search. A general entity category represents the topic or domain of entities belonging to it. In contrast, a fine-grained entity category not only specifies a topic but also limits the topic with some specific constraints. For example, the entity-category pair *Stardust: Novels* indicates that the topic (general category) of *Stardust* is *novels*. The pair *Stardust: Novels by Neil Gaiman* not only indicates the topic of *Stardust* is *novels* but also specifies

© Springer International Publishing AG, part of Springer Nature 2018
J. Pei et al. (Eds.): DASFAA 2018, LNCS 10827, pp. 861–877, 2018.
https://doi.org/10.1007/978-3-319-91452-7_55

the topic with a constraint of *by Neil Gaiman*. Obviously, fine-grained categories bring richer semantics to entities than general categories.

Entities in the most widely used knowledge base Wikipedia have been labelled with many categories (as a kind of metadata representing the types of entities/Wiki pages). To study the percentage of entities having fine-grained categories, we classify categories into general or fine-grained classes in a simple way: if the name of a category includes only one term, the category is treated as a general category (e.g., *films* and *novels*); Otherwise, it is treated as a fine-grained category (e.g., *American people* and *novels by Neil Gaiman*). We then take a survey on the Wikipedia 2008 edition, and find that about 93.5% of entities in Wikipedia are assigned with at least one fine-grained category.

Some methods have been proposed to exploit fine-grained categories of entities for improving the performance of entity search [1,8,9]. However, the power of fine-grained categories are not well exploited in these methods because practical entity search queries are often in the form of natural language questions and lack of fine-grained target categories. It is hard for users to explicitly specify such fine-grained categories. Therefore, an important problem is how to interpret high quality target categories from entity search queries. Existing approaches of using fine-grained categories [1,2,8] can be categorized into two classes: (1) The target categories are explicitly provided by users. As stated, this may not be practical because users may not know how fine-grained categories of the desired entities are defined in the underlying knowledge bases. Besides, it brings extra burdens to users and therefore affects their search experience; (2) The target categories are automatically interpreted from queries, so that they can be directly used to retrieve entities.

Automatically interpreting target categories from queries is however non-trivial. We formally define the task of interpreting fine-grained entity categories from entity search queries as: given an input of a nature language question of entity search as a query, to interpret a ranked list of fine-grained entity categories that are relevant to the query. For example, given a query *Tom Hanks movies where he plays a leading role*, we may expect to generate fine-grained categories such as {*American films, English-language films, 1980s comedy-drama films*}. The major challenge of this task is to effectively infer the relevance of a fine-grained category to a natural language query of entity search.

A number of methods have been investigated for tackling the task [2,8,10,11]. These methods can be categorized into five classes: (1) Term-centric model [2] where the relevance of a category to the query is estimated by using the text similarity between the category and the query terms; (2) Entity-centric model [10] which takes the content of document representing an entity as the contextual profile of the entity, and estimates the relevance of the entity using the relevance of it's profile. The relevance of a category is finally estimated based on the average relevance of all entities belonging to the category; (3) Type-centric model [10] which builds a profile of each category by aggregating the contexts of all the entities belonging to the category, and then ranks the categories based on the relevance of their profiles; (4) Document-centric model [8] which firstly uses standard language models to retrieve relevant documents and takes the categories of the documents as the candidate categories. Then it ranks the categories based on their frequency in the relevant documents; (5) Learning to rank approach [11] where many features from other solutions as well as those extracted from the type taxonomy are combined by a machine learning technique.

The term-centric model is the simplest approach but encounters the problem of vocabulary gap. Both entity-centric and type-centric models are based on the candidate strategy where profiles of entities or categories are required. In the candidate strategy, the relevance of a category to a query is the mean of the relevance between query terms and all entities belonging to the category [10]. The strategy will underestimate the relevance of categories with more entities because of the impacts of irrelevant entities in the category. This is the drawback of the candidate strategy, which has been verified by the experimental results of the work [16]. The document-centric model adopts a voting strategy from the top-k relevant documents of queries. When k is smaller, many relevant categories can not be found because (1) the top-k documents may not belong to the relevant categories and (2) not all documents have categories; When k is larger, the precision is very low because many popular but irrelevant categories will get more votes from the top-k documents and therefore become the top results.

We observe that entity search queries often contain terms specifying the contexts of the desired entities, as well as a topic of the desired entities. In the meanwhile, a fine-grained category often contains a topic and some contextual representation (entities belonging to the category). Topics can be widely discovered from entity search queries, entity categories, as well as entities. For example, for the query *Tom Hanks movies where he plays a leading role*, the *movies* is a topic of the desired entities. For an entity, it's topics are the headwords of categories which the entity belongs to. Based on these observations, our key idea is to interpret the fine-grained categories of entity search queries from the context perspective and the topic perspective. We propose an approach by formalizing both context-based category model and topic-based category model, to tackle the key challenge of the fine-grained category interpretation task.

In the context-based category model, we adopt the voting strategy [17] and revise it for ranking candidate categories: the relevant documents of a query are firstly retrieved, and then the entities presenting in the documents vote for their categories. By exploiting the topic of a query and that of a category, we can effectively judge whether the category is relevant to the query in term of the topic. For example, given a query *Works by Charles Rennie Mackintosh* and two categories *Buildings in Glasgow, Animals in Glasgow*, we are able to judge *Buildings in Glasgow* is relevant to the query and *Animals in Glasgow* is irrelevant, because we know the topic *Buildings* is a sub-class of the topic *Works*, while topic *Animals* is not. Therefore, we devise the topic-based category model based on the category taxonomy of Wikipedia. The context-based category model is built on an unstructured document repository, and the topic-based category model is built on a structured category taxonomy. They are complementary to each other.

Extensive experiments on two test sets (INEX-XER 2009 and SemSearch-LS) demonstrate (1) our proposed approach achieves much better empirical performance over existing solutions [2,8,10]; (2) The voting strategy based methods outperform the candidate strategy based methods; (3) Our proposed topic-based category model significantly improves the performance over both our proposed context-based category model and the existing baselines [2,8,10].

The key contributions of this work are three-fold:

- We introduce the task of interpreting fine-grained categories from entity search queries, and propose an approach by formalizing both the context-based category model and topic-based category model to tackle the task.
- We devise the context-based category model to avoid the drawback of candidate strategy based methods.
- We propose to exploit the topics of both the query and the category for inferring their relevance, based on a category taxonomy of Wikipedia.

The remainder of this paper is organized as follows. In Sect. 2, we propose the ranking model. Experimental setup is presented in Sect. 3, and followed by the experimental results and analysis given in Sect. 4. The related work is given in Sect. 5 and the conclusion is presented in Sect. 6.

2   Ranking Model

The fine-grained target category interpretation task is to retrieve a ranked list of categories in response to a query. We formulate the task in a framework using generative probabilistic model. In this framework, the relevance score of a candidate category c to a query q is denoted as $p(c|q)$. To estimate the probability $p(c|q)$, we propose an approach using the combination of context-based category model and topic-based category model as follows:

$$p(c|q) = p_c(c|q)^{1-\lambda} p_t(c|q)^{\lambda} \tag{1}$$

where $p_c(c|q)$ is the context-based relevance estimated from a context perspective, and $p_t(c|q)$ is the topic-based relevance estimated from a topic perspective. The parameter λ is to tune the weight between these two components.

2.1   Context-Based Category Model

The intuition of the context-based category model is that the more frequently do entities of a category occur in the relevant documents of q, the more relevant is the category to the query. Such a voting strategy has been widely studied and applied in the expert search [6,17]. Based on the voting strategy, the component $p_c(c|q)$ is estimated as follows:

$$p_c(c|q) = \sum_{d \in\ top(k,q)} p(c|d)p(d|q) \tag{2}$$

where $top(k,q)$ is the set of top-k relevant documents of the query q. The component $p(d|q)$ is the probability of generating the document d from the query q, and $p(c|d)$ is the probability of generating the category c from the document d.

We use the entities embedded in documents to bridge the association between documents and categories, for finding more candidate categories. The probability $p(c|d)$ is therefore estimated as:

$$p(c|d) = \sum_{e \in E(d)} p(c|e)p(e|d) \qquad (3)$$

where $E(d)$ is a set of entities embedded in the document d. The component $p(e|d)$ is the probability of generating the entity e from the document d, and the component $p(c|e)$ is the probability of generating the category c from the entity e. They are further estimated as follows:

$$p(e|d) = \frac{n(e,d)}{\sum_{e' \in E(d)} n(e',d)} \quad ; \quad p(c|e) = \frac{a(c,e)}{\sum_{c' \in C} a(c',e)} \qquad (4)$$

where $n(e,d)$ is an indicator function which equals 1 if the entity e appears in the document d, zero otherwise. The C is the entire category corpus of Wikipedia. The component $a(c,e)$ is to denote whether the entity e belongs to the category c. If e belongs to c, $a(c,e)$ equals 1; Otherwise, $a(c,e)$ equals 0.

According to the Bayes's Theorem, the probability $p(d|q)$ is rewritten as:

$$p(d|q) = \frac{p(q|d)p(d)}{p(q)} \qquad (5)$$

where the component $p(q|d)$ is the probability of generating the query q from the document d. The component $p(d)$ is the probability of the document d, which is uniformly distributed over the all documents of the entire document collection. The component $p(q)$ is the probability of the query q, which is a consistent value given the query q.

We use the standard language model to estimate the probability $p(q|d)$:

$$p(q|d) = \prod_{t \in q} \{(1 - \lambda_1)p(t|d) + \lambda_1 p(t|D)\}^{n(t,q)} \qquad (6)$$

where $p(t|d)$ is the probability of generating the term t from the document d. It is the maximum likelihood estimation of the term t in d. D is the entire document collection. The component $p(t|D)$ is the probability of generating the term t from the collection D. It is used to smooth the probability $p(t|d)$, considering that $p(t|d)$ may equal to zero. The component $n(t,q)$ represents the frequency of the term t presenting in the query q.

2.2   Topic-Based Category Model

In this study, the topic of an entity is a general category such as *films*, *people*, and *novels*. In the natural language queries, users often express their intent using a topic term, which often plays a role of query headword in the query. For example, given a query *Neil Gaiman novels*, the topic *novels* is not only the topic of desired entities but also the headword/intent of the query.

Topic Detection. In our study, we take the headword of a query as the topic of the query. The headword detection in natural language questions is a challenging task, because user's queries often do not obey the grammar of a written language [12,13]. Since the topic detection is not the contribution (focus) of this paper, we only exploit the technique of the study [12] to detect the headwords from queries and categories. The headword detection technique of [12] firstly identifies the headword-modifier pairs from abundant queries to build the instance-level headword-modifier knowledge. The headword is a query word/phrase which can represent the intent of the query, and the modifiers are other words/phrases which are to limit the headword. The instance-level headword-modifier denotes that both headword and modifier are specific instances. For example, given a query *popular smart cover iphone 5*, it identities the *smart cover* as the headword and *iphone 5* as well as *popular* as the modifiers. Since both *smart cover* and *iphone 5* are specific instances, ($smart\ cover_{[h]}$, $iphone\ 5_{[m]}$) is a instance-level headword-modifier pair. Based on a taxonomy knowledge base Probase [20], it then derives the concept-level headword-modifier knowledge from a large number of instance-level pairs. The concept-level headword-modifier denotes that both headword and modifier are concepts. For the above example, *smart cover* is an accessory and *iphone 5* is a device. Both accessory and device are concepts. Therefore, according to the pair ($smart\ cover_{[h]}$, $iphone\ 5_{[m]}$), the concept-level headword-modifier pair ($accessory_{[h]}$, $device_{[m]}$) can be created statistically from the instance-level headword-modifier pairs. Finally, each instance-level/concept-level headword-modifier pair is assigned with a score which represents the probability of that the headword and the modifier appear together in a query. Given a new query, the headword and modifiers are detected based on the knowledge of instance-level/concept-level headword-modifier pairs.

Matching. There may be a vocabulary gap between the query topic and the category topic, because users may not accurately express the topic of their desired entities. For example, users may use the query *Works by Charles Rennie Mackintosh* to retrieve all buildings designed by Charles Rennie Mackintosh, with the topic of the query as *works*. However, the topic of a relevant category can be *buildings*. A gap exists between the two topic words. To tackle the gap, we use the taxonomy of Wikipedia categories as our knowledge base to infer the probability $p_t(q|c)$, which is to capture the topic relevance between the category c and the query q.

Our approach is illustrated in Fig. 1. Two major steps of our approach are as follows:

- Step 1: We build a category profile for the query topic t_q and the category topic t_c respectively, and label the two profiles as C_q and C_c respectively. The

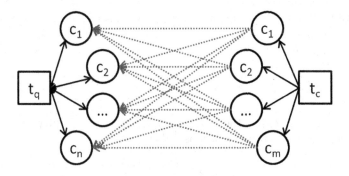

Fig. 1. An inferring approach for the relevance of two topics

profile C_q (or C_c) is a set of categories whose topic is t_q (or t_c). If t_q does not appear in the Wikipedia category corpus, we use the synonym of t_q to replace it. Noted that only one of synonyms is used in the category taxonomy of Wikipedia, because Wikipedia exploits some techniques to prevent the synonyms of same meaning existing in topics of the category taxonomy [8]. For example, given two synonyms *films* and *movies*, only *films* exists in the category taxonomy as a topic word of films/movies. The redirect information of Wikipedia is taken as our synonyms dictionary. As illustrated by the Fig. 1, we estimate the probability of generating a category $c_m \in C_c$ (or $c_n \in C_q$) from the topic t_c (or t_q). The estimation is as follows:

$$p(c_m|t_c) = \frac{a(c_m, t_c)}{\sum_{c' \in C} a(c', t_c)} \quad (7)$$

where C is the entire Wikipedia category corpus. If the topic of c' is t_c, $a(c', t_c) = 1$; Otherwise $a(c', t_c) = 0$.

- Step 2: We compute the probability of generating the category c_n from the category c_m. In the taxonomy of Wikipedia categories, each category may have some sub-categories, and also have some parents (except the root category). In other words, each category has a family tree. Therefore, we exploit the family tree to estimate the relevance of two categories. Basically, if c_n is an ancestor or a descendant of c_m, c_n will be relevant to c_m; Otherwise, they are treated as irrelevant. The distance of c_n and c_m in family tree is used to estimate the relevance:

$$p(c_n|c_m) = \frac{1}{\sigma^{D(c_n, c_m)}} \quad (8)$$

where $D(c_n, c_m)$ is the distance of c_n and c_m in the family tree of c_m. If c_n is not in the family tree of c_m, we set $D(c_n, c_m)$ as a constant value of the maximal depth of the taxonomy tree. The parameter σ is a constant value which can be learned from some validation set.

Since both the category profiles C_q and C_c may have multiple categories, each pair of two categories from C_q and C_c may be relevant to each other. The probability of generating the topic t_q from the topic t_c is then estimated as:

$$p(t_q|t_c) = \sum_{c_m \in C_c} \max_{c_n \in C_q} \{p(t_q|c_n)p(c_n|c_m)p(c_m|t_c)\} \qquad (9)$$

The component $p(t_q|c_n)$ is the probability generating the query topic t_q from the category c_n, which is estimated as follows:

$$p(t_q|c_n) = \frac{p(c_n|t_q)p(t_q)}{p(c_n)} \qquad (10)$$

where the component $p(t_q)$ is the probability of generating the topic t_q from the whole topic set, which is an uniform distribution over all the topics in Wikipedia. The component $p(c_n)$ is the probability of generating the category c_n from the whole category corpus, which is also uniformly distributed over the all categories in Wikipedia.

3 Experimental Setup

3.1 Experimental Collection

The corpus that has been used by INEX 2009 Entity Ranking track [5] is adopted by our study. It is the Wikipedia dumps of 2008 edition, which includes about 2.6 million documents. Each document represents an entity, containing some paragraphs and categories which are used to describe different aspects of the document (entity). From this document collection, 371,797 fine-grained categories are extracted, and 8,074,151 entity-to-category pairs are extracted.

Two test sets are used to evaluate the effectiveness of our model:

- INEX-XER 2009 (INEX): The INEX 2009 Entity Ranking track [5] launched an entity search test set with 55 topics whose answers are Wikipedia entities. Each topic contains five components, i.e., title, entities, categories, description, and narrative. The title component is a nature language question. Other components are to constrain or explain the question. Since our task is to interpret the fine-grained target categories from entity search queries, the title component is taken as the input of our task. One example title is *Alan Moore graphic novels adapted to film.*
- SemSearch-LS (SemSearch): This test set is used in the list search task at the 2011 Semantic Search challenge [15]. Same as [14], we use 43 out of 50 topics which have answers from Wikipedia (i.e., 7 topics without Wikipedia answers are removed). An example is *Apollo astronauts who walked on the Moon.*

3.2 Evaluation Metrics

In the above test sets, each topic/query is assigned with a list of target entities as ground truths of entity search. By using these entities, we can find a list of relevant categories to the query as the ground truths. Borrowing the idea of the work [16], we assign a relevance score to each derived ground truth, because some of them are more relevant to the query than others. The relevance of a category c to the query q, i.e., $l(c, q)$, is estimated as $l(c, q) = |E(c) \cap E(q)|$ where $E(c)$ is a set of entities which belong to c and $E(q)$ is a set of target entities of q. Note that $l(c, q)$ will be used in the evaluation metrics to distinguish different derived ground truths.

According to the work [16], we apply two metrics for evaluating the performance of different methods:

- MAP: MAP of a test set is the mean of average precision(AP) values of all queries in the set. Considering that relevant categories have different relevance levels, the average precision is defined as follows:

$$AP = \frac{\sum_{i=1}^{m} P(i)f(l(i, q))}{\sum_{j=1}^{n} l(j, q)} \tag{11}$$

  where m is the size of the list of retrieved results, n is the number of the ground truths of q, and $P(i) = \sum_{k=1}^{i} l(k, q)/i$. Besides, $f(l(i, q))$ is an indicator function equaling 1 if $l(i, q) > 0$, zero otherwise.
- Precision: $p@k$ only considers the number of relevant categories ranked within the k positions. We consider that the category c is relevant to the query q, only if $l(c, q) > 0$; Otherwise, c is irrelevant to q.

3.3 Parameter Tuning

This section describes the parameter setting used in our experiments. In our model, four parameters are as follows: (1) λ in Eq. 1, which is to tune the weight between context-based category model and topic-based category model; (2) k in Eq. 2, which is the number of documents voting for candidate categories; (3) λ_1 in Eq. 6, which is to adjust the weight of smoothing factor; (4) σ in Eq. 8, which is to magnify the distance between two categories in their family tree. In our experiments, we apply Coordinate Ascent (CA) algorithm [18] to directly optimize the mean average precision (MAP). The CA algorithm iteratively adjusts each parameter, holding all other parameters fixed. Following the work [19], we use 10-fold cross validation for each query test set, to avoid the overfitting of parameters.

3.4 Baseline Methods

Our method is compared against four baselines:

- Term-centric model (TCM) [2]: The relevance of a category to a query is estimated by using the text similarity between the query terms and the category name. The standard language model is used to compute the text similarity.

- Type-centric model (CCM) [10]: The context profile of each category is built by concatenating the contexts of all entities belonging to the category. The text similarity between the profile of a category and query terms is used to estimate the relevance of the category to the query terms.
- Entity-centric model (ECM) [10]: Instead of building profile for each category, it first models the relevance between query terms and each entity belonging to the category using it's profile. Then the arithmetic mean of the relevance of the entities is treated as the relevance of the category to query terms.
- Document-centric model (DCM) [8]: The standard language model is used to retrieve the relevant documents of a query. The categories are scored based on their occurring frequency in these documents.

A learning to rank approach is proposed in [11], which is a synthesis of the features and relevance of many other solutions. To ensure a fair comparison, we do not take it as our baseline.

4   Results and Analysis

In this section, we introduce the research questions of this paper, and conduct extensive experiments to answer them. Some labels are used to represent different models: the notation C represents the proposed context-based category model, and CT represents the combination of the context-based category model and the topic-based category model, which is exactly our solution. To measure the statistical significance, a two-tailed paired t-test is applied. The * is to denote the difference at the 0.01 level and the † is to denote the difference at the 0.05 level.

4.1   Research Question

- RQ1: Can the proposed approach CT improve significantly the overall performance of the fine-grained target category interpretation task over the baselines? (Sect. 4.2)
- RQ2: Can the voting strategy based methods outperform the candidate strategy based methods? (Sect. 4.2)
- RQ3: Can the proposed topic-based category model further improve the performance of the proposed context-based category model? (Sect. 4.2)
- RQ4: Can the topic-based category model improve the performance of the baselines? (Sect. 4.3)
- RQ5: Can our proposed approach perform better than the baselines on the different test sets? (Sect. 4.4)
- RQ6: How robust is our method with respect to parameter settings? (Sect. 4.5)

4.2   Overall Performance

Can the proposed approach CT improve significantly the overall performance of the fine-grained target category interpretation task over the baselines (RQ1)?

Table 1. Performance comparison on all queries. Significance is tested against the DCM

Models	Topics: INEX + SemSearch					
	p@1	p@2	p@3	p@5	p@10	MAP
TCM	0.340	0.361	0.337	0.278	0.214	0.239
CCM	0.289	0.294	0.275	0.239	0.236	0.255
ECM	0.268	0.309	0.289	0.272	0.243	0.267
DCM	0.474	0.459	0.405	0.398	0.318	0.329
C	$0.558^{\dagger}$	0.469	0.426	0.379	0.329	0.428^{*}
CT	$\mathbf{0.701^{*}}$	$\mathbf{0.582^{*}}$	$\mathbf{0.550^{*}}$	$\mathbf{0.505^{*}}$	$\mathbf{0.414^{*}}$	$\mathbf{0.498^{*}}$

We combine the INEX-XER 2009 topics and the SemSearch-LS topics into a larger test set (called all queries). Our proposed model (CT) is compared with the baseline models (TCM, CCM, ECM and DCM) over all queries, to test their overall performance. Table 1 reports the experimental results. We observe that the performance achieved by the model CT is much better than those achieved by the baseline models (TCM, CCM, ECM and DCM), in terms of all metrics used. Especially for the metric $p@1$, CT achieves a relative improvement of 47.9% over the state-of-the-art baseline DCM. The results give a positive answer to RQ1.

According to the results in Table 1, we can find that the model C leads to a consistent performance improvement over all the baselines. In terms of the metric MAP, the relative improvement is 30.1% over DCM, 79.1% over TCM, 67.8% over ECM, and 60.3% over CCM. In terms of other metrics, the improvements are also statistically significant. The improvement over DCM illustrates that it is a right strategy to use the entities included in retrieved documents to vote for candidate categories instead of using the retrieved documents for voting. The model C is based on the voting strategy, and the baselines CCM and ECM are both based on the candidate strategy. The results between C and CCM, as well as C and ECM can answer our second research question (RQ2): the voting strategy based methods outperform the candidate strategy based methods. Besides, RQ2 is verified further by comparing the results between DCM and ECM, as well as DCM and CCM.

To address the third research question (RQ3): can the proposed topic-based category model improve further the performance over the proposed context-based category model, we perform a group of experiments. The experimental results are shown in Table 2. The parameter k is the number of the relevant documents of a query. In Table 2, we observe that the performance of CT is much better than that of C, in terms of all k values. This illustrates that the performance improvement achieved by CT is consistent and not affected by the parameter k too much. When only the context-based category model is applied to rank the candidate categories, some popular but irrelevant categories often appear in the top results, because the entities belonging to those categories

Table 2. The impact of using the topic-based category model

k	Topics: INEX + SemSearch					
	p@1	p@2	p@3	p@5	p@10	MAP
C: with context-based category model						
10	0.485	0.448	0.388	0.355	0.321	0.404
20	0.536	0.464	0.405	0.371	0.329	0.426
30	0.557	0.469	0.426	0.379	0.329	0.428
50	0.557	0.459	0.430	0.377	0.328	0.424
100	0.515	0.469	0.416	0.367	0.319	0.416
CT: with context-based and topic-based category models						
10	0.639	0.557	0.526	0.480	0.415	0.481
20	0.691	0.572	0.546	0.511	0.431	0.497
30	0.701	0.582	0.550	0.505	0.414	0.498
50	0.711	0.593	0.550	0.505	0.421	0.496
100	0.691	0.598	0.546	0.499	0.410	0.487

often appear in the relevant documents of queries. When the topic-based category model is also applied, those irrelevant categories are depressed so that the performance is improved significantly.

4.3   Applying Topic-Based Category Model to the Baselines

Can the topic-based category model significantly improve the performance of the baselines (RQ4)? To answer this question, we use the topic-based category model to improve the baselines. The strategy of improving the baselines is to substitute our proposed context-based category model with the baselines (labeled with "+T"). For example, the label "DCM + T" represents the topic-based category model is used to improve the baseline DCM. Table 3 reports the results between the baselines and the improved baselines. Significance is tested against the baselines.

According to the results in Table 3, we observe that all baselines benefit from the usage of the topic-based category model. In terms of the metric $p@1$, the relative improvements are 26.2% on DCM, 115.3% on ECM, 82.0% on CCM, and 42.6% on TCM. The improvements of $p@1$ are very significant, especially on the ECM and CCM. This is because there are many popular but irrelevant categories appearing in the top results of ECM and CCM; The topic-based category model depresses effectively those irrelevant categories. Another interesting phenomenon of the improved baselines is the pattern of $p@k$ (i.e., $p@1 > p@2 > p@3 > p@5 > p@10$). While this ranking pattern does not always hold in the results of the baselines. The phenomenon also shows the advantage of the topic-based category model. According to the results in Table 3, we answer the fourth research question positively: the topic-based category model does significantly improve the performance of the baselines.

Table 3. Performance comparison between the baselines and the improved baselines

Models	Topics: INEX + SemSearch					
	p@1	p@2	p@3	p@5	p@10	MAP
TCM	0.340	0.361	0.337	0.278	0.214	0.239
TCM + T	0.485*	0.469*	0.423*	0.344†	0.291*	0.306*
CCM	0.289	0.294	0.275	0.239	0.236	0.255
CCM + T	0.526*	0.454*	0.423*	0.367*	0.332*	0.332*
ECM	0.268	0.309	0.289	0.272	0.243	0.267
ECM + T	0.577*	0.474*	0.454*	0.381*	0.330*	0.337*
DCM	0.474	0.459	0.405	0.398	0.318	0.329
DCM + T	0.598†	0.531	0.471†	0.425	0.358	0.408*

4.4 Breakdown by Query Subsets

Can our proposed approach perform better than the baselines on the different test sets (RQ5)? To address this question, we compare our approach (CT) with the baselines (TCM, ECM, CCM and DCM) on the two individual test sets (INEX-XER 2009 and SemSearch-LS). The comparing results are presented in Tables 4 and 5, respectively. Significance is tested against the DCM baseline.

We observe that the model CT consistently outperforms other baselines on the two test sets. According to the $p@k$ results of CT and the baselines, we find that CT achieves the best results. Take $p@1$ as an example, the performance achieved by CT is more than 1.56 times of that achieved by the best baseline (DCM) on the INEX-XER 2009; On the SemSearc-LS topics, our method achieves an improvement of 30.0% over the best baseline (DCM). Besides, the improvement of MAP over the model DCM is also significant. These improvements illustrate that the advantage of our proposed approach is consistent on the different test sets, and address the fifth research question RQ5.

Table 4. Performance comparison on the INEX-XER 2009 topics

Models	Topics: INEX-XER 2009					
	p@1	p@2	p@3	p@5	p@10	MAP
TCM	0.461	0.445	0.418	0.331	0.265	0.308
CCM	0.309	0.309	0.285	0.273	0.273	0.310
ECM	0.291	0.345	0.309	0.298	0.285	0.312
DCM	0.473	0.473	0.430	0.429	0.389	0.413
CT	**0.764***	**0.664***	**0.636***	**0.622***	**0.511***	**0.625***

Table 5. Performance comparison on the SemSearch-LS topics

Models	Topics: SemSearch-LS					
	p@1	p@2	p@3	p@5	p@10	MAP
TCM	0.143	0.250	0.230	0.210	0.148	0.150
CCM	0.262	0.274	0.195	0.188	0.260	0.182
ECM	0.238	0.262	0.262	0.238	0.188	0.207
DCM	0.476	0.440	0.373	0.357	0.236	0.219
CT	**0.619***	**0.476**	**0.437†**	**0.362**	**0.288**	**0.332***

4.5   Impacts of Parameter Settings

How robust is our approach with respect to parameter settings (RQ6)? In our model, four parameters should be investigated: λ in Eq. 1, k in Eq. 2, λ_1 in Eq. 6, and σ in Eq. 8. When investigating the impact of one parameter, we assign other parameters as default values. Therefore, we give a configuration of the four parameters: $[\lambda = 0.5; k = 30; \lambda_1 = 0.1; \sigma = 5]$. Figure 2 shows the impacts of the four parameters.

Figure 2(a) shows the impact of the parameter λ in the model CT. According to the results in Fig. 2(a), we find $\lambda = 0.3$ leads to the best performance on the different test sets. This indicates that (1) the single context-based category model performs better than the single topic-based category model, and (2)

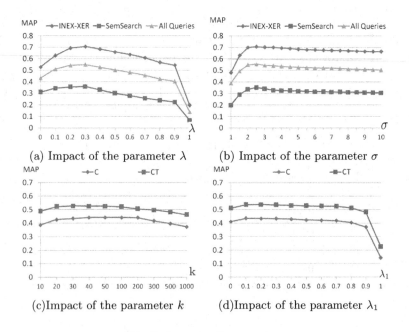

(a) Impact of the parameter λ

(b) Impact of the parameter σ

(c) Impact of the parameter k

(d) Impact of the parameter λ_1

Fig. 2. Impacts of parameters

the combination of the context-based category model and topic-based category model outperforms the any one of them. The impact of the parameter σ in the model CT is reported in Fig. 2(b). We observe that the best performance on different test sets is achieved when σ equals 2.5. This indicates that the distance of two categories in their family tree should be magnified to a suitable level.

Figure 2(c) presents the impact of the parameter k on the performance of all queries. We observe that when k is more than 30 and less than 100, the change in MAP is not obvious. But when k is less than 20 and is more than 200, the change in MAP is significant. This is because when k is less than 20, many relevant categories can not be found; When k is more than 200, some popular but irrelevant categories appear in the top results since they get more votes from the entities included in the tail documents. We present the impact of the parameter λ_1 in Fig. 2(d). When λ_1 is more than 0.1 and less than 0.9, the change in MAP is not significant. But when λ_1 is close to 1, the MAP decreases drastically. This is because the relevant documents of queries will be only based on the smoothing factor, when λ_1 equals 1.

5 Related Work

Entity search problems have been widely studied in the literature. Many techniques are proposed to exploit the fine-grained categories of entities for improving the performance of entity search [1,2,8]. In these techniques, some fine-grained target entity categories of queries are critical for fully exploiting the categories of entities. But a problem is how to find the high quality target categories for entity search queries. The INEX Entity Ranking track [5] features scenarios where the query target categories can be provided by users, in addition to query terms. In practice, it is hard for users to provide the fine-grained target entity categories of queries. This is because users may not know what fine-grained categories are in the knowledge base (supporting entity search), and don't know how to define the fine-grained entity categories for queries. Therefore, it is urgent to automatically interpret the fine-grained target entity categories from entity search queries.

It is non-trivial to automatically interpret the fine-grained target entity categories from entity search queries, because both queries and categories are short texts, and therefore their relevance is hard to be accurately estimated. The work [2] estimates the relevance of a category to a query by using the text similarity between the query terms and the category name. The method only can retrieve some relevant categories whose name includes more query terms. But if the names of relevant categories include none or less of query terms, the method will not work well. The type-centric model and the entity-centric model are proposed in [10]. The type-centric model firstly builds a contextual profile for each category by concatenating the contexts of all entities belonging to the category. Then it ranks the categories according to the text similarity between query terms and their contextual profiles. In the entity-centric model, it is first to compute the relevance of a query and each entity belonging to a category. Then the category is scored based on the arithmetic mean of the relevance of the entities. Essentially, in both the type-centric model and the entity-centric model, the relevance of a category to query terms is the mean of the relevance between query terms and all entities belonging to the category. Therefore, the two models underestimate the relevance of categories with more entities because of the impacts of irrelevant entities in the category. The document-centric model is proposed in [8]. In this model, the categories of the top-k relevant documents of queries are taken as the candidate categories. The candidate categories are ranked based on their frequency in the top-k documents. The model is not robust to the value of k. A learning to rank approach is proposed in [11], which combines many features from other solutions and those extracted from the type taxonomy by using a machine learning technique.

6   Conclusion

We define the task of interpreting fine-grained target entity categories from entity search queries, and propose an approach by formalizing both context-based category model and topic-based category model. The proposed approach achieves much better performance over the existing solutions. By exploiting the entities appearing in the relevant documents of queries to find and rank candidate categories, the context-based category model outperforms the state-of-the-art baseline DCM. The topic-based category model effectively depresses the popular but irrelevant categories from the context-based category model, and therefore improves the performance significantly when they are combined.

Acknowledgments. Yueguo Chen is supported by the National Science Foundation of China under grants No. U1711261, 61472426, 61432006, and the State Visiting Scholar Funds from the China Scholarship Council under Grant Number 201706365018. Denghao Ma is supported by the Outstanding Innovative Talents Cultivation Funded Programs 2017 of Renmin University of China and the State Scholarship Fund from China Scholarship Council under Grant Number 201706360309.

References

1. Chen, Y., Gao, L., Shi, S., Du, X., Wen, J.: Improving context and category matching for entity search. In: AAAI, pp. 16–22 (2014)
2. Balog, K., Bron, M., Rijke, R.: Query modeling for entity search based on terms, categories, and examples. ACM Trans. Inf. Syst. **29**(4), 22 (2011)
3. de Vries, A.P., Vercoustre, A.-M., Thom, J.A., Craswell, N., Lalmas, M.: Overview of the INEX 2007 entity ranking track. In: Fuhr, N., Kamps, J., Lalmas, M., Trotman, A. (eds.) INEX 2007. LNCS, vol. 4862, pp. 245–251. Springer, Heidelberg (2008). https://doi.org/10.1007/978-3-540-85902-4_22
4. Demartini, G., de Vries, A.P., Iofciu, T., Zhu, J.: Overview of the INEX 2008 entity ranking track. In: Geva, S., Kamps, J., Trotman, A. (eds.) INEX 2008. LNCS, vol. 5631, pp. 243–252. Springer, Heidelberg (2009). https://doi.org/10.1007/978-3-642-03761-0_25

5. Demartini, G., Iofciu, T., de Vries, A.P.: Overview of the INEX 2009 entity ranking track. In: Geva, S., Kamps, J., Trotman, A. (eds.) INEX 2009. LNCS, vol. 6203, pp. 254–264. Springer, Heidelberg (2010). https://doi.org/10.1007/978-3-642-14556-8_26

6. Balog, K., Rijke, M.: Combining Candidate and Document Models for Expert Search. TREC (2008)

7. Vercoustre, A.-M., Pehcevski, J., Thom, J.A.: Using Wikipedia categories and links in entity ranking. In: Fuhr, N., Kamps, J., Lalmas, M., Trotman, A. (eds.) INEX 2007. LNCS, vol. 4862, pp. 321–335. Springer, Heidelberg (2008). https://doi.org/10.1007/978-3-540-85902-4_28

8. Kaptein, R., Kamps, J.: Exploiting the category structure of Wikipedia for entity ranking. Artif. Intell. **194**, 111–129 (2013)

9. Garigliotti, D., Balog, K.: On type-aware entity retrieval. In: ICTIR, pp. 27–34 (2017)

10. Balog, K., Neumayer, R.: Hierarchical target type identification for entity-oriented queries. In: CIKM, pp. 2391–2394 (2012)

11. Garigliotti, D., Hasibi, F., Balog, K.: Target type identification for entity-bearing queries. In: SIGIR, pp. 845–848 (2017)

12. Wang, Z., Wang, H., Hu, Z.: Head, modifier, and constraint detection in short texts. In: ICDE, pp. 280–291 (2014)

13. Bendersky, M., Metzler, D., Croft, W.: Learning concept importance using a weighted dependence model. In: WSDM, pp. 31–40 (2010)

14. Balog, K., Neumayer, R.: A test collection for entity search in DBpedia. In: SIGIR, pp. 737–740 (2013)

15. Roi, B., Harry, H., Daniel, M., Peter, M., Jeffrey, P., David, R., Henry, T.: Entity search evaluation over structured web data. In: SIGIR (2011)

16. Liang, S., Rijke, M.: Formal language models for finding groups of experts. Inf. Process. Manag. **52**(4), 529–549 (2016)

17. Macdonald, C., Ounis, I.: Voting techniques for expert search. Knowl. Inf. Syst. **16**(3), 259–280 (2008)

18. Metzler, D., Bruce, W.: Linear feature-based models for information retrieval. Inf. Retr. **10**(3), 257–274 (2007)

19. Kohavi, R.: A study of cross-validation and bootstrap for accuracy estimationand model selection. In: IJCAI, pp. 1137–1145 (1995)

20. Wu, W., Li, H., Wang, H., Zhu, K.: Probase: a probabilistic taxonomy for text understanding. In: SIGMOD, pp. 481–492 (2012)

Improving Short Text Modeling
by Two-Level Attention Networks
for Sentiment Classification

Yulong Li[1], Yi Cai[1(✉)], Ho-fung Leung[2], and Qing Li[3]

[1] School of Software Engineering, South China University of Technology,
Guangzhou, China
scutliyulong@gmail.com, ycai@scut.edu.cn
[2] Department of Computer Science and Engineering,
The Chinese University of Hong Kong, Sha Tin, Hong Kong
lhf@cuhk.edu.hk
[3] Department of Computer Science, City University of Hong Kong,
Kowloon Tong, Hong Kong
itqli@cityu.edu.hk

Abstract. Understanding short texts is crucial to many applications, but it has always been challenging, due to the sparsity and ambiguity of information in short texts. In addition, sentiments expressed in those user-generated short texts are often implicit and context dependent. To address this, we propose a novel model based on two-level attention networks to identify the sentiment of short text. Our model first adopts attention mechanism to capture both local features and long-distance dependent features simultaneously, so that it is more robust against irrelevant information. Then the attention-based features are non-linearly combined with a bidirectional recurrent attention network, which enhances the expressive power of our model and automatically captures more relevant feature combinations. We evaluate the performance of our model on MR, SST-1 and SST-2 datasets. The experimental results show that our model can outperform the previous methods.

1 Introduction

A large amount of short texts are produced every day, in the forms of online reviews, instant messenger conversations, social network posts, etc. Sentiment analysis of short texts aims to identify the sentiments contained in these user-generated contents, which is an important task in many fields of natural language processing (NLP). Traditional studies typically apply the machine learning algorithms to train the classifier with the feature vectors corresponding to the labeled dataset. Recently, with the development of deep learning techniques, various neural models are proposed to automatically generate low-dimensional representations and obtain remarkable results on the sentiment classification of short texts [7].

© Springer International Publishing AG, part of Springer Nature 2018
J. Pei et al. (Eds.): DASFAA 2018, LNCS 10827, pp. 878–890, 2018.
https://doi.org/10.1007/978-3-319-91452-7_56

Table 1. a: noisy words; b: special type of sentence structure

a	negative	The film's overall mood and focus is interesting but constantly unfulfilling

The sentiment expressed by *"interesting"* varies from the key word *"unfulfilling"*, the word *"interesting"* is a noisy word to identify the right sentiment tendency

b	positive	Except this action movie, other movies of this month are not worth seeing

The sentence structure, *"except"* and *"are not worth seeing"* can produce a positive sentiment, but it hard to capture these features since their position are dispersed

Wang [22] pointed out two reasons for the sentiment classification errors of short texts as follows. First, short texts contain limited contextual information, often too sparse for a machine learning algorithm to extract useful features. Second, short texts (e.g., social network posts) do not always observe the syntax of a written language. This means traditional NLP techniques (e.g., syntactic parsing) cannot obtain good results in the short texts. Various neural network based methods are proposed to handle these problems. Some studies utilized recursive neural networks to construct a sentence-level representation vector in sentiment analysis [19]. Le and Mikolov [10] presented a paragraph vector in sentiment analysis. Tai et al. [20] put forward the tree-structured long short-term memory (LSTM) networks to improve the semantic representations.

However, all of the above methods can't handle two problems, i.e., the noisy words and some special types of sentence structures contained in short texts. As showed in Table 1a, *"but...unfulfilling"* is a key clue to the sentiment of the whole sentence. The sentiment expressed by *"interesting"* varies from the key word *"unfulfilling"*, *"interesting"* is a noisy word and expresses misleading information. To capture the sentiment information of above phrase sequence (i.e., *"but...unfulfilling"*), some LSTM-based methods are proposed. They work in a sequential order and treat each word in different positions of texts with the same computation procedure, so that all the words are treated with the same importance. But in some context like Table 1a, some words (e.g., *"unfulfilling"*) are more important than others (e.g., *"interesting"*). Some researchers proposed attention mechanism to automatically capture important semantic information in a sequence. The attention mechanism can help a model to assign attention scores to different word positions according to their importance. We adopt attention mechanism to capture important features, the model will be more robust against irrelevant information. For another example, in Table 1b, there is no explicit sentiment word shows positive sentiment. The combination, *"except"* and *"are not worth seeing"* can produce a positive sentiment. However, it is hard to capture these sentence structures (i.e., *"except"* and *"are not worth seeing"*) just by a single recurrent neural network (RNN) or convolutional neural network (CNN), since their positions are dispersed. Only using single type attention-based feature extraction is not capable to overcome such difficulty, the reason is that CNN-based feature extraction is able to learn the local features (e.g., *"except"*) from words in different positions of texts, while RNN learns the long-distance dependent features (e.g., *"are not worth seeing"*). We consider combining these two type features to enhance the representation of the short text, and applied the attention mechanism on the combined features to automatically capture more relevant feature combinations.

In pursuit of these goals, we propose a novel two-level attention model for short text sentiment classification. More specifically, at the feature extraction level, we model the local and long-distance dependent feature representations of short text by a joint CNN and RNN architectures simultaneously, and we also apply attention mechanism to capture inportant features. At the feature combination level, the attention-based features at the feature extraction level are non-linearly combined with a bidirectional recurrent attention network, which enhances the expressive power of our model and automatically captures more relevant feature combinations. Afterwards, a softmax classifier is used for sentiment classification.

To evaluate the effectiveness of the proposed model, We conduct experiments on three short text datasets and compare the proposed model with other models and baselines. The experiment results demonstrate that our model outperforms previous models.

2   Related Work

2.1   Deep Neural Networks for Sentiment Classification

Since a simple and effective approach to learn distribution representations was proposed [14], deep neural networks have achieved great improvement on sentiment classification field due to their ability of text representation learning. Kim [7] first applied neural networks for sentiment classification. The architecture is a direct application of CNN with pre-trained word embedding. Unlike word level modelings, Zhang et al. [26] use a character-level CNN for text classification and achieve competitive results. Classical models also including recursive neural network [18], recursive neural tensor network [19], recurrent neural network [13], LSTM [5] and tree-LSTMs [20] were applied into sentiment analysis currently. By utilizing syntax structures of sentences, tree-based LSTMs have been proved to be quite effective for many tasks. There are also some works that combine LSTM and CNN structures for sentence classification [8]. Tang et al. [21] use hierarchical structure in sentiment classification. They first use CNN or LSTM to get a sentence vector and then a bidirectional gated recurrent neural network to compose the sentence vectors to get a document vectors.

2.2   Attention Mechanism

The attention mechanism was proposed by Bahdanau et al. [1] in machine translation to select the reference words in original language for words in foreign language before translation. It is motivated from that, instead of decoding based on the encoding of a whole and fixed-length sentence, one can attend a specific part of the sentence. It can automatically capture the most important semantic information in a sequence, without using extra knowledge and NLP systems. Attention models have recently gained popularity in training neural networks and have been applied to various NLP tasks. Xu et al. [23] uses the attention mechanism in image caption generation to select the relevant image regions when generating words in the captions. Further uses of the attention mechanism include parsing, natural language question answering, and image question answering. The effectiveness of multiple attentions was also investigated in QA task, which shows that multiple attentions allow a model to attend different parts of the input during each pass.

3   The Proposed Model

Our model comprises three parts as shown in Fig. 2. The first part generates vector representations for each short text using attention-based CNN and RNN. The second part combines all the attention-based feature representations and generates the more suitable feature combination representation. The third part classifies short texts based on the feature combination representation.

3.1   Attention-Based Feature Extraction

A given short text of length l is represented as a sequence of m-dimensional word vectors ($e_1, e_2, \dots, e_l$), which is used by the CNN and RNN to produce the feature representations. These word vectors can be randomly initialized from a uniform distribution, or pre-trained by an embedding learning algorithm with large text corpus [7]. Compared with randomly initialization, the pre-trained word vectors can capture both semantic and syntactic information [12]. $Word2vec$[1] is a widely used pre-trained vectors. In this paper, we use it to initialization.

CNN-based Feature Extraction. CNN adopts convolutional filters to learn the local and position-invariant features from words or phrases in different places of texts [7]. For the short text ($e_1, e_2, \dots, e_l$), using a filter $W_f \in \mathbb{R}^{h \times m}$ of height h, a convolution operation on h consecutive word vectors starting from t^{th} word outputs the scalar feature

$$c_t = f(W_f \cdot E_{t:t+h-1} + b_f) \tag{1}$$

where $E_{t:t+h-1} \in \mathbb{R}^{h \times m}$ is the matrix whose i^{th} row is $e_i \in \mathbb{R}^m$ and $b_f \in \mathbb{R}$ is a bias. The symbol $\cdot$ refers to dot product and f is a non-linear function.

We perform convolution operations with k different filters of height h, and denote the resulting features as $C_t \in \mathbb{R}^k$, each of whose dimensions comes from a distinct filter. Repeating the convolution operations for each window of h consecutive words in the short text, we obtain $C_{1:l-h+1} \in \mathbb{R}^{(l-h+1) \times k}$, whose length $l - h + 1$ is shorter than the input length l. In practice, we pad the head and the tail of short text with same random vectors, as showed in Fig. 1. In such way, we are able to get $C_{1:l} \in \mathbb{R}^{l \times k}$ that is the same length as the input.

---

[1]   https://code.google.com/arhive/word2vec/.

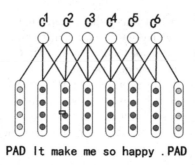

PAD It make me so happy . PAD

Fig. 1. Convolutional operation

RNN-based Feature Extraction. RNN provides a solution by incorporating memory units that allow the network to capture the long-distance dependent context information of texts. $h_t \in \mathbb{R}^s$ is the hidden state of the network at time step t. $o_t \in \mathbb{R}^k$ is the output at time step t. The formulas that govern the calculations in a RNN are as follows.

$$h_t = \sigma(Ue_t + Wh_{t-1}) \tag{2}$$

where e_t is the input at time step t. $U \in \mathbb{R}^{s \times m}$, $W \in \mathbb{R}^{s \times s}$ are weight matrices. The $\sigma(\cdot)$ denotes the logistic sigmoid function. Here h_0 is typically initialized to a zero vector.

$$o_t = softmax(Vh_t) \tag{3}$$

where $V \in \mathbb{R}^{k \times s}$ are weight matrices.

However, RNN is difficult to train because the gradients tend to either vanish or explode [2]. Recently, Long short-term memory (LSTM) [5] and gated recurrent unit (GRU) [4] are two variants of RNN to deal with the problem effectively, they have the ability to remove or add information to the cell state, carefully regulated by structures called gates.

Considering our proposed model, when classifying a particular short text, it is helpful to know the letters coming after it as well as those before, because the output at time t may not only depend on the previous elements in the sequence, but also future elements. In this paper, we use two LSTM or GRU to access both the past and future contexts for any position in a sequence, as showed in Fig. 2. The past and future features are donated as $PC_{1:l} \in \mathbb{R}^{l \times k}$ and $FC_{1:l} \in \mathbb{R}^{l \times k}$.

Attention Layer: Feature Weighting. For each of CNN-based feature and RNN-based feature, actually not all feature representations made important contributions equally to the classification. The idea of the attention mechanism is to learn to focus the attention on specific significant features [24].

We design an attention layer so as to automatically identify the parts of the input text that are relevant for sentiment classification. Each of the local,

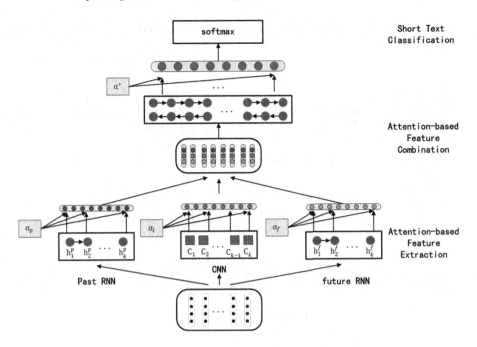

Fig. 2. The architecture of the two-level attention networks.

past and future features are then sent to the attention layer. The output of the attention layer is calculated as follows:

$$m_t = tanh(W_a f_t) \tag{4}$$

$$\alpha_t = softmax(m_t^{\mathrm{T}} \cdot m_s) \tag{5}$$

$$a = \sum_{i=1} \alpha_t f_t \tag{6}$$

where $W_a \in \mathbb{R}^{a \times k}$, the symbol $\cdot$ refers to dot product and f_t denotes C_t, PC_t or FC_t. That is, we first feed the feature annotation f_t through a one-layer MLP to get $m_t \in \mathbb{R}^{a \times 1}$ as hidden representation of f_t, then we measure the importance of the feature as the similarity of m_t^T with a feature level context vector m_s and get a normalized importance weight α_i through a softmax function. After that, we compute the feature vector a as a weighted sum of the feature annotations based on the weights.

So we can generate the attention-based local, past and future context features of $a^l \in \mathbb{R}^k$, $a^p \in \mathbb{R}^k$ and $a^f \in \mathbb{R}^k$ in the same way.

3.2 Attention-Based Feature Combination

In this section, we use the attention mechanism at feature-combination-level that is able to learn the most suitable combination form automatically for different input sequence.

Feature Combination. After generated attention-based features, Then we combine the a^l, a^p and a^f in such way: $A = [[a_1^l, a_1^p, a_1^f]^T, [a_2^l, a_2^p, a_2^f]^T, ..., [a_k^l, a_k^p, a_k^f]^T]$, where each dimension of them are combined as a vector of $[a_i^l, a_i^p, a_i^f]^T$. Figure 3 show the detailed information of the combination progress.

The combination context representations are fed into a bidirectional RNN to obtain the high-level representations $B_{1:k} \in \mathbb{R}^{k \times j}$. Bidirectional RNN extends the unidirectional RNN network by introducing a second layer, where the hidden to hidden connections flow in opposite temporal order [17]. The model is therefore able to exploit information both from the past and future [3].

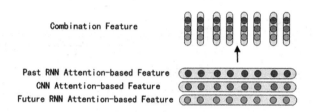

Combination Feature

Past RNN Attention-based Feature
CNN Attention-based Feature
Future RNN Attention-based Feature

Fig. 3. Combination progress

Attention Layer: Combination Weighting. We again use attention mechanism and introduce a combination level context vector n_c and use the vector to measure the importance of the combination features:

$$n_i = tanh(W_c B_i) \tag{7}$$

$$\alpha_i = softmax(n_i^T n_c) \tag{8}$$

$$a^* = \sum_{i=1} \alpha_i B_i \tag{9}$$

where $W_c \in \mathbb{R}^{p \times j}$ are weight matrices, we finally generate $a^* \in \mathbb{R}^j$ as the attention-based feature combination representation.

3.3   Sentiment Classification

In this setting, a softmax classifier are added on top of the model to produce the label for a short text S over the class space. To avoid overfitting, dropout with a masking probability p is applied on the layer.

$$\hat{p}(y|S) = softmax(W^S a^* + b^S) \tag{10}$$

where $W^S \in \mathbb{R}^{v \times j}$ are weight matrices and $b^S \in \mathbb{R}^v$ are bias vectors.

$$\hat{y} = \underset{y}{argmax}\, \hat{p}(y|S) \tag{11}$$

The cost function is the negative log-likelihood of the true class labels $\hat{y}$:

$$J(\theta) = -\frac{1}{m} \sum_{i=1}^{m} t_i \log(y_i) + \lambda \|\theta\|_F^2 \tag{12}$$

where $t \in \mathbb{R}^m$ is the one-hot represented ground truth and $y \in \mathbb{R}^m$ is the estimated probability for each class by softmax, and λ is an L2 regularization hyperparameter.

4 Experiments

To demonstrate the effective of the proposed model, we conduct experiments on three short text datasets about movie reviews, which are commonly used by the state-of-the-art models. We compare the results with previous methods.

4.1 Datasets

The detailed statistics of the datasets are summarized in Table 2. These datasets are briefly described as follows.

MR[2]. Movie reviews with one sentence per review. Classification involves detecting positive/negative reviews [16].

SST-1[3]. Stanford Sentiment Treebank is an extension of MR. The aim is to classify a review as fine-grained labels (very negative, negative, neutral, positive, very positive).

SST-2. Same as SST-1 but with neural reviews removed and binary labels (positive, negative).

In both SST-1 and SST-2 datasets, phrases (sub-sentences) are also tagged with sentiment labels. We regard phrases as individual samples during training, like [6,10]. For validating and testing, only whole sentences are considered in our experiments.

Table 2. Summary statistics of the datasets. c: Number of target classes. l: Average sentence length. V_{train}: Training set size. V_{val}: Validation set size. V_{test}: Test set size. **CV**: 10-fold cross validation.

Datasets	c	l	V_{train}	V_{val}	V_{test}
MR	2	20	10662	-	**CV**
SST-1	5	18	151525	1101	2210
SST-2	2	19	76836	872	1821

---

[2] http://www.cs.cornell.edu/people/pabo/movie-review-data/.
[3] http://nlp.stanford.edu/sentiment/.

4.2 Experimental Settings

As mentioned in Sect. 3, each recurrent unit of our model can be replaced by LSTM or GRU, and word vectors can be randomly initialized or initialized with pre-trained vectors, *word2vec*. We finally get four variants of the two-level attention networks (i.e., TA-LSTM-word2vec, TA-GRU-word2vec, TA-LSTM-rand and TA-GRU-rand). The hyper-parameter settings of the model variations may depend on the dataset being used. Only a set of commonly used hyper-parameters are introduced as follows: The proposed models are trained with backpropagation and the gradient-based optimization is performed via the *Adadelta* update rule [25]. The dimension size of the word embedding is 300, the window length of convolution filter is 4. We use the L2 regularization with the weight of 10^{-4}.

4.3 Experiment Results

Table 3 compares our models to results of the state-of-the-art methods in the sentiment classification of short texts. The best method is in bold.

Table 3. Classification accuracy (%) comparsion results on three sentiment datasets.

Group	Model	MR	SST-1	SST-2
Baseline	Naive Bayes [19]	-	41.0	81.8
	SVM [19]	-	40.7	79.4
CNN	Non-static [7]	81.5	48.0	87.2
	Multichannel [7]	81.1	47.4	88.1
Recursive	Basic [19]	-	43.2	82.4
	Matrix-vector [19]	-	44.4	82.9
	Tensor [19]	-	45.7	85.4
	Tree LSTM1 [26]	-	48.0	-
	Tree LSTM2 [20]	-	51.0	88.0
	Tree LSTM3 [9]	-	49.9	88.0
	Tree bi-LSTM [11]	79.0	-	-
Recurrent	LSTM [20]	-	46.4	84.9
	bi-LSTM [20]	-	49.1	87.5
Vector	Word vector avg [19]	-	32.7	80.1
	Paragraph vector [10]	-	48.7	87.8
TBCNNs	c-TBCNN [15]	-	50.4	86.8
	d-TBCNN [15]	-	51.4	87.9
CA-RNN	CA-LSTM [24]	-	48.5	87.7
	CA-GRU [24]	-	48.9	87.2
TA-Model	TA-LSTM-word2vec	**82.41**	**52.27**	**90.06**
	TA-GRU-word2vec	81.65	51.70	89.48
	TA-LSTM-rand	80.12	50.81	87.86
	TA-GRU-rand	79.82	51.26	88.03

For 5-class prediction in SST-1 dataset, TA-LSTM-word2vec obtains 52.27% accuracy, outperforming the state-of-the-art result, achieved by the d-TBCNN [15]. TA-GRU-word2vec is sightly worse. It achieves 51.7% accuracy. Regarding binary prediction, our TA-LSTM-word2vec model achieves 90.06% accuracy in SST-2 dataset, and 82.41% accuracy in MR dataset. It performs best among all these 2-class prediction datasets.

CA-LSTM model and CA-GRU model only use the attention mechanism on feature extraction. TA-LSTM-word2vec with two-level attention networks outperforms compared with these two models. It turns out that the two-level attention networks are able to learn the more suitable feature combination form automatically for different input sequences.

Besides, we find that our models with pre-trained vectors (i.e., TA-LSTM-word2vec and TA-GRU-word2vec), are all better than the others with randomly initialized vectors (i.e., TA-LSTM-rand and TA-GRU-rand) on all three datasets. Thus, we consider that the pre-trained vectors on large-scale corpora can solve the semantic sparsity problem to some degree.

4.4   Case Study

In order to validate that our model is able to focus on informative features and combinations in a short text, we select two sentences from datasets to show the visualization of the attention results for case study. To make the visualized results comprehensible, we make the attention module directly work on the word embedding, thus we can check whether the attention results are in line with our intuition. The visualization results are shown in Figs. 4 and 5. Note that, darker color means lower weight.

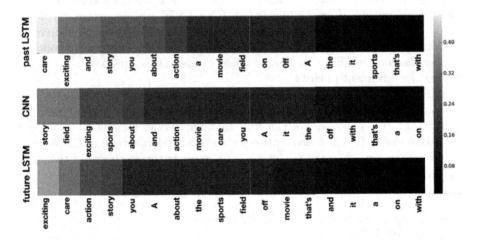

Fig. 4. The visualization of the attention results: *a sports movie with action that's exciting on the field and a story you care about off it.*

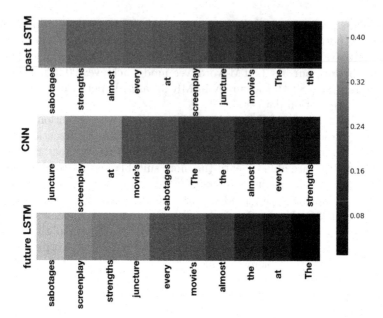

Fig. 5. The visualization of the attention results: *the screenplay sabotages the movie's strengths at almost every juncture.*

As showed in Fig. 4, "*A sports movie with action that's exciting on the field and a story you care about off it*". We observe that our model assigns different attention weights for each feature extraction layer. The words *care* and *exciting* are assigned higher weights in LSTM-based feature extraction layer, while *story* and *field* are highlighted in CNN-based feature extraction layer. Finally the model combines them and generates a positive sentiment. Moreover, the common words *A, with, on, the* and *it* are paid little attention in this context. This verifies our intuition that some noisy words makes little contribution to judging the sentiment polarity.

In Fig. 5, "*The screenplay sabotages the movie's strengths at almost every juncture*". The *sabotages* is a verbal word, unlike adjective, it expresses the implicit negative polarity. We can find that our model focuses on *sabotages* and *strengths* in LSTM-based feature extraction, and *juncture* in CNN-based feature extraction. Then the model correctly predicts the sentiment of the short text as negative. This shows that our model does well in capturing key collocation of words and sentence structure to identify the sentiment polarity.

5   Conclusions

In this paper, we propose a two-level attention model to identify the sentiment of the short text. It takes advantages of local and long-distance dependent features and combine them together. Besides, the model applies two-level attentions on feature extraction and feature combination to pay close attention to the important information to predict the final sentiment. The experimental results show that our model performs well on three benchmark datasets and achieves higher classification accuracy than the existing methods. The case study also shows that our model can reasonably pay attention to those words which are important to judging the sentiment polarity of short texts.

Acknowledgement. This work is supported by the Fundamental Research Funds for the Central Universities, SCUT (NO. 2017ZD0482015ZM136), Tiptop Scientific and Technical Innovative Youth Talents of Guangdong special support program(No. 2015TQ01X633), Science and Technology Planning Project of Guangdong Province, China (No. 2016A030310423), Science and Technology Program of Guangzhou (International Science and Technology Cooperation Program No. 201704030076 and Science and Technology Planning Major Project of Guangdong Province (No. 2015A070711001). This work presented in this paper was also partially supported by a CUHK Direct Grant for Research (Project Code EE16963).

References

1. Bahdanau, D., Cho, K., Bengio, Y.: Neural machine translation by jointly learning to align and translate. Comput. Sci. (2014)
2. Bengio, Y., Simard, P., Frasconi, P.: Learning long-term dependencies with gradient descent is difficult. IEEE Trans. Neural Netw. **5**(2), 157–166 (1994)
3. Cai, R., Zhang, X., Wang, H.: Bidirectional recurrent convolutional neural network for relation classification. In: Meeting of the Association for, Computational Linguistics, pp. 756–765 (2016)
4. Cho, K., Van Merriënboer, B., Gulcehre, C., Bahdanau, D., Bougares, F., Schwenk, H., Bengio, Y.: Learning phrase representations using RNN encoder-decoder for statistical machine translation. arXiv preprint arXiv:1406.1078 (2014)
5. Hochreiter, S., Schmidhuber, J.: Long short-term memory. Neural comput. **9**(8), 1735–1780 (1997)
6. Kalchbrenner, N., Grefenstette, E., Blunsom, P.: A convolutional neural network for modelling sentences. arXiv preprint arXiv:1404.2188 (2014)
7. Kim, Y.: Convolutional neural networks for sentence classification. Eprint Arxiv (2014)
8. Lai, S., Xu, L., Liu, K., Zhao, J.: Recurrent convolutional neural networks for text classification. AAAI **333**, 2267–2273 (2015)
9. Le, P., Zuidema, W.: Compositional distributional semantics with long short term memory. arXiv preprint arXiv:1503.02510 (2015)
10. Le, Q.V., Mikolov, T.: Distributed representations of sentences and documents. In: ICML, vol. 14, pp. 1188–1196 (2014)
11. Li, J., Luong, M.-T., Jurafsky, D., Hovy, E.: When are tree structures necessary for deep learning of representations? arXiv preprint arXiv:1503.00185 (2015)
12. Mikolov, T., Chen, K., Corrado, G., Dean, J.: Efficient estimation of word representations in vector space. arXiv preprint arXiv:1301.3781 (2013)
13. Mikolov, T., Karafit, M., Burget, L., Cernock, J., Khudanpur, S.: Recurrent neural network based language model. In: INTERSPEECH 2010, Conference of the International Speech Communication Association, Makuhari, Chiba, Japan, pp. 1045–1048, September 2010

890   Y. Li et al.

14. Mikolov, T., Sutskever, I., Chen, K., Corrado, G.S., Dean, J.: Distributed repre-
    sentations of words and phrases and their compositionality. In: Advances in Neural
    Information Processing Systems, pp. 3111–3119 (2013)
15. Mou, L., Peng, H., Li, G., Xu, Y., Zhang, L., Jin, Z.: Discriminative neural sentence
    modeling by tree-based convolution. arXiv preprint arXiv:1504.01106 (2015)
16. Pang, B., Lee, L.: Seeing stars: exploiting class relationships for sentiment catego-
    rization with respect to rating scales. In: Proceedings of the 43rd Annual Meeting
    on Association for Computational Linguistics, pp. 115–124. Association for Com-
    putational Linguistics (2005)
17. Schuster, M., Paliwal, K.K.: Bidirectional recurrent neural networks. IEEE Trans.
    Signal Process. **45**(11), 2673–2681 (1997)
18. Socher, R., Pennington, J., Huang, E.H., Ng, A.Y., Manning, C.D.: Semi-
    supervised recursive autoencoders for predicting sentiment distributions. In: Con-
    ference on Empirical Methods in Natural Language Processing, EMNLP 2011,
    27–31 July 2011, John Mcintyre Conference Centre, Edinburgh, UK, A Meeting of
    SIGDAT, A Special Interest Group of the ACL, pp. 151–161 (2011)
19. Socher, R., Perelygin, A., Wu, J.Y., Chuang, J., Manning, C.D., Ng, A.Y., Potts,
    C., et al.: Recursive deep models for semantic compositionality over a sentiment
    treebank. In: Proceedings of the Conference on Empirical Methods in Natural
    Language Processing (EMNLP), vol. 1631, p. 1642. Citeseer (2013)
20. Tai, K.S., Socher, R., Manning, C.D.: Improved semantic representations from
    tree-structured long short-term memory networks. arXiv preprint arXiv:1503.00075
    (2015)
21. Tang, D., Qin, B., Liu, T.: Document modeling with gated recurrent neural net-
    work for sentiment classification. In: Conference on Empirical Methods in Natural
    Language Processing, pp. 1422–1432 (2015)
22. Wang, H.: Understanding short texts (2013)
23. Xu, K., Ba, J., Kiros, R., Cho, K., Courville, A., Salakhutdinov, R., Zemel, R.,
    Bengio, Y.: Show, attend and tell: neural image caption generation with visual
    attention. Comput. Sci. 2048–2057 (2015)
24. Zhang, Y., Er, M.J., Wang, N., Pratama, M.: Sentiment classification using com-
    prehensive attention recurrent models. In: International Joint Conference on Neural
    Networks, (2016)
25. Zeiler, M.D.: ADADELTA: an adaptive learning rate method. Comput. Sci. (2012)
26. Zhu, X., Sobihani, P., Guo, H.: Long short-term memory over recursive structures.
    In: International Conference on Machine Learning, pp. 1604–1612 (2015)

Efficient and Scalable Mining of Frequent Subgraphs Using Distributed Graph Processing Systems

Tongtong Wang[1], Hao Huang[2], Wei Lu[1(✉)], Zhe Peng[1], and Xiaoyong Du[1]

[1] School of Information and DEKE, MOE, Renmin University of China,
Beijing, China
{wttrucer,lu-wei,pengada,duyong}@ruc.edu.cn
[2] State Key Laboratory of Software Engineering, Wuhan University, Wuhan, China
haohuang@whu.edu.cn

Abstract. Mining frequent subgraphs in large scale graph data sets helps reveal underlying knowledge. Since the mining approaches in centralized systems are often bottlenecked on calculation capacity, many parallelized solutions based on the MapReduce framework are proposed to scale out the mining process, which usually extracts frequent subgraphs in an iterative way. Nonetheless, the efficiency and scalability of these MapReduce based approaches are still bounded by the communication cost for passing the intermediate results and the unbalanced workload after a few iterations. In this paper, we propose an efficient and scalable framework for frequent subgraph mining by using distributed graph processing systems. It adopts a message-passing-free scheme among workers to reduce the communication cost, and utilizes a task scheduler to dynamically balance the workload. Experimental results on both synthetic and real-world data sets verify the efficacy of our proposed framework.

1 Introduction

Graphs are pervasively explored to model the linked data in many real-world applications such as chem-informatics [1], bio-informatics [2], image processing [3,4], and end product manufacturings [5]. Given a graph data set $\mathbb{G}$, a frequent subgraph is defined as a subgraph which is contained in at least t graphs in $\mathbb{G}$, where t is a user-defined parameter. Mining these frequent subgraphs often brings the users underlying and useful knowledge. For example, discovering frequent gene expression of mRNA and proteins in gene regulatory networks helps to reveal key gene expression, and extracting common linked parts in BOM (Bill of Material) networks helps to avoid redundant design.

So far, extensive efforts have been made to design mining approaches in centralized systems. These centralized approaches mostly use a filtering-and-verification paradigm, and can be classified into two categories, namely the Apriori-based and pattern-growth approaches. However, due to the limited computation capacity of a single node, the centralized approaches are restricted in

© Springer International Publishing AG, part of Springer Nature 2018
J. Pei et al. (Eds.): DASFAA 2018, LNCS 10827, pp. 891–907, 2018.
https://doi.org/10.1007/978-3-319-91452-7_57

terms of scaling to larger graph data sets. To avoid this flaw, some parallelized approaches, such as MRFSM [6] and FSM-H [7], have been proposed based on the MapReduce framework. Nevertheless, as these MapReduce based approaches extract frequent subgraphs in an iterative way, their efficiency and scalability are still bounded by the communication cost for passing the intermediate results and the unbalanced workload after a few iterations.

Aiming at an efficient and scalable solution for mining frequent subgraphs, we propose a framework called PFSM (**P**regel based **F**requent **S**ubgraph **M**ining) by using Pregel [8] which is a powerful distributed graph processing system. In this framework, all the subgraphs, each of which has i edges and is named i-subgraphs, can be extracted in the i-th iteration without any message passing between workers. Compared with MapReduce based approaches, our PFSM framework can preserve some intermediate results in main memory between iterations, and decrease the I/O cost. Furthermore, it adopts a task scheduler which executes a repartition method to dynamically balance the workload during the mining process. Extensive experiments on both synthetic and real-world data sets are conducted, and the results verify the efficiency and scalability of our PFSM framework.

In summary, our key contributions are as follows. (1) We propose a Pregel based framework called PFSM for frequent subgraph mining, which adopts a message-passing-free scheme to eliminate the communication cost among workers. (2) We present a dynamic workload repartition method, which helps PFSM to keep the workload balance.

The remaining sections of this paper are organized as follows. We review the related work in Sect. 2, and give the problem statement together with basic definitions in Sect. 3, following which we elaborate our PFSM framework in Sect. 4 and report the experimental results as well as our findings in Sect. 5 before concluding the paper in Sect. 6.

2   Related Work

In this section, we first review the existing approaches for frequent subgraph mining in centralized systems, followed by reviewing the parallelized approaches based on MapReduce and Pregel.

The existing *centralized approaches* for frequent subgraph mining can be categorized into two groups, namely the Apriori-based and the pattern-growth methods. AGM and FSG are two classical Apriori-based methods. They generate the subgraph candidates by adding vertices and edges, followed by checking the frequency of each candidate with a breadth-first search strategy. GSpan [9], gIndex [10], FG-Index [11] and GASTON [12] exemplify the pattern-growth methods, employing a depth-first search strategy which consumes less memory and is often more efficient. To further improve the efficiency, multi-cores technologies are used to accelerate the mining process in centralized systems [13]. Moreover, as many graph databases cannot be held into main memory, some approaches adopt novel index structures [14] or divide the graph databases into several smaller partitions [15] to enable the frequent subgraph mining over large graph databases. Nonetheless, due to the limited calculation capacity of a single categorized system, as a rule the centralized approaches are restricted in terms of scaling to larger graph data sets.

In order to scale out the calculation capacity for frequent subgraph mining, some *parallel approaches* have been proposed based on MapReduce. MG-FSM [16], MRFSM [6] and FSM-H [7] are three state-of-the-art algorithms of this type. However, since these MapReduce based approaches extract frequent subgraphs in an iterative way, they often suffer a high communication cost due to the massage passing for intermediate results, and may have a risk of unbalanced workload after a few iterations. To avoid these flaws, a few recent approaches propose to carry out the frequent subgraph mining on programming frameworks like Pregel [8], Trinity [17] and Mizan [18], which are more appropriate to handle graph data. Nevertheless, these approaches focus on frequent subgraphs mining in one single large graph [19], while in this paper, we aim to find out frequent subgraphs within a large-scale collection of moderate-sized graphs.

3   Preliminaries

In this section, we first formally define the frequent subgraph mining, and then give a brief review of the distributed graph processing systems.

3.1   Problem Definition

By abstracting the data structures from the real applications, we model each data object (like protein, mRNA, BOM, etc.) as an undirected labeled graph (a.b.a., labeled graph). Formally, a labeled graph is defined as follows:

Definition 1 (Labeled Graph). *A labeled graph g is represented by a 4-tuple (V_g, E_g, L_g, l_g), in which V_g is a set of vertices, $E_g \subseteq V_g \times V_g$ is a set of undirected edges, L_g is a set of labels, l_g is a mapping function that generates labels in L_g for vertices and edges in g. Symbols $l_g[v]$ and $l_g[u,v]$ are used to denote the label of vertex v, and the label of edge (u,v) connecting vertices u and v in g, respectively.*

Unless otherwise specified, we refer to the graphs used throughout the remainder of the paper as labeled graphs. Given two graphs s and g, **subgraph isomorphism**, formally defined in Definition 2, helps examine whether s is a subgraph of g.

Definition 2 (Subgraph Isomorphism $\subseteq$). *Given two graphs $s = (V_s, E_s, L_s, l_s)$ and $g = (V_g, E_g, L_g, l_g)$, we say s is subgraph isomorphic to g, if there exists an injective function $f: V_s \rightarrow V_g$, satisfying the following two conditions: (1) $\forall v \in V_s$, we can have $f(v) \in V_g$, and $l_s(v) = l_g(f(v))$, and (2) $\forall (u,v) \in E_s$, we can have $(f(u), f(v)) \in E_g$, and $f_s[u,v] = f_g[f(u), f(v)]$. Informally, for ease of illustration, we say s is a subgraph of g, which is denoted as $s \subseteq g$, if s is subgraph isomorphic to g.*

Example 1 (Subgraph Isomorphism). Figure 1 shows an example of three labeled graphs. According to Definition 2, we can find that $g_1 \subseteq g_2$ because there exists an injective function f: $0 \to 0$, $1 \to 1$, $2 \to 3$, $3 \to 2$, while $g_1 \not\subseteq g_3$.

Definition 3 (Embedding). *Given a subgraph s and a graph g, suppose s is a subgraph of g and f is the injective function. An embedding e is a sequence of vertices in V_g that are mapped to the vertices of V_s using f.*

As an example, Fig. 2 shows two subgraphs s_1, s_2 and their embeddings in g_2. Functionally, embedding maintains the mapping information (i.e., injective function) of two subgraph isomorphic graphs. Given a subgraph s, a graph g, and an embedding e, by utilizing the mapping information in e, we can build a unique subgraph g_e^s in g such that g_e^s being graph isomorphic to s.

Based on the above basic definition, the problem statement of frequent subgraph mining is as follows.

Problem Statement. Let D be a tremendous amount of relatively small graphs. Given a subgraph s, let $s.D$ be the posting list of graphs in D, to each of which s is subgraph isomorphic. Formally, $s.D = \{g.id \mid g \in D, s \subseteq g\}$. Given a subgraph s, s is said to be frequent if $|s.D| \geq T$, where T is called frequency threshold which is a user-defined number. Frequent Subgraph Mining is to mine every frequent subgraph s in D associated with its frequency.

3.2   The gSpan and DFS Code

gSpan [9] remains as one of the most widely used pattern-growth approaches to mine frequent subgraphs. Generally, to boost the mining efficiency, gSpan aims at (i) avoiding redundant enumeration for the same subgraph and (ii) reducing the enumeration space during the extension of frequent subgraphs. Aiming at these two targets, gSpan proposes a so-called *DFS coding technique* to transform each subgraph into an edge sequence namely *DFS code*. Technically, the DFS code is generated by performing a depth first search (DFS) on the edges of the subgraph. An edge in a DFS code is represented by a 5-tuple, $(i, j, l_s[i], l_s[i,j], l_s[j])$, where $l_s[i]$, $l_s[i,j]$, $l_s[j]$ are the labels of vertex i, edge (i,j), and vertex j in the original subgraph, respectively. Since there may exist multiple

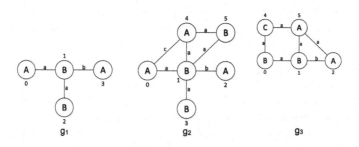

Fig. 1. Example of subgraph isomorphism testing

Subgraph	Embeddings
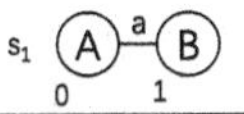 s_1 (A) $\overset{a}{\text{—}}$ (B) 0 1	[0, 1] [4, 5] [4,1]
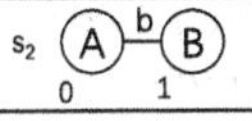 s_2 (A) $\overset{b}{\text{—}}$ (B) 0 1	[2, 1]

Fig. 2. Example of embeddings

DFS searches on a given subgraph s, s may have many different DFS codes. Therefore, a lexicographic order is proposed to compare every two DFS codes. In the following, we introduce four lemmas proposed in [9].

Lemma 1 (Minimum DFS code). *Given an i-subgraph s, let parent(s).min be its parent in (i-1)-subgraphs with the minimum DFS code. Then, s generated by parent(s).min takes the minimum DFS code.*

At first, each different 1-subgraph with its minimum DFS code is generated by just scanning the graph data set once. By introducing the minimum DFS code for the same subgraph, where the minimum DFS code can be obtained based on Lemma 1, gSpan is able to avoid redundant enumeration and restrict the extension of s in a proper way.

Lemma 2 (Redundant Enumeration Elimination). *Given a frequent subgraph s, children of s will be enumerated by extending one edge from every possible vertex of s if and only if it takes the minimum DFS code.*

Lemma 2 eliminates redundant enumeration for the same subgraphs so as to avoid many unnecessary computations. Consider the graph g_2 shown in Fig. 1 as an example. Subgraph (0,1,3) of g_2 can be extended from either (0,1) or (1,3). Since subgraph (0,1,3) that has been extended from (1,3) does not take the minimum DFS code, no enumeration is required in this case. While the same subgraph extended from (0,1) takes the minimum DFS code, and hence, enumeration of all its children will be performed by extending one edge. In this way, we can ensure that enumeration for the same subgraph is executed only once.

Lemma 3 (Enumeration Restriction). *Given a frequent subgraph s associated with the minimum DFS code, suppose L and R are the last and first discovery vertices in the depth first search. Let $P_{R \to L}$ be the path from R to L. We enumerate children of s by appending one edge in one of the following two cases: (1) backward extension: from L to a vertex in $P_{R \to L}$; (2) forward extension: from a vertex in $P_{R \to L}$ to a new vertex.*

Lemma 3 restricts the enumeration of a frequent subgraph by extending one edge from its partial vertices instead of the complete vertices. According to the restriction, gSpan can avoid unnecessary computations.

Lemma 4 (Identical Subgraph). *Given two subgraphs s_1 and s_2 of graph g, if $|E_{s_1}| = |E_{s_2}|$, and $\forall e \in E_{s_1}$, $e \in E_{s_2}$, then s_1 and s_2 are the identical subgraph.*

By introducing Lemma 4, we can verify whether two subgraphs are identical one without performing a isomorphism testing which is a time-consuming operation.

3.3 Distributed Graph Processing Systems

As discussed in Sect. 2, MapReduce is inefficient for iterative computations. In order to efficiently analyse graph data, based on Bulk Synchronous Parallel (BSP) model, a distributed graph processing framework, Pregel [8] is proposed by Google. Typically, its computation consists of an *input* phase, where a graph data set is distributed among the workers, followed by a set of iterations, called *supersteps*, separated by global synchronization points, and finally an *output* phase. In each superstep, a compute function will be called by each vertex. The compute function is related to the application and is defined by the user. It specifies the behavior of a single vertex v in a single superstep. It can read messages sent to v in the last superstep and send messages to other vertices which will be used in the next superstep. Also, this function can modify the state of the vertex and its outgoing edges.

A wide spectrum of distributed graph processing systems such as Pregel+ [20], Giraph [21] are proposed as open sourced implementations for Pregel. Among them, various optimization techniques, including message combination [21], pull-push adaptive mechanism [22], asynchronous scheduling strategies [23], etc., are designed to either reduce the communication cost, or balance the workload. Although in this paper, we implement our algorithm on top of Giraph, a widely used distributed graph processing system, other systems and their optimizations are orthogonal to our work.

4   System Overview

The state-of-the-art MapReduce based approaches are inefficient mainly due to two reasons. First, their efficiency is bounded by the expensive communication cost among workers. Note that each iteration of mining frequent subgraphs in these approaches requires to launch an individual MapReduce job. On one hand, repeated loading of source graph data across multiple jobs degrades the mining performance. On the other hand, the intermediate results generated by one job are first flushed to a persistent storage, and then loaded by the next job. Second, their scalability is bounded by the unbalanced workload. The cost of mining frequent subgraphs among different compute nodes could be severely unbalanced during the iterations, while the MapReduce based approaches are difficult to balance the workload.

To address the above two issues, we propose a Pregel based framework, namely PFSM, for frequent subgraph mining. Similarly to the existing MapReduce based approaches, PFSM extracts frequent subgraphs iteratively as well.

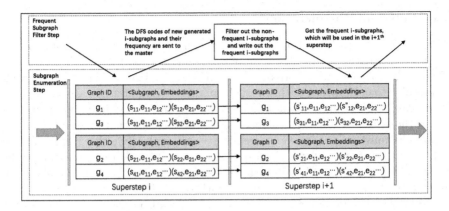

Fig. 3. An overview of PFSM

Figure 3 gives an overview of PFSM at each iteration. PFSM consists of two components, *subgraph enumeration* and *frequent subgraph filtering*. Subgraph enumeration generates frequent subgraph candidates, and frequent subgraph filtering verifies frequent subgraphs from them. The whole process that employs PFSM to mine frequent subgraphs can be elaborated as follows.

Initially, the data set D is divided into a set of partitions, which are then loaded by workers. Each worker is responsible for mining frequent subgraphs from its belonging partitions. At the i-th ($i \geq 1$) iteration, each worker sequentially scans the graphs, enumerates all possible frequent $(i + 1)$-subgraph candidates for each graph by executing the subgraph enumeration component, and sends the candidates to the master for later aggregation. Next, the master launches the frequent subgraph filtering component to collect all possible frequent $(i+1)$-subgraph candidates, count the frequency of each candidate, and output the frequent $(i + 1)$-subgraphs. The above iteration continues until no new frequent subgraphs are generated.

PFSM is able to overcome the drawbacks remaining in existing MapReduce based approaches. First, as PFSM loads the original data set once, no redundant data loading is required. Second, PFSM utilizes the in-memory pipeline computation of Pregel, and preserves the intermediate results in main memory. In this way, the I/Os to flush the intermediate results can be saved. Third, in PFSM, no communication is required among different workers to mine frequent subgraphs. For this reason, we propose a computation aware cost model to balance the workload among workers in terms of the computation. In the remainder of this section, we first present all possible frequent $(i + 1)$-subgraph enumeration in Sect. 4.1, then describe frequent i-subgraph extraction in Sect. 4.2, and finally discuss the computation-aware workload balance in Sect. 4.3.

4.1   Subgraph Enumeration

At the i-th $(i \geq 1)$ iteration, each worker launches the subgraph enumeration component separately. Intuitively, subgraph enumeration takes each graph g as the input, and generates all possible frequent $(i+1)$-subgraphs that are subgraph isomorphic to g. However, in this way, the majority computations across adjacent iterations for subgraph enumeration are redundant. To address this issue, we boost the enumeration performance based on the fact that frequent $(i + 1)$-subgraphs must be generated by frequent i-subgraphs. For this reason, at the i-th $(i \geq 1)$ iteration, for each graph g, we preserve all frequent i-subgraphs associated with their embeddings in g instead of g itself. In what follows, we assume that a subgraph s is associated with the embeddings whenever there is no ambiguity in our discussion. For ease of illustration, let F_i be all frequent i-subgraphs in D and F_i^g denote all frequent i-subgraphs that are subgraph isomorphic to g. Thus, given a graph g, subgraph enumeration aims at generating all possible frequent $(i + 1)$-subgraphs that are subgraph isomorphic to g based on all its frequent i-subgraphs. Specifically, $\forall s \in F_i^g$, we enumerate children of s by extending one edge but limited to g.

The pseudo-code of subgraph enumeration, namely $compute(msgs)$, is listed in Algorithm 1. When invoking $compute(msgs)$ function to enumerate all possible frequent $(i + 1)$-subgraphs from a graph g, we first parse the F_i^g from the $msgs$ (line 1). Since a frequent $(i+1)$-subgraphs cannot be generated based on a non-frequent i-subgraphs, we need to filter out the non-frequent subgraphs contained in F_i^g (line 3–5). Then we generate a superset of F_{i+1}^g based on F_i^g (line 6). After F_{i+1}^g is generated, it is packaged in a message and is sent to g itself, which will be used in the next iteration by g (line 11). For each new generated subgraph, we increase its frequency which will be used in the filter step (line 9–10). It is worth pointing that, since the message is only sent to g itself, there is no communication cost in $compute(msgs)$ method.

Now we modify gSpan to generate $(i + 1)$-subgraphs based on i-subgraphs. Suppose that we have extracted all 1-subgraphs associated with their embeddings in graph g. Next, given a subgraph s and all its embeddings in g, we will discuss how to identify all distinct subgraphs in g by appending one edge in g to s efficiently. Take the subgraph s_1 shown in Fig. 2 as s and g_2 shown in Fig. 1 as graph g. We first identify the vertices in path P_{R-L} for s, which are vertex 0 labeled with A and vertex 1 labeled with B. Since there only exist two vertices, we do not need to perform the backward extension. We begin from the vertex 1 of s to do the forward extension by appending a new edge. Since we have extracted all embeddings in g for s, we sequentially check the vertex v in each embedding e where v in g_e^s is mapped to vertex 1 of s. Specially, We first collect all frequent edges which start from v but are not contained in g_e^s, and append the edges one by one. For example, for embedding [0 1] emphasized in yellow of embeddings of s_{11} in Fig. 4, starting from vertex 1, we can append edges (1,2), (1,3), (1,4), (1,5), based on which four new 2-subgraphs are generated and the corresponding embeddings are highlighted in yellow in the Fig. 4 as well. Then we do the forward extension from the vertex 0 of s. For embedding [0 1], we append

---

Algorithm 1. g.compute($msgs$)

Input: $msgs$: messages received from the last superstep
1 parse F_i^g from $msgs$;
2 $F_{i+1}^g \leftarrow \phi$;
3 **for** *each subgraph s in F_i^g* **do**
4   **if** $s \notin F_i$ **then**
5   F_{i+1}^g.remove(s);

6 $F_{i+1}^g \leftarrow$ enumerate(s);
7 **if** $F_{i+1}^g = null$ **then**
8   return;

9 **for** *each s in F_{i+1}^g* **do**
10   Aggregate(s.getMinDFSCode(),1);
11 sendMessage(g,F_{i+1}^g) ;

---

one edge (0,4) that starts from vertex 0. A new 2-subgraph is generated and the corresponding embedding is emphasized in black. Similarly, we can perform the same operation for other embedding of s_1, and finally, the complete extension for s is shown in the first row of Fig. 4.

To avoid unnecessary enumeration of the same subgraphs, we need to remove redundant subgraphs with non-minimum DFS code. Take the subgraph s_{12} shown in Fig. 2 for example. By appending edge (1,0) for embedding [2 1], We can generate a subgraph s_{27} with embedding [2 1 0]. Although we can perform a graph isomorphism testing whether s has been generated or not, this operation is time-consuming. In practice, we can verify whether s is a redundant subgraph simply based on Lemma 4. For example, by constructing the subgraph from embeddings [2 1 0] of s, we find that it is identical to the subgraph constructed from embedding [0 1 2] of subgraph s_{22} shown in Fig. 4. As s is subgraph isomorphic to s_{23}, we do not need to enumerate s.

According to Lemma 1, for each subgraph s, s generated by parent(s).min takes the minimum DFS code. Hence, we rank the subgraphs in F_i^g in the ascending order of their minimum DFS codes. When a subgraph is verified as a duplicate based on Lemma 4, we can safely discard this subgraph and its embeddings since its DFS code is not minimum. After ranking and enumerating all 1-subgraphs, we can detect the other redundant subgraphs and their embeddings. For example, the subgraphs and their embeddings surrounded by dashed red lines are redundant subgraphs.

The main idea of embedding-based enumeration is outlined in Algorithm 2. At first, F_{i+1}^g and a hash set $genG$ are initialized and the $genG$ maintains all distinct g_e^s for each new generated subgraph s, which are contained in F_{i+1}^g (line 1). We then sort the subgraphs in F_i^g in ascending order of their DFS codes and collect all the distinct edges from their minimum DFS codes (line 2–3). For each subgraph s in F_i^g, we enumerate all children of it according to Lemma 3 (line 4–16). For each newly generated subgraph s', according to Lemma 4, if $g_e^{s'}$

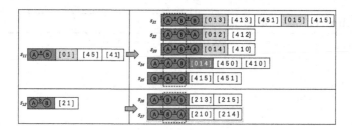

Fig. 4. Example of extension (Color figure online)

Algorithm 2. E-Extension(F_i^g)

1   $F_{i+1}^g \leftarrow \phi$; $genG \leftarrow \phi$;
2   sort subgraphs of F_i^g in the ascending rder of minimum DFS codes;
3   $E \leftarrow$ collect distinct edges from F_i^g;
4   **for** $each \in F_i^g$ **do**
5   **for** $each\ extension$ **do**
6   **for** $each\ e \in s.E$ **do**
7              $S \leftarrow \text{ext}(e, E, extension)$;
8   **for** $each\ s' \in S$ **do**
9   let e' be an embedding of $s'.E$;
10   **if** $!contain(genG, g_{e'}^{s'})$ **then**
11                     $genG \leftarrow genG \bigcup \left\{ g_{e'}^{s'} \right\}$;
12   **if** $!contain(F_{i+1}^g, s'.code)$ **then**
13                        $F_{i+1}^g \leftarrow F_{i+1}^g \bigcup \{s'\}$;
14   **else**
15                     $\bar{s} \leftarrow \text{get}(F_{i+1}^g, s')$;
16                     $\bar{s}.E \leftarrow \bar{s}.E \bigcap s'.E$;

17   **return** F_{i+1}^g;

---

in $genG$, we can verify that the associated DFS code of s' is not the minimum. we note that s' might have already been contained in F_{i+1}^g, since the same s' is generated by s but with another embedding. In this case, we merge their embeddings together (line 12–16).

4.2 Frequent Subgraph Filter

Each worker launches subgraph enumeration component to generate all possible frequent $(i + 1)$-subgraphs, each of which is denoted as a DFS code, which is a sequence of edges. Subgraphs that share the same DFS codes are graph isomorphic. Hence, in each worker, it combines the same DFS code and aggregates its frequency locally, and shuffles distinct DFS codes associated with their frequency to the master of Pregel in the cluster. The master aggregates all the new

---

Algorithm 3. master.compute(msgs)

Input: T : threshold
1 F_{i+1}^g = Aggregate.getKeys();
2 **for** *each subgraph s in F_i^g* **do**
3 | **if** *Aggregate.getValue(s) < T* **then**
4 | └ F_{i+1}^g.remove(s);

5 writeOut(F_{i+1}^g);

---

generated $(i + 1)$-subgraphs associated with their frequency, and filters out the $(i + 1)$-subgraphs with frequency less than T. Then it outputs the frequent i-subgraphs which will be used in the next superstep. The pseduo-code of frequent subgraph filter step is given in Algorithm 3.

4.3   Balance the Workload

Pregel is based on the BSP model. At each iteration, the efficiency of Pregel based approaches is decided by the slowest worker in the cluster. Therefore, balancing the workload in Pregel is of great importance to boost the mining performance. By reviewing the framework of PFSM, no communication cost is required among different workers, while the bottleneck of the mining performance relies on the subgraph enumerations. For this reason, we first propose an equal-size graph partitioning strategy to split the original graph data set, and then design a computation aware cost model to balance the workload during iterations.

Equal-Size Graph Partitioning. Initially, before proceeding any subgraph enumeration tasks, it is necessary to split the original graph data set so that the computations at the first iteration are able to be balanced. For this reason, we use an equal-size graph partitioning strategy so that the total number of edges of all graphs in each worker is balanced. To do this, we sort the graphs by their size in terms of the number of edges, and assign the graphs to the workers in a round-robin manner. As a comparison, random partitioning strategy randomly assigns the graphs to the workers. It is easy to be implemented, but it may cause load imbalance. This is because, random partitioning makes each worker process similar number of graphs, but the size of graphs may be skewed among workers.

Cost Model Based Graph Repartitioning. Consider that workload could be severely imbalanced when iterations goes on. We then propose a computation aware cost model based graph repartitioning method to balance the workload by doing dynamic data shuffling. Our cost model is motivated by the fact that the total execution time relies on the subgraph enumerations. Therefore, we target at the minimization of the total execution time by balancing the computations of subgraph enumerations among workers.

To achieve this goal, at each iteration, we collect the execution time of sub-graph enumeration for each graph, which is then accumulated to form the total execution time for subgraph enumeration of all graphs residing in each worker. When the enumeration cost among different workers is unbalanced, i.e., the gap between the fastest and slowest workers exceeds a pre-defined threshold, we do dynamic data transfer among workers. Let t_g be the execution time for g to do subgraph enumeration, W_i be the ith worker, T_i be the total execution time for graphs residing in W_i to do subgraph enumeration, and G_i be the collection of graph residing in the W_i. Formally, we can have:

$$T_i = \sum_{g \in G_i} t_g \tag{1}$$

Let T be the execution time of subgraph enumeration at each iteration. We propose a computation aware cost model, shown in Eq. 2, which approximately quantifies T.

$$T = \max_{i=1}^{m} T_i \tag{2}$$

where m is the number of workers in the cluster. Nevertheless, under Eq. 2, the execution time t_g of graph g cannot be computed before the subgraph enumeration is launched. To minimize T in practice, we make an approximation of t_g using the execution time for g to do subgraph enumeration at the previous iteration. We then propose a dynamic data transfer strategy by moving graphs from slower workers to faster workers so that the maximum T_i can be reduced. Algorithm 4 outlines our algorithm which used to generate dynamic workload transfer strategy.

The input of the algorithm is a list L_i and $L_i[i]$ is in the form of (W_i, T_i) where W_i, T_i represent a worker id and its estimated workload in the following superstep, respectively. The output is also a list L_o and the elements of it are in the form of (W_i, t_{ij}, W_j) which means the t_{ij} workload can be transport from worker W_i to worker W_j. At the beginning of our algorithm, we divide the workers into two sets according to their estimated workload. A worker is assigned to a slow set S if its workload is heavier than the average value. Otherwise a worker should be contained by the fast set F (line 1–3). In the following, for each slow worker we calculate how much workload should be transport to fast workers (line 5–18).

5   Experimental Evaluation

In this section, we first describe the experimental setup, and then verify the efficiency and scalability of our approach from the following two aspects, namely (1) the performance comparison in terms of runtime and I/O costs, and (2) the effect of load balance.

Algorithm 4. Algorithm used for generating dynamic workload transfer strategy

---

Input : L_i : the workers and their estimated workload
Output: L_o : the strategy of repartition

1 $T_{avg} = \left(\sum_1^n L_i[i].getTime()\right)/n$;
2 $S \leftarrow \{L_i[k] \mid L_i[k].getTime() > T_{avg}\}$;
3 $F \leftarrow \{L_i[k] \mid L_i[k].getTime() < T_{avg}\}$;
4 **while** *!S.empty()* **do**
5 | **for** *each* $s \in S$ **do**
6 | | **for** *each* $f \in F$ **do**
7 | | | **if** *s.getTime() < 0.1 * T_{avg}* **then**
8 | | | | S.remove(s); continue;
9 | | | $\Delta t_s = s.getTime() - T_{avg}$;
10 | | | $\Delta t_f = T_{avg} - f.getTime()$;
11 | | | **if** $\Delta t_s > \Delta t_f$ **then**
12 | | | | L_{out}.add(s.getWokerId(),Δt_f,f.getWorkerId());
13 | | | | F.remove(f); s.getTime() $\leftarrow$ s.getTime() - Δt_f ;
14 | | | **if** $\Delta t_s < \Delta t_f$ **then**
15 | | | | L_{out}.add(s.getWokerId(),Δt_s,f.getWorkerId());
16 | | | | S.remove(s); f.getTime() $\leftarrow$ f.getTime() + Δt_s ;
17 | | | **if** $\Delta t_s = \Delta t_f$ **then**
18 | | | | S.remove(s); F.remove(f);

5.1 Experimental Setup

Experimental Environment. The experimental study is conducted on a 41-node cluster, in which one node is the master and the other nodes are the workers. All these nodes are hosted on one rack, and each node has a Intel 2.4 GHz processor, 4G RAM, a 500G SATA hard disk. We install CentOS 6.5 operating system, java 1.8.0 with a 64-bit server VM, and Hadoop 0.20.203.0 on each node. We make the following changes on the default Hadoop configures, i.e., (1) the size of virtual memory for each task is 3G, and (2) each node supports one map.

Tested Approaches. We compare our proposed PFSM framework with MRFSE [24], which is a high performance parallel approach for frequent subgraph mining based on MapReduce. In both of the tested approaches, extension-based enumeration is used to generate frequent subgraph candidates.

Data Sets. We conduct the experiments over one real-world data set and several synthetic data sets. The real-world data set is extracted from the PubChem web site, which provides various interfaces for users to extract chemical structures and substances. The size of real-world the data set is 484 MB, containing one million chemical structures. Each chemical structure is represented as a graph. Each graph has 23.98 vertices, 25.76 edges, 3.5 distinct labels, and 2.0 distinct edge labels on average. The synthetic graph data sets are generated by the graph

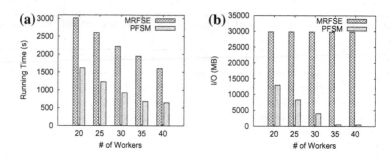

Fig. 5. Performance comparison on real-world data set.

generator provided by [11]. The size of the graphs follows a normal distribution with 5 as the variance, and the other parameters of the generator adopt the default settings.

5.2   Performance Comparison

In this experiment, we compare the performance of PFSM and MRFSE in terms of running time and I/O cost on both real-world and synthetic data sets.

Real-World Data Set. With the real-world data set, we vary the number of workers from 25 to 40, and set the frequency threshold as to $10\% \times |D|$, where $|D|$ is the cardinality of the data sets. Figure 5(a) illustrates the runtime of the tested approaches on the real-world data set, from which we can observe that the runtime of our PFSM framework is about half of MRFSE's runtime. The reason behind is twofold. (1) Since MRFSE employs one MapReduce job to mine each i-subgraphs, it has to initialize multi-MapReduce jobs to extract all the frequent subgraphs. The initialization time of MapReduce jobs takes up a large proportion of the total running time. While the PFSE only needs lightweight synchronization between adjacent supersteps. (2) MRFSE needs to flush out all the embeddings to local disk at the end of each MapReduce job and read these embeddings into main memory at the beginning of next job. This operation is time consuming and leads to great I/O cost. Although some embeddings may be flushed out to the local disk in PFSM, a large proportion of them can be maintained in main memory for the next superstep. Figure 5(b) compares the I/O cost of the two tested approaches on the real-world data set, from which we can have the following two observations, i.e., (1) the I/O cost of PFSM is much less that of MRFSE and decreases along with the increasing number of workers, and (2) the I/O cost of the MRFSE is almost invariable to the number of workers. This is because, in PFSM, more proportion of embeddings can be maintained in main memory if there are more workers. In contrast, MRFSE needs to flush out all of embeddings to the local disk, no matter how many workers are employed. Similar comparison results can be observed, when we use different frequency thresholds (varying from $5\% \times |D|$ to $15\% \times |D|$).

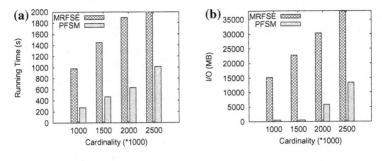

Fig. 6. Performance comparison on synthetic data set

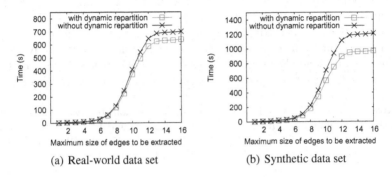

(a) Real-world data set   (b) Synthetic data set

Fig. 7. Effect of load balance

Synthetic Data Set. We evaluate the performance of PFSM and MRFSE on synthetic data sets with difference sizes, which vary from 1 million to 2.5 million. In this evaluation, the number of workers is fixed to 40, and the frequency threshold is set to $10\% \times |D|$. Figures 6(a) and (b) respectively illustrate the corresponding runtime and I/O cost of the tested approaches. From the figure, we can observe that similar to the comparison results on real-world data set, PFSM outperforms MRFSE in both runtime and I/O cost on the synthetic data sets. Moreover, the runtime and I/O cost of both approaches increase linearly with the growth of the data set size. This is because with a larger data set, each worker of the cluster needs to process more graphs in both methods.

5.3 Effect of Load Balance

In this experiment, we study the effect of our proposed dynamic repartition method for load balance in our PFSM framework on the real-world data set and synthetic data set. The size of synthetic data set is 2.5 million. To this end, we set the frequency threshold to $10\% \times |D|$, and report the efficiency performance of PFSM in Fig. 7. From the figure, we can observe that our proposed dynamic repartition method can effectively reduce the total runtime of PFSM. This is because the total runtime of PFSM is bottlenecked on the runtime of workers which have the greatest workloads. In PFSM, we enumerate the $(i+1)$-subgraph by extending frequent i-subgraph. Each worker may contain the same size of graphs at beginning. With the increasing of iterations, the frequent i-subgraphs and their embeddings contained by each worker may be skewed. The dynamic repartition method can help avoid this case by dynamically balancing the workload, and thus speeds up the mining process.

6   Conclusion

In this paper, we have investigated the problem of how to efficiently and scalably carry out frequent subgraph mining in a distributed environment. Based on the distributed graph processing system Pregel, we have proposed a mining framework called PFSM, which adopts a message-passing-free scheme among workers to reduce the communication cost, and utilizes a task scheduler to dynamically balance the workload. Experimental results on both synthetic and real-world data sets verify the efficiency and scalability of our proposed framework.

Acknowledgment. We would like to thank the anonymous reviewers for their helpful and insightful comments. This work was in part supported by the National Natural Science Foundation of China (61502504, 61732014, 61502347, U1711261), and the Technological Innovation Projects of HuBei Province (2017AAA125).

References

1. National library of medicine. http://chem.sis.nlm.nih.gov/chemidplus
2. Berman, H.M., Westbrook, J., Feng, Z., Gilliland, G., Bhat, T.N., Weissig, H., Shindyalov, I.N., Bourne, P.E.: The protein data bank. Nucleic Acids Res. **28**, 235–242 (2000)
3. Lowe, D.G.: Local feature view clustering for 3D object recognition. In: CVPR, pp. 682–688 (2001)
4. Petrakis, E.G.M., Faloutsos, C.: Similarity searching in medical image databases. IEEE Trans. Knowl. Data Eng. **9**(3), 435–447 (1997)
5. Bill of materials. https://en.wikipedia.org/wiki/Bill_of_materials
6. Lin, W., Xiao, X., Ghinita, G.: Large-scale frequent subgraph mining in mapreduce. In: ICDE, pp. 844–855 (2014)
7. Bhuiyan, M., Hasan, M.A.: An iterative mapreduce based frequent subgraph mining algorithm. IEEE Trans. Knowl. Data Eng. **27**(3), 608–620 (2015)
8. Malewicz, G., Austern, M.H., Bik, A.J.C., Dehnert, J.C., Horn, I., Leiser, N., Czajkowski, G.: Pregel: a system for large-scale graph processing. In: SIGMOD, pp. 135–146 (2010)
9. Yan, X., Han, J.: gSpan: graph-based substructure pattern mining. In: ICDM, pp. 721–724 (2002)
10. Yan, X., Yu, P.S., Han, J.: Graph indexing: a frequent structure-based approach. In: SIGMOD, pp. 335–346 (2004)
11. Cheng, J., Ke, Y., Ng, W., Lu, A.: FG-index: towards verification-free query processing on graph databases. In: SIGMOD, pp. 857–872 (2007)

12. Nijssen, S., Kok, J.N.: A quickstart in frequent structure mining can make a difference. In: KDD, pp. 647–652 (2004)
13. Han, J., Pei, J., Yin, Y.: Mining frequent patterns without candidate generation. In: SIGMOD, pp. 1–12 (2000)
14. Wang, C., Wang, W., Pei, J., Zhu, Y., Shi, B.: Scalable mining of large disk-based graph databases. In: KDD, pp. 316–325 (2004)
15. Nguyen, S.N., Orlowska, M.E., Li, X.: Graph mining based on a data partitioning approach. In: ADC, pp. 31–37 (2008)
16. Miliaraki, I., Berberich, K., Gemulla, R., Zoupanos, S.: Mind the gap: large-scale frequent sequence mining. In: SIGMOD, pp. 797–808 (2013)
17. Shao, B., Wang, H., Li, Y.: Trinity: a distributed graph engine on a memory cloud. In: SIGMOD, pp. 505–516 (2013)
18. Khayyat, Z., Awara, K., Alonazi, A., Jamjoom, H., Williams, D., Kalnis, P.: Mizan: a system for dynamic load balancing in large-scale graph processing. In: EuroSys, pp. 169–182 (2013)
19. Zhao, X., Chen, Y., Xiao, C., Ishikawa, Y., Tang, J.: Frequent subgraph mining based on pregel. Comput. J. 59(8), 1113–1128 (2016)
20. Yan, D., Cheng, J., Lu, Y., Ng, W.: Effective techniques for message reduction and load balancing in distributed graph computation. In: WWW, pp. 1307–1317 (2015)
21. Giraph - Welcome To Apache Giraph! http://giraph.apache.org/
22. Wang, Z., Gu, Y., Bao, Y., Yu, G., Yu, J.X.: Hybrid pulling/pushing for I/O-efficient distributed and iterative graph computing. In: SIGMOD, pp. 479–494 (2016)
23. Gonzalez, J.E., Low, Y., Gu, H., Bickson, D., Guestrin, C.: Powergraph: distributed graph-parallel computation on natural graphs. In: OSDI, vol. 12, no. 1, p. 2 (2012)
24. Peng, Z., Wang, T., Lu, W., Huang, H., Du, X., Zhao, F., Tung, A.K.H.: Mining frequent subgraphs from tremendous amount of small graphs using MapReduce. Knowl. Inf. Syst. 1–28 (2017)

Efficient Infrequent Itemset Mining Using Depth-First and Top-Down Lattice Traversal

Yifeng Lu[(✉)], Florian Richter, and Thomas Seidl

Database Systems and Data Mining Group, LMU Munich, Munich, Germany
{lu,richter,seidl}@dbs.ifi.lmu.de

Abstract. Frequent itemset mining is substantially studied in the past decades. In varies practical applications, frequent patterns are obvious and expected, while really interesting information might hide in obscure rarity. However, existing rare pattern mining approaches are time and memory consuming due to their apriori based candidate generation step. In this paper, we propose an efficient rare pattern extraction algorithm, which is capable of extracting the complete set of rare patterns using a top-down traversal strategy. A negative item tree is employed to accelerate the mining process. Pattern growth paradigm is used and therefore avoids expensive candidate generation.

1 Introduction

Frequent itemset (pattern) mining is a well-studied topic. In many applications, frequent patterns represent mainstream behaviors, which are usually known and expected. On the other hand, infrequent patterns can be seen as hints which help people to find out what unexpected things happened. Those interesting and unknown knowledge are important in areas such as medical or scientific research. For example, a medical database contains data about patient treatments, medications, and symptomatic consequences. Pharmacologists already know that the medications A and B might help against disease C while frequent pattern analysis would only yield this obvious information. If in rare cases, both medications $\{A, B\}$ together can lead to a lethal health condition, this result will be pruned out unless a low support parameter is given. In fact, it needs exhaustive time-intense consideration to find out patterns with low support, since the itemset lattice consists of $2^{|I|}$ different itemsets, where $|I|$ is the number of distinct items in the dataset.

The problem of rare pattern mining has not been addressed very well yet. The bottom-up traversal strategy is usually employed. All frequent patterns have to be scanned first before accessing infrequent patterns, which is inefficient when the given minimum support parameter is small. The algorithm [8] proposed to solve the above problem is based on the apriori approach. A huge number of candidates are generated during the mining process.

© Springer International Publishing AG, part of Springer Nature 2018
J. Pei et al. (Eds.): DASFAA 2018, LNCS 10827, pp. 908–915, 2018.
https://doi.org/10.1007/978-3-319-91452-7_58

2   Related Works

In general, both frequent and infrequent itemset mining algorithms can be divided into two categories depends on their itemset lattice traversing style. Breadth-first search based approaches, such as the Apriori [2] algorithm, extends the size of candidate itemsets step-wise. Depth-first search based algorithms, like FP-growth [4], avoid the explicit candidate generation step.

ARIMA [7] and FIRMA [5] follows the apriori idea with bottom-up traversing strategy employed. AfRIM [1] and Rarity [8] traverse the search space in a top-down fashion. However, the Apriori paradigm in these approaches leads to a huge number of candidates during the mining process.

To the best of our knowledge, RP-Tree algorithm [9] is the only rare pattern mining approach based on a depth-first traversing strategy. However, this algorithm makes the assumption that only patterns containing at least one infrequent item are important. Thus, infrequent patterns made up of frequent items are pruned out. The IFPMiner [3] utilizes FP-growth paradigm but only returns minimum infrequent itemsets (MII) so that the count information is lost. A detailed overview of rare itemset mining methods can be found in [6].

3   Preliminaries

Consider $I = \{i_1, i_2, \ldots, i_m\}$ to be the set of all items. Any non-empty subset $X \subseteq I$ is an *itemset*. Any itemset X with size $|X| = k$ is referred to as a k-*itemset*. A tuple $T = (tid, X)$ is called a *transaction*, where tid is the transaction identifier. For simplicity, a transaction $T = (tid, X)$ also refers to its itemset if not specified. Any non-empty itemset $Y \subseteq X$ is *contained* by a transaction $T = (tid, X)$ and we just write $Y \subseteq T$. A set of transactions establish a *transaction database* TD. Figure 1a illustrates an example transaction database.

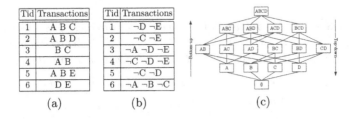

Tid	Transactions
1	A B C
2	A B D
3	B C
4	A B
5	A B E
6	D E

(a)

Tid	Transactions
1	¬D ¬E
2	¬C ¬E
3	¬A ¬D ¬E
4	¬C ¬D ¬E
5	¬C ¬D
6	¬A ¬B ¬C

(b)

(c)

Fig. 1. Example transaction database (a) and the corresponding neg-rep transaction database (b). (c) gives an simple powerset lattice with four distinct items.

Given a transaction database TD, the (absolute) *support* of an itemset X is defined as the number of transactions $T \in TD$ containing X: $X.supp = |\{T \in TD | X \subseteq T\}|$. An itemset X is *rare* if and only if: $X.supp < minSup$, where $minSup$ is a user-defined minimum support threshold. Otherwise, it is *frequent*. The main goal of this work is to extract all rare itemsets efficiently.

Itemset mining can be seen as a process of traversing through the itemset powerset lattice as shown in Fig. 1c. Algorithms can traverse in breadth-first (apriori-like) style or depth-first (pattern-growth like) style. Furthermore, different traverse directions can be combined: either bottom-up or top-down.

Given an itemset X, its support is always smaller than or equal to the support of its subset and larger than or equal to its superset. This is known as the *anti-monotonicity* property. Thus, frequent itemsets tend to be a short "subsets" while infrequent itemsets tend to be a long "supersets". In another words, infrequent itemsets tend to exist at the top part of the lattice so that a top-down traversal based approach will be more efficient [1,8]. In general, a top-down based itemset mining algorithm extracts patterns recursively:

- starting from the full itemset I,
- check k-itemsets
  - if the k-itemset is infrequent, then test its $k-1$ sub-itemsets and repeat
  - otherwise, the k-itemset and all its subsets.

The combination of top-down lattice traversal (pattern-decrease) and pattern-growth paradigm is non-trivial. In the conventional itemset notation, each item symbol expresses the *existence* of the corresponding item. For example, given $I = \{A, B, C, D, E\}$, the notation $X = \{A, B, C\}$ implies that items A, B and C exist in X. For simplicity, we refer to items and itemsets in this notation as *positive items* and *positive itemsets*. Conversely, given I, an itemset X can also be represented by those items that *do not exist in X*.

Definition 1 (Negative Item). *Given a set of items $I = \{i_1, i_2, \ldots, i_m\}$, the negative item of a positive item $i \in I$ is denoted as $\neg i$.*

Definition 2 (Negative Itemset). *Given a positive itemset $X = \{x_1, x_2, \ldots, x_n\}$, the corresponding negative itemset is denoted as $\widetilde{X} = \{\neg x_1, \neg x_2, \ldots, \neg x_n\}$. $\widetilde{\widetilde{X}} = X$.*

Definition 3 (Neg-Rep Itemset). *Given a set of items $I = \{i_1, i_2, \ldots, i_m\}$ and a positive itemset $X = \{x_1, x_2, \ldots, x_n\} \subseteq I$. The corresponding neg-rep (negative represented) itemset $\overline{X}$ is the set of negative items that X does not have, denoted as $\overline{X} = \widetilde{I \setminus X} = \{\neg i | i \in I \wedge i \notin X\}$, where "$\neg$" expresses the idea of "not exist". $\overline{\overline{X}} = X$.*

A neg-rep itemset $\overline{X}$ and its corresponding positive itemset X represent the same instance while the negative itemset $\widetilde{X}$ represents those itemsets that do not intersect with X. Converting each transaction into the neg-rep itemsets yields the *neg-rep transaction database CTD*, which contains the same itemset information in a different representation. Figure 1b gives an example of a neg-rep transaction database.

Two different supports are defined so that extracting infrequent itemset in TD is equivalent to identifying frequent itemset in CTD. Given a non-empty itemset $X = \{x_1, \ldots, x_n\}$, we have:

Definition 4 (Intersect Support). *The intersect support of X in a transaction database TD is the number of transactions that contains **all** items of X:* $X.isupp = |\{T \in TD \mid x_1 \in T \wedge x_2 \in T \wedge \cdots \wedge x_n \in T\}|$.

Definition 5 (Joint Support). *The joint support of X in a transaction database TD is the number of transactions that contains **at least one** item of X:* $X.jsupp = |\{T \in TD \mid x_1 \in T \vee x_2 \in T \vee \cdots \vee x_n \in T\}|$.

In another words, $X.isupp = |\{T \in TD | X \subseteq T\}|$ and $X.jsupp = |\{T \in TD | T \cap X \neq \emptyset\}|$. For example, given dataset in Fig. 1 and b, itemset $X = \{A, B\}$, then $X.isupp = 4$, $\widetilde{X}.jsupp = 2$. The joint support is monotonic:

Theorem 1. *Given itemsets $X_1 \subseteq X \subseteq X_2$, $X_1.jsupp \leq X.jsupp \leq X_2.jsupp$.*

Proof. Given two itemsets $X \subseteq Y$ and a transaction database TD. Let $T \in TD$. It holds $T \cap Y = T \cap (X \cup (Y \setminus X)) = (T \cap X) \cup (T \cap (Y \setminus X))$. Thus, $\{T \in TD \mid T \cap Y \neq \emptyset\} = \{T \in TD \mid T \cap X \neq \emptyset\} \cup \{T \in TD \mid T \cap (Y \setminus X) \neq \emptyset\}$. It directly follows that $|\{T \in TD \mid T \cap X \neq \emptyset\}| < |\{T \in TD \mid T \cap Y \neq \emptyset\}|$.

As the joint support of X represents the number of transactions that do not contain X, we have:

Theorem 2. *Given itemset X and dataset TD, $X.isupp + \widetilde{X}.jsupp = |TD|$*

Thus, the task of extracting rare itemsets from the transaction database is equivalent to the extraction of frequent itemsets in the neg-rep transaction database:

Lemma 1. *Given a transaction database TD. For all itemsets X with $X.isupp \leq minSup$, the joint support of the negative itemset $\widetilde{X}$ can be bound by $\widetilde{X}.jsupp \geq |TD| - minSup$.*

4 Negative Infrequent Itemset Tree Miner

In this section, we will describe our rare itemset mining algorithm: Negative Infrequent Itemset tree miner (NIIMiner). The *negative itemset tree (NI-tree)* is a prefix tree which compresses the neg-rep database. Each node $n = [\neg i, c, l]$ is a triple consisting of a negative item $\neg i$, a count value c and a successor list l. The root node $r = [is, c, l]$ stores an itemset is which is empty at the beginning. Direct successors of the root, i.e. in the list $r.l$, are called the *1st-layer* nodes. All negative items on a path from the root to any node form a negative itemset by concatenation. The count value c represents the number of occurrences of the corresponding itemset. The count of the root node represents the size of the database. The negative itemset tree is built in the same way as the well known FP-tree by scanning the whole neg-rep database. Negative items are sorted in descending order which leads to a smaller tree. The corresponding negative itemset tree for the database in Fig. 1 is shown in Fig. 2a.

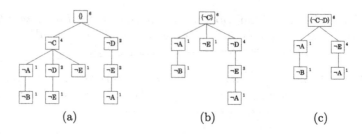

Fig. 2. Examples of (a) negative itemset tree and its corresponding de-tree by excluding (b) $\neg C$ and (c) $\neg C, \neg D$. (c) is also a de-tree of (b).

Infrequent itemsets are extracted by recursively subtracting (excluding) nodes from the NI-tree. The new tree after subtraction, called deducted tree (*de-tree*), is also an NI-tree. Items that have been subtracted so far are stored in the root node. Figure 2 illustrates two de-trees by excluding $\neg C$ and $\neg C, \neg D$ from the original NI-tree respectively. When new negative item is added to the root node, corresponding nodes in the tree are also removed. Sub-trees rooted at the excluded node will be attached and merged recursively to the node above. For example, from Fig. 2a to b, the path $-\boxed{\neg E}$ is attached to the root node and the path $-\boxed{\neg D}-\boxed{\neg E}$ is merged to the existing path. Algorithm 1 illustrates the pseudocode of the subtraction process.

Each NI-tree during the mining process corresponds to a specific itemset X whose neg-rep itemset $\overline{X}$ is stored in the root node. Other negative items, which remained in the tree, form the corresponding negative itemset $\widetilde{X}$.

Theorem 3. *Given a NI-tree root r and an itemset X with $\overline{X} = r.is$, the joint support of the negative itemset is:* $\widetilde{X}.jsupp = \sum\limits_{n \in r.l} n.c.$

Proof. Since $r.is = \overline{X}$, paths remained in the tree correspond to transactions containing at least one item in $\widetilde{X}$. Thus, the joint support of $\widetilde{X}$ is the number of transactions remained. Furthermore, the NI-tree is a prefix tree. Let $n.T$ denotes the set transactions in CTD that contribute to the count of the corresponding itemset, we have: $n'.T \subseteq n.T$, if node n' is in the subtree of node n. Thus, the count of a node summarize all counts in its subtree, $\widetilde{X}.jsupp = \sum\limits_{n \in r.l} n.c.$

For example, the NI-tree in Fig. 2c corresponds to the itemset $\{A, B, E\}$ as its neg-rep itemset $\{\neg C, \neg D\}$ is excluded. The joint support of the corresponding negative itemset $\{\neg A \neg B \neg E\}$ is $1 + 4 = 5$. The joint support is used as the stopping criteria for our mining process as mentioned in Lemma 1.

In practice, a pseudo-subtraction strategy is employed to avoid unnecessary tree construction. Firstly, only a new root node is created in the subtraction step. Related child nodes will be attached to the new root immediately. The subtraction process remains correct since the value of the joint support is not affected, as illustrated in Fig. 3a. Secondly, the subtraction process is terminated when no negative item in the 1st-layer needs to be excluded. For example, excluding $\neg D$

from the original tree will lead to the tree in Fig. 3b. The remaining $\boxed{\neg D}$ node will only be removed after the removing of node $\boxed{\neg C}$. Again, the correctness is not affected since the computation of joint-support only depends on 1st-layer nodes.

Algorithm 2 illustrates the overall procedure of NIIMiner. Negative items are excluded one by one recursively. The whole process terminated until the joint support is lower than the given threshold.

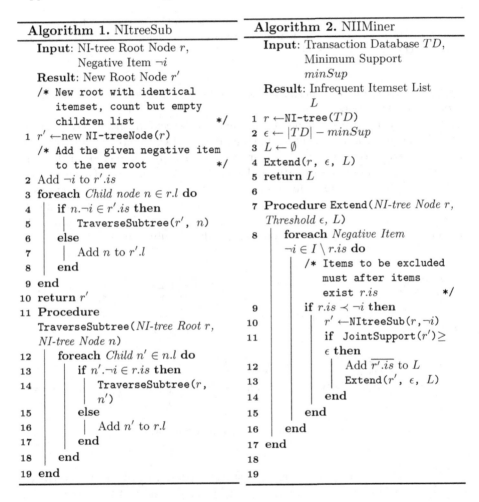

Algorithm 1. NItreeSub

Input: NI-tree Root Node r,
　　　　Negative Item $\neg i$
Result: New Root Node r'
/* New root with identical
　 itemset, count but empty
　 children list　　　　　*/
1 $r' \leftarrow$ new NI-treeNode(r)
/* Add the given negative item
　 to the new root　　　　*/
2 Add $\neg i$ to $r'.is$
3 **foreach** *Child node* $n \in r.l$ **do**
4 　 **if** $n.\neg i \in r'.is$ **then**
5 　 　 TraverseSubtree(r', n)
6 　 **else**
7 　 　 Add n to $r'.l$
8 　 **end**
9 **end**
10 **return** r'
11 **Procedure**
　 TraverseSubtree$(NI\text{-}tree\ Root\ r,$
　 $NI\text{-}tree\ Node\ n)$
12 　 **foreach** *Child* $n' \in n.l$ **do**
13 　 　 **if** $n'.\neg i \in r.is$ **then**
14 　 　 　 TraverseSubtree$(r,$
　　　　　 $n')$
15 　 　 **else**
16 　 　 　 Add n' to $r.l$
17 　 　 **end**
18 　 **end**
19 **end**

Algorithm 2. NIIMiner

Input: Transaction Database TD,
　　　　Minimum Support
　　　　$minSup$
Result: Infrequent Itemset List
　　　　L
1 $r \leftarrow$ NI-tree(TD)
2 $\epsilon \leftarrow |TD| - minSup$
3 $L \leftarrow \emptyset$
4 Extend$(r,\ \epsilon,\ L)$
5 **return** L
6
7 **Procedure** Extend$(NI\text{-}tree\ Node\ r,$
　 $Threshold\ \epsilon, L)$
8 　 **foreach** *Negative Item*
　　 $\neg i \in I \setminus r.is$ **do**
　　 /* Items to be excluded
　　　 must after items
　　　 exist $r.is$　　　　　*/
9 　 **if** $r.is \prec \neg i$ **then**
10 　 　 $r' \leftarrow$ NItreeSub$(r, \neg i)$
11 　 　 **if** JointSupport$(r') \geq$
　　　 ϵ **then**
12 　 　 　 Add $\overline{r'.is}$ to L
13 　 　 　 Extend$(r',\ \epsilon,\ L)$
14 　 　 **end**
15 　 **end**
16 　 **end**
17 **end**
18
19

5　Experimental Evaluation

We compare our NIIMiner to the Rarity [8] algorithm, which is the state-of-the-art top-down Apriori-like infrequent itemset mining approach. Other rare pattern mining algorithms, such as ARIMA [7] and AfRIM [1], are not included in our experiments since they use a breadth-first search with bottom-up traversal

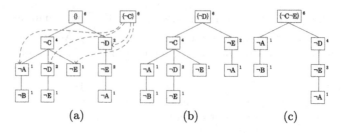

Fig. 3. (a) The pseudo-subtraction of $\neg C$ in practice, only root is created, no merging happened. (b, c) Examples of de-tree by excluding $\neg D$ and $\neg C \neg E$. Only 1st-layer nodes are checked and removed.

similar to Rarity, and their performance compared to Rarity have been conducted in detail Rarity papers.

All algorithms are implemented in Java and executed on an Intel Core i7 3.4 GHz machine running Ubuntu 16.04. Real dataset Connect-4 from UCI repository[1] is used in our experiment. Our experiments are conducted on different dataset sizes, minimum supports, and maximum itemset sizes. Given dataset size set to N and the maximum itemset size set to L, the first N transactions in a dataset and the first L items in each itemset are used. We limit the maximum itemset size since otherwise, the Rarity algorithm won't be able to finish the mining task on our machine. Experiment results are illustrated in Fig. 4.

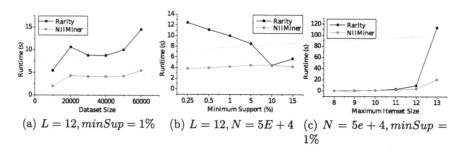

(a) $L = 12, minSup = 1\%$ (b) $L = 12, N = 5E + 4$ (c) $N = 5e + 4, minSup = 1\%$

Fig. 4. Runtime on Connect-4 dataset

Taking the advantage of the depth-first traversal over the breadth-first traversal, NIIMiner is significantly faster than Rarity under most of the settings. The only exception is under large minimum support value and small maximum itemset size settings. This is because the Connect-4 dataset has a limited number of unique items, which leads to fewer candidates during candidate generation step. Rarity also suffers from its pruning step under small minimum support settings. As shown in Fig. 4, the runtime performance of Rarity is increasing when the

---

[1] https://archive.ics.uci.edu/ml/index.php.

minimum support is decreasing, which is unusual as a top-down based approach. It costs much more time but output less rare patterns with a smaller minimum support. In real applications, extracting infrequent itemsets usually implies to find itemsets with small support rather than large support. In summary, the NIIMiner approach, with the ability to extract the complete set of rare patterns, is very efficient.

6   Conclusion and Future Works

Our novel rare itemset miner NIIMiner has proven to solve the problem of rare itemset mining in an efficient and successive manner. By utilizing the negative representations of rare itemsets to frequent itemsets, we addressed this task from its dual perspective. However, we should also notice that our NIIMiner traverses all infrequent itemsets, including those patterns with support equal to 0, known as *nonexistent patterns*. There are a huge number of nonexistent patterns, especially in a sparse dataset, which should be skipped since they are not important in many applications. NIIMiner has to spend a lot of time on traversing those patterns while Rarity only returns rare patterns that exist and could be faster than NIIMiner on sparse datasets. Further investigations are necessary to avoid the expensive traversing step on nonexistent patterns.

References

1. Adda, M., Wu, L., Feng, Y.: Rare itemset mining. In: Sixth International Conference on Machine Learning and Applications 2007, ICMLA 2007, pp. 73–80. IEEE (2007)
2. Agrawal, R., Srikant, R., et al.: Fast algorithms for mining association rules. In: Proceedings 20th International Conference Very Large Data Bases, VLDB, vol. 1215, pp. 487–499 (1994)
3. Gupta, A., Mittal, A., Bhattacharya, A.: Minimally infrequent itemset mining using pattern-growth paradigm and residual trees. In: Proceedings of the 17th International Conference on Management of Data, p. 13 (2011)
4. Han, J., Pei, J., Yin, Y.: Mining frequent patterns without candidate generation. In: Proceedings of the 2000 ACM SIGMOD International Conference on Management of Data, SIGMOD 2000, pp. 1–12. ACM, New York (2000)
5. Hoque, N., Nath, B., Bhattacharyya, D.: An efficient approach on rare association rule mining. In: Bansal, J., Singh, P., Deep, K., Pant, M., Nagar, A. (eds.) Proceedings of Seventh International Conference on Bio-Inspired Computing: Theories and Applications (BIC-TA 2012), pp. 193–203. Springer, Heidelberg (2013). https://doi.org/10.1007/978-81-322-1038-2_17
6. Koh, Y.S., Ravana, S.D.: Unsupervised rare pattern mining: a survey. ACM Trans. Knowl. Discov. Data (TKDD) 10(4), 45 (2016)
7. Szathmary, L., Napoli, A., Valtchev, P.: Towards rare itemset mining. In: 19th IEEE International Conference on Tools with Artificial Intelligence 2007, ICTAI 2007, vol. 1, pp. 305–312. IEEE (2007)
8. Troiano, L., Scibelli, G.: A time-efficient breadth-first level-wise lattice-traversal algorithm to discover rare itemsets. Data Min. Knowl. Discov. 28(3), 773–807 (2014)
9. Tsang, S., Koh, Y.S., Dobbie, G.: RP-Tree: rare pattern tree mining. In: Cuzzocrea, A., Dayal, U. (eds.) DaWaK 2011. LNCS, vol. 6862, pp. 277–288. Springer, Heidelberg (2011). https://doi.org/10.1007/978-3-642-23544-3_21

Online Subset Topic Modeling
for Interactive Documents Exploration

Linwei Li[1], Yaobo Wu[1], Yixiong Ke[1], Chaoying Liu[1], Yinan Jing[1,2(✉)],
Zhenying He[1,2(✉)], and Xiaoyang Sean Wang[1,3(✉)]

[1] Fudan University, Shanghai, China
{lwli15,ybwu13,yxke11,chaoyingliu14,jingyn,zhenying,
xywangcs}@fudan.edu.cn
[2] Shanghai Key Laboratory of Data Science, Shanghai, China
[3] Shanghai Insititute of Intelligent Electronics and Systems, Shanghai, China

Abstract. Data exploration over text databases is an important problem. In an exploration scenario, users would find something useful without previously knowing what exactly they are looking for, until the time they identify them. Therefore, labor-intensive efforts are often required, since users have to review the overview (or detail) results of ad-hoc queries and adjust the queries (*e.g.*, zoom or filter) continuously. Probabilistic topic models are often adopted as a solution to provide the overview for a given text collection, since it could discover the underlying thematic structures of unstructured text data. However, training a topic model for a selected document collection is time consuming. Moreover, frequent model retraining would be introduced by continuous query-adjusting, which leads to large amount of time wasting and therefore is unsuitable for online exploration. To remedy this problem, this paper presents STMS, an algorithm for constructing topic structures in document subsets efficiently. STMS accelerates the process of subset modeling by leveraging global precomputation and applying an efficient sampling-based inference algorithm. The experiments on real world datasets show that STMS achieves orders of magnitude speed-ups than standard topic model, while remaining comparable in terms of modeling quality.

Keywords: Subset topic modeling · OLAP · Exploratory analysis

1 Introduction

As dataset sizes increase, there is a need for modern systems which support exploration and analysis of data. This motivates us to consider exploring large-scale datasets systematically, thereby developing useful Interactive Data Exploration (*IDE*) applications. In *IDE* scenario, users can analyze "unfamiliar data" conveniently and find something useful without previously knowing what exactly they are searched for.

This paper is supported by NSFC (No. 61732004) and the Shanghai Innovation Action Project (Grant No. 16DZ1100200).

Due to lack of knowledge of given datasets, users often have to review the overview of ad-hoc queries and adjust the queries continuously. Thus, it is crucial to provide an overview for the given data collections efficiently. Data warehouse and OLAP techniques focus on this problem. By providing a rich set of OLAP operations such as *drill-down*, *roll-up*, *slice*, *pivot*, etc., OLAP systems help users carry out interactive exploration in big data to discover information of value.

Unstructured text data is ubiquitous in analytics, and it's time consuming for users to read a whole collection of text documents. Probabilistic topic models [1, 7] are techniques which can summarize the content of large text corpora. These techniques model topics as distributions of words, and represent each document as a combination of topics. If OLAP operations on a *topic* dimension is supported, then users can find whether the cuboid contains documents of their interests, and which part attracts them. Figure 1 gives an illustration.

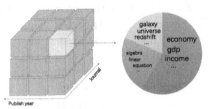

Fig. 1. Integrate topic exploration with OLAP system

To discover topics of a specified text collection, a straightforward solution is to build a topic model on the collection. However, it's very time consuming to do so. For text collection with just a moderate size (e.g. 10K documents, 8M words), training a topic model from scratch may take several minutes, which is intolerant for online users, especially when users have to explore several subsets in an *IDE* scenario. This motivates us to study how to integrate topic exploration with OLAP systems in a more efficient manner. Our works and contributions can be summarized as follows:

1. We integrate subset topic modeling into OLAP systems. By constructing topic structures in an arbitrary selected cuboid, users can do online analysis processing along a topic dimension. Importantly, we identify that global pre-computation could be used to accelerate the subset modeling as the training set is a subset in the text database.
2. To the best of our knowledge, this work is the first to study the problem of subset topic modeling. We discuss possible solutions using existing methods, and propose STMS, an algorithm which leverage global precomputation to extract topic-coherent snippets, and then construct quality topic structures efficiently in arbitrary document subsets.
3. We present evaluation of the proposed subset modeling algorithm on two real-world datasets. The experimental results show that the proposed algorithm achieves orders of magnitude speed-ups while remaining comparable performance in topic modeling to standard topic model.

2   Related Works

Latent Dirichlet Allocation [1] is a popular unsupervised method for performing probabilistic topic modeling. It has been widely used in industrial application such as information retrieval and text classification. As a generative model, LDA assumes that the words in a document are generated by first choosing a latent topic for each word, then using the topic to generate the word itself (according to the topic's multinomial distribution over words).

There are several works based on topic modeling aiming to help users understand and navigate a large document collection. Topic browser [3] implements an web-based interactive browser to display the output of topic models, FacetAtlas [2] proposes a multifaceted visualization technique which combines search technology and advanced visual analytical tools. The tools mentioned above help users understand the corpus as a whole, while some studies [4] focus on providing both a high-level summary of the corpus and links between the summary and individual documents.

However, all these methods are time consuming to prepare representation for a specific text collection. Once users select a new collection, recomputation from scratch is required, which can't meet the efficiency demand of online exploration.

Studies abound on promoting efficiency for model training. Some focused on distributed algorithms [9,12]. Some concentrate on accelerating the sampling process [8,10,12]. Some are concerned with how to run an online learning algorithm in the case of corpus growing over time [6]. These studies do improve the efficiency. However, the speed of these methods still can't meet the demand, as they also require training from scratch for a new selected subset, and the training time is proportional to the subset size.

3 Subset Topic Modeling

When given a document subset, our goal is to extract the underlying topics which outline the contents of the subset and model it. To the best of our knowledge, subset topic modeling has not been studied before. Here we will first briefly discuss possible solutions using existing methods.

1. **Vanilla LDA:** Training a LDA model from scratch for subset is a natural solution. However, it is time consuming and impractical for online analysis.
2. **Top-K Corpus-level Topics:** If the subset is selected from a text database, we can pre-train a topic model on the entire corpus. With corpus-level model, we find the top-k corpus-level topics with the highest weights in the subset.
3. **K-Means Clustering:** By representing every document with its weight distribution vectors z_m over corpus-level topics (practically better than a tf-idf vector representation for clustering task in both clustering accuracy and computation efficiency), we carry out clustering process on documents. Then we can take the word distribution of each cluster as the cluster's topic.
4. **Subset Word Distribution:** Directly using the subset word distribution as its topic is an effective solution and a reasonable baseline.

3.1 Proposed Algorithm

In this section, we propose STMS (abbr. for Subset Topic Modeling over Snippets) algorithm. The STMS algorithm first extracts topic-coherent text snippets from the subset, and then constructs topic structures over the extracted snippets. The details are as follows:

Table 1. Notations for subset topic modeling

α, β	Prior parameters for dirichlet distribution	$\overrightarrow{s_l}$	lth snippet in subset
w_{mi}	ith word in document m	z_l	Topic assigned to lth snippet
z_{mi}	Topic assigned to word w_{mi}	K	Number of topics
φ_k	Distribution over vocabulary of topic k	n_l^w	Count of word w in lth snippet
θ_m	Distribution over topics of document m	n_k^w	Count of word w in topic k
T	Number of distinct words in vocabulary	n_l	Length of lth snippet
m_k	Number of snippets assigned to topic k	n_k	Count of all words in topic k

Topic-Coherent Snippets Extraction. Aside from the topic distribution of each document, we go further to store the topic assignment z_{mi} of each word w_{mi} in each document. We use these additional information to extract topic-coherent text snippets from the subset (here a *snippet* is a concatenation of words with the same corpus-level topic assignment from different documents in the subset). We assume that words in subset with the same corpus-level topic assignment are thematically coherent. Thus, we can group them together and name the grouped words a *topic-coherent snippet*. The subset could have up to K snippets. Since K is usually quite big for large corpus (e.g. 1k topics for 10m documents), these scattered snippets can not be directly used to summarize the subset. Instead, we need to construct subset topic structures over these snippets.

Topic Modeling over Snippets. We assume that the words in a snippet are drawn from a single topic, so here we use a one layer dirichlet multinomial mixture model [11] to construct topic structures over snippets. More concretely, we assume that there are $subK$ topics in the subset, and words in a snippet are all from a same topic and drawn independently from the corresponding multinomial distribution. We list the notations for subset topic modeling in Table 1, and the generative process of the model works as follows:

1. For each of the $subK$ topics in subset, draw $\overrightarrow{\varphi_k}$ from a dirichlet prior, $\overrightarrow{\varphi_k} \sim dirichlet(\alpha)$
2. Draw topic proportion $\overrightarrow{\theta}$ in subset, $\overrightarrow{\theta} \sim dirichlet(\beta)$
3. For each of the L snippets $\overrightarrow{s_l}$:
   (a) Choose a topic $z_l \sim multinomial(\overrightarrow{\theta})$.
   (b) For each word w_i in snippet $\overrightarrow{s_l}$:
       Choose $w_i \sim p(w_i|z_l, \overrightarrow{\varphi_k})$, a multinomial probability conditioned on the topic z_l.

With the generative process, we also use collapsed gibbs sampling to do model inference, the details are demonstrated in Algorithm 1.

Algorithm 1. Inference algorithm for subset topic modeling

1: For each snippet $\overrightarrow{s_l}$ in subset, randomly assign a topic number z_l.
2: Re-scan the snippets and use collapsed gibbs sampling to re-sample each snippet's topic $z_l \sim p(z_{mi} = k | \overrightarrow{\mathbf{z}}_{m,\neg i}, \overrightarrow{\mathbf{w}}, \alpha, \beta)$.
3: Repeat step2 until convergence.
4: With topic assignments of all words, estimate the topics Φ and topic per-document allocation Θ.

The derivation of conditioned distribution $p(z_{mi} = k | \overrightarrow{\mathbf{z}}_{m,\neg i}, \overrightarrow{\mathbf{w}}, \alpha, \beta)$ is as follows:

$$p(z_l = k | \overrightarrow{\mathbf{z}}_{\neg l}, \overrightarrow{\mathbf{s}}, \alpha, \beta) \propto p(z_l = k, \overrightarrow{s_l} | \overrightarrow{\mathbf{z}}_{\neg l}, \overrightarrow{s_{\neg l}}, \alpha, \beta)$$
$$\propto p(z_l = k | \overrightarrow{z_l}, \alpha) \cdot p(\overrightarrow{s_l} | z_l = k, \overrightarrow{s_{\neg l}}, \beta) \tag{1}$$

where

$$p(z_l = k | \overrightarrow{z_l}, \alpha) = \int p(z_l = k, \overrightarrow{\theta} | \overrightarrow{z_l}, \alpha) d\overrightarrow{\theta}$$
$$= \frac{B(\overrightarrow{m} + \overrightarrow{\alpha})}{B(\overrightarrow{m_{\neg l}} + \overrightarrow{\alpha})} = \frac{m_{k,\neg l} + \alpha}{subM + subK \cdot \alpha - 1} \tag{2}$$

$$p(\overrightarrow{s_l} | z_l = k, \overrightarrow{s_{\neg l}}, \beta) = \int p(\overrightarrow{s_l}, \overrightarrow{\varphi_k} | z_l = k, \beta) d\overrightarrow{\theta}$$
$$= \frac{B(\overrightarrow{n}_k + \overrightarrow{\beta})}{B(\overrightarrow{n}_{k,\neg l} + \overrightarrow{\beta})} = \frac{\prod_{w \in \overrightarrow{s_l}}(\prod_{j=1}^{n_l^w}(n_{k,\neg l}^w + \beta + j - 1))}{\prod_{i=1}^{n_l}(n_{k,\neg l} + T\beta + i - 1)} \tag{3}$$

Computation Optimization. Equation 3 shows that the computation of sampling a latent variable z_l over $subK$ subset topics for each of the L snippets has an $O(subK \cdot n_l)$ time complexity, which leads to an $O(subK \cdot L \cdot \bar{n}_l)$ complexity for each iteration (actually the same as retraining a LDA model in the subset).

We observe that the $\prod$ operations dominate the computational costs. For product with the form: $\prod_{i=1}^{n}(x + N + i)$, where x is a real-number with decimal part and N is an integer, if the value of x is fixed for all sequential multiplications in the model inference process, we can build an index with the ith element a_i having a value of $a_i = \prod_{j=0}^{i}(x + j)$. Then the expression $\prod_{i=1}^{n}(x + N + i)$ can be calculated by a single division: $\prod_{i=1}^{n}(x + N + i) = \frac{a_{N+n}}{a_N}$, which reduces the time complexity from $O(n)$ to $O(1)$.

With this process, the computation of Eq. 3 reduces to $O(1)$ for denominator and $O(|U|)$ for numerator ($|U|$ refers to the number of unique word tokens in snippet $\overrightarrow{s_l}$), leads to a $O(L \cdot |U| \cdot subK)$ time complexity for each training iteration, which is much smaller than that of retraining a LDA model.

4   Experiments

We experiment on two real-world datasets, Enron Email dataset[1] and Wikipedia dataset[2]. Experiments are conducted on a Linux server with 2.1 GHz Intel Xeon processor and 64 GB RAM.

4.1   Perplexity

For probabilistic topic models, a natural and most widely used evaluation metric for modeling quality is the probability of held-out documents. Perplexity is such a metric used by convention in topic modeling [1]. For a test set with $subM$ documents, the perplexity is,

$$perplexity(D) = \exp\{-\frac{\sum_{m=1}^{subM} \log p(\overrightarrow{w}_d)}{\sum_{m=1}^{subM} N_m}\} \qquad (4)$$

We experiment on two datasets, Enron Email dataset containing 500K email messages with 80M words, and Wikipedia dataset containing one million wikipedia articles with 400M words. For each subset, we held out 20% of

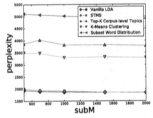

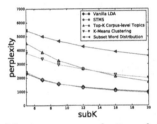

(a) Average perplexity of 50 runs on wikipedia dataset, with $K = 300, subK = 7$

(b) Average perplexity of 50 runs on wikipedia dataset, with $K = 300, subM = 1000$

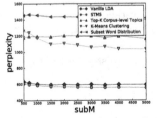

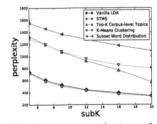

(c) Average perplexity of 50 runs on Enron Email dataset, with $K = 200, subK = 7$

(d) Average perplexity of 50 runs on Enron Email dataset, with $K = 200, subM = 1000$

Fig. 2. Comparison of the perplexity results of five methods for subset modeling

---

[1] https://snap.stanford.edu/data/email-Enron.html.

[2] https://en.wikipedia.org/wiki/Wikipedia:Database_download.

the text as test set and take the remaining 80% as training set, the results are shown in Fig. 2.

In all conditions, experimental results show that STMS's performance in modeling perplexity is comparable to the performance of standard LDA (far better than all baselines). Although STMS assumes a simpler modeling process than LDA (only one latent topic variable for per snippet), STMS is sometimes better in modeling quality. This is because with the help of corpus-level topics which are extracted from a larger collection (the corpus), STMS can have a better generalizability when it comes to held-out testset extrapolating.

4.2 Efficiency

Both vanilla LDA and STMS have satisfying modeling quality, but the computational efficiency of LDA is not competent to the scenario of subset online analysis. In this section, we test the STMS algorithm's efficiency promotion compared to standard LDA. We implement STMS algorithm in python, with core modules rewritten in cython for acceleration, and use three implementations (gibbsLDA [5], onlineLDA [6], and lightLDA [12]) of LDA as benchmarks.

Figure 3 shows the experimental results, and we can see that STMS achieves orders of magnitude speed-ups than training LDA from scratch in subset. The time required for STMS increases very slowly with the subset size. Even if the subset size is large, STMS can still give the results in a few seconds.

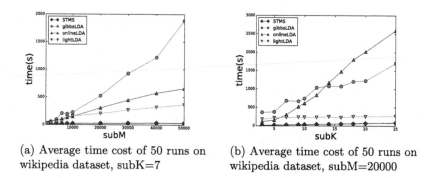

(a) Average time cost of 50 runs on wikipedia dataset, subK=7

(b) Average time cost of 50 runs on wikipedia dataset, subM=20000

Fig. 3. Efficiency promotion compared to standard LDA.

5   Conclusion

In the scenario of *IDE*, users often make duplicated efforts on training topics for given subsets of text databases. It becomes a crucial task to provide an online topic-training algorithm. In this paper, we propose STMS, an efficient subset topic modeling algorithm for constructing topic structures in document subsets. The experimental results on real world datasets show that STMS obtains orders of magnitude of speed-ups over standard LDA, while remaining comparable topic modeling quality.

References

1. Blei, D.M., Ng, A.Y., Jordan, M.I.: Latent dirichlet allocation. J. Mach. Learn. Res. **3**(Jan), 993–1022 (2003)
2. Cao, N., Sun, J., Lin, Y.R., Gotz, D., Liu, S., Qu, H.: Facetatlas: multifaceted visualization for rich text corpora. IEEE Trans. Vis. Comput. Graph. **16**(6), 1172–1181 (2010)
3. Gardner, M.J., Lutes, J., Lund, J., Hansen, J., Walker, D., Ringger, E., Seppi, K.: The topic browser: an interactive tool for browsing topic models. In: NIPS Workshop on Challenges of Data Visualization, vol. 2 (2010)
4. Görg, C., Liu, Z., Kihm, J., Choo, J., Park, H., Stasko, J.: Combining computational analyses and interactive visualization for document exploration and sensemaking in jigsaw. IEEE Trans. Vis. Comput. Graph. **19**(10), 1646–1663 (2013)
5. Griffiths, T.L., Steyvers, M.: Finding scientific topics. Proc. Nat. Acad. Sci. **101**(suppl 1), 5228–5235 (2004)
6. Hoffman, M., Bach, F.R., Blei, D.M.: Online learning for latent dirichlet allocation. In: Advances in Neural Information Processing Systems, pp. 856–864 (2010)
7. Hofmann, T.: Probabilistic latent semantic indexing. In: Proceedings of the 22nd Annual International ACM SIGIR Conference on Research and Development in Information Retrieval, pp. 50–57. ACM (1999)
8. Li, A.Q., Ahmed, A., Ravi, S., Smola, A.J.: Reducing the sampling complexity of topic models. In: Proceedings of the 20th ACM SIGKDD International Conference on Knowledge Discovery and Data Mining, pp. 891–900. ACM (2014)
9. Newman, D., Asuncion, A., Smyth, P., Welling, M.: Distributed algorithms for topic models. J. Mach. Learn. Res. **10**(Aug), 1801–1828 (2009)
10. Porteous, I., Newman, D., Ihler, A., Asuncion, A., Smyth, P., Welling, M.: Fast collapsed gibbs sampling for latent dirichlet allocation. In: Proceedings of the 14th ACM SIGKDD International Conference on Knowledge Discovery and Data Mining, pp. 569–577. ACM (2008)
11. Yin, J., Wang, J.: A dirichlet multinomial mixture model-based approach for short text clustering. In: Proceedings of the 20th ACM SIGKDD International Conference on Knowledge Discovery and Data Mining, pp. 233–242. ACM (2014)
12. Yuan, J., Gao, F., Ho, Q., Dai, W., Wei, J., Zheng, X., Xing, E.P., Liu, T.Y., Ma, W.Y.: LightLDA: big topic models on modest computer clusters. In: Proceedings of the 24th International Conference on World Wide Web, pp. 1351–1361. ACM (2015)

Main Point Generator: Summarizing with a Focus

Tong Lee Chung[1,2(✉)], Bin Xu[1,2], Yongbin Liu[3(✉)], and Chunping Ouyang[3]

[1] Department of Computer Science and Technology,
Tsinghua University, Beijing, China
`tongleechung86@gmail.com, xubin@tsinghua.edu.cn`
[2] Beijing National Research Center for Information Science and Technology
(BNRist), Beijing, China
[3] College of Computing, University of South China, Hengyang, China
`qingbinliu@163.com, ouyangcp@126.com`

Abstract. Text summarization is attracting more and more attention while deep neural network has had many successful application in NLP. One problem of such models is its inability to focus on the essentials of documents, thus generating summaries that may not be important, especially during multi-sentence summarization. In this paper, we propose Main Pointer Generator (MPG) to address the problem, where at each decoder step the whole document is taken into consideration when calculating the probability of next generated token. We experiment with CNN/Daily news corpus and results show that summaries our MPG generated follow the main theme while outperforming the original pointer generator network by about 0.5 ROUGE point.

Keywords: Text summarization · Sequence-to-sequence · Pointer
Coverage

1 Introduction

Text summarization is the task of producing a short piece of text from a long one while preserving main information [1]. *Extractive* and *abstractive* are the two general approaches to text summarization. Extractive methods focus on feature engineering for selecting words and phrases from the original text to assemble the summary. On the other hand, abstractive models may produce words that are not in the original text.

Recent advances in neural sequence to sequence [2] (encoder decoder) models, in which RNNs read and generate text with recurrent neural networks has made many advances in various NLP tasks such as machine translation, question answering as well as summarization [3,4]. Multi-sentence summarization has attracted friction recently due to introduction of new techniques and datasets [5]. Standard sequence to sequence model is the baseline of neural network summarization, which opens new possibilities to abstractive summarization [6,7].

© Springer International Publishing AG, part of Springer Nature 2018
J. Pei et al. (Eds.): DASFAA 2018, LNCS 10827, pp. 924–932, 2018.
https://doi.org/10.1007/978-3-319-91452-7_60

State-of-the-art models have been presented by several researches where some problems have been tackled when using RNNs for abstractive summarization. Attention mechanism has improved quality by showing the decoder what part of the sentence to attend to [5]. High level attributes, such as length constraint, entity-centric and source-specific are introduced to control the generation of summaries [3]. Pointer-generator with coverage mechanism model [4] handles accuracy and out-of-vocabulary problem via pointing (copying words from the document) while retaining its ability to generate new words, and uses coverage to eliminate repetition. New learning objective have been introduced to optimize ROUGE with reinforcement learning [8].

Repetition is a problem these model face where summaries tend to repeat itself. Two ways of dealing with the repetition is presented, coverage/intra-attention of words [4,8] and directly preventing generation of same n-grams [3,8]. Coverage discourages generation of the same word by using a coverage vector that sums over attention received so far during decoding showing what words have been looked at already [4], similarly intra-attention can be seen as calculating a coverage over generated words [8]. While the other solution directly prevents the generation of n-grams during decoding [3]. Though it has solved the repetition problem and improved summary quality, it might be causing another problem: the summary not being able to focus on the main points of the document. When summarizing, important words such as names, places may appear multiple times, because they are the main focus of documents such as news. Anti-repetition strategies causes the decoder to penalize them after appearing even though they are key words, and instead generates summaries that are less important. Our aim is to produce summaries that are closely related to the main theme of the document.

While the task of summarization is to preserve main information of the document, current model focus on generating summaries that do not produce false facts or do not repeat itself. Most do not consider the fact that summaries should focus on the main theme of the document. In this paper, we present Main-pointer-Generator (MPG) model, which neutralizes the negative effect of coverage and helps the decoder focus on the key parts of the document. The model, derived from pointer-generator network with coverage [4], takes into consideration a document vector at each decoding step and can generates key words of the document multiple times. We experiment on *CNN/Daily Mail* dataset that is used by various researchers and outperform state-of-the-art model and is comparable to the result of lead-3 baseline (using first three sentences as summary). We also show that the summary our model produces focuses on one theme.

2   Related Work

Summarization has been around for many decades [9], this paper focuses on multi-sentence summarization where the task is to generate summaries with multiple sentences.

Neural Abstractive Summarization. The abstractive approach needs to understand the meaning of the document, and then briefly summarize it by a highly readable human language. For this target, the RNN/LSTM models and some of their deformations are adopted to complete the neural abstractive task [10]. Recently, some researchers have used the latest neural networks model to summarize, such as the sequence-to-sequence model and attention model.

Sequence-to-Sequence Models. Most of current state-of-the-art models are based on sequence-to-sequence models which have gained many successes in machine translation [2,11]. Attention [12], pointer network [13], coverage [14] and controllable summarization [3,15] are some techniques adapted to the task of summarization.

Most similar to our work is from Fan et al. [3]. Controllable summaries lets reader decide which entity they want their summary to focus on, giving them the option on which parts to focus on. An entity marker is used to instruct the decoder where to look, where the marker can be specified by human or automatically. Such approach requires entity recognition tool to find entities. Main difference between our work is that we rely on the whole document rather than entities to find the focus.

3   Main Pointer Generator Network Model

This section describes our Main Pointer Generator Network Model (MPG) in detail. The model is derived from Pointer-Generator Network from See et al. [4] and is depicted in Figure 1. We will go through vital parts of the whole model in this section.

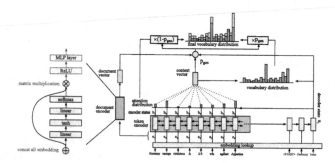

Fig. 1. Main pointer generator model. At each decoder timestep, the decoder state, context vector, and document vector is used to calculate a generation probability p_{gen}. Decoder state shows what words are already generated, the context vector shows where to attend to in the document, and the document vector helps focus on important content of the documents.

3.1   Encoding

Like most neural network model, tokens are embedded into fixed length vectors $E = (e_1, e_2, \ldots, e_n)$. Embedding flows two direction, a token encoder and a document encoder. The token encoder (a bi-directional LSTM network) produces two outputs, an output state for every input token $H = (h_1, h_2, \ldots, h_n)$ and a final output state, which is used as the first input during the decoding process. The document encoder takes in the embeddings E and generates a fixed vector representing the whole document using self attending mechanism.

The document encoder contains a bi-directional LSTM (separate from the token encoder), that generates output states $H' = (h'_1, h'_2, \ldots, h'_n)$ for each input embedding. A self attending mechanism is applied to the output states to compute a representation for the document. First attention distribution A^{doc} is calculated using $tanh$ followed by a $softmax$ operation, then multipled with the state vector H':

$$A^{doc} = \text{softmax}(W_{h2} \tanh(W_{h1} H')) \tag{1}$$

$$d' = H'^{\text{T}} A^{doc} \tag{2}$$

The multiplication gives a hidden state of the document d' with a fixed length eliminating the document size. Finally, ReLU and Multilayer Perceptron (MLP) is applied to generate a final output d with a specific size:

$$d^h = \text{ReLU}(h') \tag{3}$$

$$d = W_{output} d^h + b_{output} \tag{4}$$

W_{output} and b_{output} are both trainable parameters. By setting size of $output$, we can generate a document vector of a specific size. d is the final document vector and is calculated once for every document and shared during each decoding step.

3.2 Decoding

Now that the input document is encoded, three outputs are generated, a document vector d from document encoder, output states $H = (h_1, h_2, \ldots, h_n)$ for each token and final output state from token encoder. We now show how summary is generated at each time step. A LSTM cell takes as input a linearity of previous context vector h^*_{t-1} (described later) and embedding of previous word e'_{t-1} (previous word from reference summary when training, and previous generated word during testing) and produces a current state s_t (at step 0, inputs are final output state from token encoder and pre-defined <START> token).

$$x_t = w_{l1} h^*_{t-1} + w_{l2} e'_{t-1} + b_l \tag{5}$$

$$s_t = \text{LSTM}(x_t) \tag{6}$$

where w_{l1}, w_{l2} and b_l are trainable parameters. Attention distribution is next, which can be calculated in two ways depending on whether coverage mechanism is used. When coverage is not used, attention distribution of a token a^t is calculated using a linear operation of its encoder output state h_i and decoder state s_t followed by nonlinearity $tanh$ and $softmax$:

$$e_i^t = v^T \tanh(W_h h_i + W_s s_t + b_{attn}) \tag{7}$$

$$a^t = \text{softmax}(e^t) \tag{8}$$

W_h, W_s and b_{attn} are trainable parameters. When using coverage, a coverage vector c^t is maintained, which sums over all the previous attention distribution (at time step $t = 0$, c^0 is a zero vector because nothing is covered). Token attention distribution is calculated similar but includes c^t to the linear operation:

$$c^t = \sum_{t'=0}^{t-1} a^{t'} \tag{9}$$

$$e_i^t = v^T \tanh(W_h h_i + W_s s_t + W_c c_i^t + b_{attn}) \tag{10}$$

$$a^t = \text{softmax}(e^t) \tag{11}$$

newly included parameter W_c is trainable. The coverage vector reminds the current attention calculation which tokens were observed, thus avoiding attending to the same part of the documents. In the discussion section, we show why coverage may be the reason unimportant summary is generated. Attention is a probability over the source document telling the decoder where to look when generating words. A context vector h_t^* can now be calculated using the attention distributions and encoder states:

$$h_t^* = \sum_i a_i^t h_i \tag{12}$$

The context vector is a fixed-size representation the document based on the summary produced so far. It can be used together with the current decoder state to produce the vocabulary distribution by going through two linear and a softmax layer:

$$P_{vocab} = \text{softmax}_{vocab}(V'(V[s_t, h_t^*] + b) + b') \tag{13}$$

V, V' and b' are trainable parameters. Next token could be one with the highest probability, but this would cause misleading facts and inability to deal with out-of-vocabulary tokens. A pointer generator is then calculated to show which part of the documents it relates to and if the next token should be copied from the original document or generated by the distribution. Context vector h_t^*, current state s_t, decoder input x_t (linearity of previous word embedding and context vector), and document vector are all used to calculate a generation probability $p_{gen} \in [0, 1]$:

$$p_{gen} = \sigma(w_{h^*} h^* + w_s s_t + w_x x_t + w_d d + b_{ptr}) \tag{14}$$

where w_{h^*}, w_s, w_x, and w_d are trainable parameters, and σ is the sigmoid function. p_{gen} See et al. [4] only takes in w_{h^*}, w_s, and w_x, which means that when deciding to choose between generating and copying, only information from what have been generated is considered, the document as a whole is not considered. We adjust p_{gen} using the document vector d to focus on the main content even though appearing multiple times, thus getting the name Main Pointer Generator. d is shared during all decoding step. Using p_{gen} to re-calculate word probability by considering between copying and generating, the word probability distribution $P(w)$ from which we predict words can be calculated:

$$P(w) = p_{gen}P_{vocab}(w) + (1 - p_{gen}) \sum_{i:w_i=w} a_i^t \qquad (15)$$

3.3 Training

The loss function to train the model is the summation of negative log likelihoods of target word w^* at each time step:

$$loss_t = -\log P(w_t^*) \qquad (16)$$

$$loss = \frac{1}{T} \sum_{t=0}^{T} loss_t \qquad (17)$$

When using the coverage mechanism, a coverage loss is included to penalize repeatedly attending at the same words, making loss at each time step:

$$loss_t = -\log P(w_t^*) + \lambda \sum_i \min(a_i^t, c_i^t) \qquad (18)$$

4 Experiments and Results

We present details of our experiment in this section, which includes the hyperparameters and training of the model, the dataset used, evaluation method and results.

4.1 Model Parameters and Optimization

The model is implemented in Tensorflow and trained using a Titan X GPU. During training, we keep hyper-parameter same as See et al. [4] to show that our model produces higher scores not by fine tuning but by generating better summaries. Hidden state size is set to 256 and embedding size set to 128. Vocabulary size is 50 K. Pre-trained word embedding is not used during the training of the models. Parameters that are unique to our model are those related to the document vector. Hidden size for two non-linearity (tanh and softmax) for self attention is set to 350 and 30 respectively. Hidden size for ReLU layer is 3000. Finally, MLP layer has 256 hidden units to generate a document vector of size 256. The model is trained with AdaGrad with learning rate 0.15 and an initial accumulator value of 0.1. Gradient clipping with a maximum gradient norm of 2 is used, and do not use other forms of regularization. We follow the procedure that See et al. used by truncating the document to 400 tokens and limiting summary to 100 for training and 120 for testing. The model is trained in batches with size of 16. At test time, summaries are produced using beam search with beam size of 4.

4.2 Dataset and Evaluation

The CNN\Daily dataset is separated in training set, validation set and testing set
with 287226, 13368 and 11490 pairs of training examples respectively. Summaries
consist of 3.75 sentences on average. Our model is evaluated with the standard
ROUGE metric [16] and we report F1 score for ROUGE-1, ROUGE-2, and
ROUGE-L. We compare with the original pointer-generator + coverage model [4]
and controllable model [3], the only two models we can find that experimented
on the original text dataset. These two are the current state-of-the-art models
for multi-sentence summarization. We also compare with lead 3 baseline (first
three sentences used as summary), a hard to beat baseline. We also perform
qualitative analysis by looking at examples generated by our model and the
original pointer generator + coverage model.

4.3 Results

This section presents quantitative and qualitative experiment results.

Table 1. Results for experiment. Our model MPG + coverage has the highest ROUGE
score.

	ROUGE 1	ROUGE 2	ROUGE L
Lead-3 baseline (See et al. [4])	40.34	17.70	36.57
Pointer-generator(See et al. [4])	36.44	15.66	33.42
Pointer-generator + coverage (See et al. [4])	39.53	17.28	36.38
Controllable abstractive (Fan et al. [3])	39.75	17.29	36.54
MPG (Ours)	37.84	16.50	34.88
MPG + coverage, shared layer (ours)	39.69	17.36	36.42
MPG + coverage (Ours)	**40.12**	**17.74**	**36.82**

Table 1 show the results for our experiments. We perform experiment using
three models of our own, MPG without coverage mechanism, MPG with a shared
LSTM layer and our full MPG model. We see that coverage does produce a
boost giving a extra +2.28 on ROUGE 1, +1.24 on ROUGE 2, and +1.94 on
ROUGE L. We also see that sharing a LSTM layer between two encoders does
not perform better. But on the other hand, there are less parameters to train
faster. We can see also that by using the document vector, ROUGE score has
improved compared to the original pointer generator. When using coverage, our
model performed better with +0.59 on ROUGE 1, +0.46 on ROUGE 2, and
0.44 on ROUGE L. So we can say that using a document vector can increase
ROUGE performance by 0.5 ROUGE points. We also compare with controllable
abstractive summary with automatic control and find that our model performs
better. We can also see that our results are the only model that is closest to the
lead-3 baseline.

5 Conclusion

This paper, we investigate a problem of focusing during summarization. We analyze a state-of-the-art pointer generator + coverage model and conclude reason for such the problem. We propose a model that makes the generate summaries based of the main theme of the document. Experiment is conducted on CNN\Daily dataset and we achieve state-of-the-art results increasing ROUGE point by 0.5. This work serves as a stepping stone of neural network summarization as our technique can be integrated into other models.

Acknowledgments. This work is supported by China National High-Tech Project (863) under grant (No. 2015AA015401). Beijing Key Lab of Networked Multimedia also supports our research work. The work is supported by State Key Program of National Natural Science of China (No. 61533018), National Natural Science Foundation of China (No. 61402220), and the Philosophy and Social Science Foundation of Hunan Province (No. 16YBA323).

References

1. Nenkova, A., McKeown, K.: Automatic summarization. Found. Trends Inf. Retrieval 5(23), 103–233 (2011)
2. Sutskever, I., Vinyals, O., Le, Q.V.: Sequence to sequence learning with neural networks. In: Proceedings of the 27th International Conference on Neural Information Processing Systems - Volume 2, NIPS 2014 (2014)
3. Fan, A., Grangier, D., Auli, M.: Controllable abstractive summarization. CoRR, abs/1711.05217 (2017)
4. See, A., Liu, P.J., Manning, C.D.: Get to the point: summarization with pointer-generator networks. In: Proceedings of the 55th Annual Meeting of the Association for Computational Linguistics. Association for Computational Linguistics (2017)
5. Nallapati, R., Zhou, B., dos Santos, C.N., Gülçehre, Ç., Xiang, B.: Abstractive text summarization using sequence-to-sequence rnns and beyond. In: Proceedings of the 20th SIGNLL Conference on Computational Natural Language Learning, CoNLL 2016, Berlin, Germany, 11–12 Aug 2016
6. Chopra, S., Auli, M., Rush, A.M.: Abstractive sentence summarization with attentive recurrent neural networks. In: The 2016 Conference of the North American Chapter of the Association for Computational Linguistics: Human Language Technologies, NAACL HLT 2016, San Diego California, USA, 12–17 June 2016
7. Rush, A.M., Chopra, S., Weston, J.: A neural attention model for abstractive sentence summarization. In: Proceedings of the 2015 Conference on Empirical Methods in Natural Language Processing, EMNLP 2015, Lisbon, Portugal, 17–21 Sept 2015
8. Paulus, R., Xiong, C., Socher, R.: A deep reinforced model for abstractive summarization. CoRR, abs/1705.04304 (2017)
9. Saggion, H., Poibeau, T.: Automatic text summarization: past, present and future. In: Poibeau, T., Saggion, H., Piskorski, J., Yangarber, R. (eds.) Multi-source, Multilingual Information Extraction and Summarization. Theory and Applications of Natural Language Processing, pp. 3–21. Springer, Heidelberg (2013). https://doi.org/10.1007/978-3-642-28569-1_1

10. Shen, S., Zhao, Y., Liu, Z., Sun, M.: Neural headline generation with sentence-wise optimization. arXiv Computation and Language (2016)
11. Bahdanau, D., Cho, K., Bengio, Y.: Neural machine translation by jointly learning to align and translate. In: International Conference on Learning Representations (ICLR) (2015)
12. Vaswani, A., Shazeer, N., Parmar, N., Uszkoreit, J., Jones, L., Gomez, A.N., Kaiser, L., Polosukhin, I.: Attention is all you need. In: Guyon, I., Luxburg, U.V., Bengio, S., Wallach, H., Fergus, R., Vishwanathan, S., Garnett, R. (eds.) Advances in Neural Information Processing Systems, vol. 30, pp. 6000–6010. Curran Associates Inc. (2017)
13. Vinyals, O., Fortunato, M., Jaitly, N.: Pointer networks. In: Cortes, C., Lawrence, N.D., Lee, D.D., Sugiyama, M., Garnett, R. (eds.) Advances in Neural Information Processing Systems, vol. 28, pp. 2692–2700. Curran Associates Inc. (2015)
14. Mi, H., Sankaran, B., Wang, Z., Ittycheriah, A.: Coverage embedding models for neural machine translation. In: EMNLP (2016)
15. Kikuchi, Y., Neubig, G., Sasano, R., Takamura, H., Okumura, M.: Controlling output length in neural encoder-decoders. CoRR, abs/1609.09552 (2016)
16. Lin, C.-Y.: Rouge: a package for automatic evaluation of summaries. In: Text Summarization Branches Out: Proceedings of the ACL-04 Workshop

Correction to: Database Systems for Advanced Applications

Jian Pei, Yannis Manolopoulos, Shazia Sadiq, and Jianxin Li

Correction to:
J. Pei et al. (Eds.): *Database Systems for Advanced Applications*,
LNCS 10827, https://doi.org/10.1007/978-3-319-91452-7

The original version of this chapter titled "Free-Rider Episode Screening via Dual Partition Model" contained the following three mistakes:

1. In table 1, row 2, column 3, the average occurrence per event on STK dataset was "1.037". It should be "1,037".
2. The last model name in the legend of Figure 3 was "EIP". It should be "EDP".
3. In the experiment part the stock symbols and their companies were confused.

In the updated version these mistakes were corrected.

In the originally published version of chapters titled "BASSI: Balance and Status Combined Signed Network Embedding" and "Sample Location Selection for Efficient Distance-Aware Influence Maximization in Geo-Social Networks" the funding information in the acknowledgement section was incomplete. This has now been corrected.

The updated version of these chapters can be found at
https://doi.org/10.1007/978-3-319-91452-7_4
https://doi.org/10.1007/978-3-319-91452-7_24
https://doi.org/10.1007/978-3-319-91452-7_43

© Springer International Publishing AG, part of Springer Nature 2020
J. Pei et al. (Eds.): DASFAA 2018, LNCS 10827, p. C1, 2020.
https://doi.org/10.1007/978-3-319-91452-7_61

Author Index

Printed in the United States
By Bookmasters